SOME PHYSICAL PROPERTIES

Air (dry, at 20°C and 1 atm)

Density	1.21 kg/m³
Specific heat capacity at constant pressure	1010 J/kg·K
Ratio of specific heat capacities	1.40
Speed of sound	343 m/s
Electrical breakdown strength	3×10^6 V/m
Effective molar mass	0.0289 kg/mol

Water

Density	1000 kg/m³
Speed of sound	1460 m/s
Specific heat capacity at constant pressure	4190 J/kg·K
Heat of fusion (0°C)	333 kJ/kg
Heat of vaporization (100°C)	2260 kJ/kg
Index of refraction ($\lambda = 589$ nm)	1.33
Molar mass	0.0180 kg/mol

Earth

Mass	5.98×10^{24} kg
Mean radius	6.37×10^6 m
Free fall acceleration at the Earth's surface	9.81 m/s²
Standard atmosphere	1.01×10^5 Pa
Period of satellite at 100 km altitude	86.3 min
Radius of the geosynchronous orbit	42,200 km
Escape speed	11.2 km/s
Magnetic dipole moment	8.0×10^{22} A·m²
Mean electric field at surface	150 V/m, down

Distance to:

Moon	3.82×10^8 m
Sun	1.50×10^{11} m
Nearest star	4.04×10^{16} m
Galactic center	2.2×10^{20} m
Andromeda galaxy	2.1×10^{22} m
Edge of the observable universe	$\sim 10^{26}$ m

SUPPLEMENTS

PHYSICS, FOURTH EDITION is accompanied
by a complete supplementary package.

STUDY GUIDE (A Student's Companion to Physics)

J. RICHARD CHRISTMAN
U.S. Coast Guard Academy

Provides self-tests for conceptual understanding and problem
solving.

SOLUTIONS MANUAL

EDWARD DERRINGH
Wentworth Institute of Technology

Provides approximately 25% of the solutions to textbook problems.

LABORATORY PHYSICS, SECOND EDITION

HARRY F. MEINERS
Rensselaer Polytechnic Institute
WALTER EPPENSTEIN
Rensselaer Polytechnic Institute
KENNETH MOORE
Rensselaer Polytechnic Institute
RALPH A. OLIVA
Texas Instruments, Inc.

This laboratory manual offers a clear introduction to procedures
and instrumentation, including errors, graphing, apparatus
handling, calculators, and computers, in addition to over 70
different experiments grouped by topic.

FOR THE INSTRUCTOR

A complete supplementary package of teaching and learning
materials is available for instructors. Contact your local Wiley
representative for further information.

VOLUME TWO
EXTENDED VERSION

PHYSICS

FOURTH EDITION

VOLUME TWO
EXTENDED VERSION

PHYSICS

FOURTH EDITION

DAVID HALLIDAY

Professor of Physics, Emeritus
University of Pittsburgh

ROBERT RESNICK

Professor of Physics
Rensselaer Polytechnic Institute

KENNETH S. KRANE

Professor of Physics
Oregon State University

JOHN WILEY & SONS, INC.

New York • *Chichester* • *Brisbane* • *Toronto* • *Singapore*

Acquisitions Editor   *Clifford Mills*
Marketing Manager   *Catherine Faduska*
Production Manager   *Joe Ford*
Production Supervisor   *Lucille Buonocore*
Manufacturing Manager   *Lorraine Fumoso*
Copy Editing Manager   *Deborah Herbert*
Photo Researcher   *Jennifer Atkins*
Photo Research Manager   *Stella Kupferberg*
Illustration   *John Balbalis*
Text Design   *Karin Gerdes Kincheloe*
Cover Design Direction   *Karin Gerdes Kincheloe*
Cover Design   *Lee Goldstein*
Cover Illustration   *Roy Wiemann*

Recognizing the importance of preserving what has been written, it is a policy of
John Wiley & Sons, Inc. to have books of enduring value published in the
United States printed on acid-free paper, and we exert our best efforts to that end.

Library of Congress Cataloging-in-Publication Data

Halliday, David, 1916–
    Physics. Part Two / David Halliday, Robert Resnick, Kenneth S. Krane. —
4th ed., extended version.
        p.   cm.
    Includes index.
    ISBN 0-471-54804-9
    1. Physics.   I. Resnick, Robert, 1923–   . II. Krane, Kenneth S. III. Title.
QC21.2.H355   1992b
530—dc20
                                                    92-24917
                                                    CIP

Printed and bound by Von Hoffmann Press, Inc.

10 9 8 7 6 5 4

PREFACE TO VOLUME 2, EXTENDED VERSION

The first edition of *Physics for Students of Science and Engineering* appeared in 1960; the most recent edition (the third), called simply *Physics,* was published in 1977. The present fourth edition (1992) marks the addition of a new coauthor for the text.

The text has been updated to include new developments in physics and in its pedagogy. Based in part on our reading of the literature on these subjects, in part on the comments from numerous users of past editions, and in part on the advice of a dedicated group of reviewers of the manuscript of this edition, we have made a number of changes.

1. This volume continues the coherent treatment of energy that began in Chapters 7 and 8 and continued through the treatment of thermodynamics in Volume 1. The sign conventions for work and the handling of energy (for instance, the elimination of ill-defined terms such as "thermal energy") are consistent throughout the text.

2. Special relativity, which was treated as a Supplementary Topic in the previous edition, is integrated throughout the text. Two chapters are devoted to special relativity: one (in Volume 1) follows mechanical waves and another (in Volume 2) follows electromagnetic waves. Topics related to special relativity (for instance, relative motion, frames of reference, momentum, and energy) are treated throughout the text in chapters on kinematics, mechanics, and electromagnetism. This approach reflects our view that special relativity should be treated as part of classical physics. However, for those instructors who wish to delay special relativity until the end of the course, the material is set off in separate sections that can easily be skipped on the first reading.

3. Changes in the ordering of topics from the third edition include introducing electric potential energy before electric potential, magnetic materials before inductance, and the Biot-Savart law before Ampère's law. The linear momentum carried by electromagnetic radiation has been moved from the chapter on light (42) to that on electromagnetic waves (41), and reflection by plane mirrors is now treated in the chapter on reflection and refraction at plane surfaces (43). The previous chapter on electromagnetic oscillations has been incorporated into the chapter on inductance (38).

4. Several topics have been eliminated, including rectifiers, filters, waveguides, transmission lines, and mutual inductance. We have also eliminated use of the electric displacement vector **D** and the magnetic field intensity **H**.

5. This extended version of Volume 2 includes eight chapters (49 to 56) that discuss quantum physics and some of its applications. A new chapter (56), introducing particle physics and cosmology, has been added to those in the previous extended version, and some shuffling of topics in the atomic physics chapters (49 to 51) has occurred. Other modern applications have been "sprinkled" throughout the text: for instance, the quantized Hall effect, magnetic fields of the planets, recent tests of charge conservation, superconductivity, magnetic monopoles, and holography.

6. We have substantially increased the number of end-of-chapter problems relative to the previous edition of the extended Volume 2: there are now 1486 problems compared with 1222 previously, an increase of 22 percent. The number of end-of-chapter questions has been similarly increased from 811 to 1027 (27%). We have tried to maintain the quality and diversity of problems that have been the hallmark of previous editions of this text.

7. The number of worked examples in Volume 2 averages between six and seven per chapter, about the same as the previous edition. However, the previous edition used the worked examples to present new material (such as parallel and series combinations of resistors or capacitors), which are presented in this edition as major subsections of the text rather than as worked examples. Because we now use the worked examples (here called sample problems) only to illustrate applications of material developed in the text, this edition actually offers students far more of such examples.

8. Computational techniques are introduced through several worked examples and through a variety of end-of-chapter computer projects. Some program listings are given in an appendix to encourage students to adapt those methods to other applications.

9. We have increased and updated the references to articles in the literature that appear as footnotes throughout the text. Some references (often to articles in popular magazines such as *Scientific American*) are intended to broaden the student's background through interesting applications of a topic. In other cases, often involving items of pedagogic importance to which we wish to call the attention of students as well as instructors, we make reference to articles in journals such as the *American Journal of Physics* or *The Physics Teacher*.

10. The illustrations have been completely redone and their number in the extended Volume 2 has been increased by 26%, from 664 to 835. We have added color to many of the drawings where the additional color enhances the clarity or the pedagogy.

11. Many of the derivations, proofs, and arguments of the previous edition have been tightened up, and any assumptions or approximations have been clarified. We have thereby improved the rigor of the text without necessarily raising its level. We are concerned about indicating to students the limit of validity of a particular argument and encouraging students to consider questions such as: Does a particular result apply always or only sometimes? What happens as we go toward the quantum or the relativistic limit?

Although we have made some efforts to eliminate material from the previous edition, the additions mentioned above contribute to a text of increasing length. *It should be emphasized that few (if any) instructors will want to follow the entire text from start to finish.* We have worked to develop a text that offers a rigorous and complete introduction to physics, but the instructor is able to follow many alternate pathways through the text. The instructor who wishes to treat fewer topics in greater depth (currently called the "less is more" approach) will be able to select from among these pathways. Some sections are explicitly labeled "optional" (and are printed in smaller type), indicating that they can be skipped without loss of continuity. Depending on the course design, other sections or even entire chapters can be skipped or treated lightly. The Instructor's Guide, available as a companion volume, offers suggestions for abbreviating the coverage. In such circumstances, the curious student who desires further study can be encouraged independently to approach the omitted topics, thereby gaining a broader view of the subject. The instructor is thus provided with a wide choice of which particular reduced set of topics to cover in a course of any given length. For instructors who wish a fuller coverage, such as in courses for physics majors or honors students or in courses of length greater than one year, this text provides the additional material needed for a challenging and comprehensive experience. We hope the text will be considered a road map through physics; many roads, scenic or direct, can be taken, and all roads need not be utilized on the first journey. The eager traveler may be encouraged to return to the map to explore areas missed on previous journeys.

The text is available as separate volumes: Volume 1 (Chapters 1 to 26) covers kinematics, mechanics, and thermodynamics, and Volume 2 (Chapters 27 to 48) covers electromagnetism and optics. An extended version of Volume 2 (Chapters 27 to 56) is available with eight additional chapters which present an introduction to quantum physics and some of its applications. The following supplements are available:

Study Guide	Solutions Manual
Laboratory Manual	Instructor's Guide

A textbook contains far more contributions to the elucidation of a subject than those made by the authors alone. We have been fortunate to have the assistance of Edward Derringh (Wentworth Institute of Technology) in preparing the problem sets and J. Richard Christman (U. S. Coast Guard Academy) in preparing the Instructor's Guide and the computer projects. We have benefited from the chapter-by-chapter comments and criticisms of a dedicated team of reviewers:

Robert P. Bauman (University of Alabama)
Truman D. Black (University of Texas, Arlington)
Edmond Brown (Rensselaer Polytechnic Institute)
J. Richard Christman (U. S. Coast Guard Academy)
Sumner Davis (University of California, Berkeley)
Roger Freedman (University of California, Santa Barbara)
James B. Gerhart (University of Washington)
Richard Thompson (University of Southern California)
David Wallach (Pennsylvania State University)
Roald K. Wangsness (University of Arizona)

We are deeply indebted to these individuals for their substantial contributions to this project.

We are grateful to the staff of John Wiley & Sons for their outstanding cooperation and support, including physics editor Cliff Mills, editorial program assistant Cathy Donovan, marketing manager Cathy Faduska, illustrator John Balbalis, editorial supervisor Deborah Herbert, designer Karin Kincheloe, production supervisor Lucille Buonocore, photo researcher Jennifer Atkins, and copy editor Christina Della Bartolomea. Word processing of the manuscript for this edition was superbly done by Christina Godfrey.

May 1992

DAVID HALLIDAY
Seattle, Washington

ROBERT RESNICK
Rensselaer Polytechnic Institute
Troy, New York 12180-3590

KENNETH S. KRANE
Oregon State University
Corvallis, Oregon 97331

CONTENTS

CHAPTER 27

ELECTRIC CHARGE AND COULOMB'S LAW

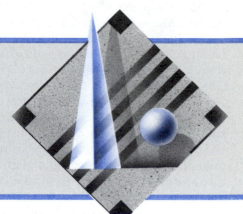

We begin here a detailed study of electromagnetism, *which will extend throughout most of the remainder of this text. Electromagnetic forces are responsible for the structure of atoms and for the binding of atoms in molecules and solids. Many properties of materials that we have studied so far are electromagnetic in their nature, such as the elasticity of solids and the surface tension of liquids. The spring force, friction, and the normal force all originate with the electromagnetic force between atoms.*

Among the examples of electromagnetism that we shall study are the force between electric charges, such as occurs between an electron and the nucleus in an atom; the motion of a charged body subject to an external electric force, such as an electron in an oscilloscope beam; the flow of electric charges through circuits and the behavior of circuit elements; the force between permanent magnets and the properties of magnetic materials; and electromagnetic radiation, which ultimately leads to the study of optics, *the nature and propagation of light.*

In this chapter, we begin with a discussion of electric charge, some properties of charged bodies, and the fundamental electric force between two charged bodies.

27-1 ELECTROMAGNETISM: A PREVIEW

The Greek philosophers, as early as 600 B.C., knew that if you rubbed a piece of amber it could pick up bits of straw. There is a direct line of development from this ancient observation to the electronic age in which we live. The strength of the connection is indicated by our word "electron," which is derived from the Greek word for amber.

The Greeks also knew that some naturally occurring "stones," which we know today as the mineral magnetite, would attract iron. From these modest origins grew the sciences of electricity and magnetism, which developed quite separately for centuries, until 1820 in fact, when Hans Christian Oersted found a connection between them: an *electric* current in a wire can deflect a *magnetic* compass needle. Oersted made this discovery while preparing a demonstration lecture for his physics students.

The new science of electromagnetism was developed further by Michael Faraday* (1791–1867), a truly gifted experimenter with a talent for physical intuition and visualization, whose collected laboratory notebooks do not

contain a single equation. James Clerk Maxwell† (1831–1879) put Faraday's ideas into mathematical form, introduced many new ideas of his own, and put electromagnetism on a sound theoretical basis. Maxwell's four equations (see Table 2 of Chapter 40) play the same role in electromagnetism as Newton's laws in classical mechanics or the laws of thermodynamics in the study of heat. We introduce and discuss Maxwell's equations individually in the chapters that follow.

Maxwell concluded that light is electromagnetic in nature and that its speed can be deduced from purely electric and magnetic measurements. Thus optics was intimately connected with electricity and magnetism. The scope of Maxwell's equations is remarkable, including the fundamental principles of all large-scale electromagnetic and optical devices such as motors, radio, television, microwave radar, microscopes, and telescopes.

---

* See "Michael Faraday," by Herbert Kondo, *Scientific American,* October 1953, p. 90. For the definitive biography, see L. Pearce Williams, *Michael Faraday* (Basic Books, 1964).
† See "James Clerk Maxwell," by James R. Newman, *Scientific American,* June 1955, p. 58.

The development of classical electromagnetism did not end with Maxwell. The English physicist Oliver Heaviside (1850–1925) and especially the Dutch physicist H. A. Lorentz (1853–1928) contributed substantially to the clarification of Maxwell's theory. Heinrich Hertz* (1857–1894) took a great step forward when, more than 20 years after Maxwell set up his theory, he produced in the laboratory electromagnetic "Maxwellian waves" of a kind that we would now call radio waves. Soon Marconi and others developed practical applications of the electromagnetic waves of Maxwell and Hertz. Albert Einstein based his relativity theory on Maxwell's equations; Einstein's 1905 paper introducing special relativity was called "On the Electrodynamics of Moving Bodies."

Present interest in electromagnetism takes two forms. On the applied or practical level, Maxwell's equations are used to study the electric and magnetic properties of new materials and to design electronic devices of increasing complexity and sophistication. On the most fundamental level, there have been efforts to combine or unify electromagnetism with the other basic forces of nature (see Section 6-1), just as the separate forces of electricity and magnetism were shown by Oersted, Faraday, and Maxwell to be part of the unified force of electromagnetism. Partial success was achieved in 1967 when Steven Weinberg and Abdus Salam independently proposed a theory, originally developed by Sheldon Glashow, that unified the electromagnetic interaction with the weak interaction, which is responsible for certain radioactive decay processes. Just as Maxwell's unification of electromagnetism gave predictions (namely, the existence of electromagnetic waves) that could be tested directly to verify the theory, the Glashow–Weinberg–Salam theory of the *electroweak* interaction gave unique predictions that could be tested experimentally. These tests have been done at high-energy particle accelerators and have verified the predictions of the electroweak theory. Glashow, Salam, and Weinberg shared the 1979 Nobel Prize for their development of this theory. Continuing theoretical efforts are underway to extend this unification to include the strong interaction, which binds nuclei together, and there are hopes eventually to include the gravitational force as well in this unification, so that one theoretical framework will include all the known fundamental interactions.

27-2 ELECTRIC CHARGE

If you walk across a carpet in dry weather, you can draw a spark by touching a metal door knob. On a grander scale, lightning is familiar to everyone. Such phenomena suggest the vast amount of *electric charge* that is stored in the familiar objects that surround us.

---

* See "Heinrich Hertz," by Philip and Emily Morrison, *Scientific American,* December 1957, p. 98.

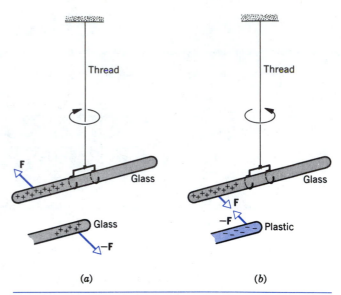

Figure 1 (*a*) Two similarly charged rods repel each other. (*b*) Two oppositely charged rods attract each other.

The electrical neutrality of most objects in our visible and tangible world conceals their content of enormous amounts of positive and negative electric charge that largely cancel each other in their external effects. Only when this electrical balance is disturbed does nature reveal to us the effects of uncompensated positive or negative charge. When we say that a body is "charged" we mean that it has a charge imbalance, even though the net charge generally represents only a tiny fraction of the total positive or negative charge contained in the body (see Sample Problem 2).

Charged bodies exert forces on each other. To show this, let us charge a glass rod by rubbing it with silk. The process of rubbing transfers a tiny amount of charge from one body to the other, thus slightly upsetting the electrical neutrality of each. If you suspend this charged rod from a thread, as in Fig. 1*a*, and if you bring a second charged glass rod nearby, the two rods repel each other. However, if you rub a plastic rod with fur it attracts the charged end of the hanging glass rod; see Fig. 1*b*.

We explain all this by saying there are two kinds of charge, one of which (the one on the glass rubbed with silk) we have come to call *positive* and the other (the one on the plastic rubbed with fur) we have come to call *negative*. These simple experiments can be summed up by saying:

Charges of the same sign repel each other, and charges of the opposite sign attract each other.

In Section 27-4, we put this rule into quantitative form, as Coulomb's law of force. We consider only charges that are either at rest with respect to each other or moving very slowly, a restriction that defines the subject of *electrostatics.*

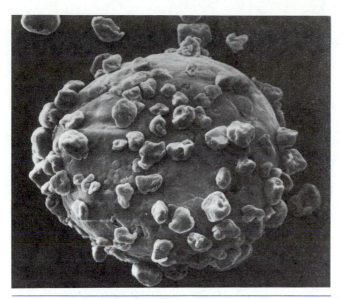

Figure 2 A carrier bead from a Xerox photocopier, covered with toner particles that stick to it by electrostatic attraction.

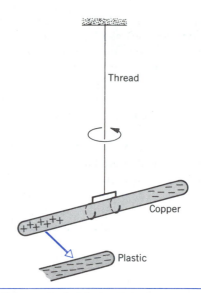

Figure 3 Either end of an isolated uncharged copper rod is attracted by a charged rod of either sign. In this case, conduction electrons in the copper rod are repelled to the far end of the copper rod, leaving the near end with a net positive charge.

The positive and negative labels for electric charge are due to Benjamin Franklin (1706–1790) who, among many other accomplishments, was a scientist of international reputation. It has even been said that Franklin's triumphs in diplomacy in France during the American War of Independence may have been made possible because he was so highly regarded as a scientist.

Electrical forces between charged bodies have many industrial applications, among them being electrostatic paint spraying and powder coating, fly-ash precipitation, nonimpact ink-jet printing, and photocopying. Figure 2, for example, shows a tiny carrier bead in a photocopying machine, covered with particles of black powder called *toner*, that stick to the carrier bead by electrostatic forces. These negatively charged toner particles are eventually attracted from their carrier beads to a positively charged latent image of the document to be copied, which is formed on a rotating drum. A charged sheet of paper then attracts the toner particles from the drum to itself, after which they are heat-fused in place to make the final copy.

27-3 CONDUCTORS AND INSULATORS

If you hold a copper rod, you cannot seem to charge it, no matter how hard you rub it or with what you rub it. However, if you fit the rod with a plastic handle, you are able to build up a charge. The explanation is that charge can flow easily through some materials, called *conductors*, of which copper is an example. In other materials, called *insulators*, charges do not flow under most circumstances; if you place charges on an insulator, such as most plastics, the charges stay where you put them. The copper

rod cannot be charged because any charges placed on it easily flow through the rod, through your body (which is also a conductor), and to the ground. The insulating handle, however, blocks the flow and allows charge to build up on the copper.

Glass, chemically pure water, and plastics are common examples of insulators. Although there are no perfect insulators, fused quartz is quite good — its insulating ability is about 10^{25} times that of copper.

Copper, metals in general, tap water, and the human body are common examples of conductors. In metals, an experiment called the *Hall effect* (see Section 34-4) shows that it is the negative charges (electrons) that are free to move. When copper atoms come together to form solid copper, their outer electrons do not remain attached to the individual atoms but become free to wander about within the rigid lattice structure formed by the positively charged ion cores. These mobile electrons are called *conduction electrons.* The positive charges in a copper rod are just as immobile as they are in a glass rod.

The experiment of Fig. 3 demonstrates the mobility of charge in a conductor. A negatively charged plastic rod attracts either end of a suspended but uncharged copper rod. The (mobile) conduction electrons in the copper rod are repelled by the negative charge on the plastic rod and move to the far end of the copper rod, leaving the near end of the copper rod with a net positive charge. A positively charged glass rod also attracts an uncharged copper rod. In this case, the conduction electrons in the copper are attracted by the positively charged glass rod to the near end of the copper rod; the far end of the copper rod is then left with a net positive charge.

This distinction between conductors and insulators becomes more quantitative when we consider the number of conduction electrons available in a given amount of material. In a typical conductor, each atom may contribute one conduction electron, and therefore there might be on the average about 10^{23} conduction electrons per cm³. In an insulator at room temperature, on the other hand, we are on the average unlikely to find even 1 conduction electron per cm³.

Intermediate between conductors and insulators are the *semiconductors* such as silicon or germanium; a typical semiconductor might contain $10^{10} - 10^{12}$ conduction electrons per cm³. One of the properties of semiconductors that makes them so useful is that the density of conduction electrons can be changed drastically by small changes in the conditions of the material, such as by introducing small quantities (less than 1 part in 10^9) of impurities or by varying the applied voltage, the temperature, or the intensity of light incident on the material.

In Chapter 32 we consider electrical conduction in various materials in more detail, and Chapter 53 of the extended text shows how quantum theory leads to a more complete understanding of electrical conduction.

27-4 COULOMB'S LAW

Charles Augustin Coulomb (1736–1806) measured electrical attractions and repulsions quantitatively and deduced the law that governs them. His apparatus, shown in Fig. 4, resembles the hanging rod of Fig. 1, except that the charges in Fig. 4 are confined to small spheres a and b.

If a and b are charged, the electric force on a tends to twist the suspension fiber. Coulomb cancelled out this twisting effect by turning the suspension head through the angle θ needed to keep the two charges at a particular separation. The angle θ is then a relative measure of the electric force acting on charge a. The device of Fig. 4 is a *torsion balance*; a similar arrangement was used later by Cavendish to measure gravitational attractions (Section 16-3).

Experiments due to Coulomb and his contemporaries showed that the electrical force exerted by one charged body on another depends directly on the product of the magnitudes of the two charges and inversely on the square of their separation.* That is,

$$F \propto \frac{q_1 q_2}{r^2} .$$

---

* In his analysis, Coulomb failed to take into account the movement of the charges on one sphere due to the other nearby charged sphere, an effect similar to that illustrated in Fig. 3. For a discussion of this point, see "Precise Calculation of the Electrostatic Force Between Charged Spheres Including Induction Effects," by Jack A. Soules, *American Journal of Physics,* December 1990, p. 1195.

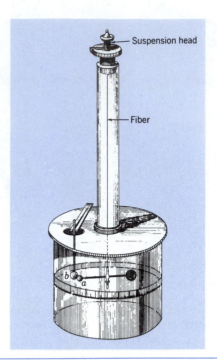

Figure 4 Coulomb's torsion balance, from his 1785 memoir to the Paris Academy of Sciences.

Here F is the magnitude of the mutual force that acts on each of the two charges a and b, q_1 and q_2 are relative measures of the charges on spheres a and b, and r is the distance between their centers. The force on each charge due to the other acts along the line connecting the charges. The two forces point in opposite directions but have equal magnitudes, even though the charges may be different.

To turn the above proportionality into an equation, let us introduce a constant of proportionality, which we represent for now as k. We thus obtain, for the force between the charges,

$$F = k \frac{q_1 q_2}{r^2} . \tag{1}$$

Equation 1, which is called *Coulomb's law,* generally holds only for charged objects whose sizes are much smaller than the distance between them. We often say that it holds only for *point charges.*†

Our belief in Coulomb's law does not rest quantitatively on Coulomb's experiments. Torsion balance measurements are difficult to make to an accuracy of better than a few percent. Such measurements could not, for example, convince us that the exponent of r in Eq. 1 is

---

† Strictly speaking, Eq. 1 should be written in terms of the absolute magnitudes of q_1 and q_2, and F then gives the magnitude of the force. The direction of the force is determined by whether the charges are of the same sign or the opposite sign. For now we ignore this detail, which will become important later in this section when we write Eq. 1 in vector form.

exactly 2 and not, say, 2.01. In Section 29-6 we show that Coulomb's law can also be deduced from an indirect experiment, which shows that, if the exponent in Eq. 1 is not exactly 2, it differs from 2 by at most 1×10^{-16}.

Coulomb's law resembles Newton's inverse square law of gravitation, $F = Gm_1m_2/r^2$, which was already more than 100 years old at the time of Coulomb's experiments. Both are inverse square laws, and the charge q plays the same role in Coulomb's law that the mass m plays in Newton's law of gravitation. One difference between the two laws is that gravitational forces, as far as we know, are always attractive, while electrostatic forces can be repulsive or attractive, depending on whether the two charges have the same or opposite signs.

There is another important difference between the two laws. In using the law of gravitation, we were able to define mass from Newton's second law, $F = ma$, and then by applying the law of gravitation to known masses we could determine the constant G. In using Coulomb's law, we take the reverse approach: we *define* the constant k to have a particular value, and we then use Coulomb's law to determine the basic unit of electric charge as the quantity of charge that produces a standard unit of force.

For example, consider the force between two equal charges of magnitude q. We could adjust q until the force has a particular value, say 1 N for a separation of $r = 1$ m, and define the resulting q as the basic unit of charge. It is, however, more precise to measure the magnetic force between two wires carrying equal currents, and therefore the fundamental SI electrical unit is the unit of current, from which the unit of charge is derived. The operational procedure for defining the SI unit of current, which is called the *ampere* (abbreviation A), is discussed in Section 35-4.

The SI unit of charge is the *coulomb* (abbreviation C), which is defined as *the amount of charge that flows in 1 second when there is a steady current of 1 ampere.* That is,

$$dq = i \, dt, \tag{2}$$

where dq (in coulombs) is the charge transferred by a current i (in amperes) during the interval dt (in seconds). For example, a wire carrying a steady current of 2 A delivers a charge of 2×10^{-6} C in a time of 10^{-6} s.

In the SI system, the constant k is expressed in the following form:

$$k = \frac{1}{4\pi\epsilon_0}. \tag{3}$$

Although the choice of this form for the constant k appears to make Coulomb's law needlessly complex, it ultimately results in a simplification of formulas of electromagnetism that are used more often than Coulomb's law.

The constant ϵ_0, which is called the *permittivity constant*, has a value that is determined by the adopted value of the speed of light, as we discuss in Chapter 41. Its value is

$$\epsilon_0 = 8.85418781762 \times 10^{-12} \ \text{C}^2/\text{N} \cdot \text{m}^2.$$

The constant k has the corresponding value (to three significant figures)

$$k = \frac{1}{4\pi\epsilon_0} = 8.99 \times 10^9 \ \text{N} \cdot \text{m}^2/\text{C}^2.$$

With this choice of the constant k, Coulomb's law can be written

$$F = \frac{1}{4\pi\epsilon_0} \frac{q_1 q_2}{r^2}. \tag{4}$$

When k has the above value, expressing q in coulombs and r in meters gives the force in newtons.

Coulomb's Law: Vector Form

So far we have considered only the magnitude of the force between two charges determined according to Coulomb's law. Force, being a vector, has directional properties as well. In the case of Coulomb's law, the direction of the force is determined by the relative sign of the two electric charges.

As illustrated in Fig. 5, suppose we have two point charges q_1 and q_2 separated by a distance r_{12}. For the moment, we assume the two charges to have the same sign, so that they repel one another. Let us consider the force *on* particle 1 exerted *by* particle 2, which we write in our usual form as $\mathbf{F}_{12}$. The position vector that locates particle 1 relative to particle 2 is $\mathbf{r}_{12}$; that is, if we were to define the origin of our coordinate system at the location of particle 2, then $\mathbf{r}_{12}$ would be the position vector of particle 1.

If the two charges have the same sign, then the force is repulsive and, as shown in Fig. 5a, $\mathbf{F}_{12}$ must be parallel to $\mathbf{r}_{12}$. If the charges have opposite signs, as in Fig. 5b, then

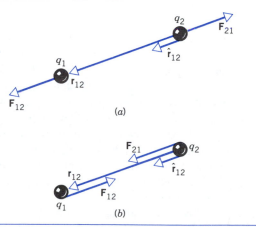

Figure 5   (a) Two point charges q_1 and q_2 of the same sign exert equal and opposite repulsive forces on one another. The vector $\mathbf{r}_{12}$ locates q_1 relative to q_2, and the unit vector $\hat{\mathbf{r}}_{12}$ points in the direction of $\mathbf{r}_{12}$. Note that $\mathbf{F}_{12}$ is parallel to $\mathbf{r}_{12}$. (b) The two charges now have opposite signs, and the force is attractive. Note that $\mathbf{F}_{12}$ is antiparallel to $\mathbf{r}_{12}$.

the force $\mathbf{F}_{12}$ is attractive and antiparallel to $\mathbf{r}_{12}$. In either case, we can represent the force as

$$\mathbf{F}_{12} = \frac{1}{4\pi\epsilon_0} \frac{q_1 q_2}{r_{12}^2} \hat{\mathbf{r}}_{12}. \tag{5}$$

Here r_{12} represents the magnitude of the vector $\mathbf{r}_{12}$, and $\hat{\mathbf{r}}_{12}$ indicates the *unit vector* in the direction of $\mathbf{r}_{12}$. That is,

$$\hat{\mathbf{r}}_{12} = \frac{\mathbf{r}_{12}}{r_{12}}. \tag{6}$$

We used a form similar to Eq. 5 to express the gravitational force (see Eqs. 2a and 2b of Chapter 16).

One other feature is apparent from Fig. 5. According to Newton's third law, the force exerted *on* particle 2 *by* particle 1, $\mathbf{F}_{21}$, is opposite to $\mathbf{F}_{12}$. This force can then be expressed in exactly the same form:

$$\mathbf{F}_{21} = \frac{1}{4\pi\epsilon_0} \frac{q_1 q_2}{r_{21}^2} \hat{\mathbf{r}}_{21}. \tag{7}$$

Here $\hat{\mathbf{r}}_{21}$ is a unit vector that points from particle 1 to particle 2; that is, it would be the unit vector in the direction of particle 2 if the origin of coordinates were at the location of particle 1.

The vector form of Coulomb's law is useful because it carries within it the directional information about $\mathbf{F}$ and whether the force is attractive or repulsive. Using the vector form is of critical importance when we consider the forces acting on an assembly of more than two charges. In this case, Eq. 5 would hold for every pair of charges, and the total force on any one charge would be found by taking the *vector* sum of the forces due to each of the other charges. For example, the force on particle 1 in an assembly would be

$$\mathbf{F}_1 = \mathbf{F}_{12} + \mathbf{F}_{13} + \mathbf{F}_{14} + \cdots, \tag{8}$$

where $\mathbf{F}_{12}$ is the force on particle 1 from particle 2, $\mathbf{F}_{13}$ is the force on particle 1 from particle 3, and so on. Equation 8 is the mathematical representation of the *principle of superposition* applied to electric forces. It permits us to calculate the force due to any pair of charges as if the other charges were not present. For instance, the force $\mathbf{F}_{13}$ that particle 3 exerts on particle 1 is completely unaffected by the presence of particle 2. The principle of superposition is not at all obvious and does not hold in many situations, particularly in the case of very strong electric forces. Only through experiment can its applicability be verified. For all situations we meet in this text, however, the principle of superposition is valid.

The significance of Coulomb's law goes far beyond the description of the forces acting between charged spheres. This law, when incorporated into the structure of quantum physics, correctly describes (1) the electrical forces that bind the electrons of an atom to its nucleus, (2) the forces that bind atoms together to form molecules, and (3) the forces that bind atoms and molecules together to form solids or liquids. Thus most of the forces of our daily

experience that are not gravitational in nature are electrical. Moreover, unlike Newton's law of gravitation, which can be considered a useful everyday approximation of the more basic general theory of relativity, Coulomb's law is an exact result for stationary charges and not an approximation from some higher law. It holds not only for ordinary objects, but also for the most fundamental "point" particles such as electrons and quarks. Coulomb's law remains valid in the quantum limit (for example, in calculating the electrostatic force between the proton and the electron in an atom of hydrogen). When charged particles move at speeds close to the speed of light, such as in a high-energy accelerator, Coulomb's law does not give a complete description of their electromagnetic interactions; instead, a more complete analysis based on Maxwell's equations must be done.

Sample Problem 1   Figure 6 shows three charged particles, held in place by forces not shown. What electrostatic force, owing to the other two charges, acts on q_1? Take $q_1 = -1.2\ \mu C$, $q_2 = +3.7\ \mu C$, $q_3 = -2.3\ \mu C$, $r_{12} = 15$ cm, $r_{13} = 10$ cm, and $\theta = 32°$.

Solution   This problem calls for the use of the superposition principle. We start by computing the magnitudes of the forces that q_2 and q_3 exert on q_1. We substitute the magnitudes of the charges into Eq. 5, disregarding their signs for the time being. We then have

$$F_{12} = \frac{1}{4\pi\epsilon_0} \frac{q_1 q_2}{r_{12}^2}$$

$$= \frac{(8.99 \times 10^9\ \text{N} \cdot \text{m}^2/\text{C}^2)(1.2 \times 10^{-6}\ \text{C})(3.7 \times 10^{-6}\ \text{C})}{(0.15\ \text{m})^2}$$

$$= 1.77\ \text{N}.$$

The charges q_1 and q_2 have opposite signs so that the force between them is attractive. Hence $\mathbf{F}_{12}$ points to the right in Fig. 6. We also have

$$F_{13} = \frac{(8.99 \times 10^9\ \text{N} \cdot \text{m}^2/\text{C}^2)(1.2 \times 10^{-6}\ \text{C})(2.3 \times 10^{-6}\ \text{C})}{(0.10\ \text{m})^2}$$

$$= 2.48\ \text{N}.$$

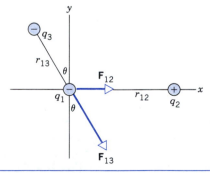

Figure 6   Sample Problem 1. The three charges exert three pairs of action–reaction forces on each other. Only the two forces acting on q_1 are shown here.

These two charges have the same (negative) sign so that the force between them is repulsive. Thus $\mathbf{F}_{13}$ points as shown in Fig. 6.

The components of the resultant force $\mathbf{F}_1$ acting on q_1 are determined by the corresponding components of Eq. 8, or

$$F_{1x} = F_{12x} + F_{13x} = F_{12} + F_{13}\sin\theta$$
$$= 1.77 \text{ N} + (2.48 \text{ N})(\sin 32°) = 3.08 \text{ N}$$

and

$$F_{1y} = F_{12y} + F_{13y} = 0 - F_{13}\cos\theta$$
$$= -(2.48 \text{ N})(\cos 32°) = -2.10 \text{ N}.$$

From these components, you can show that the magnitude of $\mathbf{F}_1$ is 3.73 N and that this vector makes an angle of $-34°$ with the x axis.

27-5 CHARGE IS QUANTIZED

In Franklin's day, electric charge was thought to be a continuous fluid, an idea that was useful for many purposes. However, we now know that fluids themselves, such as air or water, are not continuous but are made up of atoms and molecules; matter is discrete. Experiment shows that the "electrical fluid" is not continuous either but that it is made up of multiples of a certain elementary charge. That is, any charge q that can be observed and measured directly can be written

$$q = ne \qquad n = 0, \pm1, \pm2, \pm3, \ldots, \qquad (9)$$

in which e, the unit of *elementary charge*, has the experimentally determined value

$$e = 1.60217733 \times 10^{-19} \text{ C},$$

with an experimental uncertainty of about 3 parts in 10^7. The elementary charge is one of the fundamental constants of nature.

When a physical quantity such as charge exists only in discrete "packets" rather than in continuously variable amounts, we say that quantity is *quantized*. We have already seen that matter, energy, and angular momentum are quantized; charge adds one more important physical quantity to the list. Equation 9 tells us that it is possible, for example, to find a particle that carries a charge of zero, $+10e$, or $-6e$, but it is not possible to find a particle with a charge of, say, $3.57e$. Table 1 shows the charges and some other properties of the three particles that can be said to make up the material world around us.

The quantum of charge is small. For example, about 10^{19} elementary charges enter an ordinary 100-W, 120-V light bulb every second, and an equal number leave it. The graininess of electricity does not show up in large-scale phenomena, just as you cannot feel the individual molecules of water when you move your hand through it.

Since 1964, physicists have used a theory of the elementary particles according to which particles such as the proton and neutron are considered to be composite parti-

TABLE 1 SOME PROPERTIES OF THREE PARTICLES

Particle	Symbol[a]	Charge[b]	Mass[c]	Angular Momentum[d]
Electron	e^-	-1	1	$\frac{1}{2}$
Proton	p	$+1$	1836.15	$\frac{1}{2}$
Neutron	n	0	1838.68	$\frac{1}{2}$

[a] Each of the particles has an *antiparticle* with the same mass and angular momentum but the opposite charge. The antiparticles are indicated by the symbols e^+ (positive electron or positron), $\bar{p}$ (antiproton), and $\bar{n}$ (antineutron).
[b] In units of the elementary charge e.
[c] In units of the electron mass m_e.
[d] The intrinsic spin angular momentum, in units of $h/2\pi$. We introduced this concept in Section 13-6, and we give a more complete treatment in Chapter 51 of the extended version of this book.

cles made up of more fundamental units called *quarks*. An unusual feature of this theory is that the quarks are assigned fractional electric charges of $+\frac{2}{3}e$ and $-\frac{1}{3}e$. Protons and neutrons are each made up of three quarks. The proton, with its charge of $+e$, must be composed of two quarks each of charge $+\frac{2}{3}e$ and one quark of charge $-\frac{1}{3}e$. The neutron, with its net charge of 0, must include two quarks each of charge $-\frac{1}{3}e$ and one quark of charge $+\frac{2}{3}e$. Although there is firm experimental evidence for the existence of quarks within the proton and neutron, collisions involving protons or neutrons at the highest energies available in accelerators have so far failed to show evidence for the release of a free quark. Perhaps the quarks are bound so strongly in protons and neutrons that the available energy is unable to liberate one. Alternatively, it has been suggested that quarks may be required by laws governing their behavior to exist only in combinations that give electrical charges in units of e. The explanation for the failure to observe free quarks is not yet clear.

No theory has yet been developed that permits us to calculate the charge of the electron. Nor is there any definitive theory that explains why the fundamental negative charge (the electron) is exactly equal in magnitude to the fundamental positive charge (the proton). At present, we must regard the fundamental "quantum" of electric charge as a basic property of nature subject to precise measurement but whose ultimate significance is as yet beyond us.

Sample Problem 2 A penny, being electrically neutral, contains equal amounts of positive and negative charge. What is the magnitude of these equal charges?

Solution The charge q is given by NZe, in which N is the number of atoms in a penny and Ze is the magnitude of the positive and the negative charges carried by each atom.

The number N of atoms in a penny, assumed for simplicity to be made of copper, is $N_A m/M$, in which N_A is the Avogadro constant. The mass m of the coin is 3.11 g, and the mass M of 1 mol of copper (called its *molar mass*) is 63.5 g. We find

$$N = \frac{N_A m}{M} = \frac{(6.02 \times 10^{23} \text{ atoms/mol})(3.11 \text{ g})}{63.5 \text{ g/mol}}$$

$$= 2.95 \times 10^{22} \text{ atoms.}$$

Every neutral atom has a negative charge of magnitude Ze associated with its electrons and a positive charge of the same magnitude associated with its nucleus. Here e is the magnitude of the charge on the electron, which is 1.60×10^{-19} C, and Z is the atomic number of the element in question. For copper, Z is 29. The magnitude of the total negative or positive charge in a penny is then

$$q = NZe$$

$$= (2.95 \times 10^{22})(29)(1.60 \times 10^{-19} \text{ C})$$

$$= 1.37 \times 10^5 \text{ C.}$$

This is an enormous charge. By comparison, the charge that you might get by rubbing a plastic rod is perhaps 10^{-9} C, smaller by a factor of about 10^{14}. For another comparison, it would take about 38 h for a charge of 1.37×10^5 C to flow through the filament of a 100-W, 120-V light bulb. There is a lot of electric charge in ordinary matter.

Sample Problem 3   In Sample Problem 2 we saw that a copper penny contains both positive and negative charges, each of a magnitude 1.37×10^5 C. Suppose that these charges could be concentrated into two separate bundles, held 100 m apart. What attractive force would act on each bundle?

Solution   From Eq. 4 we have

$$F = \frac{1}{4\pi\epsilon_0} \frac{q^2}{r^2} = \frac{(8.99 \times 10^9 \text{ N} \cdot \text{m}^2/\text{C}^2)(1.37 \times 10^5 \text{ C})^2}{(100 \text{ m})^2}$$

$$= 1.69 \times 10^{16} \text{ N.}$$

This is about 2×10^{12} tons of force! Even if the charges were separated by one Earth diameter, the attractive force would still be about 120 tons. In all of this, we have sidestepped the problem of forming each of the separated charges into a "bundle" whose dimensions are small compared to their separation. Such bundles, if they could ever be formed, would be blasted apart by mutual Coulomb repulsion forces.

The lesson of this sample problem is that you cannot disturb the electrical neutrality of ordinary matter very much. If you try to pull out any sizable fraction of the charge contained in a body, a large Coulomb force appears automatically, tending to pull it back.

Sample Problem 4   The average distance r between the electron and the proton in the hydrogen atom is 5.3×10^{-11} m. (*a*) What is the magnitude of the average electrostatic force that acts between these two particles? (*b*) What is the magnitude of the average gravitational force that acts between these particles?

Solution   (*a*) From Eq. 4 we have, for the electrostatic force,

$$F_e = \frac{1}{4\pi\epsilon_0} \frac{q_1 q_2}{r^2} = \frac{(8.99 \times 10^9 \text{ N} \cdot \text{m}^2/\text{C}^2)(1.60 \times 10^{-19} \text{ C})^2}{(5.3 \times 10^{-11} \text{ m})^2}$$

$$= 8.2 \times 10^{-8} \text{ N.}$$

While this force may seem small (it is about equal to the weight of a speck of dust), it produces an immense effect, namely, the acceleration of the electron within the atom.

(*b*) For the gravitational force, we have

$$F_g = G \frac{m_e m_p}{r^2}$$

$$= \frac{(6.67 \times 10^{-11} \text{ N} \cdot \text{m}^2/\text{kg}^2)(9.11 \times 10^{-31} \text{ kg})(1.67 \times 10^{-27} \text{ kg})}{(5.3 \times 10^{-11} \text{ m})^2}$$

$$= 3.6 \times 10^{-47} \text{ N.}$$

We see that the gravitational force is weaker than the electrostatic force by the enormous factor of about 10^{39}. Although the gravitational force is weak, it is always attractive. Thus it can act to build up very large masses, as in the formation of stars and planets, so that large gravitational forces can develop. The electrostatic force, on the other hand, is repulsive for charges of the same sign, so that it is not possible to accumulate large concentrations of either positive or negative charge. We must always have the two together, so that they largely compensate for each other. The charges that we are accustomed to in our daily experiences are slight disturbances of this overriding balance.

Sample Problem 5   The nucleus of an iron atom has a radius of about 4×10^{-15} m and contains 26 protons. What repulsive electrostatic force acts between two protons in such a nucleus if they are separated by a distance of one radius?

Solution   From Eq. 4 we have

$$F = \frac{1}{4\pi\epsilon_0} \frac{q_p q_p}{r^2}$$

$$= \frac{(8.99 \times 10^9 \text{ N} \cdot \text{m}^2/\text{C}^2)(1.60 \times 10^{-19} \text{ C})^2}{(4 \times 10^{-15} \text{ m})^2}$$

$$= 14 \text{ N.}$$

This enormous force, more than 3 lb and acting on a single proton, must be more than balanced by the attractive nuclear force that binds the nucleus together. This force, whose range is so short that its effects cannot be felt very far outside the nucleus, is known as the "strong nuclear force" and is very well named.

27-6   CHARGE IS CONSERVED

When a glass rod is rubbed with silk, a positive charge appears on the rod. Measurement shows that a corresponding negative charge appears on the silk. This suggests that rubbing does not create charge but merely transfers it from one object to another, disturbing slightly the electrical neutrality of each. This hypothesis of the *conservation of charge* has stood up under close experimental scrutiny both for large-scale events and at the atomic and nuclear level; no exceptions have ever been found.

An interesting example of charge conservation comes about when an electron (charge $= -e$) and a positron (charge $= +e$) are brought close to each other. The two particles may simply disappear, converting all their rest energy into radiant energy. The radiant energy may appear in the form of two oppositely directed gamma rays of total energy $2m_ec^2$; thus

$$e^- + e^+ \rightarrow \gamma + \gamma.$$

The net charge is zero both before and after the event, and charge is conserved.

Certain uncharged particles, such as the neutral π meson, are permitted to decay electromagnetically into two gamma rays:

$$\pi^0 \rightarrow \gamma + \gamma.$$

This decay conserves charge, the total charge again being 0 before and after the decay. For another example, a neutron ($q = 0$) decays into a proton ($q = +e$) and an electron ($q = -e$) plus another neutral particle, a neutrino ($q = 0$). The total charge is zero, both before and after the decay, and charge is conserved. Experiments have been done to search for decays of the neutron into a proton with no electron emitted, which would violate charge conservation. No such events have been found, and the upper limit for their occurrence, relative to the charge-conserving decays, is 10^{-23}.

The decay of an electron ($q = -e$) into neutral particles, such as gamma rays (γ) or neutrinos (ν) is forbidden;

for example,

$$e^- \not\rightarrow \gamma + \nu,$$

because that decay would violate charge conservation. Attempts to observe this decay have likewise been unsuccessful, indicating that, if the decay does occur, the electron must have a lifetime of at least 10^{22} years!

Another example of charge conservation is found in the fusion of two deuterium nuclei ^{2}H (called "heavy hydrogen") to make helium. Among the possible reactions are

$$^2\text{H} + {}^2\text{H} \rightarrow {}^3\text{H} + \text{p},$$

$$^2\text{H} + {}^2\text{H} \rightarrow {}^3\text{He} + \text{n}.$$

The deuterium nucleus contains one proton and one neutron and therefore has a charge of $+e$. The nucleus of the isotope of hydrogen with mass 3, written ^{3}H and known as *tritium*, contains one proton and two neutrons, and thus also has a charge of $+e$. The first reaction therefore has a net charge of $+2e$ on each side and conserves charge. In the second reaction, the neutron is uncharged, while the nucleus of the isotope of helium with mass 3 contains two protons and one neutron and therefore has a charge of $+2e$. The second reaction thus also conserves charge. Conservation of charge explains why we never see a proton emitted when the second reaction takes place or a neutron when the first occurs.

QUESTIONS

1. You are given two metal spheres mounted on portable insulating supports. Find a way to give them equal and opposite charges. You may use a glass rod rubbed with silk but may not touch it to the spheres. Do the spheres have to be of equal size for your method to work?

2. In Question 1, find a way to give the spheres equal charges of the same sign. Again, do the spheres need to be of equal size for your method to work?

3. A charged rod attracts bits of dry cork dust, which, after touching the rod, often jump violently away from it. Explain.

4. The experiments described in Section 27-2 could be explained by postulating four kinds of charge, that is, on glass, silk, plastic, and fur. What is the argument against this?

5. A positive charge is brought very near to an uncharged insulated conductor. The conductor is grounded while the charge is kept near. Is the conductor charged positively or negatively or not at all if (a) the charge is taken away and then the ground connection is removed and (b) the ground connection is removed and then the charge is taken away?

6. A charged insulator can be discharged by passing it just above a flame. Explain how.

7. If you rub a coin briskly between your fingers, it will not seem to become charged by friction. Why?

8. If you walk briskly across a carpet, you often experience a spark upon touching a door knob. (a) What causes this? (b) How might it be prevented?

9. Why do electrostatic experiments not work well on humid days?

10. Why is it recommended that you touch the metal frame of your personal computer before installing any internal accessories?

11. An insulated rod is said to carry an electric charge. How could you verify this and determine the sign of the charge?

12. If a charged glass rod is held near one end of an insulated uncharged metal rod as in Fig. 7, electrons are drawn to one end, as shown. Why does the flow of electrons cease? After all, there is an almost inexhaustible supply of them in the metal rod.

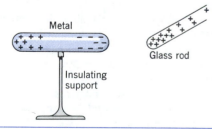

Figure 7 Questions 12 and 13.

13. In Fig. 7, does any resultant electric force act on the metal rod? Explain.

14. A person standing on an insulating stool touches a charged,

insulated conductor. Is the conductor discharged completely?

15. (a) A positively charged glass rod attracts a suspended object. Can we conclude that the object is negatively charged? (b) A positively charged glass rod repels a suspended object. Can we conclude that the object is positively charged?

16. Explain what is meant by the statement that electrostatic forces obey the principle of superposition.

17. Is the electric force that one charge exerts on another changed if other charges are brought nearby?

18. A solution of copper sulfate is a conductor. What particles serve as the charge carriers in this case?

19. If the electrons in a metal such as copper are free to move about, they must often find themselves headed toward the metal surface. Why don't they keep on going and leave the metal?

20. Would it have made any important difference if Benjamin Franklin had chosen, in effect, to call electrons positive and protons negative?

21. Coulomb's law predicts that the force exerted by one point charge on another is proportional to the product of the two charges. How might you go about testing this aspect of the law in the laboratory?

22. Explain how an atomic nucleus can be stable if it is composed of particles that are either neutral (neutrons) or carry like charges (protons).

23. An electron (charge $= -e$) circulates around a helium nucleus (charge $= +2e$) in a helium atom. Which particle exerts the larger force on the other?

24. The charge of a particle is a true characteristic of the particle, independent of its state of motion. Explain how you can test this statement by making a rigorous experimental check of whether the hydrogen atom is truly electrically neutral.

25. Earnshaw's theorem says that no particle can be in stable equilibrium under the action of electrostatic forces alone. Consider, however, point P at the center of a square of four equal positive charges, as in Fig. 8. If you put a positive test charge there it might seem to be in stable equilibrium. Every one of the four external charges pushes it toward P. Yet Earnshaw's theorem holds. Can you explain how?

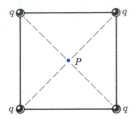

Figure 8 Question 25.

26. The quantum of charge is 1.60×10^{-19} C. Is there a corresponding quantum of mass?

27. What does it mean to say that a physical quantity is (a) quantized or (b) conserved? Give some examples.

28. In Sample Problem 4 we show that the electrical force is about 10^{39} times stronger than the gravitational force. Can you conclude from this that a galaxy, a star, or a planet must be essentially neutral electrically?

29. How do we know that electrostatic forces are not the cause of gravitational attraction, between the Earth and Moon, for example?

PROBLEMS

Section 27-4 Coulomb's Law

1. A point charge of $+3.12 \times 10^{-6}$ C is 12.3 cm distant from a second point charge of -1.48×10^{-6} C. Calculate the magnitude of the force on each charge.

2. What must be the distance between point charge $q_1 = 26.3 \mu$C and point charge $q_2 = -47.1 \mu$C in order that the attractive electrical force between them has a magnitude of 5.66 N?

3. In the return stroke of a typical lightning bolt (see Fig. 9), a current of 2.5×10^4 A flows for 20 μs. How much charge is transferred in this event?

4. Two equally charged particles, held 3.20 mm apart, are released from rest. The initial acceleration of the first particle is observed to be 7.22 m/s^2 and that of the second to be 9.16 m/s^2. The mass of the first particle is 6.31×10^{-7} kg. Find (a) the mass of the second particle and (b) the magnitude of the common charge.

5. Figure 10a shows two charges, q_1 and q_2, held a fixed distance d apart. (a) Find the strength of the electric force that acts on q_1. Assume that $q_1 = q_2 = 21.3 \mu$C and $d = 1.52$ m. (b) A third charge $q_3 = 21.3 \mu$C is brought in and placed as shown in Fig. 10b. Find the strength of the electric force on q_1 now.

Figure 9 Problem 3.

6. Two identical conducting spheres, ① and ②, carry equal amounts of charge and are fixed a distance apart large compared with their diameters. They repel each other with an electrical force of 88 mN. Suppose now that a third identical

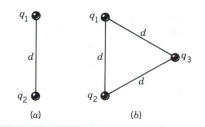

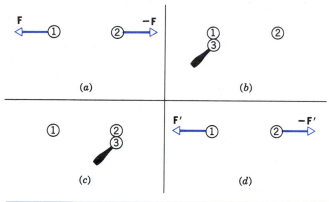

Figure 10   Problem 5.

Figure 11   Problem 6.

sphere ③, having an insulating handle and initially uncharged, is touched first to sphere ①, then to sphere ②, and finally removed. Find the force between spheres ① and ② now. See Fig. 11.

7. Three charged particles lie on a straight line and are separated by a distance d as shown in Fig. 12. Charges q_1 and q_2 are held fixed. Charge q_3, which is free to move, is found to be in equilibrium under the action of the electric forces. Find q_1 in terms of q_2.

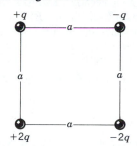

Figure 12   Problem 7.

8. In Fig. 13, find (a) the horizontal and (b) the vertical components of the resultant electric force on the charge in the lower left corner of the square. Assume that $q = 1.13$ μC and $a = 15.2$ cm. The charges are at rest.

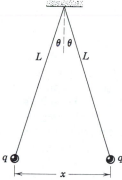

Figure 13   Problem 8.

9. Two positive charges, each 4.18 μC, and a negative charge, -6.36 μC, are fixed at the vertices of an equilateral triangle of side 13.0 cm. Find the electrical force on the negative charge.

10. Each of two small spheres is charged positively, the total charge being 52.6 μC. Each sphere is repelled from the other with a force of 1.19 N when the spheres are 1.94 m apart. Calculate the charge on each sphere.

11. Two identical conducting spheres, having charges of opposite sign, attract each other with a force of 0.108 N when separated by 50.0 cm. The spheres are suddenly connected by a thin conducting wire, which is then removed, and thereafter the spheres repel each other with a force of 0.0360 N. What were the initial charges on the spheres?

12. Two fixed charges, $+1.07$ μC and -3.28 μC, are 61.8 cm apart. Where may a third charge be located so that no net force acts on it?

13. Two *free* point charges $+q$ and $+4q$ are a distance L apart. A third charge is so placed that the entire system is in equilibrium. (a) Find the sign, magnitude, and location of the third charge. (b) Show that the equilibrium is unstable.

14. A charge Q is fixed at each of two opposite corners of a square. A charge q is placed at each of the other two corners. (a) If the resultant electrical force on Q is zero, how are Q and q related? (b) Could q be chosen to make the resultant electrical force on *every* charge zero? Explain your answer.

15. A certain charge Q is to be divided into two parts $(Q - q)$ and q. What is the relation of Q to q if the two parts, placed a given distance apart, are to have a maximum Coulomb repulsion?

16. Two similar tiny balls of mass m are hung from silk threads of length L and carry equal charges q as in Fig. 14. Assume that θ is so small that tan θ can be replaced by its approximate equal, sin θ. (a) To this approximation show that, for equilibrium,

$$x = \left(\frac{q^2 L}{2\pi\epsilon_0 mg} \right)^{1/3},$$

where x is the separation between the balls. (b) If $L = 122$ cm, $m = 11.2$ g, and $x = 4.70$ cm, what is the value of q?

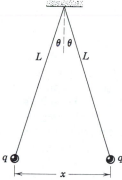

Figure 14   Problems 16, 17, and 18.

17. If the balls of Fig. 14 are conducting, (a) what happens to them after one is discharged? Explain your answer. (b) Find the new equilibrium separation.

18. Assume that each ball in Problem 16 is losing charge at the rate of 1.20 nC/s. At what instantaneous relative speed $(= dx/dt)$ do the balls approach each other initially?

19. Two equal positive point charges q are held a fixed distance $2a$ apart. A point test charge is located in a plane that is normal to the line joining these charges and midway be-

tween them. Find the radius R of the circle in this plane for which the force on the test particle has a maximum value. See Fig. 15.

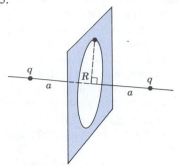

Figure 15 Problem 19.

20. Three small balls, each of mass 13.3 g, are suspended separately from a common point by silk threads, each 1.17 m long. The balls are identically charged and hang at the corners of an equilateral triangle 15.3 cm on a side. Find the charge on each ball.

21. A cube of edge a carries a point charge q at each corner. Show that the resultant electric force on any one of the charges is given by

$$F = \frac{0.262q^2}{\epsilon_0 a^2},$$

directed along the body diagonal away from the cube.

22. Two positive charges $+Q$ are held fixed a distance d apart. A particle of negative charge $-q$ and mass m is placed midway between them and then given a small displacement perpendicular to the line joining them and released. Show that the particle describes simple harmonic motion of period $(\epsilon_0 m \pi^3 d^3/qQ)^{1/2}$.

23. Calculate the period of oscillation for a particle of positive charge $+q$ displaced from the midpoint and along the line joining the charges in Problem 22.

Section 27-5 Charge Is Quantized

24. Find the total charge in coulombs of 75.0 kg of electrons.

25. In a crystal of salt, an atom of sodium transfers one of its electrons to a neighboring atom of chlorine, forming an *ionic* bond. The resulting positive sodium ion and negative chlorine ion attract each other by the electrostatic force. Calculate the force of attraction if the ions are 282 pm apart.

26. The electrostatic force between two identical ions that are separated by a distance of 5.0×10^{-10} m is 3.7×10^{-9} N. (a) Find the charge on each ion. (b) How many electrons are missing from each ion?

27. A neutron is thought to be composed of one "up" quark of charge $+\frac{2}{3}e$ and two "down" quarks each having charge $-\frac{1}{3}e$. If the down quarks are 2.6×10^{-15} m apart inside the neutron, what is the repulsive electrical force between them?

28. (a) How many electrons would have to be removed from a penny to leave it with a charge of $+1.15 \times 10^{-7}$ C? (b) To what fraction of the electrons in the penny does this correspond? See Sample Problem 2.

29. An electron is in a vacuum near the surface of the Earth. Where should a second electron be placed so that the net force on the first electron, owing to the other electron and to gravity, is zero?

30. Protons in cosmic rays strike the Earth's atmosphere at a rate, averaged over the Earth's surface, of 1500 protons/m²·s. What total current does the Earth receive from beyond its atmosphere in the form of incident cosmic ray protons?

31. Calculate the number of coulombs of positive charge in a glass of water. Assume the volume of the water to be 250 cm³.

32. In the compound CsCl (cesium chloride), the Cs atoms are situated at the corners of a cube with a Cl atom at the cube's center. The edge length of the cube is 0.40 nm; see Fig. 16. The Cs atoms are each deficient in one electron and the Cl atom carries one excess electron. (a) What is the strength of the net electric force on the Cl atom resulting from the eight Cs atoms shown? (b) Suppose that the Cs atom marked with an arrow is missing (crystal defect). What now is the net electric force on the Cl atom resulting from the seven remaining Cs atoms?

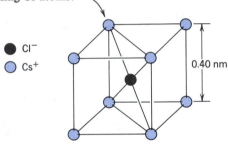

Figure 16 Problem 32.

33. (a) What equal amounts of positive charge would have to be placed on the Earth and on the Moon to neutralize their gravitational attraction? Do you need to know the Moon's distance to solve this problem? Why or why not? (b) How many metric tons of hydrogen would be needed to provide the positive charge calculated in part (a)? The molar mass of hydrogen is 1.008 g/mol.

34. Two physics students (Mary at 52.0 kg and John at 90.7 kg) are 28.0 m apart. Let each have a 0.01% imbalance in their amounts of positive and negative charge, one student being positive and the other negative. Estimate the electrostatic force of attraction between them. (*Hint*: Replace the students by spheres of water and use the result of Problem 31.)

Section 27-6 Charge Is Conserved

35. Identify the element X in the following nuclear reactions:

$$(a) \quad {}^1\mathrm{H} + {}^9\mathrm{Be} \rightarrow \mathrm{X} + \mathrm{n};$$

$$(b) \quad {}^{12}\mathrm{C} + {}^1\mathrm{H} \rightarrow \mathrm{X};$$

$$(c) \quad {}^{15}\mathrm{N} + {}^1\mathrm{H} \rightarrow {}^4\mathrm{He} + \mathrm{X}.$$

(*Hint*: See Appendix E.)

36. In the radioactive decay of ${}^{238}\mathrm{U}$ (${}^{238}\mathrm{U} \rightarrow {}^4\mathrm{He} + {}^{234}\mathrm{Th}$), the center of the emerging ${}^4\mathrm{He}$ particle is, at a certain instant, 12×10^{-15} m from the center of the residual ${}^{234}\mathrm{Th}$ nucleus. At this instant, (a) what is the force on the ${}^4\mathrm{He}$ particle and (b) what is its acceleration?

CHAPTER 28

THE ELECTRIC FIELD

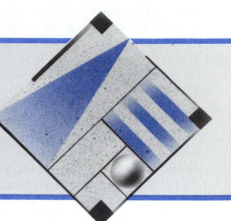

On August 25, 1989, twelve years after its launch, the spacecraft
Voyager 2 *passed close to the outer planet Neptune, a distance of 4.4×10^9 km*
from Earth. Among other discoveries, **Voyager** *reported the observation of six previously*
unknown moons of Neptune and a system of rings.

How is this information transmitted through the vast distance from **Voyager** *to Earth? The*
key to understanding this kind of communication is the **electromagnetic field.** *Electrons*
moving in electric circuits on **Voyager** *set up an electromagnetic field, and variations*
in their motion cause a disturbance in the field to travel at the speed of light. More than
4 hours later, electrons in circuits on Earth detect these changes in the field and
move accordingly.

This example involves the time-varying field set up by moving charges, while in this chapter
we are concerned with the static field of charges at rest. Nevertheless, it illustrates the
usefulness of the field concept in understanding how electromagnetic forces can act over
great distances. In later chapters we introduce the analogous magnetic field for constant
currents, and eventually we show how electromagnetic waves, such as radio waves or light,
can be regarded in terms of electromagnetic fields produced by moving charges and
varying currents.

28-1 FIELDS

The temperature has a definite value at every point in the room in which you may be sitting. You can measure the temperature at each point by putting a thermometer at that point, and you could then represent the temperature distribution throughout the room either with a mathematical function, say, $T(x,y,z)$, or else with a graph plotting the variation of T. Such a distribution of temperatures is called a *temperature field.* In a similar fashion we could measure the pressure at points throughout a fluid and so obtain a representation for the *pressure field,* describing the spatial variation of pressure. Such fields are called *scalar fields,* because the temperature T and pressure p are scalar quantities. If the temperature and pressure do not vary with time, they are also *static fields*; otherwise they are *time-varying fields* and might be represented mathematically by a function such as $T(x,y,z,t)$.

As we discussed in Section 18-5, the flow velocity in a fluid can be represented by a field of flow, which is an example of a *vector field* (see Figs. 14 – 18 of Chapter 18). Associated with every point of the fluid is a vector quan-

tity, the velocity **v** with which the fluid flows past that point. If the flow velocity remains constant in time, this vector field can also be described as a static field, represented by the mathematical function $v(x,y,z)$. Note that, even though the fluid is flowing, the *field* is static if the values at a point do not change with time.

In Section 16-7, we introduced the gravitational field **g**, defined in Eq. 19 of Chapter 16 as the gravitational force **F** per unit test mass m_0, or

$$ \mathbf{g} = \frac{\mathbf{F}}{m_0}. \tag{1} $$

This field is also a vector field and, in addition, is usually static when the distribution of mass of the gravitating body that is the source of the field remains constant. Near the surface of the Earth, and for points not too far apart, it is also a *uniform* field, meaning that **g** is the same (in direction as well as magnitude) for all points.

We can use Eq. 1 in the following way to provide an operational procedure for measuring the gravitational field. Let us use a test body of small mass m_0 and release it in the gravitational field we wish to measure. We determine its gravitational acceleration at a particular point, and Eq. 1 then tells us that the acceleration $\mathbf{F}/m_0$ is equal (in magnitude and direction) to the gravitational field $\mathbf{g}$ at that point. We specify a test body of small mass in this procedure to ensure that the test body does not disturb the mass distribution of the gravitating body and so change the very field we are trying to measure. For example, the Moon causes tides that change the distribution of mass on the Earth and so change its gravitational field; we would not want to use a test body as large as the Moon!

Before the concept of fields became widely accepted, the force between gravitating bodies was thought of as a direct and instantaneous interaction. This view, called *action at a distance,* was also used for electromagnetic forces. In the case of gravitation, it can be represented schematically as

$$\text{mass} \rightleftarrows \text{mass},$$

indicating that the two masses interact directly with one another. According to this view, the effect of a movement of one body is instantaneously transmitted to the other body. This view violates the special theory of relativity, which limits the speed at which such information can be transmitted to the speed of light c, at most. A more modern interpretation, based on the field concept and now an essential part of the general theory of relativity, can be represented as

$$\text{mass} \rightleftarrows \text{field} \rightleftarrows \text{mass},$$

in which each mass interacts not directly with the other but instead with the gravitational field established by the other. That is, the first mass sets up a field that has a certain value at every point in space; the second mass then interacts with the field at its particular location. The field plays the role of an intermediary between the two bodies. The force exerted on the second mass can be calculated from Eq. 1, given the value of the field $\mathbf{g}$ due to the first mass. The situation is completely symmetrical from the point of view of the first mass, which interacts with the gravitational field established by the second mass. Changes in the location of one mass cause variations in its gravitational field; these variations travel at the speed of light, so the field concept is consistent with the restrictions imposed by special relativity.

28-2 THE ELECTRIC FIELD E

The previous description of the gravitational field can be carried directly over to electrostatics. Coulomb's law for the force between charges encourages us to think in terms of action at a distance, represented as

$$\text{charge} \rightleftarrows \text{charge}.$$

Again introducing the field as an intermediary between the charges, we can represent the interaction as

$$\text{charge} \rightleftarrows \text{field} \rightleftarrows \text{charge}.$$

TABLE 1   SOME ELECTRIC FIELDS[a]

Location	Electric Field (N/C)
At the surface of a uranium nucleus	3×10^{21}
Within a hydrogen atom, at the electron orbit	5×10^{11}
Electric breakdown occurs in air	3×10^6
At the charged drum of a photocopier	10^5
The electron beam accelerator in a TV set	10^5
Near a charged plastic comb	10^3
In the lower atmosphere	10^2
Inside the copper wire of household circuits	10^{-2}

[a] Approximate values.

That is, the first charge sets up an *electric field,* and the second charge interacts with the electric field of the first charge. Our problem of determining the interaction between the charges is therefore reduced to two separate problems: (1) determine, by measurement or calculation, the electric field established by the first charge at every point in space, and (2) calculate the force that the field exerts on the second charge placed at a particular point in space.

In analogy with Eq. 1 for the gravitational field, we define the electric field $\mathbf{E}$ associated with a certain collection of charges in terms of the force exerted on a positive test charge q_0 at a particular point, or

$$\mathbf{E} = \frac{\mathbf{F}}{q_0}. \tag{2}$$

The direction of the vector $\mathbf{E}$ is the same as the direction of $\mathbf{F}$, because q_0 is a positive scalar.

Dimensionally, the electric field is the force per unit charge, and its SI unit is the newton/coulomb (N/C), although it is more often given, as we discuss in Chapter 30, in the equivalent unit of volt/meter (V/m). Note the similarity with the gravitational field, in which g (which is usually expressed in units of m/s^2) can also be expressed as the force per unit mass in units of newton/kilogram. Both the gravitational and electric fields can be expressed as a force divided by a property (mass or charge) of the test body. Table 1 shows some electric fields that occur in a few situations.

Figure 1 illustrates the electric field acting as the intermediary in the interaction between two charges. In Fig. 1a, charge q_1 sets up an electric field in the surrounding space, suggested by the shading of the figure. The field then acts on charge q_2, resulting in the force $\mathbf{F}_2$. From the perspective of q_1, as shown in Fig. 1b, we could just as well assert that q_2 sets up an electric field and that the force $\mathbf{F}_1$ on q_1 results from its interaction with the field of q_2. The forces are of course equal and opposite ($\mathbf{F}_1 = -\mathbf{F}_2$), even though the two electric fields may be quite different (as indicated by the difference in shading between Figs. 1a and 1b) if the charges are different.

To use Eq. 2 as an operational procedure for measuring the electric field, we must apply the same caution we did

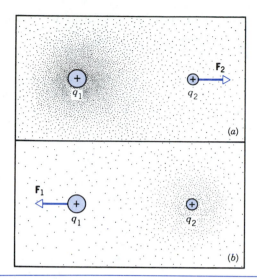

Figure 1 (a) Charge q_1 sets up an electric field that exerts a force $\mathbf{F}_2$ on charge q_2. (b) Charge q_2 sets up an electric field that exerts a force $\mathbf{F}_1$ on charge q_1. If the charges have different magnitudes, the resulting fields will be different. The forces, however, are always equal in magnitude and opposite in direction; that is, $\mathbf{F}_1 = -\mathbf{F}_2$.

in using a test mass to measure the gravitational field: the test charge should be sufficiently small so that it does not disturb the distribution of charges whose electric field we are trying to measure. That is, we should more properly write Eq. 2 as

$$\mathbf{E} = \lim_{q_0 \to 0} \frac{\mathbf{F}}{q_0} \qquad (3)$$

even though we know from Chapter 27 that this limit in actuality cannot be taken to 0 because the test charge can never be smaller than the elementary charge e. Of course, if we are *calculating* (rather than measuring) the electric field due to a specified collection of charges at fixed positions, neither the magnitude nor the sign of q_0 affects the result. As we show later in this chapter, electric fields of collections of charges can be calculated without direct reference to Eq. 3.

Sample Problem 1 A proton is placed in a uniform electric field $\mathbf{E}$. What must be the magnitude and direction of this field if the electrostatic force acting on the proton is just to balance its weight?

Solution From Eq. 2, replacing q_0 by e and F by mg, we have

$$E = \frac{F}{q_0} = \frac{mg}{e} = \frac{(1.67 \times 10^{-27} \text{ kg})(9.8 \text{ m/s}^2)}{1.60 \times 10^{-19} \text{ C}}$$

$$= 1.0 \times 10^{-7} \text{ N/C, directed up.}$$

This is a very weak field indeed. $\mathbf{E}$ must point vertically upward to float the (positively charged) proton, because $\mathbf{F} = q_0\mathbf{E}$ and $q_0 > 0$.

28-3 THE ELECTRIC FIELD OF POINT CHARGES

In this section we consider the electric field of point charges, first a single charge and then an assembly of individual charges. Later we generalize to continuous distributions of charge.

Let a positive test charge q_0 be placed a distance r from a point charge q. The magnitude of the force acting on q_0 is given by Coulomb's law,

$$F = \frac{1}{4\pi\epsilon_0} \frac{qq_0}{r^2}.$$

The magnitude of the electric field at the site of the test charge is, from Eq. 2,

$$E = \frac{F}{q_0} = \frac{1}{4\pi\epsilon_0} \frac{q}{r^2}. \qquad (4)$$

The direction of $\mathbf{E}$ is the same as the direction of $\mathbf{F}$, along a radial line from q, pointing outward if q is positive and inward if q is negative. Figure 2 shows the magnitude and direction of the electric field $\mathbf{E}$ at various points near a positive point charge. How would this figure be drawn if the charge were negative?

To find $\mathbf{E}$ for a group of N point charges, the procedure is as follows: (1) Calculate $\mathbf{E}_i$ due to each charge i at the given point *as if it were the only charge present*. (2) Add these separately calculated fields vectorially to find the resultant field $\mathbf{E}$ at the point. In equation form,

$$\mathbf{E} = \mathbf{E}_1 + \mathbf{E}_2 + \mathbf{E}_3 + \cdots$$

$$= \sum \mathbf{E}_i \quad (i = 1, 2, 3, \ldots, N). \qquad (5)$$

The sum is a vector sum, taken over all the charges. Equation 5 (like Eq. 8 of Chapter 27) is an example of the

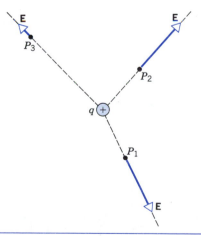

Figure 2 The electric field $\mathbf{E}$ at various points near a positive point charge q. Note that the direction of $\mathbf{E}$ is everywhere radially outward from q. The fields at P_1 and P_2, which are the same distance from q, are equal in magnitude. The field at P_3, which is twice as far from q as P_1 or P_2, has one-quarter the magnitude of the field at P_1 or P_2.

application of the *principle of superposition,* which states, in this context, that at a given point the electric fields due to separate charge distributions simply add up (vectorially) or superimpose independently. This principle may fail when the magnitudes of the fields are extremely large, but it will be valid in all situations we discuss in this text.

Sample Problem 2   In an ionized helium atom (a helium atom in which one of the two electrons has been removed), the electron and the nucleus are separated by a distance of 26.5 pm. What is the electric field due to the nucleus at the location of the electron?

Solution   We use Eq. 4, with q (the charge of the nucleus) equal to $+2e$:

$$E = \frac{1}{4\pi\epsilon_0} \frac{q}{r^2}$$

$$= \left(8.99 \times 10^9 \ \frac{\text{N} \cdot \text{m}^2}{\text{C}^2}\right) \frac{2(1.60 \times 10^{-19} \ \text{C})}{(26.5 \times 10^{-12} \ \text{m})^2}$$

$$= 4.13 \times 10^{12} \ \text{N/C}.$$

This value is 8 times the electric field that acts on an electron in hydrogen (see Table 1). The increase comes about because (1) the nuclear charge in helium is twice that in hydrogen, and (2) the orbital radius in helium is half that in hydrogen. Can you estimate the field on a similar electron in ionized uranium ($Z = 92$), from which 91 of the electrons have been removed? Such highly ionized atoms may be found in the interiors of stars.

Sample Problem 3   Figure 3 shows a charge q_1 of $+1.5 \ \mu\text{C}$ and a charge q_2 of $+2.3 \ \mu\text{C}$. The first charge is at the origin of an x axis, and the second is at a position $x = L$, where $L = 13$ cm. At what point P along the x axis is the electric field zero?

Solution   The point must lie between the charges because only in this region do the forces exerted by q_1 and by q_2 on a test charge oppose each other. If $\mathbf{E}_1$ is the electric field due to q_1 and $\mathbf{E}_2$ is that due to q_2, the magnitudes of these vectors must be equal, or

$$E_1 = E_2.$$

From Eq. 4 we then have

$$\frac{1}{4\pi\epsilon_0} \frac{q_1}{x^2} = \frac{1}{4\pi\epsilon_0} \frac{q_2}{(L - x)^2},$$

where x is the coordinate of point P. Solving for x, we obtain

$$x = \frac{L}{1 + \sqrt{q_2/q_1}} = \frac{13 \ \text{cm}}{1 + \sqrt{2.3 \ \mu\text{C}/1.5 \ \mu\text{C}}} = 5.8 \ \text{cm}.$$

This result is positive and is less than L, confirming that the zero-field point lies between the two charges, as we know it must.

The Electric Dipole

Figure 4 shows a positive and a negative charge of equal magnitude q placed a distance d apart, a configuration called an *electric dipole.* We seek to calculate the electric field $\mathbf{E}$ at point P, a distance x along the perpendicular bisector of the line joining the charges.

The positive and negative charges set up electric fields $\mathbf{E}_+$ and $\mathbf{E}_-$, respectively. The magnitudes of these two fields at P are equal, because P is equidistant from the positive and negative charges. Figure 4 also shows the directions of $\mathbf{E}_+$ and $\mathbf{E}_-$, determined by the directions of the force due to each charge alone that would act on a positive test charge at P. The total electric field at P is determined, according to Eq. 5, by the vector sum

$$\mathbf{E} = \mathbf{E}_+ + \mathbf{E}_-.$$

From Eq. 4, the magnitudes of the fields from each charge are given by

$$E_+ = E_- = \frac{1}{4\pi\epsilon_0} \frac{q}{r^2} = \frac{1}{4\pi\epsilon_0} \frac{q}{x^2 + (d/2)^2}. \quad (6)$$

Because the fields $\mathbf{E}_+$ and $\mathbf{E}_-$ have equal magnitudes and lie at equal angles θ with respect to the z direction as shown, the x component of the total field is $E_+ \sin \theta - E_- \sin \theta = 0$. The total field $\mathbf{E}$ therefore has only a z component, of magnitude

$$E = E_+ \cos \theta + E_- \cos \theta = 2E_+ \cos \theta. \quad (7)$$

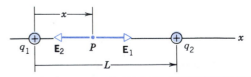

Figure 3   Sample Problem 3. At point P, the electric fields of the charges q_1 and q_2 are equal and opposite, so the net field at P is zero.

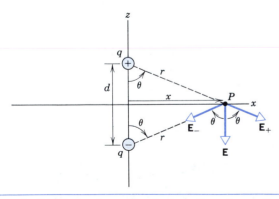

Figure 4   Positive and negative charges of equal magnitude form an electric dipole. The electric field $\mathbf{E}$ at any point is the vector sum of the fields due to the individual charges. At point P on the x axis, the field has only a z component.

From the figure we see that the angle θ is determined according to

$$\cos \theta = \frac{d/2}{\sqrt{x^2 + (d/2)^2}} .$$

Substituting this result and Eq. 6 into Eq. 7, we obtain

$$E = (2) \frac{1}{4\pi\epsilon_0} \frac{q}{x^2 + (d/2)^2} \frac{d/2}{\sqrt{x^2 + (d/2)^2}}$$

or

$$E = \frac{1}{4\pi\epsilon_0} \frac{qd}{[x^2 + (d/2)^2]^{3/2}} . \tag{8}$$

Equation 8 gives the magnitude of the electric field at P due to the dipole.

The field is proportional to the product qd, which involves the magnitudes of the dipole charges and their separation. This essential combined property of an electric dipole is called the *electric dipole moment p*, defined by

$$p = qd. \tag{9}$$

The dipole moment is a fundamental property of molecules, which often contain a negative and an equal positive charge separated by a definite distance. A molecule (*not* a crystal) of a compound such as NaCl is a good example. We can regard a molecule of NaCl as composed of a Na^+ ion (a neutral atom of sodium from which a single electron has been removed) with an electric charge of $+e$, and a Cl^- ion (a neutral atom of chlorine that has acquired an extra electron) with a charge of $-e$. The separation between Na and Cl measured for NaCl is 0.236 nm (1 nm = 10^{-9} m), and so the dipole moment is expected to be

$$p = ed = (1.60 \times 10^{-19} \text{ C})(0.236 \times 10^{-9} \text{ m})$$
$$= 3.78 \times 10^{-29} \text{ C} \cdot \text{m}.$$

The measured value is 3.00×10^{-29} C·m, indicating that the electron is not entirely removed from Na and attached to Cl. To a certain extent, the electron is shared between Na and Cl, resulting in a dipole moment somewhat smaller than expected.

Often we observe the field of an electric dipole at points P whose distance x from the dipole is very large compared with the separation d. In this case we can simplify the dipole field somewhat by making use of the binomial expansion,

$$(1 + y)^n = 1 + ny + \frac{n(n-1)}{2!} y^2 + \cdots .$$

Let us first rewrite Eq. 8 as

$$E = \frac{1}{4\pi\epsilon_0} \frac{p}{x^3} \frac{1}{[1 + (d/2x)^2]^{3/2}}$$
$$= \frac{1}{4\pi\epsilon_0} \frac{p}{x^3} \left[1 + \left(\frac{d}{2x} \right)^2 \right]^{-3/2}$$

and apply the binomial expansion to the factor in brackets, which gives

$$E = \frac{1}{4\pi\epsilon_0} \frac{p}{x^3} \left[1 + \left(-\frac{3}{2} \right) \left(\frac{d}{2x} \right)^2 + \cdots \right].$$

For this calculation it is sufficient to keep only the first term in the brackets (the 1), and so we find an expression for the magnitude of the electric field due to a dipole at distant points in its median plane:

$$E = \frac{1}{4\pi\epsilon_0} \frac{p}{x^3} . \tag{10}$$

An expression of a similar form is obtained for the field along the dipole axis (the z axis of Fig. 4); see Problem 11. A more general result for the field at any point in the xz plane can also be calculated; see Problem 12. In either case, the field at distant points varies with the distance r from the dipole as $1/r^3$. This is a characteristic result for the electric dipole field. The field varies more rapidly with distance than the $1/r^2$ dependence characteristic of a point charge. If you imagine Fig. 4 redrawn when x is very large, the angle θ approaches 90° and the fields $\mathbf{E}_+$ and $\mathbf{E}_-$ lie very nearly in opposite directions close to the x axis. The fields almost, but not quite, cancel. The $1/r^2$ variation of the fields from the individual point charges does cancel, leaving the more rapidly varying $1/r^3$ term that uniquely characterizes an electric dipole.

There are also more complicated charge distributions that give electric fields that vary as higher inverse powers of r. See Problems 13 and 14 for examples of the $1/r^4$ variation of the field of an electric *quadrupole*.

28-4   LINES OF FORCE

The concept of the electric field vector was not appreciated by Michael Faraday, who always thought in terms of *lines of force*. Although we no longer attach the same kind of reality to these lines that Faraday did, they still provide a convenient and instructive way to visualize the electric field, and we shall use them for this purpose.

Figure 5 shows the lines of force surrounding a positive point charge. You can think of this figure as an extension of Fig. 2, obtained by placing the test charge at many points around the central charge. For the purpose of the illustrations in this section, we regard a "point charge" as a small uniform sphere of charge rather than a true mathematical point. Furthermore, keep in mind as you view such drawings that they show a two-dimensional slice of a three-dimensional pattern.

Note several features of Fig. 5. (1) *The lines of force give the direction of the electric field at any point.* (In more complex patterns, in which the lines of force can be curved, it is the direction of the *tangent* to the line of force that gives the direction of **E**.) A positive test charge re-

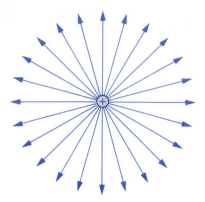

Figure 5   Lines of force surrounding a positive point charge. The direction of the force on a positive test charge, and thus the direction of the electric field at any point, is indicated by the direction of the lines. The relative spacing between the lines at any location indicates the relative strength of the field at that location. The lines are assumed to terminate on distant negative charges that are not shown.

leased at any point in the vicinity of the charge in Fig. 5 would experience a repulsive force that acts radially outward, and the test charge would move in that direction. Hence the lines of force of a positive point charge are directed radially outward. (2) *The lines of force originate on positive charges and terminate on negative charges.* The negative charges are not shown in Fig. 5, but you should imagine that the positive charge is surrounded by walls of negative charge, on which the lines of force terminate. (3) *The lines of force are drawn so that the number of lines per unit cross-sectional area (perpendicular to the lines) is proportional to the magnitude of the electric field.* Imagine an element of spherical surface of a given area close to the point charge, where many lines of force would penetrate it. As we move that area radially outward, fewer lines of force penetrate the area, because the lines of force are farther apart at large distances from the charge. This corresponds to the decrease of the electric field with increasing distance from the charge.

If the point charge of Fig. 5 were negative, the pattern of lines of force would be the same, except that all the arrows would now point inward. The force on a positive test charge would be radially inward in this case.

Figure 6 shows the lines of force for two equal positive charges. Imagine the charges to begin very far apart, where they exert negligible influence on one another and each has lines of force as shown in Fig. 5, and then to be brought together to form the pattern of Fig. 6. In the process, the lines of force that originally were between the two charges have been "pushed away" to the sides. Note that the concentration of lines is smallest in the region directly between the two charges. What does this tell us about the force on a test charge placed there? As we move far from the charges, the lines of force become nearly

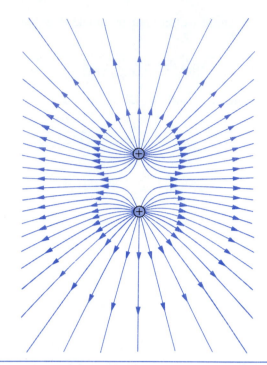

Figure 6   Lines of force surrounding two equal positive charges.

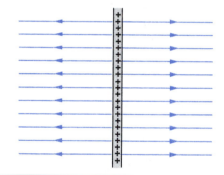

Figure 7   Lines of force close to a long line of positive charge. For a three-dimensional representation, imagine the figure rotated about an axis through the line.

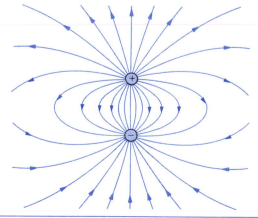

Figure 8   Lines of force surrounding positive and negative charges of equal magnitude (an electric dipole).

radial, characteristic of a single charge of magnitude equal to the total of the two charges.

Figure 6 shows that, in the regions to the left and the right of the middle of the charges, the lines of force are nearly parallel in the plane of the figure. Imagine now that the collection of two charges is extended to a long line of closely spaced positive charges, and let us consider only the region close to the middle of the line and far from either end. Figure 7 shows the resulting lines of force. Note that they are indeed parallel.

Figure 8 shows the lines of force in the case of an electric dipole, two equal charges with opposite signs. You can see here how the lines of force terminate on the negative charge. In this case the concentration of field lines is greatest in the region between the charges. What does that tell us about the electric field there? Imagine, as we did in the case of Fig. 6, that these two charges are originally far apart and are brought together. Instead of the lines of force being *repelled from* the central region, as in Fig. 6, they are *drawn into* the central region. Note the direction of the electric field along the bisector of the dipole axis, which we calculated in the previous section.

Lines of force can be made visible by applying an electric field to a suspension of tiny objects in an insulating fluid. Figure 9 shows photographs of the resulting patterns, which resemble the drawings of lines of force we have given in this section.

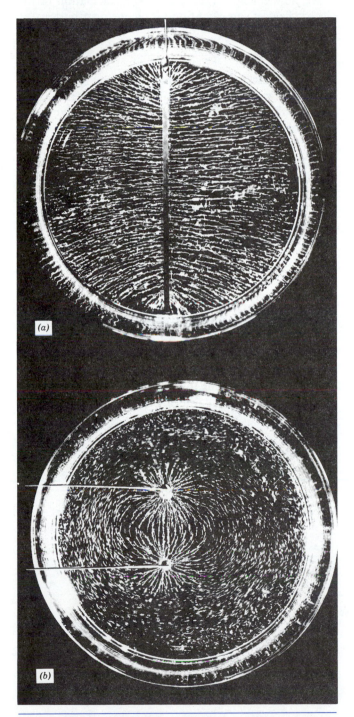

Figure 9 Photographs of the patterns of electric lines of force around (*a*) a charged plate (which produces parallel lines of force) and (*b*) two rods with equal and opposite charges (similar to the electric dipole of Fig. 8). The patterns were made visible by suspending grass seed in an insulating liquid.

Sample Problem 4   In Fig. 5, how does the magnitude of the electric field vary with the distance from the center of the charged body?

Solution   Suppose that N field lines terminate on the sphere of Fig. 5. Draw an imaginary concentric sphere of radius r. The number of lines per unit area at any point on this sphere is $N/4\pi r^2$. Because E is proportional to this quantity, we can write $E \propto 1/r^2$. Thus the electric field set up by a uniform sphere of charge varies as the inverse square of the distance from the center of the sphere, as we proved in the previous section (see Eq. 4). In much the same way, you can show that the electric field set up by the long line of charges (Fig. 7) varies as $1/r$, where r is the perpendicular distance from the axis of the line. We derive this result in the next section.

28-5   THE ELECTRIC FIELD OF CONTINUOUS CHARGE DISTRIBUTIONS

Even though electric charge is quantized (see Section 27-5), a collection of a large number of elementary charges can be regarded as a *continuous charge distribution*. The field set up by a continuous charge distribution can be computed by dividing the distribution into infinitesimal elements dq. Each element of charge establishes a field $d\mathbf{E}$ at a point P, and the resultant field at P is then found from the superposition principle by adding (that is, integrating) the field contributions due to all the charge elements, or

$$\mathbf{E} = \int d\mathbf{E}. \qquad (11)$$

The integration, like the sum in Eq. 5, is a vector operation; in the examples below, we see how such an integral is handled in three cases. Equation 11 is really a shorthand notation for separate scalar integrals over each direction; for instance, in Cartesian coordinates we have

$$E_x = \int dE_x, \quad E_y = \int dE_y, \quad \text{and} \quad E_z = \int dE_z.$$

As we discuss below, we can often simplify the calculation by arguing on the basis of symmetry that one or two of the integrals vanish or that two of them have identical values.

In calculating the electric field of a continuous charge distribution, the general strategy is to choose an arbitrary element of charge dq, find the electric field $d\mathbf{E}$ at the observation point P, and then integrate over the distribution using Eq. 11 to find the total field $\mathbf{E}$. In many cases, the charge element dq is treated as a point charge and gives a contribution to the field $d\mathbf{E}$ of magnitude given by Eq. 4, or

$$dE = \frac{1}{4\pi\epsilon_0} \frac{dq}{r^2}, \quad (12)$$

where r is the distance from the charge element dq to the point P. In other cases, we can simplify calculations by choosing dq to be an element in the form of a charge distribution that gives a known field $d\mathbf{E}$.

A continuous distribution of charge is described by its *charge density*. In a linear distribution, such as a thin filament onto which charge has been placed, an arbitrary element of length ds carries a charge dq given by

$$dq = \lambda \, ds, \quad (13)$$

where λ is the *linear charge density* (or charge per unit length) of the object. If the object is uniformly charged (that is, if the charge is distributed uniformly over the object) then λ is constant and is equal to the total charge q on the object divided by its total length L. In this case

$$dq = \frac{q}{L} \, ds \quad \text{(uniform linear charge)}. \quad (14)$$

If the charge is distributed not on a line but over a surface, the charge dq on any element of area dA is

$$dq = \sigma \, dA, \quad (15)$$

where σ is the *surface charge density* (or charge per unit area) of the object. If the charge is distributed uniformly over the surface, then σ is constant and is equal to the total charge q divided by the total area A of the surface, or

$$dq = \frac{q}{A} \, dA \quad \text{(uniform surface charge)}. \quad (16)$$

We can also consider the case in which a charge is distributed throughout a three-dimensional object, in which case the charge dq on a volume element dV is

$$dq = \rho \, dV, \quad (17)$$

where ρ is the *volume charge density* (or charge per unit volume). If the object is uniformly charged, then ρ is constant and is equal to the total charge q divided by the total volume V, or

$$dq = \frac{q}{V} \, dV \quad \text{(uniform volume charge)}. \quad (18)$$

We now consider examples of the calculation of the electric field of some continuous charge distributions.

Ring of Charge

Figure 10 shows a thin ring of radius R carrying a uniform linear charge density λ around its circumference. We may imagine the ring to be made of plastic or some other insulator, so that the charges can be regarded as fixed in place. What is the electric field at a point P, a distance z from the plane of the ring along its central axis?

Consider a differential element of the ring of length ds located at an arbitrary position on the ring in Fig. 10. It contains an element of charge given by Eq. 13, $dq = \lambda \, ds$. This element sets up a differential field $d\mathbf{E}$ at point P. From Eq. 4 we have

$$dE = \frac{1}{4\pi\epsilon_0} \frac{\lambda \, ds}{r^2} = \frac{\lambda \, ds}{4\pi\epsilon_0(z^2 + R^2)}. \quad (19)$$

Note that all charge elements that make up the ring are the same distance r from point P.

To find the resultant field at P we must add up, vectorially, all the field contributions $d\mathbf{E}$ made by the differential elements of the ring. Let us see how we can simplify this calculation by using the symmetry of the problem to eliminate certain of the integrations.

Figure 10   A uniform ring of charge. An element of the ring of length ds gives a contribution $d\mathbf{E}$ to the electric field at a point P on the axis of the ring. The total field at P is the sum of all such contributions.

In particular, we show that the electric field of the uniformly charged ring can have no x or y components. We do this by pretending such a component existed and then showing that the consequences would be unreasonable. Suppose there were an x component to the field at P; a test charge placed at P would accelerate in the x direction. Now suppose when your back was turned someone rotated the ring through 90° about the z axis. When you again look at the ring, could you tell that it had been rotated? If the ring is uniformly charged, then the physical state of the ring before the rotation is identical with that after the rotation, but a test charge now placed at P would accelerate in the y direction, because the field (and the force on the test particle) must rotate with the ring. We thus have a situation in which identical charge distributions would produce different forces on a test particle. This is an unacceptable result, and thus our original assumption must be wrong: there can be no component of the electric field perpendicular to the axis of the ring.

Another way of obtaining this result is to consider two elements of charge on the ring located at opposite ends of a diameter. The net electric field due to the two elements lies parallel to the axis, because the components perpendicular to the axis cancel one another. *All* elements around the ring can be paired in this manner, so the total field must be parallel to the z axis.

Because there is only one component to the total field (E_x and E_y being 0), the vector addition becomes a scalar addition of components parallel to the axis. The z component of $d\mathbf{E}$ is $dE \cos \theta$. From Fig. 10 we see that

$$\cos \theta = \frac{z}{r} = \frac{z}{(z^2 + R^2)^{1/2}} \cdot \qquad (20)$$

If we multiply Eqs. 19 and 20, we find

$$dE_z = dE \cos \theta = \frac{z\lambda \, ds}{4\pi\epsilon_0(z^2 + R^2)^{3/2}} \cdot \qquad (21)$$

To add the various contributions, we need add only the lengths of the elements, because all other quantities in Eq. 21 have the same value for all charge elements. Thus

$$E_z = \int dE \cos \theta = \frac{z\lambda}{4\pi\epsilon_0(z^2 + R^2)^{3/2}} \int ds \qquad (22)$$
$$= \frac{z\lambda(2\pi R)}{4\pi\epsilon_0(z^2 + R^2)^{3/2}},$$

in which the integral is simply $2\pi R$, the circumference of the ring. But $\lambda(2\pi R)$ is q, the total charge on the ring, so that we can write Eq. 22 as

$$E_z = \frac{qz}{4\pi\epsilon_0(z^2 + R^2)^{3/2}} \qquad \text{(charged ring).} \quad (23)$$

Does Eq. 23 give the correct direction for the field when z is negative? When q is negative?

For points far enough away from the ring so that

$z \gg R$, we can neglect R^2 in comparison with z^2 in the term in parentheses, in which case

$$E_z \approx \frac{1}{4\pi\epsilon_0} \frac{q}{z^2} \qquad (z \gg R), \qquad (24)$$

which (with z replaced by r) is Eq. 4, the electric field of a point charge. This should not be surprising because, at large enough distances, the ring would appear as a point charge. We note also from Eq. 23 that $E_z = 0$ for $z = 0$. This is also not surprising because a test charge at the center of the ring would be pushed or pulled equally in all directions in the plane of the ring and would experience no net force. Is this equilibrium stable or unstable?

A Disk of Charge

Figure 11 shows a circular plastic disk of radius R, carrying a uniform surface charge of density σ on its upper surface. What is the electric field at point P, a distance z from the disk along its axis?

Our plan is to divide the disk up into concentric rings and then to calculate the electric field by adding up, that is, by integrating, the contributions of the various rings. Figure 11 shows a flat ring with radius w and of width dw, its total charge being, according to Eq. 15,

$$dq = \sigma \, dA = \sigma(2\pi w)dw, \qquad (25)$$

where $dA = 2\pi w \, dw$ is the differential area of the ring.

We have already solved the problem of the electric field due to a ring of charge. Substituting dq from Eq. 25 for q in Eq. 23, and replacing R in Eq. 23 by w, we obtain

$$dE_z = \frac{z\sigma 2\pi w \, dw}{4\pi\epsilon_0(z^2 + w^2)^{3/2}} = \frac{\sigma z}{4\epsilon_0}(z^2 + w^2)^{-3/2}(2w)dw.$$

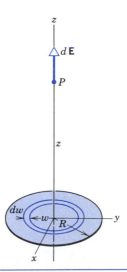

Figure 11 A disk carrying a uniform charge on its surface. The ring of radius w and width dw gives a contribution $d\mathbf{E}$ to the electric field at a point P on the axis of the disk. The total field at P is the sum of all such contributions.

We can now find E_z by integrating over the surface of the disk, that is, by integrating with respect to the variable w between the limits $w = 0$ and $w = R$. Note that z remains constant during this process. Thus

$$E_z = \int dE_z = \frac{\sigma z}{4\epsilon_0} \int_0^R (z^2 + w^2)^{-3/2}(2w)dw. \quad (26)$$

This integral is of the form $\int X^m \, dX$, in which $X = (z^2 + w^2)$, $m = -\frac{3}{2}$, and $dX = (2w)dw$. Integrating, we obtain

$$E_z = \frac{\sigma}{2\epsilon_0}\left(1 - \frac{z}{\sqrt{z^2 + R^2}}\right) \quad \text{(charged disk)} \quad (27)$$

as the final result. This equation is valid only for $z > 0$ (see Problem 28).

For $R \gg z$, the second term in the parentheses in Eq. 27 approaches zero, and this equation reduces to

$$E_z = \frac{\sigma}{2\epsilon_0} \quad \text{(infinite sheet).} \quad (28)$$

This is the electric field set up by a uniform sheet of charge of infinite extent. This is an important result which we derive in the next chapter using a different approach. Note that Eq. 28 also follows as $z \to 0$ in Eq. 27; for such nearby points the charged disk does indeed behave as if it were infinite in extent. In Problem 24 we ask you to show that Eq. 27 reduces to the field of a point charge for $z \gg R$.

Infinite Line of Charge

Figure 12 shows a section of an infinite line of charge whose linear charge density has the constant value λ. What is the field **E** at a distance y from the line?

The magnitude of the field contribution dE due to charge element $dq \, (= \lambda \, dz)$ is given, using Eq. 12, by

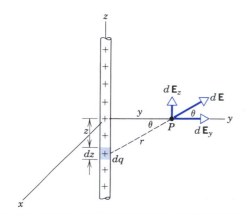

Figure 12 A uniform line of charge of great length. The element of length dz gives a contribution $d\mathbf{E}$ to the electric field at point P, whose distance y from the line is small compared with the length of the line.

$$dE = \frac{1}{4\pi\epsilon_0}\frac{dq}{r^2} = \frac{1}{4\pi\epsilon_0}\frac{\lambda \, dz}{y^2 + z^2}. \quad (29)$$

The vector $d\mathbf{E}$, as Fig. 12 shows, has the components

$$dE_y = dE \cos \theta \quad \text{and} \quad dE_z = dE \sin \theta.$$

The y and z components of the resultant vector **E** at point P are given by

$$E_y = \int dE_y = \int_{z=-\infty}^{z=+\infty} \cos \theta \, dE \quad (30a)$$

and

$$E_z = \int dE_z = \int_{z=-\infty}^{z=+\infty} \sin \theta \, dE. \quad (30b)$$

Here again we can use a symmetry argument to simplify the problem. If the line of charge were turned about the z axis, the physical situation would be unchanged, and there can thus be no component of **E** in the tangential direction at point P (the x direction of Fig. 12, perpendicular to the plane of the figure). Furthermore, if the line of charge were rotated by 180° about the y axis, thereby interchanging the portions of the line of charge along the positive and negative z directions, the physical arrangement would again be unchanged; therefore there can be no z component of the electric field (which, if it were present, would change sign upon the rotation).

Another way to show that E_z must be zero is to consider that for every charge element at positive z there is a corresponding element at negative z such that the z components of their fields cancel at P. Thus **E** points entirely in the y direction. This is strictly true only if the y axis passes through the middle of the line; however, when the line is infinitely long, we are always at its "middle" and never close to either end.

Because the contributions to E_y from the top and bottom halves of the rod are equal, we can write

$$E = E_y = 2 \int_{z=0}^{z=\infty} \cos \theta \, dE. \quad (31)$$

Note that we have changed the lower limit of integration and have introduced a compensating factor of 2. Substituting the expression for dE from Eq. 29 into Eq. 31 gives

$$E = \frac{\lambda}{2\pi\epsilon_0} \int_{z=0}^{z=\infty} \cos \theta \, \frac{dz}{y^2 + z^2}. \quad (32)$$

From Fig. 12 we see that the quantities θ and z are not independent. We can eliminate one of them, say, z, using the relation (see figure)

$$z = y \tan \theta.$$

Differentiating, we obtain

$$dz = y \sec^2 \theta \, d\theta.$$

Substituting these two expressions leads finally to

$$E = \frac{\lambda}{2\pi\epsilon_0 y} \int_{\theta=0}^{\theta=\pi/2} \cos \theta \, d\theta.$$

You should check this step carefully, noting that the limits must now be on θ and not on z. For example, as $z \rightarrow +\infty$, $\theta \rightarrow \pi/2$, as Fig. 12 shows. This equation integrates readily to

$$E = \frac{\lambda}{2\pi\epsilon_0 y}. \qquad (33)$$

This problem has *cylindrical symmetry* with respect to the z axis. At all points in the xy plane a distance r from the line of charge, the field has the value

$$E = \frac{\lambda}{2\pi\epsilon_0 r} \qquad \text{(infinite line),} \quad (34)$$

where $r = \sqrt{x^2 + y^2}$ is the distance from the line of charge to the point P at coordinates x,y.

You may wonder about the usefulness of solving a problem involving an infinite line of charge when any actual line must have a finite length (see Problem 31). However, for points close enough to finite lines and far from their ends, the equation that we have just derived yields results that are so close to the correct values that the difference can be ignored in many practical situations. It is usually unnecessary to solve exactly every geometry encountered in practical problems. Indeed, if idealizations or approximations are not made, the vast majority of significant problems of all kinds in physics and engineering cannot be solved at all.

28-6   A POINT CHARGE IN AN ELECTRIC FIELD

In the preceding sections, we have considered the first part of the charge $\rightleftarrows$ field $\rightleftarrows$ charge interaction: Given a collection of charges, what is the resulting electric field? In this section and the next, we consider the second part: What happens when we put a charged particle in a known electric field?

From Eq. 2, we know that a particle of charge q in an electric field $\mathbf{E}$ experiences a force $\mathbf{F}$ given by

$$\mathbf{F} = q\mathbf{E}.$$

To study the motion of the particle in the electric field, all we need do is use Newton's second law, $\Sigma \mathbf{F} = m\mathbf{a}$, where the resultant force on the particle includes the electric force and any other forces that may act.

As we did in our original study of Newton's laws, we can achieve a simplification if we consider the case in which the force is constant. We therefore begin by considering cases in which the electric field and the corresponding electric force are constant. Such a situation can be achieved in practice by connecting the terminals of a battery to a pair of parallel metal plates that are insulated from each other, as we discuss in the next chapter. If the distance between the plates is small compared with their dimensions, the field in the region between the plates will

be very nearly uniform, except near the edges. In the following sample problems, we assume that the field exists only in the region between the plates and drops suddenly to zero when the particle leaves that region. In reality the field decreases rapidly over a distance that is of the order of the spacing between the plates; when this distance is small, we don't make too large an error in calculating the motion of the particle if we ignore the edge effect.

Sample Problem 5   A charged drop of oil of radius $R = 2.76$ μm and density $\rho = 920$ kg/m^3 is maintained in equilibrium under the combined influence of its weight and a downward uniform electric field of magnitude $E = 1.65 \times 10^6$ N/C (Fig. 13). (a) Calculate the magnitude and sign of the charge on the drop. Express the result in terms of the elementary charge e. (b) The drop is exposed to a radioactive source that emits electrons. Two electrons strike the drop and are captured by it, changing its charge by two units. If the electric field remains at its constant value, calculate the resulting acceleration of the drop.

Solution   (a) To keep the drop in equilibrium, its weight mg must be balanced by an equal electric force of magnitude qE acting upward. Because the electric field is given as being in the downward direction, the charge q on the drop must be negative for the electric force to point in a direction opposite the field. The equilibrium condition is

$$\sum \mathbf{F} = m\mathbf{g} + q\mathbf{E} = 0.$$

Taking y components, we obtain

$$-mg + q(-E) = 0$$

or, solving for the unknown q,

$$q = -\frac{mg}{E} = -\frac{\frac{4}{3}\pi R^3 \rho g}{E}$$

$$= -\frac{\frac{4}{3}\pi(2.76 \times 10^{-6} \text{ m})^3(920 \text{ kg/m}^3)(9.8 \text{ m/s}^2)}{1.65 \times 10^6 \text{ N/C}}$$

$$= -4.8 \times 10^{-19} \text{ C}.$$

If we write q in terms of the electronic charge $-e$ as $q = n(-e)$, where n is the number of electronic charges on the drop, then

$$n = \frac{q}{-e} = \frac{-4.8 \times 10^{-19} \text{ C}}{-1.6 \times 10^{-19} \text{ C}} = 3.$$

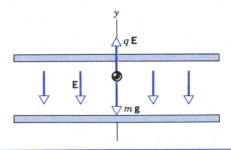

Figure 13   Sample Problem 5. A negatively charged drop is placed in a uniform electric field $\mathbf{E}$. The drop moves under the combined influence of its weight $m\mathbf{g}$ and the electric force $q\mathbf{E}$.

(b) If we add two additional electrons to the drop, its charge will become

$$q' = (n + 2)(-e) = 5(-1.6 \times 10^{-19} \text{ C}) = -8.0 \times 10^{-19} \text{ C}.$$

Newton's second law can be written

$$\sum \mathbf{F} = m\mathbf{g} + q'\mathbf{E} = m\mathbf{a}$$

and, taking y components, we obtain

$$-mg + q'(-E) = ma.$$

We can now solve for the acceleration:

$$a = -g - \frac{q'E}{m}$$

$$= -9.80 \text{ m/s}^2 - \frac{(-8.0 \times 10^{-19} \text{ C})(1.65 \times 10^6 \text{ N/C})}{\frac{4}{3}\pi(2.76 \times 10^{-6} \text{ m})^3(920 \text{ kg/m}^3)}$$

$$= -9.80 \text{ m/s}^2 + 16.3 \text{ m/s}^2 = +6.5 \text{ m/s}^2.$$

The drop accelerates in the positive y direction.

In this calculation, we have ignored the viscous drag force, which is usually quite important in this situation. We have, in effect, found the acceleration of the drop at the instant it acquired the extra two electrons. The drag force, which depends on the velocity of the drop, is initially zero if the drop starts from rest, but it increases as the drop begins to move, and so the acceleration of the drop will decrease in magnitude.

This experimental configuration forms the basis of the Millikan oil-drop experiment, which was used to measure the magnitude of the electronic charge. The experiment is discussed later in this section.

Sample Problem 6 Figure 14 shows the deflecting electrode system of an ink-jet printer. An ink drop whose mass m is 1.3×10^{-10} kg carries a charge q of -1.5×10^{-13} C and enters the deflecting plate system with a speed $v = 18$ m/s. The length L of these plates is 1.6 cm, and the electric field E between the plates is 1.4×10^6 N/C. What is the vertical deflection of the drop at the far edge of the plates? Ignore the varying electric field at the edges of the plates.

Solution Let t be the time of passage of the drop through the deflecting system. The vertical and the horizontal displacements are given by

$$y = \tfrac{1}{2}at^2 \quad \text{and} \quad L = vt,$$

respectively, in which a is the vertical acceleration of the drop.

As in the previous sample problem, we can write the y component of Newton's second law as $-mg + q(-E) = ma$. The electric force acting on the drop, $-qE$, is much greater in this case than the gravitational force mg so that the acceleration of the drop can be taken to be $-qE/m$. Eliminating t between the two equations above and substituting this value for a leads to

$$y = \frac{-qEL^2}{2mv^2}$$

$$= \frac{-(-1.5 \times 10^{-13} \text{ C})(1.4 \times 10^6 \text{ N/C})(1.6 \times 10^{-2} \text{ m})^2}{(2)(1.3 \times 10^{-10} \text{ kg})(18 \text{ m/s})^2}$$

$$= 6.4 \times 10^{-4} \text{ m} = 0.64 \text{ mm}.$$

The deflection at the paper will be larger than this because the ink drop follows a straight-line path to the paper after leaving the deflecting region, as shown in Fig. 14b. To aim the ink drops so

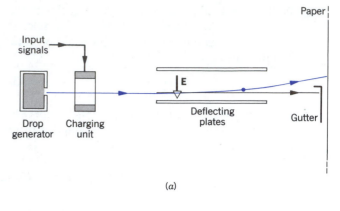

(a)

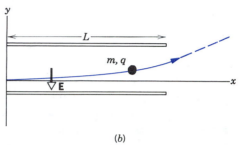

(b)

(c)

Figure 14 Sample Problem 6. (a) The essential features of an ink-jet printer. An input signal from a computer controls the charge given to the drop and thus the position at which the drop strikes the paper. A transverse force from the electric field **E** is responsible for deflecting the drop. (b) A detail of the deflecting plates. The drop moves in a parabolic path while it is between the plates, and it moves along a straight line (shown dashed) after it leaves the plates. (c) A sample of ink-jet printing, showing three enlarged letters. To print a typical letter requires about 100 drops. The drops are produced at a rate of about 100,000 per second.

that they form the characters well, it is necessary to control the charge q on the drops—to which the deflection is proportional—to within a few percent. In our treatment, we have again neglected the viscous drag forces that act on the drop; they are substantial at these high drop speeds. The analysis is the same as for the deflection of the electron beam in an electrostatic cathode ray tube.

Measuring the Elementary Charge*

Figure 15 shows a diagram of the apparatus used by the American physicist Robert A. Millikan in 1910–1913 to measure the elementary charge e. Oil droplets are introduced into chamber A by an atomizer, some of them becoming charged, either positively or negatively, in the process. Consider a drop that finds its way through a small hole in plate P_1 and drifts into chamber C. Let us assume that this drop carries a charge q, which we take to be negative.

If there is no electric field, two forces act on the drop, its weight mg and an upwardly directed viscous drag force, whose magnitude is proportional to the speed of the falling drop. The drop quickly comes to a constant terminal speed v at which these two forces are just balanced.

A downward electric field **E** is now set up in the chamber, by connecting battery B between plates P_1 and P_2. A third force, $q\mathbf{E}$, now acts on the drop. If q is negative, this force points upward, and—we assume—the drop now drifts upward, at a new terminal speed v'. In each case, the drag force points in the direction opposite to that in which the drop is moving and has a magnitude proportional to the speed of the drop. The charge q on the drop can be found from measurements of v and v'.

Millikan found that the values of q were all consistent with the relation

$$q = ne \qquad n = 0, \pm 1, \pm 2, \pm 3, \ldots,$$

in which e is the elementary charge, with a value of 1.60×10^{-19} C. Millikan's experiment is convincing

---

* For details of Millikan's experiments, see Henry A. Boorse and Lloyd Motz (eds.), *The World of the Atom* (Basic Books, 1966), Chapter 40. For the point of view of two physicists who knew Millikan as graduate students, see "Robert A. Millikan, Physics Teacher," by Alfred Romer, *The Physics Teacher,* February 1978, p. 78, and "My Work with Millikan on the Oil-Drop Experiment," by Harvey Fletcher, *Physics Today,* June 1982, p. 43.

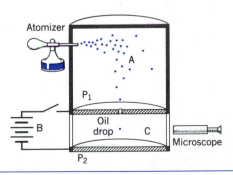

Figure 15   The Millikan oil-drop apparatus for measuring the elementary charge e. The motion of a drop is observed in chamber C, where the drop is acted on by gravity, the electric field set up by the battery B, and, if the drop is moving, a viscous drag force.

proof that charge is quantized. He was awarded the 1923 Nobel Prize in physics in part for this work. Modern measurements of the elementary charge rely on a variety of interlocking experiments, all more precise than the pioneering experiment of Millikan.

Motion in Nonuniform Electric Fields   *(Optional)*

So far we have considered only uniform fields, in which the electric field is constant in magnitude as well as in direction over the region in which the particle moves. Often, however, we encounter nonuniform fields. Once we have calculated the field, we must then solve Newton's laws in a manner appropriate for nonconstant forces, as we discussed in Chapter 6. We briefly consider an example of this procedure.

Figure 16 shows a ring of positive charge, the electric field for which is given by Eq. 23 for points on the axis. Suppose we project a positively charged particle with initial speed v_0 along the z axis toward the loop from a very large distance. What will be the subsequent motion of the particle?

We can solve this problem using the numerical technique described in Section 8-4 for a force depending on the position. We assume we are given the initial position and velocity of the particle. We can calculate the electric field at the initial position of the particle and thus determine its initial acceleration. In a small enough interval of time, we consider the acceleration to be constant, and we find the change in velocity and position in that interval as we did in Section 8-4. At the new position at the end of the first interval, we have a new electric field and a new acceleration, and we find the change in velocity and position during the second interval. Continuing in this way, we can determine the time dependence of the position and velocity of the particle.

For this calculation, we use a ring of radius $R = 3$ cm and linear charge density $\lambda = +2 \times 10^{-7}$ C/m. A proton ($q = +1.6 \times 10^{-19}$ C, $m = 1.67 \times 10^{-27}$ kg) is projected along the axis of the loop from an initial position at $z = +0.5$ m with initial velocity $v_{z0} = -7 \times 10^5$ m/s. (The negative initial velocity means that the proton is moving downward, toward the loop which lies in the xy plane.) The positively charged loop exerts a repulsive force on the positively charged proton, decreasing its speed. In Fig. 16a we plot the resulting motion in the case that the proton does not have enough initial kinetic energy to reach the plane of the loop. The proton comes instantaneously to rest at a point just above the plane of the loop and then reverses its motion as the loop now accelerates it in the positive z direction. Note that except for the region near the loop, the speed of the proton is nearly constant, because the electric field is weak at larger distances.

Figure 16b illustrates the motion in the case that the proton has more than enough initial kinetic energy to reach the plane of the loop. The repulsive force slows the proton's motion but doesn't stop it. The proton passes through the loop, with the magnitude of its velocity reaching a minimum as it passes through the loop. Once again, far from the loop the proton moves with very nearly constant velocity.

In Chapter 30 we discuss a method based on the conservation of energy, which permits v_z to be calculated directly.

A listing of the computer program that gives the solution to this problem (and to other similar one-dimensional problems) can be found in Appendix I. Problem 58 gives another example of an application of this technique. ■

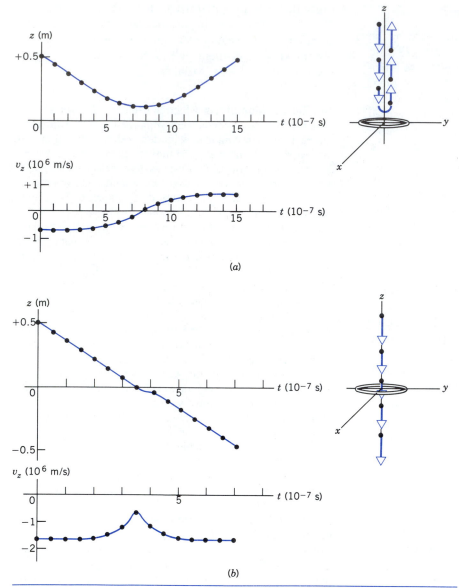

Figure 16 (a) The motion of a proton projected along the axis of a uniform positively charged ring. The position and velocity are shown. The proton comes instantaneously to rest at a time of about 8×10^{-7} s and reverses its motion. The points are the results of a numerical calculation; the curves are drawn through the points. (b) If the initial velocity of the proton is increased sufficiently, it can pass through the ring; its speed is a minimum as it passes through the center of the ring.

28-7 A DIPOLE IN AN ELECTRIC FIELD

In Section 28-3, we discussed the electric dipole, which can be represented as two equal and opposite charges $+q$ and $-q$ separated by a distance d. When we place a dipole in an *external* electric field, the force on the positive charge will be in one direction and the force on the nega-

tive charge in another direction. To account for the net effect of these forces, it is convenient to introduce the dipole moment *vector* **p**. The vector **p** has magnitude $p = qd$ and direction along the line joining the two charges pointing *from* the negative charge *toward* the positive charge. As is often the case with vectors, writing the dipole moment in vector form permits us to write the fundamental relationships involving electric dipoles in a concise form.

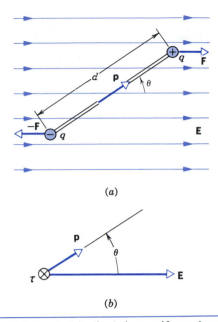

(a)

(b)

Figure 17 (a) An electric dipole in a uniform electric field. (b) The vector relationship $\boldsymbol{\tau} = \mathbf{p} \times \mathbf{E}$ between the dipole moment $\mathbf{p}$, the electric field $\mathbf{E}$, and the resultant torque $\boldsymbol{\tau}$ on the dipole. The torque points into the page.

Figure 17a shows a dipole in a uniform electric field $\mathbf{E}$. (This field is *not* that of the dipole itself but is produced by an external agent not shown in the figure.) The dipole moment $\mathbf{p}$ makes an angle θ with the direction of the field. We assume the field to be uniform, so that $\mathbf{E}$ has the same magnitude and direction at the location of $+q$ and $-q$. The forces on $+q$ and $-q$ therefore have equal magnitudes $F = qE$ but opposite directions, as shown in Fig. 17a. The net force on the dipole due to the external field is therefore zero, but there is a net torque about its center of mass that tends to rotate the dipole to bring $\mathbf{p}$ into alignment with $\mathbf{E}$. The net torque about the center of the dipole due to the two forces has a magnitude

$$\tau = F\frac{d}{2}\sin\theta + F\frac{d}{2}\sin\theta = Fd\sin\theta, \quad (35)$$

and its direction is perpendicular to the plane of the page and into the page, as indicated in Fig. 17b. We can write Eq. 35 as

$$\tau = (qE)d\sin\theta = (qd)E\sin\theta = pE\sin\theta. \quad (36)$$

Equation 36 can be written in vector form as

$$\boldsymbol{\tau} = \mathbf{p} \times \mathbf{E}, \quad (37)$$

which is consistent with the directional relationships for the cross product, as shown by the three vectors in Fig. 17b.

As is generally the case in dynamics when conservative forces act (the electrostatic force is conservative, as we discuss in Chapter 30), we can represent the system

equally well using either force equations or energy equations. Let us therefore consider the work done by the electric field in turning the dipole through an angle θ. Using the appropriate expression for work in rotational motion (Eq. 14 of Chapter 12), the work done by the external field in turning the dipole from an initial angle θ_0 to a final angle θ is

$$W = \int dW = \int_{\theta_0}^{\theta} \boldsymbol{\tau} \cdot d\boldsymbol{\theta} = \int_{\theta_0}^{\theta} -\tau \, d\theta, \quad (38)$$

where τ is the torque exerted by the external electric field. The minus sign in Eq. 38 is necessary because the torque τ tends to *decrease* θ; in vector terminology, $\boldsymbol{\tau}$ and $d\boldsymbol{\theta}$ are in opposite directions, so $\boldsymbol{\tau} \cdot d\boldsymbol{\theta} = -\tau \, d\theta$. Combining Eq. 38 with Eq. 36, we obtain

$$W = \int_{\theta_0}^{\theta} -pE\sin\theta \, d\theta = -pE\int_{\theta_0}^{\theta} \sin\theta \, d\theta$$

$$= pE(\cos\theta - \cos\theta_0). \quad (39)$$

Since the work done by the agent that produces the external field is equal to the negative of the change in potential energy of the system of field + dipole, we have

$$\Delta U \equiv U(\theta) - U(\theta_0) = -W = -pE(\cos\theta - \cos\theta_0). \quad (40)$$

We arbitrarily define the reference angle θ_0 to be 90° and choose the potential energy $U(\theta_0)$ to be zero at that angle. At any angle θ the potential energy is then

$$U = -pE\cos\theta, \quad (41)$$

which can be written in vector form as

$$U = -\mathbf{p} \cdot \mathbf{E}. \quad (42)$$

Thus U is a minimum when $\mathbf{p}$ and $\mathbf{E}$ are parallel.

The motion of a dipole in a uniform electric field can therefore be interpreted either from the perspective of force (the resultant torque on the dipole tries to rotate it into alignment with the direction of the external electric field) or energy (the potential energy of the system tends to a minimum when the dipole moment is aligned with the external field). The choice between the two is largely a matter of convenience in application to the particular problem under study.

Sample Problem 7 A molecule of water vapor (H_2O) has an electric dipole moment of magnitude $p = 6.2 \times 10^{-30}$ C·m. (This large dipole moment is responsible for many of the properties that make water such an important substance, such as its ability to act as an almost universal solvent.) Figure 18 is a representation of this molecule, showing the three nuclei and the surrounding electron clouds. The electric dipole moment $\mathbf{p}$ is represented by a vector on the axis of symmetry. The dipole moment arises because the effective center of positive charge does not coincide with the effective center of negative charge. (A contrasting case is that of a molecule of carbon dioxide, CO_2. Here the three atoms are joined in a straight line, with a carbon

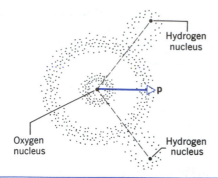

Figure 18   A molecule of H_2O, showing the three nuclei, the electron clouds, and the electric dipole moment vector **p**.

in the middle and oxygens on either side. The center of positive charge and the center of negative charge coincide at the center of mass of the molecule, and the electric dipole moment of CO_2 is zero.) (a) How far apart are the effective centers of positive and negative charge in a molecule of H_2O? (b) What is the maximum torque on a molecule of H_2O in a typical laboratory electric field of magnitude 1.5×10^4 N/C? (c) Suppose the dipole moment of a molecule of H_2O is initially pointing in a direction opposite to the field. How much work is done by the electric field in rotating the molecule into alignment with the field?

Solution   (a) There are 10 electrons and, correspondingly, 10 positive charges in this molecule. We can write, for the magnitude of the dipole moment,

$$p = qd = (10e)(d),$$

in which d is the separation we are seeking and e is the elementary charge. Thus

$$d = \frac{p}{10e} = \frac{6.2 \times 10^{-30} \text{ C} \cdot \text{m}}{(10)(1.60 \times 10^{-19} \text{ C})}$$
$$= 3.9 \times 10^{-12} \text{ m} = 3.9 \text{ pm}.$$

This is about 4% of the OH bond distance in this molecule.

(b) As Eq. 36 shows, the torque is a maximum when $\theta = 90°$. Substituting this value in that equation yields

$$\tau = pE \sin \theta = (6.2 \times 10^{-30} \text{ C} \cdot \text{m})(1.5 \times 10^4 \text{ N/C})(\sin 90°)$$
$$= 9.3 \times 10^{-26} \text{ N} \cdot \text{m}.$$

(c) The work done in rotating the dipole from $\theta_0 = 180°$ to $\theta = 0°$ is given by Eq. 39,

$$W = pE(\cos \theta - \cos \theta_0)$$
$$= pE(\cos 0° - \cos 180°)$$
$$= 2pE = (2)(6.2 \times 10^{-30} \text{ C} \cdot \text{m})(1.5 \times 10^4 \text{ N/C})$$
$$= 1.9 \times 10^{-25} \text{ J}.$$

By comparison, the average translational contribution to the internal energy ($= \frac{3}{2}kT$) of a molecule at room temperature is 6.2×10^{-21} J, which is 33,000 times larger. For the conditions of this problem, thermal agitation would overwhelm the tendency of the dipoles to align themselves with the field. That is, if we had a collection of molecules at room temperature with randomly oriented dipole moments, the application of an electric field of this magnitude would have a negligible influence on aligning the dipole moments, because of the large internal energies. If we wish to align the dipoles, we must use much stronger fields and/or much lower temperatures.

QUESTIONS

1. Name as many scalar fields and vector fields as you can.

2. (a) In the gravitational attraction between the Earth and a stone, can we say that the Earth lies in the gravitational field of the stone? (b) How is the gravitational field due to the stone related to that due to the Earth?

3. A positively charged ball hangs from a long silk thread. We wish to measure E at a point in the same horizontal plane as that of the hanging charge. To do so, we put a positive test charge q_0 at the point and measure F/q_0. Will F/q_0 be less than, equal to, or greater than E at the point in question?

4. In exploring electric fields with a test charge, we have often assumed, for convenience, that the test charge was positive. Does this really make any difference in determining the field? Illustrate in a simple case of your own devising.

5. Electric lines of force never cross. Why?

6. In Fig. 6, why do the lines of force around the edge of the figure appear, when extended backward, to radiate uniformly from the center of the figure?

7. A point charge is moving in an electric field at right angles to the lines of force. Does any force act on it?

8. In Fig. 9, why should grass seeds line up with electric lines of force? Grass seeds normally carry no electric charge. (See "Demonstration of the Electric Fields of Current-Carrying Conductors," by O. Jefimenko, *American Journal of Physics,* January 1962, p. 19.)

9. What is the origin of "static cling," a phenomenon that sometimes affects clothes as they are removed from a dryer?

10. Two point charges of unknown magnitude and sign are a distance d apart. The electric field is zero at one point between them, on the line joining them. What can you conclude about the charges?

11. Two point charges of unknown magnitude and sign are placed a distance d apart. (a) If it is possible to have $E = 0$ at any point *not* between the charges but on the line joining them, what are the necessary conditions and where is the point located? (b) Is it possible, for any arrangement of two point charges, to find *two* points (neither at infinity) at which $E = 0$? If so, under what conditions?

12. Two point charges of unknown sign and magnitude are fixed a distance d apart. Can we have $E = 0$ for off-axis points (excluding infinity)? Explain.

13. In Sample Problem 3, a charge placed at point P in Fig. 3 is

in equilibrium because no force acts on it. Is the equilibrium stable (*a*) for displacements along the line joining the charges and (*b*) for displacements at right angles to this line?

14. In Fig. 8, the force on the lower charge points up and is finite. The crowding of the lines of force, however, suggests that *E* is infinitely great at the site of this (point) charge. A charge immersed in an infinitely great field should have an infinitely great force acting on it. What is the solution to this dilemma?

15. A point charge *q* of mass *m* is released from rest in a nonuniform field. (*a*) Will it necessarily follow the line of force that passes through the release point? (*b*) Under what circumstances, if any, will a charged particle follow the electric field lines?

16. Three small spheres *x*, *y*, and *z* carry charges of equal magnitudes and with signs shown in Fig. 19. They are placed at the vertices of an isosceles triangle with the distance between *x* and *y* equal to the distance between *x* and *z*. Spheres *y* and *z* are held in place but sphere *x* is free to move on a frictionless surface. Which path will sphere *x* take when released?

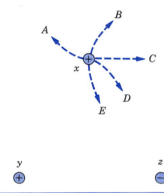

Figure 19   Question 16.

17. A positive and a negative charge of the same magnitude lie on a long straight line. What is the direction of **E** for points on this line that lie (*a*) between the charges, (*b*) outside the charges in the direction of the positive charge, (*c*) outside the charges in the direction of the negative charge, and (*d*) off the line but in the median plane of the charges?

18. In the median plane of an electric dipole, is the electric field parallel or antiparallel to the electric dipole moment **p**?

19. In what way does Eq. 10 fail to represent the lines of force of Fig. 8 if we relax the requirement that $x \gg d$?

20. (*a*) Two identical electric dipoles are placed in a straight line, as shown in Fig. 20*a*. What is the direction of the electric force on each dipole owing to the presence of the other? (*b*) Suppose that the dipoles are rearranged as in Fig. 20*b*. What now is the direction of the force?

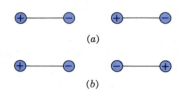

Figure 20   Question 20.

21. Compare the way *E* varies with *r* for (*a*) a point charge, (*b*) a dipole, and (*c*) a quadrupole.

22. What mathematical difficulties would you encounter if you were to calculate the electric field of a charged ring (or disk) at points *not* on the axis?

23. Equation 28 shows that *E* has the same value for all points in front of an infinite uniformly charged sheet. Is this reasonable? One might think that the field should be stronger near the sheet because the charges are so much closer.

24. Describe, in your own words, the purpose of the Millikan oil-drop experiment.

25. How does the sign of the charge on the oil drop affect the operation of the Millikan experiment?

26. Why did Millikan not try to balance electrons in his apparatus instead of oil drops?

27. You turn an electric dipole end for end in a uniform electric field. How does the work you do depend on the initial orientation of the dipole with respect to the field?

28. For what orientations of an electric dipole in a uniform electric field is the potential energy of the dipole (*a*) the greatest and (*b*) the least?

29. An electric dipole is placed in a nonuniform electric field. Is there a net force on it?

30. An electric dipole is placed at rest in a uniform external electric field, as in Fig. 17*a*, and released. Discuss its motion.

31. An electric dipole has its dipole moment **p** aligned with a uniform external electric field **E**. (*a*) Is the equilibrium stable or unstable? (*b*) Discuss the nature of the equilibrium if **p** and **E** point in opposite directions.

PROBLEMS

Section 28-2 The Electric Field **E**

1. An electron is accelerated eastward at 1.84×10^9 m/s² by an electric field. Determine the magnitude and direction of the electric field.

2. Humid air breaks down (its molecules become ionized) in an electric field of 3.0×10^6 N/C. What is the magnitude of the electric force on (*a*) an electron and (*b*) an ion (with a single electron missing) in this field?

3. An alpha particle, the nucleus of a helium atom, has a mass of 6.64×10^{-27} kg and a charge of $+2e$. What are the magnitude and direction of the electric field that will balance its weight?

4. In a uniform electric field near the surface of the Earth, a particle having a charge of -2.0×10^{-9} C is acted on by a downward electric force of 3.0×10^{-6} N. (*a*) Find the mag-

nitude of the electric field. (*b*) What is the magnitude and direction of the electric force exerted on a proton placed in this field? (*c*) What is the gravitational force on the proton? (*d*) What is the ratio of the electric force to the gravitational force in this case?

Section 28-3 The Electric Field of Point Charges

5. What is the magnitude of a point charge chosen so that the electric field 75.0 cm away has the magnitude 2.30 N/C?

6. Calculate the dipole moment of an electron and a proton 4.30 nm apart.

7. Calculate the magnitude of the electric field, due to an electric dipole of dipole moment 3.56×10^{-29} C·m, at a point 25.4 nm away along the bisector axis.

8. Find the electric field at the center of the square of Fig. 21. Assume that $q = 11.8$ nC and $a = 5.20$ cm.

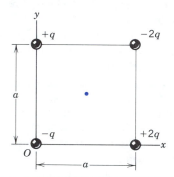

Figure 21 Problem 8.

9. A clock face has negative point charges $-q$, $-2q$, $-3q$, . . . , $-12q$ fixed at the positions of the corresponding numerals. The clock hands do not perturb the field. At what time does the hour hand point in the same direction as the electric field at the center of the dial? (*Hint:* Consider diametrically opposite charges.)

10. In Fig. 4, assume that both charges are positive. Show that E at point P in that figure, assuming $x \gg d$, is given by

$$E = \frac{1}{4\pi\epsilon_0} \frac{2q}{x^2}.$$

11. In Fig. 4, consider a point a distance z from the center of a dipole *along its axis.* (*a*) Show that, at large values of z, the electric field is given by

$$E = \frac{1}{2\pi\epsilon_0} \frac{p}{z^3}.$$

(Compare with the field at a point on the perpendicular bisector.) (*b*) What is the direction of **E**?

12. Show that the components of **E** due to a dipole are given, at distant points, by

$$E_x = \frac{1}{4\pi\epsilon_0} \frac{3pxz}{(x^2 + z^2)^{5/2}}, \qquad E_z = \frac{1}{4\pi\epsilon_0} \frac{p(2z^2 - x^2)}{(x^2 + z^2)^{5/2}},$$

where x and z are coordinates of point P in Fig. 22. Show that this general result includes the special results of Eq. 10 and Problem 11.

13. One type of electric quadrupole is formed by four charges located at the vertices of a square of side $2a$. Point P lies a

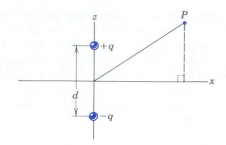

Figure 22 Problem 12.

distance x from the center of the quadrupole on a line parallel to two sides of the square as shown in Fig. 23. For $x \gg a$, show that the electric field at P is approximately given by

$$E = \frac{3(2qa^2)}{2\pi\epsilon_0 x^4}.$$

(*Hint:* Treat the quadrupole as two dipoles.)

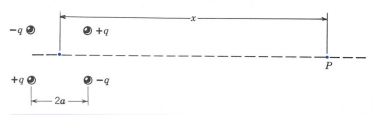

Figure 23 Problem 13.

14. Figure 24 shows one type of electric quadrupole. It consists of two dipoles whose effects at external points do not quite cancel. Show that the value of E on the axis of the quadrupole for points a distance z from its center (assume $z \gg d$) is given by

$$E = \frac{3Q}{4\pi\epsilon_0 z^4},$$

where $Q\ (= 2qd^2)$ is called the *quadrupole moment* of the charge distribution.

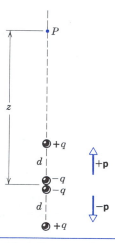

Figure 24 Problem 14.

15. Consider the ring of charge of Section 28-5. Suppose that the charge q is not distributed uniformly over the ring but that

charge q_1 is distributed uniformly over half the circumference and charge q_2 is distributed uniformly over the other half. Let $q_1 + q_2 = q$. (*a*) Find the component of the electric field at any point on the axis directed *along* the axis and compare with the uniform case. (*b*) Find the component of the electric field at any point on the axis *perpendicular* to the axis and compare with the uniform case.

Section 28-4 Lines of Force

16. Figure 25 shows field lines of an electric field; the line spacing perpendicular to the page is the same everywhere. (*a*) If the magnitude of the field at A is 40 N/C, what force does an electron at that point experience? (*b*) What is the magnitude of the field at B?

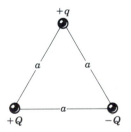

Figure 25 Problem 16.

17. Sketch qualitatively the lines of force associated with a thin, circular, uniformly charged disk of radius R. (*Hint*: Consider as limiting cases points very close to the disk, where the electric field is perpendicular to the surface, and points very far from it, where the electric field is like that of a point charge.)

18. Sketch qualitatively the lines of force associated with two separated point charges $+q$ and $-2q$.

19. Three charges are arranged in an equilateral triangle as in Fig. 26. Consider the lines of force due to $+Q$ and $-Q$, and from them identify the direction of the force that acts on $+q$ because of the presence of the other two charges. (*Hint*: See Fig. 8.)

Figure 26 Problem 19.

20. (*a*) In Fig. 27, locate the point (or points) at which the electric field is zero. (*b*) Sketch qualitatively the lines of force.

Figure 27 Problem 20.

21. Two point charges are fixed at a distance d apart (Fig. 28). Plot $E(x)$, assuming $x = 0$ at the left-hand charge. Consider both positive and negative values of x. Plot E as positive if $\mathbf{E}$ points to the right and negative if $\mathbf{E}$ points to the left. As-

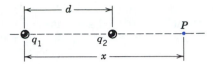

Figure 28 Problem 21.

sume $q_1 = +1.0 \times 10^{-6}$ C, $q_2 = +3.0 \times 10^{-6}$ C, and $d = 10$ cm.

22. Charges $+q$ and $-2q$ are fixed a distance d apart as in Fig. 29. (*a*) Find $\mathbf{E}$ at points A, B, and C. (*b*) Sketch roughly the electric field lines.

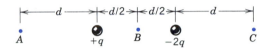

Figure 29 Problem 22.

23. Assume that the exponent in Coulomb's law is not 2 but n. Show that for $n \neq 2$ it is impossible to construct lines that will have the properties listed for lines of force in Section 28-4. For simplicity, treat an isolated point charge.

Section 28-5 The Electric Field of Continuous Charge Distributions

24. Show that Eq. 27, for the electric field of a charged disk at points on its axis, reduces to the field of a point charge for $z \gg R$.

25. At what distance along the axis of a charged disk of radius R is the electric field strength equal to one-half the value of the field at the surface of the disk at the center?

26. At what distance along the axis of a charged ring of radius R is the axial electric field strength a maximum?

27. (*a*) What total charge q must a disk of radius 2.50 cm carry in order that the electric field on the surface of the disk at its center equals the value at which air breaks down electrically, producing sparks? See Table 1. (*b*) Suppose that each atom at the surface has an effective cross-sectional area of 0.015 nm². How many atoms are at the disk's surface? (*c*) The charge in (*a*) results from some of the surface atoms carrying one excess electron. What fraction of the surface atoms must be so charged?

28. Write Eq. 27 in a form that is valid for negative as well as positive z. (*Hint*: In carrying out the integral in Eq. 26, the quantity $z/\sqrt{z^2}$ is obtained. What is the value of this quantity for $z < 0$?)

29. Measured values of the electric field E a distance z along the axis of a charged plastic disk are given below:

z (cm)	E (10^7 N/C)
0	2.043
1	1.732
2	1.442
3	1.187
4	0.972
5	0.797

Calculate (*a*) the radius of the disk and (*b*) the charge on it.

30. A thin glass rod is bent into a semicircle of radius r. A charge $+q$ is uniformly distributed along the upper half and a charge $-q$ is uniformly distributed along the lower half, as shown in Fig. 30. Find the electric field **E** at P, the center of the semicircle.

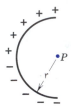

Figure 30   Problem 30.

31. A thin nonconducting rod of finite length L carries a total charge q, spread uniformly along it. Show that E at point P on the perpendicular bisector in Fig. 31 is given by

$$E = \frac{q}{2\pi\epsilon_0 y} \frac{1}{(L^2 + 4y^2)^{1/2}}.$$

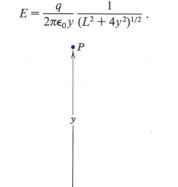

Figure 31   Problem 31.

32. An insulating rod of length L has charge $-q$ uniformly distributed along its length, as shown in Fig. 32. (a) What is the linear charge density of the rod? (b) Find the electric field at point P a distance a from the end of the rod. (c) If P were very far from the rod compared to L, the rod would look like a point charge. Show that your answer to (b) reduces to the electric field of a point charge for $a \gg L$.

Figure 32   Problem 32.

33. Sketch qualitatively the lines of force associated with three long parallel lines of charge, in a perpendicular plane. Assume that the intersections of the lines of charge with such a plane form an equilateral triangle (Fig. 33) and that each line of charge has the same linear charge density λ.

34. A "semi-infinite" insulating rod (Fig. 34) carries a constant charge per unit length of λ. Show that the electric field at the

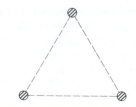

Figure 33   Problem 33.

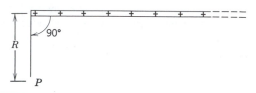

Figure 34   Problem 34.

point P makes an angle of 45° with the rod and that this result is independent of the distance R.

35. A nonconducting hemispherical cup of inner radius R has a total charge q spread uniformly over its inner surface. Find the electric field at the center of curvature. (*Hint*: Consider the cup as a stack of rings.)

Section 28-6 A Point Charge in an Electric Field

36. One defensive weapon being considered for the Strategic Defense Initiative (Star Wars) uses particle beams. For example, a proton beam striking an enemy missile could render it harmless. Such beams can be produced in "guns" using electric fields to accelerate the charged particles. (a) What acceleration would a proton experience if the electric field is 2.16×10^4 N/C? (b) What speed would the proton attain if the field acts over a distance of 1.22 cm?

37. An electron moving with a speed of 4.86×10^6 m/s is shot parallel to an electric field of strength 1030 N/C arranged so as to retard its motion. (a) How far will the electron travel in the field before coming (momentarily) to rest and (b) how much time will elapse? (c) If the electric field ends abruptly after 7.88 mm, what fraction of its initial kinetic energy will the electron lose in traversing it?

38. A uniform electric field exists in a region between two oppositely charged plates. An electron is released from rest at the surface of the negatively charged plate and strikes the surface of the opposite plate, 1.95 cm away, 14.7 ns later. (a) What is the speed of the electron as it strikes the second plate? (b) What is the magnitude of the electric field?

39. Two equal and opposite charges of magnitude 1.88×10^{-7} C are held 15.2 cm apart. (a) What are the magnitude and direction of **E** at a point midway between the charges? (b) What force (magnitude and direction) would act on an electron placed there?

40. Two point charges of magnitudes $q_1 = 2.16$ μC and $q_2 = 85.3$ nC are 11.7 cm apart. (a) Find the magnitude of the electric field that each produces at the site of the other. (b) Find the magnitude of the force on each charge.

41. In Millikan's experiment, a drop of radius 1.64 μm and density 0.851 g/cm³ is balanced when an electric field of

1.92×10^5 N/C is applied. Find the charge on the drop, in terms of e.

42. Two large parallel copper plates are 5.00 cm apart and have a uniform electric field between them as depicted in Fig. 35. An electron is released from the negative plate at the same time that a proton is released from the positive plate. Neglect the force of the particles on each other and find their distance from the positive plate when they pass each other. Does it surprise you that you need not know the electric field to solve this problem?

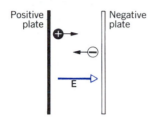

Figure 35 Problem 42.

43. In a particular early run (1911), Millikan observed that the following measured charges, among others, appeared at different times on a single drop:

6.563×10^{-19} C	13.13×10^{-19} C	19.71×10^{-19} C
8.204×10^{-19} C	16.48×10^{-19} C	22.89×10^{-19} C
11.50×10^{-19} C	18.08×10^{-19} C	26.13×10^{-19} C

What value for the quantum of charge e can be deduced from these data?

44. A uniform vertical field $\mathbf{E}$ is established in the space between two large parallel plates. A small conducting sphere of mass m is suspended in the field from a string of length L. Find the period of this pendulum when the sphere is given a charge $+q$ if the lower plate (a) is charged positively and (b) is charged negatively.

45. In Sample Problem 6, find the total deflection of the ink drop upon striking the paper 6.8 mm from the end of the deflection plates; see Fig. 14.

46. An electron is constrained to move along the axis of the ring of charge discussed in Section 28-5. Show that the electron can perform small oscillations, through the center of the ring, with a frequency given by

$$\omega = \sqrt{\frac{eq}{4\pi\epsilon_0 mR^3}} \; .$$

47. An electron is projected as in Fig. 36 at a speed of $v_0 = 5.83 \times 10^6$ m/s and at an angle of $\theta = 39.0°$; $\mathbf{E} = 1870$ N/C (directed upward), $d = 1.97$ cm, and $L = 6.20$ cm. Will the electron strike either of the plates? If it strikes a plate, which plate does it strike and at what distance from the left edge?

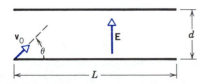

Figure 36 Problem 47.

Section 28-7 A Dipole in an Electric Field

48. An electric dipole, consisting of charges of magnitude 1.48 nC separated by 6.23 μm, is in an electric field of strength 1100 N/C. (a) What is the magnitude of the electric dipole moment? (b) What is the difference in potential energy corresponding to dipole orientations parallel and antiparallel to the field?

49. An electric dipole consists of charges $+2e$ and $-2e$ separated by 0.78 nm. It is in an electric field of strength 3.4×10^6 N/C. Calculate the magnitude of the torque on the dipole when the dipole moment is (a) parallel, (b) at a right angle, and (c) opposite to the electric field.

50. A charge $q = 3.16$ μC is 28.5 cm from a small dipole along its perpendicular bisector. The force on the charge equals 5.22×10^{-16} N. Show on a diagram (a) the direction of the force on the charge and (b) the direction of the force on the dipole. Determine (c) the magnitude of the force on the dipole and (d) the dipole moment of the dipole.

51. Find the work required to turn an electric dipole end for end in a uniform electric field $\mathbf{E}$, in terms of the dipole moment $\mathbf{p}$ and the initial angle θ_0 between $\mathbf{p}$ and $\mathbf{E}$.

52. Find the frequency of oscillation of an electric dipole, of moment p and rotational inertia I, for small amplitudes of oscillation about its equilibrium position in a uniform electric field E.

53. Consider two equal positive point charges $+q$ a distance a apart. (a) Derive an expression for dE/dz at the point midway between them, where z is the distance from the midpoint along the line joining the charges. (b) Show that the force on a small dipole placed at this point, its axis along the line joining the charges, is given by $F = p(dE/dz)$, where p is the dipole moment.

Computer Projects

54. (a) Write a computer program or design a spreadsheet to compute the components of the electric field due to a collection of point charges. Input the number of particles, their charges, and the coordinates of their positions. Then input the coordinates of the field point. Arrange the program so it returns to accept the coordinates of a new field point after it displays the field components for the previous point. For simplicity, assume that all charges are in the xy plane and the field point is also in that plane. If charge q_i has coordinates x_i and y_i then its contribution to the field at x, y is $E_{ix} = (1/4\pi\epsilon_0)q_i(x - x_i)/r_i^3$, $E_{iy} = (1/4\pi\epsilon_0)q_i(y - y_i)/r_i^3$, $E_{iz} = 0$, where $r_i = \sqrt{(x - x_i)^2 + (y - y_i)^2}$. Also have the computer calculate the magnitude of the field and the angle it makes with the x axis.

(b) Suppose two charges are located on the x axis: $q_1 = 6.0 \times 10^{-9}$ C at $x_1 = -0.030$ m and $q_2 = 3.0 \times 10^{-9}$ C at $x_2 = 0.030$ m. Use your program to calculate the electric field at the following points along the y axis: $y = 0, 0.050, 0.100, 0.150,$ and 0.200 m. Draw a diagram showing the positions of the charges and at each field point draw an arrow to represent the electric field. Its length should be proportional to the magnitude of the field there and it should make the proper angle with the x axis. You might have the program draw the vectors on the monitor screen.

(c) Now use the program to find the electric field at the following points on the y axis: $y = -0.050, -0.100, -0.150$, and -0.200 m. Draw the field vectors on the diagram. What is the relationship between the x component of the field at $y = +0.050$ m and the x component at $y = -0.050$ m? What is the relationship between the y components at these points? Do the same relationships hold for the field at other pairs of points?

55. Two charges are located on the x axis: $q_1 = -3.0 \times 10^{-9}$ C at $x_1 = -0.075$ m and $q_2 = 3.0 \times 10^{-9}$ C at $x_2 = 0.075$ m. Use the program described in the previous problem to find the electric field at the following points on the line $y = 0.030$ m: $x = -0.150, -0.100, -0.050, 0, 0.050, 0.100$, and 0.150 m. Draw a diagram showing the positions of the charges, and at each field point draw an arrow that depicts the direction and magnitude of the electric field at that point. You might program the computer to draw the arrows on the monitor screen.

   By considering the fields of the individual charges, explain qualitatively why the y component of the field is negative for field points with negative x components, zero for $x = 0$, and positive for field points with positive x coordinates. Also explain why the x component of the field reverses sign twice in the region considered. Without making a new calculation, draw field vectors at as many points as you can along the line $y = -0.030$ m.

56. (a) Two charges are located on the x axis: $q_1 = 3.0 \times 10^{-9}$ C at $x_1 = -0.075$ m and $q_2 = 6.0 \times 10^{-9}$ C at $x_2 = 0.075$ m. Use the program described previously with a trial and error technique to find the coordinates of a point where the total electric field vanishes. (b) Do the same for $q_1 = -3.0 \times 10^{-9}$ C, with q_2 and the positions of the charges as before.

57. You can use a computer to plot electric field lines. Consider charges in the xy plane and plot lines in that plane. Pick a point, with coordinates x and y. Calculate the field components E_x and E_y and magnitude E for that point. Another point on the same field line has coordinates $x + \Delta x$ and $y + \Delta y$, where $\Delta x = (E_x/E)\Delta s$, $\Delta y = (E_y/E)\Delta s$, and Δs is the distance from the first point. These expressions are approximations that are valid for Δs small. The line that joins the points is tangent to the field somewhere between them and is therefore along the field line, provided the curvature of the line between the points can be ignored. The field components and magnitude are computed for the new point and the process is repeated.

   (a) Write a computer program or design a spreadsheet to compute and plot the coordinates of points on a field line. Input the charges, their coordinates, the coordinates of the initial point on the line, and the distance Δs between adjacent points on the line. Have the computer list or plot a series of points, but have it stop when the points get far from the charges or close to any one charge. You may want to compute the coordinates of more points than are displayed. This keeps Δs small but does not generate an overwhelmingly large list.

   (b) Consider an electric dipole. Charge $q_1 = 7.1 \times 10^{-9}$ C is located at the origin and charge $q_2 = -7.1 \times 10^{-9}$ C is located on the y axis at $y = -0.40$ m. Plot four field lines. Start one at $x = 5 \times 10^{-3}$ m, $y = 5 \times 10^{-3}$ m, the second at $x = 5 \times 10^{-3}$ m, $y = -5 \times 10^{-3}$ m, the third at $x = -5 \times 10^{-3}$ m, $y = 5 \times 10^{-3}$ m, and the fourth at $x = -5 \times 10^{-3}$ m, $y = -5 \times 10^{-3}$ m. Take $\Delta s = 0.004$ m and continue plotting as long as the points are less than 2 m from the origin and greater than Δs from either charge. Draw the field line through the points.

   (c) Repeat for $q_1 = q_2 = 7.1 \times 10^{-9}$ C and all else the same. Draw four additional lines, one starting at $x = 5 \times 10^{-3}$ m, $y = -0.395$ m, the second at $x = 5 \times 10^{-3}$ m, $y = -0.405$ m, the third at $x = -5 \times 10^{-3}$ m, $y = -0.395$ m, and the fourth at $x = -5 \times 10^{-3}$ m, $y = -0.405$ m.

58. The computer program described in Appendix I can be used to investigate the motion of a particle in an electric field. Consider two particles that exert electric forces on each other. Each accelerates in response to the electric field of the other, and as their positions change the forces they exert also change.

   Two identical particles, each with charge $q = 1.9 \times 10^{-9}$ C and mass $m = 6.1 \times 10^{-15}$ kg, start with identical velocities of 3.0×10^4 m/s in the positive x direction. Initially one is at $x = 0$, $y = 6.7 \times 10^{-3}$ m and the other is at $x = 0$, $y = -6.7 \times 10^{-3}$ m. Both are in the xy plane and continue to move in that plane. Consider only the electric forces they exert on each other.

   (a) Use a computer program to plot the trajectories from time $t = 0$ to $t = 1.0 \times 10^{-6}$ s. Because the situation is symmetric you need calculate only the position and velocity of one of the charges. Use symmetry to find the position and velocity of the other at the beginning of each integration interval. Use $\Delta t = 1 \times 10^{-8}$ s for the integration interval.

   (b) Now suppose that one of the particles has charge $q = -1.9 \times 10^{-9}$ C, but all other conditions are the same. Plot the trajectories from $t = 0$ to $t = 5.0 \times 10^{-7}$ s.

CHAPTER 29

GAUSS' LAW

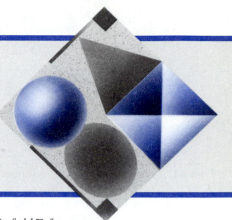

*Coulomb's law can always be used to calculate the electric field **E** for
any discrete or continuous distribution of charges at rest. The sums or integrals
might be complicated (and a computer might be needed to evaluate them numerically), but
the resulting electric field can always be found.*

*Some cases discussed in the previous chapter used simplifying arguments based on the
symmetry of the physical situation. For example, in calculating the electric field at points on
the axis of a charged circular loop, we used a symmetry argument to deduce that
components of **E** perpendicular to the axis must vanish. In this chapter we discuss an
alternative to Coulomb's law, called Gauss' law, that provides a more useful and instructive
approach to calculating the electric field in situations having certain symmetries.*

*The number of situations that can directly be analyzed using Gauss' law is small, but those
cases can be done with extraordinary ease. Although Gauss' law and Coulomb's law give
identical results in the cases in which both can be used, Gauss' law is considered a more
fundamental equation than Coulomb's law. It is fair to say that while Coulomb's law
provides the workhorse of electrostatics, Gauss' law provides the insight.*

29-1 THE FLUX OF A VECTOR FIELD

Before we discuss Gauss' law, we must first understand the concept of *flux*. The flux (symbol Φ) is a property of any vector field. The word "flux" comes from a Latin word meaning "to flow," and it is appropriate to think of the flux of a particular vector field as being a measure of the "flow" or penetration of the field vectors through an imaginary fixed surface in the field. We shall eventually consider the flux of the electric field for Gauss' law, but for now we discuss a more familiar example of a vector field, namely, the velocity field of a flowing fluid. Recall from Chapter 18 that the velocity field gives the velocity at points through which the fluid flows. The velocity field represents the fluid flow; the field itself is not flowing but is a *fixed* representation of the flow.

Figure 1 shows a field of incompressible fluid flow, which we assume for simplicity to be steady and uniform. Imagine that we place into the stream a wire bent into the shape of a square loop of area A. In Fig. 1a, the square is

placed so that its plane is perpendicular to the direction of flow. In our analysis of fluid flow (Chapter 18), we replaced the actual motion of the fluid particles by the velocity field associated with the flow. Therefore we can consider either the actual flow of material particles through the loop or the flux of the velocity field through the loop. The field concept gives us the abstraction we shall later need for Gauss' law, but of course the flow through the loop could just as well be described in terms of the fluid particles themselves.

The magnitude |Φ| of the flux of the velocity field through the loop of area A in Fig. 1a is written in terms of the volume rate of fluid flow (in units of m³/s, say) as

$$|\Phi| = vA \qquad (1)$$

in which v is the magnitude of the velocity at the location of the loop. The flux can, on the one hand, be considered as a measure of the rate at which fluid passes through the loop. In terms of the field concept (and for the purpose of introducing Gauss' law), however, it is convenient to consider it as a measure of the *number of field lines passing through the loop.*

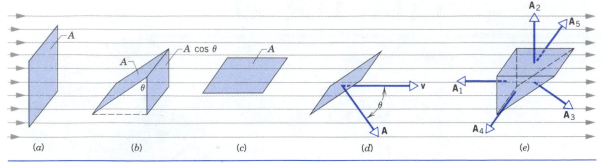

Figure 1   A wire loop of area A is immersed in a flowing stream, which we represent as a velocity field. (*a*) The loop is at right angles to the flow. (*b*) The loop is turned through an angle θ; the projection of the area perpendicular to the flow is $A \cos \theta$. (*c*) When $\theta = 90°$, none of the streamlines pass through the plane of the loop. (*d*) The area of the loop is represented by a vector $\mathbf{A}$ perpendicular to the plane of the loop. The angle between $\mathbf{A}$ and the flow velocity $\mathbf{v}$ is θ. (*e*) A closed surface made of five plane surfaces. The area $\mathbf{A}$ of each surface is represented by the outward normal.

In Fig. 1*b*, the loop has been rotated so that its plane is no longer perpendicular to the direction of the velocity. Note that the number of lines of the velocity field passing through the loop is smaller in Fig. 1*b* than in Fig. 1*a*. The projected area of the square is $A \cos \theta$, and by examining Fig. 1*b* you should convince yourself that the number of field lines passing through the inclined loop of area A is the same as the number of field lines passing through the smaller loop of area $A \cos \theta$ perpendicular to the stream. Thus the magnitude of the flux in the situation of Fig. 1*b* is

$$|\Phi| = vA \cos \theta. \qquad (2)$$

If the loop were rotated so that the fluid velocity were parallel to its surface, as in Fig. 1*c*, the flux would be zero, corresponding to $\theta = 90°$ in Eq. 2. Note that in this case no field lines pass through the loop.

Gauss' law, as we shall see, concerns the net flux through a *closed* surface. We must therefore distinguish between positive and negative flux penetrating a surface. The right side of Eq. 2 can be expressed in terms of the dot product between $\mathbf{v}$ and a vector $\mathbf{A}$ whose magnitude is the area of the surface and whose direction is perpendicular to the surface (Fig. 1*d*). However, since the normal to a surface can point either in the direction shown in Fig. 1*d* or in the reverse direction, we must have a way to specify that direction; otherwise the sign of Φ will not be clearly defined. By convention, we choose the direction of $\mathbf{A}$ to be that of the *outward normal* from a closed surface. Thus flux leaving the volume enclosed by the surface is considered positive, and flux entering the volume is considered negative. With this choice, we can then write the flux for a closed surface consisting of several individual surfaces (Fig. 1*e*, for example) as

$$\Phi = \sum \mathbf{v} \cdot \mathbf{A}, \qquad (3)$$

where $\mathbf{v}$ is the velocity vector at the surface. The sum is carried out over all the individual surfaces that make up a closed surface. The flux is a scalar quantity, because it is defined in terms of the dot product of two vectors.

Sample Problem 1   Consider the closed surface of Fig. 1*e*, which shows a volume enclosed by five surfaces (1, 2, and 3, which are parallel to the surfaces of Figs. 1*a*, 1*b*, and 1*c*, along with 4 and 5, which are parallel to the streamlines). Assuming the velocity field is uniform, so that it has the same magnitude and direction everywhere, find the total flux through the closed surface.

Solution   Using Eq. 3 we can write the total flux as the sum of the values of the flux through each of the five separate surfaces:

$$\Phi = \mathbf{v} \cdot \mathbf{A}_1 + \mathbf{v} \cdot \mathbf{A}_2 + \mathbf{v} \cdot \mathbf{A}_3 + \mathbf{v} \cdot \mathbf{A}_4 + \mathbf{v} \cdot \mathbf{A}_5.$$

Note that for surface 1 the angle between the *outward normal* $\mathbf{A}_1$ and the velocity $\mathbf{v}$ is 180°, so that the dot product $\mathbf{v} \cdot \mathbf{A}_1$ can be written $-vA_1$. The contributions from surfaces 2, 4, and 5 all vanish, because in each case (as shown in Fig. 1*e*) the vector $\mathbf{A}$ is perpendicular to $\mathbf{v}$. For surface A_3, the flux can be written $vA_3 \cos \theta$, and thus the total flux is

$$\Phi = -vA_1 + 0 + vA_3 \cos \theta + 0 + 0 = -vA_1 + vA_3 \cos \theta.$$

However, from the geometry of Fig. 1*e* we conclude that $A_3 \cos \theta = A_1$, and as a result we obtain

$$\Phi = 0.$$

That is, the total flux through the closed surface is zero.

The result of the previous sample problem should not be surprising if we remember that the velocity field is an equivalent way of representing the actual flow of material particles in the stream. Every field line that enters the closed surface of Fig. 1*e* through surface 1 leaves through surface 3. Equivalently, we can state that, for the closed surface shown in Fig. 1*e*, the net amount of fluid entering the volume enclosed by the surface is equal to the net amount of fluid leaving the volume. This is to be expected *for any closed surface* if there are within the volume no *sources* or *sinks* of fluid, that is, locations at which new fluid is created or flowing fluid is trapped. If there were a source within the volume (such as a melting ice cube that introduced additional fluid into the stream), then more fluid would leave the surface than entered it, and the total flux would be positive. If there were a sink within the volume, then more fluid would enter than would leave, and the net flux would be negative. The net positive or negative flux through the surface depends on the strength of the source or sink (that is, on the volume rate at which fluid leaves the source or enters the sink). For example, if a melting solid inside the surface released 1 cm³ of fluid per second into the stream, then we would find the net flux through the closed surface to be +1 cm³/s.

Figure 1 showed the special case of a uniform field and planar surfaces. We can easily generalize these concepts to a nonuniform field and to surfaces of arbitrary shape and orientation. Any arbitrary surface can be divided into infinitesimal elements of area dA that are approximately plane surfaces. The direction of the vector $d\mathbf{A}$ is that of the outward normal to this infinitesimal element. The field has a value $\mathbf{v}$ at the site of this element, and the net flux is found by adding the contributions of all such elements, that is, by integrating over the entire surface:

$$\Phi = \int \mathbf{v} \cdot d\mathbf{A}. \qquad (4)$$

The conclusions we derived above remain valid in this general case: if Eq. 4 is evaluated over a closed surface, then the flux is (1) *zero* if the surface encloses no sources or sinks, (2) *positive* and equal in magnitude to their strength if the surface contains only sources, or (3) *negative* and equal in magnitude to their strength if the surface contains only sinks. If the surface encloses both sources *and* sinks, the net flux can be zero, positive, or negative, depending on the relative strength of the sources and sinks.

For another example, consider the gravitational field **g** (see Section 16-7) near the Earth's surface, which (like the velocity field) is a fixed vector field. The net flux of **g** through any closed but empty container is zero. If the container encloses matter (sources of **g**), then more flux leaves the surface than enters it, and the net flux of **g** through the container is positive.

In the next section we apply similar considerations to the flux of another vector field, namely, the electric field **E**. As you might anticipate, when we discuss electrostatics the sources or sinks of the field are positive or negative charges, and the strengths of the sources or sinks are proportional to the magnitudes of the charges. Gauss' law relates the flux of the electric field through a closed surface, calculated by analogy with Eq. 4, to the net electric charge enclosed by the surface.

29-2   THE FLUX OF THE ELECTRIC FIELD

Imagine the field lines in Fig. 1 to represent an electric field of charges at rest rather than a velocity field. Even though nothing is flowing in the electrostatic case, we still use the concept of flux. The definition of electric flux is similar to that of velocity flux, with **E** replacing **v** wherever it appears. In analogy with Eq. 3, we define the flux of the electric field Φ_E as

$$\Phi_E = \sum \mathbf{E} \cdot \mathbf{A}. \qquad (5)$$

As was the case with the velocity flux, the flux Φ_E can be considered as a measure of the number of lines of the electric field that pass through the surface. The subscript E on Φ_E reminds us that we are speaking of the *electric* flux and serves to distinguish electric from magnetic flux, which we consider in Chapter 36. Equation 5 applies, as did Eq. 3, only to cases in which **E** is constant in magnitude and direction over each area **A** included in the sum.

Like the velocity flux, the flux of the electric field is a scalar. Its units are, from Eq. 5, N·m²/C.

Gauss' law deals with the flux of the electric field through a closed surface. To define Φ_E more generally, particularly in cases in which **E** is not uniform, consider Fig. 2, which shows an arbitrary closed surface immersed in a nonuniform electric field. Let us divide the surface into small squares of area ΔA, each of which is small enough so that it may be considered to be plane. Each element of area can be represented as a vector $\Delta \mathbf{A}$ whose magnitude is the area ΔA. The direction of $\Delta \mathbf{A}$ is taken as the outward-drawn normal to the surface, as in Fig. 1. Since the squares have been made very small, **E** may be taken as constant for all points on a given square.

The vectors **E** and $\Delta \mathbf{A}$ that characterize each square make an angle θ with each other. Figure 2 shows an enlarged view of three squares on the surface, marked a, b, and c. Note that at a, $\theta > 90°$ (**E** points in); at b, $\theta = 90°$ (**E** is parallel to the surface); and at c, $\theta < 90°$ (**E** points out).

A provisional definition of the total flux of the electric field over the surface is, by analogy with Eq. 5,

$$\Phi_E = \sum \mathbf{E} \cdot \Delta \mathbf{A}, \qquad (6)$$

which instructs us to add up the scalar quantity $\mathbf{E} \cdot \Delta \mathbf{A}$ for all elements of area into which the surface has been divided. For points such as a in Fig. 2 the contribution to the flux is negative; at b it is zero, and at c it is positive. Thus if **E** is everywhere outward ($\theta < 90°$), each $\mathbf{E} \cdot \Delta \mathbf{A}$ is positive, and Φ_E for the entire surface is positive. If **E** is everywhere inward ($\theta > 90°$), each $\mathbf{E} \cdot \Delta \mathbf{A}$ is negative, and Φ_E for the surface is negative. Whenever **E** is everywhere parallel to a surface ($\theta = 90°$), each $\mathbf{E} \cdot \Delta \mathbf{A}$ is zero, and Φ_E for the surface is zero.

The exact definition of electric flux is found in the

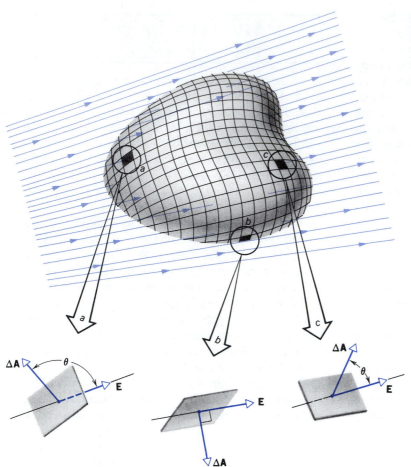

differential limit of Eq. 6. Replacing the sum over the surface by an integral over the surface yields

$$\Phi_E = \oint \mathbf{E} \cdot d\mathbf{A}. \tag{7}$$

This *surface integral* indicates that the surface in question is to be divided into infinitesimal elements of area $d\mathbf{A}$ and that the scalar quantity $\mathbf{E} \cdot d\mathbf{A}$ is to be evaluated for each element and summed over the entire surface. The circle on the integral sign indicates that the surface of integration is a *closed* surface. The flux can be calculated for any surface, whether closed or open; in Gauss' law, which we introduce in the next section, we are concerned only with closed surfaces.

Sample Problem 2 Figure 3 shows a hypothetical closed cylinder of radius R immersed in a uniform electric field **E**, the cylinder axis being parallel to the field. What is Φ_E for this closed surface?

Solution The flux Φ_E can be written as the sum of three terms, an integral over (*a*) the left cylinder cap, (*b*) the cylindrical surface, and (*c*) the right cap. Thus, from Eq. 7,

$$\Phi_E = \oint \mathbf{E} \cdot d\mathbf{A}$$

$$= \int_a \mathbf{E} \cdot d\mathbf{A} + \int_b \mathbf{E} \cdot d\mathbf{A} + \int_c \mathbf{E} \cdot d\mathbf{A}.$$

For the left cap, the angle θ for all points is 180°, **E** has a constant value, and the vectors $d\mathbf{A}$ are all parallel. Thus

$$\int_a \mathbf{E} \cdot d\mathbf{A} = \int E \, dA \cos 180° = -E \int dA = -EA,$$

where $A \,(= \pi R^2)$ is the area of the left cap. Similarly, for the right cap,

$$\int_c \mathbf{E} \cdot d\mathbf{A} = +EA,$$

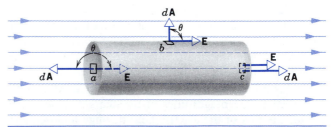

Figure 3 Sample Problem 2. A closed cylinder is immersed in a uniform electric field **E** parallel to its axis.

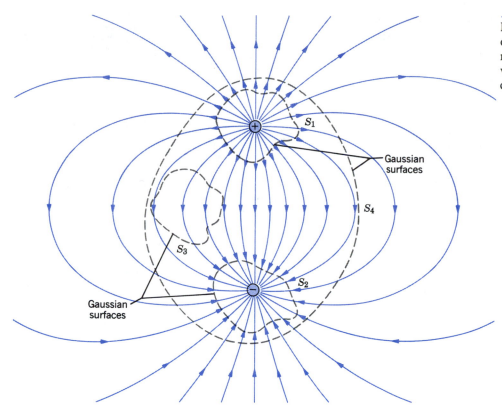

Figure 4 Two equal and opposite charges and the lines of force that represent the electric field in their vicinity. The cross sections of four closed Gaussian surfaces are shown.

the angle θ for all points being 0 here. Finally, for the cylinder wall,

$$\int_b \mathbf{E} \cdot d\mathbf{A} = 0,$$

because $\theta = 90°$; hence $\mathbf{E} \cdot d\mathbf{A} = 0$ for all points on the cylindrical surface. Thus the total flux is

$$\Phi_E = -EA + 0 + EA = 0.$$

As we shall see in the next section, this result is expected, because there are no sources or sinks of $\mathbf{E}$ (that is, charges) within the closed surface of Fig. 3. Lines of (constant) $\mathbf{E}$ enter at the left and emerge at the right, just as in Fig. 1e.

29-3 GAUSS' LAW*

Now that we have defined the flux of the electric field vector through a *closed* surface, we are ready to write Gauss' law. Let us suppose we have a collection of positive and negative charges, which establish an electric field $\mathbf{E}$

* Carl Friedrich Gauss (1777 – 1855) was a German mathematician who made substantial discoveries in number theory, geometry, and probability. He also contributed to astronomy and to measuring the size and shape of the Earth. See "Gauss," by Ian Stewart, *Scientific American*, July 1977, p. 122, for a fascinating account of the life of this remarkable mathematician.

throughout a certain region of space. We construct in that space an imaginary closed surface called a *Gaussian surface*, which may or may not enclose some of the charges. Gauss' law, which relates the total flux Φ_E through this surface to the *net* charge q enclosed by the surface, can be stated as

$$\epsilon_0 \Phi_E = q \qquad (8)$$

or

$$\epsilon_0 \oint \mathbf{E} \cdot d\mathbf{A} = q. \qquad (9)$$

We see that Gauss' law predicts that Φ_E is zero for the surface considered in Sample Problem 2, because the surface encloses no charge.

As discussed in Section 28-4, the magnitude of the electric field is proportional to the number of field lines crossing an element of area perpendicular to the field. The integral in Eq. 9 essentially counts the number of field lines passing through the surface. It is entirely reasonable that the number of field lines passing through a surface should be proportional to the net charge enclosed by the surface, as Eq. 9 requires.

The choice of the Gaussian surface is arbitrary. It is usually chosen so that the symmetry of the distribution gives, on at least part of the surface, a constant electric field, which can then be factored out of the integral of Eq. 9. In such a situation, Gauss' law can be used to evaluate the electric field.

Figure 4 shows the lines of force (and thus of electric field) of a dipole. Four closed Gaussian surfaces have been drawn, the cross sections of which are shown in the figure. On surface S_1, the electric field is everywhere outward from the surface and thus, as was the case with surface element c of Fig. 2, $\mathbf{E} \cdot d\mathbf{A}$ is everywhere positive on S_1. When we evaluate the integral of Eq. 9 over the entire closed surface, we get a positive result. Equation 9 then demands that the surface must enclose a net positive charge, as is the case. In Faraday's terminology, more lines of force leave the surface than enter it, so it must enclose a net positive charge.

On surface S_2 of Fig. 4, on the other hand, the electric field is everywhere entering the surface. Like surface element a in Fig. 2, $\mathbf{E} \cdot d\mathbf{A}$ is negative for every element of area, and the integral of Eq. 9 gives a negative value, which indicates that the surface encloses a net negative charge (as is the case). More lines of force enter the surface than leave it.

Surface S_3 encloses no charge at all, so according to Gauss' law the total flux through the surface must be zero. This is consistent with the figure, which shows that as many lines of force enter the top of the surface as leave the bottom. This is no accident; you can draw a surface in Fig. 4 of any irregular shape, and as long as it encloses neither of the charges, the number of field lines that enter the surface equals the number that leave the surface.

Surface S_4 also encloses no *net* charge, since we assumed the magnitudes of the two charges to be equal. Once again, the total flux through the surface should be zero. Some of the field lines are wholly contained within the surface and therefore don't contribute to the flux *through* the surface. However, since every field line that leaves the positive charge eventually terminates on the negative charge, every line from the positive charge that breaks the surface in the outward direction has a corresponding line that breaks the surface in the inward direction as it seeks the negative charge. The total flux is therefore zero.

Gauss' Law and Coulomb's Law

Coulomb's law can be deduced from Gauss' law and certain symmetry considerations. To do so, let us apply Gauss' law to an isolated positive point charge q as in Fig. 5. Although Gauss' law holds for any surface whatever, we choose a spherical surface of radius r centered on the charge. The advantage of this surface is that, from symmetry, $\mathbf{E}$ must be perpendicular to the surface, so the angle θ between $\mathbf{E}$ and $d\mathbf{A}$ is zero everywhere on the surface. Moreover, $\mathbf{E}$ is constant everywhere on the surface. *Constructing a Gaussian surface that takes advantage of such a symmetry is of fundamental importance in applying Gauss' law.*

In Fig. 5 both $\mathbf{E}$ and $d\mathbf{A}$ at any point on the Gaussian surface are directed radially outward, so the quantity $\mathbf{E} \cdot d\mathbf{A}$ becomes simply $E \, dA$. Gauss' law (Eq. 9) thus reduces to

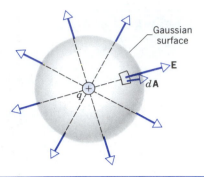

Figure 5 A spherical Gaussian surface surrounding a positive point charge q.

$$\epsilon_0 \oint \mathbf{E} \cdot d\mathbf{A} = \epsilon_0 \oint E \, dA = q.$$

Because E is constant for all points on the sphere, it can be factored from inside the integral sign, which gives

$$\epsilon_0 E \oint dA = q.$$

The integral is simply the total surface area of the sphere, $4\pi r^2$. We therefore obtain

$$\epsilon_0 E(4\pi r^2) = q$$

or

$$E = \frac{1}{4\pi\epsilon_0} \frac{q}{r^2}. \tag{10}$$

Equation 10 gives the magnitude of the electric field $\mathbf{E}$ at any point a distance r from an isolated point charge q and is identical to Eq. 4 of Chapter 28, which was obtained from Coulomb's law. Thus by choosing a Gaussian surface with the proper symmetry, we obtain Coulomb's law from Gauss' law. These two laws are totally equivalent when — as in these chapters — we apply them to problems involving charges that are either stationary or slowly moving. Gauss' law is more general in that it also covers the case of a rapidly moving charge. For such charges the electric lines of force become compressed in a plane at right angles to the direction of motion, thus losing their spherical symmetry.

Gauss' law is one of the fundamental equations of electromagnetic theory and is displayed in Table 2 of Chapter 40 as one of Maxwell's equations. Coulomb's law is not listed in that table because, as we have just proved, it can be deduced from Gauss' law and from simple assumptions about the symmetry of $\mathbf{E}$ due to a point charge.

It is interesting to note that writing the proportionality constant in Coulomb's law as $1/4\pi\epsilon_0$ permits a simpler form for Gauss' law. If we had written the Coulomb law constant simply as k, Gauss' law would have to be written as $(1/4\pi k)\Phi_E = q$. We prefer to leave the factor 4π in Coulomb's law so that it will not appear in Gauss' law or in other frequently used relations that are derived later.

29-4 A CHARGED ISOLATED CONDUCTOR

Gauss' law permits us to prove an important theorem about isolated conductors:

An excess charge placed on an isolated conductor moves entirely to the outer surface of the conductor. None of the excess charge is found within the body of the conductor. *

This might not seem unreasonable considering that like charges repel each other. You might imagine that, by moving to the surface, the added charges are getting as far away from each other as they can. We turn to Gauss' law for a quantitative proof of this qualitative speculation.

Figure 6a shows, in cross section, an isolated conductor (a lump of copper, perhaps) hanging from a thread and carrying a net positive charge q. The dashed line shows the cross section of a Gaussian surface that lies just inside the actual surface of the conductor.

The key to our proof is the realization that, under equilibrium conditions, the electric field inside the conductor must be zero. If this were not so, the field would exert a force on the conduction electrons that are present in any conductor, and internal currents would be set up. However, we know from experiment that there are no such enduring currents in an isolated conductor. Electric fields appear inside a conductor during the process of charging it, but these fields do not last long. Internal currents act quickly to redistribute the added charge in such a way that the electric fields inside the conductor vanish, the currents stop, and equilibrium (electrostatic) conditions prevail.

If **E** is zero everywhere inside the conductor, it must be zero for all points on the Gaussian surface because that surface, though close to the surface of the conductor, is definitely inside it. This means that the flux through the Gaussian surface must be zero. Gauss' law then tells us that the net charge inside the Gaussian surface must also be zero. If the added charge is not inside the Gaussian surface it can only be outside that surface, which means that *it must lie on the actual outer surface of the conductor.*

An Isolated Conductor with a Cavity

Figure 6b shows the same hanging conductor in which a cavity has been scooped out. It is perhaps reasonable to suppose that scooping out the electrically neutral material to form the cavity should not change the distribution of

---

* This statement does not apply to a wire carrying current, which cannot be considered an "isolated" conductor because it is connected to an external agent such as a battery.

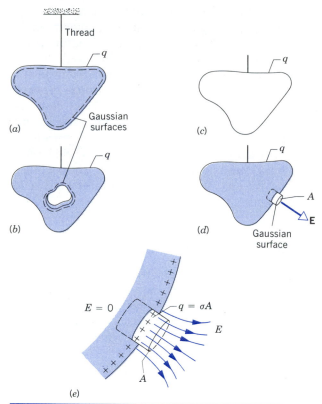

Figure 6 (a) An isolated metallic conductor carrying a charge q hangs from a thread. A Gaussian surface has been drawn just inside the surface of the conductor. (b) An internal cavity in the conductor is surrounded by a different Gaussian surface. (c) The cavity is enlarged so that it includes all of the interior of the original conductor, leaving only the charges that were on the surface. (d) A small Gaussian surface is constructed at the surface of the original conductor. (e) An enlarged view of the Gaussian surface, which encloses a charge q equal to σA. The electric field inside the conductor is zero, and the electric field just outside the conductor is perpendicular to the surface of the conductor and constant in magnitude.

charge or the pattern of the electric field that exists in Fig. 6a. Again, we turn to Gauss' law for a quantitative proof.

Draw a Gaussian surface surrounding the cavity, close to its walls but inside the conducting body. Because **E** = 0 inside the conductor, there can be no flux through this new Gaussian surface. Therefore, from Gauss' law, that surface can enclose no net charge. We conclude that there is no charge on the cavity walls; it remains on the *outer* surface of the conductor, as in Fig. 6a.

Suppose charges were placed inside the cavity. Gauss' law still requires that there be no *net* charge within the Gaussian surface, and so additional charges must be attracted to the surface of the cavity (just as charges were attracted to one end of the copper rod in Fig. 3 of Chapter 27) to make the net charge zero within the Gaussian surface.

Suppose now that, by some process, the excess charges could be "frozen" into position on the conductor surface of Fig. 6a, perhaps by embedding them in a thin plastic coating, and suppose that the conductor could then be removed completely, as in Fig. 6c. This is equivalent to enlarging the cavity of Fig. 6b until it consumes the entire conductor, leaving only the charges. The electric field pattern would not change at all; it would remain zero inside the thin shell of charge and would remain unchanged for all external points. The electric field is set up by the charges and not by the conductor. The conductor simply provides a pathway so that the charges can change their positions.

The External Electric Field

Although the excess charge on an isolated conductor moves entirely to its surface, that charge—except for an isolated spherical conductor—does not in general distribute itself uniformly over that surface. Put another way, the surface charge density σ $(= dq/dA)$ varies from point to point over the surface.

We can use Gauss' law to find a relation—at any surface point—between the surface charge density σ at that point and the electric field **E** just outside the surface at that same point. Figure 6d shows a squat cylindrical Gaussian surface, the (small) area of its two end caps being A. The end caps are parallel to the surface, one lying entirely inside the conductor and the other entirely outside. The short cylindrical walls are perpendicular to the surface of the conductor. An enlarged view of the Gaussian surface is shown in Fig. 6e.

The electric field just outside a charged isolated conductor in electrostatic equilibrium must be at right angles to the surface of the conductor. If this were not so, there would be a component of **E** lying in the surface and this component would set up surface currents that would redistribute the surface charges, violating our assumption of electrostatic equilibrium. Thus **E** is perpendicular to the surface of the conductor, and the flux through the exterior end cap of the Gaussian surface of Fig. 6e is EA. The flux through the interior end cap is zero, because $E = 0$ for all interior points of the conductor. The flux through the cylindrical walls is also zero because the lines of **E** are parallel to the surface, so they cannot pierce it. The charge q enclosed by the Gaussian surface is σA.

The total flux can then be calculated as

$$\Phi_E = \oint \mathbf{E} \cdot d\mathbf{A} = \int_{\substack{\text{outer} \\ \text{cap}}} \mathbf{E} \cdot d\mathbf{A} + \int_{\substack{\text{inner} \\ \text{cap}}} \mathbf{E} \cdot d\mathbf{A} + \int_{\substack{\text{side} \\ \text{walls}}} \mathbf{E} \cdot d\mathbf{A}$$

$$= EA + 0 + 0 = EA.$$

The electric field can now be found by using Gauss' law:

$$\epsilon_0 \Phi_E = q,$$

and substituting the values for the flux and the enclosed charge q $(= \sigma A)$, we find

$$\epsilon_0 EA = \sigma A$$

or

$$E = \frac{\sigma}{\epsilon_0}. \tag{11}$$

Compare this result with Eq. 28 of Chapter 28 (which we also derive in the next section using Gauss' law) for the electric field near a sheet of charge: $E = \sigma/2\epsilon_0$. The electric field near a conductor is *twice* the field we would expect if we considered the conductor to be a sheet of charge, even for points very close to the surface, where the immediate vicinity *does* look like a sheet of charge. How can we understand the difference between the two cases?

A sheet of charge can be constructed by spraying charges on one side of a thin layer of plastic. The charges stick where they land and are not free to move. We cannot charge a conductor in the same way. A thin layer of conducting material always has two surfaces. If we spray charge on one surface, it will travel throughout the conductor and distribute itself over all surfaces. Thus if we want to charge a thin conducting layer to a given surface charge density, we must supply enough charge to cover *both* surfaces. In effect, it takes twice as much charge to give a conducting sheet a given surface charge density as it takes to give an insulating sheet the same surface charge density.

We can understand the electric field in the case of the thin conducting sheet by referring to Fig. 7. If we regard each face of the conductor as a sheet of charge giving an electric field of $\sigma/2\epsilon_0$ (according to Eq. 28 of Chapter 28), then at point A the electric fields $\mathbf{E}_L$ from the left face and $\mathbf{E}_R$ from the right face add to give a total electric field near the conductor of $\sigma/2\epsilon_0 + \sigma/2\epsilon_0 = \sigma/\epsilon_0$. At point C, the effect is the same. At point B, however, the fields $\mathbf{E}_L$ and $\mathbf{E}_R$ are oppositely directed and sum to zero, as expected for the interior of a conductor.

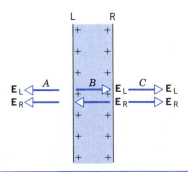

Figure 7   The electric charge near a thin conducting sheet. Note that both surfaces have charges on them. The fields $\mathbf{E}_L$ and $\mathbf{E}_R$ due, respectively, to the charges on the left and right surfaces reinforce at points A and C, and they cancel at points B in the interior of the sheet.

Sample Problem 3   The electric field just above the surface of the charged drum of a photocopying machine has a magnitude E of 2.3×10^5 N/C. What is the surface charge density on the drum if it is a conductor?

Solution   From Eq. 11 we have

$$\sigma = \epsilon_0 E = (8.85 \times 10^{-12} \text{ C}^2/\text{N} \cdot \text{m}^2)(2.3 \times 10^5 \text{ N/C})$$
$$= 2.0 \times 10^{-6} \text{ C/m}^2 = 2.0 \ \mu\text{C/m}^2.$$

Sample Problem 4   The magnitude of the average electric field normally present in the Earth's atmosphere just above the surface of the Earth is about 150 N/C, directed downward. What is the total net surface charge carried by the Earth? Assume the Earth to be a conductor.

Solution   Lines of force terminate on negative charges so that, if the Earth's electric field points downward, its average surface charge density σ must be negative. From Eq. 11 we find

$$\sigma = \epsilon_0 E = (8.85 \times 10^{-12} \text{ C}^2/\text{N} \cdot \text{m}^2)(-150 \text{ N/C})$$
$$= -1.33 \times 10^{-9} \text{ C/m}^2.$$

The Earth's total charge q is the surface charge density multiplied by $4\pi R^2$, the surface area of the (presumed spherical) Earth. Thus

$$q = \sigma 4\pi R^2$$
$$= (-1.33 \times 10^{-9} \text{ C/m}^2)(4\pi)(6.37 \times 10^6 \text{ m})^2$$
$$= -6.8 \times 10^5 \text{ C} = -680 \text{ kC}.$$

29-5   APPLICATIONS OF GAUSS' LAW

Gauss' law can be used to calculate $\mathbf{E}$ if the symmetry of the charge distribution is high. One example of this calculation, the field of a point charge, has already been discussed in connection with Eq. 10. Here we present other examples.

Infinite Line of Charge

Figure 8 shows a section of an infinite line of charge of constant linear charge density (charge per unit length) $\lambda = dq/ds$. We would like to find the electric field at a distance r from the line.

In Section 28-5 we discussed the symmetry arguments that lead us to conclude that the electric field in this case can have only a radial component. The problem therefore has cylindrical symmetry, and so as a Gaussian surface we choose a circular cylinder of radius r and length h, closed at each end by plane caps normal to the axis. E is constant over the cylindrical surface and perpendicular to the surface. The flux of $\mathbf{E}$ through this surface is $E(2\pi rh)$, where

$2\pi rh$ is the area of the surface. There is no flux through the circular caps because $\mathbf{E}$ here is parallel to the surface at every point, so that $\mathbf{E} \cdot d\mathbf{A} = 0$ everywhere on the caps.

The charge q enclosed by the Gaussian surface of Fig. 8 is λh. Gauss' law (Eq. 9) then gives

$$\epsilon_0 \oint \mathbf{E} \cdot d\mathbf{A} = q$$

$$\epsilon_0 E(2\pi rh) = \lambda h,$$

or

$$E = \frac{\lambda}{2\pi\epsilon_0 r}. \tag{12}$$

Note how much simpler is the solution using Gauss' law than that using integration methods, as in Chapter 28. Note too that the solution using Gauss' law is possible only if we choose our Gaussian surface to take full advantage of the cylindrical symmetry of the electric field set up by a long line of charge. We are free to choose any closed surface, such as a cube or a sphere (see Problem 48), for a Gaussian surface. Even though Gauss' law holds for all such surfaces, they are not all useful for the problem at hand; only the cylindrical surface of Fig. 8 is appropriate in this case.

Gauss' law has the property that it provides a useful

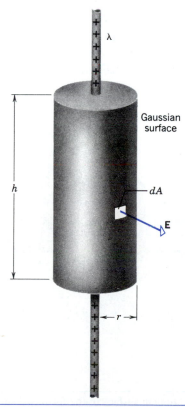

Figure 8   A Gaussian surface in the shape of a closed cylinder surrounds a portion of an infinite line of charge.

technique for calculation only in problems that have a certain degree of symmetry, but in these problems the solutions are strikingly simple.

Infinite Sheet of Charge

Figure 9 shows a portion of a thin, nonconducting, infinite sheet of charge of constant surface charge density σ (charge per unit area). We calculate the electric field at points near the sheet.

A convenient Gaussian surface is a closed cylinder of cross-sectional area A, arranged to pierce the plane as shown. From symmetry, we can conclude that **E** points at right angles to the end caps and away from the plane. Since **E** does not pierce the cylindrical surface, there is no contribution to the flux from the curved wall of the cylinder. We assume the end caps are equidistant from the sheet, and from symmetry the field has the same magnitude at the end caps. The flux through each end cap is EA and is positive for both. Gauss' law gives

$$\epsilon_0 \oint \mathbf{E} \cdot d\mathbf{A} = q$$

$$\epsilon_0(EA + EA) = \sigma A,$$

where σA is the enclosed charge. Solving for E, we obtain

$$E = \frac{\sigma}{2\epsilon_0}. \tag{13}$$

Note that E is the same for all points on each side of the sheet (and so we really didn't need to assume the end caps were equidistant from the sheet).

Although an infinite sheet of charge cannot exist physically, this derivation is still useful in that Eq. 13 gives

approximately correct results for real (not infinite) charge sheets if we consider only points that are far from the edges and whose distance from the sheet is small compared to the dimensions of the sheet.

A Spherical Shell of Charge

Figure 10 shows a cross section of a thin uniform spherical shell of charge having constant surface charge density σ and total charge q ($= 4\pi R^2\sigma$), such as we might produce by spraying charge uniformly over the surface of a spherical balloon of radius R. We use Gauss' law to establish two useful properties of this distribution, which can be summarized in the following two *shell theorems*:

  1. *A uniform spherical shell of charge behaves, for external points, as if all its charge were concentrated at its center.*

  2. *A uniform spherical shell of charge exerts no electrostatic force on a charged particle placed inside the shell.*

These two shell theorems are the electrostatic analogues of the two gravitational shell theorems presented in Chapter 16. We shall see how much simpler is our Gauss' law proof than the detailed proof of Section 16-5, in which full advantage of the spherical symmetry was not taken.

The spherical shell of Fig. 10 is surrounded by two concentric spherical Gaussian surfaces, S_1 and S_2. From a symmetry argument, we conclude that the field can have only a radial component. (Assume there were a nonradial component, and suppose someone rotated the shell through some angle about a diameter when your back was turned. When you turned back, you could use a probe of the electric field, say, a test charge, to learn that the electric field had changed direction, even though the charge distribution was the same as before the rotation. Clearly this is a contradiction. Would this symmetry argument hold if the charge were *not* uniformly distributed over the sur-

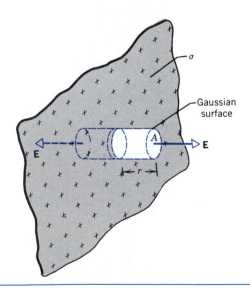

Figure 9 A Gaussian surface in the form of a small closed cylinder intersects a small portion of a sheet of positive charge. The field is perpendicular to the sheet, and so only the end caps of the Gaussian surface contribute to the flux.

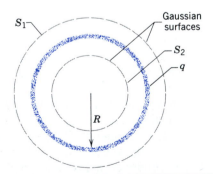

Figure 10 A cross section of a thin uniformly charged shell of total charge q. The shell is surrounded by two closed spherical Gaussian surfaces, one inside the shell and another outside the shell.

face?) Applying Gauss' law to surface S_1, for which $r > R$, gives

$$\epsilon_0 E(4\pi r^2) = q,$$

or

$$E = \frac{1}{4\pi\epsilon_0}\frac{q}{r^2} \qquad \text{(spherical shell, } r > R), \quad (14)$$

just as it did in connection with Fig. 5. Thus *the uniformly charged shell behaves like a point charge for all points outside the shell.* This proves the first shell theorem.

Applying Gauss' law to surface S_2, for which $r < R$, leads directly to

$$E = 0 \qquad \text{(spherical shell, } r < R), \quad (15)$$

because this Gaussian surface encloses no charge and because E (by another symmetry argument) has the same value everywhere on the surface. *The electric field therefore vanishes inside a uniform shell of charge;* a test charge placed anywhere in the interior would feel no electric force. This proves the second shell theorem.

These two theorems apply only in the case of a *uniformly* charged shell. If the charges were sprayed on the surface in a nonuniform manner, such that the charge density were *not* constant over the surface, these theorems would not apply. The symmetry would be lost, and as a result E could not be removed from the integral in Gauss' law. The flux would remain equal to q/ϵ_0 at exterior points and zero at interior points, but we would not be able to make such a direct connection with E as we can in the uniform case. In contrast to the uniformly charged shell, the field would *not* be zero throughout the interior.

Spherically Symmetric Charge Distribution

Figure 11 shows a cross section of a spherical distribution of charge of radius R. Here the charge is distributed throughout the spherical volume. We do not assume that the volume charge density ρ (the charge per unit volume) is a constant; however, we make the restriction that ρ at

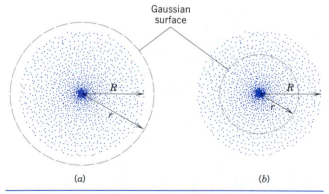

Gaussian surface

(a)   (b)

Figure 11 A cross section of a spherically symmetric charge distribution, in which the volume charge density may vary with r in this assumed nonconducting material. Closed spherical Gaussian surfaces have been drawn (*a*) outside the distribution and (*b*) within the distribution.

any point depends *only* on the distance of the point from the center, a condition called *spherical symmetry*. That is, ρ may be a function of r, but not of any angular coordinate. Let us find an expression for E for points outside (Fig. 11*a*) and inside (Fig. 11*b*) the charge distribution. Note that the object in Fig. 11 cannot be a conductor or, as we have seen, the excess charge would reside on its surface (and we could apply the shell theorems to find E).

Any spherically symmetric charge distribution, such as that of Fig. 11, can be regarded as a nest of concentric thin shells. The volume charge density ρ may vary from one shell to the next, but we make the shells so thin that we can assume ρ is constant on any particular shell. We can use the results of the previous subsection to calculate the contribution of each shell to the total electric field. The electric field from each thin shell has only a radial component, and thus the total electric field of the sphere can likewise have only a radial component. (This conclusion also follows from a symmetry argument but would not hold if the charge distribution lacked spherical symmetry, that is, if ρ depended on direction.)

Let us calculate the electric field at points that lie at a radial distance r greater than the radius R of the sphere, as shown in Fig. 11*a*. Each concentric shell, with a charge dq, contributes a radial component dE to the electric field according to Eq. 14. The total field is the total of all such components, and because all components to the field are radial, we must compute only an algebraic sum rather than a vector sum. The sum over all the shells then gives

$$E = \int dE = \int \frac{1}{4\pi\epsilon_0}\frac{dq}{r^2}$$

or, since r is constant in the integral over q,

$$E = \frac{1}{4\pi\epsilon_0}\frac{q}{r^2}, \qquad (16)$$

where q is the total charge of the sphere. Thus for points outside a spherically symmetric distribution of charge, the electric field has the value that it would have if the charge were concentrated at its center. This result is similar to the gravitational case proved in Section 16-5. Both results follow from the inverse square nature of the corresponding force laws.

We now consider the electric field for points inside the charge distribution. Figure 11*b* shows a spherical Gaussian surface of radius $r < R$. Gauss' law gives

$$\epsilon_0 \oint \mathbf{E} \cdot d\mathbf{A} = \epsilon_0 E(4\pi r^2) = q'$$

or

$$E = \frac{1}{4\pi\epsilon_0}\frac{q'}{r^2}, \qquad (17)$$

in which q' is that part of q contained *within* the sphere of radius r. According to the second shell theorem, the part of q that lies outside this sphere makes no contribution to **E** at radius r.

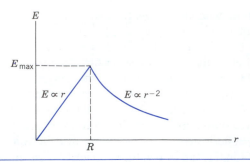

Figure 12 The variation with radius of the electric field due to a uniform spherical distribution of charge of radius R. The variation for $r > R$ applies to *any* spherically symmetric charge distribution, while that for $r < R$ applies *only* to a uniform distribution.

To continue this calculation, we must know the charge q' that is within the radius r; that is, we must know $\rho(r)$. Let us consider the special case in which the sphere is uniformly charged, so that the charge density ρ has the same value for all points within a sphere of radius R and is zero for all points outside this sphere. For points inside such a uniform sphere of charge, the fraction of charge within r is equal to the fraction of the volume within r, and so

$$\frac{q'}{q} = \frac{\frac{4}{3}\pi r^3}{\frac{4}{3}\pi R^3}$$

or

$$q' = q\left(\frac{r}{R}\right)^3,$$

where $\frac{4}{3}\pi R^3$ is the volume of the spherical charge distribution. The expression for E then becomes

$$E = \frac{1}{4\pi\epsilon_0}\frac{qr}{R^3} \qquad \text{(uniform sphere, } r < R\text{)}. \quad (18)$$

This equation becomes zero, as it should, for $r = 0$. Equation 18 applies *only* when the charge density is uniform, independent of r. Note that Eqs. 16 and 18 give the same result, as they must, for points on the surface of the charge distribution (that is, for $r = R$). Figure 12 shows the electric field for points with $r < R$ (given by Eq. 18) and for points with $r > R$ (given by Eq. 16).

Sample Problem 5 A plastic rod, whose length L is 220 cm and whose radius R is 3.6 mm, carries a negative charge q of magnitude 3.8×10^{-7} C, spread uniformly over its surface. What is the electric field near the midpoint of the rod, at a point on its surface?

Solution Although the rod is not infinitely long, for a point on its surface and near its midpoint it is effectively very long, so that we are justified in using Eq. 12. The linear charge density for the rod is

$$\lambda = \frac{q}{L} = \frac{-3.8 \times 10^{-7}\text{ C}}{2.2\text{ m}} = -1.73 \times 10^{-7}\text{ C/m}.$$

From Eq. 12 we then have

$$E = \frac{\lambda}{2\pi\epsilon_0 r}$$

$$= \frac{-1.73 \times 10^{-7}\text{ C/m}}{(2\pi)(8.85 \times 10^{-12}\text{ C}^2/\text{N}\cdot\text{m}^2)(0.0036\text{ m})}$$

$$= -8.6 \times 10^5\text{ N/C}.$$

The minus sign tells us that, because the rod is negatively charged, the direction of the electric field is radially inward, toward the axis of the rod. Sparking occurs in dry air at atmospheric pressure at an electric field strength of about 3×10^6 N/C. The field strength we calculated is lower than this by a factor of about 3.4 so that sparking should not occur.

Sample Problem 6 Figure 13a shows portions of two large sheets of charge with uniform surface charge densities of $\sigma_+ = +6.8\ \mu\text{C/m}^2$ and $\sigma_- = -4.3\ \mu\text{C/m}^2$. Find the electric field **E** (a) to the left of the sheets, (b) between the sheets, and (c) to the right of the sheets.

Solution Our strategy is to deal with each sheet separately and then to add the resulting electric fields by using the superposition principle. For the positive sheet we have, from Eq. 13,

$$E_+ = \frac{\sigma_+}{2\epsilon_0} = \frac{6.8 \times 10^{-6}\text{ C/m}^2}{(2)(8.85 \times 10^{-12}\text{ C}^2/\text{N}\cdot\text{m}^2)} = 3.84 \times 10^5\text{ N/C}.$$

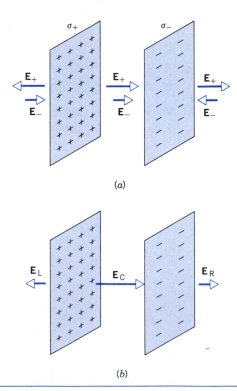

Figure 13 Sample Problem 6. (a) Two large parallel sheets of charge carry different charge distributions σ_+ and σ_-. The fields $\mathbf{E}_+$ and $\mathbf{E}_-$ would be set up by each sheet if the other were not present. (b) The net fields in the nearby regions to the left (L), center (C), and right (R) of the sheets, calculated from the vector sum of $\mathbf{E}_+$ and $\mathbf{E}_-$ in each region.

Similarly, for the negative sheet the magnitude of the field is

$$E_- = \frac{|\sigma_-|}{2\epsilon_0} = \frac{4.3 \times 10^{-6} \text{ C/m}^2}{(2)(8.85 \times 10^{-12} \text{ C}^2/\text{N} \cdot \text{m}^2)} = 2.43 \times 10^5 \text{ N/C}.$$

Figure 13*a* shows the two sets of fields calculated above, to the left of the sheets, between them, and to the right of the sheets.

The resultant fields in these three regions follow from the superposition principle. To the left of the sheets, we have (taking components of **E** in Fig. 13 to be positive if **E** points to the right and negative if **E** points to the left)

$$E_L = -E_+ + E_- = -3.84 \times 10^5 \text{ N/C} + 2.43 \times 10^5 \text{ N/C}$$
$$= -1.4 \times 10^5 \text{ N/C}.$$

The resultant (negative) electric field in this region points to the left, as Fig. 13*b* shows. To the right of the sheets, the electric field has this same magnitude but points to the right in Fig. 13*b*.

Between the sheets, the two fields add to give

$$E_C = E_+ + E_- = 3.84 \times 10^5 \text{ N/C} + 2.43 \times 10^5 \text{ N/C}$$
$$= 6.3 \times 10^5 \text{ N/C}.$$

Outside the sheets, the electric field behaves like that due to a single sheet whose surface charge density is $\sigma_+ + \sigma_-$ or $+2.5 \times 10^{-6}$ C/m^2. The field pattern of Fig. 13*b* bears this out. In Problems 22 and 23 you can investigate the case in which the two surface charge densities are equal in magnitude but opposite in sign and also the case in which they are equal in both magnitude and sign.

29-6   EXPERIMENTAL TESTS OF GAUSS' LAW AND COULOMB'S LAW

In Section 29-4, we deduced that the excess charge in a conductor must lie only on its outer surface. No charge can be within the volume of the conductor or on the surface of an empty inner cavity. This result was derived directly from Gauss' law. Therefore testing whether the charge does in fact lie entirely on the outer surface is a way of testing Gauss' law. If charge is found to be within the conductor or on an interior surface (such as the cavity in Fig. 6*b*), then Gauss' law fails. We also proved in Section 29-3 that Coulomb's law follows directly from Gauss' law. Thus if Gauss' law fails, then Coulomb's law fails. In particular, the force law might not be exactly an inverse square law. The exponent of *r* might differ from 2 by some small amount δ, so that

$$E = \frac{1}{4\pi\epsilon_0} \frac{q}{r^{2+\delta}}, \tag{19}$$

in which δ is exactly zero if Coulomb's law and Gauss' law hold.

The direct measurement of the force between two charges, described in Chapter 27, does not have the precision necessary to test whether δ is zero beyond a few percent. The observation of the charge within a conductor provides the means for a test that, as we shall see, is far more precise.

In principle, the experiment follows a procedure illustrated in Fig. 14. A charged metal ball hangs from an insulating thread and is lowered into a metal can that rests on an insulating stand. When the ball is touched to the inside of the can, the two objects form a *single conductor,* and, if Gauss' law is valid, all the charge from the ball must go to the outside of the combined conductor, as in Fig. 14*c*. When the ball is removed, it should no longer carry any charge. Touching other insulated metal objects to the inside of the can should not result in the transfer of any charge to the objects. Only on the outside of the can will it be possible to transfer charge.

Benjamin Franklin seems to have been the first to notice that there can be no charge inside an insulated metal can. In 1755 he wrote to a friend:

I electrified a silver pint cann, on an electric stand, and then lowered into it a cork-ball, of about an inch in diameter, hanging by a silk string, till the cork touched the bottom of the cann. The cork was not attracted to the inside of the cann as it would have been to the outside, and though it touched the bottom, yet when drawn out, it was not found to be electrified by that touch, as it would have been by touching the outside. The fact is singular. You require the reason; I do not know it . . . .

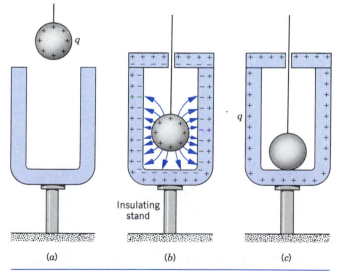

Figure 14 An arrangement conceived by Benjamin Franklin to show that the charge placed on a conductor moves to its surface. (*a*) A charged metal ball is lowered into an uncharged metal can. (*b*) The ball is inside the can and a cover is added. The lines of force between the ball and the uncharged can are shown. The ball attracts charges of the opposite sign to the inside of the can. (*c*) When the ball touches the can, they form a single conductor, and the net charge flows to the outer surface. The ball can then be removed from the can and shown to be completely uncharged, thus proving that the charge must have been transferred entirely to the can.

About 10 years later Franklin recommended this "singular fact" to the attention of his friend Joseph Priestley (1733–1804). In 1767 (about 20 years before Coulomb's experiments) Priestley checked Franklin's observation and, with remarkable insight, realized that the inverse square law of force followed from it. Thus the indirect approach is not only more accurate than the direct approach of Section 27-4 but was carried out earlier.

Priestley, reasoning by analogy with gravitation, said that the fact that no electric force acted on Franklin's cork ball when it was surrounded by a deep metal can is similar to the fact (see Section 16-5) that no gravitational force acts on a particle inside a spherical shell of matter; if gravitation obeys an inverse square law, perhaps the electrical force does also. Considering Franklin's experiment, Priestley reasoned:

> May we not infer from this that the attraction of electricity is subject to the same laws with that of gravitation and is therefore according to the squares of the distances; since it is easily demonstrated that were the earth in the form of a shell, a body in the inside of it would not be attracted to one side more than another?

Note how knowledge of one subject (gravitation) helps in understanding another (electrostatics).

Michael Faraday also carried out experiments designed to show that excess charge resides on the outside surface of a conductor. In particular, he built a large metal-covered box, which he mounted on insulating supports and charged with a powerful electrostatic generator. In Faraday's words:

> I went into the cube and lived in it, and using lighted candles, electrometers, and all other tests of electrical states, I could not find the least influence upon them . . . though all the time the outside of the cube was very powerfully charged, and large sparks and brushes were darting off from every part of its outer surface.

Coulomb's law is vitally important in physics and if δ in Eq. 19 is not zero, there are serious consequences for our

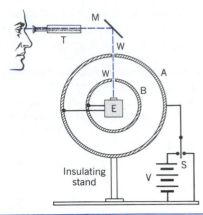

Figure 15   A modern and more precise version of the apparatus of Fig. 14, also designed to verify that charge resides only on the outside surface of a conductor. Charge is placed on sphere A by throwing switch S to the left, and the sensitive electrometer E is used to search for any charge that might move to the inner sphere B. It is expected that all the charge will remain on the outer surface (sphere A).

understanding of electromagnetism and quantum physics. The best way to measure δ is to find out *by experiment* whether an excess charge, placed on an isolated conductor, does or does not move *entirely* to its outside surface.

Modern experiments, carried out with remarkable precision, have shown that if δ in Eq. 19 is not zero it is certainly very, very small. Table 1 summarizes the results of the most important of these experiments.

Figure 15 is a drawing of the apparatus used to measure δ by Plimpton and Lawton. It consists in principle of two concentric metal shells, A and B, the former being 1.5 m in diameter. The inner shell contains a sensitive electrometer E connected so that it will indicate whether any charge moves between shells A and B. If the shells are connected electrically, any charge placed on the shell assembly should reside entirely on shell A if Gauss' law— and thus Coulomb's law—are correct as stated.

By throwing switch S to the left, a substantial charge could be placed on the sphere assembly. If any of this charge moved to shell B, it would have to pass through the electrometer and would cause a deflection, which could

TABLE 1   TESTS OF COULOMB'S INVERSE SQUARE LAW

Experimenters	Date	δ (Eq. 19)
Franklin	1755	
Priestley	1767	. . . according to the squares . . .
Robison	1769	<0.06
Cavendish	1773	<0.02
Coulomb	1785	a few percent at most
Maxwell	1873	$<5 \times 10^{-5}$
Plimpton and Lawton	1936	$<2 \times 10^{-9}$
Bartlett, Goldhagen, and Phillips	1970	$<1.3 \times 10^{-13}$
Williams, Faller, and Hill	1971	$<1.0 \times 10^{-16}$

be observed optically using telescope T, mirror M, and windows W.

However, when the switch S was thrown alternately from left to right, connecting the shell assembly either to the battery or to the ground, no effect was observed. Knowing the sensitivity of their electrometer, Plimpton and Lawton calculated that δ in Eq. 19 differs from zero by no more than 2×10^{-9}, a very small value indeed. Yet since their experiment, the limits on δ have been improved by more than seven orders of magnitude by experimenters using more detailed and precise versions of this basic apparatus.

29-7   THE NUCLEAR MODEL OF THE ATOM   (Optional)

An atom consists of negatively charged electrons bound to a core of positive charge. The positive core must have most of the atom's mass, because the total mass of the electrons of an atom typically makes up only about 1/4000 of the mass of the atom. In the early years of the 20th century, there was much speculation about the distribution of this positive charge.

According to one theory that was popular at that time, the positive charge is distributed more or less uniformly throughout the entire spherical volume of the atom. This model of the structure of the atom is called the *Thomson model* after J. J. Thomson who proposed it. (Thomson was the first to measure the charge-to-mass ratio of the electron and is therefore often credited as the discoverer of the electron.) It is also called the "plum pudding" model, because the electrons are imbedded throughout the diffuse sphere of positive charge just like raisins in a plum pudding.

One way of testing this model is to determine the electric field of the atom by probing it with a beam of positively charged projectiles that pass nearby. The particles in the beam are deflected or *scattered* by the electric field of the atom. In the following discussion, we consider only the effect on the projectile of the sphere of positive charge. We assume the projectile is both much *less* massive than the atom and much *more* massive than an electron. In this way the electrons have a negligible effect on the scattering of the projectile, and the atom can be assumed to remain at rest while the projectile is deflected.

The electric field due to a uniform sphere of positive charge was given by Eq. 16 for points outside the sphere of charge and by Eq. 18 for points inside. Let us calculate the electric field at the surface, which, as Fig. 12 shows, is the *largest* possible field that this distribution can produce. We consider a heavy atom such as gold, which has a positive charge Q of $79e$ and a radius R of about 1.0×10^{-10} m. Neglecting the electrons, the electric field at $r = R$ due to the positive charges is

$$E_{max} = \frac{1}{4\pi\epsilon_0} \frac{Q}{R^2} = \frac{(9 \times 10^9 \text{ N} \cdot \text{m}^2/\text{C}^2)(79)(1.6 \times 10^{-19} \text{ C})}{(1.0 \times 10^{-10} \text{ m})^2}$$
$$= 1.1 \times 10^{13} \text{ N/C}.$$

For the projectiles in our experiment, let us use a beam of alpha particles, which have a positive charge q of $2e$ and a mass m of 6.64×10^{-27} kg. Alpha particles are nuclei of helium atoms, which are emitted in certain radioactive decay processes. A typical kinetic energy for such a particle might be about $K = 6$ MeV or 9.6×10^{-13} J. At this energy the particle has a speed of

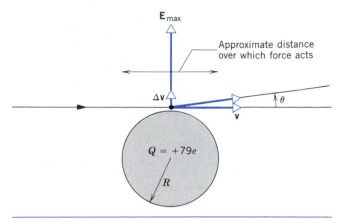

Figure 16   The scattering of a positively charged projectile passing near the surface of an atom, represented by a uniform sphere of positive charge. The electric field on the projectile causes a transverse deflection by an angle θ.

$$v = \sqrt{\frac{2K}{m}} = \sqrt{\frac{2(9.6 \times 10^{-13} \text{ J})}{6.64 \times 10^{-27} \text{ kg}}} = 1.7 \times 10^7 \text{ m/s}.$$

Note that this speed is about $0.06c$, which justifies our use of the nonrelativistic relationship between speed and kinetic energy.

Let the particle pass near the surface of the atom, where it experiences the largest electric field that this atom could exert. The corresponding force on the particle is

$$F = qE_{max} = 2(1.6 \times 10^{-19} \text{ C})(1.1 \times 10^{13} \text{ N/C}) = 3.5 \times 10^{-6} \text{ N}.$$

Figure 16 shows a schematic diagram of a scattering experiment. The actual calculation of the deflection is relatively complicated, but we can make some approximations that simplify the calculation and permit a rough estimate of the maximum deflection. Let us assume that the above force is constant and acts only during the time Δt it takes the projectile to travel a distance equal to a diameter of the atom, as indicated in Fig. 16. This time interval is

$$\Delta t = \frac{2R}{v} = \frac{2(1.0 \times 10^{-10} \text{ m})}{1.7 \times 10^7 \text{ m/s}} = 1.2 \times 10^{-17} \text{ s}.$$

The force gives the particle a transverse acceleration a, which produces a transverse velocity Δv given by

$$\Delta v = a \, \Delta t = \frac{F}{m} \Delta t = \frac{3.5 \times 10^{-6} \text{ N}}{6.64 \times 10^{-27} \text{ kg}} 1.2 \times 10^{-17} \text{ s}$$
$$= 6.6 \times 10^3 \text{ m/s}.$$

This is a small change when compared with the magnitude of the velocity of the particle (1.7×10^7 m/s). The particle will be deflected by a small angle θ that can be estimated to be about

$$\theta = \tan^{-1} \frac{\Delta v}{v} = \tan^{-1} \frac{6.6 \times 10^3 \text{ m/s}}{1.7 \times 10^7 \text{ m/s}} = 0.02°.$$

This kind of experiment was first done by Ernest Rutherford and his collaborators H. Geiger and E. Marsden at the University of Manchester in 1911. Figure 17 shows the details of the experiment they used to measure the angle of scattering. A beam of alpha particles from the radioactive source S was scattered by a thin gold foil T and observed by a detector D that could be

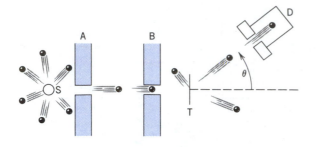

Figure 17   The experimental arrangement for studying scattering of alpha particles. The particles are emitted by a radioactive source S and fall on a thin target T (a gold foil). Scattered alpha particles are observed in a detector D that can be moved to various angles θ.

moved to any angle θ with respect to the direction of the incident beam. They determined the number of scattered particles that struck the detector per unit time at various angles.

The results of their experiment are shown schematically in Fig. 18. Although many of the particles were scattered at small angles, as our rough calculation predicts, an occasional particle, perhaps 1 in 10^4, was scattered through such a large angle that its motion is reversed. Such a result is quite surprising if one accepts the Thomson model, for which we have estimated the *maximum* deflection to be about $0.02°$. In Rutherford's words: "It was quite the most incredible event that ever happened to me in my life. It was almost as incredible as if you had fired a 15-inch shell at a piece of tissue paper and it came back and hit you."

Based on this kind of scattering experiment, Rutherford concluded that the positive charge of an atom was not diffused throughout a sphere of the same size as the atom, but instead was concentrated in a tiny region (the *nucleus*) near the center of the

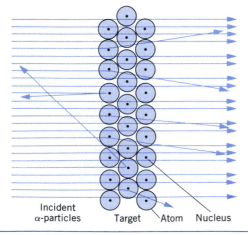

Figure 18   A schematic representation of the scattering outcomes. Most of the alpha particles pass through undeflected, but a few are deflected through small angles. An occasional particle (one in 10^4) is scattered through an angle that exceeds $90°$.

atom. In the case of a gold atom, the nucleus has a radius of about 7×10^{-15} m (7 fm), roughly 10^{-4} times smaller than the radius of the atom. That is, the nucleus occupies a volume only 10^{-12} that of the atom!

Let us calculate the maximum electric field and the corresponding force on an alpha particle that passes close to the surface of the nucleus. If we regard the nucleus as a uniform spherical ball of charge $Q = 79e$ and radius $R = 7$ fm, the maximum electric field is

$$E_{max} = \frac{1}{4\pi\epsilon_0}\frac{Q}{R^2} = \frac{(9 \times 10^9 \text{ N·m}^2/\text{C}^2)(79)(1.6 \times 10^{-19} \text{ C})}{(7.0 \times 10^{-15} \text{ m})^2}$$

$$= 2.3 \times 10^{21} \text{ N/C}.$$

This is more than eight orders of magnitude larger than the electric field that would act on a particle at the surface of a plum pudding model atom. The corresponding force is

$$F = qE_{max} = 2(1.6 \times 10^{-19} \text{ C})(2.3 \times 10^{21} \text{ N/C}) = 740 \text{ N}.$$

This is a huge force! Let us make the same simplification we did in our previous calculation and assume this force is constant and acts on the particle only during the time Δt it takes the particle to travel a distance equal to one nuclear diameter:

$$\Delta t = \frac{2R}{v} = \frac{2(7.0 \times 10^{-15} \text{ m})}{1.7 \times 10^7 \text{ m/s}} = 8.2 \times 10^{-22} \text{ s}.$$

The corresponding change in the velocity of the particle can be estimated to be

$$\Delta v = a\,\Delta t = \frac{F}{m}\Delta t = \frac{740 \text{ N}}{6.64 \times 10^{-27} \text{ kg}}\,8.2 \times 10^{-22} \text{ s}$$

$$= 9 \times 10^7 \text{ m/s}.$$

This is comparable in magnitude to the velocity itself. We conclude that a nuclear atom can produce an electric field that is sufficiently large to reverse the motion of the projectile.

We can measure the radius of the nucleus by firing alpha particles at it and measuring their deflection. The deflection can be calculated quite precisely *assuming* the projectile is always outside the charge distribution of the nucleus, in which case the electric field is given by Eq. 16. However, if we fire the projectile with enough energy, it may penetrate to the region where $r < R$, where it experiences a different electric field (given, for example, by Eq. 18 if we assume the nuclear charge distribution to be uniform) and where its deflection will therefore differ from what we would calculate by assuming the projectile is always outside the nucleus. Finding the energy at which this happens is in effect a way of measuring the radius of the nucleus of the atom. From such experiments we learn that the radius of a nucleus of an atom of mass number A is about $R_0 A^{1/3}$, where R_0 is about 1.2 fm.

Rutherford's analysis was far more detailed than the discussion we have presented here. He was able to derive a formula that gave an exact relationship between the number of scattered particles and the angle of scattering based purely on the $1/r^2$ electric field, and he tested that formula at angles from $0°$ to $180°$. His formula also depended on the *atomic number Z* of the target atoms, and so this scattering experiment provided a direct way to determine the Z of an atom. Finally, he showed that the scattering is as we have pictured it in Fig. 18: there is only a small probability to have any scattering at all, most of the projectiles passing through undeflected, and the probability to have more than one scattering of a single projectile is negligibly small. This is consistent with the small size deduced for the nucleus. The atom is mostly empty space, and there is only a very small chance of a projectile being close enough to a nucleus to experience an electric field large enough to cause a deflection. There is

practically no chance for this to happen twice to the same projectile.

This classic and painstaking series of experiments and their brilliant interpretation laid the foundation for modern atomic and nuclear physics, and Rutherford is generally credited as the founder of these fields. ∎

QUESTIONS

1. What is the basis for the statement that lines of electric force begin and end only on electric charges?

2. Positive charges are sometimes called "sources" and negative charges "sinks" of electric field. How would you justify this terminology? Are there sources and/or sinks of gravitational field?

3. By analogy with Φ_E, how would you define the flux Φ_g of a gravitational field? What is the flux of the Earth's gravitational field through the boundaries of a room, assumed to contain no matter? Through a spherical surface closely surrounding the Earth? Through a spherical surface the size of the Moon's orbit?

4. Consider the Gaussian surface that surrounds part of the charge distribution shown in Fig. 19. (a) Which of the charges contribute to the electric field at point P? (b) Would the value obtained for the flux through the surface, calculated using only the field due to q_1 and q_2, be greater than, equal to, or less than that obtained using the total field?

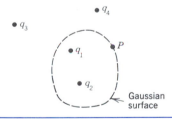

Figure 19 Question 4.

5. Suppose that an electric field in some region is found to have a constant direction but to be decreasing in strength in that direction. What do you conclude about the charge in the region? Sketch the lines of force.

6. Is it precisely true that Gauss' law states that the total number of lines of force crossing any closed surface in the outward direction is proportional to the net positive charge enclosed within the surface?

7. A point charge is placed at the center of a spherical Gaussian surface. Is Φ_E changed (a) if the surface is replaced by a cube of the same volume, (b) if the sphere is replaced by a cube of one-tenth the volume, (c) if the charge is moved off-center in the original sphere, still remaining inside, (d) if the charge is moved just outside the original sphere, (e) if a second charge is placed near, and outside, the original sphere, and (f) if a second charge is placed inside the Gaussian surface?

8. In Gauss' law,

$$\epsilon_0 \oint \mathbf{E} \cdot d\mathbf{A} = q,$$

is $\mathbf{E}$ necessarily the electric field attributable to the charge q?

9. A surface encloses an electric dipole. What can you say about Φ_E for this surface?

10. Suppose that a Gaussian surface encloses no net charge. Does Gauss' law require that $\mathbf{E}$ equal zero for all points on the surface? Is the converse of this statement true; that is, if $\mathbf{E}$ equals zero everywhere on the surface, does Gauss' law require that there be no net charge inside?

11. Is Gauss' law useful in calculating the field due to three equal charges located at the corners of an equilateral triangle? Explain why or why not.

12. A total charge Q is distributed uniformly throughout a cube of edge length a. Is the resulting electric field at an external point P, a distance r from the center C of the cube, given by $E = Q/4\pi\epsilon_0 r^2$? See Fig. 20. If not, can E be found by constructing a "concentric" cubical Gaussian surface? If not, explain why not. Can you say anything about E if $r \gg a$?

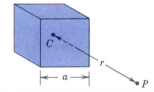

Figure 20 Question 12.

13. Is $\mathbf{E}$ necessarily zero inside a charged rubber balloon if the balloon is (a) spherical or (b) sausage shaped? For each shape, assume the charge to be distributed uniformly over the surface. How would the situation change, if at all, if the balloon has a thin layer of conducting paint on its outside surface?

14. A spherical rubber balloon carries a charge that is uniformly distributed over its surface. As the balloon is blown up, how does E vary for points (a) inside the balloon, (b) at the surface of the balloon, and (c) outside the balloon?

15. In Section 29-3 we have seen that Coulomb's law can be derived from Gauss' law. Does this necessarily mean that Gauss' law can be derived from Coulomb's law?

16. Would Gauss' law hold if the exponent in Coulomb's law were not exactly 2?

17. A large, insulated, hollow conductor carries a positive charge. A small metal ball carrying a negative charge of the same magnitude is lowered by a thread through a small opening in the top of the conductor, allowed to touch the inner surface, and then withdrawn. What is then the charge on (a) the conductor and (b) the ball?

18. Can we deduce from the argument of Section 29-4 that the electrons in the wires of a house wiring system move along the surfaces of those wires? If not, why not?

19. In Section 29-4, we assumed that E equals zero everywhere inside an isolated conductor. However, there are certainly very large electric fields inside the conductor, at points close to the electrons or to the nuclei. Does this invalidate the proof of Section 29-4? Explain.

20. Does Gauss' law, as applied in Section 29-4, require that all the conduction electrons in an insulated conductor reside on the surface?

21. A positive point charge q is located at the center of a hollow metal sphere. What charges appear on (a) the inner surface and (b) the outer surface of the sphere? (c) If you bring an (uncharged) metal object near the sphere, will it change your answers to (a) or (b) above? Will it change the way charge is distributed over the sphere?

22. If a charge $-q$ is distributed uniformly over the surface of a thin insulated spherical metal shell of radius a, there will be no electric field inside. If now a point charge $+q$ is placed at the center of the sphere, there will be no external field. This point charge can be displaced a distance $d < a$ from the center, but that gives the system a dipole moment and creates an external field. How do you account for the energy appearing in this external field?

23. How can you remove completely the excess charge from a small conducting body?

24. Explain why the spherical symmetry of Fig. 5 restricts us to a consideration of E that has only a radial component at any point. (*Hint*: Imagine other components, perhaps along the equivalent of longitude or latitude lines on the Earth's

surface. Spherical symmetry requires that these look the same from any perspective. Can you invent such field lines that satisfy this criterion?)

25. Explain why the symmetry of Fig. 8 restricts us to a consideration of E that has only a radial component at any point. Remember, in this case, that the field must not only look the same at any point along the line but must also look the same if the figure is turned end for end.

26. The *total* charge on a charged infinite rod is infinite. Why is not E also infinite? After all, according to Coulomb's law, if q is infinite, so is E.

27. Explain why the symmetry of Fig. 9 restricts us to a consideration of E that has only a component directed away from the sheet. Why, for example, could E not have a component parallel to the sheet? Remember, in this case, that the field must not only look the same at any point along the sheet in any direction but must also look the same if the sheet is rotated about a line perpendicular to the sheet.

28. The field due to an infinite sheet of charge is uniform, having the same strength at all points no matter how far from the surface charge. Explain how this can be, given the inverse square nature of Coulomb's law.

29. As you penetrate a uniform sphere of charge, E should decrease because less charge lies inside a sphere drawn through the observation point. On the other hand, E should increase because you are closer to the center of this charge. Which effect dominates, and why?

30. Given a spherically symmetric charge distribution (not of uniform radial density of charge), is E necessarily a maximum at the surface? Comment on various possibilities.

31. Does Eq. 16 hold true for Fig. 11a if (a) there is a concentric spherical cavity in the body, (b) a point charge Q is at the center of this cavity, and (c) the charge Q is inside the cavity but not at its center?

32. An atom is normally *electrically neutral*. Why then should an alpha particle be deflected by the atom under any circumstances?

PROBLEMS

Section 29-2   The Flux of the Electric Field

1. The square surface shown in Fig. 21 measures 3.2 mm on each side. It is immersed in a uniform electric field with $E = 1800$ N/C. The field lines make an angle of 65° with the

"outward pointing" normal, as shown. Calculate the flux through the surface.

2. A cube with 1.4-m edges is oriented as shown in Fig. 22 in a region of uniform electric field. Find the electric flux through the right face if the electric field, in N/C, is given by

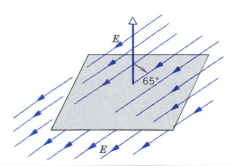

Figure 21   Problem 1.

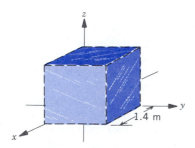

Figure 22   Problem 2.

(a) 6**i**, (b) −2**j**, and (c) −3**i** + 4**k**. (d) Calculate the total flux through the cube for each of these fields.

3. Calculate Φ_E through (a) the flat base and (b) the curved surface of a hemisphere of radius R. The field **E** is uniform and parallel to the axis of the hemisphere, and the lines of **E** enter through the flat base. (Use the outward pointing normal.)

Section 29-3 Gauss' Law

4. Charge on an originally uncharged insulated conductor is separated by holding a positively charged rod very closely nearby, as in Fig. 23. Calculate the flux for the five Gaussian surfaces shown. Assume that the induced negative charge on the conductor is equal to the positive charge q on the rod.

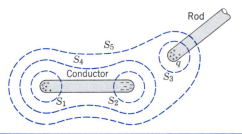

Figure 23 Problem 4.

5. A point charge of 1.84 μC is at the center of a cubical Gaussian surface 55 cm on edge. Find Φ_E through the surface.

6. The net electric flux through each face of a dice has magnitude in units of 10^3 N·m²/C equal to the number N of spots on the face (1 through 6). The flux is inward for N odd and outward for N even. What is the net charge inside the dice?

7. A point charge $+q$ is a distance $d/2$ from a square surface of side d and is directly above the center of the square as shown in Fig. 24. Find the electric flux through the square. (*Hint*: Think of the square as one face of a cube with edge d.)

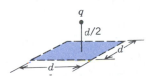

Figure 24 Problem 7.

8. A butterfly net is in a uniform electric field E as shown in Fig. 25. The rim, a circle of radius a, is aligned perpendicular to the field. Find the electric flux through the netting, relative to the outward normal.

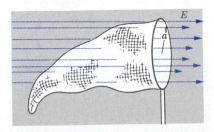

Figure 25 Problem 8.

9. It is found experimentally that the electric field in a certain region of the Earth's atmosphere is directed vertically down. At an altitude of 300 m the field is 58 N/C and at an altitude of 200 m it is 110 N/C. Find the net amount of charge contained in a cube 100 m on edge located at an altitude between 200 and 300 m. Neglect the curvature of the Earth.

10. Find the net flux through the cube of Problem 2 and Fig. 22 if the electric field is given in SI units by (a) **E** = 3y**j** and (b) **E** = −4**i** + (6 + 3y)**j**. (c) In each case, how much charge is inside the cube?

11. "Gauss' law for gravitation" is

$$\frac{1}{4\pi G}\,\Phi_g = \frac{1}{4\pi G}\oint \mathbf{g}\cdot d\mathbf{A} = -m,$$

where m is the enclosed mass and G is the universal gravitation constant. Derive Newton's law of gravitation from this. What is the significance of the minus sign?

12. A point charge q is placed at one corner of a cube of edge a. What is the flux through each of the cube faces? (*Hint*: Use Gauss' law and symmetry arguments.)

13. The electric field components in Fig. 26 are $E_x = bx^{1/2}$, $E_y = E_z = 0$, in which $b = 8830$ N/C·m$^{1/2}$. Calculate (a) the flux Φ_E through the cube and (b) the charge within the cube. Assume that $a = 13.0$ cm.

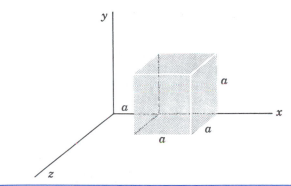

Figure 26 Problem 13.

Section 29-4 A Charged Isolated Conductor

14. A uniformly charged conducting sphere of 1.22-m radius has a surface charge density of 8.13 μC/m². (a) Find the charge on the sphere. (b) What is the total electric flux leaving the surface of the sphere? (c) Calculate the electric field at the surface of the sphere.

15. Space vehicles traveling through the Earth's radiation belts collide with trapped electrons. Since in space there is no ground, the resulting charge buildup can become significant and can damage electronic components, leading to control-circuit upsets and operational anomalies. A spherical metallic satellite 1.3 m in diameter accumulates 2.4 μC of charge in one orbital revolution. (a) Find the surface charge density. (b) Calculate the resulting electric field just outside the surface of the satellite.

16. Equation 11 ($E = \sigma/\epsilon_0$) gives the electric field at points near a charged conducting surface. Apply this equation to a conducting sphere of radius r, carrying a charge q on its surface, and show that the electric field outside the sphere is the same

as the field of a point charge at the position of the center of the sphere.

17. A conducting sphere carrying charge Q is surrounded by a spherical conducting shell. (a) What is the net charge on the inner surface of the shell? (b) Another charge q is placed outside the shell. Now what is the net charge on the inner surface of the shell? (c) If q is moved to a position between the shell and the sphere, what is the net charge on the inner surface of the shell? (d) Are your answers valid if the sphere and shell are not concentric?

18. An insulated conductor of arbitrary shape carries a net charge of $+10\,\mu C$. Inside the conductor is a hollow cavity within which is a point charge $q = +3.0\,\mu C$. What is the charge (a) on the cavity wall and (b) on the outer surface of the conductor?

19. A metal plate 8.0 cm on a side carries a total charge of $6.0\,\mu C$. (a) Using the infinite plate approximation, calculate the electric field 0.50 mm above the surface of the plate near the plate's center. (b) Estimate the field at a distance of 30 m.

Section 29-5 Applications of Gauss' Law

20. An infinite line of charge produces a field of 4.52×10^4 N/C at a distance of 1.96 m. Calculate the linear charge density.

21. (a) The drum of the photocopying machine in Sample Problem 3 has a length of 42 cm and a diameter of 12 cm. What is the total charge on the drum? (b) The manufacturer wishes to produce a desktop version of the machine. This requires reducing the size of the drum to a length of 28 cm and a diameter of 8.0 cm. The electric field at the drum surface must remain unchanged. What must be the charge on this new drum?

22. Two thin large nonconducting sheets of positive charge face each other as in Fig. 27. What is **E** at points (a) to the left of the sheets, (b) between them, and (c) to the right of the sheets? Assume the same surface charge density σ for each sheet. Consider only points not near the edges whose distance from the sheets is small compared to the dimensions of the sheet. (*Hint*: See Sample Problem 6.)

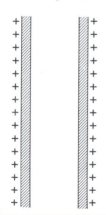

Figure 27 Problem 22.

23. Two large metal plates face each other as in Fig. 28 and carry charges with surface charge density $+\sigma$ and $-\sigma$, respectively, on their inner surfaces. Find **E** at points (a) to the left

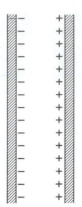

Figure 28 Problem 23.

of the sheets, (b) between them, and (c) to the right of the sheets. Consider only points not near the edges whose distances from the sheets are small compared to the dimensions of the sheet. (*Hint*: See Sample Problem 6.)

24. An electron remains stationary in an electric field directed downward in the Earth's gravitational field. If the electric field is due to charge on two large parallel conducting plates, oppositely charged and separated by 2.3 cm, what is the surface charge density, assumed to be uniform, on the plates?

25. A small sphere whose mass m is 1.12 mg carries a charge $q = 19.7$ nC. It hangs in the Earth's gravitational field from a silk thread that makes an angle $\theta = 27.4°$ with a large uniformly charged nonconducting sheet as in Fig. 29. Calculate the uniform charge density σ for the sheet.

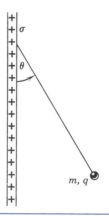

Figure 29 Problem 25.

26. Two charged concentric thin spherical shells have radii of 10.0 cm and 15.0 cm. The charge on the inner shell is 40.6 nC and that on the outer shell is 19.3 nC. Find the electric field (a) at $r = 12.0$ cm, (b) at $r = 22.0$ cm, and (c) at $r = 8.18$ cm from the center of the shells.

27. A very long straight thin wire carries -3.60 nC/m of fixed negative charge. The wire is to be surrounded by a uniform cylinder of positive charge, radius 1.50 cm, coaxial with the wire. The volume charge density ρ of the cylinder is to be selected so that the net electric field outside the cylinder is zero. Calculate the required positive charge density ρ.

28. Figure 30 shows a charge $+q$ arranged as a uniform conducting sphere of radius a and placed at the center of a spherical conducting shell of inner radius b and outer radius c. The outer shell carries a charge of $-q$. Find $E(r)$ at locations (a) within the sphere ($r < a$), (b) between the sphere and the shell ($a < r < b$), (c) inside the shell ($b < r < c$), and (d) outside the shell ($r > c$). (e) What charges appear on the inner and outer surfaces of the shell?

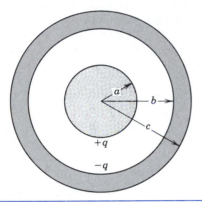

Figure 30 Problem 28.

29. A very long conducting cylinder (length L) carrying a total charge $+q$ is surrounded by a conducting cylindrical shell (also of length L) with total charge $-2q$, as shown in cross section in Fig. 31. Use Gauss' law to find (a) the electric field at points outside the conducting shell, (b) the distribution of the charge on the conducting shell, and (c) the electric field in the region between the cylinders.

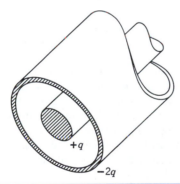

Figure 31 Problem 29.

30. Figure 32 shows a point charge $q = 126$ nC at the center of a spherical cavity of radius 3.66 cm in a piece of metal. Use Gauss' law to find the electric field (a) at point P_1, halfway from the center to the surface, and (b) at point P_2.

31. A proton orbits with a speed $v = 294$ km/s just outside a charged sphere of radius $r = 1.13$ cm. Find the charge on the sphere.

32. A large flat nonconducting surface carries a uniform charge density σ. A small circular hole of radius R has been cut in the middle of the sheet, as shown in Fig. 33. Ignore fringing of the field lines around all edges and calculate the electric field at point P, a distance z from the center of the hole along

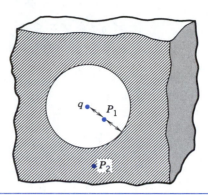

Figure 32 Problem 30.

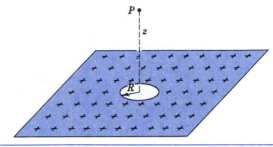

Figure 33 Problem 32.

its axis. (*Hint*: See Eq. 27 of Chapter 28 and use the principle of superposition.)

33. Figure 34 shows a section through a long, thin-walled metal tube of radius R, carrying a charge per unit length λ on its surface. Derive expressions for E for various distances r from the tube axis, considering both (a) $r > R$ and (b) $r < R$. (c) Plot your results for the range $r = 0$ to $r = 5.0$ cm, assuming that $\lambda = 2.0 \times 10^{-8}$ C/m and $R = 3.0$ cm. (*Hint*: Use cylindrical Gaussian surfaces, coaxial with the metal tube.)

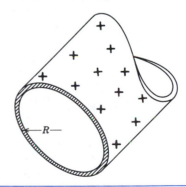

Figure 34 Problem 33.

34. Figure 35 shows a section through two long thin concentric cylinders of radii a and b. The cylinders carry equal and opposite charges per unit length λ. Using Gauss' law, prove (a) that $E = 0$ for $r < a$ and (b) that between the cylinders E is given by

$$E = \frac{1}{2\pi\epsilon_0} \frac{\lambda}{r}.$$

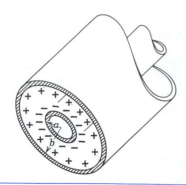

Figure 35   Problem 34.

35. In the geometry of Problem 34 a positron revolves in a circular path between and concentric with the cylinders. Find its kinetic energy, in electron-volts. Assume that $\lambda = 30$ nC/m. (Why do you not need to know the radii of the cylinders?)

36. Figure 36 shows a Geiger counter, used to detect ionizing radiation. The counter consists of a thin central wire, carrying positive charge, surrounded by a concentric circular conducting cylinder, carrying an equal negative charge. Thus a strong radial electric field is set up inside the cylinder. The cylinder contains a low-pressure inert gas. When a particle of radiation enters the tube through the cylinder walls, it ionizes a few of the gas atoms. The resulting free electrons are drawn to the positive wire. However, the electric field is so intense that, between collisions with the gas atoms, they have gained energy sufficient to ionize these atoms also. More free electrons are thereby created, and the process is repeated until the electrons reach the wire. The "avalanche" of electrons is collected by the wire, generating a signal recording the passage of the incident particle of radiation. Suppose that the radius of the central wire is 25 μm, the radius of the cylinder 1.4 cm, and the length of the tube 16 cm. The electric field at the cylinder wall is 2.9×10^4 N/C. Calculate the amount of positive charge on the central wire. (*Hint*: See Problem 34.)

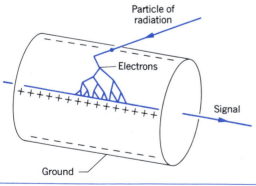

Figure 36   Problem 36.

37. Two long charged concentric cylinders have radii of 3.22 and 6.18 cm. The surface charge density on the inner cylinder is 24.7 μC/m² and that on the outer cylinder is -18.0 μC/m². Find the electric field at (*a*) $r = 4.10$ cm and (*b*) $r = 8.20$ cm.

38. An uncharged, spherical, thin, metallic shell has a point

charge q at its center. Derive expressions for the electric field (*a*) inside the shell and (*b*) outside the shell, using Gauss' law. (*c*) Has the shell any effect on the field due to q? (*d*) Has the presence of q any effect on the shell? (*e*) If a second point charge is held outside the shell, does this outside charge experience a force? (*f*) Does the inside charge experience a force? (*g*) Is there a contradiction with Newton's third law here? Why or why not?

39. A 115-keV electron is fired directly toward a large flat plastic sheet that has a surface charge density of -2.08 μC/m². From what distance must the electron be fired if it is just to fail to strike the sheet? (Ignore relativistic effects.)

40. Charged dust particles in interstellar space, each carrying one excess electron and all of the same mass, form a stable, spherical, uniform cloud. Find the mass of each particle.

41. Positive charge is distributed uniformly throughout a long, nonconducting cylindrical shell of inner radius R and outer radius $2R$. At what radial depth beneath the outer surface of the charge distribution is the electric field strength equal to one-half the surface value?

42. The spherical region $a < r < b$ carries a charge per unit volume of $\rho = A/r$, where A is a constant. At the center ($r = 0$) of the enclosed cavity is a point charge q. What should be the value of A so that the electric field in the region $a < r < b$ has constant magnitude?

43. Show that stable equilibrium under the action of electrostatic forces alone is impossible. (*Hint*: Assume that at a certain point P in an electric field $\mathbf{E}$, a charge $+q$ would be in stable equilibrium if it were placed there. Draw a spherical Gaussian surface about P, imagine how $\mathbf{E}$ must point on this surface, and apply Gauss' law to show that the assumption leads to a contradiction.) This result is known as Earnshaw's theorem.

44. A spherical region carries a uniform charge per unit volume ρ. Let $\mathbf{r}$ be the vector from the center of the sphere to a general point P within the sphere. (*a*) Show that the electric field at P is given by $\mathbf{E} = \rho\mathbf{r}/3\epsilon_0$. (*b*) A spherical cavity is created in the above sphere, as shown in Fig. 37. Using superposition concepts, show that the electric field at all points within the cavity is $\mathbf{E} = \rho\mathbf{a}/3\epsilon_0$ (uniform field), where $\mathbf{a}$ is the vector connecting the center of the sphere with the center of the cavity. Note that both these results are independent of the radii of the sphere and the cavity.

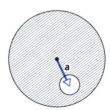

Figure 37   Problem 44.

45. Charge is distributed uniformly throughout an infinitely long cylinder of radius R. (*a*) Show that E at a distance r from the cylinder axis ($r < R$) is given by

$$E = \frac{\rho r}{2\epsilon_0},$$

where ρ is the volume charge density. (*b*) What result do you obtain for $r > R$?

46. A plane slab of thickness d has a uniform volume charge density ρ. Find the magnitude of the electric field at all points in space both (*a*) inside and (*b*) outside the slab, in terms of x, the distance measured from the median plane of the slab.

47. A solid nonconducting sphere of radius R carries a non-uniform charge distribution, the charge density being $\rho = \rho_s r/R$, where ρ_s is a constant and r is the distance from the center of the sphere. Show that (*a*) the total charge on the sphere is $Q = \pi \rho_s R^3$ and (*b*) the electric field inside the sphere is given by

$$E = \frac{1}{4\pi\epsilon_0} \frac{Q}{R^4} r^2.$$

48. Construct a *spherical* Gaussian surface centered on an infinite line of charge, calculate the flux through the sphere, and thereby show that Gauss' law is satisfied.

Section 29-7 The Nuclear Model of the Atom

49. In a 1911 paper, Ernest Rutherford said:

> In order to form some idea of the forces required to deflect an alpha particle through a large angle, consider an atom containing a point positive charge Ze at its centre and surrounded by a distribution of negative electricity, $-Ze$ uniformly distributed within a sphere of radius R. The electric field E . . . at a distance r from the center for a point *inside* the atom [is]

$$E = \frac{Ze}{4\pi\epsilon_0}\left(\frac{1}{r^2} - \frac{r}{R^3}\right).$$

Verify this equation.

50. Figure 38 shows a Thomson atom model of helium ($Z = 2$). Two electrons, at rest, are embedded inside a uniform sphere of positive charge $2e$. Find the distance d between the electrons so that the configuration is in static equilibrium.

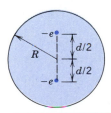

Figure 38 Problem 50.

Computer Project

51. By modifying the computer program given in Appendix I, which we used in Section 28-6 to calculate the trajectory of a particle in a nonuniform electric field, find the trajectory of a particle scattered by the electric field of another particle, as in the Rutherford scattering experiment (Section 29-7).

Choose a proton ($q = +e$, $m = 1.67 \times 10^{-27}$ kg) as the scattered particle and a gold nucleus ($Q = +79e$) as the target, which is assumed to be fixed at the origin of the xz coordinate system. Use the components E_x and E_z of the electric field of the target to find the acceleration components a_x and a_z of the proton. Take the initial position of the proton to be $z_0 = 3$ fm (the impact parameter b) when x_0 is very large and negative (say, -2000 fm), and let the proton move initially parallel to the x axis ($v_{0x} > 0$, $v_{0z} = 0$) with a speed corresponding to an initial kinetic energy K of 4.7 MeV. Choose small increments of time in doing the calculation, and tabulate x, z, v_x, v_z, $r = (x^2 + z^2)^{1/2}$, and $\phi = \tan^{-1}(z/x)$ as functions of the time t. Plot the trajectory of the particle and compare with the calculated trajectory, which can be found from Newton's laws to be given by

$$\frac{1}{r} = \frac{1}{b}\sin\phi + \frac{qQ}{8\pi\epsilon_0 b^2 K}(\cos\phi - 1).$$

To avoid errors, the time increment must be chosen to be very small. To test whether you have selected a small enough increment, run the program and examine the trajectory at times large enough that the proton is far from the gold nucleus after the scattering. The trajectory should be symmetric on either side of the point of closest approach of the projectile to the target, and the initial and final speeds should be equal.

Repeat the calculation for different values of the impact parameter b. For each value of b, determine the scattering angle $\theta = \pi - \phi$, where ϕ is evaluated as $r \to \infty$ after the scattering. Plot θ vs. b and try to determine the relationship between them.

CHAPTER 30

ELECTRIC POTENTIAL

*In Chapters 7 and 8, we learned that in some cases the energy
approach to the study of the dynamics of particles can yield not only
simplifications but also new insights. In Chapter 16, we used the energy method for
situations involving the gravitational force; we were thus able to determine such properties
as escape speeds and orbital parameters of planets and satellites.*

*One advantage of the energy method is that, although force is a vector, energy is a scalar. In
problems involving vector forces and fields, calculations requiring sums and integrals are
often complicated. For example, when we calculated the electric field in Chapter 28 for
continuous charge distributions, it was necessary to take into account the vector nature of
the field and do the integrals accordingly.*

*In this chapter, we introduce the energy method to the study of electrostatics. We begin with
the electric potential energy, a scalar that characterizes an electrostatic force just as the
gravitational potential energy characterizes a gravitational force. We then generalize to the
field of an arbitrary charge distribution and introduce the concept of electric potential. We
calculate the potential for discrete and continuous charge distributions, and we show that
the electric field and the electric potential are closely related—given one, we can find the other.*

30-1 ELECTROSTATIC AND GRAVITATIONAL FORCES

The similarity between the electrostatic and gravitational forces permits us to simplify our derivation of the electrostatic quantities by referring back to Chapter 16 for the derivation of the corresponding gravitational quantities. Note the similarities of the two force laws:

$$F = G \frac{m_1 m_2}{r^2} \qquad \text{(gravitational)}, \qquad (1a)$$

$$F = \frac{1}{4\pi\epsilon_0} \frac{q_1 q_2}{r^2} \qquad \text{(electrostatic)}, \qquad (1b)$$

which give, respectively, the gravitational force between two particles of mass m_1 and m_2 and the electrostatic force between two particles of charge q_1 and q_2, in each case separated by a distance r. The two force laws have exactly the same form: a constant (G or $1/4\pi\epsilon_0$) that gives the strength of the force, times the product of a property of the two particles (mass or charge) divided by the square of their separation. That is, both Newton's law of gravitation and Coulomb's law are *inverse square laws*.

In Section 16-7 we introduced the gravitational field strength **g**, defined at any location as the gravitation force per unit mass exerted on a test body of mass m_0 placed at that location. The electric field strength **E** was defined in Eq. 2 of Chapter 28 in a very similar fashion as the electrostatic force per unit charge exerted on a test charge q_0. Note the similarity in the mathematical definitions:

$$\mathbf{g} = \frac{\mathbf{F}}{m_0} \qquad \text{(gravitational)}, \qquad (2a)$$

$$\mathbf{E} = \frac{\mathbf{F}}{q_0} \qquad \text{(electrostatic)}. \qquad (2b)$$

In both cases, Eq. 2 gives us an operational procedure for measuring the field strength.

You will recall that the difference in potential energy ΔU when a particle moves between points a and b under the influence of a force **F** is equal to the negative of the work done by the force, or

$$\Delta U = -W_{ab}, \qquad (3)$$

where W_{ab} is the work done by the force as the particle moves from a to b. Equation 3 applies only if the force is conservative; indeed, potential energy is defined *only* for

651

conservative forces, as we discussed in Section 8-2. We can also write Eq. 3 as

$$U_b - U_a = -\int_a^b \mathbf{F} \cdot d\mathbf{s}. \qquad (4)$$

In Chapter 8 we generalized from difference in potential energy to potential energy itself by defining the potential energy to be zero at a suitable reference point. It is convenient, as in Section 16-6, to choose the reference point for potential energy to correspond to an infinite separation of the particles (where the force is zero, according to Eq. 1a), and then to define the potential energy to be zero in that condition.

The potential energy can be defined for a particular force only if the force is conservative, and we must therefore first establish the conservative nature of a force before attempting to calculate its potential energy. In Section 16-6 (see especially Fig. 13 of Chapter 16) we showed that the $1/r^2$ gravitational force is conservative, and we argued that the work done by the gravitational force when a particle moves from a to b is independent of the path taken between those locations. We can make the same argument for the electrostatic force with the same result: *the electrostatic force is conservative, and it can be represented by a potential energy.* We give the mathematical derivation in the next section.

There is one important property in which the electrostatic force differs from the gravitational force: gravitational forces are always attractive, while (depending on the relative signs of the charges) electrostatic forces can be either attractive or repulsive. As we see in the next section, this difference can affect the sign of the potential energy, but it in no way changes our argument based on the analogy between the two forces.

30-2 ELECTRIC POTENTIAL ENERGY

If you raise a stone from the surface of the Earth, the change in gravitational potential energy of the system of Earth and stone is, according to Eq. 4 of Chapter 8, the negative of the work done by the gravitational force. We can treat electrostatic situations similarly.

We have already argued in Section 30-1 by analogy with the gravitational force that the electrostatic force is conservative, and therefore *we can associate a potential energy with any system in which a charged particle is placed in an electric field and acted on by an electrostatic force.* The change in electrostatic potential energy, when a particle of charge q moves in an electric field $\mathbf{E}$, is given by Eq. 4, with the electric force $q\mathbf{E}$ substituted for the force $\mathbf{F}$:

$$U_b - U_a = -\int_a^b \mathbf{F} \cdot d\mathbf{s} = -q \int_a^b \mathbf{E} \cdot d\mathbf{s}, \qquad (5)$$

Figure 1 Two charges q_1 and q_2 separated by a distance r.

where the integral is carried out over the path of the particle from initial point a to final point b. Because the electric force is conservative, the integral is independent of path and depends only on the initial and final points a and b.

Consider two particles of charge q_1 and q_2 separated by a distance r (Fig. 1). Assume first that the charges have opposite signs, so that the force between them is attractive. If we move q_2 to the right, the electric force does negative work, the right-hand side of Eq. 5 is positive, and the potential energy of the system increases. If we release the charges from this greater separation, the separation decreases toward the original value; the potential energy of the system decreases while the kinetic energy of the system increases, in analogy with the gravitational case.

If the two charges in Fig. 1 have the same sign, moving q_2 to the *left* increases the potential energy of the system (because the electric force does negative work in this case). If we release the charges, their separation increases; the resulting decrease in potential energy is accompanied by a corresponding increase in the kinetic energy as the two charges move apart.

Let us now calculate the expression for the potential energy of the system of two point charges shown in Fig. 1. We use Eq. 5, and we assume q_2 moves toward or away from q_1 along the line connecting the two charges, which we take to be the x axis. The component E_x of the electric field due to q_1 along this line is $q_1/4\pi\epsilon_0 r^2$. This component is positive or negative according to the sign of q_1. Figure 2 shows the corresponding vector relationships. The vector $\mathbf{r}$ ($= r\mathbf{i}$, where $\mathbf{i}$ is the unit vector in the x direction) locates q_2 relative to q_1, and the vector $d\mathbf{s}$ ($= dr\mathbf{i}$) indicates the displacement of q_2. Thus $\mathbf{E} \cdot d\mathbf{s} = E_x dr$, and so, if we move q_2 from separation r_a to r_b, the change in potential energy is given by Eq. 5 as

$$U_b - U_a = -q_2 \int_{r_a}^{r_b} E_x \, dr = -\frac{1}{4\pi\epsilon_0} q_1 q_2 \int_{r_a}^{r_b} \frac{dr}{r^2}$$

$$= \frac{1}{4\pi\epsilon_0} q_1 q_2 \left(\frac{1}{r_b} - \frac{1}{r_a} \right). \qquad (6)$$

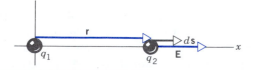

Figure 2 Charge q_2 moves relative to q_1 through a displacement $d\mathbf{s}$. The electric field due to positive charge q_1 is in the direction shown.

Equation 6 holds whether q_2 moves toward or away from q_1; in the first case, $r_b < r_a$, and in the second case, $r_b > r_a$. The equation also holds for any combination of the signs of q_1 and q_2. Furthermore, because ΔU is independent of path for a conservative force, Eq. 6 holds no matter how q_2 moves between r_a and r_b; we chose a direct radial path to simplify the calculation, but the result is valid for *any* path.

As we did in Section 16-6, we can choose a reference point a such that r_a corresponds to an infinite separation of the particles, and we define the potential energy U_a to be zero. Let r be the separation at the final point b, so that Eq. 6 reduces to

$$U(r) = \frac{1}{4\pi\epsilon_0} \frac{q_1 q_2}{r}. \tag{7}$$

Compare this result with Eq. 15 of Chapter 16 for the gravitational potential energy, which we can write as $U(r) = -Gm_1m_2/r$. If the electric force is attractive, q_1 and q_2 have opposite signs, and so the product $q_1 q_2$ is negative. In this case, the electric potential energy given by Eq. 7 is negative, as is the similarly attractive gravitational potential energy.

If the electric force is repulsive, q_1 and q_2 have the same sign, and the product $q_1 q_2$ is positive. In this case, which has no known gravitational analogue, the potential energy is positive. If we move q_2 toward q_1 from an initially infinite separation, the potential energy increases from its initial value (which we defined to be 0). If we then release q_2 from rest, it moves to larger separation, gaining kinetic energy as the system loses potential energy.

Sample Problem 1   Two protons in the nucleus of a ^{238}U atom are 6.0 fm apart. What is the potential energy associated with the electric force that acts between these two particles?

Solution   From Eq. 7, with $q_1 = q_2 = +1.60 \times 10^{-19}$ C, we obtain

$$U = \frac{1}{4\pi\epsilon_0} \frac{q_1 q_2}{r} = \frac{(8.99 \times 10^9 \text{ N}\cdot\text{m}^2/\text{C}^2)(1.60 \times 10^{-19} \text{ C})^2}{6.0 \times 10^{-15} \text{ m}}$$

$$= 3.8 \times 10^{-14} \text{ J} = 2.4 \times 10^5 \text{ eV} = 240 \text{ keV}.$$

The two protons do not fly apart because they are held together by the attractive *strong force* that binds the nucleus together. Unlike the electric force, there is no simple potential energy function that represents the strong force.

Potential Energy of a System of Charges

Suppose we have a system of point charges held in fixed positions by forces not specified. We can calculate the total potential energy of this system by applying Eq. 7 to every pair of charges in the system. For example, if we have a system of three charges as in Fig. 3, the electric potential energy of the system is

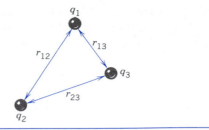

Figure 3   An assembly of three charges.

$$U = \frac{1}{4\pi\epsilon_0} \frac{q_1 q_2}{r_{12}} + \frac{1}{4\pi\epsilon_0} \frac{q_1 q_3}{r_{13}} + \frac{1}{4\pi\epsilon_0} \frac{q_2 q_3}{r_{23}}. \tag{8}$$

Note that the potential energy is a property of the system, not of any individual charge.

From this example you can immediately see the advantage of using an energy method to analyze this system: the sum involved in Eq. 8 is an *algebraic* sum of scalars. If we tried to calculate the electric field of the three charges, we would have a more complicated *vector* sum to evaluate.

There is another way to interpret the potential energy of this system. Let the three charges be initially at infinite separations from one another. We bring the first charge, q_1, in from infinity and place it at the location shown in Fig. 3. No change of potential energy is involved, because the other charges are not yet present. Bringing q_2 into position gives a potential energy $q_1 q_2/4\pi\epsilon_0 r_{12}$. Finally, bringing q_3 in from infinity to its position gives two additional terms: $q_1 q_3/4\pi\epsilon_0 r_{13}$ and $q_2 q_3/4\pi\epsilon_0 r_{23}$, which give, respectively, the potential energy of q_3 in the fields of q_1 and q_2. We can continue this process to assemble any arbitrary distribution of charge. The resulting total potential energy is independent of the order in which we assemble the charges.

When an external agent moves the charges from infinite separation to assemble a distribution such as that of Fig. 3, the agent does work in exerting a force that opposes the electrostatic force. The external agent is in effect storing energy in the system of charges. This can most easily be seen by considering the special case in which all the charges have the same sign. The charges that are already in place exert a repulsive force on new charges that are added, and the external agent must push the new charges into position. In effect, the external agent must expend energy to assemble the charge distribution. The energy is stored in the electric field of the system, and we account for it in terms of the electric potential energy of the resulting distribution. If we were suddenly to release the restraints holding the charges at their positions, they would gain kinetic energy as the system flew apart; the total kinetic energy of all the particles at infinite separation is, by conservation of energy, equal to the energy supplied by the external agent to assemble the system. If the charges had differing signs such that the total potential energy were negative, the particles would tend to move closer together if released from their positions. In this case the external agent would need to supply additional energy in the form of work to disassemble the system and move the charges to infinite separation.

We summarize this view as follows:

The electric potential energy of a system of fixed point charges is equal to the work that must be done by an external agent to assemble the system, bringing each charge in from an infinite distance. The charges are at rest in their initial positions and in their final positions.

Implicit in this definition is that we consider the reference point of potential energy to be the infinite separation of the charges, and we take the potential energy to be zero at this reference point.

For continuous charge distributions, the potential energy can be computed using a similar technique, dividing the distribution into small elements and treating each element as a point charge. We shall not consider such problems in this text.

Sample Problem 2 In the system shown in Fig. 3, assume that $r_{12} = r_{13} = r_{23} = d = 12$ cm, and that

$$q_1 = +q, \quad q_2 = -4q, \quad \text{and} \quad q_3 = +2q,$$

where $q = 150$ nC. What is the potential energy of the system?

Solution Using Eq. 8, we obtain

$$U = \frac{1}{4\pi\epsilon_0}\left(\frac{(+q)(-4q)}{d} + \frac{(+q)(+2q)}{d} + \frac{(-4q)(+2q)}{d}\right)$$

$$= -\frac{10q^2}{4\pi\epsilon_0 d}$$

$$= -\frac{(8.99 \times 10^9 \text{ N}\cdot\text{m}^2/\text{C}^2)(10)(150 \times 10^{-9} \text{ C})^2}{0.12 \text{ m}}$$

$$= -1.7 \times 10^{-2} \text{ J} = -17 \text{ mJ}.$$

The negative potential energy in this case means that negative work would be done by an external agent to assemble this structure, starting with the three charges infinitely separated and at rest. Put another way, an external agent would have to do $+17$ mJ of work to dismantle the structure completely.

30-3 ELECTRIC POTENTIAL

The force between two charged particles depends on the magnitude and sign of each charge. We have found it useful to introduce a vector quantity, the electric field, defined (see Eq. 2b) as the force per unit test charge. With this definition we can now speak of the electric field associated with a single charge.

In many applications we find it useful to work with a related scalar quantity, which is obtained from the potential energy in a similar way. This quantity is called the *electric potential* and is defined as the *potential energy per unit test charge*.

Suppose we have a collection of charges whose electric potential we wish to determine at a particular point P. We place a positive test charge q_0 an infinite distance from the collection of charges, where the electric field is zero. We then move that test charge from that infinite separation to P, and in the process the potential energy changes from 0 to U_P. The electric potential V_P at P due to the collection of charges is then defined as

$$V_P = \frac{U_P}{q_0}. \tag{9}$$

Note from Eq. 9 that potential must be a scalar, because it is calculated from the scalar quantities U and q.

Defined in this way, the potential is independent of the size of the test charge, as is the electric field defined according to Eq. 2b. (As we did in the electric field case, we assume that q_0 is a very small charge, so that it has a negligible effect on the collection of charges whose potential we wish to measure.) Equation 9 provides an operational basis for measuring the potential; as was the case with the electric field, we later establish more convenient mathematical procedures for calculating V.

Depending on the distribution of charges, the potential V_P may be positive, negative, or zero. Suppose the potential is positive at a certain point; according to Eq. 9, the potential energy at that point is positive. If we were to move a positive test charge from infinity to that point, the electric field would do negative work, which indicates that, on the average, the test charge has experienced a repulsive force. *The potential near an isolated positive charge is therefore positive.* If the potential at a point is negative, the reverse is true: as we move a positive test charge in from infinity, the electric field does positive work, and on the average the force is attractive. *The potential near an isolated negative charge is therefore negative.*

If the potential is zero at a point, no net work is done by the electric field as the test charge moves in from infinity, although the test charge may have moved through regions where it experienced attractive or repulsive electric forces. *A potential of zero at a point does not necessarily mean that the electric field is zero at that point.* Consider, for instance, a point midway between two equal and opposite charges. The potentials at that point due to the two individual charges have equal magnitudes and opposite signs, and so the total potential there is zero. However, the electric fields of the two charges have the same direction at that point, and the total electric field is certainly not zero.

Instead of making reference to a point at infinity, we often wish to find the *electric potential difference* between two points a and b in an electric field. To do so, we move a test charge q_0 from a to b. The electric potential difference is defined by an extension of Eq. 9 as

$$\Delta V = V_b - V_a = \frac{U_b - U_a}{q_0}. \qquad (10)$$

The potential at b may be greater than, less than, or the same as the potential at a, depending on the difference in potential energy between the two points or, equivalently, on the negative of the work done by the electric field as a positive test charge moves between the points. For instance, if b is at a higher potential than a $(V_b - V_a > 0)$, the electric field does *negative* work as the test charge moves from a to b.

The SI unit of potential that follows from Eq. 9 is the joule/coulomb. This combination occurs so often that a special unit, the *volt* (abbreviation V), is used to represent it; that is,

$$1 \text{ volt} = 1 \text{ joule/coulomb}.$$

The common name of "voltage" is often used for the potential at a point or the potential difference between points. When you touch the two probes of a *voltmeter* to two points in an electric circuit, you are measuring the potential difference (in volts) or voltage between those points.

Equation 10 can be written

$$\Delta U = q\,\Delta V, \qquad (11)$$

which states that when *any* charge q moves between two points whose potential difference is ΔV, the system experiences a change in potential energy ΔU given by Eq. 11. The potential difference ΔV is set up by other charges that are maintained at rest, so that the motion of charge q doesn't change the potential difference ΔV. In using Eq. 11, when ΔV is expressed in volts and q in coulombs, ΔU comes out in joules.

From Eq. 11, you can see that the *electron-volt*, which we have introduced previously as a unit of energy, follows directly from the definition of potential or potential difference. If ΔV is expressed in volts and q in units of the elementary charge e, then ΔU is expressed in electron-volts (eV). For example, consider a system in which a carbon atom from which all six electrons have been removed $(q = +6e)$ moves through a change in potential of $\Delta V = +20$ kV. The change in potential energy is

$$\Delta U = q\,\Delta V = (+6e)(+20 \text{ kV}) = +120 \text{ keV}.$$

Doing such calculations in units of eV is a great convenience when dealing with atoms or nuclei, in which the charge is easily expressed in terms of e.

Keep in mind that *potential differences* are of fundamental concern and that Eq. 9 depends on the arbitrary assignment of the value zero to the potential at the reference position (infinity); this reference potential could equally well have been chosen as any other value, say -100 V. Similarly, any other agreed-upon point could be

chosen as a reference position. In many problems the Earth is taken as a reference of potential and assigned the value zero. The location of the reference point and the value of the potential there are chosen for convenience; other choices would change the potential everywhere by the same amount but would not change the results for the potential difference.

We have already discussed that the electric field is a conservative field, and so the potential energy difference between points a and b depends only on the locations of the points and not on the path taken to move from one point to the other. Equation 10 therefore suggests that the potential difference is similarly path independent: the potential difference between any two points in an electric field is independent of the path through which the test charge moves in traveling from one point to the other.

Sample Problem 3   An alpha particle $(q = +2e)$ in a nuclear accelerator moves from one terminal at a potential of $V_a = +6.5 \times 10^6$ V to another at a potential of $V_b = 0$. (a) What is the corresponding change in the potential energy of the system? (b) Assuming the terminals and their charges do not move and that no external forces act on the system, what is the change in kinetic energy of the particle?

Solution   (a) From Eq. 11, we have

$$\begin{aligned} \Delta U = U_b - U_a &= q(V_b - V_a) \\ &= (+2)(1.6 \times 10^{-19} \text{ C})(0 - 6.5 \times 10^6 \text{ V}) \\ &= -2.1 \times 10^{-12} \text{ J}. \end{aligned}$$

(b) If no external force acts on the system, then its mechanical energy $E = U + K$ must remain constant. That is, $\Delta E = \Delta U + \Delta K = 0$, and so

$$\Delta K = -\Delta U = +2.1 \times 10^{-12} \text{ J}.$$

The alpha particle gains a kinetic energy of 2.1×10^{-12} J, in the same way that a particle falling in the Earth's gravitational field gains kinetic energy.

To see the simplifications that result, try working this problem again with the energies expressed in units of eV.

30-4   CALCULATING THE POTENTIAL FROM THE FIELD

Given the electric field **E** we can calculate the potential V, and given V we can calculate **E**. Here we discuss the calculation of V from **E**; the calculation of **E** from V is discussed in Section 30-9.

Let a and b in Fig. 4 be two points in a uniform electric field **E**, set up by an arrangement of charges not shown, and let a be a distance L in the field direction from b.

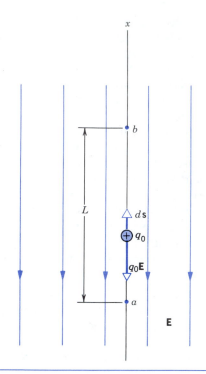

Figure 4 Test charge q_0 moves a distance L from a to b in a uniform electric field **E**.

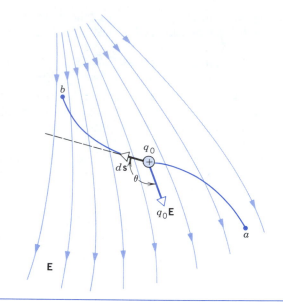

Figure 5 Test charge q_0 moves from a to b in the nonuniform electric field **E**.

Assume that a positive test charge q_0 moves from a to b along the straight line connecting them.

The electric force on the charge is $q_0\mathbf{E}$ and points in the negative x direction. As a test charge moves from a to b in the direction of $d\mathbf{s}$, the work done by the (constant) electric field is given by

$$W_{ab} = F_x\,\Delta x = (-q_0 E)(L) = -q_0 EL. \quad (12)$$

Using the definition of potential energy difference, $\Delta U = -W$, we can combine Eqs. 10 and 12 to obtain

$$V_b - V_a = \frac{U_b - U_a}{q_0} = \frac{-W_{ab}}{q_0} = EL. \quad (13)$$

This equation shows the connection between potential difference and field strength for a simple special case. Note from this equation that another SI unit for **E** is the volt/meter (V/m). You may wish to prove that a volt/meter is identical with a newton/coulomb (N/C); this latter unit was the one first presented for **E** in Section 28-2.

In Fig. 4, b has a higher potential than a. This is reasonable because the electric field does negative work on the positive test charge as it moves from a to b. Figure 4 could be used as it stands to illustrate the act of lifting a stone from a to b in the uniform gravitational field near the Earth's surface. All we need do is replace the test charge q_0 by a test mass m_0 and replace the electric field **E** by the gravitational field **g**.

What is the connection between V and **E** in the more general case in which the field is *not* uniform and in which the test body moves along a path that is *not* straight, as in Fig. 5? The electric field exerts a force $q_0\mathbf{E}$ on the test charge, as shown. An infinitesimal displacement along the path is represented by $d\mathbf{s}$. To find the total work W_{ab} done by the electric field as the test charge moves from a to b, we add up (that is, integrate) the work contributions for all the infinitesimal segments into which the path is divided. This leads to

$$W_{ab} = \int_a^b \mathbf{F}\cdot d\mathbf{s} = q_0 \int_a^b \mathbf{E}\cdot d\mathbf{s}. \quad (14)$$

Such an integral is called a *line integral*, as we discussed in Section 7-3.

With $V_b - V_a = (U_b - U_a)/q_0 = -W_{ab}/q_0$, Eq. 14 gives

$$V_b - V_a = -\int_a^b \mathbf{E}\cdot d\mathbf{s}. \quad (15)$$

It is frequently convenient to choose point a to be the reference point at ∞, where V_a is taken to be zero. We can then find the potential at any arbitrary point P using Eq. 15:

$$V_P = -\int_\infty^P \mathbf{E}\cdot d\mathbf{s}. \quad (16)$$

These two equations allow us to calculate the potential difference between any two points or the potential at any point in a known electric field **E**.

Sample Problem 4 In Fig. 6 let a test charge q_0 be moved from a to b over the path acb. Compute the potential difference between a and b.

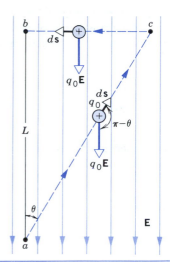

Figure 6   Sample Problem 4. A test charge q_0 moves along the path *acb* through the uniform electric field **E**.

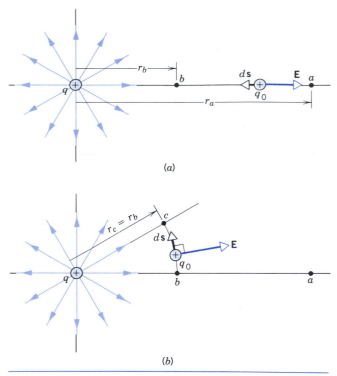

Figure 7   (*a*) A test charge q_0 moves from *a* to *b* along a radial line from a positive charge *q* that establishes an electric field **E**. (*b*) The test charge now moves from *b* to *c* along the arc of a circle centered on *q*.

Solution   For the path *ac* we have, from Eq. 15,

$$V_c - V_a = -\int_a^c \mathbf{E} \cdot d\mathbf{s} = -\int_a^c E \, ds \cos(\pi - \theta)$$

$$= E \cos \theta \int_a^c ds.$$

The integral is the length of the line *ac*, which is $L/\cos \theta$. Thus

$$V_c - V_a = E \cos \theta \, \frac{L}{\cos \theta} = EL.$$

Points *b* and *c* have the same potential because no work is done in moving a charge between them, **E** and *d***s** being at right angles for all points on the line *cb*. Thus

$$V_b - V_a = (V_b - V_c) + (V_c - V_a) = 0 + EL = EL.$$

This is the same value derived for a direct path connecting *a* and *b*, a result to be expected because the potential difference between two points is independent of path.

30-5   POTENTIAL DUE TO A POINT CHARGE

Figure 7*a* shows two points *a* and *b* near an isolated positive point charge *q*. For simplicity we assume that *a*, *b*, and *q* lie on a straight line. Let us compute the potential difference between points *a* and *b*, assuming that a positive test charge q_0 moves along a radial line from *a* to *b*.

In Fig. 7*a*, both **E** and *d***s** ($= d\mathbf{r}$) have only a radial component. Thus $\mathbf{E} \cdot d\mathbf{r} = E \, dr$, and substituting this result into Eq. 15 gives

$$V_b - V_a = -\int_a^b \mathbf{E} \cdot d\mathbf{s} = -\int_{r_a}^{r_b} E \, dr.$$

Using the expression for the electric field of a point charge, $E = q/4\pi\epsilon_0 r^2$, we obtain

$$V_b - V_a = -\frac{q}{4\pi\epsilon_0} \int_{r_a}^{r_b} \frac{dr}{r^2} = \frac{q}{4\pi\epsilon_0}\left(\frac{1}{r_b} - \frac{1}{r_a}\right). \quad (17)$$

Equation 17 gives the potential difference between points *a* and *b*. We have simplified the integration by choosing to move the test charge along a radial path, but the potential is independent of path, so Eq. 17 holds for *any* path between *a* and *b*. That is, the potential difference is a property of the points *a* and *b* themselves and not of the path *ab*.

Moreover, Eq. 17 holds for the potential difference between two points even if they do not lie on the same radial line. Figure 7*b* shows arbitrary points *a* and *c*. Because the potential difference is independent of path, we are free to choose the simplest path over which to compute the difference in potential. We choose the path *abc*, in which *ab* is radial and *bc* is along the arc of a circle centered on *q*. No work is done by the field along *bc*, because **E** is perpendicular to *d***s** everywhere on *bc*, and thus the potential difference between *a* and *c* is also given by Eq. 17.

If we wish to find the potential at any point (rather than the potential difference between two points), it is customary to choose a reference point at infinity. We choose *a* to be at infinity (that is, let $r_a \to \infty$) and define V_a to be 0 at

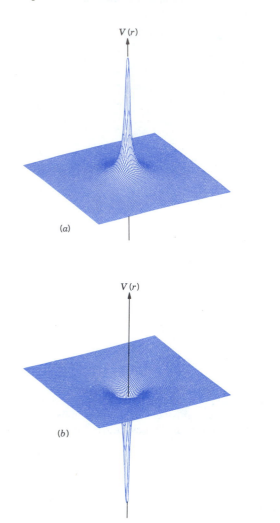

$V(r)$

(a)

$V(r)$

(b)

Figure 8   A computer-generated plot of the potential $V(r)$ in a plane near (a) a positive and (b) a negative point charge.

this position. Making these substitutions in Eq. 17 and dropping the subscript b lead to

$$V = \frac{1}{4\pi\epsilon_0} \frac{q}{r}. \qquad (18)$$

Equation 18 also is valid for any spherically symmetric distribution of total charge q, as long as r is greater than the radius of the distribution. Note that Eq. 18 could also have been obtained directly from Eq. 16.

Equation 18 shows that the potential due to a positive point charge is zero at large distances and grows to large positive values as we approach the charge. If q is negative, the potential approaches large negative values near the charge. Figure 8 shows computer-generated plots of Eq. 18 for a positive and a negative point charge. Note that these results do not depend at all on the sign of the *test charge* we used in the calculation.

Sample Problem 5   What must be the magnitude of an isolated positive point charge for the electric potential at 15 cm from the charge to be $+120$ V?

Solution   Solving Eq. 18 for q yields

$$q = V4\pi\epsilon_0 r = (120 \text{ V})(4\pi)(8.9 \times 10^{-12} \text{ C}^2/\text{N}\cdot\text{m}^2)(0.15 \text{ m})$$
$$= 2.0 \times 10^{-9} \text{ C} = 2.0 \text{ nC}.$$

This charge is comparable to charges that can be produced by friction, such as by rubbing a balloon.

Sample Problem 6   What is the electric potential at the surface of a gold nucleus? The radius is 7.0×10^{-15} m, and the atomic number Z is 79.

Solution   The nucleus, assumed spherically symmetric, behaves electrically for external points as if it were a point charge. Thus we can use Eq. 18, which gives, with $q = +79e$,

$$V = \frac{1}{4\pi\epsilon_0} \frac{q}{r} = \frac{(9.0 \times 10^9 \text{ N}\cdot\text{m}^2/\text{C}^2)(79)(1.6 \times 10^{-19} \text{ C})}{7.0 \times 10^{-15} \text{ m}}$$
$$= 1.6 \times 10^7 \text{ V}.$$

This large positive potential has no effect outside a gold *atom* because it is compensated by an equally large negative potential from the 79 atomic electrons of gold.

30-6   POTENTIAL DUE TO A COLLECTION OF POINT CHARGES

The potential at any point due to a group of N point charges is found by (1) calculating the potential V_i due to each charge, as if the other charges were not present, and (2) adding the quantities so obtained:

$$V = V_1 + V_2 + \cdots + V_N,$$

or, using Eq. 18,

$$V = \sum_{i=1}^{N} V_i = \frac{1}{4\pi\epsilon_0} \sum_i \frac{q_i}{r_i}, \qquad (19)$$

where q_i is the value (magnitude and sign) of the ith charge and r_i is the distance of this charge from the point in question. Once again, we see the benefit gained by using the potential, which is a *scalar*: the sum used to calculate V is an *algebraic sum* and not a vector sum like the one used to calculate **E** for a group of point charges (see Eq. 5 of Chapter 28). This is an important computational advantage of using potential rather than electric field.

The potential at a point due to one of the charges is not affected by the presence of the other charges. To find the total potential, we add the potentials due to each of the charges as if it were the only one present. This is the *principle of superposition*, which applies to potential as well as to electric field.

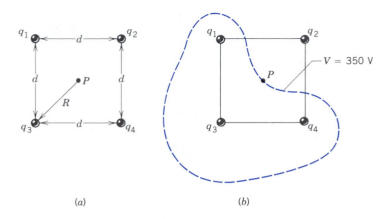

(a)   (b)

Sample Problem 7 Calculate the potential at point P, located at the center of the square of point charges shown in Fig. 9a. Assume that $d = 1.3$ m and that the charges are

$$q_1 = +12 \text{ nC}, \quad q_3 = +31 \text{ nC},$$
$$q_2 = -24 \text{ nC}, \quad q_4 = +17 \text{ nC}.$$

Solution From Eq. 19 we have

$$V_P = \sum_i V_i = \frac{1}{4\pi\epsilon_0} \frac{q_1 + q_2 + q_3 + q_4}{R}.$$

The distance R of each charge from the center of the square is $d/\sqrt{2}$ or 0.919 m, so that

$$V_P = \frac{(8.99 \times 10^9 \text{ N} \cdot \text{m}^2/\text{C}^2)(12 - 24 + 31 + 17) \times 10^{-9} \text{ C}}{0.919 \text{ m}}$$
$$= 3.5 \times 10^2 \text{ V}.$$

Close to any of the three positive charges in Fig. 9a, the potential can have very large positive values. Close to the single negative charge in that figure, the potential can have large negative values. There must then be other points within the boundaries of the square that have the same potential as that at point P. The dashed line in Fig. 9b connects other points in the plane that have this same value of the potential. As we discuss later in Section 30-8, such *equipotential surfaces* provide a useful way of visualizing the potentials of various charge distributions.

Potential Due to a Dipole

Two equal charges of opposite sign, $\pm q$, separated by a distance d, constitute an electric dipole; see Section 28-3. The electric dipole moment $\mathbf{p}$ has the magnitude qd and points from the negative charge to the positive charge. Here we derive an expression for the electric potential V due to a dipole.

A point P is specified by giving the quantities r and θ in Fig. 10. From symmetry, it is clear that the potential does not change as point P rotates about the z axis, r and θ being fixed. (Equivalently, consider what would happen if the dipole were rotated about the z axis: there would be no change in the physical situation.) Thus if we find V for points in the plane of Fig. 10, we have found V for all

points in space. Applying Eq. 19 gives

$$V_P = \sum_i V_i = V_1 + V_2$$
$$= \frac{1}{4\pi\epsilon_0}\left(\frac{q}{r_1} + \frac{-q}{r_2}\right) = \frac{q}{4\pi\epsilon_0}\frac{r_2 - r_1}{r_1 r_2}, \quad (20)$$

which is an exact relationship.

For naturally occurring dipoles, such as many molecules, the observation point P is located very far from the dipole, such that $r \gg d$. Under this condition, we can deduce from Fig. 10 that

$$r_2 - r_1 \approx d \cos\theta \quad \text{and} \quad r_1 r_2 \approx r^2,$$

and the potential reduces to

$$V \approx \frac{q}{4\pi\epsilon_0}\frac{d \cos\theta}{r^2} = \frac{1}{4\pi\epsilon_0}\frac{p \cos\theta}{r^2}. \quad (21)$$

Note that $V = 0$ everywhere in the equatorial plane ($\theta = 90°$). This means that the electric field of the dipole does no work when a test charge moves from infinity along a line in the midplane of the dipole (for instance, the

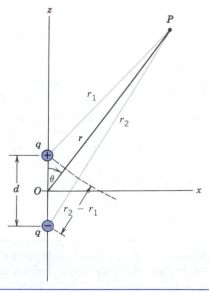

Figure 10 A point P in the field of an electric dipole.

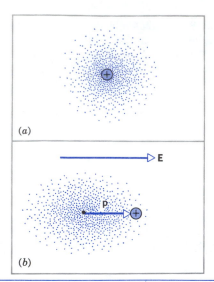

(a)

(b)

Figure 11   (a) An atom is represented by its positively charged nucleus and its diffuse negatively charged electron cloud. The centers of positive and negative charge coincide. (b) When the atom is placed in an external electric field, the positive and negative charges experience forces in opposite directions, and the centers of the positive and negative charges no longer coincide. The atom acquires an induced dipole moment.

x axis in Fig. 10). For a given r, the potential has its greatest positive value for $\theta = 0°$ and its greatest negative value for $\theta = 180°$. Note that V does not depend separately on q and d but only on their product p.

Although certain molecules, such as water, do have permanent electric dipole moments (see Fig. 18 of Chapter 28), individual atoms and many other molecules do not. However, dipole moments may be induced by placing *any* atom or molecule in an external electric field. The action of the field, as Fig. 11 shows, is to separate the centers of positive and negative charge. We say that the atom becomes *polarized* and acquires an *induced electric dipole moment*. Induced dipole moments disappear when the electric field is removed.

Electric dipoles are important in situations other than atomic and molecular ones. Radio and TV antennas are often in the form of a metal wire or rod in which electrons surge back and forth periodically. At a certain time one end of the wire or rod is negative and the other end positive. Half a cycle later the polarity of the ends is exactly reversed. This is an *oscillating* electric dipole. It is so named because its dipole moment changes in a periodic way with time.

Sample Problem 8   An *electric quadrupole* consists of two electric dipoles so arranged that they almost, but not quite, cancel each other in their electric effects at distant points (see Fig. 12). Calculate $V(r)$ for points on the axis of this quadrupole.

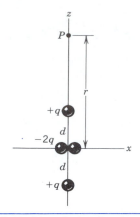

Figure 12   Sample Problem 8. An electric quadrupole, consisting of two oppositely directed electric dipoles.

Solution   Applying Eq. 19 to Fig. 12 yields

$$V = \sum_i V_i = \frac{1}{4\pi\epsilon_0}\left(\frac{q}{r-d} + \frac{-2q}{r} + \frac{q}{r+d}\right)$$

$$= \frac{1}{4\pi\epsilon_0}\frac{2qd^2}{r(r^2-d^2)} = \frac{1}{4\pi\epsilon_0}\frac{2dq^2}{r^3(1-d^2/r^2)}.$$

Because $d \ll r$, we can neglect d^2/r^2 compared with 1, in which case the potential becomes

$$V = \frac{1}{4\pi\epsilon_0}\frac{Q}{r^3}, \tag{22}$$

where $Q \ (= 2qd^2)$ is the *electric quadrupole moment* of the charge assembly of Fig. 12. Note that V varies (1) as $1/r$ for a point charge (see Eq. 18), (2) as $1/r^2$ for a dipole (see Eq. 21), and (3) as $1/r^3$ for a quadrupole (see Eq. 22).

Note too that (1) a dipole is two equal and opposite charges that do not quite coincide in space so that their electric effects at distant points do not quite cancel, and (2) a quadrupole is two equal and opposite dipoles that do not quite coincide in space so that their electric effects at distant points again do not quite cancel. We can continue to construct more complex assemblies of electric charges. This process turns out to be useful, because the electric potential of *any* charge distribution can be represented as a series of terms in increasing powers of $1/r$. The $1/r$ part, called the *monopole* term, depends on the net charge of the distribution, and the succeeding terms ($1/r^2$, the *dipole* term; $1/r^3$, the *quadrupole* term; and so on) indicate how the charge is distributed. This type of analysis is called an *expansion in multipoles*.

30-7   THE ELECTRIC POTENTIAL OF CONTINUOUS CHARGE DISTRIBUTIONS

To calculate the electric potential of a continuous charge distribution, we follow the same method we used in Section 28-5 to calculate the electric field of a continuous

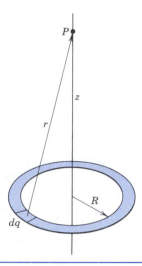

Figure 13   A uniformly charged ring. To find the potential at P, we calculate the total effect of all charge elements such as dq.

charge distribution. The calculation is simpler in the case of the potential, because the potential is a scalar, and it is therefore not necessary to take into account the different directions of the contributions from the different elements of charge.

In analogy with Section 28-5, we assume we have either a line of charge with linear charge density λ, a surface of charge with surface charge density σ, or a volume of charge with volume charge density ρ. We divide the object into small elements of charge dq, where

$$dq = \lambda \, ds, \quad dq = \sigma \, dA, \quad \text{or} \quad dq = \rho \, dv,$$

according to the geometry of the problem.*

Each element dq can be treated as a point charge, with a contribution dV to the potential calculated according to Eq. 18, which becomes

$$dV = \frac{1}{4\pi\epsilon_0} \frac{dq}{r}. \tag{23}$$

To find the potential due to the entire distribution, it is necessary to integrate the individual contributions of all the elements, or

$$V = \int dV = \frac{1}{4\pi\epsilon_0} \int \frac{dq}{r}. \tag{24}$$

In many problems, the object is uniformly charged, so the charge density is uniform and comes out of the integral.

As an example, let us find the electric potential at point P, a distance z along the axis of a uniform ring of radius R and total charge q (Fig. 13). Consider a charge element dq on the ring. The potential dV due to this element is given

---

* We write the volume element as dv, so it will not be confused with the differential element of potential dV.

by Eq. 23. However, all such elements of the ring are the same distance r from point P, and so, as we integrate over the ring, r remains constant and can be taken out of the integral. The remaining integral, $\int dq$, gives simply the total charge q on the ring. The potential at point P can thus be expressed as

$$V = \frac{1}{4\pi\epsilon_0} \frac{q}{\sqrt{R^2 + z^2}} \qquad \text{(ring of charge),} \quad (25)$$

since $r = \sqrt{R^2 + z^2}$.

Sample Problem 9   Calculate the potential at a point on the axis of a circular plastic disk of radius R, one surface of which carries a uniform charge density σ.

Solution   The disk is shown in Fig. 14. Consider a charge element dq consisting of a circular ring of radius w and width dw, for which

$$dq = \sigma(2\pi w)(dw),$$

where $(2\pi w)(dw)$ is the surface area of the ring. The contribution of this ring to the potential at P is given by Eq. 25:

$$dV = \frac{1}{4\pi\epsilon_0} \frac{dq}{r} = \frac{1}{4\pi\epsilon_0} \frac{\sigma 2\pi w \, dw}{\sqrt{w^2 + z^2}}.$$

The potential V is found by integrating over all the rings into which the disk can be divided, or

$$V = \int dV = \frac{\sigma}{2\epsilon_0} \int_0^R (w^2 + z^2)^{-1/2} w \, dw,$$

which gives

$$V = \frac{\sigma}{2\epsilon_0} (\sqrt{R^2 + z^2} - z) \quad \text{(uniformly charged disk).} \quad (26)$$

This general result is valid for all positive values of z. In the

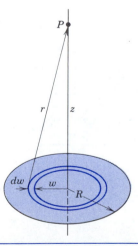

Figure 14   Sample Problem 9. A plastic disk of radius R carries a uniform charge density σ on one surface. The element of charge dq is a uniformly charged ring.

special case of $z \gg R$, the quantity $\sqrt{R^2 + z^2}$ can be approximated as

$$\sqrt{R^2 + z^2} = z\left(1 + \frac{R^2}{z^2}\right)^{1/2} = z\left(1 + \frac{1}{2}\frac{R^2}{z^2} + \cdots\right) \approx z + \frac{R^2}{2z},$$

in which the quantity in parentheses in the second member of this equation has been expanded by the binomial theorem. Using this approximation, Eq. 26 becomes

$$V \approx \frac{\sigma}{2\epsilon_0}\left(z + \frac{R^2}{2z} - z\right) = \frac{\sigma\pi R^2}{4\pi\epsilon_0 z} = \frac{1}{4\pi\epsilon_0}\frac{q}{z},$$

where $q\ (= \sigma\pi R^2)$ is the total charge on the disk. This limiting result is expected because the disk behaves like a point charge for $z \gg R$.

30-8   EQUIPOTENTIAL SURFACES

The lines of force (or, equivalently, lines to which the electric field is tangent) provide a convenient way of visualizing the field due to any charge distribution. We can make a similar graphical representation based on the electric potential. In this method, we draw a family of surfaces connecting points having the same value of the electric potential. These surfaces are called *equipotential surfaces*.

Consider first a uniform electric field **E**, for which the lines of force are shown in Fig. 15*a*. As we derived in Eq. 4, the potential difference between any two points (such as A and B_1 in Fig. 15*a*) separated by a distance L along the direction of the field has magnitude equal to EL. That is, the work done by the electric field as a positive test charge q_0 moves from A to B_1 is $q_0 EL$. If we then move the test charge perpendicular to the field, such as from B_1 to B_2 or B_3, no work is done by the electric field (because $\mathbf{E} \cdot d\mathbf{s} = 0$), and the potential difference between B_1 and B_2 or B_3 is

zero. In fact, all points on the line that contains B_1, B_2, and B_3 have the same potential. If we were to extend this drawing of a uniform field to three dimensions, the points having a given value of potential form a planar surface: *for a uniform electric field, the equipotential surfaces are planes.* Figure 15*a* shows (in cross section) a family of planar equipotential surfaces. The magnitude of the difference in potential between any point on one plane and any point on a neighboring plane is EL, where L is the (constant) spacing between the planes.

The potential of a point charge depends on the radial distance from the charge (Eq. 18). Thus all points at a given radius have the same potential, and *the equipotential surfaces of a point charge form a family of concentric spheres*, shown in cross section in Fig. 15*b* as concentric circles. The circles have been drawn so that the potential difference between any equipotential surface and its neighbor has the same value (that is, $\Delta V_{AB} = \Delta V_{BC} = \Delta V_{CD}$); the equipotential surfaces of a point charge are *not* equally spaced, in contrast to Fig. 15*a*. For a dipole, the equipotential surfaces are more complicated (Fig. 15*c*).

When a test charge moves along an equipotential surface, no work is done on it by the electric field. This follows directly from Eq. 10, for if $\Delta V = 0$, then $\Delta U = 0$, and the work W is correspondingly equal to 0. Furthermore, because of the path independence of potential, this result holds for any two points on the equipotential surface, even if the path between them does not lie entirely on the equipotential surface.

Figure 16 shows an arbitrary family of equipotential surfaces. The work done by the field when a charge moves along paths 1 or 2 is zero because both these paths begin and end on the same equipotential surface. Along paths 3 and 4 the work is not zero but has the same value for these two paths because the initial and the final potentials are identical; paths 3 and 4 connect the same pair of equipotential surfaces.

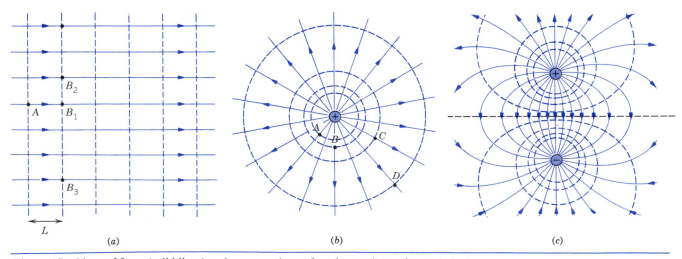

(a)   (b)   (c)

Figure 15   Lines of force (solid lines) and cross sections of equipotential surfaces (dashed lines) for (*a*) a uniform field, (*b*) a positive point charge, and (*c*) an electric dipole.

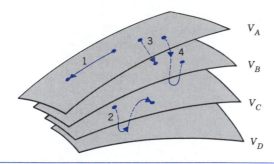

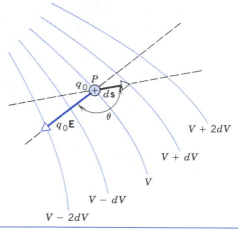

Figure 16   Portions of four equipotential surfaces. Four different paths for moving a test particle are shown.

Figure 17   A test charge q_0 is moved from one equipotential surface to another through the displacement $d\mathbf{s}$.

From an examination of Fig. 15, we see that the equipotential surfaces are always at right angles to the lines of force and thus to **E**. If **E** were *not* at right angles to the equipotential surface, **E** would have a component lying in that surface. This component would exert a force on a test charge, and thus work would be done on a test charge as it moves about on the equipotential surface. But, according to Eq. 10, work cannot be done if the surface is truly an equipotential. Therefore **E** must be at right angles to the surface. In the next section, we consider the calculation of **E** from V, which again emphasizes that **E** must be perpendicular to the equipotential surface.

30-9   CALCULATING THE FIELD FROM THE POTENTIAL

The potential V and the field **E** are equivalent descriptions for electrostatics. Equation 16, $V = -\int \mathbf{E} \cdot d\mathbf{s}$, suggests how to calculate V from **E**. Now we consider how to calculate **E** if we know V throughout a certain region.

We already have determined how to solve this problem graphically. If **E** is known at every point in space, the lines of force can be drawn; then a family of equipotentials can be sketched in by drawing surfaces perpendicular to the lines of force. These equipotentials describe the behavior of V. Conversely, if V is given as a function of position, a set of equipotential surfaces can be drawn. The lines of force can then be found by drawing lines perpendicular to the equipotential surfaces, thus describing the behavior of **E**. Here we seek the mathematical equivalent of this second graphical process, finding **E** from V. See Fig. 15 for examples of lines of force and the corresponding equipotentials.

Figure 17 shows a cross section of a family of equipotential surfaces, differing in potential by the amount dV. The figure shows that **E** at a typical point P is at right angles to the equipotential surface through P.

Let a test charge q_0 move from P through the displacement $d\mathbf{s}$ to the equipotential surface marked $V + dV$. The work that is done by the electric field is $-q_0 dV$. From another point of view we can calculate the work done on the test charge by the electric field according to

$$dW = \mathbf{F} \cdot d\mathbf{s},$$

where $\mathbf{F} (= q_0 \mathbf{E})$ is the force exerted on the charge by the electric field. The work done by the field can thus be written

$$dW = q_0 \mathbf{E} \cdot d\mathbf{s} = q_0 E \, ds \cos \theta.$$

These two expressions for the work must be equal, which gives

$$-q_0 \, dV = q_0 E \, ds \cos \theta$$

or

$$E \cos \theta = -\frac{dV}{ds}.$$

Now $E \cos \theta$, which we call E_s, is the component of **E** in the direction of $d\mathbf{s}$ in Fig. 17. We therefore obtain

$$E_s = -\frac{dV}{ds}. \qquad (27)$$

In words, this equation states: *the negative of the rate of change of the potential with position in any direction is the component of* **E** *in that direction*. The minus sign implies that **E** points in the direction of *decreasing V*, as in Fig. 17. It is clear from Eq. 27 that an appropriate unit for **E** is the volt/meter (V/m).

There will be one direction $d\mathbf{s}$ for which the quantity $-dV/ds$ is a maximum. From Eq. 27, we see that E_s will also be a maximum for this direction and will in fact be E itself. Thus

$$E = -\left(\frac{dV}{ds}\right)_{\text{max}}. \qquad (28)$$

The maximum value of dV/ds at a given point is called the *potential gradient* at that point. The direction $d\mathbf{s}$ for which dV/ds has its maximum value is always at right angles to the equipotential surface, corresponding to the direction of **E** in Fig. 17. Consider again the equipotential surfaces for the uniform field (Fig. 15a), and imagine the field lines to be removed from that figure. Suppose a test charge were located at point A, and that you were to move the test charge a fixed distance ds in any direction and determine the resulting change in potential (such as, by measuring the work done on the test charge). From Fig. 15a it is quite clear that, for a given magnitude of ds, the maximum change in potential will occur when you move the charge as far as possible from the first equipotential plane and as close as possible to the next one. This will occur only if you move the charge perpendicular to the plane, which then indicates that the electric field must be perpendicular to the equipotential plane. By carrying out this procedure for many points, you could draw a "map" of the electric field for any set of equipotential surfaces.

If we take the direction ds to be, in turn, in the directions of the x, y, and z axes, we can find the three components of **E** at any point, from Eq. 27:

$$E_x = -\frac{\partial V}{\partial x}, \quad E_y = -\frac{\partial V}{\partial y}, \quad \text{and} \quad E_z = -\frac{\partial V}{\partial z}. \quad (29)$$

Thus if V is known for all points of space, that is, if the function $V(x, y, z)$ is known, the components of **E**, and thus **E** itself, can be found by taking derivatives.*

We therefore have two methods for calculating **E** for continuous charge distributions. One is based on integrating Coulomb's law (see Eqs. 11 and 12 of Chapter 28), and the other is based on differentiating V (see Eq. 29). In practice, the second method is often less difficult.

Sample Problem 10   Using Eq. 26 for the potential on the axis of a uniformly charged disk, derive an expression for the electric field at axial points.

Solution   From symmetry, **E** must lie along the axis of the disk (the z axis). Using Eq. 29, we have

$$E_z = -\frac{\partial V}{\partial z} = -\frac{\sigma}{2\epsilon_0} \frac{d}{dz}[(z^2 + R^2)^{1/2} - z]$$

$$= \frac{\sigma}{2\epsilon_0}\left(1 - \frac{z}{\sqrt{z^2 + R^2}}\right).$$

This is the same expression that we derived in Section 28-5 by direct integration, using Coulomb's law; compare with Eq. 27 of that chapter.

Sample Problem 11   Figure 18 shows a (distant) point P in the field of a dipole located at the origin of an xz coordinate system. Calculate **E** as a function of position.

---

* The symbol $\partial V/\partial x$ denotes a *partial derivative*. In taking this derivative of the function $V(x, y, z)$, the quantity x is to be viewed as a variable and y and z are to be regarded as constants. Similar considerations hold for $\partial V/\partial y$ and $\partial V/\partial z$.

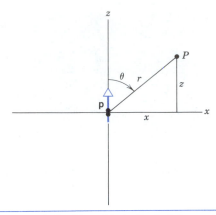

Figure 18   Sample Problem 11. A dipole is located at the origin of the xz system.

Solution   From symmetry, **E** at points in the plane of Fig. 18 lies in this plane and can be expressed in terms of its components E_x and E_z, E_y being zero. Let us first express the potential in rectangular coordinates rather than polar coordinates, making use of

$$r = (x^2 + z^2)^{1/2} \quad \text{and} \quad \cos\theta = \frac{z}{(x^2 + z^2)^{1/2}}.$$

V is given by Eq. 21,

$$V = \frac{1}{4\pi\epsilon_0}\frac{p\cos\theta}{r^2}.$$

Substituting for r^2 and $\cos\theta$, we obtain

$$V = \frac{p}{4\pi\epsilon_0}\frac{z}{(x^2 + z^2)^{3/2}}.$$

We find E_z from Eq. 29, recalling that x is to be treated as a constant in this calculation,

$$E_z = -\frac{\partial V}{\partial z} = -\frac{p}{4\pi\epsilon_0}\frac{(x^2 + z^2)^{3/2} - z[\frac{3}{2}(x^2 + z^2)^{1/2}](2z)}{(x^2 + z^2)^3}$$

$$= -\frac{p}{4\pi\epsilon_0}\frac{x^2 - 2z^2}{(x^2 + z^2)^{5/2}}. \quad (30)$$

Putting $x = 0$ describes distant points along the dipole axis (that is, the z axis), and the expression for E_z reduces to

$$E_z = \frac{1}{4\pi\epsilon_0}\frac{2p}{z^3}.$$

This result agrees exactly with that found in Chapter 28 (see Problem 11 of Chapter 28) for the field along the dipole axis. Note that along the z axis, $E_x = 0$ from symmetry.

Putting $z = 0$ in Eq. 30 gives E_z for distant points in the median plane of the dipole:

$$E_z = -\frac{1}{4\pi\epsilon_0}\frac{p}{x^3},$$

which agrees exactly with the result found in Eq. 10 of Chapter 28, for, again from symmetry, E_x equals zero in the median plane. The minus sign in this equation indicates that **E** points in the negative z direction.

The component E_x is also found from Eq. 29, recalling that z is to be taken as a constant during this calculation:

$$E_x = -\frac{\partial V}{\partial x} = -\frac{pz}{4\pi\epsilon_0}\left(-\frac{3}{2}\right)(x^2 + z^2)^{-5/2}(2x)$$

$$= \frac{3p}{4\pi\epsilon_0}\frac{xz}{(x^2 + z^2)^{5/2}}. \tag{31}$$

As expected, E_x vanishes both on the dipole axis ($x = 0$) and in the median plane ($z = 0$).

30-10   AN ISOLATED CONDUCTOR

In Section 29-4, we used Gauss' law to prove an important theorem about isolated conductors: an excess charge placed on an isolated conductor moves entirely to the outer surface of the conductor. In equilibrium, none of the charge is found inside the body of the conductor or on any interior surfaces, even when the conductor has internal cavities (provided that there is no net charge within any of the cavities).

This property of conductors can be stated equivalently in the language of potential:

An excess charge placed on an isolated conductor distributes itself on the surface so that all points of the conductor — whether on the surface or inside — come to the same potential.

This property holds true even if the conductor has internal cavities, whether or not they contain a net charge.

The proof of this statement is based on the experimental observation that, in the steady-state situation, internal currents do not exist in a conductor. If two points within a conductor were at different potentials, then free charges (presumably negatively charged electrons) would move from regions of low potential to regions of high potential. Such movement of charges would contradict the observation of no currents in the steady state. Therefore internal points cannot be at different potentials.

We can also prove this statement based on Eq. 15. We learned in Section 29-4 that the electric field vanishes in a conductor. If $\mathbf{E} = 0$ *everywhere* within a conductor, then the integral $\int \mathbf{E} \cdot d\mathbf{s}$ vanishes on any path between any pair of endpoints a and b within the conductor. Thus $V_b - V_a = 0$ for all possible pairs of points, and the potential has a constant value.

We also deduced in Section 29-4 that the electric field near the surface of a conductor is perpendicular to its surface. This is consistent with the surface of the conductor being an equipotential; as we showed in Section 30-9, the electric field is always perpendicular to equipotential surfaces.

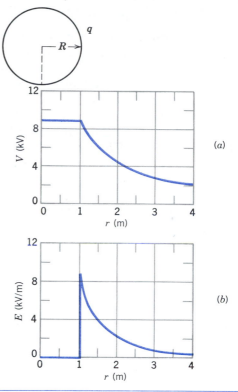

Figure 19   (a) The potential and (b) the electric field for a spherical shell having a uniform charge.

Figure 19 shows the variation of the potential with radial distance for an isolated spherical conducting shell of radius 1.0 m carrying a charge of 1.0 μC. For points outside the shell, $V(r)$ can be calculated from Eq. 16 because the charge q behaves, for external points, as if it were concentrated at the center of the sphere. Equation 16 gives the potential as we approach from outside, up to the surface of the sphere. Now suppose there is a tiny hole in the surface, just sufficient to allow us to push a test charge into the interior. No additional electrical force acts on the test charge from inside, so its potential does not change. As Fig. 19a shows, the potential everywhere inside is equal to that on the surface.

Figure 19b shows the electric field for this same spherical shell. Note that $\mathbf{E} = 0$ everywhere inside. We can obtain Fig. 19b from Fig. 19a by differentiating, according to Eq. 28; we can obtain Fig. 19a from Fig. 19b by integrating, according to Eq. 16.

Figure 19 would hold without change if the conductor were a *solid* conducting sphere rather than a spherical shell as we assumed. However, compare Fig. 19b (conducting shell or sphere) with Fig. 12 of Chapter 29, which described a *nonconducting* sphere. The difference comes about because the charge on the conducting shell or sphere lies entirely on the surface, but for the nonconducting sphere it can be spread throughout the volume.

A Conductor in an External Electric Field

All points of a conductor must be at the same potential whether or not the conductor carries a net charge. Furthermore, this is true even if the electric field that gives rise to the potential is externally imposed and not a result of a net charge on the conductor.

Figure 20 shows an uncharged conductor placed in an external electric field. The field was uniform before the conductor was placed in it. The free conducting electrons of the conductor move in response to the field, with the negative charges tending to accumulate on one side of the conductor and the positive charges on the other. As shown in Fig. 20, the field lines, which must start or end on free charges, are distorted from their previously uniform configuration. The equipotential surfaces are plane sheets in the uniform regions far from the conductor, and near the conductor they gradually assume the shape of its surface, which as we have discussed must be an equipotential surface.

If the surface charges on the conductor could somehow be frozen in space and the conductor removed, the field lines would be unchanged. In particular, in the region formerly occupied by the conductor, the charges give rise

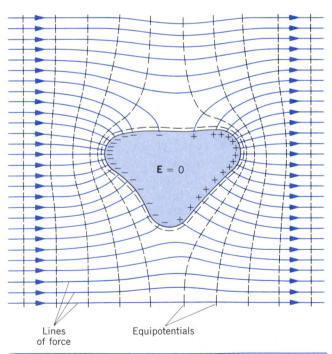

Figure 20 An uncharged conductor is placed in an external electric field. The conduction electrons distribute themselves on the surface to produce a charge distribution as shown, reducing the field inside the conductor to zero. Note the distortion of the lines of force (solid lines) and the equipotentials (dashed lines) when the conductor is placed in the previously uniform field.

to a uniform field that points to the left in Fig. 20 and exactly cancels the original uniform field to give zero field in the conductor's interior. Outside that region, the surface charges give a field that combines vectorially with the original uniform field to give the resultant shown.

A pattern of field lines such as that drawn in Fig. 20 can be made visible by surrounding the conductor with a suspension of small particles that align with the field lines (see Fig. 9 of Chapter 28). Alternatively, the equipotential surfaces can be mapped with a pair of electronic probes by fixing one probe and using the other to locate all points at a potential difference of zero relative to the first point.

Corona Discharge *(Optional)*

Although the surface charge is distributed uniformly on a spherical conductor, this will *not* be the case on conductors of arbitrary shape.* Near sharp points or edges, the surface charge density—and thus the electric field just outside the surface—can reach very high values.

To see qualitatively how this occurs, consider two conducting spheres of different radii connected by a fine wire (Fig. 21). Let the entire assembly be raised to some arbitrary potential V. The (equal) potentials of the two spheres, using Eq. 18, are

$$V = \frac{1}{4\pi\epsilon_0} \frac{q_1}{R_1} = \frac{1}{4\pi\epsilon_0} \frac{q_2}{R_2},$$

which yields

$$\frac{q_1}{q_2} = \frac{R_1}{R_2}. \tag{32}$$

Note that Eq. 18, which we originally derived for a point charge, holds for any spherically symmetric charge distribution. We assume the spheres to be so far apart that the charge on one doesn't affect the distribution of charge on the other.

The ratio of the surface charge densities of the two spheres is

$$\frac{\sigma_1}{\sigma_2} = \frac{q_1/4\pi R_1^2}{q_2/4\pi R_2^2} = \frac{q_1 R_2^2}{q_2 R_1^2}.$$

Combining this result with Eq. 32 gives

$$\frac{\sigma_1}{\sigma_2} = \frac{R_2}{R_1}. \tag{33}$$

Equation 33 suggests that the smaller sphere has the larger surface charge density. In the geometry shown in Fig. 21, this implies that the electric field close to the smaller sphere is greater than the electric field near the larger sphere. The smaller the radius of the sphere, the larger the electric field near its surface.

Near a sharp conductor (that is, one of very small radius) the electric field may be large enough to ionize molecules in the surrounding air; as a result the normally nonconducting air can conduct and carry charge away from the conductor. Such an effect is called a *corona discharge*. Electrostatic paint sprayers use a corona discharge to transfer charge to droplets of paint,

---

* See "The Lightning-rod Fallacy," by Richard H. Price and Ronald J. Crowley, *American Journal of Physics*, September 1985, p. 843, for a careful discussion of this phenomenon.

Figure 21 Two conducting spheres connected by a long fine wire.

which are then accelerated by an electric field. Photocopy machines based on the xerography process use a wire to produce a corona discharge that transfers charge to a selenium-covered surface; the charge is neutralized on regions where light strikes the surface, and the remaining charged areas attract a fine black powder that forms the image. ∎

30-11 THE ELECTROSTATIC ACCELERATOR *(Optional)*

Many studies of nuclei involve nuclear reactions, which occur when a beam of particles is incident on a target. One method that is used to accelerate particles for nuclear reactions is based on an electrostatic technique. A particle of positive charge q "falls" through a negative change in potential ΔV and therefore experiences a negative change in its potential energy, $\Delta U = q\,\Delta V$, according to Eq. 11. The corresponding increase in the kinetic energy of the particle is $\Delta K = -\Delta U$, and, assuming the particle starts from rest, its final kinetic energy is

$$K = -q\,\Delta V. \qquad (34)$$

For ionized atoms, q is normally positive (although there is an important application of Eq. 34 that makes use of negative ions and positive potential differences). To obtain the highest energy possible for the beam, we would like to have the largest difference in potential. For applications of interest in nuclear physics, particles with kinetic energies of millions of electron-volts (MeV) are required to overcome the Coulomb force of repulsion

between the incident and target particles. Kinetic energies of MeV require potential differences of millions of volts.

An electrostatic device that can produce such large potential differences is illustrated in Fig. 22. A small conducting sphere of radius r and carrying charge q is located inside a larger shell of radius R that carries charge Q. A conducting path is momentarily established between the two conductors, and the charge q then moves entirely to the outer conductor, no matter how much charge Q is already residing there (see also Fig. 14 of Chapter 29 and the accompanying discussion in Section 29-6). If there is a convenient mechanism for replenishing the charge q on the inner sphere from an external supply, the charge Q on the outer sphere and its potential can, in principle, be increased without limit. In practice, the terminal potential is limited by sparking that occurs through air (Fig. 23).

This well-known principle of electrostatics was first applied to accelerating nuclear particles by Robert J. Van de Graaff in the early 1930s, and the accelerator has become known as a *Van de Graaff accelerator*. Potentials of several million volts were easily achieved, the limiting potential coming from the leakage of charge through the insulating supports or breakdown of air (or the high-pressure insulating gas) surrounding the high-voltage terminal.

Figure 24 shows the basic design of the Van de Graaff accelerator. Charge is sprayed from a sharp tip (called a corona point) at A onto a moving belt made of insulating material (often rubber). The belt carries the charge into the high-voltage terminal, where it is removed by another corona point B and travels to the outer conductor. Inside the terminal is a source of positive ions, for example, nuclei of hydrogen (protons) or helium (alpha particles). The ions "fall" from the high potential, gaining a kinetic energy of several MeV in the process. The terminal is enclosed in a tank that contains insulating gas to prevent sparking.

A clever variation of this basic design makes use of the same high voltage to accelerate ions twice, thereby gaining an additional increase in kinetic energy. A source of *negative* ions, made

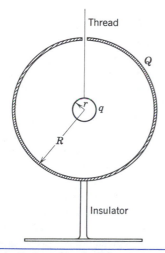

Figure 22 A small charged sphere is suspended inside a larger charged spherical shell.

Figure 23 An electrostatic generator, with a potential of 2.7 million volts, causing sparking due to conduction through air.

by adding an electron to a neutral atom, is located outside the terminal. These negative ions "fall toward" the positive potential of the terminal. Inside the high-voltage terminal, the beam passes through a chamber consisting of a gas or thin foil, which is designed to remove or strip several electrons from the negative ions, turning them into positive ions which then "fall from" the

positive potential. Such "tandem" Van de Graaff accelerators currently use a terminal voltage of 25 million volts to accelerate ions such as carbon or oxygen to kinetic energies in excess of 100 MeV.

Sample Problem 12   Calculate the potential difference between the two spheres illustrated in Fig. 22.

Solution   The potential difference $V(R) - V(r)$ has two contributions: one from the small sphere and one from the large spherical shell. These can be calculated independently and added algebraically. Let us first consider the large shell. Figure 19a shows that the potential at all interior points has the same value as the potential on the surface. Thus the contribution of the large shell to the difference $V(R) - V(r)$ is 0.

All that remains then is to evaluate the difference considering only the small sphere. For all points external to the small sphere, we can treat it as a point charge, and the potential difference can be found from Eq. 19:

$$V(R) - V(r) = \frac{q}{4\pi\epsilon_0}\left(\frac{1}{R} - \frac{1}{r}\right).$$

This expression gives the difference in potential between the inner sphere and the outer shell. Note that this is *independent of the charge Q on the outer shell.* If q is positive, the difference will always be negative, indicating that the outer shell will always be at a lower potential. If positive charge is permitted to flow between the spheres, it will always flow from higher to lower potential, that is, from the inner to the outer sphere, no matter how much charge already resides on the outer spherical shell. ∎

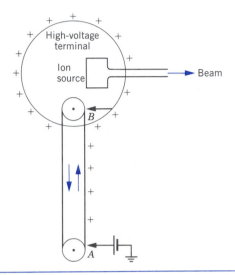

Figure 24   Diagram of Van de Graaff accelerator. Positive charge is sprayed onto the moving belt at *A* and is removed from the belt at *B*, where it flows to the terminal, which becomes charged to a potential *V*. Positively charged ions are repelled from the terminal to form the accelerator beam.

QUESTIONS

1. Are we free to call the potential of the Earth + 100 V instead of zero? What effect would such an assumption have on measured values of (*a*) potentials and (*b*) potential differences?

2. What would happen to you if you were on an insulated stand and your potential was increased by 10 kV with respect to the Earth?

3. Why is the electron-volt often a more convenient unit of energy than the joule?

4. How would a proton-volt compare with an electron-volt? The mass of a proton is 1840 times that of an electron.

5. Do electrons tend to go to regions of high potential or of low potential?

6. Does the amount of work per unit charge required to transfer electric charge from one point to another in an electrostatic field depend on the amount of charge transferred?

7. Distinguish between potential difference and difference of potential energy. Give examples of statements in which each term is used properly.

8. Estimate the combined energy of all the electrons striking the screen of a cathode ray oscilloscope in 1 second.

9. Why is it possible to shield a room against electrical forces but not against gravitational forces?

10. Suppose that the Earth has a net charge that is not zero. Why is it still possible to adopt the Earth as a standard reference point of potential and to assign the potential $V = 0$ to it?

11. Can there be a potential difference between two conductors that carry like charges of the same magnitude?

12. Give examples of situations in which the potential of a charged body has a sign opposite to that of its charge.

13. Can two different equipotential surfaces intersect?

14. An electrical worker was accidentally electrocuted and a newspaper account reported: "He accidentally touched a high-voltage cable and 20,000 V of electricity surged through his body." Criticize this statement.

15. Advice to mountaineers caught in lightning and thunderstorms is (*a*) get rapidly off peaks and ridges and (*b*) put both feet together and crouch in the open, only the feet touching the ground. What is the basis of this good advice?

16. If E equals zero at a given point, must *V* equal zero for that point? Give some examples to prove your answer.

17. If you know E only at a given point, can you calculate *V* at that point? If not, what further information do you need?

18. In Fig. 16, is the electric field E greater at the left or at the right of the figure?

19. Is the uniformly charged, nonconducting disk of Sample Problem 9 a surface of constant potential? Explain.

20. We have seen that, inside a hollow conductor, you are shielded from the fields of outside charges. If you are *outside* a hollow conductor that contains charges, are you shielded from the fields of these charges? Explain why or why not.

21. If the surface of a charged conductor is an equipotential, does that mean that charge is distributed uniformly over that surface? If the electric field is constant in magnitude over the surface of a charged conductor, does *that* mean that the charge is distributed uniformly?

22. In Section 30-10 we were reminded that charge delivered to the *inside* of an isolated conductor is transferred *entirely* to the outer surface of the conductor, no matter how much charge is already there. Can you keep this up forever? If not, what stops you?

23. Why can an isolated atom not have a permanent electric dipole moment?

24. Ions and electrons act like condensation centers; water droplets form around them in air. Explain why.

25. If V equals a constant throughout a given region of space, what can you say about **E** in that region?

26. In Chapter 16 we saw that the gravitational field strength is zero inside a spherical shell of matter. The electrical field strength is zero not only inside an isolated charged spherical conductor but inside an isolated conductor of any shape. Is the gravitational field strength inside, say, a cubical shell of matter zero? If not, in what respect is the analogy not complete?

27. How can you ensure that the electric potential in a given region of space will have a constant value?

28. Devise an arrangement of three point charges, separated by finite distances, that has zero electric potential energy.

29. A charge is placed on an insulated conductor in the form of a perfect cube. What will be the relative charge density at various points on the cube (surfaces, edges, corners)? What will happen to the charge if the cube is in air?

30. We have seen (Section 30-10) that the potential inside a conductor is the same as that on its surface. (a) What if the conductor is irregularly shaped and has an irregularly shaped cavity inside? (b) What if the cavity has a small "worm hole" connecting it to the outside? (c) What if the cavity is closed but has a point charge suspended within it? Discuss the potential within the conducting material and at different points within the cavities.

31. An isolated conducting spherical shell carries a negative charge. What will happen if a positively charged metal object is placed in contact with the shell interior? Discuss the three cases in which the positive charge is (a) less than, (b) equal to, and (c) greater than the negative charge in magnitude.

32. An uncharged metal sphere suspended by a silk thread is placed in a uniform external electric field. What is the magnitude of the electric field for points inside the sphere? Is your answer changed if the sphere carries a charge?

PROBLEMS

Section 30-2 Electric Potential Energy

1. In the quark model of fundamental particles, a proton is composed of three quarks: two "up" quarks, each having charge $+\frac{2}{3}e$, and one "down" quark, having charge $-\frac{1}{3}e$. Suppose that the three quarks are equidistant from each other. Take the distance to be 1.32×10^{-15} m and calculate (a) the potential energy of the interaction between the two "up" quarks and (b) the total electrical potential energy of the system.

2. Derive an expression for the work required by an external agent to put the four charges together as indicated in Fig. 25. Each side of the square has length a.

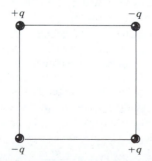

Figure 25 Problem 2.

3. A decade before Einstein published his theory of relativity, J. J. Thomson proposed that the electron might be made up of small parts and that its mass is due to the electrical interaction of the parts. Furthermore, he suggested that the energy equals mc^2. Make a rough estimate of the electron mass in the following way: assume that the electron is composed of three identical parts that are brought in from infinity and placed at the vertices of an equilateral triangle having sides equal to the classical radius of the electron, 2.82×10^{-15} m. (a) Find the total electrical potential energy of this arrangement. (b) Divide by c^2 and compare your result to the accepted electron mass (9.11×10^{-31} kg). The result improves if more parts are assumed. Today, the electron is thought to be a single, indivisible particle.

4. The charges shown in Fig. 26 are fixed in space. Find the value of the distance x so that the electrical potential energy of the system is zero.

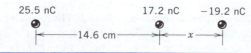

Figure 26 Problem 4.

5. Figure 27 shows an idealized representation of a ^{238}U nucleus ($Z = 92$) on the verge of fission. Calculate (a) the repulsive force acting on each fragment and (b) the mutual electric potential energy of the two fragments. Assume that the fragments are equal in size and charge, spherical, and just touching. The radius of the initially spherical ^{238}U nucleus is 8.0 fm. Assume that the material out of which nuclei are made has a constant density.

Figure 27   Problem 5.

Section 30-3 Electric Potential

6. Two parallel, flat, conducting surfaces of spacing $d = 1.0$ cm have a potential difference ΔV of 10.3 kV. An electron is projected from one plate directly toward the second. What is the initial velocity of the electron if it comes to rest just at the surface of the second plate? Ignore relativistic effects.

7. In a typical lightning flash the potential difference between discharge points is about 1.0×10^9 V and the quantity of charge transferred is about 30 C. (a) How much energy is released? (b) If all the energy released could be used to accelerate a 1200 kg automobile from rest, what would be its final speed? (c) If it could be used to melt ice, how much ice would it melt at 0°C?

8. The electric potential difference between discharge points during a particular thunderstorm is 1.23×10^9 V. What is the magnitude of the change in the electrical potential energy of an electron that moves between these points? Give your answer in (a) joules and (b) electron-volts.

9. (a) Through what potential difference must an electron fall, according to Newtonian mechanics, to acquire a speed v equal to the speed c of light? (b) Newtonian mechanics fails as $v \to c$. Therefore, using the correct relativistic expression for the kinetic energy (see Eq. 27 of Chapter 21)

$$K = mc^2 \left[\frac{1}{\sqrt{1 - (v/c)^2}} - 1 \right]$$

in place of the Newtonian expression $K = \frac{1}{2}mv^2$, determine the actual electron speed acquired in falling through the potential difference computed in (a). Express this speed as an appropriate fraction of the speed of light.

10. An electron is projected with an initial speed of 3.44×10^5 m/s directly toward a proton that is essentially at rest. If the electron is initially a great distance from the proton, at what distance from the proton is its speed instantaneously equal to twice its initial value?

11. A particle of charge q is kept in a fixed position at a point P and a second particle of mass m, having the same charge q, is initially held at rest a distance r_1 from P. The second particle is then released and is repelled from the first one. Determine its speed at the instant it is a distance r_2 from P. Let $q = 3.1$ μC, $m = 18$ mg, $r_1 = 0.90$ mm, and $r_2 = 2.5$ mm.

12. Calculate (a) the electric potential established by the nucleus

of a hydrogen atom at the average distance of the circulating electron ($r = 5.29 \times 10^{-11}$ m), (b) the electric potential energy of the atom when the electron is at this radius, and (c) the kinetic energy of the electron, assuming it to be moving in a circular orbit of this radius centered on the nucleus. (d) How much energy is required to ionize the hydrogen atom? Express all energies in electron-volts.

13. A particle of (positive) charge Q is assumed to have a fixed position at P. A second particle of mass m and (negative) charge $-q$ moves at constant speed in a circle of radius r_1, centered at P. Derive an expression for the work W that must be done by an external agent on the second particle in order to increase the radius of the circle of motion, centered at P, to r_2.

14. In the rectangle shown in Fig. 28, the sides have lengths 5.0 cm and 15 cm, $q_1 = -5.0$ μC and $q_2 = +2.0$ μC. (a) What are the electric potentials at corner B and at corner A? (b) How much external work is required to move a third charge $q_3 = +3.0$ μC from B to A along a diagonal of the rectangle? (c) In this process, is the external work converted into electrostatic potential energy or vice versa? Explain.

Figure 28   Problem 14.

15. Three charges of $+122$ mC each are placed on the corners of an equilateral triangle, 1.72 m on a side. If energy is supplied at the rate of 831 W, how many days would be required to move one of the charges onto the midpoint of the line joining the other two?

Section 30-4 Calculating the Potential from the Field

16. An infinite sheet of charge has a charge density $\sigma = 0.12$ μC/m^2. How far apart are the equipotential surfaces whose potentials differ by 48 V?

17. Two large parallel conducting plates are 12.0 cm apart and carry equal but opposite charges on their facing surfaces. An electron placed midway between the two plates experiences a force of 3.90×10^{-15} N. (a) Find the electric field at the position of the electron. (b) What is the potential difference between the plates?

18. In the Millikan oil-drop experiment (see Section 28-6), an electric field of 1.92×10^5 N/C is maintained at balance across two plates separated by 1.50 cm. Find the potential difference between the plates.

19. A Geiger counter has a metal cylinder 2.10 cm in diameter along whose axis is stretched a wire 1.34×10^{-4} cm in diameter. If 855 V is applied between them, find the electric field at the surface of (a) the wire and (b) the cylinder. (*Hint*: Use the result of Problem 36, Chapter 29.)

20. The electric field inside a nonconducting sphere of radius R, containing uniform charge density, is radially directed and has magnitude

$$E(r) = \frac{qr}{4\pi\epsilon_0 R^3},$$

where q is the total charge in the sphere and r is the distance from the center of the sphere. (*a*) Find the potential $V(r)$ inside the sphere, taking $V = 0$ at $r = 0$. (*b*) What is the difference in electric potential between a point on the surface and the center of the sphere? If q is positive, which point is at the higher potential? (*c*) Show that the potential at a distance r from the center, where $r < R$, is given by

$$V = \frac{q(3R^2 - r^2)}{8\pi\epsilon_0 R^3},$$

where the zero of potential is taken at $r = \infty$. Why does this result differ from that of part (*a*)?

Section 30-5 Potential Due to a Point Charge

21. A gold nucleus contains a positive charge equal to that of 79 protons and has a radius of 7.0 fm; see Sample Problem 6. An alpha particle (which consists of two protons and two neutrons) has a kinetic energy K at points far from the nucleus and is traveling directly toward it. The alpha particle just touches the surface of the nucleus where its velocity is reversed in direction. (*a*) Calculate K. (*b*) The actual alpha particle energy used in the experiment of Rutherford and his collaborators that led to the discovery of the concept of the atomic nucleus was 5.0 MeV. What do you conclude?

22. Compute the escape speed for an electron from the surface of a uniformly charged sphere of radius 1.22 cm and total charge 1.76×10^{-15} C. Neglect gravitational forces.

23. A point charge has $q = +1.16\ \mu C$. Consider point A, which is 2.06 m distant, and point B, which is 1.17 m distant in a direction diametrically opposite, as in Fig. 29*a*. (*a*) Find the potential difference $V_A - V_B$. (*b*) Repeat if points A and B are located as in Fig. 29*b*.

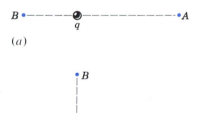

Figure 29 Problem 23.

24. Much of the material comprising Saturn's rings (see Fig. 30) is in the form of tiny dust particles having radii on the order of 1.0 μm. These grains are in a region containing a dilute ionized gas, and they pick up excess electrons. If the electric potential at the surface of a grain is -400 V, how many excess electrons has it picked up?

25. As a space shuttle moves through the dilute ionized gas of the Earth's ionosphere, its potential is typically changed by -1.0 V before it completes one revolution. By assuming that the shuttle is a sphere of radius 10 m, estimate the amount of charge it collects.

26. A particle of mass m, charge $q > 0$, and initial kinetic energy K is projected (from "infinity") toward a heavy nucleus of charge Q, assumed to have a fixed position in our reference frame. (*a*) If the aim is "perfect," how close to the center of the nucleus is the particle when it comes instantaneously to

Figure 30 Problem 24.

rest? (*b*) With a particular imperfect aim the particle's closest approach to the nucleus is twice the distance determined in part (*a*). Determine the speed of the particle at this closest distance of approach. Assume that the particle does not reach the surface of the nucleus.

27. A spherical drop of water carrying a charge of 32.0 pC has a potential of 512 V at its surface. (*a*) What is the radius of the drop? (*b*) If two such drops of the same charge and radius combine to form a single spherical drop, what is the potential at the surface of the new drop so formed?

28. Suppose that the negative charge in a copper one-cent coin were removed to a very large distance from the Earth — perhaps to a distant galaxy — and that the positive charge were distributed uniformly over the Earth's surface. By how much would the electric potential at the surface of the Earth change? (See Sample Problem 2 in Chapter 27.)

29. An electric field of approximately 100 V/m is often observed near the surface of the Earth. If this field were the same over the entire surface, what would be the electric potential of a point on the surface? See Sample Problem 6.

Section 30-6 Potential Due to a Collection of Point Charges

30. The ammonia molecule NH_3 has a permanent electric dipole moment equal to 1.47 D, where D is the debye unit with a value of 3.34×10^{-30} C·m. Calculate the electric potential due to an ammonia molecule at a point 52.0 nm away along the axis of the dipole.

31. (*a*) For Fig. 31, derive an expression for $V_A - V_B$. (*b*) Does your result reduce to the expected answer when $d = 0$? When $a = 0$? When $q = 0$?

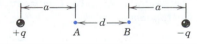

Figure 31 Problem 31.

32. In Fig. 32, locate the points, if any, (*a*) where $V = 0$ and (*b*) where $\mathbf{E} = 0$. Consider only points on the axis.

Figure 32   Problem 32.

33. A point charge $q_1 = +6e$ is fixed at the origin of a rectangular coordinate system, and a second point charge $q_2 = -10e$ is fixed at $x = 9.60$ nm, $y = 0$. The locus of all points in the xy plane with $V = 0$ is a circle centered on the x axis, as shown in Fig. 33. Find (*a*) the location x_c of the center of the circle and (*b*) the radius R of the circle. (*c*) Is the $V = 5$ V equipotential also a circle?

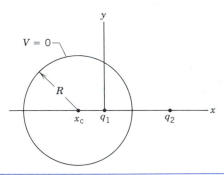

Figure 33   Problem 33.

34. Two charges $q = +2.13\ \mu$C are fixed in space a distance $d = 1.96$ cm apart, as shown in Fig. 34. (*a*) What is the electric potential at point C? (*b*) You bring a third charge $Q = +1.91\ \mu$C slowly from infinity to C. How much work must you do? (*c*) What is the potential energy U of the configuration when the third charge is in place?

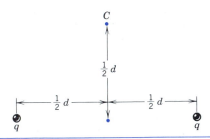

Figure 34   Problem 34.

35. For the charge configuration of Fig. 35 show that $V(r)$ for points on the vertical axis, assuming $r \gg d$, is given by

$$V = \frac{1}{4\pi\epsilon_0}\frac{q}{r}\left(1 + \frac{2d}{r}\right).$$

(*Hint*: The charge configuration can be viewed as the sum of an isolated charge and a dipole.)

Section 30-7 The Electric Potential of Continuous Charge Distributions

36. Figure 36 shows, edge-on, an "infinite" sheet of positive charge density σ. (*a*) How much work is done by the electric field of the sheet as a small positive test charge q_0 is moved from an initial position on the sheet to a final position located a perpendicular distance z from the sheet? (*b*) Use the result from (*a*) to show that the electric potential of an infinite sheet of charge can be written

$$V = V_0 - (\sigma/2\epsilon_0)z,$$

where V_0 is the potential at the surface of the sheet.

37. An electric charge of -9.12 nC is uniformly distributed around a ring of radius 1.48 m that lies in the yz plane with its center at the origin. A particle carrying a charge of -5.93 pC is located on the x axis at $x = 3.07$ m. Calculate the work done by an external agent in moving the point charge to the origin.

38. A total amount of positive charge Q is spread onto a nonconducting flat circular annulus of inner radius a and outer radius b. The charge is distributed so that the charge density (charge per unit area) is given by $\sigma = k/r^3$, where r is the distance from the center of the annulus to any point on it. Show that the potential at the center of the annulus is given by

$$V = \frac{Q}{8\pi\epsilon_0}\left(\frac{a+b}{ab}\right).$$

Section 30-8 Equipotential Surfaces

39. Two line charges are parallel to the z axis. One, of charge per unit length $+\lambda$, is a distance a to the right of this axis. The other, of charge per unit length $-\lambda$, is a distance a to the left of this axis (the lines and the z axis being in the same plane). Sketch some of the equipotential surfaces.

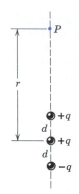

Figure 35   Problem 35.

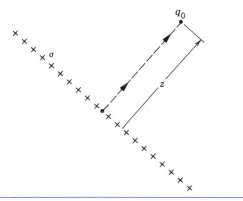

Figure 36   Problem 36.

40. In moving from A to B along an electric field line, the electric field does 3.94×10^{-19} J of work on an electron in the field illustrated in Fig. 37. What are the differences in the electric potential (a) $V_B - V_A$, (b) $V_C - V_A$, and (c) $V_C - V_B$?

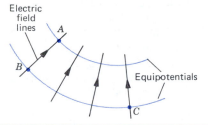

Figure 37   Problem 40.

41. Consider a point charge with $q = 1.5 \times 10^{-8}$ C. (a) What is the radius of an equipotential surface having a potential of 30 V? (b) Are surfaces whose potentials differ by a constant amount (1.0 V, say) evenly spaced?

42. In Fig. 38 sketch qualitatively (a) the lines of force and (b) the intersections of the equipotential surfaces with the plane of the figure. (*Hint:* Consider the behavior close to each point charge and at considerable distances from the pair of charges.)

Figure 38   Problem 42.

43. Three long parallel lines of charge have the relative linear charge densities shown in Fig. 39. Sketch some lines of force and the intersections of some equipotential surfaces with the plane of this figure.

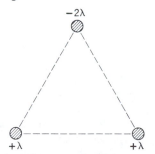

Figure 39   Problem 43.

Section 30-9 Calculating the Field from the Potential

44. Suppose that the electric potential varies along the x axis as shown in the graph of Fig. 40. Of the intervals shown (ignore the behavior at the end points of the intervals), determine the intervals in which E_x has (a) its greatest absolute value and (b) its least. (c) Plot E_x versus x.

45. Two large parallel metal plates are 1.48 cm apart and carry equal but opposite charges on their facing surfaces. The negative plate is grounded and its potential is taken to be zero. If the potential halfway between the plates is $+5.52$ V, what is the electric field in this region?

46. From Eq. 25 derive an expression for E at axial points of a uniformly charged ring.

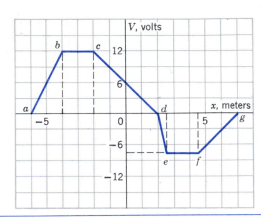

Figure 40   Problem 44.

47. Calculate the radial potential gradient, in V/m, at the surface of a gold nucleus. See Sample Problem 6.

48. Problem 49 in Chapter 29 deals with Rutherford's calculation of the electric field a distance r from the center of an atom. He also gave the electric potential as

$$V = \frac{Ze}{4\pi\epsilon_0}\left(\frac{1}{r} - \frac{3}{2R} + \frac{r^2}{2R^3}\right).$$

(a) Show how the expression for the electric field given in Problem 49 of Chapter 29 follows from this expression for V. (b) Why does this expression for V not go to zero as $r \to \infty$?

49. The electric potential V in the space between the plates of a particular, and now obsolete, vacuum tube is given by $V = 1530x^2$, where V is in volts if x, the distance from one of the plates, is in meters. Calculate the magnitude and direction of the electric field at $x = 1.28$ cm.

50. A charge per unit length λ is distributed uniformly along a straight-line segment of length L. (a) Determine the potential (chosen to be zero at infinity) at a point P a distance y from one end of the charged segment and in line with it (see Fig. 41). (b) Use the result of (a) to compute the component of the electric field at P in the y direction (along the line). (c) Determine the component of the electric field at P in a direction perpendicular to the straight line.

Figure 41   Problem 50.

51. On a thin rod of length L lying along the x axis with one end at the origin ($x = 0$), as in Fig. 42, there is distributed a charge per unit length given by $\lambda = kx$, where k is a constant. (a) Taking the electrostatic potential at infinity to be zero, find V at the point P on the y axis. (b) Determine the vertical

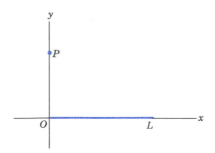

Figure 42 Problem 51.

component, E_y, of the electric field at P from the result of part (a) and also by direct calculation. (c) Why cannot E_x, the horizontal component of the electric field at P, be found using the result of part (a)? (d) At what distance from the rod along the y axis is the potential equal to one-half the value at the left end of the rod?

Section 30-10 An Isolated Conductor

52. A thin conducting spherical shell of outer radius 20 cm carries a charge of $+3.0$ μC. Sketch (a) the magnitude of the electric field $\mathbf{E}$ and (b) the potential V versus the distance r from the center of the shell.

53. Consider two widely separated conducting spheres, 1 and 2, the second having twice the diameter of the first. The smaller sphere initially has a positive charge q and the larger one is initially uncharged. You now connect the spheres with a long thin wire. (a) How are the final potentials V_1 and V_2 of the spheres related? (b) Find the final charges q_1 and q_2 on the spheres in terms of q.

54. If the Earth had a net charge equivalent to 1 electron/m^2 of surface area (a very artificial assumption), (a) what would be the Earth's potential? (b) What would be the electric field due to the Earth just outside its surface?

55. A charge of 15 nC can be produced by simple rubbing. To what potential would such a charge raise an isolated conducting sphere of 16-cm radius?

56. Find (a) the charge and (b) the charge density on the surface of a conducting sphere of radius 15.2 cm whose potential is 215 V.

57. Consider the Earth to be a spherical conductor of radius 6370 km and to be initially uncharged. A metal sphere, having a radius of 13 cm and carrying a charge of -6.2 nC is *earthed*, that is, put into electrical contact with the Earth. Show that this process effectively discharges the sphere, by calculating the fraction of the excess electrons originally present on the sphere that remain after the sphere is earthed.

58. Two conducting spheres, one of radius 5.88 cm and the other of radius 12.2 cm, each have a charge of 28.6 nC and are very far apart. If the spheres are subsequently connected by a conducting wire, find (a) the final charge on and (b) the potential of each sphere.

59. Consider a thin, isolated, conducting, spherical shell that is uniformly charged to a constant charge density σ (C/m^2). How much work does it take to move a small positive test charge q_0 (a) from the surface of the shell to the interior, through a small hole, (b) from one point on the surface to another, regardless of path, (c) from point to point inside the shell, and (d) from any point P outside the shell over any

path, whether or not it pierces the shell, back to P? (e) For the conditions given, does it matter whether or not the shell is conducting?

60. Two identical conducting spheres of radius 15.0 cm are separated by a distance of 10.0 m. What is the charge on each sphere if the potential of one is $+1500$ V and the other is -1500 V? What assumptions have you made?

61. The metal object in Fig. 43 is a figure of revolution about the horizontal axis. If it is charged negatively, sketch roughly a few equipotentials and lines of force. Use physical reasoning rather than mathematical analysis.

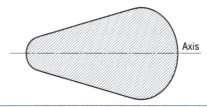

Figure 43 Problem 61.

62. A copper sphere whose radius is 1.08 cm has a very thin surface coating of nickel. Some of the nickel atoms are radioactive, each atom emitting an electron as it decays. Half of these electrons enter the copper sphere, each depositing 100 keV of energy there. The other half of the electrons escape, each carrying away a charge of $-e$. The nickel coating has an activity of 10.0 mCi ($=$ 10.0 millicuries $=$ 3.70×10^8 radioactive decays per second). The sphere is hung from a long, nonconducting string and insulated from its surroundings. How long will it take for the potential of the sphere to increase by 1000 V?

63. A charged metal sphere of radius 16.2 cm has a net charge of 31.5 nC. (a) Find the electric potential at the sphere's surface. (b) At what distance from the sphere's surface has the electric potential decreased by 550 V?

Section 30-11 The Electrostatic Accelerator

64. (a) How much charge is required to raise an isolated metallic sphere of 1.0-m radius to a potential of 1.0 MV? Repeat for a sphere of 1.0-cm radius. (b) Why use a large sphere in an electrostatic accelerator when the same potential can be achieved using a smaller charge with a small sphere? (*Hint:* Calculate the charge densities.)

65. Let the potential difference between the high-potential inner shell of a Van de Graaff accelerator and the point at which charges are sprayed onto the moving belt be 3.41 MV. If the belt transfers charge to the shell at the rate of 2.83 mC/s, what minimum power must be provided to drive the belt?

66. The high-voltage electrode of an electrostatic accelerator is a charged spherical metal shell having a potential $V = +9.15$ MV. (a) Electrical breakdown occurs in the gas in this machine at a field $E = 100$ MV/m. To prevent such breakdown, what restriction must be made on the radius r of the shell? (b) A long moving rubber belt transfers charge to the shell at 320 $\mu C/s$, the potential of the shell remaining constant because of leakage. What minimum power is required to transfer the charge? (c) The belt is of width $w = 48.5$ cm and travels at speed $v = 33.0$ m/s. What is the surface charge density on the belt?

Computer Projects

67. Charge $q_1 = -1.2 \times 10^{-9}$ C is at the origin and charge $q_2 = 2.5 \times 10^{-9}$ C is at $x = 0$, $y = 0.5$ m in the xy plane. Write a computer program or design a spreadsheet to calculate the electric potential due to these charges at any point in the xy plane. You should be able to input the coordinates of the point, then the computer will display the potential. It should then return to accept the coordinates of another point. Take the zero of potential to be far from both charges.

(a) Use the program to plot the 5-V equipotential surface in the xy plane. On a piece of graph paper draw axes that run from -5 m to $+5$ m in both the x and y directions. Mark the positions of the charges. First set $x = 0$ and try various values of y until you find two that differ by less than 0.005 m and straddle $V = 5$ V. Avoid the positions of the charges. Take the average position of the two points to be a point on the surface. Since the surface is closed you should find two points on it with the same x coordinate. When you have found them mark them on the graph. Then go on to $x = 0.25$ m. Continue to increment x by 0.25 m until you are beyond the equipotential surface — that is, until no point is found. Complete the diagram by marking points on the surface for negative values of x. Since the surface is symmetric about $x = 0$ you do not need to compute the points. Draw the surface through the points you have marked.

(b) Now draw the 3-V equipotential surface in the xy plane. Be careful here. For some values of x there are 4 points for which $V = 3$ V. There are in fact two 3-V equipotential surfaces.

68. The magnitude of an electric field is given by $E = |dV/ds|$, where ds is the (infinitesimal) distance between the equipotential surfaces for V and $V + dV$. E can be approximated by $|\Delta V/\Delta s|$ for two surfaces separated by a finite distance Δs. Consider the charge configuration of the previous problem and use your computer program to plot the 6-V equipotential surface in the neighborhood of the point where it crosses the positive x axis. If you did not work the previous problem also plot the 5-V equipotential surface in that region. The most efficient plan is to set $y = -0.1$, 0, and $+0.1$ m, in turn, and for each value of y search for two closely spaced values of x that straddle the equipotential surface. Draw a perpendicular line from one surface to the other and measure Δs, then calculate $E = |\Delta V/\Delta s|$, with $\Delta V = 1$ V, E in V/m, and Δs in meters. Check the accuracy of your result by using Coulomb's law to calculate the magnitude of the electric field at the point on the x axis halfway between the equipotential surfaces.

CHAPTER 31

CAPACITORS AND DIELECTRICS

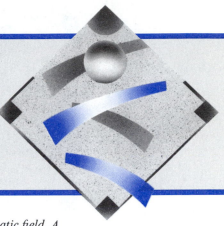

A capacitor* is a device that stores energy in an electrostatic field. A flash bulb, for example, requires a short burst of electric energy that exceeds what a battery can generally provide. We can draw energy relatively slowly (over several seconds) from the battery into a capacitor which releases the energy rapidly (within milliseconds) through the bulb. Much larger capacitors are used to provide intense laser pulses in attempts to induce thermonuclear fusion in tiny pellets of hydrogen. In this case the power level is about 10^{14} W, about 200 times the entire generating capability of the United States, but it lasts for only about 10^{-9} s.

Capacitors are also used to produce electric fields, such as the parallel-plate device that deflects beams of charged particles, as illustrated in Figs. 13–15 of Chapter 28. In this chapter we consider the electrostatic field and the stored energy of capacitors.

Capacitors have other important functions in electronic circuits, especially for time-varying voltages and currents. For transmitting and receiving radio and TV signals, capacitors are fundamental components of electromagnetic oscillators, as we discuss in Chapter 39.

31-1 CAPACITANCE

Figure 1 shows a generalized capacitor, consisting of two conductors a and b of arbitrary shape. No matter what their geometry, these conductors are called *plates*. We assume that they are totally isolated from their surroundings. We further assume, for the time being, that the conductors exist in a vacuum.

A capacitor is said to be *charged* if its plates carry equal and opposite charges $+q$ and $-q$, respectively. Note that q is *not* the net charge on the capacitor, which is zero. In our discussion of capacitors, we let q represent the absolute value of the charge on either plate; that is, q represents a magnitude only, and the sign of the charge on a given plate must be specified.

We can charge a capacitor by connecting the two plates to opposite terminals of a battery. Because the plates are conductors, they are equipotentials, and the potential difference of the battery appears across the plates. In charging the capacitor, the battery transfers equal and opposite charges to the two plates. For convenience, we represent

* See "Capacitors," by Donald M. Trotter, Jr., *Scientific American,* July 1988, p. 86.

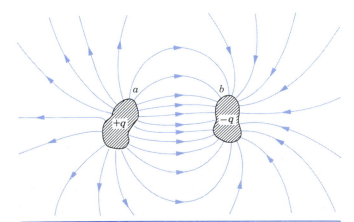

Figure 1 Two conductors, isolated from one another and from their surroundings, form a capacitor. When the capacitor is charged, the conductors carry equal but opposite charges of magnitude q. The two conductors are called *plates* no matter what their shape.

the *magnitude* of the potential difference between the plates by V.

As we show in the next section, there is a direct proportionality between the magnitude of the charge q on a

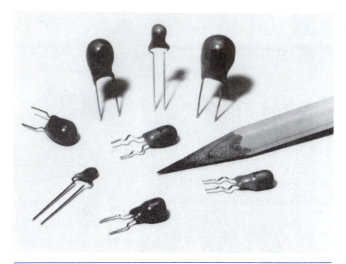

Figure 2 An assortment of capacitors that might be found in electronic circuits.

capacitor and the potential difference V between its plates. That is, we can write

$$q = CV \qquad (1)$$

in which C, the constant of proportionality, is called the *capacitance* of the capacitor. In the next section, we also show that C depends on the shapes and relative positions of the plates, and we calculate the actual dependence of C on these variables in three important special cases. C also depends on the material that fills the space between the plates (see Section 31-5); for the present, however, we assume that space to be a vacuum.

The SI unit of capacitance that follows from Eq. 1 is the coulomb/volt, which is given the name *farad* (abbreviation F):

1 farad = 1 coulomb/volt.

The unit is named in honor of Michael Faraday who, among his other contributions, developed the concept of capacitance. The submultiples of the farad, the *microfarad* (1 μF = 10^{-6} F) and the *picofarad* (1 pF = 10^{-12} F), are more convenient units in practice. Figure 2 shows some capacitors in the microfarad or picofarad range that might be found in electronic or computing equipment.

Sample Problem 1 A storage capacitor on a random access memory (RAM) chip has a capacitance of 55 fF. If it is charged to 5.3 V, how many excess electrons are there on its negative plate?

Solution If the negative plate has N excess electrons, it carries a net charge of magnitude $q = Ne$. Using Eq. 1, we obtain

$$N = \frac{q}{e} = \frac{CV}{e} = \frac{(55 \times 10^{-15} \text{ F})(5.3 \text{ V})}{1.60 \times 10^{-19} \text{ C}} = 1.8 \times 10^6 \text{ electrons.}$$

For electrons, this is a very small number. A speck of household dust, so tiny that it essentially never settles, contains about 10^{17} electrons (and the same number of protons).

Analogy with Fluid Flow *(Optional)*

In situations involving electric circuits, it is often useful to draw analogies between the movement of electric charge and the movement of material particles such as occurs in fluid flow. In the case of a capacitor, an analogy can be made between a capacitor carrying a charge q and a rigid container of volume v (we use v rather than V for volume so as not to confuse it with potential difference) containing n moles of an ideal gas. The gas pressure p is directly proportional to n for a fixed temperature, according to the ideal gas law (Eq. 7 of Chapter 23)

$$n = \left(\frac{v}{RT}\right) p.$$

For the capacitor (Eq. 1)

$$q = (C)V.$$

Comparison shows that the capacitance C of the capacitor is analogous to the volume v of the container, assuming a fixed temperature for the gas. In fact, the word "capacitor" brings to mind the word "capacity," in the same sense that the volume of a container for gas has a certain "capacity."

We can force more gas into the container by imposing a higher pressure, just as we can force more charge into the capacitor by imposing a higher voltage. Note that any amount of charge can be put on the capacitor, and any mass of gas can be put in the container, up to certain limits. These correspond to electrical breakdown ("arcing over") for the capacitor and to rupture of the walls for the container. ∎

31-2 CALCULATING THE CAPACITANCE

Our task here is to calculate the capacitance of a capacitor once we know its geometry. Because we consider a number of different geometries, it seems wise to develop a general plan to simplify the work. In brief our plan is as follows: (1) assume a charge q on the plates; (2) calculate the electric field $\mathbf{E}$ between the plates in terms of this charge, using Gauss' law; (3) knowing $\mathbf{E}$, calculate the potential difference V between the plates from Eq. 15 of Chapter 30; (4) calculate C from $C = q/V$ (Eq. 1).

Before we start, we can simplify the calculation of both the electric field and the potential difference by making certain assumptions. We discuss each in turn.

Calculating the Electric Field

The electric field is related to the charge on the plates by Gauss' law, or

$$\epsilon_0 \oint \mathbf{E} \cdot d\mathbf{A} = q. \qquad (2)$$

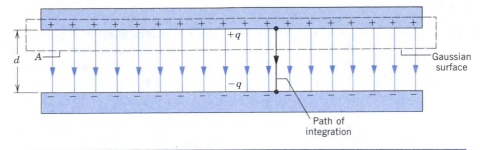

Figure 3   A charged parallel-plate capacitor in cross section. A Gaussian surface has been drawn enclosing the charge on the positive plate. The vertical line shows the path of integration used in Eq. 5.

Here q is the charge contained within the Gaussian surface, and the integral is carried out over that surface. We consider only cases in which, whenever flux passes through the Gaussian surface, the electric field $\mathbf{E}$ has a constant magnitude E, and the vectors $\mathbf{E}$ and $d\mathbf{A}$ are parallel. Equation 2 then reduces to

$$\epsilon_0 EA = q, \tag{3}$$

in which A is the area of that part of the Gaussian surface through which flux passes. For convenience, we draw the Gaussian surface so that it completely encloses the charge on the positive plate; see Fig. 3 for an example.

Calculating the Potential Difference

The potential difference between the plates is related to the electric field $\mathbf{E}$ by Eq. 15 of Chapter 30,

$$V_f - V_i = -\int_i^f \mathbf{E} \cdot d\mathbf{s}, \tag{4}$$

in which the integral is evaluated along any path that starts on one plate and ends on the other. We always choose a path that follows an electric field line from the positive plate to the negative plate, as shown in Fig. 3. For this path, the vectors $\mathbf{E}$ and $d\mathbf{s}$ point in the same direction, so that the quantity $V_f - V_i$ is negative. Since we are looking for V, the *absolute value* of the potential difference between the plates, we can set $V_f - V_i = -V$. We can recast Eq. 4 as

$$V = \int_+^- E\, ds, \tag{5}$$

in which the $+$ and the $-$ signs remind us that our path of integration starts on the positive plate and ends on the negative plate.

The electric field between the plates of a capacitor is the sum of the fields due to the two plates: $\mathbf{E} = \mathbf{E}_+ + \mathbf{E}_-$, where $\mathbf{E}_+$ is the field due to the charges on the positive plate and $\mathbf{E}_-$ is the field due to the charges on the negative plate. By Gauss' law, E_+ and E_- must each be proportional to q, so E is proportional to q, and by Eq. 5 V is also proportional to q. That is, if we double q (the charge on

each plate), E and V are likewise doubled. Because V is proportional to q, the ratio q/V is a constant and independent of q. We define this ratio to be the capacitance C, according to Eq. 1.

We are now ready to apply Eqs. 3 and 5 to some particular cases.

A Parallel-Plate Capacitor

We assume, as Fig. 3 suggests, that the plates of this capacitor are so large and so close together that we can neglect the "fringing" of the electric field at the edges of the plates. We take $\mathbf{E}$ to be constant throughout the volume between the plates.

Let us draw a Gaussian surface that includes the charge q on the positive plate, as Fig. 3 shows. The electric field can then be found from Eq. 3: $E = q/\epsilon_0 A$, where A is the area of the plates. Equation 5 then yields

$$V = \int_+^- E\, ds = \frac{q}{\epsilon_0 A} \int_0^d ds = \frac{qd}{\epsilon_0 A}. \tag{6}$$

In Eq. 6, E is constant and can be removed from the integral; the second integral above is simply the plate separation d.

Note in Eq. 6 that V is equal to a constant times q. According to Eq. 1, this constant is just $1/C$, and so

$$C = \epsilon_0 \frac{A}{d} \qquad \text{(parallel-plate capacitor)}. \tag{7}$$

The capacitance does indeed depend only on geometrical factors, namely, the plate area A and plate separation d.

As an aside we point out that Eq. 7 suggests one reason why we wrote the electrostatic constant in Coulomb's law in the form $1/4\pi\epsilon_0$. If we had not done so, Eq. 7—which is used more often in practice than is Coulomb's law—would have been less simple in form. We note further that Eq. 7 suggests units for the permittivity constant ϵ_0 that are more appropriate for problems involving capacitors, namely,

$$\epsilon_0 = 8.85 \times 10^{-12} \text{ F/m} = 8.85 \text{ pF/m}.$$

We have previously expressed this constant as

$$\epsilon_0 = 8.85 \times 10^{-12} \text{ C}^2/\text{N} \cdot \text{m}^2,$$

involving units that are useful when dealing with problems that involve Coulomb's law. The two sets of units are equivalent.

A Cylindrical Capacitor

Figure 4 shows, in cross section, a cylindrical capacitor of length L formed by two coaxial cylinders of radii a and b. We assume that $L \gg b$ so that we can neglect the "fringing" of the electric field that occurs at the ends of the cylinders.

As a Gaussian surface, we choose a cylinder of length L and radius r, closed by end caps. Equation 3 yields

$$q = \epsilon_0 EA = \epsilon_0 E(2\pi rL)$$

in which $2\pi rL$ is the area of the curved part of the Gaussian surface. Solving for E gives

$$E = \frac{q}{2\pi \epsilon_0 Lr}. \tag{8}$$

Substitution of this result into Eq. 5 yields

$$V = \int_+^- E \, ds = \frac{q}{2\pi \epsilon_0 L} \int_a^b \frac{dr}{r} = \frac{q}{2\pi \epsilon_0 L} \ln\left(\frac{b}{a}\right). \tag{9}$$

From the relation $C = q/V$, we then have

$$C = 2\pi \epsilon_0 \frac{L}{\ln(b/a)} \quad \text{(cylindrical capacitor).} \tag{10}$$

We see that the capacitance of the cylindrical capacitor, like that of a parallel-plate capacitor, depends only on geometrical factors, in this case L, b, and a.

A Spherical Capacitor

Figure 4 can also represent a central cross section of a capacitor that consists of two concentric spherical shells of radii a and b. As a Gaussian surface we draw a sphere of radius r. Applying Eq. 3 to this surface yields

$$q = \epsilon_0 EA = \epsilon_0 E(4\pi r^2),$$

in which $4\pi r^2$ is the area of the spherical Gaussian surface. We solve this equation for E, obtaining

$$E = \frac{1}{4\pi \epsilon_0} \frac{q}{r^2}, \tag{11}$$

which we recognize as the expression for the electric field due to a uniform spherical charge distribution.

If we substitute this expression into Eq. 5, we find

$$V = \int_+^- E \, ds = \frac{q}{4\pi \epsilon_0} \int_a^b \frac{dr}{r^2} = \frac{q}{4\pi \epsilon_0} \left(\frac{1}{a} - \frac{1}{b}\right)$$

$$= \frac{q}{4\pi \epsilon_0} \frac{b - a}{ab}. \tag{12}$$

Substituting Eq. 12 into Eq. 1 and solving for C, we obtain

$$C = 4\pi \epsilon_0 \frac{ab}{b - a} \quad \text{(spherical capacitor).} \tag{13}$$

An Isolated Sphere

We can assign a capacitance to a single isolated conductor by assuming that the "missing plate" is a conducting sphere of infinite radius. After all, the field lines that leave the surface of a charged isolated conductor must end somewhere; the walls of the room in which the conductor is housed can serve effectively as our sphere of infinite radius.

If we let $b \to \infty$ in Eq. 13 and substitute R for a, we find

$$C = 4\pi \epsilon_0 R \quad \text{(isolated sphere).} \tag{14}$$

Comparing Eqs. 7, 10, 13, and 14, we note that C is always expressed as ϵ_0 times a quantity with the dimension of length. The units for ϵ_0 (F/m) are consistent with this relationship.

Sample Problem 2   The plates of a parallel-plate capacitor are separated by a distance $d = 1.0$ mm. What must be the plate area if the capacitance is to be 1.0 F?

Solution   From Eq. 7 we have

$$A = \frac{Cd}{\epsilon_0} = \frac{(1.0 \text{ F})(1.0 \times 10^{-3} \text{ m})}{8.85 \times 10^{-12} \text{ F/m}} = 1.1 \times 10^8 \text{ m}^2.$$

This is the area of a square more than 10 km on edge. The farad is indeed a large unit. Modern technology, however, has permitted the construction of 1-F capacitors of very modest size. These "Supercaps" are used as backup voltage sources for computers; they can maintain the computer memory for up to 30 days in case of power failure.

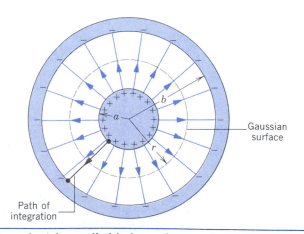

Figure 4   A long cylindrical capacitor seen in cross section. A cylindrical Gaussian surface has been drawn enclosing the inner conductor. The path of integration used in evaluating Eq. 5 is shown. The same figure could illustrate a cross section through the center of a spherical capacitor.

Sample Problem 3   The space between the conductors of a long coaxial cable, used to transmit TV signals, has an inner radius $a = 0.15$ mm and an outer radius $b = 2.1$ mm. What is the capacitance per unit length of this cable?

Solution   From Eq. 10 we have

$$\frac{C}{L} = \frac{2\pi\epsilon_0}{\ln(b/a)} = \frac{(2\pi)(8.85 \text{ pF/m})}{\ln(2.1 \text{ mm}/0.15 \text{ mm})} = 21 \text{ pF/m}.$$

Sample Problem 4   What is the capacitance of the Earth, viewed as an isolated conducting sphere of radius 6370 km?

Solution   From Eq. 14 we have

$$C = 4\pi\epsilon_0 R = (4\pi)(8.85 \times 10^{-12} \text{ F/m})(6.37 \times 10^6 \text{ m})$$
$$= 7.1 \times 10^{-4} \text{ F} = 710 \text{ }\mu\text{F}.$$

A tiny 1-F Supercap has a capacitance that is about 1400 times larger than that of the Earth.

31-3   CAPACITORS IN SERIES AND PARALLEL

In analyzing electric circuits, it is often desirable to know the *equivalent capacitance* of two or more capacitors that are connected in a certain way. By "equivalent capacitance" we mean the capacitance of a single capacitor that can be substituted for the combination with no change in the operation of the rest of the circuit. In an electric circuit, a capacitor is indicated by the symbol ─┤├─, which looks like a parallel-plate capacitor but represents any type of capacitor.

Capacitors Connected in Parallel

Figure 5a shows two capacitors connected *in parallel*. There are three properties that characterize a parallel connection of circuit elements. (1) In traveling from a to b, we can take any of several (two, in this case) *parallel* paths, each of which goes through *only one* of the parallel elements. (2) When a battery of potential difference V is connected across the combination (that is, the leads of the

battery are connected to points a and b in Fig. 5), the same potential difference V appears across each element of the parallel connection. The wires and capacitor plates are conductors and therefore equipotentials. The potential at a appears on the wires connected to a and on the two left-hand capacitor plates; similarly, the potential at b appears on all the wires connected to b and on the two right-hand capacitor plates. (3) The total charge that is delivered by the battery to the combination is shared among the elements.

With these principles in mind, we can now find the equivalent capacitance C_{eq} that gives the same total capacitance between points a and b, as indicated in Fig. 5b. We assume a battery of potential difference V to be connected between points a and b. For each capacitor, we can write (using Eq. 1)

$$q_1 = C_1 V \quad \text{and} \quad q_2 = C_2 V. \tag{15}$$

In writing these equations, we have used the same value of the potential difference across the capacitors, in accordance with the second characteristic of a parallel connection stated previously. The battery extracts charge q from one side of the circuit and moves it to the other side. This charge is shared among the two elements according to the third characteristic, such that the sum of the charges on the two capacitors equals the total charge:

$$q = q_1 + q_2. \tag{16}$$

If the parallel combination were replaced with a single capacitor C_{eq} and connected to the same battery, the requirement that the circuit operate in identical fashion means that the same charge q must be transferred by the battery. That is, for the equivalent capacitor,

$$q = C_{eq} V. \tag{17}$$

Substituting Eq. 16 into Eq. 17, and then putting Eqs. 15 into the result, we obtain

$$C_{eq} V = C_1 V + C_2 V,$$

or

$$C_{eq} = C_1 + C_2. \tag{18}$$

If we have more than two capacitors in parallel, we can first replace C_1 and C_2 with their equivalent C_{12} determined according to Eq. 18. We then find the equivalent capacitance of C_{12} and the next parallel capacitor C_3. Continuing this process, we can extend Eq. 18 to any number of capacitors connected in parallel:

$$C_{eq} = \sum_n C_n \qquad \text{(parallel combination).} \tag{19}$$

That is, to find the equivalent capacitance of a parallel combination, simply add the individual capacitances. Note that the equivalent capacitance is always larger than the largest capacitance in the parallel combination. The parallel combination can store more charge than any one of the individual capacitors.

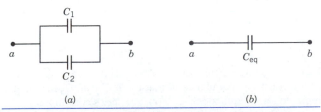

Figure 5   (*a*) Two capacitors in parallel. (*b*) The equivalent capacitance that can replace the parallel combination.

Capacitors Connected in Series

Figure 6 shows two capacitors connected *in series*. There are three properties that distinguish a series connection of circuit elements. (1) If we attempt to travel from a to b, we must pass through *all* the circuit elements *in succession*. (2) When a battery is connected across the combination, the potential difference V of the battery equals the sum of the potential differences across each of the elements. (3) The charge q delivered to each element of the series combination has the same value.

To understand this last property, note the region of Fig. 6 enclosed by the dashed line. Let us assume the battery puts a charge $-q$ on the left-hand plate of C_1. Since a capacitor carries equal and opposite charges on its plates, a charge $+q$ appears on the right-hand plate of C_1. But the H-shaped conductor enclosed by the dashed line is electrically isolated from the rest of the circuit; initially it carries no net charge, and no charge can be transferred to it. If a charge $+q$ appears on the right-hand plate of C_1, then a charge $-q$ must appear on the left-hand plate of C_2. That is, $n\ (=q/e)$ electrons move from the right-hand plate of C_1 to the left-hand plate of C_2. If there were more than two capacitors in series, a similar argument can be made across the entire line of capacitors, the result being that the left-hand plate of *every* capacitor in the series connection carries a charge q of one sign, and the right-hand plate of *every* capacitor in the series connection carries a charge of equal magnitude q and opposite sign.

For the individual capacitors we can write, using Eq. 1,

$$V_1 = \frac{q}{C_1} \quad \text{and} \quad V_2 = \frac{q}{C_2}, \qquad (20)$$

with the same charge q on each capacitor, but different potential differences across each. According to the second property of a series connection, we have

$$V = V_1 + V_2. \qquad (21)$$

We seek the equivalent capacitance C_{eq} that can replace the combination, such that the battery would move the same amount of charge:

$$V = \frac{q}{C_{eq}}. \qquad (22)$$

Substituting Eq. 21 into Eq. 22 and then using Eqs. 20, we obtain

$$\frac{q}{C_{eq}} = \frac{q}{C_1} + \frac{q}{C_2},$$

or

$$\frac{1}{C_{eq}} = \frac{1}{C_1} + \frac{1}{C_2}. \qquad (23)$$

If we have several capacitors in series, we can use Eq. 23 to find the equivalent capacitance C_{12} of the first two. We then find the equivalent capacitance of C_{12} and the next capacitor in series, C_3. Continuing in this way, we find the

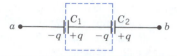

Figure 6 A series combination of two capacitors.

equivalent capacitance of any number of capacitors in series,

$$\frac{1}{C_{eq}} = \sum_n \frac{1}{C_n} \qquad \text{(series combination).} \qquad (24)$$

That is, to find the equivalent capacitance of a series combination, take the reciprocal of the sum of the reciprocals of the individual capacitances. Note that the equivalent capacitance of the series combination is always smaller than the smallest individual capacitance in the series.

Occasionally, capacitors are connected in ways that are not immediately identifiable as series or parallel combinations. As Sample Problem 5 shows, such combinations can often (but not always) be broken down into smaller units that can be analyzed as series or parallel connections.

Sample Problem 5   (a) Find the equivalent capacitance of the combination shown in Fig. 7a. Assume

$$C_1 = 12.0\ \mu\text{F}, \quad C_2 = 5.3\ \mu\text{F}, \quad \text{and} \quad C_3 = 4.5\ \mu\text{F}.$$

(b) A potential difference $V = 12.5$ V is applied to the terminals in Fig. 7a. What is the charge on C_1?

Solution   (a) Capacitors C_1 and C_2 are in parallel. From Eq. 18, their equivalent capacitance is

$$C_{12} = C_1 + C_2 = 12.0\ \mu\text{F} + 5.3\ \mu\text{F} = 17.3\ \mu\text{F}.$$

As Fig. 7b shows, C_{12} and C_3 are in series. From Eq. 23, the final equivalent combination (see Fig. 7c) is found from

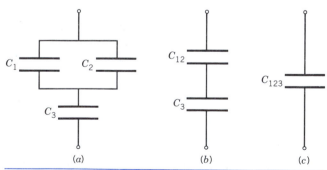

(a) (b) (c)

Figure 7 Sample Problem 5. (a) A combination of three capacitors. (b) The parallel combination of C_1 and C_2 has been replaced by its equivalent, C_{12}. (c) The series combination of C_{12} and C_3 has been replaced by its equivalent, C_{123}.

$$\frac{1}{C_{123}} = \frac{1}{C_{12}} + \frac{1}{C_3} = \frac{1}{17.3 \ \mu F} + \frac{1}{4.5 \ \mu F} = 0.280 \ \mu F^{-1},$$

or

$$C_{123} = \frac{1}{0.280 \ \mu F^{-1}} = 3.57 \ \mu F.$$

(*b*) We treat the equivalent capacitors C_{12} and C_{123} exactly as we would real capacitors of the same capacitance. The charge on C_{123} in Fig. 7*c* is then

$$q_{123} = C_{123} V = (3.57 \ \mu F)(12.5 \ V) = 44.6 \ \mu C.$$

This same charge exists on each capacitor in the series combination of Fig. 7*b*. The potential difference across C_{12} in that figure is then

$$V_{12} = \frac{q_{12}}{C_{12}} = \frac{44.6 \ \mu C}{17.3 \ \mu F} = 2.58 \ V.$$

This same potential difference appears across C_1 in Fig. 7*a*, so that

$$q_1 = C_1 V_1 = (12 \ \mu F)(2.68 \ V)$$
$$= 31 \ \mu C.$$

31-4 ENERGY STORAGE IN AN ELECTRIC FIELD

As we pointed out in the introduction to this chapter, an important use of capacitors is to store electrostatic energy in applications ranging from flash lamps to laser systems (see Fig. 8), both of which depend for their operation on the charging and discharging of capacitors.

In Section 30-2 we showed that any charge configuration has a certain *electric potential energy U*, equal to the work *W* (which may be positive or negative) that is done by an external agent that assembles the charge configuration from its individual components, originally assumed to be infinitely far apart and at rest. This potential energy is similar to that of mechanical systems, such as a compressed spring or the Earth–Moon system.

For a simple example, work is done when two equal and opposite charges are separated. This energy is stored as electric potential energy in the system, and it can be recovered as kinetic energy if the charges are allowed to come together again. Similarly, a charged capacitor has stored in it an electrical potential energy *U* equal to the work *W* done by the external agent as the capacitor is charged. This energy can be recovered if the capacitor is allowed to discharge. Alternatively, we can visualize the work of charging by imagining that an external agent pulls electrons from the positive plate and pushes them onto the negative plate, thereby bringing about the charge separation. Normally, the work of charging is done by a battery, at the expense of its store of chemical energy.

Suppose that at a time *t* a charge q' has already been transferred from one plate to the other. The potential

Figure 8 This bank of 10,000 capacitors at the Lawrence Livermore National Laboratory stores 60 MJ of electric energy and releases it in 1 ms to flashlamps that drive a system of lasers. The installation is part of the Nova project, which is attempting to produce sustained nuclear fusion reactions.

difference V' between the plates at that moment is $V' = q'/C$. If an increment of charge dq' is now transferred, the resulting small change dU in the electric potential energy is, according to Eq. 10 of Chapter 30 ($\Delta V = \Delta U / q_0$),

$$dU = V' \ dq' = \frac{q'}{C} \ dq'.$$

If this process is continued until a total charge *q* has been transferred, the total potential energy is

$$U = \int dU = \int_0^q \frac{q'}{C} \ dq' \tag{25}$$

or

$$U = \frac{q^2}{2C}. \tag{26}$$

From the relation $q = CV$ we can also write this as

$$U = \tfrac{1}{2} C V^2. \tag{27}$$

It is reasonable to suppose that the energy stored in a capacitor resides in the electric field between its plates, just as the energy carried by an electromagnetic wave can be regarded as residing in its electric field. As q or V in Eqs. 26 and 27 increase, for example, so does the electric field E; when q and V are zero, so is E.

In a parallel-plate capacitor, neglecting fringing, the electric field has the same value for all points between the plates. It follows that the *energy density u*, which is the stored energy per unit volume, should also be the same everywhere between the plates; u is given by the stored energy U divided by the volume Ad, or

$$u = \frac{U}{Ad} = \frac{\frac{1}{2}CV^2}{Ad}.$$

Substituting the relation $C = \epsilon_0 A/d$ (Eq. 7) leads to

$$u = \frac{\epsilon_0}{2}\left(\frac{V}{d}\right)^2.$$

However, V/d is the electric field E, so that

$$u = \tfrac{1}{2}\epsilon_0 E^2. \tag{28}$$

Although we derived this equation for the special case of a parallel-plate capacitor, it is true in general. *If an electric field* **E** *exists at any point in space (a vacuum), we can think of that point as the site of stored energy in amount, per unit volume, of* $\tfrac{1}{2}\epsilon_0 E^2$.

In general, E varies with location, so u is a function of the coordinates. For the special case of the parallel-plate capacitor, E and u do not vary with location in the region between the plates.

Sample Problem 6   A 3.55-μF capacitor C_1 is charged to a potential difference $V_0 = 6.30$ V, using a battery. The charging battery is then removed, and the capacitor is connected as in Fig. 9 to an uncharged 8.95-μF capacitor C_2. After the switch S is closed, charge flows from C_1 to C_2 until an equilibrium is established, with both capacitors at the same potential difference V. (*a*) What is this common potential difference? (*b*) What is the energy stored in the electric field before and after the switch S in Fig. 9 is thrown?

Solution   (*a*) The original charge q_0 is now shared by two capacitors, or

$$q_0 = q_1 + q_2.$$

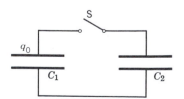

Figure 9   Sample Problem 6. Capacitor C_1 has previously been charged to a potential difference V_0 by a battery that has been removed. When the switch S is closed, the initial charge q_0 on C_1 is shared with C_2.

Applying the relation $q = CV$ to each term yields

$$C_1 V_0 = C_1 V + C_2 V,$$

or

$$V = V_0 \frac{C_1}{C_1 + C_2} = \frac{(6.30\text{ V})(3.55\ \mu\text{F})}{3.55\ \mu\text{F} + 8.95\ \mu\text{F}} = 1.79\text{ V}.$$

If we know the battery voltage V_0 and the value of C_1, we can determine an unknown capacitance C_2 by measuring the value of V in an arrangement similar to that of Fig. 9.

(*b*) The initial stored energy is

$$U_i = \tfrac{1}{2}C_1 V_0^2 = \tfrac{1}{2}(3.55 \times 10^{-6}\text{ F})(6.30\text{ V})^2$$
$$= 7.05 \times 10^{-5}\text{ J} = 70.5\ \mu\text{J}.$$

The final energy is

$$U_f = \tfrac{1}{2}C_1 V^2 + \tfrac{1}{2}C_2 V^2 = \tfrac{1}{2}(C_1 + C_2)V^2$$
$$= \tfrac{1}{2}(3.55 \times 10^{-6}\text{ F} + 8.95 \times 10^{-6}\text{ F})(1.79\text{ V})^2$$
$$= 2.00 \times 10^{-5}\text{ J} = 20.0\ \mu\text{J}.$$

We conclude that $U_f < U_i$, by about 72%. This is not a violation of energy conservation. The "missing" energy appears as thermal energy in the connecting wires, as we discuss in the next chapter.*

Sample Problem 7   An isolated conducting sphere whose radius R is 6.85 cm carries a charge $q = 1.25$ nC. (*a*) How much energy is stored in the electric field of this charged conductor? (*b*) What is the energy density at the surface of the sphere? (*c*) What is the radius R_0 of a spherical surface such that one-half of the stored potential energy lies within it?

Solution   (*a*) From Eqs. 26 and 14 we have

$$U = \frac{q^2}{2C} = \frac{q^2}{8\pi\epsilon_0 R} = \frac{(1.25 \times 10^{-9}\text{ C})^2}{(8\pi)(8.85 \times 10^{-12}\text{ F/m})(0.0685\text{ m})}$$
$$= 1.03 \times 10^{-7}\text{ J} = 103\text{ nJ}.$$

(*b*) From Eq. 28,

$$u = \tfrac{1}{2}\epsilon_0 E^2,$$

so that we must first find E at the surface of the sphere. This is given by

$$E = \frac{1}{4\pi\epsilon_0}\frac{q}{R^2}.$$

The energy density is then

$$u = \tfrac{1}{2}\epsilon_0 E^2 = \frac{q^2}{32\pi^2 \epsilon_0 R^4}$$
$$= \frac{(1.25 \times 10^{-9}\text{ C})^2}{(32\pi^2)(8.85 \times 10^{-12}\text{ C}^2/\text{N}\cdot\text{m}^2)(0.0685\text{ m})^4}$$
$$= 2.54 \times 10^{-5}\text{ J/m}^3 = 25.4\ \mu\text{J/m}^3.$$

(*c*) The energy that lies in a spherical shell between radii r and $r + dr$ is

$$dU = (u)(4\pi r^2)(dr),$$

---

* Some slight amount of energy is also radiated away. For a critical discussion, see "Two-Capacitor Problem: A More Realistic View," by R. A. Powell, *American Journal of Physics*, May 1979, p. 460.

where $(4\pi r^2)(dr)$ is the volume of the spherical shell. Using the result of part (b) for the energy density evaluated at a radius r, we obtain

$$dU = \frac{q^2}{32\pi^2\epsilon_0 r^4} 4\pi r^2 \, dr = \frac{q^2}{8\pi\epsilon_0} \frac{dr}{r^2}.$$

The condition given for this problem is

$$\int_R^{R_0} dU = \frac{1}{2} \int_R^{\infty} dU$$

or, using the result obtained above for dU and canceling constant factors from both sides,

$$\int_R^{R_0} \frac{dr}{r^2} = \frac{1}{2} \int_R^{\infty} \frac{dr}{r^2},$$

which becomes

$$\frac{1}{R} - \frac{1}{R_0} = \frac{1}{2R}.$$

Solving for R_0 yields

$$R_0 = 2R = (2)(6.85 \text{ cm}) = 13.7 \text{ cm}.$$

Half the stored energy is contained within a spherical surface whose radius is twice the radius of the conducting sphere.

31-5   CAPACITOR WITH DIELECTRIC

Up to this point we have calculated the capacitance assuming that there is no material in the space between the plates of the capacitor. The presence of material alters the capacitance of the capacitor and (possibly) the electric field between its plates. In this section we discuss the effect of filling the region between the plates with one of a number of insulating substances known as a *dielectrics*.

Michael Faraday in 1837 first investigated the effect of filling the space between capacitor plates with dielectrics. Faraday constructed two identical capacitors, filling one with dielectric and the other with air under normal conditions. When both capacitors were charged to the *same potential difference*, Faraday's experiments showed that *the charge on the capacitor with the dielectric was greater than that on the other.*

Since q is larger for the same V with the dielectric present, it follows from the relation $C = q/V$ that the capacitance of a capacitor *increases* if a dielectric is placed between the plates. (We assume, unless stated otherwise, that the dielectric completely fills the space between the plates.) The dimensionless factor by which the capacitance increases, relative to its value C_0 when no dielectric is present, is called the *dielectric constant* κ_e:

$$\kappa_e = C/C_0. \tag{29}$$

The dielectric constant is a fundamental property of the dielectric material and is independent of the size or shape of the conductor. Table 1 shows the dielectric constants of

TABLE 1   SOME PROPERTIES OF DIELECTRICS[a]

Material	Dielectric Constant κ_e	Dielectric Strength (kV/mm)
Vacuum	1 (exact)	∞
Air (1 atm)	1.00059	3
Polystyrene	2.6	24
Paper	3.5	16
Transformer oil	4.5	12
Pyrex	4.7	14
Mica	5.4	160
Porcelain	6.5	4
Silicon	12	
Water (25°C)	78.5	
Water (20°C)	80.4	
Titania ceramic	130	
Strontium titanate	310	8

[a] Measured at room temperature.

various materials. Note that, for most practical applications, air and vacuum are equivalent in their dielectric effects.

Figure 10 provides some insight into Faraday's experiments. The battery B initially charges the capacitor with charge q, and the battery remains connected to ensure that the potential difference V and the electric field **E** between the plates remain constant. After a dielectric slab is inserted, the charge increases by a factor of κ_e to a value of $\kappa_e q$. The additional charge $(\kappa_e - 1)q$ is moved from the negative to the positive plate by the battery as the dielectric slab is inserted.

Alternatively, as in Fig. 11, we can disconnect the battery after the capacitor is charged to charge q. As we now insert the dielectric slab, the charge remains constant (because there is no path for charge transfer), but the potential difference changes. In this case, we find that the potential difference *decreases* by a factor κ_e from V to V/κ_e after the dielectric is inserted. The electric field also decreases by the factor κ_e. We expect this decrease in V on the basis of the expression $q = CV$; if q is constant, then

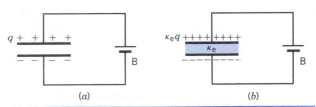

Figure 10   (a) An originally uncharged, empty capacitor is charged by a battery B. In a circuit, a battery is indicated by the symbol $\dashv\vdash$, the longer side indicating the more positive terminal. The battery maintains a constant potential difference between its terminals. (b) The battery remains connected as the region between the capacitor plates is filled with a dielectric. In this case, the potential difference remains constant while the charge on the capacitor increases.

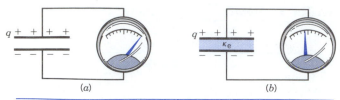

Figure 11 (*a*) An originally uncharged, empty capacitor is charged by a battery, which is then removed. The voltmeter shows the potential difference between the plates. (*b*) The region between the plates is filled with a dielectric. The charge remains constant, but the potential difference decreases.

the increase in C by the factor κ_e must be compensated by an equivalent decrease in V by the same factor.

If the purpose of a capacitor is to store charge, then its ability is enhanced by the dielectric, which permits it to store a factor κ_e more charge for the same potential difference. However, the presence of the dielectric also limits the potential difference that can be maintained across the plates. If this limit is exceeded, the dielectric material breaks down, resulting in a conducting path between the plates. Every dielectric material has a characteristic *dielectric strength*, which is the maximum value of the electric field that it can tolerate without breakdown. Some of these values are shown in Table 1.

For a parallel-plate capacitor filled with dielectric, the capacitance is

$$C = \frac{\kappa_e \epsilon_0 A}{d}. \tag{30}$$

Equation 7 is a special case of this result with $\kappa_e = 1$, corresponding to a vacuum between the plates. The capacitance of *any* capacitor is increased by a factor of κ_e when the entire space where the electric field exists is completely filled with a dielectric. We can similarly correct Eqs. 10, 13, or 14 for the presence of a dielectric filling the region between the plates.

The replacement of ϵ_0 with $\kappa_e \epsilon_0$ accounts for the effect on the capacitance of filling the capacitor with dielectric. This same change can be used to modify any of the equations of electrostatics to account for the presence of a dielectric that fills the entire space. For a point charge q imbedded in a dielectric, the electric field is (see Eq. 4 of Chapter 28)

$$E = \frac{1}{4\pi\kappa_e\epsilon_0} \frac{q}{r^2}. \tag{31}$$

Equation 31 gives the *total* field in the dielectric. The field due to the point charge is still given by Coulomb's law (without the factor κ_e), but the dielectric itself produces another electric field, which combines with the field of the point charge to give Eq. 31.

In a similar manner, the electric field near the surface of an isolated, charged conductor immersed in a dielectric is

$$E = \frac{\sigma}{\kappa_e\epsilon_0}. \tag{32}$$

The conductor gives a contribution σ/ϵ_0 to the field, and the dielectric gives an additional contribution so that the total field is given by Eq. 32. In both Eq. 31 and Eq. 32, the presence of the dielectric causes ϵ_0 to be replaced with $\kappa_e\epsilon_0$. Note that the effect of this replacement is to weaken the electric field. In the next section, we discuss how the microscopic properties of the dielectric account for this reduction.

Sample Problem 8 A parallel-plate capacitor whose capacitance C_0 is 13.5 pF has a potential difference $V = 12.5$ V between its plates. The charging battery is now disconnected and a porcelain slab ($\kappa_e = 6.5$) is slipped between the plates as in Fig. 11*b*. What is the stored energy of the unit, both before and after the slab is introduced?

Solution The initial stored energy is given by Eq. 27 as

$$U_i = \tfrac{1}{2}C_0V^2 = \tfrac{1}{2}(13.5 \times 10^{-12} \text{ F})(12.5 \text{ V})^2$$
$$= 1.055 \times 10^{-9} \text{ J} = 1055 \text{ pJ}.$$

We can write the final energy from Eq. 26 in the form

$$U_f = \frac{q^2}{2C}$$

because, from the conditions of the problem statement, q (but not V) remains constant as the slab is introduced. After the slab is in place, the capacitance increases to $\kappa_e C_0$ so that

$$U_f = \frac{q^2}{2\kappa_e C_0} = \frac{U_i}{\kappa_e} = \frac{1055 \text{ pJ}}{6.5} = 162 \text{ pJ}.$$

The energy after the slab is introduced is smaller by a factor of $1/\kappa_e$.

The "missing" energy, in principle, would be apparent to the person who introduced the slab. The capacitor would exert a force on the slab and would do work on it, in amount

$$W = U_i - U_f = 1055 \text{ pJ} - 162 \text{ pJ} = 893 \text{ pJ}.$$

If the slab were introduced with no restraint and if there were no friction, the slab would oscillate back and forth between the plates. The system consisting of capacitor + slab has a constant energy of 1055 pJ; the energy shuttles back and forth between kinetic energy of the moving slab and stored energy of the electric field. At the instant the oscillating slab filled the space between the plates, its kinetic energy would be 893 pJ.

31-6 DIELECTRICS: AN ATOMIC VIEW

We now seek to understand, in atomic terms, what happens when we place a dielectric in an electric field. There are two possibilities. The molecules of some dielectrics, like water (see Fig. 18 of Chapter 28), have permanent electric dipole moments. In such materials (called *polar dielectrics*) the electric dipole moments **p** tend to align themselves with an external electric field, as in Fig. 12.

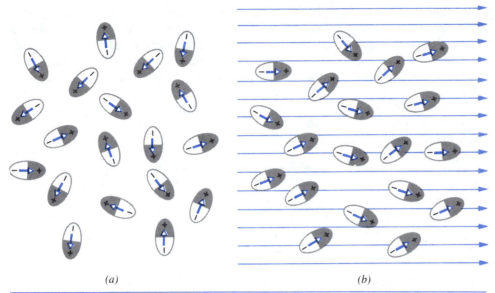

Figure 12 (*a*) A collection of molecules with permanent electric dipole moments. When there is no external electric field, the molecules are randomly oriented. (*b*) An external electric field produces a partial alignment of the dipoles. Thermal agitation prevents complete alignment.

Because the molecules are in constant thermal agitation, the degree of alignment is not complete but increases as the applied electric field increases or as the temperature decreases. In the absence of an applied field, the dipoles are randomly oriented.

In *nonpolar dielectrics,* the molecules do not have permanent electric dipole moments but can acquire them by *induction* when placed in an electric field. In Section 30-6 (see Fig. 11 of Chapter 30), we saw that the external electric field tends to separate the negative and the positive charge in the atom or molecule. This *induced electric dipole moment* is present only when the electric field is present. It is proportional to the electric field (for normal field strengths) and is created already lined up with the electric field as Fig. 11 of Chapter 30 suggests. Polar dielectrics can also acquire induced electric dipole moments in external fields.

Let us use a parallel-plate capacitor, carrying a fixed charge q and not connected to a battery, to provide a uniform external electric field $\mathbf{E}_0$ into which we place a dielectric slab (Fig. 13*a*). The overall effect of alignment and induction is to separate the center of positive charge of the entire slab slightly from the center of negative charge. Although the slab as a whole remains electrically

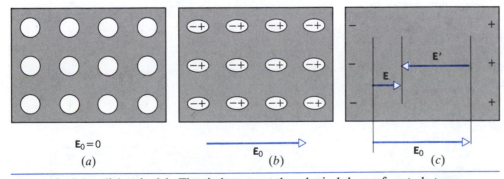

Figure 13 (*a*) A dielectric slab. The circles suggest the spherical shape of neutral atoms within the slab. (*b*) An external electric field $\mathbf{E}_0$ separates the positive and negative charges of the atom. An element of volume in the interior of the slab contains no net charge, but there is a net induced surface charge on the slab, negative on the left side and positive on the right side. (*c*) The net induced surface charges set up an induced electric field $\mathbf{E}'$, which is opposite in direction to the applied field $\mathbf{E}_0$. In the interior of the slab, the net field $\mathbf{E}$ is the vector sum of $\mathbf{E}_0$ and $\mathbf{E}'$.

neutral, it becomes *polarized*, as Fig. 13b suggests. The net effect is a buildup of positive charge on the right face of the slab and negative charge on the left face; within the slab no excess charge appears in any given volume element.

Since the slab as a whole remains neutral, the positive *induced surface charge* must be equal in magnitude to the negative induced surface charge. Note that in this process electrons in the dielectric are displaced from their equilibrium positions by distances that are considerably less than an atomic diameter. There is no transfer of charge over macroscopic distances such as occurs when a current is set up in a conductor.

Figure 13c shows that the induced surface charges always appear in such a way that the electric field **E**′ set up *by them* opposes the external electric field **E**₀. The *resultant* field **E** in the dielectric is the vector sum of **E**₀ and **E**′. It points in the same direction as **E**₀ but is smaller. *If we place a dielectric in an electric field, induced surface charges appear which tend to weaken the original field within the dielectric.*

This weakening of the electric field reveals itself in Fig. 11 as a reduction in potential difference between the plates of a charged isolated capacitor when a dielectric is introduced between the plates. The relation $V = Ed$ for a parallel-plate capacitor (see Eq. 6) holds whether or not dielectric is present and shows that the reduction in V described in Fig. 11 is directly connected to the reduction in E described in Fig. 13. Both E and V are reduced by the factor κ_e. (Note that this holds only when the battery is no longer connected. If the battery remained connected, V would be constant but q would increase. The increased electric field from this additional charge on the capacitor would be opposed by the field **E**′ in the dielectric, and the result would be a constant E.)

Induced charge is the explanation of the attraction to a charged rod of uncharged bits of nonconducting material such as paper. Figure 14 shows a bit of paper in the field of a charged rod. Surface charges appear on the paper as shown. The negatively charged end of the paper is pulled toward the rod, and the positively charged end is repelled. These two forces do not have the same magnitude because the negative end, being closer to the rod, is in a stronger field and experiences a stronger force. The net effect is an attraction. If a dielectric object is placed in a *uniform* electric field, induced surface charges appear but the object experiences no net force.

In Sample Problem 8 we pointed out that, if we insert a

dielectric slab into a parallel-plate capacitor carrying a fixed charge q, a force acts on the slab drawing it into the capacitor. This force is provided by the electrostatic attraction between the charges $\pm q$ on the capacitor plates and the induced surface charges $\mp q'$ on the dielectric slab. When the slab is only part way into the capacitor, neither q nor q' is uniformly distributed. (See Question 26.)

31-7 DIELECTRICS AND GAUSS' LAW

So far our use of Gauss' law has been confined to situations in which no dielectric was present. Now let us apply this law to a parallel-plate capacitor filled with a material of dielectric constant κ_e.

Figure 15 shows the capacitor both with and without the dielectric. We assume that the charge q on the plates is the same in each case. Gaussian surfaces have been drawn as in Fig. 3.

If no dielectric is present (Fig. 15a), Gauss' law gives

$$\epsilon_0 \oint \mathbf{E} \cdot d\mathbf{A} = \epsilon_0 E_0 A = q$$

or

$$E_0 = \frac{q}{\epsilon_0 A} . \tag{33}$$

If the dielectric is present (Fig. 15b), Gauss' law gives

$$\epsilon_0 \oint \mathbf{E} \cdot d\mathbf{A} = \epsilon_0 E A = q - q'$$

or

$$E = \frac{q}{\epsilon_0 A} - \frac{q'}{\epsilon_0 A} , \tag{34}$$

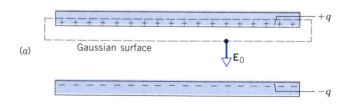

Figure 15 (a) A parallel-plate capacitor. (b) A dielectric slab is inserted, while the charge q on the plates remains constant. Induced charged q' appears on the surface of the dielectric slab.

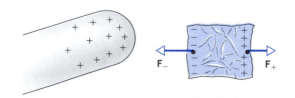

Figure 14 A charged rod attracts an uncharged bit of paper because unbalanced forces act on the induced surface charges.

in which $-q'$, the *induced surface charge*, must be distinguished from q, the *free charge* on the plates. These two charges $+q$ and $-q'$, both of which lie within the Gaussian surface, are opposite in sign; the *net* charge within the Gaussian surface is $q + (-q') = q - q'$.

The dielectric reduces the electric field by the factor κ_e, and so

$$E = \frac{E_0}{\kappa_e} = \frac{q}{\kappa_e \epsilon_0 A}. \tag{35}$$

Inserting this in Eq. 34 yields

$$\frac{q}{\kappa_e \epsilon_0 A} = \frac{q}{\epsilon_0 A} - \frac{q'}{\epsilon_0 A}$$

or

$$q' = q \left(1 - \frac{1}{\kappa_e}\right). \tag{36}$$

This shows that the induced surface charge q' is always less in magnitude than the free charge q and is equal to zero if no dielectric is present, that is, if $\kappa_e = 1$.

Now we write Gauss' law for the case of Fig. 15b in the form

$$\epsilon_0 \oint \mathbf{E} \cdot d\mathbf{A} = q - q', \tag{37}$$

$q - q'$ again being the net charge within the Gaussian surface. Substituting from Eq. 36 for q' leads, after some rearrangement, to

$$\epsilon_0 \oint \kappa_e \mathbf{E} \cdot d\mathbf{A} = q. \tag{38}$$

This important relation, although derived for a parallel-plate capacitor, is true generally and is the form in which Gauss' law is usually written when dielectrics are present. Note the following:

1. The flux integral now deals with $\kappa_e \mathbf{E}$ instead of $\mathbf{E}$. This is consistent with the *reduction* of E in a dielectric by the factor κ_e, because $\kappa_e \mathbf{E}$ (dielectric present) equals $\mathbf{E}_0$ (no dielectric). For generality, we allow for the possibility that κ_e is not constant by putting it inside the integral.

2. The charge q contained within the Gaussian surface is taken to be the *free charge only*. Induced surface charge is deliberately omitted on the right side of Eq. 38, having

been taken into account by the introduction of κ_e on the left side. Equations 37 and 38 are completely equivalent formulations.

Sample Problem 9   Figure 16 shows a parallel-plate capacitor of plate area A and plate separation d. A potential difference V_0 is applied between the plates. The battery is then disconnected, and a dielectric slab of thickness b and dielectric constant κ_e is placed between the plates as shown. Assume

$$A = 115 \text{ cm}^2, \quad d = 1.24 \text{ cm}, \quad b = 0.78 \text{ cm},$$

$$\kappa_e = 2.61, \quad V_0 = 85.5 \text{ V}.$$

(*a*) What is the capacitance C_0 before the slab is inserted? (*b*) What free charge appears on the plates? (*c*) What is the electric field E_0 in the gaps between the plates and the dielectric slab? (*d*) Calculate the electric field E in the dielectric slab. (*e*) What is the potential difference between the plates after the slab has been introduced? (*f*) What is the capacitance with the slab in place?

Solution   (*a*) From Eq. 7 we have

$$C_0 = \frac{\epsilon_0 A}{d} = \frac{(8.55 \times 10^{-12} \text{ F/m})(115 \times 10^{-4} \text{ m}^2)}{1.24 \times 10^{-2} \text{ m}}$$

$$= 8.21 \times 10^{-12} \text{ F} = 8.21 \text{ pF}.$$

(*b*) The free charge on the plates can be found from Eq. 1,

$$q = C_0 V_0 = (8.21 \times 10^{-12} \text{ F})(85.5 \text{ V})$$

$$= 7.02 \times 10^{-10} \text{ C} = 702 \text{ pC}.$$

Because the charging battery was disconnected before the slab was introduced, the free charge remains unchanged as the slab is put into place.

(*c*) Let us apply Gauss' law in the form given in Eq. 38 to the upper Gaussian surface in Fig. 16, which encloses only the free charge on the upper capacitor plate. We have

$$\epsilon_0 \oint \kappa_e \mathbf{E} \cdot d\mathbf{A} = \epsilon_0 (1) E_0 A = q$$

or

$$E_0 = \frac{q}{\epsilon_0 A} = \frac{7.02 \times 10^{-10} \text{ C}}{(8.85 \times 10^{-12} \text{ F/m})(115 \times 10^{-4} \text{ m}^2)}$$

$$= 6900 \text{ V/m} = 6.90 \text{ kV/m}.$$

Note that we put $\kappa_e = 1$ in this equation because the Gaussian

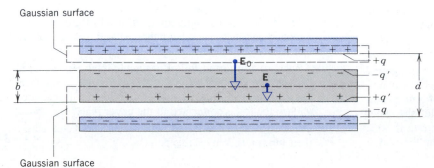

Gaussian surface

Gaussian surface

Figure 16   Sample Problem 9. A parallel-plate capacitor contains a dielectric that only partially fills the space between the plates.

surface over which Gauss' law was integrated does not pass through any dielectric. Note too that the value of E_0 remains unchanged as the slab is introduced. It depends only on the free charge on the plates.

(*d*) Again we apply Eq. 38, this time to the lower Gaussian surface in Fig. 16 and including only the free charge $-q$. We find

$$\epsilon_0 \oint \kappa_e \mathbf{E} \cdot d\mathbf{A} = -\epsilon_0 \kappa_e EA = -q$$

or

$$E = \frac{q}{\kappa_e \epsilon_0 A} = \frac{E_0}{\kappa_e} = \frac{6.90 \text{ kV/m}}{2.61} = 2.64 \text{ kV/m.}$$

The minus sign appears when we evaluate the dot product $\mathbf{E} \cdot d\mathbf{A}$ because $\mathbf{E}$ and $d\mathbf{A}$ are in opposite directions, $d\mathbf{A}$ always being in the direction of the *outward* normal to the closed Gaussian surface.

(*e*) To find the potential difference, we use Eq. 6,

$$V = \int_+^- E \, ds = E_0(d - b) + Eb$$

$$= (6900 \text{ V/m})(0.0124 \text{ m} - 0.0078 \text{ m})$$

$$+ (2640 \text{ V/m})(0.0078 \text{ m})$$

$$= 52.3 \text{ V.}$$

TABLE 2   SUMMARY OF RESULTS FOR SAMPLE PROBLEM 9

Quantity		No Slab	Partial Slab	Full Slab
C	pF	8.21	13.4	21.4
q	pC	702	702	702
q'	pC	—	433	433
V	V	85.5	52.3	32.8
E_0	kV/m	6.90	6.90	6.90[a]
E	kV/m	—	2.64	2.64

[a] Assumes that a very narrow gap is present.

This contrasts with the original applied potential difference of 85.5 V.

(*f*) From Eq. 1, the capacitance with the slab in place is

$$C = \frac{q}{V} = \frac{7.02 \times 10^{-10} \text{ C}}{52.3 \text{ V}}$$

$$= 1.34 \times 10^{-11} \text{ F} = 13.4 \text{ pF.}$$

Table 2 summarizes the results of this sample problem and also includes the results that would have followed if the dielectric slab had completely filled the space between the plates.

QUESTIONS

1. A capacitor is connected across a battery. (*a*) Why does each plate receive a charge of exactly the same magnitude? (*b*) Is this true even if the plates are of different sizes?

2. You are given two capacitors, C_1 and C_2, in which $C_1 > C_2$. How could things be arranged so that C_2 could hold more charge than C_1?

3. The relation $\sigma \propto 1/R$, in which σ is the surface charge density and R is the radius of curvature (see Eq. 33 of Chapter 30) suggests that the charge placed on an isolated conductor concentrates on points and avoids flat surfaces, where $R = \infty$. How do we reconcile this with Fig. 3, in which the charge is definitely on the flat surface of either plate?

4. In connection with Eq. 1 ($q = CV$) we said that C is a constant. Yet we pointed out (see Eq. 7) that it depends on the geometry (and also, as we saw later, on the medium). If C is indeed a constant, with respect to what variables does it remain constant?

5. In Fig. 1, suppose that a and b are nonconductors, the charge being distributed arbitrarily over their surfaces. (*a*) Would Eq. 1 ($q = CV$) hold, with C independent of the charge arrangements? (*b*) How would you define V in this case?

6. You are given a parallel-plate capacitor with square plates of area A and separation d, in a vacuum. What is the qualitative effect of each of the following on its capacitance? (*a*) Reduce d. (*b*) Put a slab of copper between the plates, touching neither plate. (*c*) Double the area of both plates. (*d*) Double the area of one plate only. (*e*) Slide the plates parallel to each other so that the area of overlap is 50%. (*f*) Double the potential difference between the plates.

(*g*) Tilt one plate so that the separation remains d at one end but is $\frac{1}{2}d$ at the other.

7. You have two isolated conductors, each of which has a certain capacitance; see Fig. 17. If you join these conductors by a fine wire, how do you calculate the capacitance of the combination? In joining them with the wire, have you connected them in series or parallel?

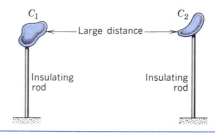

Figure 17   Question 7.

8. The capacitance of a conductor is affected by the presence of a second conductor that is uncharged and isolated electrically. Why?

9. A sheet of aluminum foil of negligible thickness is placed between the plates of a capacitor as in Fig. 18. What effect has it on the capacitance if (*a*) the foil is electrically insulated and (*b*) the foil is connected to the upper plate?

10. Capacitors often are stored with a wire connected across their terminals. Why is this done?

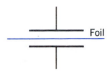

Figure 18 Question 9.

11. If you were not to neglect the fringing of the electric field lines in a parallel-plate capacitor, would you calculate a higher or a lower capacitance?

12. Two circular copper disks are facing each other a certain distance apart. In what ways could you reduce the capacitance of this combination?

13. Would you expect the dielectric constant of a material to vary with temperature? If so, how? Does whether or not the molecules have permanent dipole moments matter here?

14. Discuss similarities and differences when (a) a dielectric slab and (b) a conducting slab are inserted between the plates of a parallel-plate capacitor. Assume the slab thicknesses to be one-half the plate separation.

15. An oil-filled, parallel-plate capacitor has been designed to have a capacitance C and to operate safely at or below a certain maximum potential difference V_m without arcing over. However, the designer did not do a good job and the capacitor occasionally arcs over. What can be done to redesign the capacitor, keeping C and V_m unchanged and using the same dielectric?

16. Show that the dielectric constant of a conductor can be taken to be infinitely great.

17. For a given potential difference does a capacitor store more or less charge with a dielectric than it does without a dielectric (vacuum)? Explain in terms of the microscopic picture of the situation.

18. An electric field can polarize gases in several ways: by distorting the electron clouds of molecules; by orienting polar molecules; by bending or stretching the bonds in polar molecules. How does this differ from polarization of molecules in liquids and solids?

19. A dielectric object in a nonuniform electric field experiences a net force. Why is there no net force if the field is uniform?

20. A stream of tap water can be deflected if a charged rod is brought close to the stream. Explain carefully how this happens.

21. Water has a high dielectric constant (see Table 1). Why isn't it used ordinarily as a dielectric material in capacitors?

22. Figure 19 shows an actual 1-F capacitor available for use in student laboratories. It is only a few centimeters in diameter. Considering the result of Sample Problem 2, how can such a capacitor be constructed?

23. A dielectric slab is inserted in one end of a charged parallel-plate capacitor (the plates being horizontal and the charging battery having been disconnected) and then released. Describe what happens. Neglect friction.

24. A parallel-plate capacitor is charged by using a battery, which is then disconnected. A dielectric slab is then slipped between the plates. Describe qualitatively what happens to

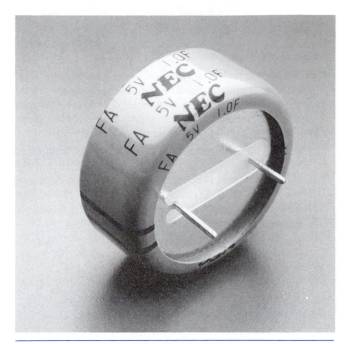

Figure 19 Question 22.

the charge, the capacitance, the potential difference, the electric field, and the stored energy.

25. While a parallel-plate capacitor remains connected to a battery, a dielectric slab is slipped between the plates. Describe qualitatively what happens to the charge, the capacitance, the potential difference, the electric field, and the stored energy. Is work required to insert the slab?

26. Imagine a dielectric slab, of width equal to the plate separation, inserted only halfway into a parallel-plate capacitor carrying a fixed charge q. Sketch qualitatively the distribution of the charge q on the plates and the induced charge q' on the slab.

27. Two identical capacitors are connected as shown in Fig. 20. A dielectric slab is slipped between the plates of one capacitor, the battery remaining connected. Describe qualitatively what happens to the charge, the capacitance, the potential difference, the electric field, and the stored energy for each capacitor.

28. In this chapter we have assumed electrostatic conditions; that is, the potential difference V between the capacitor plates remains constant. Suppose, however, that, as it often does in practice, V varies sinusoidally with time with an angular frequency ω. Would you expect the dielectric constant κ_e to vary with ω?

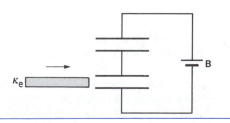

Figure 20 Question 27.

PROBLEMS

Section 31-1 Capacitance

1. An electrometer is a device used to measure static charge. Unknown charge is placed on the plates of a capacitor and the potential difference is measured. What minimum charge can be measured by an electrometer with a capacitance of 50 pF and a voltage sensitivity of 0.15 V?

2. The two metal objects in Fig. 21 have net charges of $+73.0$ pC and -73.0 pC, and this results in a 19.2-V potential difference between them. (a) What is the capacitance of the system? (b) If the charges are changed to $+210$ pC and -210 pC, what does the capacitance become? (c) What does the potential difference become?

Figure 21 Problem 2.

3. The capacitor in Fig. 22 has a capacitance of 26.0 μF and is initially uncharged. The battery supplies 125 V. After switch S has been closed for a long time, how much charge will have passed through the battery B?

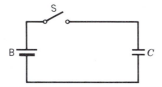

Figure 22 Problem 3.

Section 31-2 Calculating the Capacitance

4. A parallel-plate capacitor has circular plates of 8.22-cm radius and 1.31-mm separation. (a) Calculate the capacitance. (b) What charge will appear on the plates if a potential difference of 116 V is applied?

5. The plate and cathode of a vacuum tube diode are in the form of two concentric cylinders with the cathode as the central cylinder. The cathode diameter is 1.62 mm and the plate diameter 18.3 mm with both elements having a length of 2.38 cm. Calculate the capacitance of the diode.

6. Two sheets of aluminum foil have a separation of 1.20 mm, a capacitance of 9.70 pF, and are charged to 13.0 V. (a) Calculate the plate area. (b) The separation is now decreased by 0.10 mm with the charge held constant. Find the new capacitance. (c) By how much does the potential difference change? Explain how a microphone might be constructed using this principle.

7. The plates of a spherical capacitor have radii 38.0 mm and 40.0 mm. (a) Calculate the capacitance. (b) What must be the plate area of a parallel-plate capacitor with the same plate separation and capacitance?

8. Suppose that the two spherical shells of a spherical capacitor have their radii approximately equal. Under these conditions the device approximates a parallel-plate capacitor with $b - a = d$. Show that Eq. 13 for the spherical capacitor does indeed reduce to Eq. 7 for the parallel-plate capacitor in this case.

9. In Section 31-2 the capacitance of a cylindrical capacitor was calculated. Using the approximation (see Appendix H) that $\ln(1 + x) \approx x$ when $x \ll 1$, show that the capacitance approaches that of a parallel-plate capacitor when the spacing between the two cylinders is small.

10. A capacitor is to be designed to operate, with constant capacitance, in an environment of fluctuating temperature. As shown in Fig. 23, the capacitor is a parallel-plate type with plastic "spacers" to keep the plates aligned. (a) Show that the rate of change of capacitance C with temperature T is given by

$$\frac{dC}{dT} = C\left(\frac{1}{A}\frac{dA}{dT} - \frac{1}{x}\frac{dx}{dT}\right),$$

where A is the plate area and x the plate separation. (b) If the plates are aluminum, what should be the coefficient of thermal expansion of the spacers in order that the capacitance not vary with temperature? (Ignore the effect of the spacers on the capacitance.)

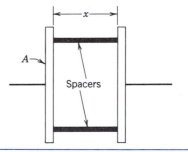

Figure 23 Problem 10.

Section 31-3 Capacitors in Series and Parallel

11. How many 1.00-μF capacitors must be connected in parallel to store a charge of 1.00 C with a potential of 110 V across the capacitors?

12. In Fig. 24 find the equivalent capacitance of the combination. Assume that $C_1 = 10.3$ μF, $C_2 = 4.80$ μF, and $C_3 = 3.90$ μF.

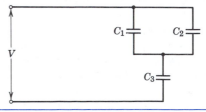

Figure 24 Problems 12, 19, and 36.

13. In Fig. 25 find the equivalent capacitance of the combination. Assume that $C_1 = 10.3\ \mu F$, $C_2 = 4.80\ \mu F$, and $C_3 = 3.90\ \mu F$.

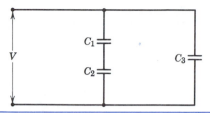

Figure 25 Problem 13.

14. Each of the uncharged capacitors in Fig. 26 has a capacitance of 25.0 μF. A potential difference of 4200 V is established when the switch S is closed. How much charge then passes through the meter A?

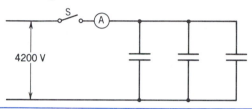

Figure 26 Problem 14.

15. A 6.0-μF capacitor is connected in series with a 4.0-μF capacitor; a potential difference of 200 V is applied across the pair. (*a*) Calculate the equivalent capacitance. (*b*) What is the charge on each capacitor? (*c*) What is the potential difference across each capacitor?

16. Work Problem 15 for the same two capacitors connected in parallel.

17. (*a*) Three capacitors are connected in parallel. Each has plate area A and plate spacing d. What must be the spacing of a single capacitor of plate area A if its capacitance equals that of the parallel combination? (*b*) What must be the spacing if the three capacitors are connected in series?

18. In Fig. 27 a variable air capacitor of the type used in tuning radios is shown. Alternate plates are connected together, one group being fixed in position, the other group being capable of rotation. Consider a pile of n plates of alternate polarity, each having an area A and separated from adjacent plates by a distance d. Show that this capacitor has a maximum capacitance of

$$C = \frac{(n-1)\epsilon_0 A}{d}.$$

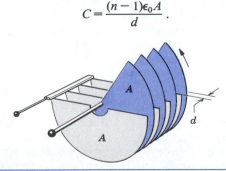

Figure 27 Problem 18.

19. In Fig. 24 suppose that capacitor C_3 breaks down electrically, becoming equivalent to a conducting path. What *changes* in (*a*) the charge and (*b*) the potential difference occur for capacitor C_1? Assume that $V = 115$ V.

20. You have several 2.0-μF capacitors, each capable of withstanding 200 V without breakdown. How would you assemble a combination having an equivalent capacitance of (*a*) 0.40 μF or of (*b*) 1.2 μF, each combination capable of withstanding 1000 V?

21. Figure 28 shows two capacitors in series, the rigid center section of length b being movable vertically. Show that the equivalent capacitance of the series combination is independent of the position of the center section and is given by

$$C = \frac{\epsilon_0 A}{a - b}.$$

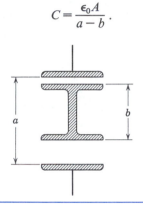

Figure 28 Problem 21.

22. A 108-pF capacitor is charged to a potential difference of 52.4 V, the charging battery then being disconnected. The capacitor is then connected in parallel with a second (initially uncharged) capacitor. The measured potential difference drops to 35.8 V. Find the capacitance of this second capacitor.

23. In Fig. 29, capacitors $C_1 = 1.16\ \mu F$ and $C_2 = 3.22\ \mu F$ are each charged to a potential $V = 96.6$ V but with opposite polarity, so that points a and c are on the side of the respective positive plates of C_1 and C_2, and points b and d are on the side of the respective negative plates. Switches S_1 and S_2 are now closed. (*a*) What is the potential difference between points e and f? (*b*) What is the charge on C_1? (*c*) What is the charge on C_2?

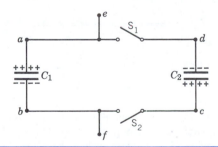

Figure 29 Problem 23.

24. When switch S is thrown to the left in Fig. 30, the plates of the capacitor C_1 acquire a potential difference V_0. C_2 and C_3

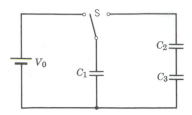

Figure 30   Problem 24.

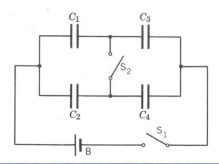

Figure 33   Problem 27.

are initially uncharged. The switch is now thrown to the right. What are the final charges q_1, q_2, q_3 on the corresponding capacitors?

25. Figure 31 shows two identical capacitors of capacitance C in a circuit with two (ideal) diodes D. A 100-V battery is connected to the input terminals, (a) first with terminal a positive and (b) later with terminal b positive. In each case, what is the potential difference across the output terminals? (An ideal diode has the property that positive charge flows through it only in the direction of the arrow and negative charge flows through it only in the opposite direction.)

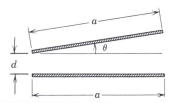

Figure 31   Problem 25.

26. A capacitor has square plates, each of side a, making an angle θ with each other as shown in Fig. 32. Show that for small θ the capacitance is given by

$$C = \frac{\epsilon_0 a^2}{d}\left(1 - \frac{a\theta}{2d}\right).$$

(*Hint*: The capacitor may be divided into differential strips that are effectively in parallel.)

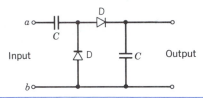

Figure 32   Problem 26.

27. In Fig. 33 the battery B supplies 12 V. (a) Find the charge on each capacitor when switch S_1 is closed and (b) when (later) switch S_2 is also closed. Take $C_1 = 1.0\ \mu\text{F}$, $C_2 = 2.0\ \mu\text{F}$, $C_3 = 3.0\ \mu\text{F}$, and $C_4 = 4.0\ \mu\text{F}$.

28. Find the equivalent capacitance between points x and y in Fig. 34. Assume that $C_2 = 10\ \mu\text{F}$ and that the other capacitors are all $4.0\ \mu\text{F}$. (*Hint*: Apply a potential difference V between x and y and write down all the relationships that

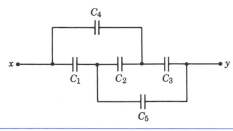

Figure 34   Problem 28.

involve the charges and potential differences for the separate capacitors.)

Section 31-4 Energy Storage in an Electric Field

29. How much energy is stored in 2.0 m³ of air due to the "fair weather" electric field of strength 150 V/m?

30. Attempts to build a controlled thermonuclear fusion reactor, which, if successful, could provide the world with a vast supply of energy from heavy hydrogen in seawater, usually involve huge electric currents for short periods of time in magnetic field windings. For example, ZT-40 at the Los Alamos National Laboratory has rooms full of capacitors. One of the capacitor banks provides 61.0 mF at 10.0 kV. Calculate the stored energy (a) in joules and (b) in kW·h.

31. A parallel-plate air capacitor having area 42.0 cm² and spacing of 1.30 mm is charged to a potential difference of 625 V. Find (a) the capacitance, (b) the magnitude of the charge on each plate, (c) the stored energy, (d) the electric field between the plates, and (e) the energy density between the plates.

32. Two capacitors, $2.12\ \mu\text{F}$ and $3.88\ \mu\text{F}$, are connected in series across a 328-V potential difference. Calculate the total energy stored in the capacitors.

33. An isolated metal sphere whose diameter is 12.6 cm has a potential of 8150 V. Calculate the energy density in the electric field near the surface of the sphere.

34. A parallel-connected bank of 2100 5.0-μF capacitors is used to store electric energy. What does it cost to charge this bank to 55 kV, assuming a rate of 3.0 ¢/kW·h?

35. One capacitor is charged until its stored energy is 4.0 J, the charging battery then being removed. A second uncharged capacitor is then connected to it in parallel. (a) If the charge distributes equally, what is now the total energy stored in the electric fields? (b) Where did the excess energy go?

36. In Fig. 24 find (a) the charge, (b) the potential difference, and (c) the stored energy for each capacitor. Assume the numerical values of Problem 12, with $V = 112$ V.

37. A parallel-plate capacitor has plates of area A and separation d and is charged to a potential difference V. The charging battery is then disconnected and the plates are pulled apart until their separation is $2d$. Derive expressions in terms of A, d, and V for (a) the new potential difference, (b) the initial and the final stored energy, and (c) the work required to separate the plates.

38. A cylindrical capacitor has radii a and b as in Fig. 4. Show that half the stored electric potential energy lies within a cylinder whose radius is

$$r = \sqrt{ab}.$$

39. (a) Calculate the energy density of the electric field at a distance r from an electron (presumed to be a particle) at rest. (b) Assume now that the electron is not a point but a sphere of radius R over whose surface the electron charge is uniformly distributed. Determine the energy associated with the external electric field in vacuum of the electron as a function of R. (c) If you now associate this energy with the mass of the electron, you can, using $E_0 = mc^2$, calculate a value for R. Evaluate this radius numerically; it is often called the *classical radius* of the electron.

40. Show that the plates of a parallel-plate capacitor attract each other with a force given by

$$F = \frac{q^2}{2\epsilon_0 A}.$$

Prove this by calculating the work necessary to increase the plate separation from x to $x + dx$, the charge q remaining constant.

41. Using the result of Problem 40, show that the force per unit area (the *electrostatic stress*) acting on either capacitor plate is given by $\frac{1}{2}\epsilon_0 E^2$. Actually, this result is true, in general, for a conductor of *any* shape with an electric field **E** at its surface.

42. A soap bubble of radius R_0 is slowly given a charge q. Because of mutual repulsion of the surface charges, the radius increases slightly to R. The air pressure inside the bubble drops, because of the expansion, to $p(V_0/V)$, where p is the atmospheric pressure, V_0 is the initial volume, and V is the final volume. Show that

$$q^2 = 32\pi^2\epsilon_0 pR(R^3 - R_0^3).$$

(*Hint*: Consider the forces acting on a small area of the charged bubble. These are due to (i) gas pressure, (ii) atmospheric pressure, (iii) electrostatic stress; see Problem 41.)

Section 31-5 Capacitor with Dielectric

43. An air-filled parallel-plate capacitor has a capacitance of 1.32 pF. The separation of the plates is doubled and wax inserted between them. The new capacitance is 2.57 pF. Find the dielectric constant of the wax.

44. Given a 7.40-pF air capacitor, you are asked to design a capacitor to store up to 6.61 μJ with a maximum potential difference of 630 V. What dielectric in Table 1 will you use to fill the gap in the air capacitor if you do not allow for a margin of error?

45. For making a parallel-plate capacitor you have available two plates of copper, a sheet of mica (thickness = 0.10 mm, $\kappa_e = 5.4$), a sheet of glass (thickness = 0.20 mm, $\kappa_e = 7.0$), and a slab of paraffin (thickness = 1.0 cm, $\kappa_e = 2.0$). To obtain the largest capacitance, which sheet should you place between the copper plates?

46. A parallel-plate air capacitor has a capacitance of 51.3 pF. (a) If its plates each have an area of 0.350 m^2, what is their separation? (b) If the region between the plates is now filled with material having a dielectric constant of 5.60, what is the capacitance?

47. A coaxial cable used in a transmission line responds as a "distributed" capacitance to the circuit feeding it. Calculate the capacitance of 1.00 km for a cable having an inner radius of 0.110 mm and an outer radius of 0.588 mm. Assume that the space between the conductors is filled with polystyrene.

48. A certain substance has a dielectric constant of 2.80 and a dielectric strength of 18.2 MV/m. If it is used as the dielectric material in a parallel-plate capacitor, what minimum area may the plates of the capacitor have in order that the capacitance be 68.4 nF and that the capacitor be able to withstand a potential difference of 4.13 kV?

49. You are asked to construct a capacitor having a capacitance near 1.0 nF and a breakdown potential in excess of 10 kV. You think of using the sides of a tall drinking glass (Pyrex), lining the inside and outside with aluminum foil (neglect the ends). What are (a) the capacitance and (b) the breakdown potential? You use a glass 15 cm tall with an inner radius of 3.6 cm and an outer radius of 3.8 cm.

50. You have been assigned to design a transportable capacitor that can store 250 kJ of energy. You select a parallel-plate type with dielectric. (a) What is the minimum capacitor volume achievable using a dielectric selected from those listed in Table 1 that have values of dielectric strength? (b) Modern high-performance capacitors that can store 250 kJ have volumes of 0.087 m^3. Assuming that the dielectric used has the same dielectric strength as in (a), what must be its dielectric constant?

51. A slab of copper of thickness b is thrust into a parallel-plate capacitor as shown in Fig. 35. (a) What is the capacitance after the slab is introduced? (b) If a charge q is maintained on the plates, find the ratio of the stored energy before to that after the slab is inserted. (c) How much work is done on the slab as it is inserted? Is the slab pulled in or do you have to push it in?

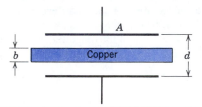

Figure 35 Problem 51.

52. Reconsider Problem 51 assuming that the potential difference V rather than the charge is held constant.

53. A cylindrical ionization chamber has a central wire anode of

radius 0.180 mm and a coaxial cylindrical cathode of radius 11.0 mm. It is filled with a gas whose dielectric strength is 2.20 MV/m. Find the largest potential difference that should be applied between anode and cathode if the gas is to avoid electric breakdown before radiation penetrates the chamber window.

54. A parallel-plate capacitor is filled with two dielectrics as in Fig. 36. Show that the capacitance is given by

$$C = \frac{\epsilon_0 A}{d}\left(\frac{\kappa_{e1} + \kappa_{e2}}{2}\right).$$

Check this formula for all the limiting cases that you can think of. (*Hint*: Can you justify regarding this arrangement as two capacitors in parallel?)

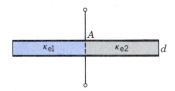

Figure 36   Problem 54.

55. A parallel-plate capacitor is filled with two dielectrics as in Fig. 37. Show that the capacitance is given by

$$C = \frac{2\epsilon_0 A}{d}\left(\frac{\kappa_{e1}\kappa_{e2}}{\kappa_{e1} + \kappa_{e2}}\right).$$

Check this formula for all the limiting cases that you can think of. (*Hint*: Can you justify regarding this arrangement as two capacitors in series?)

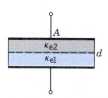

Figure 37   Problem 55.

56. What is the capacitance of the capacitor in Fig. 38?

Section 31-7   Dielectrics and Gauss' Law

57. A parallel-plate capacitor has a capacitance of 112 pF, a plate area of 96.5 cm², and a mica dielectric ($\kappa_e = 5.40$). At a 55.0-V potential difference, calculate (*a*) the electric field strength in the mica, (*b*) the magnitude of the free charge on the plates, and (*c*) the magnitude of the induced surface charge.

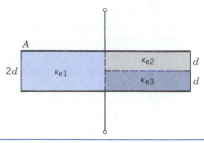

Figure 38   Problem 56.

58. In Sample Problem 9, suppose that the battery remains connected during the time that the dielectric slab is being introduced. Calculate (*a*) the capacitance, (*b*) the charge on the capacitor plates, (*c*) the electric field in the gap, and (*d*) the electric field in the slab, after the slab is introduced.

59. Two parallel plates of area 110 cm² are each given equal but opposite charges of 890 nC. The electric field within the dielectric material filling the space between the plates is 1.40 MV/m. (*a*) Calculate the dielectric constant of the material. (*b*) Determine the magnitude of the charge induced on each dielectric surface.

60. In the capacitor of Sample Problem 9 (Fig. 16), (*a*) what fraction of the energy is stored in the air gaps? (*b*) What fraction is stored in the slab?

61. A parallel-plate capacitor has plates of area 0.118 m² and a separation of 1.22 cm. A battery charges the plates to a potential difference of 120 V and is then disconnected. A dielectric slab of thickness 4.30 mm and dielectric constant 4.80 is then placed symmetrically between the plates. (*a*) Find the capacitance before the slab is inserted. (*b*) What is the capacitance with the slab in place? (*c*) What is the free charge q before and after the slab is inserted? (*d*) Determine the electric field in the space between the plates and dielectric. (*e*) What is the electric field in the dielectric? (*f*) With the slab in place what is the potential difference across the plates? (*g*) How much external work is involved in the process of inserting the slab?

62. A dielectric slab of thickness b is inserted between the plates of a parallel-plate capacitor of plate separation d. Show that the capacitance is given by

$$C = \frac{\kappa_e \epsilon_0 A}{\kappa_e d - b(\kappa_e - 1)}.$$

(*Hint*: Derive the formula following the pattern of Sample Problem 9.) Does this formula predict the correct numerical result of Sample Problem 9? Verify that the formula gives reasonable results for the special cases of $b = 0$, $\kappa_e = 1$, and $b = d$.

CHAPTER 32

CURRENT AND RESISTANCE

The previous five chapters dealt with electrostatics, *that is, with charges at rest. With this chapter we begin our study of* electric currents, *that is, of charges in motion.*

Examples of electric currents abound, ranging from the large currents that constitute lightning strokes to the tiny nerve currents that regulate our muscular activity. We are familiar with currents resulting from charges flowing through solid conductors (household wiring, light bulbs), semiconductors (integrated circuits), gases (fluorescent lamps), liquids (automobile batteries), and even evacuated spaces (TV picture tubes).

On a global scale, charged particles trapped in the Van Allen radiation belts surge back and forth above the atmosphere between the north and the south magnetic poles. On the scale of the solar system, enormous currents of protons, electrons, and ions travel radially outward from the Sun as the solar wind. On the galactic scale, cosmic rays, which are largely energetic protons, stream through the galaxy.

32-1 ELECTRIC CURRENT

The free electrons in an isolated metallic conductor, such as the length of wire illustrated in Fig. 1a, are in random motion like the molecules of a gas confined to a container. They have no net directed motion along the wire. If we pass a hypothetical plane through the wire, the rate at which electrons cross that plane in one direction is equal to the rate at which they cross in the other direction; the *net* rate is zero. (Here we assume our observation time is long enough so that the small statistical fluctuations in the number of electrons crossing the plane average to zero. In some cases, the fluctuations can be important. For example, they contribute to the electrical noise in circuits.)

Whether the conductor of Fig. 1a is charged or uncharged, there is no net flow of charge in its interior. In the absence of an externally applied field, no electric field exists within the volume of the conductor or parallel to its surface. Even though an abundance of conduction electrons is available, there is no force on the electrons and no net flow of charge.

In Fig. 1b, a battery has been connected across the ends of the conductor. If the battery maintains a potential difference V and the wire has length L, then an electric field

of magnitude V/L is established in the conductor. This electric field **E** acts on the electrons and gives them a net motion in the direction opposite to **E**. If the battery could maintain the potential difference, then the charges would continue to circulate indefinitely. In reality, a battery can maintain the current only as long as it is able to convert chemical energy to electrical energy; eventually the battery's source of energy is exhausted, and the potential difference cannot be maintained.

The existence of an electric field inside a conductor does not contradict Section 29-4, in which we asserted that **E** equals zero inside a conductor. In that section, which dealt with a state in which all net motion of charge had stopped (electrostatics), we assumed that the conductor was insulated and that no potential difference was deliberately maintained between any two points on it, as by a battery. In this chapter, which deals with charges in motion, we relax this restriction.

If a *net* charge dq passes through any surface in a time interval dt, we say that an *electric current i* has been established, where

$$i = dq/dt. \qquad (1)$$

For current in a wire, we take dq to be the charge that passes through a cross section in the time dt.

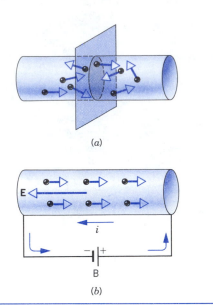

(a)

(b)

Figure 1 (*a*) In an isolated conductor, the electrons are in random motion. The net flow of charge across an arbitrary plane is zero. (*b*) A battery B connected across the conductor sets up an electric field **E**, and the electrons acquire a net motion due to the field.

Note that we require a *net* charge dq to flow for a current to be established. In Fig. 1*a*, equal numbers of electrons are flowing in both directions across the plane; even though there may be a considerable number of electrons flowing across the plane, the current is zero. For another example, the flow of water through a garden hose does not give rise to an electric current according to our definition because the electrically neutral molecules flowing across any surface carry equal positive and negative charges; thus the net flow of charge is zero.

The SI unit of current is the *ampere* (abbreviation A). According to Eq. 1, we have

$$1 \text{ ampere} = 1 \text{ coulomb/second.}$$

You will recall from Section 27-4 that Eq. 1 provides the definition of the coulomb, because the ampere is a SI base unit (see Appendix A). The determination of this fundamental quantity is discussed in Section 35-4.

The net charge that passes through the surface in any time interval is found by integrating the current:

$$q = \int i \, dt. \qquad (2)$$

If the current is constant in time, then the charge q that flows in time t determines the current i according to

$$i = q/t. \qquad (3)$$

In this chapter we consider only currents that are constant in time; currents that vary with time are considered in Chapter 33. Although there are many different kinds of currents (some of which are mentioned in the introduc-

tion), in this chapter we restrict our discussion to electrons moving through solid conductors.

We assume that, under steady conditions, charge does not collect at or drain away from any point in our idealized wire. In the language of Section 18-2, there are no sources or sinks of charge in the wire. When we made this assumption in our study of incompressible fluids, we concluded that the rate at which the fluid flows past any cross section of a pipe is the same even if the cross section varies. The fluid flows faster where the pipe is smaller and slower where it is larger, but the volume rate of flow, measured perhaps in liters/second, remains constant. In the same way, *the electric current i is the same for all cross sections of a conductor, even though the cross-sectional area may be different at different points.*

Although in metals the charge carriers are electrons, in electrolytes or in gaseous conductors (plasmas) they may also be positive or negative ions, or both. We need a convention for labeling the direction of current because charges of opposite sign move in opposite directions in a given field. A positive charge moving in one direction is equivalent in nearly all external effects to a negative charge moving in the opposite direction. Hence, for simplicity and algebraic consistency, we adopt the following convention:

The direction of current is the direction that positive charges would move, even if the actual charge carriers are negative.

If the charge carriers are negative, they simply move opposite to the direction of the current arrow (see Fig. 1*b*).

Under most circumstances, we analyze electric circuits based on an assumed direction for the current, without taking into account whether the actual charge carriers are positive or negative. In rare cases (see, for example, the Hall effect in Section 34-4) we must take into account the sign of the charge carriers.

Even though we assign it a direction, current is a scalar and not a vector. The arrow that we draw to indicate the direction of the current merely shows the sense of the charge flow through the wire and is *not* to be taken as a vector. Current does not obey the laws of vector addition, as can be seen from Fig. 2. The current i_1 in wire 1 divides

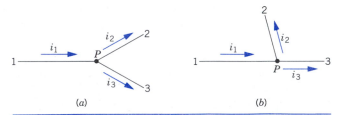

(a)   (b)

Figure 2 (*a*) At point *P*, the current i_1 divides into currents i_2 and i_3, such that $i_1 = i_2 + i_3$. (*b*) Changing the direction of the wires does not change the way the currents add, illustrating that currents add like scalars, not like vectors.

into two branches i_2 and i_3 in wires 2 and 3, such that $i_1 = i_2 + i_3$. Changing the directions of the wires does not change the way the currents are added, as it would if they added like vectors.

32-2   CURRENT DENSITY

The current i is a characteristic of a particular conductor. It is a macroscopic quantity, like the mass of an object, the volume of an object, or the length of a rod. A related microscopic quantity is the *current density* $\mathbf{j}$. It is a vector and is characteristic of a point inside a conductor rather than of the conductor as a whole. If the current is distributed uniformly across a conductor of cross-sectional area A, as in Fig. 3, the magnitude of the current density for all points on that cross section is

$$j = i/A. \qquad (4)$$

The vector $\mathbf{j}$ at any point is oriented in the direction that a positive charge carrier would move at that point. An electron at that point moves in the direction $-\mathbf{j}$. In Fig. 3, $\mathbf{j}$ is a constant vector and points to the left; the electrons drift to the right.

In general, for a particular surface (which need not be plane) that cuts across a conductor, i is the flux of the vector $\mathbf{j}$ over that surface, or

$$i = \int \mathbf{j} \cdot d\mathbf{A}, \qquad (5)$$

where $d\mathbf{A}$ is an element of surface area and the integral is done over the surface in question. The vector $d\mathbf{A}$ is taken to be perpendicular to the surface element such that $\mathbf{j} \cdot d\mathbf{A}$ is positive, giving a positive current i. Equation 4 (written as $i = jA$) is a special case of Eq. 5 in which the surface of integration is a plane cross section of the conductor and in which $\mathbf{j}$ is constant over this surface and at right angles to it. However, we may apply Eq. 5 to *any* surface through which we wish to know the current. Equation 5 shows clearly that i is a scalar because the integrand $\mathbf{j} \cdot d\mathbf{A}$ is a scalar.

The electric field exerts a force ($= -e\mathbf{E}$) on the electrons in a conductor but this force does not produce a *net* acceleration because the electrons keep colliding with the atoms or ions that make up the conductor. This array of ions, coupled together by strong springlike forces of electromagnetic origin, is called the *lattice* (see Fig. 11 of Chapter 14). The overall effect of the collisions is to transfer kinetic energy from the accelerating electrons into vibrational energy of the lattice. The electrons acquire a constant average *drift speed* v_d in the direction $-\mathbf{E}$. There is a close analogy to a ball falling in a uniform gravitational field $\mathbf{g}$ at a constant terminal speed through a viscous fluid. The gravitational force ($m\mathbf{g}$) acting on the falling ball does not increase the ball's kinetic energy (which is constant); instead, energy is transferred to the fluid by molecular collisions and produces a small rise in temperature.

We can compute the drift speed v_d of charge carriers in a conductor from the current density j. Figure 3 shows the conduction electrons in a wire moving to the right at an assumed constant drift speed v_d. The number of conduction electrons in a length L of the wire is nAL, where n is the number of conduction electrons per unit volume and AL is the volume of the length L of the wire. A charge of magnitude

$$q = (nAL)e$$

passes out of this segment of the wire, through its right end, in a time t given by

$$t = \frac{L}{v_d}.$$

The current i is

$$i = \frac{q}{t} = \frac{nALe}{L/v_d} = nAev_d.$$

Solving for v_d and recalling that $j = i/A$ (Eq. 4), we obtain

$$v_d = \frac{i}{nAe} = \frac{j}{ne}. \qquad (6)$$

Since both v_d and j are vectors, we can rewrite Eq. 6 as a vector equation. We follow our adopted convention for positive current density, which means we must take the direction of $\mathbf{j}$ to be opposite to that of $\mathbf{v_d}$. The vector equivalent of Eq. 6 is therefore

$$\mathbf{j} = -ne\mathbf{v_d}. \qquad (7)$$

Figure 3 shows that, for electrons, these vectors are indeed in opposite directions.

As the following sample problems illustrate, the drift speed in typical conductors is quite small, often of the order of cm/s. In contrast, the random thermal motion of conduction electrons in a metal takes place with typical speeds of 10^6 m/s.

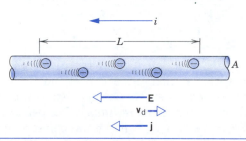

Figure 3   The electric field causes electrons to drift to the right. The conventional current (the hypothetical direction of flow of positive charge) is to the left. The current density $\mathbf{j}$ is likewise drawn as if the charge carriers were positive, so that $\mathbf{j}$ and $\mathbf{E}$ are in the same direction.

Sample Problem 1   One end of an aluminum wire whose diameter is 2.5 mm is welded to one end of a copper wire whose diameter is 1.8 mm. The composite wire carries a steady current i of 1.3 A. What is the current density in each wire?

Solution   We may take the current density as (a different) constant within each wire except for points near the junction. The current density is given by Eq. 4,

$$j = \frac{i}{A}.$$

The cross-sectional area A of the aluminum wire is

$$A_{Al} = \frac{1}{4}\pi d^2 = (\pi/4)(2.5 \times 10^{-3} \text{ m})^2 = 4.91 \times 10^{-6} \text{ m}^2$$

so that

$$j_{Al} = \frac{1.3 \text{ A}}{4.91 \times 10^{-6} \text{ m}^2} = 2.6 \times 10^5 \text{ A/m}^2 = 26 \text{ A/cm}^2.$$

As you can verify, the cross-sectional area of the copper wire is $2.54 \times 10^{-6} \text{ m}^2$, so that

$$j_{Cu} = \frac{1.3 \text{ A}}{2.54 \times 10^{-6} \text{ m}^2} = 5.1 \times 10^5 \text{ A/m}^2 = 51 \text{ A/cm}^2.$$

The fact that the wires are of different materials does not enter here.

Sample Problem 2   What is the drift speed of the conduction electrons in the copper wire of Sample Problem 1?

Solution   The drift speed is given by Eq. 6,

$$v_d = \frac{j}{ne}.$$

In copper, there is very nearly one conduction electron per atom on the average. The number n of electrons per unit volume is therefore the same as the number of atoms per unit volume and is given by

$$\frac{n}{N_A} = \frac{\rho_m}{M} \quad \text{or} \quad \frac{\text{atoms/m}^3}{\text{atoms/mol}} = \frac{\text{mass/m}^3}{\text{mass/mol}}.$$

Here ρ_m is the (mass) density of copper, N_A is the Avogadro constant, and M is the molar mass of copper.* Thus

$$n = \frac{N_A \rho_m}{M} = \frac{(6.02 \times 10^{23} \text{ electrons/mol})(8.96 \times 10^3 \text{ kg/m}^3)}{63.5 \times 10^{-3} \text{ kg/mol}}$$

$$= 8.49 \times 10^{28} \text{ electrons/m}^3.$$

We then have

$$v_d = \frac{5.1 \times 10^5 \text{ A/m}^2}{(8.49 \times 10^{28} \text{ electrons/m}^3)(1.60 \times 10^{-19} \text{ C/electron})}$$

$$= 3.8 \times 10^{-5} \text{ m/s} = 14 \text{ cm/h}.$$

You should be able to show that for the aluminum wire, $v_d = 2.7 \times 10^{-5}$ m/s = 9.7 cm/h. Can you explain, in physical terms, why the drift speed is smaller in aluminum than in copper, even though the two wires carry the same current?

---

* We use the subscript m to make it clear that the density referred to here is a mass density (kg/m³), not a charge density (C/m³).

If the electrons drift at such a low speed, why do electrical effects seem to occur immediately after a switch is thrown, such as when you turn on the room lights? Confusion on this point results from not distinguishing between the drift speed of the electrons and the speed at which *changes* in the electric field configuration travel along wires. This latter speed approaches that of light. Similarly, when you turn the valve on your garden hose, with the hose full of water, a pressure wave travels along the hose at the speed of sound in water. The speed at which the water moves through the hose—measured perhaps with a dye marker—is much lower.

Sample Problem 3   A strip of silicon, of width $w = 3.2$ mm and thickness $d = 250 \ \mu$m, carries a current i of 190 mA. The silicon is an *n-type semiconductor*, having been "doped" with a controlled amount of phosphorus impurity. The doping has the effect of greatly increasing n, the number of charge carriers (electrons, in this case) per unit volume, as compared with the value for pure silicon. In this case, $n = 8.0 \times 10^{21}$ m^{-3}. (*a*) What is the current density in the strip? (*b*) What is the drift speed?

Solution   (*a*) From Eq. 4,

$$j = \frac{i}{wd} = \frac{190 \times 10^{-3} \text{ A}}{(3.2 \times 10^{-3} \text{ m})(250 \times 10^{-6} \text{ m})}$$

$$= 2.4 \times 10^5 \text{ A/m}^2.$$

(*b*) From Eq. 6,

$$v_d = \frac{j}{ne} = \frac{2.4 \times 10^5 \text{ A/m}^2}{(8.0 \times 10^{21} \text{ m}^{-3})(1.60 \times 10^{-19} \text{ C})}$$

$$= 190 \text{ m/s}.$$

The drift speed (190 m/s) of the electrons in this silicon semiconductor is much greater than the drift speed (3.8×10^{-5} m/s) of the conduction electrons in the metallic copper conductor of Sample Problem 2, even though the current densities are similar. The number of charge carriers in this semiconductor (8.0×10^{21} m^{-3}) is much smaller than the number of charge carriers in the copper conductor (8.49×10^{28} m^{-3}). The smaller number of charge carriers must drift faster in the semiconductor if they are to establish the same current density that the greater number of charge carriers establish in copper.

32-3   RESISTANCE, RESISTIVITY, AND CONDUCTIVITY

If we apply the same potential difference between the ends of geometrically similar rods of copper and of wood, very different currents result. The characteristic of the conductor that enters here is its *resistance*. We determine the resistance of a conductor between two points by applying a potential difference V between those points and measuring the current i that results. The resistance R is then

$$R = V/i. \tag{8}$$

If V is in volts and i in amperes, the resistance R is in volts/ampere, which is given the name of *ohms* (abbreviation Ω), such that

$$1 \text{ ohm} = 1 \text{ volt/ampere.}$$

A conductor whose function in a circuit is to provide a specified resistance is called a *resistor* (symbol -⋁⋁⋁-).

The flow of charge through a conductor is often compared with the flow of water through a pipe as a result of a difference in pressure between the ends of the pipe, established perhaps by a pump. The pressure difference is analogous to the potential difference between the ends of a conductor, established perhaps by a battery. The rate of flow of water (liters/second, say) is analogous to the rate of flow of charge (coulombs/second, or amperes). The rate of flow of water for a given pressure difference is determined by the nature of the pipe: its length, cross section, and solid interior impediments (for instance, gravel in the pipe). These characteristics of the pipe are analogous to the resistance of a conductor.

The ohm is not a SI base unit (see Appendix A); no primary standard of the ohm is kept and maintained. However, resistance is such an important quantity in science and technology that a *practical reference standard* is maintained at the National Institute of Standards and Technology. Since January 1, 1990, this *representation of the ohm* (as it is known) has been based on the *quantum Hall effect* (see Section 34-4), a precise and highly reproducible quantum phenomenon that is independent of the properties of any particular material.

Related to resistance is the *resistivity* ρ, which is a characteristic of a material rather than of a particular specimen of a material; it is defined by

$$\rho = \frac{E}{j}. \qquad (9)$$

The units of ρ are those of E (V/m) divided by j (A/m^2), which are equivalent to $\Omega \cdot m$. Figure 3 indicates that E and j are vectors, and we can write Eq. 9 in vector form as

$$\mathbf{E} = \rho \mathbf{j}. \qquad (10)$$

Equations 9 and 10 are valid only for *isotropic* materials, whose electrical properties are the same in all directions.

The resistivity of copper is 1.7×10^{-8} $\Omega \cdot m$; that of fused quartz is about 10^{16} $\Omega \cdot m$. Few physical properties are measurable over such a range of values. Table 1 lists resistivities for some common materials.

Some substances cannot readily be classified as conductors or insulators. Plastics generally have large resistivities that would lead us to classify them with the insulators. For example, household electrical wiring normally uses plastic for insulation. However, by doping plastics with certain chemicals, their conductivity can match that of copper.*

---

* See "Plastics that Conduct Electricity," by Richard B. Kaner and Alan G. MacDiarmid, *Scientific American*, February 1988, p. 106.

TABLE 1   RESISTIVITY OF SOME MATERIALS AT ROOM TEMPERATURE (20°C)

Material	Resistivity, ρ ($\Omega \cdot m$)	Temperature Coefficient of Resistivity $\overline{\alpha}$ (per C°)
Typical Metals		
Silver	1.62×10^{-8}	4.1×10^{-3}
Copper	1.69×10^{-8}	4.3×10^{-3}
Aluminum	2.75×10^{-8}	4.4×10^{-3}
Tungsten	5.25×10^{-8}	4.5×10^{-3}
Iron	9.68×10^{-8}	6.5×10^{-3}
Platinum	10.6×10^{-8}	3.9×10^{-3}
Manganin[a]	48.2×10^{-8}	0.002×10^{-3}
Typical Semiconductors		
Silicon pure	2.5×10^{3}	-70×10^{-3}
Silicon n-type[b]	8.7×10^{-4}	
Silicon p-type[c]	2.8×10^{-3}	
Typical Insulators		
Glass	10^{10}–10^{14}	
Polystyrene	$> 10^{14}$	
Fused quartz	$\approx 10^{16}$	

[a] An alloy specifically designed to have a small value of α.
[b] Pure silicon "doped" with phosphorus impurities to a charge carrier density of 10^{23} m^{-3}.
[c] Pure silicon "doped" with aluminum impurities to a charge carrier density of 10^{23} m^{-3}.

Sometimes we prefer to speak of the *conductivity* σ of a material rather than its resistivity. These are reciprocal quantities, related by

$$\sigma = 1/\rho. \qquad (11)$$

The SI units of σ are $(\Omega \cdot m)^{-1}$. Equation 10 can be written in terms of the conductivity as

$$\mathbf{j} = \sigma \mathbf{E}. \qquad (12)$$

If we know the resistivity ρ of a material, we should be able to calculate the resistance R of a particular piece of the material. Consider a cylindrical conductor, of cross-sectional area A and length L carrying a steady current i with a potential difference V between its ends (see Fig. 4). If the cylinder cross sections at each end are equipotential surfaces, the electric field and the current density are constant for all points in the cylinder and have the values

$$E = \frac{V}{L} \quad \text{and} \quad j = \frac{i}{A}.$$

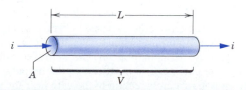

Figure 4   A potential difference V is applied across a cylindrical conductor of length L and cross-sectional area A, establishing a current i.

The resistivity ρ is

$$\rho = \frac{E}{j} = \frac{V/L}{i/A} .$$

But V/i is the resistance R, which leads to

$$R = \rho \frac{L}{A} . \tag{13}$$

We stress that Eq. 13 applies only to a homogeneous, isotropic conductor of uniform cross section subject to a uniform electric field.

Sample Problem 4 A rectangular block of iron has dimensions $1.2 \text{ cm} \times 1.2 \text{ cm} \times 15 \text{ cm}$. (a) What is the resistance of the block measured between the two square ends? (b) What is the resistance between two opposing rectangular faces? The resistivity of iron at room temperature is $9.68 \times 10^{-8} \ \Omega \cdot \text{m}$.

Solution (a) The area of a square end is $(1.2 \times 10^{-2} \text{ m})^2$ or $1.44 \times 10^{-4} \text{ m}^2$. From Eq. 13,

$$R = \frac{\rho L}{A} = \frac{(9.68 \times 10^{-8} \ \Omega \cdot \text{m})(0.15 \text{ m})}{1.44 \times 10^{-4} \text{ m}^2}$$

$$= 1.0 \times 10^{-4} \ \Omega = 100 \ \mu\Omega.$$

(b) The area of a rectangular face is $(1.2 \times 10^{-2} \text{ m})(0.15 \text{ m})$ or $1.80 \times 10^{-3} \text{ m}^2$. From Eq. 13,

$$R = \frac{\rho L}{A} = \frac{(9.68 \times 10^{-8} \ \Omega \cdot \text{m})(1.2 \times 10^{-2} \text{ m})}{1.80 \times 10^{-3} \text{ m}^2}$$

$$= 6.5 \times 10^{-7} \ \Omega = 0.65 \ \mu\Omega.$$

We assume in each case that the potential difference is applied to the block in such a way that the surfaces between which the resistance is desired are equipotentials. Otherwise, Eq. 13 would not be valid.

Microscopic and Macroscopic Quantities *(Optional)*
V, i, and R are *macroscopic* quantities, applying to a particular body or extended region. The corresponding *microscopic* quantities are **E**, **j**, and ρ (or σ); they have values at every point in a body. The macroscopic quantities are related by Eq. 8 ($V = iR$) and the microscopic quantities by Eqs. 9, 10, and 12.

The macroscopic quantities can be found by integrating over the microscopic quantities, using relations already given, namely,

$$i = \int \mathbf{j} \cdot d\mathbf{A}$$

and

$$V_{ab} = -V_{ba} = \int_a^b \mathbf{E} \cdot d\mathbf{s}.$$

The current integral is a surface integral, carried out over any cross section of the conductor. The field integral is a line integral carried out along an arbitrary line drawn along the conductor, connecting any two equipotential surfaces, identified by a and b. For a long wire connected to a battery, equipotential surface a might be chosen as a cross section of the wire near the positive battery terminal, and b might be a cross section near the negative terminal.

We can express the resistance of a conductor between a and b in microscopic terms by dividing the two equations:

$$R = \frac{V_{ab}}{i} = \frac{\displaystyle\int_a^b \mathbf{E} \cdot d\mathbf{s}}{\displaystyle\int \mathbf{j} \cdot d\mathbf{A}} .$$

If the conductor is a long cylinder of cross section A and length L, and if points a and b are its ends, the above equation for R reduces to

$$R = \frac{EL}{jA} = \rho \frac{L}{A} ,$$

which is Eq. 13.

The macroscopic quantities V, i, and R are of primary interest when we are making electrical measurements on real conducting objects. They are the quantities whose values are indicated on meters. The microscopic quantities **E**, **j**, and ρ are of primary importance when we are concerned with the fundamental behavior of matter (rather than of specimens of matter), as we usually are in the research area of *solid state* (or *condensed matter*) physics. Section 32-5 accordingly deals with an atomic view of the *resistivity* of a metal and not of the *resistance* of a metallic specimen. The microscopic quantities are also important when we are interested in the interior behavior of irregularly shaped conducting objects. ∎

Temperature Variation of Resistivity *(Optional)*
Figure 5 shows a summary of some experimental measurements of the resistivity of copper at different temperatures. For practical use of this information, it would be helpful to express it in the form of an equation. Over a limited range of temperature, the relationship between resistivity and temperature is nearly linear. We can fit a straight line to any selected region of Fig. 5, using two points to determine the slope of the line. Choosing a reference point, such as that labeled T_0, ρ_0 in the figure, we can

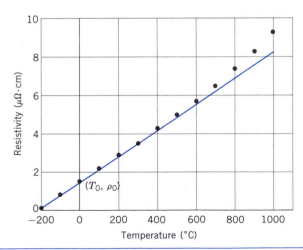

Figure 5 The dots show selected measurements of the resistivity of copper at different temperatures. Over any given range of temperature, the variation in the resistivity with T can be approximated by a straight line; for example, the line shown fits the data from about $-100°\text{C}$ to $400°\text{C}$.

express the resistivity ρ at an arbitrary temperature T from the empirical equation of the straight line in Fig. 5, which is

$$\rho - \rho_0 = \rho_0 \overline{\alpha}(T - T_0). \qquad (14)$$

[This expression is very similar to that for linear thermal expansion ($\Delta L = \alpha L \, \Delta T$), which we introduced in Section 22-5.] We have written the slope of this line as $\rho_0 \overline{\alpha}$. If we solve Eq. 14 for $\overline{\alpha}$, we obtain

$$\overline{\alpha} = \frac{1}{\rho_0} \frac{\rho - \rho_0}{T - T_0} . \qquad (15)$$

The quantity $\overline{\alpha}$ is the *mean (or average) temperature coefficient of resistivity* over the region of temperature between the two points used to determine the slope of the line. We can define a more general temperature coefficient of resistivity as

$$\alpha = \frac{1}{\rho} \frac{d\rho}{dT} , \qquad (16)$$

which is the fractional change in resistivity $d\rho/\rho$ per change in temperature dT. That is, α gives the dependence of resistivity on temperature *at a particular temperature*, while $\overline{\alpha}$ gives the average dependence *over a particular interval*. The coefficient α is in general dependent on temperature.

For most practical purposes, Eq. 14 gives results that are within the acceptable range of accuracy. Typical values of $\overline{\alpha}$ are given in Table 1. For more precise work, such as the use of the platinum resistance thermometer to measure temperature (see Section 22-3), the linear approximation is not sufficient. In this case we can add terms in $(T - T_0)^2$ and $(T - T_0)^3$ to the right side of Eq. 14 to improve the precision. The coefficients of these additional terms must be determined empirically, in analogy with the coefficient $\overline{\alpha}$ in Eq. 14. ■

32-4   OHM'S LAW

Let us select a particular sample of conducting material, apply a uniform potential difference across it, and measure the resulting current. We repeat the measurement for various values of the potential difference and plot the results, as in Fig. 6a. The experimental points clearly fall along a straight line, which indicates that the ratio V/i (the inverse of the slope of the line) is a constant. The resistance of this device is a constant, independent of the potential difference across it or the current through it. Note that the line extends to negative potential differences and currents.

In this case, we say that the material obeys *Ohm's law*:

A conducting device obeys Ohm's law if the resistance between any pair of points is independent of the magnitude and polarity of the applied potential difference.

A material or a circuit element that obeys Ohm's law is called *ohmic.*

Modern electronic circuits also depend on devices that do *not* obey Ohm's law. An example of the current–voltage relationship for a nonohmic device (a *pn* junction

diode) is shown in Fig. 6b. Note that the current does not increase linearly with the voltage, and also note that the device behaves very differently for negative potential differences than it does for positive ones.

We stress that the relationship $V = iR$ is *not* a statement of Ohm's law. A conductor obeys Ohm's law only if its V versus i graph is linear, that is, if R is independent of V and i. The relationship $R = V/i$ remains as the general definition of the resistance of a conductor whether or not the conductor obeys Ohm's law.

The microscopic equivalent of the relationship $V = iR$ is Eq. 10, $\mathbf{E} = \rho \mathbf{j}$. A conducting material is said to obey Ohm's law if a plot of E versus j is linear, that is, if the resistivity ρ is independent of E and j. Ohm's law is a specific property of certain materials and is not a general law of electromagnetism, for example, like Gauss' law.

Analogy Between Current and Heat Flow   *(Optional)*
A close analogy exists between the flow of charge established by a potential difference and the flow of heat established by a temperature difference. Consider a thin electrically conducting slab of thickness Δx and area A. Let a potential difference ΔV be maintained between opposing faces. The current i is given by Eqs. 8 ($i = V/R$) and 13 ($R = \rho L/A$), or

$$i = \frac{V_a - V_b}{R} = \frac{(V_a - V_b)A}{\rho L} = -\frac{(V_b - V_a)A}{\rho \Delta x} .$$

In the limiting case of a slab of thickness dx this becomes

$$i = -\rho^{-1} A \frac{dV}{dx}$$

or, replacing the inverse of the resistivity by the conductivity σ,

$$\frac{dq}{dt} = -\sigma A \frac{dV}{dx} . \qquad (17)$$

The minus sign in Eq. 17 indicates that positive charge flows in the direction of decreasing V; that is, dq/dt is positive when dV/dx is negative.

The analogous heat flow equation (see Section 25-7) is

$$\frac{dQ}{dt} = -kA \frac{dT}{dx} , \qquad (18)$$

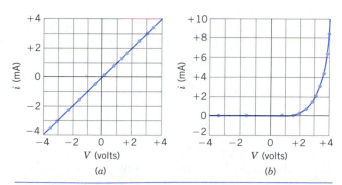

Figure 6   (a) A current–voltage plot for a material that obeys Ohm's law, in this case a 1000-Ω resistor. (b) A current–voltage plot for a material that does not obey Ohm's law, in this case a *pn* junction diode.

which shows that k, the thermal conductivity, corresponds to σ, and dT/dx, the temperature gradient, corresponds to dV/dx, the potential gradient. For pure metals there is more than a formal mathematical analogy between Eqs. 17 and 18. Both heat energy and charge are carried by the free electrons in such metals; empirically, a good electrical conductor (silver, say) is also a good heat conductor, and the electrical conductivity σ is directly related to the thermal conductivity k. ■

32-5 OHM'S LAW: A MICROSCOPIC VIEW

As we discussed previously, Ohm's law is not a fundamental law of electromagnetism because it depends on the properties of the conducting medium. The law is very simple in form, and it is curious that many materials obey it so well, whereas other materials do not obey it at all. Let us see if we can understand why metals obey Ohm's law, which we shall write (see Eq. 10) in the microscopic form $\mathbf{E} = \rho\mathbf{j}$.

In a metal the valence electrons are not attached to individual atoms but are free to move about within the lattice and are called *conduction electrons*. In copper there is one such electron per atom, the other 28 remaining bound to the copper nuclei to form ionic cores.

The theory of electrical conduction in metals is often based on the *free-electron model*, in which (as a first approximation) the conduction electrons are assumed to move freely throughout the conducting material, somewhat like molecules of gas in a container. In fact, the assembly of conduction electrons is sometimes called an *electron gas*. As we shall see, however, we cannot neglect the effect of the ion cores on this "gas."

The classical Maxwellian velocity distribution (see Section 24-3) for the electron gas would suggest that the conduction electrons have a broad distribution of velocities from zero to infinity, with a well-defined average. However, in considering the electrons we cannot ignore quantum mechanics, which gives a very different view. In the quantum distribution (see Fig. 16 of Chapter 24) the electrons that readily contribute to electrical conduction are concentrated in a very narrow interval of kinetic energies and therefore of speeds. To a very good approximation, we can assume that the electrons move with a uniform average speed. In the case of copper, this speed is about $\bar{v} = 1.6 \times 10^6$ m/s. Furthermore, whereas the Maxwellian average speed depends strongly on the temperature, the effective speed obtained from the quantum distribution is nearly independent of temperature.

In the absence of an electric field, the electrons move randomly, again like the molecules of gas in a container. Occasionally, an electron collides with an ionic core of the lattice, suffering a sudden change in direction in the process. As we did in the case of collisions of gas molecules, we can associate a mean free path λ and a mean free time τ to the average distance and time between collisions. (Col-

lisions between the electrons themselves are rare and do not affect the electrical properties of the conductor.)

In an ideal metallic crystal (containing no defects or impurities) at 0 K, electron–lattice collisions would not occur, according to the predictions of quantum physics; that is, $\lambda \to \infty$ as $T \to 0$ K for ideal crystals. Collisions take place in actual crystals because (1) the ionic cores at any temperature T are vibrating about their equilibrium positions in a random way; (2) impurities, that is, foreign atoms, may be present; and (3) the crystal may contain lattice imperfections, such as missing atoms and displaced atoms. Consequently, the resistivity of a metal can be increased by (1) raising its temperature, (2) adding small amounts of impurities, and (3) straining it severely, as by drawing it through a die, to increase the number of lattice imperfections.

When we apply an electric field to a metal, the electrons modify their random motion in such a way that they drift slowly, in the opposite direction to that of the field, with an average drift speed v_d. This drift speed is very much less (by a factor of something like 10^{10}; see Sample Problem 2) than the effective average speed $\bar{v}$. Figure 7 suggests the relationship between these two speeds. The solid lines suggest a possible random path followed by an electron in the absence of an applied field; the electron proceeds from x to y, making six collisions on the way. The dashed lines show how this same event *might* have occurred if an electric field $\mathbf{E}$ had been applied. Note that the electron drifts steadily to the right, ending at y' rather than at y. In preparing Fig. 7, it has been assumed that the drift speed v_d is $0.02\bar{v}$; actually, it is more like $10^{-10}\bar{v}$, so that the "drift" exhibited in the figure is greatly exaggerated.

We can calculate the drift speed v_d in terms of the applied electric field E and of $\bar{v}$ and λ. When a field is applied to an electron in the metal, it experiences a force eE,

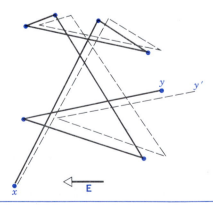

Figure 7   The solid line segments show an electron moving from x to y, making six collisions en route. The dashed lines show what its path *might* have been in the presence of an applied electric field $\mathbf{E}$. Note the gradual but steady drift in the direction of $-\mathbf{E}$. (Actually, the dashed lines should be slightly curved to represent the parabolic paths followed by the electrons between collisions.)

which imparts to it an acceleration a given by Newton's second law,

$$a = \frac{eE}{m}.$$

Consider an electron that has just collided with an ion core. The collision, in general, momentarily destroys the tendency to drift, and the electron has a truly random direction after the collision. During the time interval to the next collision, the electron's speed changes, on the average, by an amount $a(\lambda/\bar{v})$ or $a\tau$, where τ is the mean time between collisions. We identify this with the drift speed v_d, or*

$$v_d = a\tau = \frac{eE\tau}{m}. \tag{19}$$

We may also express v_d in terms of the current density (Eq. 6), which gives

$$v_d = \frac{j}{ne} = \frac{eE\tau}{m}.$$

Combining this with Eq. 9 ($\rho = E/j$), we finally obtain

$$\rho = \frac{m}{ne^2\tau}. \tag{20}$$

Note that m, n, and e in this equation are constants. Thus Eq. 20 can be taken as a statement that metals obey Ohm's law if we can show that τ is a constant. In particular, we must show that τ does not depend on the applied electric field E. In this case ρ does not depend on E, which (see Section 32-4) is the criterion that a material obey Ohm's law. The quantity τ depends on the speed distribution of the conduction electrons. We have seen that this distribution is affected only very slightly by the application of even a relatively large electric field, since $\bar{v}$ is of the order of 10^6 m/s, and v_d (see Sample Problem 2) is only of the order of 10^{-4} m/s, a ratio of 10^{10}. Whatever the value of τ is (for copper at 20°C, say) in the absence of a field, it remains essentially unchanged when the field is applied. Thus the right side of Eq. 20 is independent of E (which

---

* It may be tempting to write Eq. 19 as $v_d = \frac{1}{2}a\tau$, reasoning that $a\tau$ is the electron's *final* velocity, and thus that its *average* velocity is half that value. The extra factor of $\frac{1}{2}$ would be correct if we followed a typical electron, taking its drift speed to be the average of its velocity over its mean time τ between collisions. However, the drift speed is proportional to the current density j and must be calculated from the average velocity of *all* the electrons taken at one instant of time. For each electron, the velocity at any time is at, where t is the time since the last collision for that electron. Since the acceleration a is the same for all electrons, the average value of at at a given instant is $a\tau$, where τ is the average time since the last collision, which is the same as the mean time between collisions. For a discussion of this point, see *Electricity and Magnetism*, 2nd ed., by Edward Purcell (McGraw-Hill, 1985), Section 4.4. See also "Drift Speed and Collision Time," by Donald E. Tilley, *American Journal of Physics*, June 1976, p. 597.

means that ρ is independent of E), and the material obeys Ohm's law.

Sample Problem 5   (a) What is the mean free time τ between collisions for the conduction electrons in copper? (b) What is the mean free path λ for these collisions? Assume an effective speed $\bar{v}$ of 1.6×10^6 m/s.

Solution   (a) From Eq. 20 we have

$$\tau = \frac{m}{ne^2\rho}$$

$$= \frac{9.11 \times 10^{-31} \text{ kg}}{(8.49 \times 10^{28} \text{ m}^{-3})(1.60 \times 10^{-19} \text{ C})^2(1.69 \times 10^{-8} \ \Omega \cdot \text{m})}$$

$$= 2.48 \times 10^{-14} \text{ s}.$$

The value of n, the number of conduction electrons per unit volume in copper, was obtained from Sample Problem 2; the value of ρ comes from Table 1.

(b) We define the mean free path from

$$\lambda = \tau\bar{v} = (2.48 \times 10^{-14} \text{ s})(1.6 \times 10^6 \text{ m/s})$$

$$= 4.0 \times 10^{-8} \text{ m} = 40 \text{ nm}.$$

This is about 150 times the distance between nearest-neighbor ions in a copper lattice. A full treatment based on quantum physics reveals that we cannot view a "collision" as a direct interaction between an electron and an ion. Rather, it is an interaction between an electron and the thermal vibrations of the lattice, lattice imperfections, or lattice impurity atoms. An electron can pass very freely through an "ideal" lattice, that is, a geometrically "perfect" lattice close to the absolute zero of temperature. Mean free paths as large as 10 cm have been observed under such conditions.

32-6   ENERGY TRANSFERS IN AN ELECTRIC CIRCUIT

Figure 8 shows a circuit consisting of a battery B connected to a "black box." A steady current i exists in the connecting wires, and a steady potential difference V_{ab}

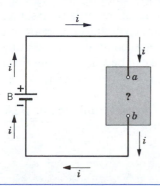

Figure 8   A battery B sets up a current i in a circuit containing a "black box," that is, a box whose contents are unknown.

exists between the terminals a and b. The box might contain a resistor, a motor, or a storage battery, among other things.

Terminal a, connected to the positive battery terminal, is at a higher potential than terminal b. The potential energy of a charge dq that moves through the box from a to b decreases by $dq\, V_{ab}$ (see Section 30-3). The conservation-of-energy principle tells us that this energy is transferred in the box from electric energy to some other form. What that other form will be depends on what is in the box. In a time dt the energy dU transferred inside the box is then

$$dU = dq\, V_{ab} = i\, dt\, V_{ab}.$$

We find the *rate* of energy transfer or power P according to

$$P = \frac{dU}{dt} = iV_{ab}. \tag{21}$$

If the device in the box is a motor, the energy appears largely as mechanical work done by the motor; if the device is a storage battery that is being charged, the energy appears largely as stored chemical energy in this second battery.

If the device is a resistor, the energy appears in the resistor as internal energy (associated with atomic motion and observed, perhaps, as an increase in temperature). To see this, consider a stone of mass m that falls through a height h. It decreases its gravitational potential energy by mgh. If the stone falls in a vacuum or—for practical purposes—in air, this energy is transformed into kinetic energy of the stone. If the stone falls into the depths of the ocean, however, its speed eventually becomes constant, which means that the kinetic energy no longer increases. The potential energy that is steadily being made available as the stone falls then appears as internal energy in the stone and the surrounding water. It is the viscous, frictionlike drag of the water on the surface of the stone that stops the stone from accelerating, and it is at this surface that the transformation to internal energy occurs.

The course of an electron through the resistor is much like that of the stone through water. On average, the electrons travel with a constant drift speed v_d and thus do not gain kinetic energy. They lose electric energy through collisions with atoms of the resistor. As a result, the amplitudes of the atomic vibrations increase; on a macroscopic scale this can correspond to a temperature increase. Subsequently, there can be a flow of energy out of the resistor as heat, if the environment is at a lower temperature than the resistor.

For a resistor we can combine Eqs. 8 ($R = V/i$) and 21 and obtain either

$$P = i^2 R \tag{22}$$

or

$$P = \frac{V^2}{R}. \tag{23}$$

Note that Eq. 21 applies to electrical energy transfer of *all* kinds; Eqs. 22 and 23 apply only to the transfer of electrical energy to internal energy in a resistor. Equations 22 and 23 are known as *Joule's law*, and the corresponding energy transferred to the resistor or its surroundings is called *Joule heating*. This law is a particular way of writing the conservation-of-energy principle for the special case in which electrical energy is transferred into internal energy in a resistor.

The unit of power that follows from Eq. 21 is the volt·ampere. We can show the volt·ampere to be equivalent to the watt as a unit of power by using the definitions of the volt (joule/coulomb) and ampere (coulomb/second):

$$1 \text{ volt·ampere} = 1\,\frac{\text{joule}}{\text{coulomb}} \cdot \frac{\text{coulomb}}{\text{second}}$$

$$= 1\,\frac{\text{joule}}{\text{second}} = 1 \text{ watt.}$$

We previously introduced the watt as a unit of power in Section 7-5.

Sample Problem 6   You are given a length of heating wire made of a nickel–chromium–iron alloy called Nichrome; it has a resistance R of 72 Ω. It is to be connected across a 120-V line. Under which circumstances will the wire dissipate more heat: (*a*) its entire length is connected across the line, or (*b*) the wire is cut in half and the two halves are connected in parallel across the line?

Solution   (*a*) The power P dissipated by the entire wire is, from Eq. 23,

$$P = \frac{V^2}{R} = \frac{(120 \text{ V})^2}{72 \text{ Ω}} = 200 \text{ W.}$$

(*b*) The power for a wire of half length (and thus half resistance) is

$$P' = \frac{V^2}{\frac{1}{2}R} = \frac{(120 \text{ V})^2}{36 \text{ Ω}} = 400 \text{ W.}$$

There are two halves so that the power obtained from both of them is 800 W, or four times that for the single wire. This would seem to suggest that you could buy a heating wire, cut it in half, and reconnect it to obtain four times the heat output. Why is this not such a good idea?

32-7 SEMICONDUCTORS   *(Optional)*

A class of materials called *semiconductors* is intermediate between conductors and insulators in its ability to conduct electricity. Among the elements, silicon and germaniun are common examples of room-temperature semiconductors. One important property of semiconductors is that their ability to conduct can be changed dramatically by external factors, such as by changes in the temperature, applied voltage, or incident light. You can see from Table 1 that, although pure silicon is a relatively poor conductor, a low concentration of impurity atoms (added to pure silicon at a level of one impurity atom per 10^6 silicon atoms) can change the conductivity by six or seven orders of magnitude. You can also see that the conductivity of silicon is at least an order of magnitude more sensitive to changes in temperature than that of a typical conductor. Because of these properties, semiconductors have found wide applications in such devices as switching and control circuits, and they are now essential components of integrated circuits and computer memories.

To describe the properties of conductors, insulators, and semiconductors in microscopic detail requires the application of the principles of quantum physics. However, we can gain a qualitative understanding of the differences between conductors, insulators, and semiconductors by referring to Fig. 9, which shows energy states that might typically represent electrons in conductors, semiconductors, and insulators. The electrons have permitted energies that are discrete or *quantized* (see Section 8-8), but which group together in *bands*. Within the bands, the permitted energy states, which are so close together that they are virtually continuous, may be *occupied* (electrons having the permitted energy) or *unoccupied* (no electrons having that energy). Between the bands there is an *energy gap*, which contains no states that an individual electron may occupy. An electron may jump from an occupied state to any unoccupied one. At ordinary temperatures, the internal energy distribution provides the source of the energy needed for electrons to jump to higher states.

Figure 9*a* illustrates the energy bands that represent a conductor. The valence band, which is the highest band occupied by electrons, is only partially occupied, so that electrons have many empty states to which they can easily jump. An applied electric field can encourage electrons to make these small jumps and contribute to a current in the material. This ease of movement of the electrons is what makes the material a conductor.

Figure 9*b* shows bands that might characterize a semiconductor, such as silicon. At very low temperature, the valence band is completely occupied, and the upper (conduction) band is completely empty. At ordinary temperatures, there is a small probability that an electron from one of the occupied states in the lower band has enough energy to jump across the gap to one of the empty states in the upper band. The probability for such a

jump depends on the energy distribution, which according to Eq. 27 of Chapter 24 includes the factor $e^{-\Delta E/kT}$, where ΔE is the energy gap. Taking $\Delta E = 0.7$ eV (typical for silicon) and $kT = 0.025$ eV at room temperature, the exponential factor is 7×10^{-13}. Although this is a small number, there are so many electrons available in a piece of silicon (about 10^{23} per gram) that a reasonable number (perhaps 10^{11} per gram) are in the upper band. In this band they can move easily from occupied to empty states and contribute to the ability of a semiconductor to transport electric charge. (In the process of jumping to the conduction band, electrons leave vacancies or *holes* in the valence band. Other electrons in the valence band can jump to those vacancies, thereby also contributing to the conductivity.)

Another difference between conductors and semiconductors is in their temperature coefficients of resistivity. Metals are kept from being perfect conductors by deviations from the perfect lattice structure, such as might be caused by the presence of impurities or defects in the lattice. The vibration of the ion cores about their equilibrium lattice positions is a major contributor to the resistivity of metals. Since this effect increases with temperature, the resistivity of metals increases with temperature. The same effect of course also occurs in semiconductors, but it is overwhelmed by a much greater effect that *decreases* the resistivity with increasing temperature. As the temperature increases, more electrons acquire enough energy to be excited across the energy gap into the conduction band, thereby increasing the conductivity and decreasing the resistivity. As Table 1 shows, silicon (in contrast to the metals listed) has a *negative* temperature coefficient of resistivity.

Figure 9*c* shows typical energy bands for an insulator, such as sodium chloride. The band structure is very similar to that of a semiconductor, with the valence band occupied and the conduction band empty. The major difference is in the size of the energy gap, which might be typically 2 eV or more in the case of an insulator (compared with perhaps 0.7 eV in a semiconductor). This relatively small difference makes an enormous difference in the exponential factor that gives the probability of an electron acquiring enough energy to jump across the gap. For an insulator at room temperature, the factor $e^{-\Delta E/kT}$ is typically 2×10^{-35}, so that in a gram of material (10^{23} atoms) there is a

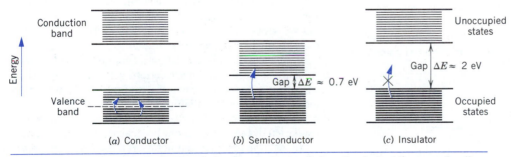

Figure 9 (*a*) Energy bands characteristic of a conductor. Below the dashed line, nearly all energy states are occupied, while nearly all states above that line are empty. Electrons can easily jump from occupied states to empty ones, as suggested by the arrows. (*b*) In a semiconductor, the dividing line between filled and empty states occurs in the gap. The electrical conductivity is determined in part by the number of electrons that jump to occupy states in the conduction band. (*c*) The energy bands in an insulator resemble those in a semiconductor; the major difference is in the size of the energy gap. At ordinary temperatures, there is no probability for an electron to jump to the empty states in the conduction band.

negligible probability at ordinary temperatures of *even one electron* being in the conduction band where it could move freely. In insulators, therefore, all electrons are confined to the valence band, where they have no empty states to enter and thus are not at all free to travel throughout the material.

Note that the principal difference between semiconductors and insulators is in the relationship between the gap energy and kT. At very low temperature, a semiconductor becomes an insulator, while at a high enough temperature (which is, however, above the point at which the material would be vaporized), an insulator could become a semiconductor.

We consider more details of the application of quantum theory to the structure of semiconductors in Chapter 53 of the extended text. ■

32-8   SUPERCONDUCTIVITY   *(Optional)*

As we reduce the temperature of a conductor, the resistivity grows smaller, as Fig. 5 suggests. What happens as we approach the absolute zero of the temperature scale?

The part of the resistivity due to scattering of electrons by atoms vibrating from their equilibrium lattice positions decreases as the temperature decreases, because the amplitude of the vibration decreases with temperature. According to quantum theory, the atoms retain a certain minimum vibrational motion, even at the absolute zero of temperature. Furthermore, the contributions of defects and impurities to the resistivity remain as T falls to 0. We therefore expect the resistivity to decrease with decreasing temperature, but to remain finite at the lowest temperatures. Many materials do in fact show this type of behavior.

Quite a different kind of behavior was discovered in 1911 by the Dutch physicist Kammerlingh Onnes, who was studying the resistivity of mercury at low temperature. He discovered that, below a temperature of about 4 K, mercury suddenly lost all resistivity and became a *perfect* conductor, called a *superconductor*. This was not a gradual change, as Eq. 14 and Fig. 5 suggest, but a sudden transition, as indicated by Fig. 10. The resistivity of a superconductor is not just very small; it is zero! If a current is established in a superconducting material, it should persist forever, even with no electric field present.

The availability of superconducting materials immediately suggests a number of applications. (1) Energy can be transported and stored in electrical wires without resistive losses. That is, a power company can produce electrical energy when the demand is light, perhaps overnight, and store the current in a supercon-

ducting ring. Electrical power can then be released during peak demand times the following day. Such a ring now operates in Tacoma, Washington, to store 5 MW of power. In smaller laboratory test rings, currents have been stored for several years with no reduction. (2) Superconducting electromagnets can produce larger magnetic fields than conventional electromagnets. As we discuss in Chapter 35, a current-carrying wire gives rise to a magnetic field in the surrounding space, just as an electric charge sets up an electric field. With superconducting wires, larger currents and therefore larger magnetic fields can be produced. Applications of this technology include magnetically levitated trains and bending magnets for beams of particles in large accelerators such as Fermilab. (3) Superconducting components in electronic circuits would generate no Joule heating and would permit further miniaturization of circuits. The next generation of mainframe computers may employ superconducting components.

Progress in applying this exciting technology proceeded slowly in the 75 years following Kammerlingh Onnes' discovery for one reason: the elements and compounds that displayed superconductivity did so only at very low temperatures, in most cases below 20 K. To achieve such temperatures, the superconducting material is generally immersed in a bath of liquid helium at 4 K. The liquid helium is costly and so, while there have been many scientific applications of superconductivity, commercial applications have been held back by the high cost of liquid helium.

Beginning in 1986 a series of ceramic materials was discovered which remained superconducting at relatively high temperatures. The first of these kept its superconductivity to a temperature of 90 K. While this is still a low temperature by ordinary standards, it marks an important step: it can be maintained in a bath of liquid nitrogen (77 K), which costs about an order of magnitude less than liquid helium, thereby opening commercial possibilities that had not been feasible with liquid-helium-cooled materials.*

Superconductivity should not be regarded merely as an improvement in the conductivity of materials that are already good conductors. The best room-temperature conductors (copper, silver, and gold) do not show any superconductivity at all.

An understanding of this distinction can be found in the microscopic basis of superconductivity. Ordinary materials are good conductors if they have free electrons that can move easily through the lattice. Atoms of copper, silver, and gold have a single weakly bound valence electron that can be contributed to the electron gas that permeates the lattice. According to one theory, superconductors depend on the motion of highly correlated *pairs* of electrons. Since electrons generally don't like to form pairs, a special circumstance is required: two electrons each interact strongly with the lattice and thus in effect with each other. The situation is somewhat like two boats on a lake, where the wake left by the motion of one boat causes the other to move, even though the first boat did not exert a force directly on the second boat. Thus a good ordinary conductor depends on hav-

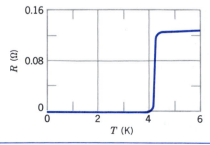

Figure 10   The resistivity of mercury drops to zero at a temperature of about 4 K. Mercury is a solid at this low temperature.

---

* See "The New Superconductors: Prospects for Applications," by Alan M. Wolsky, Robert F. Giese, and Edward J. Daniels, *Scientific American*, February 1989, p. 60, and "Superconductors Beyond 1-2-3," by Robert J. Cava, *Scientific American*, August 1990, p. 42.

ing electrons that interact *weakly* with the lattice, while a superconductor seems to require electrons that interact *strongly* with the lattice.

More details about superconductors and the application of quantum theory to understanding their properties can be found in Chapter 53 of the extended text. ■

QUESTIONS

1. Name other physical quantities that, like current, are scalars having a sense represented by an arrow in a diagram.

2. In our convention for the direction of current arrows (a) would it have been more convenient, or even possible, to have assumed all charge carriers to be negative? (b) Would it have been more convenient, or even possible, to have labeled the electron as positive, the proton as negative, and so on?

3. What experimental evidence can you give to show that the electric charges in current electricity and those in electrostatics are identical?

4. Explain in your own words why we can have $E \neq 0$ inside a conductor in this chapter, whereas we took $E = 0$ for granted in Section 29-4.

5. A current i enters one corner of a square sheet of copper and leaves at the opposite corner. Sketch arrows at various points within the square to represent the relative values of the current density j. Intuitive guesses rather than detailed mathematical analyses are called for.

6. Can you see any logic behind the assignment of gauge numbers to household wire? See Problem 6. If not, then why is this system used?

7. A potential difference V is applied to a copper wire of diameter d and length L. What is the effect on the electron drift speed of (a) doubling V, (b) doubling L, and (c) doubling d?

8. Why is it not possible to measure the drift speed for electrons by timing their travel along a conductor?

9. Describe briefly some possible designs of variable resistors.

10. A potential difference V is applied to a circular cylinder of carbon by clamping it between circular copper electrodes, as in Fig. 11. Discuss the difficulty of calculating the resistance of the carbon cylinder using the relation $R = \rho L/A$.

Copper

Carbon

Copper

Figure 11 Question 10.

11. You are given a cube of aluminum and access to two battery terminals. How would you connect the terminals to the cube to ensure (a) a maximum and (b) a minimum resistance?

12. How would you measure the resistance of a pretzel-shaped block of metal? Give specific details to clarify the concept.

13. Sliding across the seat of an automobile can generate potentials of several thousand volts. Why isn't the slider electrocuted?

14. Discuss the difficulties of testing whether the filament of a light bulb obeys Ohm's law.

15. Will the drift velocity of electrons in a current-carrying metal conductor change when the temperature of the conductor is increased? Explain.

16. Explain why the momentum that conduction electrons transfer to the ions in a metal conductor does not give rise to a resultant force on the conductor.

17. List in tabular form similarities and differences between the flow of charge along a conductor, the flow of water through a horizontal pipe, and the conduction of heat through a slab. Consider such ideas as what causes the flow, what opposes it, what particles (if any) participate, and the units in which the flow may be measured.

18. How does the relation $V = iR$ apply to resistors that do *not* obey Ohm's law?

19. A cow and a man are standing in a meadow when lightning strikes the ground nearby. Why is the cow more likely to be killed than the man? The responsible phenomenon is called "step voltage."

20. The lines in Fig. 7 should be curved slightly. Why?

21. A fuse in an electrical circuit is a wire that is designed to melt, and thereby open the circuit, if the current exceeds a predetermined value. What are some characteristics of ideal fuse wire?

22. Why does an incandescent light bulb grow dimmer with use?

23. The character and quality of our daily lives are influenced greatly by devices that do not obey Ohm's law. What can you say in support of this claim?

24. From a student's paper: "The relationship $R = V/i$ tells us that the resistance of a conductor is directly proportional to the potential difference applied to it." What do you think of this proposition?

25. Carbon has a negative temperature coefficient of resistivity. This means that its resistivity drops as its temperature increases. Would its resistivity disappear entirely at some high enough temperature?

26. What special characteristics must heating wire have?

27. Equation 22 ($P = i^2 R$) seems to suggest that the rate of increase of internal energy in a resistor is reduced if the resistance is made less; Eq. 23 ($P = V^2/R$) seems to suggest just the opposite. How do you reconcile this apparent paradox?

28. Why do electric power companies reduce voltage during times of heavy demand? What is being saved?

29. Is the filament resistance lower or higher in a 500-W light

bulb than in a 100-W bulb? Both bulbs are designed to operate at 120 V.

30. Five wires of the same length and diameter are connected in turn between two points maintained at constant potential difference. Will internal energy be developed at a faster rate in the wire of (a) the smallest or (b) the largest resistance?

31. Why is it better to send 10 MW of electric power long distances at 10 kV rather than at 220 V?

PROBLEMS

Section 32-2 Current Density

1. A current of 4.82 A exists in a 12.4-Ω resistor for 4.60 min. (a) How much charge and (b) how many electrons pass through any cross section of the resistor in this time?

2. The current in the electron beam of a typical video display terminal is 200 μA. How many electrons strike the screen each minute?

3. Suppose that we have 2.10×10^8 doubly charged positive ions per cubic centimeter, all moving north with a speed of 1.40×10^5 m/s. (a) Calculate the current density, in magnitude and direction. (b) Can you calculate the total current in this ion beam? If not, what additional information is needed?

4. A small but measurable current of 123 pA exists in a copper wire whose diameter is 2.46 mm. Calculate (a) the current density and (b) the electron drift speed. See Sample Problem 2.

5. Suppose that the material composing a fuse (see Question 21) melts once the current density rises to 440 A/cm². What diameter of cylindrical wire should be used for the fuse to limit the current to 0.552 A?

6. The (United States) National Electric Code, which sets maximum safe currents for rubber-insulated copper wires of various diameters, is given (in part) below. Plot the safe current density as a function of diameter. Which wire gauge has the maximum safe current density?

Gauge[a]	4	6	8	10	12	14	16	18
Diameter (mils)[b]	204	162	129	102	81	64	51	40
Safe current (A)	70	50	35	25	20	15	6	3

[a] A way of identifying the wire diameter.
[b] 1 mil = 10^{-3} in.

7. A current is established in a gas discharge tube when a sufficiently high potential difference is applied across the two electrodes in the tube. The gas ionizes; electrons move toward the positive terminal and singly charged positive ions toward the negative terminal. What are the magnitude and direction of the current in a hydrogen discharge tube in which 3.1×10^{18} electrons and 1.1×10^{18} protons move past a cross-sectional area of the tube each second?

8. A *pn* junction is formed from two different semiconducting materials in the form of identical cylinders with radius 0.165 mm, as depicted in Fig. 12. In one application 3.50×10^{15} electrons per second flow across the junction from the *n* to the *p* side while 2.25×10^{15} holes per second flow from the *p* to the *n* side. (A hole acts like a particle with charge $+1.6 \times 10^{-19}$ C.) Find (a) the total current and (b) the current density.

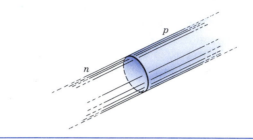

Figure 12   Problem 8.

9. You are given an isolated conducting sphere of 13-cm radius. One wire carries a current of 1.0000020 A into it. Another wire carries a current of 1.0000000 A out of it. How long would it take for the sphere to increase in potential by 980 V?

10. The belt of an electrostatic accelerator is 52.0 cm wide and travels at 28.0 m/s. The belt carries charge into the sphere at a rate corresponding to 95.0 μA. Compute the surface charge density on the belt. See Section 30-11.

11. Near the Earth, the density of protons in the solar wind is 8.70 cm^{-3} and their speed is 470 km/s. (a) Find the current density of these protons. (b) If the Earth's magnetic field did not deflect them, the protons would strike the Earth. What total current would the Earth receive?

12. In a hypothetical fusion research lab, high-temperature helium gas is completely ionized, each helium atom being separated into two free electrons and the remaining positively charged nucleus (alpha particle). An applied electric field causes the alpha particles to drift to the east at 25 m/s while the electrons drift to the west at 88 m/s. The alpha particle density is $2.8 \times 10^{15} \text{ cm}^{-3}$. Calculate the net current density; specify the current direction.

13. How long does it take electrons to get from a car battery to the starting motor? Assume that the current is 115 A and the electrons travel through copper wire with cross-sectional area 31.2 mm² and length 85.5 cm. See Sample Problem 2.

14. A steady beam of alpha particles ($q = 2e$) traveling with kinetic energy 22.4 MeV carries a current of 250 nA. (a) If the beam is directed perpendicular to a plane surface, how many alpha particles strike the surface in 2.90 s? (b) At any instant, how many alpha particles are there in a given 18.0-cm length of the beam? (c) Through what potential difference was it necessary to accelerate each alpha particle from rest to bring it to an energy of 22.4 MeV?

15. In the two intersecting storage rings of circumference 950 m at CERN, protons of kinetic energy 28.0 GeV form beams of current 30.0 A each. (a) Find the total charge carried by the protons in each ring. Assume that the protons travel at the speed of light. (*b*) A beam is deflected out of a ring onto a 43.5-kg copper block. By how much does the temperature of the block rise?

16. (*a*) The current density across a cylindrical conductor of radius *R* varies according to the equation

$$j = j_0(1 - r/R),$$

where *r* is the distance from the axis. Thus the current density is a maximum j_0 at the axis $r = 0$ and decreases linearly to zero at the surface $r = R$. Calculate the current in terms of j_0 and the conductor's cross-sectional area $A = \pi R^2$. (*b*) Suppose that, instead, the current density is a maximum j_0 at the surface and decreases linearly to zero at the axis, so that

$$j = j_0 r/R.$$

Calculate the current. Why is the result different from (*a*)?

Section 32-3 Resistance, Resistivity, and Conductivity

17. A steel trolley-car rail has a cross-sectional area of 56 cm². What is the resistance of 11 km of rail? The resistivity of the steel is $3.0 \times 10^{-7} \ \Omega \cdot m$.

18. A human being can be electrocuted if a current as small as 50 mA passes near the heart. An electrician working with sweaty hands makes good contact with two conductors being held one in each hand. If the electrician's resistance is 1800 Ω, what might the fatal voltage be? (Electricians often work with "live" wires.)

19. A wire 4.0 m long and 6.0 mm in diameter has a resistance of 15 mΩ. A potential difference of 23 V is applied between the ends. (*a*) What is the current in the wire? (*b*) Calculate the current density. (*c*) Calculate the resistivity of the wire material. Can you identify the material? See Table 1.

20. A fluid with resistivity $9.40 \ \Omega \cdot m$ seeps into the space between the plates of a 110-pF parallel-plate air capacitor. When the space is completely filled, what is the resistance between the plates?

21. Show that if changes in the dimensions of a conductor with changing temperature can be ignored, then the resistance varies with temperature according to $R - R_0 = \bar{\alpha} R_0 (T - T_0)$.

22. From the slope of the line in Fig. 5, estimate the average temperature coefficient of resistivity for copper at room temperature and compare with the value given in Table 1.

23. (*a*) At what temperature would the resistance of a copper conductor be double its resistance at 20°C? (Use 20°C as the reference point in Eq. 14; compare your answer with Fig. 5.) (*b*) Does this same temperature hold for all copper conductors, regardless of shape or size?

24. The copper windings of a motor have a resistance of 50 Ω at 20°C when the motor is idle. After running for several hours the resistance rises to 58 Ω. What is the temperature of the windings? Ignore changes in the dimensions of the windings. See Table 1.

25. A 4.0-cm-long caterpillar crawls in the direction of electron drift along a 5.2-mm-diameter bare copper wire that carries a current of 12 A. (*a*) Find the potential difference between the two ends of the caterpillar. (*b*) Is its tail positive or nega-

tive compared to its head? (*c*) How much time could it take the caterpillar to crawl 1.0 cm and still keep up with the drifting electrons in the wire?

26. A coil is formed by winding 250 turns of insulated gauge 8 copper wire (see Problem 6) in a single layer on a cylindrical form whose radius is 12.2 cm. Find the resistance of the coil. Neglect the thickness of the insulation. See Table 1.

27. A wire with a resistance of 6.0 Ω is drawn out through a die so that its new length is three times its original length. Find the resistance of the longer wire, assuming that the resistivity and density of the material are not changed during the drawing process.

28. What must be the diameter of an iron wire if it is to have the same resistance as a copper wire 1.19 mm in diameter, both wires being the same length?

29. Two conductors are made of the same material and have the same length. Conductor A is a solid wire of diameter *D*. Conductor B is a hollow tube of outside diameter 2*D* and inside diameter *D*. Find the resistance ratio, R_A/R_B, measured between their ends.

30. A copper wire and an iron wire of the same length have the same potential difference applied to them. (*a*) What must be the ratio of their radii if the current is to be the same? (*b*) Can the current density be made the same by suitable choices of the radii?

31. An electrical cable consists of 125 strands of fine wire, each having 2.65-μΩ resistance. The same potential difference is applied between the ends of each strand and results in a total current of 750 mA. (*a*) What is the current in each strand? (*b*) What is the applied potential difference? (*c*) What is the resistance of the cable?

32. A common flashlight bulb is rated at 310 mA and 2.90 V, the values of the current and voltage under operating conditions. If the resistance of the bulb filament when cold ($T_0 = 20°C$) is 1.12 Ω, calculate the temperature of the filament when the bulb is on. The filament is made of tungsten. Assume that Eq. 14 holds over the temperature range encountered.

33. When 115 V is applied across a 9.66-m-long wire, the current density is 1.42 A/cm². Calculate the conductivity of the wire material.

34. A block in the shape of a rectangular solid has a cross-sectional area of 3.50 cm², a length of 15.8 cm, and a resistance of 935 Ω. The material of which the block is made has 5.33×10^{22} conduction electrons/m³. A potential difference of 35.8 V is maintained between its ends. (*a*) Find the current in the block. (*b*) Assuming that the current density is uniform, what is its value? Calculate (*c*) the drift velocity of the conduction electrons and (*d*) the electric field in the block.

35. Copper and aluminum are being considered for a high-voltage transmission line that must carry a current of 62.3 A. The resistance per unit length is to be 0.152 Ω/km. Compute for each choice of cable material (*a*) the current density and (*b*) the mass of 1.00 m of the cable. The densities of copper and aluminum are 8960 and 2700 kg/m³, respectively.

36. In the lower atmosphere of the Earth there are negative and positive ions, created by radioactive elements in the soil and

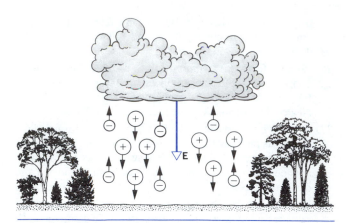

Figure 13   Problem 36.

Figure 14   Problem 40.

cosmic rays from space. In a certain region, the atmospheric electric field strength is 120 V/m, directed vertically down. Due to this field, singly charged positive ions, 620 per cm³, drift downward, and singly charged negative ions, 550 per cm³, drift upward; see Fig. 13. The measured conductivity is $2.70 \times 10^{-14}/\Omega \cdot m$. Calculate (*a*) the ion drift speed, assumed the same for positive and negative ions, and (*b*) the current density.

37. A rod of a certain metal is 1.6 m long and 5.5 mm in diameter. The resistance between its ends (at 20°C) is $1.09 \times 10^{-3}\ \Omega$. A round disk is formed of this same material, 2.14 cm in diameter and 1.35 mm thick. (*a*) What is the material? (*b*) What is the resistance between the opposing round faces, assuming equipotential sufaces?

38. When a metal rod is heated, not only its resistance but also its length and its cross-sectional area change. The relation $R = \rho L/A$ suggests that all three factors should be taken into account in measuring ρ at various temperatures. (*a*) If the temperature changes by 1.0 C°, what fractional changes in R, L, and A occur for a copper conductor? (*b*) What conclusion do you draw? The coefficient of linear expansion is $1.7 \times 10^{-5}/C°$.

39. It is desired to make a long cylindrical conductor whose temperature coefficient of resistivity at 20°C will be close to zero. If such a conductor is made by assembling alternate disks of iron and carbon, find the ratio of the thickness of a carbon disk to that of an iron disk. (For carbon, $\rho = 3500 \times 10^{-8}\ \Omega \cdot m$ and $\alpha = -0.50 \times 10^{-3}/C°$.)

40. A resistor is in the shape of a truncated right circular cone (Fig. 14). The end radii are a and b, and the altitude is L. If the taper is small, we may assume that the current density is uniform across any cross section. (*a*) Calculate the resistance of this object. (*b*) Show that your answer reduces to $\rho L/A$ for the special case of zero taper ($a = b$).

Section 32-4 Ohm's Law

41. For a hypothetical electronic device, the potential difference V in volts, measured across the device, is related to the current i in mA by $V = 3.55\ i^2$. (*a*) Find the resistance when the current is 2.40 mA. (*b*) At what value of the current is the resistance equal to 16.0 Ω?

42. Using data from Fig. 6*b*, plot the resistance of the *pn* junction diode as a function of applied potential difference.

Section 32-5 Ohm's Law: A Microscopic View

43. Calculate the mean free time between collisions for conduction electrons in aluminum at 20°C. Each atom of aluminum contributes three conduction electrons. Take needed data from Table 1 and Appendix D. See also Sample Problem 2.

44. Show that, according to the free-electron model of electrical conduction in metals and classical physics, the resistivity of metals should be proportional to $\sqrt{T}$, where T is absolute temperature. (*Hint*: Treat the electrons as an ideal gas.)

Section 32-6 Energy Transfers in an Electric Circuit

45. A student's 9.0-V, 7.5-W portable radio was left on from 9:00 p.m. until 3:00 a.m. How much charge passed through the wires?

46. The headlights of a moving car draw 9.7 A from the 12-V alternator, which is driven by the engine. Assume the alternator is 82% efficient and calculate the horsepower the engine must supply to run the lights.

47. A space heater, operating from a 120-V line, has a hot resistance of 14.0 Ω. (*a*) At what rate is electrical energy transfered into internal energy? (*b*) At 5.22¢/kW·h, what does it cost to operate the device for 6 h 25 min?

48. The National Board of Fire Underwriters has fixed safe current-carrying capacities for various sizes and types of wire. For #10 rubber-coated copper wire (diameter = 0.10 in.) the maximum safe current is 25 A. At this current, find (*a*) the current density, (*b*) the electric field, (*c*) the potential difference for 1000 ft of wire, and (*d*) the rate at which internal energy is developed for 1000 ft of wire.

49. A 100-W light bulb is plugged into a standard 120-V outlet. (*a*) How much does it cost per month (31 days) to leave the light turned on? Assume electric energy cost 6¢/kW·h. (*b*) What is the resistance of the bulb? (*c*) What is the current in the bulb? (*d*) Is the resistance different when the bulb is turned off?

50. A Nichrome heater dissipates 500 W when the applied potential difference is 110 V and the wire temperature is 800°C. How much power would it dissipate if the wire temperature were held at 200°C by immersion in a bath of

cooling oil? The applied potential difference remains the same; α for Nichrome at 800°C is $4.0 \times 10^{-4}/\text{C}°$.

51. An electron linear accelerator produces a pulsed beam of electrons. The pulse current is 485 mA and the pulse duration is 95.0 ns. (*a*) How many electrons are accelerated per pulse? (*b*) Find the average current for a machine operating at 520 pulses/s. (*c*) If the electrons are accelerated to an energy of 47.7 MeV, what are the values of average and peak power outputs of the accelerator?

52. A cylindrical resistor of radius 5.12 mm and length 1.96 cm is made of material that has a resistivity of 3.50×10^{-5} $\Omega \cdot \text{m}$. What are (*a*) the current density and (*b*) the potential difference when the power dissipation is 1.55 W?

53. A heating element is made by maintaining a potential difference of 75 V along the length of a Nichrome wire with a 2.6 mm² cross section and a resistivity of $5.0 \times 10^{-7} \Omega \cdot \text{m}$. (*a*) If the element dissipates 4.8 kW, what is its length? (*b*) If a potential difference of 110 V is used to obtain the same power output, what should the length be?

54. A coil of current-carrying Nichrome wire is immersed in a liquid contained in a calorimeter. When the potential difference across the coil is 12 V and the current through the coil is 5.2 A, the liquid boils at a steady rate, evaporating at the rate of 21 mg/s. Calculate the heat of vaporization of the liquid.

55. A resistance coil, wired to an external battery, is placed inside an adiabatic cylinder fitted with a frictionless piston and containing an ideal gas. A current $i = 240$ mA flows through the coil, which has a resistance $R = 550$ Ω. At what speed v must the piston, mass $m = 11.8$ kg, move upward in order that the temperature of the gas remains unchanged? See Fig. 15.

56. An electric immersion heater normally takes 93.5 min to bring cold water in a well-insulated container to a certain temperature, after which a thermostat switches the heater off. One day the line voltage is reduced by 6.20% because of a laboratory overload. How long will it now take to heat the water? Assume that the resistance of the heating element is the same for each of these two modes of operation.

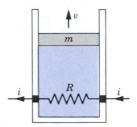

Figure 15 Problem 55.

57. Two isolated conducting spheres, each of radius 14.0 cm, are charged to potentials of 240 and 440 V and are then connected by a fine wire. Calculate the internal energy developed in the wire.

58. The current carried by the electron beam in a particular cathode ray tube is 4.14 mA. The speed of the electrons is 2.82×10^7 m/s and the beam travels a distance of 31.5 cm in reaching the screen. (*a*) How many electrons are in the beam at any instant? (*b*) Find the power dissipated at the screen. (Ignore relativistic effects.)

59. A 420-W immersion heater is placed in a pot containing 2.10 liters of water at 18.5°C. (*a*) How long will it take to bring the water to boiling temperature, assuming that 77.0% of the available energy is absorbed by the water? (*b*) How much longer will it take to boil half the water away?

60. A 32-μF capacitor is connected across a programmed power supply. During the interval from $t = 0$ to $t = 3$ s the output voltage of the supply is given by $V(t) = 6 + 4t - 2t^2$ volts. At $t = 0.50$ s find (*a*) the charge on the capacitor, (*b*) the current into the capacitor, and (*c*) the power output from the power supply.

61. A potential difference V is applied to a wire of cross-sectional area A, length L, and conductivity σ. You want to change the applied potential difference and draw out the wire so the power dissipated is increased by a factor of 30 and the current is increased by a factor of 4. What should be the new values of the (*a*) length and (*b*) cross-sectional area?

CHAPTER 33

DC CIRCUITS

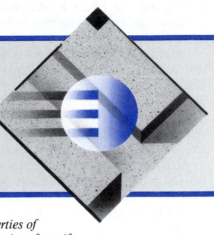

In the previous chapter, we discussed some general properties of current and resistance. In this chapter, we begin to study the behavior of specific electric circuits that include resistive elements, which may be individual resistors or may be the internal resistances of circuit elements such as batteries or wires.

*We confine our study in this chapter to **direct current** (DC) circuits, in which the direction of the current does not change with time. In DC circuits that contain only batteries and resistors, the magnitude of the current does not vary with time, while in DC circuits containing capacitors, the magnitude of the current may be time dependent. Alternating current (AC) circuits, in which the current changes direction periodically, are considered in Chapter 39.*

33-1 ELECTROMOTIVE FORCE

An external energy source is required by most electrical circuits to move charge through the circuit. The circuit therefore must include a device that maintains a potential difference between two points in the circuit, just as a circulating fluid requires an analogous device (a pump) that maintains a *pressure* difference between two points.

Any device that performs this task in an electrical circuit is called a source (or a seat) of *electromotive force* (symbol $\mathscr{E}$; abbreviation emf). It is sometimes useful to consider a seat of emf as a mechanism that creates a "hill" of potential and moves charge "uphill," from which it flows "downhill" through the rest of the circuit. A common seat of emf is the ordinary battery; another is the electric generator found in power plants. Solar cells are sources of emf used both in spacecraft and in pocket calculators. Other less commonly found sources of emf are fuel cells (used to power the space shuttle) and thermopiles. Biological systems, including the human heart, also function as seats of emf.

Figure 1a shows a seat of emf $\mathscr{E}$, which we can consider to be a battery, connected to a resistor R. The seat of emf maintains its upper terminal at a high potential and its lower terminal at a low potential, as indicated by the + and − signs. In the external circuit, positive charge carriers would be driven in the direction shown by the arrows marked i. In other words, a clockwise current is set up in the circuit of Fig. 1a.

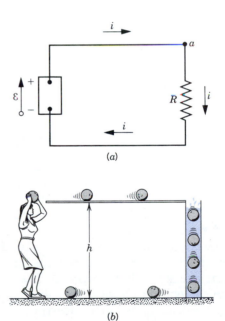

Figure 1 (*a*) A simple electric circuit, in which the emf $\mathscr{E}$ does work on the charge carriers and maintains a steady current through the resistor. (*b*) A gravitational analogue, in which work done by the person maintains a steady flow of bowling balls through the viscous medium.

An emf is represented by an arrow that is placed next to the seat and points in the direction in which the emf, acting alone, would cause a positive charge carrier to move in the external circuit. We draw a small circle on the tail of an emf arrow so that we will not confuse it with a current arrow.

A seat of emf must be able to do work on charge carriers that enter it. In its interior, the seat acts to move positive charges from a point of low potential (the negative terminal) through the seat to a point of high potential (the positive terminal). The charges then move through the external circuit, dissipating energy in the process, and return to the negative terminal, from which the emf raises them to the positive terminal again, and the cycle continues. (Note that, in accordance with our usual convention, we analyze the circuit as if positive charge were flowing. The actual motion of the electrons is in the opposite direction.)

When a steady current has been established in the circuit of Fig. 1a, a charge *dq* passes through *any* cross section of the circuit in time *dt*. In particular, this charge enters the seat of emf $\mathcal{E}$ at its low-potential end and leaves at its high-potential end. The seat must do an amount of work *dW* on the (positive) charge carriers to force them to go to the point of higher potential. The emf $\mathcal{E}$ of the seat is defined as the work per unit charge, or

$$\mathcal{E} = dW/dq. \tag{1}$$

The unit of emf is the joule/coulomb, which is the *volt* (abbreviation V):

$$1 \text{ volt} = 1 \text{ joule/coulomb.}$$

Note from Eq. 1 that the electromotive force is not actually a force; that is, we do not measure it in newtons. The name originates from the early history of the subject.

The work done by a seat of emf on charge carriers in its interior must be derived from a source of energy within the seat. The energy source may be chemical (as in a battery or a fuel cell), mechanical (a generator), thermal (a thermopile), or radiant (a solar cell). We can describe a seat of emf as a device in which some other form of energy is changed into electrical energy. The energy provided by the source of emf in Fig. 1a is stored in the electric and the magnetic* fields that surround the circuit. This stored energy does not increase because it is converted to internal energy in the resistor and dissipated as Joule heating, at the same rate at which it is supplied. The electric and magnetic fields play the role of intermediary in the energy transfer process, acting as storage reservoirs.

Figure 1b shows a gravitational analogue of Fig. 1a. In the top figure the seat of emf does work on the charge carriers. This energy, stored in passage as electromagnetic field energy, appears eventually as internal energy in resistor *R*. In the lower figure the person, in lifting the bowl-

ing balls from the floor to the shelf, does work on them. This energy is stored, in passage, as gravitational field energy. The balls roll slowly and uniformly along the shelf, dropping from the right end into a cylinder full of viscous oil. They sink to the bottom at an essentially constant speed, are removed by a trapdoor mechanism not shown, and roll back along the floor to the left. The energy put into the system by the person appears eventually as internal energy in the viscous fluid, resulting in a temperature rise. The energy supplied by the person comes from internal (chemical) energy. The circulation of charges in Fig. 1a stops eventually if the source of emf runs out of energy; the circulation of bowling balls in Fig. 1b stops eventually if the person runs out of energy.

Figure 2a shows a circuit containing two ideal (resistanceless) batteries A and B, a resistor of resistance *R*, and an ideal electric motor M employed in lifting a weight. The batteries are connected so that they tend to send charges around the circuit in opposite directions; the actual direction of the current is determined by battery B, which has the larger emf. Figure 2b shows the energy transfers in this circuit. The chemical energy in battery B is steadily depleted, the energy appearing in the three forms shown on the right. Battery A is being charged while battery B is being discharged. Again, the electric and magnetic fields that surround the circuit act as an intermediary.

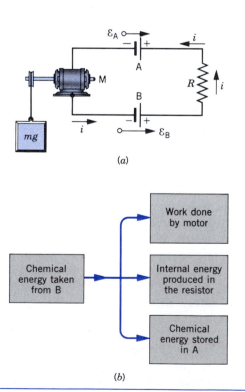

(a)

(b)

Figure 2   (a) $\mathcal{E}_B > \mathcal{E}_A$, so that battery B determines the direction of the current in this single-loop circuit. (b) Energy transfers in this circuit.

---

* A current in a wire is surrounded by a magnetic field, and this field, like the electric field, can also be viewed as a site of stored energy (see Section 38-4).

Reversibility *(Optional)*

It is part of the definition of an ideal emf that the energy transfer process be *reversible,* at least in principle. Recall that a reversible process is one that passes through equilibrium states; its course can be reversed by making an infinitesimal change in the environment of the system (see Section 26-1). A battery, for example, can be either charged or discharged; a generator can be driven mechanically, producing electrical energy, or it can be operated backward as a motor. The (reversible) energy transfers here are

$$\text{electrical} \rightleftarrows \text{chemical}$$

and

$$\text{electrical} \rightleftarrows \text{mechanical}.$$

The energy transfer from electrical energy to internal energy is not reversible. We can easily raise the temperature of a conductor by supplying electric energy to it, but it is *not* possible to set up a current in a closed copper loop by raising its temperature uniformly. Because of this lack of reversibility, we do not associate an emf with the Joule effect, that is, with energy transfers associated with Joule heating in the wires or circuit elements. ■

33-2 CALCULATING THE CURRENT IN A SINGLE LOOP

Consider a single-loop circuit, such as that of Fig. 1*a,* containing one seat of emf $\mathcal{E}$ and one resistor R. In a time dt an amount of energy given by $i^2R\,dt$ appears in the resistor as internal energy (see Eq. 22 of Chapter 32). During this same time a charge $dq\,(= i\,dt)$ moves through the seat of emf, and the seat does work on this charge (see Eq. 1) given by

$$dW = \mathcal{E}\,dq = \mathcal{E}i\,dt.$$

From the conservation of energy principle, the work done by the seat must equal the internal energy deposited in the resistor, or

$$\mathcal{E}i\,dt = i^2R\,dt.$$

Solving for i, we obtain

$$i = \mathcal{E}/R. \qquad (2)$$

We can also derive Eq. 2 by considering that, if electric potential is to have any meaning, a given point can have only one value of potential at any given time. If we start at any point in the circuit of Fig. 1*a* and go around the circuit in either direction, adding up algebraically the changes in potential that we encounter, we must find the same potential when we return to our starting point. We summarize this rule as follows:

The algebraic sum of the changes in potential encountered in a complete traversal of any closed circuit is zero.

This statement is called *Kirchhoff's second rule*; for brevity we call it the *loop rule.* This rule is a particular way of

stating the law of conservation of energy for a charge carrier traveling a closed circuit.

In Fig. 1*a* let us start at point *a,* whose potential is V_a, and traverse the circuit clockwise. (The numerical value of V_a is not important because, as in most electric circuit situations, we are concerned here with *differences* of potential.) In going through the resistor, there is a change in potential of $-iR$. The minus sign shows that the top of the resistor is higher in potential than the bottom, which must be true, because positive charge carriers move of their own accord from high to low potential. As we traverse the battery from bottom to top, there is an *increase* of potential equal to $+\mathcal{E}$, because the battery does (positive) work on the charge carriers; that is, it moves them from a point of low potential to one of high potential. Adding the algebraic sum of the changes in potential to the initial potential V_a must yield the identical final value V_a, or

$$V_a - iR + \mathcal{E} = V_a.$$

We write this as

$$-iR + \mathcal{E} = 0,$$

which is independent of the value of V_a and which asserts explicitly that the algebraic sum of the potential changes for a complete circuit traversal is zero. This relation leads directly to Eq. 2.

These two ways to find the current in single-loop circuits, one based on the conservation of energy and the other on the concept of potential, are completely equivalent because potential differences are defined in terms of work and energy (see Section 30-3).

To prepare for the study of more complex circuits, let us examine the rules for finding potential differences; these rules follow from the previous discussion. They are not meant to be memorized but rather to be so thoroughly understood that it becomes trivial to re-derive them on each application.

1. If a resistor is traversed in the direction of the current, the change in potential is $-iR$; in the opposite direction it is $+iR$.

2. If a seat of emf is traversed in the direction of the emf (the direction of the arrow, or from the negative terminal to the positive terminal), the change in potential is $+\mathcal{E}$; in the opposite direction it is $-\mathcal{E}$.

Finally, keep in mind that we are always referring to the direction of the current as the direction of flow of positive charges, opposite to the actual direction of flow of the electrons.

Internal Resistance of a Seat of emf

Figure 3*a* shows a single-loop circuit, which emphasizes that all seats of emf have an intrinsic *internal resistance r.* This resistance cannot be removed — although we would usually like to do so — because it is an inherent part of the

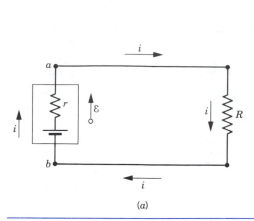

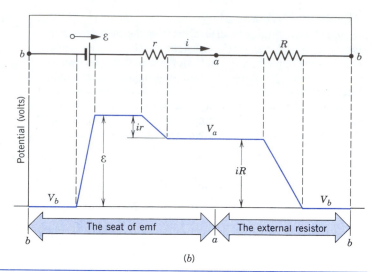

Figure 3 (a) A single-loop circuit, containing a seat of emf having internal resistance r. (b) The circuit is drawn with the components along a straight line along the top. The potential changes encountered in traveling clockwise around the circuit starting at point b are shown at the bottom.

device. The figure shows the internal resistance r and the emf separately, although they actually occupy the same region of space.

We can apply the loop rule starting at any point in the circuit. Starting at b and going around clockwise, we obtain

$$V_b + \mathcal{E} - ir - iR = V_b$$

or

$$+\mathcal{E} - ir - iR = 0.$$

Compare these equations with Fig. 3b, which shows the changes in potential graphically. In writing these equations, note that we traversed r and R in the direction of the current and $\mathcal{E}$ in the direction of the emf. The same equation follows if we start at any other point in the circuit or if we traverse the circuit in a counterclockwise direction. Solving for i gives

$$i = \frac{\mathcal{E}}{R + r}. \qquad (3)$$

Note that the internal resistance r *reduces* the current that the emf can supply to the external circuit.

33-3 POTENTIAL DIFFERENCES

We often want to find the potential difference between two points in a circuit. In Fig. 3a, for example, how does the potential difference $V_{ab} (= V_a - V_b)$ between points b and a depend on the fixed circuit parameters $\mathcal{E}$, r, and R? To find their relationship, let us start at point b and traverse the circuit counterclockwise to point a, passing

through resistor R. If V_a and V_b are the potentials at a and b, respectively, we have

$$V_b + iR = V_a$$

because we experience an increase in potential in traversing a resistor in the direction opposite to the current. We rewrite this relation in terms of V_{ab}, the potential difference between a and b, as

$$V_{ab} = V_a - V_b = +iR,$$

which tells us that V_{ab} has magnitude iR and that point a is more positive than point b. Combining this last equation with Eq. 3 yields

$$V_{ab} = \mathcal{E}\frac{R}{R + r}. \qquad (4)$$

In summary, to find the potential difference between any two points in a circuit, start at one point, travel through the circuit to the other, and add algebraically the changes in potential encountered. This algebraic sum is the potential difference between the points. This procedure is similar to that for finding the current in a closed loop, except that here the potential differences are added over part of a loop and not over the whole loop.

You can travel *any* path through the circuit between the two points, and you get the same value of the potential difference, because *path independence* is an essential part of our concept of potential. The potential difference between any two points can have only one value; we must obtain the same result for all paths that connect those points. (Similarly, if we consider two points on the side of a hill, the measured difference in gravitational potential between them is the same no matter what path is followed in going from one to the other.) In Fig. 3a let us recalculate

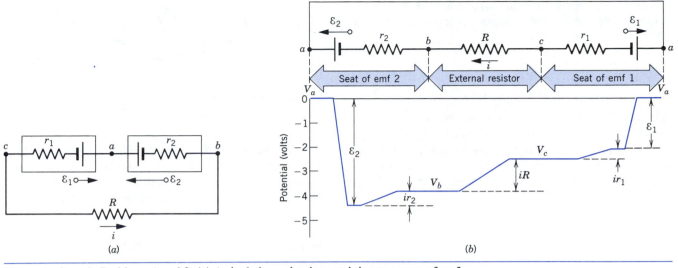

Figure 4   Sample Problems 1 and 2. (*a*) A single-loop circuit containing two seats of emf.
(*b*) The changes in potential encountered in traveling clockwise around the circuit starting at
point *a*.

V_{ab}, using a path starting at *a* and going counterclockwise
through the seat of emf. (This is equivalent to starting at *a*
in Fig. 3*b* and moving to the left toward point *b*.) We have

$$V_a + ir - \mathcal{E} = V_b$$

or

$$V_{ab} = V_a - V_b = +\mathcal{E} - ir.$$

Combining this result with Eq. 3 leads to Eq. 4.

The quantity V_{ab} is the potential difference across the
battery terminals. We see from Eq. 4 that V_{ab} is equal to $\mathcal{E}$
only if either the battery has no internal resistance ($r = 0$)
or the external circuit is open ($R = \infty$).

Sample Problem 1   What is the current in the circuit of Fig. 4*a*?
The emfs and the resistors have the following values:

$$\mathcal{E}_1 = 2.1 \text{ V}, \qquad \mathcal{E}_2 = 4.4 \text{ V},$$

$$r_1 = 1.8 \text{ }\Omega, \qquad r_2 = 2.3 \text{ }\Omega, \qquad R = 5.5 \text{ }\Omega.$$

Solution   The two emfs are connected so that they oppose each
other but $\mathcal{E}_2$, because it is larger than $\mathcal{E}_1$, controls the direction
of the current in the circuit, which is counterclockwise. The loop
rule, applied clockwise from point *a*, yields

$$-\mathcal{E}_2 + ir_2 + iR + ir_1 + \mathcal{E}_1 = 0.$$

Check that this same equation results by going around counterclockwise or by starting at some point other than *a*. Also, compare this equation term by term with Fig. 4*b*, which shows the
potential changes graphically.

Solving for the current *i*, we obtain

$$i = \frac{\mathcal{E}_2 - \mathcal{E}_1}{R + r_1 + r_2} = \frac{4.4 \text{ V} - 2.1 \text{ V}}{5.5 \text{ }\Omega + 1.8 \text{ }\Omega + 2.3 \text{ }\Omega} = 0.24 \text{ A}.$$

It is not necessary to know the direction of the current in ad-

vance. To show this, let us assume that the current in Fig. 4*a* is
clockwise, that is, opposite to the direction of the current arrow
in Fig. 4*a*. The loop rule would then yield (going clockwise from
a)

$$-\mathcal{E}_2 - ir_2 - iR - ir_1 + \mathcal{E}_1 = 0$$

or

$$i = -\frac{\mathcal{E}_2 - \mathcal{E}_1}{R + r_1 + r_2}.$$

Substituting numerical values yields $i = -0.24$ A for the
current. The minus sign is a signal that the current is in the
opposite direction from that which we have assumed.

In more complex circuits involving many loops and branches,
it is often impossible to know in advance the actual directions for
the currents in all parts of the circuit. However, the current
directions for each branch may be chosen arbitrarily. If you get
an answer with a positive sign for a particular current, you have
chosen its direction correctly; if you get a negative sign, the
current is opposite in direction to that chosen. In either case, the
numerical value is correct.

Sample Problem 2   (*a*) What is the potential difference between points *a* and *b* in Fig. 4*a*? (*b*) What is the potential difference between points *a* and *c* in Fig. 4*a*?

Solution   (*a*) This potential difference is the terminal potential
difference of battery 2, which includes emf $\mathcal{E}_2$ and internal resistance r_2. Let us start at point *b* and traverse the circuit counterclockwise to point *a*, passing directly through the seat of emf.
We find

$$V_b - ir_2 + \mathcal{E}_2 = V_a$$

or

$$V_a - V_b = -ir_2 + \mathcal{E}_2 = -(0.24 \text{ A})(2.3 \text{ }\Omega) + 4.4 \text{ V} = +3.8 \text{ V}.$$

We see that *a* is more positive than *b* and the potential difference
between them (3.8 V) is *smaller* than the emf (4.4 V); see Fig.
4*b*.

We can verify this result by starting at point b in Fig. 4a and traversing the circuit clockwise to point a. For this different path we find

$$V_b + iR + ir_1 + \mathcal{E}_1 = V_a$$

or

$$V_a - V_b = iR + ir_1 + \mathcal{E}_1$$
$$= (0.24 \text{ A})(5.5 \text{ }\Omega + 1.8 \text{ }\Omega) + 2.1 \text{ V} = +3.8 \text{ V},$$

exactly as before. The potential difference between two points has the same value for all paths connecting those points.

(b) Note that the potential difference between a and c is the terminal potential difference of battery 1, consisting of emf $\mathcal{E}_1$ and internal resistance r_1. Let us start at c and traverse the circuit clockwise to point a. We find

$$V_c + ir_1 + \mathcal{E}_1 = V_a$$

or

$$V_a - V_c = ir_1 + \mathcal{E}_1 = (0.24 \text{ A})(1.8 \text{ }\Omega) + 2.1 \text{ V} = +2.5 \text{ V}.$$

This tells us that a is at a higher potential than c. The terminal potential difference (2.5 V) is in this case *larger* than the emf (2.1 V); see Fig. 4b. Charge is being forced through $\mathcal{E}_1$ in a direction opposite to that in which it would send charge if it were acting by itself; if $\mathcal{E}_1$ were a storage battery it would be charging at the expense of $\mathcal{E}_2$.

33-4   RESISTORS IN SERIES AND PARALLEL

Just as was the case with capacitors (see Section 31-3), resistors often occur in circuits in various combinations. In analyzing such circuits, it is helpful to replace the combination of resistors with a single *equivalent resistance* R_{eq} whose value is chosen such that the operation of the circuit is unchanged. We consider two ways that resistors can be combined.

Resistors Connected in Parallel

Recall our definition of a parallel combination of circuit elements in Section 31-3: we can travel through the combination by crossing *only one* of the elements, the same potential difference V appears across each element, and the flow of charge is shared among the elements.

Figure 5 shows two resistors connected in parallel. We seek the equivalent resistance between points a and b. Let us assume we connect a battery (or other source of emf) that maintains a potential difference V between points a and b. The potential difference across each resistor is V. The current through each of the resistors is, from Eq. 2,

$$i_1 = V/R_1 \quad \text{and} \quad i_2 = V/R_2. \tag{5}$$

According to the properties of a parallel circuit, the total current i must be shared among the branches, so

$$i = i_1 + i_2. \tag{6}$$

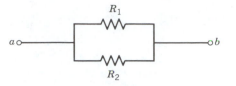

Figure 5   Two resistors in parallel.

If we were to replace the parallel combination by a single equivalent resistance R_{eq}, the same total current i must flow (because the replacement must not change the operation of the circuit). The current is then

$$i = V/R_{eq}. \tag{7}$$

Substituting Eqs. 5 and 7 into Eq. 6, we obtain

$$\frac{V}{R_{eq}} = \frac{V}{R_1} + \frac{V}{R_2}$$

or

$$\frac{1}{R_{eq}} = \frac{1}{R_1} + \frac{1}{R_2}. \tag{8}$$

To find the equivalent resistance of a parallel combination of more than two resistors, we first find the equivalent resistance R_{12} of R_1 and R_2 using Eq. 8. We then find the equivalent resistance of R_{12} and the next parallel resistance, R_3, again using Eq. 8. Continuing in this way, we obtain a general expression for the equivalent resistance of a parallel combination of any number of resistors,

$$\frac{1}{R_{eq}} = \sum_n \frac{1}{R_n} \quad \text{(parallel combination).} \tag{9}$$

That is, to find the equivalent resistance of a parallel combination, add the reciprocals of the individual resistances and take the reciprocal of the resulting sum. Note that R_{eq} is always *smaller than* the smallest resistance in the parallel combination — by adding more paths for the current, we get more current for the same potential difference.

In the special case of two resistors in parallel, Eq. 8 can be written

$$R_{eq} = \frac{R_1 R_2}{R_1 + R_2}, \tag{10}$$

or as the product of the two resistances divided by their sum.

Resistors Connected in Series

Figure 6 shows two resistors connected in series. Recall the properties of a series combination of circuit elements (see Section 31-3): to travel through the combination, we must travel through *all* the elements in succession, a battery connected across the combination gives (in general) a different potential drop across each element, and the same current is maintained in each element.

Suppose a battery of potential difference V is connected

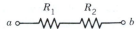

Figure 6   Two resistors in series.

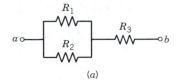

(a)

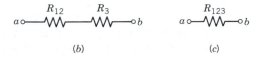

(b)   (c)

Figure 7   Sample Problem 3. (a) The parallel combination of R_1 and R_2 is in series with R_3. (b) The parallel combination of R_1 and R_2 has been replaced by its equivalent resistance, R_{12}. (c) The series combination of R_{12} and R_3 has been replaced by its equivalent resistance, R_{123}.

across points a and b in Fig. 6. A current i is set up in the combination and in each of the resistors. The potential differences across the resistors are

$$V_1 = iR_1 \quad \text{and} \quad V_2 = iR_2. \tag{11}$$

The sum of these potential differences must give the potential difference across points a and b maintained by the battery, or

$$V = V_1 + V_2. \tag{12}$$

If we replaced the combination by its equivalent resistance R_{eq}, the same current i would be established, and so

$$V = iR_{eq}. \tag{13}$$

Combining Eqs. 11, 12, and 13, we obtain

$$iR_{eq} = iR_1 + iR_2,$$

or

$$R_{eq} = R_1 + R_2. \tag{14}$$

Extending this result to a series combination of any number of resistors, we obtain

$$R_{eq} = \sum_n R_n \quad \text{(series combination)}. \tag{15}$$

That is, to find the equivalent resistance of a series combination, find the algebraic sum of the individual resistors. Note that the equivalent resistance of a series combination is always *larger* than the largest resistance in the series—adding more resistors in series means we get less current for the same potential difference.

Comparing these results with Eqs. 19 and 24 of Chapter 31 for the series and parallel combinations of capacitors, we see that resistors in parallel add like capacitors in series, and resistors in series add like capacitors in parallel. This has to do with the different way the two quantities are defined, resistance being potential/current and capacitance being charge/potential.

Occasionally, resistors may appear in combinations that are neither parallel nor series. In such a case, the equivalent resistance can sometimes be found by breaking the problem into smaller units that can be regarded as series or parallel connections.

Sample Problem 3   (a) Find the equivalent resistance of the combination shown in Fig. 7a, using the values $R_1 = 4.6\ \Omega$, $R_2 = 3.5\ \Omega$, and $R_3 = 2.8\ \Omega$. (b) What is the value of the current through R_1 when a 12.0-V battery is connected across points a and b?

Solution   (a) We first find the equivalent resistance R_{12} of the parallel combination of R_1 and R_2. Using Eq. 10 we obtain

$$R_{12} = \frac{R_1 R_2}{R_1 + R_2} = \frac{(4.6\ \Omega)(3.5\ \Omega)}{4.6\ \Omega + 3.5\ \Omega} = 2.0\ \Omega.$$

R_{12} and R_3 are in series, as shown in Fig. 7b. Using Eq. 14, we can find the equivalent resistance R_{123} of this series combination, which is the equivalent resistance of the entire original combination:

$$R_{123} = R_{12} + R_3 = 2.0\ \Omega + 2.8\ \Omega = 4.8\ \Omega.$$

(b) With a 12.0-V battery connected across points a and b in Fig. 7c, the resulting current is

$$i = \frac{V}{R_{123}} = \frac{12.0\ \text{V}}{4.8\ \Omega} = 2.5\ \text{A}.$$

With this current in the series combination in Fig. 7b, the potential difference across R_{12} is

$$V_{12} = iR_{12} = (2.5\ \text{A})(2.0\ \Omega) = 5.0\ \text{V}.$$

In a parallel combination, the same potential difference appears across each element (and across their combination). The potential difference across R_1 (and R_2) is therefore 5.0 V, and the current through R_1 is

$$i_1 = \frac{V_{12}}{R_1} = \frac{5.0\ \text{V}}{4.6\ \Omega} = 1.1\ \text{A}.$$

Sample Problem 4   Figure 8a shows a cube made of 12 resistors, each of resistance R. Find R_{12}, the equivalent resistance across a cube edge.

Solution   Although this problem at first looks hopeless to divide into series and parallel subunits, the symmetry of the connections suggests a way to do so. The key is the realization that, from considerations of symmetry alone, points 3 and 6 must be at the same potential. So must points 4 and 5.

If two points in a circuit have the same potential, the currents in the circuit do not change if you connect these points by a wire. There is no current in the wire because there is no potential difference between its ends. Points 3 and 6 may therefore be connected by a wire, and similarly points 4 and 5 may be connected.

This allows us to redraw the cube as in Fig. 8b. From this point, it is simply a matter of reducing the circuit between the input terminals to a single resistor, using the rules for resistors in series and in parallel. In Fig. 8c, we make a start by replacing five parallel combinations of two resistors by their equivalents, each of resistance $\frac{1}{2}R$.

In Fig. 8d, we have added the three resistors that are in series in the right-hand loop, obtaining a single equivalent resistance of 2R. In Fig. 8e, we have replaced the two resistors that now form the right-hand loop by a single equivalent resistor $\frac{2}{3}R$. In so doing, it is useful to recall that the equivalent resistance of two resistors in parallel is equal to their product divided by their sum (see Eq. 10).

In Fig. 8f, we have added the three series resistors of Fig. 8e, obtaining $\frac{7}{5}R$, and in Fig. 8g we have reduced this parallel combination to the single equivalent resistance that we seek, namely,

$$R_{12} = \tfrac{7}{12}R.$$

You can also use these methods to find R_{13}, the equivalent resistance of a cube across a face diagonal, and R_{17}, the equivalent resistance across a body diagonal (see Problem 29).

33-5   MULTILOOP CIRCUITS

Figure 9 shows a circuit containing more than one loop. For simplicity, we have neglected the internal resistances of the batteries. When we analyze such circuits, it is helpful to consider their *junctions* and *branches*. A junction in a multiloop circuit such as that of Fig. 9 is a point in the circuit at which three or more wire segments meet. There are two junctions in the circuit of Fig. 9, at b and d. (Points a and c in Fig. 9 are *not* junctions, because only two wire segments meet at those points.)

A branch is any circuit path that starts on one junction and proceeds along the circuit to the next junction. There are three branches in the circuit of Fig. 9; that is, there are three paths that connect junctions b and d—the left branch bad, the right branch bcd, and the central branch bd.

In single-loop circuits, such as those of Figs. 3 and 4, there is only one current to determine. In multiloop circuits, however, each branch has its own individual current, which must be determined in our analysis of the circuit. In the circuit of Fig. 9, the three (unknown) currents are labeled i_1 (for branch bad), i_2 (for branch bcd), and i_3 (for branch bd). The directions of the currents have been chosen arbitrarily. If you look carefully, you will note that i_3 must point in a direction opposite to the one we have shown. We have deliberately drawn it in wrong to show how the formal mathematical procedures always correct such incorrect guesses.

Note that we cannot analyze the circuit of Fig. 9 in terms of series or parallel collections of resistors. Review the criteria that defined series and parallel combinations, and you will conclude that it is not possible to consider

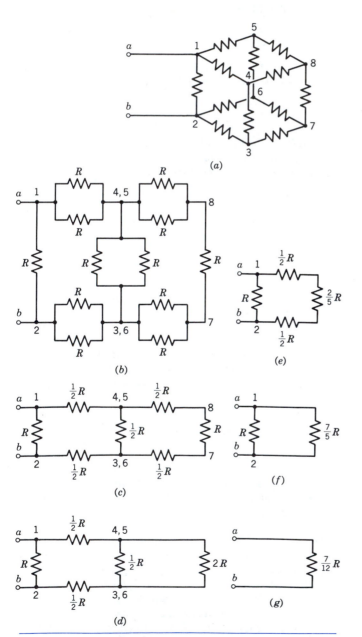

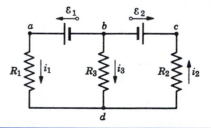

Figure 8   Sample Problem 4. (a) A cube formed of 12 identical resistors. (b)–(g) The step-by-step reduction of the cube to a single equivalent resistance.

Figure 9   A two-loop circuit. Given the emfs and resistances, we would like to find the three currents.

any combination of R_1, R_2, and R_3 as being in series or parallel.

The three currents i_1, i_2, and i_3 carry charge either toward junction d or away from it. Charge does not collect at junction d because the circuit is in a steady-state condition; charge must be removed from the junction by the currents at the same rate that charge is brought into the junction. At junction d in Fig. 9, the total rate at which charge enters the junction is given by $i_1 + i_3$, and the rate at which charge leaves is given by i_2. Equating the currents entering and leaving the junction, we obtain

$$i_1 + i_3 = i_2. \tag{16}$$

This equation suggests a general principle for the solution of multiloop circuits:

At any junction the sum of currents leaving the junction (those with arrows pointing away from the junction) equals the sum of currents entering the junction (those with arrows pointing toward the junction).

This *junction rule*, which is also known as *Kirchhoff's first rule*, is simply a statement of the conservation of charge. Our basic tools for analyzing circuits are (1) the conservation of energy (the loop rule — see Section 33-2) and (2) the conservation of charge (the junction rule).

For the circuit of Fig. 9, the junction rule yields only one relationship among the three unknowns. Applying the rule at junction b leads to exactly the same equation, as you can easily verify. To solve for the three unknowns, we need two more independent equations; they can be found from the loop rule.

In single-loop circuits there is only one conducting loop around which to apply the loop rule, and the current is the same in all parts of this loop. In multiloop circuits there is more than one loop, and the current in general is *not* the same in all parts of any given loop.

If we traverse the left loop of Fig. 9 in a counterclockwise direction starting and ending at point b, the loop rule gives

$$\mathscr{E}_1 - i_1 R_1 + i_3 R_3 = 0. \tag{17}$$

The right loop gives (again going counterclockwise from b)

$$-i_3 R_3 - i_2 R_2 - \mathscr{E}_2 = 0. \tag{18}$$

These two equations, together with the relation derived earlier with the junction rule (Eq. 16), are the three simultaneous equations needed to solve for the unknowns i_1, i_2, and i_3. Solving (you should supply the missing steps), we find

$$i_1 = \frac{\mathscr{E}_1(R_2 + R_3) - \mathscr{E}_2 R_3}{R_1 R_2 + R_2 R_3 + R_1 R_3}, \tag{19}$$

$$i_2 = \frac{\mathscr{E}_1 R_3 - \mathscr{E}_2(R_1 + R_3)}{R_1 R_2 + R_2 R_3 + R_1 R_3}, \tag{20}$$

$$i_3 = \frac{-\mathscr{E}_1 R_2 - \mathscr{E}_2 R_1}{R_1 R_2 + R_2 R_3 + R_1 R_3}. \tag{21}$$

Equation 21 shows that no matter what numerical values are given to the emfs and to the resistances, the current i_3 always has a negative value. This means that it always points up in Fig. 9 rather than down, as we assumed. The currents i_1 and i_2 may be in either direction, depending on the numerical values of the emfs and the resistances.

To check these results, verify that Eqs. 19–21 reduce to sensible conclusions in special cases. For $R_3 = \infty$, for example, we find

$$i_1 = i_2 = \frac{\mathscr{E}_1 - \mathscr{E}_2}{R_1 + R_2} \quad \text{and} \quad i_3 = 0.$$

What do these equations reduce to for $R_2 = \infty$?

The loop theorem can be applied to the large loop consisting of the entire circuit *abcda* of Fig. 9. This fact might suggest that there are more equations than we need, for there are only three unknowns, and we already have three equations written in terms of them. However, the loop rule yields for this loop

$$-i_1 R_1 - i_2 R_2 - \mathscr{E}_2 + \mathscr{E}_1 = 0,$$

which is nothing more than the sum of Eqs. 17 and 18. The large loop does not yield another *independent* equation. For multiloop circuits the number of independent equations must equal the number of branches (or the number of different currents). The number of independent *junction* equations is one less than the number of junctions (one equation in the case of the circuit of Fig. 9, which has two junctions). The remaining equations must be *loop* equations.

Sample Problem 5 Figure 10 shows a circuit whose elements have the following values:

$$\mathscr{E}_1 = 2.1 \text{ V}, \qquad \mathscr{E}_2 = 6.3 \text{ V},$$
$$R_1 = 1.7 \ \Omega, \qquad R_2 = 3.5 \ \Omega.$$

Find the currents in the three branches of the circuit.

Solution Let us draw and label the currents as shown in the figure, choosing the current directions arbitrarily. Applying the junction rule at a, we find

$$i_1 + i_2 = i_3. \tag{22}$$

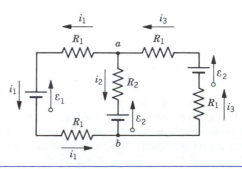

Figure 10 Sample Problems 5 and 6. A two-loop circuit.

Now let us start at point a and traverse the left-hand loop in a counterclockwise direction. We find

$$-i_1R_1 - \mathcal{E}_1 - i_1R_1 + \mathcal{E}_2 + i_2R_2 = 0$$

or

$$2i_1R_1 - i_2R_2 = \mathcal{E}_2 - \mathcal{E}_1. \qquad (23)$$

If we traverse the right-hand loop in a clockwise direction from point a, we find

$$+i_3R_1 - \mathcal{E}_2 + i_3R_1 + \mathcal{E}_2 + i_2R_2 = 0$$

or

$$i_2R_2 + 2i_3R_1 = 0. \qquad (24)$$

Equations 22, 23, and 24 are three independent simultaneous equations involving the three variables i_1, i_2, and i_3. We can solve these equations for these variables, obtaining, after a little algebra,

$$i_1 = \frac{(\mathcal{E}_2 - \mathcal{E}_1)(2R_1 + R_2)}{4R_1(R_1 + R_2)}$$

$$= \frac{(6.3 \text{ V} - 2.1 \text{ V})(2 \times 1.7 \ \Omega + 3.5 \ \Omega)}{(4)(1.7 \ \Omega)(1.7 \ \Omega + 3.5 \ \Omega)} = 0.82 \text{ A},$$

$$i_2 = -\frac{\mathcal{E}_2 - \mathcal{E}_1}{2(R_1 + R_2)}$$

$$= -\frac{6.3 \text{ V} - 2.1 \text{ V}}{(2)(1.7 \ \Omega + 3.5 \ \Omega)} = -0.40 \text{ A},$$

$$i_3 = \frac{(\mathcal{E}_2 - \mathcal{E}_1)R_2}{4R_1(R_1 + R_2)}$$

$$= \frac{(6.3 \text{ V} - 2.1 \text{ V})(3.5 \ \Omega)}{(4)(1.7 \ \Omega)(1.7 \ \Omega + 3.5 \ \Omega)} = 0.42 \text{ A}.$$

The signs of the currents tell us that we have guessed correctly about the directions of i_1 and i_3 but that we are wrong about the direction of i_2; it should point up—and not down—in the central branch of the circuit of Fig. 10.

Note that, having discovered that current i_2 is pointing in the wrong direction, we do not need to change it in Fig. 10. We can leave it in the figure as it is, as long as we remember to substitute a negative numerical value for i_2 in all subsequent calculations involving that current.

Sample Problem 6   What is the potential difference between points a and b in the circuit of Fig. 10?

Solution   For the potential difference between a and b, we have, traversing branch ab in Fig. 10 and assuming the current directions shown,

$$V_a - i_2R_2 - \mathcal{E}_2 = V_b,$$

or

$$V_a - V_b = \mathcal{E}_2 + i_2R_2$$
$$= 6.3 \text{ V} + (-0.40 \text{ A})(3.5 \ \Omega) = +4.9 \text{ V}.$$

The positive sign tells us that a is more positive in potential than b. We should expect this result from looking at the circuit diagram, because all three batteries have their positive terminals on the top side of the figure.

33-6   MEASURING INSTRUMENTS

Several electrical measuring instruments involve circuits that can be analyzed by the methods of this chapter. We discuss three of them.

The Ammeter

An instrument used to measure currents is called an *ammeter*. To measure the current in a wire, you usually have to break or cut the wire and insert the ammeter so that the current to be measured passes through the meter; see Fig. 11.

It is essential that the resistance R_A of the ammeter be very small (ideally zero) compared to other resistances in the circuit. Otherwise, the very presence of the meter would change the current to be measured. In the single-loop circuit of Fig. 11, the required condition, assuming that the voltmeter is not connected, is

$$R_A \ll r + R_1 + R_2.$$

An ammeter can also be used as an *ohmmeter* to measure an unknown resistance; see Problem 40.

The Voltmeter

A meter to measure potential differences is called a *voltmeter*. To find the potential difference between any two points in the circuit, the voltmeter terminals are connected at those points, without breaking the circuit; see Fig. 11.

It is essential that the resistance R_V of a voltmeter be very large (ideally infinite) compared to any circuit element across which the voltmeter is connected. Otherwise, significant current would pass through the meter, changing the current through the circuit element in parallel with

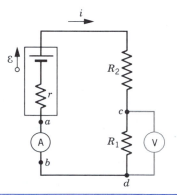

Figure 11   A single-loop circuit, illustrating the connection of an ammeter A, which measures the current i, and a voltmeter V, which measures the potential difference between points c and d.

the meter and consequently changing the potential difference being measured. In Fig. 11, the required condition is

$$R_V \gg R_1.$$

Often a single unit is packaged so that, by external switching, it can serve as either an ammeter, a voltmeter, or an ohmmeter. Such a versatile unit is called a *multimeter*. Its output readings may take the form of a pointer moving over a scale or of a digital display.

The Potentiometer

A *potentiometer* is a device for measuring an unknown emf $\mathcal{E}_x$ by comparing it with a known standard emf $\mathcal{E}_s$. Figure 12 shows its basic elements. The resistor that extends from a to e is a carefully made precision resistor with a sliding contact shown positioned at d. The resistance R in the figure is the resistance between points a and d.

When using the instrument, $\mathcal{E}_s$ is first placed in the position $\mathcal{E}$, and the sliding contact is adjusted until the current i is zero as noted on the sensitive ammeter A. The potentiometer is then said to be *balanced*, the value of R at balance being R_s. In this balance condition we have, considering the loop *abcda*,

$$\mathcal{E}_s = i_0 R_s. \qquad (25)$$

Because $i = 0$ in branch *abcd*, the internal resistance r of the standard source of emf (or of the ammeter) does not enter.

The process is now repeated with $\mathcal{E}_x$ substituted for $\mathcal{E}_s$, the potentiometer being balanced once more. The current i_0 remains unchanged (because $i = 0$), and the new balance condition is

$$\mathcal{E}_x = i_0 R_x. \qquad (26)$$

From Eqs. 25 and 26 we then have

$$\mathcal{E}_x = \mathcal{E}_s \frac{R_x}{R_s}. \qquad (27)$$

The unknown emf can be found in terms of the known

emf by making two adjustments of the precision resistor. Note that this result is independent of the value of $\mathcal{E}_0$.

In the past, the potentiometer served as a secondary voltage standard that enabled a researcher in any laboratory to determine an unknown emf by comparing it with that of a carefully calibrated *standard cell* (an electrochemical device similar to a battery). Today, the volt is defined in terms of a more precise quantum standard that is relatively easy to duplicate in the laboratory—the quantized voltage steps of a sandwich consisting of two superconductors separated by a thin insulating layer, called a *Josephson junction*.*

The potentiometer is an example of a *null instrument*, which permits precision measurement by adjusting the value of a circuit element until a meter reads zero (null). In this case, the null reading permits us to measure $\mathcal{E}_x$ when no current passes through it, and so our measurement is independent of the internal resistance r of the source of emf. Another null device is the Wheatstone bridge; see Problem 46.

33-7   *RC* CIRCUITS

The preceding sections dealt with circuits containing only resistors, in which the currents did not vary with time. Here we introduce the capacitor as a circuit element, which leads us to the study of time-varying currents.

Suppose we charge the capacitor in Fig. 13 by throwing switch S to position a. (Later we consider the connection to position b.) What current is set up in the resulting

---

* Brian Josephson, a British physicist, was a 22-year-old graduate student when he discovered the properties of this junction, for which he was honored with the 1973 Nobel Prize in physics.

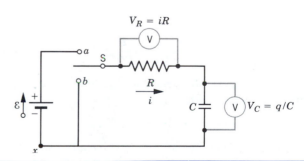

Figure 13   When switch S is connected to a, the capacitor C is charged by emf $\mathcal{E}$ through the resistor R. After the capacitor is charged, the switch is moved to b, and the capacitor discharges through R. A voltmeter connected across R measures the potential difference $V_R (= iR)$ across the resistor and thus determines the current i. A voltmeter connected across the capacitor measures the potential difference $V_C (= q/C)$ across the capacitor and thus determines the charge q.

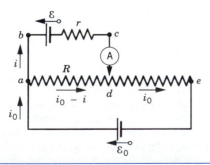

Figure 12   The basic elements of a potentiometer, used to compare emfs.

single-loop circuit? Let us apply conservation of energy principles.

In time dt a charge $dq\,(=i\,dt)$ moves through any cross section of the circuit. The work $(=\mathcal{E}\,dq$; see Eq. 1) done by the seat of emf must equal the internal energy $(=i^2R\,dt)$ produced in the resistor during time dt, plus the increase dU in the amount of energy $U\,(=q^2/2C$; see Eq. 26 of Chapter 31) that is stored in the capacitor. Conservation of energy gives

$$\mathcal{E}\,dq = i^2R\,dt + d\left(\frac{q^2}{2C}\right)$$

or

$$\mathcal{E}\,dq = i^2R\,dt + \frac{q}{C}\,dq.$$

Dividing by dt yields

$$\mathcal{E}\frac{dq}{dt} = i^2R + \frac{q}{C}\frac{dq}{dt}.$$

Since q is the charge on the upper plate, positive i means positive dq/dt. With $i = dq/dt$, this equation becomes

$$\mathcal{E} = iR + \frac{q}{C}. \tag{28}$$

Equation 28 also follows from the loop theorem, as it must, since the loop theorem was derived from the conservation of energy principle. Starting from point x and going around the circuit clockwise, we experience an increase in potential in going through the seat of emf and decreases in potential in going through the resistor and the capacitor, or

$$\mathcal{E} - iR - \frac{q}{C} = 0,$$

which is identical to Eq. 28.

To solve Eq. 28, we first substitute dq/dt for i, which gives

$$\mathcal{E} = R\frac{dq}{dt} + \frac{q}{C}. \tag{29}$$

We can rewrite Eq. 29 as

$$\frac{dq}{q - \mathcal{E}C} = -\frac{dt}{RC}. \tag{30}$$

Integrating this result in the case that $q = 0$ at $t = 0$, we obtain (after solving for q),

$$q = C\mathcal{E}(1 - e^{-t/RC}). \tag{31}$$

We can check that this function $q(t)$ is really a solution of Eq. 29 by substituting it into that equation and seeing whether an identity results. Differentiating Eq. 31 with respect to time yields

$$i = \frac{dq}{dt} = \frac{\mathcal{E}}{R}e^{-t/RC}. \tag{32}$$

Substituting q (Eq. 31) and dq/dt (Eq. 32) into Eq. 29 yields an identity, as you should verify. Equation 31 is therefore a solution of Eq. 29.

In the laboratory we can determine i and q conveniently by measuring quantities that are proportional to them, namely, the potential difference $V_R\,(=iR)$ across the resistor and the potential difference $V_C\,(=q/C)$ across the capacitor. Such measurements can be accomplished rather easily, as illustrated in Fig. 13, by connecting voltmeters (or oscilloscope probes) across the resistor and the capacitor. Figure 14 shows the resulting plots of V_R and V_C. Note the following: (1) At $t = 0$, $V_R = \mathcal{E}$ (the full potential difference appears across R), and $V_C = 0$ (the capacitor is not charged). (2) As $t \to \infty$, $V_C \to \mathcal{E}$ (the capacitor becomes fully charged), and $V_R \to 0$ (the current stops flowing). (3) At all times, $V_R + V_C = \mathcal{E}$, as Eq. 29 requires.

The quantity RC in Eqs. 31 and 32 has the dimensions of time (because the exponent must be dimensionless) and is called the *capacitive time constant* τ_C of the circuit:

$$\tau_C = RC. \tag{33}$$

It is the time at which the charge on the capacitor has increased to within a factor of $1 - e^{-1}\,(\approx 63\%)$ of its final value $C\mathcal{E}$. To show this, we put $t = \tau_C = RC$ in Eq. 31 to obtain

$$q = C\mathcal{E}(1 - e^{-1}) = 0.63C\mathcal{E}.$$

Figure 14a shows that if a resistance is included in a circuit with a charging capacitor, the increase of the charge of the capacitor toward its limiting value is *delayed* by a time characterized by the time constant RC. With no resistor present ($RC = 0$), the charge would rise immediately to its limiting value. Although we have shown that this time delay follows from an application of the loop theorem to RC circuits, it is important to develop a physical understanding of the causes of the delay.

When switch S in Fig. 13 is closed on a, the charge on the capacitor is initially zero, so the potential difference across the capacitor is initially zero. At this time, Eq. 28

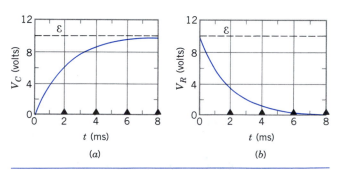

Figure 14 (*a*) As indicated by the potential difference V_C, during the charging process the charge on the capacitor increases with time, and V_C approaches the value of the emf $\mathcal{E}$. The time is measured after the switch is closed on a at $t = 0$. (*b*) The potential difference across the resistor decreases with time, approaching 0 at later times because the current falls to zero once the capacitor is fully charged. The curves have been drawn for $\mathcal{E} = 10$ V, $R = 2000\ \Omega$, and $C = 1\ \mu$F. The filled triangles represent successive time constants.

shows that $\mathcal{E} = iR$, and so $i = \mathcal{E}/R$ at $t = 0$. Because of this current, charge flows to the capacitor and the potential difference across the capacitor increases with time. Equation 28 now shows that, because the emf $\mathcal{E}$ is a constant, any increase in the potential difference across the capacitor must be balanced by a corresponding *decrease* in the potential difference across the resistor, with a similar decrease in the current. This decrease in the current means that the charge on the capacitor increases more slowly. This process continues until the current decreases to zero, at which time there is no potential drop across the resistor. The entire potential difference of the emf now appears across the capacitor, which is fully charged $(q = C\mathcal{E})$. Unless changes are made in the circuit, there is no further flow of charge. Review the derivations of Eqs. 31 and 32 and study Fig. 14 with the qualitative arguments of this paragraph in mind.

Sample Problem 7   A resistor R ($=6.2$ MΩ) and a capacitor C ($=2.4$ μF) are connected in series, and a 12-V battery of negligible internal resistance is connected across their combination. (*a*) What is the capacitive time constant of this circuit? (*b*) At what time after the battery is connected does the potential difference across the capacitor equal 5.6 V?

Solution   (*a*) From Eq. 33,

$$\tau_C = RC = (6.2 \times 10^6 \ \Omega)(2.4 \times 10^{-6} \ \text{F}) = 15 \ \text{s}.$$

(*b*) The potential difference across the capacitor is $V_C = q/C$, which according to Eq. 31 can be written

$$V_C = \frac{q}{C} = \mathcal{E}(1 - e^{-t/RC}).$$

Solving for t, we obtain (using $\tau_C = RC$)

$$t = -\tau_C \ln\left(1 - \frac{V_C}{\mathcal{E}}\right)$$

$$= -(15 \ \text{s}) \ln\left(1 - \frac{5.6 \ \text{V}}{12 \ \text{V}}\right) = 9.4 \ \text{s}.$$

As we found above, after a time τ_C ($=15$ s), the potential difference across the capacitor is $0.63\mathcal{E} = 7.6$ V. It is reasonable that in a shorter time of 9.4 s, the potential difference across the capacitor reaches only the smaller value of 5.6 V.

Discharging a Capacitor

Assume now that the switch S in Fig. 13 has been in position *a* for a time that is much greater than RC. For all practical purposes, the capacitor is fully charged, and no charge is flowing. The switch S is then thrown to position *b*. How do the charge of the capacitor and the current vary with time?

With the switch S closed on *b*, the capacitor discharges through the resistor. There is no emf in the circuit and Eq. 28 for the circuit, with $\mathcal{E} = 0$, becomes simply

$$iR + \frac{q}{C} = 0. \tag{34}$$

Putting $i = dq/dt$ allows us to write the equation of the circuit (compare Eq. 29)

$$R\frac{dq}{dt} + \frac{q}{C} = 0. \tag{35}$$

The solution is, as you may readily derive by integration (after writing $dq/q = -dt/RC$) and verify by substitution,

$$q = q_0 e^{-t/\tau_C}, \tag{36}$$

q_0 being the initial charge on the capacitor ($= \mathcal{E}C$, in our case). The capacitive time constant τ_C ($=RC$) appears in this expression for a discharging capacitor as well as in that for a charging capacitor (Eq. 31). We see that at a time such that $t = \tau_C = RC$, the capacitor charge is reduced to $q_0 e^{-1}$, which is about 37% of the initial charge q_0.

Differentiating Eq. 36, we find the current during discharge,

$$i = \frac{dq}{dt} = -\frac{q_0}{RC} e^{-t/\tau_C}. \tag{37}$$

The negative sign shows that the current is in the direction opposite to that shown in Fig. 13. This is as it should be, since the capacitor is discharging rather than charging. Since $q_0 = C\mathcal{E}$, we can write Eq. 37 as

$$i = -\frac{\mathcal{E}}{R} e^{-t/\tau_C}. \tag{38}$$

The initial current, found by setting $t = 0$ in Eq. 38, is $-\mathcal{E}/R$. This is reasonable because the initial potential difference across the resistor is $\mathcal{E}$.

The potential differences across R and C, which are, respectively, proportional to i and q, can again be measured as indicated in Fig. 13. Typical results are shown in Fig. 15. Note that, as suggested by Eq. 36, V_C ($=q/C$) falls

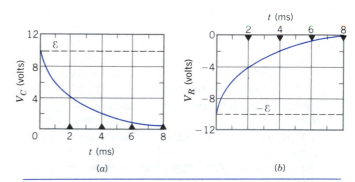

Figure 15   (*a*) After the capacitor has become fully charged, the switch in Fig. 13 is thrown from *a* to *b*, which we take to define a new $t = 0$. The potential difference across the capacitor decreases exponentially to zero as the capacitor discharges. (*b*) When the switch is initially moved to *b*, the potential difference across the resistor is negative compared with its value during the charging process shown in Fig. 14. As the capacitor discharges, the magnitude of the current falls exponentially to zero, and the potential drop across the resistor also approaches zero.

exponentially from its maximum value, which occurs at $t = 0$, whereas V_R ($= iR$) is negative and rises exponentially to zero. Note also that $V_C + V_R = 0$, as required by Eq. 34.

Sample Problem 8   A capacitor C discharges through a resistor R. (*a*) After how many time constants does its charge fall to one-half its initial value? (*b*) After how many time constants does the stored energy drop to half its initial value?

Solution   (*a*) The charge on the capacitor varies according to Eq. 36,

$$q = q_0 e^{-t/\tau_C},$$

in which q_0 is the initial charge. We seek the time t at which $q = \frac{1}{2}q_0$, or

$$\tfrac{1}{2}q_0 = q_0 e^{-t/\tau_C}.$$

Canceling q_0 and taking the natural logarithm of each side, we find

$$-\ln 2 = -\frac{t}{\tau_C}$$

or

$$t = (\ln 2)\tau_C = 0.69\tau_C.$$

The charge drops to half its initial value after 0.69 time constants.

(*b*) The energy of the capacitor is

$$U = \frac{q^2}{2C} = \frac{q_0^2}{2C} e^{-2t/\tau_c} = U_0 e^{-2t/\tau_c},$$

in which U_0 is the initial stored energy. The time at which $U = \frac{1}{2}U_0$ is found from

$$\tfrac{1}{2}U_0 = U_0 e^{-2t/\tau_c}.$$

Canceling U_0 and taking the logarithm of each side, we obtain

$$-\ln 2 = -2t/\tau_C$$

or

$$t = \tau_C \frac{\ln 2}{2} = 0.35\tau_C.$$

The stored energy drops to half its initial value after 0.35 time constants have elapsed. This remains true no matter what the initial scored energy may be. The time ($0.69\tau_C$) needed for the *charge* to fall to half its initial value is greater than the time ($0.35\tau_C$) needed for the *energy* to fall to half its initial value. Why?

QUESTIONS

1. Does the direction of the emf provided by a battery depend on the direction of current flow through the battery?

2. In Fig. 2, discuss what changes would occur if we increased the mass m by such an amount that the "motor" reversed direction and became a "generator," that is, a seat of emf.

3. Discuss in detail the statement that the energy method and the loop rule method for solving circuits are perfectly equivalent.

4. Devise a method for measuring the emf and the internal resistance of a battery.

5. What is the origin of the internal resistance of a battery? Does this depend on the age or size of the battery?

6. The current passing through a battery of emf $\mathscr{E}$ and internal resistance r is decreased by some external means. Does the potential difference between the terminals of the battery necessarily decrease or increase? Explain.

7. How could you calculate V_{ab} in Fig. 3*a* by following a path from *a* to *b* that does not lie in the conducting circuit?

8. A 25-W, 120-V bulb glows at normal brightness when connected across a bank of batteries. A 500-W, 120-V bulb glows only dimly when connected across the same bank. How could this happen?

9. Under what circumstances can the terminal potential difference of a battery exceed its emf?

10. Automobiles generally use a 12-V electrical system. Years ago a 6-V system was used. Why the change? Why not 24 V?

11. The loop rule is based on the conservation of energy principle and the junction rule on the conservation of charge principle. Explain just how these rules are based on these principles.

12. Under what circumstances would you want to connect batteries in parallel? In series?

13. Compare and contrast the formulas for the equivalent values of series and parallel combinations of (*a*) capacitors and (*b*) resistors.

14. Under what circumstances would you want to connect resistors in parallel? In series?

15. What is the difference between an emf and a potential difference?

16. Referring to Fig. 9, use a qualitative argument to convince yourself that i_3 is drawn in the wrong direction.

17. Explain in your own words why the resistance of an ammeter should be very small whereas that of a voltmeter should be very large.

18. Do the junction and loop rules apply to a circuit containing a capacitor?

19. Show that the product RC in Eqs. 31 and 32 has the dimensions of time, that is, that 1 second = 1 ohm × 1 farad.

20. A capacitor, resistor, and battery are connected in series. The charge that the capacitor stores is unaffected by the resistance of the resistor. What purpose, then, is served by the resistor?

21. Explain why, in Sample Problem 8, the energy falls to half its initial value more rapidly than does the charge.

22. The light flash in a camera is produced by the discharge of a capacitor across the lamp. Why don't we just connect the photoflash lamp directly across the power supply used to charge the capacitor?

23. Does the time required for the charge on a capacitor in an *RC* circuit to build up to a given fraction of its final value

depend on the value of the applied emf? Does the time required for the charge to change by a given amount depend on the applied emf?

24. A capacitor is connected across the terminals of a battery. Does the charge that eventually appears on the capacitor plates depend on the value of the internal resistance of the battery?

25. Devise a method whereby an *RC* circuit can be used to measure very high resistances.

26. In Fig. 13 suppose that switch S is closed on *a*. Explain why, in view of the fact that the negative terminal of the battery is not connected to resistance *R*, the current in *R* should be $\mathcal{E}/R$, as Eq. 32 predicts.

27. In Fig. 13 suppose that switch S is closed on *a*. Why does the charge on capacitor *C* not rise instantaneously to $q = C\mathcal{E}$? After all, the positive battery terminal is connected to one plate of the capacitor and the negative terminal to the other.

PROBLEMS

Section 33-1 Electromotive Force

1. A 5.12-A current is set up in an external circuit by a 6.00-V storage battery for 5.75 min. By how much is the chemical energy of the battery reduced?

2. (*a*) How much work does a 12.0-V seat of emf do on an electron as it passes through from the positive to the negative terminal? (*b*) If 3.40×10^{18} electrons pass through each second, what is the power output of the seat?

3. A certain 12-V car battery carries an initial charge of 125 A·h. Assuming that the potential across the terminals stays constant until the battery is completely discharged, for how long can it deliver energy at the rate of 110 W?

4. A standard flashlight battery can deliver about 2.0 W·h of energy before it runs down. (*a*) If a battery costs 80 cents, what is the cost of operating a 100-W lamp for 8.0 h using batteries? (*b*) What is the cost if power provided by an electric utility company, at 12 cents per kW·h, is used?

Section 33-3 Potential Differences

5. In Fig. 16 the potential at point *P* is 100 V. What is the potential at point *Q*?

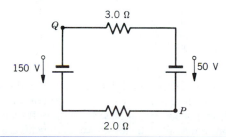

Figure 16   Problem 5.

6. A gasoline gauge for an automobile is shown schematically in Fig. 17. The indicator (on the dashboard) has a resistance of 10 Ω. The tank unit is simply a float connected to a resistor that has a resistance of 140 Ω when the tank is empty, 20 Ω when it is full, and varies linearly with the volume of gasoline. Find the current in the circuit when the tank is (*a*) empty, (*b*) half full, and (*c*) full.

7. (*a*) In Fig. 18 what value must *R* have if the current in the circuit is to be 50 mA? Take $\mathcal{E}_1 = 2.0$ V, $\mathcal{E}_2 = 3.0$ V, and $r_1 = r_2 = 3.0$ Ω. (*b*) What is the rate at which internal energy appears in *R*?

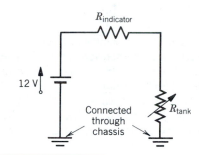

Figure 17   Problem 6.

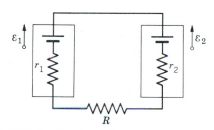

Figure 18   Problem 7.

8. The current in a single-loop circuit is 5.0 A. When an additional resistance of 2.0 Ω is inserted in series, the current drops to 4.0 A. What was the resistance in the original circuit?

9. The section of circuit *AB* (see Fig. 19) absorbs 53.0 W of power when a current $i = 1.20$ A passes through it in the indicated direction. (*a*) Find the potential difference between *A* and *B*. (*b*) If the element *C* does not have an internal resistance, what is its emf? (*c*) Which terminal, left or right, is positive?

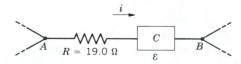

Figure 19   Problem 9.

10. Internal energy is to be generated in a 108-mΩ resistor at the rate of 9.88 W by connecting it to a battery whose emf is

1.50 V. (*a*) What is the internal resistance of the battery? (*b*) What potential difference exists across the resistor?

11. The starting motor of an automobile is turning slowly and the mechanic has to decide whether to replace the motor, the cable, or the battery. The manufacturer's manual says that the 12-V battery can have no more than 0.020 Ω internal resistance, the motor no more than 0.200 Ω resistance, and the cable no more than 0.040 Ω resistance. The mechanic turns on the motor and measures 11.4 V across the battery, 3.0 V across the cable, and a current of 50 A. Which part is defective?

12. Two batteries having the same emf $\mathscr{E}$ but different internal resistances r_1 and r_2 $(r_1 > r_2)$ are connected in series to an external resistance R. (*a*) Find the value of R that makes the potential difference zero between the terminals of one battery. (*b*) Which battery is it?

13. A solar cell generates a potential difference of 0.10 V when a 500-Ω resistor is connected across it and a potential difference of 0.16 V when a 1000-Ω resistor is substituted. What are (*a*) the internal resistance and (*b*) the emf of the solar cell? (*c*) The area of the cell is 5.0 cm² and the intensity of light striking it is 2.0 mW/cm². What is the efficiency of the cell for converting light energy to internal energy in the 1000-Ω external resistor?

14. (*a*) In the circuit of Fig. 3*a* show that the power delivered to R as internal energy is a maximum when R is equal to the internal resistance r of the battery. (*b*) Show that this maximum power is $P = \mathscr{E}^2/4r$.

15. A battery of emf $\mathscr{E} = 2.0$ V and internal resistance $r = 0.50$ Ω is driving a motor. The motor is lifting a 2.0-N object at constant speed $v = 0.50$ m/s. Assuming no power losses, find (*a*) the current i in the circuit and (*b*) the potential difference V across the terminals of the motor. (*c*) Discuss the fact that there are two solutions to this problem.

Section 33-4 Resistors in Series and Parallel

16. Four 18-Ω resistors are connected in parallel across a 27-V battery. What is the current through the battery?

17. By using only two resistors—singly, in series, or in parallel—you are able to obtain resistances of 3.0, 4.0, 12, and 16 Ω. What are the separate resistances of the resistors?

18. In Fig. 20, find the equivalent resistance between points (*a*) *A* and *B*, (*b*) *A* and *C*, and (*c*) *B* and *C*.

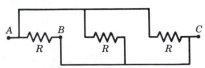

Figure 20   Problem 18.

19. A circuit containing five resistors connected to a 12-V battery is shown in Fig. 21. Find the potential drop across the 5.0-Ω resistor.

20. A 120-V power line is protected by a 15-A fuse. What is the maximum number of 500-W lamps that can be simultaneously operated in parallel on this line?

21. Two resistors R_1 and R_2 may be connected either in series or parallel across a (resistanceless) battery with emf $\mathscr{E}$. We de-

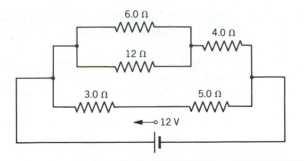

Figure 21   Problem 19.

sire the internal energy transfer rate for the parallel combination to be five times that for the series combination. If $R_1 = 100$ Ω, what is R_2?

22. You are given a number of 10-Ω resistors, each capable of dissipating only 1.0 W. What is the minimum number of such resistors that you need to combine in series or parallel combinations to make a 10-Ω resistor capable of dissipating at least 5.0 W?

23. A three-way 120-V lamp bulb, rated for 100-200-300 W, burns out a filament. Afterward, the bulb operates at the same intensity on its lowest and its highest switch positions but does not operate at all on the middle position. (*a*) How are the two filaments wired inside the bulb? (*b*) Calculate the resistances of the filaments.

24. (*a*) In Fig. 22 find the equivalent resistance of the network shown. (*b*) Calculate the current in each resistor. Put $R_1 = 112$ Ω, $R_2 = 42.0$ Ω, $R_3 = 61.6$ Ω, $R_4 = 75.0$ Ω, and $\mathscr{E} = 6.22$ V.

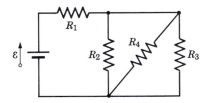

Figure 22   Problem 24.

25. Conducting rails *A* and *B*, having equal lengths of 42.6 m and cross-sectional area of 91.0 cm², are connected in series. A potential of 630 V is applied across the terminal points of the connected rails. The resistances of the rails are 76.2 and 35.0 μΩ. Determine (*a*) the resistivities of the rails, (*b*) the current density in each rail, (*c*) the electric field strength in each rail, and (*d*) the potential difference across each rail.

26. In the circuit of Fig. 23, $\mathscr{E}$, R_1, and R_2 have constant values but R can be varied. Find an expression for R that results in the maximum heating in that resistor.

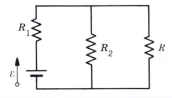

Figure 23   Problem 26.

27. In Fig. 24, find the equivalent resistance between points (*a*) *F* and *H* and (*b*) *F* and *G*.

Figure 24   Problem 27.

28. Find the equivalent resistance between points *x* and *y* shown in Fig. 25. Four of the resistors have equal resistance *R*, as shown; the "middle" resistor has value $r \neq R$. (Compare with Problem 28 of Chapter 31.)

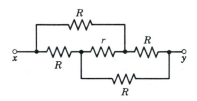

Figure 25   Problem 28.

29. Twelve resistors, each of resistance *R* ohms, form a cube (see Fig. 8*a*). (*a*) Find R_{13}, the equivalent resistance of a face diagonal. (*b*) Find R_{17}, the equivalent resistance of a body diagonal. See Sample Problem 4.

Section 33-5 Multiloop Circuits

30. In Fig. 26 find (*a*) the current in each resistor, and (*b*) the potential difference between *a* and *b*. Put $\mathcal{E}_1 = 6.0$ V, $\mathcal{E}_2 = 5.0$ V, $\mathcal{E}_3 = 4.0$ V, $R_1 = 100$ Ω, and $R_2 = 50$ Ω.

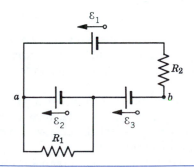

Figure 26   Problem 30.

31. Two light bulbs, one of resistance R_1 and the other of resistance $R_2 (< R_1)$ are connected (*a*) in parallel and (*b*) in series. Which bulb is brighter in each case?

32. In Fig. 9 calculate the potential difference $V_c - V_d$ between points *c* and *d* by as many paths as possible. Assume that $\mathcal{E}_1 = 4.22$ V, $\mathcal{E}_2 = 1.13$ V, $R_1 = 9.77$ Ω, $R_2 = 11.6$ Ω, and $R_3 = 5.40$ Ω.

33. What current, in terms of $\mathcal{E}$ and *R*, does the ammeter A in Fig. 27 read? Assume that A has zero resistance.

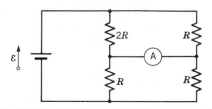

Figure 27   Problem 33.

34. When the lights of an automobile are switched on, an ammeter in series with them reads 10.0 A and a voltmeter connected across them reads 12.0 V. See Fig. 28. When the electric starting motor is turned on, the ammeter reading drops to 8.00 A and the lights dim somewhat. If the internal resistance of the battery is 50.0 mΩ and that of the ammeter is negligible, what are (*a*) the emf of the battery and (*b*) the current through the starting motor when the lights are on?

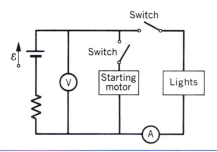

Figure 28   Problem 34.

35. Figure 29 shows a battery connected across a uniform resistor R_0. A sliding contact can move across the resistor from *x* = 0 at the left to *x* = 10 cm at the right. Find an expression for the power dissipated in the resistor *R* as a function of *x*. Plot the function for $\mathcal{E} = 50$ V, $R = 2000$ Ω, and $R_0 = 100$ Ω.

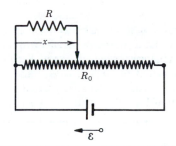

Figure 29   Problem 35.

36. You are given two batteries of emf values $\mathcal{E}_1$ and $\mathcal{E}_2$ and internal resistances r_1 and r_2. They may be connected either in (*a*) parallel or (*b*) series and are used to establish a current

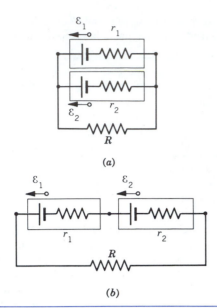

(a)

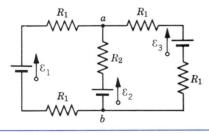

(b)

Figure 30   Problem 36.

in a resistor R, as shown in Fig. 30. Derive expressions for the current in R for both methods of connection.

37. (a) Calculate the current through each source of emf in Fig. 31. (b) Calculate $V_b - V_a$. Assume that $R_1 = 1.20\ \Omega$, $R_2 = 2.30\ \Omega$, $\mathcal{E}_1 = 2.00$ V, $\mathcal{E}_2 = 3.80$ V, and $\mathcal{E}_3 = 5.00$ V.

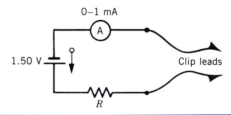

Figure 31   Problem 37.

38. A battery of emf $\mathcal{E}_1$ and internal resistance $r_1 = 140\ \Omega$ is used to operate a device with resistance $R = 34\ \Omega$. However, the emf $\mathcal{E}_1$ fluctuates between 25 and 27 V; therefore the current through R fluctuates also. To stabilize the current through R, a second battery, with internal resistance $r_2 = 0.11\ \Omega$, is introduced in parallel with the first battery. This second battery is stable in its emf. See Fig. 32. Find the change in current through R as $\mathcal{E}_1$ varies (a) before and (b) after the second battery is inserted into the circuit. (c) What should be the value of $\mathcal{E}_2$ so that the average

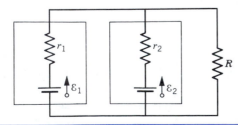

Figure 32   Problem 38.

current through R, calculated with $\mathcal{E}_1 = 26$ V (its average value), is not changed by the introduction of the second battery?

39. In Fig. 33 imagine an ammeter inserted in the branch containing R_3. (a) What will it read, assuming $\mathcal{E} = 5.0$ V, $R_1 = 2.0\ \Omega$, $R_2 = 4.0\ \Omega$, and $R_3 = 6.0\ \Omega$? (b) The ammeter and the source of emf are now physically interchanged. Show that the ammeter reading remains unchanged.

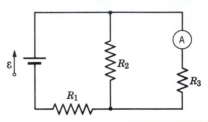

Figure 33   Problem 39.

Section 33-6 Measuring Instruments

40. A simple ohmmeter is made by connecting a 1.50-V flashlight battery in series with a resistor R and a 1.00-mA ammeter, as shown in Fig. 34. R is adjusted so that when the circuit terminals are shorted together the meter deflects to its full-scale value of 1.00 mA. What external resistance across the terminals results in a deflection of (a) 10%, (b) 50%, and (c) 90% of full scale? (d) If the ammeter has a resistance of 18.5 Ω and the internal resistance of the battery is negligible, what is the value of R?

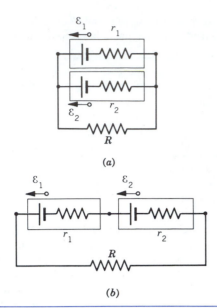

Figure 34   Problem 40.

41. In Fig. 11 assume that $\mathcal{E} = 5.0$ V, $r = 2.0\ \Omega$, $R_1 = 5.0\ \Omega$, and $R_2 = 4.0\ \Omega$. If $R_A = 0.10\ \Omega$, what percent error is made in reading the current? Assume that the voltmeter is not present.

42. In Fig. 11 assume that $\mathcal{E} = 3.0$ V, $r = 100\ \Omega$, $R_1 = 250\ \Omega$, and $R_2 = 300\ \Omega$. If $R_V = 5.0$ kΩ, what percent error is made in reading the potential difference across R_1? Ignore the presence of the ammeter.

43. A voltmeter (resistance R_V) and an ammeter (resistance R_A) are connected to measure a resistance R, as in Fig. 35a. The resistance is given by $R = V/i$, where V is the voltmeter reading and i is the current in the resistor R. Some of the current registered by the ammeter (i') goes through the voltmeter so that the ratio of the meter readings ($= V/i'$) gives only an *apparent* resistance reading R'. Show that R and R' are related by

$$\frac{1}{R} = \frac{1}{R'} - \frac{1}{R_V}.$$

Note that as $R_V \to \infty$, $R' \to R$.

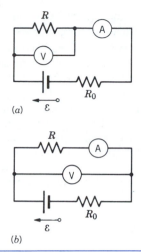

(a)

(b)

Figure 35 Problems 43, 44, and 45.

44. If meters are used to measure resistance, they may also be connected as they are in Fig. 35b. Again the ratio of the meter readings gives only an apparent resistance R'. Show that R' is related to R by

$$R = R' - R_A,$$

in which R_A is the ammeter resistance. Note that as $R_A \to 0$, $R' \to R$.

45. In Fig. 35 the ammeter and voltmeter resistances are 3.00 Ω and 300 Ω, respectively. (a) If $R = 85.0$ Ω, what will the meters read for the two different connections? (b) What apparent resistance R' will be computed in each case? Take $\mathcal{E} = 12.0$ V and $R_0 = 100$ Ω.

46. In Fig. 36 R_s is to be adjusted in value until points a and b are brought to exactly the same potential. (One tests for this condition by momentarily connecting a sensitive ammeter between a and b; if these points are at the same potential, the ammeter will not deflect.) Show that when this adjustment is made, the following relation holds:

$$R_x = R_s(R_2/R_1).$$

An unknown resistance (R_x) can be measured in terms of a standard (R_s) using this device, which is called a Wheatstone bridge.

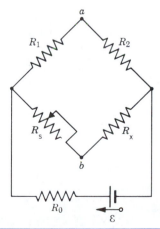

Figure 36 Problems 46 and 47.

47. If points a and b in Fig. 36 are connected by a wire of resistance r, show that the current in the wire is

$$i = \frac{\mathcal{E}(R_s - R_x)}{(R + 2r)(R_s + R_x) + 2R_sR_x},$$

where $\mathcal{E}$ is the emf of the battery. Assume that R_1 and R_2 are equal ($R_1 = R_2 = R$) and that R_0 equals zero. Is this formula consistent with the result of Problem 46?

Section 33-7 RC Circuits

48. In an RC series circuit $\mathcal{E} = 11.0$ V, $R = 1.42$ MΩ, and $C = 1.80$ μF. (a) Calculate the time constant. (b) Find the maximum charge that will appear on the capacitor during charging. (c) How long does it take for the charge to build up to 15.5 μC?

49. How many time constants must elapse before a capacitor in an RC circuit is charged to within 1.00% of its equilibrium charge?

50. A 15.2-kΩ resistor and a capacitor are connected in series and a 13.0-V potential is suddenly applied. The potential across the capacitor rises to 5.00 V in 1.28 μs. (a) Calculate the time constant. (b) Find the capacitance of the capacitor.

51. An RC circuit is discharged by closing a switch at time $t = 0$. The initial potential difference across the capacitor is 100 V. If the potential difference has decreased to 1.06 V after 10.0 s, (a) calculate the time constant of the circuit. (b) What will be the potential difference at $t = 17$ s?

52. A controller on an electronic arcade game consists of a variable resistor connected across the plates of a 220-nF capacitor. The capacitor is charged to 5.00 V, then discharged through the resistor. The time for the potential difference across the plates to decrease to 800 mV is measured by an internal clock. If the range of discharge times that can be handled is from 10.0 μs to 6.00 ms, what should be the range of the resistance of the resistor?

53. Figure 37 shows the circuit of a flashing lamp, like those attached to barrels at highway construction sites. The fluorescent lamp L is connected in parallel across the capacitor C of an RC circuit. Current passes through the lamp only when the potential across it reaches the breakdown voltage V_L; in this event, the capacitor discharges through the lamp and it flashes for a very short time. Suppose that two flashes per second are needed. Using a lamp with breakdown voltage $V_L = 72$ V, a 95-V battery, and a 0.15-μF capacitor, what should be the resistance R of the resistor?

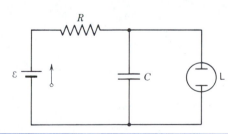

Figure 37 Problem 53.

54. A 1.0-μF capacitor with an initial stored energy of 0.50 J is discharged through a 1.0-MΩ resistor. (a) What is the initial charge on the capacitor? (b) What is the current through the

resistor when the discharge starts? (*c*) Determine V_C, the voltage across the capacitor, and V_R, the voltage across the resistor, as functions of time. (*d*) Express the rate of generation of internal energy in the resistor as a function of time.

55. A 3.0-MΩ resistor and a 1.0-μF capacitor are connected in a single-loop circuit with a seat of emf with $\mathcal{E} = 4.0$ V. At 1.0 s after the connection is made, what are the rates at which (*a*) the charge on the capacitor is increasing, (*b*) energy is being stored in the capacitor, (*c*) internal energy is appearing in the resistor, and (*d*) energy is being delivered by the seat of emf?

56. (*a*) Carry out the missing steps to obtain Eq. 31 from Eq. 30.

(*b*) In a similar manner, obtain Eq. 36 from Eq. 35. Note that $q = q_0$ (capacitor charged) at $t = 0$.

57. Prove that when switch S in Fig. 13 is thrown from *a* to *b*, all the energy stored in the capacitor is transformed into internal energy in the resistor. Assume that the capacitor is fully charged before the switch is thrown.

58. An initially uncharged capacitor C is fully charged by a constant emf $\mathcal{E}$ in series with a resistor R. (*a*) Show that the final energy stored in the capacitor is half the energy supplied by the emf. (*b*) By direct integration of i^2R over the charging time, show that the internal energy dissipated by the resistor is also half the energy supplied by the emf.

CHAPTER 34

THE MAGNETIC FIELD

The science of magnetism *had its origin in ancient times. It grew from the observation that certain naturally occurring stones would attract one another and would also attract small bits of one metal, iron, but not other metals, such as gold or silver. The word "magnetism" comes from the name of the district (Magnesia) in Asia Minor, one of the locations where these stones were found.*

Today we have put that discovery to great practical use, from small "refrigerator" magnets to magnetic recording tape and computer disks. The magnetism of individual atomic nuclei is used by physicians to make images of organs deep within the body. Spacecraft have measured the magnetism of the Earth and the other planets to learn about their internal structure.

In this chapter we begin our study of magnetism by considering the magnetic field and its effects on a moving electric charge. In the next chapter, we consider the production of magnetic fields by electric currents. In later chapters, we continue to explore the close relationship between electricity and magnetism, which are linked together under the common designation electromagnetism.

34-1 THE MAGNETIC FIELD B

Just as in ancient times, small bits of iron are still used to reveal the presence of magnetic effects. Figure 1 shows the distribution of iron fillings in the space near a small *permanent magnet,* in this case a short iron bar. Figure 2 shows a corresponding distribution for a current-carrying wire.

We describe the space around a permanent magnet or a current-carrying conductor as the location of a *magnetic field,* just as we described the space around a charged object as the location of an electric field. The magnitude and direction of the magnetic field, which we define in the next section, are indicated by the vector **B**.* Figure 3 shows an electromagnet, which might be used to produce large magnetic fields in the laboratory.

---

* There is not general agreement on the naming of field vectors in magnetism. **B** may be called the *magnetic induction* or *magnetic flux density,* while another field vector, denoted by **H**, may be called the magnetic field. We regard **B** as the more fundamental quantity and therefore call it the magnetic field.

In electrostatics, we represented the relation between electric field and electric charge symbolically by

$$\text{electric charge} \rightleftarrows \mathbf{E} \rightleftarrows \text{electric charge.} \quad (1)$$

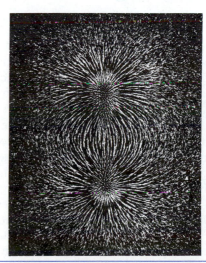

Figure 1 Iron filings sprinkled on a sheet of paper covering a bar magnet. The distribution of the filings suggests the pattern of lines of the magnetic field.

Figure 2 Iron filings on a sheet of paper through which passes a wire carrying a current. The pattern suggests the lines of the magnetic field.

Figure 3 A laboratory electromagnet, consisting of two coils C about 1 m in diameter and two iron pole pieces P, all supported in a rigid frame F. A large magnetic field, in this case horizontal, is established in the few-centimeter gap between the pole pieces.

That is, electric charges set up an electric field, which in turn can exert a force of electric origin on other charges.

It is tempting to try to exploit the symmetry between electric and magnetic fields by writing

$$\text{magnetic charge} \rightleftarrows \mathbf{B} \rightleftarrows \text{magnetic charge.} \quad (2)$$

However, individual magnetic charges (called *magnetic monopoles*; see Section 37-1) either do not exist or are so exceedingly rare that such a relationship is of no practical value. The more useful relationship is

$$\text{moving electric charge} \rightleftarrows \mathbf{B} \rightleftarrows \text{moving electric charge,} \quad (3)$$

which we can also write as

$$\text{electric current} \rightleftarrows \mathbf{B} \rightleftarrows \text{electric current.} \quad (4)$$

A moving electric charge or an electric current sets up a magnetic field, which can then exert a magnetic force on other *moving* charges or currents. There is indeed a symmetry between Eq. 1 for the electric field and Eq. 3 or 4 for the magnetic field.

Another similarity between $\mathbf{E}$ and $\mathbf{B}$ is that we represent both with field lines. As was the case with electric field lines, the lines of $\mathbf{B}$ are drawn so that the tangent to any line gives the direction of $\mathbf{B}$ at that point, and the number of lines crossing any particular area at right angles gives a measure of the magnitude of $\mathbf{B}$. That is, the lines are close together where B is large, and the lines are far apart where B is small. However, there is one very important difference between the two cases: the electric force on a charged particle is always parallel to the lines of $\mathbf{E}$ but, as we shall see, the magnetic force on a moving charged particle is always perpendicular to the lines of $\mathbf{B}$. A difference of this sort is suggested by a comparison of Eqs. 1 and 3: Eq. 1

involves only one vector ($\mathbf{E}$), while Eq. 3 involves two vectors ($\mathbf{B}$ and $\mathbf{v}$). The magnetic force on a moving charge is thus more complex than the electric force on a static charge. Another difference, as we shall see, is that the lines of $\mathbf{E}$ always begin and end at charges, while the lines of $\mathbf{B}$ always form closed loops.

34-2 THE MAGNETIC FORCE ON A MOVING CHARGE

Our goal in this chapter is to establish a set of procedures for determining if a magnetic field is present in a certain region of space (such as between the poles of the electromagnet of Fig. 3) and to study its effects in terms of the force of magnetic origin exerted on objects, such as moving charges, that are present in that region. In the next chapter we consider the source of the $\mathbf{B}$ field and the calculation of its magnitude and direction.

Let us therefore consider a set of measurements that, at least in principle, could be done to study the magnetic force that may act on an electric charge. (In these experiments, we consider only electric or magnetic forces; we assume that the experiments are carried out in an environment where other forces, such as gravity, may be neglected.)

1. We first test for the presence of an *electric* force by placing a small test charge at rest at various locations. Later we can subtract the electric force (if any) from the total force, which presumably leaves only the magnetic force. We assume this has been done, so that from now on we can ignore any electric force that acts on the charge.

2. Next we project the test charge q through a particular point P with a velocity **v**. We find that the magnetic force **F**, if it is present, always acts sideways, that is, at right angles to the direction of **v**. We can repeat the experiment by projecting the charge through P in different directions; we find that, no matter what the direction of **v**, the magnetic force is always at right angles to that direction.

3. As we vary the direction of **v** through point P, we also find that the magnitude of F changes from zero when **v** has a certain direction to a maximum when it is at right angles to that direction. At intermediate angles, the magnitude of F varies as the sine of the angle ϕ that the velocity vector makes with that particular direction. (Note that there are actually two directions of **v** for which **F** is zero; these directions are opposite to each other, that is, $\phi = 0°$ or $180°$.)

4. As we vary the magnitude of the velocity, we find that the magnitude of F varies in direct proportion.

5. We also find that F is proportional to the magnitude of the test charge q, and that **F** reverses direction when q changes sign.

We now define the magnetic field **B** in the following way, based on these observations: the direction of **B** at point P is the same as one of the directions of **v** (to be specified shortly) in which the force is zero, and the magnitude of **B** is determined from the magnitude $F_\perp$ of the maximum force exerted when the test charge is projected perpendicular to the direction of **B**; that is,

$$B = \frac{F_\perp}{qv}. \tag{5}$$

At arbitrary angles, our observations are summarized by the formula

$$F = qvB \sin \phi, \tag{6}$$

where ϕ is the smaller angle between **v** and **B**. Because F, v, and B are vectors, Eq. 6 can be written as a vector product:

$$\mathbf{F} = q\mathbf{v} \times \mathbf{B}. \tag{7}$$

By writing $\mathbf{v} \times \mathbf{B}$ instead of $\mathbf{B} \times \mathbf{v}$ in Eq. 7, we have specified which of the two possible directions of **B** that we want to use.

Figure 4 shows the geometrical relationship among the vectors **F**, **v**, and **B**. Note that, as is always the case for a vector product, **F** is perpendicular to the plane formed by **v** and **B**. Thus **F** is always perpendicular to **v**, and the magnetic force is always a sideways deflecting force. Note also that **F** vanishes when **v** is either parallel or antiparallel to the direction of **B** (in which case $\phi = 0°$ or $180°$, and $\mathbf{v} \times \mathbf{B} = 0$), and that **F** has its maximum magnitude, equal to qvB, when **v** is at right angles to **B**.

Because the magnetic force is always perpendicular to **v**, it cannot change the magnitude of **v**, only its direction. Equivalently, the force is always at right angles to the displacement of the particle and can do no work on it.

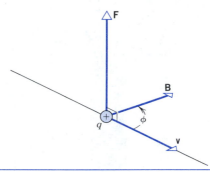

Figure 4 A particle with a positive charge q moving with velocity **v** through a magnetic field **B** experiences a magnetic deflecting force **F**.

Thus a constant magnetic field cannot change the kinetic energy of a moving charged particle. (In Chapter 36 we consider time-varying magnetic fields, which *can* change the kinetic energy of a particle. In this chapter, we deal only with magnetic fields that do not vary with time.)

Equation 7, which serves as the definition of the magnetic field **B**, indicates both its magnitude and its direction. We define the electric field similarly through an equation, $\mathbf{F} = q\mathbf{E}$, so that by measuring the electric force we can determine the magnitude *and* direction of the electric field. Magnetic fields cannot be determined quite so simply with a single measurement. As Fig. 4 suggests, measuring **F** for a single **v** is not sufficient to determine **B**, because the direction of **F** does not indicate the direction of **B**. We must first find the direction of **B** (for example, by finding the directions of **v** for which there is no force), and then a single additional measurement can determine its magnitude.

The SI unit of **B** is the *tesla* (abbreviation T). It follows from Eq. 5 that

$$1 \text{ tesla} = 1 \frac{\text{newton}}{\text{coulomb} \cdot \text{meter/second}} = 1 \frac{\text{newton}}{\text{ampere} \cdot \text{meter}}.$$

An earlier (non-SI) unit for **B**, still in common use, is the *gauss,* related to the tesla by

$$1 \text{ tesla} = 10^4 \text{ gauss}.$$

Table 1 gives some typical values of magnetic fields.

TABLE 1 TYPICAL VALUES OF SOME MAGNETIC FIELDS[a]

Location	Magnetic Field (T)
At the surface of a neutron star (calculated)	10^8
Near a superconducting magnet	5
Near a large electromagnet	1
Near a small bar magnet	10^{-2}
At the surface of the Earth	10^{-4}
In interstellar space	10^{-10}
In a magnetically shielded room	10^{-14}

[a] Approximate values.

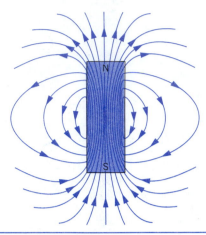

Figure 5   The magnetic field lines for a bar magnet. The lines form closed loops, leaving the magnet at its north pole and entering at its south pole.

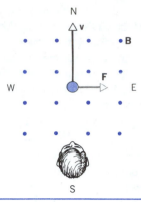

Figure 6   Sample Problem 1. A view from above of a student sitting in a room in which a vertically upward magnetic field deflects a moving proton toward the east. (The dots, which represent the points of arrows, symbolize vectors pointing out of the page.)

Figure 5 (see also Fig. 1) shows the lines of **B** of a bar magnet. Note that the lines of **B** pass through the magnet, forming closed loops. From the clustering of field lines outside the magnet near its ends, we infer that the magnetic field has its greatest magnitude there. These ends are called the *poles* of the magnet, with the designations *north* and *south* given to the poles from which the lines respectively emerge and enter.

Opposite magnetic poles attract one another (thus the north pole of one bar magnet attracts the south pole of another), and like magnetic poles repel one another. An ordinary magnetic compass is nothing but a suspended magnet, whose north end points in the general direction of geographic north. Thus the Earth's magnetic pole in the Arctic region must be a *south* magnetic pole, and the pole in Antarctica must be a *north* magnetic pole. Near the equator the magnetic field lines are nearly parallel to the surface and directed from geographic south to north (as you can deduce from turning Fig. 5 upside down).

Sample Problem 1   A uniform magnetic field **B**, with magnitude 1.2 mT, points vertically upward throughout the volume of the room in which you are sitting. A 5.3-MeV proton moves horizontally from south to north through a certain point in the room. What magnetic deflecting force acts on the proton as it passes through this point? The proton mass is 1.67×10^{-27} kg.

Solution   The magnetic deflecting force depends on the speed of the proton, which we can find from $K = \frac{1}{2}mv^2$. Solving for v, we find

$$v = \sqrt{\frac{2K}{m}} = \sqrt{\frac{(2)(5.3 \text{ MeV})(1.60 \times 10^{-13} \text{ J/MeV})}{1.67 \times 10^{-27} \text{ kg}}}$$

$$= 3.2 \times 10^7 \text{ m/s}.$$

Equation 6 then yields

$$F = qvB \sin \phi$$
$$= (1.60 \times 10^{-19} \text{ C})(3.2 \times 10^7 \text{m/s})(1.2 \times 10^{-3} \text{ T})(\sin 90°)$$
$$= 6.1 \times 10^{-15} \text{ N}.$$

This may seem like a small force, but it acts on a particle of small mass, producing a large acceleration, namely,

$$a = \frac{F}{m} = \frac{6.1 \times 10^{-15} \text{ N}}{1.67 \times 10^{-27} \text{ kg}} = 3.7 \times 10^{12} \text{ m/s}^2.$$

It remains to find the direction of **F** when, as in Fig. 6, **v** points horizontally from south to north, and **B** points vertically up. Using Eq. 7 and the right-hand rule for the direction of vector products (see Section 3-5), we conclude that the deflecting force **F** must point horizontally from west to east, as Fig. 6 shows.

If the charge of the particle had been negative, the magnetic deflecting force would have pointed in the opposite direction, that is, horizontally from east to west. This is predicted automatically by Eq. 7, if we substitute $-q$ for q.

In this calculation, we used the (approximate) classical expression ($K = \frac{1}{2}mv^2$) for the kinetic energy of the proton rather than the (exact) relativistic expression (see Eq. 25 of Chapter 7). The criterion for safely using the classical expression is $K \ll mc^2$, where mc^2 is the rest energy of the particle. In this case $K = 5.3$ MeV, and the rest energy of a proton (see Appendix F) is 938 MeV. This proton passes the test, and we were justified in using the classical $K = \frac{1}{2}mv^2$ formula for the kinetic energy. In dealing with energetic particles, we must always be alert to this point.

The Lorentz Force

If both an electric field **E** and a magnetic field **B** act on a charged particle, the total force on it can be expressed as

$$\mathbf{F} = q\mathbf{E} + q\mathbf{v} \times \mathbf{B}. \tag{8}$$

This force is called the *Lorentz force*. The Lorentz force is not a new kind of force: it is merely the sum of the electric

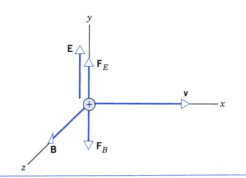

Figure 7   A positively charged particle, moving through a region in which there are electric and magnetic fields perpendicular to one another, experiences opposite electric and magnetic forces $\mathbf{F}_E$ and $\mathbf{F}_B$.

and magnetic forces that may simultaneously act on a charged particle. The electric part of this force acts on any charged particle, whether at rest or in motion; the magnetic part acts only on moving charged particles.

One common application of the Lorentz force occurs when a beam of charged particles passes through a region in which the **E** and **B** fields are perpendicular to each other and to the velocity of the particles. If **E**, **B**, and **v** are oriented as shown in Fig. 7, then the electric force $\mathbf{F}_E = q\mathbf{E}$ is in the opposite direction to the magnetic force $\mathbf{F}_B = q\mathbf{v} \times \mathbf{B}$. We can adjust the magnetic and electric fields until the magnitudes of the forces are equal, in which case the Lorentz force is zero. In scalar terms,

$$qE = qvB \tag{9}$$

or

$$v = \frac{E}{B} . \tag{10}$$

The crossed **E** and **B** fields therefore serve as a *velocity selector*: only particles with speed $v = E/B$ pass through the region unaffected by the two fields, while particles with other velocities are deflected. This value of v is independent of the charge or mass of the particles.

Beams of charged particles are often prepared using methods that give a distribution of velocities (for example, a thermal distribution such as that of Fig. 11 of Chapter 24). Using a velocity selector we can isolate particles with a chosen speed from the beam. This principle was applied in 1897 by J. J. Thomson in his discovery of the electron and the measurement of its charge-to-mass ratio. Figure 8 shows a modern version of his apparatus. Thomson first measured the vertical deflection y of the beam when only the electric field was present. From Sample Problem 6 of Chapter 28, the deflection is

$$y = -\frac{qEL^2}{2mv^2} . \tag{11}$$

In this expression, as in Fig. 8, we take the positive y direction to be upward, and E is the *magnitude* of the electric field. The deflection y of a negatively charged particle is positive in Eq. 11 and Fig. 8.

Then the magnetic field was turned on and adjusted until the beam deflection was zero (equivalent to that measured with no fields present). In this case $v = E/B$, and solving for the charge-to-mass ratio with $q = -e$ gives

$$\frac{e}{m} = \frac{2yE}{B^2 L^2} . \tag{12}$$

Thomson's value for e/m (expressed in modern units) was 1.7×10^{11} C/kg, in good agreement with the current value of $1.75881962 \times 10^{11}$ C/kg.

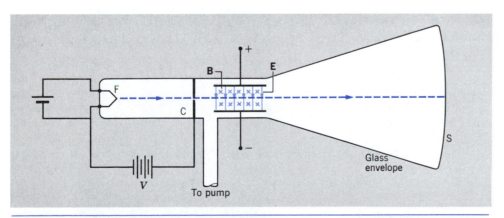

Figure 8   A modern version of J. J. Thomson's apparatus for measuring the charge-to-mass ratio of the electron. The filament F produces a beam of electrons with a distribution of speeds. The electric field E is set up by connecting a battery across the plate terminals. The magnetic field **B** is set up by means of current-carrying coils (not shown). The beam makes a visible spot where it strikes the screen S. (The crosses, which represent the tails of arrows, symbolize **B** vectors pointing into the page.)

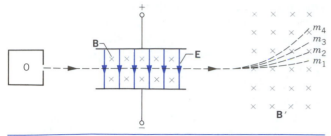

Figure 9 Schematic diagram of a mass spectrometer. A beam of ionized atoms having a mixture of different masses leaves an oven O and enters a region of perpendicular **E** and **B** fields. Only those atoms with speeds $v = E/B$ pass undeflected through the region. Another magnetic field **B′** deflects the atoms along circular paths whose radii are determined by the masses of the atoms.

Another application of the velocity selector is in the mass spectrometer, a device for separating ions by mass (see Section 1-5). In this case a beam of ions, perhaps including species of differing masses, may be obtained from a vapor of the material heated in an oven (see Fig. 9). A velocity selector passes only ions of a particular speed, and when the resulting beam is then passed through another magnetic field, the paths of the particles are circular arcs (as we show in the next section) whose radii are determined by the momentum of the particles. Since all the particles have the same speed, the radius of the path is determined by the mass, and each different mass component in the beam follows a path of a different radius. These atoms can be collected and measured or else formed into a beam for subsequent experiments. See Problems 17 and 22–24 for other details on separating ions by their mass.

Since **B** is perpendicular to **v**, the magnitude of the magnetic force can be written $|q|vB$, and Newton's second law with a centripetal acceleration of v^2/r gives

$$|q|vB = m\frac{v^2}{r} \qquad (13)$$

or

$$r = \frac{mv}{|q|B} = \frac{p}{|q|B}. \qquad (14)$$

Thus the radius of the path is determined by the momentum p of the particles, their charge, and the strength of the magnetic field. If the source of the electrons in Fig. 10 had projected them with a smaller speed, they would have moved in a circle of smaller radius.

The angular velocity of the circular motion is

$$\omega = \frac{v}{r} = \frac{|q|B}{m}, \qquad (15)$$

and the corresponding frequency is

$$v = \frac{\omega}{2\pi} = \frac{|q|B}{2\pi m}. \qquad (16)$$

Note that the frequency associated with the circular motion does not depend on the speed of the particle (as long as $v \ll c$, as we discuss below). Thus, if electrons in Fig. 10 were projected with a smaller speed, they would require the same time to complete the smaller circle that the faster electrons require to complete the larger circle. The frequency given by Eq. 16 is called the *cyclotron frequency*, because particles circulate at this frequency in a cyclotron. The frequency is characteristic of a particular particle moving in a particular magnetic field, just as the oscillating pendulum or the mass–spring system has its characteristic frequency.

34-3   CIRCULATING CHARGES

Figure 10 shows a beam of electrons traveling through an evacuated chamber in which there is a uniform magnetic field **B** out of the plane of the figure. The magnetic deflecting force is the only important force that acts on the electrons. The beam clearly follows a circular path in the plane of the figure. Let us see how we can understand this behavior.

The magnetic deflecting force has two properties that affect the trajectories of charged particles: (1) it does not change the speed of the particles, and (2) it always acts perpendicular to the velocity of the particles. These are exactly the characteristics we require for a particle to move in a circle at constant speed, as in the case of the electrons in Fig. 10.

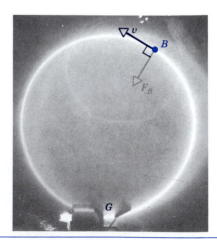

Figure 10 Electrons circulating in a chamber containing a gas at low pressure. The beam is made visible by collisions with the atoms of the gas. A uniform magnetic field **B**, pointing out of the plane of the figure at right angles to it, fills the chamber. The magnetic force **F**$_B$ is directed radially inward.

Figure 11   A cyclotron accelerator. The magnets are in the large chambers at the top and bottom. The beam is visible as it emerges from the accelerator because, like the beam of electrons in Fig. 10, it ionizes air molecules in collisions.

The Cyclotron

The cyclotron (Fig. 11) is an accelerator that produces beams of energetic charged particles, which might be used in nuclear reaction experiments. Figure 12 shows a schematic view of a cyclotron. It consists of two hollow metal

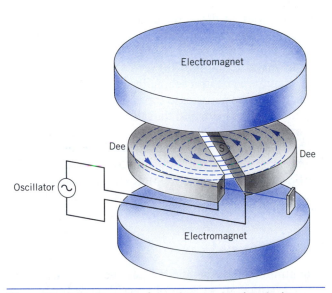

Figure 12   The elements of a cyclotron, showing the ion source S and the dees. The electromagnets provide a uniform vertical magnetic field. The particles spiral outward within the hollow dees, gaining energy every time they cross the gap between the dees.

D-shaped objects called *dees*. The dees are made of conducting material such as sheets of copper and are open along their straight edges. They are connected to an electric oscillator, which establishes an oscillating potential difference between the dees. A magnetic field is perpendicular to the plane of the dees. At the center of the instrument is a source that emits the ions we wish to accelerate.

When the ions are in the gap between the dees, they are accelerated by the potential difference between the dees. They then enter one of the dees, where they feel no electric field (the electric field inside a conductor being zero), but the magnetic field (which is not shielded by the copper dees) bends their path into a semicircle. When the particles next enter the gap, the oscillator has reversed the direction of the electric field, and the particles are again accelerated as they cross the gap. Moving with greater speed, they travel a path of greater radius, as required by Eq. 14. However, according to Eq. 16, *it takes them exactly the same amount of time to travel the larger semicircle*; this is the critical characteristic of the operation of the cyclotron. The frequency of the electric oscillator must be adjusted to be equal to the cyclotron frequency (determined by the magnetic field and the charge and mass of the particle to be accelerated); this equality of frequencies is called the *resonance condition*. If the resonance condition is satisfied, particles continue to accelerate in the gap and "coast" around the semicircles, gaining a small increment of energy in each circuit, until they are deflected out of the accelerator.

The final speed of the particles is determined by the radius R at which the particles leave the accelerator. From Eq. 14,

$$v = \frac{|q|BR}{m},\qquad(17)$$

and the corresponding (nonrelativistic) kinetic energy of the particles is

$$K = \tfrac{1}{2}mv^2 = \frac{q^2B^2R^2}{2m}.\qquad(18)$$

Typical cyclotrons produce beams of protons with maximum energies in the range of 10 MeV. For a given mass, ions with higher electric charges emerge with energies that increase with the square of the charge.

It is somewhat surprising that the energy in Eq. 18 depends on the magnetic field, which does not accelerate the particles, but does not depend on the electric potential difference that causes the acceleration. A larger potential difference gives the particles a larger "kick" with each cycle; the radius increases more quickly, and the particles make fewer cycles before leaving the accelerator. With a smaller potential difference, the particles make more cycles but get a smaller "kick" each time. Thus the energy of the particles is independent of the potential difference.

The Synchrotron

In principle, we should be able to increase the energy of the beam of particles in a cyclotron by increasing the radius. However, above about 50 MeV, the resonance condition is lost. To understand this effect we must return to Eq. 14, in which we used the classical momentum mv. Even at a proton kinetic energy of 50 MeV, $v/c = 0.3$; thus the classical expression mv should not be used. The expression $r = p/|q|B$, however, is correct, if we use the relativistic expression for the momentum, $p = mv/\sqrt{1 - v^2/c^2}$ (see Eq. 22 of Chapter 9), and so Eq. 16 becomes

$$v = \frac{|q|B\sqrt{1 - v^2/c^2}}{2\pi m}.\qquad(19)$$

In this case, the frequency ν is no longer constant (as it was in Eq. 16) but now depends on the speed v. The resonance between the circulating frequency and the oscillator frequency no longer occurs.

This difficulty can be relieved by adjusting the magnetic field so that it increases at larger radii. Cyclotrons operating on this principle include the 500-MeV proton accelerators in the nuclear physics laboratories near Vancouver, Canada and Zurich, Switzerland. Continued increase in the energy is limited by the cost of building larger magnets; to reach an energy of 500 GeV would require a magnet with an area of about 1000 acres!

Higher energies are reached using an accelerator of a different design, called a *synchrotron*. One example is the 1000-GeV proton synchrotron at the Fermi National Ac-

Figure 13   A view along the Fermilab tunnel. The accelerated beam passes through many individual magnet sections, of rectangular cross section and of length about 2 m, several of which can be seen here.

celerator Laboratory near Chicago (Fig. 13; see also Fig. 19 of Chapter 10). Instead of a single magnet, a synchrotron uses many individual magnets (about 3000 in Fermilab) along the circumference of a circle; each magnet bends the beam through a small angle (0.1°). At a gap in the ring, an electric field accelerates the particles. Particles are accelerated in bursts, and both the frequency of the accelerating potential and the strength of the magnetic field are varied as the particles are accelerated, thereby maintaining the resonance at all energies and keeping the orbital radius constant. In the Fermilab accelerator, the protons make about 400,000 revolutions around the 4-mile circumference in reaching their full energy. It takes about 10 s for the particles to travel this distance at speeds near the speed of light, and thus the accelerator produces one burst every 10 s.

There are presently plans to build an even larger synchrotron, the Superconducting Supercollider (SSC). The SSC ring will be about 20 times the size and produce particles with 20 times the energy of the Fermilab accelerator.*

The Magnetic Mirror

A nonuniform magnetic field can be used to trap a charged particle in a region of space. Figure 14 shows a

---

* See "The Superconducting Supercollider," by J. David Jackson, Maury Tigner, and Stanley Wojcicki, *Scientific American*, March 1986, p. 66.

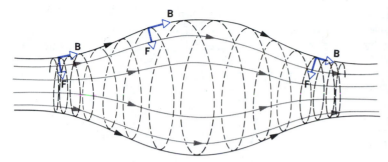

Figure 14   A charged particle spiraling in a nonuniform magnetic field. The field is greater at the left and right ends of the region than it is at the center. Particles can be trapped, spiraling back and forth between the strong-field regions at the ends. Note that the magnetic force vectors at each end of this "magnetic bottle" have components that point toward the center; it is these force components that serve to confine the particles.

schematic view of the operation of such a magnetic mirror. The charged particles tend to move in circles about the field direction. Suppose they also are drifting laterally, say to the right in Fig. 14. The motion is therefore that of a helix, like a coiled spring. The field increases near the ends of the "magnetic bottle," and the force has a small component pointing toward the center of the region, which reverses the direction of the motion of the particles and causes them to spiral in the opposite direction, until they are eventually reflected from the opposite end. The particles continue to travel back and forth, confined to the space between the two high-field regions. Such an arrangement is used to confine the hot ionized gases (called *plasmas*) that are used in research into controlled thermonuclear fusion.

A similar phenomenon occurs in the Earth's magnetic field, as shown in Fig. 15. Electrons and protons are trapped in different regions of the Earth's field and spiral back and forth between the high-field regions near the poles in a time of a few seconds. These fast particles are responsible for the so-called Van Allen radiation belts that surround the Earth.

Sample Problem 2   A particular cyclotron is designed with dees of radius $R = 75$ cm and with magnets that can provide a field of 1.5 T. (*a*) To what frequency should the oscillator be set if deuterons are to be accelerated? (*b*) What is the maximum energy of deuterons that can be obtained?

Solution   (*a*) A deuteron is a nucleus of heavy hydrogen, with a charge $q = +e$ and a mass of 3.34×10^{-27} kg, about twice the mass of ordinary hydrogen. Using Eq. 16 we can find the frequency:

$$\nu = \frac{|q|B}{2\pi m} = \frac{(1.60 \times 10^{-19} \text{ C})(1.5 \text{ T})}{2\pi(3.34 \times 10^{-27} \text{ kg})}$$

$$= 1.1 \times 10^7 \text{ Hz} = 11 \text{ MHz}.$$

(*b*) The maximum energy occurs for deuterons that emerge at the maximum radius R. According to Eq. 18,

$$K = \frac{q^2 B^2 R^2}{2m} = \frac{(1.60 \times 10^{-19} \text{ C})^2 (1.5 \text{ T})^2 (0.75 \text{ m})^2}{2(3.34 \times 10^{-27} \text{ kg})}$$

$$= 4.85 \times 10^{-12} \text{ J} = 30 \text{ MeV}.$$

Deuterons of this energy have a range in air of a few meters, as suggested by Fig. 11.

Numerical Calculation of Path   *(Optional)*

Consider a particle of positive charge q and mass m that passes through the origin moving with speed v_0 in the x direction at $t = 0$ (Fig. 16). A uniform field B_0 is parallel to the z direction. What is the path of the particle?

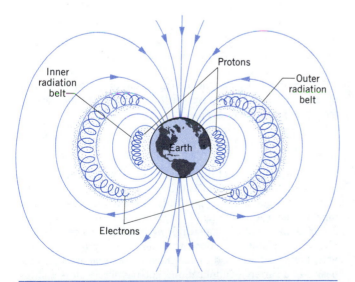

Figure 15   The Earth's magnetic field, showing protons and electrons trapped in the Van Allen radiation belts.

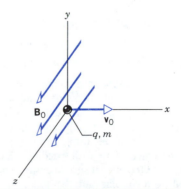

Figure 16   A particle of charge q and mass m passes through the origin with velocity $\mathbf{v}_0$ in the x direction in a region in which there is a uniform field $\mathbf{B}_0$ in the z direction.

There are three methods by which this problem can be solved: (1) use Eq. 14 to find the path, knowing that it must be a circle; (2) use Eq. 7 to find the components of the force on the particle and then solve Newton's laws analytically to obtain $x(t)$, $y(t)$, and $z(t)$; and (3) solve Newton's laws numerically. To demonstrate a general technique that can be applied even when the field is not uniform, we choose method 3. Methods 1 and 2 are considered in Problems 34 and 35.

We begin by writing the components of the force, using Eq. 7 and the expression for the components of the cross product (Eq. 17 of Chapter 3):

$$\mathbf{F} = q(\mathbf{v} \times \mathbf{B}) = q(v_y B_z - v_z B_y)\mathbf{i} + q(v_z B_x - v_x B_z)\mathbf{j} + q(v_x B_y - v_y B_x)\mathbf{k},$$

or, with $B_x = B_y = 0$ and $B_z = B_0$,

$$F_x = q(v_y B_z - v_z B_y) = q v_y B_0,$$
$$F_y = q(v_z B_x - v_x B_z) = -q v_x B_0,$$
$$F_z = q(v_x B_y - v_y B_x) = 0.$$

With no force in the z direction, there can be no acceleration in that direction. The initial velocity has no z component, and thus $v_z = 0$ at all times. The motion is therefore confined to the xy plane. Considering only the x and y motions, Newton's second law then becomes

x component: $\qquad F_x = q v_y B_0 = m \dfrac{dv_x}{dt}$,

y component: $\qquad F_y = -q v_x B_0 = m \dfrac{dv_y}{dt}$.

We solve these equations numerically, as we did in Sections 6-6, 6-7, and 8-4. The motion is divided into intervals of time δt that are small enough that the acceleration can be taken as approximately constant during the interval. We rewrite the above equations in a form that gives the increments of velocity δv_x and δv_y obtained in the interval δt:

$$\delta v_x = (q B_0/m) v_y\, \delta t,$$
$$\delta v_y = -(q B_0/m) v_x\, \delta t.$$

Beginning with the first interval ($t = 0$ to $t = \delta t$), in which $v_x = v_0$ and $v_y = 0$, we find the increments of velocity and then use the formulas of constant acceleration to find the position and velocity at the end of the interval:

$$v_x = v_{0x} + \delta v_x$$
$$v_y = v_{0y} + \delta v_y$$
$$x = x_0 + \bar{v}_x\, \delta t = x_0 + \tfrac{1}{2}(v_{0x} + v_x)\delta t$$
$$y = y_0 + \bar{v}_y\, \delta t = y_0 + \tfrac{1}{2}(v_{0y} + v_y)\delta t$$

where $\bar{v}_x$ and $\bar{v}_y$ are the components of the average velocity in the interval. Continuing through the second and succeeding intervals, we can find x and y at any future time. Appendix I gives a computer program in the BASIC language that does the calculation. Figure 17a shows the resulting motion, calculated for an alpha particle moving initially with speed $v_0 = 3.0 \times 10^6$ m/s in a field $B_0 = 0.15$ T. Of course it should come as no surprise that the motion follows a circular path.

The advantage of this method is that it can easily be adapted to cases in which the field is not uniform. In such cases, the motion

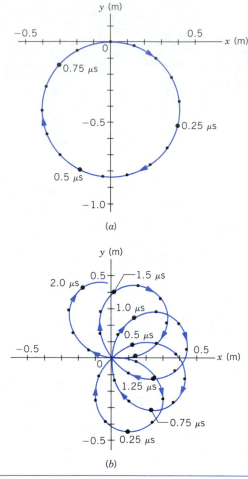

Figure 17 (*a*) The path of the particle is a circle if the field is uniform. The small dots show the positions calculated at intervals of 0.05 μs. (*b*) The path of the particle in the case of a particular nonuniform field.

is not circular, so that method 1 cannot be used, and Newton's laws may have no obvious analytical solution, so that method 2 may not be possible. Method 3 can be used no matter what the nature of the field.

For example, suppose the field again has only a z component in the xy plane but increases with the distance of the particle from the origin according to

$$B = B_0 \left(1 + \frac{\sqrt{x^2 + y^2}}{R} \right),$$

where R is the radius of the particle's path in the previous case (corresponding to the field B_0). Only one minor change in the computer program (see Appendix I) is necessary, and the resulting motion is shown in Fig. 17b. This beautiful and symmetrical flower-shaped pattern is a surprising result of this calculation. Similar calculations are done to design the nonuniform magnetic fields that are used to confine and focus charged particle beams in a variety of applications, such as accelerators and fusion reactors. ∎

34-4   THE HALL EFFECT

In 1879, Edwin H. Hall* conducted an experiment that permitted direct measurement of the sign and the number density (number per unit volume) of charge carriers in a conductor. The *Hall effect* plays a critical role in our understanding of electrical conduction in metals and semiconductors.

Consider a flat strip of material of width w carrying a current i, as shown in Fig. 18. The direction of the current i is the conventional one, opposite to the direction of motion of the electrons. A uniform magnetic field **B** is established perpendicular to the plane of the strip, such as by placing the strip between the poles of an electromagnet. The charge carriers (electrons, for instance) experience a magnetic deflecting force $\mathbf{F} = q\mathbf{v} \times \mathbf{B}$, as shown in the figure, and move to the right side of the strip. Note that positive charges moving in the direction of i experience a deflecting force in the *same* direction.

The buildup of charge along the right side of the strip (and a corresponding deficiency of charge of that sign on the opposite side of the strip), which is the Hall effect, produces an electric field **E** across the strip, as shown in Fig. 18b. Equivalently, a potential difference $V = E/w$, called the *Hall potential difference* (or Hall voltage), exists across the strip. We can measure V by connecting the leads of a voltmeter to points x and y of Fig. 18. As we show below, the sign of V gives the sign of the charge carriers, and the magnitude of V gives their density (number per unit volume). If the charge carriers are electrons, for example, an excess of negative charges builds up on the right side of the strip, and point y is at a lower potential than point x. This may seem like an obvious conclusion in the case of metals; however, you should keep in mind that Hall's work was done nearly 20 years before Thomson's discovery of the electron, and the nature of electrical conduction in metals was not at all obvious at that time.

Let us assume that conduction in the material is due to charge carriers of a particular sign (positive or negative) moving with drift velocity $\mathbf{v_d}$. As the charge carriers drift, they are deflected to the right in Fig. 18 by the magnetic force. As the charges collect on the right side, they set up an electric field that acts inside the conductor to oppose the sideways motion of additional charge carriers. Eventually, an equilibrium is reached, and the Hall voltage reaches its maximum; the sideways magnetic force $(q\mathbf{v_d} \times \mathbf{B})$ is then balanced by the sideways electric force

---

* At the time of his discovery, Hall was a 24-year-old graduate student at the Johns Hopkins University. His research supervisor was Professor Henry A. Rowland, who had a few years earlier shown that a moving electric charge produced the same magnetic effect as an electric current. See "Rowland's Physics," by John D. Miller, *Physics Today,* July 1976, p. 39.

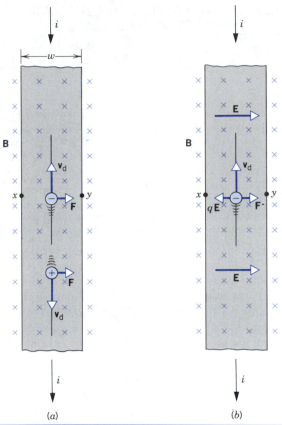

Figure 18   A strip of copper immersed in a magnetic field **B** carries a current i. (*a*) The situation just after the magnetic field has been turned on, and (*b*) the situation at equilibrium, which quickly follows. Note that negative charges pile up on the right side of the strip, leaving uncompensated positive charges on the left. Point x is at a higher potential than point y.

($q\mathbf{E}$). In vector terms, the Lorentz force on the charge carriers under these circumstances is zero:

$$q\mathbf{E} + q\mathbf{v_d} \times \mathbf{B} = 0, \qquad (20)$$

or

$$\mathbf{E} = -\mathbf{v_d} \times \mathbf{B}. \qquad (21)$$

Since $\mathbf{v_d}$ and **B** are at right angles, we can write Eq. 21 in terms of magnitudes as

$$E = v_d B. \qquad (22)$$

From Eq. 6 of Chapter 32 we can write the drift speed as $v_d = j/ne$, where j is the current density in the strip and n is the density of charge carriers. The current density j is the current i per unit cross-sectional area A of the strip. If t is the thickness of the strip, then its cross-sectional area A can be written as wt. Substituting V/w for the electric field E, we obtain

$$\frac{V}{w} = v_d B = \frac{j}{ne}B = \frac{i}{wtne}B$$

TABLE 2 HALL EFFECT RESULTS FOR SELECTED MATERIALS

Material	$n\ (10^{28}/\text{m}^3)$	Sign of V	Number per per atom[a]
Na	2.5	−	0.99
K	1.5	−	1.1
Cu	11	−	1.3
Ag	7.4	−	1.3
Al	21	−	3.5
Sb	0.31	−	0.09
Be	2.6	+	2.2
Zn	19	+	2.9
Si (pure)	1.5×10^{-12}	−	3×10^{-13}
Si (typical *n*-type)	10^{-7}	−	2×10^{-8}

[a] The number of charge carriers per atom of the material as determined from the number per unit volume and the density and molar mass of the material.

or, solving for the density of charge carriers,

$$n = \frac{iB}{etV}. \tag{23}$$

From a measurement of the magnitude of the Hall potential difference *V* we can find the number density of the charge carriers. Table 2 shows a summary of Hall effect data for several metals and semiconductors. For some monovalent metals (Na, K, Cu, Ag) the Hall effect indicates that each atom contributes approximately one free electron to the conduction. For other metals, the number of electrons can be greater than one per atom (Al) or less than one per atom (Sb). For some metals (Be, Zn), the Hall potential difference shows that the charge carriers have a *positive* sign. In this case the conduction is dominated by *holes*, unoccupied energy levels in the valence band (see Section 32-7 and Chapter 53 of the extended text). The holes correspond to the absence of an electron and thus behave like positive charge carriers moving through the material. For some materials, semiconductors in particular, there may be substantial contributions from both electrons and holes, and the simple interpretation of the Hall effect in terms of free conduction by one type of charge carrier is not sufficient. In this case we must use more detailed calculations based on quantum theory.

Sample Problem 3 A strip of copper 150 μm thick is placed in a magnetic field $B = 0.65$ T perpendicular to the plane of the strip, and a current $i = 23$ A is set up in the strip. What Hall potential difference *V* would appear across the width of the strip if there were one charge carrier per atom?

Solution In Sample Problem 2 of Chapter 32, we calculated the number of charge carriers per unit volume for copper, assuming that each atom contributes one electron, and we found

$$n = 8.49 \times 10^{28} \text{ electrons/m}^3.$$

From Eq. 23 then,

$$V = \frac{iB}{net} = \frac{(23 \text{ A})(0.65 \text{ T})}{(8.49 \times 10^{28} \text{ m}^{-3})(1.60 \times 10^{-19} \text{ C})(150 \times 10^{-6} \text{ m})}$$

$$= 7.3 \times 10^{-6} \text{ V} = 7.3 \ \mu\text{V}.$$

This potential difference, though small, is readily measurable.

The Quantized Hall Effect* *(Optional)*

Let us rewrite Eq. 23 as

$$\frac{V}{i} = \frac{1}{etn} B. \tag{24}$$

The quantity on the left has the dimension of resistance (voltage divided by current), although it is not a resistance in the conventional sense. It is commonly called the *Hall resistance*. We can determine the Hall resistance by measuring the Hall voltage *V* in a material carrying a current *i*.

Equation 24 shows that the Hall resistance is expected to increase linearly with the magnetic field *B* for a particular sample of material (in which *n* and *t* are constants). A plot of the Hall resistance against *B* should be a straight line.

In experiments done in 1980, German physicist Klaus von Klitzing discovered that, at high magnetic fields and low temperatures (about 1 K), the Hall resistance did not increase linearly with the field; instead, the plot showed a series of "stair steps," as shown in Fig. 19. This effect has become known as the *quantized Hall effect*, and von Klitzing was awarded the 1985 Nobel Prize in physics for his discovery.

The explanation for this effect involves the circular paths in which electrons are forced to move by the field. Quantum mechanics prevents the electron orbits from overlapping. As the field increases, the orbital radius decreases, permitting more orbits to bunch together on one side of the material. Because the orbital motion of electrons is quantized (only certain orbits being allowed), the changes in orbital motion occur suddenly, corresponding to the steps in Fig. 19. A natural unit of resistance

---

* See "The Quantized Hall Effect," by Bertrand I. Halperin, *Scientific American,* April 1986, p. 52.

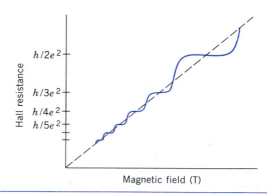

Figure 19 The quantized Hall effect. The dashed line shows the expected classical behavior. The steps show the quantum behavior.

corresponding to orbital motion is h/e^2, where h is Planck's constant, and the steps in Fig. 19 occur at Hall resistances of $h/2e^2$, $h/3e^2$, $h/4e^2$, and so on.

The quantized Hall resistance h/e^2 has the value $25812.806 \, \Omega$ and is known to a precision of less than 1 part in 10^{10}, so the quantized Hall effect has provided a new standard for resistance. This standard, which can be duplicated exactly in laboratories around the world, became the new representation for the ohm in 1990. ■

34-5   THE MAGNETIC FORCE ON A CURRENT

A current is a collection of moving charges. Because a magnetic field exerts a sideways force on a moving charge, it should also exert a sideways force on a wire carrying a current. That is, a sideways force is exerted on the conduction electrons in the wire, but since the electrons cannot escape sideways, the force must be transmitted to the wire itself. Figure 20 shows a wire that passes through a region in which a magnetic field **B** exists. When the wire carries no current (Fig. 20a), it experiences no deflection. When a current is carried by the wire, it deflects (Fig. 20b); when the current is reversed (Fig. 20c), the deflection reverses. The deflection also reverses when the field **B** is reversed.

To understand this effect, we consider the individual charges flowing in a wire (Fig. 21). We use the free-electron model (Section 32-5) for current in a wire, assuming the electrons to move with a constant velocity, the drift velocity v_d. The actual direction of motion of the electrons is of course opposite to the direction we take for the current i in the wire.

The wire passes through a region in which a uniform field **B** exists. The sideways force on each electron (of charge $q = -e$) due to the magnetic field is $-ev_d \times \mathbf{B}$. Let us consider the total sideways force on a segment of the wire of length L. The same force (magnitude and direction) acts on each electron in the segment, and the total force **F** on the segment is therefore equal to the number N of electrons times the force on each electron:

$$\mathbf{F} = -Nev_d \times \mathbf{B}. \qquad (25)$$

How many electrons are contained in that segment of wire? If n is the number density (number per unit volume) of electrons, then the total number N of electrons in the segment is nAL, where A is the cross-sectional area of the wire. Substituting into Eq. 25, we obtain

$$\mathbf{F} = -nALev_d \times \mathbf{B}. \qquad (26)$$

Equation 6 of Chapter 32 ($v_d = i/nAe$) permits us to write Eq. 26 in terms of the current i. To preserve the vector relationship of Eq. 26, we define the vector **L** to be equal in magnitude to the length of the segment and to point in the direction of the current (opposite to the direction of electron flow). The vectors v_d and **L** have opposite directions, and we can write the scalar relationship $nALev_d = iL$ using vectors as

$$-nALev_d = i\mathbf{L}. \qquad (27)$$

Substituting Eq. 27 into Eq. 26, we obtain an expression for the force on the segment:

$$\mathbf{F} = i\mathbf{L} \times \mathbf{B}. \qquad (28)$$

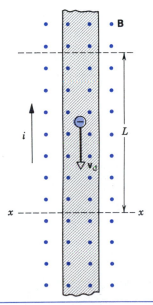

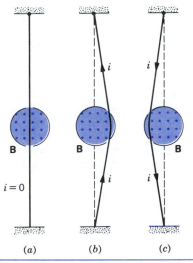

Figure 20   A flexible wire passes between the poles of a magnet. (a) There is no current in the wire. (b) A current is established in the wire. (c) The current is reversed.

Figure 21   A close-up view of a length L of the wire of Fig. 20b. The current direction is upward, which means that electrons drift downward. A magnetic field emerges from the plane of the figure, so that the wire is deflected to the right.

Equation 28 is similar to Eq. 7 ($\mathbf{F} = q\mathbf{v} \times \mathbf{B}$), in that either can be taken to be the defining equation for the magnetic field. Figure 22 shows the vector relationship between $\mathbf{F}$, $\mathbf{L}$, and $\mathbf{B}$; compare with Fig. 4 to see the similarities between Eqs. 28 and 7.

If the segment is perpendicular to the direction of the field, the magnitude of the force can be written

$$F = iLB. \qquad (29)$$

If the wire is not straight or the field is not uniform, we can imagine the wire to be broken into small segments of length $d\mathbf{s}$; we make the segments small enough that they are approximately straight and the field is approximately uniform. The force on each segment can then be written

$$d\mathbf{F} = i \, d\mathbf{s} \times \mathbf{B}. \qquad (30)$$

We can find the total force on the segment of length L by doing a suitable integration over the length.

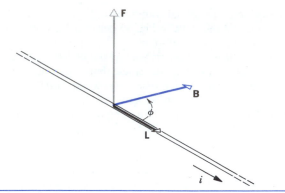

Figure 22 A directed wire segment $\mathbf{L}$ makes an angle ϕ with a magnetic field. Compare carefully with Fig. 4.

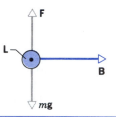

Figure 23 Sample Problem 4. A wire (shown in cross section) can be made to "float" in a magnetic field, with the upward magnetic force $\mathbf{F}$ balancing the downward pull of gravity. The current in the wire emerges from the paper.

Sample Problem 4 A straight, horizontal segment of copper wire carries a current $i = 28$ A. What are the magnitude and direction of the magnetic field needed to "float" the wire, that is, to balance its weight? Its linear mass density is 46.6 g/m.

Solution Figure 23 shows the arrangement. For a length L of wire we have (see Eq. 29)

$$mg = iLB,$$

or

$$B = \frac{(m/L)g}{i} = \frac{(46.6 \times 10^{-3} \text{ kg/m})(9.8 \text{ m/s}^2)}{28 \text{ A}}$$

$$= 1.6 \times 10^{-2} \text{ T} = 16 \text{ mT}.$$

This is about 400 times the strength of the Earth's magnetic field.

Sample Problem 5 Figure 24 shows a wire segment, placed in a uniform magnetic field $\mathbf{B}$ that points out of the plane of the figure. If the segment carries a current i, what resultant magnetic force $\mathbf{F}$ acts on it?

Solution According to Eq. 29, the magnetic force that acts on each straight section has the magnitude

$$F_1 = F_3 = iLB$$

and points down, as shown by the arrows in the figure. The force $d\mathbf{F}$ that acts on a segment of the arc of length $ds = R \, d\theta$ has magnitude

$$dF = iB \, ds = iB(R \, d\theta)$$

and direction radially toward O, the center of the arc. Note that only the downward component ($d\mathbf{F} \sin \theta$) of this force element

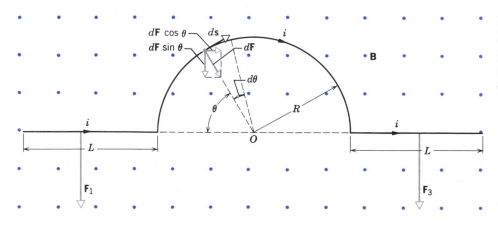

Figure 24 Sample Problem 5. A wire segment carrying a current i is immersed in a magnetic field. The resultant force on the wire is directed downward.

is effective. The horizontal component ($d\mathbf{F} \cos \theta$) is canceled by an oppositely directed horizontal component due to a symmetrically located segment on the opposite side of the arc.

The total force on the central arc points down and is given by

$$F_2 = \int_0^\pi dF \sin\theta = \int_0^\pi (iBR\, d\theta) \sin\theta$$

$$= iBR \int_0^\pi \sin\theta\, d\theta = 2iBR.$$

The resultant force on the entire wire is then

$$F = F_1 + F_2 + F_3 = iLB + 2iBR + iLB$$

$$= iB(2L + 2R).$$

Note that this force is the same as the force that would act on a straight wire of length $2L + 2R$. This would be true no matter what the shape of the central segment, shown as a semicircle in Fig. 24. Can you convince yourself that this is so?

Sample Problem 6   A rectangular loop of wire (Fig. 25), consisting of nine turns and having width $a = 0.103$ m and length $b = 0.685$ m is attached to one pan of a balance. A portion of the loop passes through a region in which there is a uniform magnetic field of magnitude B perpendicular to the plane of the loop, as shown in Fig. 25. The apparatus is carefully adjusted so that the weight of the loop is balanced by an equal weight (not shown)

on the opposite pan . A current $i = 0.224$ A is set up in the wire, and it is found that to restore the balance to its previous equilibrium condition a mass of $m = 13.7$ g must be added to the right-hand pan of the balance. Find the magnitude and direction of the magnetic field.

Solution   Whether the field is into or out of the plane of the page of Fig. 25, the forces on the two lower portions of the long sides of the loop cancel. We therefore consider only the force F on the bottom of the loop, which has magnitude iaB for each of the nine segments of the bottom of the loop that pass through the field. Since it was necessary to add weight to the same pan from which the loop was suspended, the magnetic force on the bottom segment must point upward; the upward magnetic force F is balanced by the additional weight mg on that side. For the force to be upward, the magnetic field must point *into* the plane of the paper (check this with the right-hand rule for vector products). The equilibrium condition is

$$mg = F = 9(iaB)$$

or

$$B = \frac{mg}{9ia} = \frac{(0.0137 \text{ kg})(9.80 \text{ m/s}^2)}{9(0.224 \text{ A})(0.103 \text{ m})} = 0.647 \text{ T}.$$

A device operating on this general principle can be used to provide accurate measurements of magnetic fields.

34-6   TORQUE ON A CURRENT LOOP

When a loop of wire carrying a current is placed in a magnetic field, that loop can experience a torque that tends to rotate it about a particular axis (which for generality we can take through the center of mass of the loop). This principle is the basis of operation of electric motors as well as of galvanometers on which analog current and voltage meters are based. In this section we consider this torque.

Figure 26 shows a rectangular loop of wire in a uniform magnetic field **B**. For simplicity, only the loop itself is shown; we assume that the wires that bring current to and from the loop are twisted together so that there is no net magnetic force on them. We also assume that the loop is suspended in such a way that it is free to rotate about any axis.

The uniform field **B** is in the y direction of the coordinate system of Fig. 26. The loop is oriented so that the z axis lies in its plane. In this orientation, sides 1 and 3 of the loop are perpendicular to **B**. (In the next section, we consider the more general case in which the loop has an arbitrary orientation.) The plane of the loop is indicated by a unit vector **n** that is perpendicular to the plane; the direction of **n** is determined by using the right-hand rule, so that if the fingers of your right hand indicate the direction of the current in the loop, the thumb gives the direction of **n**. The vector **n** makes an angle θ with **B**.

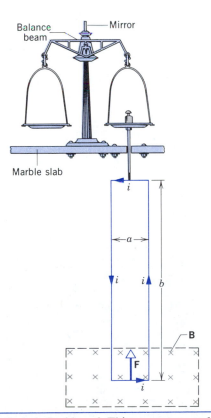

Figure 25   Sample Problem 6. This apparatus can be used to measure **B**. A light beam reflected from the mirror on the balance beam provides a sensitive indication of the deflection.

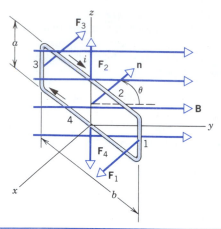

Figure 26   A rectangular loop of wire carrying a current i is placed in a uniform magnetic field. The unit vector **n** is normal to the plane of the loop and makes an angle θ with the field. A torque acts to rotate the loop about the z axis so that **n** aligns with **B**.

The net force on the loop can be determined using Eq. 28, $\mathbf{F} = i\mathbf{L} \times \mathbf{B}$, to calculate the force on each of its four sides. (If the sides of the loop were not straight, it would be necessary to use Eq. 30 to find the magnetic force on it.) As indicated by Fig. 22, the force on each segment must be perpendicular both to **B** and to the direction of the current in the segment. Thus the magnitude of the force $\mathbf{F}_2$ on side 2 (of length b) is

$$F_2 = ibB \sin (90° - \theta) = ibB \cos \theta. \qquad (31)$$

This force points in the positive z direction. The force $\mathbf{F}_4$ on side 4 has magnitude

$$F_4 = ibB \sin (90° + \theta) = ibB \cos \theta, \qquad (32)$$

and points in the negative z direction. These forces are equal and opposite, and so contribute nothing to the net force on the loop. Furthermore, they have the same line of action, so the net torque due to these two forces is also zero.

The forces $\mathbf{F}_1$ and $\mathbf{F}_3$ have a common magnitude of iaB. They are oppositely directed parallel and antiparallel to the x axis in Fig. 26, so they also contribute nothing to the net force on the loop. The sum of all four forces gives a resultant of zero, and we conclude that the center of mass of the loop does not accelerate under the influence of the net magnetic force.

The torques due to forces $\mathbf{F}_1$ and $\mathbf{F}_3$ do *not* cancel, however, because they do not have the same line of action. These two forces tend to rotate the loop about an axis parallel to the z axis. *The direction of the rotation tends to bring* **n** *into alignment with* **B**. That is, in the situation shown in Fig. 26, the loop would rotate clockwise as viewed from the positive z axis, thereby reducing the angle θ. If the current in the loop were reversed, **n** would have the opposite direction, and the loop would again

rotate through the angle (equal to $\pi - \theta$ in Fig. 26) necessary to bring **n** into alignment with **B**.

The forces $\mathbf{F}_1$ and $\mathbf{F}_3$ have moment arms about the z axis of $(b/2)\sin \theta$, and so the total torque on the loop is

$$\tau = 2(iaB)(b/2)\sin \theta = iabB \sin \theta, \qquad (33)$$

where the factor of 2 enters because both forces contribute equally to the torque. Note that if **n** is already parallel to **B** (so that $\theta = 0$) there is no torque.

Equation 33 gives the torque on a single rectangular loop in the field. If we have a coil of N turns (such as might be found in a motor or a galvanometer), Eq. 33 gives the torque on each turn, and the total torque on the coil would be

$$\tau = NiAB \sin \theta, \qquad (34)$$

where we have substituted A, the area of the rectangular loop, for the product ab.

Equation 34 can be shown to hold in general, *for all plane loops of area A, whether they are rectangular or not.* We generalize this result in the next section.

Sample Problem 7   Analog voltmeters and ammeters, in which the reading is displayed by the deflection of a pointer over a scale, work by measuring the torque exerted by a magnetic field on a current loop. Figure 27 shows the rudiments of a *galvanometer*, on which both analog ammeters and analog voltmeters are based. The coil is 2.1 cm high and 1.2 cm wide; it has 250 turns and is mounted so that it can rotate about its axis in a uniform radial magnetic field with $B = 0.23$ T. A spring provides a countertorque that balances the magnetic torque, resulting in a steady angular deflection ϕ corresponding to a given steady current i in the coil. If a current of $100\ \mu A$ produces an angular deflection of

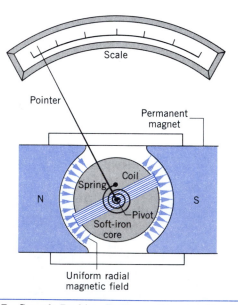

Figure 27   Sample Problem 7. The rudiments of a galvanometer. Depending on the external circuit, this device can act as either a voltmeter or an ammeter.

28° (=0.49 rad), what must be the torsional constant κ of the spring?

Solution   Setting the magnetic torque (Eq. 34) equal to the restoring torque $\kappa\phi$ of the spring yields

$$\tau = NiAB \sin \theta = \kappa\phi,$$

in which ϕ is the angular deflection of the pointer and A (=2.52×10^{-4} m^2) is the area of the coil. Note that the normal to the plane of the coil (that is, the pointer) is always at right angles to the (radial) magnetic field so that $\theta = 90°$ for all pointer positions.

Solving for κ, we find

$$\kappa = \frac{NiAB \sin \theta}{\phi}$$

$$= \frac{(250)(100 \times 10^{-6}\ \text{A})(2.52 \times 10^{-4}\ \text{m}^2)(0.23\ \text{T})(\sin 90°)}{0.49\ \text{rad}}$$

$$= 3.0 \times 10^{-6}\ \text{N} \cdot \text{m/rad}.$$

Many modern ammeters and voltmeters are of the digital, direct-reading type and operate in a way that does not involve a moving coil.

34-7   THE MAGNETIC DIPOLE

In Section 28-7 we considered the effect of an *electric* field **E** on an electric dipole, which we pictured as two equal and opposite charges separated by a distance. Defining the electric dipole moment **p** in a particular way, we found (see Eq. 37 of Chapter 28) that the electric field exerted a torque on the electric dipole that tended to rotate the dipole so that **p** aligned with **E**. This statement appears very similar to one made at the end of the previous section about the effect of a magnetic field on a current loop: the torque on the loop tends to rotate it so that the normal vector **n** aligns with **B**. This similarity suggests that we can use equations similar to those for the electric dipole to analyze the effect of a magnetic field on a current loop. We are encouraged to make this analogy by the similarity between the electric field lines of an electric dipole (see Figs. 8 and 9b of Chapter 28) and the magnetic field lines of a bar magnet, which is an example of a *magnetic dipole* (see Figs. 1 and 5 of this chapter).

The torque on an electric dipole is (Eq. 37 of Chapter 28)

$$\tau = \mathbf{p} \times \mathbf{E}, \tag{35}$$

which can also be written in terms of magnitudes as $\tau = pE \sin \theta$, where θ is the angle between **p** and **E**. Equation 34 of this chapter gives the torque on a coil of current-carrying wire as $\tau = NiAB \sin \theta$. The similarity of these two expressions is striking. Let us, by analogy with the electric case, define a vector $\boldsymbol{\mu}$, the *magnetic dipole moment*, to have magnitude

$$\mu = NiA \tag{36}$$

and direction parallel to **n** (Fig. 26). That is, with the fingers of the right hand in the direction of the current, the thumb gives the direction of $\boldsymbol{\mu}$. We can therefore write Eq. 34 as $\tau = \mu B \sin \theta$ or, in vector form, as

$$\tau = \boldsymbol{\mu} \times \mathbf{B}. \tag{37}$$

Although we have not proved it in general, Eq. 37 gives the most general description of the torque exerted on *any* planar current loop in a uniform magnetic field **B**. It holds no matter what the shape of the loop or the angle between its plane and the field.

We can continue the analogy between electric and magnetic fields by considering the work done to change the orientation of a magnetic dipole in a magnetic field and relating that work to the potential energy of a magnetic dipole in a magnetic field. We can write the potential energy as

$$U = -\mu B \cos \theta = -\boldsymbol{\mu} \cdot \mathbf{B}, \tag{38}$$

for a magnetic dipole whose moment μ makes an angle θ with **B**. This equation is similar to the corresponding expression for an electric dipole, $U = -\mathbf{p} \cdot \mathbf{E}$ (Eq. 42 of Chapter 28).

The magnetic force, like all forces that depend on velocity, is in general *not* conservative and therefore cannot generally be represented by a potential energy. In this special case, in which the torque on a dipole depends on its position relative to the field, it *is* possible to define a potential energy for the *system* consisting of the dipole in the field. Note that the potential energy is not characteristic of the field alone, but of the dipole *in* the field. In general, we cannot define a scalar "magnetic potential energy" of a point charge or "magnetic potential" of the field itself such as we did for electric fields in Chapter 30.

A great variety of physical systems have magnetic dipole moments: the Earth, bar magnets, current loops, atoms, nuclei, and elementary particles. Table 3 gives some typical values; more details on magnetic dipole moments are given in Chapter 37.

Note that Eq. 38 suggests units for μ of energy divided by magnetic field, or J/T. Equation 36 suggests units of current times area, or A · m^2. You can show that these two

TABLE 3   SELECTED VALUES OF MAGNETIC DIPOLE MOMENTS

System	μ (J/T)
Nucleus of nitrogen atom	2.0×10^{-28}
Proton	1.4×10^{-26}
Electron	9.3×10^{-24}
Nitrogen atom	2.8×10^{-23}
Typical small coil[a]	5.4×10^{-6}
Small bar magnet	5
Superconducting coil	400
The Earth	8.0×10^{22}

[a] That of Sample Problem 8, for instance.

units are equivalent, and the choice between them is one of convenience. As indicated by the example of nitrogen, nuclear magnetic dipole moments are typically three to six orders of magnitude smaller than atomic magnetic dipole moments. Several conclusions follow immediately from this observation. (1) Electrons cannot be constituents of the nucleus; otherwise nuclear magnetic dipoole moments would typically have magnitudes about the same as that of the electron. (2) Ordinary magnetic effects in materials are determined by *atomic* magnetism, rather than the much weaker *nuclear* magnetism. (3) To exert a particular torque necessary to align nuclear dipoles requires a magnetic field about three to six orders of magnitude larger than that necessary to align atomic dipoles.

Sample Problem 8   (*a*) What is the magnetic dipole moment of the coil of Sample Problem 7, assuming that it carries a current of 85 μA? (*b*) The magnetic dipole moment of the coil is lined up with an external magnetic field whose strength is 0.85 T. How

much work would be done by an external agent to rotate the coil through 180°?

Solution   (*a*) The magnitude of the magnetic dipole moment of the coil, whose area A is 2.52×10^{-4} m², is

$$\mu = NiA$$
$$= (250)(85 \times 10^{-6} \text{ A})(2.52 \times 10^{-4} \text{ m}^2)$$
$$= 5.36 \times 10^{-6} \text{ A} \cdot \text{m}^2 = 5.36 \times 10^{-6} \text{ J/T}.$$

The direction of μ, as inspection of Fig. 27 shows, must be that of the pointer. You can verify this by showing that, if we assume μ to be in the pointer direction, the torque predicted by Eq. 37 would indeed move the pointer clockwise across the scale.

(*b*) The external work is equal to the increase in potential energy of the system, which is

$$W = \Delta U = -\mu B \cos 180° - (-\mu B \cos 0°) = 2\mu B$$
$$= 2(5.36 \times 10^{-6} \text{ J/T})(0.85 \text{ T}) = 9.1 \times 10^{-6} \text{ J} = 9.1 \ \mu\text{J}.$$

This is about equal to the work needed to lift an aspirin tablet through a vertical height of about 3 mm.

QUESTIONS

1. Of the three vectors in the equation $\mathbf{F} = q\mathbf{v} \times \mathbf{B}$, which pairs are always at right angles? Which may have any angle between them?

2. Why do we not simply define the direction of the magnetic field **B** to be the direction of the magnetic force that acts on a moving charge?

3. Imagine that you are sitting in a room with your back to one wall and that an electron beam, traveling horizontally from the back wall to the front wall, is deflected to your right. What is the direction of the uniform magnetic field that exists in the room?

4. How could we rule out that the forces between two magnets are electrostatic forces?

5. If an electron is not deflected in passing through a certain region of space, can we be sure that there is no magnetic field in that region?

6. If a moving electron is deflected sideways in passing through a certain region of space, can we be sure that a magnetic field exists in that region?

7. A beam of electrons can be deflected either by an electric field or by a magnetic field. Is one method better than the other? In any sense easier?

8. Electric fields can be represented by maps of equipotential surfaces. Can the same be done for magnetic fields? Explain.

9. Is a magnetic force conservative or nonconservative? Justify your answer. Could we define a magnetic potential energy as we defined electric or gravitational potential energy?

10. A charged particle passes through a magnetic field and is deflected. This means that a force acted on it and changed its momentum. Where there is a force there must be a reaction force. On what object does it act?

11. In the Thomson experiment we neglected the deflections

produced by the gravitational field and magnetic field of the Earth. What errors are thereby introduced?

12. Imagine the room in which you are seated to be filled with a uniform magnetic field pointing vertically downward. At the center of the room two electrons are suddenly projected horizontally with the same initial speed but in opposite directions. (*a*) Describe their motions. (*b*) Describe their motions if one particle is an electron and one a positron, that is, a positively charged electron. (The electrons will gradually slow down as they collide with molecules of the air in the room.)

13. Figure 28 shows the tracks of two electrons (e⁻) and a positron (e⁺) in a bubble chamber. A magnetic field fills the chamber, perpendicular to the plane of the figure. Why are

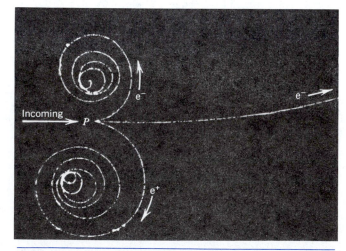

Figure 28   Question 13.

the tracks spirals and not circles? What can you tell about the particles from their tracks? What is the direction of the magnetic field?

14. What are the primary functions of (a) the electric field and (b) the magnetic field in the cyclotron?

15. In a given magnetic field, would a proton or an electron, traveling at the same speed, have the greater frequency of revolution? Consider relativistic effects.

16. What central fact makes the operation of a conventional cyclotron possible? Ignore relativistic considerations.

17. A bare copper wire emerges from one wall of a room, crosses the room, and disappears into the opposite wall. You are told that there is a steady current in the wire. How can you find its direction? Describe as many ways as you can think of. You may use any reasonable piece of equipment, but you may not cut the wire.

18. Discuss the possibility of using the Hall effect to measure the strength B of a magnetic field.

19. (a) In measuring Hall potential differences, why must we be careful that points x and y in Fig. 18 are exactly opposite each other? (b) If one of the contacts is movable, what procedure might we follow in adjusting it to make sure that the two points are properly located?

20. In Section 34-5, we state that a magnetic field **B** exerts a sideways force on the conduction electrons in, say, a copper wire carrying a current i. We have tacitly assumed that this same force acts on the conductor itself. Are there some missing steps in this argument? If so, supply them.

21. A straight copper wire carrying a current i is immersed in a magnetic field **B**, at right angles to it. We know that **B** exerts a sideways force on the free (or conduction) electrons. Does it do so on the bound electrons? After all, they are not at rest. Discuss.

22. Does Eq. 28 ($\mathbf{F} = i\mathbf{L} \times \mathbf{B}$) hold for a straight wire whose cross section varies irregularly along its length (a "lumpy" wire)?

23. A current in a magnetic field experiences a force. Therefore it should be possible to pump conducting liquids by sending a current through the liquid (in an appropriate direction) and letting it pass through a magnetic field. Design such a pump. This principle is used to pump liquid sodium (a

conductor, but highly corrosive) in some nuclear reactors, where it is used as a coolant. What advantages would such a pump have?

24. A uniform magnetic field fills a certain cubical region of space. Can an electron be fired into this cube from the outside in such a way that it will travel in a closed circular path inside the cube?

25. A conductor, even though it is carrying a current, has zero net charge. Why then does a magnetic field exert a force on it?

26. You wish to modify a galvanometer (see Sample Problem 7) to make it into (a) an ammeter and (b) a voltmeter. What do you need to do in each case?

27. A rectangular current loop is in an arbitrary orientation in an external magnetic field. How much work is required to rotate the loop about an axis perpendicular to its plane?

28. Equation 37 ($\boldsymbol{\tau} = \boldsymbol{\mu} \times \mathbf{B}$) shows that there is no torque on a current loop in an external magnetic field if the angle between the axis of the loop and the field is (a) 0° or (b) 180°. Discuss the nature of the equilibrium (that is, whether it is stable, neutral, or unstable) for these two positions.

29. In Sample Problem 8 we showed that the work required to turn a current loop end-for-end in an external magnetic field is $2\mu B$. Does this result hold no matter what the original orientation of the loop was?

30. Imagine that the room in which you are seated is filled with a uniform magnetic field pointing vertically upward. A circular loop of wire has its plane horizontal. For what direction of current in the loop, as viewed from above, will the loop be in stable equilibrium with respect to forces and torques of magnetic origin?

31. The torque exerted by a magnetic field on a magnetic dipole can be used to measure the strength of that magnetic field. For an accurate measurement, does it matter whether the dipole moment is small or not? Recall that, in the case of measurement of an electric field, the test charge was to be as small as possible so as not to disturb the source of the field.

32. You are given a frictionless sphere the size of a Ping-Pong ball and told that it contains a magnetic dipole. What experiments would you carry out to find the magnitude and the direction of its magnetic dipole moment?

PROBLEMS

Section 34-2 The Magnetic Force on a Moving Charge

1. Four particles follow the paths shown in Fig. 29 as they pass through the magnetic field there. What can one conclude about the charge of each particle?

2. An electron in a TV camera tube is moving at 7.2×10^6 m/s in a magnetic field of strength 83 mT. (a) Without knowing the direction of the field, what could be the greatest and least magnitudes of the force the electron could feel due to the field? (b) At one point the acceleration of the electron is 4.9×10^{16} m/s². What is the angle between the electron's velocity and the magnetic field?

3. An electric field of 1.5 kV/m and a magnetic field of 0.44 T act on a moving electron to produce no force. (a) Calculate

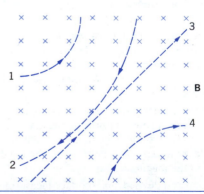

Figure 29 Problem 1.

the minimum electron speed v. (b) Draw the vectors **E**, **B**, and **v**.

4. A proton traveling at 23.0° with respect to a magnetic field of strength 2.63 mT experiences a magnetic force of 6.48×10^{-17} N. Calculate (a) the speed and (b) the kinetic energy in eV of the proton.

5. A cosmic ray proton impinges on the Earth near the equator with a vertical velocity of 2.8×10^7 m/s. Assume that the horizontal component of the Earth's magnetic field at the equator is 30 μT. Calculate the ratio of the magnetic force on the proton to the gravitational force on it.

6. An electron is accelerated through a potential difference of 1.0 kV and directed into a region between two parallel plates separated by 20 mm with a potential difference of 100 V between them. If the electron enters moving perpendicular to the electric field between the plates, what magnetic field is necessary perpendicular to both the electron path and the electric field so that the electron travels in a straight line?

7. An electron in a uniform magnetic field has a velocity **v** = $40\mathbf{i} + 35\mathbf{j}$ km/s. It experiences a force **F** = $-4.2\mathbf{i} + 4.8\mathbf{j}$ fN. If $B_x = 0$, calculate the magnetic field.

8. An ion source is producing ions of ^{6}Li (mass = 6.01 u) each carrying a net charge of $+e$. The ions are accelerated by a potential difference of 10.8 kV and pass horizontally into a region in which there is a vertical magnetic field $B = 1.22$ T. Calculate the strength of the horizontal electric field to be set up over the same region that will allow the ^{6}Li ions to pass through undeflected.

9. The electrons in the beam of a television tube have a kinetic energy of 12.0 keV. The tube is oriented so that the electrons move horizontally from magnetic south to magnetic north. The vertical component of the Earth's magnetic field points down and has a magnitude of 55.0 μT. (a) In what direction will the beam deflect? (b) What is the acceleration of a given electron due to the magnetic field? (c) How far will the beam deflect in moving 20.0 cm through the television tube?

10. An electron has an initial velocity $12.0\mathbf{j} + 15.0\mathbf{k}$ km/s and a constant acceleration of $(2.00 \times 10^{12}$ m/s$^2)\mathbf{i}$ in a region in which uniform electric and magnetic fields are present. If **B** = $400\mathbf{i}$ μT, find the electric field **E**.

Section 34-3 Circulating Charges

11. (a) In a magnetic field with $B = 0.50$ T, for what path radius will an electron circulate at 0.10 the speed of light? (b) What will be its kinetic energy in eV? Ignore the small relativistic effects.

12. A 1.22-keV electron is circulating in a plane at right angles to a uniform magnetic field. The orbit radius is 24.7 cm. Calculate (a) the speed of the electron, (b) the magnetic field, (c) the frequency of revolution, and (d) the period of the motion.

13. An electron is accelerated from rest by a potential difference of 350 V. It then enters a uniform magnetic field of magnitude 200 mT, its velocity being at right angles to this field. Calculate (a) the speed of the electron and (b) the radius of its path in the magnetic field.

14. S. A. Goudsmit devised a method for measuring accurately the masses of heavy ions by timing their period of revolution in a known magnetic field. A singly charged ion of iodine makes 7.00 rev in a field of 45.0 mT in 1.29 ms. Calculate its mass, in atomic mass units. Actually, the mass measurements are carried out to much greater accuracy than these approximate data suggest.

15. An alpha particle ($q = +2e$, $m = 4.0$ u) travels in a circular path of radius 4.5 cm in a magnetic field with $B = 1.2$ T. Calculate (a) its speed, (b) its period of revolution, (c) its kinetic energy in eV, and (d) the potential difference through which it would have to be accelerated to achieve this energy.

16. A beam of electrons whose kinetic energy is K emerges from a thin-foil "window" at the end of an accelerator tube. There is a metal plate a distance d from this window and at right angles to the direction of the emerging beam. See Fig. 30. (a) Show that we can prevent the beam from hitting the plate if we apply a magnetic field B such that

$$B \geq \sqrt{\frac{2mK}{e^2d^2}},$$

in which m and e are the electron mass and charge. (b) How should **B** be oriented?

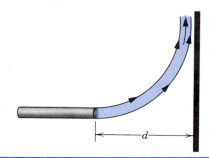

Figure 30 Problem 16.

17. Bainbridge's mass spectrometer, shown in Fig. 31, separates ions having the same velocity. The ions, after entering through slits S_1 and S_2, pass through a velocity selector composed of an electric field produced by the charged plates P and P', and a magnetic field **B** perpendicular to the electric field and the ion path. Those ions that pass undeviated through the crossed **E** and **B** fields enter into a region where a second magnetic field **B'** exists, and are bent into circular paths. A photographic plate registers their arrival. Show that $q/m = E/rBB'$, where r is the radius of the circular orbit.

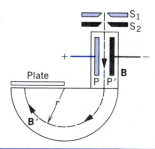

Figure 31 Problem 17.

18. A physicist is designing a cyclotron to accelerate protons to 0.100c. The magnet used will produce a field of 1.40 T.

Calculate (*a*) the radius of the cyclotron and (*b*) the corresponding oscillator frequency. Relativity considerations are not significant.

19. In a nuclear experiment a proton with kinetic energy K_p moves in a uniform magnetic field in a circular path. What energy must (*a*) an alpha particle and (*b*) a deuteron have if they are to circulate in the same orbit? (For a deuteron, $q = +e$, $m = 2.0$ u; for an alpha particle, $q = +2e$, $m = 4.0$ u.)

20. A proton, a deuteron, and an alpha particle, accelerated through the same potential difference V, enter a region of uniform magnetic field, moving at right angles to **B**. (*a*) Find their kinetic energies. If the radius of the proton's circular path is r_p, what are the radii of (*b*) the deuteron and (*c*) the alpha particle paths, in terms of r_p?

21. A proton, a deuteron, and an alpha particle with the same kinetic energy enter a region of uniform magnetic field, moving at right angles to **B**. The proton moves in a circle of radius r_p. In terms of r_p, what are the radii of (*a*) the deuteron path and (*b*) the alpha particle path?

22. Figure 32 shows an arrangement used to measure the masses of ions. An ion of mass m and charge $+q$ is produced essentially at rest in source S, a chamber in which a gas discharge is taking place. The ion is accelerated by potential difference V and allowed to enter a magnetic field **B**. In the field it moves in a semicircle, striking a photographic plate at distance x from the entry slit. Show that the ion mass m is given by

$$m = \frac{B^2 q}{8V} x^2.$$

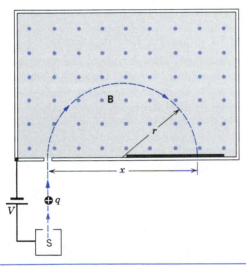

Figure 32 Problem 22.

23. Two types of singly ionized atoms having the same charge q and mass differing by a small amount Δm are introduced into the mass spectrometer described in Problem 22. (*a*) Calculate the difference in mass in terms of V, q, m (of either), B, and the distance Δx between the spots on the photographic plate. (*b*) Calculate Δx for a beam of singly ionized chlorine atoms of masses 35.0 and 37.0 u if $V = 7.33$ kV and $B = 520$ mT.

24. In a mass spectrometer (see Problem 22) used for commercial purposes, uranium ions of mass 238 u and charge $+2e$ are separated from related species. The ions are first accelerated through a potential difference of 105 kV and then pass into a magnetic field, where they travel a 180° arc of radius 97.3 cm. They are then collected in a cup after passing through a slit of width 1.20 mm and a height of 1.14 cm. (*a*) What is the magnitude of the (perpendicular) magnetic field in the separator? If the machine is designed to separate out 90.0 mg of material per hour, calculate (*b*) the current of the desired ions in the machine and (*c*) the internal energy dissipated in the cup in 1.00 h.

25. A neutral particle is at rest in a uniform magnetic field of magnitude B. At time $t = 0$ it decays into two charged particles each of mass m. (*a*) If the charge of one of the particles is $+q$, what is the charge of the other? (*b*) The two particles move off in separate paths both of which lie in the plane perpendicular to **B**. At a later time the particles collide. Express the time from decay until collision in terms of m, B, and q.

26. A deuteron in a cyclotron is moving in a magnetic field with an orbit radius of 50 cm. Because of a grazing collision with a target, the deuteron breaks up, with a negligible loss of kinetic energy, into a proton and a neutron. Discuss the subsequent motions of each. Assume that the deuteron energy is shared equally by the proton and neutron at breakup.

27. (*a*) What speed would a proton need to circle the Earth at the equator, if the Earth's magnetic field is everywhere horizontal there and directed along longitudinal lines? Relativistic effects must be taken into account. Take the magnitude of the Earth's magnetic field to be 41 μT at the equator. (*b*) Draw the velocity and magnetic field vectors corresponding to this situation.

28. Compute the radius of the path of a 10.0-MeV electron moving perpendicular to a uniform 2.20-T magnetic field. Use both the (*a*) classical and (*b*) relativistic formulas. (*c*) Calculate the true period of the circular motion. Is the result independent of the speed of the electron?

29. Ionization measurements show that a particular nuclear particle carries a double charge ($= 2e$) and is moving with a speed of 0.710c. It follows a circular path of radius 4.72 m in a magnetic field of 1.33 T. Find the mass of the particle and identify it.

30. The proton synchrotron at Fermilab accelerates protons to a kinetic energy of 500 GeV. At this energy, calculate (*a*) the speed parameter and (*b*) the magnetic field at the proton orbit that has a radius of curvature of 750 m. (The proton has a rest energy of 938 MeV.)

31. A 22.5-eV positron (positively charged electron) is projected into a uniform magnetic field $B = 455$ μT with its velocity vector making an angle of 65.5° with **B**. Find (*a*) the period, (*b*) the pitch p, and (*c*) the radius r of the helical path. See Fig. 33.

32. In Bohr's theory of the hydrogen atom the electron can be thought of as moving in a circular orbit of radius r about the proton. Suppose that such an atom is placed in a magnetic field, with the plane of the orbit at right angles to **B**. (*a*) If the electron is circulating clockwise, as viewed by an observer sighting along **B**, will the angular frequency increase or decrease? (*b*) What if the electron is circulating counterclock-

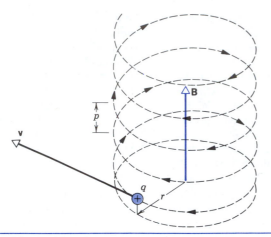

Figure 33 Problem 31.

wise? Assume that the orbit radius does not change. [*Hint*: The centripetal force is now partially electric ($\mathbf{F}_E$) and partially magnetic ($\mathbf{F}_B$) in origin.] (*c*) Show that the change in frequency of revolution caused by the magnetic field is given approximately by

$$\Delta \nu = \pm \frac{Be}{4\pi m}.$$

Such frequency shifts were observed by Zeeman in 1896. (*Hint*: Calculate the frequency of revolution without the magnetic field and also with it. Subtract, bearing in mind that because the effect of the magnetic field is very small, some — but not all — terms containing B can be set equal to zero with little error.)

33. Estimate the total path length traveled by a deuteron in a cyclotron during the acceleration process. Assume an accelerating potential between the dees of 80 kV, a dee radius of 53 cm, and an oscillator frequency of 12 MHz.

34. Consider a particle of mass m and charge q moving in the xy plane under the influence of a uniform magnetic field $\mathbf{B}$ pointing in the $+z$ direction. Write expressions for the coordinates $x(t)$ and $y(t)$ of the particle as functions of time t, assuming that the particle moves in a circle of radius R centered at the origin of coordinates.

35. Consider the particle of Problem 34, but this time *prove* (rather than *assuming*) that the particle moves in a circular path by solving Newton's law analytically. (*Hint*: Solve the expression for F_y to find v_x and substitute into the expression for F_x to obtain an equation that can be solved for v_y. Do the same for v_x by substituting into the F_y equation. Finally, obtain $x(t)$ and $y(t)$ from v_x and v_y.)

Section 34-4 The Hall Effect

36. In a Hall effect experiment, a current of 3.2 A lengthwise in a conductor 1.2 cm wide, 4.0 cm long, and 9.5 μm thick produces a transverse Hall voltage (across the width) of 40 μV when a magnetic field of 1.4 T is passed perpendicularly through the thin conductor. From these data, find (*a*) the drift velocity of the charge carriers and (*b*) the number density of charge carriers. From Table 2, identify the conductor. (*c*) Show on a diagram the polarity of the Hall voltage with a given current and magnetic field direction, assuming charge carriers are (negative) electrons.

37. Show that, in terms of the Hall electric field E and the current density j, the number of charge carriers per unit volume is given by

$$n = \frac{jB}{eE}.$$

38. (*a*) Show that the ratio of the Hall electric field E to the electric field E_c responsible for the current is

$$\frac{E}{E_c} = \frac{B}{ne\rho},$$

where ρ is the resistivity of the material. (*b*) Compute the ratio numerically for Sample Problem 3. See Table 1 in Chapter 32.

39. A metal strip 6.5 cm long, 0.88 cm wide, and 0.76 mm thick moves with constant velocity **v** through a magnetic field $B = 1.2$ mT perpendicular to the strip, as shown in Fig. 34. A potential difference of 3.9 μV is measured between points x and y across the strip. Calculate the speed v.

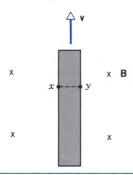

Figure 34 Problem 39.

Section 34-5 The Magnetic Force on a Current

40. A horizontal conductor in a power line carries a current of 5.12 kA from south to north. The Earth's magnetic field in the vicinity of the line is 58.0 μT and is directed toward the north and inclined downward at 70.0° to the horizontal. Find the magnitude and direction of the magnetic force on 100 m of the conductor due to the Earth's field.

41. A wire of length 62.0 cm and mass 13.0 g is suspended by a pair of flexible leads in a magnetic field of 440 mT. Find the magnitude and direction of the current in the wire required to remove the tension in the supporting leads. See Fig. 35.

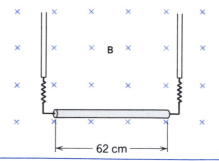

Figure 35 Problem 41.

42. A metal wire of mass m slides without friction on two horizontal rails spaced a distance d apart, as in Fig. 36. The track lies in a vertical uniform magnetic field **B**. A constant

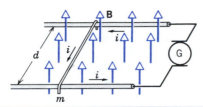

Figure 36 Problem 42.

current i flows from generator G along one rail, across the wire, and back down the other rail. Find the velocity (speed and direction) of the wire as a function of time, assuming it to be at rest at $t = 0$.

43. Consider the possibility of a new design for an electric train. The engine is driven by the force due to the vertical component of the Earth's magnetic field on a conducting axle. Current is passed down one rail, into a conducting wheel, through the axle, through another conducting wheel, and then back to the source via the other rail. (a) What current is needed to provide a modest 10-kN force? Take the vertical component of the Earth's field to be $10 \, \mu T$ and the length of the axle to be 3.0 m. (b) How much power would be lost for each ohm of resistance in the rails? (c) Is such a train totally unrealistic or just marginally unrealistic?

44. Figure 37 shows a wire of arbitrary shape carrying a current i between points a and b. The wire lies in a plane at right angles to a uniform magnetic field **B**. Prove that the force on the wire is the same as that on a straight wire carrying a current i directly from a to b. (Hint: Replace the wire by a series of "steps" parallel and perpendicular to the straight line joining a and b.)

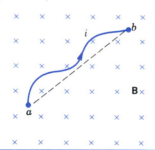

Figure 37 Problem 44.

45. A U-shaped wire of mass m and length L is immersed with its two ends in mercury (Fig. 38). The wire is in a homogeneous magnetic field **B**. If a charge, that is, a current pulse $q = \int i \, dt$, is sent through the wire, the wire will jump up. Calculate, from the height h that the wire reaches, the size of the charge or current pulse, assuming that the time of the current pulse is very small in comparison with the time of flight. Make use of the fact that impulse of force equals $\int F \, dt$, which equals mv. (Hint: Relate $\int i \, dt$ to $\int F \, dt$.) Evaluate q for $B = 0.12$ T, $m = 13$ g, $L = 20$ cm, and $h = 3.1$ m.

46. A 1.15-kg copper rod rests on two horizontal rails 95.0 cm apart and carries a current of 53.2 A from one rail to the other. The coefficient of static friction is 0.58. Find the smallest magnetic field (not necessarily vertical) that would cause the bar to slide.

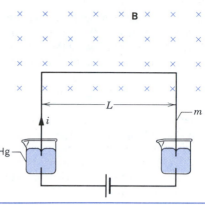

Figure 38 Problem 45.

47. A long, rigid conductor, lying along the x axis, carries a current of 5.0 A in the $-x$ direction. A magnetic field **B** is present, given by $\mathbf{B} = 3\mathbf{i} + 8x^2\mathbf{j}$, with x in meters and **B** in mT. Calculate the force on the 2.0-m segment of the conductor that lies between $x = 1.2$ m and $x = 3.2$ m.

Section 34-6 Torque on a Current Loop

48. Figure 39 shows a rectangular, 20-turn loop of wire, 12 cm by 5.0 cm. It carries a current of 0.10 A and is hinged at one side. It is mounted with its plane at an angle of 33° to the direction of a uniform magnetic field of 0.50 T. Calculate the torque about the hinge line acting on the loop.

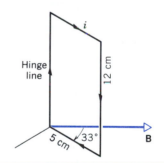

Figure 39 Problem 48.

49. A single-turn current loop, carrying a current of 4.00 A, is in the shape of a right triangle with sides 50 cm, 120 cm, and 130 cm. The loop is in a uniform magnetic field of magnitude 75.0 mT whose direction is parallel to the current in the 130-cm side of the loop. (a) Find the magnetic force on each of the three sides of the loop. (b) Show that the total magnetic force on the loop is zero.

50. A stationary, circular wall clock has a face with a radius of 15 cm. Six turns of wire are wound around its perimeter; the wire carries a current 2.0 A in the clockwise direction. The clock is located where there is a constant, uniform external magnetic field of 70 mT (but the clock still keeps perfect time). At exactly 1:00 p.m., the hour hand of the clock points in the direction of the external magnetic field. (a) After how many minutes will the minute hand point in the direction of the torque on the winding due to the magnetic field? (b) What is the magnitude of this torque?

51. A length L of wire carries a current i. Show that if the wire is formed into a circular coil, the maximum torque in a given magnetic field is developed when the coil has one turn only and the maximum torque has the magnitude

$$\tau = \frac{1}{4\pi} L^2 i B.$$

52. Prove that the relation $\tau = NiAB \sin \theta$ holds for closed loops of arbitrary shape and not only for rectangular loops as in Fig. 26. (*Hint:* Replace the loop of arbitrary shape by an assembly of adjacent long, thin, approximately rectangular loops that are nearly equivalent to it as far as the distribution of current is concerned.)

53. Figure 40 shows a wire ring of radius a at right angles to the general direction of a radially symmetric diverging magnetic field. The magnetic field at the ring is everywhere of the same magnitude B, and its direction at the ring is everywhere at an angle θ with a normal to the plane of the ring. The twisted lead wires have no effect on the problem. Find the magnitude and direction of the force the field exerts on the ring if the ring carries a current i as shown in the figure.

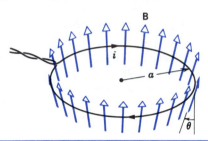

Figure 40 Problem 53.

54. A certain galvanometer has a resistance of 75.3 Ω; its needle experiences a full-scale deflection when a current of 1.62 mA passes through its coil. (*a*) Determine the value of the auxiliary resistance required to convert the galvanometer into a voltmeter that reads 1.00 V at full-scale deflection. How is it to be connected? (*b*) Determine the value of the auxiliary resistance required to convert the galvanometer into an ammeter that reads 50.0 mA at full-scale deflection. How is it to be connected?

55. Figure 41 shows a wooden cylinder with a mass $m = 262$ g and a length $L = 12.7$ cm, with $N = 13$ turns of wire wrapped around it longitudinally, so that the plane of the wire loop contains the axis of the cylinder. What is the least current through the loop that will prevent the cylinder from rolling down a plane inclined at an angle θ to the horizontal, in the presence of a vertical, uniform magnetic field of 477 mT, if the plane of the windings is parallel to the inclined plane?

Section 34-7 The Magnetic Dipole

56. A circular coil of 160 turns has a radius of 1.93 cm. (*a*) Calculate the current that results in a magnetic moment of 2.33 A·m². (*b*) Find the maximum torque that the coil, carrying this current, can experience in a uniform 34.6-mT magnetic field.

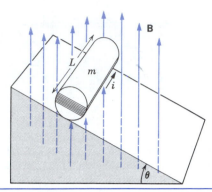

Figure 41 Problem 55.

57. The magnetic dipole moment of the Earth is 8.0×10^{22} J/T. Assume that this is produced by charges flowing in the molten outer core of the Earth. If the radius of the circular path is 3500 km, calculate the required current.

58. A circular wire loop whose radius is 16.0 cm carries a current of 2.58 A. It is placed so that the normal to its plane makes an angle of 41.0° with a uniform magnetic field of 1.20 T. (*a*) Calculate the magnetic dipole moment of the loop. (*b*) Find the torque on the loop.

59. Two concentric circular loops, radii 20.0 and 30.0 cm, in the xy plane each carry a clockwise current of 7.00 A, as shown in Fig. 42. (*a*) Find the net magnetic moment of this system. (*b*) Repeat if the current in the outer loop is reversed.

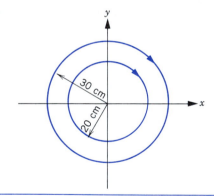

Figure 42 Problem 59.

60. A circular loop of wire having a radius of 8.0 cm carries a current of 0.20 A. A unit vector parallel to the dipole moment μ of the loop is given by $0.60\mathbf{i} - 0.80\mathbf{j}$. If the loop is located in a magnetic field given in T by $\mathbf{B} = 0.25\mathbf{i} + 0.30\mathbf{k}$, find (*a*) the torque on the loop and (*b*) the magnetic potential energy of the loop.

Computer Projects

61. A particle of charge $q = 1.6 \times 10^{-19}$ C and mass $m = 1.7 \times 10^{-27}$ kg moves in a uniform magnetic field of 1.1 T in the positive z direction. At time $t = 0$ it is at the origin and has a velocity of 6.0×10^5 m/s in the positive x direction. (*a*) Use the computer program given in Appendix I to plot the position of the particle from $t = 0$ to $t = 6.5 \times 10^{-8}$ s.

Use $\Delta t = 3 \times 10^{-11}$ s for the integration interval. Also have the computer calculate and display the speed of the particle when it displays its position. (*b*) Is the speed constant to within computational accuracy? If the first two significant figures of the computed speed are not constant reduce the value of Δt and try again. (*c*) Measure the radius of the orbit and compare the result with mv/qB.

62. The magnetic field in the neighborhood of the origin is in the positive z direction and its magnitude in tesla is given by $B = 50r$, where r is the distance in meters from the z axis. A particle of charge 1.6×10^{-19} C and mass 1.7×10^{-27} kg is to be injected into the field with a velocity of 6.0×10^5 m/s in the negative y direction from a point on the x axis. If the initial distance from the z axis obeys $mv^2/r = qvB$ then the orbit will be circular. (*a*) What is this distance? (*b*) Use a computer program to plot the orbit from $t = 0$, when the particle is injected, to $t = 1.2 \times 10^{-7}$ s. Take the initial coordinates to be $x = R$ and $y = 0$, where R is the value of r you found in part (*a*). Take the integration interval to be $\Delta t = 5 \times 10^{-11}$ s. Also have the computer calculate the speed of the particle for every displayed point. Is the speed constant? If the first two significant figures of the computed speed are not constant, reduce the value of Δt. Is the orbit circular? (*c*) Now start the particle at $x = 0.5R$, $y = 0$ and plot the orbit for the same time interval. Is it circular? Is the speed constant?

63. (*a*) Consider a magnetic field in the positive z direction, with magnitude in tesla given by $B = 7.0 \times 10^{-3}/x$. A particle with charge $q = 1.6 \times 10^{-19}$ C is initially at $x = 5.0 \times 10^{-2}$ m, $y = 0$ and is moving in the positive y direction with a speed of 7.0×10^5 m/s. Use a computer program to plot the orbit from $t = 0$ to $t = 2.5 \times 10^{-6}$ s. Use $\Delta t = 2.5 \times 10^{-10}$ s for the integration interval. Also have the computer calculate the speed for every point displayed. Is the speed constant? If the first two significant figures of the speed are not constant, reduce the values of Δt. Is the orbit circular? (*b*) Now suppose the charge starts at the same point but with a velocity of 7.0×10^5 m/s in the negative y direction. Use the program to plot its orbit from $t = 0$ to $t = 1.0 \times 10^{-6}$ s. Use an integration interval of $\Delta t = 8 \times 10^{-11}$ s. Check the constancy of the speed to see if Δt needs adjustment. (*c*) Notice that in both cases the charge drifts in the negative y direction as it spirals in the field. Use your knowledge of motion in a uniform field to explain qualitatively the shapes of the two orbits. (*d*) How can you change the initial conditions so the charge drifts in the *positive y* direction?

64. A uniform 1.2-T magnetic field is in the positive z direction and a uniform electric field is in the negative x direction. A particle with charge $q = 1.6 \times 10^{-19}$ C and mass $m = 1.7 \times 10^{-26}$ kg starts at the origin with velocity 5.0×10^4 m/s, in the positive y direction. For each of the following electric field magnitudes use a computer program to plot the orbit from $t = 0$ to $t = 1.0 \times 10^{-6}$ s. Use $\Delta t = 1 \times 10^{-9}$ s for the integration interval. (*a*) 1.0×10^4 V/m. (*b*) 3.0×10^4 V/m. (*c*) 6.0×10^4 V/m. (*d*) 9.0×10^4 V/m.

CHAPTER 35

AMPÈRE'S LAW

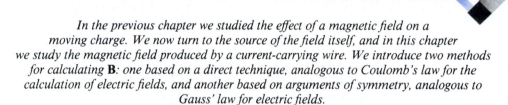

In the previous chapter we studied the effect of a magnetic field on a
moving charge. We now turn to the source of the field itself, and in this chapter
we study the magnetic field produced by a current-carrying wire. We introduce two methods
for calculating **B**: one based on a direct technique, analogous to Coulomb's law for the
calculation of electric fields, and another based on arguments of symmetry, analogous to
Gauss' law for electric fields.

In analogy with our previous study of the electric fields of some simple charge distributions,
we investigate in this chapter the magnetic fields produced by some simple current
distributions: straight wires and circular loops. We discuss the magnetic dipole field, which
is similar to the electric dipole field. Finally, we show that the relationship between electric
and magnetic fields is deeper than merely the similarity of equations; the relationship
extends to the transformation of the fields into one another when charge or current
distributions are viewed from different inertial frames.

35-1 THE BIOT–SAVART LAW

The discovery that currents produce magnetic fields was
made by Hans Christian Oersted in 1820. Oersted observed that, as illustrated in Fig. 1, when a compass is
placed near a straight current-carrying wire, the needle
always aligns perpendicular to the wire (neglecting the
influence of the Earth's magnetic field on the compass).
This was the first experimental link between electricity
and magnetism, and it provided the beginning of the development of a formal theory of electromagnetism. In
modern terms, we analyze Oersted's experiment by saying that the current in the wire sets up a magnetic field,
which exerts a torque on the compass needle and aligns it
with the field.

We now develop a procedure for calculating the magnetic field due to a specified current distribution. Before
considering the magnetic field, let us first review the analogous procedure for calculating electric fields.

Figure 2 shows two charge distributions q_1 and q_2 of
arbitrary size and shape. We consider charge elements dq_1

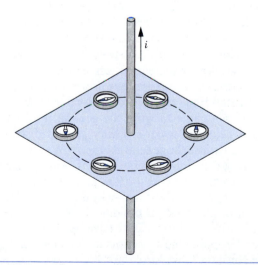

Figure 1 Oersted's experiment. The direction of the compass
needle is always perpendicular to the direction of the current
in the wire.

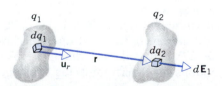

Figure 2 Two arbitrary charge distributions q_1 and q_2. An
element of charge dq_1 sets up an electric field $d\mathbf{E}_1$ at the location of dq_2.

761

and dq_2 in the two distributions. The electric field $d\mathbf{E}_1$ set up by dq_1 at the location of dq_2 is given by

$$d\mathbf{E}_1 = \frac{1}{4\pi\epsilon_0}\frac{dq_1}{r^2}\mathbf{u}_r = \frac{1}{4\pi\epsilon_0}\frac{dq_1}{r^3}\mathbf{r}, \qquad (1)$$

where $\mathbf{r}$ is the vector from dq_1 to dq_2 (Fig. 2), r is its magnitude, and $\mathbf{u}_r (= \mathbf{r}/r)$ is a unit vector in the direction of $\mathbf{r}$. To find the total electric field $\mathbf{E}_1$ acting at dq_2 due to the entire distribution q_1, we integrate over q_1:

$$\mathbf{E}_1 = \frac{1}{4\pi\epsilon_0}\int\frac{dq_1}{r^2}\mathbf{u}_r = \frac{1}{4\pi\epsilon_0}\int\frac{dq_1}{r^3}\mathbf{r}. \qquad (2)$$

The force $d\mathbf{F}_{21}$ acting on dq_2 due to the charge distribution q_1 can then be written

$$d\mathbf{F}_{21} = \mathbf{E}_1\,dq_2. \qquad (3)$$

Equations 1 or 2 (for the electric field of a charge distribution) and 3 (giving the force due to that distribution acting on another charge) together can be taken to be a form of Coulomb's law for finding the electrostatic force between charges.

In the case of magnetic fields, we seek the force between *current* elements (Fig. 3). That is, we consider two currents i_1 and i_2 and their corresponding current elements $i_1\,d\mathbf{s}_1$ and $i_2\,d\mathbf{s}_2$. We assume, based on our results from the previous chapter, that the relative directions of the current elements (specified by the vectors $d\mathbf{s}_1$ and $d\mathbf{s}_2$) will be important and that the force between the currents may involve cross products of vectors. Coulomb's law for the force between charges was developed as a statement of experimental results; an analogous magnetic force law was proposed in 1820 by French physicist André-Marie Ampère* soon after he learned of Oersted's results. The magnetic force $d\mathbf{F}_{21}$ exerted on current element 2 by i_1 can be written, using Eq. 30 of Chapter 34,

$$d\mathbf{F}_{21} = i_2\,d\mathbf{s}_2 \times \mathbf{B}_1, \qquad (4)$$

where the magnetic field $\mathbf{B}_1$ at the location of the current element $i_2\,d\mathbf{s}_2$ is due to the entire current i_1. The contribution $d\mathbf{B}_1$ of each current element of i_1 to the total field $\mathbf{B}_1$ is given by

$$d\mathbf{B}_1 = k\frac{i_1\,d\mathbf{s}_1 \times \mathbf{u}_r}{r^2} = k\frac{i_1\,d\mathbf{s}_1 \times \mathbf{r}}{r^3}, \qquad (5)$$

where $\mathbf{r}$ is the vector from current element 1 to current element 2, and $\mathbf{u}_r$ is the unit vector in the direction of $\mathbf{r}$. Equations 4 and 5 together give the magnetic force between current elements in a manner analogous to Eqs. 1 and 3 for charge elements.

An undetermined constant k is included in Eq. 5, just as we included a similar constant in Coulomb's law (see Eq. 1 of Chapter 27). You will recall that in electrostatics we had two options for determining the constant in Cou-

* See "André Marie Ampère," by L. Pearce Williams, *Scientific American*, January 1989, p. 90.

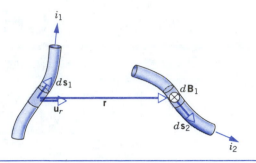

Figure 3 Two arbitrary current distributions i_1 and i_2. The element of current in the length $d\mathbf{s}_1$ of one wire sets up a magnetic field $d\mathbf{B}_1$ at the location of an element of current in the length $d\mathbf{s}_2$ of the other wire.

lomb's law: (1) set the constant equal to a convenient value, and use the force law to determine by experiment the unit of electric charge, or (2) define the unit of charge and then determine the constant by experiment. We chose option 2, defining the unit of charge in terms of the unit of current. In the case of the constant in the magnetic force law we choose option 1: set the constant equal to a convenient value and use the force law to define the unit of current, the ampere. The constant k in SI units is defined to have the exact value 10^{-7} tesla·meter/ampere (T·m/A). However, as was the case in electrostatics, we find it convenient to write the constant in a different form:

$$k = \frac{\mu_0}{4\pi} = 10^{-7}\ \text{T·m/A}, \qquad (6)$$

where the constant μ_0, called the *permeability constant*, has the exact value

$$\mu_0 = 4\pi \times 10^{-7}\ \text{T·m/A}.$$

The permeability constant μ_0 plays a role in calculating magnetic fields similar to that of the permittivity constant ϵ_0 in calculating electric fields. The two constants are not independent of one another; as we show in Chapter 41, they are linked through the speed of light c, such that $c = 1/\sqrt{\mu_0\epsilon_0}$. We are therefore *not* free to choose both constants arbitrarily; we can set one arbitrarily, but then the other is determined by the accepted value of c.

We are now able to write the general results for the magnetic field due to an arbitrary current distribution. Figure 4 illustrates the general geometry. We are no longer considering the force between two current elements; instead, we calculate the field $d\mathbf{B}$ at point P due to a single element of current $i\,d\mathbf{s}$. If we are interested in calculating the effect of that field on moving charges or currents at point P, we use the formulas we developed in the previous chapter. Dropping the subscripts in Eq. 5 and using Eq. 6 for the constant k, we have

$$d\mathbf{B} = \frac{\mu_0}{4\pi}\frac{i\,d\mathbf{s} \times \mathbf{u}_r}{r^2} = \frac{\mu_0}{4\pi}\frac{i\,d\mathbf{s} \times \mathbf{r}}{r^3}. \qquad (7)$$

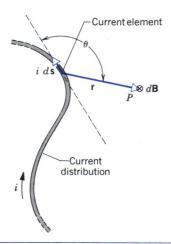

Figure 4   The element $i\,d\mathbf{s}$ of an arbitrary current distribution sets up a magnetic field $d\mathbf{B}$ into the plane of the page at the point P.

This result is known as the *Biot–Savart law*. The direction of $d\mathbf{B}$ is the same as the direction of $d\mathbf{s} \times \mathbf{u}_r$ (or $d\mathbf{s} \times \mathbf{r}$), into the plane of the paper in Fig. 4.

We can express the magnitude of $d\mathbf{B}$ from the Biot–Savart law as

$$dB = \frac{\mu_0}{4\pi} \frac{i\,ds\,\sin\theta}{r^2}, \tag{8}$$

where θ is the angle between $d\mathbf{s}$ (which is in the direction of i) and $\mathbf{r}$, as shown in Fig. 4.

To find the total field $\mathbf{B}$ due to the entire current distribution, we must integrate over all current elements $i\,d\mathbf{s}$:

$$\mathbf{B} = \int d\mathbf{B} = \frac{\mu_0}{4\pi} \int \frac{i\,d\mathbf{s} \times \mathbf{u}_r}{r^2} = \frac{\mu_0}{4\pi} \int \frac{i\,d\mathbf{s} \times \mathbf{r}}{r^3}. \tag{9}$$

Just as we did in Chapter 28 for electric fields, we may have to take into account in computing this integral that not all the elements $d\mathbf{B}$ are in the same direction (see Section 28-5 for examples of this kind of vector integral in the case of electric fields).

35-2   APPLICATIONS OF THE BIOT–SAVART LAW

A Long Straight Wire

We illustrate the law of Biot and Savart by applying it to find $\mathbf{B}$ due to a current i in a long straight wire. Figure 5 shows a typical current element $i\,d\mathbf{s}$. The magnitude of the contribution $d\mathbf{B}$ of this element to the magnetic field at P is found from Eq. 8,

$$dB = \frac{\mu_0 i}{4\pi} \frac{ds\,\sin\theta}{r^2}.$$

We choose x to be the variable of integration that runs along the wire, and so the length of the current element is

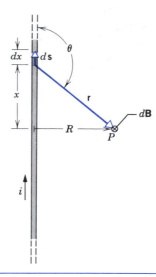

Figure 5   The magnetic field $d\mathbf{B}$ established by a current element in a long straight wire points into the page at P.

dx. The directions of the contributions $d\mathbf{B}$ at point P for all elements are the same, namely, into the plane of the figure at right angles to the page. This is the direction of the vector product $d\mathbf{s} \times \mathbf{r}$. We can thus evaluate a scalar integral rather than the vector integral of Eq. 9, and B can be written

$$B = \int dB = \frac{\mu_0 i}{4\pi} \int_{x=-\infty}^{x=+\infty} \frac{\sin\theta\,dx}{r^2}. \tag{10}$$

Now x, θ, and r are not independent, being related (see Fig. 5) by

$$r = \sqrt{x^2 + R^2}$$

and

$$\sin\theta = \sin(\pi - \theta) = \frac{R}{\sqrt{x^2 + R^2}},$$

so that Eq. 10 becomes

$$B = \frac{\mu_0 i}{4\pi} \int_{-\infty}^{+\infty} \frac{R\,dx}{(x^2 + R^2)^{3/2}} = \frac{\mu_0 i}{4\pi R} \frac{x}{(x^2 + R^2)^{1/2}}\Big|_{x=-\infty}^{x=+\infty}$$

or

$$B = \frac{\mu_0 i}{2\pi R}. \tag{11}$$

This problem reminds us of its electrostatic equivalent. We derived an expression for $\mathbf{E}$ due to a long charged rod by integration methods, using Coulomb's law (Section 28-5). We also solved the same problem using Gauss' law (Section 29-5). Later in this chapter, we consider a law of magnetic fields, Ampère's law, which is similar to Gauss' law in that it simplifies magnetic field calculations in cases (such as this one) that have a high degree of symmetry.

A Circular Current Loop

Figure 6 shows a circular loop of radius R carrying a current i. Let us calculate $\mathbf{B}$ at a point P on the axis a distance z from the center of the loop.

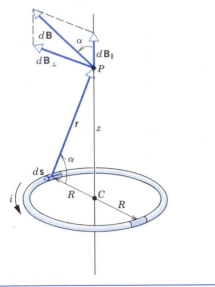

Figure 6 A circular loop of current. The element $i\,d\mathbf{s}$ of the loop sets up a field $d\mathbf{B}$ at a point P on the axis of the loop.

The angle θ between the current element $i\,d\mathbf{s}$ and $\mathbf{r}$ is $90°$. From the Biot–Savart law, we know that the vector $d\mathbf{B}$ for this element is at right angles to the plane formed by $i\,d\mathbf{s}$ and $\mathbf{r}$ and thus lies at right angles to $\mathbf{r}$, as the figure shows.

Let us resolve $d\mathbf{B}$ into two components, one, $d\mathbf{B}_\parallel$, along the axis of the loop and another, $d\mathbf{B}_\perp$, at right angles to the axis. Only $d\mathbf{B}_\parallel$ contributes to the total magnetic field $\mathbf{B}$ at point P. This follows because the components $d\mathbf{B}_\parallel$ for all current elements lie on the axis and add directly; however, the components $d\mathbf{B}_\perp$ point in different directions perpendicular to the axis, and the sum of all $d\mathbf{B}_\perp$ for the complete loop is zero, from symmetry. (A diametrically opposite current element, indicated in Fig. 6, produces the same $d\mathbf{B}_\parallel$ but the opposite $d\mathbf{B}_\perp$.) We can therefore replace the vector integral over all $d\mathbf{B}$ with a scalar integral over the parallel components only:

$$B = \int dB_\parallel. \tag{12}$$

For the current element in Fig. 6 the Biot–Savart law (Eq. 8) gives

$$dB = \frac{\mu_0 i}{4\pi} \frac{ds \sin 90°}{r^2}. \tag{13}$$

We also have

$$dB_\parallel = dB \cos \alpha,$$

which, combined with Eq. 13, gives

$$dB_\parallel = \frac{\mu_0 i \cos \alpha \, ds}{4\pi r^2}. \tag{14}$$

Figure 6 shows that r and α are not independent of each

other. Let us express each in terms of z, the distance from the center of the loop to the point P. The relationships are

$$r = \sqrt{R^2 + z^2}$$

and

$$\cos \alpha = \frac{R}{r} = \frac{R}{\sqrt{R^2 + z^2}}.$$

Substituting these values into Eq. 14 for $dB_\parallel$ gives

$$dB_\parallel = \frac{\mu_0 i R}{4\pi (R^2 + z^2)^{3/2}} \, ds.$$

Note that i, R, and z have the same values for all current elements. Integrating this equation, we obtain

$$B = \int dB_\parallel = \frac{\mu_0 i R}{4\pi (R^2 + z^2)^{3/2}} \int ds$$

or, noting that $\int ds$ is simply the circumference of the loop ($= 2\pi R$),

$$B = \frac{\mu_0 i R^2}{2(R^2 + z^2)^{3/2}}. \tag{15}$$

At the center of the loop ($z = 0$), Eq. 15 reduces to

$$B = \frac{\mu_0 i}{2R}. \tag{16}$$

The magnitude of the magnetic field on the axis of a circular current loop is given by Eq. 15. The field has its largest value in the plane of the loop (Eq. 16) and decreases as the distance z increases. The direction of the field is determined by the right-hand rule: grasp the wire in the right hand, with the thumb in the direction of the current, and the fingers curl in the direction of the magnetic field.

If $z \gg R$, so that points close to the loop are not considered, Eq. 15 reduces to

$$B = \frac{\mu_0 i R^2}{2z^3}.$$

For a tightly wound coil of N identical circular loops, the total field is N times this value, or (substituting the area $A = \pi R^2$ of the loop)

$$B = \frac{\mu_0}{2\pi} \frac{N i A}{z^3} = \frac{\mu_0}{2\pi} \frac{\mu}{z^3}, \tag{17}$$

where μ is the *magnetic dipole moment* (see Section 34-7) of the current loop. This reminds us of the result derived in Problem 11 of Chapter 28 [$E = (1/2\pi\epsilon_0)(p/z^3)$], which is the formula for the *electric* field on the axis of an *electric* dipole. Problem 33 gives an example of the calculation of the magnetic field at distant points perpendicular to the axis of a magnetic dipole.

We have shown in two ways that we can regard a current loop as a magnetic dipole: it experiences a torque given by $\tau = \mu \times \mathbf{B}$ when we place it in an *external* magnetic field (Eq. 37 of Chapter 34); it generates its own

TABLE 1   SOME DIPOLE EQUATIONS

Property	Electric Dipole	Magnetic Dipole
Torque in an external field	$\tau = \mathbf{p} \times \mathbf{E}$	$\tau = \mu \times \mathbf{B}$
Energy in an external field	$U = -\mathbf{p} \cdot \mathbf{E}$	$U = -\mu \cdot \mathbf{B}$
Field at distant points along axis	$E = \dfrac{1}{2\pi\epsilon_0} \dfrac{p}{z^3}$	$B = \dfrac{\mu_0}{2\pi} \dfrac{\mu}{z^3}$
Field at distant points along perpendicular bisector	$E = \dfrac{1}{4\pi\epsilon_0} \dfrac{p}{x^3}$	$B = \dfrac{\mu_0}{4\pi} \dfrac{\mu}{x^3}$

magnetic field given, for points on the axis, by Eq. 17. Table 1 summarizes some properties of electric and magnetic dipoles.

Sample Problem 1   Two long parallel wires a distance $2d$ apart carry equal currents i in opposite directions, as shown in Fig. 7a. Derive an expression for the magnetic field B at a point P on the line connecting the wires and a distance x from the point midway between them.

Solution   Study of Fig. 7a shows that $\mathbf{B}_1$ due to the current i_1 and $\mathbf{B}_2$ due to the current i_2 point in the same direction at P. Each is given by Eq. 11 ($B = \mu_0 i / 2\pi R$) so that

$$B = B_1 + B_2 = \frac{\mu_0 i}{2\pi(d+x)} + \frac{\mu_0 i}{2\pi(d-x)} = \frac{\mu_0 i d}{\pi(d^2 - x^2)}.$$

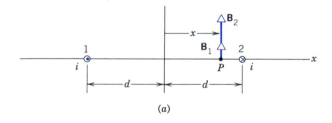

(a)

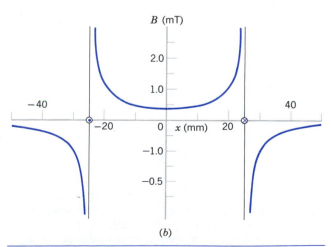

(b)

Figure 7   Sample Problem 1. (*a*) The magnetic fields at point P due to the currents in wires 1 and 2. (*b*) The resultant field at P, calculated for $i = 25$ A and $d = 25$ mm.

Inspection of this result shows that (1) B is symmetrical about $x = 0$; (2) B has its minimum value ($= \mu_0 i / \pi d$) at $x = 0$; and (3) $B \to \infty$ as $x \to \pm d$. This last conclusion is not correct, because Eq. 11 cannot be applied to points inside the wires. In reality (see Sample Problem 5, for example) the field due to each wire would vanish at the center of that wire.

You should show that our result for the combined field remains valid at points where $|x| > d$. Figure 7b shows the variation of B with x for $i = 25$ A and $d = 25$ mm.

Sample Problem 2   In the Bohr model of the hydrogen atom, the electron circulates around the nucleus in a path of radius 5.29×10^{-11} m at a frequency v of 6.63×10^{15} Hz (or rev/s). (*a*) What value of B is set up at the center of the orbit? (*b*) What is the equivalent magnetic dipole moment?

Solution   (*a*) The current is the rate at which charge passes any point on the orbit and is given by

$$i = ev = (1.60 \times 10^{-19}\ \text{C})(6.63 \times 10^{15}\ \text{Hz}) = 1.06 \times 10^{-3}\ \text{A}.$$

The magnetic field B at the center of the orbit is given by Eq. 16,

$$B = \frac{\mu_0 i}{2R} = \frac{(4\pi \times 10^{-7}\ \text{T}\cdot\text{m/A})(1.06 \times 10^{-3}\ \text{A})}{2(5.29 \times 10^{-11}\ \text{m})} = 12.6\ \text{T}.$$

(*b*) From Eq. 36 of Chapter 34 with N (the number of loops) $= 1$, we have

$$\mu = iA = (1.06 \times 10^{-3}\ \text{A})(\pi)(5.29 \times 10^{-11}\ \text{m})^2$$
$$= 9.31 \times 10^{-24}\ \text{A}\cdot\text{m}^2.$$

Sample Problem 3   Figure 8 shows a flat strip of copper of width a and negligible thickness carrying a current i. Find the magnetic field $\mathbf{B}$ at point P, at a distance R from the center of the strip along its perpendicular bisector.

Solution   Let us subdivide the strip into long infinitesimal filaments of width dx, each of which may be treated as a wire carrying a current di given by $i(dx/a)$. The field contribution dB at point P in Fig. 8 is given, for the element shown, by the differential form of Eq. 11, or

$$dB = \frac{\mu_0}{2\pi} \frac{di}{r} = \frac{\mu_0}{2\pi} \frac{i(dx/a)}{R \sec \theta},$$

in which $r = R/\cos \theta = R \sec \theta$. Note that the vector $d\mathbf{B}$ is at right angles to the line marked r.

Only the horizontal component of $d\mathbf{B}$, namely, $dB \cos \theta$, is effective; the vertical component is canceled by the contribution of a symmetrically located filament on the other side of the origin (the second shaded strip in Fig. 8). Thus B at point P is given by the (scalar) integral

$$B = \int dB \cos \theta = \int \frac{\mu_0 i (dx/a)}{2\pi R \sec \theta} \cos \theta$$

$$= \frac{\mu_0 i}{2\pi a R} \int \frac{dx}{\sec^2 \theta}.$$

The variables x and θ are not independent, being related by

$$x = R \tan \theta$$

or

$$dx = R \sec^2 \theta \, d\theta.$$

The limits on θ are $\pm \alpha$, where $\alpha = \tan^{-1}(a/2R)$. Substituting for dx in the expression for B, we find

$$B = \frac{\mu_0 i}{2\pi a R} \int \frac{R \sec^2 \theta \, d\theta}{\sec^2 \theta}$$

$$= \frac{\mu_0 i}{2\pi a} \int_{-\alpha}^{+\alpha} d\theta = \frac{\mu_0 i}{\pi a} \alpha = \frac{\mu_0 i}{\pi a} \tan^{-1} \frac{a}{2R}.$$

This is the general result for the magnetic field due to the strip.

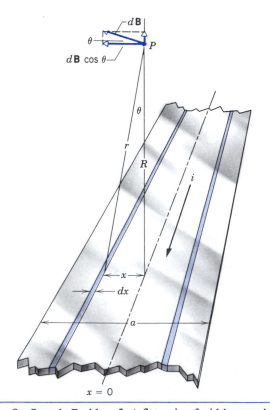

Figure 8   Sample Problem 3. A flat strip of width a carries a current i.

At points far from the strip, α is a small angle, for which $\alpha \approx \tan \alpha = a/2R$. Thus we have, as an approximate result,

$$B \approx \frac{\mu_0 i}{\pi a} \left(\frac{a}{2R} \right) = \frac{\mu_0}{2\pi} \frac{i}{R}.$$

This result is expected because at distant points the strip cannot be distinguished from a thin wire (see Eq. 11).

35-3   LINES OF B

Figure 9 shows lines representing the magnetic field $\mathbf{B}$ near a long straight wire. Note the increase in the spacing of the lines with increasing distance from the wire. This represents the $1/r$ decrease in B predicted by Eq. 11.

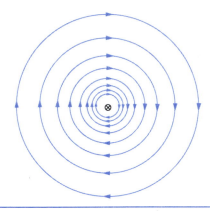

Figure 9   The lines of the magnetic field are concentric circles for a long straight current-carrying wire. Their direction is given by the right-hand rule.

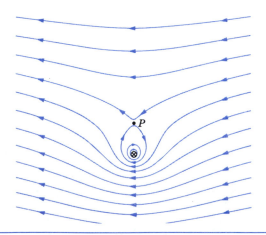

Figure 10   A long straight wire carrying a current into the page is immersed in a uniform external magnetic field. The magnetic field lines shown represent the resultant field formed by combining at each point the vectors representing the original uniform field and the field established by the current in the wire.

Figure 10 shows the resultant magnetic lines associated with a current in a wire that is oriented at right angles to a uniform *external* field $\mathbf{B_e}$ directed to the left. At any point the total resultant magnetic field $\mathbf{B_t}$ is the vector sum of $\mathbf{B_e}$ and $\mathbf{B_i}$, where $\mathbf{B_i}$ is the magnetic field set up by the current in the wire. The fields $\mathbf{B_e}$ and $\mathbf{B_i}$ tend to cancel above the wire and to reinforce each other below the wire. At point P in Fig. 10, $\mathbf{B_e}$ and $\mathbf{B_i}$ cancel exactly, and $\mathbf{B_t} = 0$. Very near the wire the field is represented by circular lines, and $\mathbf{B_t} \approx \mathbf{B_i}$.

To Michael Faraday, who originated the concept, the magnetic field lines represented the action of mechanical forces, somewhat like stretched rubber bands. Using Faraday's interpretation, we can readily see that the wire in Fig. 10 is pulled upward by the "tension" in the field lines. This concept is only of limited usefulness, and today we use lines of $\mathbf{B}$ largely for purposes of visualization. For quantitative calculations we use the field vectors, and we would describe the magnetic force on the wire in Fig. 10 using the relation $\mathbf{F} = i\mathbf{L} \times \mathbf{B}$.

In applying this relation to Fig. 10, we recall that the force on the wire is caused by the *external field* in which the wire is immersed; that is, it is $\mathbf{B_e}$, which points to the left. Since $\mathbf{L}$ points into the page, the magnetic force on the wire ($= i\mathbf{L} \times \mathbf{B_e}$) does indeed point up. It is important to use only the *external* field in such calculations because the field set up by the current in the wire cannot exert a force on the wire, just as the gravitational field of the Earth cannot exert a force on the Earth itself but only on another body. In Fig. 9, for example, there is no magnetic force on the wire because no external magnetic field is present.

35-4   TWO PARALLEL CONDUCTORS

Soon after Oersted's discovery that a current-carrying conductor would deflect the needle of a magnetic compass, Ampère concluded that two such conductors would attract each other with a force of magnetic origin.

We analyze the magnetic interaction of two currents in a manner similar to that of our analysis of the electric interaction between two charges:

$$\text{charge} \rightleftarrows \mathbf{E} \rightleftarrows \text{charge}.$$

That is, one charge sets up an electric field, and the other charge interacts with the field at its particular location. We use a similar procedure for the magnetic interaction:

$$\text{current} \rightleftarrows \mathbf{B} \rightleftarrows \text{current}.$$

Here a current sets up a magnetic field, and the other current then interacts with that field.

Wire 1 in Fig. 11, carrying a current i_1, produces a magnetic field $\mathbf{B_1}$ whose magnitude at the site of the sec-

ond wire is, according to Eq. 11,

$$B_1 = \frac{\mu_0 i_1}{2\pi d}.$$

The right-hand rule shows that the direction of $\mathbf{B_1}$ at wire 2 is down, as shown in the figure.

Wire 2, which carries a current i_2, can thus be considered to be immersed in an *external* magnetic field $\mathbf{B_1}$. A length L of this wire experiences a sideways magnetic force $\mathbf{F_{21}} = i_2\mathbf{L} \times \mathbf{B_1}$ of magnitude

$$F_{21} = i_2 L B_1 = \frac{\mu_0 L i_1 i_2}{2\pi d}. \tag{18}$$

The vector rule for the cross product shows that $\mathbf{F_{21}}$ lies in the plane of the wires and points toward wire 1 in Fig. 11.

We could equally well have started with wire 2 by first computing the magnetic field $\mathbf{B_2}$ produced by wire 2 at the site of wire 1 and then finding the force $\mathbf{F_{12}}$ exerted on a length L of wire 1 by the field of wire 2. This force on wire 1 would, for parallel currents, point toward wire 2 in Fig. 11. The forces that the two wires exert on each other are equal in magnitude and opposite in direction; they form an action–reaction pair according to Newton's third law.

If the currents in Fig. 11 were antiparallel, we would find that the forces on the wires would have the opposite directions: the wires would repel one another. The general rule is:

Parallel currents attract, and antiparallel currents repel.

This rule is in a sense opposite to the rule for electric charges, in that like (parallel) currents attract, but like (same sign) charges repel.

The force between long parallel wires is used to define the ampere. Given two long parallel wires of negligible circular cross section separated in vacuum by a distance of 1 meter, the ampere is defined as the current in each wire that would produce a force of 2×10^{-7} newtons per meter of length.

Primary measurements of current can be made with a current balance, shown schematically in Fig. 12. It consists of a carefully wound coil of wire placed between two other coils; the outer coils are fastened to a table, while the

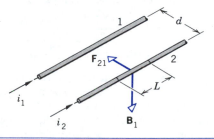

Figure 11 Two parallel wires carrying currents in the same direction attract each other. The field $\mathbf{B_1}$ at wire 2 is that due to the current in wire 1.

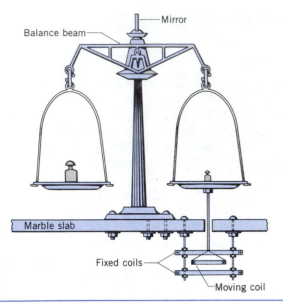

Figure 12 A current balance.

inner coil hangs from the arm of a balance. All three coils carry the same current.

Just like the parallel wires of Fig. 11, the coils exert forces on one another, which can be measured by loading weights on the balance pan. The current can be determined from this measured force and the dimensions of the coils. This arrangement using coils is more practical than the long parallel wires of Fig. 11. Current balance measurements are used to calibrate other more convenient secondary standards for measuring current.

Sample Problem 4   A long horizontal rigidly supported wire carries a current i_a of 96 A. Directly above it and parallel to it is a fine wire that carries a current i_b of 23 A and weighs 0.073 N/m. How far above the lower wire should this second wire be strung if we hope to support it by magnetic repulsion?

Solution   To provide a repulsion, the two currents must point in opposite directions. For equilibrium, the magnetic force per unit length must equal the weight per unit length and must be oppositely directed. Solving Eq. 18 for d yields

$$d = \frac{\mu_0 i_a i_b}{2\pi(F/L)} = \frac{(4\pi \times 10^{-7}\ \text{T·m/A})(96\ \text{A})(23\ \text{A})}{2\pi(0.073\ \text{N/m})}$$
$$= 6.0 \times 10^{-3}\ \text{m} = 6.0\ \text{mm}.$$

We assume that the wire diameters are much smaller than their separation. This assumption is necessary because in deriving Eq. 18 we tacitly assumed that the magnetic field produced by one wire is uniform for all points within the second wire.

Is the equilibrium of the suspended wire stable or unstable against vertical displacements? This can be tested by displacing the wire vertically and examining how the forces on the wire change. Is the equilibrium stable or unstable against horizontal displacements?

Suppose that the fine wire is suspended *below* the rigidly supported wire. How may it be made to "float"? Is the equilibrium stable or unstable against vertical displacements? Against horizontal displacements?

35-5 AMPÈRE'S LAW

Coulomb's law can be considered a fundamental law of electrostatics; we can use it to calculate the electric field associated with any distribution of electric charges. In Chapter 29, however, we showed that Gauss' law permitted us to solve a certain class of problems, those containing a high degree of symmetry, with ease and elegance. Furthermore, we showed that Gauss' law contained within it Coulomb's law for the electric field of a point charge. We consider Gauss' law to be more basic than Coulomb's law, and Gauss' law is one of the four fundamental (Maxwell) equations of electromagnetism.

The situation in magnetism is similar. Using the Biot–Savart law, we can calculate the magnetic field of any distribution of currents, just as we used Eq. 2 (which is equivalent to Coulomb's law) to calculate the electric field of any distribution of charges. A more fundamental approach to magnetic fields uses a law that (like Gauss' law for electric fields) takes advantage of the symmetry present in certain problems to simplify the calculation of **B**. This law is considered more fundamental than the Biot–Savart law and leads to another of the four Maxwell equations.

This new result is called *Ampère's law* and is written

$$\oint \mathbf{B} \cdot d\mathbf{s} = \mu_0 i. \tag{19}$$

You will recall that, in using Gauss' law, we first constructed an imaginary closed surface (a Gaussian surface) that enclosed a certain amount of charge. In using Ampère's law we construct an imaginary closed curve (called an *Ampèrian loop*), as indicated in Fig. 13. The left side of Eq. 19 tells us to divide the curve into small segments of length $d\mathbf{s}$. As we travel around the loop (our direction of travel determining the direction of $d\mathbf{s}$), we evaluate the quantity $\mathbf{B} \cdot d\mathbf{s}$ and add (integrate) all such quantities around the loop.

The integral on the left of Eq. 19 is called a *line integral*. (Previously we used line integrals in Chapter 7 to calculate work and in Chapter 30 to calculate potential difference.) The circle superimposed on the integral sign reminds us that the line integral is to be evaluated around a *closed* path. Letting θ represent the angle between $d\mathbf{s}$ and **B**, we can write the line integral as

$$\oint \mathbf{B} \cdot d\mathbf{s} = \oint B\, ds \cos \theta. \tag{20}$$

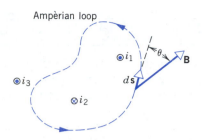

Figure 13 Ampère's law applied to an arbitrary loop that encloses two wires but excludes a third. Note the directions of the currents.

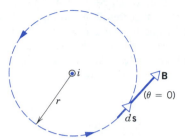

Figure 14 A circular Ampèrian loop is used to find the magnetic field set up by a current in a long straight wire. The wire is perpendicular to the page, and the direction of the current is out of the page.

The right side of Eq. 19 is the total current "enclosed" by the loop; that is, it is the total current carried by wires that pierce the surface bounded by the loop. As for charges in the case of Gauss' law, currents outside the loop are not included. Figure 13 shows three wires carrying current. The magnetic field **B** is the net effect of the currents in all wires. However, in the evaluation of the right side of Eq. 19, we include only the currents i_1 and i_2, because the wire carrying i_3 does not pass through the surface enclosed by the loop. The two wires that pass through the loop carry currents in opposite direction. A right-hand rule is used to assign signs to currents: with the fingers of your right hand in the direction in which the loop is traveled, currents in the direction of your thumb (such as i_1) are taken to be positive, while currents in the opposite direction (such as i_2) are taken to be negative. The net current i in the case of Fig. 13 is thus $i = i_1 - i_2$.

The magnetic field **B** at points on the loop and within the loop certainly depends on the current i_3; however, the *integral* of **B**·d**s** around the loop does *not* depend on currents such as i_3 that do not penetrate the surface enclosed by the loop. This is reasonable, because **B**·d**s** for the field established by i_1 or i_2 always has the same sign as we travel around the loop; however, **B**·d**s** for the field due only to i_3 changes sign as we travel around the loop, and in fact the positive and negative contributions exactly cancel one another.

Note that including the arbitrary constant of 4π in the Biot–Savart law reduces the constant that appears in Ampère's law to simply μ_0. (A similar simplification of Gauss' law was obtained by including the constant 4π in Coulomb's law.)

We were able to use Gauss' law to calculate electric fields only in cases having a high degree of symmetry. In those cases, we argued that E was constant and could be brought out of the integral. We choose Ampèrian loops in a similar manner, so that B is constant and can be brought out of the integral.

By way of illustration, let us use Ampère's law to find the magnetic field at a distance r from a long straight wire, a problem we have solved already using the Biot–Savart law. As illustrated in Fig. 14, we choose as our Ampèrian

path a circle of radius r. From the symmetry of the problem, **B** can depend only on r (and not, for instance, on the angular coordinate around the circle). By choosing a path that is everywhere the same distance from the wire, we know that B is constant around the path.

We know from Oersted's experiments that **B** has only a tangential component. Thus the angle θ is zero, and the line integral becomes

$$\oint B\, ds \cos\theta = B \oint ds = B(2\pi r).$$

Note that the integral of ds around the path is simply the length of the path, or $2\pi r$ in the case of the circle. The right side of Ampère's law is simply $\mu_0 i$ (taken as positive, in accordance with the right-hand rule). Ampère's law gives

$$B(2\pi r) = \mu_0 i$$

or

$$B = \frac{\mu_0 i}{2\pi r}.$$

This is identical with Eq. 11, a result we obtained (with considerably more effort) using the Biot–Savart law.

Sample Problem 5 Derive an expression for **B** at a distance r from the center of a long cylindrical wire of radius R, where $r < R$. The wire carries a current i, distributed uniformly over the cross section of the wire.

Solution Figure 15 shows a circular Ampèrian loop inside the wire. Symmetry suggests that **B** is constant in magnitude along the loop and tangent to it as shown. Ampère's law gives

$$B(2\pi r) = \mu_0 i \frac{\pi r^2}{\pi R^2},$$

where the right side includes only the fraction of the current that passes through the surface enclosed by the path of integration. Solving for B yields

$$B = \frac{\mu_0 i r}{2\pi R^2}.$$

At the surface of the wire ($r = R$) this equation reduces to the same expression as that found by putting $r = R$ in Eq. 11 ($B = \mu_0 i / 2\pi R$). That is, both expressions give the same result for the field at the surface of the wire. Figure 16 shows how the field depends on r both inside and outside the wire.

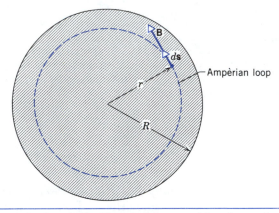

Figure 15   Sample Problem 5. A long straight wire carries a current that is emerging from the page and is uniformly distributed over the circular cross section of the wire. A circular Ampèrian loop is drawn inside the wire.

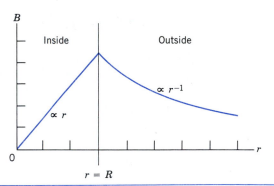

Figure 16   The magnetic field calculated for the wire shown in Fig. 15. Note that the largest field occurs at the surface of the wire.

35-6   SOLENOIDS AND TOROIDS

Two classes of practical devices based on windings of current loops are *solenoids* and *toroids*. A solenoid is often used to establish a uniform magnetic field, just as a parallel-plate capacitor establishes a uniform electric field. In door bells and loudspeakers, a solenoid often provides the magnetic field that accelerates a magnetic material. Toroids are also used to establish large fields.

Solenoids

A solenoid is a long wire wound in a close-packed helix and carrying a current i. We assume that the helix is very long compared with its diameter. What is the magnetic field **B** that is set up by the solenoid?

Figure 17 shows, for the sake of illustration only, a section of a "stretched-out" solenoid. For points close to a single turn of the solenoid, the observer is not aware that the wire is bent in an arc. The wire behaves magnetically almost like a long straight wire, and the lines of **B** due to this single turn are almost concentric circles.

The solenoid field is the vector sum of the fields set up by all the turns that make up the solenoid. Figure 17 suggests that the fields tend to cancel between adjacent wires. It also suggests that, at points inside the solenoid and reasonably far from the wires, **B** is parallel to the solenoid axis. In the limiting case of tightly packed square wires, the solenoid becomes essentially a cylindrical current sheet, and the requirements of symmetry then make it rigorously true that **B** is parallel to the axis of the solenoid. In the following we assume this to be the case.

For points such as P in Fig. 17 the field set up by the upper part of the solenoid turns (marked $\odot$, because the current is out of the page) points to the left and tends to cancel the field set up by the lower part of the solenoid turns (marked $\otimes$, because the current is into the page),

Figure 17   A section of a solenoid that has been stretched out for this illustration. The magnetic field lines are shown.

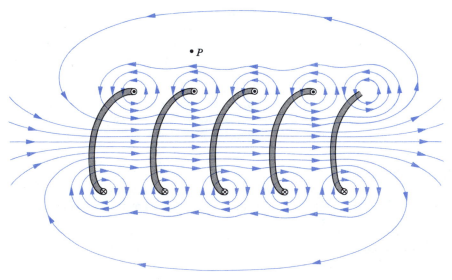

which points to the right. As the solenoid becomes more and more ideal, that is, as it approaches the configuration of an infinitely long cylindrical current sheet, the field **B** at outside points approaches zero. Taking the external field to be zero is a good assumption for a practical solenoid if its length is much greater than its diameter and if we consider only external points near the central region of the solenoid, that is, away from the ends. Figure 18 shows the lines of **B** for a real solenoid, which is far from ideal in that the length is not much greater than the diameter. Even here the spacing of the lines of **B** in the central plane shows that the external field is much weaker than the internal field.

Let us apply Ampère's law,

$$\oint \mathbf{B} \cdot d\mathbf{s} = \mu_0 i,$$

to the rectangular path *abcd* in the ideal solenoid of Fig. 19. We write the integral $\oint \mathbf{B} \cdot d\mathbf{s}$ as the sum of four integrals, one for each path segment:

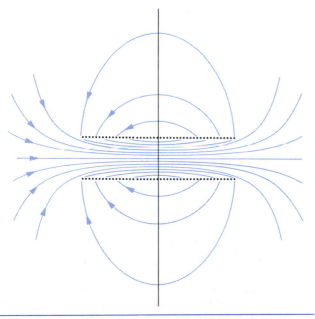

Figure 18 Magnetic field lines for a solenoid of finite length. Note that the field is stronger (indicated by the greater density of field lines) inside the solenoid than it is outside.

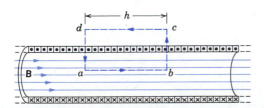

Figure 19 An Ampèrian loop (the rectangle *abcd*) is used to calculate the magnetic field of this long idealized solenoid.

$$\oint \mathbf{B} \cdot d\mathbf{s} = \int_a^b \mathbf{B} \cdot d\mathbf{s} + \int_b^c \mathbf{B} \cdot d\mathbf{s} + \int_c^d \mathbf{B} \cdot d\mathbf{s} + \int_d^a \mathbf{B} \cdot d\mathbf{s}. \quad (21)$$

The first integral on the right is *Bh*, where *B* is the magnitude of **B** inside the solenoid and *h* is the arbitrary length of the path from *a* to *b*. Note that path *ab*, though parallel to the solenoid axis, need not coincide with it. It will turn out that *B* inside the solenoid is constant over its cross section and independent of the distance from the axis (as suggested by the equal spacing of the lines of **B** in Fig. 18 near the center of the solenoid).

The second and fourth integrals in Eq. 21 are zero because for every element of these paths **B** is either at right angles to the path (for points inside the solenoid) or is zero (for points outside). In either case, $\mathbf{B} \cdot d\mathbf{s}$ is zero, and the integrals vanish. The third integral, which includes the part of the rectangle that lies outside the solenoid, is zero because we have taken **B** as zero for all external points for an ideal solenoid.

For the entire rectangular path, $\oint \mathbf{B} \cdot d\mathbf{s}$ has the value *Bh*. The net current *i* that passes through the rectangular Ampèrian loop is not the same as the current i_0 in the solenoid because the windings pass through the loop more than once. Let *n* be the number of turns per unit length; then the total current, which is out of the page inside the rectangular Ampèrian loop of Fig. 19, is

$$i = i_0 nh.$$

Ampère's law then becomes

$$Bh = \mu_0 i_0 nh$$

or

$$B = \mu_0 i_0 n. \quad (22)$$

Equation 22 shows that the magnetic field inside a solenoid depends only on the current i_0 and the number of turns per unit length *n*.

Although we derived Eq. 22 for an infinitely long ideal solenoid, it holds quite well for actual solenoids at internal points near the center of the solenoid. For an ideal solenoid, Eq. 22 suggests that *B* does not depend on the diameter or the length of the solenoid and that *B* is constant over the solenoid cross section. A solenoid is a practical way to set up a uniform magnetic field.

Toroids

Figure 20 shows a toroid, which we may consider to be a solenoid bent into the shape of a doughnut. Let us find the magnetic field at interior points using Ampère's law and certain considerations of symmetry.

From symmetry, the lines of **B** form concentric circles inside the toroid, as shown in the figure. Let us choose a concentric circle of radius *r* as an Ampèrian loop and traverse it in the clockwise direction. Ampère's law yields

$$B(2\pi r) = \mu_0 i_0 N,$$

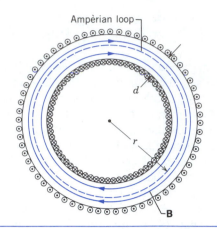

Figure 20 A toroid. The interior field can be found using the circular Ampèrian loop shown.

where i_0 is the current in the toroid windings and N is the total number of turns. This gives

$$B = \frac{\mu_0 i_0 N}{2\pi r} . \qquad (23)$$

In contrast to the solenoid, B is not constant over the cross section of a toroid. You should be able to show, from Ampère's law, that $B = 0$ for points outside an ideal toroid.

Close inspection of Eq. 23 justifies our earlier statement that a toroid is "a solenoid bent into the shape of a doughnut." The denominator in Eq. 23, $2\pi r$, is the central circumference of the toroid, and $N/2\pi r$ is just n, the number of turns per unit length. With this substitution, Eq. 23 reduces to $B = \mu_0 i_0 n$, the equation for the magnetic field in the central region of a solenoid.

The direction of the magnetic field within a toroid (or a solenoid) follows from the right--hand rule: curl the fingers of your right hand in the direction of the current; your extended right thumb then points in the direction of the magnetic field.

Toroids form the central feature of the *tokamak*, a device showing promise as the basis for a fusion power reactor. We discuss its mode of operation in Chapter 55 of the extended version of this book.

Sample Problem 6 A solenoid has a length $L = 1.23$ m and an inner diameter $d = 3.55$ cm. It has five layers of windings of 850 turns each and carries a current $i_0 = 5.57$ A. What is B at its center?

Solution From Eq. 22

$$B = \mu_0 i_0 n = (4\pi \times 10^{-7} \text{ T·m/A})(5.57 \text{ A}) \left(\frac{5 \times 850 \text{ turns}}{1.23 \text{ m}} \right)$$

$$= 2.42 \times 10^{-2} \text{ T} = 24.2 \text{ mT}.$$

Note that Eq. 22 applies even if the solenoid has more than one layer of windings because the diameter of the windings does not enter into the equation.

The Field Outside a Solenoid   *(Optional)*

We have so far neglected the field outside the solenoid, but even for an ideal solenoid, the field at points outside the winding is not zero. Figure 21 shows an Ampèrian path in the shape of a circle of radius r. Because the solenoid windings are helical, one turn of the winding pierces the surface enclosed by the circle. The product $\mathbf{B} \cdot d\mathbf{s}$ for this path depends on the tangential component of the field B_t, and thus Ampère's law gives

$$B_t(2\pi r) = \mu_0 i_0$$

or

$$B_t = \frac{\mu_0 i_0}{2\pi r} , \qquad (24)$$

which is the same field (in magnitude *and* direction) that would be set up by a straight wire. Note that the windings, in addition to carrying current around the surface of the solenoid, also carry current from left to right in Fig. 21, and in this respect the solenoid behaves like a straight wire at points outside the windings.

The tangential field is much smaller than the interior field (Eq. 22), as we can see by taking the ratio

$$\frac{B_t}{B} = \frac{\mu_0 i_0 / 2\pi r}{\mu_0 i_0 n} = \frac{1}{2\pi r n} .$$

Suppose the solenoid consists of one layer of turns in which the wires are touching one another, as in Fig. 19. Every interval along the solenoid of length equal to the diameter D of the wire contains one turn, and thus the number of turns per unit length n must be $1/D$. The ratio thus becomes

$$\frac{B_t}{B} = \frac{D}{2\pi r} . \qquad (25)$$

For a typical wire, $D = 0.1$ mm. The distance r to exterior points must be at least as large as the radius of the solenoid, which might be a few centimeters. Thus $B_t/B \leq 0.001$, and the tangential exterior field is indeed negligible compared with the interior field along the axis. We are therefore safe in neglecting the exterior field.

By drawing an Ampèrian circle similar to that of Fig. 21 but with radius smaller than that of the solenoid, you should be able to show that the tangential component of the interior field is zero. ∎

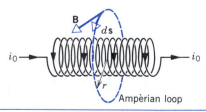

Figure 21 A circular Ampèrian loop of radius r is used to find the tangential field external to a solenoid.

35-7 ELECTROMAGNETISM AND FRAMES OF REFERENCE (Optional)

Figure 22a shows a particle carrying a positive charge q at rest near a long straight wire that carries a current i. We view the system from a frame of reference S in which the wire is at rest. Inside the wire are negative electrons moving with the drift velocity v_d and positive ion cores at rest. In any given length of wire, the number of electrons equals the number of ion cores, and the net charge is zero. The electrons can instantaneously be considered as a line of negative charge, which sets up an electric field at the location of q according to Eq. 33 of Chapter 28:

$$E = \frac{\lambda_-}{2\pi\epsilon_0 r},$$

where λ_- is the linear charge density of electrons (a negative number). The positive ion cores also set up an electric field given by a similar expression, depending on the linear charge density λ_+ of positive ions. Because the charge densities are of equal magnitude and opposite sign, $\lambda_+ + \lambda_- = 0$, and the net electric field that acts on the particle is zero.

There is a nonzero magnetic field on the particle, but because the particle is at rest, there is no magnetic force. Therefore no net force of electromagnetic origin acts on the particle in this frame of reference.

Now let us consider the situation from the perspective of a frame of reference S' moving parallel to the wire with velocity v_d (the drift velocity of the electrons). Figure 22b shows the situation in this frame of reference, in which the electrons are at rest and the ion cores move to the right with velocity v_d. Clearly, in this case the particle, being in motion, experiences a magnetic force F_B as shown in the figure.

Observers in different inertial frames must agree that if there is no acceleration in the S frame, there must also be no accelera-tion in the S' frame. The particle must therefore experience no net force in S', and so there must be another force in addition to F_B that acts on the particle to give a net force of zero.

This additional force that acts in the S' frame must be of electric origin. Consider in Fig. 22a a length L of the wire. We can imagine that length of the wire to consist of two measuring rods, a positively charged rod (the ions) at rest and a negatively charged rod (the electrons) in motion. The two rods have the same length (in S) and contain the same number of charges. When we transform those rods into S', we find that the rod of negative charge has a greater length in S'. In S, this moving rod has its *contracted length*, according to the relativistic effect of length contraction we considered in Section 21-3. In S', it is at rest and has its *proper length*, which is longer than the contracted length in S. The negative linear charge density λ'_- in S' is smaller in magnitude than that in S (that is, $|\lambda'_-| < |\lambda_-|$), because the same amount of charge is spread over a greater length in S'.

For the positive charges, the situation is opposite. In S, the positive charges are at rest, and the rod of positive charge has its proper length. In S', it is in motion and has a shorter contracted length. The linear density λ'_+ of positive charge in S' is greater than that in S ($\lambda'_+ > \lambda_+$), because the same amount of charge is spread over a shorter length. We therefore have the following relationships for the charge densities:

$$\text{in } S: \qquad \lambda_+ = |\lambda_-|,$$
$$\text{in } S': \qquad \lambda'_+ > |\lambda'_-|.$$

The charge q experiences the electric fields due to a line of positive charge and a line of negative charge. In S', these fields do not cancel, because the linear charge densities are different. The electric field at q in S' is therefore that due to a net linear density of positive charge, and q is repelled from the wire. The electric force F_E on q therefore opposes the magnetic force F_B, as shown in Fig. 22b. A detailed calculation* shows that the resulting electric force is exactly equal to the magnetic force, and the net force in S' is zero. Thus the particle experiences no acceleration in either reference frame. We can extend this result to situations other than the special case we considered here in which S' moves at velocity v_d with respect to S. In other frames of reference, the electric force and the magnetic force have values different from their values in S'; however, in every frame they are equal and opposite to one another and the net force on the particle is zero in *every* frame of reference.

This is a remarkable result. According to special relativity, electric and magnetic fields do not have separate existences. A field that is purely electric or purely magnetic in one frame of reference has both electric and magnetic components in another frame. Using relativistic transformation equations, we can easily pass back and forth from one frame to another, and we can often solve difficult problems by choosing a frame of reference in which the fields have a simpler character and then transforming the result back to the original frame. Special relativity can be of great practical value in solving such problems, because the techniques of special relativity may turn out to be simpler than the classical techniques.

In mathematical language, we say that the laws of electromagnetism (Maxwell's equations) are invariant with respect to the

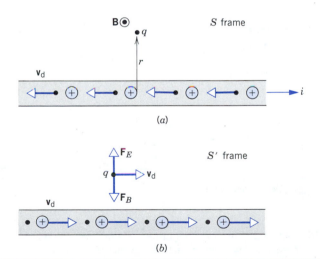

(a)

(b)

Figure 22 (a) A particle of charge q is at rest in equilibrium near a wire carrying a current i. The situation is viewed from a reference frame S at rest with respect to the particle. (b) The same situation viewed from a frame S' that is moving with the drift velocity of the electrons in the wire. The particle is also in equilibrium in this frame under the influence of the two forces F_E and F_B.

---

* See, for example, R. Resnick, *Introduction to Special Relativity* (Wiley, 1968), Chapter 4.

Lorentz transformation. Recall our discussion in Section 3-6 about *invariant* physical laws: we write down the law in one frame of reference, transform to another frame, and obtain a law of exactly the same mathematical form. For example, Gauss' law, one of the four Maxwell equations, has exactly the same form in every frame of reference.

Einstein's words are direct and to the point: "The force acting on a body in motion in a magnetic field is nothing else but an electric field." (In fact, Einstein's original 1905 paper, in which he first presented the ideas of special relativity, was titled "On the Electrodynamics of Moving Bodies.") In this context, we can regard magnetism as a relativistic effect, depending on the velocity of the charge relative to the observer. However, unlike other relativistic effects, it has substantial observable consequences at speeds far smaller than the speed of light. ∎

QUESTIONS

1. A beam of 20-MeV protons emerges from a cyclotron. Do these particles cause a magnetic field?

2. Discuss analogies and differences between Coulomb's law and the Biot–Savart law.

3. Consider a magnetic field line. Is the magnitude of **B** constant or variable along such a line? Can you give an example of each case?

4. In electronics, wires that carry equal but opposite currents are often twisted together to reduce their magnetic effect at distant points. Why is this effective?

5. Consider two charges, first (a) of the same sign and then (b) of opposite signs, that are moving along separated parallel paths with the same velocity. Compare the directions of the mutual electric and magnetic forces in each case.

6. Is there any way to set up a magnetic field other than by causing charges to move?

7. Give details of three ways in which you can measure the magnetic field **B** at a point P, a perpendicular distance r from a long straight wire carrying a constant current i. Base them on (a) projecting a particle of charge q through point P with velocity **v**, parallel to the wire; (b) measuring the force per unit length exerted on a second wire, parallel to the first wire and carrying a current i'; (c) measuring the torque exerted on a small magnetic dipole located a perpendicular distance r from the wire.

8. How might you measure the magnetic dipole moment of a compass needle?

9. A circular loop of wire lies on the floor of the room in which you are sitting. It carries a constant current i in a clockwise sense, as viewed from above. What is the direction of the magnetic dipole moment of this current loop?

10. Is **B** uniform for all points within a circular loop of wire carrying a current? Explain.

11. In Fig. 10, explain the relation between the figure and the equation $\mathbf{F} = i\mathbf{L} \times \mathbf{B}$.

12. Two long parallel conductors carry equal currents i in the same direction. Sketch roughly the resultant lines of **B** due to the action of both currents. Does your figure suggest an attraction between the wires?

13. A current is sent through a vertical spring from whose lower end a weight is hanging. What will happen?

14. Equation 11 ($B = \mu_0 i/2\pi R$) suggests that a strong magnetic field is set up at points near a long wire carrying a current. Since there is a current i and a magnetic field **B**, why is there not a force on the wire in accord with the equation $\mathbf{F} = i\mathbf{L} \times \mathbf{B}$?

15. Two long straight wires pass near one another at right angles. If the wires are free to move, describe what happens when currents are sent through both of them.

16. Two fixed wires cross each other perpendicularly so that they do not actually touch but are close to each other, as shown in Fig. 23. Equal currents i exist in each wire in the directions indicated. In what region(s) will there be some points of zero net magnetic field?

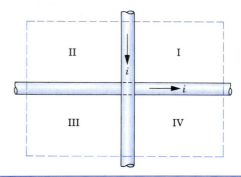

Figure 23   Question 16.

17. A messy loop of limp wire is placed on a frictionless table and anchored at points a and b as shown in Fig. 24. If a current i is now passed through the wire, will it try to form a circular loop or will it try to bunch up further?

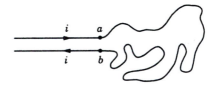

Figure 24   Question 17.

18. Can the path of integration around which we apply Ampère's law pass through a conductor?

19. Suppose we set up a path of integration around a cable that contains 12 wires with different currents (some in opposite directions) in each wire. How do we calculate i in Ampère's law in such a case?

20. Apply Ampère's law qualitatively to the three paths shown in Fig. 25.

21. Discuss analogies and differences between Gauss' law and Ampère's law.

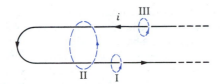

Figure 25 Question 20.

22. Does it necessarily follow from symmetry arguments alone that the lines of **B** around a long straight wire carrying a current i must be concentric circles?

23. A steady longitudinal uniform current is set up in a long copper tube. Is there a magnetic field (a) inside and/or (b) outside the tube?

24. A very long conductor has a square cross section and contains a coaxial cavity also with a square cross section. Current is distributed uniformly over the material cross section of the conductor. Is the magnetic field in the cavity equal to zero? Justify your answer.

25. A long straight wire of radius R carries a steady current i. How does the magnetic field generated by this current depend on R? Consider points both outside and inside the wire.

26. A long straight wire carries a constant current i. What does Ampère's law require for (a) a loop that encloses the wire but is not circular, (b) a loop that does not enclose the wire, and (c) a loop that encloses the wire but does not all lie in one plane?

27. Two long solenoids are nested on the same axis, as in Fig. 26. They carry identical currents but in opposite directions. If

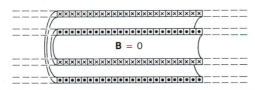

Figure 26 Question 27.

there is no magnetic field inside the inner solenoid, what can you say about n, the number of turns per unit length, for the two solenoids? Which one, if either, has the larger value?

28. The magnetic field at the center of a circular current loop has the value $B = \mu_0 i/2R$; see Eq. 16. However, the *electric* field at the center of a ring of charge is *zero*. Why this difference?

29. A steady current is set up in a cubical network of resistive wires, as in Fig. 27. Use symmetry arguments to show that the magnetic field at the center of the cube is zero.

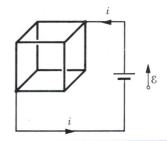

Figure 27 Question 29.

30. As an exercise in vector representation, contrast and compare Fig. 16 of Chapter 18, which deals with fluid flow, with Fig. 9 of this chapter, which deals with the magnetic field. How strong an analogy can you make?

31. Does Eq. 22 ($B = \mu_0 i_0 n$) hold for a solenoid of square cross section?

32. A toroid is described as a solenoid bent into the shape of a doughnut. The magnetic field outside an ideal solenoid is not zero. What can you say about the strength of the magnetic field outside an ideal toroid?

33. Drifting electrons constitute the current in a wire and a magnetic field is associated with this current. What current and magnetic field would be measured by an observer moving along the wire at the electron drift velocity?

PROBLEMS

Section 35-2 Applications of the Biot–Savart Law

1. A #10 bare copper wire (2.6 mm in diameter) can carry a current of 50 A without overheating. For this current, what is the magnetic field at the surface of the wire?

2. A surveyor is using a magnetic compass 6.3 m below a power line in which there is a steady current of 120 A. Will this interfere seriously with the compass reading? The horizontal component of Earth's magnetic field at the site is 21 μT (= 0.21 gauss).

3. The 25-kV electron gun in a TV tube fires an electron beam 0.22 mm in diameter at the screen, 5.6×10^{14} electrons arriving each second. Calculate the magnetic field produced by the beam at a point 1.5 mm from the axis of the beam.

4. At a location in the Philippines, the Earth's magnetic field has a value of 39.0 μT and is horizontal and due north. The net field is zero 8.13 cm above a long straight horizontal wire

that carries a steady current. (a) Calculate the current and (b) find its direction.

5. A long straight wire carries a current of 48.8 A. An electron, traveling at 1.08×10^7 m/s, is 5.20 cm from the wire. Calculate the force that acts on the electron if the electron velocity is directed (a) toward the wire, (b) parallel to the current, and (c) at right angles to the directions defined by (a) and (b).

6. A straight conductor carrying a current i is split into identical semicircular turns as shown in Fig. 28. What is the mag-

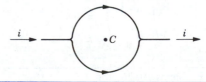

Figure 28 Problem 6.

netic field strength at the center C of the circular loop so formed?

7. Two long parallel wires are 8.10 cm apart. What equal currents must flow in the wires if the magnetic field halfway between them is to have a magnitude of 296 μT?

8. Two long straight parallel wires, separated by 0.75 cm, are perpendicular to the plane of the page as shown in Fig. 29. Wire W_1 carries a current of 6.6 A into the page. What must be the current (magnitude and direction) in wire W_2 for the resultant magnetic field at point P to be zero?

Figure 29   Problem 8.

9. Figure 30a shows a length of wire carrying a current i and bent into a circular coil of one turn. In Fig. 30b the same length of wire has been bent more sharply, to give a double loop of smaller radius. (a) If B_a and B_b are magnitudes of the magnetic fields at the centers of the two loops, what is the ratio B_b/B_a? (b) What is the ratio of their dipole moments, μ_b/μ_a?

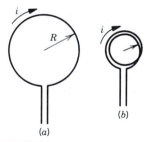

Figure 30   Problem 9.

10. Figure 31 shows an arrangement known as a *Helmholtz coil*. It consists of two circular coaxial coils each of N turns and radius R, separated by a distance R. They carry equal currents i in the same direction. Find the magnetic field at P, midway between the coils.

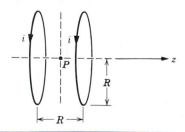

Figure 31   Problems 10, 26, and 27.

11. A student makes an electromagnet by winding 320 turns of wire around a wooden cylinder of diameter 4.80 cm. The coil is connected to a battery producing a current of 4.20 A in the wire. (a) What is the magnetic moment of this device? (b) At what axial distance $z \gg d$ will the magnetic field of this dipole be 5.0 μT (approximately one-tenth the Earth's magnetic field)?

12. A long hairpin is formed by bending a piece of wire as shown in Fig. 32. If the wire carries a current $i = 11.5$ A, (a) what are the magnitude and direction of **B** at point a? (b) At point b, very far from a? Take $R = 5.20$ mm.

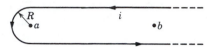

Figure 32   Problem 12.

13. A wire carrying current i has the configuration shown in Fig. 33. Two semi-infinite straight sections, each tangent to the same circle, are connected by a circular arc, of angle θ, along the circumference of the circle, with all sections lying in the same plane. What must θ be in order for B to be zero at the center of the circle?

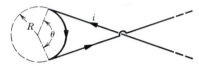

Figure 33   Problem 13.

14. A straight section of wire of length L carries a current i. (a) Show that the magnetic field associated with this segment at P, a perpendicular distance D from one end of the wire (see Fig. 34), is given by

$$B = \frac{\mu_0 i}{4\pi D} \frac{L}{(L^2 + D^2)^{1/2}}.$$

(b) Show that the magnetic field is zero at point Q, along the line of the wire.

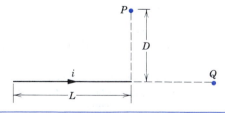

Figure 34   Problem 14.

15. Consider the circuit of Fig. 35. The curved segments are arcs of circles of radii a and b. The straight segments are along the radii. Find the magnetic field **B** at P, assuming a current i in the circuit.

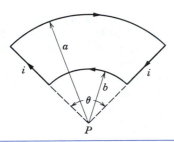

Figure 35 Problem 15.

16. A straight wire segment of length L carries a current i. Show that the magnetic field **B** associated with this segment, at a distance R from the segment along a perpendicular bisector (see Fig. 36), is given in magnitude by

$$B = \frac{\mu_0 i}{2\pi R} \frac{L}{(L^2 + 4R^2)^{1/2}}.$$

Show that this expression reduces to an expected result as $L \to \infty$.

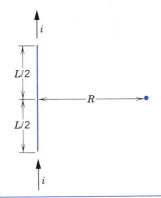

Figure 36 Problem 16.

17. Show that B at the center of a rectangular loop of wire of length L and width W, carrying a current i, is given by

$$B = \frac{2\mu_0 i}{\pi} \frac{(L^2 + W^2)^{1/2}}{LW}.$$

Show that this reduces to a result consistent with Sample Problem 1 for $L \gg W$.

18. A square loop of wire of edge a carries a current i. (a) Show that B for a point on the axis of the loop and a distance z from its center is given by

$$B(z) = \frac{4\mu_0 i a^2}{\pi(4z^2 + a^2)(4z^2 + 2a^2)^{1/2}}.$$

(b) To what does this reduce at the center of the loop?

19. The magnetic field B for various points on the axis of a square current loop of side a is given in Problem 18. (a) Show that the axial field for this loop for $z \gg a$ is that of a magnetic dipole (see Eq. 17). (b) Find the magnetic dipole moment of this loop.

20. You are given a length L of wire in which a current i may be established. The wire may be formed into a circle or a square. Show that the square yields the greater value for B at the central point.

21. Figure 37 shows a cross section of a long, thin ribbon of width w that is carrying a uniformly distributed total current i into the page. Calculate the magnitude and the direction of the magnetic field **B** at a point P in the plane of the ribbon at a distance d from its edge. (*Hint*: Imagine the ribbon to be constructed from many long, thin, parallel wires.)

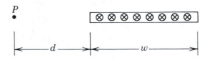

Figure 37 Problem 21.

22. Two long straight parallel wires 12.2 cm apart each carry a current of 115 A. Figure 38 shows a cross section, with the wires running perpendicular to the page and point P lying on the perpendicular bisector of d. Find the magnitude and direction of the magnetic field at P when the current in the left-hand wire is out of the page and the current in the right-hand wire is (a) out of the page and (b) into the page.

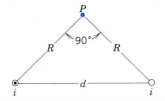

Figure 38 Problem 22.

23. In Fig. 7a assume that both currents are in the same direction, out of the plane of the figure. Show that the magnetic field in the plane defined by the wires is

$$B = \frac{\mu_0 i x}{\pi(x^2 - d^2)}.$$

Assume $i = 25$ A and $d = 2.5$ cm in Fig. 7a and plot B for the range -2.5 cm $< x < +2.5$ cm. Assume that the wire diameters are negligible.

24. Two long wires a distance d apart carry equal antiparallel currents i, as in Fig. 39. (a) Show that the magnetic field strength at point P, which is equidistant from the wires, is given by

$$B = \frac{2\mu_0 i d}{\pi(4R^2 + d^2)}.$$

(b) In what direction does **B** point?

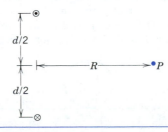

Figure 39 Problem 24.

25. You are given a closed circuit with radii a and b, as shown in Fig. 40, carrying current i. Find the magnetic dipole moment of the circuit.

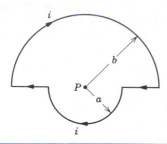

Figure 40   Problem 25.

26. Two 300-turn coils each carry a current i. They are arranged a distance apart equal to their radius, as in Fig. 31. For $R = 5.0$ cm and $i = 50$ A, plot B as a function of distance z along the common axis over the range $z = -5$ cm to $z = +5$ cm, taking $z = 0$ at the midpoint P. Such coils provide an especially uniform field B near point P. (*Hint*: See Eq. 15.)

27. In Problem 10 (Fig. 31) let the separation of the coils be a variable s (not necessarily equal to the coil radius R). (a) Show that the first derivative of the magnetic field (dB/dz) vanishes at the midpoint P regardless of the value of s. Why would you expect this to be true from symmetry? (b) Show that the second derivative of the magnetic field (d^2B/dz^2) also vanishes at P if $s = R$. This accounts for the uniformity of B near P for this particular coil separation.

28. A circular loop of radius 12 cm carries a current of 13 A. A second loop of radius 0.82 cm, having 50 turns and a current of 1.3 A is at the center of the first loop. (a) What magnetic field does the large loop set up at its center? (b) Calculate the torque that acts on the small loop. Assume that the planes of the two loops are at right angles and that the magnetic field due to the large loop is essentially uniform throughout the volume occupied by the small loop.

29. (a) A wire in the form of a regular polygon of n sides is just enclosed by a circle of radius a. If the current in this wire is i, show that the magnetic field **B** at the center of the circle is given in magnitude by

$$B = \frac{\mu_0 n i}{2\pi a} \tan (\pi/n).$$

(b) Show that as $n \to \infty$ this result approaches that of a circular loop. (c) Find the dipole moment of the polygon.

30. (a) A long wire is bent into the shape shown in Fig. 41, without cross contact at P. The radius of the circular section is R. Determine the magnitude and direction of **B** at the center C of the circular portion when the current i is as indicated. (b) The circular part of the wire is rotated without distortion about its (dashed) diameter perpendicular to the

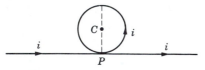

Figure 41   Problem 30.

straight portion of the wire. The magnetic moment associated with the circular loop is now in the direction of the current in the straight part of the wire. Determine **B** at C in this case.

31. (a) Calculate **B** at point P in Fig. 42. (b) Is the field strength at P greater or less than at the center of the square?

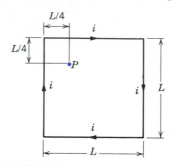

Figure 42   Problem 31.

32. A thin plastic disk of radius R has a charge q uniformly distributed over its surface. If the disk rotates at an angular frequency ω about its axis, show that (a) the magnetic field at the center of the disk is

$$B = \frac{\mu_0 \omega q}{2\pi R}$$

and (b) the magnetic dipole moment of the disk is

$$\mu = \frac{\omega q R^2}{4}.$$

(*Hint*: The rotating disk is equivalent to an array of current loops.)

33. Consider the rectangular loop carrying current i shown in Fig. 43. Point P is located a distance x from the center of the loop. Find an expression for the magnetic field at P due to the current loop, assuming that P is very far away. Verify that your expression agrees with the appropriate entry in Table 1, with $\mu = iab$. (*Hint*: Opposite sides of the rectangle can be treated together, but consider carefully the directions of **B** due to each side.)

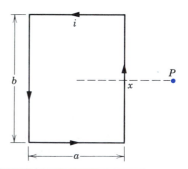

Figure 43   Problem 33.

Section 35-4   Two Parallel Conductors

34. Figure 44 shows five long parallel wires in the xy plane. Each wire carries a current $i = 3.22$ A in the positive x direction. The separation between adjacent wires is $d = 8.30$ cm. Find

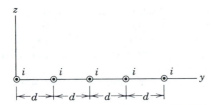

Figure 44 Problem 34.

the magnetic force per meter, magnitude and direction, exerted on each of these five wires.

35. Four long copper wires are parallel to each other and arranged in a square; see Fig. 45. They carry equal currents i out of the page, as shown. Calculate the force per meter on any one wire; give magnitude and direction. Assume that $i = 18.7$ A and $a = 24.5$ cm. (In the case of parallel motion of charged particles in a plasma, this is known as the pinch effect.)

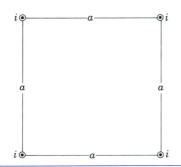

Figure 45 Problem 35.

36. Figure 46 shows a long wire carrying a current i_1. The rectangular loop carries a current i_2. Calculate the resultant force acting on the loop. Assume that $a = 1.10$ cm, $b = 9.20$ cm, $L = 32.3$ cm, $i_1 = 28.6$ A, and $i_2 = 21.8$ A.

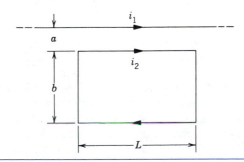

Figure 46 Problem 36.

37. Figure 47 shows an idealized schematic of an "electromagnetic rail gun," designed to fire projectiles at speeds up to 10 km/s. (The feasibility of these devices as defenses against ballistic missiles is being studied.) The projectile P sits between and in contact with two parallel rails along which it can slide. A generator G provides a current that flows up one rail, across the projectile, and back down the other rail.

(a) Let w be the distance between the rails, r the radius of the rails (presumed circular), and i the current. Show that the force on the projectile is to the right and given approximately by

$$F = \frac{1}{2} \left(\frac{i^2 \mu_0}{\pi} \right) \ln \left(\frac{w + r}{r} \right).$$

(b) If the projectile (in this case a test slug) starts from the left end of the rail at rest, find the speed v at which it is expelled at the right. Assume that $i = 450$ kA, $w = 12$ mm, $r = 6.7$ cm, $L = 4.0$ m, and that the mass of the slug is $m = 10$ g.

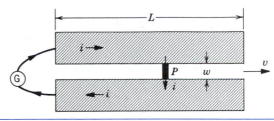

Figure 47 Problem 37.

38. In Sample Problem 4, suppose that the upper wire is displaced downward a small distance and then released. Show that the resulting motion of the wire is simple harmonic with the same frequency of oscillation as a simple pendulum of length d.

Section 35-5 Ampère's Law

39. Each of the indicated eight conductors in Fig. 48 carries 2.0 A of current into or out of the page. Two paths are indicated for the line integral $\oint \mathbf{B} \cdot d\mathbf{s}$. What is the value of the integral for (a) the dotted path and (b) the dashed path?

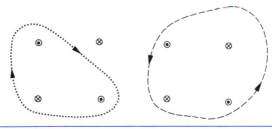

Figure 48 Problem 39.

40. Eight wires cut the page perpendicularly at the points shown in Fig. 49. A wire labeled with the integer k ($k = 1, 2, \ldots, 8$) bears the current ki_0. For those with odd k, the current is out

Figure 49 Problem 40.

of the page; for those with even k it is into the page. Evaluate $\oint \mathbf{B} \cdot d\mathbf{s}$ along the closed loop in the direction shown.

41. In a certain region there is a uniform current density of 15 A/m² in the positive z direction. What is the value of $\oint \mathbf{B} \cdot d\mathbf{s}$ when the line integral is taken along the three straight-line segments from $(4d, 0, 0)$ to $(4d, 3d, 0)$ to $(0, 0, 0)$ to $(4d, 0, 0)$, where $d = 23$ cm?

42. Consider a long cylindrical wire of radius R carrying a current i distributed uniformly over the cross section. At what *two* distances from the axis of the wire is the magnetic field strength, due to the current, equal to one-half the value at the surface?

43. Show that a uniform magnetic field $\mathbf{B}$ cannot drop abruptly to zero as one moves at right angles to it, as suggested by the horizontal arrow through point a in Fig. 50. (*Hint*: Apply Ampère's law to the rectangular path shown by the dashed lines.) In actual magnets "fringing" of the lines of $\mathbf{B}$ always occurs, which means that $\mathbf{B}$ approaches zero in a gradual manner. Modify the $\mathbf{B}$ lines in the figure to indicate a more realistic situation.

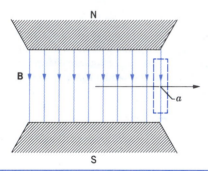

Figure 50   Problem 43.

44. Figure 51 shows a cross section of a hollow cylindrical conductor of radii a and b, carrying a uniformly distributed current i. (*a*) Using the circular Ampèrian loop shown, verify that $B(r)$ for the range $b < r < a$ is given by

$$B(r) = \frac{\mu_0 i}{2\pi(a^2 - b^2)} \frac{r^2 - b^2}{r}.$$

(*b*) Test this formula for the special cases of $r = a$, $r = b$, and $b = 0$. (*c*) Assume $a = 2.0$ cm, $b = 1.8$ cm, and $i = 100$ A and plot $B(r)$ for the range $0 < r < 6$ cm.

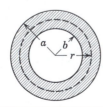

Figure 51   Problem 44.

45. Figure 52 shows a cross section of a long conductor of a type called a coaxial cable of radii a, b, and c. Equal but antipar-

allel uniformly distributed currents i exist in the two conductors. Derive expressions for $B(r)$ in the ranges (*a*) $r < c$, (*b*) $c < r < b$, (*c*) $b < r < a$, and (*d*) $r > a$. (*e*) Test these expressions for all the special cases that occur to you. (*f*) Assume $a = 2.0$ cm, $b = 1.8$ cm, $c = 0.40$ cm, and $i = 120$ A and plot $B(r)$ over the range $0 < r < 3$ cm.

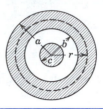

Figure 52   Problem 45.

46. A conductor consists of an infinite number of adjacent wires, each infinitely long and carrying a current i_0. Show that the lines of $\mathbf{B}$ are as represented in Fig. 53 and that B for all points above and below the infinite current sheet is given by

$$B = \tfrac{1}{2}\mu_0 n i_0,$$

where n is the number of wires per unit length. Derive both by direct application of Ampère's law and by considering the problem as a limiting case of Sample Problem 3.

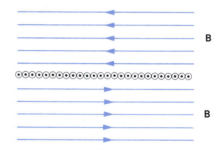

Figure 53   Problem 46.

47. The current density inside a long, solid, cylindrical wire of radius a is in the direction of the axis and varies linearly with radial distance r from the axis according to $j = j_0 r/a$. Find the magnetic field inside the wire. Express your answer in terms of the total current i carried by the wire.

48. Figure 54 shows a cross section of a long cylindrical conductor of radius a containing a long cylindrical hole of radius b. The axes of the two cylinders are parallel and are a distance d apart. A current i is uniformly distributed over the cross-hatched area in the figure. (*a*) Use superposition ideas to show that the magnetic field at the center of the hole is

$$B = \frac{\mu_0 i d}{2\pi(a^2 - b^2)}.$$

(*b*) Discuss the two special cases $b = 0$ and $d = 0$. (*c*) Can you use Ampère's law to show that the magnetic field in the hole is uniform? (*Hint*: Regard the cylindrical hole as filled with two equal currents moving in opposite directions, thus canceling each other. Assume that each of these currents has

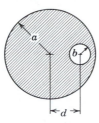

Figure 54 Problem 48.

the same current density as that in the actual conductor. Thus we superimpose the fields due to two complete cylinders of current, of radii a and b, each cylinder having the same current density.)

49. A long circular pipe, with an outside radius of R, carries a (uniformly distributed) current of i_0 (into the paper as shown in Fig. 55). A wire runs parallel to the pipe at a distance $3R$ from center to center. Calculate the magnitude and direction of the current in the wire that would cause the resultant magnetic field at the point P to have the same magnitude, but the opposite direction, as the resultant field at the center of the pipe.

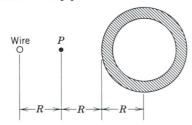

Figure 55 Problem 49.

Section 35-6 Solenoids and Toroids

50. A solenoid 95.6 cm long has a radius of 1.90 cm, a winding of 1230 turns, and carries a current of 3.58 A. Calculate the strength of the magnetic field inside the solenoid.

51. A solenoid 1.33 m long and 2.60 cm in diameter carries a current of 17.8 A. The magnetic field inside the solenoid is 22.4 mT. Find the length of the wire forming the solenoid.

52. A toroid having a square cross section, 5.20 cm on edge, and an inner radius of 16.2 cm has 535 turns and carries a current of 813 mA. Calculate the magnetic field inside the toroid at (a) the inner radius and (b) the outer radius of the toroid.

53. A long solenoid has 100 turns per centimeter. An electron moves within the solenoid in a circle of radius 2.30 cm perpendicular to the solenoid axis. The speed of the electron is $0.0460c$ (c = speed of light). Find the current in the solenoid.

54. A long solenoid with 115 turns/cm and a radius of 7.20 cm carries a current of 1.94 mA. A current of 6.30 A flows in a straight conductor along the axis of the solenoid. (a) At what radial distance from the axis will the direction of the resulting magnetic field be at 40.0° from the axial direction? (b) What is the magnitude of the magnetic field?

55. An interesting (and frustrating) effect occurs when one attempts to confine a collection of electrons and positive ions (a plasma) in the magnetic field of a toroid. Particles whose motion is perpendicular to the **B** field will not execute circular paths because the field strength varies with radial distance from the axis of the toroid. This effect, which is shown (exaggerated) in Fig. 56, causes particles of opposite sign to drift in opposite directions parallel to the axis of the toroid. (a) What is the sign of the charge on the particle whose path is sketched in the figure? (b) If the particle path has a radius of curvature of 11 cm when its radial distance from the axis of the toroid is 125 cm, what will be the radius of curvature when the particle is 110 cm from the axis?

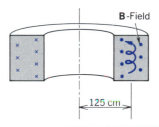

Figure 56 Problem 55.

56. Derive the solenoid equation (Eq. 22) starting from the expression for the field on the axis of a circular loop (Eq. 15). (*Hint*: Subdivide the solenoid into a series of current loops of infinitesimal thickness and integrate. See Fig. 17.)

CHAPTER 36

FARADAY'S LAW OF INDUCTION

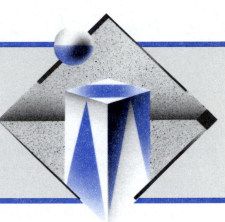

*We can often anticipate the outcome of an experiment by considering
how it is related by symmetry to other experiments. For example, a current
loop in a magnetic field experiences a torque (due to the field) that rotates the loop.
Consider a similar situation: a loop of wire in which there is no current is placed in a
magnetic field, and a torque applied by an external agent rotates the loop. We find that a
current appears in the loop! For a loop of wire in a magnetic field, a current produces a
torque, and a torque produces a current. This is an example of the symmetry of nature.*

The appearance of current in the loop is one example of the application of **Faraday's law of
induction,** *which is the subject of this chapter. Faraday's law, which is one of the four
Maxwell equations, was deduced from a number of simple and direct experiments, which
can easily be done in the laboratory and which serve directly to demonstrate Faraday's law.*

36-1 FARADAY'S EXPERIMENTS

Faraday's law of induction was discovered through experiments carried out by Michael Faraday in England in 1831 and by Joseph Henry in the United States at about the same time.* Even though Faraday published his results first, which gives him priority of discovery, the SI unit of inductance (see Chapter 38) is called the *henry* (abbreviation H). On the other hand, the SI unit of capacitance is, as we have seen, called the *farad* (abbreviation F). In Chapter 38, where we discuss oscillations in capacitive–inductive circuits, we see how appropriate it is to link the names of these two talented contemporaries in a single context.

---

* In addition to their independent simultaneous discovery of the law of induction, Faraday and Henry had several other similarities in their lives. Both were apprentices at an early age. Faraday, at age 14, was apprenticed to a London bookbinder. Henry, at age 13, was apprenticed to a watchmaker in Albany, New York. In later years Faraday was appointed director of the Royal Institution in London, whose founding was due in large part to an American, Benjamin Thompson (Count Rumford). Henry, on the other hand, became secretary of the Smithsonian Institution in Washington, DC, which was founded by an endowment from an Englishman, James Smithson.

Figure 1 shows a coil of wire as a part of a circuit containing an ammeter. Normally, we would expect the ammeter to show no current in the circuit because there seems to be no electromotive force. However, if we push a bar magnet toward the coil, with its north pole facing the coil, a remarkable thing happens. *While the magnet is moving,* the ammeter deflects, showing that a current has been set up in the coil. If we hold the magnet stationary with respect to the coil, the ammeter does not deflect. If we move the magnet *away* from the coil, the meter again deflects, but in the opposite direction, which means that the current in the coil is in the opposite direction. If we use the south pole end of a magnet instead of the north pole

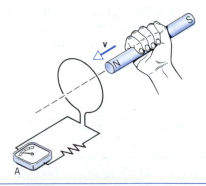

Figure 1 The ammeter A deflects when the magnet is moving with respect to the coil.

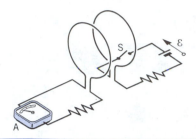

Figure 2   The ammeter A deflects momentarily when switch S is closed or opened. No physical motion of the coils is involved.

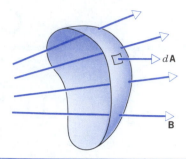

Figure 3   The magnetic field **B** through an area A gives a magnetic flux through the surface. The element of area $d\mathbf{A}$ is represented by a vector.

end, the experiment works as described but the deflections are reversed. The faster the magnet is moved, the greater is the reading of the meter. Further experimentation shows that *what matters is the relative motion of the magnet and the coil.* It makes no difference whether we move the magnet toward the coil or the coil toward the magnet.

The current that appears in this experiment is called an *induced current* and is said to be set up by an *induced electromotive force.* Note that there are no batteries anywhere in the circuit. Faraday was able to deduce from experiments like this the law that gives the magnitude and direction of the induced emfs. Such emfs are very important in practice. The chances are good that the lights in the room in which you are reading this book are operated from an induced emf produced in a commercial electric generator.

In another experiment, the apparatus of Fig. 2 is used. The coils are placed close together but at rest with respect to each other. When we close the switch S, thus setting up a steady current in the right-hand coil, the meter deflects momentarily; when we open the switch, thus interrupting this current, the meter again deflects momentarily, but in the opposite direction. None of the apparatus is physically moving in this experiment.

Experiment shows that there is an induced emf in the left coil of Fig. 2 whenever the current in the right coil is *changing.* It is the *rate at which the current is changing and not the size of the current* that is significant.

The common feature of these two experiments is *motion* or *change.* It is the *moving* magnet or the *changing* current that is responsible for the induced emfs. In the next section, we give the mathematical basis for these effects.

36-2   FARADAY'S LAW OF INDUCTION

Imagine that there are lines of magnetic field coming from the bar magnet of Fig. 1 and from the right-hand current loop in Fig. 2. Some of those field lines pass through the

left-hand coil in both figures. As the magnet is moved in the situation of Fig. 1, or as the switch is opened or closed in Fig. 2, the number of lines of the magnetic field passing through the left-hand coil changes. As Faraday's experiments showed, and as Faraday's technique of field lines helps us visualize, *it is the change in the number of field lines passing through a circuit loop that induces the emf in the loop.* Specifically, it is the *rate of change* in the number of field lines passing through the loop that determines the induced emf.

To make this statement quantitative, we introduce the *magnetic flux* Φ_B. Like the electric flux (see Section 29-2), the magnetic flux can be considered to be a measure of the number of field lines passing through a surface. In analogy with the electric flux (see Eq. 7 of Chapter 29), the magnetic flux through *any* surface is defined as

$$\Phi_B = \int \mathbf{B} \cdot d\mathbf{A}. \tag{1}$$

Here $d\mathbf{A}$ is an element of area of the surface (shown in Fig. 3), and the integration is carried out over the entire surface through which we wish to calculate the flux (for example, the surface enclosed by the left-hand loop in Fig. 1). If the magnetic field has a constant magnitude and direction over a planar area A, the flux can be written

$$\Phi_B = BA \cos \theta, \tag{2}$$

where θ is the angle between the normal to the surface and the direction of the field.

The SI unit of magnetic flux is the tesla·meter², which is given the name of *weber* (abbreviation Wb). That is,

$$1 \text{ weber} = 1 \text{ tesla·meter}^2.$$

Inverting this relationship, we see that the tesla is equivalent to the weber/meter², which was the unit used for magnetic fields before the tesla was adopted as the SI unit.

In terms of the magnetic flux, the emf induced in a circuit is given by *Faraday's law of induction:*

The induced emf in a circuit is equal to the negative of the rate at which the magnetic flux through the circuit is changing with time.

In mathematical terms, Faraday's law is

$$\mathscr{E} = -\frac{d\Phi_B}{dt}. \tag{3}$$

where $\mathscr{E}$ is the induced emf. If the rate of change of flux is in units of webers per second, the emf has units of volts. The minus sign in Eq. 3 is very important, because it tells us the direction of the induced emf. We consider this sign in detail in the next section.

If the coil consists of N turns, then an induced emf appears in every turn, and the total induced emf in the circuit is the sum of the individual values, just as in the case of batteries connected in series. If the coil is so tightly wound that each turn may be considered to occupy the same region of space and therefore to experience the same change of flux, then the total induced emf is

$$\mathscr{E} = -N\frac{d\Phi_B}{dt}. \tag{4}$$

There are many ways of changing the flux through a loop: moving a magnet relative to the loop (as in Fig. 1), changing the current in a nearby circuit (as in Fig. 2 and also as in a transformer), moving the loop in a nonuniform field, rotating the loop in a fixed magnetic field such that the angle θ in Eq. 2 changes (as in a generator), or changing the size or shape of the loop. In each of these methods, an emf is induced in the loop.

Finally, we note that, even though Eq. 3 is known as Faraday's law, it was not written in that form by Faraday, who was untrained in mathematics. In fact, Faraday's three-volume published work on electromagnetism, a landmark achievement in the development of physics and chemistry, contains not a single equation!

Sample Problem 1   The long solenoid S of Fig. 4 has 220 turns/cm and carries a current $i = 1.5$ A; its diameter d is 3.2 cm. At its center we place a 130-turn close-packed coil C of diameter $d_C = 2.1$ cm. The current in the solenoid is increased from zero to 1.5 A at a steady rate over a period of 0.16 s. What is the absolute value (that is, the magnitude without regard for sign) of the induced emf that appears in the central coil while the current in the solenoid is being changed?

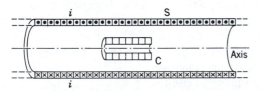

Figure 4   Sample Problem 1. A coil C is located inside a solenoid S. The solenoid carries a current that emerges from the page at the top and enters at the bottom, as indicated by the dots and crosses. When the current in the solenoid is changing, an induced emf appears in the coil.

Solution   The absolute value of the final flux through each turn of this coil is given by Eq. 2 with $\theta = 0$,

$$\Phi_B = BA.$$

The magnetic field B at the center of the solenoid is given by Eq. 22 of Chapter 35, or

$$B = \mu_0 in = (4\pi \times 10^{-7}\ \text{T·m/A})(1.5\ \text{A})$$
$$\times (220\ \text{turns/cm})(100\ \text{cm/m})$$
$$= 4.15 \times 10^{-2}\ \text{T}.$$

In terms of its diameter d_C, the area of the central coil (not of the solenoid) is given by $\frac{1}{4}\pi d_C^2$, which works out to be $3.46 \times 10^{-4}\ \text{m}^2$. The absolute value of the final flux through each turn of the coil is then

$$\Phi_B = (4.15 \times 10^{-2}\ \text{T})(3.46 \times 10^{-4}\ \text{m}^2)$$
$$= 1.44 \times 10^{-5}\ \text{Wb} = 14.4\ \mu\text{Wb}.$$

The induced emf follows from Faraday's law (Eq. 4), in which we ignore the minus sign because we are seeking only the absolute value of the emf:

$$\mathscr{E} = \frac{N\,\Delta\Phi_B}{\Delta t}$$

in which N is the number of turns in the inner coil C. The change $\Delta\Phi_B$ in the flux through each turn of the central coil is thus 14.4 μWb. This change occurs in 0.16 s, giving for the magnitude of the induced emf

$$\mathscr{E} = \frac{N\,\Delta\Phi_B}{\Delta t} = \frac{(130)(14.4 \times 10^{-6}\ \text{Wb})}{0.16\ \text{s}}$$
$$= 1.2 \times 10^{-2}\ \text{V} = 12\ \text{mV}.$$

We shall explain in the next section how to find the *direction* of the induced emf. For now, we can predict its direction by the following argument. Suppose an increase in the flux from the outer coil caused a current in the inner coil that produced a magnetic field in the same direction as the original field. This would in turn increase the flux through the area enclosed by the outer coil, which should similarly cause its current to increase, thereby increasing again the current in the inner coil, and so on. Is this a reasonable outcome?

36-3   LENZ' LAW

Thus far we have not specified the directions of the induced emfs. Although we can find these directions from a formal analysis of Faraday's law, we prefer to find them from the conservation-of-energy principle. In mechanics the energy principle often allows us to draw conclusions about mechanical systems without analyzing them in detail. We use the same approach here. The rule for determining the direction of the induced current was proposed in 1834 by Heinrich Friedrich Lenz (1804–1865) and is known as *Lenz' law*:

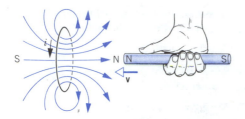

Figure 5   When the magnet is pushed toward the loop, the induced current i has the direction shown, setting up a magnetic field that opposes the motion of the magnet. This illustrates the application of Lenz' law.

The induced current in a closed conducting loop appears in such a direction that it opposes the change that produced it.

The minus sign in Faraday's law suggests this opposition.

Lenz' law refers to induced *currents,* which means that it applies only to closed conducting circuits. If the circuit is open, we can usually think in terms of what would happen if it *were* closed and in this way find the direction of the induced emf.

Consider the first of Faraday's experiments described in Section 36-1. Figure 5 shows the north pole of a magnet and a cross section of a nearby conducting loop. As we push the magnet toward the loop (or the loop toward the magnet) an induced current is set up in the loop. What is its direction?

A current loop sets up a magnetic field at distant points like that of a magnetic dipole, one face of the loop being a north pole, the opposite face being a south pole. The north pole, as for bar magnets, is that face *from* which the lines of **B** emerge. If, as Lenz' law predicts, the loop in Fig. 5 is to oppose the motion of the magnet toward it, the face of the loop *toward* the magnet must become a north pole. The two north poles—one of the current loop and one of the magnet—repel each other. The right-hand rule applied to the loop shows that for the magnetic field set up by the loop to emerge from the right face of the loop, the induced current must be as shown. The current is counterclockwise as we sight along the magnet toward the loop.

When we push the magnet toward the loop (or the loop toward the magnet), an induced current appears. In terms of Lenz' law this pushing is the "change" that produces the induced current, and, according to this law, the induced current opposes the "push." If we pull the magnet away from the coil, the induced current opposes the "pull" by creating a *south* pole on the right-hand face of the loop of Fig. 5. To make the right-hand face a south pole, the current must be opposite to that shown in Fig. 5. Whether we pull or push the magnet, its motion is automatically opposed.

The agent that causes the magnet to move, either toward the coil or away from it, always experiences a resisting force and is thus required to do work. From the

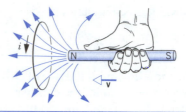

Figure 6   Another view of the operation of Lenz' law. When the magnet is pushed toward the loop, the magnetic flux through the loop is increased. The induced current through the loop sets up a magnetic field that opposes the increase in flux.

conservation-of-energy principle this work done on the system must exactly equal the internal (Joule) energy produced in the coil, since these are the only two energy transfers that occur in the system. If the magnet is moved more rapidly, the agent does work at a greater rate and the rate of production of internal energy increases correspondingly. If we cut the loop and then perform the experiment, there is no induced current, no internal energy change, no force on the magnet, and no work required to move it. There is still an emf in the loop, but, like a battery connected to an open circuit, it does not set up a current.

If the current in Fig. 5 were in the *opposite* direction to that shown, as the magnet moves toward the loop, the face of the loop toward the magnet would be a south pole, which would *pull* the bar magnet toward the loop. We would only need to push the magnet slightly to start the process and then the action would be self-perpetuating. The magnet would accelerate toward the loop, increasing its kinetic energy all the time. At the same time internal energy would appear in the loop at a rate that would increase with time. This would indeed be a something-for-nothing situation! Needless to say, it does not occur.

Let us apply Lenz' law to Fig. 5 in a different way. Figure 6 shows the lines of **B** for the bar magnet.* From this point of view the "change" is the increase in Φ_B through the loop caused by bringing the magnet nearer. The induced current opposes this change by setting up a field that tends to oppose the increase in flux caused by the moving magnet. Thus the field due to the induced current must point from left to right through the plane of the coil, in agreement with our earlier conclusion.

It is not significant here that the induced field opposes the *field* of the magnet but rather that it opposes the *change,* which in this case is the *increase* in Φ_B through the loop. If we withdraw the magnet, we reduce Φ_B through the loop. The induced field must now oppose this decrease in Φ_B (that is, the change) by *reinforcing* the magnetic field. In each case the induced field opposes the change that gives rise to it.

---

* There are two magnetic fields in this problem—one connected with the current loop and one with the bar magnet. You must always be certain which one is meant.

We can now obtain the direction of the current in the small coil C of Sample Problem 1. The field of the solenoid S points to the right in Fig. 4 and is increasing. The current in C must oppose this increase in flux through C and so must set up a field that opposes the field of S. The current in C is therefore in a direction opposite to that in S. If the current in S were *decreasing* instead of increasing, a similar argument shows that the induced current in C would have the same direction as the current in S.

Eddy Currents

When the magnetic flux through a large piece of conducting material changes, induced currents appear in the material (Fig. 7). These currents are called *eddy currents*. In some cases, the eddy currents may produce undesirable effects. For example, they increase the internal energy and thus can increase the temperature of the material. For this reason, materials that are subject to changing magnetic fields are often *laminated* or constructed in many small layers insulated from one another. Instead of one large loop, the eddy currents follow many smaller loops, thereby increasing the total length of their paths and the corresponding resistance; the resistive heating $\mathcal{E}^2/R$ is smaller, and the increase in internal energy is smaller. On the other hand, eddy-current heating can be used to advantage, as in an *induction furnace*, in which a sample of material can be heated using a rapidly changing magnetic field. Induction furnaces are used in cases in which it is not possible to make thermal contact with the material to be heated, such as when it is enclosed in a vacuum chamber.

Eddy currents are real currents and produce the same effects as real currents. In particular, a force $\mathbf{F} = i\mathbf{L} \times \mathbf{B}$ is exerted on the part of the eddy-current path in Fig. 7 that passes through the field. This force is transmitted to the material, and Lenz' law can be used to show (see Question 26) that the force opposes the motion of the conductor.

This gives rise to a form of *magnetic braking,* in which magnetic fields applied to a rotating wheel or a moving track produce forces that decelerate the motion. Such a brake has no moving parts or mechanical linkages and is not subject to the frictional wear of ordinary mechanical brakes. Moreover, it is most efficient at high speed (because the magnetic force increases with the relative speed), where the wear on mechanical brakes would be greatest.

36-4 MOTIONAL EMF

The example of Fig. 6, although easy to understand qualitatively, does not lend itself to quantitative calculations. Consider then Fig. 8, which shows a rectangular loop of wire of width D, one end of which is in a uniform field **B** pointing at right angles to the plane of the loop. This field **B** may be produced, for example, in the gap of a large electromagnet. The dashed lines show the assumed limits of the magnetic field. The loop is pulled to the right at a constant speed v.

The situation described by Fig. 8 does not differ in any essential detail from that of Fig. 6. In each case a conducting loop and a magnet are in relative motion; in each case the flux of the field of the magnet through the loop is being caused to change with time. The important difference between the two arrangements is that the situation of Fig. 8 permits easier calculations.

The external agent (the hand in Fig. 8) pulls the loop to the right at constant speed v by exerting a force F. We wish to calculate the mechanical power $P = Fv$ expended by the external agent or, equivalently, the rate at which it does work on the loop, and to compare that result with the rate at which the induced current in the loop produces internal energy.

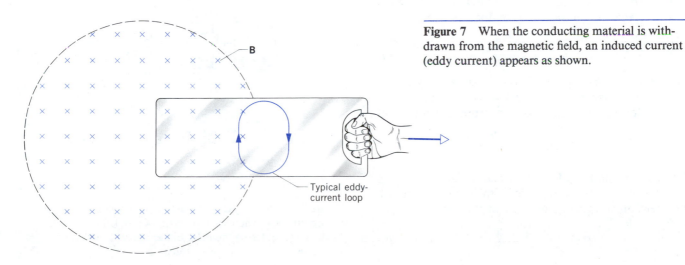

Figure 7 When the conducting material is withdrawn from the magnetic field, an induced current (eddy current) appears as shown.

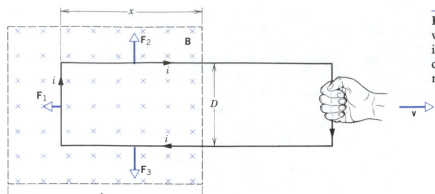

Figure 8 When the closed conducting loop is withdrawn from the field, an induced current i is produced in the loop. Internal energy is produced by the current at the same rate at which mechanical work is done on the loop.

The flux Φ_B enclosed by the loop in Fig. 8 is

$$\Phi_B = BDx,$$

where Dx is the area of that part of the loop in which B is not zero. We find the emf $\mathcal{E}$ from Faraday's law:

$$\mathcal{E} = -\frac{d\Phi_B}{dt} = -\frac{d}{dt}(BDx) = -BD\frac{dx}{dt} = BDv, \quad (5)$$

where we have set $-dx/dt$ equal to the speed v at which the loop is pulled out of the magnetic field, since x is decreasing. Note that the only dimension of the loop that enters into Eq. 5 is the length D of the left end conductor. As we shall see later, the induced emf in Fig. 8 may be regarded as localized here. An induced emf such as this, produced by the relative motion of a conductor and the source of a magnetic field, is sometimes called a *motional emf*.

The emf BDv sets up a current in the loop given by

$$i = \frac{\mathcal{E}}{R} = \frac{BDv}{R}, \quad (6)$$

where R is the loop resistance. From Lenz' law, this current (and thus $\mathcal{E}$) must be clockwise in Fig. 8; it opposes the "change" (the decrease in Φ_B) by setting up a field that is parallel to the external field within the loop.

The current in the loop gives rise to magnetic forces $\mathbf{F}_1$, $\mathbf{F}_2$, and $\mathbf{F}_3$ that act on the three conductors, according to Eq. 28 of Chapter 34,

$$\mathbf{F} = i\mathbf{L} \times \mathbf{B}. \quad (7)$$

Because $\mathbf{F}_2$ and $\mathbf{F}_3$ are equal and opposite, they cancel each other; $\mathbf{F}_1$, which is the force that opposes our effort to move the loop, is given in magnitude from Eqs. 6 and 7 as

$$F_1 = iDB \sin 90° = \frac{B^2 D^2 v}{R}. \quad (8)$$

The agent that pulls the loop must exert a force F equal in magnitude to F_1, if the loop is to move at constant speed. The agent must therefore do work at the steady rate of

$$P = F_1 v = \frac{B^2 D^2 v^2}{R}. \quad (9)$$

We can also compute the rate at which energy is dissipated in the loop as a result of Joule heating by the induced current. This is given by

$$P = i^2 R = \left(\frac{BDv}{R}\right)^2 R = \frac{B^2 D^2 v^2}{R}, \quad (10)$$

which agrees precisely with Eq. 9 for the rate at which mechanical work is done on the loop. The work done by the external agent is eventually dissipated as Joule heating of the loop.

Figure 9 shows a side view of the loop in the field. In Fig. 9a the loop is stationary; in Fig. 9b we are moving it to the right; in Fig. 9c we are moving it to the left. The lines of $\mathbf{B}$ in these figures represent the *resultant field* produced by the vector addition of the field due to the magnet and the field due to the induced current, if any, in the loop. Ac-

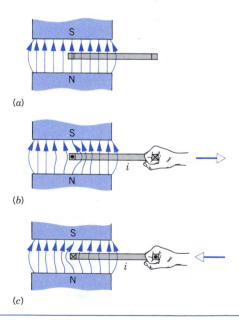

Figure 9 Magnetic field lines acting on a conducting loop in a magnetic field when the loop is (a) at rest, (b) leaving the field, and (c) entering the field. Either attempt to move the loop gives rise to an opposing force.

cording to Faraday's view, in which we regard the magnetic field lines as stretched rubber bands (see Section 35-3), the magnetic field lines in Fig. 9 suggest convincingly that the agent moving the coil always experiences an opposing force.

Sample Problem 2 Figure 10*a* shows a rectangular loop of resistance R, width D, and length a being pulled at constant speed v through a region of thickness d in which a uniform magnetic field **B** is set up by a magnet. As functions of the position x of the right-hand edge of the loop, plot (*a*) the flux Φ_B through the loop, (*b*) the induced emf $\mathcal{E}$, and (*c*) the rate P of production of internal energy in the loop. Use $D = 4$ cm, $a = 10$ cm, $d = 15$ cm, $R = 16$ Ω, $B = 2.0$ T, and $v = 1.0$ m/s.

Solution (*a*) The flux Φ_B is zero when the loop is not in the field; it is BDa when the loop is entirely in the field; it is BDx when the loop is entering the field and $BD[a - (x - d)]$ when the loop is leaving the field. These conclusions, which you should verify, are shown graphically in Fig. 10*b*.

(*b*) The induced emf $\mathcal{E}$ is given by $\mathcal{E} = -d\Phi_B/dt$, which we can write as

$$\mathcal{E} = -\frac{d\Phi_B}{dt} = -\frac{d\Phi_B}{dx}\frac{dx}{dt} = -\frac{d\Phi_B}{dx} v,$$

where $d\Phi_B/dx$ is the slope of the curve of Fig. 10*b*. The emf $\mathcal{E}$ is plotted as a function of x in Fig. 10*c*. Using the same type of reasoning as that used for Fig. 8, we deduce from Lenz' law that when the loop is entering the field, the emf $\mathcal{E}$ acts counterclockwise as seen from above. Note that there is no emf when the loop is entirely in the magnetic field because the flux Φ_B through the loop is not changing with time, as Fig. 10*b* shows.

(*c*) The rate of internal energy production is given by $P = \mathcal{E}^2/R$. It may be calculated by squaring the ordinate of the curve of Fig. 10*c* and dividing by R. The result is plotted in Fig. 10*d*.

If the fringing of the magnetic field, which cannot be avoided in practice (see Problem 43 of Chapter 35), is taken into account, the sharp bends and corners in Fig. 10 will be replaced by smooth curves. What changes would occur in the curves of Fig. 10 if the loop were cut so that it no longer formed a closed conducting path?

Sample Problem 3 A copper rod of length R rotates at angular frequency ω in a uniform magnetic field **B** as shown in Fig. 11. Find the emf $\mathcal{E}$ developed between the two ends of the rod. (We might measure this emf by placing a conducting rail along the dashed circle in the figure and connecting a voltmeter between the rail and point O.)

Solution If a wire of length dr is moved at velocity **v** at right angles to a field **B**, a motional emf $d\mathcal{E}$ will be developed (see Eq. 5) given by

$$d\mathcal{E} = Bv\, dr.$$

The rod of Fig. 11 may be divided into elements of length dr, the linear speed v of each element being ωr. Each element is perpendicular to **B** and is also moving in a direction at right angles to **B** so that, since the emf's $d\mathcal{E}$ of each element are "in series,"

$$\mathcal{E} = \int d\mathcal{E} = \int_0^R Bv\, dr = \int_0^R B\omega r\, dr = \tfrac{1}{2}B\omega R^2.$$

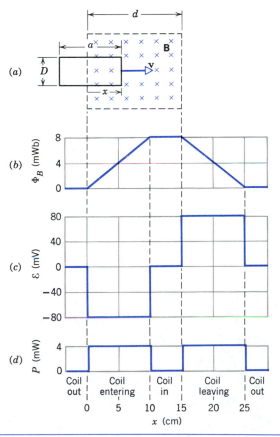

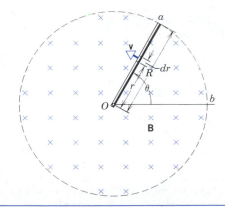

Figure 10 Sample Problem 2. (*a*) A closed conducting loop is pulled at constant speed completely through a region in which there is a uniform magnetic field **B**. (*b*) The magnetic flux through the loop as a function of the coordinate x of the right side of the loop. (*c*) The induced emf as a function of x. (*d*) The rate at which internal energy appears in the loop as it is moved.

Figure 11 Sample Problem 3. A copper rod rotates in a uniform magnetic field.

For a second approach, consider that at any instant the flux enclosed by the sector aOb in Fig. 11 is given by

$$\Phi_B = BA = B(\tfrac{1}{2}R^2\theta),$$

where $\tfrac{1}{2}R^2\theta$ is the area of the sector. Differentiating gives

$$\frac{d\Phi_B}{dt} = \tfrac{1}{2}BR^2\frac{d\theta}{dt} = \tfrac{1}{2}B\omega R^2.$$

From Faraday's law, this is precisely the magnitude of $\mathcal{E}$ and agrees with the previous result.

36-5 INDUCED ELECTRIC FIELDS

Suppose we place a loop of conducting wire in an external magnetic field (as in Fig. 12a). The field, which we assume to have a uniform strength over the area of the loop, may be established by an external electromagnet. By varying the current in the electromagnet, we can vary the strength of the magnetic field.

As **B** is varied, the magnetic flux through the loop varies with time, and from Faraday's and Lenz' laws we can calculate the magnitude and direction of the induced emf and the induced current in the loop. Before the field began changing, there was no current in the loop; while the field is changing, charges flow in the loop. For charges to begin moving, they must be accelerated by an electric field. This *induced electric field* occurs with a changing magnetic field, according to Faraday's law.

The induced electric field is just as real as any that might be set up by static charges; for instance, it exerts a force $q_0\mathbf{E}$ on a test charge. Moreover, the presence of the electric field has nothing to do with the presence of the loop of wire; if we were to remove the loop completely, the electric field would still be present. We could fill the space with a "gas" of electrons or ionized atoms; these particles would experience the same induced electric field **E**.

Let us therefore replace the loop of wire with a circular path of arbitrary radius r (Fig. 12b). The path, which we take in a plane perpendicular to the direction of **B**, encloses a region of space in which the magnetic field is changing at a rate $d\mathbf{B}/dt$. We assume that the rate $d\mathbf{B}/dt$ is

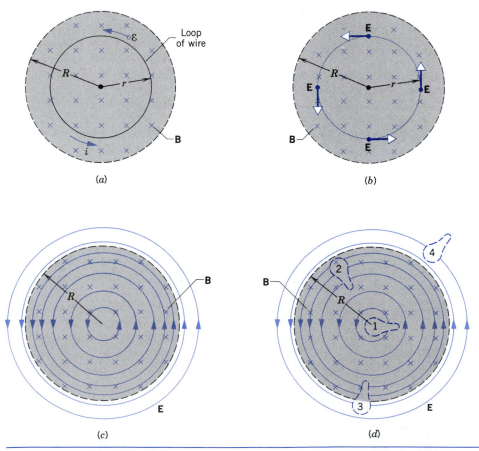

(a)   (b)

(c)   (d)

Figure 12 (a) If the magnetic field increases at a steady rate, a constant current appears, as shown, in the loop of wire of radius r. (b) Induced electric fields exist in the region, even when the ring is removed. (c) The complete picture of the induced electric fields, displayed as lines of force. (d) Four similar closed paths around which an emf may be calculated.

the same at every point in the area enclosed by the path. The circular path encloses a flux Φ_B which is changing at a rate $d\Phi_B/dt$ owing to the variation in the magnetic field. An induced emf appears around the path, and therefore there is an induced electric field at all points around the circle. From symmetry, we conclude that **E** must have the same magnitude at all points around the circle, there being no preferred direction in this space. Furthermore, **E** can have no radial component, a conclusion that follows from Gauss' law: construct an imaginary cylindrical Gaussian surface perpendicular to the plane of Fig. 12*b*. If there were a radial component to **E**, there would be a net *electric* flux into or out of the surface, which would require that the surface enclose a net electric charge. Since there is no such charge, the electric flux must be zero and the radial component of **E** must be zero. Thus the induced electric field is tangential, and the electric field lines are concentric circles, as in Fig. 12*c*.

Consider a test charge q_0 moving around the circular path of Fig. 12*b*. The work W done on the charge by the induced electric field in one revolution is $\mathscr{E}q_0$. Equivalently, we can express the work as the electric force q_0E times the displacement $2\pi r$ covered in one revolution. Setting these two expressions for W equal to one another and canceling the factor q_0, we obtain

$$\mathscr{E} = E(2\pi r). \tag{11}$$

The right side of Eq. 11 can be expressed as a line integral of **E** around the circle, which can be written in more general cases (for instance, when **E** is not constant or when the chosen path is not a circle) as

$$\mathscr{E} = \oint \mathbf{E} \cdot d\mathbf{s}. \tag{12}$$

Note that Eq. 12 reduces directly to Eq. 11 in our special case of a circular path with constant tangential E.

Replacing the emf by Eq. 12, we can write Faraday's law of induction ($\mathscr{E} = -d\Phi_B/dt$) as

$$\oint \mathbf{E} \cdot d\mathbf{s} = -\frac{d\Phi_B}{dt}. \tag{13}$$

It is in this form that Faraday's law appears as one of the four basic Maxwell equations of electromagnetism. In this form, it is apparent that Faraday's law implies that a changing magnetic field produces an electric field.

In Fig. 12, we have assumed that the magnetic field is increasing; that is, both dB/dt and $d\Phi_B/dt$ are positive. By Lenz' law, the induced emf opposes this change, and thus the induced currents create a magnetic field that points out of the plane of the figure. Since the currents must be counterclockwise, the lines of induced electric field **E** (which is responsible for the current) must also be counterclockwise. If, on the other hand, the magnetic field were decreasing ($dB/dt < 0$), the lines of induced electric field would be clockwise, such that the induced current again opposes the change in Φ_B.

Faraday's law in the form of Eq. 13 can be applied to paths of any geometry, not only the special circular path we chose in Fig. 12*b*. Figure 12*d* shows four such paths, all having the same shape and area but located in different positions in the changing field. For paths 1 and 2, the induced emf is the same because these paths lie entirely within the changing magnetic field and thus have the same value of $d\Phi_B/dt$. However, even though the emf $\mathscr{E}$ ($=\oint \mathbf{E} \cdot d\mathbf{s}$) is the same for these two paths, the distribution of electric field vectors around the paths is different, as indicated by the lines of the electric field. For path 3, the emf is smaller because both Φ_B and $d\Phi_B/dt$ are smaller, and for path 4 the induced emf is zero, even though the electric field is not zero at any pont along the path.

The induced electric fields that are set up by the induction process are not associated with charges but with a changing magnetic flux. Although both kinds of electric fields exert forces on charges, there is a difference between them. The simplest evidence for this difference is that lines of **E** associated with a changing magnetic flux can form closed loops (see Fig. 12); lines of **E** associated with charges do not form closed loops but are always drawn to start on a positive charge and end on a negative charge.

Equation 15 of Chapter 30, which defined the potential difference between two points *a* and *b*, is

$$V_b - V_a = \frac{-W_{ab}}{q_0} = -\int_a^b \mathbf{E} \cdot d\mathbf{s}. \tag{14}$$

If potential is to have any useful meaning, this integral (and W_{ab}) must have the same value for every path connecting *a* and *b*. This proved to be true for every case examined in earlier chapters.

An interesting special case comes up if *a* and *b* are the same point. The path connecting them is now a closed loop; V_a must be identical with V_b, and Eq. 14 reduces to

$$\oint \mathbf{E} \cdot d\mathbf{s} = 0. \tag{15}$$

However, when changing magnetic flux is present, $\oint \mathbf{E} \cdot d\mathbf{s}$ is *not* zero but is, according to Faraday's law (see Eq. 13), $-d\Phi_B/dt$. Electric fields associated with stationary charges are *conservative*, but those associated with changing magnetic fields are *nonconservative*; see Section 8-2. The (nonconservative) electric fields produced by induction cannot be described by an electric potential.

A similar argument can be given in the case of magnetic fields produced by currents in wires. The lines of **B** also form closed loops (see Fig. 9 of Chapter 35), and consequently magnetic potential has no meaning in such cases.

Sample Problem 4 In Fig. 12*b*, assume that $R = 8.5$ cm and that $dB/dt = 0.13$ T/s. (*a*) What is the magnitude of the electric field **E** for $r = 5.2$ cm? (*b*) What is the magnitude of the induced electric field for $r = 12.5$ cm?

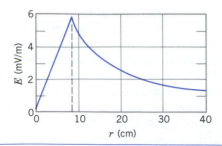

Figure 13 The induced electric field found in Sample Problem 4.

Solution (a) From Faraday's law (Eq. 13) we have

$$E(2\pi r) = -\frac{d\Phi_B}{dt}.$$

We note that $r < R$. The flux Φ_B through a closed path of radius r is then

$$\Phi_B = B(\pi r^2),$$

so that

$$E(2\pi r) = -(\pi r^2)\frac{dB}{dt}.$$

Solving for E and taking magnitudes, we find

$$E = \frac{1}{2}\left|\frac{dB}{dt}\right| r. \tag{16}$$

Note that the induced electric field E depends on dB/dt but not on B. For $r = 5.2$ cm, we have, for the magnitude of $\mathbf{E}$,

$$E = \frac{1}{2}\left|\frac{dB}{dt}\right| r = \frac{1}{2}(0.13 \text{ T/s})(5.2 \times 10^{-2} \text{ m})$$

$$= 0.0034 \text{ V/m} = 3.4 \text{ mV/m}.$$

(b) In this case we have $r > R$ so that the entire flux of the magnet passes through the circular path. Thus

$$\Phi_B = B(\pi R^2).$$

From Faraday's law (Eq. 13) we then find

$$E(2\pi r) = -\frac{d\Phi_B}{dt} = -(\pi R^2)\frac{dB}{dt}.$$

Solving for E and again taking magnitudes, we find

$$E = \frac{1}{2}\left|\frac{dB}{dt}\right|\frac{R^2}{r}. \tag{17}$$

An electric field is induced in this case even at points that are well outside the (changing) magnetic field, an important result that makes transformers (see Section 39-5) possible. For $r = 12.5$ cm, Eq. 17 gives

$$E = \frac{1}{2}(0.13 \text{ T/s})\frac{(8.5 \times 10^{-2} \text{ m})^2}{12.5 \times 10^{-2} \text{ m}}$$

$$= 3.8 \times 10^{-3} \text{ V/m} = 3.8 \text{ mV/m}.$$

Equations 16 and 17 yield the same result, as they must, for $r = R$. Figure 13 shows a plot of $E(r)$ based on these two equations.

36-6 THE BETATRON*

The betatron is a device for accelerating electrons (also known as beta particles) to high speeds using induced electric fields produced by changing magnetic fields. Such high-energy electrons can be used for basic research in physics as well as for producing x rays for applied research in industry and for medical purposes such as cancer therapy. The betatron provides an excellent illustration of the "reality" of induced electric fields. Typically, betatrons can produce energies of 100 MeV, in which case the electrons are highly relativistic ($v = 0.999987c$). Betatrons can produce enormous currents, in the range of 10^3-10^5 A. They are, however, pulsed machines, producing pulses of typical width μs or less separated by time intervals in the range of $0.01-1$ s.

Figure 14 shows a cross section through the inner structure of a betatron. It consists of a large electromagnet M, the field of which (indicated by the field lines) can be varied by changing the current in coils C. The electrons circulate in the evacuated ceramic doughnut-shaped tube marked D. Their orbit is at right angles to the plane of the figure, emerging from the left and entering at the right.

The magnetic field has several functions: (1) it guides the electrons in a circular path; (2) the changing magnetic field produces an induced electric field that accelerates the electrons in their path; (3) it maintains a constant radius of the path of the electrons; (4) it introduces electrons into the orbit and then removes them from the orbit after they have attained their full energy; and (5) it provides a restoring force that tends to resist any tendency of the electrons to leave their orbit, either vertically or radially. It is remarkable that the magnetic field is capable of performing all these operations.

The coils carry an alternating current and produce the magnetic field shown in Fig. 15. For electrons to circulate in the direction shown in Fig. 14 (counterclockwise as viewed from above), the magnetic field must be pointing upward (taken as positive in Fig. 15). Furthermore, the changing field must have positive slope ($dB/dt > 0$ so that $d\Phi_B/dt > 0$) if the electrons are to be accelerated (rather than decelerated) during the cycle. Thus only the first quarter-cycle of Fig. 15 is useful for the operation of the betatron; the electrons are injected at $t = 0$ and extracted at $t = T/4$. For the remaining three-quarters of a cycle, the device produces no beam.

---

* For a review of developments and applications of betatrons and similar devices, see "Ultra-high-current Electron Induction Accelerators," by Chris A. Kapetanakos and Phillip Sprangle, *Physics Today*, February 1985, p. 58.

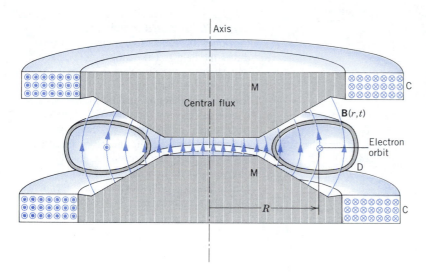

Figure 14   A cross section of a betatron, showing the orbit of the accelerating electrons and a "snapshot" of the time-varying magnetic field at a certain moment during the cycle. The magnetic field is produced by the coils C and is shaped by the magnetic pole pieces M.

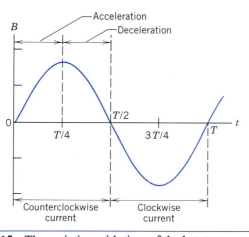

Figure 15   The variation with time of the betatron magnetic field B during one cycle.

Sample Problem 5   In a 100-MeV betatron, the orbit radius R is 84 cm. The magnetic field in the region enclosed by the orbit rises periodically (60 times per second) from zero to a maximum average value $B_{av,m} = 0.80$ T in an accelerating interval of one-fourth of a period, or 4.2 ms. (*a*) How much energy does the electron gain in one average trip around its orbit in this changing flux? (*b*) What is the *average* speed of an electron during its acceleration cycle?

Solution   (*a*) The central flux rises during the accelerating interval from zero to a maximum of

$$\Phi_B = (B_{av,m})(\pi R^2)$$
$$= (0.80 \text{ T})(\pi)(0.84 \text{ m})^2 = 1.8 \text{ Wb}.$$

The average value of $d\Phi_B/dt$ during the accelerating interval is then

$$\left(\frac{d\Phi_B}{dt}\right)_{av} = \frac{\Delta\Phi_B}{\Delta t} = \frac{1.8 \text{ Wb}}{4.2 \times 10^{-3} \text{ s}} = 430 \text{ Wb/s}.$$

From Faraday's law (Eq. 3) this is also the average emf in volts. Thus the electron increases its energy by an average of 430 eV per revolution in this changing flux. To achieve its full final energy of 100 MeV, it has to make about 230,000 revolutions in its orbit, a total path length of about 1200 km.

(*b*) The length of the acceleration cycle is given as 4.2 ms, and the path length is calculated above to be 1200 km. The average speed is then

$$\bar{v} = \frac{1200 \times 10^3 \text{ m}}{4.2 \times 10^{-3} \text{ s}} = 2.86 \times 10^8 \text{ m/s}.$$

This is 95% of the speed of light. The actual speed of the fully accelerated electron, when it has reached its final energy of 100 MeV, is 99.9987% of the speed of light.

36-7   INDUCTION AND RELATIVE MOTION   *(Optional)*

In Section 35-7, we discussed that the classification of electromagnetic effects into purely electric or purely magnetic was dependent on the reference frame of the observer. That is, what appears to be a magnetic field in one frame of reference can appear as a mixture of electric and magnetic fields in another frame of reference. Since motional emf is determined by the velocity of the object moving through the magnetic field, it clearly depends on the reference frame of the observer. Other observers in different inertial frames will measure different velocities and different magnetic field strengths. It is therefore essential in calculating induced emfs and currents to specify the reference frame of the observer.

Figure 16*a* shows a closed loop which an external agent (not shown) causes to move at velocity **v** with respect to a magnet that provides a uniform field **B** over a region. An observer *S* is at rest with respect to the magnet used to establish the field **B**. The induced emf in this case is a *motional emf* because the conducting loop is moving with respect to this observer.

Consider a positive charge carrier at the center of the left end of the loop. To observer *S*, this charge q is constrained to move through the field **B** with velocity **v** to the right along with the

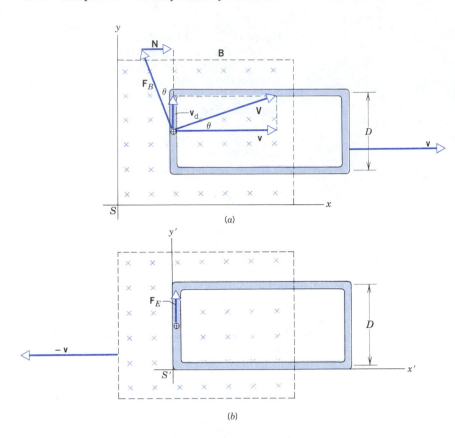

(a)

(b)

Figure 16 A closed conducting loop is in motion with respect to a magnet that produces the field **B**. (*a*) An observer *S*, fixed with respect to the magnet, sees the loop moving to the right and observes a magnetic force $F_B \cos \theta$ acting upward on the positive charge carriers. (*b*) An observer *S'*, fixed with respect to the loop, sees the magnet moving toward the left and observes an *electric* force acting upward on the positive charge carriers. In both figures there are internal forces of collision (not shown) that keep the charge carriers from accelerating.

loop, and it experiences a magnetic force given by $\mathbf{F} = q\mathbf{v} \times \mathbf{B}$ (not shown in Fig. 16*a*). This force causes the carriers to move upward (in the *y* direction) along the conductor; eventually, they acquire the drift velocity $\mathbf{v}_d$, as shown in Fig. 16*a*.

The resultant equilibrium velocity of the carriers is now **V**, the vector sum of **v** and $\mathbf{v}_d$. In this situation the magnetic force $\mathbf{F}_B$ is

$$\mathbf{F}_B = q\mathbf{V} \times \mathbf{B} \qquad (18)$$

acting (as usual) at right angles to the resultant velocity **V** of the carrier, as shown in Fig. 16*a*.

Acting alone, $\mathbf{F}_B$ would tend to push the carriers through the left wall of the conductor. Because this does not happen the conductor wall must exert a normal force **N** on the carriers (see Fig. 16*a*) of magnitude such that $\mathbf{v}_d$ lies parallel to the axis of the wire; in other words, **N** exactly cancels the horizontal component of $\mathbf{F}_B$, leaving only the component $F_B \cos \theta$ that lies along the direction of the conductor. This latter component of force on the carrier is also canceled out in this case by the average impulsive force $\overline{F}_i$ associated with the internal collisions that the carrier experiences as it drifts with (constant) speed v_d through the wire.

The kinetic energy of the charge carrier as it drifts through the wire remains constant. This is consistent with the fact that the resultant force acting on the charge carrier ($= \mathbf{F}_B + \overline{\mathbf{F}}_i + \mathbf{N}$) is zero. The work done by $\mathbf{F}_B$ is zero because magnetic forces, acting at right angles to the velocity of a moving charge, can do no work on that charge. Thus the (negative) work done on the carrier by the average internal collision force $\overline{\mathbf{F}}_i$ must be exactly canceled by the (positive) work done on the carrier by the force **N**. Ultimately, **N** is supplied by the agent that pulls the loop through the magnetic field, and the mechanical energy expended

by this agent appears as internal energy in the loop, as we have seen in Section 36-4.

Let us then calculate the work dW done on the carrier in time dt by the force **N**; it is

$$dW = N(v\,dt) \qquad (19)$$

in which $v\,dt$ is the distance that the loop (and the carrier) has moved to the right in Fig. 16*a* in time dt. We can write for N (see Eq. 18 and Fig. 16*a*)

$$N = F_B \sin \theta = (qVB)(v_d/V) = qBv_d. \qquad (20)$$

Substituting Eq. 20 into Eq. 19 yields

$$dW = (qBv_d)(v\,dt)$$
$$= (qBv)(v_d\,dt) = qBv\,ds \qquad (21)$$

in which $ds \, (= v_d\,dt)$ is the distance the carrier drifts along the conductor in time dt.

The work done on the carrier as it makes a complete circuit of the loop is found by integrating Eq. 21 around the loop and is

$$W = \oint dW = qBvD. \qquad (22)$$

This follows because work contributions for the top and the bottom of the loops are opposite in sign and cancel, and no work is done in those portions of the loop that lie outside the magnetic field.

An agent that does work on charge carriers, thus establishing a current in a closed conducting loop, can be viewed as an emf. Using Eq. 22, we find

$$\mathcal{E} = \frac{W}{q} = \frac{qBvD}{q} = BDv, \qquad (23)$$

which is the same result that we derived from Faraday's law of induction; see Eq. 5. Thus a motional emf is intimately connected with the sideways deflection of a charged particle moving through a magnetic field.

We now consider how the situation of Fig. 16a would appear to an observer S' who is *at rest with respect to the loop*. To this observer, the magnet is moving to the left in Fig. 16b with velocity $-\mathbf{v}$, and the charge q does not move in the x' direction with the loop but drifts clockwise around the loop. S' measures an emf $\mathscr{E}'$ which is accounted for, at the microscopic level, by postulating that an electric field $\mathbf{E}'$ is induced in the loop by the action of the moving magnet. The emf $\mathscr{E}'$ is related to $\mathbf{E}'$ by Eq. 12,

$$\mathscr{E}' = \oint \mathbf{E}' \cdot d\mathbf{s}.$$

The induced field $\mathbf{E}'$, which has the same origin as the induced fields that we discussed in Section 36-5, exerts a force $q\mathbf{E}'$ on the charge carrier.

The induced field $\mathbf{E}'$ that produces the current exists only in the left end of the loop. (As we carry out the integral of Eq. 12 around the loop, contributions to the integral from the x' component of $\mathbf{E}'$ cancel on the top and bottom sides, while there is no contribution from parts of the loop that are not in the magnetic field.) Using Eq. 12 we then obtain

$$\mathscr{E}' = E'D. \tag{24}$$

For motion at speeds small compared with the speed of light, the emfs given by Eqs. 23 and 24 must be identical, because the relative motion of the loop and the magnet is identical in the two cases shown in Fig. 16. Equating these relations yields

$$E'D = BDv,$$

or

$$E' = vB. \tag{25}$$

In Fig. 16b the vector $\mathbf{E}'$ points upward along the axis of the left end of the conducting loop because this is the direction in which positive charges are observed to drift. The directions of $\mathbf{v}$ and $\mathbf{B}$ are clearly shown in this figure. We see then that Eq. 25 is consistent with the more general vector relation

$$\mathbf{E}' = \mathbf{v} \times \mathbf{B}. \tag{26}$$

We have not proved Eq. 26 except for the special case of Fig. 16; nevertheless it is true in general, no matter what the angle between $\mathbf{v}$ and $\mathbf{B}$.

We interpret Eq. 26 in the following way. Observer S fixed with respect to the magnet is aware only of a magnetic field. To this observer, the force arises from the motion of the charges through $\mathbf{B}$. Observer S' fixed on the charge carrier is aware of an electric field $\mathbf{E}'$ also and attributes the force on the charge (at rest initially with respect to S') to the electric field. S says the force is of purely magnetic origin, while S' says the force is of purely

electric origin. From the point of view of S, the induced emf is given by $\oint(\mathbf{v} \times \mathbf{B}) \cdot d\mathbf{s}$. From the point of view of S', the same induced emf is given by $\oint \mathbf{E}' \cdot d\mathbf{s}$, where $\mathbf{E}'$ is the (induced) electric field vector that S' observes at points along the circuit.

For a third observer S'', relative to whom both the magnet and the loop are moving, the force tending to move charges around the loop is neither purely electric nor purely magnetic, but a bit of each. In summary, in the equation

$$\mathbf{F}/q = \mathbf{E} + \mathbf{v} \times \mathbf{B},$$

different observers form different assessments of $\mathbf{E}$, $\mathbf{B}$, and $\mathbf{v}$ but, when these are combined, all observers form the same assessment of $\mathbf{F}/q$, and all obtain the same value for the induced emf in the loop (which depends only on the *relative* motion). That is, the total force (and, hence, the total acceleration) is the same for all observers, but each observer forms a different estimate of the separate electric and magnetic forces contributing to the same total force.

The essential point is that what seems like a magnetic field to one observer may seem like a mixture of an electric field and a magnetic field to a second observer in a different inertial reference frame. Both observers agree, however, on the overall measurable result, in the case of Fig. 16, the current in the loop. We are forced to conclude that magnetic and electric fields are *not* independent of each other and have no separate unique existence; they depend on the inertial frame, as we also concluded in Section 35-7.

All the results of this section assume that the relative speed between S and S' is small compared with the speed of light c. If v is comparable to c, the appropriate set of relativistic transformations must be applied. In this case, we would find that the induced emfs measured by S and S' would no longer be equal, and that the induced electric field is not given by Eq. 26. However, if we are careful to define all quantities in the proper relativistic manner, we find again that the basic laws of electromagnetism, including Faraday's law, hold in all inertial reference frames.* Indeed, such considerations led Einstein to the special theory of relativity; in the language of special relativity, we say that Maxwell's equations are invariant with respect to the Lorentz transformation. ■

---

* For a careful discussion of motional emfs in the case of velocities that are not necessarily small compared with c, see "Application of Special Relativity to a Simple System in which a Motional emf Exists," by Murray D. Sirkis, *American Journal of Physics*, June 1986, p. 538. Further considerations of the relativistic transformation of electric and magnetic fields can be found in *Introduction to Special Relativity*, by Robert Resnick (Wiley, 1968), Chapter 4.

QUESTIONS

1. Show that 1 volt = 1 weber/second.

2. Are induced emfs and currents different in any way from emfs and currents provided by a battery connected to a conducting loop?

3. Is the size of the voltage induced in a coil through which a magnet moves affected by the strength of the magnet? If so, explain how.

4. Explain in your own words the difference between a mag-

netic field **B** and the flux of a magnetic field Φ_B. Are they vectors or scalars? In what units may each be expressed? How are these units related? Are either or both (or neither) properties of a given point in space?

5. Can a charged particle at rest be set in motion by the action of a magnetic field? If not, why not? If so, how? Consider both static and time-varying fields.

6. Account qualitatively for the configurations of the lines of **B** in Fig. 9a–c.

7. In Faraday's law of induction, does the induced emf depend on the resistance of the circuit? If so, how?

8. You drop a bar magnet along the axis of a long copper tube. Describe the motion of the magnet and the energy interchanges involved. Neglect air resistance.

9. You are playing with a metal loop, moving it back and forth in a magnetic field, as in Fig. 9. How can you tell, without detailed inspection, whether or not the loop has a narrow saw cut across it, rendering it nonconducting?

10. Figure 17 shows an inclined wooden track that passes, for part of its length, through a strong magnetic field. You roll a copper disk down the track. Describe the motion of the disk as it rolls from the top of the track to the bottom.

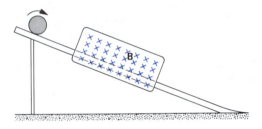

Figure 17   Question 10.

11. Figure 18 shows a copper ring, hung from a ceiling by two threads. Describe in detail how you might most effectively use a bar magnet to get this ring to swing back and forth.

Figure 18   Question 11.

12. Is an emf induced in a long solenoid by a bar magnet that moves inside it along the solenoid axis? Explain your answer.

13. Two conducting loops face each other a distance d apart, as shown in Fig. 19. An observer sights along their common

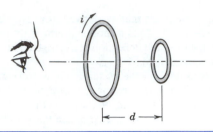

Figure 19   Question 13.

axis from left to right. If a clockwise current i is suddenly established in the larger loop, by a battery not shown, (a) what is the direction of the induced current in the smaller loop? (b) What is the direction of the force (if any) that acts on the smaller loop?

14. What is the direction of the induced emf in coil Y of Fig. 20 (a) when coil Y is moved toward coil X? (b) When the current in coil X is decreased, without any change in the relative positions of the coils?

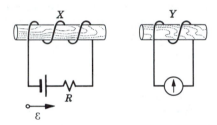

Figure 20   Question 14.

15. The north pole of a magnet is moved away from a copper ring, as in Fig. 21. In the part of the ring farthest from the reader, which way does the current point?

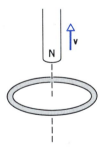

Figure 21   Question 15.

16. A circular loop moves with constant velocity through regions where uniform magnetic fields of the same magnitude are directed into or out of the plane of the page, as indicated in Fig. 22. At which of the seven indicated positions will the emf be (a) clockwise, (b) counterclockwise, and (c) zero?

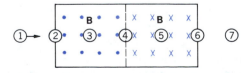

Figure 22   Question 16.

17. A short solenoid carrying a steady current is moving toward a conducting loop as in Fig. 23. What is the direction of the induced current in the loop as one sights toward it as shown?

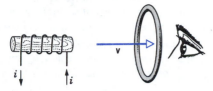

Figure 23   Question 17.

18. The resistance R in the left-hand circuit of Fig. 24 is being increased at a steady rate. What is the direction of the induced current in the right-hand circuit?

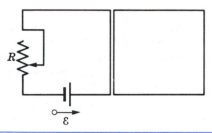

Figure 24   Question 18.

19. What is the direction of the induced current through resistor R in Fig. 25 (a) immediately after switch S is closed, (b) some time after switch S is closed, and (c) immediately after switch S is opened? (d) When switch S is held closed, from which end of the longer coil do field lines emerge? This is the effective north pole of the coil. (e) How do the conduction electrons in the coil containing R know about the flux within the long coil? What really gets them moving?

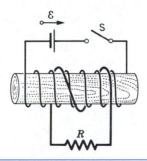

Figure 25   Question 19.

20. Can an induced current ever establish a magnetic field **B** that is in the same direction as the magnetic field inducing the current? Justify your answer.

21. How can you summarize in one statement all the ways of determining the direction of an induced emf?

22. The loop of wire shown in Fig. 26 rotates with constant angular speed about the x axis. A uniform magnetic field **B**,

whose direction is that of the positive y axis, is present. For what portions of the rotation is the induced current in the loop (a) from P to Q, (b) from Q to P, and (c) zero? Repeat if the direction of rotation is reversed from that shown in the figure.

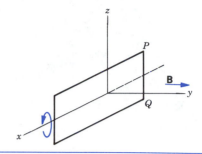

Figure 26   Question 22.

23. In Fig. 27 the straight movable wire segment is moving to the right with constant velocity **v**. An induced current appears in the direction shown. What is the direction of the uniform magnetic field (assumed constant and perpendicular to the page) in region A?

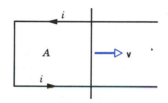

Figure 27   Question 23.

24. A conducting loop, shown in Fig. 28, is removed from the permanent magnet by pulling it vertically upward. (a) What is the direction of the induced current? (b) Is a force required to remove the loop? (Ignore the weight of the loop.) (c) Does the total amount of internal energy produced depend on the time taken to remove it?

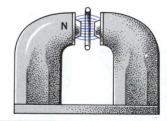

Figure 28   Question 24.

25. A plane closed loop is placed in a uniform magnetic field. In what ways can the loop be moved without inducing an emf? Consider motions both of translation and rotation.

26. A strip of copper is mounted as a pendulum about O in Fig. 29. It is free to swing through a magnetic field normal to the page. If the strip has slots cut in it as shown, it can swing freely through the field. If a strip without slots is substituted, the motion is strongly damped (*magnetic damping*). Ex-

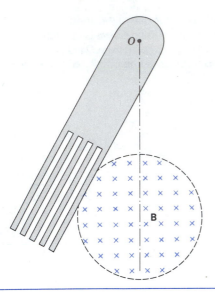

Figure 29   Question 26.

plain the observations. (*Hint*: Use Lenz' law; consider the paths that the charge carriers in the strip must follow if they are to oppose the motion.)

27. Consider a conducting sheet lying in a plane perpendicular to a magnetic field **B**, as shown in Fig. 30. (*a*) If **B** suddenly changes, the full change in **B** is not immediately detected at points near *P* (*electromagnetic shielding*). Explain. (*b*) If the resistivity of the sheet is zero, the change is never detected at *P*. Explain. (*c*) If **B** changes periodically at high frequency and the conductor is made of material with a low resistivity, the region near *P* is almost completely shielded from the changes in flux. Explain. (*d*) Why is such a conductor not useful as a shield from static magnetic fields?

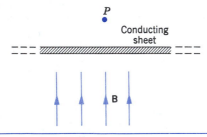

Figure 30   Question 27.

28. (*a*) In Fig. 12*b*, need the circle of radius *r* be a conducting loop in order that **E** and $\mathscr{E}$ be present? (*b*) If the circle of radius *r* were not concentric (moved slightly to the left, say), would $\mathscr{E}$ change? Would the configuration of **E** around the circle change? (*c*) For a concentric circle of radius *r*, with *r* > *R*, does an emf exist? Do electric fields exist?

29. A copper ring and a wooden ring of the same dimensions are placed so that there is the same changing magnetic flux through each. Compare the induced electric fields in the two rings.

30. An airliner is cruising in level flight over Alaska, where Earth's magnetic field has a large downward component. Which of its wingtips (right or left) has more electrons than the other?

31. In Fig. 12*d* how can the induced emfs around paths 1 and 2 be identical? The induced electric fields are much weaker near path 1 than near path 2, as the spacing of the lines of force shows. See also Fig. 13.

32. A cyclotron (see Section 34-3) is a so-called *resonance device*. Does a betatron depend on resonance? Discuss.

33. Show that, in the betatron of Fig. 14, the directions of the lines of **B** are correctly drawn to be consistent with the direction of circulation shown for the electrons.

34. In the betatron of Fig. 14 you want to increase the orbit radius by suddenly imposing an additional central flux $\Delta\Phi_B$ (set up by suddenly establishing a current in an auxiliary coil not shown). Should the lines of **B** associated with this flux increment be in the same direction as the lines shown in the figure or in the opposite direction? Assume that the magnetic field at the orbit position remains relatively unchanged by this flux increment.

35. In the betatron of Fig. 14, why is the iron core of the magnet made of laminated sheets rather than of solid metal as for the cyclotron of Section 34-3?

36. In Fig. 16*a* we can see that a force ($F_B \cos\theta$) acts on the charge carriers in the left branch of the loop. However, if there is to be a continuous current in the loop, and there is, a force of some sort must act on charge carriers in the other three branches of the loop to maintain the same drift speed v_d in these branches. What is its source? (*Hint*: Consider that the left branch of the loop was the only conducting element, the other three being nonconducting. Would not positive charge pile up at the top of the left half and negative charge at the bottom?)

PROBLEMS

Section 36-2 Faraday's Law of Induction

1. At a certain location in the northern hemisphere, the Earth's magnetic field has a magnitude of 42 μT and points downward at 57° to the vertical. Calculate the flux through a horizontal surface of area 2.5 m²; see Fig. 31.

2. A circular UHF television antenna has a diameter of 11.2 cm. The magnetic field of a TV signal is normal to the plane of the loop and, at one instant of time, its magnitude is changing at the rate 157 mT/s. The field is uniform. Find the emf in the antenna.

3. In Fig. 32 the magnetic flux through the loop shown increases according to the relation

$$\Phi_B = 6t^2 + 7t,$$

where Φ_B is in milliwebers and *t* is in seconds. (*a*) What is the

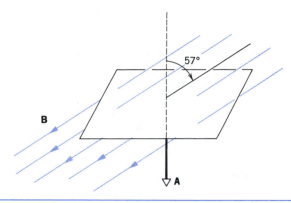

Figure 31 Problem 1.

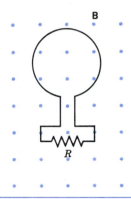

Figure 32 Problems 3 and 11.

absolute value of the emf induced in the loop when $t = 2.0$ s? (b) What is the direction of the current through the resistor?

4. The magnetic field through a one-turn loop of wire 16 cm in radius and 8.5 Ω in resistance changes with time as shown in Fig. 33. Calculate the emf in the loop as a function of time. Consider the time intervals (a) $t = 0$ to $t = 2$ s; (b) $t = 2$ s to $t = 4$ s; (c) $t = 4$ s to $t = 8$ s. The (uniform) magnetic field is at right angles to the plane of the loop.

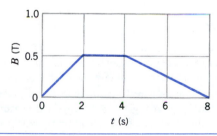

Figure 33 Problem 4.

5. A uniform magnetic field is normal to the plane of a circular loop 10.4 cm in diameter made of copper wire (diameter = 2.50 mm). (a) Calculate the resistance of the wire. (See Table 1 in Chapter 32.) (b) At what rate must the magnetic field change with time if an induced current of 9.66 A is to appear in the loop?

6. A loop antenna of area A and resistance R is perpendicular to a uniform magnetic field **B**. The field drops linearly to zero in a time interval Δt. Find an expression for the total internal energy dissipated in the loop.

7. Suppose that the current in the solenoid of Sample Problem 1 now changes, not as in that sample problem, but according to $i = 3.0t + 1.0t^2$, where i is in amperes and t is given in seconds. (a) Plot the induced emf in the coil from $t = 0$ to $t = 4$ s. (b) The resistance of the coil is 0.15 Ω. What is the current in the coil at $t = 2.0$ s?

8. In Fig. 34 a 120-turn coil of radius 1.8 cm and resistance 5.3 Ω is placed outside a solenoid like that of Sample Problem 1. If the current in the solenoid is changed as in that sample problem, (a) what current appears in the coil while the solenoid current is being changed? (b) How do the conduction electrons in the coil "get the message" from the solenoid that they should move to establish a current? After all, the magnetic flux is entirely confined to the interior of the solenoid.

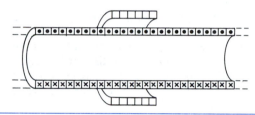

Figure 34 Problem 8.

9. You are given 52.5 cm of copper wire (diameter = 1.10 mm). It is formed into a circular loop and placed at right angles to a uniform magnetic field that is increasing with time at the constant rate of 9.82 mT/s. At what rate is internal energy generated in the loop?

10. A closed loop of wire consists of a pair of identical semicircles, radius 3.7 cm, lying in mutually perpendicular planes. The loop was formed by folding a circular loop along a diameter until the two halves became perpendicular. A uniform magnetic field **B** of magnitude 76 mT is directed perpendicular to the fold diameter and makes angles of 62° and 28° with the planes of the semicircles, as shown in Fig. 35. The magnetic field is reduced at a uniform rate to zero during a time interval of 4.5 ms. Determine the induced emf.

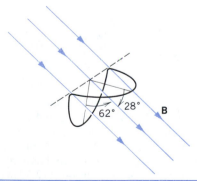

Figure 35 Problem 10.

11. In Fig. 32 let the flux for the loop be $\Phi_B(0)$ at time $t = 0$. Then let the magnetic field **B** vary in a continuous but un-

specified way, in both magnitude and direction, so that at time t the flux is represented by $\Phi_B(t)$. (a) Show that the net charge $q(t)$ that has passed through resistor R in time t is

$$q(t) = \frac{1}{R}[\Phi_B(0) - \Phi_B(t)],$$

independent of the way **B** has changed. (b) If $\Phi_B(t) = \Phi_B(0)$ in a particular case we have $q(t) = 0$. Is the induced current necessarily zero throughout the time interval from 0 to t?

12. Around a cylindrical core of cross-sectional area 12.2 cm² are wrapped 125 turns of insulated copper wire. The two terminals are connected to a resistor. The total resistance in the circuit is 13.3 Ω. An externally applied uniform longitudinal magnetic field in the core changes from 1.57 T in one direction to 1.57 T in the opposite direction in 2.88 ms. How much charge flows through the circuit? (*Hint*: See Problem 11.)

13. A uniform magnetic field **B** is changing in magnitude at a constant rate dB/dt. You are given a mass m of copper which is to be drawn into a wire of radius r and formed into a circular loop of radius R. Show that the induced current in the loop does not depend on the size of the wire or of the loop and, assuming **B** perpendicular to the loop, is given by

$$i = \frac{m}{4\pi\rho\delta}\frac{dB}{dt},$$

where ρ is the resistivity and δ the density of copper.

14. A square wire loop with 2.3-m sides is perpendicular to a uniform magnetic field, with half the area of the loop in the field, as shown in Fig. 36. The loop contains a 2.0-V battery with negligible internal resistance. If the magnitude of the field varies with time according to $B = 0.042 - 0.87t$, with B in tesla and t in seconds, what is the total emf in the circuit?

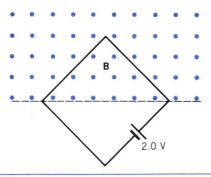

Figure 36   Problem 14.

15. A wire is bent into three circular segments of radius $r = 10.4$ cm as shown in Fig. 37. Each segment is a quadrant of a circle, *ab* lying in the xy plane, *bc* lying in the yz plane, and *ca* lying in the zx plane. (a) If a uniform magnetic field **B** points in the positive x direction, find the emf developed in the wire when B increases at the rate of 3.32 mT/s. (b) What is the direction of the emf in the segment *bc*?

16. For the situation shown in Fig. 38, $a = 12$ cm, $b = 16$ cm. The current in the long straight wire is given by $i = 4.5t^2 - 10t$, where i is in amperes and t is in seconds. Find the emf in the square loop at $t = 3.0$ s.

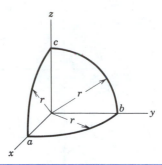

Figure 37   Problem 15.

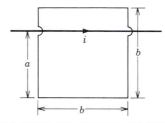

Figure 38   Problem 16.

17. In Fig. 39, the square has sides of length 2.0 cm. A magnetic field points out of the page; its magnitude is given by $B = 4t^2y$, where B is in tesla, t is in seconds, and y is in meters. Determine the emf around the square at $t = 2.5$ s and give its direction.

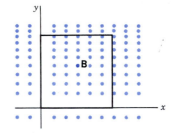

Figure 39   Problem 17.

Section 36-4 Motional emf

18. An automobile having a radio antenna 110 cm long travels at 90 km/h in a region where Earth's magnetic field is 55 μT. Find the maximum possible value of the induced emf.

19. A circular loop of wire 10 cm in diameter is placed with its normal making an angle of 30° with the direction of a uniform 0.50-T magnetic field. The loop is "wobbled" so that its normal rotates in a cone about the field direction at the constant rate of 100 rev/min; the angle between the normal and the field direction ($=30°$) remains unchanged during the process. What emf appears in the loop?

20. Figure 40 shows a conducting rod of length L being pulled along horizontal, frictionless, conducting rails at a constant velocity **v**. A uniform vertical magnetic field **B** fills the region in which the rod moves. Assume that $L = 10.8$ cm, $v = 4.86$ m/s, and $B = 1.18$ T. (a) Find the induced emf in the rod. (b) Calculate the current in the conducting loop. Assume that the resistance of the rod is 415 mΩ and that the

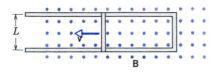

Figure 40   Problem 20.

resistance of the rails is negligibly small. (*c*) At what rate is internal energy being generated in the rod? (*d*) Find the force that must be applied by an external agent to the rod to maintain its motion. (*e*) At what rate does this force do work on the rod? Compare this answer with the answer to (*c*).

21. In Fig. 41 a conducting rod of mass *m* and length *L* slides without friction on two long horizontal rails. A uniform vertical magnetic field **B** fills the region in which the rod is free to move. The generator G supplies a constant current *i* that flows down one rail, across the rod, and back to the generator along the other rail. Find the velocity of the rod as a function of time, assuming it to be at rest at $t = 0$.

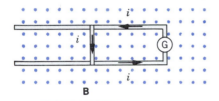

Figure 41   Problems 21 and 22.

22. In Problem 21 (see Fig. 41) the constant-current generator G is replaced by a battery that supplies a constant emf $\mathcal{E}$. (*a*) Show that the velocity of the rod now approaches a constant terminal value **v** and give its magnitude and direction. (*b*) What is the current in the rod when this terminal velocity is reached? (*c*) Analyze both this situation and that of Problem 21 from the point of view of energy transfers.

23. A circular loop made of a stretched conducting elastic material has a 1.23-m radius. It is placed with its plane at right angles to a uniform 785-mT magnetic field. When released, the radius of the loop starts to decrease at an instantaneous rate of 7.50 cm/s. Calculate the emf induced in the loop at that instant.

24. Figure 42 shows two parallel loops of wire having a common axis. The smaller loop (radius *r*) is above the larger loop

Figure 42   Problem 24.

(radius *R*), by a distance $x \gg R$. Consequently the magnetic field, due to the current *i* in the larger loop, is nearly constant throughout the smaller loop and equal to the value on the axis. Suppose that *x* is increasing at the constant rate $dx/dt = v$. (*a*) Determine the magnetic flux across the area bounded by the smaller loop as a function of *x*. (*b*) Compute the emf generated in the smaller loop. (*c*) Determine the direction of the induced current flowing in the smaller loop.

25. A small bar magnet is pulled rapidly through a conducting loop, along its axis. Sketch qualitatively (*a*) the induced current and (*b*) the rate of internal energy production as a function of the position of the center of the magnet. Assume that the north pole of the magnet enters the loop first and that the magnet moves at constant speed. Plot the induced current as positive if it is clockwise as viewed along the path of the magnet.

26. A stiff wire bent into a semicircle of radius *a* is rotated with a frequency *v* in a uniform magnetic field, as suggested in Fig. 43. What are (*a*) the frequency and (*b*) the amplitude of the emf induced in the loop?

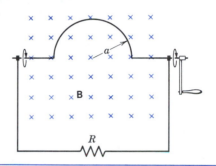

Figure 43   Problem 26.

27. A rectangular loop of *N* turns and of length *a* and width *b* is rotated at a frequency *v* in a uniform magnetic field **B**, as in Fig. 44. (*a*) Show that an induced emf given by

$$\mathcal{E} = 2\pi v NabB \sin 2\pi vt = \mathcal{E}_0 \sin 2\pi vt$$

appears in the loop. This is the principle of the commercial alternating current generator. (*b*) Design a loop that will produce an emf with $\mathcal{E}_0 = 150$ V when rotated at 60 rev/s in a magnetic field of 0.50 T.

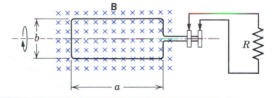

Figure 44   Problem 27.

28. A conducting wire of fixed length *L* can be wound into *N* circular turns and used as the armature of a generator. To get the largest emf, what value of *N* would you choose?

29. The armature of a motor has 97 turns each of area 190 cm² and rotates in a uniform magnetic field of 0.33 T. A potential difference of 24 V is applied. If no load is attached and

friction is neglected, find the rotational speed at equilibrium.

30. A generator consists of 100 turns of wire formed into a rectangular loop 50 cm by 30 cm, placed entirely in a uniform magnetic field with magnitude $B = 3.5$ T. What is the maximum value of the emf produced when the loop is spun at 1000 revolutions per minute about an axis perpendicular to **B**?

31. At a certain place, the Earth's magnetic field has magnitude $B = 59$ μT and is inclined downward at an angle of 70° to the horizontal. A flat horizontal circular coil of wire with a radius of 13 cm has 950 turns and a total resistance of 85 Ω. It is connected to a galvanometer with 140 Ω resistance. The coil is flipped through a half revolution about a diameter, so it is again horizontal. How much charge flows through the galvanometer during the flip? (*Hint*: See Problem 11.)

32. In the arrangement of Sample Problem 3 put $B = 1.2$ T and $R = 5.3$ cm. If $\mathscr{E} = 1.4$ V, what acceleration will a point at the end of the rotating rod experience?

33. Figure 45 shows a rod of length L caused to move at constant speed v along horizontal conducting rails. In this case the magnetic field in which the rod moves is not uniform but is provided by a current i in a long parallel wire. Assume that $v = 4.86$ m/s, $a = 10.2$ mm, $L = 9.83$ cm, and $i = 110$ A. (*a*) Calculate the emf induced in the rod. (*b*) What is the current in the conducting loop? Assume that the resistance of the rod is 415 mΩ and that the resistance of the rails is negligible. (*c*) At what rate is internal energy being generated in the rod? (*d*) What force must be applied to the rod by an external agent to maintain its motion? (*e*) At what rate does this external agent do work on the rod? Compare this answer to (*c*).

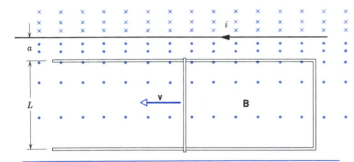

Figure 45   Problem 33.

34. Two straight conducting rails form an angle θ where their ends are joined. A conducting bar in contact with the rails and forming an isosceles triangle with them starts at the vertex at time $t = 0$ and moves with constant velocity **v** to the right, as shown in Fig. 46. A magnetic field **B** points out of the page. (*a*) Find the emf induced as a function of time. (*b*) If $\theta = 110°$, $B = 352$ mT, and $v = 5.21$ m/s, when is the induced emf equal to 56.8 V?

35. A rectangular loop of wire with length a, width b, and resistance R is placed near an infinitely long wire carrying current i, as shown in Fig. 47. The distance from the long wire to the loop is D. Find (*a*) the magnitude of the magnetic flux through the loop and (*b*) the current in the loop as it moves away from the long wire with speed v.

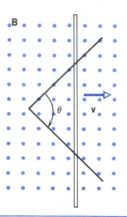

Figure 46   Problem 34.

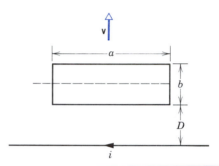

Figure 47   Problem 35.

36. Figure 48 shows a "homopolar generator," a device with a solid conducting disk as rotor. This machine can produce a greater emf than one using wire loop rotors, since it can spin at a much higher angular speed before centrifugal forces disrupt the rotor. (*a*) Show that the emf produced is given by

$$\mathscr{E} = \pi v B R^2$$

where v is the spin frequency, R the rotor radius, and B the uniform magnetic field perpendicular to the rotor. (*b*) Find the torque that must be provided by the motor spinning the rotor when the output current is i.

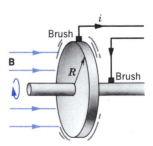

Figure 48   Problem 36.

37. A rod with length L, mass m, and resistance R slides without friction down parallel conducting rails of negligible resistance, as in Fig. 49. The rails are connected together at the bottom as shown, forming a conducting loop with the rod as the top member. The plane of the rails makes an angle θ with the horizontal and a uniform vertical magnetic field **B** exists

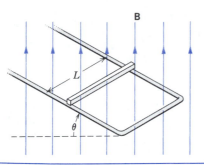

Figure 49 Problem 37.

throughout the region. (*a*) Show that the rod acquires a steady-state terminal velocity whose magnitude is

$$v = \frac{mgR}{B^2L^2} \frac{\sin \theta}{\cos^2 \theta}.$$

(*b*) Show that the rate at which internal energy is being generated in the rod is equal to the rate at which the rod is losing gravitational potential energy. (*c*) Discuss the situation if **B** were directed down instead of up.

38. A wire whose cross-sectional area is 1.2 mm² and whose resistivity is 1.7×10^{-8} Ω·m is bent into a circular arc of radius $r = 24$ cm as shown in Fig. 50. An additional straight length of this wire, *OP*, is free to pivot about *O* and makes sliding contact with the arc at *P*. Finally, another straight length of this wire, *OQ*, completes the circuit. The entire arrangement is located in a magnetic field $B = 0.15$ T directed out of the plane of the figure. The straight wire *OP* starts from rest with $\theta = 0$ and has a constant angular acceleration of 12 rad/s². (*a*) Find the resistance of the loop *OPQO* as a function of θ. (*b*) Find the magnetic flux through the loop as a function of θ. (*c*) For what value of θ is the induced current in the loop a maximum? (*d*) What is the maximum value of the induced current in the loop?

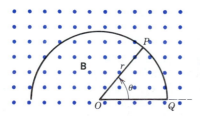

Figure 50 Problem 38.

39. An electromagnetic "eddy current" brake consists of a disk of conductivity σ and thickness *t* rotating about an axis through its center with a magnetic field **B** applied perpendicular to the plane of the disk over a small area a^2 (see Fig. 51). If the area a^2 is at a distance *r* from the axis, find an approximate expression for the torque tending to slow down the disk at the instant its angular velocity equals ω.

Figure 51 Problem 39.

Section 36-5 Induced Electric Fields

40. A long solenoid has a diameter of 12.6 cm. When a current *i* is passed through its windings, a uniform magnetic field $B = 28.6$ mT is produced in its interior. By decreasing *i*, the field is caused to decrease at the rate 6.51 mT/s. Calculate the magnitude of the induced electric field (*a*) 2.20 cm and (*b*) 8.20 cm from the axis of the solenoid.

41. Figure 52 shows two circular regions R_1 and R_2 with radii $r_1 = 21.2$ cm and $r_2 = 32.3$ cm, respectively. In R_1 there is a uniform magnetic field $B_1 = 48.6$ mT into the page and in R_2 there is a uniform magnetic field $B_2 = 77.2$ mT out of the page (ignore any fringing of these fields). Both fields are decreasing at the rate 8.50 mT/s. Calculate the integral $\oint \mathbf{E} \cdot d\mathbf{s}$ for each of the three indicated paths.

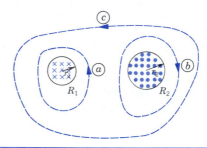

Figure 52 Problem 41.

42. Figure 53 shows a uniform magnetic field **B** confined to a cylindrical volume of radius *R*. **B** is decreasing in magnitude at a constant rate of 10.7 mT/s. What is the instantaneous acceleration (direction and magnitude) experienced by an electron placed at *a*, at *b*, and at *c*? Assume $r = 4.82$ cm. (The necessary fringing of the field beyond *R* will not change your answer as long as there is axial symmetry about the perpendicular axis through *b*.)

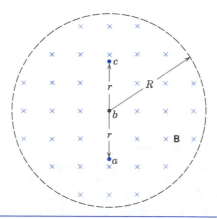

Figure 53 Problem 42.

43. Prove that the electric field **E** in a charged parallel-plate capacitor cannot drop abruptly to zero as one moves at right angles to it, as suggested by the arrow in Fig. 54 (see point *a*). In actual capacitors fringing of the lines of force always occurs, which means that **E** approaches zero in a continuous and gradual way; compare with Problem 43, Chapter 35. (*Hint*: Apply Faraday's law to the rectangular path shown by the dashed lines.)

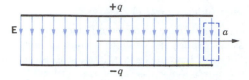

Figure 54   Problem 43.

44. Early in 1981 the Francis Bitter National Magnet Laboratory at M.I.T. commenced operation of a 3.3-cm diameter cylindrical magnet, which produces a 30-T field, then the world's largest steady-state field. The field can be varied sinusoidally between the limits of 29.6 T and 30.0 T at a frequency of 15 Hz. When this is done, what is the maximum value of the induced electric field at a radial distance of 1.6 cm from the axis? This magnet is described in *Physics Today*, August 1984.

45. A uniform magnetic field **B** fills a cylindrical volume of radius R. A metal rod of length L is placed as shown in Fig. 55. If B is changing at the rate dB/dt, show that the emf that is produced by the changing magnetic field and that acts between the ends of the rod is given by

$$\mathcal{E} = \frac{dB}{dt} \frac{L}{2} \sqrt{R^2 - \left(\frac{L}{2}\right)^2}.$$

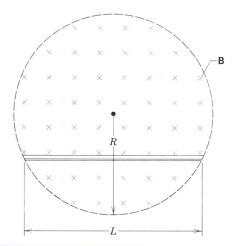

Figure 55   Problem 45.

Section 36-6 The Betatron

46. Figure 56a shows a top view of the electron orbit in a betatron. Electrons are accelerated in a circular orbit in the xy plane and then withdrawn to strike the target T. The magnetic field **B** is along the z axis (the positive z axis is out of the page). The magnetic field B_z along this axis varies sinusoidally as shown in Fig. 56b. Recall that the magnetic field

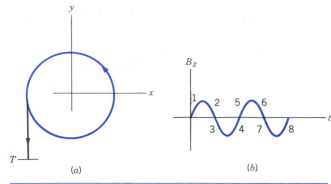

Figure 56   Problem 46.

must (*i*) guide the electrons in their circular path and (*ii*) generate the electric field that accelerates the electrons. Which quarter cycle(s) in Fig. 56b are suitable (*a*) according to (*i*), (*b*) according to (*ii*), and (*c*) for operation of the betatron?

47. In a certain betatron, the radius of the electron orbit is 32 cm and the magnetic field at the orbit is given by $B_{orb} = 0.28 \sin 120\pi t$, where t is in seconds and B_{orb} is in tesla. In the betatron, the average value B_{av} of the field enclosed by the electron orbit is equal to twice the value B_{orb} at the electron orbit. (*a*) Calculate the induced electric field felt by the electrons at $t = 0$. (*b*) Find the acceleration of the electrons at this instant. Ignore relativistic effects.

48. Some measurements of the maximum magnetic field as a function of radius for a betatron are:

r (cm)	B (tesla)	r (cm)	B (tesla)
0	0.950	81.2	0.409
10.2	0.950	83.7	0.400
68.2	0.950	88.9	0.381
73.2	0.528	91.4	0.372
75.2	0.451	93.5	0.360
77.3	0.428	95.5	0.340

Show by graphical analysis that the relation $B_{av} = 2B_{orb}$ mentioned in Problem 47 as essential to betatron operation is satisfied at the orbit radius $R = 84$ cm. (*Hint*: Note that

$$B_{av} = \frac{1}{\pi R^2} \int_0^R B(r) 2\pi r \, dr$$

and evaluate the integral graphically.)

Section 36-7 Induction and Relative Motion

49. (*a*) Estimate θ in Fig. 16. Recall that $v_d = 4 \times 10^{-2}$ cm/s in a typical case. Assume $v = 15$ cm/s. (*b*) It is clear that θ will be small. However, must we have $\theta \neq 0$ for the arguments presented in connection with this figure to be valid?

CHAPTER 37

MAGNETIC PROPERTIES OF MATTER

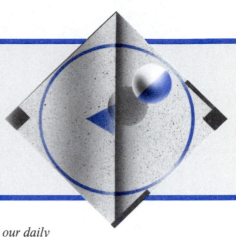

Magnetic materials play increasingly important roles in our daily lives. Materials such as iron, which are permanent magnets at ordinary temperatures, are commonly used in electric motors and generators as well as in certain types of loudspeakers. Other materials can be "magnetized" and "demagnetized" with relative ease; these materials have found wide use for storing information in such applications as magnetic recording tape (used in audio tape recorders and VCRs), computer disks, and credit cards. Still other materials are analogous to dielectrics in that they acquire an induced magnetic field in response to an external magnetic field; the induced field vanishes when the external field is removed.

In this chapter we consider the internal structure of materials that is responsible for their magnetic properties. We consider a magnetic form of Gauss' law, which takes into account the apparent nonexistence of isolated magnetic poles. We show that the behavior of different magnetic materials can be understood in terms of the magnetic dipole moments of individual atoms. A complete understanding of magnetic properties requires methods of quantum mechanics that are beyond the level of this text, but a qualitative understanding can be achieved based on principles discussed in this chapter.

37-1 GAUSS' LAW FOR MAGNETISM

Figure 1a shows the electric field associated with an insulating rod having equal quantities of positive and negative charge placed on opposite ends. This is an example of an electric dipole. Figure 1b shows the analogous case of the magnetic dipole, such as the familiar bar magnet, with a north pole at one end and a south pole at the other end. At this level the electric and magnetic cases look quite similar. (Compare Fig. 9b of Chapter 28 with Fig. 1 of Chapter 34 to see another illustration of this similarity.) In fact, we might be led to postulate the existence of individual magnetic poles analogous to electric charges; such poles, if they existed, would produce magnetic fields (similar to electric fields produced by charges) proportional to the strength of the pole and inversely proportional to the square of the distance from the pole. As we shall see, this hypothesis is not consistent with experiment.

Let us cut the objects of Fig. 1 in half and separate the two pieces. Figure 2 shows that the electric and magnetic cases are no longer similar. In the electric case, we have two objects that, if separated by a sufficiently large dis-

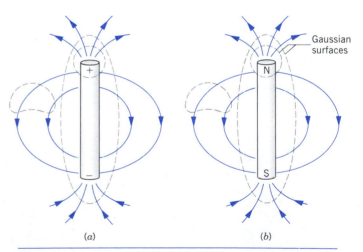

Figure 1 (a) An electric dipole, consisting of an insulating rod with a positive charge at one end and a negative charge at the other. Several Gaussian surfaces are shown. (b) A magnetic dipole, consisting of a bar magnet with a north pole at one end and a south pole at the other.

tance, could be regarded as point charges of opposite polarities, each producing a field characteristic of a point charge. In the magnetic case, however, we obtain not iso-

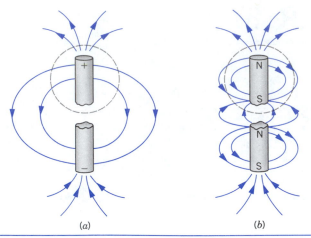

(a)　　　　　　　(b)

Figure 2 (a) When the electric dipole of Fig. 1a is cut in half, the positive charge is isolated on one piece and the negative charge on the other. (b) When the magnetic dipole of Fig. 1b is cut in half, a new pair of north and south poles appears. Note the difference in the field patterns.

lated north and south poles but instead a pair of magnets, each with its own north and south poles.

This appears to be an important difference between electric and magnetic dipoles: an electric dipole can be separated into its constituent single charges (or "poles"), but a magnetic dipole cannot. Each time we try to divide a magnetic dipole into separate north and south poles, we create a new pair of poles. It's a bit like cutting a piece of string with two ends to try to make two pieces of string each with only one end!

This effect occurs microscopically, down to the level of individual atoms. As we discuss in the next section, each atom behaves like a magnetic dipole having a north and a south pole, and as far as we yet know the dipole, rather than the single isolated pole, appears to be the smallest fundamental unit of magnetic structure.

This difference between electric and magnetic fields has a mathematical expression in the form of Gauss' law. In Fig. 1a, the flux of the electric field through the different Gaussian surfaces depends on the net charge enclosed by each surface. If the surface encloses no charge at all, or no net charge (that is, equal quantities of positive and negative charge, such as the entire dipole), the flux of the electric field vector through the surface is zero. If the surface cuts through the dipole, so that it encloses a net charge q, the flux Φ_E of the electric field is given by Gauss' law:

$$\Phi_E = \oint \mathbf{E} \cdot d\mathbf{A} = q/\epsilon_0. \qquad (1)$$

We can similarly construct Gaussian surfaces for the magnetic field, as in Fig. 1b. If the Gaussian surface contains no net "magnetic charge," the flux Φ_B of the magnetic field through the surface is zero. However, as we have seen, even those Gaussian surfaces that cut through the bar magnet enclose no net magnetic charge, because every cut through the magnet gives a piece having both a north and a south pole. The magnetic form of Gauss' law is written

$$\Phi_B = \oint \mathbf{B} \cdot d\mathbf{A} = 0. \qquad (2)$$

The net flux of the magnetic field through any closed surface is zero.

Figure 3 shows a more detailed representation of the magnetic fields of a bar magnet and a solenoid, both of which can be considered as magnetic dipoles. Note in Fig. 3a that lines of **B** enter the Gaussian surface inside the magnet and leave it outside the magnet. The total inward flux equals the total outward flux, and the net flux Φ_B for the surface is zero. The same is true for the Gaussian surface through the solenoid shown in Fig. 3b. In neither case is there a single point from which the lines of **B** originate or to which they converge; that is, *there is no isolated magnetic charge.*

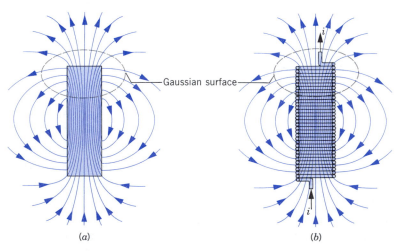

(a)　　　　　　　(b)

Figure 3 Lines of **B** for (a) a bar magnet and (b) a short solenoid. In each case, the north pole is at the top of the figure. The dashed lines represent Gaussian surfaces.

Magnetic Monopoles

We showed in Chapter 29 that Gauss' law for electric fields is equivalent to Coulomb's law, which is based on the experimental observation of the force between point charges. Gauss' law for magnetism is also based on an experimental observation, the failure to observe isolated magnetic poles, such as a single north pole or south pole.

The existence of isolated magnetic charges was proposed in 1931 by theoretical physicist Paul Dirac on the basis of arguments using quantum mechanics and symmetry. He called those charges *magnetic monopoles* and derived some basic properties expected of them, including the magnitude of the "magnetic charge" (analogous to the electronic charge e). Following Dirac's prediction, searches for magnetic monopoles were made using large particle accelerators as well as by examining samples of terrestrial and extraterrestrial matter. None of these early searches turned up any evidence for the existence of magnetic monopoles.

Recent attempts to unify the laws of physics, bringing together the strong, weak, and electromagnetic forces into a single framework, have reawakened interest in magnetic monopoles. These theories predict the existence of extremely massive magnetic monopoles, roughly 10^{16} times the mass of the proton. This is certainly far too massive to be made in any accelerator on Earth; in fact, the only known conditions under which such monopoles could have been made would have occurred in the hot, dense matter of the early universe. Searches for magnetic monopoles continue to be made, but convincing evidence for their existence has not yet been obtained.* For the present, we assume either that monopoles do not exist, so that Eq. 2 is exactly and universally valid, or else that if they do exist they are so exceedingly rare that Eq. 2 is a highly accurate approximation. Equation 2 then assumes a fundamental role as a description of the behavior of magnetic fields in nature, and it is included as one of the four Maxwell equations of electromagnetism.

37-2 ATOMIC AND NUCLEAR MAGNETISM

The differences in microscopic behavior between electric and magnetic fields can best be appreciated by looking at the fundamental atomic and nuclear structure that produces the fields. Consider the dielectric medium shown in Fig. 13 of Chapter 31. The medium consists of electric

---

* See "Searches for Magnetic Monopoles and Fractional Electric Charges," by Susan B. Felch, *The Physics Teacher,* March 1984, p. 142. See also "Superheavy Magnetic Monopoles," by Richard A. Carrigan, Jr. and W. Peter Trower, *Scientific American,* April 1982, p. 106.

dipoles that are aligned in an external electric field. These dipoles produce an induced electric field in the medium. If we cut the medium in half, assuming we don't cut any of the dipoles, we get two similar dielectric media; each has an induced positive charge on one end and an induced negative charge on the other end. We can keep dividing the material until we reach the level of a single atom or molecule, which has a negative charge on one end and a positive charge on the other end. With one final cut we can divide and separate the positive and negative charges.

The magnetic medium appears macroscopically to behave similarly. Figure 4 represents a magnetic medium as a collection of magnetic dipoles. If we cut the medium in half without cutting any of the dipoles, each of the two halves has a north pole at one end and a south pole at the other. We can continue cutting only until we reach the level of a single atom. Here we discover that the magnetic dipole consists not of two individual and opposite charges, as in the electric case, but instead is a tiny current loop, in which the current corresponds, for instance, to the circulation of the electron in the atom. Just as in the case of the current loops we considered in Section 34-7, the atomic current has an associated magnetic dipole moment. There is no way to divide this dipole into separate poles, so the dipole is the smallest fundamental unit of magnetism.

Let us consider a simple model in which an electron moves in a circular orbit in an atom. The magnetic dipole moment μ of this current loop is, according to Eq. 36 of Chapter 34,

$$\mu = iA, \tag{3}$$

where i is the effective current associated with the circulation of the electron and A is the area enclosed by the orbit. The current is just the charge e divided by the period T for one orbit,

$$i = \frac{e}{T} = \frac{e}{2\pi r/v}, \tag{4}$$

where r is the radius of the orbit and v is the tangential

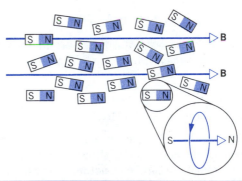

Figure 4 A magnetic material can be regarded as a collection of magnetic dipole moments, each with a north and a south pole. Microscopically, each dipole is actually a current loop that cannot be split into individual poles.

speed of the electron. The magnetic moment can then be written

$$\mu = iA = \left(\frac{ev}{2\pi r}\right)(\pi r^2) = \frac{erv}{2}. \qquad (5)$$

In terms of the angular momentum $l = mvr$, this is

$$\mu_l = \frac{e}{2m} l. \qquad (6)$$

We write this as μ_l to remind us that this contribution to the magnetic dipole moment of an atom depends on the orbital angular momentum l.

We generalize Eq. 6 by noting that (1) both μ and l are vectors, so Eq. 6 should more properly be written in vector form, and (2) the circulation of *all* the electrons in an atom contributes to the magnetic dipole moment. We therefore obtain

$$\mu_L = -\frac{e}{2m} \mathbf{L}, \qquad (7)$$

where μ_L is the total orbital magnetic dipole moment of the atom and $\mathbf{L} = \Sigma \, \mathbf{l}_i$ is the total orbital angular momentum of all the electrons in the atom. The negative sign appears because, owing to the electron's negative electric charge, the current has the opposite sense to the electronic motion, so that the magnetic moment vector is in a direction opposite to that of the angular momentum.

From quantum mechanics (see Section 13-6) we learn that any particular component of the angular momentum $\mathbf{L}$ is *quantized* in integral multiples of $h/2\pi$, where h is the Planck constant. That is, the orbital angular momentum can take values of $h/2\pi$, $2(h/2\pi)$, $3(h/2\pi)$, and so forth, but never, for example, $1.8(h/2\pi)$ or $4.2(h/2\pi)$. A natural unit in which to measure atomic magnetic dipole moments, called the *Bohr magneton* μ_B, is obtained by letting L have magnitude $h/2\pi$ in Eq. 7, in which case the magnetic moment has the value of one Bohr magneton μ_B, or

$$\mu_B = \frac{e}{2m}\frac{h}{2\pi} = \frac{eh}{4\pi m}. \qquad (8)$$

Substituting the appropriate values for the charge and mass of the electron into Eq. 8, we find

$$\mu_B = 9.27 \times 10^{-24} \text{ J/T}.$$

Magnetic moments associated with the orbital motion of electrons in atoms typically have magnitudes on the order of μ_B.

Intrinsic Magnetic Moments

Experiments in the 1920s, done by passing beams of atoms through magnetic fields, showed that the above model of the magnetic dipole structure of the atom was not sufficient to explain the observed properties. It was necessary to introduce another kind of magnetic moment for the electron, called the *intrinsic magnetic moment* μ_s.

Associated with this intrinsic magnetic moment, by

TABLE 1 SPINS AND MAGNETIC MOMENTS OF SOME PARTICLES

Particle	s (units of $h/2\pi$)	μ_s (units of μ_B)
Electron	$\frac{1}{2}$	$-1.001\ 159\ 652\ 193$
Proton	$\frac{1}{2}$	$+0.001\ 521\ 032\ 202$
Neutron	$\frac{1}{2}$	$-0.001\ 041\ 875\ 63$
Deuteron (^{2}H)	1	$+0.000\ 466\ 975\ 448$
Alpha	0	0
Photon	1	0

analogy with Eq. 6, is an intrinsic angular momentum s. The vector relationship between the intrinsic magnetic moment and the intrinsic angular moment can be written

$$\mu_s = -\frac{e}{m} \mathbf{s}. \qquad (9)$$

For an electron, the component of the intrinsic angular momentum $\mathbf{s}$ on any chosen axis is predicted by quantum theory and confirmed by experiment to have the value $\frac{1}{2}(h/2\pi)$,* as illustrated in Fig. 19 of Chapter 13. It is sometimes convenient to picture the intrinsic magnetic moment by considering the electron to be a ball of charge, spinning on its axis. (Hence the intrinsic angular momentum s is also known as "spin.") However, this picture is not strictly correct because, as far as we know, the electron is a point particle with a radius of zero.

By analogy with Eq. 7, we can define the *total* intrinsic magnetic dipole moment vector μ_S of an atom to be

$$\mu_S = -\frac{e}{m} \mathbf{S}, \qquad (10)$$

where $\mathbf{S} = \Sigma \, \mathbf{s}_i$ is the total spin of the electrons in the atom.

The orbital angular momentum and the orbital magnetic dipole moment of a *single* electron are properties of its particular state of motion. The intrinsic angular momentum (spin) and the intrinsic magnetic dipole moment of a single electron are fundamental characteristics of the electron itself, along with its mass and electric charge. In fact, every elementary particle has a certain intrinsic angular momentum and a corresponding intrinsic magnetic dipole moment. Table 1 gives some examples of these values. Note the incredible precision of these measured values, the uncertainties for which are in the last one or two digits. The values for the neutron and proton have respective precisions of 1 part in 10^7 and 10^8, while that for the electron has a precision of 1 part in 10^{11}, making it the most precise measurement ever done!

---

* The apparent difference of a factor of 2 between Eq. 6 and Eq. 9 comes about because the basic unit of s for the electron is $\frac{1}{2}(h/2\pi)$, while the basic unit for l is $h/2\pi$. In both cases, the fundamental unit of μ is μ_B. Actually, the difference between Eq. 6 and Eq. 9 is not *exactly* 2; it is predicted for the electron to be 2.002 319 304 386, a result that has been verified experimentally to all 12 decimal places.

The magnetic properties of a material are determined by the total magnetic dipole moment of its atoms, obtained from the vector sum of the orbital part, Eq. 7, and the spin part, Eq. 10. In a complex atom containing many electrons, the sums necessary to determine **L** and **S** may be very complicated. In many cases, however, the electrons couple pairwise so that the total **L** and **S** are zero. Materials made from these atoms are virtually nonmagnetic, except for a very weak induced effect called *diamagnetism*, which we consider in Section 37-4. In other atoms, either **L** or **S** (or both) may be nonzero; these atoms are responsible for the induced magnetic field in certain materials that is analogous to the induced *electric* field in a dielectric material. Such materials are called *paramagnetic*. The most familiar type of magnetism is *ferromagnetism*, in which, owing to the interactions among the atoms, the magnetic effects persist in the material even when the external magnetic field is removed. In Section 37-4 we discuss how our simple model of atomic magnetism helps us to understand ferromagnetic behavior as well.

Nuclear Magnetism

The nucleus, which is composed of protons and neutrons in orbital motion under the influence of their mutual forces, has a magnetic moment with two parts: an orbital part, due to the motion of the protons (neutrons, being uncharged, do not contribute to the orbital magnetic moment even though they may have orbital angular momentum), and an intrinsic part, due to the intrinsic magnetic moments of the protons and neutrons. (It may seem surprising that the uncharged neutron has a nonzero intrinsic magnetic moment. If the neutron were truly an elementary particle with no electric charge, it would indeed have no magnetic dipole moment. The nonzero magnetic dipole moment of the neutron is a clue to its internal structure and can be fairly well accounted for by considering the neutron to be composed of three charged quarks.)

Nuclei have orbital and spin magnetic dipole moments that can be expressed in the form of Eqs. 7 and 10. However, the mass that appears in these equations (the electron mass) must be replaced by the proton or neutron mass, which is about 1800 times the electron mass. Typical nuclear magnetic dipole moments are smaller than atomic dipole moments by a factor of the order of 10^{-3}, and their contribution to the magnetic properties of materials is usually negligible.

The effects of nuclear magnetism become important in the case of *nuclear magnetic resonance*, in which the nucleus is subject to electromagnetic radiation of a precisely defined frequency corresponding to that necessary to cause the nuclear magnetic moment to change direction. We can align the nuclear magnetic moments in a sample of material by a static magnetic field; the direction of the dipoles reverses when they absorb the time-varying elec-

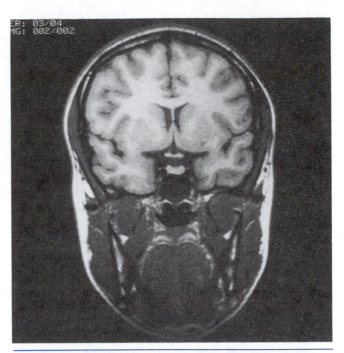

Figure 5 A cross section of a human head, taken by magnetic resonance imaging (MRI) techniques. It shows detail not visible on x-ray images and involves no radiation health risk to the patient.

tromagnetic radiation. The absorption of this radiation can easily be detected. This effect is the basis of magnetic resonance imaging (MRI), a diagnostic technique in which images of organs of the body can be obtained using radiation far less dangerous to the body than x rays (Fig. 5).

Sample Problem 1 A neutron is in a magnetic field of strength $B = 1.5$ T. The spin of the neutron is initially parallel to the direction of **B**. How much external work must be done to reverse the direction of the spin of the neutron?

Solution The energy of interaction of a magnetic dipole with a magnetic field was given by Eq. 38 of Chapter 34, $U = -\boldsymbol{\mu} \cdot \mathbf{B}$. Table 1 shows that the magnetic moment of the neutron, like that of the electron, is negative, meaning that, along any chosen axis, the component of the vector representing its spin magnetic moment is always opposite to that representing its spin. When the spin is parallel to the field, as in the initial state of this problem, $\boldsymbol{\mu}$ is opposite to the field and so the initial energy U_i is

$$U_i = -\boldsymbol{\mu} \cdot \mathbf{B} = |\mu|B,$$

because the angle between $\boldsymbol{\mu}$ and **B** is 180°. We write this in terms of the magnitude of $\boldsymbol{\mu}$ because we have already taken its sign into account in the dot product. When the spin changes direction (called a "spin flip"), the magnetic moment becomes *parallel* to **B**, and the final energy is

$$U_f = -\boldsymbol{\mu} \cdot \mathbf{B} = -|\mu|B.$$

The external work done on the system is equal to the change in energy, or

$$W = U_f - U_i = -|\mu|B - |\mu|B = -2|\mu|B$$
$$= -2(0.00104 \, \mu_B)[(9.27 \times 10^{-24} \, \text{J/T})/\mu_B](1.5 \, \text{T})$$
$$= -2.9 \times 10^{-26} \, \text{J} = -0.18 \, \mu\text{eV}.$$

Because the environment does *negative* work on the system, the system does *positive* work on its environment. This energy might be transmitted to the environment in the form of electromagnetic radiation, which would be in the radio-frequency range of the spectrum and would have a frequency of 44 MHz, slightly below the tuning range of an FM radio.

37-3 MAGNETIZATION

In Chapter 31 we considered the effect of filling the space between capacitor plates with a dielectric medium, and we found that inserting the dielectric while keeping the charge on the plates constant reduced the electric field between the plates. That is, if E_0 is the electric field without the dielectric, then the field E with the dielectric is given by Eq. 35 of Chapter 31, which we write in vector form as

$$\mathbf{E} = \mathbf{E}_0/\kappa_e. \tag{11}$$

The effect of the dielectric is characterized by the dielectric constant κ_e.

Consider instead a magnetic medium composed of atoms having magnetic dipole moments μ_i. These dipoles in general point in various directions in space. Let us compute the net dipole moment μ of a volume V of the material by taking the *vector* sum of all the dipoles in that volume: $\mu = \Sigma \, \mu_i$. We then define the *magnetization* $\mathbf{M}$ of the medium to be the net dipole moment per unit volume, or

$$\mathbf{M} = \frac{\mu}{V} = \frac{\Sigma \mu_i}{V}. \tag{12}$$

For the magnetization to be considered a microscopic quantity, Eq. 12 should be written as the limit as the volume approaches zero. This permits us to consider a material as having a *uniform* magnetization.

Suppose such a material is placed in a uniform field $\mathbf{B}_0$. This applied field "magnetizes" the material and aligns the dipoles. The aligned dipoles produce a magnetic field of their own, in analogy with the *electric* field produced by the electric dipoles in a dielectric medium (see Section 31-6). At any point in space, the net magnetic field $\mathbf{B}$ is the sum of the applied field $\mathbf{B}_0$ and that produced by the dipoles, which we call $\mathbf{B}_M$, so that

$$\mathbf{B} = \mathbf{B}_0 + \mathbf{B}_M. \tag{13}$$

The field $\mathbf{B}_M$ can include contributions from permanent dipoles in paramagnetic materials (analogous to polar dielectrics) and induced dipoles in all materials (as in nonpolar dielectrics).

The magnetization field $\mathbf{B}_M$ is related to the magnetization $\mathbf{M}$, which (as defined in Eq. 12) is also determined by the dipoles in the material. In weak fields, $\mathbf{M}$ is proportional to the applied field $\mathbf{B}_0$. However, $\mathbf{B}_M$ is in general difficult to calculate unless the magnetization is uniform and the geometry has a high degree of symmetry. As an example of such a case, we consider a long (ideal) solenoid of circular cross section filled with a magnetic material (Fig. 6). In this case, the applied field is uniform throughout the interior; both $\mathbf{B}_0$ and $\mathbf{M}$ are parallel to the axis, and it can be shown that $\mathbf{B}_M = \mu_0 \mathbf{M}$ in the interior of the solenoid. (You should check the dimensions and show that $\mu_0 \mathbf{M}$ has the same dimensions as $\mathbf{B}$.)

We can therefore write the net field as

$$\mathbf{B} = \mathbf{B}_0 + \mu_0 \mathbf{M} \tag{14}$$

as illustrated in Fig. 6b. In weak fields, $\mathbf{M}$ increases linearly with the applied field $\mathbf{B}_0$, and so $\mathbf{B}$ must be proportional to $\mathbf{B}_0$. In this case, we can write

$$\mathbf{B} = \kappa_m \mathbf{B}_0, \tag{15}$$

where κ_m is the *permeability constant* of the material, which is defined relative to a vacuum, for which $\kappa_m = 1$. Permeability constants of most common materials (excepting ferromagnets) have values very close to 1, as we discuss in the next section. For materials other than ferromagnets, the permeability constant may depend on such properties as the temperature and density of the material, but not on the field $\mathbf{B}_0$. Under ordinary circumstances, Eq. 15 describes a linear relationship with the net field $\mathbf{B}$ increasing linearly as the applied field increases. For ferromagnets, on the other hand, we can regard Eq. 15 as defining a particular κ_m that depends on the applied field $\mathbf{B}_0$, so that Eq. 15 is no longer linear.*

Combining Eqs. 14 and 15, we can write the magnetization induced by the applied field as

$$\mu_0 \mathbf{M} = (\kappa_m - 1)\mathbf{B}_0. \tag{16}$$

The quantity $\kappa_m - 1$ is typically of order 10^{-3} to 10^{-6} for most nonferromagnetic materials, and so the contribution of the magnetization $\mu_0 \mathbf{M}$ to the total field is generally far smaller than $\mathbf{B}_0$. This is in great contrast to the case of electric fields, in which κ_e has values for typical materi-

---

* There is, as always, an analogy here between electric and magnetic fields. There are dielectric materials, called *ferroelectrics*, in which the relationship between $\mathbf{E}$ and $\mathbf{E}_0$ is nonlinear; that is, κ_e is dependent on the applied field $\mathbf{E}_0$. From such materials we can construct quasipermanent electric dipoles, called *electrets*, which are analogous to permanent magnets. Most dielectric materials in common use are linear, whereas the most commonly useful magnetic materials are nonlinear.

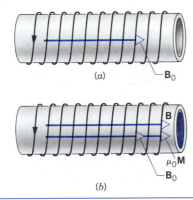

Figure 6 (*a*) In an empty solenoid, the current establishes a field $\mathbf{B}_0$. (*b*) When the solenoid is filled with magnetic material, the total field $\mathbf{B}$ includes contributions $\mathbf{B}_0$ from the current and $\mu_0 \mathbf{M}$ from the magnetic material.

als in the range of 3–100. The net electric field is modified substantially by the dielectric medium, while the magnetic medium has only a very small effect on the magnetic field for nonferromagnets.

Sample Problem 2 The magnetic field in the interior of a certain solenoid has the value 6.5×10^{-4} T when the solenoid is empty. When it is filled with iron, the field becomes 1.4 T. (*a*) Find the relative permeability under these conditions. (*b*) Find the average magnetic moment of an iron atom under these conditions.

Solution (*a*) From Eq. 15, we have (taking magnitudes only)

$$\kappa_m = \frac{B}{B_0} = \frac{1.4 \text{ T}}{6.5 \times 10^{-4} \text{ T}} = 2300.$$

(*b*) Using Eq. 14, we obtain

$$M = \frac{B - B_0}{\mu_0} = \frac{1.4 \text{ T} - 6.5 \times 10^{-4} \text{ T}}{4\pi \times 10^{-7} \text{ T} \cdot \text{m/A}} = 1.11 \times 10^6 \text{ A/m}.$$

Note that the units of M can also be expressed as $A \cdot m^2/m^3$. This represents the magnetic moment per unit volume of the iron. To find the magnetic moment per atom, we need the number density n of atoms (the number of atoms per unit volume):

$$n = \frac{\text{atoms}}{\text{volume}} = \frac{\text{mass}}{\text{volume}} \frac{\text{atoms}}{\text{mass}}$$

$$= \frac{\text{mass}}{\text{volume}} \frac{\text{atoms/mole}}{\text{mass/mole}} = \rho \frac{N_A}{m}.$$

Here ρ is the density of iron, N_A is the Avogadro constant, and m is the molar mass of iron. Putting in the values, we obtain

$$n = (7.85 \times 10^3 \text{ kg/m}^3) \frac{6.02 \times 10^{23} \text{ atoms/mole}}{0.0559 \text{ kg/mole}}$$

$$= 8.45 \times 10^{28} \text{ atoms/m}^3.$$

The average magnetic moment per atom is

$$\mu = \frac{M}{n} = \frac{1.11 \times 10^6 \text{ A/m}}{8.45 \times 10^{28}/\text{m}^3} = 1.31 \times 10^{-23} \text{ J/T} = 1.4 \,\mu_B.$$

This result is quite consistent with what we expect for an atomic magnetic moment. The calculation suggests that each atom of the sample of iron is contributing its full magnetic dipole moment to the magnetization of the material, a situation that characterizes ferromagnets.

37-4 MAGNETIC MATERIALS

We are now in a position to understand some characteristics of three types of magnetic materials. As we shall see, these classifications depend in part on the magnetic dipole moments of the atoms of the material and in part on the interactions among the atoms.

Paramagnetism

Paramagnetism occurs in materials whose atoms have permanent magnetic dipole moments; it makes no difference whether these dipole moments are of the orbital or spin types.

In a sample of a paramagnetic material with no applied field, the atomic dipole moments initially are randomly oriented in space (Fig. 7*a*). The magnetization, computed according to Eq. 12, is zero, because the random directions of the μ_i cause the vector sum to vanish, just as the randomly directed velocities of the molecules in a sample of a gas sum to give zero for the center-of-mass velocity of the entire sample.

When an external magnetic field is applied to the material (perhaps by placing it within the windings of a solenoid), the resulting torque on the dipoles tends to align them with the field (Fig. 7*b*). The vector sum of the individual dipole moments is no longer zero. The field inside the material now has two components: the applied field $\mathbf{B}_0$ and the induced field $\mu_0 \mathbf{M}$ from the magnetization of the dipoles. Note that these two fields are parallel; the dipoles enhance the applied field, in contrast to the electrical case in which the dipole field opposed the applied

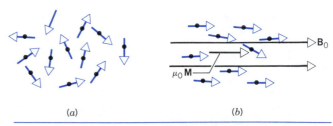

Figure 7 (*a*) In an unmagnetized sample, the atomic magnetic moments are randomly oriented. (*b*) When an external field $\mathbf{B}_0$ is applied, the dipoles rotate into alignment with the field, and the vector sum of the atomic dipole moments gives a contribution $\mu_0 \mathbf{M}$ to the field in the material.

**TABLE 2   RELATIVE PERMEABILITY OF
SOME PARAMAGNETIC MATERIALS
AT ROOM TEMPERATURE**

Material	$\kappa_m - 1$
Gd_2O_3	1.2×10^{-2}
$CuCl_2$	3.5×10^{-4}
Chromium	3.3×10^{-4}
Tungsten	6.8×10^{-5}
Aluminum	2.2×10^{-5}
Magnesium	1.2×10^{-5}
Oxygen (1 atm)	1.9×10^{-6}
Air (1 atm)	3.6×10^{-7}

field and reduced the total electric field in the material (see Fig. 13 of Chapter 31). The ratio between $\mu_0 \mathbf{M}$ and $\mathbf{B}_0$ is determined, according to Eq. 16, by $\kappa_m - 1$, which is small and positive for paramagnetic materials. Table 2 shows some representative values.

The thermal motion of the atoms tends to disturb the alignment of the dipoles, and consequently the magnetization decreases with increasing temperature. The relationship between M and the temperature T was discovered to be an inverse one by Pierre Curie in 1895 and is written

$$M = C\frac{B_0}{T}, \tag{17}$$

which is known as *Curie's law,* the constant C being known as the Curie constant.

Because the magnetization of a particular sample depends on the vector sum of its atomic magnetic dipoles, the magnetization reaches its maximum value when all the dipoles are parallel. If there are N such dipoles in the volume V, the maximum value of μ is $N\mu_i$, which occurs when all N magnetic dipoles μ_i are parallel. In this case

$$M_{max} = \frac{N}{V}\mu_i. \tag{18}$$

When the magnetization reaches this *saturation* value, increases in the applied field $\mathbf{B}_0$ have no further effect on the magnetization. Curie's law, which requires that M increase linearly with B_0, is valid only when the magnetization is far from saturation, that is, when B_0/T is small. Figure 8 shows the measured magnetization M, as a fraction of the maximum value M_{max}, as a function of B_0/T for various temperatures for the paramagnetic salt chrome alum, $CrK(SO_4)_2 \cdot 12H_2O$. (It is the chromium ions in this salt that are responsible for the paramagnetism.) Note the approach to saturation, and note that Curie's law is valid only at small values of B_0/T (corresponding to small applied fields or high temperatures). A complete treatment using quantum statistical mechanics gives an excellent fit to the data.

When the external magnetic field is removed from a paramagnetic sample, the thermal motion causes the directions of the magnetic dipole moments to become random again; the magnetic forces between atoms are too weak to hold the alignment and prevent the randomization. This effect can be used to achieve cooling in a process known as *adiabatic demagnetization.* A sample is magnetized at constant temperature. The dipoles move into a state of minimum energy in full or partial alignment with the applied field, and in doing so they must give up energy to the surrounding material. This energy flows as heat to the thermal reservoir of the environment. Now the sample is thermally isolated from its environment and is demagnetized adiabatically. When the dipoles become randomized, the increase in their magnetic energy must be compensated by a corresponding decrease in the internal energy of the system (since heat cannot flow to or from the isolated system in an adiabatic process). The temperature of the sample must therefore decrease. The lowest temperature that can be reached is determined by the residual field caused by the dipoles. The demagnetization of atomic magnetic dipoles can be used to achieve temperatures on the order of 0.001 K, while the demagnetization of the much smaller nuclear magnetic dipoles permits temperatures in the range of 10^{-6} K to be obtained.

Diamagnetism

In 1847, Michael Faraday discovered that a specimen of bismuth was *repelled* by a strong magnet. He called such substances diamagnetic. (In contrast, paramagnetic substances are always *attracted* by a magnet.) Diamagnetism occurs in all materials. However, it is generally a much weaker effect than paramagnetism, and therefore it can most easily be observed only in materials that are not paramagnetic. Such materials might be those having atomic magnetic dipole moments of zero, perhaps originating from atoms having several electrons with their orbital and spin magnetic moments adding vectorially to zero.

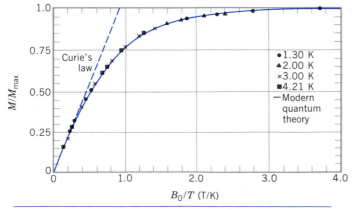

Figure 8   For a paramagnetic material, the ratio of the magnetization M to its saturation value M_{max} varies with B_0/T.

Diamagnetism is analogous to the effect of induced electric fields in electrostatics. An uncharged bit of material such as paper is attracted to a charged rod of either polarity. The molecules of the paper do not have permanent electric dipole moments but acquire *induced* dipole moments from the action of the electric field, and these induced moments can then be attracted by the field (see Fig. 14 of Chapter 31).

In diamagnetic materials, atoms having no *permanent* magnetic dipole moments acquire *induced* dipole moments when they are placed in an external magnetic field. Consider the orbiting electrons in an atom to behave like current loops. When an external field $\mathbf{B}_0$ is applied, the flux through the loop changes. By Lenz' law, the motion must change in such a way that an induced field opposes this increase in flux. A calculation based on circular orbits (see Problem 25) shows that the change in motion is accomplished by a slight speeding up or slowing down of the orbital motion, such that the circular frequency associated with the orbital motion changes by

$$\Delta\omega = \pm \frac{eB_0}{2m}, \tag{19}$$

where B_0 is the magnitude of the applied field and m is the mass of an electron. This change in the orbital frequency in effect changes the orbital magnetic moment of an electron (see Eq. 5 and Sample Problem 4).

If we were to bring a single atom of a material such as bismuth near the north pole of a magnet, the field (which points away from the pole) tends to increase the flux through the current loop that represents the circulating electron. According to Lenz' law, there must be an induced field pointing in the opposite direction (toward the pole). The induced north pole is on the side of the loop toward the magnet, and the two north poles repel one another.

This effect occurs no matter what the sense of rotation of the original orbit, so the magnetization in a diamagnetic material opposes the applied field. The ratio of the magnetization contribution to the field $\mu_0 M$ to the applied field B_0, given by $\kappa_m - 1$ according to Eq. 16, amounts to about -10^{-6} to -10^{-5} for typical diamagnetic materials. Table 3 shows some diamagnetic materials and their permeability constants.

Ferromagnetism

Ferromagnetism, like paramagnetism, occurs in materials in which the atoms have permanent magnetic dipole moments. What distinguishes ferromagnetic materials from paramagnetic materials is that in ferromagnetic materials there is a strong interaction between neighboring atomic dipole moments that keeps them aligned even when the external magnetic field is removed. Whether or not this occurs depends on the strength of the atomic dipoles and, because the dipole field changes with dis-

TABLE 3 RELATIVE PERMEABILITY OF SOME DIAMAGNETIC SUBSTANCES AT ROOM TEMPERATURE

Substance	$\kappa_m - 1$
Mercury	-3.2×10^{-5}
Silver	-2.6×10^{-5}
Bismuth	-1.7×10^{-5}
Ethyl alcohol	-1.3×10^{-5}
Copper	-9.7×10^{-6}
Carbon dioxide (1 atm)	-1.1×10^{-8}
Nitrogen (1 atm)	-5.4×10^{-9}

tance, on the separation between the atoms of the material. Certain atoms might be ferromagnetic in one kind of material but not in another, because their spacing is different. Familiar ferromagnetic materials at room temperature include the elements iron, cobalt, and nickel. Less familiar ferromagnetic elements, some of which show their ferromagnetism only at temperatures much below room temperature, are the elements of the rare earths, such as gadolinium or dysprosium. Compounds and alloys also may be ferromagnetic; for example, CrO_2, the basic ingredient of magnetic tape, is ferromagnetic even though neither of the elements chromium or oxygen is ferromagnetic at room temperature.

We can decrease the effectiveness of the coupling between neighboring atoms that causes ferromagnetism by increasing the temperature of a substance. The temperature at which a ferromagnetic material becomes paramagnetic is called its *Curie temperature.* The Curie temperature of iron, for instance, is 770°C; above this temperature, iron is paramagnetic. The Curie temperature of gadolinium metal is 16°C; at room temperature, gadolinium is paramagnetic, while at temperatures below 16°C, gadolinium becomes ferromagnetic.

The enhancement of the applied field in ferromagnets is considerable. The total magnetic field $\mathbf{B}$ inside a ferromagnet may be 10^3 or 10^4 times the applied field $\mathbf{B}_0$. The permeability κ_m of a ferromagnetic material is not a constant; neither the field $\mathbf{B}$ nor the magnetization $\mathbf{M}$ increases linearly with $\mathbf{B}_0$, even at small values of $\mathbf{B}_0$.

Let us insert a ferromagnetic material such as iron into the solenoid of Fig. 6b. We assume that the current is initially zero and that the iron is unmagnetized, so that initially both B_0 and M are zero. We increase B_0 by increasing the current in the solenoid. The magnetization increases rapidly toward a saturation value as indicated in Fig. 9 by the segment *ab*. Now we decrease the current to zero. The magnetization does not retrace its original path, but instead the iron remains magnetized (at point *c*) even when the applied field B_0 is zero. If we then reverse the direction of the current in the solenoid, we reach a saturated magnetization in the opposite direction (point *d*), and returning the current to zero we find that the sample retains a permanent magnetization at point *e*. We can

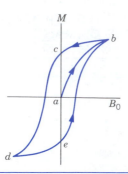

Figure 9   The variation of the magnetization of a sample of ferromagnetic material as the applied field is changed. The loop *bcdeb* is called a *hysteresis curve.*

then increase the current again to return to the saturated magnetization in the original direction (point *b*). The path *bcdeb* can be repeatedly followed.

The behavior shown in Fig. 9 is called *hysteresis.* At points *c* and *e*, the iron is magnetized, even though there is no current in the solenoid. Furthermore, the iron "remembers" how it became magnetized, a negative current producing a magnetization different from a positive one. This "memory" is essential to the operation of magnetic storage of information, such as on cassette tapes or computer disks.

The approach of a ferromagnet to saturation occurs through a mechanism different from that of a paramagnet (which we described by means of the rotation of individual magnetic dipoles into alignment with the applied field). A material such as iron is composed of a large number of microscopic crystals. Within each crystal are *magnetic domains,* regions of roughly 0.01 mm in size in which the coupling of atomic magnetic dipoles produces

essentially perfect alignment of all the atoms. Figure 10 shows a pattern of domains in a single crystal of ferromagnetic nickel. There are many domains, each with its dipoles pointing in a different direction, and the net result of adding these dipole moments in an unmagnetized ferromagnet gives a magnetization of zero.

When the ferromagnet is placed in an external field, two effects may occur: (1) dipoles outside the walls of domains that are aligned with the field can rotate into alignment, in effect allowing such domains to grow at the expense of neighboring domains; and (2) the dipoles of nonaligned domains may swing entirely into alignment with the applied field. In either case, there are now more dipoles aligned with the field, and the material has a large magnetization. When the field is removed, the domain walls do not move completely back to their former positions, and the material retains a magnetization in the direction of the applied field.

Sample Problem 3   A paramagnetic substance is composed of atoms with a magnetic dipole moment of 3.3 μ_B. It is placed in a magnetic field of strength 5.2 T. To what temperature must the substance be cooled so that the magnetic energy of each atom would be as large as the mean translational kinetic energy per atom?

Solution   The magnetic energy of a dipole in an external field is $U = -\mu \cdot B$, and the mean translational kinetic energy per atom is $(3/2)kT$ (see Section 23-4). These are equal in magnitude when the temperature is

$$T = \frac{\mu B}{(3/2)k} = \frac{(3.3)(9.27 \times 10^{-24}\ \text{J/T})(5.2\ \text{T})}{(1.5)(1.38 \times 10^{-23}\ \text{J/K})} = 7.7\ \text{K}.$$

Sample Problem 4   Calculate the change in magnetic moment of a circulating electron in an applied field B_0 of 2.0 T acting perpendicular to the plane of the orbit. Take $r = 5.29 \times 10^{-11}$ m for the radius of the orbit, corresponding to the normal state of an atom of hydrogen.

Solution   We can write Eq. 5 as

$$\mu = \tfrac{1}{2}erv = \tfrac{1}{2}er^2\omega,$$

using $v = r\omega$. The change $\Delta\mu$ in magnetic moment corresponding to a change in the angular frequency is then

$$\Delta\mu = \tfrac{1}{2}er^2\Delta\omega = \tfrac{1}{2}er^2\left(\pm\frac{eB_0}{2m}\right) = \pm\frac{e^2B_0r^2}{4m}$$

$$= \pm\frac{(1.6 \times 10^{-19}\ \text{C})^2(2.0\ \text{T})(5.29 \times 10^{-11}\ \text{m})^2}{4(9.1 \times 10^{-31}\ \text{kg})}$$

$$= \pm 3.9 \times 10^{-29}\ \text{J/T},$$

where we have used Eq. 19 for $\Delta\omega$.

Compared with the value of the magnetic moment, $\mu_B = 9.27 \times 10^{-24}$ J/T, we see that this effect amounts to only about 4×10^{-6} of the magnetic moment. This is consistent with the order of magnitude expected for diamagnetic effects (Table 3).

Figure 10   Domain patterns for a single crystal of nickel. The white lines, which show the boundaries of the domains, are produced by iron oxide powder sprinkled on the surface. The arrows illustrate the orientation of the magnetic dipoles within the domains.

37-5 THE MAGNETISM OF THE PLANETS *(Optional)*

Although magnetic compasses had already been in use as navigational instruments for several centuries, the explanation for their behavior was not well understood until 1600, when Sir William Gilbert, later physician to Queen Elizabeth I, proposed that the Earth is a huge magnet, with a magnetic pole near each geographic pole. Subsequent researchers have carefully mapped the Earth's magnetic field, and interplanetary spacecraft have studied the magnetic fields of other planets.

The Earth's field can be considered roughly that of a magnetic dipole, with moment $\mu = 8.0 \times 10^{22}$ J/T. The field at the surface has a magnitude that ranges from about 30 μT near the equator to about 60 μT near the poles. (For a dipole, we expect the magnetic field on the axis to be twice the field at the same distance along the bisector; see Table 1 of Chapter 35.) The axis of the dipole makes an angle of about 11.5° with the Earth's rotational axis (which itself makes an angle of 23.5° with the normal to the plane of the Earth's orbit about the Sun, as shown in Fig. 11). What we commonly call the north magnetic pole, which is located in northern Canada, is in fact the south pole of the Earth's dipole, as we have defined it by the converging of the magnetic field lines. The south magnetic pole, which is located in Antarctica, is represented by the north pole of a dipole, because the lines of **B** emerge from it. Put another way, when you use a magnetic compass to tell direction, the end of the compass that points toward the north is a true north pole of the suspended magnet in your compass; it is attracted toward a true south pole, which is near the north geographical pole of Earth.

The Earth's magnetic field has practical importance not only in navigation but also in prospecting and in communications. It has therefore been studied extensively for many years, on the surface by measuring its magnitude and direction and above its

Figure 12   The spectacular aurora borealis, also known as the "northern lights."

surface by using orbiting satellites. Among its other effects are the Van Allen radiation belts that surround Earth (see Fig. 15 of Chapter 34) and the so-called "northern lights," the brilliant display of the aurora* (Fig. 12).

Because we find magnetized rocks in the ground, it is tempting to suggest a core of permanently magnetized rocks as the source of the Earth's magnetic field. However, this cannot be correct, because the temperature of the core is several thousand degrees, far above the Curie temperature of iron. Iron in the Earth's core therefore cannot be ferromagnetic.

Furthermore, from measurements over the past few hundred years we know that the north magnetic pole migrates relative to the north geographic pole, and from the geologic record we know that the poles reverse on a time scale of several hundred thousand years. (Moreover, as we discuss later, some planets in the solar system that have compositions similar to Earth's have no magnetic field, whereas other planets that certainly contain no magnetic material have very large fields.) Such observations are difficult to explain based on the assumption of a permanently magnetized core.

The exact source of the Earth's magnetism is not completely understood, but it probably involves some sort of *dynamo* effect. The outer core contains minerals in a liquid state, which easily conduct electricity. A small initial magnetic field causes currents to flow in this moving conductor, by Faraday's law of induction. These currents may enhance the magnetic field, and this enhanced field is what we observe as the Earth's field. However, we know from our study of induction that a conductor moving in a magnetic field experiences a braking force. The source of the energy needed to overcome the braking force and keep the core moving is not yet understood.

The Earth contains a record of changes in both the direction and the magnitude of the field. Ancient pottery samples, for example, contain tiny iron particles, which became magnetized in the Earth's field as the pottery was cooled after its firing. From the strength of the magnetization of the particles, we can deduce the intensity of the Earth's field at the time and place of the firing. A geological record of similar origin is preserved in the

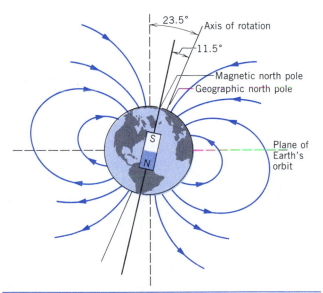

Figure 11   A simplified representation of the Earth's magnetic field near its surface. Note that the magnetic north pole is actually a south pole of the dipole that represents the Earth's field. The magnetic axis lies roughly halfway between the axis of rotation and the normal to the plane of the Earth's orbit (vertical dashed line).

---

* See "The Dynamic Aurora," by Syun-Ichi Akasofu, *Scientific American,* May 1989, p. 90.

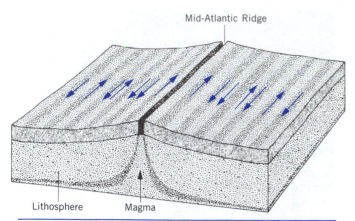

Figure 13   As molten material emerges through a ridge in the ocean floor and cools, it preserves a record of the direction of the Earth's magnetic field at that time (arrows). Each segment might represent a time of 100,000 to 1,000,000 years.

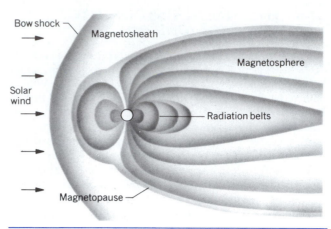

Figure 14   The magnetic field far from the Earth's surface shows the influence of the dipole field as well as that of the solar wind. The long magnetic tail stretches downstream for several thousand Earth diameters.

ocean floor (Fig. 13). As molten magma oozes from a ridge and solidifies, the iron particles become magnetized. The direction of magnetization of the particles shows the direction of the Earth's field. From the patterns of magnetization, we can deduce that the Earth's poles have reversed fairly regularly over geologic history. This reversal occurs about every 100,000–1,000,000 years and has become more frequent in recent times. The reasons for these reversals and their accelerating rate are not known but presumably involve the dynamo effect in some way.*

As we move away from the Earth, its field decreases, and we begin to observe modifications resulting from the *solar wind,* a stream of charged particles coming from the Sun (Fig. 14). As a result, a long tail associated with the Earth's field extends for many thousands of Earth diameters. Because the Sun has such a large effect on the Earth's magnetic field, even at distances of a few Earth radii, it can influence phenomena that involve the Earth's field, such as radio communication and the aurora.

In recent years, interplanetary space probes have been able to measure the direction and magnitude of the magnetic fields of the planets. These observations support the dynamo mechanism as the source of these fields. Table 4 shows values of the magnetic dipole moments and surface magnetic fields of the planets.

Venus, whose core is similar to Earth's, has no field because its rotation is too slow (once every 244 Earth days) to sustain the dynamo effect. Mars, whose rotational period is nearly the same as Earth's, has no field because its core is presumably too small, a fact deduced from the measured mean density of Mars. The outer planets (Jupiter and beyond) are composed mostly of hydrogen and helium, which ordinarily are not expected to be magnetic; however, at the high pressures and temperatures near the center of these planets, hydrogen and helium can behave like metals, in particular showing large electrical conductivity and permitting the dynamo effect.

Figure 15 shows the alignment of the rotational axis and magnetic field axis of Jupiter and Uranus; compare these with the Earth shown in Fig. 11. Note that the rotational axis of Uranus is nearly parallel to the plane of its orbit, in contrast to the other planets. Notice also that the magnetic axis of Uranus is badly

TABLE 4   MAGNETIC FIELDS IN THE SOLAR SYSTEM

Planet	μ (A·m²)	B at Surface (μT)
Mercury	5×10^{19}	0.35
Venus	$<10^{19}$	<0.01
Earth	8.0×10^{22}	30
Mars	$<2 \times 10^{18}$	<0.01
Jupiter	1.6×10^{27}	430
Saturn	4.7×10^{25}	20
Uranus	4.0×10^{24}	10–100
Neptune	2.2×10^{24}	10–100

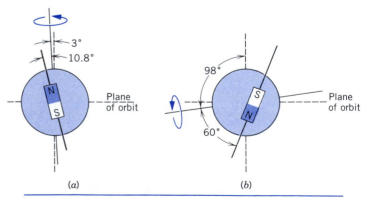

Figure 15   (*a*) The alignment of the magnetic dipole axis of Jupiter relative to its axis of rotation and the plane of its orbit. Note that, in contrast to the Earth, the north magnetic pole is a true north pole of the dipole field. (*b*) The alignment of the magnetic dipole axis of Uranus.

* See "The Evolution of the Earth's Magnetic Field," by Jeremy Bloxham and David Gubbins, *Scientific American,* December 1989, p. 68; and "The Source of the Earth's Magnetic Field," by Charles R. Carrigan and David Gubbins, *Scientific American,* February 1979, p. 118.

misaligned with its rotational axis and that the dipole is displaced from the center of the planet. A similar situation occurs for the planet Neptune. Unfortunately, our observational information on the planets is limited to that gathered from space flights that were in the neighborhood of the planet only for a day or so. If we could examine their other physical properties and their geologic records, we would learn a great deal more about the origin of planetary magnetism.† ■

Sample Problem 5 A measurement of the horizontal component B_h of the Earth's field at the location of Tucson, Arizona gave a value of 26 μT. By suspending a small magnet like a compass that is free to swing in a vertical plane, it is possible to measure the angle between the field direction and the horizontal plane, called the *inclination* or the *dip angle* ϕ_i. The dip angle at Tucson was measured to be 59°. Find the magnitude of the field and its vertical component at that location.

Solution As Fig. 16 shows, the magnitude of the field can be found from

---

† See "Magnetic Fields in the Cosmos," by E. N. Parker, *Scientific American,* August 1983, p. 44; and "Uranus," by Andrew P. Ingersoll, *Scientific American,* January 1987, p. 38.

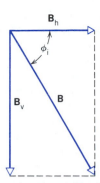

Figure 16 Sample Problem 5. The horizontal and vertical components of the Earth's magnetic field near Tucson, Arizona. The angle ϕ_i is the dip angle.

$$B = \frac{B_h}{\cos \phi_i} = \frac{26 \, \mu\text{T}}{\cos 59°} = 50 \, \mu\text{T}.$$

The vertical component is given by

$$B_v = B_h \tan \phi_i = (26 \, \mu\text{T})(\tan 59°) = 43 \, \mu\text{T}.$$

As expected for a dipole field (see Fig. 11), measured values of the dip angle range from 0° near the equator (actually, the *magnetic* equator) to 90° near the poles.

QUESTIONS

1. Two iron bars are identical in appearance. One is a magnet and one is not. How can you tell them apart? You are not permitted to suspend either bar as a compass needle or to use any other apparatus.

2. Two iron bars always attract, no matter the combination in which their ends are brought near each other. Can you conclude that one of the bars must be unmagnetized?

3. How are these phenomena similar and different? (*a*) A charged rod can attract small pieces of uncharged insulators. (*b*) A permanent magnet can attract any nonmagnetized sample of ferromagnetic material.

4. How can you determine the polarity of an unlabeled magnet?

5. Show that, classically, a spinning positive charge will have a spin magnetic moment that points in the same direction as its spin angular momentum.

6. The neutron, which has no charge, has a magnetic dipole moment. Is this possible on the basis of classical electromagnetism, or does this evidence alone indicate that classical electromagnetism has broken down?

7. Must all permanent magnets have identifiable north and south poles? Consider geometries other than the bar or horseshoe magnet.

8. Consider these two situations: (*a*) a (hypothetical) magnetic monopole is pulled through a single-turn conducting loop along its axis, at a constant speed; (*b*) a short bar magnet (a magnetic dipole) is similarly pulled. Compare qualitatively the net amounts of charge transferred through any cross

section of the loop during these two processes. Experiments designed to detect possible magnetic monopoles exploit such differences.

9. A certain short iron rod is found, by test, to have a north pole at each end. You sprinkle iron filings over the rod. Where (in the simplest case) will they cling? Make a rough sketch of what the lines of **B** must look like, both inside and outside the rod.

10. Starting with **A** and **B** in the positions and orientations shown in Fig. 17, with **A** fixed but **B** free to rotate, what happens (*a*) if **A** is an electric dipole and **B** is a magnetic dipole; (*b*) if **A** and **B** are both magnetic dipoles; (*c*) if **A** and **B** are both electric dipoles? Answer the same questions if **B** is fixed and **A** is free to rotate.

Figure 17 Question 10.

11. You are a manufacturer of compasses. (*a*) Describe ways in which you might magnetize the needles. (*b*) The end of the needle that points north is usually painted a characteristic color. Without suspending the needle in the Earth's field, how might you find out which end of the needle to paint? (*c*) Is the painted end a north or a south magnetic pole?

12. Would you expect the magnetization at saturation for a

paramagnetic substance to be very much different from that for a saturated ferromagnetic substance of about the same size? Why or why not?

13. Can you give a reason for the fact that ferromagnetic materials become purely paramagnetic at depths greater than about 20 km below the Earth's surface?

14. It is desired to demagnetize a sample of ferromagnetic material that retains the magnetism acquired when placed in an external field. Must the temperature of the sample be raised to the melting temperature to accomplish this?

15. The magnetization induced in a given diamagnetic sphere by a given external magnetic field does not vary with temperature, in sharp contrast to the situation in paramagnetism. Explain this behavior in terms of the description that we have given of the origin of diamagnetism.

16. Explain why a magnet attracts an unmagnetized iron object such as a nail.

17. Does any net force or torque act on (a) an unmagnetized iron bar or (b) a permanent bar magnet when placed in a uniform magnetic field?

18. A nail is placed at rest on a frictionless tabletop near a strong magnet. It is released and attracted to the magnet. What is the source of the kinetic energy that it has just before it strikes the magnet?

19. Superconductors are said to be perfectly diamagnetic. Explain.

20. Explain why a small bar magnet that is placed vertically above a bowl made of superconducting lead needs no contact forces to support it.

21. Compare the magnetization curves for a paramagnetic substance (see Fig. 8) and for a ferromagnetic substance (see Fig. 9). What would a similar curve for a diamagnetic substance look like?

22. Why do iron filings line up with a magnetic field? After all, they are not intrinsically magnetized.

23. The Earth's magnetic field can be represented closely by that of a magnetic dipole located at or near the center of the Earth. The Earth's magnetic poles can be thought of as (a) the points where the axis of this dipole passes through the Earth's surface or (b) the points on the Earth's surface where a dip needle would point vertically. Are these necessarily the same points?

24. A "friend" borrows your favorite compass and paints the entire needle red. When you discover this you are lost in a cave and have with you two flashlights, a few meters of wire, and (of course) this book. How might you discover which end of your compass needle is the north-seeking end?

25. How can you magnetize an iron bar if the Earth is the only magnet around?

26. How would you go about shielding a certain volume of space from constant external magnetic fields? If you think it can't be done, explain why.

27. Cosmic rays are charged particles that strike our atmosphere from some external source. We find that more low-energy cosmic rays reach the Earth near the north and south magnetic poles than at the (magnetic) equator. Why is this so?

28. How might the magnetic dipole moment of the Earth be measured?

29. Give three reasons for believing that the flux Φ_B of the Earth's magnetic field is greater through the boundaries of Alaska than through those of Texas.

30. Aurorae are most frequently observed, not at the north and south magnetic poles, but at magnetic latitudes about 23° away from these poles (passing through Hudson Bay, for example, in the northern geomagnetic hemisphere). Can you think of any reason, however qualitative, why the auroral activity should not be strongest at the poles themselves?

31. Can you think of a mechanism by which a magnetic storm, that is, a strong perturbation of the Earth's magnetic field, can interfere with radio communication?

PROBLEMS

Section 37-1 Gauss' Law for Magnetism

1. The magnetic flux through each of five faces of a dice is given by $\Phi_B = \pm N$ Wb, where $N (= 1$ to 5) is the number of spots on the face. The flux is positive (outward) for N even and negative (inward) for N odd. What is the flux through the sixth face of the dice?

2. A Gaussian surface in the shape of a right circular cylinder has a radius of 13 cm and a length of 80 cm. Through one end there is an inward magnetic flux of 25 μWb. At the other end there is a uniform magnetic field of 1.6 mT, normal to the surface and directed outward. Calculate the net magnetic flux through the curved surface.

3. Figure 18 shows four arrangements of pairs of small compass needles, set up in a space in which there is no external magnetic field. Identify the equilibrium in each case as stable or unstable. For each pair consider only the torque acting on one needle due to the magnetic field set up by the other. Explain your answers.

4. A simple bar magnet hangs from a string as in Fig. 19. A

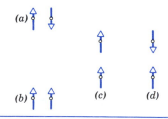

Figure 18   Problem 3.

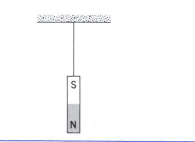

Figure 19   Problem 4.

uniform magnetic field **B** directed horizontally is then established. Sketch the resulting orientation of the string and the magnet.

5. Two wires, parallel to the z axis and a distance $4r$ apart, carry equal currents i in opposite directions, as shown in Fig. 20. A circular cylinder of radius r and length L has its axis on the z axis, midway between the wires. Use Gauss' law for magnetism to calculate the net outward magnetic flux through the half of the cylindrical surface above the x axis. (*Hint*: Find the flux through that portion of the xz plane that is within the cylinder.)

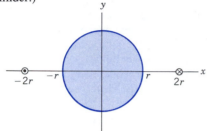

Figure 20 Problem 5.

Section 37-2 Atomic and Nuclear Magnetism

6. Using the values of spin angular momentum s and spin magnetic moment μ_s given in Table 1 for the free electron, numerically verify Eq. 9.

7. In the lowest energy state of the hydrogen atom the most probable distance between the single orbiting electron and the central proton is 5.29×10^{-11} m. Calculate (*a*) the electric field and (*b*) the magnetic field set up by the proton at this distance, measured along the proton's axis of spin. See Table 1 for the magnetic moment of the proton.

8. Suppose that the hydrogen nuclei (protons) in 1.50 g of water could all be aligned. Calculate the magnetic field that would be produced 5.33 m from the sample along the alignment axis.

9. A charge q is distributed uniformly around a thin ring of radius r. The ring is rotating about an axis through its center and perpendicular to its plane at an angular speed ω. (*a*) Show that the magnetic moment due to the rotating charge is

$$\mu = \tfrac{1}{2}q\omega r^2.$$

(*b*) If L is the angular momentum of the ring, show that $\mu/L = q/2m$.

10. Assume that the electron is a small sphere of radius R, its charge and mass being spread uniformly throughout its volume. Such an electron has a "spin" angular momentum L and a magnetic moment μ. Show that $e/m = 2\mu/L$. Is this prediction in agreement with experiment? (*Hint*: The spherical electron must be divided into infinitesimal current loops and an expression for the magnetic moment found by integration. This model of the electron is too mechanistic to be in the spirit of quantum physics.)

Section 37-3 Magnetization

11. A magnet in the shape of a cylindrical rod has a length of 4.8 cm and a diameter of 1.1 cm. It has a uniform magnetization of 5.3 kA/m. Calculate its magnetic dipole moment.

12. The dipole moment associated with an atom of iron in an iron bar is 2.22 μ_B. Assume that all the atoms in the bar, which is 4.86 cm long and has a cross-sectional area of 1.31 cm², have their dipole moments aligned. (*a*) What is the dipole moment of the bar? (*b*) What torque must be exerted to hold this magnet at right angles to an external field of 1.53 T?

13. A solenoid with 16 turns/cm carries a current of 1.3 A. (*a*) By how much does the magnetic field inside the solenoid increase when a close-fitting chromium rod is inserted? (*b*) Find the magnetization of the rod. (See Table 2.)

14. An electron with kinetic energy K_e travels in a circular path that is perpendicular to a uniform magnetic field, subject only to the force of the field. (*a*) Show that the magnetic dipole moment due to its orbital motion has magnitude $\mu = K_e/B$ and that it is in the direction opposite to that of **B**. (*b*) What is the magnitude and direction of the magnetic dipole moment of a positive ion with kinetic energy K_i under the same circumstances? (*c*) An ionized gas consists of 5.28×10^{21} electrons/m³ and the same number of ions/m³. Take the average electron kinetic energy to be 6.21×10^{-20} J and the average ion kinetic energy to be 7.58×10^{-21} J. Calculate the magnetization of the gas for a magnetic field of 1.18 T.

Section 37-4 Magnetic Materials

15. A 0.50-T magnetic field is applied to a paramagnetic gas whose atoms have an intrinsic magnetic dipole moment of 1.2×10^{-23} J/T. At what temperature will the mean kinetic energy of translation of the gas atoms be equal to the energy required to reverse such a dipole end for end in this magnetic field?

16. Measurements in mines and boreholes indicate that the temperature in the Earth increases with depth at the average rate of 30 C°/km. Assuming a surface temperature of 20°C, at what depth does iron cease to be ferromagnetic? (The Curie temperature of iron varies very little with pressure.)

17. A sample of the paramagnetic salt to which the magnetization curve of Fig. 8 applies is held at room temperature (300 K). At what applied magnetic field would the degree of magnetic saturation of the sample be (*a*) 50%? (*b*) 90%? (*c*) Are these fields attainable in the laboratory?

18. A sample of the paramagnetic salt to which the magnetization curve of Fig. 8 applies is immersed in a magnetic field of 1.8 T. At what temperature would the degree of magnetic saturation of the sample be (*a*) 50% and (*b*) 90%?

19. The paramagnetic salt to which the magnetization curve of Fig. 8 applies is to be tested to see whether it obeys Curie's law. The sample is placed in a 0.50-T magnetic field that remains constant throughout the experiment. The magnetization M is then measured at temperatures ranging from 10 to 300 K. Would it be found that Curie's law is valid under these conditions?

20. A paramagnetic substance is (weakly) attracted to a pole of a magnet. Figure 21 shows a model of this phenomenon. The "paramagnetic substance" is a current loop L, which is placed on the axis of a bar magnet nearer to its north pole than its south pole. Because of the torque $\tau = \mu \times B$ exerted on the loop by the **B** field of the bar magnet, the magnetic dipole moment μ of the loop will align itself to be parallel to

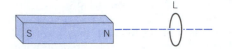

Figure 21 Problems 20 and 21.

B. (*a*) Make a sketch showing the **B** field lines due to the bar magnet. (*b*) Show the direction of the current *i* in the loop. (*c*) Using $d\mathbf{F} = i\,d\mathbf{s} \times \mathbf{B}$ show from (*a*) and (*b*) that the net force on L is toward the north pole of the bar magnet.

21. A diamagnetic substance is (weakly) repelled by a pole of a magnet. Figure 21 shows a model of this phenomenon. The "diamagnetic substance" is a current loop L that is placed on the axis of a bar magnet nearer to its north pole than its south pole. Because the substance is diamagnetic the magnetic moment μ of the loop will align itself to be antiparallel to the **B** field of the bar magnet. (*a*) Make a sketch showing the **B** field lines due to the bar magnet. (*b*) Show the direction of the current *i* in the loop. (*c*) Using $d\mathbf{F} = i\,d\mathbf{s} \times \mathbf{B}$, show from (*a*) and (*b*) that the net force on L is away from the north pole of the bar magnet.

22. The saturation magnetization of the ferromagnetic metal nickel is 511 kA/m. Calculate the magnetic moment of a single nickel atom. (Obtain needed data from Appendix D.)

23. The coupling mentioned in Section 37-4 as being responsible for ferromagnetism is *not* the mutual magnetic interaction energy between two elementary magnetic dipoles. To show this, calculate (*a*) the magnetic field a distance of 10 nm away along the dipole axis from an atom with magnetic dipole moment 1.5×10^{-23} J/T (cobalt) and (*b*) the minimum energy required to turn a second identical dipole end for end in this field. Compare with the results of Sample Problem 3. What do you conclude?

24. Consider a solid containing *N* atoms per unit volume, each atom having a magnetic dipole moment μ. Suppose the direction of μ can be only parallel or antiparallel to an externally applied magnetic field **B** (this will be the case if μ is due to the spin of a single electron). According to statistical mechanics, it can be shown that the probability of an atom being in a state with energy *U* is proportional to $e^{-U/kT}$ where *T* is the temperature and *k* is Boltzmann's constant (Boltzmann distribution; see Section 24-6). Thus, since $U = -\mu \cdot \mathbf{B}$, the fraction of atoms whose dipole moment is parallel to **B** is proportional to $e^{\mu B/kT}$ and the fraction of atoms whose dipole moment is antiparallel to **B** is proportional to $e^{-\mu B/kT}$. (*a*) Show that the magnetization of this solid is $M = N\mu \tanh (\mu B/kT)$. Here tanh is the hyperbolic tangent function: $\tanh x = (e^x - e^{-x})/(e^x + e^{-x})$. (*b*) Show that (*a*) reduces to $M = N\mu^2 B/kT$ for $\mu B \ll kT$. (*c*) Show that (*a*) reduces to $M = N\mu$ for $\mu B \gg kT$. (*d*) Show that (*b*) and (*c*) agree qualitatively with Fig. 8.

25. Consider an atom in which an electron moves in a circular orbit with radius *r* and angular frequency ω_0. A magnetic field is applied perpendicular to the plane of the orbit. As a result of the magnetic force, the electron circulates in an orbit with the same radius *r* but with a new angular frequency $\omega = \omega_0 + \Delta\omega$. (*a*) Show that, when the field is applied, the change in the centripetal acceleration of the electron is $2r\omega_0\,\Delta\omega$. (*b*) Assuming that the change in

centripetal acceleration is entirely due to the magnetic force, derive Eq. 19.

Section 37-5 The Magnetism of the Planets

26. In Sample Problem 5 the vertical component of the Earth's magnetic field in Tucson, Arizona, was found to be 43 μT. Assume this is the average value for all of Arizona, which has an area of 295,000 square kilometers, and calculate the net magnetic flux through the rest of the Earth's surface (the entire surface excluding Arizona). Is the flux outward or inward?

27. The Earth has a magnetic dipole moment of 8.0×10^{22} J/T. (*a*) What current would have to be set up in a single turn of wire going around the Earth at its magnetic equator if we wished to set up such a dipole? (*b*) Could such an arrangement be used to cancel out the Earth's magnetism at points in space well above the Earth's surface? (*c*) On the Earth's surface?

28. The magnetic dipole moment of the Earth is 8.0×10^{22} J/T. (*a*) If the origin of this magnetism were a magnetized iron sphere at the center of the Earth, what would be its radius? (*b*) What fraction of the volume of the Earth would the sphere occupy? The density of the Earth's inner core is 14 g/cm^3. The magnetic dipole moment of an iron atom is 2.1×10^{-23} J/T.

29. The magnetic field of the Earth can be approximated as a dipole magnetic field, with horizontal and vertical components, at a point a distance *r* from the Earth's center, given by

$$B_{\rm h} = \frac{\mu_0 \mu}{4\pi r^3} \cos L_{\rm m}, \quad B_{\rm v} = \frac{\mu_0 \mu}{2\pi r^3} \sin L_{\rm m},$$

where $L_{\rm m}$ is the *magnetic latitude* (latitude measured from the magnetic equator toward the north or south magnetic pole). The magnetic dipole moment μ is 8.0×10^{22} A·m^2. (*a*) Show that the strength at latitude $L_{\rm m}$ is given by

$$B = \frac{\mu_0 \mu}{4\pi r^3} \sqrt{1 + 3 \sin^2 L_{\rm m}}.$$

(*b*) Show that the inclination $\phi_{\rm i}$ of the magnetic field is related to the magnetic latitude $L_{\rm m}$ by

$$\tan \phi_{\rm i} = 2 \tan L_{\rm m}.$$

30. Use the results displayed in Problem 29 to predict the value of the Earth's magnetic field (magnitude and inclination) at (*a*) the magnetic equator; (*b*) a point at magnetic latitude 60°; and (*c*) the north magnetic pole.

31. Find the altitude above the Earth's surface where the Earth's magnetic field has a magnitude one-half the surface value at the same magnetic latitude. (Use the dipole field approximation given in Problem 29.)

32. Using the dipole field approximation to the Earth's magnetic field (see Problem 29), calculate the maximum strength of the magnetic field at the core–mantle boundary, which is 2900 km below the Earth's surface.

33. Use the properties of the dipole field displayed in Problem 29 to calculate the magnitude and inclination angle of the Earth's magnetic field at the north geographic pole. (*Hint:* The angle between the magnetic axis and the rotational axis of the Earth is 11.5°.) Why do the calculated values probably not agree with the measured values?

CHAPTER 38

INDUCTANCE

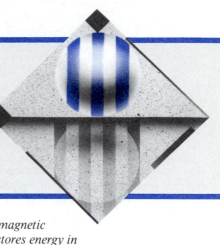

An inductor is a circuit element that stores energy in the magnetic field surrounding its current-carrying wires, just as a capacitor stores energy in the electric field between its charged plates. Previously we used the ideal parallel-plate capacitor as a convenient representation of any capacitor; in this chapter we similarly use the ideal solenoid to represent an inductor.

In Chapter 31 we showed that a capacitor is characterized by the value of its capacitance, which can be calculated from the geometry of its construction and which then describes the behavior of the capacitor in an electrical circuit. In this chapter, we show that an inductor is characterized by its inductance, *which depends on the geometry of its construction and which describes its behavior in a circuit.*

When a circuit contains both an inductor and a capacitor, the energy stored in the circuit can oscillate back and forth between them, just as the energy in a mechanical oscillator can oscillate between kinetic and potential. Such circuits, which behave as electromagnetic oscillators, are discussed at the end of this chapter.

38-1 INDUCTANCE

Capacitance is defined by Eq. 1 of Chapter 31,

$$V_C = \frac{1}{C} q. \tag{1}$$

This equation, which is ultimately based on Coulomb's law, asserts that the potential difference V_C across a capacitor is proportional to the charge q stored in the capacitor; the proportionality constant, C^{-1}, gives the (inverse of the) capacitance. We regard the quantities in Eq. 1 as being magnitudes only; the sign of the potential difference is such that the plate with the positive charge has the higher potential.

The *inductance L* of a circuit element (such as a solenoid) is defined by a similar relationship,

$$\mathcal{E}_L = L \frac{di}{dt}, \tag{2}$$

where all quantities are again taken to be magnitudes only. This equation, which we later show to be based on Faraday's law, asserts that a time-varying current through the inductor gives rise to an emf $\mathcal{E}_L$ across the inductor, and that the emf $\mathcal{E}_L$ is proportional to the rate of change of the current. The proportionality constant L gives the

inductance. Like the capacitance C, the inductance L is always taken to be a positive quantity.

Equation 2 shows that the SI unit of inductance is the volt·second/ampere. This combination of units has been given the special name of the *henry* (abbreviation H), so that

$$1 \text{ henry} = 1 \text{ volt·second/ampere.}$$

This unit is named after Joseph Henry (1797–1878), an American physicist and a contemporary of Faraday. In an electrical circuit diagram, an inductor is represented by the symbol —\$\text{\textperiodcentered}\$000\$\text{\textperiodcentered}\$—, which resembles the shape of a solenoid.

To find the relationship between the sign of $\mathcal{E}_L$ and the sign of di/dt, we use Lenz' law. Figure 1 shows an ideal solenoid in which a steady current i has been established (perhaps by a battery, not shown in the figure). Let us

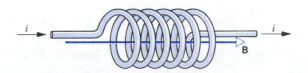

Figure 1 An arbitrary inductor, represented as a solenoid. The current i establishes a magnetic field **B**.

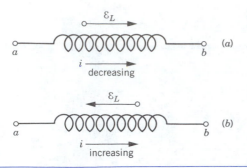

Figure 2 (a) A *decreasing* current induces in the inductor an emf that opposes the decrease in current. (b) An *increasing* current induces in the inductor an emf that opposes the increase.

suddenly *decrease* the (battery) emf in the circuit. The current i at once starts to *decrease*. This decrease in current is the change which, according to Lenz' law, the inductance must oppose. To oppose the falling current, the induced emf must provide an additional current in the *same* direction as i.

If instead we suddenly *increase* the (battery) emf, the current i starts at once to *increase*. Now Lenz' law shows that the increase in current is opposed by the inductance through an additional current in a direction *opposite* to i.

In each case, the induced emf acts to oppose the *change* in the current. Figure 2 summarizes the relationship between the sign of di/dt and the sign of $\mathcal{E}_L$. In Fig. 2a, V_b is greater than V_a, such that $V_b - V_a = |L\, di/dt|$. Since i is decreasing, di/dt is negative, so we can write this as

$$V_b - V_a = -L\, di/dt. \tag{3}$$

In Fig. 2b, di/dt is positive and V_a is greater than V_b, so Eq. 3 applies in this case as well. Equation 3 is particularly useful when we use the loop theorem to analyze circuits containing inductors.

38-2   CALCULATING THE INDUCTANCE

We calculated the capacitance of an arbitrary charged conductor (free of dielectric substance) by using Coulomb's law in the form of Gauss' law to find the electric field $\mathbf{E}$ in terms of the charge q stored in the capacitor; by writing the potential difference as

$$\Delta V = -\int \mathbf{E} \cdot d\mathbf{s} \tag{4}$$

we can then substitute for $\mathbf{E}$ and deduce the dependence of ΔV on q, and Eq. 1 then gives the capacitance. We demonstrated this technique in the examples of Section 31-2.

We adopt a similar technique to calculate inductance.

This technique is based on Faraday's law. We first determine the magnetic field $\mathbf{B}$ for the geometry of a particular inductor (which for the time being we assume contains no magnetic material). This enables the magnetic flux Φ_B through each turn of the coil to be obtained. We assume that the flux has the same value for each of the N turns of the coil. The product $N\Phi_B$ is known as the number of *flux linkages* of the inductor. The emf can be found from Faraday's law

$$\mathcal{E}_L = -\frac{d(N\Phi_B)}{dt}. \tag{5}$$

Equations 2 and 5 relate the emf in an inductor to the current (Eq. 2) or to a property that is proportional to the current (Φ_B in Eq. 5). Comparing the two equations (and taking the magnitude of all quantities), we find

$$L\frac{di}{dt} = \frac{d(N\Phi_B)}{dt}.$$

Integrating with respect to the time, we find

$$Li = N\Phi_B,$$

or

$$L = \frac{N\Phi_B}{i}. \tag{6}$$

Equation 6, which is based on Faraday's law, permits the inductance to be found directly from the number of flux linkages. Note that, since Φ_B is proportional to the current i, the ratio in Eq. 6 is *independent* of i and (like the capacitance) depends only on the geometry of the device.

The Inductance of a Solenoid

Let us apply Eq. 6 to calculate L for a section of length l of a long solenoid of cross-sectional area A; we assume the section is near the center of the solenoid so that edge effects need not be considered. In Section 35-6, the magnetic field B inside a solenoid carrying a current i was shown to be

$$B = \mu_0 ni, \tag{7}$$

where n is the number of turns per unit length. The number of flux linkages in the length l is

$$N\Phi_B = (nl)(BA),$$

which becomes, after substituting for B,

$$N\Phi_B = \mu_0 n^2 liA. \tag{8}$$

Equation 6 then gives the inductance directly:

$$L = \frac{N\Phi_B}{i} = \frac{\mu_0 n^2 liA}{i} = \mu_0 n^2 lA. \tag{9}$$

The inductance per unit length of the solenoid can be written

$$\frac{L}{l} = \mu_0 n^2 A. \tag{10}$$

This expression involves only geometrical factors—the cross-sectional area and the number of turns per unit length. The proportionality to n^2 is expected; if we double the number of turns per unit length, not only is the number N of turns doubled, but the flux Φ_B through *each turn* is doubled, and the number of flux linkages increases by a factor of 4, as does the inductance.

Equations 9 and 10 are valid for a solenoid of length very much greater than its radius. We have neglected the spreading of the magnetic field lines near the end of a solenoid, just as we neglected the fringing of the electric field near the edges of the plates of a capacitor.

The Inductance of a Toroid

We now calculate the inductance of a toroid of rectangular cross section, as shown in Fig. 3. The magnetic field B in a toroid was given by Eq. 23 of Chapter 35,

$$B = \frac{\mu_0 i N}{2\pi r}, \tag{11}$$

where N is the total number of turns of the toroid. Note that the magnetic field is not constant inside the toroid but varies with the radius r.

The flux Φ_B through the cross section of the toroid is

$$\Phi_B = \int \mathbf{B} \cdot d\mathbf{A} = \int_a^b B(h\,dr) = \int_a^b \frac{\mu_0 i N}{2\pi r} h\,dr$$

$$= \frac{\mu_0 i N h}{2\pi} \int_a^b \frac{dr}{r} = \frac{\mu_0 i N h}{2\pi} \ln \frac{b}{a},$$

where $h\,dr$ is the area of the elementary strip between the dashed lines shown in Fig. 3. The inductance can then be found directly from Eq. 6:

$$L = \frac{N\Phi_B}{i} = \frac{\mu_0 N^2 h}{2\pi} \ln \frac{b}{a}. \tag{12}$$

Once again, L depends only on geometrical factors.

Inductors with Magnetic Materials

In Section 31-5, we showed that the capacitance C of a capacitor filled with a dielectric substance is increased by a factor κ_e, the dielectric constant, relative to the capacitance C_0 when no dielectric is present:

$$C = \kappa_e C_0. \tag{13}$$

We were able to convert equations derived for empty capacitors to account for the case with the dielectric by replacing the permittivity constant ϵ_0 with the product $\kappa_e \epsilon_0$.

When a magnetic field $\mathbf{B}_0$ acts on a magnetic substance, the total field $\mathbf{B}$ (including the applied field $\mathbf{B}_0$ plus the field due to the dipoles of the material) can be written

$$\mathbf{B} = \kappa_m \mathbf{B}_0 \tag{14}$$

as we showed in Section 37-3. Here κ_m is the permeability

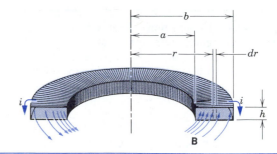

Figure 3   A cross section of a toroid, showing the current in the windings and the magnetic field in the interior.

constant of the material. Since the applied field $\mathbf{B}_0$ includes the factor μ_0, we can account for the effect of the magnetic material by replacing μ_0 with the quantity $\kappa_m \mu_0$, in analogy with the similar substitution made in the case of capacitors containing dielectrics.

In the case of an inductor, the field $\mathbf{B}_0$ would appear in the inductor if no magnetic material were present. The field $\mathbf{B}$ appears in the inductor when it is filled with magnetic material. In the expressions for inductance, we can account for the presence of a magnetic material filling the inductor by substituting $\kappa_m \mu_0$ for μ_0, or, in analogy with Eq. 13,

$$L = \kappa_m L_0, \tag{15}$$

where L is the inductance with the magnetic material present and L_0 is the inductance of the empty inductor. Thus a solenoid filled with a magnetic substance of permeability constant κ_m has an inductance given by

$$L = \kappa_m \mu_0 n^2 l A, \tag{16}$$

which we find by substituting $\kappa_m \mu_0$ for μ_0 in Eq. 9.

Because the permeability constants of paramagnetic or diamagnetic substances do not differ substantially from 1, the inductances of inductors filled with such substances are nearly equal to their values when empty, and no major change in the properties of the inductor is obtained by filling the inductor with a paramagnetic or a diamagnetic material. In the case of a ferromagnetic material, however, substantial changes can occur. Although the permeability constant is not defined in general for ferromagnetic materials (because the total field does not increase in linear proportion to the applied field), under particular circumstances B can be several thousand times B_0. Thus the "effective" permeability constant for a ferromagnet can have values in the range of 10^3 to 10^4, and the inductance of an inductor filled with ferromagnetic material (that is, one in which the windings are made on a core of a material such as iron) can be greater than the inductance of a similar set of windings on an empty core by a factor of 10^3 to 10^4. Ferromagnetic cores provide the means to obtain large inductances, just as dielectric materials in capacitors permit large capacitances to be obtained.

Sample Problem 1   A section of a solenoid of length $l = 12$ cm and having a circular cross section of diameter $d = 1.6$ cm carries a steady current of $i = 3.80$ A. The section contains 75 turns along its length. (*a*) What is the inductance of the solenoid when the core is empty? (*b*) The current is reduced at a constant rate to 3.20 A in a time of 15 s. What is the resulting emf developed by the solenoid, and in what direction does it act?

Solution   (*a*) The inductance of the solenoid is found from Eq. 9:

$$L = \mu_0 n^2 l A$$
$$= (4\pi \times 10^{-7} \text{ H/m})(75 \text{ turns}/0.12 \text{ m})^2 (0.12 \text{ m})(\pi)(0.008 \text{ m})^2$$
$$= 1.2 \times 10^{-5} \text{ H} = 12 \, \mu\text{H}.$$

Note that we have expressed μ_0 in units of H/m. An inductance can always be expressed as μ_0 times a quantity with the dimension of length. A similar situation holds for capacitance; see Section 31-2.

(*b*) The rate at which the current changes is

$$\frac{di}{dt} = \frac{3.20 \text{ A} - 3.80 \text{ A}}{15 \text{ s}} = -0.040 \text{ A/s},$$

and the corresponding emf has magnitude given by Eq. 2:

$$\mathcal{E}_L = |L \, di/dt| = (12 \, \mu\text{H})(0.040 \text{ A/s}) = 0.48 \, \mu\text{V}.$$

Because the current is decreasing, the induced emf must act in the same direction as the current, so that the induced emf opposes the decreases in the current.

Sample Problem 2   The core of the solenoid of Sample Problem 1 is filled with iron while the current is held constant at 3.20 A. The magnetization of the iron is saturated such that $B = 1.4$ T. What is the resulting inductance?

Solution   The "effective" permeability constant of the core subject to this particular applied field is determined from

$$\kappa_m = \frac{B}{B_0} = \frac{B}{\mu_0 n i}$$
$$= \frac{1.4 \text{ T}}{(4\pi \times 10^{-7} \text{ T} \cdot \text{m/A})(75 \text{ turns}/0.12 \text{ m})(3.20 \text{ A})} = 557.$$

The inductance is given by Eq. 15 as

$$L = \kappa_m L_0 = (557)(12 \, \mu\text{H}) = 6.7 \text{ mH}.$$

38-3   *LR* CIRCUITS

In Section 33-7 we saw that if we suddenly introduce an emf $\mathcal{E}$, perhaps by using a battery, into a single-loop circuit containing a resistor R and a capacitor C, the charge does not build up immediately to its final equilibrium value $C\mathcal{E}$ but approaches it exponentially, as described by Eq. 31 of Chapter 33:

$$q = C\mathcal{E}(1 - e^{-t/\tau_C}). \tag{17}$$

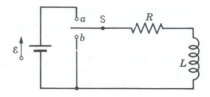

Figure 4   An *LR* circuit.

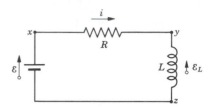

Figure 5   The *LR* circuit of Fig. 4 when the switch is closed on *a*.

The rate at which charge builds up is determined by the *capacitive time constant* τ_C, defined by

$$\tau_C = RC. \tag{18}$$

If in this same circuit the battery emf $\mathcal{E}$ is suddenly removed when the capacitor has stored a charge q_0, the charge does not immediately fall to zero but approaches it exponentially, as described by Eq. 36 of Chapter 33,

$$q = q_0 e^{-t/\tau_C}. \tag{19}$$

The same time constant τ_C describes the rise and the fall of the charge on the capacitor.

A similar rise (or fall) of the current occurs if we suddenly introduce an emf $\mathcal{E}$ into (or remove it from) a single-loop circuit containing a resistor R and an inductor L. When the switch S in Fig. 4 is closed on a, the current in the resistor starts to rise. If the inductor were not present, the current would rise rapidly to a steady value $\mathcal{E}/R$. Because of the inductor, however, an induced emf $\mathcal{E}_L$ appears in the circuit and, from Lenz' law, this emf opposes the *rise* of current, which means that it opposes the battery emf $\mathcal{E}$ in polarity. Thus the current in the resistor depends on the sum of two emfs, a constant one $\mathcal{E}$ due to the battery and a variable one $\mathcal{E}_L$ of the opposite sign due to the inductance. As long as this second emf is present, the current in the resistor is less than $\mathcal{E}/R$.

As time goes on, the current increases less rapidly, and the induced emf $\mathcal{E}_L$, which is proportional to di/dt, becomes smaller. The current in the circuit approaches the value $\mathcal{E}/R$ exponentially, as we prove below.

Now let us analyze this circuit quantitatively. When the switch S in Fig. 4 is thrown to a, the circuit reduces to that of Fig. 5. Let us apply the loop theorem, starting at x in Fig. 5 and going clockwise around the loop. For the direction of current shown, x is higher in potential than y, which means that we encounter a change in potential of $V_y - V_x = -iR$ as we traverse the resistor. Point y is higher in potential than point z because, for an increasing current, the induced emf opposes the *rise* of the current by pointing as shown. Thus as we traverse the inductor from y to z we encounter a change in potential of $V_z - V_y = -L(di/dt)$, according to Eq. 3. Finally, we encounter a rise in potential of $+\mathcal{E}$ in traversing the battery from z to x. The loop theorem gives

$$-iR - L\frac{di}{dt} + \mathcal{E} = 0$$

or

$$L\frac{di}{dt} + iR = \mathcal{E}. \tag{20}$$

To solve Eq. 20, we must find the function $i(t)$ such that when it and its first derivative are substituted in Eq. 20 the equation is satisfied.

Although there are formal rules for solving equations such as Eq. 20, it is also possible to solve it by direct integration (see Problem 20). It is even simpler in this case to try to guess at the solution, guided by physical reasoning and by previous experience. We can test the proposed solution by substituting it into Eq. 20 and seeing whether the resulting equation reduces to an identity.

In this case we guess at a solution similar to that for the buildup of charge on a capacitor in an *RC* circuit (Eq. 17). We also require on physical grounds that the solution $i(t)$ have two mathematical properties. (1) The initial current must be zero; that is, $i(0) = 0$. The current builds up from the value of zero just after the switch is closed. (2) The current must approach the value $\mathcal{E}/R$ as t becomes large. This second requirement follows from the expectation that the change in current gradually decreases, and when di/dt dies away the influence of the inductor on the circuit disappears. We therefore try as a solution the function

$$i(t) = \frac{\mathcal{E}}{R}(1 - e^{-t/\tau_L}). \tag{21}$$

Note that this mathematical form has the two properties $i(0) = 0$ and $i \to \mathcal{E}/R$ as $t \to \infty$. The time constant τ_L must be determined by substituting $i(t)$ and its derivative di/dt into Eq. 20. Differentiating Eq. 21, we obtain

$$\frac{di}{dt} = \frac{\mathcal{E}}{R}\frac{1}{\tau_L}e^{-t/\tau_L}. \tag{22}$$

Doing the substitutions and the necessary algebra, we find that Eq. 20 is satisfied if

$$\tau_L = \frac{L}{R}. \tag{23}$$

τ_L is called the *inductive time constant*. In analogy with the capacitive time constant $\tau_C = RC$, it indicates how rapidly the current in an *LR* circuit approaches the steady value.

To show that the quantity $\tau_L = L/R$ has the dimension of time, we have

$$[\tau_L] = \frac{[L]}{[R]} = \frac{\text{henry}}{\text{ohm}} = \frac{\text{volt} \cdot \text{second/ampere}}{\text{ohm}}$$

$$= \left(\frac{\text{volt}}{\text{ampere} \cdot \text{ohm}}\right)\text{second} = \text{second},$$

where the quantity in parentheses equals 1 because 1 ohm = 1 volt/ampere (as in $R = V/i$).

The physical significance of τ_L follows from Eq. 21. If we put $t = \tau_L$ into this equation, it reduces to

$$i = \frac{\mathcal{E}}{R}(1 - e^{-1}) = (1 - 0.37)\frac{\mathcal{E}}{R} = 0.63\frac{\mathcal{E}}{R}.$$

The time constant τ_L is that time at which the current in the circuit is less than its final steady value $\mathcal{E}/R$ by a factor of $1/e$ (about 37%).

The complete solution for the current in an *LR* circuit can be written

$$i(t) = \frac{\mathcal{E}}{R}(1 - e^{-tR/L}). \tag{24}$$

Figure 6 shows the potential drop $V_R\ [= |V_y - V_x| = i(t)R]$ across the resistor R and the potential drop $V_L\ [= |V_z - V_y| = L(di/dt)]$ across the ideal inductor.

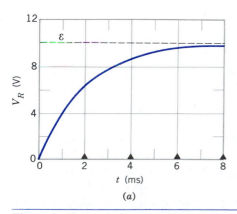

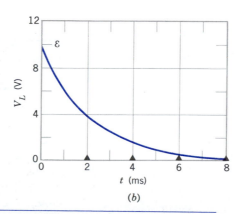

(a)   (b)

Figure 6   The variation with time of (a) V_R, the potential difference across the resistor in the circuit of Fig. 5, and (b) V_L, the potential difference across the inductor in that circuit. The curves are drawn for $R = 2000\ \Omega$, $L = 4.0$ H, and $\mathcal{E} = 10$ V. The inductive time constant τ_L is 2 ms; successive intervals equal to τ_L are marked by the triangles along the horizontal axis.

If the switch S in Fig. 4 is thrown to b when the current in the circuit has some value i_0, the effect is to remove the battery from the circuit. The equation that governs the subsequent decay of the current in the circuit can be found by putting $\mathcal{E} = 0$ in Eq. 20, which gives

$$L\frac{di}{dt} + iR = 0. \qquad (25)$$

By direct substitution or by integration, it can be shown that the solution to this equation is

$$i(t) = i_0 e^{-t/\tau_L}, \qquad (26)$$

where i_0 is the current at $t = 0$ (which now means the time at which the switch is thrown to b). The decay of the current occurs with the same exponential time constant $\tau_L = L/R$ as does the rise in the current. Note the similarity with Eq. 19 for the decay of the charge on a capacitor.

Throwing the switch in Fig. 4 back and forth between a and b can be accomplished electronically by removing the battery from Fig. 5 and replacing it with a generator that produces a *square wave*, of the form shown in Fig. 7a. This waveform oscillates back and forth between the values $\mathcal{E}$ and 0 in a fixed time interval, which we choose to be much greater than τ_L.

If we connect the terminals of an oscilloscope across the resistor (points x and y in Fig. 5), the waveform displayed is that of the current in the circuit, which is identical in form to V_R, as shown in Fig. 7b. The current builds up to its maximum value $\mathcal{E}/R$ when the applied emf has the value $\mathcal{E}$, and it decays exponentially to zero (according to Eq. 26) when the applied emf is zero.

If we connect the oscilloscope terminals across the inductor (points y and z in Fig. 5), the waveform displayed is that of the *derivative* of the current, which has the same form as V_L, as shown in Fig. 7c. According to Eq. 22, this form is

$$V_L = L\frac{di}{dt} = \mathcal{E}e^{-t/\tau_L} \qquad (27)$$

when the applied emf has the value $\mathcal{E}$. When the applied emf is zero, differentiating Eq. 26 shows that

$$V_L = L\frac{di}{dt} = -\mathcal{E}e^{-t/\tau_L},$$

since $\mathcal{E} = i_0 R$ in this case. We see that this result is just the negative of Eq. 27. This agrees with the alternating series of positive and negative exponentials shown in Fig. 7c.

Note that adding the curves of Fig. 7b and 7c gives Fig. 7a. That is, $V_R + V_L = \mathcal{E}$, which must be true according to the loop theorem.

Sample Problem 3   A solenoid has an inductance of 53 mH and a resistance of 0.37 Ω. If it is connected to a battery, how long will it take for the current to reach one-half its final equilibrium value?

Solution   The equilibrium value of the current, which is reached at $t \rightarrow \infty$, is $\mathcal{E}/R$ from Eq. 24. If the current has half this value at a particular time t_0, this equation becomes

$$\frac{1}{2}\frac{\mathcal{E}}{R} = \frac{\mathcal{E}}{R}(1 - e^{-t_0/\tau_L}),$$

or

$$e^{-t_0/\tau_L} = \frac{1}{2}.$$

Solving for t_0 by rearranging and taking the (natural) logarithm of each side, we find

$$t_0 = \tau_L \ln 2 = \frac{L}{R}\ln 2 = \frac{53 \times 10^{-3}\text{ H}}{0.37\ \Omega}\ln 2 = 0.10\text{ s}.$$

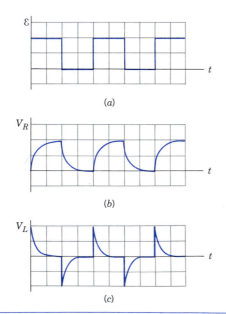

(a)

(b)

(c)

Figure 7   (a) A source of emf varying as a square wave is applied to the circuit of Fig. 5. (b) The potential difference across the resistor. (c) The potential difference across the inductor.

38-4   ENERGY STORAGE IN A MAGNETIC FIELD

When a stone is lifted from the Earth, the external work done is stored as potential energy of the Earth–stone system. We can regard the process of separating the two objects as a way of storing energy in the gravitational field. When the stone is released, the energy can be recovered in the form of kinetic energy as the stone and Earth move closer together. In a similar manner, the work done in separating two charges of different signs is stored as the energy of the electric field of the charges; that energy can be recovered by allowing the charges to move together.

We can also consider the energy stored in the (gravitational or electric) field surrounding an isolated body, such as the Earth or a single charge. We regard the energy stored in that field as representing the energy expended in assembling the body from its constituent mass or charge elements, assumed initially to be at rest at infinite separations.

Energy can similarly be stored in a magnetic field. For example, consider two long, rigid, parallel wires carrying current in the same direction. The wires attract each other, and the work done in separating them is stored in the magnetic field surrounding them. We can recover that additional stored magnetic energy by letting the wires move back to their original positions.

We also regard energy as stored in the magnetic field of an isolated wire, in analogy with the energy of the electric field of an isolated charge. Before considering this subject in general it is helpful to consider the energy stored in the magnetic field of an inductor, just as we introduced energy storage in an electric field in Section 31-4 by considering the electric energy stored in a capacitor.

Figure 5 shows a source of emf $\mathcal{E}$ connected to a resistor R and an inductor L. The loop theorem applied to this circuit gives

$$\mathcal{E} = iR + L\frac{di}{dt},$$

as we already found in Eq. 20. Recall that the loop theorem is basically an expression of the principle of conservation of energy for single-loop circuits. Multiplying each side of this expression by i, we obtain

$$\mathcal{E}i = i^2R + Li\frac{di}{dt}, \tag{28}$$

which has the following physical interpretation in terms of work and energy:

1. If a charge dq passes through the seat of emf $\mathcal{E}$ in Fig. 5 in a time dt, the seat does work on it in the amount $\mathcal{E}\,dq$. The *rate* of doing work is $(\mathcal{E}\,dq)/dt$ or $\mathcal{E}i$. Thus the left side of Eq. 28 is the *rate at which the seat of emf delivers energy to the circuit.*

2. The second term in Eq. 28, i^2R, is the *rate at which energy is dissipated in the resistor.* This energy appears as the internal energy associated with atomic motions in the resistor.

3. Energy delivered to the circuit but not dissipated in the resistor must, by our hypothesis, be stored in the magnetic field. Since Eq. 28 represents a statement of the conservation of energy for LR circuits, the last term must represent the *rate at which energy is stored in the magnetic field.*

Let U_B represent the energy stored in the magnetic field; then the rate at which energy is stored is dU_B/dt. Equating the rate of energy storage to the last term of Eq. 28, we obtain

$$\frac{dU_B}{dt} = Li\frac{di}{dt} \tag{29}$$

or

$$dU_B = Li\,di. \tag{30}$$

Suppose we start with no current in the inductor ($i = 0$) and no stored energy in its magnetic field. We gradually increase the current to the final value i. The energy U_B stored in the magnetic field can be found by integrating Eq. 30, which gives

$$\int_0^{U_B} dU_B = \int_0^i Li\,di$$

or

$$U_B = \tfrac{1}{2}Li^2, \tag{31}$$

which represents the total stored magnetic energy in an inductance L carrying a current i.

If the switch in Fig. 4 is thrown from a to b after a current i is established, the stored energy in the inductor dissipates through Joule heating in the resistor. The current in this case is given by Eq. 26.

An analogous situation holds in charging and discharging a capacitor. When the capacitor has accumulated a charge q, the energy stored in the electric field is

$$U_E = \frac{1}{2}\frac{q^2}{C}.$$

We derived this expression in Section 31-4 by setting the stored energy equal to the work that must be done in setting up the field. The capacitor can discharge through a resistor, in which case the stored energy is again dissipated through Joule heating.

The necessity to dissipate the energy stored in an inductor is the reason that a "make before break" switch is needed in the circuit of Fig. 4. In this type of switch, the connection to b is made before the connection to a is broken. If such a switch were *not* used, the circuit would be momentarily open when the switch was thrown from a to b, in which case the current would be interrupted; the energy stored in the inductor would dissipate suddenly as a spark across the switch terminals.

Sample Problem 4   A coil has an inductance of 53 mH and resistance of 0.35 Ω. (*a*) If a 12-V emf is applied, how much energy is stored in the magnetic field after the current has built up to its maximum value? (*b*) In terms of τ_L, how long does it take for the stored energy to reach half of its maximum value?

Solution   (*a*) From Eq. 21 the maximum current is

$$i_m = \frac{\mathcal{E}}{R} = \frac{12\text{ V}}{0.35\ \Omega} = 34.3\text{ A}.$$

Substituting this current into Eq. 31, we find the stored energy:

$$U_B = \tfrac{1}{2}Li_m^2 = \tfrac{1}{2}(53 \times 10^{-3}\text{ H})(34.3\text{ A})^2$$
$$= 31\text{ J}.$$

(b) Let i be the current at the instant the stored energy has half its maximum value. Then

$$\tfrac{1}{2}Li^2 = (\tfrac{1}{2})\tfrac{1}{2}Li_m^2$$

or

$$i = i_m/\sqrt{2}.$$

But i is given by Eq. 21 and i_m (see above) is $\mathcal{E}/R$, so that

$$\frac{\mathcal{E}}{R}(1 - e^{-t/\tau_L}) = \frac{\mathcal{E}}{\sqrt{2}R}.$$

This can be written

$$e^{-t/\tau_L} = 1 - 1/\sqrt{2} = 0.293,$$

which yields

$$-\frac{t}{\tau_L} = \ln 0.293 = -1.23$$

or

$$t = 1.23\tau_L.$$

The stored energy reaches half its maximum value after 1.23 time constants.

Sample Problem 5   A 3.56-H inductor is placed in series with a 12.8-Ω resistor, an emf of 3.24 V being suddenly applied to the combination. At 0.278 s (which is one inductive time constant) after the contact is made, find (a) the rate P at which energy is being delivered by the battery, (b) the rate P_R at which internal energy appears in the resistor, and (c) the rate P_B at which energy is stored in the magnetic field.

Solution   (a) The current is given by Eq. 21. At $t = \tau_L$, we obtain

$$i = \frac{\mathcal{E}}{R}(1 - e^{-t/\tau_L}) = \frac{3.24\text{ V}}{12.8\ \Omega}(1 - e^{-1}) = 0.1600\text{ A}.$$

The rate P at which the battery delivers energy is then

$$P = \mathcal{E}i = (3.24\text{ V})(0.1600\text{ A}) = 0.5184\text{ W}.$$

(b) The rate P_R at which energy is dissipated in the resistor is given by

$$P_R = i^2R = (0.1600\text{ A})^2(12.8\ \Omega) = 0.3277\text{ W}.$$

(c) The rate $P_B (= dU_B/dt)$ at which energy is being stored in the magnetic field is given by Eq. 29. Using Eq. 22 with $t = \tau_L$, we obtain

$$\frac{di}{dt} = \frac{\mathcal{E}}{L}e^{-t/\tau_L} = \frac{3.24\text{ V}}{3.56\text{ H}}e^{-1} = 0.3348\text{ A/s}.$$

From Eq. 29 the desired rate is then

$$P_B = \frac{dU_B}{dt} = Li\frac{di}{dt}$$

$$= (3.56\text{ H})(0.1600\text{ A})(0.3348\text{ A/s}) = 0.1907\text{ W}.$$

Note that, as required by energy conservation,

$$P = P_R + P_B,$$

or

$$P = 0.3277\text{ W} + 0.1907\text{ W} = 0.5184\text{ W}.$$

Energy Density and the Magnetic Field

We now derive an expression for the *energy density* (energy per unit volume) u_B in a magnetic field. Consider a very long solenoid of cross-sectional area A whose interior contains no material. A portion of length l far from either end encloses a volume Al. The magnetic energy stored in this portion of the solenoid must lie entirely within this volume because the magnetic field outside the solenoid is essentially zero. Moreover, the stored energy must be uniformly distributed throughout the volume of the solenoid because the magnetic field is uniform everywhere inside. Thus we can write the energy density as

$$u_B = \frac{U_B}{Al}$$

or, since

$$U_B = \tfrac{1}{2}Li^2,$$

we have

$$u_B = \frac{\tfrac{1}{2}Li^2}{Al}.$$

To express this in terms of the magnetic field, we can solve Eq. 7 ($B = \mu_0 in$) for i and substitute in this equation. We can also substitute for L using the relation $L = \mu_0 n^2 lA$ (Eq. 9). Doing so yields finally

$$u_B = \frac{1}{2\mu_0}B^2. \tag{32}$$

This equation gives the energy density stored at any point (in a vacuum or in a nonmagnetic substance) where the magnetic field is **B**. The equation is true for all magnetic field configurations, even though we derived it by considering a special case, the solenoid. Equation 32 is to be compared with Eq. 28 of Chapter 31,

$$u_E = \tfrac{1}{2}\epsilon_0 E^2, \tag{33}$$

which gives the energy density (in a vacuum) at any point in an electric field. Note that both u_B and u_E are proportional to the square of the appropriate field quantity, B or E.

The solenoid plays a role for magnetic fields similar to that of the parallel-plate capacitor for electric fields. In each case we have a simple device that can be used for setting up a uniform field throughout a well-defined region of space and for deducing, in a simple way, properties of these fields.

Sample Problem 6   A long coaxial cable (Fig. 8) consists of two concentric cylindrical conductors with radii a and b, where $b \gg a$. Its central conductor carries a steady current i, and the outer conductor provides the return path. (a) Calculate the energy stored in the magnetic field for a length l of such a cable. (b) What is the inductance of a length l of the cable?

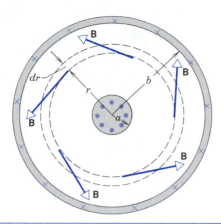

Figure 8 Sample Problem 6. Cross section of a coaxial cable, which carries steady equal but opposite currents in its inner and outer conductors. In the region between the conductors the lines of **B** form circles.

Solution (a) In the space between the two conductors Ampère's law,

$$\oint \mathbf{B} \cdot d\mathbf{s} = \mu_0 i,$$

leads to

$$B(2\pi r) = \mu_0 i$$

or

$$B = \frac{\mu_0 i}{2\pi r}.$$

Ampère's law shows further that the magnetic field is zero for points outside the outer conductor (why?). The outer conductor is so thin that we can neglect the magnetic energy stored in that conductor. We similarly assume that the inner conductor is so small that the magnetic energy in its volume is negligible. We therefore consider the stored magnetic energy to reside entirely in the space between the conductors.

The energy density for points between the conductors, from Eq. 32, is

$$u_B = \frac{1}{2\mu_0} B^2 = \frac{1}{2\mu_0} \left(\frac{\mu_0 i}{2\pi r} \right)^2 = \frac{\mu_0 i^2}{8\pi^2 r^2}.$$

Consider a volume element dV consisting of a cylindrical shell whose radii are r and $r + dr$ and whose length (perpendicular to the plane of Fig. 8) is l. The energy dU_B contained in it is

$$dU_B = u_B dV = \frac{\mu_0 i^2}{8\pi^2 r^2} (2\pi r l)(dr) = \frac{\mu_0 i^2 l}{4\pi} \frac{dr}{r}.$$

The total stored magnetic energy is found by integration:

$$U_B = \int dU_B = \frac{\mu_0 i^2 l}{4\pi} \int_a^b \frac{dr}{r} = \frac{\mu_0 i^2 l}{4\pi} \ln \frac{b}{a}.$$

(b) We can find the inductance L from Eq. 31 ($U_B = \frac{1}{2} L i^2$), which leads to

$$L = \frac{2U_B}{i^2} = \frac{\mu_0 l}{2\pi} \ln \frac{b}{a}.$$

You should also derive this expression directly from the definition of inductance, using the procedures of Section 38-2 (see Problem 15).

Sample Problem 7   Compare the energy required to set up, in a cube 10 cm on edge, (a) a uniform electric field of 1.0×10^5 V/m and (b) a uniform magnetic field of 1.0 T. Both these fields would be judged reasonably large but they are readily available in the laboratory.

Solution   (a) In the electric case we have, where V_0 is the volume of the cube,

$$U_E = u_E V_0 = \frac{1}{2} \epsilon_0 E^2 V_0$$
$$= (0.5)(8.9 \times 10^{-12} \text{ C}^2/\text{N} \cdot \text{m}^2)(10^5 \text{ V/m})^2(0.1 \text{ m})^3$$
$$= 4.5 \times 10^{-5} \text{ J}.$$

(b) In the magnetic case, from Eq. 32 we have

$$U_B = u_B V_0 = \frac{B^2}{2\mu_0} V_0 = \frac{(1.0 \text{ T})^2(0.1 \text{ m})^3}{(2)(4\pi \times 10^{-7} \text{ T} \cdot \text{m/A})}$$
$$= 400 \text{ J}.$$

In terms of fields normally available in the laboratory, much larger amounts of energy can be stored in a magnetic field than in an electric one, the ratio being about 10^7 in this example. Conversely, much more energy is required to set up a magnetic field of reasonable laboratory magnitude than is required to set up an electric field of similarly reasonable magnitude.

38-5   ELECTROMAGNETIC OSCILLATIONS: QUALITATIVE

We now turn to a study of the properties of circuits that contain both a capacitor C and an inductor L. Such a circuit forms an *electromagnetic oscillator*, in which the current varies sinusoidally with time, much as the displacement of a mechanical oscillator varies with time. In fact, as we shall see, there are several analogies between electromagnetic and mechanical oscillators. These analogies help us understand electromagnetic oscillators based on our previous study of mechanical oscillators (Chapter 15).

For the time being, we assume the circuit to include no resistance. The circuit *with* resistance, which we consider in Section 38-7, is analogous to the damped oscillator we discussed in Section 15-8. We also assume that no source of emf is present in the circuit; oscillating circuits with emf present, which we also consider in Section 38-7, are analogous to forced mechanical oscillators such as we discussed in Section 15-9.

With no source of emf present, the energy in the circuit comes from the energy initially stored in one or both of the components. Let us assume the capacitor C is charged (from some external source that doesn't concern us) so that it contains a charge q_m, at which time it is removed from the external source and connected to the inductor L.

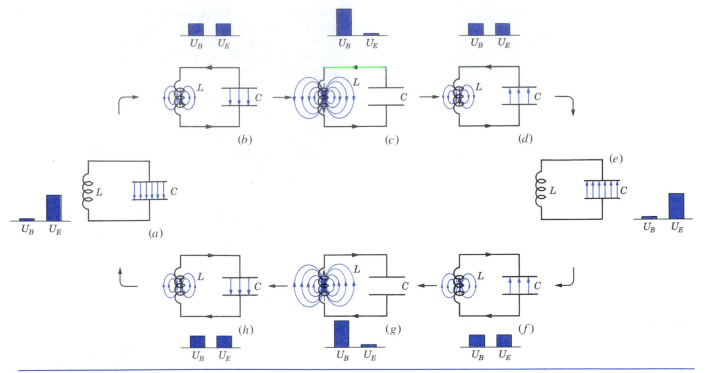

Figure 9 Eight stages in a single cycle of oscillation of a resistanceless *LC* circuit. The bar graphs show the stored magnetic and electric energies. The arrow through the inductor shows the current.

The *LC* circuit is shown in Fig. 9*a*. At first, the energy U_E stored in the capacitor is

$$U_E = \frac{1}{2}\frac{q_m^2}{C}, \qquad (34)$$

while the energy $U_B = \frac{1}{2}Li^2$ stored in the inductor is initially zero, because the current is zero.

The capacitor now starts to discharge through the inductor, positive charge carriers moving counterclockwise, as shown in Fig. 9*b*. A current $i = dq/dt$ now flows through the inductor, increasing its stored energy from zero. At the same time, the discharging of the capacitor reduces its stored energy. If the circuit is free of resistance, no energy is dissipated, and the decrease in the energy stored in the capacitor is exactly compensated by an increase in the energy stored in the inductor, such that the total energy remains constant. In effect, the electric field decreases and the magnetic field increases, energy being transferred from one to the other.

At a time corresponding to Fig. 9*c*, the capacitor is fully discharged, and the energy stored in the capacitor is zero. The current in the inductor has reached its maximum value, and all the energy in the circuit is stored in the magnetic field of the inductor. Note that, even though $q = 0$ at this instant, dq/dt differs from zero because charge is flowing.

The current in the inductor continues to transport charge from the top plate of the capacitor to the bottom plate, as in Fig. 9*d*; energy is now flowing from the induc-

tor back into the capacitor as its electric field builds up again. Eventually (see Fig. 9*e*) all the energy has been transferred back to the capacitor, which is now fully charged but in the opposite sense of Fig. 9*a*. The situation continues as the capacitor now discharges until the energy is completely back with the inductor, the magnetic field and the corresponding energy having their maximum values (Fig. 9*g*). Finally, the current in the inductor charges the capacitor once again until the capacitor is fully charged and the circuit is back in its original condition (Fig. 9*a*). The process then begins again, and the cycle repeats indefinitely. In the absence of resistance, which would cause energy to be dissipated, the charge and current return to their same maximum values in each cycle.

The oscillation of the *LC* circuit takes place with a definite frequency v (measured in Hz) corresponding to an angular frequency ω ($= 2\pi v$ and measured in rad/s). As we discuss in the next section, ω is determined by L and C. By suitable choices of L and C, we can build oscillating circuits with frequencies that range from below audio frequencies (10 Hz) to above microwave frequencies (10 GHz).

To determine the charge q as a function of the time, we can measure the variable potential difference $V_C(t)$ that exists across the capacitor C, which is related to the charge q by

$$V_C = \frac{1}{C}q.$$

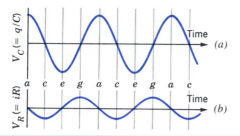

Figure 10 (*a*) The potential difference across the capacitor in the circuit of Fig. 9 as a function of time. This quantity is proportional to the charge on the capacitor. (*b*) The potential difference across a small resistor inserted into the circuit of Fig. 9. This quantity is proportional to the current in the circuit. The letters indicate the corresponding stages in the oscillation of Fig. 9.

We can determine the current by inserting into the circuit a resistor R so small that its effect on the circuit is negligible. The potential difference $V_R(t)$ across R is proportional to the current, according to

$$V_R = iR.$$

If we were to display $V_C(t)$ and $V_R(t)$, such as on the screen of an oscilloscope, the result might be similar to that shown in Fig. 10.

Sample Problem 8 A 1.5-μF capacitor is charged to 57 V. The charging battery is then disconnected and a 12-mH coil is connected across the capacitor, so that LC oscillations occur. What is the maximum current in the coil? Assume that the circuit contains no resistance.

Solution From the conservation-of-energy principle, the maximum stored energy in the capacitor must equal the maximum stored energy in the inductor. Using Eqs. 31 and 34, we obtain

$$\frac{q_m^2}{2C} = \tfrac{1}{2}Li_m^2,$$

where i_m is the maximum current and q_m is the maximum charge. Note that the maximum current and maximum charge do not occur at the same time but one-fourth of a cycle apart; see Figs. 9 and 10. Solving for i_m and substituting CV for q_m, we find

$$i_m = V\sqrt{\frac{C}{L}} = (57 \text{ V})\sqrt{\frac{1.5 \times 10^{-6} \text{ F}}{12 \times 10^{-3} \text{ H}}} = 0.64 \text{ A}.$$

Analogy to Simple Harmonic Motion

Figure 6 of Chapter 8 shows that in an oscillating block–spring system, as in an oscillating LC circuit, two kinds of energy occur. One is potential energy of the compressed or extended spring; the other is kinetic energy of the moving block. These are given by the familiar formulas in the first column of Table 1. The table suggests that a capacitor is in some way like a spring, an inductor is like a massive

TABLE 1 ENERGY IN OSCILLATING SYSTEMS

Mechanical		Electromagnetic	
Spring	$U_s = \tfrac{1}{2}kx^2$	Capacitor	$U_E = \tfrac{1}{2}C^{-1}q^2$
Block	$K = \tfrac{1}{2}mv^2$	Inductor	$U_B = \tfrac{1}{2}Li^2$
	$v = dx/dt$		$i = dq/dt$

object (the block), and certain electromagnetic quantities "correspond" to certain mechanical ones, namely,

$$q \text{ corresponds to } x, \qquad i \text{ corresponds to } v,$$
$$1/C \text{ corresponds to } k, \qquad L \text{ corresponds to } m. \tag{35}$$

Comparison of Fig. 9, which shows the oscillations of a resistanceless LC circuit, with Fig. 6 of Chapter 8, which shows the oscillations in a frictionless block–spring system, indicates how close the correspondence is. Note how v and i correspond in the two figures; also x and q. Note also how in each case the energy alternates between two forms, magnetic and electric for the LC system, and kinetic and potential for the block–spring system.

In Section 15-3 we saw that the natural angular frequency of a mechanical simple harmonic oscillator is

$$\omega = 2\pi v = \sqrt{\frac{k}{m}} .$$

The correspondence between the two systems suggests that to find the frequency of oscillation of a (resistanceless) LC circuit, k should be replaced by $1/C$ and m by L, which gives

$$\omega = 2\pi v = \sqrt{\frac{1}{LC}} . \tag{36}$$

This formula can also be derived from a rigorous analysis of the electromagnetic oscillation, as shown in the next section.

38-6 ELECTROMAGNETIC OSCILLATIONS: QUANTITATIVE

We now derive an expression for the frequency of oscillation of a (resistanceless) LC circuit using the conservation-of-energy principle. The total energy U present at any instant in an oscillating LC circuit is

$$U = U_B + U_E = \frac{1}{2}Li^2 + \frac{1}{2}\frac{q^2}{C}, \tag{37}$$

which indicates that at any arbitrary time the energy is stored partly in the magnetic field of the inductor and partly in the electric field of the capacitor. If we assume the circuit resistance to be zero, no energy is dissipated and U remains constant with time, even though i and q

vary. In more formal language, dU/dt must be zero. This leads to

$$\frac{dU}{dt} = \frac{d}{dt}\left(\frac{1}{2}Li^2 + \frac{1}{2}\frac{q^2}{C}\right) = Li\frac{di}{dt} + \frac{q}{C}\frac{dq}{dt} = 0. \quad (38)$$

We let q represent the charge on a particular plate of the capacitor (for instance, the upper plate in Fig. 9), and i then represents the rate at which charge flows into that plate (so that $i > 0$ when positive charge flows into the plate). In this case

$$i = \frac{dq}{dt} \quad \text{and} \quad \frac{di}{dt} = \frac{d^2q}{dt^2},$$

and substituting into Eq. 38 we obtain

$$\frac{d^2q}{dt^2} + \frac{1}{LC}q = 0. \quad (39)$$

Equation 39 describes the oscillations of a (resistanceless) LC circuit. To solve it, note the similarity of Eq. 4 of Chapter 15,

$$\frac{d^2x}{dt^2} + \frac{k}{m}x = 0, \quad (40)$$

which describes the mechanical oscillation of a particle on a spring. Fundamentally, it is by comparing these two equations that the correspondences of Eq. 35 arise.

The solution of Eq. 40 obtained in Chapter 15 was

$$x = x_m \cos(\omega t + \phi),$$

where x_m is the amplitude of the motion and ϕ is an arbitrary phase constant. Since q corresponds to x, we can write the solution of Eq. 39 as

$$q = q_m \cos(\omega t + \phi), \quad (41)$$

where ω is the still unknown angular frequency of the electromagnetic oscillations.

We can test whether Eq. 41 is indeed a solution of Eq. 39 by substituting it and its second derivative in that equation. To find the second derivative, we write

$$\frac{dq}{dt} = i = -\omega q_m \sin(\omega t + \phi) \quad (42)$$

and

$$\frac{d^2q}{dt^2} = -\omega^2 q_m \cos(\omega t + \phi). \quad (43)$$

Substituting q and d^2q/dt^2 into Eq. 39 yields

$$-L\omega^2 q_m \cos(\omega t + \phi) + \frac{1}{C}q_m \cos(\omega t + \phi) = 0.$$

Canceling $q_m \cos(\omega t + \phi)$ and rearranging leads to

$$\omega = \sqrt{\frac{1}{LC}}. \quad (44)$$

Thus, if ω is given the value $1/\sqrt{LC}$, Eq. 41 is indeed a solution of Eq. 39. This expression for ω agrees with Eq. 36, which we arrived at by the correspondence between mechanical and electromagnetic oscillations.

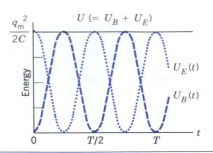

Figure 11   The stored magnetic energy and electric energy and their sum in an LC circuit as functions of the time. $T (= 2\pi/\omega)$ is the period of the oscillation.

The phase constant ϕ in Eq. 41 is determined by the conditions at $t = 0$. If the initial condition is as represented by Fig. 9a, then we put $\phi = 0$ in order that Eq. 41 may predict $q = q_m$ at $t = 0$. What initial physical condition is implied by $\phi = 90°$? $180°$? $270°$? Which of the states shown in Fig. 9 correspond to these choices of ϕ?

The stored electric energy in the LC circuit, using Eq. 41, is

$$U_E = \frac{1}{2}\frac{q^2}{C} = \frac{q_m^2}{2C}\cos^2(\omega t + \phi), \quad (45)$$

and the magnetic energy, using Eq. 42, is

$$U_B = \tfrac{1}{2}Li^2 = \tfrac{1}{2}L\omega^2 q_m^2 \sin^2(\omega t + \phi).$$

Substituting Eq. 44 for ω into this last equation yields

$$U_B = \frac{q_m^2}{2C}\sin^2(\omega t + \phi). \quad (46)$$

Figure 11 shows plots of $U_E(t)$ and $U_B(t)$ for the case of $\phi = 0$. Note that (1) the maximum values of U_E and U_B are the same ($= q_m^2/2C$); (2) the sum of U_E and U_B is a constant ($= q_m^2/2C$); (3) when U_E has its maximum value, U_B is zero and conversely; and (4) U_B and U_E each reach their maximum value *twice* during each cycle. This analysis supports the qualitative analysis of Section 38-5. Compare this discussion with that given in Section 15-4 for the energy transfers in a mechanical simple harmonic oscillator.

Sample Problem 9   (a) In an oscillating LC circuit, what value of charge, expressed in terms of the maximum charge, is present on the capacitor when the energy is shared equally between the electric and the magnetic field? (b) At what time t will this condition occur, assuming the capacitor to be fully charged initially? Assume that $L = 12$ mH and $C = 1.7$ μF.

Solution   (a) The stored energy U_E and the *maximum* stored energy U_m in the capacitor are, respectively,

$$U_E = \frac{q^2}{2C} \quad \text{and} \quad U_m = \frac{q_m^2}{2C}.$$

Substituting $U_E = \frac{1}{2}U_m$ yields

$$\frac{q^2}{2C} = \frac{1}{2}\frac{q_m^2}{2C}$$

or

$$q = \frac{q_m}{\sqrt{2}}.$$

(b) Since $\phi = 0$ in Eq. 41 because $q = q_m$ at $t = 0$, we have

$$q = q_m \cos \omega t = \frac{q_m}{\sqrt{2}},$$

which leads to

$$\omega t = \cos^{-1}\frac{1}{\sqrt{2}} = \frac{\pi}{4}$$

or, using $\omega = 1/\sqrt{LC}$,

$$t = \frac{\pi}{4\omega} = \frac{\pi\sqrt{LC}}{4} = \frac{\pi\sqrt{(12 \times 10^{-3}\text{ H})(1.7 \times 10^{-6}\text{ F})}}{4}$$
$$= 1.1 \times 10^{-4}\text{ s}.$$

38-7 DAMPED AND FORCED OSCILLATIONS

A resistance R is always present in any real LC circuit. When we take this resistance into account, we find that the total electromagnetic energy U is not constant but decreases with time as it is dissipated as internal energy in the resistor. As we shall see, the analogy with the damped block–spring oscillator of Section 15-8 is exact. As before, we have

$$U = U_B + U_E = \frac{1}{2}Li^2 + \frac{q^2}{2C}. \tag{47}$$

U is no longer constant but rather

$$\frac{dU}{dt} = -i^2R, \tag{48}$$

the minus sign signifying that the stored energy U decreases with time, being converted to internal energy in the resistor at the rate i^2R. Differentiating Eq. 47 and combining the result with Eq. 48, we have

$$Li\frac{di}{dt} + \frac{q}{C}\frac{dq}{dt} = -i^2R.$$

Substituting dq/dt for i and d^2q/dt^2 for di/dt and dividing by i, we obtain

$$L\frac{d^2q}{dt^2} + R\frac{dq}{dt} + \frac{1}{C}q = 0, \tag{49}$$

which describes the damped LC oscillations. If we put $R = 0$, Eq. 49 reduces, as it must, to Eq. 39, which describes the undamped LC oscillations.

We state without proof that the general solution of Eq. 49 can be written in the form

$$q = q_m e^{-Rt/2L} \cos(\omega't + \phi), \tag{50}$$

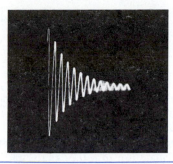

Figure 12 Photograph of an oscilloscope trace showing the oscillation of an LC circuit. The oscillation decreases in amplitude because energy is dissipated in the resistance of the circuit.

in which

$$\omega' = \sqrt{\omega^2 - (R/2L)^2}. \tag{51}$$

Using the analogies of Eq. 35, we see that Eq. 50 is the exact equivalent of Eq. 38 of Chapter 15, the equation for the displacement as a function of time in damped simple harmonic motion. Comparing Eq. 51 with Eq. 39 of Chapter 15, we see that the resistance R corresponds to the damping constant b of the damped mechanical oscillator.

Figure 12 shows the current in a damped LC circuit as a function of the time. (Compare Fig. 19 of Chapter 15.) The current oscillates sinusoidally with frequency ω', and the current amplitude decreases exponentially with time. The frequency ω' is strictly less than the frequency ω $(= 1/\sqrt{LC})$ of the undamped oscillations, but for most cases of interest we can put $\omega' = \omega$ with negligible error.

Sample Problem 10   A circuit has $L = 12$ mH, $C = 1.6\ \mu$F, and $R = 1.5\ \Omega$. (a) After what time t will the amplitude of the charge oscillations drop to one-half of its initial value? (b) To how many periods of oscillation does this correspond?

Solution   (a) This will occur when the amplitude factor $e^{-Rt/2L}$ in Eq. 50 has the value 1/2, or

$$e^{-Rt/2L} = \tfrac{1}{2}.$$

Taking the natural logarithm of each side gives

$$-Rt/2L = \ln\tfrac{1}{2} = -\ln 2,$$

or, solving for t,

$$t = \frac{2L}{R}\ln 2 = \frac{(2)(12 \times 10^{-3}\text{ H})}{1.5\ \Omega}\ln 2 = 0.0111\text{ s}.$$

(b) The number of oscillations is the elapsed time divided by the period, which is related to the angular frequency ω by $T = 2\pi/\omega$. The angular frequency is

$$\omega = \frac{1}{\sqrt{LC}} = \frac{1}{\sqrt{(12 \times 10^{-3}\text{ H})(1.6 \times 10^{-6}\text{ F})}} = 7220\text{ rad/s}.$$

The period of oscillation is then

$$T = \frac{2\pi}{\omega} = \frac{2\pi}{7220\text{ rad/s}} = 8.70 \times 10^{-4}\text{ s}.$$

The elapsed time, expressed in terms of the period of oscillation, is then

$$\frac{t}{T} = \frac{0.0111 \text{ s}}{8.70 \times 10^{-4} \text{ s}} \approx 13.$$

The amplitude drops to one-half after about 13 cycles of oscillation. By comparison, the damping in this example is less severe than that shown in Fig. 12, where the amplitude drops to one-half in about three cycles.

In this sample problem, we have used ω rather than ω'. From Eq. 51, we calculate $\omega - \omega' = 0.27$ rad/s, and so we make a negligible error in using ω.

Forced Oscillations and Resonance

Consider a damped LC circuit containing a resistance R. If the damping is small, the circuit oscillates at the frequency $\omega = 1/\sqrt{LC}$, which we call the *natural frequency* of the system.

Suppose now that we drive the circuit with a time-varying emf given by

$$\mathcal{E} = \mathcal{E}_m \cos \omega''t \tag{52}$$

using an external generator. Here ω'', which can be varied at will, is the frequency of this external source. We describe such oscillations as *forced*. When the emf described by Eq. 52 is first applied, time-varying transient currents appear in the circuit. Our interest, however, is in the sinusoidal currents that exist in the circuit after these initial transients have died away. Whatever the natural frequency ω may be, *these oscillations of charge, current, or potential difference in the circuit must occur at the external driving frequency ω''*.

Figure 13 compares the electromagnetic oscillating system with a corresponding mechanical system. A vibrator V, which imposes an external alternating force, corresponds to generator V, which imposes an external alter-

nating emf. Other quantities "correspond" as before (see Table 1): displacement to charge and velocity to current. The inductance L, which opposes changes in current, corresponds to the mass (inertia) m, which opposes changes in velocity. The spring constant k and the inverse capacitance C^{-1} represent the "stiffness" of their systems, giving, respectively, the response (displacement) of the spring to the force and the response (charge) of the capacitor to the emf.

In Chapter 39, we derive the solution for the current in the circuit of Fig. 13a, which we can write in the form

$$i = i_m \sin (\omega''t - \phi). \tag{53}$$

The current amplitude i_m in Eq. 53 is a measure of the response of the circuit of Fig. 13a to the driving emf. It is reasonable to suppose (from experience in pushing swings, for example) that i_m is large when the driving frequency ω'' is close to the natural frequency ω of the system. In other words, we expect that a plot of i_m versus ω'' exhibits a maximum when

$$\omega'' = \omega = 1/\sqrt{LC}, \tag{54}$$

which we call the *resonance* condition.

Figure 14 shows three plots of i_m as a function of the ratio ω''/ω, each plot corresponding to a different value of the resistance R. We see that each of these peaks does indeed have a maximum value when the resonance condition of Eq. 54 is satisfied. Note that as R is decreased, the resonance peak becomes sharper, as shown by the three horizontal arrows drawn at the half-maximum level of each curve.

Figure 14 suggests the common experience of tuning a radio set. In turning the tuning knob, we are adjusting the natural frequency ω of an internal LC circuit to match the

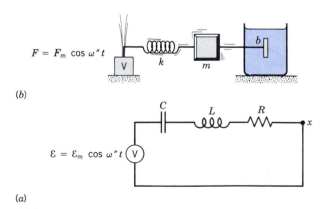

(b)

(a)

Figure 13   (a) Electromagnetic oscillations of a circuit are driven at an angular frequency ω''. (b) Mechanical oscillations of a spring system are driven at an angular frequency ω''. Corresponding elements of the two systems are drawn opposite each other.

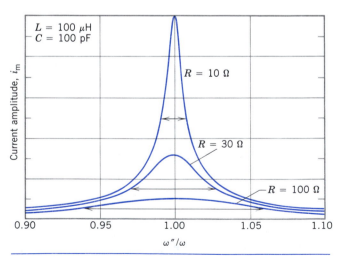

Figure 14   Resonance curves for the forced oscillating circuit of Fig. 13a. The three curves correspond to different values of the resistance of the circuit. The horizontal arrows indicate the width or "sharpness" of each resonance.

driving frequency ω'' of the signal transmitted by the antenna of the broadcasting station; we are looking for resonance. In a metropolitan area, where there are many signals whose frequencies are often close together, sharpness of tuning becomes important.

Figure 14 is similar to Fig. 20 of Chapter 15, which shows resonance peaks for the forced oscillations of a mechanical oscillator such as that of Fig. 13*b*. In this case also, the maximum response occurs when $\omega'' = \omega$, and

the resonance peaks become sharper as the damping factor (the coefficient *b*) is reduced. Note that the curves of Fig. 14 and of Fig. 20 of Chapter 15 are not exactly alike. The former is a plot of current amplitude, while the latter is a plot of displacement amplitude. The mechanical variable that corresponds to current is not displacement but velocity. Nevertheless, both sets of curves illustrate the resonance phenomenon.

QUESTIONS

1. Show that the dimensions of the two expressions for L, $N\Phi_B/i$ (Eq. 6) and $\mathcal{E}_L/(di/dt)$ (Eq. 2), are the same.

2. If the flux passing through each turn of a coil is the same, the inductance of the coil may be calculated from $L = N\Phi_B/i$ (Eq. 6). How might one compute L for a coil for which this assumption is not valid?

3. Give examples of how the flux linked by a coil can change due to stretching or compression of the coil.

4. You want to wind a coil so that it has resistance but essentially no inductance. How would you do it?

5. A long cylinder is wound from left to right with one layer of wire, giving it n turns per unit length with an inductance of L_1, as in Fig. 15*a*. If the winding is now continued, in the same *sense* but returning from right to left, as in Fig. 15*b*, so as to give a second layer also of n turns per unit length, then what is the value of the inductance? Explain.

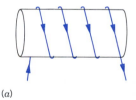

(*a*)

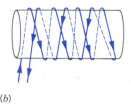

(*b*)

Figure 15   Question 5.

6. Is the inductance per unit length for a real solenoid near its center the same as, less than, or greater than the inductance per unit length near its ends? Justify your answer.

7. Explain why the inductance of a coaxial cable is expected to increase when the radius of the outer conductor is increased, the radius of the inner conductor remaining fixed.

8. You are given a length l of copper wire. How would you arrange it to obtain the maximum inductance?

9. Explain how a long straight wire can show induction effects. How would you go about looking for them?

10. A steady current is set up in a coil with a very large inductive time constant. When the current is interrupted with a switch, a heavy arc tends to appear at the switch blades. Explain why. (*Note*: Interrupting currents in highly inductive circuits can be destructive and dangerous.)

11. Suppose that you connect an ideal (that is, essentially resistanceless) coil across an ideal (again, essentially resistanceless) battery. You might think that, because there is no resistance in the circuit, the current would jump at once to a very large value. On the other hand, you might think that, because the inductive time constant ($= L/R$) is very large, the current would rise very slowly, if at all. What actually happens?

12. In an *LR* circuit like that of Fig. 5, can the induced emf ever be larger than the battery emf?

13. In an *LR* circuit like that of Fig. 5, is the current in the resistor always the same as the current in the inductor?

14. In the circuit of Fig. 4, the induced emf is a maximum at the instant the switch is closed on *a*. How can this be since there is no current in the inductor at this instant?

15. Does the time required for the current in a particular *LR* circuit to build up to a given fraction of its equilibrium value depend on the value of the applied constant emf?

16. If the current in a source of emf is in the direction of the emf, the energy of the source decreases; if a current is in a direction opposite to the emf (as in charging a battery), the energy of the source increases. Do these statements apply to the inductor in Figs. 2*a* and 2*b*?

17. Can the back emf in an inductor be in the same sense as the emf of the source, which gives the inductor its magnetic energy?

18. The switch in Fig. 4, having been closed on *a* for a "long" time, is thrown to *b*. What happens to the energy that is stored in the inductor?

19. A coil has a (measured) inductance L and a (measured) resistance R. Is its inductive time constant necessarily given by $\tau_L = L/R$? Bear in mind that we derived that equation (see Fig. 4) for a situation in which the inductive and resistive elements are physically separated. Discuss.

20. Figure 6*a* in this chapter and Fig. 14*b* in Chapter 33 are plots of $V_R(t)$ for, respectively, an *LR* circuit and an *RC* circuit. Why are these two curves so different? Account for each in terms of physical processes going on in the appropriate circuit.

21. Two solenoids, *A* and *B*, have the same diameter and length and contain only one layer of copper windings, with adjacent turns touching, insulation thickness being negligible. Solenoid *A* contains many turns of fine wire and solenoid *B* contains fewer turns of heavier wire. (*a*) Which solenoid has the larger inductance? (*b*) Which solenoid has the larger inductive time constant? Justify your answers.

22. Can you make an argument based on the manipulation of bar magnets to suggest that energy may be stored in a magnetic field?

23. Draw all the formal analogies you can think of between a parallel-plate capacitor (for electric fields) and a long solenoid (for magnetic fields).

24. In each of the following operations energy is expended. Some of this energy is returnable (can be reconverted) into electrical energy that can be made to do useful work and some becomes unavailable for useful work or is wasted in other ways. In which case will there be the *least* fraction of returnable electrical energy? (*a*) Charging a capacitor; (*b*) charging a storage battery; (*c*) sending a current through a resistor; (*d*) setting up a magnetic field; and (*e*) moving a conductor in a magnetic field.

25. The current in a solenoid is reversed. What changes does this make in the magnetic field **B** and the energy density u_B at various points along the solenoid axis?

26. Commercial devices such as motors and generators that are involved in the transformation of energy between electrical and mechanical forms involve magnetic rather than electrostatic fields. Why should this be so?

27. Why doesn't the *LC* circuit of Fig. 9 simply stop oscillating when the capacitor has been completely discharged?

28. How might you start an *LC* circuit into oscillation with its initial condition being represented by Fig. 9*c*? Devise a switching scheme to bring this about.

29. The lower curve *b* in Fig. 10 is proportional to the derivative of the upper curve *a*. Explain why.

30. In an oscillating *LC* circuit, assumed resistanceless, what determines (*a*) the frequency and (*b*) the amplitude of the oscillations?

31. In connection with Figs. 9*c* and 9*g*, explain how there can be a current in the inductor even though there is no charge on the capacitor.

32. In Fig. 9, what changes are required if the oscillations are to proceed counterclockwise around the figure?

33. In Fig. 9, what phase constants ϕ in Eq. 41 would permit the eight circuit situations shown to serve in turn as initial conditions?

34. What constructional difficulties would you encounter if you tried to build an *LC* circuit of the type shown in Fig. 9 to oscillate (*a*) at 0.01 Hz or (*b*) at 10 GHz?

35. Two inductors L_1 and L_2 and two capacitors C_1 and C_2 can be connected in series according to the arrangement in Fig. 16*a* or 16*b*. Are the frequencies of the two oscillating circuits equal? Consider the two cases (*a*) $C_1 = C_2$, $L_1 = L_2$ and (*b*) $C_1 \neq C_2$, $L_1 \neq L_2$.

36. In the mechanical analogy to the oscillating *LC* circuit, what mechanical quantity corresponds to the potential difference?

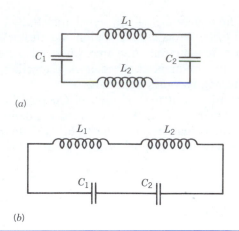

(*a*)

(*b*)

Figure 16   Question 35.

37. In comparing the electromagnetic oscillating system to a mechanical oscillating system, to what mechanical properties are the following electromagnetic properties analogous: capacitance, resistance, charge, electric field energy, magnetic field energy, inductance, and current?

38. Two springs are joined and connected to an object with mass *m*, the arrangement being free to oscillate on a horizontal frictionless surface as in Fig. 17. Sketch the electromagnetic analog of this mechanical oscillating system.

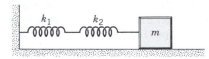

Figure 17   Question 38.

39. Explain why it is not possible to have (*a*) a real *LC* circuit without resistance, (*b*) a real inductor without inherent capacitance, or (*c*) a real capacitor without inherent inductance. Discuss the practical validity of the *LC* circuit of Fig. 9, in which each of the above realities is ignored.

40. All practical *LC* circuits must contain some resistance. However, one can buy a packaged audio oscillator in which the output maintains a constant amplitude indefinitely and does not decay, as it does in Fig. 12. How can this happen?

41. What would a resonance curve for $R = 0$ look like if plotted in Fig. 14?

42. Can you see any physical reason for assuming that *R* is "small" in Eqs. 50 and 51? (*Hint*: Consider what might happen if the damping *R* were so large that Eq. 50 would not even go through one cycle of oscillation before *q* was reduced essentially to zero. Could this happen? If so, what do you imagine Fig. 12 would look like?)

43. What is the difference between free, damped, and forced oscillating circuits?

44. Tabulate as many mechanical or electrical systems as you can think of that possess a natural frequency, along with the formula for that frequency if given in the text.

45. In an oscillatory radio receiver circuit, is it desirable to have a low or a high *Q*-factor? Explain. (See Problem 71.)

PROBLEMS

Section 38-2 Calculating the Inductance

1. The inductance of a close-packed coil of 400 turns is 8.0 mH. Calculate the magnetic flux through the coil when the current is 5.0 mA.

2. A circular coil has a 10.3-cm radius and consists of 34 closely wound turns of wire. An externally produced magnetic field of 2.62 mT is perpendicular to the coil. (*a*) If no current is in the coil, what is the number of flux linkages? (*b*) When the current in the coil is 3.77 A in a certain direction, the net flux through the coil is found to vanish. Find the inductance of the coil.

3. A solenoid is wound with a single layer of insulated copper wire (diameter, 2.52 mm). It is 4.10 cm in diameter and 2.0 m long. What is the inductance per meter for the solenoid near its center? Assume that adjacent wires touch and that insulation thickness is negligible.

4. At a given instant the current and the induced emf in an inductor are as indicated in Fig. 18. (*a*) Is the current increasing or decreasing? (*b*) The emf is 17 V, and the rate of change of the current is 25 kA/s; what is the value of the inductance?

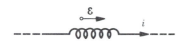

Figure 18   Problem 4.

5. The inductance of a closely wound *N*-turn coil is such that an emf of 3.0 mV is induced when the current changes at the rate 5.0 A/s. A steady current of 8.0 A produces a magnetic flux of 40 μWb through each turn. (*a*) Calculate the inductance of the coil. (*b*) How many turns does the coil have?

6. A toroid having a 5.20-cm square cross section and an inside radius of 15.3 cm has 536 turns of wire and carries a current of 810 mA. Calculate the magnetic flux through a cross section.

7. A solenoid 126 cm long is formed from 1870 windings carrying a current of 4.36 A. The core of the solenoid is filled with iron, and the effective permeability constant is 968. Calculate the inductance of the solenoid, assuming that it can be treated as ideal, with a diameter of 5.45 cm.

8. The current *i* through a 4.6-H inductor varies with time *t* as shown on the graph of Fig. 19. Calculate the induced emf during the time intervals (*a*) *t* = 0 to *t* = 2 ms, (*b*) *t* = 2 ms

to *t* = 5 ms, and (*c*) *t* = 5 ms to *t* = 6 ms. (Ignore the behavior at the ends of the intervals.)

9. A long thin solenoid can be bent into a ring to form a toroid. Show that if the solenoid is long and thin enough, the equation for the inductance of a toroid (Eq. 12) is equivalent to that for a solenoid of the appropriate length (Eq. 9).

10. Two inductors L_1 and L_2 are connected in series and are separated by a large distance. (*a*) Show that the equivalent inductance is given by

$$L_{eq} = L_1 + L_2.$$

(*b*) Why must their separation be large for this relationship to hold?

11. Two inductors L_1 and L_2 are connected in parallel and separated by a large distance. (*a*) Show that the equivalent inductance is given from

$$\frac{1}{L_{eq}} = \frac{1}{L_1} + \frac{1}{L_2}.$$

(*b*) Why must their separation be large for this relationship to hold?

12. A wide copper strip of width *W* is bent into a piece of slender tubing of radius *R* with two plane extensions, as shown in Fig. 20. A current *i* flows through the strip, distributed uniformly over its width. In this way a "one-turn solenoid" has been formed. (*a*) Derive an expression for the magnitude of the magnetic field **B** in the tubular part (far away from the edges). (*Hint*: Assume that the field outside this one-turn solenoid is negligibly small.) (*b*) Find also the inductance of this one-turn solenoid, neglecting the two plane extensions.

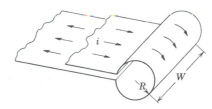

Figure 20   Problem 12.

13. Two long parallel wires, each of radius *a*, whose centers are a distance *d* apart carry equal currents in opposite directions. Show that, neglecting the flux within the wires themselves, the inductance of a length *l* of such a pair of wires is given by

$$L = \frac{\mu_0 l}{\pi} \ln \frac{d - a}{a}.$$

See Sample Problem 1, Chapter 35. (*Hint*: Calculate the flux through a rectangle of which the wires form two opposite sides.)

14. Two long, parallel copper wires (diameter = 2.60 mm) carry currents of 11.3 A in opposite directions. (*a*) If their centers are 21.8 mm apart, calculate the flux per meter of wire that exists in the space between the axes of the wires. (*b*) What fraction of this flux lies inside the wires, and therefore, what is the fractional error made in ignoring this flux in

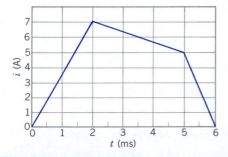

Figure 19   Problem 8.

calculating the inductance of two parallel wires? See Problem 13. (c) Repeat the calculations of (a) for parallel currents.

15. Find the inductance of the coaxial cable of Fig. 8 directly from Eq. 6. (*Hint*: Calculate the flux through a rectangular surface, perpendicular to **B**, of length *l* and width *b − a*.)

Section 38-3 LR Circuits

16. The current in an *LR* circuit builds up to one-third of its steady-state value in 5.22 s. Calculate the inductive time constant.

17. The current in an *LR* circuit drops from 1.16 A to 10.2 mA in the 1.50 s immediately following removal of the battery from the circuit. If *L* is 9.44 H, find the resistance *R* in the circuit.

18. (a) Consider the *LR* circuit of Fig. 4. In terms of the battery emf $\mathcal{E}$, what is the induced emf $\mathcal{E}_L$ when the switch has just been closed on *a*? (b) What is $\mathcal{E}_L$ after two time constants? (c) After how many time constants will $\mathcal{E}_L$ be just one-half of the battery emf $\mathcal{E}$?

19. The number of flux linkages through a certain coil of 745-mΩ resistance is 26.2 mWb when there is a current of 5.48 A in it. (a) Calculate the inductance of the coil. (b) If a 6.00-V battery is suddenly connected across the coil, how long will it take for the current to rise from 0 to 2.53 A?

20. (a) Show that Eq. 20 can be written

$$\frac{di}{i - \mathcal{E}/R} = -\frac{R}{L}\,dt.$$

(b) Integrate this equation to obtain Eq. 21.

21. Suppose the emf of the battery in the circuit of Fig. 5 varies with time *t* so the current is given by $i(t) = 3.0 + 5.0t$, where *i* is in amperes and *t* is in seconds. Take $R = 4.0\ \Omega$, $L = 6.0$ H, and find an expression for the battery emf as a function of time. (*Hint*: Apply the loop rule.)

22. At $t = 0$ a battery is connected to an inductor and resistor connected in series. The table below gives the measured potential difference, in volts, across the inductor as a function of time, in ms, following the connection of the battery. Deduce (a) the emf of the battery and (b) the time constant of the circuit.

t (ms)	V_L (V)	*t* (ms)	V_L (V)
1.0	18.2	5.0	5.98
2.0	13.8	6.0	4.53
3.0	10.4	7.0	3.43
4.0	7.90	8.0	2.60

23. A 45-V potential difference is suddenly applied to a coil with $L = 50$ mH and $R = 180\ \Omega$. At what rate is the current increasing after 1.2 ms?

24. A wooden toroidal core with a square cross section has an inner radius of 10 cm and an outer radius of 12 cm. It is wound with one layer of wire (diameter, 0.96 mm; resistance per unit length 21 mΩ/m). Calculate (a) the inductance and (b) the inductive time constant. Ignore the thickness of the insulation.

25. In Fig. 21, $\mathcal{E} = 100$ V, $R_1 = 10\ \Omega$, $R_2 = 20\ \Omega$, $R_3 = 30\ \Omega$, and $L = 2.0$ H. Find the values of i_1 and i_2 (a) immediately

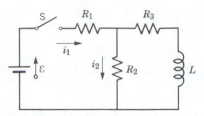

Figure 21 Problem 25.

after switch S is closed; (b) a long time later; (c) immediately after switch S is opened again; (d) a long time later.

26. In the circuit shown in Fig. 22, $\mathcal{E} = 10$ V, $R_1 = 5.0\ \Omega$, $R_2 = 10\ \Omega$, and $L = 5.0$ H. For the two separate conditions (I) switch S just closed and (II) switch S closed for a long time, calculate (a) the current i_1 through R_1, (b) the current i_2 through R_2, (c) the current *i* through the switch, (d) the potential difference across R_2, (e) the potential difference across *L*, and (f) di_2/dt.

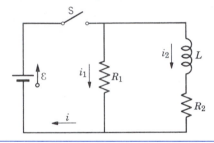

Figure 22 Problem 26.

27. Show that the inductive time constant τ_L can also be defined as the time that would be required for the current in an *LR* circuit to reach its equilibrium value *if it continued to increase at its initial rate*.

28. In Fig. 23 the component in the upper branch is an ideal 3.0-A fuse. It has zero resistance as long as the current through it remains less than 3.0 A. If the current reaches 3.0 A, it "blows" and thereafter it has infinite resistance. Switch S is closed at time $t = 0$. (a) When does the fuse blow? (b) Sketch a graph of the current *i* through the inductor as a function of time. Mark the time at which the fuse blows.

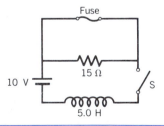

Figure 23 Problem 28.

Section 38-4 Energy Storage in a Magnetic Field

29. The magnetic energy stored in a certain inductor is 25.3 mJ when the current is 62.0 mA. (a) Calculate the inductance.

(*b*) What current is required for the magnetic energy to be four times as much?

30. A 92-mH toroidal inductor encloses a volume of 0.022 m³. If the average energy density in the toroid is 71 J/m³, calculate the current.

31. Find the magnetic energy density at the center of a circulating electron in the hydrogen atom (see Sample Problem 2, Chapter 35).

32. A solenoid 85.3 cm long has a cross-sectional area of 17.2 cm². There are 950 turns of wire carrying a current of 6.57 A. (*a*) Calculate the magnetic field energy density inside the solenoid. (*b*) Find the total energy stored in the magnetic field inside the solenoid. (Neglect end effects.)

33. What must be the magnitude of a uniform electric field if it is to have the same energy density as that possessed by a 0.50-T magnetic field?

34. The magnetic field in the interstellar space of our galaxy has a magnitude of about 100 pT. (*a*) Calculate the corresponding energy density, in eV/cm³. (*b*) How much energy is stored in this field in a cube 10 light-years on edge? (For scale, note that the nearest star, other than the Sun, is 4.3 light-years distant and the "radius" of our galaxy is about 80,000 light-years.)

35. The coil of a superconducting electromagnet used for nuclear magnetic resonance investigations has an inductance of 152 H and carries a current of 32 A. The coil is immersed in liquid helium, which has a latent heat of vaporization of 85 J/mol. (*a*) Calculate the energy in the magnetic field of the coil. (*b*) Find the mass of helium that is boiled off if the superconductor is quenched and thereby suddenly develops a finite resistance.

36. Suppose that the inductive time constant for the circuit of Fig. 5 is 37.5 ms and the current in the circuit is zero at time $t = 0$. At what time does the rate at which energy is dissipated in the resistor equal the rate at which energy is being stored in the inductor?

37. A coil is connected in series with a 10.4-kΩ resistor. When a 55.0-V battery is applied to the two, the current reaches a value of 1.96 mA after 5.20 ms. (*a*) Find the inductance of the coil. (*b*) How much energy is stored in the coil at this same moment?

38. For the circuit of Fig. 5, assume that $\mathcal{E} = 12.2$ V, $R = 7.34$ Ω, and $L = 5.48$ H. The battery is connected at time $t = 0$. (*a*) How much energy is delivered by the battery during the first 2.00 s? (*b*) How much of this energy is stored in the magnetic field of the inductor? (*c*) How much has appeared in the resistor?

39. (*a*) Find an expression for the energy density as a function of the radial distance r for a toroid of rectangular cross section. (*b*) Integrating the energy density over the volume of the toroid, calculate the total energy stored in the field of the toroid. (*c*) Using Eq. 12, evaluate the energy stored in the toroid directly from the inductance and compare with (*b*).

40. A length of copper wire carries a current of 10 A, uniformly distributed. Calculate (*a*) the magnetic energy density and (*b*) the electric energy density at the surface of the wire. The wire diameter is 2.5 mm and its resistance per unit length is 3.3 Ω/km.

41. The magnetic field at the Earth's surface has a strength of

about 60 μT. Assuming this to be relatively constant over radial distances small compared with the radius of the Earth and neglecting the variations near the magnetic poles, calculate the energy stored in a shell between the Earth's surface and 16 km above the surface.

42. Prove that, after switch S in Fig. 4 is thrown from a to b, all the energy stored in the inductor ultimately appears as internal energy in the resistor.

43. A long wire carries a current i uniformly distributed over a cross section of the wire. (*a*) Show that the magnetic energy of a length l stored *within* the wire equals $\mu_0 i^2 l/16\pi$. (Why does it not depend on the wire diameter?) (*b*) Show that the inductance for a length l of the wire associated with the flux *inside* the wire is $\mu_0 l/8\pi$.

Section 38-5 Electromagnetic Oscillations: Qualitative

44. What is the capacitance of an LC circuit if the maximum charge on the capacitor is 1.63 μC and the total energy is 142 μJ?

45. A 1.48-mH inductor in an LC circuit stores a maximum energy of 11.2 μJ. What is the peak current?

46. In an oscillating LC circuit $L = 1.13$ mH and $C = 3.88$ μF. The maximum charge on the capacitor is 2.94 μC. Find the maximum current.

47. LC oscillators have been used in circuits connected to loudspeakers to create some of the sounds of "electronic music." What inductance must be used with a 6.7-μF capacitor to produce a frequency of 10 kHz, near the upper end of the audible range of frequencies?

48. You are given a 10.0-mH inductor and two capacitors, of 5.00-μF and 2.00-μF capacitance. List the resonant frequencies that can be generated by connecting these elements in various combinations.

49. Consider the circuit shown in Fig. 24. With switch S_1 closed and the other two switches open, the circuit has a time constant τ_C. With switch S_2 closed and the other two switches open, the circuit has a time constant τ_L. With switch S_3 closed and the other two switches open, the circuit oscillates with a period T. Show that $T = 2\pi\sqrt{\tau_C\tau_L}$.

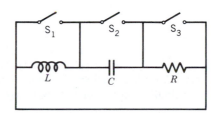

Figure 24 Problem 49.

50. A 485-g body oscillates on a spring that, when extended 2.10 mm from equilibrium, has a restoring force of 8.13 N. (*a*) Calculate the angular frequency of oscillation. (*b*) What is its period of oscillation? (*c*) What is the capacitance of the analogous LC system if L is chosen to be 5.20 H?

Section 38-6 Electromagnetic Oscillations: Quantitative

51. For a certain LC circuit the total energy is converted from electrical energy in the capacitor to magnetic energy in the

inductor in 1.52 μs. (a) What is the period of oscillation? (b) What is the frequency of oscillation? (c) How long after the magnetic energy is a maximum will it be a maximum again?

52. In an *LC* circuit with $L = 52.2$ mH and $C = 4.21$ μF, the current is initially a maximum. How long will it take before the capacitor is fully charged for the first time?

53. An oscillating *LC* circuit is designed to operate at a peak current of 31 mA. The inductance of 42 mH is fixed and the frequency is varied by changing C. (a) If the capacitor has a maximum peak voltage of 50 V, can the circuit safely operate at a frequency of 1.0 MHz? (b) What is the maximum safe operating frequency? (c) What is the minimum capacitance?

54. An oscillating *LC* circuit consisting of a 1.13-nF capacitor and a 3.17-mH coil has a peak potential drop of 2.87 V. Find (a) the maximum charge on the capacitor, (b) the peak current in the circuit, and (c) the maximum energy stored in the magnetic field of the coil.

55. An *LC* circuit has an inductance of 3.0 mH and a capacitance of 10 μF. Calculate (a) the angular frequency and (b) the period of oscillation. (c) At time $t = 0$ the capacitor is charged to 200 μC, and the current is zero. Sketch roughly the charge on the capacitor as a function of time.

56. In the circuit shown in Fig. 25 the switch has been in position *a* for a long time. It is now thrown to *b*. (a) Calculate the frequency of the resulting oscillating current. (b) What will be the amplitude of the current oscillations?

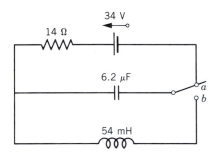

Figure 25 Problem 56.

57. (a) In an oscillating *LC* circuit, in terms of the maximum charge on the capacitor, what value of charge is present when the energy in the electric field is one-half that in the magnetic field? (b) What fraction of a period must elapse following the time the capacitor is fully charged for this condition to arise?

58. An inductor is connected across a capacitor whose capacitance can be varied by turning a knob. We wish to make the frequency of the *LC* oscillations vary linearly with the angle of rotation of the knob, going from 200 to 400 kHz as the knob turns through 180°. If $L = 1.0$ mH, plot C as a function of angle for the 180° rotation.

59. A variable capacitor with a range from 10 to 365 pF is used with a coil to form a variable-frequency *LC* circuit to tune the input to a radio. (a) What ratio of maximum to minimum frequencies may be tuned with such a capacitor? (b) If this capacitor is to tune from 0.54 to 1.60 MHz, the ratio computed in (a) is too large. By adding a capacitor in parallel to the variable capacitor this range may be adjusted. How large should this capacitor be and what inductance should be chosen in order to tune the desired range of frequencies?

60. In an *LC* circuit $L = 24.8$ mH and $C = 7.73$ μF. At time $t = 0$ the current is 9.16 mA, the charge on the capacitor is 3.83 μC, and the capacitor is charging. (a) What is the total energy in the circuit? (b) What is the maximum charge on the capacitor? (c) What is the maximum current? (d) If the charge on the capacitor is given by $q = q_m \cos(\omega t + \phi)$, what is the phase angle ϕ? (e) Suppose the data are the same, except that the capacitor is discharging at $t = 0$. What then is the phase angle ϕ?

61. In an oscillating *LC* circuit $L = 3.0$ mH and $C = 2.7$ μF. At $t = 0$ the charge on the capacitor is zero and the current is 2.0 A. (a) What is the maximum charge that will appear on the capacitor? (b) In terms of the period T of oscillation, how much time will elapse after $t = 0$ until the energy stored in the capacitor will be increasing at its greatest rate? (c) What is this greatest rate at which energy flows into the capacitor?

62. The resonant frequency of a series circuit containing inductance L_1 and capacitance C_1 is ω_0. A second series circuit, containing inductance L_2 and capacitance C_2, has the same resonant frequency. In terms of ω_0, what is the resonant frequency of a series circuit containing all four of these elements? Neglect resistance. (*Hint*: Use the formulas for equivalent capacitance and equivalent inductance.)

63. Three identical inductors L and two identical capacitors C are connected in a two-loop circuit as shown in Fig. 26. (a) Suppose the currents are as shown in Fig. 26a. What is the current in the middle inductor? Write down the loop equations and show that they are satisfied provided that the current oscillates with angular frequency $\omega = 1/\sqrt{LC}$. (b) Now suppose the currents are as shown in Fig. 26b. What is the current in the middle inductor? Write down the loop equations and show that they are satisfied provided the current oscillates with angular frequency $\omega = 1/\sqrt{3LC}$. (c) In view of the fact that the circuit can oscillate at two different frequencies, show that it is not possible to replace this two-loop circuit by an equivalent single-loop *LC* circuit.

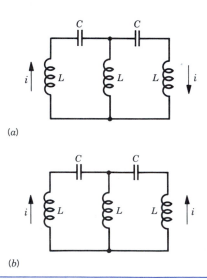

Figure 26 Problem 63.

64. In Fig. 27 the 900-μF capacitor is initially charged to 100 V and the 100-μF capacitor is uncharged. Describe in detail how one might charge the 100-μF capacitor to 300 V by manipulating switches S_1 and S_2.

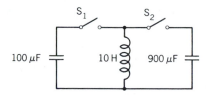

Figure 27 Problem 64.

Section 38-7 Damped and Forced Oscillations

65. In a damped LC circuit, find the time required for the maximum energy present in the capacitor during one oscillation to fall to one-half of its initial value. Assume $q = q_m$ at $t = 0$.

66. A single-loop circuit consists of a 7.22-Ω resistor, a 12.3-H inductor, and a 3.18-μF capacitor. Initially, the capacitor has a charge of 6.31 μC and the current is zero. Calculate the charge on the capacitor N complete cycles later for $N = 5$, 10, and 100.

67. How much resistance R should be connected to an inductor $L = 220$ mH and capacitor $C = 12$ μF in series in order that

the maximum charge on the capacitor decay to 99% of its initial value in 50 cycles?

68. (*a*) By direct substitution of Eq. 50 into Eq. 49, show that $\omega' = \sqrt{\omega^2 - (R/2L)^2}$. (*b*) By what fraction does the frequency of oscillation shift when the resistance is increased from 0 to 100 Ω in a circuit with $L = 4.4$ H and $C = 7.3$ μF?

69. A circuit has $L = 12.6$ mH and $C = 1.15$ μF. How much resistance must be inserted in the circuit to reduce the (undamped) resonant frequency by 0.01%?

70. Suppose that in a damped LC circuit the amplitude of the charge oscillations drops to one-half its initial value after n cycles. Show that the fractional reduction in the frequency of resonance, caused by the presence of the resistor, is given to a close approximation by

$$\frac{\omega - \omega'}{\omega} = \frac{0.0061}{n^2},$$

which is independent of L, C, or R.

71. In a damped LC circuit show that the fraction of the energy lost per cycle of oscillation, $\Delta U/U$, is given to a close approximation by $2\pi R/\omega L$. The quantity $\omega L/R$ is often called the Q of the circuit (for "quality"). A "high-Q" circuit has low resistance and a low fractional energy loss per cycle $(= 2\pi/Q)$.

CHAPTER 39

ALTERNATING CURRENT CIRCUITS

*Circuits involving alternating currents (commonly abbreviated AC)
are used in electric power distribution systems, in radio, television, and other
communication devices, and in a wide variety of electric motors. The designation
"alternating" means that the current changes direction, alternating periodically from one
direction to the other. Generally we work with currents that vary sinusoidally with time;
however, as we have seen previously in the case of wave motion, complex waveforms can be
viewed as combinations of sinusoidal waves (through Fourier analysis), and by analogy we
can understand the behavior of circuits having currents of arbitrary time dependence by first
understanding the behavior of circuits having currents that vary sinusoidally with time.*

*In this chapter we study the behavior of simple circuits containing resistors, inductors, and
capacitors when a sinusoidally varying source of emf is present.*

39-1 ALTERNATING CURRENTS

Previously we discussed the current produced when emfs
that vary with time in some different ways are applied to
circuits containing individual or combined elements of
resistance R, inductance L, and capacitance C. In Chapter
33, we discussed the steady currents resulting from the
application of steady emfs to purely resistive networks. In
Section 33-7, we discussed the response of a single-loop
RC circuit to the sudden application of an emf, and in
Section 38-3 the LR circuit was similarly considered. Sections 38-5 and 38-6 discussed the behavior of an LC circuit with no source of emf and the behavior of an RLC
circuit to a sinusoidal emf at or near resonance.

Here we consider the alternating current in a single-
loop RLC circuit that results when it is driven by a source
of emf that varies with time as

$$\mathcal{E} = \mathcal{E}_m \sin \omega t, \tag{1}$$

where $\mathcal{E}_m$ is the amplitude of the varying emf. The angular
frequency ω (in rad/s) is related to the frequency v (in Hz)
according to $\omega = 2\pi v$.

One possible way of producing a sinusoidally alternating emf is indicated in Fig. 1. As the coil rotates in a
uniform magnetic field, a sinusoidal emf is induced according to Faraday's law (see Section 36-4). This is a simple example of an AC *generator,* a more complex version
of which might be found in a commercial power plant. In
a circuit, the symbol for a source of alternating emf, such
as that of Fig. 1, is ──(∿)── .

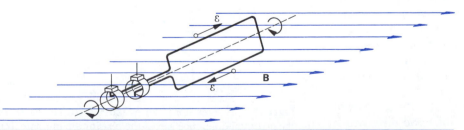

Figure 1 The basic principle of an alternating current generator is a conducting loop rotated
in an external magnetic field. The alternating emf appears across the two rings in contact
with the ends of the loop.

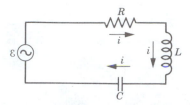

Figure 2 A single-loop circuit, consisting of a resistor, an inductor, and a capacitor. A generator supplies a source of alternating emf that establishes an alternating current.

Our goal in this chapter is to understand the result of applying an alternating emf, of the form of Eq. 1, to a circuit containing resistive, inductive, and capacitive elements. There are many ways these elements can be connected in a circuit; as an example of the analysis of AC circuits, we consider in this chapter the series *RLC* circuit shown in Fig. 2, in which a resistor *R*, inductor *L*, and capacitor *C* are connected in series across an alternating emf of the form of Eq. 1.

For a short time after the emf is initially applied to the circuit, the current varies erratically with time. These variations, called *transients,* rapidly die away, after which we find that *the current varies sinusoidally with the same angular frequency as the source of emf.* We assume that we examine the circuit after it has settled into this condition, in which the current can be written

$$i = i_m \sin (\omega t - \phi), \tag{2}$$

where i_m is the *current amplitude* (the maximum magnitude of the current) and ϕ is a phase constant or phase angle that indicates the phase relationship between $\mathscr{E}$ and i. (Note that we have assumed a phase constant of 0 in Eq. 1 for the emf. Note also that we write the phase constant in Eq. 2 with a minus sign; this choice is customary in discussing the phase relationship between the current and the emf.) The angular frequency ω in Eq. 2 is the same as that in Eq. 1.

We assume that $\mathscr{E}_m$, ω, R, L, and C are known. The goal of our calculation is to find i_m and ϕ, so that Eq. 2 completely characterizes the current. We use a general method for the series *RLC* circuit; a similar procedure can be used to analyze more complicated circuits (containing elements in various series and parallel combinations). It can also be applied to nonsinusoidal emfs, because more complicated emfs can be written in terms of sinusoidal emfs using the techniques of Fourier analysis (see Section 19-7), and the resultant current can similarly be considered to be the superposition of many terms of the form of Eq. 2. Understanding the series *RLC* circuit driven by a sinusoidal emf is therefore essential to understanding time-dependent behavior in all circuits.

In this chapter we are not specifically concerned with the phenomenon of resonance, which we discussed in Section 38-6. The angular frequency ω is completely arbitrary and is not necessarily close to the natural angular

frequency of oscillation of the circuit. Our general derivation of Eq. 2 in the next two sections includes resonance as a special case, but it remains a general result valid for any ω.

39-2 THREE SEPARATE ELEMENTS

Before analyzing the circuit of Fig. 2, it is helpful to discuss the response of each of the three elements separately to an alternating current of the form of Eq. 2. We assume that we deal with ideal elements; for instance, the inductor has only inductance and no resistance or capacitance.

A Resistive Element

Figure 3*a* shows a resistor in a section of a circuit in which a current *i* (given by Eq. 2) has been established by means not shown in the figure. Defining V_R $(= V_a - V_b)$ as the potential difference across the resistor, we can write

$$V_R = iR = i_m R \sin (\omega t - \phi). \tag{3}$$

Comparison of Eqs. 2 and 3 shows that the time-varying quantities V_R and i are *in phase:* they reach their maxi-

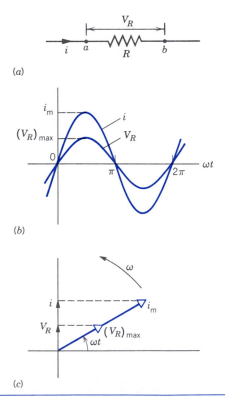

Figure 3 (*a*) A resistor in an AC circuit. (*b*) The current and the potential difference across the resistor are in phase. (*c*) A phasor diagram representing the current and potential difference.

mum values at the same time. This phase relationship is illustrated in Fig. 3b.

Figure 3c shows another way of looking at the situation. It is called a *phasor diagram,* in which the phasors, represented by the open arrows, rotate counterclockwise with an angular frequency ω about the origin. The phasors have the following properties. (1) The length of a phasor is proportional to the *maximum* value of the alternating quantity involved: for the potential difference, $(V_R)_{max} = i_m R$ from Eq. 3, and for the current, i_m from Eq. 2. (2) The projection of a phasor on the vertical axis gives the *instantaneous* value of the alternating quantity involved. The arrows on the vertical axis represent the time-varying quantities V_R and i, as in Eqs. 2 and 3, respectively. That V_R and i are in phase follows from the fact that their phasors lie along the same line in Fig. 3c.

The phasor diagram is very similar to Fig. 14 of Chapter 15, in which we made the connection between uniform circular motion and simple harmonic motion. You may recall that the projection on any axis of the position of a particle moving in uniform circular motion gives a displacement that varies sinusoidally, in analogy with simple harmonic motion. Here as the phasors rotate, their projections on the vertical axis give a sinusoidally varying current or voltage. Follow the rotation of the phasors in Fig. 3c and convince yourself that this phasor diagram completely and correctly describes Eqs. 2 and 3.

An Inductive Element

Figure 4a shows a portion of a circuit containing only an inductive element. The potential difference V_L ($= V_a - V_b$) across the inductor is related to the current by Eq. 3 of Chapter 38:

$$V_L = L\frac{di}{dt} = Li_m\omega \cos(\omega t - \phi), \qquad (4)$$

using Eq. 2 for the current. The trigonometric identity $\cos\theta = \sin(\theta + \pi/2)$ allows us to write Eq. 4 as

$$V_L = Li_m\omega \sin(\omega t - \phi + \pi/2). \qquad (5)$$

Comparison of Eqs. 2 and 5 shows that the time-varying quantities V_L and i are not in phase; they are one-quarter cycle out of phase, with V_L ahead of i (or i behind V_L). It is commonly said that the current *lags* the potential difference by 90° in an inductor. We show this in Fig. 4b, which is a plot of Eqs. 2 and 5. Note that, as time goes on, i reaches its maximum *after* V_L does, by one-quarter cycle.

This phase relationship between i and V_L is indicated in the phasor diagram of Fig. 4c. As the phasors rotate counterclockwise, it is clear that the i phasor follows (that is, *lags*) the V_L phasor by one-quarter cycle.

In analyzing AC circuits, it is convenient to define the *inductive reactance* X_L:

$$X_L = \omega L, \qquad (6)$$

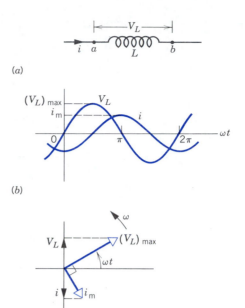

(a)

(b)

(c)

Figure 4   (*a*) An inductor in an AC circuit. (*b*) The current lags the potential difference across the inductor by 90°. (*c*) A phasor diagram representing the current and potential difference.

in terms of which we can rewrite Eq. 5 as

$$V_L = i_m X_L \sin(\omega t - \phi + \pi/2). \qquad (7)$$

Comparing Eqs. 3 and 7, we see that the SI unit for X_L must be the same as that of R, namely, the ohm. This can be seen directly by comparing Eq. 6 with the expression for the inductive time constant, $\tau_L = L/R$. Even though both are measured in ohms, a reactance is not the same as a resistance.

The maximum value of V_L is, from Eq. 7,

$$(V_L)_{max} = i_m X_L. \qquad (8)$$

A Capacitive Element

Figure 5a shows a portion of a circuit containing only a capacitive element. Again, a current i given by Eq. 2 has been established by means not shown.* Let the charge on the left-hand plate be q, so that a positive current into that

---

* It may at first be difficult to think of a capacitor as a part of a current-carrying circuit; clearly charge does not flow through the capacitor. It may be helpful to consider the flow of charge in this way: the current i brings charge q to the left-hand plate of the capacitor, so a charge $-q$ must flow to the right-hand plate from whatever circuit is beyond the capacitor to the right. This flow of charge $-q$ from right to left is entirely equivalent to a flow of charge $+q$ from left to right, which is identical to the current on the left side of the capacitor. Thus a current on one side of the capacitor can appear on the other side, even though there is no conducting path between the two plates!

(a)

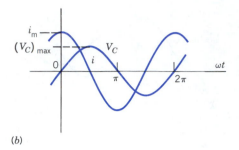

(b)

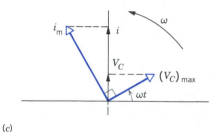

(c)

Figure 5 (a) A capacitor in an AC circuit. (b) The current leads the potential difference across the capacitor by 90°. (c) A phasor diagram representing the current and potential difference.

plate gives an increase in q; that is, $i = dq/dt$ implies $dq > 0$ when $i > 0$. The potential difference V_C ($= V_a - V_b$) across the capacitor is given by

$$V_C = \frac{q}{C} = \frac{\int i\, dt}{C}. \tag{9}$$

Integrating the current i given by Eq. 2, we find

$$V_C = -\frac{i_m}{\omega C} \cos(\omega t - \phi)$$

$$= \frac{i_m}{\omega C} \sin(\omega t - \phi - \pi/2), \tag{10}$$

where we have used the trigonometric identity $\cos\theta = -\sin(\theta - \pi/2)$.

Comparing Eqs. 2 and 10, we see that i and V_C are 90° out of phase, with i ahead of V_C. Figure 5b shows i and V_C plotted as functions of the time; note that i reaches its maximum one-quarter cycle or 90° before V_C. Equivalently, we may say that the current *leads* the potential difference by 90° in a capacitor.

The phase relationship is shown in the phasor diagram of Fig. 5c. As the phasors rotate counterclockwise, it is clear that the i phasor leads the V_C phasor by one-quarter cycle.

In analogy with the inductive reactance, it is convenient to define the *capacitive reactance X_C*:

$$X_C = \frac{1}{\omega C}, \tag{11}$$

in terms of which we can rewrite Eq. 10 as

$$V_C = i_m X_C \sin(\omega t - \phi - \pi/2). \tag{12}$$

Comparing Eqs. 3 and 12, we see that the unit of X_C must also be the ohm. This conclusion also follows by comparing Eq. 11 with the expression $\tau_C = RC$ for the capacitive time constant.

The maximum value of V_C is, from Eq. 12,

$$(V_C)_{max} = i_m X_C. \tag{13}$$

Table 1 summarizes the results derived for the three individual circuit elements.

Sample Problem 1 In Fig. 4a let $L = 230$ mH, $v = 60$ Hz, and $(V_L)_{max} = 36$ V. (a) Find the inductive reactance X_L. (b) Find the current amplitude in the circuit.

Solution (a) From Eq. 6

$$X_L = \omega L = 2\pi v L = (2\pi)(60 \text{ Hz})(230 \times 10^{-3} \text{ H})$$

$$= 87\ \Omega.$$

(b) From Eq. 8, the current amplitude is

$$i_m = \frac{(V_L)_{max}}{X_L} = \frac{36 \text{ V}}{87\ \Omega} = 0.41 \text{ A}.$$

TABLE 1 PHASE AND AMPLITUDE RELATIONS FOR ALTERNATING CURRENTS AND VOLTAGES

Circuit Element	Symbol	Impedance[a]	Phase of the Current	Amplitude Relation
Resistor	R	R	In phase with V_R	$(V_R)_{max} = i_m R$
Inductor	L	X_L	Lags V_L by 90°	$(V_L)_{max} = i_m X_L$
Capacitor	C	X_C	Leads V_C by 90°	$(V_C)_{max} = i_m X_C$

Many students have remembered the phase relations from: "ELI the ICE man." Here L and C stand for inductance and capacitance; E stands for voltage and I for current. Thus in an inductive circuit (ELI) the current (I) lags the voltage (E).

[a] *Impedance* is a general term that includes both resistance and reactance.

We see that, although a reactance is not a resistance, the inductive reactance plays the same role for an inductor that the resistance does for a resistor. Note that, if you doubled the frequency, the inductive reactance would double and the current amplitude would be cut in half. We can also understand this physically: to get the same value of V_L, you must change the current at the same rate ($V_L = L \, di/dt$). If the frequency doubles, you cut the time of change in half so that the maximum current is also cut in half. To sum up: for inductors, the higher the frequency, the higher the reactance.

Sample Problem 2   In Fig. 5a let $C = 15 \, \mu$F, $\nu = 60$ Hz, and $(V_C)_{max} = 36$ V. (a) Find the capacitive reactance X_C. (b) Find the current amplitude in this circuit.

Solution   (a) From Eq. 11, we have

$$X_C = \frac{1}{\omega C} = \frac{1}{2\pi\nu C}$$

$$= \frac{1}{(2\pi)(60 \text{ Hz})(15 \times 10^{-6} \text{ F})} = 177 \, \Omega.$$

(b) From Eq. 13, we have for the current amplitude

$$i_m = \frac{(V_C)_{max}}{X_C} = \frac{36 \text{ V}}{177 \, \Omega} = 0.20 \text{ A}.$$

Note that, if you doubled the frequency, the capacitive reactance would drop to half its value and the current amplitude would double. We can understand this physically: to get the same value of V_C you must deliver the same charge to the capacitor ($V_C = q/C$). If the frequency doubles, then you have only half the time to deliver this charge so that the maximum current must double. To sum up: for capacitors, the higher the frequency, the lower the reactance.

39-3   THE SINGLE-LOOP *RLC* CIRCUIT

Having finished our analysis of separate R, L, and C elements, we now return to the analysis of the circuit of Fig. 2, in which all three elements are present. The emf is given by Eq. 1

$$\mathcal{E} = \mathcal{E}_m \sin \omega t,$$

and the current in the circuit has the form of Eq. 2,

$$i = i_m \sin (\omega t - \phi).$$

Our goal is to determine i_m and ϕ.

We start by applying the loop theorem to the circuit of Fig. 2, obtaining $\mathcal{E} - V_R - V_L - V_C = 0$, or

$$\mathcal{E} = V_R + V_L + V_C. \tag{14}$$

Equation 14 can be solved for the current amplitude i_m and phase ϕ using a variety of techniques: a trigonometric analysis, a graphical analysis using phasors, and a differential analysis.

Trigonometric Analysis

We have already obtained relationships between the potential difference across each element and the current through the element. Let us therefore substitute Eqs. 3, 7, and 12 into Eq. 14, from which we obtain

$$\mathcal{E}_m \sin \omega t = i_m R \sin (\omega t - \phi)$$
$$+ i_m X_L \sin (\omega t - \phi + \pi/2) \tag{15}$$
$$+ i_m X_C \sin (\omega t - \phi - \pi/2),$$

in which we have substituted Eq. 1 for the emf. Using trigonometric identities, Eq. 15 can be written

$$\mathcal{E}_m \sin \omega t = i_m R \sin (\omega t - \phi) + i_m X_L \cos (\omega t - \phi)$$
$$- i_m X_C \cos (\omega t - \phi)$$
$$= i_m [R \sin (\omega t - \phi) \tag{16}$$
$$+ (X_L - X_C) \cos (\omega t - \phi)].$$

Following additional trigonometric manipulations (see Problem 18), Eq. 16 can be reduced to

$$\mathcal{E}_m \sin \omega t = i_m \sqrt{R^2 + (X_L - X_C)^2} \sin \omega t \tag{17}$$

provided we choose

$$\tan \phi = \frac{X_L - X_C}{R} = \frac{\omega L - 1/\omega C}{R}. \tag{18}$$

The current amplitude is found directly from Eq. 17:

$$i_m = \frac{\mathcal{E}_m}{\sqrt{R^2 + (X_L - X_C)^2}} = \frac{\mathcal{E}_m}{\sqrt{R^2 + (\omega L - 1/\omega C)^2}}. \tag{19}$$

This completes the analysis of the series *RLC* circuit, because we have accomplished our goal of expressing the current amplitude i_m and phase ϕ in terms of the parameters of the circuit ($\mathcal{E}_m$, ω, R, L, and C). Note that the phase ϕ does not depend on the amplitude $\mathcal{E}_m$ of the applied emf; changing $\mathcal{E}_m$ changes i_m but not ϕ: the *scale* of the result changes but not its *nature*.

The quantity in the denominator of Eq. 19 is called the *impedance Z* of the series *RLC* circuit:

$$Z = \sqrt{R^2 + (X_L - X_C)^2}, \tag{20}$$

and so Eq. 19 can be written

$$i_m = \frac{\mathcal{E}_m}{Z}, \tag{21}$$

which reminds us of the relation $i = \mathcal{E}/R$ for single-loop resistive networks with steady emfs. The SI unit of impedance is evidently the ohm.

Equation 19 gives the current amplitude in Eq. 53 of Chapter 38, and Fig. 14 of Chapter 38 is a plot of Eq. 19. The current i_m has its maximum value when the impedance Z has its minimum value R, which occurs when $X_L = X_C$, or

$$\omega L = 1/\omega C,$$

so that

$$\omega = 1/\sqrt{LC}, \tag{22}$$

which is the *resonance* condition given in Eq. 54 of Chapter 38. Although Eq. 19 is a general result valid for any driving frequency, it includes the resonance condition as a special case.

Graphical Analysis

It is instructive to use a phasor diagram to analyze the series *RLC* circuit. Figure 6*a* shows a phasor representing the current. It has length i_m, and its projection on the vertical axis is $i_m \sin(\omega t - \phi)$, which is the time-varying current i. In Fig. 6*b* we have drawn phasors representing the individual potential differences across *R*, *L*, and *C*.

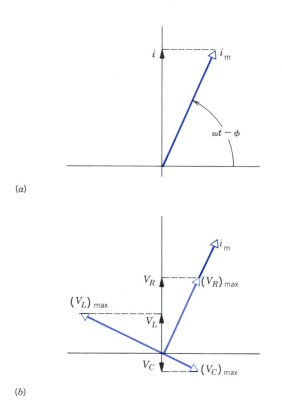

(a)

(b)

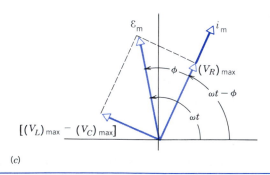

(c)

Figure 6 (*a*) A phasor representing the alternating current in the *RLC* circuit of Fig. 2. (*b*) Phasors representing the potential differences across the resistor, capacitor, and inductor. Note their phase differences with respect to the current. (*c*) A phasor representing the alternating emf has been added.

Note their maximum values and time-varying projections on the vertical axis. Be sure to note that the phases are in agreement with our conclusions from Section 39-2: V_R is in phase with the current, V_L *leads* the current by 90°, and V_C *lags* the current by 90°.

In accordance with Eq. 14, the *algebraic* sum of the (instantaneous) projections of V_R, V_L, and V_C on the vertical axis gives the (instantaneous) value of $\mathscr{E}$. On the other hand, we assert that the *vector* sum of the phasor amplitudes $(V_R)_{max}$, $(V_L)_{max}$, and $(V_C)_{max}$ yields a phasor whose amplitude is the $\mathscr{E}_m$ of Eq. 1. The projection of $\mathscr{E}_m$ on the vertical axis is the time-varying $\mathscr{E}$ of Eq. 1; that is, it is $V_R + V_L + V_C$ as Eq. 14 asserts. In vector operations, the (algebraic) sum of the projections of any number of vectors on a given straight line is equal to the projection on that line of the (vector) sum of those vectors.

In Fig. 6*c*, we have first formed the vector sum of $(V_L)_{max}$ and $(V_C)_{max}$, which is the phasor $(V_L)_{max} - (V_C)_{max}$. Next we form the vector sum of this phasor with $(V_R)_{max}$. Because these two phasors are at right angles, the amplitude of their sum, which is the amplitude of the phasor $\mathscr{E}_m$, is

$$\mathscr{E}_m = \sqrt{[(V_R)_{max}]^2 + [(V_L)_{max} - (V_C)_{max}]^2}$$
$$= \sqrt{(i_m R)^2 + (i_m X_L - i_m X_C)^2}$$
$$= i_m \sqrt{R^2 + (X_L - X_C)^2} \qquad (23)$$

using Eqs. 3, 8, and 13 to replace the phasor amplitudes. Equation 23 is identical with Eq. 19, which we obtained from the trigonometric analysis.

As shown in Fig. 6*c*, ϕ is the angle between the i_m and $\mathscr{E}_m$ phasors, and we see from the figure that

$$\tan \phi = \frac{(V_L)_{max} - (V_C)_{max}}{(V_R)_{max}}$$
$$= \frac{i_m(X_L - X_C)}{i_m R}$$
$$= \frac{X_L - X_C}{R}, \qquad (24)$$

which is identical with Eq. 18.

We drew Fig. 6*b* arbitrarily with $X_L > X_C$; that is, we assumed the circuit of Fig. 2 to be more inductive than capacitive. For this assumption, i_m *lags* $\mathscr{E}_m$ (although not by so much as one-quarter cycle as it did in the purely inductive element shown in Fig. 4). The phase angle ϕ in Eq. 23 (and thus in Eq. 2) is positive but less than +90°.

If, on the other hand, we had $X_C > X_L$, the circuit would be more capacitive than inductive and i_m would lead $\mathscr{E}_m$ (although not by as much as one-quarter cycle, as it did in the purely capacitive element shown in Fig. 5). Consistent with this change from lagging to leading, the angle ϕ in Eq. 23 (and thus in Eq. 2) would automatically become negative.

Another way of interpreting the resonance condition makes use of the phasor diagram of Fig. 6. At resonance,

$X_L = X_C$ and, according to Eq. 23, $\phi = 0$. In this case, the phasors $(V_L)_{max}$ and $(V_C)_{max}$ in Fig. 6 are equal and opposite, and so i_m is in phase with $\mathcal{E}_m$.

Once again, keep in mind that, while the techniques we have demonstrated here are valid for *any* AC circuit, the results hold *only* for the series *RLC* circuit. Furthermore, remember that we are examining the circuit only in the steady-state situation, after the short-lived transient variations have died away.

Sample Problem 3   In Fig. 2 let $R = 160\ \Omega$, $C = 15\ \mu F$, $L = 230$ mH, $v = 60$ Hz, and $\mathcal{E}_m = 36$ V. Find (*a*) the inductive reactance X_L, (*b*) the capacitive reactance X_C, (*c*) the impedance Z for the circuit, (*d*) the current amplitude i_m, and (*e*) the phase constant ϕ.

Solution   (*a*) $X_L = 87\ \Omega$, as in Sample Problem 1.

(*b*) $X_C = 177\ \Omega$, as in Sample Problem 2. Note that $X_C > X_L$ so that the circuit is more capacitive than inductive.

(*c*) From Eq. 20,

$$Z = \sqrt{R^2 + (X_L - X_C)^2}$$
$$= \sqrt{(160\ \Omega)^2 + (87\ \Omega - 177\ \Omega)^2} = 184\ \Omega.$$

(*d*) From Eq. 21,

$$i_m = \frac{\mathcal{E}_m}{Z} = \frac{36\ \text{V}}{184\ \Omega} = 0.196\ \text{A}.$$

(*e*) From Eq. 18 we have

$$\tan \phi = \frac{X_L - X_C}{R} = \frac{87\ \Omega - 177\ \Omega}{160\ \Omega} = -0.563.$$

Thus we have

$$\phi = \tan^{-1}(-0.563) = -29.4°.$$

A negative phase constant is appropriate for a capacitive load, as can be inferred from Table 1 and Fig. 6.

Sample Problem 4   (*a*) What is the resonance frequency in Hz of the circuit of Sample Problem 3? (*b*) What is the current amplitude at resonance?

Solution   (*a*) From Eq. 22,

$$\omega = \frac{1}{\sqrt{LC}} = \frac{1}{\sqrt{(0.23\ \text{H})(15 \times 10^{-6}\ \text{F})}} = 538\ \text{rad/s}.$$

Then

$$v = \frac{\omega}{2\pi} = 86\ \text{Hz}.$$

(*b*) At resonance, $X_L = X_C$, and so $Z = R$. From Eq. 21,

$$i_m = \frac{\mathcal{E}_m}{R} = \frac{36\ \text{V}}{160\ \Omega} = 0.23\ \text{A}.$$

The 60-Hz frequency of Sample Problem 3 is fairly close to resonance.

Differential Analysis   *(Optional)*

With $V_C = q/C$ and $V_L = L\, di/dt$, Eq. 14 can be written

$$\mathcal{E} = iR + L\frac{di}{dt} + \frac{q}{C}, \tag{25}$$

or, using $i = dq/dt$,

$$L\frac{d^2q}{dt^2} + R\frac{dq}{dt} + \frac{1}{C}q = \mathcal{E}_m \sin \omega t. \tag{26}$$

This equation is in the same form as that for the forced mechanical oscillator discussed in Section 15-9 (see Eq. 41 of Chapter 15). Making the analogies

$$x \to q, \quad m \to L, \quad b \to R, \quad \text{and} \quad k \to 1/C,$$

which we also used in Sections 38-5–38-7, we can immediately adapt the result given in Eq. 42 of Chapter 15 for the forced, damped mechanical oscillator to the driven, damped (that is, resistive) electromagnetic oscillator:

$$q = -\frac{\mathcal{E}_m}{\omega Z}\cos(\omega t - \phi), \tag{27}$$

where, as you should show, ωZ is G as defined by Eq. 43 of Chapter 15. Differentiating Eq. 27 to find the current, we obtain Eq. 2, $i = i_m \sin(\omega t - \phi)$, with $i_m = \mathcal{E}_m/Z$. You should also show that the phase ϕ given by Eq. 44 of Chapter 15 reduces to Eq. 18 when we replace the mechanical quantities by their electromagnetic analogues.

Seeking analogies, such as we have done here between mechanical and electromagnetic resonance, is a useful technique that not only provides insight into new phenomena but also saves work in their analysis, because we can adapt mathematical results obtained for one system to the analysis of another. We recognize the common characteristics of the two systems: a sinusoidal driving element; an inertial element, which resists changes in motion (m, which resists changes in v, and L, which resists changes in i); a dissipative element (b and R, each part of terms linear in the rate of change of the coordinate); and a restoring element (k and $1/C$, each part of terms linear in the coordinate). Common features of both solutions are: a stable sinusoidal oscillation at the driving frequency after an initial period of rapidly decaying transients; a phase difference between the driver and oscillating coordinate that is independent of the driving amplitude; and resonance at a particular frequency whose value is determined only by the inertial and restoring elements. ∎

39-4   POWER IN AC CIRCUITS

In an electrical circuit, energy is supplied by the source of emf, stored by the capacitive and inductive elements, and dissipated in resistive elements. Conservation of energy requires that, at any particular time, the rate at which energy is supplied by the source of emf must equal the rate at which it is stored in the capacitive and inductive elements plus the rate at which it is dissipated in the resistive elements. (We assume ideal capacitive and inductive elements that have no internal resistance.)

Let us consider a resistor as an isolated element (as in

Fig. 3) in an AC circuit in which the current is given by Eq. 2. (We examine the circuit in its steady state, a sufficiently long time after the source of emf has been connected to the circuit.) Just as in a DC circuit, the rate of energy dissipation (Joule heating) in a resistor in an AC circuit is given by

$$P = i^2R = i_m^2 R \sin^2(\omega t - \phi). \tag{28}$$

The energy dissipated in the resistor fluctuates with time, as does the energy stored in the inductive or capacitive elements. In most cases involving alternating currents, it is of no interest how the power varies during each cycle; the main interest is the *average* power dissipated during any particular cycle. The *average* energy stored in the inductive or capacitive elements remains constant over any complete cycle; in effect, energy is transferred from the source of emf to the resistive elements, where it is dissipated.

For example, the commercial power company supplies an AC source of emf to your home that varies with a frequency of $\nu = 60$ Hz. You are charged for the *average* power you consume; the power company is not concerned with whether you are operating a purely resistive device, in which the maximum power is dissipated in phase with the source of emf, or a partially capacitive or inductive device such as a motor, in which the current maximum (and therefore the power maximum) might occur out of phase with the emf. If the power company measured your energy use in a time smaller than $\frac{1}{60}$ s, they would notice variations in the rate at which you use energy, but in measuring over a time longer than $\frac{1}{60}$ s only the average rate of energy consumption becomes important.

We write the average power $\overline{P}$ by taking the average value of Eq. 28. The average value of the $\sin^2$ over any whole number of cycles is $\frac{1}{2}$, independent of the phase constant. The average power is then

$$\overline{P} = \tfrac{1}{2}i_m^2 R, \tag{29}$$

which we can also write as

$$\overline{P} = (i_m/\sqrt{2})^2 R. \tag{30}$$

The quantity $i_m/\sqrt{2}$ is equal to the *root-mean-square* (rms) value of the current:

$$i_{rms} = \frac{i_m}{\sqrt{2}}. \tag{31}$$

It is the result you would obtain if you first squared the current, then took its average (or mean) over a whole number of cycles, and then took the square root. (We defined the rms molecular speed in the same way in Chapter 24.) It is convenient to write the power in terms of rms values, because AC current and voltage meters are designed to report rms values. The common 120 V of household wiring is a rms value; the peak voltage is $\mathscr{E}_m = \sqrt{2}\mathscr{E}_{rms} = \sqrt{2}(120$ V$) = 170$ V.

In terms of i_{rms}, Eq. 31 can be written

$$\overline{P} = i_{rms}^2 R. \tag{32}$$

Equation 32 is similar to the expression $P = i^2R$, which describes the power dissipated in a resistor in a DC circuit. If we replace DC currents and voltages with the rms values of AC currents and voltages, DC expressions for power dissipation can be used to obtain the average AC power dissipation.

So far we have been considering the power dissipated in an isolated resistive element in an AC circuit. Let us now consider a full AC circuit from the standpoint of power dissipation. For this purpose we again choose the series *RLC* circuit as an example.

The work dW done by a source of emf $\mathscr{E}$ on a charge dq is given by $dW = \mathscr{E}\,dq$. The power $P\,(=dW/dt)$ is then $\mathscr{E}\,dq/dt = \mathscr{E}i$, or, using Eqs. 1 and 2,

$$P = \mathscr{E}i = \mathscr{E}_m i_m \sin\omega t \sin(\omega t - \phi). \tag{33}$$

We are seldom interested in this instantaneous power, which is usually a rapidly fluctuating function of the time. To find the *average* power, let us first use a trigonometric identity to expand the factor $\sin(\omega t - \phi)$:

$$\begin{aligned} P &= \mathscr{E}_m i_m \sin\omega t\,(\sin\omega t \cos\phi - \cos\omega t \sin\phi) \\ &= \mathscr{E}_m i_m(\sin^2\omega t \cos\phi - \sin\omega t \cos\omega t \sin\phi). \end{aligned} \tag{34}$$

When we now average over a complete cycle, the $\sin^2\omega t$ term gives the value $\frac{1}{2}$, while the $\sin\omega t \cos\omega t$ term gives 0, as you should show (see Problem 22). The average power is then

$$\overline{P} = \tfrac{1}{2}\mathscr{E}_m i_m \cos\phi. \tag{35}$$

Replacing both $\mathscr{E}_m$ and i_m with their rms values ($\mathscr{E}_{rms} = \mathscr{E}_m/\sqrt{2}$ and $i_{rms} = i_m/\sqrt{2}$), we can write Eq. 35 as

$$\overline{P} = \mathscr{E}_{rms} i_{rms} \cos\phi. \tag{36}$$

The quantity $\cos\phi$ in Eq. 36 is called the *power factor* of the AC circuit. Let us evaluate the power factor for the series *RLC* circuit. From Eq. 18, $\tan\phi = (X_L - X_C)/R$, we can show that

$$\cos\phi = \frac{R}{\sqrt{R^2 + (X_L - X_C)^2}} = \frac{R}{Z}. \tag{37}$$

According to Eq. 36, the power delivered to the circuit by the source of emf is maximum when $\cos\phi = 1$, which occurs when the circuit is purely resistive and contains no capacitors or inductors, or at resonance when $X_L = X_C$ so that $Z = R$. In this case the average power is

$$\overline{P} = \mathscr{E}_{rms} i_{rms} \qquad \text{(resistive load)}. \tag{38}$$

If the load is strongly inductive, as it often is in the case of motors, compressors, and the like, the power delivered to the load can be maximized by increasing the capacitance of the circuit. Power companies often place capacitors throughout their transmission system to bring this about.

Sample Problem 5   Consider again the circuit of Fig. 2, using the same parameters that we used in Sample Problem 3, namely, $R = 160 \ \Omega$, $C = 15 \ \mu F$, $L = 230$ mH, $v = 60$ Hz, and $\mathcal{E}_m = 36$ V. Find (a) the rms emf, (b) the rms current, (c) the power factor, and (d) the average power dissipated in the resistor.

Solution   (a)

$$\mathcal{E}_{rms} = \mathcal{E}_m/\sqrt{2} = 36 \ V/\sqrt{2} = 25.5 \ V.$$

(b) In Sample Problem 3 we found $i_m = 0.196$ A. We then have

$$i_{rms} = i_m/\sqrt{2} = (0.196 \ A)/\sqrt{2} = 0.139 \ A.$$

(c) In Sample Problem 3 we found that the phase constant ϕ was $-29.4°$. Thus

$$\text{power factor} = \cos(-29.4°) = 0.871.$$

(d) From Eq. 32 we have

$$\overline{P} = i_{rms}^2 R = (0.139 \ A)^2(160 \ \Omega) = 3.1 \ W.$$

Alternatively, Eq. 36 yields

$$\overline{P} = \mathcal{E}_{rms} i_{rms} \cos \phi$$
$$= (25.5 \ V)(0.139 \ A)(0.871) = 3.1 \ W,$$

in full agreement. That is, the average power dissipated in the resistor equals the average power supplied by the emf. In effect, energy is transferred from the emf to the resistive load, where it is dissipated. Note that, to get agreement of these results to two significant figures, we had to use three significant figures for the currents and voltages. Aside from such numerical rounding errors, Eqs. 32 and 36 give identical results.

39-5   THE TRANSFORMER   (Optional)

In DC circuits the power dissipation in a resistive load is given by Eq. 21 of Chapter 32 ($P = iV$). For a given power requirement, we have our choice of a relatively large current i and a relatively small potential difference V or just the reverse, provided that their product remains constant. In the same way, for purely resistive AC circuits (in which the power factor, $\cos \phi$ in Eq. 36, is equal to 1), the average power dissipation is given by Eq. 38 ($\overline{P} = i_{rms} \mathcal{E}_{rms}$) and we have the same choice as to the relative values of i_{rms} and $\mathcal{E}_{rms}$.

In electric power distribution systems it is desirable, both for reasons of safety and the efficient design of equipment, to have relatively low voltages at both the generating end (the electric power plant) and the receiving end (the home or factory). For example, no one wants an electric toaster or a child's electric train to operate at, say, 10 kV.

On the other hand, in the transmission of electric energy from the generating plant to the consumer, we want the *lowest* practical current (and thus the *largest* practical potential difference) so as to minimize the $i^2 R$ energy dissipation in the transmission line. Values such as $\mathcal{E}_{rms} = 350$ kV are typical. Thus there is a fundamental mismatch between the requirements for efficient transmission on the one hand and efficient and safe generation and consumption on the other hand.

To overcome this problem, we need a device that can, as design considerations require, raise (or lower) the potential dif-

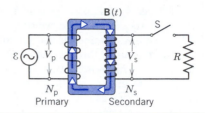

Figure 7   An ideal transformer, showing two coils wound on an iron core.

ference in a circuit, keeping the product $i_{rms} \mathcal{E}_{rms}$ essentially constant. The alternating current *transformer* of Fig. 7 is such a device. Operating on the basis of Faraday's law of induction, the transformer has no direct current counterpart of equivalent simplicity, which is why DC distribution systems, strongly advocated by Edison, have now been essentially totally replaced by AC systems, strongly advocated by Tesla and others.*

In Fig. 7 two coils are shown wound around an iron core. The *primary* winding, of N_p turns, is connected to an alternating current generator whose emf $\mathcal{E}$ is given by $\mathcal{E} = \mathcal{E}_m \sin \omega t$. The *secondary* winding, of N_s turns, is an open circuit as long as switch S is open, which we assume for the present. Thus there is no current in the secondary winding. We assume further that we can neglect all dissipative elements, such as the resistances of the primary and secondary windings. Actually, well-designed, high-capacity transformers can have energy losses as low as 1% so that our assumption of an ideal transformer is not unreasonable.

For the above conditions the primary winding is a pure inductance, as in Fig. 4a. The (very small) primary current, called the *magnetizing current* $i_{mag}(t)$, lags the primary potential difference $V_p(t)$ by 90°; the power factor ($=\cos \phi$ in Eq. 36) is zero, so no power is delivered from the generator to the transformer.

However, the small alternating primary current $i_{mag}(t)$ induces an alternating magnetic flux $\Phi_B(t)$ in the iron core, and we assume that all this flux links the turns of the secondary windings. (That is, we assume that all the magnetic field lines form closed loops within the iron core and none "escape" into the surroundings.) From Faraday's law of induction the emf *per turn* $\mathcal{E}_T$ (equal to $-d\Phi_B/dt$) is the same for both the primary and secondary windings, because the primary and secondary fluxes are equal. On a rms basis, we can write

$$\left(\frac{d\Phi_B}{dt}\right)_{primary} = \left(\frac{d\Phi_B}{dt}\right)_{secondary} \tag{39}$$

or

$$(\mathcal{E}_T)_{rms, \ primary} = (\mathcal{E}_T)_{rms, \ secondary}. \tag{40}$$

For each winding, the emf per turn equals the potential difference divided by the number of turns in the winding; Eq. 40 can then be written

$$\frac{V_p}{N_p} = \frac{V_s}{N_s}. \tag{41}$$

Here V_p and V_s refer to rms quantities. Solving for V_s, we obtain

$$V_s = V_p(N_s/N_p). \tag{42}$$

---

* See "The Transformer," by John W. Coltman, *Scientific American,* January 1988, p. 86.

If $N_s > N_p$ (in which case $V_s > V_p$), we speak of a *step-up transformer;* if $N_s < N_p$, we speak of a *step-down transformer.*

In all of the above we have assumed an open circuit secondary so that no power is transmitted through the transformer. If we now close switch S in Fig. 7, however, we have a more practical situation in which the secondary winding is connected with a resistive load R. In the general case, the load would also contain inductive and capacitive elements, but we confine ourselves to this special case of a purely resistive load.

Several things happen when we close switch S. (1) A rms current i_s appears in the secondary circuit, with a corresponding average power dissipation $i_s^2 R (= V_s^2/R)$ in the resistive load. (2) The alternating secondary current induces its own alternating magnetic flux in the iron core, and this flux induces (from Faraday's law and Lenz' law) an opposing emf in the primary windings. (3) V_p, however, cannot change in response to this opposing emf because it must always equal the emf that is provided by the generator; closing switch S cannot change this fact. (4) To ensure this, a new alternating current i_p must appear in the primary circuit, its magnitude and phase constant being just that needed to cancel the opposing emf generated in the primary windings by i_s.

Rather than analyze the above rather complex process in detail, we take advantage of the overall view provided by the conservation of energy principle. For an ideal transformer with a resistive load this tells us that

$$i_p V_p = i_s V_s. \tag{43}$$

Because Eq. 42 holds whether or not the switch S of Fig. 7 is closed, we then have

$$i_s = i_p(N_p/N_s) \tag{44}$$

as the transformation relation for currents.

Finally, knowing that $i_s = V_s/R$, we can use Eqs. 42 and 44 to obtain

$$i_p = \frac{V_p}{(N_p/N_s)^2 R}, \tag{45}$$

which tells us that, from the point of view of the primary circuit, the equivalent resistance of the load is not R but

$$R_{eq} = (N_p/N_s)^2 R. \tag{46}$$

Equation 46 suggests still another function for the transformer. We have seen that, for maximum transfer of energy from a seat of emf to a resistive load, the resistance of the generator and the resistance of the load must be equal. (See Problem 14

of Chapter 33.) The same relation holds for AC circuits except that the *impedance* (rather than the resistance) of the generator must be matched to that of the load. It often happens—as when we wish to connect a speaker to an amplifier—that this condition is far from met, the amplifier being of high impedance and the speaker of low impedance. We can match the impedances of the two devices by coupling them through a transformer with a suitable turns ratio. ■

Sample Problem 6   A transformer on a utility pole operates at $V_p = 8.5$ kV on the primary side and supplies electric energy to a number of nearby houses at $V_s = 120$ V, both quantities being rms values. The rate of average energy consumption in the houses served by the transformer at a given time is 78 kW. Assume an ideal transformer, a resistive load, and a power factor of unity. (*a*) What is the turns ratio N_p/N_s of this step-down transformer? (*b*) What are the rms currents in the primary and secondary windings of the transformer? (*c*) What is the equivalent resistive load in the secondary circuit? (*d*) What is the equivalent resistive load in the primary circuit?

Solution   (*a*) From Eq. 42 we have

$$\frac{N_p}{N_s} = \frac{V_p}{V_s} = \frac{8.5 \times 10^3 \text{ V}}{120 \text{ V}} = 70.8.$$

(*b*) From Eq. 38,

$$i_p = \frac{\overline{P}}{V_p} = \frac{78 \times 10^3 \text{ W}}{8.5 \times 10^3 \text{ V}} = 9.18 \text{ A}$$

and

$$i_s = \frac{\overline{P}}{V_s} = \frac{78 \times 10^3 \text{ W}}{120 \text{ V}} = 650 \text{ A}.$$

(*c*) In the secondary circuit,

$$R_s = \frac{V_s}{i_s} = \frac{120 \text{ V}}{650 \text{ A}} = 0.185 \text{ } \Omega.$$

(*d*) Here we have

$$R_p = \frac{V_p}{i_p} = \frac{8.5 \times 10^3 \text{ V}}{9.18 \text{ A}} = 930 \text{ } \Omega.$$

We can verify this from Eq. 46, which we write as

$$R_p = (N_p/N_s)^2 R_s = (70.8)^2(0.185 \text{ } \Omega) = 930 \text{ } \Omega.$$

QUESTIONS

1. In the relation $\omega = 2\pi\nu$ when using SI units we measure ω in radians per second and ν in hertz or cycles per second. The radian is a measure of angle. What connection do angles have with alternating current?

2. If the output of an AC generator such as that in Fig. 1 is connected to an *RLC* circuit such as that of Fig. 2, what is the ultimate source of the energy dissipated in the resistor?

3. Why would power distribution systems be less effective without alternating current?

4. In the circuit of Fig. 2, why is it safe to assume that (*a*) the alternating current of Eq. 2 has the same angular frequency ω as the alternating emf of Eq. 1, and (*b*) that the phase angle ϕ in Eq. 2 does not vary with time? What would happen if either of these (true) statements were false?

5. How does a phasor differ from a vector? We know, for example, that emfs, potential differences, and currents are not vectors. How then can we justify constructions such as Fig. 6?

6. In the purely resistive circuit element of Fig. 3, does the maximum value i_m of the alternating current vary with the angular frequency of the applied emf?

7. Would any of the discussion of Section 39-3 be invalid if the phasor diagrams were to rotate clockwise, rather than counterclockwise as we assumed?

8. Suppose that, in a series *RLC* circuit, the frequency of the applied voltage is changed continuously from a very low value to a very high value. How does the phase constant change?

9. Could the alternating current *resistance* of a device depend on the frequency?

10. From the analysis of an *RLC* circuit we can determine the behavior of an *RL* circuit (no capacitor) by putting $C = \infty$, whereas we put $L = 0$ to determine the behavior of an *RC* circuit (no inductor). Explain this difference.

11. During World War II, at a large research laboratory in this country, an alternating current generator was located a mile or so from the laboratory building it served. A technician increased the speed of the generator to compensate for what he called "the loss of frequency along the transmission line" connecting the generator with the laboratory building. Comment on this procedure.

12. As the speed of the blades of a rotating fan is increased from zero, a series of stationary patterns can be observed when the blades are illuminated by light from an alternating current source. The effect is more pronounced when a fluorescent tube or neon lamp is used than it is with a tungsten filament lamp. Explain these observations.

13. Assume that in Fig. 2 we let $\omega \to 0$. Does Eq. 19 approach an expected value? What is this value? Discuss.

14. Discuss in your own words what it means to say that an alternating current "leads" or "lags" an alternating emf.

15. If, as we stated in Section 39-3, a given circuit is "more inductive than capacitive," that is, that $X_L > X_C$, (a) does this mean, for a fixed angular frequency, that L is relatively "large" and C is relatively "small," or L and C are both relatively "large"? (b) For fixed values of L and C does this mean that ω is relatively "large" or relatively "small"?

16. How could you determine, in a series *RLC* circuit, whether the circuit frequency is above or below resonance?

17. Criticize this statement: "If $X_L > X_C$, then we must have $L > 1/C$."

18. How, if at all, must Kirchhoff's rules (the loop and junction rules) for direct current circuits be modified when applied to alternating current circuits?

19. Do the loop rule and the junction rule apply to multiloop AC circuits as well as to multiloop DC circuits?

20. In Sample Problem 5 what would be the effect on $\overline{P}$ if you increased (a) R, (b) C, and (c) L? (d) How would ϕ in Eq. 36 change in these three cases?

21. If $R = 0$ in the circuit of Fig. 2, there can be no power dissipation in the circuit. However, an alternating emf and an alternating current are still present. Discuss the energy flow in the circuit under these conditions.

22. Is there an rms *power* of an alternating current circuit?

23. Do commercial power station engineers like to have a low power factor or a high one, or does it make any difference to them? Between what values can the power factor range? What determines the power factor; is it characteristic of the generator, of the transmission line, of the circuit to which the transmission line is connected, or some combination of these?

24. Can the instantaneous power delivered by a source of alternating current ever be negative? Can the power factor ever be negative? If so, explain the meaning of these negative values.

25. In a series *RLC* circuit the emf is leading the current for a particular frequency of operation. You now lower the frequency slightly. Does the total impedance of the circuit increase, decrease, or stay the same?

26. If you know the power factor ($= \cos \phi$ in Eq. 36) for a given *RLC* circuit, can you tell whether or not the applied alternating emf is leading or lagging the current? If so, how? If not, why not?

27. What is the permissible range of values of the phase angle ϕ in Eq. 2? Of the power factor in Eq. 36?

28. Why is it useful to use the rms notation for alternating currents and voltages?

29. You want to reduce your electric bill. Do you hope for a small or a large power factor or does it make any difference? If it does, is there anything you can do about it? Discuss.

30. In Eq. 36 is ϕ the phase angle between $\mathscr{E}(t)$ and $i(t)$ or between $\mathscr{E}_{rms}$ and i_{rms}? Explain.

31. A doorbell transformer is designed for a primary rms input of 120 V and a secondary rms output of 6 V. What would happen if the primary and secondary connections were accidentally interchanged during installation? Would you have to wait for someone to push the doorbell to find out? Discuss.

32. You are given a transformer enclosed in a wooden box, its primary and secondary terminals being available at two opposite faces of the box. How could you find its turns ratio without opening the box?

33. In the transformer of Fig. 7, with the secondary an open circuit, what is the phase relationship between (a) the applied emf and the primary current, (b) the applied emf and the magnetic field in the transformer core, and (c) the primary current and the magnetic field in the transformer core?

34. What are some applications of a step-up transformer? Of a step-down transformer?

35. What determines which winding of a transformer is the primary and which the secondary? Can a transformer have a single primary and two secondaries? A single secondary and two primaries?

36. Instead of the 120-V, 60-Hz current typical of the United States, Europe uses 240-V, 50-Hz alternating currents. While on vacation in Europe, you would like to use some of your American appliances, such as a clock, an electric razor, and a hair dryer. Can you do so simply by plugging in a 2:1 step-up transformer? Explain why this apparently simple step may or may not suffice.

PROBLEMS

Section 39-2 Three Separate Elements

1. Let Eq. 1 describe the effective emf available at an ordinary 60-Hz AC outlet. To what angular frequency ω does this correspond? How does the utility company establish this frequency?

2. A 45.2-mH inductor has a reactance of 1.28 kΩ. (a) Find the frequency. (b) What is the capacitance of a capacitor with the same reactance at that frequency? (c) If the frequency is doubled, what are the reactances of the inductor and capacitor?

3. (a) At what angular frequency would a 6.23-mH inductor and a 11.4-μF capacitor have the same reactance? (b) What would this reactance be? (c) Show that this frequency would be equal to the natural frequency of free LC oscillations.

4. The output of an AC generator is $\mathcal{E} = \mathcal{E}_m \sin \omega t$, with $\mathcal{E}_m = 25.0$ V and $\omega = 377$ rad/s. It is connected to a 12.7-H inductor. (a) What is the maximum value of the current? (b) When the current is a maximum, what is the emf of the generator? (c) When the emf of the generator is -13.8 V and increasing in magnitude, what is the current? (d) For the conditions of part (c), is the generator supplying energy to or taking energy from the rest of the circuit?

5. The AC generator of Problem 4 is connected to a 4.15-μF capacitor. (a) What is the maximum value of the current? (b) When the current is a maximum, what is the emf of the generator? (c) When the emf of the generator is -13.8 V and increasing in magnitude, what is the current? (d) For the conditions of part (c), is the generator supplying energy to or taking energy from the rest of the circuit?

6. The output of an AC generator is given by $\mathcal{E} = \mathcal{E}_m \sin (\omega t - \pi/4)$, where $\mathcal{E}_m = 31.4$ V and $\omega = 350$ rad/s. The current is given by $i(t) = i_m \sin (\omega t - 3\pi/4)$, where $i_m = 622$ mA. (a) At what time, after $t = 0$, does the generator emf first reach a maximum? (b) At what time, after $t = 0$, does the current first reach a maximum? (c) The circuit contains a single element other than the generator. Is it a capacitor, an inductor, or a resistor? Justify your answer. (d) What is the value of the capacitance, inductance, or resistance, as the case may be?

7. Repeat the previous problem except that now $i = i_m \sin (\omega t + \pi/4)$.

8. A three-phase generator G produces electrical power that is transmitted by means of three wires as shown in Fig. 8. The potentials (relative to a common reference level) of these wires are $V_1 = V_m \sin \omega t$, $V_2 = V_m \sin (\omega t - 120°)$, and $V_3 = V_m \sin (\omega t - 240°)$. Some industrial equipment (for example, motors) has three terminals and is designed to be connected directly to these three wires. To use a more conventional two-terminal device (for example, a light bulb), one connects it to any two of the three wires. Show that the

potential difference between *any two* of the wires (a) oscillates sinusoidally with angular frequency ω and (b) has amplitude $V_m \sqrt{3}$.

Section 39-3 The Single-Loop RLC Circuit

9. Redraw (roughly) Figs. 6b and 6c for the cases of $X_C > X_L$ and $X_C = X_L$.

10. (a) Recalculate all the quantities asked for in Sample Problem 3 for $C = 70$ μF, the other parameters in that sample problem remaining unchanged. (b) Draw to scale a phasor diagram like that of Fig. 6c for this new situation and compare the two diagrams closely.

11. Consider the resonance curves of Fig. 14, Chapter 38. (a) Show that for frequencies above resonance the circuit is predominantly inductive and for frequencies below resonance it is predominantly capacitive. (b) How does the circuit behave at resonance? (c) Sketch a phasor diagram like that of Fig. 6c for conditions at a frequency higher than resonance, at resonance, and lower than resonance.

12. Verify mathematically that the following geometrical construction correctly gives both the impedance Z and the phase constant ϕ. Referring to Fig. 9, (1) draw an arrow in the $+y$ direction of magnitude X_C, (2) draw an arrow in the $-y$ direction of magnitude X_L, and (3) draw an arrow of magnitude R in the $+x$ direction. Then the magnitude of the "resultant" of these arrows is Z and the angle (measured below the $+x$ axis) of this resultant is ϕ.

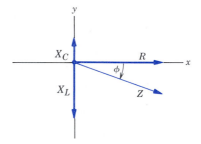

Figure 9 Problem 12.

13. Can the amplitude of the voltage across an inductor be greater than the amplitude of the generator emf in an RLC circuit? Consider a circuit with $\mathcal{E}_m = 10$ V, $R = 9.6$ Ω, $L = 1.2$ H, and $C = 1.3$ μF. Find the amplitude of the voltage across the inductor at resonance.

14. A coil of inductance 88.3 mH and unknown resistance and a 937-nF capacitor are connected in series with an oscillator of frequency 941 Hz. The phase angle ϕ between the applied emf and current is 75.0°. Find the resistance of the coil.

15. When the generator emf in Sample Problem 3 is a maximum, what is the voltage across (a) the generator, (b) the resistor, (c) the capacitor, and (d) the inductor? (e) By summing these with appropriate signs, verify that the loop rule is satisfied.

16. A resistor–inductor–capacitor combination, R_1, L_1, C_1, has a resonant frequency that is just the same as that of a

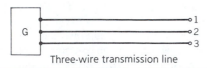

Figure 8 Problem 8.

different combination, R_2, L_2, C_2. You now connect the two combinations in series. Show that this new circuit also has the same resonant frequency as the separate individual circuits.

17. For a certain *RLC* circuit the maximum generator emf is 125 V and the maximum current is 3.20 A. If the current leads the generator emf by 56.3°, (*a*) what is the impedance and (*b*) what is the resistance of the circuit? (*c*) Is the circuit predominantly capacitive or inductive?

18. Use Eq. 18 to obtain relationships for sin ϕ and cos ϕ in terms of R, X_L, and X_C. Then substitute those expressions into Eq. 16 to obtain Eq. 17.

19. In a certain *RLC* circuit, operating at 60 Hz, the maximum voltage across the inductor is twice the maximum voltage across the resistor, while the maximum voltage across the capacitor is the same as the maximum voltage across the resistor. (*a*) By what phase angle does the current lag the generator emf? (*b*) If the maximum generator emf is 34.4 V, what should be the resistance of the circuit to obtain a maximum current of 320 mA?

20. An *RLC* circuit has $R = 5.12\,\Omega$, $C = 19.3\,\mu\text{F}$, $L = 988\,\text{mH}$, and $\mathcal{E}_m = 31.3\,\text{V}$. (*a*) At what angular frequency ω will the current have its maximum value, as in the resonance curves of Fig. 14 in Chapter 38? (*b*) What is this maximum value? (*c*) At what two angular frequencies ω_1 and ω_2 will the current amplitude have one-half of this maximum value? (*d*) Find the fractional width $[= (\omega_1 - \omega_2)/\omega]$ of the resonance curve.

21. (*a*) Show that the fractional width of the resonance curves of Fig. 14 in Chapter 38 is given, to a close approximation, by

$$\frac{\Delta\omega}{\omega} = \frac{\sqrt{3}R}{\omega L},$$

in which ω is the resonant frequency and $\Delta\omega$ is the width of the resonance peak at $i = \frac{1}{2}i_m$. Note (see Problem 71 of Chapter 38) that this expression may be written as $\sqrt{3}/Q$, which shows clearly that a "high-Q" circuit has a sharp resonance peak, that is, a small value of $\Delta\omega/\omega$. (*b*) Use this result to check part (*d*) of Problem 20.

Section 39-4 Power in AC Circuits

22. Show that $\overline{\sin^2 \omega t} = \frac{1}{2}$ and $\overline{\sin \omega t \cos \omega t} = 0$, where the averages are taken over one or more complete cycles.

23. An electric motor connected to a 120-V, 60-Hz power outlet does mechanical work at the rate of 0.10 hp (1 hp = 746 W). If it draws an rms current of 650 mA, what is its resistance, in terms of power transfer? Would this be the same as the resistance of its coils, as measured with an ohmmeter with the motor disconnected from the power outlet?

24. Show that the average power delivered to an *RLC* circuit can also be written

$$\overline{P} = \mathcal{E}^2_{\text{rms}} R/Z^2.$$

Show that this expression gives reasonable results for a purely resistive circuit, for an *RLC* circuit at resonance, for a purely capacitive circuit, and for a purely inductive circuit.

25. Calculate the average power dissipated in Sample Problem 3 assuming (*a*) that the inductor is removed from the circuit and (*b*) that the capacitor is removed.

26. An air conditioner connected to a 120-V rms AC line is equivalent to a 12.2-Ω resistance and a 2.30-Ω inductive reactance in series. (*a*) Calculate the impedance of the air conditioner. (*b*) Find the average power supplied to the appliance. (*c*) What is the value of the rms current?

27. A high-impedance AC voltmeter is connected in turn across the inductor, the capacitor, and the resistor in a series circuit having an AC source of 100 V (rms) and gives the same reading in volts in each case. What is this reading?

28. A farmer runs a water pump at 3.8 A rms. The connecting line from the transformer is 1.2 km long and consists of two copper wires each 1.8 mm in diameter. The temperature is 5.4°C. How much power is lost in transmission through the line?

29. In Fig. 10 show that the power dissipated in the resistor R is a maximum when $R = r$, in which r is the internal resistance of the AC generator. In the text we have tacitly assumed, up to this point, that $r = 0$. Compare with the DC situation.

Figure 10 Problems 29 and 43.

30. Consider the FM antenna circuit shown in Fig. 11, with $L = 8.22\,\mu\text{H}$, $C = 0.270\,\text{pF}$, and $R = 74.7\,\Omega$. The radio signal induces an alternating emf in the antenna with $\mathcal{E}_{\text{rms}} = 9.13\,\mu\text{V}$. Find (*a*) the frequency of the incoming waves for which the antenna is "in tune," (*b*) the rms current in the antenna, and (*c*) the rms potential difference across the capacitor.

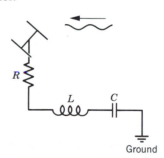

Figure 11 Problem 30.

31. Figure 12 shows an AC generator connected to a "black box" through a pair of terminals. The box contains an *RLC* circuit, possibly even a multiloop circuit, whose elements and arrangements we do not know. Measurements outside the box reveal that

$$\mathcal{E}(t) = (75\,\text{V})\sin \omega t$$

and

$$i(t) = (1.2\,\text{A})\sin (\omega t + 42°).$$

(*a*) What is the power factor? (*b*) Does the current lead or lag

Figure 12 Problem 31.

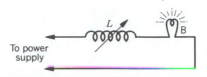

Figure 14 Problem 36.

the emf? (c) Is the circuit in the box largely inductive or largely capacitive in nature? (d) Is the circuit in the box in resonance? (e) Must there be a capacitor in the box? An inductor? A resistor? (f) What average power is delivered to the box by the generator? (g) Why don't you need to know the angular frequency ω to answer all these questions?

32. In an *RLC* circuit such as that of Fig. 2 assume that $R = 5.0\ \Omega$, $L = 60$ mH, $v = 60$ Hz, and $\mathcal{E}_m = 30$ V. For what values of the capacitance would the average power dissipated in the resistor be (a) a maximum and (b) a minimum? (c) What are these maximum and minimum powers? (d) What are the corresponding phase angles? (e) What are the corresponding power factors?

33. In Fig. 13, $R = 15.0\ \Omega$, $C = 4.72\ \mu$F, and $L = 25.3$ mH. The generator provides a sinusoidal voltage of 75.0 V (rms) and frequency $v = 550$ Hz. (a) Calculate the rms current amplitude. (b) Find the rms voltages V_{ab}, V_{bc}, V_{cd}, V_{bd}, V_{ad}. (c) What average power is dissipated by each of the three circuit elements?

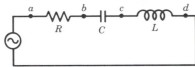

Figure 13 Problem 33.

34. In an *RLC* circuit, $R = 16.0\ \Omega$, $C = 31.2\ \mu$F, $L = 9.20$ mH, $\mathcal{E} = \mathcal{E}_m \sin \omega t$ with $\mathcal{E}_m = 45.0$ V, and $\omega = 3000$ rad/s. For time $t = 0.442$ ms find (a) the rate at which energy is being supplied by the generator, (b) the rate at which energy is being stored in the capacitor, (c) the rate at which energy is being stored in the inductor, and (d) the rate at which energy is being dissipated in the resistor. (e) What is the meaning of a negative result for any of parts (a), (b), and (c)? (f) Show that the results of parts (b), (c), and (d) sum to the result of part (a).

35. For an *RLC* circuit show that in one cycle with period T (a) the energy stored in the capacitor does not change; (b) the energy stored in the inductor does not change; (c) the generator supplies energy $(\frac{1}{2}T)\mathcal{E}_m i_m \cos \phi$; and (d) the resistor dissipates energy $(\frac{1}{2}T)Ri_m^2$. (e) Show that the quantities found in (c) and (d) are equal.

36. A typical "light dimmer" used to dim the stage lights in a theater consists of a variable inductor L connected in series with the light bulb B as shown in Fig. 14. The power supply is 120 V (rms) at 60.0 Hz; the light bulb is marked "120 V, 1000 W." (a) What maximum inductance L is required if the power in the light bulb is to be varied by a factor of five? Assume that the resistance of the light bulb is independent of its temperature. (b) Could one use a variable resistor instead of an inductor? If so, what maximum resistance is required? Why isn't this done?

37. The AC generator in Fig. 15 supplies 170 V (max) at 60 Hz. With the switch open as in the diagram, the resulting current leads the generator emf by 20°. With the switch in position 1 the current lags the generator emf by 10°. When the switch is in position 2 the maximum current is 2.82 A. Find the values of R, L, and C.

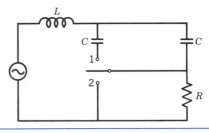

Figure 15 Problem 37.

Section 39-5 The Transformer

38. A generator supplies 150 V to the primary coil of a transformer of 65 turns. If the secondary coil has 780 turns, what is the secondary voltage?

39. A transformer has 500 primary turns and 10 secondary turns. (a) If V_p for the primary is 120 V (rms), what is V_s for the secondary, assumed an open circuit? (b) If the secondary is now connected to a resistive load of 15 Ω, what are the currents in the primary and secondary windings?

40. Figure 16 shows an "autotransformer." It consists of a single coil (with an iron core). Three "taps" are provided. Between taps T_1 and T_2 there are 200 turns and between taps T_2 and T_3 there are 800 turns. Any two taps can be considered the "primary terminals" and any two taps can be considered the "secondary terminals." List all the ratios by which the primary voltage may be changed to a secondary voltage.

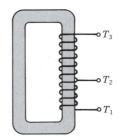

Figure 16 Problem 40.

41. In Fig. 7 show that $i_p(t)$ in the primary circuit remains unchanged if a resistance R' [$= R(N_p/N_s)^2$] is connected di-

rectly across the generator, the transformer and the secondary circuit being removed. That is,

$$i_p(t) = \frac{\mathcal{E}(t)}{R'}.$$

In this sense we see that a transformer not only "transforms" potential differences and currents but also resistances. In the more general case, in which the secondary load in Fig. 7 contains capacitive and inductive elements as well as resistive, we say that a transformer transforms *impedances.*

42. An electrical engineer designs an ideal transformer to run an x-ray machine at a peak potential of 74 kV and 270-mA rms current. The transformer operates from a 220-V rms power supply. However, resistance in the wires connecting the power supply to the transformer was ignored. Upon installation, it is realized that the supply wires have a resistance of 0.62 Ω. By how much must the supply voltage be increased in order to maintain the same operating parameters at the transformer?

43. In Fig. 10 let the rectangular box on the left represent the (high-impedance) output of an audio amplifier, with $r = 1000 \, \Omega$. Let $R = 10 \, \Omega$ represent the (low-impedance) coil of a loudspeaker. We learned that a transformer can be used to "transform" resistances, making them behave electrically as if they were larger or smaller than they actually are. Sketch the primary and secondary coils of a transformer to be introduced between the "amplifier" and the "speaker" in Fig. 10 to "match the impedances." What must be the turns ratio?

Computer Projects

44. A resistor R is connected in series with an inductor L and an emf $\mathcal{E}$. The current i obeys $L \, di/dt = -Ri + \mathcal{E}$, an equation that has the same mathematical form as Newton's second law for one-dimensional motion. The current replaces the velocity, $-Ri + \mathcal{E}$ replaces the force and L replaces the mass. You can use a computer program described in Section 6-6 to find the current as a function of time. The net charge that has passed any point on the circuit replaces the coordinate in Newton's second law. You may omit it from the program or retain it and take its initial value to be 0. (a) Take $R = 100 \, \Omega$, $L = 3.0 \times 10^{-2}$ H, and $\mathcal{E} = 15 \sin(1.0 \times 10^4 t)$, where $\mathcal{E}$ is in volts and t is in seconds. The current is zero at $t = 0$. Use the program to plot $i(t)$ from $t = 0$ to $t = 2.5 \times 10^{-3}$ s. Use $\Delta t = 2.0 \times 10^{-6}$ s for the integration interval. Notice the transients: the current is not sinusoidal at first but eventually becomes sinusoidal. On the part of the graph for which the current is most nearly sinusoidal mark the times when the emf has its maximum value. Does the current lead or lag the emf? By what time interval? The phase difference in radians is 2π times this time interval divided by the period. Calculate its value. (b) Repeat for an inductance of 0.10 H. How does the change in inductance

affect the current amplitude? How does it affect the time interval between a maximum of the emf and the nearest maximum of the current?

45. A resistor R, an inductor L, and a capacitor C are connected in series with an emf $\mathcal{E}$. The charge q on the capacitor obeys $L \, d^2q/dt^2 = \mathcal{E} - Ri - q/C$. This equation is mathematically identical to Newton's second law for one-dimensional motion: q replaces the coordinate, $i \, (= dq/dt)$ replaces the velocity, L replaces the mass, and $\mathcal{E} - Ri - q/C$ replaces the force. Use a computer program described in Chapter 6 to find the charge and current as functions of time. (a) Take $R = 100 \, \Omega$, $L = 3.0 \times 10^{-2}$ H, $C = 3.0 \times 10^{-6}$ F, and $\mathcal{E} = 15 \sin(1.0 \times 10^4 t)$, where $\mathcal{E}$ is in volts and t is in seconds. Take both the current and charge to be zero at time $t = 0$ and use the program to plot q and i from $t = 0$ to $t = 2.5 \times 10^{-3}$ s. Use $\Delta t = 2 \times 10^{-6}$ s for the integration interval. Notice that the transients die out and the current and charge become sinusoidal after a time. On the graph measure the current amplitude, then use $\mathcal{E}_m = i_m Z$ to calculate the impedance Z of the circuit. Compare your result with $Z = \sqrt{(X_L - X_C)^2 + R^2}$. Does the current lead or lag the emf? What is the time interval between a maximum of the emf and the nearest maximum of the current? (b) Repeat the calculation and answer the questions for an emf given by $15 \sin(2.0 \times 10^4 t)$. (c) Repeat the calculation and answer the questions for an emf given by $15 \sin \omega_0 t$, where ω_0 is the natural angular frequency of the circuit. Extend the range of the graph so about 3 cycles are plotted.

46. Consider the circuit described in the previous problem and use the values given in part (a) of that problem. (a) Modify the computer program to calculate and display the power supplied by the seat of emf ($\mathcal{E}i$), the rate of energy dissipation in the resistor (i^2R), the rate of energy storage in the electric field of the capacitor (iq/C), and the rate of energy storage in the magnetic field of the inductor ($Li \, di/dt = \mathcal{E}i - i^2R - iq/C$). Use the program to plot these quantities as functions of time from $t = 0$ to $t = 2.5 \times 10^{-3}$ s. (b) Identify the time intervals during which the seat of emf is supplying energy to the circuit and the intervals during which it is removing energy. After transients have died away do you think the seat of emf supplies more energy than it removes, removes more energy than it supplies, or supplies and removes the same energy? (c) Identify the time intervals during which energy is being transferred from the circuit to the magnetic field of the inductor and the intervals during which it is being transferred from the magnetic field to the circuit. Before the transients die out is there a net of flow of energy into or out of the inductor? What happens after the transients die out? (d) Identify the time intervals during which energy is being transferred from the circuit to the electric field of the capacitor. Before the transients die out is there a net flow of energy into or out of the capacitor? What happens after the transients die out?

CHAPTER 40

MAXWELL'S EQUATIONS

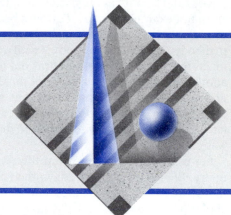

*In classical mechanics and in thermodynamics, we tried to obtain the
smallest, most compact set of equations or laws that enable us to analyze the
behavior of physical systems. In classical mechanics, Newton's three laws of motion provide
the framework. In thermodynamics, the three laws (numbered zero, one, and two) are used
to interpret a wide variety of experiments.*

*The basic equations of electromagnetism, which we have treated individually in previous
chapters, are known as* Maxwell's equations, *after Scottish physicist James Clerk Maxwell
(1831–1879), who was the first to make the equations part of a comprehensive and
symmetrical theory of electromagnetism. In this chapter, we summarize Maxwell's
equations and show that an argument based on symmetry leads to an important missing
term in one of our previous equations. In the next chapter, we show how these equations,
including the additional term, are essential in understanding electromagnetic waves, thereby
bringing optics, radio and TV transmission, microwave ovens, and magnetically levitated
trains all into the realm of electromagnetism.*

40-1 THE BASIC EQUATIONS OF ELECTROMAGNETISM

In this chapter we seek to identify a basic set of equations
for electromagnetism. We shall take several steps to accomplish this objective. First, we display in Table 1 a
tentative set of equations. These equations have been derived in the previous 13 chapters. Keep in mind that each
of these four equations is a statement about a different set
of experimental results. After studying this table, we shall
conclude from an argument based on symmetry that
these equations are not yet complete and that there may
be (and indeed *is*) a missing term in one of them.

The missing term proves to be no trifling correction: it
completes the description of electromagnetism and establishes optics as an integral part of electromagnetism. In
particular, it allows us to predict that the speed of light c
(and of all electromagnetic waves) in free space is related
to purely electric and magnetic quantities by

$$c = \frac{1}{\sqrt{\epsilon_0 \mu_0}} . \tag{1}$$

This relationship, along with additional predictions of the
electromagnetic equations, was later verified by experiment for light, radio waves, and other electromagnetic
waves.

We have seen how the principle of symmetry permeates

TABLE 1   TENTATIVE[a] BASIC EQUATIONS OF ELECTROMAGNETISM

Symbol	Name	Equation	Section Reference
I	Gauss' law for electricity	$\oint \mathbf{E} \cdot d\mathbf{A} = q/\epsilon_0$	29-3
II	Gauss' law for magnetism	$\oint \mathbf{B} \cdot d\mathbf{A} = 0$	37-1
III	Faraday's law of induction	$\oint \mathbf{E} \cdot d\mathbf{s} = -d\Phi_B/dt$	36-2
IV	Ampère's law	$\oint \mathbf{B} \cdot d\mathbf{s} = \mu_0 i$	35-5

[a] "Tentative" suggests, as we shall see later, that Eq. IV is not yet complete and requires an additional term; see Table 2.

859

physics and how it has often led to new insights or discoveries. For example, if body A attracts body B with a force of magnitude F, then we might expect from symmetry that body B should attract body A with a force of the same magnitude. This expectation turns out to be correct. For another example, the symmetry of the theory describing ordinary negatively charged electrons suggests that the electron should have a positively charged counterpart; the later discovery of the positron showed that this prediction was correct.

Let us examine Table 1 from the standpoint of symmetry. We ignore any lack of symmetry in the equations that arises from ϵ_0 and μ_0; these constants result from our choice of unit systems and play no role in considerations of symmetry. (There are, in fact, systems of units in which $\epsilon_0 = \mu_0 = 1$.)

With this in mind we see that the left sides of the equations in Table 1 are completely symmetrical, in pairs. Equations I and II are surface integrals of $\mathbf{E}$ and $\mathbf{B}$, respectively, over closed surfaces. Equations III and IV are line integrals of $\mathbf{E}$ and $\mathbf{B}$, respectively, around closed loops.

The right sides of these equations, on the other hand, are *not* symmetrical. There are two kinds of asymmetries:

1. The first asymmetry, which is not really the concern of this chapter, deals with the apparent fact that there are no isolated centers of magnetic charge (magnetic monopoles; see Section 37-1) analogous to isolated centers of electric charge (electrons, for instance). Thus we account for the q on the right side of Eq. I and for the 0 on the right side of Eq. II. In the same way, the term $i\,(=dq/dt)$, representing the current of electric charges, appears on the right side of Eq. IV, but there is no corresponding term representing a current of magnetic charges on the right of Eq. III. The desire for symmetry in these equations has led to the prediction that magnetic monopoles should exist. Despite many experimental searches for monopoles, there is as yet no confirmation of their existence. Later in this chapter we discuss how to symmetrize Maxwell's equations if magnetic monopoles are proved to exist.

2. The second asymmetry, which is more significant for the discussions of this chapter, is equally prominent. On the right side of Eq. III we find the term $-d\Phi_B/dt$. This equation, also known as Faraday's law of induction, can be loosely interpreted by saying:

If you change a magnetic field $(d\Phi_B/dt)$, you produce an electric field $(\oint \mathbf{E} \cdot d\mathbf{s})$.

We learned this in Section 36-1 where we showed that if you push a bar magnet through a closed conducting loop, you do indeed induce an electric field, and thus a current, in that loop.

From the principle of symmetry we are entitled to suspect that the analogous relation holds, that is:

If you change an electric field $(d\Phi_E/dt)$, you produce a magnetic field $(\oint \mathbf{B} \cdot d\mathbf{s})$.

This supposition, which we discuss more fully in the next section, provides us with the missing term in Eq. IV and turns out to meet the test of experiment.

40-2 INDUCED MAGNETIC FIELDS AND THE DISPLACEMENT CURRENT

Here we discuss in detail the evidence for the supposition of the previous section: namely, a changing electric field induces a magnetic field. Although we are guided primarily by considerations of symmetry, we also find direct experimental verification.

Figure 1*a* shows a circular parallel-plate capacitor. A current i enters the left-hand plate (which we assume to carry a positive charge), and an equal current i leaves the right-hand plate. An Ampèrian loop surrounds the wire in Fig. 1*a* and forms the boundary for a surface that is pierced by the wire. The current in the wire sets up a magnetic field; in Section 35-5 we saw that the magnetic field and the current are related by Ampère's law,

$$\oint \mathbf{B} \cdot d\mathbf{s} = \mu_0 i. \qquad (2)$$

That is, the line integral of the magnetic field around the

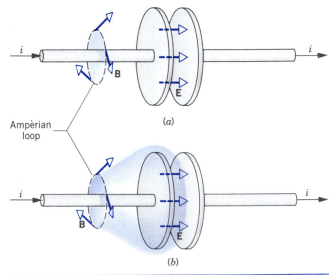

Figure 1 (*a*) An Ampèrian loop encloses a surface through which passes a wire carrying a current. (*b*) The same Ampèrian loop encloses a surface that passes between the capacitor plates. No conduction current passes through the surface.

loop is proportional to the total current that passes through the surface bounded by the loop.

In Fig. 1*b*, we have kept the same loop but have stretched the surface bounded by the loop so that it encloses the entire left-hand capacitor plate. Since the loop has not changed (nor has the magnetic field), the left side of Ampère's law gives the same result, but the right side gives a very different result, namely, zero, because no conducting wires pass through the surface. We appear to have a violation of Ampère's law!

To restore Ampère's law so that it correctly describes the situation of Fig. 1*b*, we rely on the conclusion given in the previous section based on symmetry: *a magnetic field is set up by a changing electric field.* Let us consider the situation of Fig. 1 in more detail. As charge is transported into the capacitor, the electric field in its interior changes at a certain rate dE/dt. The electric field lines pass through the surface of Fig. 1*b*; we account for the passage of field lines through this surface in terms of the electric flux Φ_E, and a changing electric field must give a correspondingly changing electric flux, $d\Phi_E/dt$.

To describe this new effect quantitatively, we are guided by analogy with Faraday's law of induction,

$$\oint \mathbf{E} \cdot d\mathbf{s} = -\frac{d\Phi_B}{dt}, \tag{3}$$

which asserts that an electric field (left side) is produced by a changing magnetic field (right side). For the symmetrical counterpart we write*

$$\oint \mathbf{B} \cdot d\mathbf{s} = \mu_0\epsilon_0 \frac{d\Phi_E}{dt}. \tag{4}$$

Equation 4 asserts that a magnetic field (left term) can be produced by a changing electric field (right term).

The situation shown in Fig. 1*a* is described by Ampère's law in the form of Eq. 1, while the situation of Fig. 1*b* is described by Eq. 4. In the first case, it is the current through the surface that gives the magnetic field, while in the second case, it is the changing electric flux through the surface that gives the magnetic field. In general, we must account for *both* ways of producing a magnetic field: (*a*) by a current and (*b*) by a changing electric flux, and so we must modify Ampère's law to read

$$\oint \mathbf{B} \cdot d\mathbf{s} = \mu_0 i + \mu_0\epsilon_0 \frac{d\Phi_E}{dt}. \tag{5}$$

Maxwell is responsible for this important generalization of Ampère's law. It is a central and vital contribution, as we have pointed out earlier.

In Chapter 35 we assumed that no changing electric fields were present so that the term $d\Phi_E/dt$ in Eq. 5 was

---

* Our system of units requires that we insert the constants ϵ_0 and μ_0 in Eq. 4. In some unit systems they would not appear.

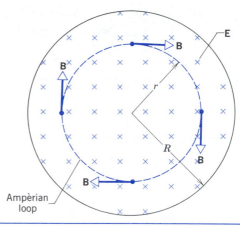

Figure 2   The induced magnetic field **B**, shown at four points, produced by the changing electric field **E** inside the capacitor of Fig. 1. The electric field is increasing in magnitude. Compare with Fig. 12 of Chapter 36.

---

zero. In the discussion of Fig. 1*b* we assumed that there were no conduction currents in the space containing the electric field. Thus the term i in Eq. 5 is zero in that case. We see now that each of these situations is a special case. If there were fine wires connecting the two plates in Fig. 1*b*, there would be contributions from both terms in Eq. 5.†

An alternative way of interpreting Eq. 5 is suggested by Fig. 2, which shows the electric field in the region between the capacitor plates of Fig. 1. We now take our Ampèrian loop to be a circular path in this region. On the right side of Eq. 5, the term i is zero, but the term $d\Phi_E/dt$ is not zero. In fact, the flux through the surface is positive if the field lines are as shown, and the flux is increasing (corresponding to the electric field increasing) as positive charge is transported into the left-hand plate of Fig. 1. The line integral of **B** around the loop must also be positive, and the directions of **B** must be as shown in Fig. 2.

Figure 2 suggests a beautiful example of the symmetry of nature. A changing *magnetic* field induces an *electric* field (Faraday's law); now we see that a changing *electric* field induces a *magnetic* field. Carefully compare Fig. 2

---

† There is a third way of setting up a magnetic field: the use of magnetic materials. For example, Eq. 5 does not account for the entire field in a solenoid wound on an iron core. The effect of the magnetic material can be included by adding a third term to Eq. 5, which can then be written

$$\oint \mathbf{B} \cdot d\mathbf{s} = \mu_0 i + \mu_0\epsilon_0 \frac{d\Phi_E}{dt} + \mu_0 i_M,$$

where i_M is the *magnetization current*, which can be regarded as the additional current that must flow through the empty solenoid to produce the same field that the current i produces when the magnetic material is present. We assume that no magnetic materials are present, so that this term need not be included.

with Fig. 12 of Chapter 36, which illustrates the production of an electric field by a changing magnetic field. In each case the appropriate flux Φ_B or Φ_E is *increasing*. However, experiment shows that the lines of **E** in Fig. 12 of Chapter 36 are *counterclockwise,* whereas those of **B** in Fig. 2 are *clockwise.* This difference requires that the minus sign of Eq. 3 be omitted from Eq. 4.

Sample Problem 1   A parallel-plate capacitor with circular plates is being charged as in Fig. 2. (*a*) Derive an expression for the induced magnetic field at various radii r in the region between the plates. Consider both $r \leq R$ and $r \geq R$. (*b*) Find B at $r = R$ for $dE/dt = 10^{12}$ V/m·s and $R = 5.0$ cm.

Solution   (*a*) From Eq. 4,

$$\oint \mathbf{B} \cdot d\mathbf{s} = \mu_0 \epsilon_0 \frac{d\Phi_E}{dt},$$

we can write, for $r \leq R$,

$$(B)(2\pi r) = \mu_0 \epsilon_0 \frac{d}{dt}[(E)(\pi r^2)] = \mu_0 \epsilon_0 \pi r^2 \frac{dE}{dt}.$$

Solving for B yields

$$B = \tfrac{1}{2}\mu_0 \epsilon_0 r \frac{dE}{dt} \qquad (r \leq R).$$

For $r \geq R$, Eq. 4 yields

$$(B)(2\pi r) = \mu_0 \epsilon_0 \frac{d}{dt}[(E)(\pi R^2)] = \mu_0 \epsilon_0 \pi R^2 \frac{dE}{dt},$$

or

$$B = \frac{\mu_0 \epsilon_0 R^2}{2r} \frac{dE}{dt} \qquad (r \geq R).$$

(*b*) At $r = R$ the two equations for B reduce to the same expression, or

$$
\begin{aligned}
B &= \tfrac{1}{2}\mu_0 \epsilon_0 R \frac{dE}{dt} \\
&= \tfrac{1}{2}(4\pi \times 10^{-7} \text{ T·m/A})(8.9 \times 10^{-12} \text{ C}^2/\text{N·m}^2) \\
&\quad \times (5.0 \times 10^{-2} \text{ m})(10^{12} \text{ V/m·s}) \\
&= 2.8 \times 10^{-7} \text{ T} = 280 \text{ nT}.
\end{aligned}
$$

This shows that the induced magnetic fields in this example are so small that they can scarcely be measured with simple apparatus, in sharp contrast to induced *electric* fields (Faraday's law), which can be demonstrated easily. This experimental difference is in part due to the fact that induced emfs can easily be multiplied by using a coil of many turns. No technique of comparable simplicity exists for magnetic fields. In experiments involving oscillations at very high frequencies, dE/dt can be very large, resulting in significantly larger values of the induced magnetic field.

Displacement Current

Equation 5 shows that the term $\epsilon_0 \, d\Phi_E/dt$ has the dimensions of a current. Even though no motion of charge is involved, there are advantages in giving this term the

name *displacement current.** The displacement current i_d is defined according to

$$i_d = \epsilon_0 \frac{d\Phi_E}{dt}. \tag{6}$$

Thus we can say that a magnetic field can be set up either by a conduction current i or by a displacement current i_d, and we can rewrite Eq. 5 as

$$\oint \mathbf{B} \cdot d\mathbf{s} = \mu_0(i + i_d). \tag{7}$$

The concept of displacement current permits us to retain the notion that *current is continuous,* a principle established for steady conduction currents in Section 32-1. In Fig. 1*b*, for example, a conduction current i enters the positive plate and leaves the negative plate. The *conduction* current is *not* continuous across the capacitor gap because no charge is transported across this gap. However, the displacement current i_d in the gap proves to be exactly equal to i, thus retaining the concept of the continuity of current.

Let us calculate the displacement current i_d in the capacitor gap of Fig. 1*b*. The charge q on the plates is related to the electric field E in the gap by Eq. 3 of Chapter 31,

$$q = \epsilon_0 EA.$$

Differentiating gives

$$i = \frac{dq}{dt} = \epsilon_0 \frac{d(EA)}{dt}.$$

The quantity EA is the electric flux Φ_E, and thus

$$i = \epsilon_0 \frac{d\Phi_E}{dt}.$$

Comparison with Eq. 6 shows

$$i = i_d.$$

Thus the displacement current in the gap equals the conduction current in the wires, which shows that the current is continuous.

When the capacitor is fully charged, the conduction current drops to zero (no current flows in the wires). The electric field between the plates becomes constant; thus $dE/dt = 0$, and so the displacement current also drops to zero.

The displacement current i_d, given by Eq. 6, has a direction as well as a magnitude. The direction of the conduction current i is that of the conduction current density vector **j**. Similarly, the direction of the displacement current i_d is that of the displacement current density vector **j**$_d$, which, as we deduce from Eq. 6, is just $\epsilon_0(d\mathbf{E}/dt)$.

---

* The word "displacement" was introduced for historical reasons. It has nothing to do with our previous use of displacement to indicate the position of a particle.

The right-hand rule applied to $\mathbf{j}_d$ gives the direction of the associated magnetic field, just as it does for the conduction current density $\mathbf{j}$.

Sample Problem 2 What is the displacement current for the situation of Sample Problem 1?

Solution From Eq. 6, the definition of displacement current,

$$i_d = \epsilon_0 \frac{d\Phi_E}{dt} = \epsilon_0 \frac{d}{dt}[(E)(\pi R^2)] = \epsilon_0 \pi R^2 \frac{dE}{dt}$$

$$= (8.9 \times 10^{-12} \ \text{C}^2/\text{N} \cdot \text{m}^2)(\pi)(5 \times 10^{-2} \ \text{m})^2(10^{12} \ \text{V/m} \cdot \text{s})$$

$$= 0.070 \ \text{A} = 70 \ \text{mA}.$$

This is a reasonably large current, yet we determined in Sample Problem 1 that it produces a magnetic field of only 280 nT. A current of 70 mA flowing in a thin wire would produce a large magnetic field near the surface of the wire, easily detectable by a compass needle.

The difference is *not* caused by the fact that one current is a conduction current and the other is a displacement current. Under the same conditions, both kinds of current are equally effective in generating a magnetic field. The difference arises because the conduction current, in this case, is confined to a thin wire but the displacement current is spread out over an area equal to the surface area of the capacitor plates. Thus the capacitor behaves like a "fat wire" of radius 5 cm, carrying a (displacement) current of 70 mA. Its largest magnetic effect, which occurs at the capacitor edge, is much smaller than would be the case at the surface of a thin wire. (See also Problem 12.)

40-3 MAXWELL'S EQUATION

Equation 5 completes our presentation of the ba-ua-tions of electromagnetism, called *Maxwell's equations.* They are summarized in Table 2, which replaces the "tentative" set of Table 1, the difference between the two sets being the "missing" displacement current term in Eq. IV of Table 1. Also in Table 2 we list the crucial experiments that led to each of Maxwell's equations. This list of experiments reminds us that Maxwell's equations were not mere theoretical speculations but were developed to explain the results of laboratory experiments.

Maxwell described his theory of electromagnetism in a lengthy *Treatise on Electricity and Magnetism,* published in 1873, just six years before his death. The *Treatise* does not contain the four equations in the form in which we have presented them. It was British physicist Oliver Heaviside (1850–1925), described as "an unemployed, largely self-educated former telegrapher," who pointed out the symmetry between $\mathbf{E}$ and $\mathbf{B}$ in the equations and cast the four equations in the form in which we know them today.

Let us consider some features of these remarkable equations.

1. *Symmetry.* The inclusion of the displacement current term in Eq. IV of Table 2 certainly makes Eqs. III and IV look more similar, thereby improving the symmetry of the set of equations. They are still not completely symmet-

TABLE 2 BASIC EQUATIONS OF ELECTROMAGNETISM (MAXWELL'S EQUATIONS)[a]

Number	Name	Equation	Describes	Crucial Experiment	Chapter Reference
I	Gauss' law for electricity	$\oint \mathbf{E} \cdot d\mathbf{A} = q/\epsilon_0$	Charge and the electric field	(a) Like charges repel and unlike charges attract, as the inverse square of their separation. (b) A charge on an insulated conductor moves to its outer surface.	29
II	Gauss' law for magnetism	$\oint \mathbf{B} \cdot d\mathbf{A} = 0$	The magnetic field	It has thus far not been possible to verify the existence of a magnetic monopole.	37
III	Faraday's law of induction	$\oint \mathbf{E} \cdot d\mathbf{s} = -d\Phi_B/dt$	The electrical effect of a changing magnetic field	A bar magnet, thrust through a closed loop of wire, will set up a current in the loop.	36
IV	Ampère's law (as extended by Maxwell)	$\oint \mathbf{B} \cdot d\mathbf{s} = \mu_0 i$ $+\mu_0 \epsilon_0 \ d\Phi_E/dt$	The magnetic effect of a current or a changing electric field	(a) A current in a wire sets up a magnetic field near the wire.	35
				(b) The speed of light can be calculated from purely electromagnetic measurements.	41

[a] Written on the assumption that no dielectric or magnetic material is present.

ric, however. A completely symmetric set would result if the existence of individual magnetic charges (monopoles) were confirmed. If such magnetic charges were discovered, experiments with them would be possible. By analogy with our previous development of electromagnetism, two experiments come to mind. One experiment, similar to Coulomb's original experiment, would be to measure the force between monopoles to determine whether it obeyed an inverse-square law. If so, then Eq. II could be written $\oint \mathbf{B} \cdot d\mathbf{A} = \mu_0 q_{\mathrm{m}}$. This form of Gauss' law for magnetism would assert that the flux of the magnetic field through any closed surface is proportional to the net magnetic charge q_{m} enclosed by the surface. In this case Eqs. I and II would become more symmetric.

The second experiment, similar to that of Oersted, would be to show that a current of magnetic charges produces an electric field. In this case we would add to the right side of Eq. III a term involving $i_{\mathrm{m}} = dq_{\mathrm{m}}/dt$, the current of magnetic charges. With this addition, Eqs. III and IV would become more symmetric.

So far there is no conclusive evidence for magnetic monopoles, so the above experiments remain speculations, and the set of equations in Table 2 is our best description of the properties of electric and magnetic fields. However, note how easily a major discovery such as the magnetic monopole could be incorporated into the basic equations of electromagnetism.

2. *Electromagnetic waves.* The four equations of Table 1 were of course known long before Maxwell's time (he was born in the year that Faraday discovered the law of induction). Taken together, they suggest no new effects beyond the original experiments they represent. It is only when the displacement current is added that new physics emerges. This new physics includes the prediction of the existence of electromagnetic waves, which were discovered experimentally by Heinrich Hertz in 1888, 15 years after Maxwell's *Treatise* was published. In the next chapter, we show how electromagnetic waves, which can transport energy and momentum through empty space by means of electromagnetic fields, follow from Maxwell's equations.

3. *Electromagnetism and relativity.* We have already suggested in the introduction to this chapter that Maxwell's equations are for electromagnetism what Newton's laws are for mechanics. There is, however, an important difference. Einstein's theory of relativity was presented in 1905, more than 30 years after Maxwell's work and more than 200 years after Newton's. Relativity necessitated major changes in Newton's laws for motion at speeds near that of light, but *no changes whatever were required in Maxwell's equations.* Maxwell's equations are totally consistent with the special theory of relativity, and in fact Einstein's theory grew out of his thinking about Maxwell's equations. In the language of physics, we say that Maxwell's equations are invariant under a Lorentz trans-

formation, but Newton's laws are not. (See Section 35-7 for a discussion of the relativistic transformation of **E** and **B** fields.)

40-4   MAXWELL'S EQUATIONS AND CAVITY OSCILLATIONS   *(Optional)*

There are many situations involving electromagnetic fields that we can use as a demonstration of Maxwell's equations. We delay until Chapter 41 any discussion of tests involving electromagnetic waves. Here we discuss a resonant cavity, which can be considered to be an electromagnetic oscillator with distributed elements.

By way of analogy, consider the *acoustic* resonant cavity of Fig. 3. (An organ pipe, closed at both ends, is an example of such an acoustic resonator.) In a simple oscillator, such as a block on a spring or an LC circuit, we can "lump" the stored energy into separate items: the kinetic energy of the block and the potential energy of the spring, or the stored magnetic energy of the inductor and the stored electric energy of the capacitor. In the acoustic resonator, such a division is not possible. Every tiny element of the gas within the tube has both potential and kinetic energy; such a system is said to have *distributed* elements. The electromagnetic resonant cavity likewise has distributed elements.

One characteristic of a distributed system is that it has a large number of resonant modes (the lumped system by contrast having few, often just one). Figure 3 shows the fundamental mode of the acoustic cavity. It illustrates a series of "snapshots" of the pressure and velocity variations throughout one cycle. Note that the pressure and the velocity vary with time and with location along the tube. There is a pressure antinode at each end of a closed pipe. Where the pressure variation is greatest, the velocity is zero (Figs. 3a and 3e), in analogy with the block–spring system at its maximum displacement. When the pressure is uniform, the velocities have their maximum values (Figs. 3c and 3g).

As shown by the bar graphs accompanying each "snapshot" of Fig. 3, the energy of the resonator oscillates between the kinetic energy of the moving gas and the potential energy associated with the compression and rarefaction of the gas. The energy may be all potential (Figs. 3a and 3e), all kinetic (Figs. 3c and 3g), or a mixture of both.

By analogy with the acoustic cavity, we can consider a cylindrical electromagnetic resonant cavity. Instead of pressure and velocity, we describe the state of the resonator by its electric and magnetic fields. Imagine the ends of the cavity to be a parallel-plate capacitor. To start the field oscillations, we connect a sinusoidally varying source of emf. This gives rise to a changing electric field in the cavity. As was the case in Fig. 2, the changing electric field causes a magnetic field, and thus within the cavity there are magnetic and electric fields that vary with location and with time.

Like the acoustic resonator, the electromagnetic resonator stores its energy in two forms; in this case they are the energies associated with the electric field and the magnetic field. Every element of volume of the cavity contributes to both kinds of energy, and thus the electromagnetic cavity has distributed elements.

Figure 4 shows, in similar fashion to Fig. 3, a series of "snapshots" of the cavity illustrating the electric and magnetic fields at

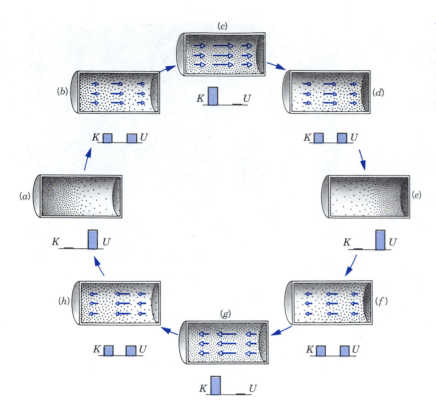

Figure 3 Eight stages in a cycle of oscillation of a cylindrical acoustic resonant cavity (such as a closed organ pipe). The bar graphs below each figure show the kinetic energy K and the potential energy U. The arrows represent the directed velocities of small volume elements of the gas.

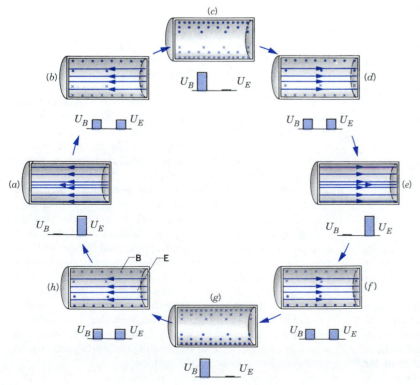

Figure 4 Eight stages in a cycle of oscillation of a cylindrical electromagnetic resonant cavity. The bar graphs below each figure show the stored electric energy U_E and magnetic energy U_B. The lines of **B** are circles concentric with the axis, and the lines of **E** are parallel to the axis. Compare with Fig. 3; both figures are examples of oscillations involving distributed elements.

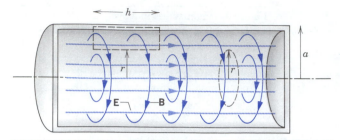

Figure 5 A more detailed representation of a cylindrical electromagnetic resonant cavity at the instant of Fig. 4d. The dashed rectangle is used to apply Faraday's law, and the dashed circle is used for Ampère's law.

Figure 6 The interior of the 2-mile Stanford Linear Accelerator. The large vertical cylinder is one of the several hundred electromagnetic resonant cavities (klystrons) that supply the electric fields needed to accelerate the electrons. Each klystron produces a peak power of 67 MW.

various times during one cycle of oscillation of the fundamental mode. Note the oscillation of the energy between the two forms, corresponding to the electric and magnetic energy densities,

$$u_E = \tfrac{1}{2}\epsilon_0 E^2 \quad \text{and} \quad u_B = \frac{1}{2\mu_0} B^2.$$

Integrating over the volume of the cavity, we can find the total energy in each of the two forms.

Figure 5 shows a more detailed representation of the electric and magnetic fields at one particular instant of the oscillation, corresponding to Fig. 4d. Note from Fig. 4d that the magnetic field is decreasing, and the electric field is increasing. Let us apply Faraday's law,

$$\oint \mathbf{E} \cdot d\mathbf{s} = -\frac{d\Phi_B}{dt},$$

to the dashed rectangle of dimensions h and $a - r$. There is a definite magnetic flux Φ_B through this rectangular area, and this flux is decreasing with time because **B** is decreasing.

For a cavity made of conducting material, we can set **E** to zero for the upper leg of the integration path, which lies inside the cavity wall. Also, on the two side legs **E** and $d\mathbf{s}$ are at right angles, so $\mathbf{E} \cdot d\mathbf{s} = 0$ on that part of the rectangular path. The only contribution to the line integral of **E** around the perimeter of the rectangle comes from the lower segment, and so

$$\oint \mathbf{E} \cdot d\mathbf{s} = hE(r),$$

in which $E(r)$ is the value of E at a radius r from the axis of the cavity. Inserting this result for the line integral into Faraday's law, we obtain

$$E(r) = -\frac{1}{h}\frac{d\Phi_B}{dt}. \tag{8}$$

Equation 8 shows that $E(r)$ depends on the rate at which Φ_B through the path shown is changing with time and that it has its maximum magnitude when $d\Phi_B/dt$ is a maximum. This occurs when **B** is zero, that is, when **B** is changing its direction; recall that a sine or cosine is changing most rapidly (it has the steepest slope) at the instant it crosses the axis between positive and negative values. The electric field pattern in the cavity has its *maximum* value when the magnetic field is *zero* everywhere, consistent with Figs. 4a and 4e and with the concept of the interchange of energy between electric and magnetic fields. You can show, by applying Lenz' law, that the electric field in Fig. 5

indeed points to the *right,* as shown, if the magnetic field is *decreasing.*

Let us apply Ampère's law in the form

$$\oint \mathbf{B} \cdot d\mathbf{s} = \mu_0 i + \mu_0 \epsilon_0 \frac{d\Phi_E}{dt},$$

to the dashed circular path of radius r shown in the figure. No charge is transported through the area bounded by the circular path, so the conduction current i is zero. The line integral on the left is $(B)(2\pi r)$, and so the equation reduces to

$$B(r) = \frac{\mu_0 \epsilon_0}{2\pi r}\frac{d\Phi_E}{dt}. \tag{9}$$

Equation 9 shows that the magnetic field $B(r)$ is proportional to the rate at which the electric flux Φ_E through the ring is changing with time. The field $B(r)$ has its maximum value when $d\Phi_E/dt$ is at its maximum; this occurs when $\mathbf{E} = 0$, that is, when **E** is reversing its direction. Thus we see that **B** has its *maximum* value when **E** is *zero* for all points in the cavity. This is consistent with Figs. 4c and 4g and with the concept of the interchange of energy between electric and magnetic forms. A comparison with Fig. 2, which like Fig. 5 corresponds to an increasing electric field, shows that the lines of **B** are indeed clockwise, as viewed along the direction of the electric field.

Comparison of Eqs. 8 and 9 suggests the complete interdependence of **B** and **E** in the cavity. As the magnetic field changes with time, it induces the electric field in a way described by Faraday's law. The electric field, which also changes with time, induces the magnetic field in a way described by Maxwell's extension of Ampère's law. The oscillations, once established, sustain each other and would continue indefinitely were it not for losses due to production of internal energy in the conducting cavity walls or leakage of energy from openings that might be present in the walls. In Chapter 41 we show that a similar interplay of **B** and **E** occurs not only in standing electromagnetic waves in cavities but also in traveling electromagnetic waves, such as radio waves or visible light.

In a resonant acoustic cavity, such as an organ pipe, we provide a source of energy (for example, by directing a stream of air

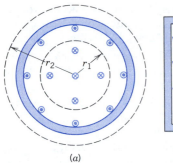

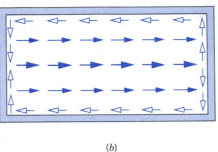

Figure 7 Sample Problem 3. Cross sections of the cavity of Figs. 4 and 5, showing (*a*) the conduction current coming up the walls and the displacement current going down the cavity volume, and (*b*) the displacement current (solid arrowheads) in the volume of the cavity and the conduction current (open arrowheads) in the walls. The arrows represent current densities. Note that the total current (conduction + displacement) is continuous; that is, it is possible to form closed current loops.

(*a*)   (*b*)

against a sharp edge), allow the standing wave to be established in the cavity with a frequency determined by the geometry of the cavity, and arrange for a portion of the energy of the wave to leave the pipe, where it is heard by the listener. In an electromagnetic cavity, the sequence of events is similar. The oscillations must be stimulated externally, such as by a current. A standing electromagnetic wave is established, whose frequency depends on the dimensions of the cylindrical cavity. A portion of the wave is then permitted to leave the cavity. A common use of such resonant cavities is in accelerators that produce beams of charged particles with high energies. Figure 6 shows the interior of the 2-mile electron accelerator at Stanford, in which a series of hundreds of resonant cavities (called *klystrons*) feeds electromagnetic waves into the accelerator. The electrons travel along the straight 2-mile path, subject to a sequence of accelerating electric fields, which boost the energies of the electrons to nearly 50 GeV. ∎

Sample Problem 3 In Fig. 5 analyze the currents (both conduction and displacement) that occur in the cavity (both in its conducting walls and within its volume). Show the relationship between these currents and the electric and magnetic fields and also show that, considering both conduction and displacement currents together, it is reasonable to conclude that current is continuous around closed loops.

Solution Figure 7 shows two views of the cavity, at an instant corresponding to that of Fig. 5. For simplicity, we do not show the **E** and **B** fields; the arrows represent currents. Because *E* is increasing in Figs. 5 and 7, the positive charge on the left end cap must be increasing. Thus there must be conduction currents in

the walls pointing from right to left in Fig. 7*b*. These currents are also shown by the dots (representing the tips of arrows) near the cavity walls in Fig. 7*a*.

Bearing in mind that $\epsilon_0 \, d\Phi_E/dt$ is a displacement current, we can write Eq. 9 as

$$B(r) = \frac{\mu_0}{2\pi r} \left(\epsilon_0 \frac{d\Phi_E}{dt} \right) = \frac{\mu_0}{2\pi r} i_d.$$

This equation stresses that **B** in the cavity is associated with a displacement current; compare Eq. 11 of Chapter 35, $B = \mu_0 i/2\pi r$. Applying the right-hand rule in Fig. 5 shows that the displacement current i_d must be directed into the plane of Fig. 7*a* if it is to be associated with the clockwise lines of **B** that are present.

The displacement current is represented in Fig. 7*b* by arrows that point to the right and in Fig. 7*a* by crosses that represent arrows entering the page. Figure 7 shows that the current is continuous, directed up the walls as a conduction current and then back down through the volume of the cavity as a displacement current. Applying Ampère's law as extended by Maxwell,

$$\oint \mathbf{B} \cdot d\mathbf{s} = \mu_0 (i_d + i), \qquad (10)$$

to the circular path of radius r_1 in Fig. 7*a*, we see that **B** at that path is due entirely to the displacement current, the conduction current *i* *within the path* being zero.

For the path of radius r_2, the *net* current enclosed is zero because the conduction current in the walls is exactly equal and opposite to the displacement current in the cavity volume. Since *i* equals i_d in magnitude, but is oppositely directed, it follows from Eq. 10 that *B* must be zero for all points outside the cavity, in agreement with observation.

QUESTIONS

1. In your own words explain why Faraday's law of induction (see Table 2) can be interpreted by saying "a changing magnetic field generates an electric field."

2. If a uniform flux Φ_E through a plane circular ring decreases with time, is the induced magnetic field (as viewed along the direction of **E**) clockwise or counterclockwise?

3. If (as is true) there are unit systems in which ϵ_0 and μ_0 do not appear, how can Eq. 1 be true?

4. Compare Tables 1 and 2. Is it enough to rely on the principle of symmetry alone or do we really need experimental verification for the "missing" term in Eq. IV?

5. Why is it so easy to show that "a changing magnetic field produces an electric field" but so hard to show in a simple way that "a changing electric field produces a magnetic field"?

6. In Fig. 2 consider a circle with $r > R$. How can a magnetic

field be induced around this circle, as Sample Problem 1 shows? After all, there is no electric field at the location of this circle and $dE/dt = 0$ here.

7. In Fig. 2, **E** is into the figure and is increasing in magnitude. Find the direction of **B** if, instead, (a) **E** is into the figure and decreasing, (b) **E** is out of the figure and increasing, (c) **E** is out of the figure and decreasing, and (d) **E** remains constant.

8. In Fig. 9c of Chapter 38, a displacement current is needed to maintain continuity of current in the capacitor. How can one exist, considering that there is no charge on the capacitor?

9. (a) In Fig. 2 what is the direction of the displacement current i_d? In this same figure, can you find a rule relating the directions (b) of **B** and **E** and (c) of **B** and $d\mathbf{E}/dt$?

10. What advantages are there in calling the term $\epsilon_0 d\Phi_E/dt$ in Eq. IV, Table 2, a displacement current?

11. Can a displacement current be measured with an ammeter? Explain.

12. Why are the magnetic fields of conduction currents in wires so easy to detect but the magnetic effects of displacement current in capacitors so hard to detect?

13. In Table 2 there are three kinds of apparent lack of symmetry in Maxwell's equations. (a) The quantities ϵ_0 and/or μ_0 appear in I and IV but not in II and III. (b) There is a minus sign in III but no minus sign in IV. (c) There are missing "magnetic pole terms" in II and III. Which of these represent genuine lack of symmetry? If magnetic monopoles were discovered, how would you rewrite these equations to include them? (*Hint*: Let q_m be the magnetic pole strength, analogous to the quantum of charge e; what SI units would q_m have?)

14. Maxwell's equations as displayed in Table 2 are written on the assumption that no dielectric materials are present. How should the equations be written if this restriction is removed?

15. List as many (a) lumped and (b) distributed mechanical oscillating systems as you can.

16. A coil has a measured inductance L. In a practical case it also has a capacitance C, adjacent windings behaving as "plates." The coil can be made to oscillate at a certain frequency without attaching it to an external capacitor. Is this a case of distributed elements? Do you suppose that it can oscillate at more than one frequency? Discuss.

17. Can a given circuit element (a capacitor, say) behave like a "lumped" element under some circumstances and like a "distributed" element under others?

18. Are oscillating systems (mechanical, say) *either* lumped or distributed? That is, is there no middle ground? (a) Consider a lumped system such as an idealized block–spring arrangement. How might you change it physically to make it more distributed? (b) Consider a distributed system such as a vibrating string. How might you change it physically to make it more lumped?

19. Discuss the periodic flow of energy, if any, from point to point in an acoustic resonant cavity.

20. An air-filled acoustic resonant cavity and an electromagnetic resonant cavity of the same size have resonant frequencies that are in the ratio of 10^6 or so. Which has the higher frequency and why?

21. Electromagnetic cavities are often silver-plated on the inside. Why?

22. At what parts of the cycle will (a) the conduction current and (b) the displacement current in the cavity of Fig. 4 be zero?

23. Discuss the time variation during one complete cycle of the charges that appear at various points on the inner walls of the oscillating electromagnetic cavity of Fig. 4.

24. Would you expect that the arrangement of the magnetic and electric fields in Fig. 5 is the only possible arrangement? If there are other arrangements, would you expect them to have higher or lower frequencies than that shown in Fig. 5?

25. In connection with Fig. 7, in what sense can the end caps be considered as capacitor plates? In what sense can the cylindrical walls be considered as an inductor? (*Note*: Figure 7 is clearly a case of distributed elements but there must be a smooth transition between distributed and lumped elements.)

26. (a) In Fig. 5 is it possible to apply Faraday's law usefully to the dashed circle? (b) Is it possible to apply Ampère's law usefully to the dashed rectangle? Discuss.

PROBLEMS

Section 40-1 The Basic Equations of Electromagnetism

1. By substituting numerical values of ϵ_0 and μ_0 used in previous chapters, verify the numerical value of the speed of light from Eq. 1 and show that the equation is dimensionally correct.

2. (a) Show that $\sqrt{\mu_0/\epsilon_0} = 377\ \Omega$ (called the "impedance of free space"). (b) Show that the angular frequency of ordinary 60 Hz AC is 377 rad/s. (c) Compare (a) with (b). Do you think that this coincidence is the reason that 60 Hz was originally chosen as the frequency for AC generators? Recall that, in Europe, 50 Hz is used.

Section 40-2 Induced Magnetic Fields and the Displacement Current

3. For the situation of Sample Problem 1, where is the induced magnetic field equal to one-half of its maximum value?

4. Prove that the displacement current in a parallel-plate capacitor can be written

$$i_d = C\frac{dV}{dt}.$$

5. You are given a 1.0-μF parallel-plate capacitor. How would you establish an (instantaneous) displacement current of 1.0 mA in the space between its plates?

6. In Sample Problem 1 show that the *displacement current density* j_d is given, for $r < R$, by

$$j_d = \epsilon_0 \frac{dE}{dt}.$$

7. A parallel-plate capacitor has square plates 1.22 m on a side as in Fig. 8. There is a charging current of 1.84 A flowing into (and out of) the capacitor. (*a*) What is the displacement current through the region between the plates? (*b*) What is dE/dt in this region? (*c*) What is the displacement current through the square dashed path between the plates? (*d*) What is $\oint \mathbf{B} \cdot d\mathbf{s}$ around this square dashed path?

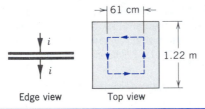

Edge view   Top view

Figure 8   Problem 7.

8. Figure 9 shows the plates P_1 and P_2 of a circular parallel-plate capacitor of radius R. They are connected as shown to long straight wires in which a constant conduction current i exists. Also shown are three hypothetical circles of radius r, two of them outside the capacitor and one between the plates. Show that the magnetic field at the circumference of each of these circles is given by

$$B = \frac{\mu_0 i}{2\pi r}.$$

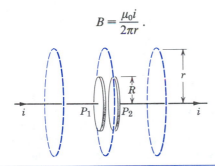

Figure 9   Problem 8.

9. A uniform electric field collapses to zero from an initial strength of 0.60 MV/m in a time of 15 μs in the manner

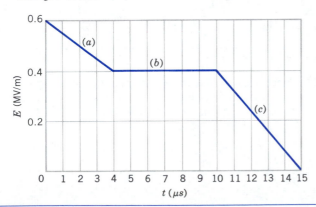

Figure 10   Problem 9.

shown in Fig. 10. Calculate the displacement current, through a 1.9-m² region perpendicular to the field, during each of the time intervals (*a*), (*b*), and (*c*) shown on the graph. (Ignore the behavior at the ends of the intervals.)

10. In Sample Problem 1 show that the expressions derived for $B(r)$ can be written

$$B(r) = \frac{\mu_0 i_d}{2\pi r} \qquad (r \geq R),$$

$$B(r) = \frac{\mu_0 i_d r}{2\pi R^2} \qquad (r \leq R).$$

Note that these expressions are of just the same form as those derived in Chapter 35 except that the conduction current i has been replaced by the displacement current i_d.

11. A parallel-plate capacitor with circular plates 21.6 cm in diameter is being charged as in Fig. 2. The displacement current density throughout the region is uniform, into the paper in the diagram, and has a value of 1.87 mA/cm². (*a*) Calculate the magnetic field B at a distance $r = 53.0$ mm from the axis of symmetry of the region. (*b*) Calculate dE/dt in this region.

12. In 1929 M. R. Van Cauwenberghe succeeded in measuring directly, for the first time, the displacement current i_d between the plates of a parallel-plate capacitor to which an alternating potential difference was applied, as suggested by Fig. 2. He used circular plates whose effective radius was 40.0 cm and whose capacitance was 100 pF. The applied potential difference had a maximum value V_m of 174 kV at a frequency of 50.0 Hz. (*a*) What maximum displacement current was present between the plates? (*b*) Why was the applied potential difference chosen to be as high as it is? (The delicacy of these measurements is such that they were only performed in a direct manner more than 60 years after Maxwell enunciated the concept of displacement current! The experiment is described in *Journal de Physique*, No. 8, 1929.)

13. Suppose that a circular-plate capacitor has a radius R of 32.1 mm and a plate separation of 4.80 mm. A sinusoidal potential difference with a maximum value of 162 V and a frequency of 60.0 Hz is applied between the plates. Find the maximum value of the induced magnetic field at $r = R$.

14. The capacitor in Fig. 11 consisting of two circular plates with radius $R = 18.2$ cm is connected to a source of emf $\mathcal{E} = \mathcal{E}_m \sin \omega t$, where $\mathcal{E}_m = 225$ V and $\omega = 128$ rad/s. The maximum value of the displacement current is $i_d = 7.63$ μA. Neglect fringing of the electric field at the edges of the plates. (*a*) What is the maximum value of the current i? (*b*) What is the maximum value of $d\Phi_E/dt$, where Φ_E is the electric flux through the region between the plates? (*c*) What

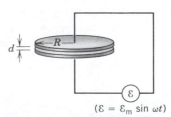

$(\mathcal{E} = \mathcal{E}_m \sin \omega t)$

Figure 11   Problem 14.

is the separation d between the plates? (*d*) Find the maximum value of the magnitude of **B** between the plates at a distance $r = 11.0$ cm from the center.

Section 40-3 Maxwell's Equations

15. Collect and tabulate expressions for the following four quantities, considering both $r < R$ and $r > R$. Place the derivations side by side and study them as interesting applications of Maxwell's equations to problems having cylindrical symmetry. (*a*) $B(r)$ for a current i in a long wire of radius R. (*b*) $E(r)$ for a long uniform cylinder of charge of radius R. (*c*) $B(r)$ for a parallel-plate capacitor, with circular plates of radius R, in which E is changing at a constant rate. (*d*) $E(r)$ for a cylindrical region of radius R in which a uniform magnetic field B is changing at a constant rate.

16. A long cylindrical conducting rod with radius R is centered on the x axis as shown in Fig. 12. A narrow saw cut is made in the rod at $x = b$. A conduction current i, increasing with time and given by $i = \alpha t$, flows toward the right in the rod; α is a (positive) proportionality constant. At $t = 0$ there is no charge on the cut faces near $x = b$. (*a*) Find the magnitude of the charge on these faces, as a function of time. (*b*) Use Eq. I in Table 2 to find E in the gap as a function of time. (*c*) Sketch the lines of **B** for $r < R$, where r is the distance from the x axis. (*d*) Use Eq. IV in Table 2 to find $B(r)$ in the gap for $r < R$. (*e*) Compare the above answer with $B(r)$ in the rod for $r < R$.

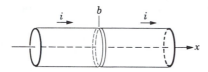

Figure 12   Problem 16.

17. Two adjacent closed paths *abefa* and *bcdeb* share the common edge *be* as shown in Fig. 13. (*a*) We may apply $\oint \mathbf{E} \cdot d\mathbf{s} = -d\Phi_B/dt$ (Eq. III of Table 2) to each of these two closed paths separately. Show that, from this alone, Eq. III is *automatically* satisfied for the composite path *abcdefa*. (*b*) Repeat using Eq. IV. (*c*) This relation is called a "self-consistency" property; why must each of Maxwell's equations be self-consistent?

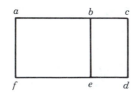

Figure 13   Problem 17.

18. Two adjacent closed parallelepipeds share a common face as shown in Fig. 14. (*a*) We may apply $\oint \mathbf{E} \cdot d\mathbf{A} = q/\epsilon_0$ (Eq. I in Table 2) to each of these two closed surfaces separately.

Show that, from this alone, Eq. I is *automatically* satisfied for the composite closed surface. (*b*) Repeat using Eq. II. See Problem 17.

Figure 14   Problem 18.

Section 40-4 Maxwell's Equations and Cavity Oscillations

19. What would be the dimensions of a cylindrical electromagnetic resonant cavity (like that described in the text) operating, in the fundamental mode, at 60 Hz, the frequency of household alternating current? (The angular frequency is given by $\omega = 2.41c/a$, where a is the radius of the cavity, in meters.)

20. A cylindrical electromagnetic cavity 4.8 cm in diameter and 7.3 cm long is oscillating in the mode shown in Fig. 4. (*a*) Assume that, for points on the axis of the cavity, $E_m = 13$ kV/m. The frequency of oscillation is 2.4 GHz. For such axial points, what is the maximum rate $(dE/dt)_m$ at which E changes? (*b*) Assume that the average value of $(dE/dt)_m$, for all points over a cross section of the cavity, is one-half the value found above for axial points. On this assumption, what is the maximum value of B at the cylindrical surface of the cavity?

21. In microscopic terms the principle of continuity of current may be expressed as

$$\oint (\mathbf{j} + \mathbf{j_d}) \cdot d\mathbf{A} = 0,$$

in which **j** is the conduction current density and $\mathbf{j_d}$ is the displacement current density. The integral is to be taken over any closed surface; the equation essentially says that whatever current flows into the enclosed volume must also flow out. (*a*) Apply this equation to the surface shown by the dashed lines in Fig. 15 shortly after switch S is closed. (*b*) Apply it to various surfaces that may be drawn in the cavity of Fig. 7, including some that cut the cavity walls.

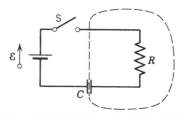

Figure 15   Problem 21.

22. Sketch diagrams like those shown in Fig. 4 showing a cycle of oscillation of a cylindrical electromagnetic resonant cavity operating, not in the fundamental mode as in that figure, but in the first overtone.

CHAPTER 41

ELECTROMAGNETIC WAVES

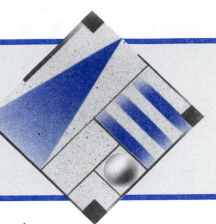

Maxwell's equations, the topic of the previous chapter, not only summarize the properties of electric and magnetic fields in a compact manner; the equations also lead to entirely new phenomena. Perhaps the supreme achievement of Maxwell's theory was the prediction of the existence of electromagnetic waves and the realization that light could be understood as a type of electromagnetic wave.

In this chapter, we show how the equations for electromagnetic waves follow from Maxwell's equations, and we discuss the properties of the resulting waves. Our description of electromagnetic waves uses many of the terms we used previously in our study of mechanical waves in Chapters 19 and 20; we consider sinusoidal waves, and we describe them in such familiar terms as amplitude, frequency, wavelength, and phase velocity. Here we consider electromagnetic waves in general terms, and in the next chapter we consider the properties of light waves in more detail. These two chapters form a bridge to the study of optics in the chapters that follow.

41-1 THE ELECTROMAGNETIC SPECTRUM*

In Maxwell's time, light and the adjoining infrared and ultraviolet radiations were the only known types of electromagnetic radiations. Today the electromagnetic spectrum, shown in Fig. 1, includes a broad range of different kinds of radiations from a variety of sources. From Maxwell's theory we conclude that, even though these radiations differ greatly in their properties, in their means of production, and in the ways we observe them, they share other features in common: they all can be described in

---

* The word *spectrum* comes from a Latin word meaning "form" or "appearance." Other familiar words from the same root include "spectacle" and "species." Newton introduced the word to describe the rainbow-like image that resulted when a beam of sunlight passed through a glass prism. Today we speak of the *electromagnetic spectrum* to indicate the many different kinds of electromagnetic radiation, classified according to their frequency or wavelength on a scale from small to large. We also speak of the *political spectrum,* which similarly indicates the broad range of political views on a scale from ultraconservative to ultraliberal.

terms of electric and magnetic fields, and they all travel through vacuum with the same speed (the speed of light). In fact, from the fundamental point of view, they differ *only* in wavelength or frequency. The names given to the various regions of the spectrum in Fig. 1 have to do only with the way the different types of waves are produced or observed; they have nothing to do with any fundamental property of the waves. Other than the difference in their wavelengths, there is no experimental way to distinguish a wave in the visible region from one in the infrared region; the waves have identical forms and identical mathematical descriptions. There are no gaps in the spectrum, nor are there sharp boundaries between the various categories. (Certain regions of the spectrum are assigned by law for commercial or other uses, such as TV, AM, or FM broadcasting.)

Let us consider some of these types of electromagnetic radiation in more detail.

1. *Light.* The visible region of the spectrum is the one most familiar to us, because as a species we have adapted receptors (eyes) that are sensitive to the most intense electromagnetic radiation emitted by the Sun, the closest extraterrestrial source. The limits of the wavelength of the visible region are from about 400 nm (violet) to about 700 nm (red).

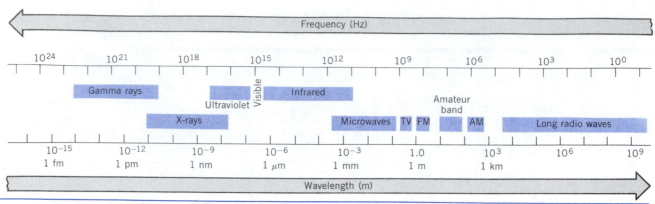

Figure 1 The electromagnetic spectrum. Note that both the wavelength and frequency scales are logarithmic.

Light is often emitted when the outer (or valence) electrons in atoms change their state of motion; for this reason, such transitions in the state of the electron are called *optical transitions.* The color of the light tells us something about the atoms or the object from which it was emitted. The study of the light emitted from the Sun and from distant stars gives information about their composition.

2. *Infrared.* Infrared radiation, which has wavelengths longer than the visible (from 0.7 μm to about 1 mm), is commonly emitted by atoms or molecules when they change their rotational or vibrational motion. Often this change occurs as a change in the internal energy of the emitting object and is observed as a change in the internal energy of the object that detects the radiation. In this case, infrared radiation is an important means of heat transfer and is sometimes called *heat radiation.* The warmth you feel when you place your hand near a glowing light bulb is primarily a result of the infrared radiation emitted from the bulb and absorbed by your hand. All objects emit electromagnetic radiation (called "thermal radiation;" see Chapter 49 of the extended text) because of their temperature. Objects of temperatures in the range we normally encounter (say, 3 K to 3000 K) emit their most intense thermal radiation in the infrared region of the spectrum. Mapping the infrared radiation from space has given us information that supplements that obtained from the visible radiation (Fig. 2).

3. *Microwaves.* Microwaves can be regarded as short radio waves, with typical wavelengths in the range 1 mm to 1 m. They are commonly produced by electromagnetic oscillators in electric circuits, as in the case of microwave ovens. Microwaves are often used to transmit telephone conversations; Fig. 3 shows a microwave station that serves to relay telephone calls. Microwaves also reach us from extraterrestrial sources. The most abundant component is the *microwave background radiation,* which is believed to be the electromagnetic radiation associated with the "Big Bang" fireball that marked the birth of the universe some 10^{10} years ago; as the universe expanded and

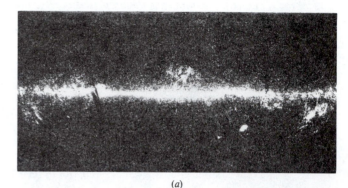

(a)

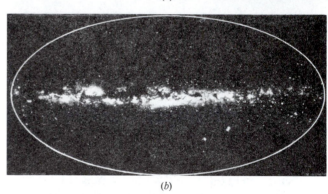

(b)

Figure 2 (a) Infrared image of our Milky Way Galaxy taken by the IRAS satellite. (b) Visible-light image of the Milky Way. Parts of the visible image, especially those near the center of the galaxy, are obscured by dust clouds, which do not affect the infrared image. The two large objects below the galaxy and right of center are the Large and Small Magellanic Clouds, which are companion galaxies to the Milky Way.

cooled, the wavelength of this radiation was stretched until it is now in the microwave region, with a peak wavelength of about 1 mm. Neutral hydrogen atoms, which populate the regions between the stars in our galaxy, are another common extraterrestrial source of microwaves, emitting radiation with a wavelength of 21 cm.

4. *Radio waves.* Radio waves have wavelengths longer than 1 m. They are produced from terrestrial sources

Figure 3 A microwave relay station, which receives and then re-transmits signals that carry long-distance telephone calls.

Figure 4   One of the 27 25-m diameter radiotelescope antenna dishes at the Very Large Array near Socorro, New Mexico. The 27 dishes are arranged on a Y-shaped railroad track, each leg of which is 10 miles long. This arrangement is equivalent to a single dish 20 miles in diameter.

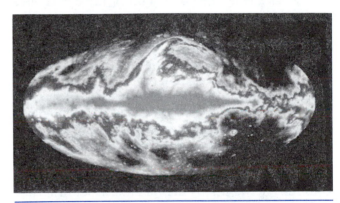

Figure 5   A radio image of the Milky Way Galaxy. (Compare with Fig. 2.) This image was taken at a wavelength of 73 cm. This radiation mostly originates from high-energy electrons that are deflected by magnetic fields in the galaxy. Note the intense emissions out of the plane of the Galaxy, which do not appear in Fig. 2.

through electrons oscillating in wires of electric circuits. By carefully choosing the geometry of these circuits, as in an antenna, we can control the distribution in space of the emitted radiation (if the antenna acts as a transmitter) or the sensitivity of the detector (if the antenna acts as a receiver). Traveling outward at the speed of light, the expanding wavefront of TV signals transmitted on Earth since about 1950 has now reached approximately 400 stars, carrying information to their inhabitants, if any, about our civilization.

Radio waves reach us from extraterrestrial sources, the Sun being a major source that often interferes with radio or TV reception on Earth. Jupiter is also an active source of radio emissions. Mapping the radio emissions from extraterrestrial sources, known as *radio astronomy,* has provided information about the universe that is often not obtainable using optical telescopes. Furthermore, because the Earth's atmosphere does not absorb strongly at

radio wavelengths, radio astronomy provides certain advantages over optical, infrared, or microwave astronomy on Earth. Figure 4 shows an example of a radiotelescope, and Fig. 5 gives a typical result of the observation of our galaxy at radio wavelengths.

One of the most startling discoveries of radio astronomy was the existence of pulsed sources of radio waves, first observed in 1968. These objects, known as *pulsars,* emit very short bursts of radio waves separated in time by intervals of the order of seconds. This time interval between pulses is extremely stable, varying by less than 10^{-9} s. Pulsars are believed to originate from rotating neutron stars, in which electrons trapped by the magnetic field experience large centripetal accelerations owing to the rotation. The highly directional radio emissions sweep by the Earth like a searchlight beacon as the star rotates. Pulsars have been observed over the full range of the spectrum, including visible and x-ray wavelengths.

5. *Ultraviolet.* The radiations of wavelengths shorter than the visible begin with the ultraviolet (1 nm to 400 nm), which can be produced in atomic transitions of the outer electrons as well as in radiation from thermal sources such as the Sun. Because our atmosphere absorbs strongly at ultraviolet wavelengths, little of this radiation from the Sun reaches the ground. However, the principal agent of this absorption is atmospheric ozone, which has been depleted in recent years as a result of chemical reactions with fluorocarbons released from aerosol sprays, refrigeration equipment, and other sources. Brief exposure to ultraviolet radiation causes common sunburn, but long-term exposure can lead to more serious effects, including skin cancer. Ultraviolet astronomy is done using observatories carried into Earth orbit by satellites.

6. *X rays.* X rays (typical wavelengths 0.01 nm to 10 nm) can be produced with discrete wavelengths in individual transitions among the inner (most tightly bound) electrons of an atom, and they can also be produced when charged particles (such as electrons) are decelerated. X-ray wavelengths correspond roughly to the spacing between the atoms of solids; therefore scattering of x rays from materials is a useful way of studying their structure. X rays can easily penetrate soft tissue but are stopped by bone and other solid matter; for this reason they have found wide use in medical diagnosis.

X-ray astronomy, like ultraviolet astronomy, is done with orbiting observatories. Most stars, such as the Sun, are not strong x-ray emitters; however, in certain systems consisting of two nearby stars orbiting about their common center of mass (called a *binary* system), material from one star can be heated and accelerated as it falls into the other, emitting x rays in the process. Although confirming evidence is not yet available, it is believed that the more massive member of certain x-ray binaries may be a black hole.

7. *Gamma rays.* Gamma rays are electromagnetic radiations with the shortest wavelengths (less than 10 pm). They are the most penetrating of electromagnetic radiations, and exposure to intense gamma radiation can have a harmful effect on the human body. These radiations can be emitted in transitions of an atomic nucleus from one state to another and can also occur in the decays of certain elementary particles; for example, a neutral pion can decay into two gamma rays according to

$$\pi^0 \rightarrow \gamma + \gamma,$$

and an electron and a positron (the antiparticle of the electron) can mutually annihilate into two gamma rays:

$$e^- + e^+ \rightarrow \gamma + \gamma.$$

In general, each such process emits gamma rays of a unique wavelength. In gamma-ray astronomy, detection of such radiations (and measurement of their wavelength) serves as evidence of particular nuclear processes in the universe.

From the above descriptions, you can see that there are both natural and artificial sources of all types of electromagnetic radiations, and you can also see that the study of electromagnetic radiations at all wavelengths has in recent years been used to provide a more accurate picture of the structure and evolution of the universe.

In describing the emission of electromagnetic radiation as a wave phenomenon, we are concentrating on one particular aspect. We consider the atoms of the system that emits the radiation to behave cooperatively; for example, the participation of the electrons from many atoms is necessary for the emission of light from the hot filament of a light bulb. As an alternative, we can study the emission of electromagnetic radiation by a single atom. In this case, we focus our attention on one bundle of electromagnetic energy (called a *quantum*), and we generally observe the radiation not as a smoothly varying wave but as a concentrated bundle of electromagnetic energy. Some experiments seem inconsistent with the wave interpretation and can be explained only in terms of particles or quanta of electromagnetic radiation. In this chapter, we emphasize the wave aspects and ignore these particle aspects. In Chapter 49 of the extended version of this text we consider the particle aspects, which are complementary to the wave aspects in forming a complete understanding of electromagnetic radiation.

41-2 GENERATING AN ELECTROMAGNETIC WAVE

An electric charge at rest sets up a pattern of electric field lines. A charge in motion at constant speed sets up a pattern of magnetic field lines, in addition to the electric field lines. Once a steady condition has been reached (that is, after the charge is in motion and the fields are established in space), there is an energy density in space associated with the electric and magnetic fields, but the energy density remains constant in time. No signal, other than evidence of its presence, is transported from the charge to distant points; there is no transport of energy or momentum, and there is no electromagnetic radiation.

If, on the other hand, you were to wiggle the charge back and forth, you could send signals to a distant friend who had the necessary equipment to detect changes in the electric and magnetic fields. With a pre-arranged code, you could send information by wiggling the charge at a certain rate or in a certain direction. In this case, you would be signaling by means of an electromagnetic wave. To produce this wave, it is necessary to accelerate the

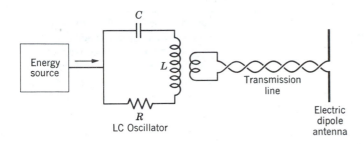

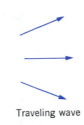

Figure 6   An arrangement for generating a traveling electromagnetic wave, in this case a shortwave radio wave.

charge. That is, *static charges and charges in motion at constant velocity do not radiate; accelerated charges radiate.* Put another way, the uniform motion of the charge is a current that does not change with time, and the accelerated motion of the charge is correspondingly a current that varies with time; thus we can equivalently regard radiation as being produced by time-varying currents.

In the laboratory, a convenient way of generating an electromagnetic wave is to cause currents in wires to vary with time. We assume for simplicity a sinusoidal time variation. Figure 6 shows a circuit that might be used for this purpose. It consists of an oscillating *RLC* circuit, with an external source that restores the energy that is dissipated in the circuit or carried away by the radiation. The current in the circuit varies sinusoidally with the resonant circular frequency ω, which is approximately $1/\sqrt{LC}$ if the resistive losses are small (see Section 38–7). The oscillator is coupled through a transformer to a *transmission line*, which serves to carry the current to an *antenna*. (Coaxial cables, which carry TV signals to many homes, are common examples of transmission lines.)

The geometry of the antenna determines the geometrical properties of the radiated electric and magnetic fields. We assume a *dipole antenna,* which, as Fig. 6 shows, can be considered simply as two straight conductors. Charges surge back and forth in these two conductors at the frequency ω, driven by the oscillator. We can regard the antenna as an oscillating electric dipole, in which one branch carries an instantaneous charge q, and the other branch carries $-q$. The charge q varies sinusoidally with

time and changes sign every half cycle. The charges are certainly accelerated as they move back and forth in the antenna, and as a result the antenna is a source of *electric dipole radiation.* At any point in space there are electric and magnetic fields that vary sinusoidally with time.*

Figure 7 shows a series of "snapshots" that give a schematic picture of how the radiation field is formed. Only the electric field is shown; the corresponding magnetic field can be inferred from the current in the conductors using the right-hand rule. Figure 8 gives a more complete view of the electromagnetic wave that might be generated by the antenna. The figure is a slice through the *xy* plane; to obtain a more complete picture of the field, we must imagine the figure to be rotated about the *y* axis. We assume that we observe the field at distances from the dipole that are large compared with its dimensions and compared with the wavelength of the radiation; the field observed under these conditions is called the *radiation field.* At smaller distances, we would observe the more complicated *near field,* which we do not discuss here. Note that the field "breaks away" from the antenna and forms closed loops, in contrast to the static field of an electric dipole, in which the field lines always start on positive charges and end on negative charges.

---

* Most of the radiations we encounter, from radio waves to light to x rays and gamma rays, are of the dipole type. Radio and TV antennas are generally designed to transmit dipole radiation. Individual atoms and nuclei can often be considered as oscillating dipoles from the standpoint of emitting radiation.

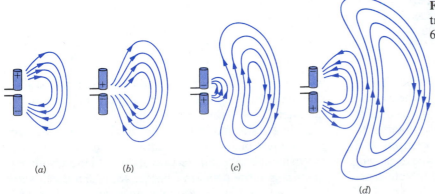

Figure 7   Successive stages in the emission of a traveling wave, such as from the antenna of Fig. 6. Only the electric field patterns are shown.

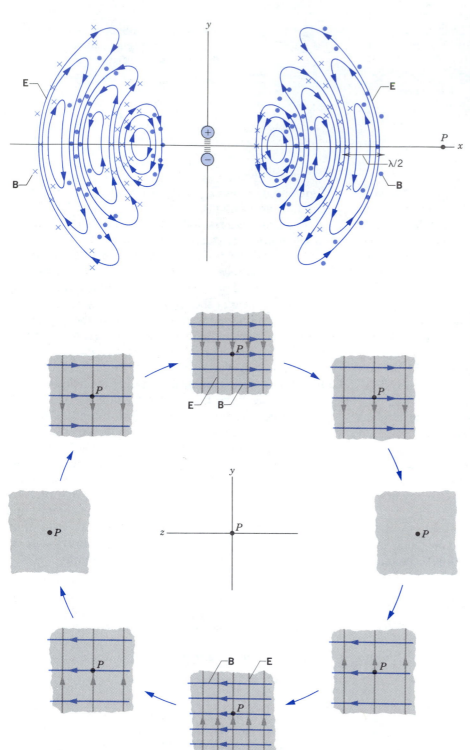

Figure 8   The **E** and **B** fields radiated from an electric dipole. The fields are shown at distances that are large compared with the dimensions of the dipole. A distant observer at point *P* records a plane wave moving in the *x* direction.

Figure 9   Eight cyclical "snapshots" of the plane electromagnetic wave radiated from the oscillating dipole of Fig. 8 observed at point *P*. The direction of travel of the wave (*x* direction in Fig. 8) is out of the plane of the page. Lines of **E** are vertical, and lines of **B** are horizontal.

An alternative view of the radiation field is given in Fig. 9, which represents a series of "snapshots" of the electric and magnetic fields sweeping past an observer located at point *P* on the *x* axis of Fig. 8. We assume the observer to be located so far from the dipole that the wavefronts can be regarded as planes. As is always the case, the density of field lines indicates the strength of the field. Note especially that (1) **E** and **B** are in phase (they both reach their

maxima at the same instant, and they both are zero at the same instant), and (2) **E** and **B** are perpendicular to one another. These conclusions follow from an analysis of traveling electromagnetic waves in free space using Maxwell's equations, which is summarized in Section 41–3.

An additional characteristic of this radiation, which we discuss in more detail in Chapter 48, is that it is *linearly polarized*; that is, the **E** vector everywhere points along the same line, in this case the y direction. This remains true at all points on the x axis and at all times. This direction of polarization is determined by the direction of the axis of the dipole. Light emitted by a disordered collection of atoms, such as the filament of an ordinary light bulb, is unpolarized; in effect, the individual atomic dipoles are randomly oriented in space. In a laser, the atoms are stimulated to emit radiation with their dipole axes aligned; laser light is therefore polarized.

41-3   TRAVELING WAVES AND MAXWELL'S EQUATIONS

The preceding discussion has given us a qualitative picture of one type of electromagnetic traveling wave. In this section we consider the mathematical description of the wave, which we show to be consistent with Maxwell's equations. In doing so, we also show that the speed of such waves in empty space is the same as the speed of light, which leads us to conclude that light is itself an electromagnetic wave.

Suppose the observer in Fig. 8 is at such a great distance from the oscillating dipole that the wavefronts passing point P (shown in Fig. 9) are planes. The lines of **E** are parallel to the y axis, and the lines of **B** are parallel to the z axis. We write the **E** and **B** fields in the usual mathematical form of a sinusoidal traveling wave (see Section 19-3):

$$E(x,t) = E_m \sin (kx - \omega t), \tag{1}$$

$$B(x,t) = B_m \sin (kx - \omega t). \tag{2}$$

Here ω is the angular frequency associated with the oscillating dipole, and the wave number k has its usual meaning of $2\pi/\lambda$. If the wave propagates with phase speed c, then ω and k are related according to $c = \omega/k$. Figure 10 represents the sinusoidal oscillation of the **E** and **B** fields as a function of x at a particular instant of time.

The amplitudes E_m and B_m will later be shown to be related to one another. Note that in writing these equations for the magnitudes of **E** and **B** we have assumed that E and B are in phase; that is, the phase constants in Eqs. 1 and 2 have the same value (which we have taken to be zero). Later we show that this choice follows from Maxwell's equations.

Figure 11 shows a three-dimensional "snapshot" of a plane wave traveling along the x direction. It represents a different way of showing the same wave illustrated in Fig. 10. Let us consider the wave as it passes through the thin rectangular box at point P in Fig. 11. In Fig. 12 we have redrawn two sections through the three-dimensional wave. Figure 12a shows a section parallel to the xy plane; the lines of **E** lie in this section, while the lines of **B** are perpendicular to it. Figure 12b shows a section parallel to the xz plane; here the lines of **B** lie in the section, and the lines of **E** are perpendicular.

As the wave passes over the fixed rectangle in Fig. 12a, the magnetic flux through the rectangle changes, which must give rise to an induced electric field around the rectangle, according to Faraday's law of induction. This induced electric field is simply the electric field associated with the traveling wave.

To see this in more detail, let us apply Lenz' law to the induction process. The flux Φ_B for the shaded rectangle of

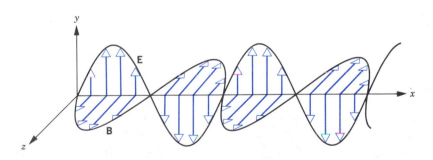

Figure 10   A linearly polarized, sinusoidally varying plane wave propagating in the x direction. The figure represents a snapshot at a particular time.

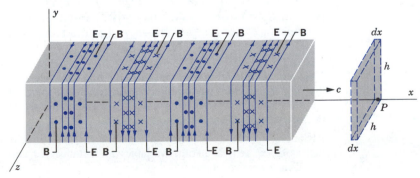

Figure 11   Another representation of the plane wave of Fig. 10. Energy is transported through a hypothetical thin rectangular box at P. Note that at all points of the wave, the vector $\mathbf{E} \times \mathbf{B}$ points in the direction in which the wave is moving.

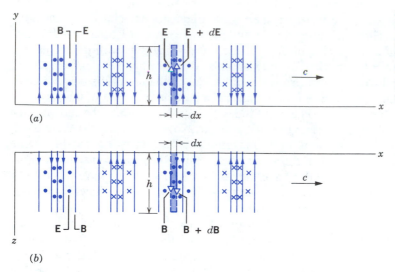

Figure 12   (*a*) The wave of Fig. 11 viewed in the *xy* plane. As the wave sweeps past, the magnetic flux through the shaded rectangle changes, inducing an electric field. (*b*) The wave of Fig. 11 viewed in the *xz* plane. As the wave sweeps past, the electric flux through the shaded rectangle changes, inducing a magnetic field.

Fig. 12*a* is *decreasing* with time, because the wave is moving through the rectangle to the right, and a region of weaker magnetic field is moving into the rectangle. The induced field acts to oppose this change, which means that if we imagine the boundary of the shaded rectangle to be a conducting loop, a *counterclockwise* induced current would appear in it. This current would induce a field **B** that, within the rectangle, would point out of the page, thus opposing the decrease in Φ_B. There is of course no conducting loop, but the net induced electric field would be consistent with this explanation, because the larger field $E + dE$ on the right side of the loop would give rise to a net counterclockwise current. Thus the electric field configuration in Fig. 12*a* is consistent with the concept that it is induced by the changing magnetic field.

In similar fashion, as the wave passes over the shaded rectangle in Fig. 12*b*, the electric flux through the rectangle changes, thereby giving rise to an induced magnetic field. (This effect depends on the displacement current term in Eq. IV of Table 2 in Chapter 40, and you can now see its importance in Maxwell's modified form of Ampère's law.) The induced magnetic field is simply the magnetic field associated with the traveling wave.

You can see that the variations in **E** and **B** are intimately connected with one another: a varying **E** field gives rise to a varying **B** field, which in turn gives rise to a varying **E** field, and so on. In this way the electric and magnetic fields of the wave sustain one another through empty space, and no medium is required for the wave to propagate.

Mathematical Description

For a more detailed analysis, let us apply Faraday's law of induction,

$$\oint \mathbf{E} \cdot d\mathbf{s} = -\frac{d\Phi_B}{dt}, \tag{3}$$

going counterclockwise around the shaded rectangle of Fig. 12*a*. There is no contribution to the integral from the top or bottom of the rectangle because **E** and *d***s** are at right angles here. The integral then becomes

$$\oint \mathbf{E} \cdot d\mathbf{s} = (E + dE)h - Eh = dE\, h.$$

The flux Φ_B for the rectangle is*

$$\Phi_B = (B)(dx\, h),$$

where B is the magnitude of **B** at the rectangular strip and $dx\, h$ is the area of the strip. Differentiating gives

$$\frac{d\Phi_B}{dt} = h\, dx\, \frac{dB}{dt}.$$

From Eq. 3 we then have

$$dE\, h = -h\, dx\, \frac{dB}{dt},$$

or

$$\frac{dE}{dx} = -\frac{dB}{dt}. \tag{4}$$

Actually, both B and E are functions of x and t; see Eqs. 1 and 2. In evaluating dE/dx, we assume that t is constant because Fig. 12*a* is an "instantaneous snapshot." Also, in evaluating dB/dt we assume that x is constant since what is required is the time rate of change of B at a particular place, the strip in Fig. 12*a*. The derivatives under these circumstances are *partial derivatives*,† and a somewhat

---

* We use a right-hand rule for the sign of the flux: if the fingers of the right hand point in the direction in which we integrate around the path, then the thumb indicates the direction in which the field through the enclosed area gives a positive flux.

† In taking a partial derivative with respect to a certain variable, such as $\partial E/\partial x$, we treat all other variables (for instance, y, z, and t) as if they were constants.

different notation is used for them; see, for example, Sections 19–4 and 19–5. In this notation Eq. 4 becomes

$$\frac{\partial E}{\partial x} = -\frac{\partial B}{\partial t}. \tag{5}$$

The minus sign in this equation is appropriate and necessary, for, although E is increasing with x at the site of the shaded rectangle in Fig. 12a, B is decreasing with t. Since $E(x,t)$ and $B(x,t)$ are known (see Eqs. 1 and 2), Eq. 5 reduces to

$$kE_{\rm m} \cos (kx - \omega t) = \omega B_{\rm m} \cos (kx - \omega t).$$

If we had used different phase constants in Eqs. 1 and 2, the cosine terms in this equation would be out of phase, and the two sides could not be equal at all x and t. Equation 5, which follows directly from applying Maxwell's equations, shows that E and B must be in phase.

Eliminating the cosine term, we obtain

$$\frac{E_{\rm m}}{B_{\rm m}} = \frac{\omega}{k} = c. \tag{6}$$

The ratio of the amplitudes of the electric and the magnetic components of the wave is the speed c of the wave. From Eqs. 1 and 2 we see that the ratio of the amplitudes is the same as the ratio of the instantaneous values, or

$$E = cB. \tag{7}$$

This important result will be useful in later sections.

We now turn our attention to Fig. 12b, in which the electric flux Φ_E for the shaded rectangle is decreasing with time as the wave moves through it. According to Maxwell's modified form of Ampère's law (with $i = 0$, because there is no conduction current in a traveling electromagnetic wave),

$$\oint \mathbf{B} \cdot d\mathbf{s} = \mu_0 \epsilon_0 \frac{d\Phi_E}{dt}, \tag{8}$$

this changing flux induces a magnetic field at points around the periphery of the rectangle.

Comparison of the shaded rectangles in Fig. 12 shows that for each the appropriate flux, Φ_B or Φ_E, is *decreasing* with time. However, if we proceed counterclockwise around the upper and lower shaded rectangles, we see that $\oint \mathbf{E} \cdot d\mathbf{s}$ is *positive*, whereas $\oint \mathbf{B} \cdot d\mathbf{s}$ is *negative*, as we show below. This is as it should be. Comparing Fig. 12b of Chapter 36 with Fig. 2 of Chapter 40, we note that although the fluxes Φ_B and Φ_E in those figures are changing with time in the same way (both are increasing), the lines of the induced **E** and **B** fields circulate in opposite directions.

The integral in Eq. 8, evaluated by proceeding counterclockwise around the shaded rectangle of Fig. 12b, is

$$\oint \mathbf{B} \cdot d\mathbf{s} = -(B + dB)h + Bh = -h \, dB,$$

where B is the magnitude of **B** at the left edge of the strip and $B + dB$ is its magnitude at the right edge.

The flux Φ_E through the rectangle of Fig. 12b is

$$\Phi_E = (E)(h \, dx).$$

Differentiating gives

$$\frac{d\Phi_E}{dt} = h \, dx \frac{dE}{dt}.$$

Thus we can write Eq. 8 as

$$-h \, dB = \mu_0 \epsilon_0 \left(h \, dx \frac{dE}{dt} \right)$$

or, substituting partial derivatives,

$$-\frac{\partial B}{\partial x} = \mu_0 \epsilon_0 \frac{\partial E}{\partial t}. \tag{9}$$

Again, the minus sign in this equation is appropriate and necessary, for, although B is increasing with x at the site of the shaded rectangle in Fig. 12b, E is decreasing with t.

Combining this equation with Eqs. 1 and 2, we find

$$-kB_{\rm m} \cos (kx - \omega t) = -\mu_0 \epsilon_0 \omega E_{\rm m} \cos (kx - \omega t),$$

or

$$\frac{E_{\rm m}}{B_{\rm m}} = \frac{k}{\mu_0 \epsilon_0 \omega} = \frac{1}{\mu_0 \epsilon_0 c}. \tag{10}$$

Eliminating $E_{\rm m}/B_{\rm m}$ between Eqs. 6 and 10 gives

$$c = \frac{1}{\sqrt{\mu_0 \epsilon_0}}. \tag{11}$$

Substituting numerical values, we obtain

$$c = \frac{1}{\sqrt{(4\pi \times 10^{-7} \, \text{T} \cdot \text{m/A})(8.9 \times 10^{-12} \, \text{C}^2/\text{N} \cdot \text{m}^2)}}$$
$$= 3.0 \times 10^8 \, \text{m/s},$$

which is the speed of light in free space! This emergence of the speed of light from purely electromagnetic considerations is a crowning achievement of Maxwell's electromagnetic theory. Maxwell made this prediction before radio waves were known and before it was realized that light was electromagnetic in nature. His prediction led to the concept of the electromagnetic spectrum and to the discovery of radio waves by Heinrich Hertz in 1890. It permitted optics to be discussed as a branch of electromagnetism and allowed its fundamental laws to be derived from Maxwell's equations.

Because μ_0 is defined to be exactly $4\pi \times 10^{-7}$ H/m, and the speed of light is now given the exact value of 299,792,458 m/s, Eq. 11 permits us to obtain a defined value of ϵ_0:

$$\epsilon_0 = \frac{1}{c^2 \mu_0} = 8.85418782 \times 10^{-12} \, \text{C}^2/\text{N} \cdot \text{m}^2.$$

Curiously, Maxwell himself did not view the propagation of electromagnetic waves and electromagnetic phenomena in general, in anything like the terms suggested by, say, Fig. 11. Like all physicists of his day he believed firmly that space was permeated by a subtle substance called the *ether* and that electromagnetic phenomena could be accounted for in terms of rotating vortices in this ether.

It is a tribute to Maxwell's genius that, with such mechanical models in his mind, he was able to deduce the laws of electromagnetism that bear his name. These laws, as we have pointed out, not only required no change when Einstein's special theory of relativity came on the scene three decades later but, indeed, were strongly confirmed by that theory. Today, as discussed in Chapter 21, we no longer find it necessary to invoke the ether concept to explain the propagation of electromagnetic waves.

41-4 ENERGY TRANSPORT AND THE POYNTING VECTOR

Like any form of wave, an electromagnetic wave can transport energy from one location to another. Light from a bulb and radiant heat from a fire are common examples of energy flowing by means of electromagnetic waves.

The energy flow in an electromagnetic wave is commonly measured in terms of the rate of energy flow per unit area (or, equivalently, electromagnetic power per unit area). We describe the magnitude and direction of the energy flow in terms of a vector called the *Poynting vector*[*] **S**, defined from

$$S = \frac{1}{\mu_0} \mathbf{E} \times \mathbf{B}. \tag{12}$$

The vectors **E** and **B** refer to the fields of a wave at a particular point in space, and **S** indicates the Poynting vector at that point. Note that, according to our usual rules for the cross product of two vectors, **S** must be perpendicular to the plane formed by **E** and **B**, and the direction of **S** is determined by the right-hand rule. Check these directional relationships with the plane wave shown in Figs. 10 and 11; note that even though the directions of **E** and **B** may change, their cross product always points in the positive *x* direction, which is the direction of travel of the wave.

An electromagnetic wave can be uniquely specified by giving its **E** field and its direction of travel (which is the same as the direction of **S**). It is not necessary to give **B**, because the magnitude of **B** is determined from the magnitude of **E** using Eq. 7, and the direction of **B** can be found from the directions of **E** and **S** based on Eq. 12.

[*] The Poynting vector is named for John Henry Poynting (1852–1914), who first discussed its properties. Poynting was a British physicist who was known for his studies of electromagnetism and gravitation.

The dimension of *B* is the same as the dimension of *E/c*. Using this result and the dimensions of *E* and μ_0, you can show that the dimension of **S** is power per unit area. Its SI unit is watts/meter².

For the plane electromagnetic wave of Fig. 10, Eq. 12 reduces to

$$S = \frac{1}{\mu_0} EB, \tag{13}$$

which can also be written, using Eq. 7,

$$S = \frac{1}{\mu_0 c} E^2 \quad \text{or} \quad S = \frac{c}{\mu_0} B^2, \tag{14}$$

where *S*, *E*, and *B* are instantaneous values at the observation point. Let us show that these results are consistent with our previous results for the energy density associated with electric and magnetic fields in the special case of a plane wave. Consider the electromagnetic energy in the rectangular box of Fig. 11 as the wave passes through it. At any instant, the electromagnetic energy in the box is

$$dU = dU_E + dU_B = (u_E + u_B)(A\,dx), \tag{15}$$

where *A dx* is the volume of the box, and u_E and u_B are, respectively, the electric and magnetic energy densities. Using Eq. 28 of Chapter 31 for u_E and Eq. 32 of Chapter 38 for u_B, we obtain

$$dU = \left(\frac{1}{2} \epsilon_0 E^2 + \frac{1}{2\mu_0} B^2 \right) A\,dx. \tag{16}$$

Equation 7 (*E = cB*) can be used to eliminate *one E* in the first term and *one B* in the second term, which gives

$$dU = \left[\frac{1}{2} \epsilon_0 E(cB) + \frac{1}{2\mu_0} B \left(\frac{E}{c} \right) \right] A\,dx$$

$$= \frac{(\mu_0 \epsilon_0 c^2 + 1)(EBA\,dx)}{2\mu_0 c}.$$

From Eq. 11, however, $\mu_0 \epsilon_0 c^2 = 1$, so that

$$dU = \frac{EBA\,dx}{\mu_0 c}. \tag{17}$$

This energy *dU* passes through the box in a time *dt* equal to *dx/c*. The magnitude of *S*, given in terms of energy flow per unit time per unit area, is

$$S = \frac{dU}{dt\,A} = \frac{EBA\,dx}{(\mu_0 c)(dx/c)A} = \frac{1}{\mu_0} EB,$$

in agreement with Eq. 13.

This expression relates the magnitudes of *E*, *B*, and *S* at a particular instant of time. The frequencies of many electromagnetic waves (light waves, for instance) are so great that *E* and *B* fluctuate too rapidly for their time variation to be measured directly. In many experiments, therefore, we are more interested in knowing the *time average* of *S*, taken over one or more cycles of the wave. The time average $\overline{S}$ is also known as the *intensity I* of the wave.

From Eq. 14 and Eq. 1, we obtain

$$I = \bar{S} = \frac{1}{\mu_0 c}\,\overline{E^2} = \frac{1}{\mu_0 c}\,E_m^2\,\overline{\sin^2(kx - \omega t)}.$$

The time average of the $\sin^2$ over any whole number of cycles is $\frac{1}{2}$, and so

$$I = \bar{S} = \frac{1}{2\mu_0 c}\,E_m^2 = \frac{1}{2\mu_0}\,E_m B_m. \qquad (18)$$

The intensity may also be expressed in terms of the rms (root-mean-square) magnitudes of the fields. Recalling that $E_m = \sqrt{2}E_{rms}$, we obtain

$$I = \bar{S} = \frac{1}{\mu_0 c}\,E_{rms}^2 = \frac{1}{\mu_0}\,E_{rms} B_{rms}. \qquad (19)$$

Sample Problem 1 An observer is 1.8 m from a light source (of dimensions much smaller than 1.8 m) whose power output P is 250 W. Calculate the rms values of the electric and magnetic fields at the position of the observer. Assume that the source radiates uniformly in all directions.

Solution The intensity of the light at a distance r from the source is given by

$$I = \frac{P}{4\pi r^2},$$

where $4\pi r^2$ is the area of a sphere of radius r centered on the source. The intensity is also given by Eq. 19, so that

$$I = \frac{P}{4\pi r^2} = \frac{1}{\mu_0 c}\,E_{rms}^2.$$

The rms electric field is

$$E_{rms} = \sqrt{\frac{P\mu_0 c}{4\pi r^2}}$$
$$= \sqrt{\frac{(250\ \text{W})(4\pi \times 10^{-7}\ \text{H/m})(3.00 \times 10^8\ \text{m/s})}{(4\pi)(1.8\ \text{m})^2}}$$
$$= 48\ \text{V/m}.$$

The rms value of the magnetic field follows from Eq. 7 and is

$$B_{rms} = \frac{E_{rms}}{c} = \frac{48\ \text{V/m}}{3.00 \times 10^8\ \text{m/s}}$$
$$= 1.6 \times 10^{-7}\ \text{T} = 0.16\ \mu\text{T}.$$

Note that E_{rms} (= 48 V/m) is appreciable as judged by ordinary laboratory standards but that B_{rms} (= 0.16 μT) is quite small. This helps to explain why most instruments used for the detection and measurements of electromagnetic waves respond to the electric component of the wave. It is wrong, however, to say that the electric component of an electromagnetic wave is "stronger" than the magnetic component. You cannot compare quantities that are measured in different units. As we have seen, the electric and the magnetic components are on an absolutely equal basis as far as the propagation of the wave is concerned. Their average energies, which *can* be compared, are exactly equal.

41-5 MOMENTUM AND PRESSURE OF RADIATION *(Optional)*

Besides carrying energy, electromagnetic waves may also transport linear momentum. In other words, it is possible to exert a pressure (a *radiation pressure**) on an object by shining a light on it. Such forces must be small in relation to forces of our daily experience because we do not ordinarily notice them. We do not, after all, fall over backward when we raise a window shade in a dark room and let sunlight shine on us. Radiation pressure effects are, however, important in the life cycles of stars because of the incredibly high temperatures (2×10^7 K for our Sun) that we associate with stellar interiors. The first measurements of radiation pressure were made in 1901–1903 by Nichols and Hull in the United States and by Lebedev in Russia, about 30 years after the existence of such effects had been predicted theoretically by Maxwell.

Let a parallel beam of light fall on an object for a time t, the incident light being *entirely absorbed* by the object. The electric field of the light causes charges (electrons) in the material to move in a direction transverse to the direction of the beam. The force $q\mathbf{v} \times \mathbf{B}$ on these moving charges due to the *magnetic* field of the light is in the direction of the beam. The absorption of the light correspondingly transfers momentum in the beam direction to particles of the absorber. If energy U is absorbed, the momentum p delivered to the object during this time is given, according to Maxwell's prediction, by

$$p = \frac{U}{c} \qquad \text{(total absorption)}, \qquad (20)$$

where c is the speed of light. The direction of $\mathbf{p}$ is the direction of the incident beam. Later in this section we give a rigorous derivation of this result.

If the light energy U is *entirely reflected*, the momentum delivered will be twice that given above, or

$$p = \frac{2U}{c} \qquad \text{(total reflection)}. \qquad (21)$$

In the same way, twice as much momentum is delivered to an object when a perfectly elastic tennis ball is bounced from it as when it is struck by a perfectly inelastic ball (a lump of putty, say) of the same mass and speed. If the light energy U is partly reflected and partly absorbed, the delivered momentum is between U/c and $2U/c$.

Nichols and Hull in 1903 measured radiation pressures and verified Eq. 21 using a torsion balance technique. They allowed light to fall on mirror M as in Fig. 13; the radiation pressure caused the balance arm to turn through a measured angle θ, twisting the torsion fiber F. Given the torsion constant of the fiber, the experimenters could calculate a numerical value for this pressure. Nichols and Hull measured the intensity of their light beam by allowing it to fall on a blackened metal disk of known absorptivity and measuring the resulting temperature rise of this disk. In a particular run these experimenters measured a radiation pressure of 7.01×10^{-6} N/m^2; for their light

---

* See "Radiation Pressure," by G. E. Henry, *Scientific American,* June 1957, p. 99; see also "The Pressure of Laser Light," by Arthur Ashkin, *Scientific American,* February 1972, p. 63.

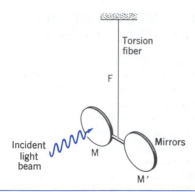

Figure 13 The arrangement of Nichols and Hull for measuring radiation pressure. Many details of this delicate experiment are omitted from the drawing.

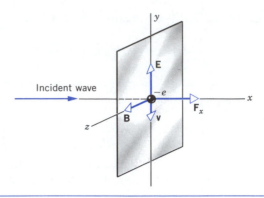

Figure 14 An incident plane light wave falls on an electron in a thin resistive sheet. Instantaneous values of **E**, **B**, the electron velocity **v**, and the radiation force $\mathbf{F}_x$ are shown.

beam, the value predicted, using Eq. 21, was 7.05×10^{-6} N/m², in excellent agreement. Assuming a mirror area of 1 cm², this represents a force on the mirror of only 7×10^{-10} N, a remarkably small force.

The success of the experiment of Nichols and Hull was the result in large part of the care they took to eliminate spurious deflecting effects caused by changes in the speed distribution of the molecules in the gas surrounding the mirror. These changes were brought about by the small rise in the temperature of the mirror as it absorbed light energy from the incident beam. This "radiometer effect" is responsible for the spinning action of the familiar toy radiometers when placed in a beam of sunlight. In a perfect vacuum such effects would not occur, but in the best vacuums available in 1903 radiometer effects were present and had to be taken specifically into account in the design of the experiment.

Sample Problem 2   A beam of light with an intensity $I (= \overline{S})$ of 12 W/cm² falls perpendicularly on a perfectly reflecting plane mirror of 1.5-cm² area. What force acts on the mirror?

Solution   From Newton's second law, the average force on the mirror is given by

$$F = \frac{\Delta p}{\Delta t},$$

where Δp is the momentum transferred to the mirror in time Δt. From Eq. 21 we have

$$\Delta p = \frac{2\,\Delta U}{c} = \frac{2\overline{S}A\,\Delta t}{c}.$$

We have then for the force

$$F = \frac{2\overline{S}A}{c} = \frac{(2)(12 \times 10^4 \text{ W/m}^2)(1.5 \times 10^{-4} \text{ m}^2)}{3.00 \times 10^8 \text{ m/s}}$$

$$= 1.2 \times 10^{-7} \text{ N}.$$

This is a very small force, about equal to the weight of a very small grain of table salt. Note that the pressure exerted by the radiation, which we can define in the usual way as force per unit area or F/A, is given by $2\overline{S}/c$.

Let us now derive Eq. 20 in the particular case of a plane electromagnetic wave traveling in the x direction and falling on a large thin sheet of a material of high resistivity as in Fig. 14. A small part of the incident energy is absorbed within the sheet, but most of it is transmitted if the sheet is thin enough. (Some of the incident energy is also reflected, but the reflected wave is of such low intensity that we can ignore it in the derivation that follows.)

The incident wave vectors **E** and **B** vary with time at the sheet as

$$\mathbf{E} = \mathbf{E_m} \sin \omega t \qquad (22)$$

and

$$\mathbf{B} = \mathbf{B_m} \sin \omega t, \qquad (23)$$

where **E** is parallel to the $\pm y$ axis and **B** is parallel to the $\pm z$ axis.

In Section 32-5 we saw that the effect of a (constant) electric force $(= -eE)$ on a conduction electron in a metal was to make it move with a (constant) drift speed v_d. The electron behaves as if it is immersed in a viscous fluid, the electric force acting on it being counterbalanced by a "viscous" force, which may be taken as proportional to the electron speed. Thus for a constant field E, after equilibrium is established,

$$eE = bv_d, \qquad (24)$$

where b is a resistive damping coefficient. The electron equilibrium speed, dropping the subscript d, is thus

$$v = \frac{eE}{b}. \qquad (25)$$

If the applied electric field varies with time and if the variation is slow enough, the electron speed can continually readjust itself to the changing value of E so that its speed continues to be given essentially by its equilibrium value (Eq. 25) at all times. These readjustments are more rapidly made in a medium of greater viscosity, just as a stone falling in air reaches a constant equilibrium rate of descent only relatively slowly but one falling in a viscous oil does so quite rapidly. We assume that the sheet in Fig. 14 is so viscous, that is, that its resistivity is so high, that Eq. 25 remains true even for the rapid oscillations of E in the incident light beam.

As the electron vibrates parallel to the y axis, it experiences a *second* force due to the *magnetic* component of the wave. This force $\mathbf{F}_x (= -e\mathbf{v} \times \mathbf{B})$ points in the x direction, being at right angles to the plane formed by **v** and **B**, that is, the *yz* plane. The instantaneous magnitude of F_x is given by

$$F_x = evB = \frac{e^2EB}{b} . \tag{26}$$

F_x always points in the positive *x* direction because **v** and **B** reverse their directions simultaneously; this force is, in fact, the mechanism by which the radiation pressure acts on the sheet of Fig. 14.

From Newton's second law, F_x is the rate dp_e/dt at which the incident wave delivers momentum to each electron in the sheet, or

$$\frac{dp_e}{dt} = \frac{e^2EB}{b} . \tag{27}$$

Momentum is delivered at this rate to every electron in the sheet and thus to the sheet itself. It remains to relate the momentum transfer to the sheet to the absorption of energy within the sheet.

The electric field component of the incident wave does work on each oscillating electron at an instantaneous rate given by

$$\frac{dU_e}{dt} = F_E v = (eE)\left(\frac{eE}{b}\right) = \frac{e^2E^2}{b} .$$

Note that the magnetic force F_x, always being at right angles to the velocity **v**, does no work on the oscillating electron. Equation 7 shows that for a plane wave in free space *B* and *E* are related by

$$E = Bc.$$

Substituting above for *one* of the *E*'s leads to

$$\frac{dU_e}{dt} = \frac{e^2EBc}{b} . \tag{28}$$

This equation represents the rate, per electron, at which energy is absorbed from the incident wave.

Comparing Eqs. 27 and 28 shows that

$$\frac{dp_e}{dt} = \frac{1}{c}\frac{dU_e}{dt} .$$

Integrating yields

$$\int \frac{dp_e}{dt} \, dt = \frac{1}{c} \int \frac{dU_e}{dt} \, dt,$$

or

$$p_e = \frac{U_e}{c} , \tag{29}$$

where p_e is the momentum delivered to a single electron in any given time *t* and U_e is the energy absorbed by that electron in the same time interval. Multiplying each side by the number of free electrons in the sheet leads to Eq. 20.

Although we derived Eq. 29 for a particular kind of absorber, no characteristics of the absorber—for example, the resistive damping coefficient *b*—remain in the final expression. This is as it should be because Eq. 29 is a general property of radiation absorbed by *any* material. ■

QUESTIONS

1. Electromagnetic waves reach us from the farthest depths of space. From the information they carry, can we tell what the universe is like at the present moment? At any selected time in the past?

2. If you are asked on an examination what fraction of the electromagnetic spectrum lies in the visible range, what would you reply?

3. List several ways in which radio waves differ from visible light waves. In what ways are they the same?

4. How would you characterize electromagnetic radiation that has a frequency of 10 kHz? 10^{20} Hz? A wavelength of 500 nm? 10 km? 0.50 nm?

5. What determines the desirable length and orientation of the "rabbit ears" antenna on a TV set?

6. How does a microwave oven cook food? You can boil water in a plastic bag in such an oven. How can this happen?

7. Speaking loosely we can say that the electric and the magnetic components of a traveling electromagnetic wave "feed on each other." What does this mean?

8. "Displacement currents are present in a traveling electromagnetic wave and we may associate the magnetic field component of the wave with these currents." Is this statement true? Discuss it in detail.

9. Can an electromagnetic wave be deflected by a magnetic field? By an electric field?

10. Why is Maxwell's modification of Ampère's law (that is, the term $\mu_0\epsilon_0 \, d\Phi_E/dt$ in Table 2, Chapter 40) needed to understand the propagation of electromagnetic waves?

11. Is it conceivable that electromagnetic theory might some day be able to predict the value of *c* (3×10^8 m/s), not in terms of μ_0 and ϵ_0, but directly and numerically without recourse to any measurement?

12. If you were to calculate the Poynting vector for various points in and around a transformer, what would you expect the field pattern to look like? Assume that an alternating potential difference has been applied to the primary windings and that a resistive load is connected across the secondary.

13. Name two historic experiments, in addition to the radiation pressure measurements of Nichols and Hull, in which a torsion balance was used. Both are described in this book, one in Volume 1 and one in Volume 2.

14. In Section 41-5 we stated that the force on the mirror in the radiation pressure experiment of Nichols and Hull (see Fig. 13) was about 7×10^{-10} N. Identify an object whose weight at the Earth's surface is about this magnitude.

15. Can an object absorb light energy without having linear momentum transferred to it? If so, give an example. If not, explain why.

16. When you turn on a flashlight does it experience any force associated with the emission of the light?

17. We associated energy and linear momentum with electromagnetic waves. Is angular momentum present also?

18. What is the relation, if any, between the intensity *I* of an electromagnetic wave and the magnitude *S* of its Poynting vector?

19. As you recline in a beach chair in the Sun, why are you so conscious of the thermal energy delivered to you but totally unresponsive to the linear momentum delivered from the same source? Is it true that when you catch a hard-pitched baseball, you are conscious of the energy delivered but not of the momentum?

20. When a parallel beam of light falls on an object, the momen-

tum transfers are given by Eqs. 20 and 21. Do these equations still hold if the light source is moving rapidly toward or away from the object at, perhaps, a speed of $0.1c$?

21. Radiation pressure is believed responsible for setting an upper limit (of about $100M_{Sun}$) to the mass of a star. Explain.

PROBLEMS

Section 41-1 The Electromagnetic Spectrum

1. Show that the frequency and wavelength markings on Fig. 1 obey the relation $\nu\lambda = c$.

2. Project Seafarer was an ambitious program to construct an enormous antenna, buried underground on a site about 4000 square miles in area. Its purpose was to transmit signals to submarines while they were deeply submerged. If the effective wavelength was 1.0×10^4 Earth radii, what would be (a) the frequency and (b) the period of the radiations emitted? Ordinarily, electromagnetic radiations do not penetrate very far into conductors such as seawater. Can you think of any reason why such ELF (extremely low frequency) radiations should penetrate more effectively? Think of the limiting case of zero frequency. (Why not transmit signals at zero frequency?)

3. (a) The wavelength of the most energetic x rays produced when electrons accelerated to 18 GeV in the Stanford Linear Accelerator slam into a solid target is 0.067 fm. What is the frequency of these x rays? (b) A VLF (very low frequency) radio wave has a frequency of only 30 Hz. What is its wavelength?

4. The radiation from a certain HeNe laser, although centered on 632.8 nm, has a finite "linewidth" of 0.010 nm. Calculate the linewidth in frequency units.

Section 41-2 Generating an Electromagnetic Wave

5. What inductance is required with a 17-pF capacitor in order to construct an oscillator capable of generating 550-nm (i.e., visible) electromagnetic waves? Comment on your answer.

6. Figure 15 shows an *LC* oscillator connected by a transmission line to an antenna of a *magnetic* dipole type. Compare with Fig. 6, which shows a similar arrangement but with an *electric* dipole type of antenna. (a) What is the basis for the names of these two antenna types? (b) Draw figures corresponding to Figs. 8 and 9 to describe the electromagnetic wave that sweeps past the observer at point *P* in Fig. 15.

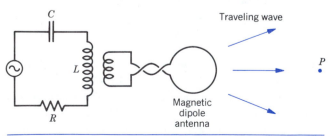

Figure 15 Problem 6.

Section 41-3 Traveling Waves and Maxwell's Equations

7. A certain plane electromagnetic wave has a maximum electric field of 321 μV/m. Find the maximum magnetic field.

8. The electric field associated with a plane electromagnetic wave is given by $E_x = 0$, $E_y = 0$, $E_z = E_0 \sin k(x - ct)$, where $E_0 = 2.34 \times 10^{-4}$ V/m and $k = 9.72 \times 10^6$ m^{-1}. The wave is propagating in the $+x$ direction. (a) Write expressions for the components of the magnetic field of the wave. (b) Find the wavelength of the wave.

9. Start from Eqs. 5 and 9 and show that $E(x, t)$ and $B(x, t)$, the electric and magnetic field components of a plane traveling electromagnetic wave, must satisfy the "wave equations"

$$\frac{\partial^2 E}{\partial t^2} = c^2 \frac{\partial^2 E}{\partial x^2}$$

and

$$\frac{\partial^2 B}{\partial t^2} = c^2 \frac{\partial^2 B}{\partial x^2}.$$

10. (a) Show that Eqs. 1 and 2 satisfy the wave equations displayed in Problem 9. (b) Show that any expressions of the form

$$E = E_m f(kx \pm \omega t)$$

and

$$B = B_m f(kx \pm \omega t),$$

where $f(kx \pm \omega t)$ denotes an arbitrary function, also satisfy these wave equations.

Section 41-4 Energy Transport and the Poynting Vector

11. Currently operating neodymium-glass lasers can provide 100 TW of power in 1.0-ns pulses at a wavelength of 0.26 μm. How much energy is contained in a single pulse?

12. Show, by finding the direction of the Poynting vector **S**, that the directions of the electric and magnetic fields at all points in Figs. 8, 9, 10, 11, and 12 are consistent at all times with the assumed directions of propagation.

13. Our closest stellar neighbor, α-Centauri, is 4.30 light-years away. It has been suggested that TV programs from our planet have reached this star and may have been viewed by the hypothetical inhabitants of a hypothetical planet orbiting this star. A TV station on Earth has a power output of 960 kW. Find the intensity of its signal at α-Centauri.

14. A plane electromagnetic wave is traveling in the negative y direction. At a particular position and time, the magnetic field is along the positive z axis and has a magnitude of 28 nT. What are the direction and magnitude of the electric field at that position and at that time?

15. The intensity of direct solar radiation that was unabsorbed by the atmosphere on a particular summer day is 130 W/m². How close would you have to stand to a 1.0-kW electric heater to feel the same intensity? Assume that the heater radiates uniformly in all directions.

16. (*a*) Show that in a plane traveling electromagnetic wave the average intensity, that is, the average rate of energy transport per unit area, is given by

$$\overline{S} = \frac{cB_m^2}{2\mu_0}.$$

(*b*) What is the average intensity of a plane traveling electromagnetic wave if B_m, the maximum value of its magnetic field component, is 1.0×10^{-4} T (= 1.0 gauss)?

17. You walk 162 m directly toward a street lamp and find that the intensity increases to 1.50 times the intensity at your original position. (*a*) How far from the lamp were you first standing? (The lamp radiates uniformly in all directions.) (*b*) Can you find the power output of the lamp? If not, explain why.

18. Prove that, for any point in an electromagnetic wave such as that of Fig. 10, the density of the energy stored in the electric field equals that of the energy stored in the magnetic field.

19. The maximum electric field at a distance of 11.2 m from a point light source is 1.96 V/m. Calculate (*a*) the maximum value of the magnetic field, (*b*) the intensity, and (*c*) the power output of the source.

20. Sunlight strikes the Earth, just outside its atmosphere, with an intensity of 1.38 kW/m². Calculate (*a*) E_m and (*b*) B_m for sunlight, assuming it to be a plane wave.

21. A cube of edge a has its edges parallel to the x, y, and z axes of a rectangular coordinate system. A uniform electric field **E** is parallel to the y axis and a uniform magnetic field **B** is parallel to the x axis. Calculate (*a*) the rate at which, according to the Poynting vector point of view, energy may be said to pass through each face of the cube and (*b*) the net rate at which the energy stored in the cube may be said to change.

22. The radiation emitted by a laser is not exactly a parallel beam; rather, the beam spreads out in the form of a cone with circular cross section. The angle θ of the cone (see Fig. 16) is called the *full-angle beam divergence*. A 3.85-kW argon laser, radiating at 514.5 nm, is aimed at the Moon in a ranging experiment; the laser has a full-angle beam divergence of 0.880 μrad. Find the intensity of the beam at the Moon's surface.

Figure 16 Problem 22.

23. A HeNe laser, radiating at 632.8 nm, has a power output of 3.10 mW and a full-angle beam divergence (see Problem 22) of 172 μrad. (*a*) Find the intensity of the beam 38.2 m from the laser. (*b*) What would be the power output of an isotropic source that provides this same intensity at the same distance?

24. Frank D. Drake, an active investigator in the SETI (Search for Extra-Terrestrial Intelligence) program, has said that the large radio telescope in Arecibo, Puerto Rico, "can detect a signal which lays down on the entire surface of the Earth a power of only one picowatt." See Fig. 17. (*a*) What is the power actually received by the Arecibo antenna for such a signal? The antenna diameter is 305 m. (*b*) What would be the power output of a source at the center of our galaxy that could provide such a signal? The galactic center is 2.3×10^4 ly away. Take the source as radiating uniformly in all directions.

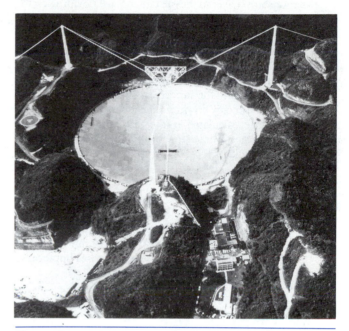

Figure 17 Problem 24.

25. An airplane flying at a distance of 11.3 km from a radio transmitter receives a signal of 7.83 μW/m². Calculate (*a*) the amplitude of the electric field at the airplane due to this signal; (*b*) the amplitude of the magnetic field at the airplane; (*c*) the total power radiated by the transmitter, assuming the transmitter to radiate uniformly in all directions.

26. During a test, a NATO surveillance radar system, operating at 12 GHz with 183 kW of output power, attempts to detect an incoming "enemy" aircraft at 88.2 km. The target aircraft is designed to have a very small effective area for reflection of radar waves of 0.222 m². Assume that the radar beam spreads out isotropically into the forward hemisphere both upon transmission and reflection and ignore absorption in the atmosphere. For the reflected beam as received back at the radar site, calculate (*a*) the intensity, (*b*) the maximum value of the electric field vector, and (*c*) the rms value of the magnetic field.

27. The average intensity of sunlight, falling at normal incidence just outside the Earth's atmosphere, varies during the year due to the changing Earth–Sun distance. Show that the fractional yearly variation is given by $\Delta I/I = 4e$ approximately, where e is the eccentricity of the Earth's elliptical orbit around the Sun.

28. A copper wire (diameter = 2.48 mm; resistance 1.00 Ω per 300 m) carries a current of 25.0 A. Calculate (*a*) the electric field, (*b*) the magnetic field, and (*c*) the Poynting vector magnitude for a point on the surface of the wire.

29. Consider the possibility of standing electromagnetic waves:

$$E = E_m(\sin \omega t)(\sin kx),$$

$$B = B_m(\cos \omega t)(\cos kx).$$

(*a*) Show that these satisfy Eqs. 5 and 9 if E_m is suitably related to B_m and ω is suitably related to k. What are these relationships? (*b*) Find the (instantaneous) Poynting vector. (*c*) Show that the time-average power flow across any area is zero. (*d*) Describe the flow of energy in this situation.

30. Figure 18 shows a cylindrical resistor of length *l*, radius *a*, and resistivity ρ, carrying a current *i*. (*a*) Show that the Poynting vector **S** at the surface of the resistor is everywhere directed normal to the surface, as shown. (*b*) Show that the rate at which energy flows into the resistor through its cylindrical surface, calculated by integrating the Poynting vector over this surface, is equal to the rate at which internal energy is produced; that is,

$$\int \mathbf{S} \cdot d\mathbf{A} = i^2 R,$$

where *d***A** is an element of area of the cylindrical surface. This suggests that, according to the Poynting vector point of view, the energy that appears in a resistor as internal energy does not enter it through the connecting wires but through the space around the wires and the resistor.

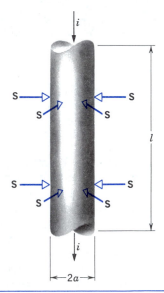

Figure 18   Problem 30.

31. A coaxial cable (inner radius *a*, outer radius *b*) is used as a transmission line between a battery $\mathscr{E}$ and a resistor *R*, as shown in Fig. 19. (*a*) Calculate *E*, *B* for $a < r < b$. (*b*) Calculate the Poynting vector *S* for $a < r < b$. (*c*) By suitably integrating the Poynting vector, show that the total power flowing across the annular cross section $a < r < b$ is $\mathscr{E}^2/R$. Is this reasonable? (*d*) Show that the direction of **S** is always from the battery to the resistor, no matter which way the battery is connected.

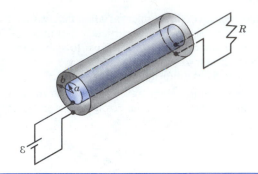

Figure 19   Problem 31.

32. Figure 20 shows a parallel-plate capacitor being charged. (*a*) Show that the Poynting vector **S** points everywhere radially into the cylindrical volume. (*b*) Show that the rate at which energy flows into this volume, calculated by integrating the Poynting vector over the cylindrical boundary of this volume, is equal to the rate at which the stored electrostatic energy increases; that is,

$$\int \mathbf{S} \cdot d\mathbf{A} = Ad \frac{d}{dt}\left(\frac{1}{2}\epsilon_0 E^2\right),$$

where *Ad* is the volume of the capacitor and $\frac{1}{2}\epsilon_0 E^2$ is the energy density for all points within that volume. This analysis shows that, according to the Poynting vector point of view, the energy stored in a capacitor does not enter it through the wires but through the space around the wires and the plates. (*Hint*: To find **S** we must first find **B**, which is the magnetic field set up by the displacement current during the charging process; see Fig. 2 of Chapter 40. Ignore fringing of the lines of **E**.)

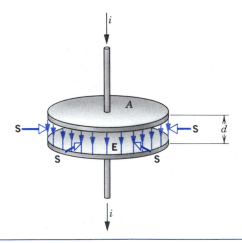

Figure 20   Problem 32.

Section 41-5   Momentum and Pressure of Radiation

33. Suppose that you lie in the Sun for 2.5 h, exposing an area of 1.3 m² at 90° to the Sun's rays of intensity 1.1 kW/m². Assuming complete absorption of the rays, how much momentum is delivered to your body?

34. Show (*a*) that the force *F* exerted by a laser beam of intensity *I* on a perfectly reflecting object of area *A* normal to

the beam is given by $F = 2IA/c$ and (b) that the pressure $P = 2I/c$.

35. High-power lasers are used to compress gas plasmas by radiation pressure. The reflectivity of a plasma is unity if the electron density is high enough. A laser generating pulses of radiation of peak power 1.5 GW is focused onto 1.3 mm² of high-electron-density plasma. Find the pressure exerted on the plasma.

36. (a) Show that the average intensity of the solar radiation that falls normally on a surface just outside the Earth's atmosphere is 1.38 kW/m². (b) What radiation pressure is exerted on this surface, assuming complete absorption? (c) How does this pressure compare with the Earth's sea-level atmospheric pressure, which is 101 kPa?

37. Radiation from the Sun striking the Earth has an intensity of 1.38 kW/m². (a) Assuming that the Earth behaves like a flat disk at right angles to the Sun's rays and that all the incident energy is absorbed, calculate the force on the Earth due to radiation pressure. (b) Compare it with the force due to the Sun's gravitational attraction by calculating the ratio $F_{\text{rad}}/F_{\text{grav}}$.

38. Calculate the radiation pressure 1.50 m away from a 500-W light bulb. Assume that the surface on which the pressure is exerted faces the bulb and is perfectly absorbing and that the bulb radiates uniformly in all directions.

39. A plane electromagnetic wave, with wavelength 3.18 m, travels in free space in the $+x$ direction with its electric vector $\mathbf{E}$, of amplitude 288 V/m, directed along the y axis. (a) What is the frequency of the wave? (b) What is the direction and amplitude of the magnetic field associated with the wave? (c) If $E = E_m \sin(kx - \omega t)$, what are the values of k and ω? (d) Find the intensity of the wave. (e) If the wave falls upon a perfectly absorbing sheet of area 1.85 m², at what rate would momentum be delivered to the sheet and what is the radiation pressure exerted on the sheet?

40. Show that the vector $c\epsilon_0\mathbf{E}\times\mathbf{B}$ has the dimensions of momentum/(area·time), whereas $(1/\mu_0)\,\mathbf{E}\times\mathbf{B}$ has the dimensions of energy/(area·time). (The vector $c\epsilon_0\mathbf{E}\times\mathbf{B}$ may be used for computing momentum flow in the same manner that $\mathbf{S} = (1/\mu_0)\,\mathbf{E}\times\mathbf{B}$ is used to compute energy flow.)

41. Radiation of intensity I is normally incident on an object that absorbs a fraction f of it and reflects the rest. What is the radiation pressure?

42. Prove, for a plane wave at normal incidence on a plane surface, that the radiation pressure on the surface is equal to the energy density in the beam outside the surface. This relation holds no matter what fraction of the incident energy is reflected.

43. Prove, for a stream of bullets striking a plane surface at normal incidence, that the "pressure" is twice the kinetic energy density in the stream above the surface; assume that the bullets are completely absorbed by the surface. Contrast this with the behavior of light; see Problem 42.

44. A small spaceship whose mass, with occupant, is 1500 kg is drifting in outer space, where the gravitational field is negli-

gible. If the astronaut turns on a 10.0-kW laser beam, what speed would the ship attain in one day because of the reaction force associated with the momentum carried away by the beam?

45. A helium–neon laser of the type often found in physics laboratories has a beam power output of 5.00 mW at a wavelength of 633 nm. The beam is focused by a lens to a circular spot whose effective diameter may be taken to be 2.10 wavelengths. Calculate (a) the intensity of the focused beam, (b) the radiation pressure exerted on a tiny, perfectly absorbing sphere whose diameter is that of the focal spot, (c) the force exerted on this sphere, and (d) the acceleration imparted to it. Assume a sphere density of 4.88 g/cm³.

46. A laser has a power output of 4.6 W and a beam diameter of 2.6 mm. If it is aimed vertically upward, what is the height H of a perfectly reflecting cylinder that can be made to "hover" by the radiation pressure exerted by the beam? Assume that the density of the cylinder is 1.2 g/cm³. See Fig. 21.

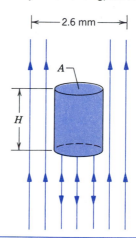

Figure 21 Problem 46.

47. It has been proposed that a spaceship might be propelled in the solar system by radiation pressure, using a large sail made of foil. How large must the sail be if the radiation force is to be equal in magnitude to the Sun's gravitational attraction? Assume that the mass of the ship + sail is 1650 kg, that the sail is perfectly reflecting, and that the sail is oriented at right angles to the Sun's rays. See Appendix C for needed data.

48. Verify the value of the radiation force on the Sun yacht *Diana*, described in Problem 9 of Chapter 5.

49. A particle in the solar system is under the combined influence of the Sun's gravitational attraction and the radiation force due to the Sun's rays. Assume that the particle is a sphere of density 1.00 g/cm³ and that all the incident light is absorbed. (a) Show that all particles with radius less than some critical radius R_0 will be blown out of the solar system. (b) Calculate R_0. Note that R_0 does not depend on the distance from the particle to the Sun.

CHAPTER 42

THE NATURE
AND PROPAGATION
OF LIGHT

There is nothing in its fundamental nature that distinguishes light from any other electromagnetic wave. The descriptions of electromagnetic waves given in the previous chapter apply equally well to light waves. What distinguishes light from other electromagnetic waves is that we have receptors (eyes) that are sensitive to electromagnetic radiation only in a narrow range of wavelengths from about 400 nm (violet) to about 700 nm (red).

In this chapter, we discuss some of the characteristics of light waves, including the sources of visible radiation, the speed of propagation in vacuum and in matter, and the Doppler effect for light that occurs when the source and the observer are in relative motion. In later chapters, we deal with optics, which continues our study of the propagation of light. This chapter serves as a bridge between our previous discussion of electromagnetic waves in general and the coming discussions of optics. However, you should keep in mind that much of what we cover in this chapter applies equally well to other kinds of electromagnetic waves.

42-1 VISIBLE LIGHT

We may operationally define visible light to be electromagnetic radiation to which the eye is sensitive. The sensitivity of individual observers may vary, but typical humans can observe radiation in the range in wavelength of 400 nm to 700 nm (corresponding to a range in fre-

quency of 7×10^{14} Hz to 4×10^{14} Hz. Within that range, the sensitivity to different wavelengths is not at all constant. Figure 1 shows a representation of the variation in the sensitivity of a standard observer to radiations of differing wavelength but constant radiant intensity over the visible region of the spectrum. The greatest sensitivity occurs near 555 nm, corresponding to light of a yellow-green color. The limits of the visible region are not well defined, because the sensitivity curve approaches the axis asymptotically at both long and short wavelengths. The limits corresponding to a sensitivity equal to 1% of that of the peak are 430 nm (violet) and 690 nm (red). Keep in mind that Fig. 1 applies only to a standard human observer; the eyes of animals may have different sensitivities, and electronic devices may have broader or narrower sensitivity curves.* (Compare the range of *visible* wave-

---

* The assignment of color to the various regions of the visible spectrum is quite arbitrary, because color is primarily a psychological label rather than a physical quality. Just as there is no fundamental physical distinction between light and other electromagnetic waves, there is no fundamental physical distinction between blue light and red light. For more on the perception of color, see "The Retinex Theory of Color Vision," by Edwin H. Land, *Scientific American*, December 1977, p. 108, and *Eye, Brain, and Vision*, by David H. Hubel (Scientific American Library Series, 1988), Chapter 8.

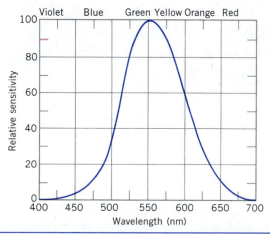

Figure 1 The relative sensitivity of the human eye as a function of wavelength.

lengths, less than a factor of 2, with the range of *audible* wavelengths or frequencies, which Fig. 4 of Chapter 20 shows to be about a factor of 100 at the 1% limit.)

Sources of visible light depend ultimately on the motion of electrons. Electrons in atoms can be raised from their lowest energy state to higher states by various means, such as by heating the substance or by passing an electric current through it. When the electrons eventually drop back to their lowest levels, the atoms emit radiation that may be in the visible region of the spectrum. Emission of visible light is particularly likely when the outer (valence) electrons are the ones making the transitions.

The most familiar source of visible light is the Sun. Its surface emits radiation across the entire electromagnetic spectrum, but its most intense radiation is in the region we define as visible, and the Sun's radiant intensity peaks at a wavelength of about 550 nm, corresponding precisely to the peak in the sensitivity of our standard observer (Fig. 1). This suggests that, through natural selection, our eyes evolved in such a way that their sensitivity matched the Sun's spectrum.

All objects emit electromagnetic radiation, called *thermal radiation*, because of their temperature. Objects such as the Sun, whose thermal radiation is visible, are called *incandescent*. Other common incandescent objects are the filaments of ordinary light bulbs and the glowing embers in a charcoal fire. Incandescence is normally associated with hot objects; typically, temperatures in excess of 1000°C are required.

It is also possible for light to be emitted from cool objects; this phenomenon is called *luminescence*. Examples include common fluorescent lamps, lightning, glowing watch and clock dials, and television receivers. In the case of a fluorescent lamp, an electric current passed through the gas in the tube causes the electrons to move to higher energy states; when the electrons return to their original energy states, they give up their excess energy in the form of ultraviolet radiation. This radiation is absorbed by atoms of the coating on the inside of the glass tube, which then emit visible light. In the case of glowing clock dials, it is *incident* light that causes the excitation.

Luminescent objects can be put into two categories depending on the duration of light emission after the source of excitation is removed. Objects in which the emission of light ceases immediately (within 10^{-8} s) after the excitation is removed are called *fluorescent*, for example, the fluorescent lamp. Objects that continue to glow longer than 10^{-8} s after the source of the excitation is removed (such as the clock dial) are called *phosphorescent*, and the material that causes this effect is called a *phosphor* (Fig. 2).

Luminescence can have a variety of causes. When the energy that excites the atoms originates from a chemical reaction, it is called *chemiluminescence*. Often the effect occurs in living things, such as in fireflies and many marine organisms, in which case it is called *bioluminescence*

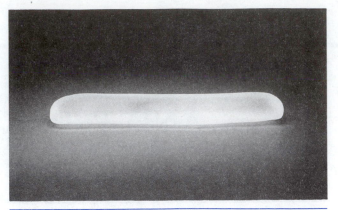

Figure 2 A phosphorescent material, a crystal of sodium borate, emits visible light when it absorbs ultraviolet radiation.

Figure 3 The dots of light are glow-worms in a cave in New Zealand. The light attracts insects, which are trapped and serve as food for the glow-worm larvae.

(Fig. 3). Light can also be emitted when certain crystals, for example, sugar, are crushed; the effect, called *triboluminescence*, can be observed in a dark room by crunching

Wintergreen Life-Savers™ between the teeth. Other causes of luminescence include electric currents (as in lightning or light-emitting diodes) and the impact of high-energy particles (as in the aurora borealis).

42-2 THE SPEED OF LIGHT

According to Maxwell's theory, all electromagnetic waves travel through empty space with the same speed. We call this speed "the speed of light," even though it applies to all electromagnetic radiations, not just light. This speed is one of the fundamental constants of nature. Its precise knowledge is important in relating frequency and wavelength of electromagnetic waves (according to $c = \lambda\nu$). In the early 1900s, the precision of measured wavelengths exceeded that of the speed of light as it was known at that time, and as a result the frequency of electromagnetic waves could not be calculated with great precision. (Frequencies were generally expressed in units of inverse length for this reason.) Today, the measurement of frequency (and thus of time intervals) can be done with a precision that exceeds that of wavelengths; as a result, the meter is no longer a primary standard (see Section 1-4).

Until the 17th century, it was generally believed that light propagated instantaneously; that is, the speed of light was thought to be infinite. Galileo discussed this question in his famous work, *Dialogue Concerning Two New Sciences*, published in 1638. He presented his arguments in the form of a dialogue between several characters, including Simplicio (representing the scientifically ignorant) and Sagredo (representing the voice of reason and probably Galileo himself):

SIMPLICIO: Everyday experience shows that the propagation of light is instantaneous; for when we see a piece of artillery fired, at a great distance, the flash reaches our eyes without a lapse of time, but the sound reaches the ear only after a noticeable interval.

SAGREDO: Well, Simplicio, the only thing I am able to infer from this familiar bit of experience is that sound, in reaching our ear, travels more slowly than light; it does not inform me whether the coming of the light is instantaneous or whether, although extremely rapid, it still occupies time.

Galileo then goes on to describe an experiment (which he actually carried out) to measure the speed of light. He and an assistant stood facing one another at night, separated by a distance of about a mile, each carrying a lantern that could be covered or uncovered at will. Galileo started by uncovering his lantern, and the assistant was to uncover his lantern when he saw the light from Galileo's. Galileo

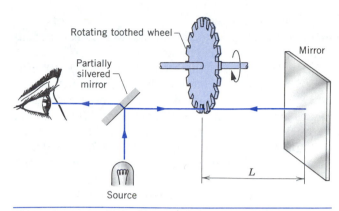

Figure 4 A schematic diagram of Fizeau's apparatus for measuring the speed of light.

then tried to measure the time interval between the instant at which he uncovered his lantern and the instant at which the light from his assistant's lantern reached him. Although Galileo was not able to determine a value for the speed of light (the round-trip time for a separation of 1 mile being only 11 μs, several orders of magnitude smaller than human reaction times), he is credited with the first attempt at measuring the speed of light.

In 1676, Ole Roemer, a Danish astronomer working in Paris, used astronomical observations to deduce that the speed of light is finite. His conclusion was based on a discrepancy between the predicted and observed times of eclipses of Jupiter's innermost moon, Io (see Problem 6). About 50 years later James Bradley, an English astronomer, used a different technique based on starlight to obtain a value of 3×10^8 m/s.

The next major improvement in measuring the speed of light did not come for more than a century. In 1849, the French physicist Hippolyte Louis Fizeau (1819–1896) used a mechanical arrangement, illustrated in Fig. 4. In essence, a light beam was made to travel a long round-trip path (of length $L = 8630$ m each way), passing through a rotating toothed wheel in each direction. The rotating wheel chops the beam going toward the mirror into short pulses. If, during the time the pulse travels the round trip to the mirror and back, the wheel rotates so that a tooth is now blocking the light path, the observer does not see the light pulse. When this occurs, the time $2L/c$ it takes the light beam to make the round trip between the wheel and the mirror must equal the time θ/ω it takes for the wheel to rotate at angular speed ω through the angle θ between the center of a tooth and the center of a gap. That is,

$$\frac{2L}{c} = \frac{\theta}{\omega}$$

or

$$c = \frac{2L\omega}{\theta}. \tag{1}$$

Chopped beams are used in similar ways to measure the

speeds of neutrons and other particles. (A variant of this method was used to verify the Maxwell speed distribution; see Fig. 12 of Chapter 24.)

Fizeau's result using this method was 3.133×10^8 m/s. Other experimenters, among whom was the U.S. physicist Albert A. Michelson, used similar mechanical techniques throughout the late 19th and early 20th centuries. Michelson's work was noteworthy for its care and precision, and he was awarded the 1907 Nobel Prize in physics for his research using optical techniques to make precise measurements. As a result of these investigations, the uncertainty in c was reduced to about 1000 m/s.

The development of electronic techniques, especially as applied to microwaves, permitted a new class of measurements to be done in the 1950s. These measurements gave results that agreed with Michelson's and had similar limits of uncertainty.

The breakthrough in measurements of the speed of light came in the 1970s with the application of lasers. By measuring the frequency and wavelength directly, the speed of light could be obtained from $c = \lambda v$. Refinements of this technique have resulted in values of c with uncertainties smaller than 1 m/s. Table 1 summarizes some of the measurements of c we have discussed.* Note the reduction in the limit of uncertainty over the years.

The precision of measuring frequency (about 1 part in

---

* For references to some of these measurements, see "Resource Letter RMSL-1: Recent Measurements of the Speed of Light and the Redefinition of the Meter," by Harry E. Bates, *American Journal of Physics*, August 1988, p. 682.

10^{13}) has far exceeded that of measuring wavelength (about 1 part in 10^9). As a result, we now define the speed of light to have the exact value

$$c = 299{,}792{,}458 \text{ m/s},$$

and the second is defined based on measurements of frequency, so that the meter is now a secondary standard, defined in terms of the second and the value of c.

The Speed of Light in Matter

When we refer to "the speed of light," we usually mean the speed *in vacuum*. We have discussed in Chapter 41 the propagation of electromagnetic radiation, which takes place through the coupling between its electric and magnetic fields. In dielectric materials, we have seen in Section 31-7 that the electric field is altered by a factor of κ_e, the dielectric constant of the material. A convenient way of modifying equations for electric fields in vacuum to account for the presence of dielectric materials is, as shown in Section 31-5, to replace the permittivity constant ϵ_0 with the quantity $\kappa_e \epsilon_0$.

We also must account for the effect of the *magnetic* properties of the medium on the magnetic field of the propagating electromagnetic wave. As we discussed in Section 37-3, magnetic materials are characterized by a relative permeability constant κ_m, and in analogy with the electric field we can modify the magnetic field equations in matter by replacing the permeability constant μ_0 with the quantity $\kappa_m \mu_0$.

TABLE 1 SPEED OF ELECTROMAGNETIC RADIATION IN FREE SPACE
(SOME SELECTED MEASUREMENTS)

Date	Experimenter	Country	Method	Speed (km/s)	Uncertainty (km/s)
1600 (?)	Galileo	Italy	Lanterns and shutters	"Extraordinarily rapid"	
1676	Roemer	France	Moons of Jupiter	"Finite"	
1729	Bradley	England	Aberration of starlight	304,000	
1849	Fizeau	France	Toothed wheel	313,300	
1862	Foucault	France	Rotating mirror	298,000	500
1880	Michelson	United States	Rotating mirror	299,910	50
1906	Rosa and Dorsey	United States	Electromagnetic theory	299,781	10
1923	Mercier	France	Standing waves on wires	299,782	15
1926	Michelson	United States	Rotating mirror	299,796	4
1950	Bergstrand	Sweden	Geodimeter	299,792.7	0.25
1950	Essen	England	Microwave cavity	299,792.5	3
1950	Bol and Hansen	United States	Microwave cavity	299,789.3	0.4
1951	Aslakson	United States	Shoran radar	299,794.2	1.9
1952	Rank et al.	United States	Molecular spectra	299,766	7
1952	Froome	England	Microwave interferometer	299,792.6	0.7
1958	Froome	England	Microwave interferometer	299,792.50	0.10
1967	Grosse	Germany	Geodimeter	299,792.5	0.05
1973	Evenson et al.	United States	Laser techniques	299,792.4574	0.0012
1978	Woods et al.	England	Laser techniques	299,792.4588	0.0002
1987	Jennings et al.	United States	Laser techniques	299,792.4586	0.0003

TABLE 2 SPEED OF LIGHT IN SELECTED MATERIALS[a]

Material	Speed of Light (10^8 m/s)
Vacuum	3.00
Air	3.00
Water	2.26
Sugar solution (50%)	2.11
Crown glass	1.97
Diamond	1.24

[a] For yellow light ($\lambda = 589$ nm).

Making these substitutions, we can therefore modify Eq. 11 of Chapter 41 to give the speed of light in matter:

$$v = \frac{1}{\sqrt{\kappa_m \kappa_e}} \frac{1}{\sqrt{\mu_0 \epsilon_0}} = \frac{c}{\sqrt{\kappa_m \kappa_e}}. \tag{2}$$

Materials that transmit light are normally nonferromagnetic, and therefore κ_m differs from 1 typically by no more than 10^{-4} (see Tables 2 and 3 of Chapter 37). It is therefore the dielectric constant κ_e that determines the speed of light in a material. However, the dielectric constants that are listed in Table 1 of Chapter 31 cannot be used in Eq. 2, because those values are characteristic of *static* situations. Recall that the dielectric constant is in effect a measure of the response of the dipoles (permanent or induced) to an applied electric field. If the applied field varies at high frequency, the dipoles may not have time to respond, and we cannot use the static dielectric constants in the case of a rapidly varying **E** field. At the frequencies characteristic of a light wave (10^{15} Hz), the field oscillates too rapidly for the dipoles to follow completely. Furthermore, κ_e in Eq. 2 varies with frequency, so that the speed of light in matter depends on the wavelength or frequency of the light.

Table 2 shows values of the speed of light in various materials.

Sample Problem 1 The speed of light of yellow color ($\lambda = 589$ nm) in water is 2.26×10^8 m/s. What is the effective dielectric constant for water at this frequency?

Solution We use Eq. 2 and assume that, to sufficient accuracy for this calculation, $\kappa_m = 1$. Solving Eq. 2 for κ_e, we obtain

$$\kappa_e = \left(\frac{c}{v}\right)^2 = \left(\frac{3.00 \times 10^8 \text{ m/s}}{2.26 \times 10^8 \text{ m/s}}\right)^2 = 1.76.$$

This is very different from the *static* dielectric constant for water, which has a value of about 80 at room temperature, suggesting the difficulty that the dipole moments of water molecules have in following the variation of the electric field at this frequency. In general, dielectric constants at high frequency are smaller than the corresponding static values, which means that the high-frequency induced electric field is smaller than the static induced electric field.

Propagation of Light in Matter *(Optional)*

The mechanism responsible for the propagation of light in matter is scattering (in effect, absorption of the incident light by the atoms or molecules of the medium and reemission of the scattered light). The phases of the scattered waves traveling transverse to the direction of the incident light cause nearly complete destructive interference in the transverse directions. The scattered waves traveling parallel to the direction of the incident light are not in phase with the incident light; as a result of the interference between the two waves, the phase of their combination differs from the phase of the incident wave. We observe this change in phase as a change in speed.

The electric field of the incident light causes the electrons in an atom to oscillate with the frequency of the incident light. It is reasonable to expect that the phase of the reemitted wave depends on the frequency of the atomic oscillation and therefore on the frequency of the original wave. When the incident and scattered waves interfere, the phase of their combination depends on their phase difference and hence on the frequency. As a result, the speed of light in a material depends on the frequency or wavelength. This phenomenon, which is called *dispersion*, is discussed in Chapter 43.

In a typical solid, the distance over which the original light is absorbed and reemitted is of the order of micrometers, and in air it is of the order of millimeters. In effect, the light that we see from the Sun comes to our eyes not directly from the Sun but from the molecules of air a few millimeters in front of our eyes. ■

42-3 THE DOPPLER EFFECT FOR LIGHT

In Section 20-7 we showed that if a source of *sound* is moving toward an observer at a speed u, the frequency v heard by the observer is (see Eq. 39 of Chapter 20, which we have rearranged and in which we have substituted u for v_S)

$$v = v_0 \frac{1}{1 - u/v} \quad \text{(sound wave, observer fixed, source approaching).} \tag{3}$$

In this equation v_0 is the frequency heard when the source is at rest, and v is the speed of sound. This change in frequency due to relative motion is called the *Doppler effect*.

If the source is at rest in the transmitting medium but the observer is moving toward the source at speed u, the observed frequency (see Eq. 36 of Chapter 20, in which u has been substituted for v_O) is

$$v = v_0(1 + u/v) \quad \text{(sound wave, source fixed, observer approaching).} \tag{4}$$

For identical values of the relative separation speed u of the source and the observer, the frequencies predicted by Eqs. 3 and 4 are different. This is not surprising, because a source of sound moving through a medium in which the observer is at rest is physically different from an observer moving through that medium with the source at rest, as we see by comparing Figs. 12 and 13 of Chapter 20 and as we demonstrated in Sample Problem 5 of Chapter 20.

We might be tempted to apply Eqs. 3 and 4 to light, substituting c, the speed of light, for v, the speed of sound. For light, as contrasted with sound, however, it has proved impossible to identify a medium of transmission relative to which the source and the observer are moving. This means that "source approaching observer" and "observer approaching source" are physically identical situations and must exhibit *exactly the same* Doppler-shifted frequency. As applied to light, either Eq. 3 or Eq. 4 or both must be incorrect. As we show in the next section, the Doppler effect predicted by the theory of relativity is

$$
\begin{aligned}
v &= v_0 \frac{1 + u/c}{\sqrt{1 - u^2/c^2}} \\
&= v_0 \sqrt{\frac{1 + u/c}{1 - u/c}} \quad \text{(light wave, source and observer approaching).} \quad (5) \\
&= v_0 \frac{\sqrt{1 - u^2/c^2}}{1 - u/c}
\end{aligned}
$$

Equation 5 applies only in the special case in which the direction of propagation of the light is the same as the direction of the relative motion of S and S'. In the next section, we derive a more general result valid for any direction. We can modify Eqs. 3, 4, and 5 if the source and the observer are *separating from* each other by replacing u with $-u$.

Equations 3, 4, and 5 give similar results if the ratio u/c is small, as we can see by expanding the equations using the binomial theorem, substituting c for v, which gives

$$
\text{Eq. 3:} \qquad v = v_0 \left[1 + \frac{u}{c} + \left(\frac{u}{c} \right)^2 + \cdots \right], \quad (6)
$$

$$
\text{Eq. 4:} \qquad v = v_0 \left(1 + \frac{u}{c} \right), \quad (7)
$$

$$
\text{Eq. 5:} \qquad v = v_0 \left[1 + \frac{u}{c} + \frac{1}{2} \left(\frac{u}{c} \right)^2 + \cdots \right]. \quad (8)
$$

The ratio u/c for most light sources, even those of atomic dimensions, is small. The terms in u^2/c^2 (and higher-order terms) in such cases are negligibly small, and the first-order term u/c gives a reasonable estimate of the Doppler shift.

Under nearly all circumstances the differences among these three equations are not important. Nevertheless, it is of extreme interest to carry out at least one experiment precisely enough to serve as a test of Eq. 5 and thus, in part, of the theory of relativity.

The classic experimental test was done in 1938 by H. E. Ives and G. R. Stilwell. They sent a beam of hydrogen atoms, generated in a gas discharge, down a tube at speed u, as in Fig. 5. They could observe light emitted by these atoms in a direction opposite to **u** (atom 1, for example) using a mirror, and also in a direction parallel to **u** (atom 2, for example). With a precise spectrograph, they could photograph a particular characteristic spectral line of this light, obtaining, on a frequency scale, the lines marked v_1' and v_2' in Fig. 5*b*. It is also possible to photograph, on the same photographic plate, a line corresponding to light emitted from atoms *at rest*; such a line appears as v in Fig. 5*b*. A fundamental measured quantity in this experiment is $\Delta v/v$, defined from

$$
\frac{\Delta v}{v} = \frac{\Delta v_2 - \Delta v_1}{v}, \quad (9)
$$

(see Fig. 5*b*). It measures the extent to which the frequency of the light from resting atoms fails to lie halfway between the frequencies v_1' and v_2'. Table 3 shows that the measured results agree with the formula predicted by the theory of relativity (Eq. 5) and not with the classical formula borrowed from the theory of sound propagation in a material medium (Eq. 3).

Ives and Stilwell did not present their experimental results as evidence for the support of Einstein's theory of

Figure 5   Apparatus used in the Ives–Stilwell experiment.

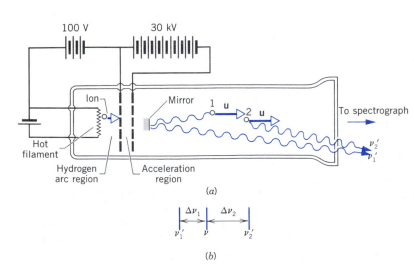

(a)

(b)

TABLE 3   THE IVES–STILWELL EXPERIMENT

Speed u (10^6 m/s)	$\Delta v/v$, 10^{-5}		
	Classical	Relativistic	Experiment
0.865	1.67	0.835	0.762
1.01	2.26	1.13	1.1
1.15	2.90	1.45	1.42
1.33	3.94	1.97	1.9

relativity but rather gave them an alternative theoretical explanation. Modern observers, looking not only at their excellent experiment but at the whole range of experimental evidence, now give the Ives–Stilwell experiment the interpretation we have described, as a test of the relativistic Doppler effect.

Sample Problem 2   A distant quasar is moving away from Earth at a speed u. An astronomer is searching for a certain spectral line in the light from the quasar. That line, emitted by atomic hydrogen, is observed using hydrogen discharge tubes on Earth to have a wavelength of $\lambda_0 = 121.6$ nm. The astronomer finds the hydrogen spectral line emitted by the quasar at a wavelength of $\lambda = 460.9$ nm. Assuming the quasar to be moving radially away from Earth, what is its speed relative to the Earth?

Solution   We use Eq. 5, which we rewrite in terms of wavelength and substitute $-u$ for u because the source and observer are separating:

$$\frac{c}{\lambda} = \frac{c}{\lambda_0} \frac{1 - u/c}{\sqrt{1 - u^2/c^2}}$$

or

$$\lambda = \lambda_0 \sqrt{\frac{1 + u/c}{1 - u/c}}. \qquad (10)$$

Solving for the speed, we find

$$\frac{u}{c} = \frac{(\lambda/\lambda_0)^2 - 1}{(\lambda/\lambda_0)^2 + 1}. \qquad (11)$$

With $\lambda/\lambda_0 = 460.9$ nm/121.6 nm $= 3.79$, we obtain

$$\frac{u}{c} = \frac{(3.79)^2 - 1}{(3.79)^2 + 1} = 0.87.$$

The quasar is moving away from Earth at 87% of the speed of light. This calculation determines only the radial or line-of-sight component of the relative velocity.

The Doppler effect causes the wavelengths of light from objects receding from Earth to be lengthened or shifted toward the red (long-wavelength) end of the visible spectrum. Hence it is known as the *red shift*. Figure 6 shows an example of a red-shifted spectrum, from which it is possible to determine the speed of the galaxy relative to the Earth. Evidence from many such observations shows that all distant objects are moving away from us, and that there is a direct (linear) relationship between the speed of the object and its distance from Earth: the more distant the object, the faster it moves away from us. This linear behavior, deduced from measurements of the red shift, is the primary evidence for the expansion of the universe.

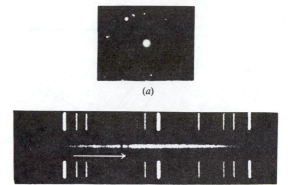

(a)

(b)

Figure 6   (a) A galaxy in the constellation Corona Borealis. (b) The central streak shows the wavelength spectrum of the light emitted by this galaxy. The two dark vertical bands show absorption lines associated with calcium, which is present in the galaxy. The line spectra above and below are recorded from a laboratory source to provide a wavelength calibration. The horizontal arrow shows how far the calcium lines are displaced from where they would be expected to appear if they were emitted by a source at rest in the laboratory. From this Doppler shift, the recessional speed of the galaxy is deduced to be about 21,000 km/s.

42-4   DERIVATION OF THE RELATIVISTIC DOPPLER EFFECT   *(Optional)*

In this section, we use Einstein's two postulates, along with the equations of the Lorentz transformation, to derive the equation for the relativistic Doppler effect.

In Chapter 21, we illustrated through numerous examples the way that the equations of the Lorentz transformation can be used to relate measurements made by one inertial observer S to those of another observer S' who moves at constant velocity with respect to S. Here we compare the results of the two observers when they measure the same light wave. As in Chapter 21, we assume the relative motion between S and S' takes place in the common xx' direction with speed u.

We consider the case of a train of plane electromagnetic waves that travel at speed c' in the S' frame. The source of the plane waves is at rest according to S', who measures wave number k' ($= 2\pi/\lambda'$) and angular frequency ω' ($= 2\pi v'$); these are of course related by $c' = \omega'/k'$. If the wave traveled along the x' direction, the variation with space and time of the electric field of the wave in S' would be given by a sinusoidal expression of the form

$$\mathbf{E}' = \mathbf{E}'_m \sin{(k'x' - \omega't')},$$

and if the wave traveled along the y' direction, the electric field would be of the form

$$\mathbf{E}' = \mathbf{E}'_m \sin{(k'y' - \omega't')}.$$

Similar expressions are obtained for $\mathbf{B}'$.

Let us now consider a more general case, in which the wave travels parallel to the $x'y'$ plane in a direction that makes an

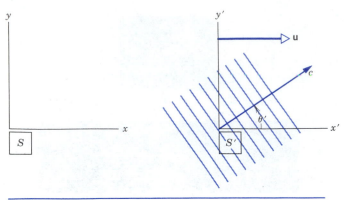

Figure 7 A source at rest in S' emits plane wavefronts that travel in a direction at an angle θ' with respect to the x' axis. The frame of S' (including the source) moves at velocity **u** relative to S.

angle of θ' with the x' axis (Fig. 7). In this case the sinusoidally varying part of the fields can be shown to be given by

$$\sin (k'x' \cos \theta' + k'y' \sin \theta' - \omega't').$$

Note that this reduces to the previous expressions for a wave that travels in the x' direction ($\theta' = 0$) or in the y' direction ($\theta' = 90°$). It is more convenient to express the sinusoidal variation as

$$\sin 2\pi \left(\frac{x' \cos \theta' + y' \sin \theta'}{\lambda'} - v't' \right), \quad (12)$$

where $c' = \lambda'v'$.

We now wish to observe this wave from the laboratory frame of reference S, relative to which S' (including the source of the waves) moves at speed u in the x direction. How is the wave observed in the S frame related to the wave observed in S'?

Let us first see what Einstein's postulates tell us about the form of the wave in the S frame. The first postulate demands that, if the wave satisfies a wave equation (see, for example, Eq. 25 of Chapter 19) in S', then it also must satisfy a wave equation in S. That is, in the S frame, the variation of the wave must be of the form

$$\sin 2\pi \left(\frac{x \cos \theta + y \sin \theta}{\lambda} - vt \right), \quad (13)$$

where $c = \lambda v$. The second postulate requires that the phase velocity in S be equal to the phase velocity in S'; that is, $c = c'$.

We now proceed by applying the Lorentz transformation. From Table 2 of Chapter 21, we obtain the Lorentz transformation equations for x', y', and t'. We substitute those expressions into Eq. 12, which gives

$$\sin 2\pi \left(\frac{\gamma(x - ut) \cos \theta' + y \sin \theta'}{\lambda'} - v'\gamma(t - ux/c^2) \right),$$

where $\gamma = 1/\sqrt{1 - u^2/c^2}$. After some algebraic manipulation, this becomes

$$\sin 2\pi \left(\frac{\gamma(\cos \theta' + u/c)}{\lambda'} x + \frac{\sin \theta'}{\lambda'} y \right.$$
$$\left. - \gamma v'[1 + (u/c)\cos \theta']t \right). \quad (14)$$

Consistent with the first postulate, Eq. 14 is indeed in the same

form as Eq. 13, if the coefficients of x, y, and t in the two expressions are equal. That is,

$$\frac{\cos \theta}{\lambda} = \frac{\gamma(\cos \theta' + u/c)}{\lambda'}, \quad (15)$$

$$\frac{\sin \theta}{\lambda} = \frac{\sin \theta'}{\lambda'}, \quad (16)$$

$$v = \gamma v'[1 + (u/c)\cos \theta']. \quad (17)$$

Since we are seeking a result of a measurement in the S frame, we eliminate the unknown angle θ' from Eqs. 15 and 17 and solve for v, which gives

$$v = v_0 \frac{\sqrt{1 - u^2/c^2}}{1 - (u/c)\cos \theta}. \quad (18)$$

We have replaced the frequency v' with the frequency v_0, to remind us that it is measured in a frame of reference in which the source is at rest. It can therefore be considered a proper frequency, analogous to the proper time. We shall return to this point later.

Equation 18 is the relativistic expression for the Doppler effect, written for the case in which the source and observer are moving *toward* one another; in this case, the observer in S measures a *higher* frequency. Note that Eq. 18 reduces to Eq. 5 if we put $\theta = 0$. For motion of the source *away from* the observer, we substitute $-u$ for u, in which case the observer in S would measure a *lower* frequency.

Equations 15–17 also permit us to relate the directions of propagation θ and θ' as seen from two different reference frames. This relativistic effect is called *aberration*. (See Problem 22.) That is, from the reference frame of S in Fig. 7, the light wave propagates with a different wavelength λ (the Doppler shift) and in a different direction θ (aberration).

Sample Problem 3 An Earth satellite is orbiting from west to east at an altitude of $h = 153$ km in a circular orbit above the equator (see Fig. 8). A tracking ship is located at the equator on the Prime Meridian at 0° longitude (just off the west coast of Africa). The satellite emits radio waves at a frequency of 122.450 MHz. To what frequency should the ship tune its receiver when the satellite is (a) directly overhead; (b) at 10° longitude west of the ship; and (c) at 10° longitude east of the ship?

Solution (a) We let the S' frame be moving with the satellite at the instant it is overhead; the S frame is that of the ship below. The frequency v_0 observed in the S' frame (the satellite) is 122.450 MHz. The relative speed u between the frames is determined by the orbital speed of the satellite at an altitude h or at a radius $R = R_E + h$, where R_E is the radius of the Earth. The gravitational acceleration at a radius R is MG/R^2, which must supply the centripetal acceleration u^2/R necessary for a circular orbit. Thus

$$\frac{MG}{R^2} = \frac{u^2}{R}$$

or

$$u = \sqrt{\frac{MG}{R}} = \sqrt{\frac{MG}{R_E + h}}$$

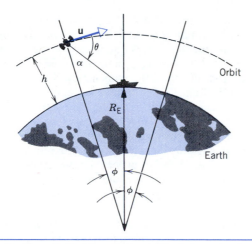

Figure 8   Sample Problem 3. A satellite is in a circular orbit at an altitude h above the surface of the Earth. A ship on the surface observes signals beamed by the satellite.

$$= \sqrt{\frac{(5.98 \times 10^{24} \text{ kg})(6.67 \times 10^{-11} \text{ N} \cdot \text{m}^2/\text{kg}^2)}{6370 \text{ km} + 153 \text{ km}}}$$

$$= 7.82 \times 10^3 \text{ m/s} = 2.61 \times 10^{-5} c.$$

When the satellite is directly overhead, the Doppler shift is obtained from Eq. 18 with $\theta = 90°$:

$$\nu = \nu_0 \sqrt{1 - u^2/c^2}.$$

With $u/c = 2.61 \times 10^{-5}$, we have $u^2/c^2 = 6.8 \times 10^{-10}$. The quantity under the radical differs from 1 by only a few parts in 10^{10}, so that, to the desired precision,

$$\nu \approx \nu_0 = 122.450 \text{ MHz}.$$

(b) When the satellite is not overhead, it is necessary to calculate the angle θ between the velocity of the satellite and the direct line to the tracking ship (see Fig. 8). We can find the angle α ($= \pi/2 - \theta$) by applying the law of sines to the triangle formed by the satellite, the ship, and the center of the Earth:

$$\frac{\sin \alpha}{R_E} = \frac{\sin (\pi - \alpha - \phi)}{R_E + h}.$$

Solving, we find

$$\tan \alpha = \frac{R_E \sin \phi}{h + R_E(1 - \cos \phi)}$$

$$= \frac{(6370 \text{ km})(\sin 10°)}{153 \text{ km} + (6370 \text{ km})(1 - \cos 10°)} = 4.428,$$

or

$$\alpha = \tan^{-1} 4.428 = 77.3°$$

and so

$$\theta = \pi/2 - \alpha = 90° - 77.3° = 12.7°.$$

We can calculate the Doppler-shifted frequency using Eq. 18, neglecting the Lorentz factor γ, which we showed in part (a) did not differ significantly from 1. The remaining Doppler shift is

$$\nu \approx \frac{\nu_0}{1 - (u/c)\cos \theta} = \frac{122.450 \text{ MHz}}{1 - (2.61 \times 10^{-5})(\cos 12.7°)}$$

$$= 122.453 \text{ MHz}.$$

(c) After the satellite passes overhead and moves east of the tracking station, its motion becomes away from the observer, and we can calculate the Doppler shift by making the substitution $u \rightarrow -u$. The result is

$$\nu \approx \frac{\nu_0}{1 + (u/c)\cos \theta} = \frac{122.450 \text{ MHz}}{1 + (2.61 \times 10^{-5})(\cos 12.7°)}$$

$$= 122.447 \text{ MHz}.$$

We see that the frequency detected on Earth varies from 122.453 MHz (satellite approaching) to 122.450 MHz (satellite overhead) to 122.447 MHz (satellite receding). A measurement of the Doppler-shifted frequency is thus sufficient to locate the satellite. ∎

42-5   CONSEQUENCES OF THE RELATIVISTIC DOPPLER EFFECT   *(Optional)*

We have already discussed two very important and commonly observed consequences of the relativistic Doppler effect: the motional red shift of distant objects in the universe (see Sample Problem 2) and the frequency shift that can be used to track satellites (see Sample Problem 3). Here we consider two additional consequences, the transverse Doppler effect and the twin paradox.

Transverse Doppler Effect

An important difference between the classical and relativistic formulas for the Doppler effect arises when we consider the case of $\theta = 90°$, in which the relative motion of the source and observer is at right angles to the direction of propagation of the wavefronts. Carrying through the classical analysis of Chapter 20, we would find that there is no (classical) Doppler shift in this case. The relativistic expression (Eq. 18), on the other hand, predicts that the observer measures a frequency of

$$\nu = \nu_0 \sqrt{1 - u^2/c^2}. \tag{19}$$

Equation 19 is known as the *transverse Doppler effect* and is a purely relativistic effect with no classical counterpart. Note that the observed frequency ν is always lower than the frequency ν_0 emitted by the source.

If we expand Eq. 19 using the binomial theorem, we obtain

$$\nu = \nu_0 \left(1 - \frac{1}{2} \frac{u^2}{c^2} + \cdots \right). \tag{20}$$

Comparing Eq. 20 with Eqs. 6–8, we see that the transverse Doppler shift contains no term proportional to the first power of u/c. Both the longitudinal relativistic Doppler effect and the classical Doppler effect contain such a term. The absence of such a term in Eq. 20 is consistent with the failure of the classical theory to predict such an effect.

A particularly striking confirmation of the transverse Doppler effect resulted from experiments done in 1963 by Walter Kundig. A source of gamma rays was placed at the center of the rotor of a centrifuge. On the rim of the centrifuge was placed a foil that absorbed gamma rays emitted by the source. The absorption depends on the frequency of the radiation that reaches the foil, and a detector behind the foil measured the absorption as the

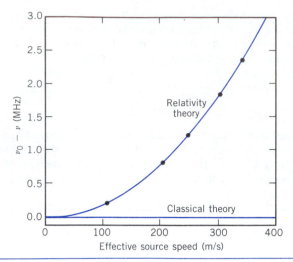

Figure 9   The results of Kundig's experiment for the transverse Doppler effect agree with relativity theory and disagree with classical theory, which predicts no effect.

rotor speed was varied. When the centrifuge is rotating, the foil on the rim is in transverse motion relative to the source at the center, and the radiation that reaches the foil is subject to the transverse Doppler shift. Even though the tangential speed of the absorber was only a few hundred meters/second, corresponding to a value of u^2/c^2 of about 10^{-12}, this sensitive experiment was able to obtain clear evidence of the transverse Doppler shift. Figure 9 shows a summary of Kundig's results, which agree with the relativistic formula and disagree with the classical formula, which predicts no transverse effect.

The transverse Doppler shift can also be interpreted as an effect of time dilation. The source of waves in S' can be regarded as a clock, ticking at a rate determined by the period $T_0 = 1/v_0$, a proper time in S'. The observer in S measures a longer (dilated) period T and thus a smaller frequency $v = 1/T$. The confirmation of the transverse Doppler effect can therefore be taken as another confirmation of relativistic time dilation.

The Twin Paradox Revisited

The Doppler effect permits a reanalysis of the twin paradox, which was discussed in Section 21-7, in a way that reveals unam-

biguously which twin is "really" moving. Assume Fred and Ethel each have identical clocks that were previously calibrated to keep Earth time. The clocks can be used, in their respective frames, to record the passage of time in that frame in units of Earth years (but of course the years appear to be of different durations if the frames are in relative motion). Let us suppose that Ethel in her spaceship is moving away from Fred and his space platform at a relative speed $u = 0.6c$ toward a star whose distance from the platform is measured (by Fred) to be 12 light-years. (Assume the star is at rest with respect to Fred.) According to Fred, Ethel's outward journey takes a time of (12 light-years)/$0.6c$ = 20 years, and the return journey at the same speed takes an equal time. Fred therefore measures the passage of 40 years on his clock, and he ages 40 years during Ethel's journey. In Ethel's frame of reference, the distance to the star is contracted by the factor of $\sqrt{1 - u^2/c^2} = 0.8$, and thus the contracted distance to the star is (12 light-years)(0.8) = 9.6 light-years, according to Ethel. Watching the scenery of space sail by at a speed of $0.6c$, Ethel arrives at the star after the passage of (9.6 light-years)/$0.6c$ = 16 years on her clock, and she measures an equal interval for the return trip. Ethel therefore ages only 32 years during the round trip.

Suppose Fred sends Ethel a pulse of light once each year (on their birthday, perhaps). The frequency of the light signal transmitted by Fred is (as measured by Fred) $v_0 = 1$ y^{-1}, but the Doppler-shifted frequency as observed by Ethel is, according to Eq. 5,

$$v = (1 \text{ y}^{-1}) \sqrt{\frac{1 - 0.6}{1 + 0.6}} = 0.5 \text{ y}^{-1}$$

during the outbound journey. Ethel thus receives, during the outward journey that she measures to be 16 years long, a total of (0.5 y^{-1})(16 y) = 8 signals. During her return trip, the Doppler-shifted frequency becomes 2 y^{-1}, which we obtain by substituting $-u$ for u in the above calculation. The number of signals she receives during the return trip is thus (2 y^{-1})(16 y) = 32 signals. Thus Ethel, who ages only 32 years by her clock during the entire round-trip journey, receives a total of 8 + 32 = 40 signals from Fred, showing that Fred has celebrated 40 birthdays during the trip that Ethel measures to be 32 years long. Ethel, the traveler, is the younger upon their reunion.

In Problem 28, you are asked to carry out a similar analysis if it is Ethel who is sending the signals. You should of course find the same result, with both twins agreeing that Ethel is the younger. ∎

QUESTIONS

1. How might an eye-sensitivity curve like that of Fig. 1 be measured?

2. Why are danger signals in red, when the eye is most sensitive to yellow-green?

3. Comment on this definition of the limits of the spectrum of visible light given by a physiologist: "The limits of the visible spectrum occur when the eye is no better adapted than any other organ of the body to serve as a detector."

4. In connection with Fig. 1, (*a*) do you think it possible that the wavelength of maximum sensitivity could vary if the

intensity of the light is changed? (*b*) What might the curve of Fig. 1 look like for a group of color-blind people who could not, for example, distinguish red from green?

5. Suppose that human eyes were insensitive to visible light but were very sensitive to infrared light. What environmental changes would be needed if you were to (*a*) walk down a long corridor and (*b*) drive a car? Would the phenomenon of color exist? How might traffic lights have to be modified?

6. What feature of light corresponds to loudness in sound?

7. How could Galileo have tested experimentally that reaction

times were an overwhelming source of error in his attempt to measure the speed of light, described in Section 42-2?

8. Can you think of any "everyday" observation (that is, without experimental apparatus) to show that the speed of light is not infinite? Think of lightning flashes, possible discrepancies between the predicted time of sunrise and the observed time, radio communications between Earth and astronauts in orbiting spaceships, and so on.

9. Comment on this statement: Because of the way the meter is defined, it is no longer possible to measure the speed of light.

10. Is the fact that many stars appear white evidence that electromagnetic waves of all colors travel through a vacuum at the same speed?

11. It has been suggested that the speed of light may change slightly in value as time goes on. Can you find any evidence for this in Table 1?

12. In a vacuum, does the speed of light depend on (a) the wavelength, (b) the frequency, (c) the intensity, (d) the speed of the source, or (e) the speed of the observer?

13. Atoms are mostly empty space. However, the speed of light passing through a transparent solid made up of such atoms is often considerably less than the speed of light in free space. How can this be?

14. Is the Doppler effect simply a time-dilation effect and nothing more, or is there something else to it?

15. One member of a binary star system emits visible light. Show on a simple graph how the Doppler frequency shift on Earth varies with time.

16. Can a galaxy be so distant that its recession speed equals c? If so, how can we see the galaxy? That is, will its light ever reach us?

17. Gamma rays are electromagnetic radiation emitted from radioactive nuclei. In free space, do they travel with the same speed as visible light? Does their speed depend on the speed of the nucleus that emits them?

18. Perhaps the simplest astronomical observation that you can make is this: When the Sun sets, the sky becomes dark. This is true and seems obvious but an argument can be made that it should not be so. Consider: "Assuming an infinite universe, uniformly populated by stars more or less like our Sun, we can say that a straight line projected from the observer in any direction will eventually hit a star. The *distances R* of most of these stars will be very great indeed so that the stars illuminate the observer only very weakly, the illumination varying as $1/R^2$. On the other hand, the *number* of distant stars located within a spherical shell whose radii are R and $R + dR$ increases as R^2 (assuming that dR is constant). Can you prove this last statement? These two effects seem to cancel precisely. Thus the night sky should be virtually infinitely bright, the observer being illuminated by an infinity of suns." Can you see any flaw in this argument (usually called *Olber's paradox*)? Think of the finite speed of light, the large scale of the universe, the expanding universe and the associated red shift, the finite lifetime of stars, and so on. (See "The Dark Sky Paradox," by E. R. Harrison, *American Journal of Physics,* February 1977, p. 119, for an excellent historical review and a lucid explanation.)

PROBLEMS

Section 42-1 Visible Light

1. (a) At what wavelengths does the eye of a standard observer have half its maximum sensitivity? (b) What are the wavelength, the frequency, and the period of the light for which the eye is the most sensitive?

2. How many complete vibrations are contained in the wavetrain of light of wavelength 520 nm emitted by an atom for a time of 430 ps?

Section 42-2 The Speed of Light

3. (a) Suppose that we were able to establish radio communication with the hypothetical inhabitants of a hypothetical planet orbiting our nearest star, α-Centauri, which is 4.34 light-years from us. How long would it take to receive a reply to a message? (b) Repeat for the Great Nebula in Andromeda, one of our closest extragalactic neighbors but 2.2×10^6 light-years distant. What do these considerations lead you to conclude about the nature of our possible communication with extragalactic peoples?

4. (a) How long does it take a radio signal to travel 150 km from a transmitter to a receiving antenna? (b) We see a full Moon by reflected sunlight. How much earlier did the light that enters our eye leave the Sun? (c) What is the round-trip travel time for light between Earth and a spaceship orbiting Saturn, 1.3×10^9 km distant? (d) The Crab nebula, which is about 6500 light-years distant, is thought to be the result of a supernova explosion recorded by Chinese astronomers in A.D. 1054. In approximately what year did the explosion actually occur?

5. The uncertainty of the distance to the Moon, as measured by the reflection of laser light from reflectors placed on the Moon by *Apollo 11* astronauts, is about 2 cm. This uncertainty is associated with the measurement of the elapsed time; what uncertainty in this time is implied?

6. In 1676, Ole Roemer deduced that the speed of light is finite by observing the time of the eclipse of one of Jupiter's satellites, Io (see Fig. 10). Based on the known orbital properties of Io, it was predicted to emerge from Jupiter's shadow at a particular time, corresponding to the Earth at position x in its orbit. When the Earth was actually at position y, Io emerged from Jupiter's shadow about 10 min late. Roemer concluded that the discrepancy must be due to the additional time necessary for light from Io to travel the additional distance of the radius of the Earth's orbit. What value can be calculated for the speed of light from this observation? (These observations can also be interpreted in terms of the Doppler effect of light. See "The Doppler Interpretation of Roemer's Method," by V. M. Babovicić, D. M. Davidović, and B. A. Anićin, *American Journal of Physics,* June 1991, p. 515.)

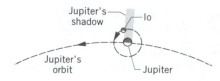

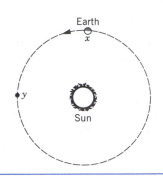

Figure 10   Problem 6.

7. Consider a star located on a line through the Sun, drawn perpendicular to the plane of the Earth's orbit about the Sun. The star's distance is much greater than the diameter of the Earth's orbit. Show that, due to the finite speed of light, a telescope through which the star is seen must be tilted at an angle $\alpha = 20.5''$ to the perpendicular, in the direction the Earth is moving; see Fig. 11. This phenomenon, called *aberration*, is noticeable and was first explained by James Bradley in 1729. (See Problem 22 for a description of aberration based on relativity.)

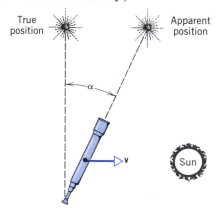

Figure 11   Problem 7.

Section 42-3 The Doppler Effect for Light

8. Show that, for speeds $u \ll c$, the Doppler shift can be written in the approximate form

$$\frac{\Delta\lambda}{\lambda} = \frac{u}{c},$$

where $\Delta\lambda$ is the change in wavelength.

9. Show that, for the 21.1-cm line so much used by radio-astronomers, a Doppler frequency shift in kHz can be converted to a radial velocity in km/s by multiplying by 0.211, provided that $u \ll c$.

10. A spaceship, moving away from Earth at a speed of $0.892c$, reports back by transmitting on a frequency (measured in the spaceship frame) of 100 MHz. To what frequency must Earth receivers be tuned to receive these signals?

11. A rocketship is receding from Earth at a speed of $0.20c$. A light in the rocketship appears blue to passengers on the ship. What color would it appear to be to an observer on Earth? (See Fig. 1.)

12. The "red shift" of radiation from a distant galaxy consists of the light H_γ, known to have a wavelength of 434 nm when observed in the laboratory, appearing to have a wavelength of 462 nm. (*a*) What is the speed of the galaxy in the line of sight relative to the Earth? (*b*) Is it approaching or receding?

13. In the spectrum of quasar 3C9, some of the familiar hydrogen lines appear but they are shifted so far forward toward the red that their wavelengths are observed to be three times as large as that observed in the light from hydrogen atoms at rest in the laboratory. (*a*) Show that the classical Doppler equation, which assumes that light behaves like sound, gives a velocity of recession greater than c. (*b*) Assuming that the relative motion of 3C9 and the Earth is entirely one of recession, find the recession speed predicted by the relativistic Doppler equation.

14. With what speed would you have to go through a red light in order to have it appear green? Take 620 nm as the wavelength of red light and 540 nm as the wavelength for green light.

15. Calculate the Doppler wavelength shifts expected for light of wavelength 553 nm emitted from the edge of the Sun's disk at the equator due to the Sun's rotation. See Appendix C for needed data.

16. Hydrogen molecules at 700 K emit light of frequency 457 THz. (*a*) Determine the change in frequency of the light observed due to the motion of a molecule moving toward an observer with the root-mean-square speed. (*b*) Find the frequency shift if such light originated from hydrogen atoms instead of molecules.

17. In the experiment of Ives and Stilwell the speed u of the hydrogen atoms in a particular run was 8.65×10^5 m/s. Calculate $\Delta v/v$, on the assumptions that (*a*) Eq. 6 is correct and (*b*) that Eq. 8 is correct; compare your results with those given in Table 3 for this speed. Retain the first three terms only in Eqs. 6 and 8.

18. Microwaves, which travel with the speed of light, are reflected from a distant airplane approaching the wave source. It is found that when the reflected waves are beat against the waves radiating from the source the beat frequency is 990 Hz. If the microwaves are 12.0 cm in wavelength, what is the approach speed of the airplane? (*Hint*: See Problem 65 in Chapter 20.)

19. A, on Earth, signals with a flashlight every 6 min. B is on a space station that is stationary with respect to Earth. C is on a rocket traveling from A to B with a constant velocity of $0.60c$ relative to A; see Fig. 12. (*a*) At what intervals does B receive signals from A? (*b*) At what intervals does C receive

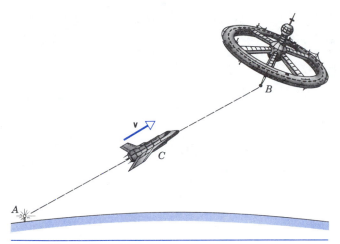

Figure 12   Problem 19.

signals from A? (c) If C flashes a light every time a flash is received from A, at what intervals does B receive C's flashes?

20. A radar transmitter T is fixed to a reference frame S' that is moving to the right with speed u relative to reference frame S (see Fig. 13). A mechanical timer (essentially a clock) in frame S', having a period τ_0 (measured in S') causes transmitter T to emit radar pulses, which travel at the speed of light and are received by R, a receiver fixed in frame S. (a) What would be the period τ of the timer relative to observer A, who is fixed in frame S? (b) Show that the receiver R would observe the time interval between pulses arriving from T, not as τ or as τ_0, but as

$$\tau_R = \tau_0 \sqrt{\frac{c+u}{c-u}}.$$

(c) Explain why the observer at R measures a different period for the transmitter than does observer A, who is in the same reference frame. (*Hint*: A clock and a radar pulse are not the same.)

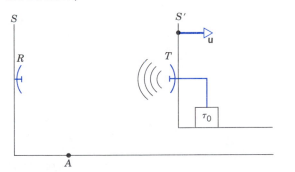

Figure 13   Problem 20.

Section 42-4 Derivation of the Relativistic Doppler Effect

21. An Earth satellite, transmitting on a frequency of 40 MHz, passes directly over a radio receiving station at an altitude of 400 km and at a speed of 2.8×10^4 km/h. Plot the change in frequency, attributable to the Doppler effect, as a function of time, counting $t = 0$ as the instant the satellite is over the station. (*Hint*: The speed in the Doppler formula is not the actual speed of the satellite but its component in the direc-

tion of the station. Neglect the curvature of the Earth and of the satellite orbit.)

22. (a) By combining Eqs. 15 and 16, show that the relationship between θ and θ' can be written

$$\tan \theta = \frac{\sin \theta' \sqrt{1 - u^2/c^2}}{\cos \theta' + u/c}.$$

This relativistic effect is called *aberration*. It gives the angle of emission according to S when the emission angle is θ' according to S' (see Fig. 7). Does this equation reduce to the expected result when $u = 0$? (b) Without doing any further calculation, invert this expression to give the angle θ' observed by S' when the emission angle θ is observed by S.

23. A source of light, at rest in the S' frame, emits radiation uniformly in all directions. (a) Show that the fraction of light emitted into a cone of half-angle θ' is given by

$$f = \tfrac{1}{2}(1 - \cos \theta').$$

Calculate f for $\theta' = 30°$. (b) The source is viewed from frame S, the relative velocity of the two frames being $0.80c$. Find the value of θ (in frame S) to which this value of f corresponds, using the aberration formula; see Problem 22. Repeat the calculation for $u/c = 0.90$ and for $u/c = 0.990$. Can you see why this aberration phenomenon is often referred to as the "headlight effect"?

24. A radioactive nucleus moves with a uniform velocity of $0.050c$ in the laboratory frame. It decays by emitting a gamma ray. Find the direction of propagation of the gamma ray in the laboratory frame. Assume that the gamma ray is emitted (a) parallel to the direction of motion of the nucleus, as seen in the frame in which the nucleus is at rest, (b) at $45°$ to this direction, and (c) at $90°$ to this direction.

Section 42-5 Consequences of the Relativistic Doppler Effect

25. Give the Doppler wavelength shift $\lambda - \lambda_0$, if any, for the sodium D_2 line (589.00 nm) emitted from a source moving in a circle with constant speed $0.122c$ as measured by an observer fixed at the center of the circle.

26. A source of light moves at right angles to the line of sight to a detector. The speed of the source is $0.662c$. At what speed must an identical source move at $75.0°$ to the line of sight if the Doppler shifts as recorded by the detector for the two sources are equal?

27. A source of radio waves, rest frequency 188 MHz, is moving at a speed of $0.717c$ transverse to the line of sight to a detector. At what angle to the line of sight must a second source, rest frequency 162 MHz, move also at $0.717c$ if the frequencies of the two sources as received at the detector are to be equal?

28. Consider again the twin paradox. Suppose now that Ethel sends a birthday signal to Fred once each year (according to her clock). (a) At what rate does Fred receive the signals during Ethel's outward journey? (b) How many signals does Fred receive during Ethel's outward journey? (*Hint*: According to Fred's clock, when does the signal arrive showing Ethel's arrival at the distant star?) (c) At what rate does Fred receive signals during Ethel's return journey? (d) What is the total number of Ethel's birthday signals that Fred receives during her journey?

CHAPTER 43

REFLECTION AND REFRACTION AT PLANE SURFACES

Optics *refers to the study of the properties of light and its propagation through various materials. Traditional applications of optics include corrective lenses for vision and image formation by telescopes and microscopes. Modern applications include information storage and retrieval, such as in CD players or supermarket bar-code scanners, and signal transmission through optical fiber cables, which can carry a greater density of information than copper wires and are lighter in weight and less susceptible to electronic interference.*

In this chapter and the next, we consider cases in which light travels in straight lines and encounters objects whose size is much larger than the wavelength of the light. This is the realm of geometrical optics, *which includes the study of the properties of mirrors and lenses. The passage of light through very narrow slits or around very narrow barriers, whose dimensions may be comparable to the wavelength of the light, is a part of* physical optics *(or* wave optics*), which we begin to consider in Chapter 45.*

43-1 GEOMETRICAL OPTICS AND WAVE OPTICS

In our description of wave motion in Chapter 19, we used the *ray* as a convenient way to represent the motion of a train of waves; the ray is perpendicular to the wavefronts and indicates the direction of travel of the wave. A ray is a convenient geometrical construction that, as we see in the next chapter, is often helpful in studying the optical behavior of a system such as a lens. A ray is not a physical entity, however, and it is not possible to isolate one.

Consider a train of plane light waves of wavelength λ incident on a barrier in which there is a slit of width a. As suggested by Fig. 1a, if $\lambda \ll a$, the waves pass through the slit, and the barrier forms a sharp "shadow." As we make the slit smaller, we find that the light flares out into what was formerly the shadow of the barrier, as shown in Fig. 1b. This phenomenon, which is known as *diffraction*, occurs when the size of a slit (or an obstacle) in the path of the wave is comparable to the wavelength. We consider diffraction in detail in Chapter 46. Note (Fig. 1c) that diffraction becomes more pronounced as the slit width becomes smaller; thus an attempt to isolate a single ray will be futile.

Diffraction is not the exclusive property of light waves. In fact, phenomena we study for light (reflection, refraction, interference, diffraction, and polarization) can occur for other kinds of wave motion, even for mechanical waves. For example, Fig. 2 shows that diffraction can occur for water waves. For another example, when you shout through an open doorway, the sound waves are diffracted (the wavelength being comparable to the size of the doorway opening), and a friend can hear you even though she cannot see you (light waves having too small a wavelength to be diffracted noticeably upon passing through the opening).

If a is a measure of the smallest transverse dimension of a slit or obstacle, then the effects of diffraction can be ignored if the ratio of a/λ is large enough. In this case, the light appears to travel in straight-line paths that we can represent as rays. This is the condition for *geometrical optics,* also known as *ray optics.* When a light beam encounters such obstacles as mirrors, lenses, or prisms whose lateral size is much greater than the wavelength of light, we are safe in using the equations of geometrical optics.

If the condition for geometrical optics is not met, we cannot describe the behavior of light by rays but must take its wave nature specifically into account. In this case we

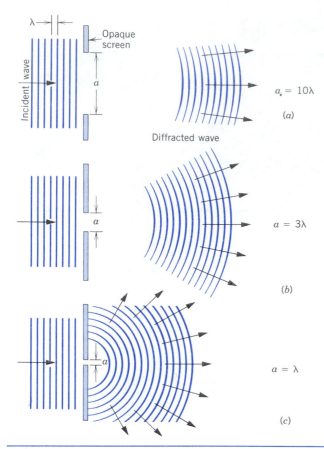

Figure 1   An attempt to isolate a ray by reducing the slit width fails because of diffraction, which becomes more pronounced for a fixed wavelength λ as the slit width a is reduced.

are in the realm of *physical optics* or *wave optics,* which includes geometrical optics as a limiting case, much as relativistic mechanics includes classical mechanics as a limiting case. We begin our discussion of wave optics in Chapter 45.

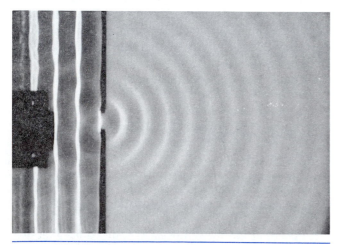

Figure 2   Diffraction of water waves at a slit in a ripple tank. Note that the slit width is about the same size as the wavelength. Compare with Fig. 1*c*.

43-2   REFLECTION AND REFRACTION

When you look at a pane of window glass, you of course notice that light reaches you from the other side of the glass, and a friend standing on the other side of the glass is able to see you. If you look carefully, however, you may also see your reflection in the glass. If you were to shine a flashlight on the glass, your friend would see the beam of light, but you might also see some of the light reflected back toward you.

In general, these two effects can occur whenever a beam of light travels from one medium (the air, for instance) to another (the glass). Part of the beam may be reflected back into the first medium, and part may be transmitted into the second medium. Figure 3 illustrates these two effects. Note that the beam of light may be bent or *refracted* as it enters the second medium.*

Geometrical optics includes the study of reflection and refraction. In this section we summarize the laws of reflection and refraction; later in this chapter we derive these laws and give examples of their applications when the boundary between the two media is a plane. Cases in which the boundary is curved, such as in spherical mirrors or lenses, are discussed in the next chapter.

In Fig. 3, the beams are represented by rays. The rays, which are drawn as straight lines perpendicular to the (plane) wavefronts, indicate the direction of motion of the wavefronts. Note the three rays shown in Fig. 3: the original or *incident* ray, the *reflected* ray, and the *refracted* ray, which changes direction as it enters the second medium.

At the point where the incident ray strikes the surface, we draw a line normal (perpendicular) to the surface, and we define three angles measured with respect to the normal: the *angle of incidence* θ_1, the *angle of reflection* θ'_1, and the *angle of refraction* θ_2. (The subscripts on the angles indicate the medium through which the ray travels. In our case, the ray is incident from medium 1, the air, and enters medium 2, the glass.) The plane formed by the incident ray and the normal is called the *plane of incidence*; it is the plane of the page in Fig. 3.

From experiment, we deduce the following laws governing reflection and refraction:

The Law of Reflection   The reflected ray lies in the plane of incidence, and

$$\theta'_1 = \theta_1. \tag{1}$$

The Law of Refraction   The refracted ray lies in the plane of incidence, and

$$n_1 \sin \theta_1 = n_2 \sin \theta_2. \tag{2}$$

* *Refracted* comes from the Latin for "broken"; the same root occurs in the word "fracture." If you dip a slanted pencil part way into a bowl of water, the pencil appears to be "broken."

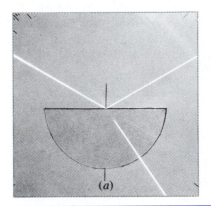
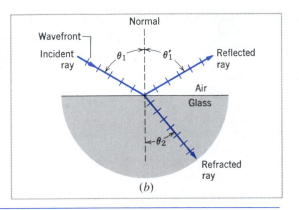

Figure 3   (*a*) A photograph showing the reflection and refraction of a light beam incident on a plane glass surface. (*b*) A representation using rays. The angles of incidence θ_1, reflection θ_1', and refraction θ_2 are marked. Note that the angles are measured between the normal to the surface and the appropriate ray.

Equation 2 is called Snell's law. Here n_1 and n_2 are dimensionless constants called the *index of refraction* of medium 1 and medium 2. The index of refraction n of a medium is the ratio between the speed of light c in vacuum and the speed of light v in that medium:

$$n = \frac{c}{v}. \tag{3}$$

We discussed the speed of light in various materials in Section 42-2. It is fair to say that refraction occurs because the speed of light changes from one medium to another. We develop this idea further in Section 43-5.

   Table 1 shows some examples of the index of refraction of various materials. Note that, for most purposes, air can be regarded as equivalent to a vacuum in its refraction of light. The index of refraction of a material generally varies with the wavelength of the light (see Fig. 4). Refraction can thus be used to analyze a beam of light into its constituent wavelengths, such as occurs in a rainbow.

Reflection and Refraction of Electromagnetic Waves (Optional)

The laws of reflection and refraction hold for all regions of the electromagnetic spectrum, not just for light. In fact, Eqs. 1 and 2 can be derived from Maxwell's equations, which makes them generally applicable to electromagnetic waves. Experimental evi-

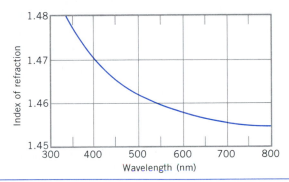

Figure 4   The index of refraction of fused quartz as a function of wavelength.

dence for this general applicability includes the reflection of microwaves or radio waves from the ionosphere and the refraction of x rays by crystals.

   We normally think of highly polished or smooth surfaces as "good" reflectors, but other surfaces may reflect as well, for example, a sheet of paper. The reflection by the paper (which is called a *diffuse reflection*) scatters the light more or less in all directions. It is largely by diffuse reflections that we see nonluminous objects around us. The difference between diffuse and specular (mirrorlike) reflection depends on the roughness of the surface: a reflected beam is formed only if the typical dimensions

TABLE 1   SOME INDICES OF REFRACTION[a]

Medium	Index	Medium	Index
Vacuum (exactly)	1.00000	Typical crown glass	1.52
Air (STP)	1.00029	Sodium chloride	1.54
Water (20°C)	1.33	Polystyrene	1.55
Acetone	1.36	Carbon disulfide	1.63
Ethyl alcohol	1.36	Heavy flint glass	1.65
Sugar solution (30%)	1.38	Sapphire	1.77
Fused quartz	1.46	Heaviest flint glass	1.89
Sugar solution (80%)	1.49	Diamond	2.42

[a] For a wavelength of 589 nm (yellow sodium light).

of the surface irregularities of the reflector are substantially less than the wavelength of the incident light. Thus the classification of the reflective properties of a surface depends on the wavelength of the radiation that strikes the surface. The bottom of a cast-iron skillet, for example, may be a good reflector for microwaves of wavelength 0.5 cm but is not a good reflector for visible light.

Maxwell's equations permit us to calculate how the incident energy is divided between the reflected and refracted beams. Figure 5 shows the theoretical prediction for (a) a light beam in air falling on a glass–air interface, and (b) a light beam in glass falling on a glass–air interface. Figure 5a shows that for angles of incidence up to about 60°, less than 10% of the light energy is reflected. At grazing incidence (that is, at angles of incidence near 90°), the surface becomes an excellent reflector. Another example of this effect is the high reflecting power of a wet road when light from automobile headlights strikes the road near grazing incidence.

Figure 5b shows clearly that at a certain critical angle (41.8° in this case), *all* the light is reflected. We consider this phenomenon, called *total internal reflection,* in Section 43-6. ■

Sample Problem 1 Figure 6 shows an incident ray *i* striking a plane mirror *MM′* at angle of incidence *θ*. Mirror *M′M″* is perpendicular to *MM′*. Trace this ray through its subsequent reflections.

Solution The reflected ray *r* makes an angle *θ* with the normal at *b* and falls as an incident ray on mirror *M′M″*. Its angle of incidence *θ′* on this mirror is *π/2 − θ*. A second reflected ray *r′* makes an angle *θ′* with the normal erected at *b′*. Rays *i* and *r′* are antiparallel for any value of *θ*. To see this, note that

$$\phi = \pi - 2\theta' = \pi - 2\left(\frac{\pi}{2} - \theta\right) = 2\theta.$$

Two lines are parallel if their opposite interior angles for an intersecting line (*φ* and *2θ*) are equal.

Repeat the problem if the angle between the mirrors is 120° rather than 90°.

The three-dimensional analogue of Fig. 6 is the *corner reflector,* which consists of three perpendicular plane mirrors joined like the positive sections of the coordinate planes of an *xyz* system. A corner reflector has the property that, for *any* direction of incidence, an incident ray is reflected back in the opposite direction. Highway reflectors use this principle, so that light from the headlights of an oncoming car is reflected back toward the car, no matter when the direction of approach of the car or the angle of the headlights above the road. Corner reflectors were placed on the Moon by the Apollo astronauts; timing a reflected laser beam from Earth permits precise determination of the Earth–Moon separation.

Sample Problem 2 A light beam in air is incident on the plane surface of a block of quartz and makes an angle of 30° with the normal. The beam contains two wavelengths, 400 and 500 nm. The indices of refraction for quartz at these wavelengths are 1.4702 and 1.4624, respectively. What is the angle between the two refracted beams in the quartz?

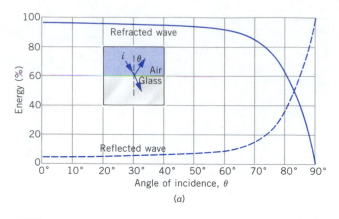

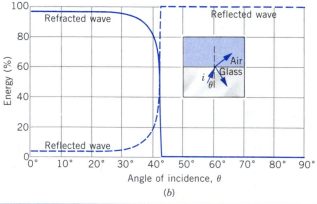

Figure 5 (a) The percentage of energy reflected and refracted when a wave in air is incident on glass (n = 1.50). (b) The same for a wave in glass incident on air, showing total internal reflection.

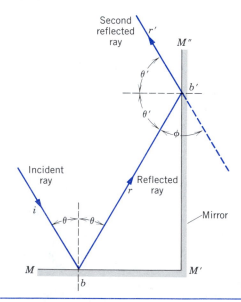

Figure 6 Sample Problem 1. A two-dimensional corner reflector.

Solution From Eq. 2 we have, for the 400-nm beam (taking $n_1 = 1$ for air)

$$\sin \theta_1 = n_2 \sin \theta_2,$$

or

$$\sin 30° = (1.4702) \sin \theta_2,$$

which leads to

$$\theta_2 = 19.88°.$$

For the 500-nm beam we have

$$\sin 30° = (1.4624) \sin \theta_2,$$

or

$$\theta_2 = 19.99°.$$

The angle $\Delta \theta_2$ between the beams is 0.11°, with the shorter wavelength component bent through the larger angle, that is, having the smaller angle of refraction. The difference in angle decreases as the angle of incidence decreases, becoming 0.018° for $\theta_1 = 5°$. In optical instruments using lenses, the variation in the refraction angle with wavelength leads to a distortion called *chromatic aberration*. Using small angles of incidence reduces the distortion due to chromatic aberration.

Sample Problem 3 A light beam in air is incident on one face of a glass prism as in Fig. 7. The angle θ is chosen so that the emerging ray also makes an angle θ with the normal to the other face. Derive an expression for the index of refraction of the prism material, taking $n = 1$ for air.

Solution Note that $\angle bad + \alpha = \pi/2$ and that $\angle bad + \phi/2 = \pi/2$, where ϕ is the prism angle. Therefore

$$\alpha = \tfrac{1}{2}\phi. \tag{4}$$

The deviation angle ψ is the sum of the two opposite interior angles in triangle *aed*, or

$$\psi = 2(\theta - \alpha).$$

Substituting $\tfrac{1}{2}\phi$ for α and solving for θ yield

$$\theta = \tfrac{1}{2}(\psi + \phi). \tag{5}$$

At point a, θ is the angle of incidence and α the angle of refraction. The law of refraction (see Eq. 2) is

$$\sin \theta = n \sin \alpha,$$

in which n is the index of refraction of the glass.

From Eqs. 4 and 5 this yields

$$\sin \frac{\psi + \phi}{2} = n \sin \frac{\phi}{2}$$

or

$$n = \frac{\sin (\psi + \phi)/2}{\sin (\phi/2)}, \tag{6}$$

which is the desired relation. This equation holds only for θ chosen so that the light ray passes symmetrically through the prism. In this case ψ is called the *angle of minimum deviation*; if θ is either increased or decreased, a larger deviation angle results.

43-3 DERIVING THE LAW OF REFLECTION

The law of reflection can be derived in several different ways. We discuss two of these derivations here.

Huygens' Principle

The Dutch physicist Christiaan Huygens* put forth a simple theory of light in 1678. This theory assumes that light is a wave, but it says nothing about the nature of the wave. (In particular, since Maxwell's theory of electromagnetism would not appear for nearly two centuries, Huygens' theory gives no hint of the electromagnetic character of light.) Huygens did not know whether light was a transverse wave or a longitudinal one; he did not know the wavelengths of visible light; he had little knowledge of the

---

* Christiaan Huygens (1629–1695) was a scientist of remarkable depth and influence. In addition to the wave theory of light, his accomplishments included improvements in telescope design that permitted him to deduce the shape of the rings of Saturn, the development of the pendulum clock, and contributions to the theory of rotating bodies (including the first recognition of the existence of centripetal acceleration) and colliding objects (including the principle of conservation of momentum).

Figure 7 Sample Problem 3.

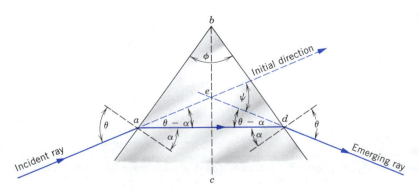

speed of light. Nevertheless, his theory was a useful guide to experiment for many years and remains useful today for pedagogic and certain other practical purposes. We must not expect it to yield the same wealth of detailed information that Maxwell's more complete electromagnetic theory does.

Huygens' theory is based on a geometrical construction that allows us to tell where a given wavefront will be at any time in the future if we know its present position. Huygens' principle can be stated as follows:

All points on a wavefront can be considered as point sources for the production of spherical secondary wavelets. After a time t the new position of a wavefront is the surface tangent to these secondary wavelets.

Consider a trivial example. Given a wavefront (*ab* in Fig. 8) in a plane wave in free space, where will the wavefront be a time *t* later? Following Huygens' principle, we let several points on this plane (the dots in Fig. 8) serve as centers for secondary spherical wavelets. In a time *t* the radius of these spherical waves is *ct*, where *c* is the speed of light in free space. We represent the plane tangent to these spheres at time *t* by *de*. As we expect, it is parallel to plane *ab* and a perpendicular distance *ct* from it. Thus plane wavefronts are propagated as planes and with speed *c*. Note that the Huygens method involves a three-dimensional construction and that Fig. 8 is the intersection of this construction with the plane of the page.

We might expect that, contrary to observation, a wave should be radiated backward as well as forward from the dots in Fig. 8. This result is avoided by assuming that the intensity of the spherical wavelets is not uniform in all directions but varies continuously from a maximum in the forward direction to a minimum of zero in the back direction. This is suggested by the shading of the circular arcs in Fig. 8. Huygens' method can be applied quantitatively to *all* wave phenomena; see Problem 24. The method was put on a firm mathematical footing two centuries after Huygens by Gustav Kirchhoff (1824 – 1887), who proved that the intensity of the wavelets varies with direction as described above.

Now we show how the law of reflection follows from Huygens' principle. Figure 9*a* shows three wavefronts in a plane wave falling on a plane mirror *MM'*. For convenience the wavefronts are chosen to be one wavelength apart. Note that θ_1, the angle between the wavefronts and the mirror, is the same as the angle between the incident ray and the normal to the mirror. In other words, θ_1 is the *angle of incidence*. The three wavefronts are related to each other by the Huygens construction, as in Fig. 8.

Let us regard point *a* in the wavefront in Fig. 9*b* as a source of a Huygens wavelet, which expands after a time λ/c to include point *b* on the surface of the mirror. Light from point *p* in this same wavefront cannot move beyond the mirror but must expand upward as a spherical Huy-

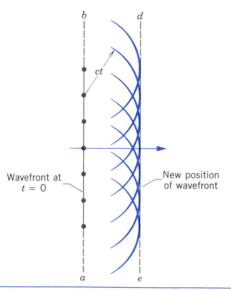

Figure 8 The propagation of a plane wave in free space is described by the Huygens construction. Note that the ray (horizontal arrow) representing the wave is perpendicular to the wavefronts.

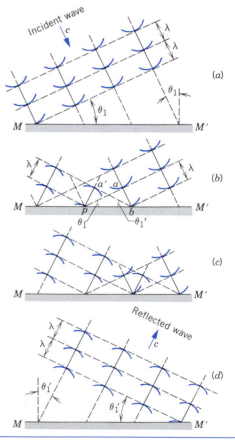

Figure 9 The reflection of a plane wave from a plane mirror as analyzed by the Huygens construction.

gens wavelet. Setting a compass to radius λ and swinging an arc about p provides a semicircle to which the reflected wavefront must be tangent. Since point b must lie on the new wavefront, this tangent must pass through b. Note that the angle θ_1' between the wavefront and the mirror is the same as the angle between the reflected ray and the normal to the mirror. In other words, θ_1' is the *angle of reflection*.

Consider right triangles abp and $a'bp$. They have side bp in common, and side $ab\ (=\lambda)$ is equal to side $a'p$. The two right triangles are thus congruent and we may conclude that

$$\theta_1 = \theta_1',$$

verifying the law of reflection. If you recall that the Huygens construction is three dimensional and that the arcs shown represent segments of spherical surfaces, you will be able to convince yourself that the reflected ray lies in the plane formed by the incident ray and the normal to the mirror, that is, the plane of Fig. 9. This is also a requirement of the law of reflection.

Figures 9c and 9d show how the process continues until all three incident wavefronts have been reflected.

Fermat's Principle

In 1650 Pierre Fermat* discovered a remarkable principle, which we can express in these terms:

A light ray traveling from one fixed point to another fixed point follows a path such that, compared with nearby paths, the time required is either a minimum or a maximum or remains unchanged (that is, stationary).

We can readily derive the law of reflection from this principle. Figure 10 shows two fixed points A and B and a reflecting ray APB connecting them. (We assume that ray APB lies in the plane of the figure; see Problem 25.) The total length L of this ray is

$$L = \sqrt{a^2 + x^2} + \sqrt{b^2 + (d - x)^2},$$

where x locates the point P at which the ray touches the mirror.

According to Fermat's principle, P will have a position such that the time of travel $t = L/c$ of the light must be a minimum (or a maximum or must remain unchanged),

* Pierre Fermat (1601–1665) was a French mathematician who is remembered for his development of analytic geometry and his many contributions to number theory. Perhaps his most challenging result is known as *Fermat's last theorem*: the equation $x^n + y^n = z^n$, in which x, y, z, and n are positive integers, has no solution for $n > 2$. Despite Fermat's claim of a proof of this theorem (a proof which he failed to publish), it has eluded mathematicians for more than 300 years.

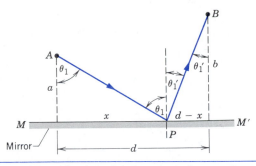

Figure 10   The reflection of a plane wave from a plane mirror as analyzed by using Fermat's principle. A ray from A passes through B after reflection at P.

which occurs when $dt/dx = 0$. Taking this derivative yields

$$\frac{dt}{dx} = \frac{1}{c}\frac{dL}{dx}$$

$$= \frac{1}{2c}(a^2 + x^2)^{-1/2}(2x)$$

$$+ \frac{1}{2c}[b^2 + (d - x)^2]^{-1/2}(2)(d - x)(-1) = 0,$$

which we can rewrite as

$$\frac{x}{\sqrt{a^2 + x^2}} = \frac{d - x}{\sqrt{b^2 + (d - x)^2}}.$$

(In evaluating the derivative, note that we hold the endpoints fixed and vary the path by allowing x to vary.) Comparison with Fig. 10 shows that we can rewrite this as

$$\sin \theta_1 = \sin \theta_1',$$

or

$$\theta_1 = \theta_1',$$

which is the law of reflection.

43-4   IMAGE FORMATION BY PLANE MIRRORS

Perhaps our most familiar optical experience is looking into a mirror. Figure 11 shows a point source of light O, which we call the *object*, placed a distance o in front of a plane mirror. The light that falls on the mirror is represented by rays emanating from O.† At the point at which

† In our previous discussion of reflection in this chapter, we assumed an incident *plane* wave; the incident rays are parallel to each other in that case. Here we have a *point* source, and the rays striking the mirror are *diverging* from that point source. We can regard the point source as a source of spherical waves, and the rays radiating from the source are perpendicular to the spherical wavefronts.

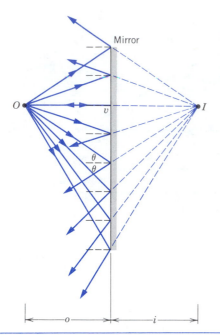

Figure 11   A point object *O* forms a virtual image *I* in a plane mirror. The rays *appear* to diverge from *I* but no light is actually present at that point.

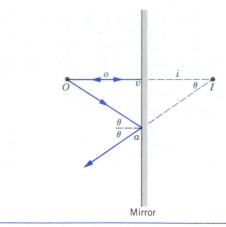

Figure 12   Two rays from Fig. 11. Ray *Oa* makes an arbitrary angle θ with the normal to the surface of the mirror.

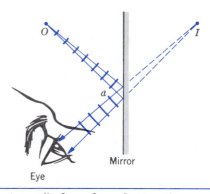

Figure 13   A pencil of rays from *O* enters the eye after reflection from the mirror. Only a small portion of the mirror near *a* is effective. The small arcs represent portions of the spherical wavefronts. The light appears to come from *I*.

each ray strikes the mirror we construct a reflected ray. If we extend the reflected rays backward, they intersect at a point *I*, which we call the *image* of the object *O*. The image is the same distance behind the mirror that the object *O* is in front of it, which we prove below.

Images may be *real* or *virtual*. In a real image light actually passes through the image point; in a virtual image the light *behaves* as though it diverges from the image point, although, in fact, it does not pass through this point; see Fig. 11. Images of diverging light in plane mirrors are always virtual. We know from daily experience how "real" such a virtual image *appears to be* and how definite is its location in the space behind the mirror, even though this space may, in fact, be occupied by a brick wall.

Figure 12 shows two rays from Fig. 11. One strikes the mirror at *v*, along a perpendicular line. The other strikes it at an arbitrary point *a*, making an angle of incidence θ with the normal at that point. Elementary geometry shows that the angles *aOv* and *aIv* are also equal to θ. Thus the right triangles *aOv* and *aIv* are congruent and so

$$o = -i, \qquad (7)$$

in which we introduce the minus sign to show that *I* and *O* are on opposite sides of the mirror. Equation 7 does not involve θ, which means that *all* rays from *O* striking the mirror pass through *I* when extended backward, as we have seen in Fig. 11. Other than assuming that the mirror is truly plane and that the conditions for geometrical optics hold, we have made no approximations in deriving

Eq. 7. A point object produces a point image in a plane mirror, with $o = -i$, no matter how large the angle θ in Fig. 12.

Because of the finite diameter of the pupil of the eye, only rays that lie fairly close together can enter the eye after reflection at a mirror. For the eye position shown in Fig. 13, only a small patch of the mirror near point *a* is effective in forming the image; the rest of the mirror may be covered up or removed. If we move our eye to another location, a different patch of the mirror will be effective; the location of the virtual image *I* will remain unchanged, however, as long as the object remains fixed.

If the object is an extended source such as the head of a person, a virtual image is also formed. We can consider an extended source to be an array of point sources, each of which produces spherical waves. From Eq. 7, every object point of the source has a corresponding image point that lies an equal distance directly behind the plane of the mirror. Thus the image reproduces the object point by point. Most of us prove this every day by looking into a mirror.

Image Reversal

As Fig. 14*a* shows, the image of a left hand appears to be a right hand. We interpret this appearance as a reversal of left and right. That is, if you raise your left hand, then your mirror image raises its right hand. It is often then asked: Why does a mirror reverse left and right but not also reverse up and down?

Figure 14*b* illustrates the way a mirror reverses the image of a three-dimensional object, represented simply as a set of three mutually perpendicular arrows. Note that the arrows parallel to the plane of the mirror (arrows *x* and *y*) are identical with their mirror images. Only the *z* arrow has its direction changed by the reflection. It is therefore more accurate to say that a mirror reverses front to back rather than left to right. The transformation of a left hand to a right hand is accomplished, in a sense, by exchanging the front and back of the hand.

Note also that the object can be considered to represent a conventional right-handed coordinate system (*x* "crossed into" *y* points in the *z* direction), while the image is a left-handed coordinate system (*x* "crossed into" *y* points in the *negative z* direction). Such reversals apply to physical objects as well; for example, the image of a screw with right-handed threads is a screw with left-handed threads.

If we knew for a fact that all humans were right-handed, then we could surely tell the difference between a physical situation and its mirror image; the "real" person would be using the right hand, while the image would use the left. However, if humans were ambidextrous, we could not use this feature to distinguish between the real world and the looking-glass world. The same distinction has been applied to the laws of physics: if the laws of physics have perfect left–right symmetry, then the mirror image of an experiment would also be a physically possible experiment. If, however, the laws lack that symmetry, then the outcome of certain mirror-image experiments would not be physically possible. In 1956, it was discovered that the so-called weak interaction, which causes certain radioactive decays, lacks this symmetry, which is called *parity* (see Section 3-6). This experiment provided the first fundamental basis for a distinction between our world and its mirror image.*

Sample Problem 4 Find the minimum length *h* of a mirror that is needed for a person of height *H* to see his entire reflection.

Solution Figure 15 shows the person's foot *f*, eyes *e*, and top of head *t*. For him to see his entire height, a ray of light (*tae*) must leave the top of his head, reflect from the mirror at *a*, and enter his eyes, while another ray (*fce*) must leave his feet, reflect from

---

* For some amusing discussions about symmetry and the distinctions between objects and their mirror images, see *The Ambidextrous Universe,* by Martin Gardner (Scribner's, 1979), and *Reality's Mirror,* by Bryan Bunch (Wiley, 1989).

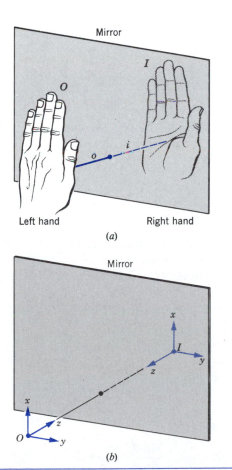

Figure 14 (*a*) The object *O* is a left hand; its image *I* is a right hand. (*b*) Study of a reflected three-arrow object shows that a mirror interchanges front and back, rather than left and right.

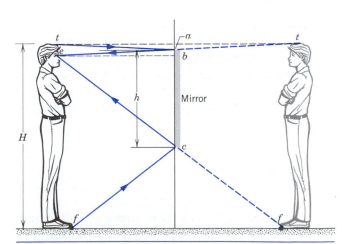

Figure 15 Sample Problem 4.

the mirror at *c*, and enter his eyes. The person will see a full-height reflection (including the virtual images of points *t* and *f*) if the length of the mirror is at least *ac*.

From the geometry of Fig. 15, we see that

$$ab = \tfrac{1}{2}te \quad \text{and} \quad bc = \tfrac{1}{2}ef,$$

where point *b* is at the same height as the eyes. Thus

$$ac = ab + bc = \tfrac{1}{2}te + \tfrac{1}{2}ef = \tfrac{1}{2}tf.$$

With *h* = *ac* and *H* = *tf*, we obtain

$$h = \tfrac{1}{2}H.$$

The person can see his entire image if the mirror is at least half his height. Portions of the mirror below *c* show reflections of the floor in front of his feet, while those above *t* show the person what is above his head. Note that the distance of the person from the mirror makes no difference in this calculation, which remains valid for any object distance from a plane mirror.

43-5 DERIVING THE LAW OF REFRACTION

In analogy with Section 43-3, here we use Huygens' principle and Fermat's principle to derive the law of refraction (Eq. 2).

Huygens' Principle

Figure 16 shows four stages in the refraction of three successive wavefronts in a plane wave falling on an interface between air (medium 1) and glass (medium 2). For convenience, we assume that the incident wavefronts are separated by λ_1, the wavelength as measured in medium 1. Let the speed of light in air be v_1 and that in glass be v_2. We assume that

$$v_2 < v_1. \tag{8}$$

This assumption about the speeds is vital to the derivation that follows.

The wavefronts in Fig. 16*a* are related to each other by the Huygens construction of Fig. 8. As in Fig. 9, θ_1 is the angle of incidence. In Fig. 16*b* consider the time ($= \lambda_1/v_1$) during which a Huygens wavelet from point *e* moves to include point *d*. Light from point *h*, traveling through glass at a reduced speed (recall the assumption of Eq. 8) moves a shorter distance

$$\lambda_2 = \lambda_1 \frac{v_2}{v_1} \tag{9}$$

during this time. This follows from $v = \lambda \nu$ and $\nu_1 = \nu_2$. The refracted wavefront must be tangent to an arc of this radius centered on *h*. Since *d* lies on the new wavefront, the tangent must pass through this point, as shown. Note that θ_2, the angle between the refracted wavefront and the air–glass interface, is the same as the angle between the

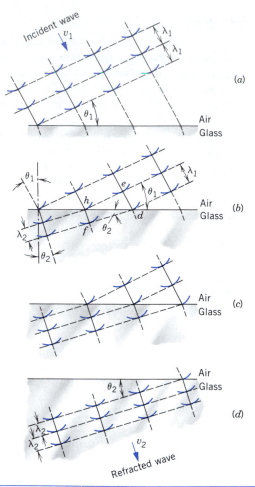

Figure 16 The refraction of a plane wave at a plane interface as described by the Huygens construction. For simplicity the reflected wave is not shown. Note the change in wavelength of the refracted wave.

refracted ray and the normal to this interface. In other words, θ_2 is the *angle of refraction*. Note too that the wavelength in glass (λ_2) is less than the wavelength in air (λ_1).

For the right triangles *hde* and *hdf* we may write

$$\sin \theta_1 = \frac{\lambda_1}{hd} \quad \text{(for } hde\text{)}$$

and

$$\sin \theta_2 = \frac{\lambda_2}{hd} \quad \text{(for } hdf\text{)}.$$

Dividing and using Eq. 9 yields

$$\frac{\sin \theta_1}{\sin \theta_2} = \frac{\lambda_1}{\lambda_2} = \frac{v_1}{v_2}. \tag{10}$$

Introducing a common factor of *c* allows us to rewrite Eq. 10 as

$$\frac{c}{v_1} \sin \theta_1 = \frac{c}{v_2} \sin \theta_2. \tag{11}$$

According to Eq. 3, $c/v_1 = n_1$ and $c/v_2 = n_2$, so that Eq. 11 becomes

$$n_1 \sin \theta_1 = n_2 \sin \theta_2, \qquad (12)$$

which is the law of refraction.

If one medium is a vacuum, Eq. 9 becomes

$$\lambda_n = \lambda \frac{v}{c} = \frac{\lambda}{n}, \qquad (13)$$

where λ_n is the wavelength of light in a medium of index n and λ is the wavelength in vacuum. In passing from one medium to another, both the speed of light and its wavelength are reduced by the same factor, but the frequency of the light is unchanged.

The application of Huygens' principle to refraction requires that if a light ray is bent toward the normal in passing from air to an optically dense medium, then the speed of light in that optically dense medium (glass, say) must be *less* than that in air; see Eq. 8. This requirement holds for all wave theories of light. In the early particle theory of light put forward by Newton, the explanation of refraction required that the speed of light in the medium in which light is bent toward the normal (the optically denser medium) be *greater* than that in air. The denser medium was thought to exert attractive forces on the light "corpuscles" as they neared the surface, speeding them up and changing their direction to cause them to make a smaller angle with the normal.

An experimental comparison of the speed of light in air and in an optically denser medium is therefore critical in deciding between the wave and corpuscular theories of light. Such a measurement was first carried out by Foucault in 1850; he showed conclusively that light travels more slowly in water than in air, thus ruling out the corpuscular theory of Newton.

Fermat's Principle

To prove the law of refraction from Fermat's principle, consider Fig. 17, which shows two fixed points A and B in two different media and a refracting ray APB connecting them. The time t for the ray to travel from A to B is given by

$$t = \frac{L_1}{v_1} + \frac{L_2}{v_2}.$$

Using the relation $n = c/v$ we can write this as

$$t = \frac{n_1 L_1 + n_2 L_2}{c} = \frac{L}{c}, \qquad (14)$$

where L is the *optical path length*, defined as

$$L = n_1 L_1 + n_2 L_2. \qquad (15)$$

For any light ray traveling through successive media, the optical path length is the sum of the products of the geometrical path length of each segment and the index of refraction of that medium. Equation 13 ($\lambda_n = \lambda/n$) shows

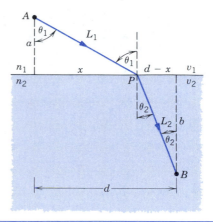

Figure 17   The refraction of a plane wave at a plane interface as analyzed using Fermat's principle. A ray from A passes through B after refraction at P.

that the optical path length is equal to the length that this same number of waves would have if the medium were a vacuum. Do not confuse the optical path length with the geometrical path length, which is $L_1 + L_2$ for the ray of Fig. 17.

Fermat's principle requires that the time t for the light to travel the path APB must be a minimum (or a maximum or must remain unchanged), which in turn requires that x be chosen so that $dt/dx = 0$. The optical path length in Fig. 17 is

$$L = n_1 L_1 + n_2 L_2 = n_1 \sqrt{a^2 + x^2} + n_2 \sqrt{b^2 + (d-x)^2}.$$

Substituting this result into Eq. 14 and differentiating, we obtain

$$\frac{dt}{dx} = \frac{1}{c} \frac{dL}{dx}$$

$$= \frac{n_1}{2c} (a^2 + x^2)^{-1/2}(2x)$$

$$+ \frac{n_2}{2c} [b^2 + (d-x)^2]^{-1/2} (2)(d-x)(-1) = 0,$$

which we can write as

$$n_1 \frac{x}{\sqrt{a^2 + x^2}} = n_2 \frac{d-x}{\sqrt{b^2 + (d-x)^2}}.$$

Comparison with Fig. 17 shows that we can write this as

$$n_1 \sin \theta_1 = n_2 \sin \theta_2,$$

which is the law of refraction.

Sample Problem 5   Red light of wavelength 632 nm in free space is incident, at an angle of $\theta_1 = 39°$ with respect to the normal, on a glass microscope slide of thickness $d = 0.78$ mm and index of refraction $n = 1.52$ (Fig. 18). Find (*a*) the wavelength in the glass and (*b*) the optical path length of the light in traveling through the glass.

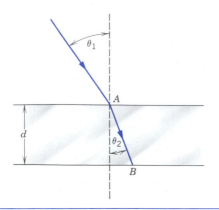

Figure 18   Sample Problem 5.

Solution   (*a*) We can find the wavelength in the glass using Eq. 13, which gives

$$\lambda_n = \frac{\lambda}{n} = \frac{632 \text{ nm}}{1.52} = 416 \text{ nm}.$$

(*b*) The angle of refraction is found from Eq. 12,

$$\sin \theta_2 = \frac{\sin \theta_1}{n} = \frac{\sin 39°}{1.52} = 0.414,$$

or

$$\theta_2 = 24.5°,$$

and the actual length of the path through the glass is

$$AB = \frac{d}{\cos \theta_2} = \frac{0.78 \text{ mm}}{\cos 24.5°} = 0.856 \text{ mm}.$$

The optical path is

$$L = n(AB) = 1.52(0.856 \text{ mm}) = 1.30 \text{ mm}.$$

43-6   TOTAL INTERNAL REFLECTION

Figure 19 shows rays from a point source in glass falling on a glass–air interface. As the angle of incidence θ is increased, we reach a situation (see ray *e*) at which the refracted ray points along the surface, the angle of refrac-

tion being 90°. For angles of incidence larger than this *critical angle* θ_c, there is no refracted ray, and we speak of *total internal reflection*.

We find the critical angle by putting $\theta_2 = 90°$ in the law of refraction (see Eq. 2):

$$n_1 \sin \theta_c = n_2 \sin 90°,$$

or

$$\theta_c = \sin^{-1} \frac{n_2}{n_1}. \tag{16}$$

For glass in air, $\theta_c = \sin^{-1}(1.00/1.50) = 41.8°$. Figure 5 indicates that the energy of the reflected wave becomes 100% when the angle of incidence exceeds 41.8°.

The sine of an angle cannot exceed unity so that we must have $n_2 < n_1$. This tells us that total internal reflection cannot occur when the incident light is in the medium of lower index of refraction. The word *total* means just that; the reflection occurs with no loss of intensity. In ordinary reflection from a mirror, by way of contrast, there is an intensity loss of about 4%.

Total internal reflection makes possible fiber optical devices by means of which physicians can visually inspect many internal body sites; see Fig. 20. In these devices, a bundle of fibers transmits an image that can be inspected visually outside the body.* Optical fibers are also used for telephone communications and, because of their light weight and freedom from electromagnetic interference, for carrying signals on aircraft. Figure 21 shows light emerging from an optical fiber.†

As Fig. 22 shows, the fiber consists of a central core that is graded smoothly into an outer cladding layer of a material of lower index of refraction. Only those rays that are internally reflected can be propagated along the fiber. To reduce attenuation of the signal as it passes along the fiber, materials of extreme purity have been developed. If sea-

---

* See "Optical Fibers in Medicine," by Abraham Katzir, *Scientific American,* May 1989, p. 120.
† See "Lightwaves and Telecommunication," by Stewart E. Miller, *American Scientist,* January–February 1984, p. 66, and "Light-Wave Communications," by W. S. Boyle, *Scientific American,* August 1977, p. 40.

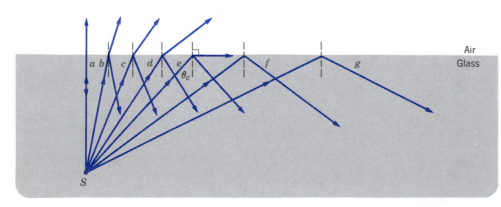

Figure 19   Total internal reflection of light from a point source S occurs for all angles of incidence greater than the critical angle θ_c. At the critical angle, the refracted ray points along the air–glass interface.

water were as transparent as the glass from which optical fibers are made, it would be possible to see the sea bottom by reflected sunlight at a depth of several miles.

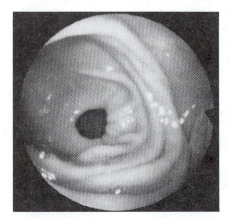

Figure 20   Fiber optic image of the passage from the stomach into the small intestine.

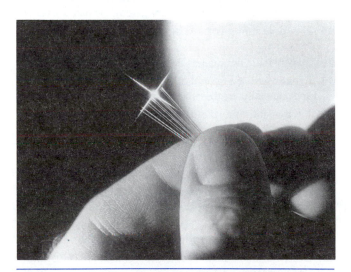

Figure 21   Light is transmitted through an optical fiber.

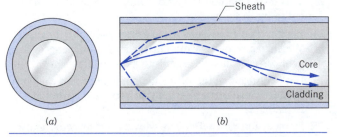

Figure 22   (*a*) An optical fiber shown in cross section. The diameter of the fiber is about the same as that of a human hair. (*b*) A transverse view, showing propagation by total internal reflection. The core, the cladding (of lower index than the core), and the protective sheath are shown.

Sample Problem 6   Figure 23*a* shows a triangular prism of glass, a ray incident normal to one face being totally reflected. If θ_1 is 45°, what can you conclude about the index of refraction n of the glass?

Solution   The angle θ_1 must be equal to or greater than the critical angle θ_c, where θ_c is given by Eq. 16:

$$\theta_c = \sin^{-1} \frac{n_2}{n_1} = \sin^{-1} \frac{1}{n},$$

or

$$n = \frac{1}{\sin \theta_c},$$

in which the index of refraction of air ($= n_2$) is set equal to unity. Since total internal reflection occurs, θ_c must be less than 45°, and so

$$n > \frac{1}{\sin 45°} = 1.41.$$

Thus the index of refraction of the glass must exceed 1.41. If n were less than 1.41, the ray shown in Fig. 23*a* would be partially refracted into the air, instead of totally reflected back into the glass.

Sample Problem 7   What happens if the prism in Sample Problem 6 (assume that $n_1 = 1.50$) is immersed in water ($n_2 = 1.33$)? See Fig. 23*b*.

Solution   The new critical angle, given by Eq. 16, is

$$\theta_c = \sin^{-1} \frac{n_2}{n_1} = \sin^{-1} \frac{1.33}{1.50} = 62.5°.$$

The actual angle of incidence ($= 45°$) is less than this so that we do *not* have total internal reflection.

There is a reflected ray r, with an angle of reflection of 45°, as Fig. 23*b* shows. There is also a refracted ray r', with an angle of refraction given by

$$n_1 \sin \theta_1 = n_2 \sin \theta_2$$

$$(1.50)(\sin 45°) = (1.33) \sin \theta_2,$$

which yields $\theta_2 = 52.9°$. Show that as $n_2 \to n_1$, $\theta_c \to 90°$.

How does the incident ray i in Figs. 23*a* and 23*b* determine whether there is air or water beyond the glass? That is, how does it "know" whether to be totally reflected or partially refracted? The traveling wave in glass establishes

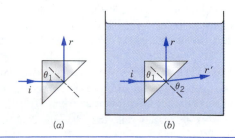

Figure 23   (*a*) Sample Problem 6. (*b*) Sample Problem 7.

electric and magnetic fields that are strongly decreasing exponential functions of distance that penetrate a few wavelengths into the next medium. These fields are not those associated with a traveling wave but can be regarded as "sampling" the medium beyond the boundary. We can demonstrate this penetration by placing a second glass prism near the first, as in Fig. 24. In sampling medium 2 (the air), the fields also sense the second prism; although waves are forbidden by the law of refraction from appearing in the air gap between the prisms, they are not forbidden from propagating in the second prism. Note from Fig. 24 that the light ray appears in the second prism but not in the air gap. This situation is called *frustrated total internal reflection* and is a general property of waves. (It can be done, for instance, with microwaves.) In quantum mechanics the wavelike properties of material particles permit a similar effect called *barrier penetration*: a particle

Figure 24   Frustrated total internal reflection. The thicker the air gap, the smaller the intensity of the light in the second prism (indicated by the width of the rays). Note that light waves do not appear *in* the gap.

can pass from one allowed region to another allowed region by penetrating a region in which it is forbidden. We consider barrier penetration in Chapter 50 of the extended text.

QUESTIONS

1. Describe what your immediate environment would be like if all objects were totally absorbing of light. Sitting in a chair in a room, could you see anything? If a cat entered the room could you see it?

2. Can you think of a simple test or observation to prove that the law of reflection is the same for all wavelengths, under conditions in which geometrical optics prevail?

3. A street light, viewed by reflection across a body of water in which there are ripples, appears very elongated in the line of vision but not sideways. Explain.

4. Shortwave broadcasts from Europe are heard in the United States even though the path is not a straight line. Explain how.

5. The travel time of signals from satellites to receiving stations on Earth varies with the frequency of the signal. Why?

6. By what percentage does the speed of blue light in fused quartz differ from that of red light?

7. Can (a) reflection phenomena or (b) refraction phenomena be used to determine the wavelength of light?

8. How can one determine the indices of refraction of the media in Table 1 relative to water, given the data in that table?

9. Would you expect sound waves to obey the laws of reflection and refraction obeyed by light waves? Discuss the propagation of spherical and cylindrical waves using Huygens' principle. Does Huygens' principle apply to sound waves in air?

10. If Huygens' principle predicts the laws of reflection and refraction, why is it necessary or desirable to view light as an electromagnetic wave, with all its attendant complexity?

11. A light beam is broadened upon entering water. Explain.

12. What is a plausible explanation for the observation that a street appears darker when wet than when dry?

13. Sound waves are largely reflected when incident on water from air. Why?

14. Is it correct to say that there is no interaction between visible light and a transparent medium through which it passes?

15. How does atmospheric refraction affect the apparent time of sunset?

16. Stars twinkle but planets do not. Why?

17. Explain why the far end of a pool filled to a uniform depth appears shallower than the end near the observer at its edge.

18. It is a bright sunny day and you want to create a rainbow in your back yard using a garden hose. Exactly how do you go about it? Incidentally, why can't you walk under, or go to the end of, a rainbow?

19. Is it possible, by using one or more prisms, to recombine into white light the color spectrum formed when white light passes through a single prism? If yes, explain how.

20. You are given a cube of glass. How can you find the speed of light (from a sodium light source) in this cube? ·

21. Describe and explain what a fish sees as it looks in various directions above its horizon.

22. How did Foucault's measurement of the speed of light in water decide between the wave and particle theories of light?

23. Why does a diamond "sparkle" more than a glass imitation cut to the same shape?

24. Light has (a) a wavelength, (b) a frequency, and (c) a speed. Which, if any, of these quantities remains unchanged when light passes from a vacuum into a slab of glass?

25. Is it plausible that the wavelength of light should change in passing from air into glass but that its frequency should not? Explain.

26. In reflection and refraction why do the reflected and refracted rays lie in the plane defined by the incident ray and the normal to the surface? Can you think of any exceptions?

27. What causes mirages? Does it have anything to do with the fact that the index of refraction of air is not constant but

varies with its density? See "Mirages," by Alistair B. Fraser and William B. Mach, *Scientific American,* January 1976, p. 102.

28. Can a virtual image be photographed by exposing a film at the location of the image? Explain.

29. At night, in a lighted room, you blow a smoke ring toward a window pane. If you focus your eyes on the ring as it approaches the pane it will seem to go right through the glass into the darkness beyond. What is the explanation of this illusion?

30. In driving a car you sometimes see vehicles such as ambulances with letters printed on them in such a way that they read in the normal fashion when you look through the rear-view mirror. Print your name so that it may be so read.

31. We have seen that a single reflection in a plane mirror reverses right and left. When we drive down a highway, for example, the letters on the highway signs are reversed as seen through the rear-view mirror. And yet, as seen through this same mirror, you still seem to be driving down the right lane. Why does the mirror reverse the signs and not the lanes? Or does it? Discuss.

32. We all know that when we look into a mirror right and left are reversed. Our right hand will seem to be a left hand; if we part our hair on the left it will seem to be parted on the right, and so on. Can you think of a system of mirrors that would let us see ourselves as others see us? If so, draw it and prove your point by drawing some typical rays.

33. Devise a system of plane mirrors that will let you see the back of your head. Trace the rays to prove your point.

34. Design a periscope, taking advantage of total internal reflection. What are the advantages compared with silvered mirrors?

35. What characteristics must a material have in order to serve as an efficient "light pipe"?

36. A certain toothbrush has a red plastic handle into which rows of nylon bristles are set. The tops of the bristles (but not their sides) appear red. Explain.

37. Why are optical fibers more effective carriers of information than, say, microwaves or cables? Think of the frequencies involved.

38. What does "optical path length" mean? Can the optical path length ever be less than the geometrical path length? Ever greater?

39. A solution of copper sulfate appears blue when we view it through transmitted light. Does this mean that a copper sulfate solution absorbs blue light selectively? Discuss.

PROBLEMS

Section 43-2 Reflection and Refraction

1. In Fig. 25 find the angles (*a*) θ_1 and (*b*) θ_2.

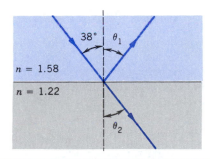

Figure 25   Problem 1.

2. Light in vacuum is incident on the surface of a glass slab. In the vacuum the beam makes an angle of 32.5° with the normal to the surface, while in the glass it makes an angle of 21.0° with the normal. Find the index of refraction of the glass.

3. The speed of yellow sodium light in a certain liquid is measured to be 1.92×10^8 m/s. Find the index of refraction of this liquid with respect to air, for sodium light.

4. Find the speed in fused quartz of light of wavelength 550 nm. (See Fig. 4.)

5. When an electron moves through a medium at a speed exceeding the speed of light in that medium, it radiates electromagnetic waves (the Cerenkov effect). What minimum speed must an electron have in a liquid of index of refraction 1.54 in order to radiate?

6. A laser beam travels along the axis of a straight section of pipeline 1.61 km long. The pipe normally contains air at standard temperature and pressure, but it may also be evacuated. In which case would the travel time for the beam be greater and by how much?

7. When the rectangular metal tank in Fig. 26 is filled to the top with an unknown liquid, an observer with eyes level with the top of the tank can just see the corner *E*. Find the index of refraction of the liquid.

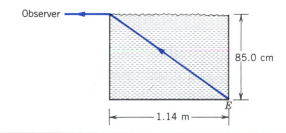

Figure 26   Problem 7.

8. Ocean waves moving at a speed of 4.0 m/s are approaching a beach at an angle of 30° to the normal, as shown in Fig. 27. Suppose the water depth changes abruptly and the wave speed drops to 3.0 m/s. Close to the beach, what is the angle θ between the direction of wave motion and the normal? (Assume the same law of refraction as for light.) Explain why

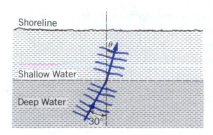

Figure 27   Problem 8.

most waves come in normal to a shore even though at large distances they approach at a variety of angles.

9. A ray of light goes through an equilateral prism in the position of minimum deviation. The total deviation is 37°. What is the index of refraction of the prism? See Sample Problem 3.

10. Two perpendicular mirrors form the sides of a vessel filled with water, as shown in Fig. 28. A light ray is incident from above, normal to the water surface. (*a*) Show that the emerging ray is parallel to the incident ray. Assume that there are two reflections at the mirror surfaces. (*b*) Repeat the analysis for the case of oblique incidence, the ray lying in the plane of the figure.

Figure 28   Problem 10.

11. In Fig. 7 (Sample Problem 3) show by graphical ray tracing, using a protractor, that if θ for the incident ray is either increased or decreased, the deviation angle ψ is increased.

12. Light from a laser enters a glass block at A and emerges at B; see Fig. 29. The glass block has a length $L = 54.7$ cm and an index of refraction $n = 1.63$. The angle of incidence is $\theta = 24.0°$. Find the time needed for light to pass through the block.

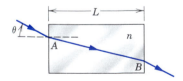

Figure 29   Problem 12.

13. A diver beneath the surface of water in a lake looks up at 27° from the vertical to see a life ring floating on the surface. Through the center of the ring can be seen the top of a smokestack known to be 98 m high. How far is the base of the smokestack from the life ring?

14. A bottom-weighted 200-cm-long vertical pole extends from the bottom of a swimming pool to a point 64 cm above the water. Sunlight is incident at 55° above the horizon. Find the length of the shadow of the pole on the level bottom of the pool.

15. Prove that a ray of light incident on the surface of a sheet of plate glass of thickness t emerges from the opposite face parallel to its initial direction but displaced sideways, as in Fig. 30. (*a*) Show that, for small angles of incidence θ, this displacement is given by

$$x = t\theta \frac{n-1}{n},$$

where n is the index of refraction and θ is measured in radians. (*b*) Calculate the displacement at a 10° angle of incidence through a 1.0-cm-thick sheet of crown glass.

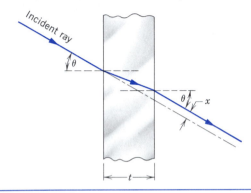

Figure 30   Problem 15.

16. A glass prism with an apex angle of 60° has $n = 1.60$. (*a*) What is the smallest angle of incidence for which a ray can enter one face of the prism and emerge from the other? (*b*) What angle of incidence would be required for the ray to pass through the prism symmetrically? See Sample Problem 3.

17. A coin lies at the bottom of a pool with depth d and index of refraction n, as shown in Fig. 31. Show that light rays that are close to the normal appear to come from a point $d_{app} = d/n$ below the surface. This distance is the *apparent depth* of the pool.

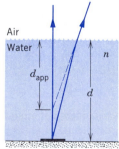

Figure 31   Problem 17.

18. The apparent depth of a pool depends on the angle of viewing. Suppose that you place a coin at the bottom of a swimming pool filled with water ($n = 1.33$) to a depth of 2.16 m. Find the apparent depth of the coin below the surface when viewed (*a*) at near normal incidence and (*b*) by rays that leave the coin making an angle of 35.0° with the normal to the bottom of the pool. See Problem 17.

19. A layer of water ($n = 1.33$) 20 mm thick floats on a layer of carbon tetrachloride ($n = 1.46$) 41 mm thick. How far below the water surface, viewed at near normal incidence, does the bottom of the tank seem to be?

20. The index of refraction of the Earth's atmosphere decreases monotonically with height from its surface value (about 1.00029) to the value in space (about 1.00000) at the top of the atmosphere. This continuous (or graded) variation can be approximated by considering the atmosphere to be composed of three (or more) plane parallel layers in each of which the index of refraction is constant. Thus, in Fig. 32, $n_3 > n_2 > n_1 > 1.00000$. Consider a ray of light from a star S that strikes the top of the atmosphere at an angle θ with the vertical. (*a*) Show that the apparent direction θ_3 of the star with the vertical as seen by an observer at the Earth's surface is obtained from

$$\sin \theta_3 = \frac{1}{n_3} \sin \theta.$$

(*Hint*: Apply the law of refraction to successive pairs of layers of the atmosphere; ignore the curvature of the Earth.) (*b*) Calculate the shift in position of a star observed to be $50°$ from the vertical. (The very small effects due to atmospheric refraction can be most important; for example, they must be taken into account in using navigation satellites to obtain accurate fixes of position on the Earth.)

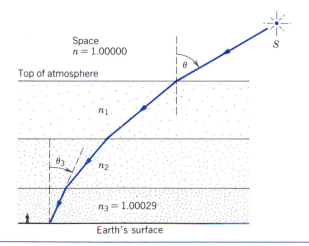

Figure 32 Problem 20.

21. You stand at one end of a long airport runway. A vertical temperature gradient in the air has resulted in the index of refraction of the air above the runway to vary with height y according to $n = n_0(1 + ay)$, where n_0 is the index of refraction at the runway surface and $a = 1.5 \times 10^{-6}$ m^{-1}. Your eyes are at a height $h = 1.7$ m above the runway. Beyond what horizontal distance d can you not see the runway? See Fig. 33 and Problem 20.

22. A *corner reflector,* much used in optical, microwave, and other applications, consists of three plane mirrors fastened together as the corner of a cube. It has the property that an incident ray is returned, after three reflections, with its direction exactly reversed. Prove this result.

23. Muons (mass $= 106$ MeV/c^2) and neutral pions (mass $=$

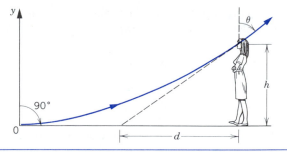

Figure 33 Problem 21.

135 MeV/c^2), each with momentum 145 MeV/c, pass through a transparent material. Find the range of index of refraction of the material so that only the muons emit Cerenkov radiation. (See Problem 5.)

Section 43-3 Deriving the Law of Reflection

24. One end of a stick is dragged through water at a speed v that is greater than the speed u of water waves. Applying Huygens' construction to the water waves, show that a conical wavefront is set up and that its half-angle α is given by

$$\sin \alpha = u/v.$$

This is familiar as the bow wave of a ship or the shock wave caused by an object moving through air with a speed exceeding that of sound, as in Fig. 14 of Chapter 20.

25. Using Fermat's principle, prove that the reflected ray, the incident ray, and the normal lie in one plane.

Section 43-4 Image Formation by Plane Mirrors

26. A small object is 10 cm in front of a plane mirror. If you stand behind the object, 30 cm from the mirror, and look at its image, for what distance must you focus your eyes?

27. You are standing in front of a large plane mirror, contemplating your image. If you move toward the mirror at speed v, at what speed does your image move toward you? Report this speed both (*a*) in your own reference frame and (*b*) in the reference frame of the room in which the mirror is at rest.

28. Figure 34 shows (top view) that Bernie B is walking directly toward the center of a vertical mirror M. How close to the mirror will he be when Sarah S is just able to see him? Take $d = 3.0$ m.

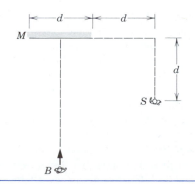

Figure 34 Problem 28.

29. Prove that if a plane mirror is rotated through an angle α, the reflected beam is rotated through an angle 2α. Show that this result is reasonable for $\alpha = 45°$.

30. In Fig. 13 you rotate the mirror 30° counterclockwise about its bottom edge, leaving the point object O in place. Is the image point displaced? If so, where is it? Can the eye still see the image without being moved? Sketch a figure showing the new situation.

31. A small object O is placed one-third of the way between two parallel plane mirrors as in Fig. 35. Trace approriate bundles of rays for viewing the four images that lie closest to the object.

Figure 35   Problem 31.

32. Two plane mirrors make an angle of 90° with each other. What is the largest number of images of an object placed between them that can be seen by a properly placed eye? The object does not lie on the mirror bisector.

33. Figure 36 shows a small light bulb suspended 250 cm above the surface of the water in a swimming pool. The water is 186 cm deep and the bottom of the pool is a large mirror. Where is the image of the light bulb when viewed from near normal incidence?

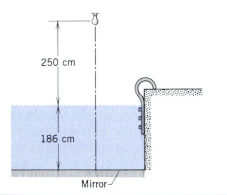

Figure 36   Problem 33.

34. A point object is 10 cm away from a plane mirror while the eye of an observer (pupil diameter 5.0 mm) is 24 cm away. Assuming both the eye and the point to be on the same line perpendicular to the surface, find the area of the mirror used in observing the reflection of the point.

35. You put a point source of light S a distance d in front of a screen A. How is the intensity at the center of the screen changed if you put a mirror M a distance d behind the source, as in Fig. 37? (*Hint*: Recall the variation of intensity with distance from a point source of light.)

36. Solve Problem 32 if the angle between the mirrors is (*a*) 45°, (*b*) 60°, and (*c*) 120°, the object always being placed on the bisector of the mirrors.

Figure 37   Problem 35.

37. How many images of yourself can you see in a room whose ceiling and two adjacent walls are mirrors? Explain.

Section 43-5 Deriving the Law of Refraction

38. The wavelength of yellow sodium light in air is 589 nm. (*a*) What is its frequency? (*b*) What is its wavelength in glass whose index of refraction is 1.53? (*c*) From the results of (*a*) and (*b*) find its speed in this glass.

39. Light of wavelength 612 nm in a vacuum travels 1.57 μm in a medium of index of refraction 1.51. Find (*a*) the wavelength in the medium, (*b*) the optical path length, and (*c*) the phase difference after moving that distance, with respect to light traveling the same distance in a vacuum.

Section 43-6 Total Internal Reflection

40. Prove that the optical path lengths for reflection and refraction in Figs. 10 and 17 are each a minimum when compared with other nearby paths connecting the same two points. (*Hint*: Examine the quantity d^2L/dx^2.)

41. Two materials, A and B, have indices of refraction of 1.667 and 1.586, respectively. (*a*) Find the critical angle for total internal reflection at an interface between the two materials. (*b*) In which direction must an incident ray be propagating if it is to be totally reflected?

42. A ray of light is incident normally on the face ab of a glass prism ($n = 1.52$), as shown in Fig. 38. (*a*) Assuming that the prism is immersed in air, find the largest value for the angle ϕ so that the ray is totally reflected at face ac. (*b*) Find ϕ if the prism is immersed in water.

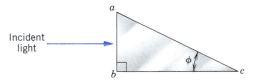

Figure 38   Problem 42.

43. A drop of liquid may be placed on a semicircular slab of glass as in Fig. 39. (*a*) Show how to determine the index of refraction of the liquid by observing total internal reflection. The index of refraction of the glass is unknown and must also be determined. Is the range of indices of refraction that can be measured in this way restricted in any sense? (*b*) In reality, how practical is this method?

44. A fish is 1.8 m below the surface of a smooth lake. At what angle above the horizontal must it look to see the light from a small camp fire burning at the water's edge 92 m away?

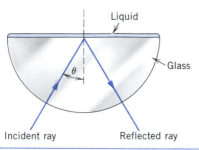

Figure 39 Problem 43.

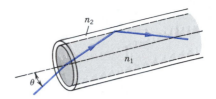

Figure 41 Problem 48.

45. A point source of light is 82.0 cm below the surface of a body of water. Find the diameter of the largest circle at the surface through which light can emerge from the water.

46. A light ray falls on a square glass slab as in Fig. 40. What must be the minimum index of refraction of the glass if total internal reflection occurs at the vertical face?

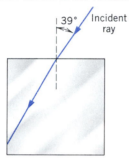

Figure 40 Problem 46.

47. A point source of light is placed a distance h below the surface of a large deep lake. (a) Show that the fraction f of the light energy that escapes directly from the water surface is independent of h and is given by

$$f = \tfrac{1}{2}(1 - \sqrt{1 - 1/n^2}),$$

where n is the index of refraction of water. (*Note:* Absorption within the water and reflection at the surface (except where it is total) have been neglected.) (b) Evaluate this fraction numerically.

48. A particular optical fiber consists of a nongraded glass core (index of refraction n_1) surrounded by a cladding (index of refraction $n_2 < n_1$). Suppose a beam of light enters the fiber from air at an angle θ with the fiber axis as shown in Fig. 41. (a) Show that the greatest possible value of θ for which a ray can be propagated down the fiber is given by

$$\theta = \sin^{-1} \sqrt{n_1^2 - n_2^2}.$$

(b) Assume the glass and coating indices of refraction are 1.58 and 1.53, respectively, and calculate the value of this angle.

49. In an optical fiber (see Problem 48), different rays travel different paths along the fiber, leading to different travel times. This causes a light pulse to spread out as it travels

along the fiber, resulting in information loss. The delay time should be minimized in designing a fiber. Consider a ray that travels a distance L along a fiber axis and another that is reflected, at the critical angle, as it travels to the same destination as the first. (a) Show that the difference Δt in the times of arrival is given by

$$\Delta t = \frac{L}{c} \frac{n_1}{n_2} (n_1 - n_2),$$

where n_1 is the index of refraction of the core and n_2 is the index of refraction of the cladding. (b) Evaluate Δt for the fiber of Problem 48, with $L = 350$ km.

50. A light ray of given wavelength, initially in air, strikes a 90° prism at P (see Fig. 42) and is refracted there and at Q to such an extent that it just grazes the right-hand prism surface at Q. (a) Determine the index of refraction of the prism for this wavelength in terms of the angle of incidence θ_1 that gives rise to this situation. (b) Give a numerical upper bound for the index of refraction of the prism. Show, by ray diagrams, what happens if the angle of incidence at P is (c) slightly greater or (d) slightly less than θ_1.

Figure 42 Problem 50.

51. A plane wave of white light traveling in fused quartz strikes a plane surface of the quartz, making an angle of incidence θ. Is it possible for the internally reflected beam to appear (a) bluish or (b) reddish? (c) Roughly what value of θ must be used? (*Hint:* White light will appear bluish if wavelengths corresponding to red are removed.)

52. A glass cube has a small spot at its center. (a) What parts of the cube face must be covered to prevent the spot from being seen, no matter what the direction of viewing? (b) What fraction of the cube surface must be so covered? Assume a cube edge of 12.6 mm and an index of refraction of 1.52. (Neglect the subsequent behavior of an internally reflected ray.)

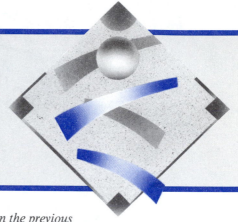

CHAPTER 44

SPHERICAL MIRRORS AND LENSES

*Reflection and refraction at plane surfaces, considered in the previous
chapter, are of limited usefulness in optical instruments. For one reason, they
are unable to change diverging light into converging light; diverging light, such as from a
point source, remains diverging light after reflection from a plane mirror or refraction across
a plane boudary.*

*If the mirror or the refracting surface is curved, plane wavefronts can be changed into curved
wavefronts, which can then converge to a point or appear to diverge from a point. Diverging
light can even be turned into converging light and focused to form an image, such as in a
camera, a telescope, or the human eye. Using combinations of mirrors and lenses, we can
make tiny objects appear large or distant objects appear close.*

*In this chapter, we analyze the formation of images by spherical lenses and mirrors.
Through either algebraic or graphical methods, we can find the image and determine its
size relative to the original object. Examples including the microscope and the telescope
show how these principles can be used to design optical systems that extend the range of
human vision to the very small or the very distant.*

44-1 SPHERICAL MIRRORS

In Section 43-4, we discussed the formation of an image by a plane mirror. We discovered that a plane mirror forms an image that appears to be behind the mirror; that is, when we observe the image, the light appears to come from a point behind the mirror. We called this a *virtual* image, and we found that the image was the same size as the object and that it was located at a (negative) distance i behind the mirror equal in magnitude to the distance o of the object in front of the mirror, as illustrated in Fig. 1a.

Suppose that, instead of making the mirror flat, we give it a slight curvature. In particular, we consider mirrors that have a spherical shape. Figures 1b and 1c show the effect in two different cases. In the first case (Fig. 1b), the mirror is *concave* (meaning "hollow," like a cave) with respect to the location of the object. Note that, in comparison with the plane mirror, the image is (1) magnified (that is, larger than the object) and (2) located at a greater distance behind the mirror (that is, i has a larger negative value). Such mirrors are commonly used for shaving or applying make-up, when magnification is desirable, even though the field of view may be reduced. Figure 1b applies

only when the distance of the object from the mirror is small (less than $r/2$, as we shall see).

In the second case (Fig. 1c), the mirror is *convex* with respect to the location of the object. Note that the image is (1) reduced in size and (2) closer to the mirror, compared with the plane mirror. Examples of such mirrors are right-hand side-view mirrors in automobiles and surveillance mirrors used in retail stores. The field of view is wider than that of a plane mirror.

Suppose the spherical mirrors in Fig. 1 were flexible. If we were to bend either one to make it more planar, the image would approach the location and size of the image in a plane mirror. We can therefore consider a plane mirror to be a special case of a spherical mirror, in which the radius of the sphere becomes infinitely large. Our equations describing the spherical mirror should reduce to the plane mirror equation ($o = -i$) as the radius tends toward infinity.

The Mirror Equation

At the end of this section, we derive the equation that relates the *object distance o* and the *image distance i* for a spherical mirror. We consider the special case in which

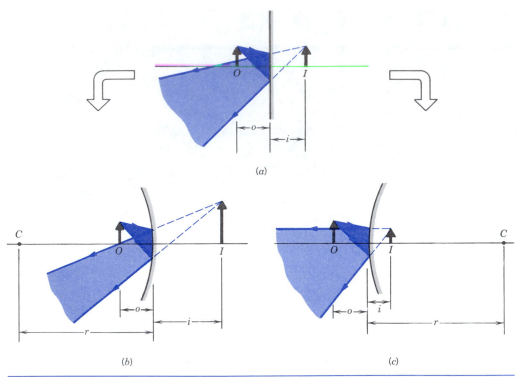

(a)

(b)   (c)

Figure 1 (*a*) An object *O* forms a virtual image *I* in a plane mirror. (*b*) If the mirror is bent so it becomes *concave*, the image moves away from the mirror and becomes larger. (*c*) If the plane mirror is bent so it becomes *convex*, the image moves closer to the mirror and becomes smaller. Point *C* is called the *center of curvature* of the mirror; it is the center of the spherical surface of which the mirror is a part.

the rays of light from the object make small angles with the axis of the mirror. Such rays are called *paraxial* rays. Put another way, the dimensions of the mirror are small compared with its radius of curvature. Our description would not apply to a fully illuminated mirror in the shape of an entire hemisphere.

The mirror equation relates the three distances in Fig. 1: *o, i,* and the radius of curvature *r* of the mirror. This relationship is given by the spherical mirror equation,

$$\frac{1}{o} + \frac{1}{i} = \frac{2}{r}. \tag{1}$$

It is convenient to define the *focal length f* of the mirror to be just half the radius of curvature, or

$$f = r/2. \tag{2}$$

In terms of the focal length, the mirror equation can be written

$$\frac{1}{o} + \frac{1}{i} = \frac{1}{f}. \tag{3}$$

Figure 2 shows parallel light rays incident on the mirror. Parallel light can be obtained from an object at a great distance from the mirror, such that the wavefronts from the object are essentially planes. In practice, parallel light

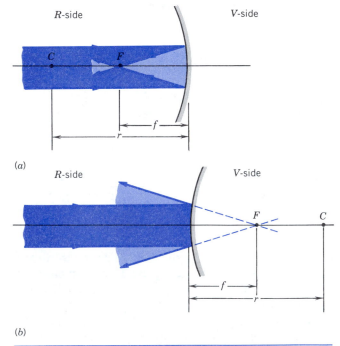

(a)

(b)

Figure 2 (*a*) In a concave mirror, incident parallel light is brought to a real focus at *F* on the *R*-side of the mirror. (*b*) In a convex mirror, incident parallel light appears to diverge from a virtual focus at *F* on the *V*-side of the mirror.

can be obtained using another mirror or a lens. Parallel light is brought to a focus at a point F called the *focal point*. This point is distance f from the mirror. Equation 3 shows that if $o = \infty$, corresponding to the object at a very great distance from the mirror, then $i = f$.

Equations 1 and 3 can be used to find the *location* of the image; we should also like to know its size, compared with the size of the object. For this purpose we define the *lateral magnification m* as

$$|m| = \frac{\text{lateral size of image}}{\text{lateral size of object}} . \tag{4}$$

The sign of m is defined so that $m > 0$ if the image is upright or erect with respect to the object, and $m < 0$ if the image is inverted with respect to the object. As we derive later in this section, the lateral magnification is given by

$$m = -\frac{i}{o} . \tag{5}$$

Sign Conventions

Figure 2 suggests the sign conventions that must be considered in using Eqs. 1 and 3. The side of the mirror from which the light is incident is called the *R*-side, because it is on this side that a *real image* will be formed. Real images are those that are formed by converging light; equivalently, we can say that real images are those that can be viewed on a screen placed at the position of the image. On the *R*-side of the mirror, i, o, r, and f are taken to be positive.

The region behind the mirror is called the *V*-side, because on this side *virtual images* can be formed. Virtual images are those formed by diverging light that cannot be shown on a screen. On the *V*-side, o, i, r, and f are taken to be negative.

According to these sign conventions, in Fig. 1*b* the object distance o is positive (because the object is on the *R*-side of the mirror) and the image distance i is negative (because the image is on the *V*-side). The center of curvature C is on the *R*-side, so the radius of curvature r is positive. In Fig. 1*c*, o is positive and i is negative, as in Fig. 1*b*, but r is negative, because C is on the *V*-side.

Figure 3 shows the image distances for three different object distances as an object is moved toward a concave mirror. In Fig. 3*a*, the object and image distances are positive, because the object and the image both appear on the *R*-side of the mirror. In Fig. 3*b*, the object is at the focal point. With $o = f$, Eq. 1 gives $i = \infty$. This is consistent with parallel light emerging from the mirror. In Fig. 3*c*, the object distance remains positive but is now smaller than f. In this case, Eq. 1 gives a negative value for i; that is, a virtual image forms on the *V*-side, as shown.

In Fig. 3*a*, the lateral magnification m determined according to Eq. 5 is negative, because o and i are both positive. The image is therefore inverted. (It is also en-

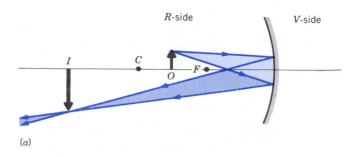

(a)

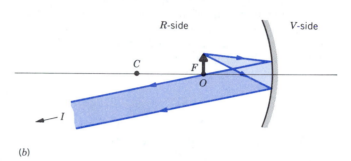

(b)

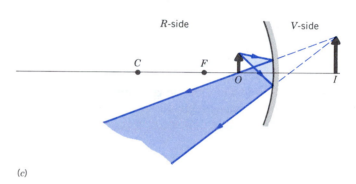

(c)

Figure 3   An object is moved successively closer to a concave mirror, from (*a*) just beyond the focal point to (*b*) the focal point and then (*c*) within the focal point. In the process, the image moves from (*a*) its position on the *R*-side to (*b*) infinity, and then (*c*) reappears on the *V*-side.

larged, because i happens to be greater than o in the case illustrated.) In Fig. 3*c*, o and i have opposite signs, so m is positive and the image is correspondingly erect, as illustrated.

Figure 4 suggests one possible arrangement in which the object is considered to be on the *V*-side of the mirror, so that o is negative. Converging light (produced by another optical device, such as a lens or a mirror, that is not shown) is incident on the mirror. If the mirror were not present, the light would converge to an image at the location O shown. This location defines the position of a *virtual object*, and the distance between this location and the mirror is the (negative) object distance. The image distance is positive. The magnification is positive, and thus the image is erect, as shown. Can you predict the outcome

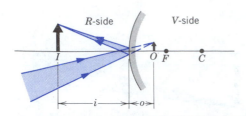

Figure 4   Converging light (from mirrors or lenses not illustrated) is incident on a convex mirror. The *virtual object* at *O* shows the location where the light would be focused if the mirror were not present. Of course, no light is present on the *V*-side of the mirror. A real image *I* is formed. This arrangement produces a real image only if the magnitude of the object distance is less than the focal length, but in a similar situation a convex lens *always* forms a real image.

if the mirror in Fig. 4 were made concave instead of convex? How would the resulting image distance compare in magnitude with the object distance? Would the image be erect or inverted?

Ray Tracing

It is a good idea to check the results of algebraic computations obtained from the mirror equation against a graphical method for locating the image. This method is called *ray tracing*. As suggested in Figs. 1–4, bundles of rays either converge to a real image or diverge from a virtual

image. If we can draw these rays as they reflect from the mirror, we can locate the image.

We can simplify this procedure by drawing a few basic rays, the intersection of which serves to locate the image. These rays, which are shown in Fig. 5, are selected from an infinite number of possible rays for convenience in locating the image. These rays need not necessarily exist in actuality (part of the mirror might be covered by an aperture, for example), but they nevertheless can be used to find the image (which is complete even if some of the rays are blocked). The rays are:

1. *A ray parallel to the axis*, which is reflected to pass through the focal point (in the case of a converging mirror, Fig. 5*a*) or to appear to come from the focal point (in the case of a diverging mirror, Fig. 5*c*).

2. *A ray passing through the focal point* (converging mirror, Fig. 5*a*) or appearing to do so (diverging mirror, Fig. 5*c*), which is reflected to be parallel to the axis.

3. *A ray passing through the center of curvature C*, which is reflected back along its original path (Figs. 5*b* and 5*d*).

4. *A ray striking the vertex of the mirror* (the point where the axis intersects the mirror), which is reflected at an equal angle on the opposite side of the axis (Figs. 5*b* and 5*d*).

Any two of these four rays can be used to locate the image, as indicated in Fig. 5.

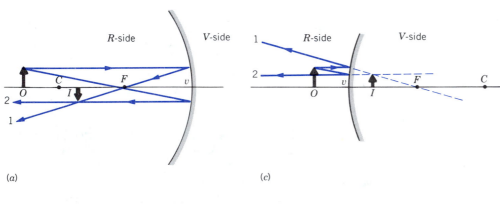

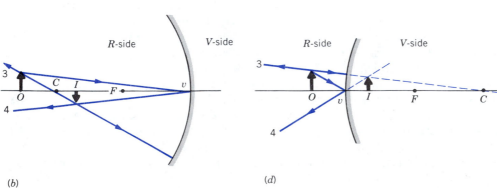

(a)

(b)

(c)

(d)

Figure 5   (*a,b*) Four rays that may be used in graphical constructions to locate the image of an object in a concave mirror. Note that the image is real and inverted. (*c,d*) Four similar rays drawn in the case of a convex mirror. The image is virtual and upright.

Sample Problem 1   In the situation shown in Figs. 5*a* and 5*b*, suppose $f = 12$ cm and $o = 30$ cm. Find the position of the image and the lateral magnification.

Solution   Solving Eq. 1 for $1/i$, we obtain

$$\frac{1}{i} = \frac{1}{f} - \frac{1}{o} = \frac{1}{12 \text{ cm}} - \frac{1}{30 \text{ cm}},$$

or

$$i = 20 \text{ cm}.$$

This is consistent with Figs. 5*a* and 5*b*.

Using Eq. 5, the magnification is found to be

$$m = -\frac{i}{o} = -\frac{20 \text{ cm}}{30 \text{ cm}} = -0.67.$$

The image is 2/3 the size of the object and (as indicated by the minus sign) is inverted. These are consistent with Figs. 5*a* and 5*b*.

Sample Problem 2   A convex mirror has a radius of curvature of 22 cm. An object is placed 14 cm from the mirror. Locate and describe the image using (*a*) graphical and (*b*) algebraic methods.

Solution   (*a*) Figure 6 shows the object and the mirror. Rays 1, 2, and 3 are drawn to locate the image. The image is virtual, erect, and located on the *V*-side of the mirror; the image distance is in magnitude about half the object distance, and the image is about half as tall as the object.

(*b*) According to our sign convention, the radius is negative if the center of curvature is located on the *V*-side of the mirror. Using Eq. 1, we have

$$\frac{1}{o} + \frac{1}{i} = \frac{2}{r}$$

or

$$\frac{1}{+14 \text{ cm}} + \frac{1}{i} = \frac{2}{-22 \text{ cm}},$$

which yields

$$i = -6.2 \text{ cm}.$$

This value is consistent with the result of our graphical construction.

The lateral magnification is, from Eq. 5,

$$m = -\frac{i}{o} = -\frac{-6.2 \text{ cm}}{+14 \text{ cm}} = +0.44,$$

which is also consistent with the result obtained graphically. Note that $m > 0$, indicating that the image is upright.

Derivation of Mirror Equations

Figure 7 shows a point object O on the axis of a concave spherical mirror whose radius of curvature is r. A ray from O that makes an arbitrary angle α with the axis intersects the axis at I after reflection from the mirror at a. A ray that leaves O along the axis is reflected back along itself at v and also passes through I. Thus I is the image of O; it is a *real* image because light actually passes through I. Let us find the location of I.

A useful theorem is that the exterior angle of a triangle is equal to the sum of the two opposite interior angles. Applying this to triangles OaC and OaI in Fig. 7 yields

$$\beta = \alpha + \theta$$

and

$$\gamma = \alpha + 2\theta.$$

Eliminating θ between these equations leads to

$$\alpha + \gamma = 2\beta. \tag{6}$$

In radian measure we can write angles α, β, and γ as

$$\alpha \approx \frac{av}{vO} = \frac{av}{o},$$

$$\beta = \frac{av}{vC} = \frac{av}{r}, \tag{7}$$

$$\gamma \approx \frac{av}{vI} = \frac{av}{i}.$$

Note that only the equation for β is exact, because the center of curvature of arc av is at C and not at O or I. However, the equations for α and for γ are approximately

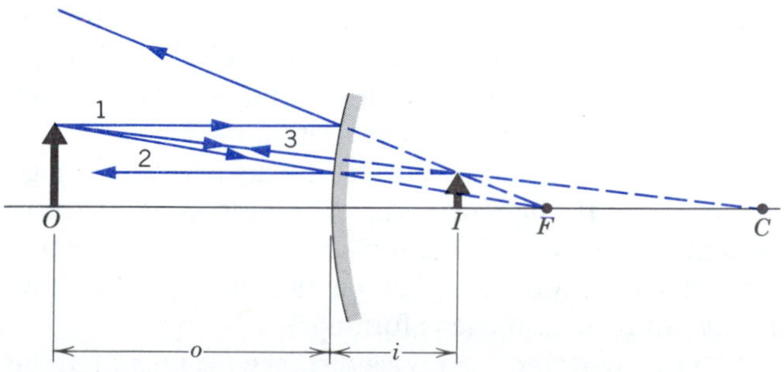

Figure 6   Sample Problem 2.

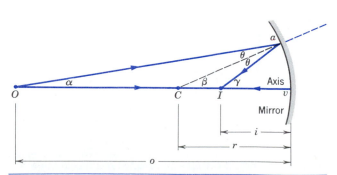

Figure 7   A point object O forms a real point image I after reflection from a concave mirror.

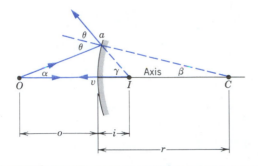

Figure 8 A point object O forms a virtual point image I after reflection from a convex mirror. Compare with Fig. 7.

correct if these angles are sufficiently small. *In all that follows we assume that the rays diverging from the object make only a small angle α with the axis of the mirror.* We call such rays, which lie close to the mirror axis, *paraxial* rays. We did not find it necessary to make such an assumption for plane mirrors. Substituting Eqs. 7 into Eq. 6 and canceling av yield Eq. 1,

$$\frac{1}{o} + \frac{1}{i} = \frac{2}{r},\qquad(1)$$

which is the equation we set out to prove.

Figure 8 shows a point object on the axis of a convex mirror. The angles are labeled similarly to those of Fig. 7. We can carry out a similar analysis to that given previously, which again yields Eq. 1, *provided* we follow the sign convention that i and r are taken to be negative in Fig. 8. This derivation is left as an exercise (see Problem 6).

Significantly, Eq. 1 does not contain α (or β, γ, or θ), so that it holds for all rays that strike the mirror provided that they are sufficiently paraxial. In an actual case, the rays can be made as paraxial as one likes by putting a circular diaphragm in front of the mirror, centered about the vertex v; this will impose a certain maximum value of α.

To derive the equation for lateral magnification (Eq. 5), consider Fig. 9, which shows a ray (avb) that originates on the tip of the object, reflects from the mirror at point v, and passes through the tip of the image. The law of reflec-

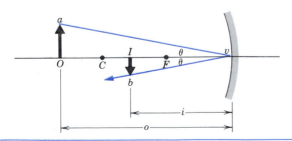

Figure 9 An object O forms an inverted real image I in a concave mirror.

tion demands that this ray make equal angles θ with the mirror axis as shown. For the two similar right triangles aOv and bIv we can write

$$\frac{Ib}{Oa} = \frac{vI}{vO}.$$

The quantity on the left (apart from a question of sign) is the *lateral magnification m* of the mirror given by Eq. 4. Since we want to represent an *inverted* image by a *negative* magnification, we arbitrarily define m for this case as $-(Ib/Oa)$. Since $vI = i$ and $vO = o$, we have at once the result previously given as Eq. 5,

$$m = -\frac{i}{o}.\qquad(5)$$

This equation gives the magnification for spherical and plane mirrors under all circumstances. For a plane mirror, $o = -i$ and the predicted magnification is $+1$, which, in agreement with experience, indicates an *erect* image the same size as the object.

Images in spherical mirrors suffer from several distortions that arise because the assumption of paraxial rays is never completely justified. In general, a point source does not result in a point image; see Problem 2. Apart from this, distortion arises because the magnification varies somewhat with distance from the mirror axis, Eq. 5 being strictly correct only for paraxial rays. Finally, we must always keep in mind that geometrical optics is itself only a special case of physical optics; the effects of diffraction (see Chapter 46) can also distort or defocus the image.

44-2 SPHERICAL REFRACTING SURFACES

In Fig. 10a, light from a point object O falls on a convex spherical refracting surface of radius of curvature r. The surface separates two media; the index of refraction of the medium containing the incident light is n_1, while that of the medium containing the refracted light is n_2. Such a diagram might represent light that is incident on a small region of a glass sphere; note that the real image is formed within the glass (medium 2). Although we do not often encounter images of this type, understanding the images produced by spherical refracting surfaces is essential in the discussion of thin lenses in Section 44-3.

In Fig. 10b, a concave surface forms a virtual image when $n_1 < n_2$, the light in medium 2 diverging as if it came from the image point I. Figure 10c shows a surface that is again concave with respect to the incident light, but now $n_1 > n_2$ and a real image is formed.

As we prove later, the image distance i is related to the

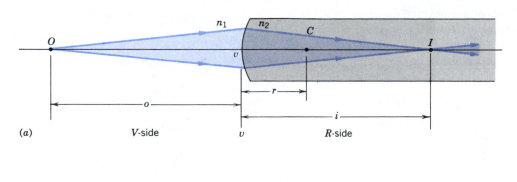

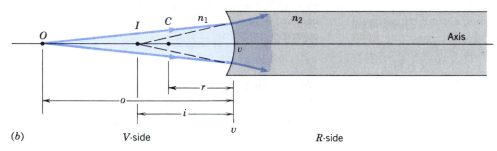

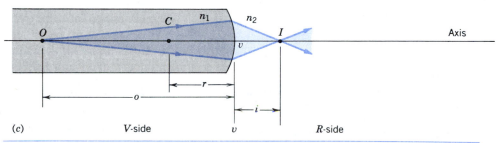

Figure 10   (*a*) A real image of a point object is formed by refraction at a convex spherical boundary between two media; in this case, $n_2 > n_1$. (*b*) A virtual image is formed by refraction at a concave spherical boundary when $n_2 > n_1$. (*c*) The same as (*b*), except that $n_2 < n_1$.

object distance o, radius of curvature r, and two indices of refraction according to

$$\frac{n_1}{o} + \frac{n_2}{i} = \frac{n_2 - n_1}{r}.$$
$$(8)$$

This single equation, with appropriate sign conventions, is sufficient to analyze both convex and concave surfaces. The only restriction, as was the case in our discussion of spherical mirrors, is that the rays must be paraxial.

The sign conventions to be used with Eq. 8 are summarized in Fig. 11. If a real image is to be formed by converging light from the surface, it must appear on the side of the surface opposite to the incident light. This side is called the *R*-side. Virtual images, as shown in Fig. 10*b*, are formed on the same side as the incident light, which we call the *V*-side. The radius of curvature is taken as positive if the center of curvature *C* is located on the *R*-side (as in Fig. 10*a*) and negative if *C* is located on the *V*-side (as in

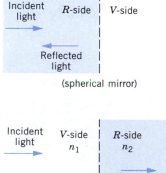

Figure 11   Real images are formed on the same side as the incident light for mirrors but on the opposite side for refracting surfaces and for lenses.

Figs. 10*b* and 10*c*). Object distances are positive for real objects (on the *V*-side), while image distances are positive for real images (on the *R*-side). The image distance *i* is positive in Figs. 10*a* and 10*c*, and *i* is negative in Fig. 10*b*.

Sample Problem 3   Locate the image for the geometry shown in Fig. 10*a*, assuming the radius of curvature *r* to be 11 cm, $n_1 = 1.0$, and $n_2 = 1.9$. Let the object be 19 cm to the left of the vertex *v*.

Solution   From Eq. 8,

$$\frac{n_1}{o} + \frac{n_2}{i} = \frac{n_2 - n_1}{r},$$

we have

$$\frac{1.0}{+19 \text{ cm}} + \frac{1.9}{i} = \frac{1.9 - 1.0}{+11 \text{ cm}}.$$

Note that *r* is positive because the center of curvature *C* of the surface in Fig. 10*a* lies on the *R*-side. If we solve the above equation for *i*, we find

$$i = +65 \text{ cm}.$$

This result agrees with Fig. 10*a* and is consistent with the sign conventions. The light actually passes through the image point *I* so the image is real, as indicated by the positive sign that we found for *i*. Remember also that n_1 (= 1.0 in this case) always refers to the medium on the side of the surface from which the light comes.

Sample Problem 4   A fish is swimming along a horizontal diameter and is 10 cm from the side of a spherical bowl of radius 15 cm (Fig. 12). Take the index of refraction of water to be $n_1 = 1.33$ and find the location of the fish according to an observer outside the bowl. Assume the glass bowl is so thin that refraction due to the glass can be ignored.

Solution   According to the sign conventions, in the geometry of Fig. 12 we take *o* to be positive because the object is on the *V*-side

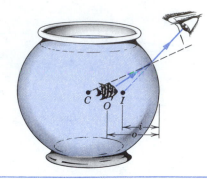

Figure 12   Sample Problem 4. Note that the ray from *O* is bent away from the normal (indicated by the dashed line), in accordance with Snell's law.

of the spherical surface, and we take *r* to be negative because *C* is on the *V*-side. We use Eq. 8 with $n_2 = 1$, which we solve to give

$$\frac{n_2}{i} = \frac{n_2 - n_1}{r} - \frac{n_1}{o} = \frac{1 - 1.33}{-15 \text{ cm}} - \frac{1.33}{10 \text{ cm}}.$$

Solving, we find

$$i = -6.4 \text{ cm}.$$

That is, the fish appears closer to the side of the bowl than it really is.

Derivation of Refracting Surface Formula (Eq. 8)

Figure 13 shows a point source *O* near a spherical refracting surface of radius *r*. A ray from *O* strikes the surface at *v* at normal incidence and passes undeviated into medium 2 through the center of curvature *C*. This ray establishes a convenient axis for our calculation. A second ray, which makes a small but arbitrary angle α with the axis and strikes the refracting surface at *a*, is refracted according to

$$n_1 \sin \theta_1 = n_2 \sin \theta_2.$$

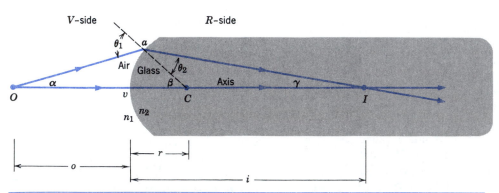

Figure 13   A point object *O* forms a real point image *I* after refraction at a spherical convex surface between two media.

The refracted ray intersects the first ray at I, thereby locating the image of O.

As in the derivation of the mirror equation, we use the theorem that the exterior angle of a triangle is equal to the sum of the two opposite interior angles. Applying this to triangles COa and ICa yields

$$\theta_1 = \alpha + \beta \quad \text{and} \quad \beta = \theta_2 + \gamma. \quad (9)$$

As we did in Section 44-1, we assume all rays are paraxial, so that all angles (β, γ, θ_1, θ_2) are small, and the sine of each angle can be replaced by the angle itself. This permits us to write the law of refraction as

$$n_1\theta_1 \approx n_2\theta_2. \quad (10)$$

Combining Eqs. 9 and 10 leads, after rearrangement, to

$$n_1\alpha + n_2\gamma = (n_2 - n_1)\beta. \quad (11)$$

In radian measure the angles α, β, and γ in Fig. 13 are

$$\alpha \approx \frac{av}{o}, \quad \beta = \frac{av}{r}, \quad \text{and} \quad \gamma \approx \frac{av}{i}. \quad (12)$$

Only the second of these equations is exact. The other two are approximate because I and O are *not* the centers of circles of which av is an arc. However, for paraxial rays (α small enough) the inaccuracies in Eqs. 12 can be made as small as desired.

Substituting Eqs. 12 into Eq. 11 leads directly to Eq. 8.

44-3 THIN LENSES

There are many common examples of the refraction of light by a lens. The lenses in our eyes focus light on the retina, while the corrective lenses of eyeglasses or contact lenses compensate for deficiencies in our vision. The multi-element lens of a camera focuses light on the film. In this section we consider the properties of such lenses.

In most refraction situations there is more than one refracting surface. This is true even for a contact lens, where the light passes first from air into glass and then from glass into the eye. We consider here only the special case of a *thin lens*; that is, the thickness of the lens is small compared with the object distance o, the image distance i, or the radii of curvature r_1 and r_2 of either of the two refracting surfaces. For such a lens—as we shall prove later in this section—these quantities are related by

$$\frac{1}{o} + \frac{1}{i} = \frac{1}{f} \quad (13)$$

in which the focal length f of the lens is given by

$$\frac{1}{f} = (n - 1)\left(\frac{1}{r_1} - \frac{1}{r_2}\right). \quad (14)$$

Equations 13 and 14 are approximations that hold only for thin lenses and paraxial rays. Note that Eq. 13 is the same equation that we used for spherical mirrors. Equation 14 is often called the *lens maker's equation*; it relates the focal length of the lens to the index of refraction n of the lens material and the radii of curvature of the two surfaces.

In Eq. 14, r_1 is the radius of curvature of the lens surface on which the light first falls and r_2 is that of the second surface. Equation 14 is used in cases in which a lens of index of refraction n is immersed in air. If the lens is immersed in a medium for which the index of refraction is not unity, Eq. 14 still holds if we replace n in that formula by $n_{\text{lens}}/n_{\text{medium}}$.

The lateral magnification of a thin lens is given by the same formula as that of a spherical mirror,

$$m = -\frac{i}{o}. \quad (15)$$

Later in this section we derive this result.

Sign Conventions

The sign conventions for o, i, r_1, and r_2 are similar to those for spherical mirrors and refracting surfaces; see Fig. 11. Figure 14 illustrates these sign conventions. As before, we have an R-side and a V-side.

1. The radii of curvature r_1 (referring to the first surface struck by the light) and r_2 (referring to the second surface struck by the light) are positive if the corresponding centers of curvature are on the R-side. The radii are negative if the corresponding centers of curvature are on the V-side. In Fig. 14a, the center of curvature C_1 is on the R-side, so r_1 is positive, while C_2 is on the V-side, so r_2 is negative. Inspection of Eq. 13 shows that, when $r_1 > 0$ and $r_2 < 0$, the focal length f is always positive. Such a lens is called a *converging* lens; a lens that is thicker at the center than at the edges, when immersed in a medium of index of refraction lower than that of the lens, is always a converging lens.

In Fig. 14b, C_1 is on the V-side, while C_2 is on the R-side. Hence r_1 is negative and r_2 is positive. In this case, Eq. 13 shows that f is always negative. Such a lens is called a *diverging* lens; a lens that is thinner at the center than at the edges, when immersed in a medium of lower index of refraction, is always a diverging lens.

2. The object distance o is positive if the object is real and lies on the V-side of the lens, as in both Figs. 14a and 14b. Light from a real object is diverging when it strikes the lens. It is also possible to have converging light strike the lens, as in Fig. 14c. In this case, if the lens were not present, the converging light would form an image at O on the R-side of the lens; we take this image as a virtual object, and we take o as negative in this case.

3. The image distance i is positive if the (real) image lies on the R-side of the lens, as in Figs. 14a and 14c, while i is negative if the (virtual) image lies on the V-side of the lens, as in Fig. 14b.

4. According to Eq. 15, the magnification is negative when both i and o are positive, as in Fig. 14a, corresponding to an *inverted* image. In the case of an erect image, as in Figs. 14b and 14c, the magnification is positive, because o and i have opposite signs. In the case shown in Fig.

14b, o is positive and i is negative, while in Fig. 14c, o is negative and i is positive.

A useful representation of the sign conventions for both spherical mirrors and thin lenses can be obtained if we write the mirror equation (Eq. 3) and the lens equation (Eq. 13) in this form:

$$\frac{1}{o/|f|} + \frac{1}{i/|f|} = \pm 1, \qquad (16)$$

which is obtained by multiplying both equations by $|f|$, the absolute value of the focal length f. On the right side of Eq. 16, we choose $+1$ for a converging lens or a concave mirror and -1 for a diverging lens or a convex mirror. See Problem 21.

Figure 15 is a graphical representation of Eq. 16, with

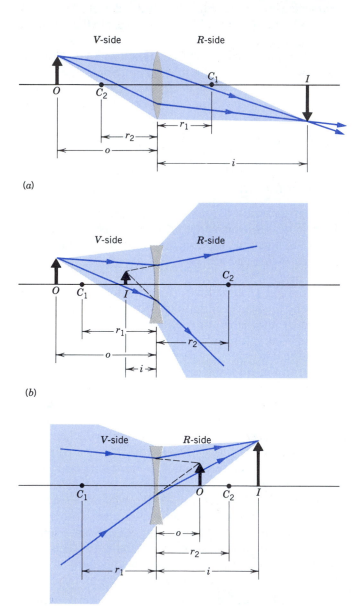

(a)

(b)

(c)

Figure 14 (a) A real, inverted image is formed by a converging lens. Such a lens has a positive focal length and is thicker at the center than at the edges. (b) A virtual, erect image is formed by a diverging lens. Such a lens has a negative focal length and is thinner at the center than at the edges. (c) Converging light gives a virtual object at O. A real, erect image is formed at I by this diverging lens.

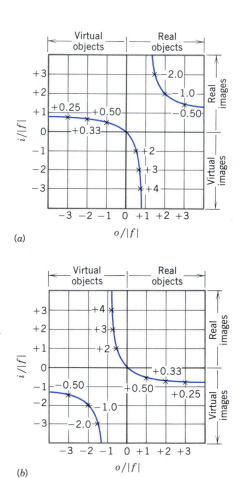

(a)

(b)

Figure 15 (a) A representation of $i/|f|$ and $o/|f|$ for concave mirrors and converging lenses. Note that (lower left quadrant) a virtual object cannot produce a virtual image. The numbers near the crosses are the magnifications (see Eq. 15), positive values indicating erect images and negative values indicating inverted images. (b) The same for convex mirrors and diverging lenses. Note that (upper right quadrant) a real object cannot produce a real image. See "Image Formation in Lenses and Mirrors, a Complete Representation," by Albert A. Bartlett, *The Physics Teacher*, May 1976, p. 296.

converging lenses and concave mirrors represented in Fig. 15a and diverging lenses and convex mirrors in Fig. 15b. Each graph contains two branches of a hyperbola, one with positive magnification and one with negative magnification. These two graphs neatly summarize all possible applications of Eqs. 3 and 5 (for spherical mirrors) and Eqs. 13 and 15 (for thin lenses).

In contrast with a spherical mirror or a spherical refracting surface, a lens has *two* focal points. In a thin lens, the two focal points are located at equal distances f from the lens on either side of the lens.

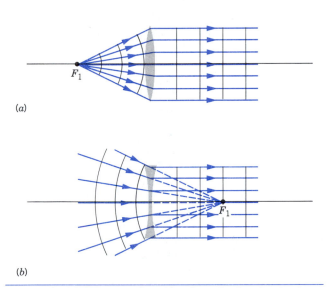

(a)

(b)

Figure 16 (a) When a point object is at the first focal point F_1 of a converging lens, parallel light emerges from the lens. (b) In the case of a diverging lens, a virtual point object gives parallel light.

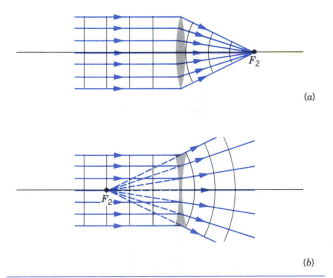

(a)

(b)

Figure 17 (a) When parallel light is incident on a converging lens, the light is brought to a focus at the second focal point F_2. (b) When parallel light is incident on a diverging lens, it appears to emerge from the second focal point.

When a point object is located at the *first focal point F_1*, parallel light emerges from the lens, as shown in Fig. 16a. In the case of a diverging lens (Fig. 16b), the point object is a virtual object. Converging light, which would have been focused at F_1 if the lens were not there, is defocused into parallel light by the diverging lens. The *second focal point F_2* is the point where parallel light is (or appears to be) focused by the lens, as shown in Fig. 17. Note from comparing Figs. 16 and 17 that the locations of the first and second focal points in a converging lens are opposite to those in a diverging lens.

In Figs. 16 and 17, all rays contain the same number of wavelengths; that is, they have the same *optical path length* (see Section 43-5). Note how the different geometrical lengths of the paths of the rays through the lens (where the speed of light is smaller than it is in air) changes the spherical wavefronts into planes or the plane wavefronts into spherical ones.

Ray Tracing

As was the case with spherical mirrors, it is helpful to locate the image formed by a thin lens using a graphical method with a few basic rays. Figure 18 shows three rays that can be used:

1. A ray (ray 1 in Fig. 18) passing through (or, when extended, appearing to pass through) the first focal point F_1 emerges from the lens parallel to the axis.

2. A ray (ray 2 in Fig. 18) parallel to the axis passes through (or, when extended, appears to pass through) the second focal point F_2.

3. A ray (ray 3 in Fig. 18) falling on the lens at its center passes through the lens undeflected, because near its center the lens behaves like a flat piece of glass with parallel sides, which doesn't change the direction of the ray.

Any two of these rays can be used to locate the image; the third is available as a check. Note from Fig. 18 that for all three rays, we consider the refraction to take place in a plane at the location of the lens. This can be done only for a thin lens.

Sample Problem 5 The lenses of Fig. 14 have radii of curvature of magnitude 42 cm and are made of glass with $n = 1.65$. Compute their focal lengths.

Solution Since C_1 lies on the R-side of the lens in Fig. 14a, r_1 is positive ($= +42$ cm). Since C_2 lies on the V-side, r_2 is negative ($= -42$ cm). Substituting in Eq. 14 yields

$$\frac{1}{f} = (n-1)\left(\frac{1}{r_1} - \frac{1}{r_2}\right) = (1.65 - 1)\left(\frac{1}{+42 \text{ cm}} - \frac{1}{-42 \text{ cm}}\right)$$

or

$$f = +32 \text{ cm}.$$

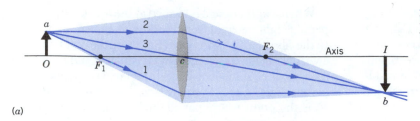

Figure 18 Three rays that can be used to locate the image formed by a thin lens.

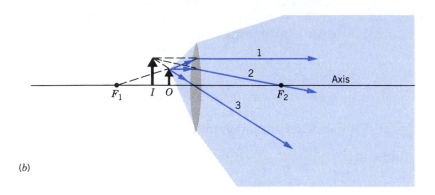

(a)

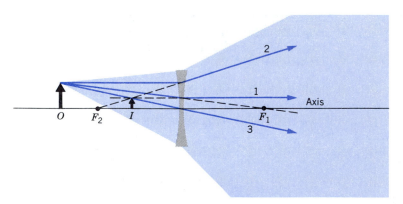

(b)

A positive focal length indicates that, in agreement with what we have said, parallel incident light converges after refraction to form a real focus.

In Figs. 14b and 14c, C_1 lies on the V-side of the lens so that r_1 is negative (= −42 cm). Since r_2 is positive (= +42 cm), Eq. 14 yields

$$\frac{1}{f} = (n-1)\left(\frac{1}{r_1} - \frac{1}{r_2}\right) = (1.65 - 1)\left(\frac{1}{-42 \text{ cm}} - \frac{1}{+42 \text{ cm}}\right)$$

or

$$f = -32 \text{ cm}.$$

Sample Problem 6 An object is 38 cm in front of a diverging lens of focal length −24 cm. Find the location and lateral magnification of the image using (a) graphical and (b) algebraic techniques.

Solution (a) The ray diagram is shown in Fig. 19. Ray 1 is headed toward F_1 when it strikes the lens; it emerges parallel to

the axis. Ray 2, originally parallel to the axis, emerges as if it came from F_2. Ray 3 passes undeviated through the center of the lens. All three rays appear to come from the tip of the image I. Note that only two of these rays would have been sufficient to locate the image; however, it is helpful to draw a third ray to reduce the chance of making an error.

The graphical construction suggests that the image is virtual, erect, located at about 2/3 of a focal length from the lens on the V-side, and about 1/3 the height of the object.

(b) Using Eq. 13, we have

$$\frac{1}{+38 \text{ cm}} + \frac{1}{i} = \frac{1}{-24 \text{ cm}}$$

or

$$i = -15 \text{ cm},$$

consistent with the graphical result. The magnification is

$$m = -\frac{i}{o} = -\frac{-15 \text{ cm}}{+38 \text{ cm}} = +0.39,$$

also consistent with the graphical result.

Figure 19 Sample Problem 6.

Derivation of the Thin Lens Formulas (Eqs. 13 and 14)

Our plan is to consider each lens surface separately, using the image formed by the first surface as an object for the second.

Figure 20a shows such a thick glass "lens" of length L whose surfaces are ground to radii r_1 and r_2. A point object O is placed near the left surface as shown. A ray leaving O along the axis is not deflected on entering or leaving the lens.

A second ray leaving O at an angle α with the axis strikes the surface at point a, is refracted, and strikes the second surface at point b. The ray is again refracted and crosses the axis at I, which, being the intersection of two rays from O, is the image of point O, formed after refraction at two surfaces.

Figure 20b shows the first surface, which forms a virtual image of O at I'. To locate I', we use Eq. 8, with $n_1 = 1$ and $n_2 = n$:

$$\frac{1}{o} + \frac{n}{i'} = \frac{n-1}{r_1},$$

or, taking into account that i' is negative,

$$\frac{1}{o} - \frac{n}{|i'|} = \frac{n-1}{r_1}. \tag{17}$$

Figure 20c shows the second surface. Unless an observer at point b were aware of the existence of the first surface, we would think that the light striking that point originated at point I' in Fig. 20b and that the region to the left of the surface was filled with glass. Thus the (virtual) image I' formed by the first surface serves as a real object O' for the second surface. The distance of this object from the second surface is

$$o' = |i'| + L. \tag{18}$$

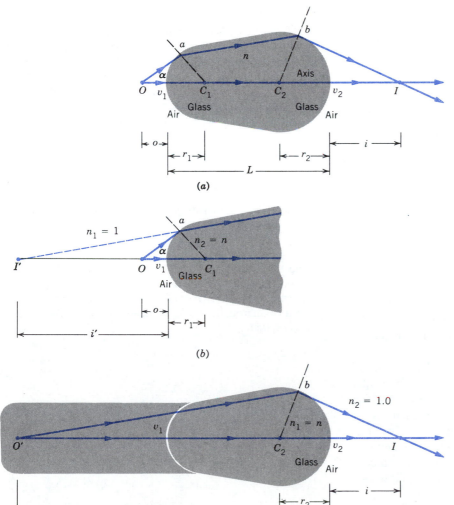

Figure 20   (a) Two rays from O form a real image at I after refraction at two spherical surfaces, the first surface being converging and the second diverging. (b) The first surface and (c) the second surface, shown separately. The vertical scale has been greatly exaggerated for clarity.

In applying Eq. 8 to the second surface, we insert $n_1 = n$ and $n_2 = 1$ because the object behaves as if it were imbedded in glass. If we use Eq. 18, Eq. 8 becomes

$$\frac{n}{|i'| + L} + \frac{1}{i} = \frac{1 - n}{r_2}. \qquad (19)$$

Let us now assume that the thickness L of the "lens" in Fig. 20a is so small that we can neglect it in comparison with other linear quantities in this figure (such as o, i, o', i', r_1, and r_2). In all that follows we make this *thin lens approximation*. Putting $L = 0$ in Eq. 19 leads to

$$\frac{n}{|i'|} + \frac{1}{i} = -\frac{n - 1}{r_2}. \qquad (20)$$

Adding Eqs. 17 and 20 gives

$$\frac{1}{o} + \frac{1}{i} = (n - 1)\left(\frac{1}{r_1} - \frac{1}{r_2}\right). \qquad (21)$$

Defining the right side of Eq. 21 to be $1/f$ leads directly to Eqs. 13 and 14, completing the derivation.

To derive Eq. 15 for the lateral magnification, we refer to Fig. 18a. Right triangles acO and bcI are similar, because angles acO and bcI are equal. For the corresponding sides of the similar triangles, we obtain

$$\frac{bI}{aO} = \frac{cI}{cO}. \qquad (22)$$

The right side of this expression is just i/o, while the left side is $-m$, the minus sign indicating that the image is inverted. With these substitutions, Eq. 22 reduces directly to Eq. 15.

44-4 COMPOUND OPTICAL SYSTEMS

A single mirror or lens is seldom a useful optical device. In such instruments as binoculars, telescopes, microscopes, and cameras, images are formed by a combination of several lenses or mirrors. In this section we consider the images formed by systems containing several optical elements.

The analysis of the formation of images by compound optical systems is straightforward. We merely consider the elements one at a time, as if the others were not present, taking the *image* formed by one element as the *object* for the next. We apply the previously derived formulas for the spherical mirror (Eqs. 3 and 5) or thin lens (Eqs. 13 and 15), taking careful account of the sign conventions in each case. In particular, note the following:

When diverging *light from the image formed by one element strikes the next element, we treat that image as a* real object *for the next element. When* converg-ing *light from the image formed by one element strikes the next element, we treat that image as a* virtual object *for the next element.*

Sample Problem 7 Two identical converging lenses of focal lengths $f = f' = +15$ cm are separated by a distance d of 6 cm, as shown in Fig. 21. A luminous source is placed a distance of $o = 10$ cm from the first lens. Locate the final image.

Solution We begin by locating the image using a ray diagram, as shown in Fig. 21. Rays 2 and 3 from the object O are refracted as shown by the first lens; extended backward they show the location of the (virtual) image I produced by the first lens. This image then acts as the object O' of the second lens, and rays $2'$ and $3'$ give the position of the final image I', which is inverted and real.

Applying Eq. 13, we can find the position of the first image:

$$\frac{1}{+10 \text{ cm}} + \frac{1}{i} = \frac{1}{+15 \text{ cm}}$$

or

$$i = -30 \text{ cm}.$$

That is, the image is virtual and formed 30 cm from the lens on its V-side. Treating this image as the object O' for the second lens, the object distance o' is $|i| + d = 30$ cm $+ 6$ cm $= 36$ cm. Note that, although I is a *virtual* image for the first lens, O' is a *real* object for the second lens, because *diverging* light leaves the object and strikes the second lens.

Applying Eq. 13 once again, we have

$$\frac{1}{+36 \text{ cm}} + \frac{1}{i'} = \frac{1}{+15 \text{ cm}}$$

or

$$i' = +26 \text{ cm},$$

corresponding to a real image on the R-side of the second lens, as shown in Fig. 21.

Sample Problem 8 The optical system shown in Fig. 22 consists of two lenses, of focal lengths $f = +12$ cm and $f' = -32$ cm, separated by a distance of $d = 22$ cm. A luminous object is placed 18 cm from the first lens. Locate the final image produced by this system.

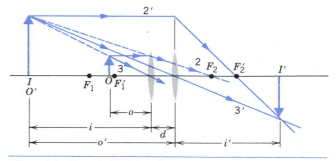

Figure 21 Sample Problem 7.

Solution A graphical construction using a ray diagram is given in Fig. 22. The real image I of the first lens would form to the right of the second lens. Because light forming this image is converging as it strikes the second lens, we treat it as a virtual object O' for the second lens.

For the first lens, Eq. 13 gives

$$\frac{1}{+18 \text{ cm}} + \frac{1}{i} = \frac{1}{+12 \text{ cm}}$$

or

$$i = +36 \text{ cm}.$$

The real image would be formed 36 cm from the first lens, as shown. The distance o' from the virtual object O' to the second lens has magnitude $i - d$ or 36 cm − 22 cm = 14 cm. Because O' is a virtual object, we take the distance o' to be negative. Now Eq. 13 gives

$$\frac{1}{-14 \text{ cm}} + \frac{1}{i'} = \frac{1}{-32 \text{ cm}}$$

or

$$i' = +25 \text{ cm}.$$

The real image I' forms on the R-side of the second lens.

Sample Problem 9 The object in Fig. 22 has a height h of 2.4 cm. Find the height of the image.

Solution We seek the lateral magnification of the compound system. Once again, we treat the compound system as two separate systems, and the total lateral magnification m_t of the combined system is the product of the lateral magnifications m and m' of the individual lenses:

$$m_t = mm' = \left(-\frac{i}{o}\right)\left(-\frac{i'}{o'}\right) = \left(-\frac{+36 \text{ cm}}{+18 \text{ cm}}\right)\left(-\frac{+25 \text{ cm}}{-14 \text{ cm}}\right)$$

$$= -3.57,$$

where we have used the values of the object and image distances found in Sample Problem 8. The height h_f of the final image is

$$h_f = m_t h = (-3.57)(2.4 \text{ cm}) = -8.6 \text{ cm}.$$

The minus sign reminds us that the final image is inverted with respect to the original object.

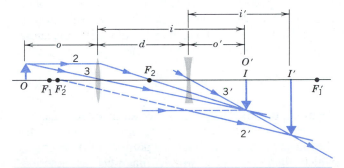

Figure 22 Sample Problem 8.

44-5 OPTICAL INSTRUMENTS

The human eye is a remarkably effective organ, but its range can be extended in many ways by optical instruments such as eyeglasses or contact lenses, simple magnifiers, motion picture projectors, cameras (including TV cameras), microscopes, and telescopes. In many cases these devices extend the scope of our vision beyond the visible range; satellite-borne infrared cameras and x-ray microscopes are examples.

In almost all cases of modern sophisticated optical instruments, the mirror and thin lens formulas hold only as approximations. In typical laboratory microscopes the lens can by no means be considered "thin." In most optical instruments lenses are compound; that is, they are made of several components. Figure 23, for example, shows the components of a typical zoom lens, commonly used in TV cameras to provide a 20:1 range in focal lengths.

In this section we consider optical devices that are designed to produce an enlarged image: we want something to *appear* larger than it appears to the unaided eye. The lateral magnification is an incomplete measure of the apparent size of an image produced by an optical system. An optical system might produce an enlarged image ($|m| > 1$) but may place that image so much farther from us than the object that it would actually appear to the observer to be smaller than the object. Even though the lateral magnification may be greater than unity, and thus the image size greater than the object size, the net result is not what the observer would call a "magnified" image.

The Simple Magnifier

Figure 24 represents the formation of an image by a human eye. The size of the image on the retina is determined by the angle θ subtended by the object. For small objects located at relatively large distances from the eye, the angle θ can be approximated as

$$\theta \approx \frac{h}{d}, \tag{23}$$

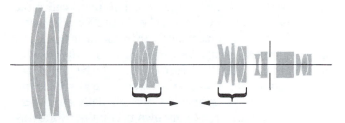

Figure 23 The components of a zoom lens in a TV camera. The central sections of the lens system move as shown. None of the lenses is "thin," and the paraxial approximation is not imposed.

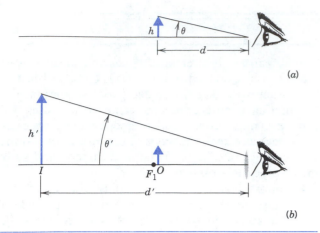

Figure 24 (a) An object of height h at a distance d from the eye subtends an angle θ. (b) When the object is viewed through a lens used as a simple magnifier, the image I of height h' is at a distance d' and subtends an angle θ' at the eye.

where h is the size of the object and d is its distance from the eye.

In Fig. 24b, the observer is viewing the object through a lens that forms an image of lateral size h' at a distance d' from the eye. The apparent angular size of the image to the observer is, again for small angles,

$$\theta' \approx \frac{h'}{d'} . \tag{24}$$

The image viewed through the lens will *appear* larger than the original object to the observer if it subtends a larger solid angle than the object subtends. It is therefore not the lateral magnification m $(= h'/h)$ that is important in measuring the apparent size of the image; it is the *angular magnification* m_θ, defined as

$$m_\theta = \frac{\theta'}{\theta} . \tag{25}$$

In effect, m_θ is the ratio of the size of the two images on the retina, one with the lens and one without.

The normal human eye can focus a sharp image of an object on the retina if the object O is located anywhere from infinity (the stars, say) to a certain point called the *near point P_n*, which we take to be about 25 cm from the eye. If you view an object closer than the near point, the perceived retinal image becomes fuzzy. The location of the near point normally varies with age. We have all heard stories about people who claim not to need glasses but who read their newspapers at arm's length; their near points are receding! Find your own near point by moving this page closer to your eyes, considered separately, until you reach a position at which the image begins to become indistinct.

We take as our basis for comparison the angular size

that an object would appear to have if it were placed at the near point. Thus

$$\theta = \frac{h}{25 \text{ cm}} . \tag{26}$$

If we place the object so that it is just inside the first focal point of a converging lens, as in Fig. 24b, a virtual image is formed far away from the lens. The lateral magnification m has magnitude i/o, and the distance d' to the image is i. The lateral size of the image is, taking magnitudes of all quantities,

$$h' = mh = \frac{i}{o} h \tag{27}$$

and the angular size is

$$\theta' = \frac{h'}{d'} = \frac{(i/o)h}{i} = \frac{h}{o} \approx \frac{h}{f} , \tag{28}$$

where the last step can be taken because we assumed the object to be placed close to the focal point. The angular magnification is

$$m_\theta = \frac{\theta'}{\theta} = \frac{h/f}{h/25 \text{ cm}}$$

or

$$m_\theta = \frac{25 \text{ cm}}{f} . \tag{29}$$

Equation 29 gives the angular magnification of the *simple magnifier*, which uses only one lens. The ordinary "magnifying glass," used by stamp collectors and actors portraying Sherlock Holmes, is in reality a simple magnifier. To obtain large angular magnification, we want f as small as possible. In practice, an angular magnification of about 10 is the best we can do before lens aberrations begin to distort the image. More sophisticated magnifiers, such as the compound microscope discussed next, can have appreciably greater angular magnifications.

Compound Microscope

Figure 25 shows a thin lens version of a compound microscope, used for viewing small objects that are very close to the objective lens of the instrument. The object O, of height h, is placed just outside the first focal point F_1 of the objective lens, whose focal length is f_{ob}. A real, inverted image I of height h' is formed by the objective, the lateral magnification being given by Eq. 15, or

$$m = -\frac{h'}{h} = -\frac{s \tan \theta}{f_{ob} \tan \theta} = -\frac{s}{f_{ob}} . \tag{30}$$

As usual, the minus sign indicates an inverted image.

The distance s (called the *tube length*) is chosen so that the image I falls on the first focal point F_1' of the eyepiece, which then acts as a simple magnifier as described previously. Parallel rays enter the eye, and a final image I'

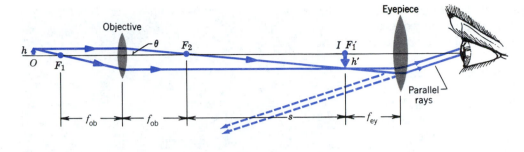

Figure 25   A thin lens version of a compound microscope (not drawn to scale).

forms at infinity. The final magnification M is the product of the linear magnification m for the objective lens (Eq. 30) and the angular magnification m_θ of the eyepiece (Eq. 29), or

$$M = mm_\theta = -\frac{s}{f_{ob}}\frac{25\text{ cm}}{f_{ey}}. \tag{31}$$

Refracting Telescope

Like microscopes, telescopes come in a large variety of forms. The form we describe here is the simple refracting telescope consisting of an objective lens and an eyepiece, both represented in Fig. 26 by thin lenses. In practice, just as in microscopes, each lens may be a compound lens system.

At first glance it may seem that the lens arrangements for telescopes and for microscopes are similar. However, telescopes are designed to view large objects, such as galaxies, stars, and planets, at large distances, whereas microscopes are designed for just the opposite purpose. Note also that in Fig. 26 the second focal point of the objective F_2 coincides with the first focal point of the eyepiece F'_1, but in Fig. 25 these points are separated by the tube length s.

In Fig. 26 parallel rays from a distant object strike the objective lens, making an angle θ_{ob} with the telescope axis and forming a real, inverted image at the common focal point F_2, F'_1. This image acts as an object for the eyepiece and a (still inverted) virtual image is formed at infinity. The rays defining the image make an angle θ_{ey} with the telescope axis.

The angular magnification m_θ of the telescope is θ_{ey}/θ_{ob}. For paraxial rays (rays close to the axis) we can write $\theta_{ob} = h'/f_{ob}$ and $\theta_{ey} = h'/f_{ey}$, which gives

$$m_\theta = -\frac{f_{ob}}{f_{ey}}, \tag{32}$$

the minus sign indicating an inverted final image.

Magnification is only one of the design factors of an astronomical telescope and is indeed easily achieved. A good telescope needs *light-gathering power*, which determines how bright the image is. This is important when viewing faint objects such as distant galaxies and is accomplished by making the objective lens diameter as large as possible. *Field of view* is another important parameter. An instrument designed for galactic observation (narrow field of view) must be quite different from one designed for the observation of meteors (wide field of view). The telescope designer must also take account of lens and mirror aberrations including *spherical aberration* (that is, lenses and mirrors with truly spherical surfaces do not form sharp images) and *chromatic aberration* (that is, for simple lenses the index of refraction and thus the focal length vary with wavelength so that fuzzy images are

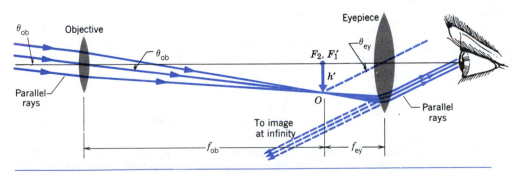

Figure 26   A thin lens version of a refracting telescope (not drawn to scale).

formed, displaying unnatural colors). The effects of diffraction (see Section 46-4) limit the ability of any optical instrument to distinguish between two objects (stars, say) whose angular separation is small.

To build refracting telescopes of larger diameters (for better light-gathering efficiency), we must also make the lenses thicker, which increases the distortions and aberrations caused by the lens. The largest refracting telescopes, which were built around the end of the 19th century, have lenses about 1 m in diameter. Reflecting telescopes, in which the objective element is a mirror rather than a lens, do not suffer from these distortions, because the light reflects from the front surface of the mirror. The largest single reflecting telescopes have diameters around 5–6 m, and thus have about 25–36 times the light-gathering capability of the largest refracting telescopes. Even larger reflecting telescopes can be constructed by combining the light from many individual mirrors into a single image.

Earth-bound optical telescopes are limited in their ability to form sharp images by atmospheric distortion; the natural turbulence in the atmosphere distorts the (nearly) plane wavefronts that reach the Earth from distant objects. One cure for this problem has been obtained through the development of *adaptive optics*; by sensing the atmospheric distortion, the shape of a flexible mirror can be modified to compensate for the distortion and thus produce a sharp image. An alternative way to eliminate the effects of the atmosphere is to place the telescope

Figure 27   The Hubble Space Telescope.

above the atmosphere. Figure 27 shows the Hubble Space Telescope, a reflecting telescope that was launched into Earth orbit by a space shuttle in 1990.

QUESTIONS

1. In many city buses a convex mirror is suspended over the door, in full view of the driver. Why not a plane or concave mirror?

2. Dentists and dental hygienists use a small mirror with a long handle attached to examine your teeth. Is the mirror concave, convex, or plane, and why?

3. Under what conditions will a spherical mirror, which may be concave or convex, form (*a*) a real image, (*b*) an inverted image, and (*c*) an image smaller than the object?

4. Can a virtual image be projected onto a screen?

5. You are looking at a dog through a glass window pane. Where is the image of the dog? Is it real or virtual? Is it upright or inverted? What is the magnification? (*Hint*: Think of the window pane as the limiting case of a thin lens in which the radii of curvature have been allowed to become infinitely large.)

6. In some cars the right (passenger) side mirror bears the notation: "Objects in the mirror are closer than they appear." What feature of the mirror requires this warning? What advantages does the mirror have to compensate for this disadvantage? Do cars viewed in this mirror appear to be

moving faster or slower than they would if viewed in a plane mirror?

7. We have all seen TV pictures of a baseball game shot from a camera located somewhere behind second base. The pitcher and the batter are about 60 ft apart but they look much closer on the TV screen. Why are the images viewed through a telephoto lens foreshortened in this way?

8. An unsymmetrical thin lens forms an image of a point object on its axis. Is the image location changed if the lens is reversed?

9. Why has a lens two focal points and a mirror only one?

10. Under what conditions will a thin lens, which may be converging or diverging, form (*a*) a real image, (*b*) an inverted image, and (*c*) an image smaller than the object?

11. A diver wants to use an air-filled plastic bag as a converging lens for underwater visibility. Sketch a suitable cross section for the bag.

12. In connection with Fig. 17*a*, all rays originating on the same wavefront in the incident wave have the same optical path to the image point. Discuss this in connection with Fermat's principle (see Chapter 43).

13. What is the significance of the origin of coordinates in Figs. 15a and 15b?

14. Why does chromatic aberration occur in simple lenses but not in mirrors?

15. Consider various lens aberrations. Is it possible in principle to make a lens free of all aberrations (for example, by grinding the surfaces) when focusing monochromatic light?

16. A concave mirror and a converging lens have the same focal length in air. Do they have the same focal length when immersed in water? If not, which has the greater focal length?

17. Under what conditions will a thin lens have a lateral magnification (a) of −1 and (b) of +1?

18. How does the focal length of a thin glass lens for blue light compare with that for red light, assuming the lens is (a) diverging and (b) converging?

19. Does the focal length of a lens depend on the medium in which the lens is immersed? Is it possible for a given lens to act as a converging lens in one medium and a diverging lens in another medium?

20. Are the following statements true for a glass lens in air? (a) A lens that is thicker at the center than at the edges is a converging lens for parallel light. (b) A lens that is thicker at the edges than at the center is a diverging lens for parallel light.

21. Under what conditions would the lateral magnification ($m = -i/o$) for lenses and mirrors become infinite? Is there any practical significance to such a condition?

22. Is the focal length of a spherical mirror affected by the medium in which it is immersed? Of a thin lens? Why the difference, if any?

23. Why is the magnification of a simple magnifier (see the derivation leading to Eq. 29) defined in terms of angles rather than image/object size?

24. Ordinary spectacles do not magnify but a simple magnifier does. What then is the function of spectacles?

25. The "*f-number*" of a camera lens (see Problem 39) is its focal length divided by its aperture (effective diameter). Why is this useful to know in photography? How can the *f*-number of the lens be changed? How is exposure time related to *f*-number?

26. A magnifying glass of small focal length allows finer detail to be examined than does one of long focal length. Explain.

27. Estimate the greatest distance at which the human eye can read the headlines of a newspaper.

28. Does it matter whether (a) an astronomical telescope, (b) a compound microscope, (c) a simple magnifier, (d) a camera, including a TV camera, or (e) a projector, including a slide projector, produces upright or inverted images? What about real or virtual images?

29. The unaided human eye produces a real but inverted image on the retina. (a) Why then don't we perceive objects such as people and trees as upside down? (b) We don't, of course, but suppose that we wore special glasses so that we did. If you then turned this book upside down, could you read this question with the same facility that you do now?

30. Which of the following—a converging lens, a diverging lens, a concave mirror, a convex mirror, or a plane mirror —is used: (a) As a magnifying glass? (b) As the reflector in the lamphouse of a slide projector? (c) As the objective of a reflecting telescope? (d) In a kaleidoscope? (e) As the eyepiece of opera glasses? (f) To obtain a wider rear view from the driver's seat in a car?

31. What properties of a lens would make it a good burning glass (a lens that, aimed at the Sun, will quickly ignite paper or twigs placed behind it)?

32. In William Golding's *Lord of the Flies* the character Piggy uses his glasses to focus the Sun's rays and kindle a fire. Later, the boys abuse Piggy and break his glasses. He is unable to identify them at close range because he is near-sighted. Find the flaw in this narrative. (*Boston Globe,* December 17, 1985, Letters.)

33. Explain the function of the objective lens of a microscope. Why use an objective lens at all? Why not just use a very powerful magnifier?

34. Why do astronomers use optical telescopes in looking at the sky? After all, the stars are so far away that they still appear to be points of light, without any detail discernible.

35. A watchmaker uses diverging eyeglasses for driving, no glasses for reading, and converging glasses in occupational work. Is the watchmaker nearsighted or farsighted? Explain. (See Problem 38.)

36. Why are all recent large astronomical telescopes of the reflecting rather than the refracting variety? Think of mechanical mounting problems for lenses and mirrors, the difficulty of shaping (that is, "figuring") the various optical surfaces involved, problems with small flaws in the optical glass blanks used to make lenses and mirrors, and so on.

37. Explain why (a) ultraviolet light is sometimes used to illuminate objects under a microscope, (b) blue filters are sometimes used to photograph a star seen through a telescope, and (c) infrared light is often used to get greater clarity in landscape photographs.

PROBLEMS

Section 44-1 Spherical Mirrors

1. A concave shaving mirror has a radius of curvature of 35 cm. It is positioned so that the image of a man's face is 2.7 times the size of his face. How far is the mirror from the man's face?

2. Redraw Fig. 28 on a large sheet of paper and trace carefully

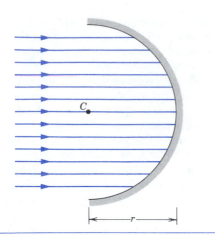

Figure 28   Problem 2.

the reflected rays, using the law of reflection. Is a point focus formed? Discuss.

3. Fill in the table below, each column of which refers to a spherical or plane mirror and a real object. Check your results by ray tracing. Distances are in centimeters; if a number has no plus or minus sign in front of it, find the correct sign.

4. (a) A luminous point is moving at speed v_o toward a spherical mirror, along its axis. Show that the speed at which the image of this point object is moving is given by

$$v_i = -\left(\frac{r}{2o-r}\right)^2 v_o.$$

(b) Assume that the mirror is concave, with $r = 15$ cm and that $v_0 = 5.0$ cm/s. Find the speed of the image if the object is far outside the focal point ($o = 75$ cm). (c) If it is close to the focal point ($o = 7.7$ cm). (d) If it is very close to the mirror ($o = 0.15$ cm).

5. A short linear object of length L lies on the axis of a spherical mirror, a distance o from the mirror. (a) Show that its image will have a length L', where

$$L' = L\left(\frac{f}{o-f}\right)^2.$$

(b) Show that the *longitudinal magnification* $m'\ (= L'/L)$ is equal to m^2, where m is the lateral magnification.

6. Repeat the derivation leading to Eq. 1 using the geometry of Fig. 8 for the convex mirror, and show that Eq. 1 is valid in this case only if i and r are taken to be negative.

Section 44-2 Spherical Refracting Surfaces

7. Figure 29 shows the cross section of a hollow glass tube of internal radius r, external radius R, and index of refraction n. (a) Convince yourself that the ray ABC shown defines the apparent internal radius r^* as seen from the side. (b) Show that $r^* = nr$, independent of R.

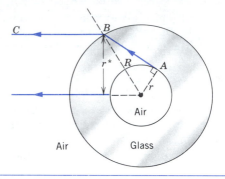

Figure 29   Problem 7.

8. Fill in the table at the top of the next page, each column of which refers to a spherical surface separating two media with different indices of refraction. Distances are measured in centimeters. The object is real in all cases. Draw a figure for each situation and construct the appropriate rays graphically. Assume a point object.

9. A parallel beam of light from a laser falls on a solid transparent sphere of index of refraction n, as shown in Fig. 30. (a) Show that the beam cannot be brought to a focus at the back of the sphere unless the beam width is small compared with the radius of the sphere. (b) If the condition in (a) is

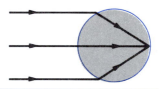

Figure 30   Problem 9.

satisfied, what is the index of refraction of the sphere? (c) What index of refraction, if any, will focus the beam at the center of the sphere?

10. A narrow parallel incident beam of light falls, from the left, on a solid glass sphere at normal incidence. The radius of the

TABLE FOR PROBLEM 3

	a	b	c	d	e	f	g	h
Type	Concave						Convex	
f (cm)	20		+20			20		
r (cm)				−40			40	
i (cm)				−10			4.0	
o (cm)	+10	+10	+30	+60				+24
m		+1.0		−0.50		+0.10		0.50
Real image?		No						
Upright image?								No

TABLE FOR PROBLEM 8

	a	b	c	d	e	f	g	h
n_1	1.0	1.0	1.0	1.0	1.5	1.5	1.5	1.5
n_2	1.5	1.5	1.5		1.0	1.0	1.0	
o (cm)	+10	+10	+10		+20	+10		+70	+100
i (cm)		−13	+600	−20	−6.0	−7.5		+600
r (cm)	+30		+30	−20		−30	+30	−30
Real image?								

sphere is R and its index of refraction is $n < 2$. Find the distance of the image from the right edge of the sphere.

Section 44-3 Thin Lenses

11. An object is 20 cm to the left of a thin diverging lens having a focal length of −30 cm. Where is the image formed? Obtain the image position both by calculation and also from a ray diagram.

12. A double-convex lens is to be made of glass with an index of refraction of 1.5. One surface is to have twice the radius of curvature of the other and the focal length is to be 60 mm. Find the radii.

13. Suppose that you focus an image of the Sun on a screen, using a thin lens whose focal length is 27 cm. Find the diameter of the image. (See Appendix C for needed data on the Sun.)

14. A lens is made of glass having an index of refraction of 1.5. One side of the lens is flat and the other convex with a radius of curvature of 20 cm. (a) Find the focal length of the lens. (b) If an object is placed 40 cm to the left of the lens, where will the image be located?

15. Show that the focal length f for a thin lens whose index of refraction is n and which is immersed in a fluid whose index of refraction is n' is given by

$$\frac{1}{f} = \frac{n - n'}{n'} \left(\frac{1}{r_1} - \frac{1}{r_2} \right).$$

16. An object is placed at the center of curvature of a double-concave lens, both of whose radii of curvature have the same magnitude. (a) Find the image distance in terms of the radius of curvature r and the index of refraction n of the

glass. (b) Describe the nature of the image. (c) Verify your result with a ray diagram.

17. You have a supply of flat glass disks ($n = 1.5$) and a lens-grinding machine that can be set to grind radii of curvature of either 40 cm or 60 cm. You are asked to prepare a set of six lenses like those shown in Fig. 31. What will be the focal length of each lens? (*Note*: Where you have a choice of radii of curvature, select the smaller one.)

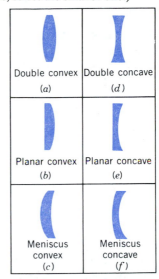

Figure 31 Problem 17.

18. To the extent possible, fill in the table below, each column of which refers to a thin lens. If a quantity cannot be calculated,

TABLE FOR PROBLEM 18

	a	b	c	d	e	f	g	h	i
Type	Converging								
f (cm)	10	+10	10	10					
r_1 (cm)					+30	−30	−30		
r_2 (cm)					−30	+30	−60		
i (cm)									
o (cm)	+20	+5.0	+5.0	+5.0	+10	+10	+10	+10	+10
n					1.5	1.5	1.5		
m			>1	<1				0.50	0.50
Real image?									Yes
Upright image?								Yes	

write "X." Distances are in centimeters; if a number (except in row *n*) has no plus sign or minus sign in front of it, find the correct sign. Draw a figure for each situation and construct the appropriate rays graphically. The object is real in all cases.

19. The formula

$$\frac{1}{o} + \frac{1}{i} = \frac{1}{f}$$

is called the *Gaussian* form of the thin lens formula. Another form of this formula, the *Newtonian* form, is obtained by considering the distance x from the object to the first focal point and the distance x' from the second focal point to the image. Show that

$$xx' = f^2.$$

20. Reproduce Fig. 15 from first principles, that is, from Eq. 13. How do you know: (*a*) That the lens is diverging or converging? (*b*) That the image is real or virtual? (*c*) That the object is real or virtual? (*d*) That the lateral magnification is >1 or <1?

21. Show that Eq. 16 is correct.

22. An illuminated arrow forms a real inverted image of itself at a distance $d = 40.0$ cm, measured along the optic axis of a lens; see Fig. 32. The image is just half the size of the object. (*a*) What kind of lens must be used to produce this image? (*b*) How far from the object must the lens be placed? (*c*) What is the focal length of the lens?

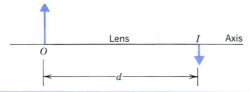

Figure 32 Problem 22.

23. An illuminated slide is mounted 44 cm from a screen. How far from the slide must a lens of focal length 11 cm be placed in order to focus an image on the screen?

24. Show that the distance between a real object and its real image formed by a thin converging lens is always greater than or equal to four times the focal length of the lens.

25. A luminous object and a screen are a fixed distance D apart. (*a*) Show that a converging lens of focal length f will form a real image on the screen for two positions that are separated by

$$d = \sqrt{D(D - 4f)}.$$

(*b*) Show that the ratio of the two image sizes for these two positions is

$$\left(\frac{D - d}{D + d}\right)^2.$$

Section 44-4 Compound Optical Systems

26. Two converging lenses, with focal lengths f_1 and f_2, are positioned a distance $f_1 + f_2$ apart, as shown in Fig. 33. Arrangements like this are called *beam expanders* and are often used to increase the diameters of light beams from lasers. (*a*) If W_1 is the incident beam width, show that the width of the

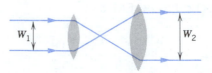

Figure 33 Problem 26.

emerging beam is $W_2 = (f_2/f_1) W_1$. (*b*) Show how a combination of one diverging and one converging lens can also be arranged as a beam expander. Incident rays parallel to the axis should exit parallel to the axis. (*c*) Calculate the ratio of the intensity of the beam emerging from the beam expander to the intensity of the laser beam.

27. A converging lens with a focal length of +20 cm is located 10 cm to the left of a diverging lens having a focal length of −15 cm. If a real object is located 40 cm to the left of the first lens, locate and describe completely the image formed.

28. A thin flat plate of partially reflecting glass is a distance b from a convex mirror. A point source of light S is placed a distance a in front of the plate (see Fig. 34) so that its image in the partially reflecting plate coincides with its image in the mirror. If $b = 7.50$ cm and the focal length of the mirror is $f = -28.2$ cm, find a and draw the ray diagram.

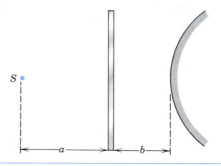

Figure 34 Problem 28.

29. (*a*) Show that a thin converging lens of focal length f followed by a thin diverging lens of focal length $-f$ will bring parallel light to a focus beyond the second lens provided that the separation of the lenses L satisfies $0 < L < f$. (*b*) Does this property change if the lenses are interchanged? (*c*) What happens when $L = 0$?

30. An upright object is placed a distance in front of a converging lens equal to twice the focal length f_1 of the lens. On the other side of the lens is a converging mirror of focal length f_2 separated from the lens by a distance $2(f_1 + f_2)$; see Fig. 35.

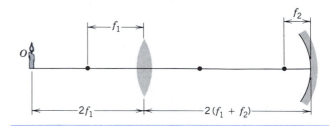

Figure 35 Problem 30.

(a) Find the location, nature, and relative size of the final image, as seen by an eye looking toward the mirror through the lens. (b) Draw the appropriate ray diagram.

31. An object is placed 1.12 m in front of a converging lens, of focal length 58.0 cm, which is 1.97 m in front of a plane mirror. (a) Where is the final image, measured from the lens, that would be seen by an eye looking toward the mirror through the lens? (b) Is the final image real or virtual? (c) Is the final image upright or inverted? (d) What is the lateral magnification?

32. An object is 20.0 cm to the left of a lens with a focal length of +10.0 cm. A second lens of focal length +12.5 cm is 30.0 cm to the right of the first lens. (a) Using the image formed by the first lens as the object for the second, find the location and relative size of the final image. (b) Verify your conclusions by drawing the lens system to scale and constructing a ray diagram. (c) Describe the final image.

33. Two thin lenses of focal lengths f_1 and f_2 are in contact. Show that they are equivalent to a single thin lens with a focal length given by

$$f = \frac{f_1 f_2}{f_1 + f_2}.$$

34. The *power P* of a lens is defined by $P = 1/f$, where f is the focal length. The unit of power is the *diopter,* where 1 diopter = 1/meter. (a) Why is this a reasonable definition to use for lens power? (b) Show that the net power of two lenses in contact is given by $P = P_1 + P_2$, where P_1 and P_2 are the powers of the separate lenses. (*Hint*: See Problem 33.)

Section 44-5 Optical Instruments

35. The angular magnification of an astronomical telescope in normal adjustment is 36, and the diameter of the objective lens is 72 mm. What is the minimum diameter of the eyepiece required to collect all the light entering the objective from a distant point source on the axis of the instrument?

36. A microscope of the type shown in Fig. 25 has a focal length for the objective lens of 4.2 cm and for the eyepiece lens of 7.7 cm. The distance between the lenses is 25 cm. (a) What is the distance s in Fig. 25? (b) To reproduce the conditions of Fig. 25 how far beyond F_1 in that figure should the object be placed? (c) What is the lateral magnification m of the objective? (d) What is the angular magnification m_θ of the eyepiece? (e) What is the overall magnification M of the microscope?

37. Figure 36a suggests a normal human eye. Parallel rays entering a relaxed eye gazing at infinity produce a real, inverted image on the retina. The eye thus acts as a converging lens. Most of the refraction occurs at the outer surface of the eye, the *cornea*. Assume a focal length f for the eye of 2.50 cm. In Fig. 36b the object is moved in to a distance $o = 36.0$ cm from the eye. To form an image on the retina the effective focal length of the eye must be reduced to f'. This is done by the action of the ciliary muscles that change the shape of the lens and thus the effective focal length of the eye. (a) Find f' from the above data. (b) Would the effective radii of curvature of the lens become larger or smaller in the transition from Fig. 36a to 36b? (In the figure the structure of the eye is only roughly suggested and Fig. 36b is not to scale.)

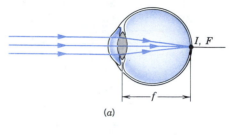

(a)

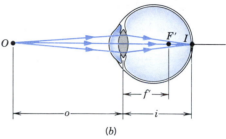

(b)

Figure 36 Problem 37.

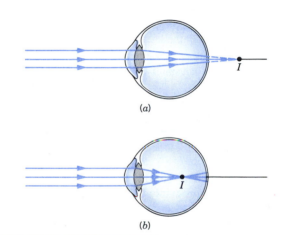

(a)

(b)

Figure 37 Problem 38.

38. In an eye that is *farsighted* the eye focuses parallel rays so that the image would form behind the retina, as in Fig. 37a. In an eye that is *nearsighted* the image is formed in front of the retina, as in Fig. 37b. (a) How would you design a corrective lens for each eye defect? Make a ray diagram for each case. (b) If you need spectacles only for reading, are you nearsighted or farsighted? (c) What is the function of bifocal spectacles, in which the upper parts and lower parts have different focal lengths?

39. Figure 38 shows an idealized camera focused on an object at infinity. A real, inverted image I is formed on the film, the image distance i being equal to the (fixed) focal length f (= 5.0 cm, say) of the lens system. In Fig. 38b the object O is closer to the camera, the object distance o being, say, 100 cm. To focus an image I on the film, we must extend the lens away from the camera (why?). (a) Find i' in Fig. 38b. (b) By how much must the lens be moved? Note that the camera differs from the eye (see Problem 37) in this respect. In the camera, f remains constant and the image distance i must be adjusted by moving the lens. For the eye the image

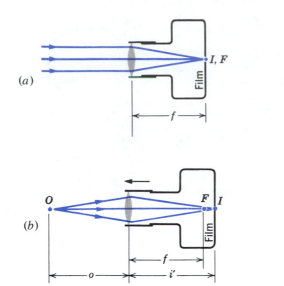

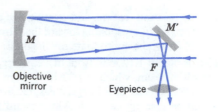

Figure 39   Problem 42.

Figure 38   Problem 39.

distance i remains constant and the focal length f is adjusted by distorting the lens. Compare Fig. 36 and Fig. 38 carefully.

40. The focal length of a small camera is 50 mm and the focusing range extends from 1.2 m out to infinity. Find the range of movement necessary between lens and film.

41. In a compound microscope, the object is 12.0 mm from the objective lens. The lenses are 285 mm apart and the intermediate image is 48.0 mm from the eyepiece. What magnification is produced?

42. Isaac Newton, having convinced himself (erroneously as it turned out) that chromatic aberration was an inherent property of refracting telescopes, invented the reflecting telescope, shown schematically in Fig. 39. He presented his second model of this telescope, which has a magnifying power of 38, to the Royal Society, which still has it. In Fig. 39 incident light falls, closely parallel to the telescope axis, on the objective mirror M. After reflection from small mirror M' (the figure is not to scale), the rays form a real, inverted image in the focal plane through F. This image is then viewed through an eyepiece. (a) Show that the angular magnification m_θ is also given by Eq. 32, or

$$m_\theta = -f_{ob}/f_{ey}$$

where f_{ob} is the focal length of the objective mirror and f_{ey} that of the eyepiece. (b) The 200-in. mirror in the reflecting telescope at Mt. Palomar in California has a focal length of 16.8 m. Estimate the size of the image formed in the focal plane of this mirror when the object is a meter stick 2.0 km away. Assume parallel incident rays. (c) The mirror of a different reflecting astronomical telescope has an effective radius of curvature ("effective" because such mirrors are ground to a parabolic rather than a spherical shape, to eliminate spherical aberration defects) of 10 m. To give an angular magnification of 200, what must be the focal length of the eyepiece?

43. A photographer stands 44.5 m from a railroad track, the line of vision being perpendicular to the tracks. A train passes at 135 km/h and the photographer takes a picture. Using a camera with focal length 3.6 cm, find the maximum exposure time so that the blurring of the image on the film does not exceed 0.75 mm.

CHAPTER 45

INTERFERENCE

The previous two chapters dealt with geometrical optics, in which the light encounters obstacles or apertures (lenses, for instance) of dimensions much larger than the wavelength of the light. You may wish to review Section 43-1, where we discussed the limit of validity of geometrical optics.

In this chapter and the next, we discuss the phenomena of interference *and* diffraction, *in which light encounters obstacles or apertures whose size is comparable to its wavelength. This is the realm of* physical optics *(also known as wave optics), which differs from geometrical optics in that physical or wave optics involves effects that depend on the wave nature of light. In fact, it is from interference and diffraction experiments that we obtain proof that light behaves (at least in these circumstances) like a wave rather than a stream of particles (as Newton believed).*

Although we deal only with light waves in this chapter, all other kinds of waves (such as sound waves and water waves) also can experience interference and diffraction. For example, in the placement of loudspeakers in a room, it is necessary to consider the interference and diffraction of sound waves. The principles we develop for light waves apply equally to other kinds of waves.

45-1 DOUBLE-SLIT INTERFERENCE

When otherwise identical waves from two sources overlap at a point in space, the combined wave intensity at that point can be greater or less than the intensity of either of the two waves. We call this effect *interference*. The interference can be either *constructive,* when the net intensity is greater than the individual intensities, or *destructive,* when the net intensity is less than the individual intensities. As we discuss later, whether the interference is constructive or destructive depends on the relative phase of the two waves.

Although any number of waves can in principle interfere, we consider here the interference of only two waves. We assume that the sources of the waves each emit at only a single wavelength. (The case of sources that emit waves of several wavelengths can be handled by considering the separate interferences of the individual component wavelengths.)

We also assume, for the time being, that the phase relationship between the two waves does not change with time. Such waves are said to be *coherent.* When coherent waves interfere, the intensity of the combined wave at any point in space does not change with time. Coherence, which is a necessary condition for interference to occur, is discussed in the next section.

Two different light sources cannot in general be made coherent, because the emission of light by the atoms of one source is independent of that of the other. The peaks and valleys of the waves from the two sources do not maintain a definite phase relationship, and the waves are said to be *incoherent.* To do interference experiments with light, it is usually necessary to divide the light from a single source into two components and to treat each component as if it were emitted from an independent source of light. If we do this properly, the two components can be made to interfere. Later we consider several schemes to create this division of the light wave; here we consider the technique of passing a light wave through two narrow openings or slits. The widths of the slits must be of the same order as

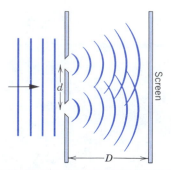

Figure 1 A train of plane light waves (for example, from a laser) is incident on a barrier into which are cut two narrow slits separated by a distance d. The widths of the slits are small compared with the wavelength, so that the waves passing through the slits spread out (diffract) and illuminate the screen.

the wavelength of the light. Thus we are clearly out of the realm in which geometrical optics applies (see Fig. 1 of Chapter 43).

Figure 1 shows a barrier into which two narrow parallel slits have been cut. A train of plane light waves, such as might be obtained from a laser, is incident on the slits. Portions of each incident wavefront pass through the slits, and so the slits can be considered as two sources of coherent light waves. (The spreading of light as it passes through the slits, illustrated in Fig. 1, is called *diffraction* and is discussed in the next chapter. For now, we regard the slits as so narrow that each can be considered as a line of point sources of light, each point source emitting spherical Huygens wavelets as discussed in Section 43-3.) Note that the two waves can overlap and interfere where they strike the screen. To simplify the analysis, we assume that the distance D between the slits and the screen is very much greater than the slit separation d. (Alternatively, we can place a lens between the slit and the screen to focus the emerging light on the screen, as we discuss later.)

When we view the screen, we see an alternating series of bright and dark bands, or *interference fringes,* corresponding respectively to maxima and minima in the intensity of the light, as shown in Fig. 2. Figure 3 shows a pattern of maxima and minima in the intensity of interfering water waves in a ripple tank. The interference of light waves and water waves to produce these maxima and minima can be understood based on similar analyses.

To analyze the interference pattern, we consider waves from each slit that combine at an arbitrary point P on the screen C in Fig. 4. The point P is at distances of r_1 and r_2 from the narrow slits S_1 and S_2, respectively. The line S_2b is drawn so that the lines PS_2 and Pb have equal lengths. If d, the slit spacing, is much smaller than the distance D between the slits and the screen (the ratio d/D in the figure being exaggerated for clarity), S_2b is then almost perpendicular to both r_1 and r_2. This means that angle S_1S_2b is almost equal to angle PaO, both angles being marked θ in the figure; equivalently, the lines r_1 and r_2 may be taken as nearly parallel.

Figure 2 The interference pattern, consisting of bright and dark bands or fringes, that would appear on the screen of Fig. 1.

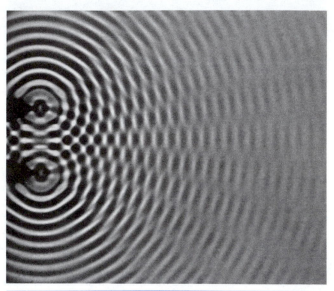

Figure 3 An interference pattern produced by water waves in a ripple tank. Two vibrating prongs create two patterns of circular ripples, which overlap to give a pattern of maxima and minima in the waves. The right-hand edge of this photograph plays the role of the screen in Fig. 1. Note that, along this "screen," there is an alternating pattern of maxima and minima, as in Fig. 2.

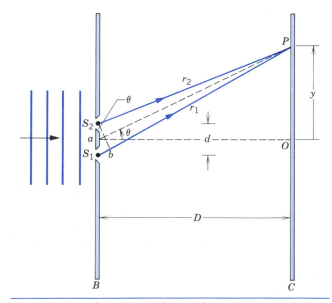

Figure 4 Rays from S_1 and S_2 combine at P. In actuality, $D \gg d$; the figure has been distorted for clarity. Point a is the midpoint between the slits.

The two rays arriving at P in Fig. 4 from S_1 and S_2 are in phase at the source slits; both are derived from the same wavefront in the incident plane wave. Because the rays travel different optical path lengths, they arrive at P with a phase difference. The number of wavelengths contained in the path difference S_1b determines the type of interference at P.

To have a *maximum* at P, the two rays must arrive in phase, and so S_1b ($= d \sin \theta$) must contain a whole number of wavelengths, or

$$S_1b = m\lambda \qquad m = 0, 1, 2, \ldots ,$$

which we can write as

$$d \sin \theta = m\lambda \qquad m = 0, 1, 2, \ldots \quad \text{(maxima).} \quad (1)$$

Note that each maximum above O in Fig. 4 has a symmetrically located maximum below O; these correspond to using $m = -1, -2, \ldots$ in Eq. 1. The central maximum is described by $m = 0$.

For a *minimum* at P, the two rays must differ in phase by an odd multiple of π, for which S_1b ($= d \sin \theta$) must contain a half-integral number of wavelengths, or

$$d \sin \theta = (m + \tfrac{1}{2})\lambda \quad m = 0, 1, 2, \ldots \quad \text{(minima).} \quad (2)$$

Negative values of m locate the minima on the lower half of the screen.

It is mathematically simpler to deal with plane waves that are incident on the double slit and that emerge from it. However, plane wavefronts do not form an image on a screen at any finite distance D from the slits. We therefore often use a lens, as shown in Fig. 5, to focus parallel rays from the slits onto the screen. Light focused at P must have struck the lens parallel to the line Px drawn from P

through the center of the lens. Under these conditions, rays r_1 and r_2 are strictly parallel, even though the requirement $D \gg d$ may not be met.

If a lens is used between the slits and the screen, it may seem that a phase difference should develop between the rays beyond the plane represented by S_2b, the geometrical path lengths between this plane and P being clearly different. In Section 44-3, however, we saw that for parallel rays focused by a lens the *optical* path lengths are equal. Two rays with equal optical path lengths contain the same number of wavelengths, so no phase difference occurs as a result of the light passing through the lens.

Young's Double-Slit Experiment

An interference experiment of the kind described above was first done in 1801 by Thomas Young.* Young's experiment provided the first conclusive proof of the wave nature of light. Because, as indicated by Eqs. 1 and 2, the spacing of the interference fringes depends on the wavelength, Young's experiments provided the first direct measurement of the wavelength of light.

There were of course no lasers in Young's time, so he created a source of coherent light by allowing sunlight to fall on a narrow opening S_0, as shown in Fig. 6. The spreading wavelets from S_0 give coherent wavefronts that pass through the two openings. Young used pinholes rather than slits for his experiments, and as a result the interference pattern was more complicated than that of Fig. 2. Nevertheless, his conclusions regarding the wave nature of light were unambiguous. Even when done with a laser, the double-slit experiment is often known as Young's experiment.

Sample Problem 1   The double-slit arrangement in Fig. 4 is illuminated with light from a mercury vapor lamp filtered so that only the strong green line ($\lambda = 546$ nm) is visible. The slits are 0.12 mm apart, and the screen on which the interference pattern appears is 55 cm away. What is the angular position of the first minimum? Of the tenth maximum?

Solution   At the first minimum we put $m = 0$ in Eq. 2, or

$$\sin \theta = \frac{(m + \tfrac{1}{2})\lambda}{d} = \frac{(\tfrac{1}{2})(546 \times 10^{-9} \text{ m})}{0.12 \times 10^{-3} \text{ m}} = 0.0023.$$

This value for $\sin \theta$ is so small that we can take it to be the value of θ, expressed in radians; expressed in degrees it is 0.13°.

---

* Thomas Young (1773–1829) was originally trained as a physician. His interest in sense perception and vision led him to physics and the study of light. Among his other scientific accomplishments were studies of surface tension and elasticity, which was recognized with the naming of the elastic modulus, now known as Young's modulus, in his honor. He was also noted for his interest in hieroglyphics and contributed to the deciphering of the Rosetta stone, which provided the first understanding of ancient Egyptian languages.

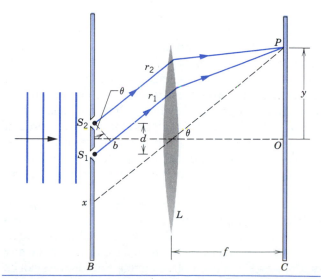

Figure 5   A lens is used to produce the interference fringes. Compare with Fig. 4. In actuality, $f \gg d$; the figure is again distorted for clarity.

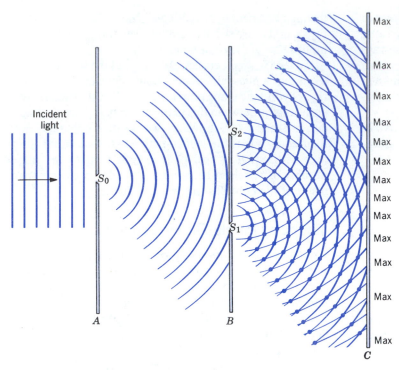

Figure 6   In Young's interference experiment, light diffracted from pinhole S_0 falls on pinholes S_1 and S_2 in screen B. Light diffracted from these two pinholes overlaps on screen C, producing the interference pattern.

At the tenth maximum (not counting the central maximum) we must put $m = 10$ in Eq. 1. Doing so and calculating as before we find an angular position of 2.6°. For these conditions we see that the angular spread of the first dozen or so fringes is small.

Sample Problem 2   What is the linear distance on screen C between the adjacent maxima m and $m + 1$ of Sample Problem 1?

Solution   If θ is small enough, we can use the approximation

$$\sin \theta \approx \tan \theta \approx \theta.$$

From Fig. 4 we see that

$$\tan \theta = \frac{y}{D}.$$

Substituting this into Eq. 1 for $\sin \theta$ leads to

$$y = m \frac{\lambda D}{d} \qquad m = 0, 1, 2, \ldots \quad \text{(maxima)}.$$

The positions of any two adjacent maxima are given by

$$y_m = m \frac{\lambda D}{d}$$

and

$$y_{m+1} = (m + 1) \frac{\lambda D}{d}.$$

We find their separation Δy by subtracting:

$$\Delta y = y_{m+1} - y_m = \frac{\lambda D}{d}$$

$$= \frac{(546 \times 10^{-9} \text{ m})(55 \times 10^{-2} \text{ m})}{0.12 \times 10^{-3} \text{ m}} = 2.5 \text{ mm}.$$

As long as θ in Fig. 4 is small, the separation of the interference fringes is independent of m; that is, the fringes are evenly spaced, as shown in Fig. 2. If the incident light contains more than one wavelength, the separate interference patterns, which have different fringe spacings, are superimposed.

45-2   COHERENCE

In deriving Eq. 1, we determined that a maximum would appear on the screen in Fig. 4 whenever the *path difference* in the waves traveling to a point P on the screen from the two slits S_1 and S_2 was equal to a whole number of wavelengths. Another description uses the *phase difference* ϕ between the two waves from S_1 and S_2. At certain points on the screen, where the phase difference is 0, $\pm 2\pi$, $\pm 4\pi$, . . . , the two waves are in phase and a maximum in the intensity occurs. At other points, the phase difference is $\pm \pi$, $\pm 3\pi$, $\pm 5\pi$, . . . , and the two waves are out of phase; at those points a minimum of intensity occurs. At still other points on the screen, the phase difference may have other values that cannot be expressed as an integer multiple of π.

For an interference pattern to occur at all, *the phase difference at points on the screen must not change with time.* If this occurs, we say that the beams from S_1 and S_2 are completely *coherent*. Coherence can occur for any type of waves. For example, coherent sound waves can be obtained by driving two different loudspeakers from the same audio oscillator, and coherent radio waves similarly result when two different antennas are connected to the same electromagnetic oscillator.

Suppose instead that slits S_1 and S_2 are replaced by two completely independent light sources, such as two fine incandescent wires placed side by side. No interference fringes appear on screen C but only a relatively uniform illumination. We can interpret this if we make the reasonable assumption that for completely independent light sources the phase difference between the two beams arriving at any point on the screen varies with time in a random way. At a certain instant conditions may be right for cancellation, and a short time later (perhaps 10^{-8} s) they may be right for reinforcement. The eye cannot follow these rapid variations and sees only a uniform illumination. The intensity at any point is equal to the sum of the intensities that each source S_1 and S_2 produces separately at that point. Under these conditions the two beams emerging from S_1 and S_2 are said to be completely *incoherent*.

To find the intensity resulting from the overlap of completely *coherent* light beams, we (1) add the wave amplitudes, taking the (constant) phase difference properly into account, and then (2) square the resultant amplitude to obtain a quantity proportional to the resultant intensity. For completely *incoherent* light beams, on the other hand, we (1) square the individual amplitudes to obtain quantities proportional to the individual intensities and then (2) add the individual intensities to obtain the resultant intensity. These procedures agree with the experimental facts that for completely independent light sources the resultant intensity at every point is always greater than the intensity produced at that point by either light source acting alone, while for coherent sources the intensity at some points may be less than that produced by either source alone.

Under what experimental conditions are coherent or incoherent beams produced? Consider a parallel beam of microwave radiation emerging from an antenna connected by a coaxial cable to an oscillator based on an electromagnetic resonant cavity. The cavity oscillations (see Section 40-4) are completely periodic with time and produce, at the antenna, a completely periodic variation of **E** and **B** with time. The radiated wave at large enough distances from the antenna is well represented by Fig. 10

of Chapter 41. Note that (1) the wave has essentially infinite extent in time, including both future times ($t > 0$, say) and past times ($t < 0$); see Fig. 7*a*. At any point, as the wave passes by, the wave disturbance (**E** or **B**) varies with time in a perfectly periodic way. (2) The wavefronts at points far removed from the antenna are parallel planes of essentially infinite extent at right angles to the propagation direction. At any instant of time the wave disturbance varies with distance along the propagation direction in a perfectly periodic way.

Two beams generated from a single traveling wave like that of Fig. 10 of Chapter 41 are completely coherent. One way to generate two such beams is to put an opaque screen containing two slits in the path of the beam. The waves emerging from the slits always have a constant phase difference at any point in the region in which they overlap, and interference fringes are produced. Coherent radio beams can also be readily established, as can coherent elastic waves in solids, liquids, and gases. The two prongs of the vibrating tapper in Fig. 3, for example, generate two coherent waves in the water of the ripple tank.

If we turn instead to common sources of visible light, such as incandescent wires or an electric discharge passing through a gas, we become aware of a fundamental difference. In both of these sources the fundamental light emission processes occur in individual atoms, and these atoms do not act together in a cooperative (that is, *coherent*) way. The act of light emission by a single atom takes, in a typical case, about 10^{-8} s, and the emitted light is properly described as a finite *wavetrain* (Fig. 7*b*) rather than as an infinite wave (Fig. 7*a*). For emission times such as these, the wavetrains are a few meters long. For actual light sources, such as low-pressure gas discharge tubes, the wavetrains are typically of the order of centimeters long. This is the limit of distances over which light from such sources remains coherent.

Interference effects from ordinary light sources can be produced by putting a very narrow slit (S_0 in Fig. 6) directly in front of the source. This ensures that the wavetrains that strike slits S_1 and S_2 in screen B in Fig. 6 originate from the same small region of the source. If the path lengths from all points within slit S_0 to S_1 and S_2 are nearly equal, light passing through the double slits is in phase, and a stationary interference pattern is produced

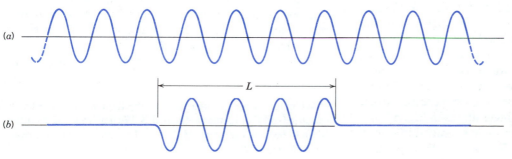

Figure 7   (*a*) A section of an infinite wave. (*b*) A wavetrain of finite length L.

on the screen C. If slit S_0 is made so wide that there are points within S_0 for which the path lengths to S_1 and S_2 differ by one-half wavelength, the light passing through the double slits from sources at such points would be out of phase; this light contributes maxima on the screen where there were previously minima and minima where there were previously maxima. The effect on the screen becomes an incoherent superposition of the light from effectively many sources, thereby washing out the interference pattern.

If the slit S_0 is small enough, a given constant phase difference is maintained at any point on the screen C between beams passing through the slits S_1 and S_2. We can regard this light as coherent, within a distance characterized by the length of its wavetrain.

If the width of slit S_0 in Fig. 6 is gradually increased, it is observed experimentally that the maxima of the interference fringes become reduced in intensity and that the intensity in the fringe minima is no longer strictly zero. In other words, the fringes become less distinct. If S_0 is opened extremely wide, the lowering of the maximum intensity and the raising of the minimum intensity are such that the fringes disappear, leaving only a uniform illumination. Under these conditions we say that the beams from S_1 and S_2 pass continuously from a condition of complete coherence to one of complete incoherence. When not at either of these two limits, the beams are said to be partially coherent.

Partial coherence can also be demonstrated in two beams that are produced using a partially silvered mirror, which reflects part of the incident light and transmits the rest. The two beams so produced can be made to traverse paths of different lengths before they are recombined. If the path difference is small compared with the average length of a wavetrain, the interference fringes are sharply defined and go essentially to zero at their minima. If the path difference is deliberately made longer, the two beams begin to lose their coherence and the fringes become less distinct. Finally, when the path difference is larger than the average length of a wavetrain, the fringes disappear altogether. In this way, it is possible to go gradually from complete coherence, through partial coherence, to complete incoherence.

Before 1960, it was not possible to construct a source of visible light that gave an infinite wave such as that of Fig. 7a. In the sources of visible light that were previously available, the atoms did not behave cooperatively, and the light was not coherent. We now have highly coherent sources of visible light: the familiar *laser,* which stands for *l*ight *a*mplification through *s*timulated *e*mission of *r*adiation. Using a laser beam, we can do double-slit interference in the geometry shown in Fig. 1 by merely illuminating a double slit with the laser. It is not necessary to use the diffracted light from the single slit, as in Fig. 6.

The coherence of the laser beam has resulted in a number of practical applications. In many of these applica-

tions, the beam from the laser is split into two beams (using a partially silvered mirror). The two beams travel different paths and are then made to recombine, where they interfere. Because the coherence length of light from lasers can be tens or hundreds of kilometers, interference patterns are produced for large path differences between the two beams. One application of this coherence is in holography (see Section 47-5), in which one beam is reflected from an object and the interference pattern between the direct and reflected beams is stored on photographic film, which can be used to reconstruct a three-dimensional image of the object. Changes in the path length of one of the beams can be easily detected over large distances through changes in the interference pattern; laser interferometers based on this principle are used to track the movement of geologic plates over the Earth's surface. Other applications involve the Doppler shift of a beam reflected from a moving object; when the two beams recombine, a pattern of beats is produced. Figure 20 of Chapter 2 showed an application of this effect to the measurement of the free-fall acceleration. Other applications that rely on the coherence of the laser beam include communication over long distances using optical signals.

45-3 INTENSITY IN DOUBLE-SLIT INTERFERENCE

Equations 1 and 2 give the locations of the maxima and minima of the interference pattern. They do not, however, indicate how the intensity varies between the maxima and minima. In this section we derive an expression for the intensity I at any point P located by the angle θ in Fig. 4.

Let us assume that the electric field components* of the two waves in Fig. 4 vary with time at point P as

$$E_1 = E_0 \sin \omega t \qquad (3)$$

and

$$E_2 = E_0 \sin (\omega t + \phi), \qquad (4)$$

where $\omega \, (= 2\pi\nu)$ is the angular frequency of the waves and ϕ is the phase difference between them. Note that ϕ depends on the location of point P, which is described by the angle θ in Fig. 4. We assume that the slits are so narrow that the diffracted light from each slit illuminates the central portion of the screen uniformly. This means that near the center of the screen E_0 is independent of the position of P, that is, of the value of θ.

If the slit separation d is much smaller than the distance

---

* We could choose to characterize the light wave either by its electric field **E** or its magnetic field **B**. We generally use **E** rather than **B**, because the effects of **B** on the human eye and on various light detectors are exceedingly small.

D to the screen, the E vectors from the two interfering waves are nearly parallel, and we can replace the vector sum of the E fields with the scalar sum of their components,

$$E = E_1 + E_2, \qquad (5)$$

which, as we prove later in this section, can be written

$$E = E_\theta \sin(\omega t + \beta), \qquad (6)$$

where the phase β is

$$\beta = \tfrac{1}{2}\phi \qquad (7)$$

and the amplitude is

$$E_\theta = 2E_0 \cos \beta. \qquad (8)$$

The amplitude E_θ of the resultant wave disturbance, which determines the intensity of the interference fringes, depends on β, which in turn depends on the value of θ, that is, on the location of point P in Fig. 4. The maximum possible value of the amplitude E_θ is $2E_0$, equal to twice the amplitude E_0 of the combining waves, corresponding to complete reinforcement.

In Section 41-4, we showed that the intensity I of an electromagnetic wave is proportional to the square of its electric field amplitude E_m (see Eq. 18 of Chapter 41):

$$I = \frac{1}{2\mu_0 c} E_m^2. \qquad (9)$$

The ratio of the intensities of two light waves can therefore be expressed as the ratio of the squares of the amplitudes of their electric fields. If I_θ is the intensity of the resultant wave at P, and I_0 is the intensity that each single wave acting alone would produce, then

$$\frac{I_\theta}{I_0} = \left(\frac{E_\theta}{E_0}\right)^2. \qquad (10)$$

Combining Eqs. 8 and 10, we obtain

$$I_\theta = 4I_0 \cos^2 \beta. \qquad (11)$$

Note that the intensity of the resultant wave at any point P varies from zero [for a point at which $\phi\,(=2\beta) = \pi$, say] to four times the intensity I_0 of each individual wave [for a point at which $\phi\,(=2\beta) = 0$, say]. Let us compute I_θ as a function of the angle θ in Fig. 4.

The phase difference ϕ in Eq. 4 is associated with the path difference $S_1 b$ in Fig. 4. If $S_1 b$ is $\tfrac{1}{2}\lambda$, ϕ is π; if $S_1 b$ is λ, ϕ is 2π, and so forth. In general,

$$\frac{\text{phase difference}}{2\pi} = \frac{\text{path difference}}{\lambda}.$$

Letting ϕ be the phase difference and recalling that the path difference in Fig. 4 is $d \sin \theta$, we can write this as

$$\phi = \frac{2\pi}{\lambda}(d \sin \theta),$$

or, using Eq. 7,

$$\beta = \tfrac{1}{2}\phi = \frac{\pi d}{\lambda} \sin \theta. \qquad (12)$$

The intensity at any θ can therefore be written

$$I_\theta = 4I_0 \cos^2 \tfrac{1}{2}\phi \qquad (13)$$

or

$$I_\theta = 4I_0 \cos^2\left(\frac{\pi d \sin \theta}{\lambda}\right). \qquad (14)$$

From Eq. 13, we see that intensity maxima occur where $\cos^2 \tfrac{1}{2}\phi = 1$, or

$$\phi = 2m\pi \qquad m = 0, \pm 1, \pm 2, \ldots.$$

Using Eq. 12, we can write this as

$$d \sin \theta = m\lambda \qquad m = 0, \pm 1, \pm 2, \ldots \quad \text{(maxima)},$$

which is the same as Eq. 1. Intensity minima occur, according to Eq. 13, where $\cos^2 \tfrac{1}{2}\phi = 0$, or

$$\phi = (2m + 1)\pi \qquad m = 0, \pm 1, \pm 2, \ldots,$$

which we write using Eq. 12 as

$$d \sin \theta = (m + \tfrac{1}{2})\lambda \qquad m = 0, \pm 1, \pm 2, \ldots \quad \text{(minima)},$$

in agreement with Eq. 2.

Figure 8 shows the intensity pattern for double-slit interference. The horizontal solid line is I_0; this describes the (uniform) intensity pattern on the screen if one of the slits is covered up. If the two sources were incoherent, the intensity would be uniform over the screen and would be $2I_0$, indicated by the horizontal dashed line of Fig. 8. For coherent sources we expect the energy to be merely redis-

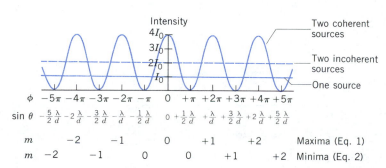

Figure 8   The intensity pattern for double-slit interference, assuming that the two interfering waves illuminate this region of the screen uniformly; that is, I_0 is independent of position.

tributed over the screen, because energy is neither created nor destroyed by the interference process. Thus the *average* intensity in the interference pattern should be $2I_0$, as for incoherent sources. This follows at once if, in Eq. 13, we substitute for the cosine-squared term the value $\frac{1}{2}$, which always results when we average the square of a sine or a cosine term over one or more half-cycles.

Adding Wave Disturbances

We now derive Eqs. 6–8 for the combined electric field of the light in double-slit interference. This derivation can be done algebraically, using the methods of Section 19-8. However, the algebraic method becomes extremely difficult when we wish to add more than two wave disturbances, as we do in later chapters. We therefore use a graphical method, which proves to be convenient in more complicated situations. This method is based on rotating *phasors* and is similar to that used in the analysis of alternating current circuits in Chapter 39.

A sinusoidal wave disturbance such as that of Eq. 3 can be represented graphically using a rotating phasor. In Fig. 9a a phasor of magnitude E_0 is allowed to rotate about the origin in a counterclockwise direction with an angular frequency ω. The alternating wave disturbance E_1 (Eq. 3) is represented by the projection of this phasor on the vertical axis.

A second wave disturbance E_2, given by Eq. 4, which has the same amplitude E_0 but a phase difference ϕ with respect to E_1, can be represented graphically (Fig. 9b) as the projection on the vertical axis of a second phasor of the same magnitude E_0 which makes an angle ϕ with the first phasor. The sum E of E_1 and E_2 is the sum of the projections of the two phasors on the vertical axis. This is revealed more clearly if we redraw the phasors, as in Fig. 9c, placing the foot of one arrow at the head of the other, maintaining the proper phase difference, and letting the whole assembly rotate counterclockwise about the origin.

In Fig. 9c, E can also be regarded as the projection on the vertical axis of a phasor of length E_θ, which is the vector sum of the two phasors of magnitude E_0. From that

figure, we see that the projection can be written

$$E = E_\theta \sin(\omega t + \beta),$$

in agreement with Eq. 6. Note that the (algebraic) sum of the projections of the two phasors is equal to the projection of the (vector) sum of the two phasors.

In most problems in optics we are concerned only with the *amplitude* E_θ of the resultant wave disturbance and not with its time variation. This is because the eye and other common measuring instruments respond to the resultant intensity of the light (that is, to the square of the amplitude) and cannot detect the rapid time variations that characterize visible light. For sodium light ($\lambda = 589$ nm), for example, the frequency ν ($= \omega/2\pi$) is 5.1×10^{14} Hz. Often, then, we need not consider the rotation of the phasors but can confine our attention to finding the amplitude of the resultant phasor.

In Fig. 9c the three phasors form an isosceles triangle whose sides have lengths E_0, E_0, and E_θ. In any triangle, an exterior angle (ϕ in this case) is equal to the sum of the two opposite interior angles (β and β), and so

$$\beta = \tfrac{1}{2}\phi.$$

It is also clear from Fig. 9c that the length of the base of this triangle is

$$E_\theta = 2E_0 \cos \beta.$$

These results are identical with Eqs. 7 and 8.

In a more general case we might want to find the resultant of more than two sinusoidally varying wave disturbances. The general procedure is the following:

1. Construct a series of phasors representing the functions to be added. Draw them end to end, maintaining the proper phase relationships between adjacent phasors.

2. Construct the sum of this phasor array, analogous to a sum of vectors. The length of the resulting phasor gives the amplitude of the electric field. The angle between it and the first phasor is the phase of the resultant with respect to this first phasor. The projection of this phasor on the vertical axis gives the time variation of the resultant wave disturbance.

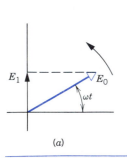
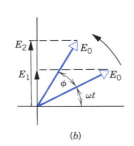
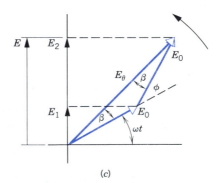

Figure 9 (a) A time-varying wave E_1 is represented by a rotating vector or phasor. (b) Two waves E_1 and E_2 differing in phase by ϕ. (c) Another way of drawing (b).

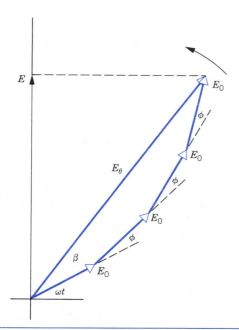

Figure 10   Sample Problem 3. Four waves are added graphically, using the method of phasors.

Sample Problem 3   Find graphically the resultant $E(t)$ of the following wave disturbances:

$$E_1 = E_0 \sin \omega t,$$

$$E_2 = E_0 \sin (\omega t + 15°),$$

$$E_3 = E_0 \sin (\omega t + 30°),$$

$$E_4 = E_0 \sin (\omega t + 45°).$$

Solution   Figure 10 shows the assembly of four phasors that represent these functions. The phase angle ϕ between successive phasors is 15°. We find by graphical measurement with a ruler and a protractor that the amplitude E_θ is 3.8 times as long as E_0 and that the resultant wave makes a phase angle β of 22.5° with respect to E_1. In other words,

$$E(t) = E_1 + E_2 + E_3 + E_4$$
$$= 3.8E_0 \sin (\omega t + 22.5°).$$

Check this result by direct trigonometric calculation or by geometric calculation from the phasor diagram of Fig. 10.

45-4   INTERFERENCE FROM THIN FILMS

The colors that we see when sunlight falls on a soap bubble, an oil slick, or a ruby-throated hummingbird are caused by the interference of light waves reflected from the front and back surfaces of thin transparent films. The film thickness is typically of the order of magnitude of the wavelength of light. Thin films deposited on optical com-

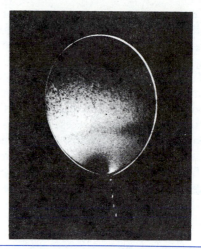

Figure 11   A soapy water film on a wire loop, viewed by reflected light. The black segment at the top is not a torn film. It occurs because the film, by drainage, is so thin there that destructive interference occurs between light reflected from the front and back surfaces of the film.

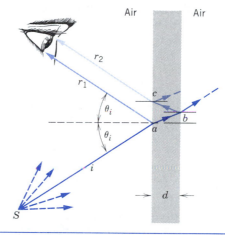

Figure 12   A thin film is viewed by light reflected from a source S. Waves reflected from the front and back surfaces enter the eye as shown, and the intensity of the resultant light wave is determined by the phase difference between the combining waves. The medium on either side of the film is assumed to be air.

---

ponents, such as camera lenses, can reduce reflection and enhance the intensity of the transmitted light. Thin coatings on windows can enhance the reflectivity for infrared radiation while having less effect on the visible radiation. In this way it is possible to reduce the heating effect of sunlight on a building.

Depending on its thickness, a thin film can be perfectly reflecting or perfectly transmitting for light of a given wavelength, as shown in Fig. 11. These effects result from constructive or destructive interference.

Figure 12 shows a transparent film of uniform thickness d illuminated by monochromatic light of wavelength λ from a point source S. The eye is positioned so that a particular incident ray i from the source enters the eye as ray r_1 after reflection from the front surface of the film at *a*. The incident ray also enters the film at *a* as a refracted ray and is reflected from the back surface of the film at *b*; it then emerges from the front surface of the film at *c* and also enters the eye, as ray r_2. The geometry of Fig. 12 is such that rays r_1 and r_2 are parallel. Having originated in the same point source, they are also coherent and so are capable of interfering. Because these two rays have traveled over paths of different lengths, have traversed different media, and—as we shall see—have suffered different kinds of reflections at *a* and *b*, there is a phase difference between them. The intensity perceived by the eye, as the parallel rays from the region *ac* of the film enter it, is determined by this phase difference.

For near-normal incidence ($\theta_i \approx 0$ in Fig. 12) the geometrical path difference for the two rays from *S* is close to $2d$. We might expect the resultant wave reflected from the film near *a* to be an interference maximum if the distance $2d$ is an integral number of wavelengths. This statement must be modified for two reasons.

First, the wavelength must refer to the wavelength λ_n of the light in the film and not to its wavelength λ in air; that is, we are concerned with *optical* path lengths rather than *geometrical* path lengths. The wavelengths λ and λ_n are related by Eq. 13 of Chapter 43,

$$\lambda_n = \lambda/n, \qquad (15)$$

where *n* is the index of refraction of the film.

To bring out the second point, let us assume that the film is so thin that $2d$ is very much less than one wavelength. The phase difference between the two waves would be close to zero on our assumption, and we would expect such a film to appear bright on reflection. However, it appears dark. This is clear from Fig. 11, in which the action of gravity produces a wedge-shaped film, extremely thin at its top edge. As drainage continues, the dark area increases in size. To explain this and many similar phenomena, one or the other of the two rays of Fig. 12 must suffer an abrupt phase change of π ($=180°$) when it is reflected at the air–film interface. As it turns out, only the ray reflected from the front surface suffers this phase change. The other ray is not changed abruptly in phase, either on transmission through the front surface or on reflection at the back surface.

In Section 19-9 we discussed phase changes on reflection for transverse waves in strings. To extend these ideas, consider the composite string of Fig. 13, which consists of two parts with different masses per unit length, stretched to a given tension. A pulse in the heavier string moves to the right in Fig. 13*a*, approaching the junction. Later there will be reflected and transmitted pulses, the reflected pulse being *in phase* with the incident pulse. In Fig. 13*b* the situation is reversed, the incident pulse now being in the lighter string. In this case the reflected pulse differs in phase from the incident pulse by π ($=180°$). In each case the transmitted pulse is in phase with the incident pulse.

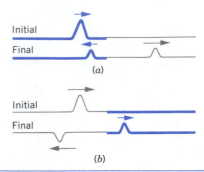

Figure 13   Phase changes on reflection at a junction between two strings of different linear mass densities. The wave speed is greater in the lighter string. (*a*) The incident pulse is in the heavier string. (*b*) The incident pulse is in the lighter string.

Figure 13*a* suggests a light wave in glass, say, approaching a surface beyond which there is a less optically dense medium (one of lower index of refraction) such as air. Figure 13*b* suggests a light wave in air approaching glass. To sum up the optical situation, when reflection occurs from an interface beyond which the medium has a *lower* index of refraction, the reflected wave undergoes *no phase change*; when the medium beyond the interface has a *higher* index, there is a phase change of π.* The transmitted wave does not experience a change of phase in either case.

We are now able to take into account both factors that determine the nature of the interference, namely, differences in optical path length and phase changes on reflection. For the two rays of Fig. 12 to combine to give a *maximum* intensity, assuming normal incidence, we must have

$$2d = (m + \tfrac{1}{2})\lambda_n \qquad m = 0, 1, 2, \ldots .$$

The term $\tfrac{1}{2}\lambda_n$ is introduced because upon reflection there is a phase change of $180°$, equivalent to half a wavelength. Substituting λ/n for λ_n yields finally

$$2dn = (m + \tfrac{1}{2})\lambda \qquad m = 0, 1, 2, \ldots \quad \text{(maxima).} \quad (16)$$

The conditions for a *minimum* intensity are

$$2dn = m\lambda \qquad m = 0, 1, 2, \ldots \quad \text{(minima).} \quad (17)$$

These equations hold when the index of refraction of the film is either greater or less than the indices of the media on *each* side of the film. Only in these cases will there be a relative phase change of $180°$ for reflections at the two surfaces. A water film in air and an air film in the space between two glass plates provide examples of cases to

---

* These statements, which can be proved rigorously from Maxwell's equations (see also Section 45-5), must be modified for light falling on a less dense medium at an angle such that total internal reflection occurs. They must also be modified for reflection from metallic surfaces.

which Eqs. 16 and 17 apply. Sample Problem 5 provides a case in which they do not apply.

If the film thickness is not uniform, as in Fig. 11 where the film is wedge shaped, constructive interference occurs in certain parts of the film and destructive interference occurs in others. Bands of maximum and of minimum intensity appear, called *fringes of constant thickness*. The width and spacing of the fringes depend on the variation of the film thickness d. If the film is illuminated with white light rather than monochromatic light, the light reflected from various parts of the film is modified by the various constructive or destructive interferences that occur. This accounts for the brilliant colors of soap bubbles and oil slicks.

Only if the film is "thin" (d being no more than a few wavelengths of light) is it possible to obtain these types of fringes, that is, fringes that appear localized on the film and are associated with a variable film thickness. For very thick films (say $d \approx 1$ cm), the path difference between the two rays of Fig. 12 is many wavelengths, and the phase difference at a given point on the film changes rapidly as we move even a small distance away from a. For "thin" films, however, the phase difference at a also holds for reasonably nearby points; there is a characteristic "patch brightness" for any point on the film, as Fig. 11 shows. Interference fringes can be produced for thick films; they are not localized on the film but are at infinity (see Section 45-6).

Sample Problem 4 A water film ($n = 1.33$) in air is 320 nm thick. If it is illuminated with white light at normal incidence, what color will it appear to be in reflected light?

Solution Solving Eq. 16 for λ, we obtain

$$\lambda = \frac{2dn}{m + \frac{1}{2}} = \frac{(2)(320 \text{ nm})(1.33)}{m + \frac{1}{2}} = \frac{851 \text{ nm}}{m + \frac{1}{2}} \quad \text{(maxima)}.$$

From Eq. 17 the minima are given by

$$\lambda = \frac{851 \text{ nm}}{m} \quad \text{(minima)}.$$

Maxima and minima occur for the following wavelengths:

m	0 (max)	1 (min)	1 (max)	2 (min)	2 (max)
λ (nm)	1702	851	567	426	340

Only the maximum corresponding to $m = 1$ lies in the visible region (between about 400 and 700 nm); light of wavelength 567 nm appears yellow-green. If white light is used to illuminate the film, the yellow-green component is enhanced when viewed by reflection. What is the color of the light transmitted through the film?

Sample Problem 5 Lenses are often coated with thin films of transparent substances such as MgF_2 ($n = 1.38$) to reduce the

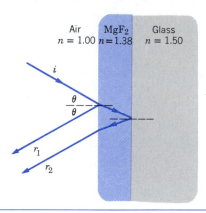

Figure 14 Sample Problem 5. Unwanted reflections from glass can be suppressed (at a chosen wavelength) by coating the glass with a film of the proper thickness.

reflection from the glass surface. How thick a coating is needed to produce a minimum reflection at the center of the visible spectrum ($\lambda = 550$ nm)?

Solution We assume that the light strikes the lens at near-normal incidence (θ is exaggerated for clarity in Fig. 14), and we seek destructive interference between rays r_1 and r_2. Equation 17 does not apply because in this case a phase change of 180° is associated with *each* ray, for at *both* the upper and lower surfaces of the MgF_2 film the reflection is from a medium of greater index of refraction.

There is no *net* change in phase produced by the two reflections, which means that the optical path difference for destructive interference is $(m + \frac{1}{2})\lambda_n$ (compare Eq. 16), leading to

$$2dn = (m + \tfrac{1}{2})\lambda \qquad m = 0, 1, 2, \ldots \quad \text{(minima)}.$$

Solving for d and putting $m = 0$ we obtain

$$d = \frac{(m + \frac{1}{2})\lambda}{2n} = \frac{\lambda}{4n} = \frac{550 \text{ nm}}{(4)(1.38)} = 100 \text{ nm}.$$

Sample Problem 6 Figure 15 shows a plano-convex lens of radius of curvature R resting on an accurately plane glass plate and illuminated from above by light of wavelength λ. Figure 16 shows that circular interference fringes (called *Newton's rings*) appear, associated with the variable thickness air film between the lens and the plate. Find the radii of the circular interference maxima.

Solution Here it is the ray from the *bottom* of the (air) film rather than from the top that undergoes a phase change of 180°, for it is the one reflected from a medium of higher refractive index. The condition for a maximum remains unchanged (Eq. 16), however, and is

$$2d = (m + \tfrac{1}{2})\lambda \qquad m = 0, 1, 2, \ldots, \qquad (18)$$

assuming $n = 1$ for the air film. From Fig. 15 we can write

$$d = R - \sqrt{R^2 - r^2} = R - R\left[1 - \left(\frac{r}{R}\right)^2\right]^{1/2}.$$

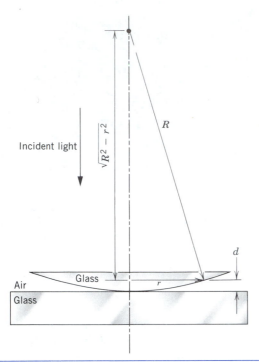

Figure 15 Sample Problem 6. The apparatus for observing Newton's rings.

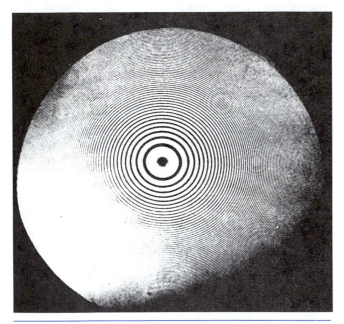

Figure 16 Circular interference fringes (Newton's rings) observed with the apparatus of Fig. 15.

If $r/R \ll 1$, we can expand the square bracket by the binomial theorem, keeping only two terms, or

$$d = R - R\left[1 - \frac{1}{2}\left(\frac{r}{R}\right)^2 + \cdots\right] \approx \frac{r^2}{2R}.$$

Combining with Eq. 18 yields

$$r = \sqrt{(m + \tfrac{1}{2})\lambda R} \qquad m = 0, 1, 2, \ldots \text{ (maxima)},$$

which gives the radii of the bright rings. If white light is used, each spectrum component produces its own set of circular fringes, and the sets all overlap.

Note that $r > 0$ for $m = 0$. That is, the first bright ring is at $r > 0$, and consequently the center must be dark, as shown in Fig. 16. This observation can be taken as experimental evidence for the 180° phase change upon reflection used to obtain Eq. 18.

45-5 OPTICAL REVERSIBILITY AND PHASE CHANGES ON REFLECTION *(Optional)*

G. G. Stokes (1819–1903) used the principle of optical reversibility to investigate the reflection of light at an interface between two media. The principle states that if there is no absorption of light, a light ray that is reflected or refracted will retrace its original path if its direction is reversed. This reminds us that any mechanical system can run backward as well as forward, provided there is no dissipation of energy such as by friction.

Figure 17*a* shows a wave of amplitude E reflected and refracted at a surface separating media 1 and 2, where $n_2 > n_1$. The amplitude of the reflected wave is $r_{12}E$, in which r_{12} is an *amplitude reflection coefficient*. The amplitude of the refracted wave is $t_{12}E$, where t_{12} is an *amplitude transmission coefficient*.

The sign of the coefficient indicates the relative phase of the reflected or transmitted component. If we consider only the possibility of phase changes of 0 or 180°, then if $r_{12} = +0.5$, for example, we have a reduction in amplitude on reflection by one-half and no change in phase. For $r_{12} = -0.5$ we have a phase change of 180° because

$$E \sin (\omega t + 180°) = -E \sin \omega t.$$

In Fig. 17*b*, the rays indicated by $r_{12}E$ and $t_{12}E$ have been reversed in direction. Ray $r_{12}E$, identified by the single arrows in the figure, is reflected and refracted, producing the rays of amplitudes $r_{12}^2 E$ and $r_{12}t_{12}E$. Ray $t_{12}E$, identified by the triple arrows, is also reflected and refracted, producing the rays of amplitudes $t_{12}t_{21}E$ and $t_{12}r_{21}E$ as shown. Note that r_{12} describes a ray in medium 1 reflected from medium 2, and r_{21} describes a ray in medium 2 reflected from medium 1. Similarly, t_{12} describes a ray that passes from medium 1 to medium 2; t_{21} describes a ray that passes from medium 2 to medium 1.

Based on the reversibility principle, we conclude that the two rays in the upper left of Fig. 17*b* must be equivalent to the incident ray of Fig. 17*a*, reversed; the two rays in the lower left of Fig. 17*b* must cancel. This second requirement leads to

$$r_{12}t_{12}E + t_{12}r_{21}E = 0,$$

or

$$r_{12} = -r_{21}.$$

This result tells us that if we compare a wave reflected from medium 1 with one reflected from medium 2, they behave differently in that one or the other undergoes a phase change of 180°.

Experiment shows that the ray reflected from the more optically dense medium suffers the change in phase of 180°. This can be demonstrated using the setup shown in Fig. 18, which is called the Lloyd's mirror experiment. Interference occurs on the screen at an arbitrary point P as a result of the overlap of the direct and

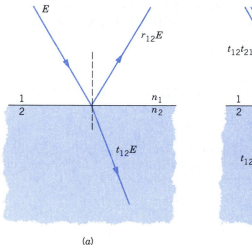
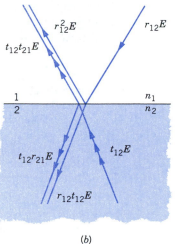

Figure 17   (*a*) A ray is reflected and refracted at an interface. (*b*) The optically reversed situation; the two rays in the lower left must cancel.

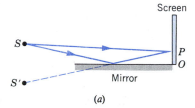

(*a*)

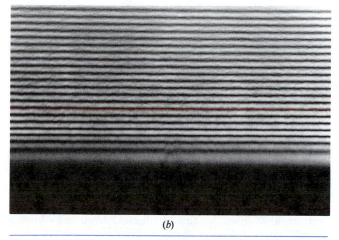

(*b*)

Figure 18   (*a*) The experimental setup for Lloyd's mirror. Fringes appear on the screen as a result of interference between the direct and reflected beams. (*b*) Fringes observed in the Lloyd's mirror experiment.

reflected beams. We can analyze this experiment as two-source interference, in which one of the sources (S') is the *virtual image* of S in the plane mirror. However, there is one important difference between the apparatus of Fig. 18 and the double-slit experiment: the light from the virtual source S' has been reflected from the mirror and has undergone a phase change of 180°. As a result of this phase change, the lower edge of the screen (at O) shows a dark fringe, instead of the bright fringe that appears at the corresponding point (the center of the screen) in the double-slit experiment. Put another way, the appearance of the dark fringe at O

shows that one of the interfering beams has been shifted in phase by 180°. Since there is nothing to change the phase of the direct beam SP, it must be the reflected beam that experiences the change in phase. This shows that reflection from a more optically dense medium involves a change in phase of 180°.   ■

45-6   MICHELSON'S INTERFEROMETER*

An *interferometer* is a device that can be used to measure lengths or changes in length with great accuracy by means of interference fringes. We describe the form originally built by A. A. Michelson (1852–1931) in 1881.

Consider light that leaves point P on extended source S (Fig. 19) and falls on half-silvered mirror M (sometimes called a *beam splitter*). This mirror has a silver coating just thick enough to transmit half the incident light and to reflect half; in the figure we have assumed for convenience that this mirror has negligible thickness. At M the light divides into two waves. One proceeds by transmission toward mirror M_1; the other proceeds by reflection toward M_2. The waves are reflected at each of these mirrors and are sent back along their directions of incidence, each wave eventually entering the eye E. Because the waves are coherent, being derived from the same point on the source, they interfere.

If the mirrors M_1 and M_2 are exactly perpendicular to each other, the effect is that of light from an extended source S falling on a uniformly thick slab of air, between glass, whose thickness is equal to $d_2 - d_1$. Interference

---

* See "Michelson: America's First Nobel Prize Winner in Science," by R. S. Shankland, *The Physics Teacher,* January 1977, p. 19. See also "Michelson and his Interferometer," by R. S. Shankland, *Physics Today,* April 1974, p. 36.

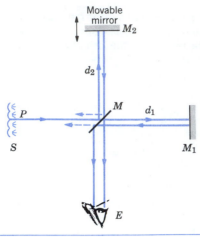

Figure 19   Michelson's interferometer, showing the path of a ray originating at point P of an extended source S. The ray from P splits at M; the two rays are reflected from mirrors M_1 and M_2 and then recombine at M. Mirror M_2 can be moved to change the path difference between the combining rays.

fringes appear, caused by small changes in the angle of incidence of the light from different points on the extended source as it strikes the equivalent air film. For *thick* films a path difference of one wavelength can be brought about by a very small change in the angle of incidence.

If M_2 is moved backward or forward, the effect is to change the thickness of the equivalent air film. Suppose that the center of the (circular) fringe pattern appears bright and that M_2 is moved just enough to cause the first bright circular fringe to move to the center of the pattern. The path of the light beam traveling back and forth to M_2 has been changed by one wavelength. This means (because the light passes twice through the equivalent air film) that the mirror must have moved through a distance of $\frac{1}{2}\lambda$.

The interferometer is used to measure changes in length by counting the number of interference fringes that pass the field of view as mirror M_2 is moved. Length measurements made in this way can be accurate if large numbers of fringes are counted.

Michelson measured the length of the standard meter, kept in Paris, in terms of the wavelength of a certain monochromatic red light emitted from a light source containing cadmium. He showed that the standard meter was equivalent to 1,553,163.5 wavelengths of the red cadmium light. For this work he received the Nobel prize in 1907. Michelson's work laid the foundation for the eventual abandonment (in 1961) of the meter bar as a standard of length and for the redefinition of the meter in terms of the wavelength of light. In 1983, as we have seen, even this wavelength standard was not precise enough to meet the

growing requirements of science and technology and was replaced by a new standard based on a defined value for the speed of light.

Sample Problem 7   Yellow light ($\lambda = 589.00$ nm) illuminates a Michelson interferometer. How many bright fringes will be counted as the mirror is moved through 1.0000 cm?

Solution   Each fringe corresponds to a movement of the mirror through one-half wavelength. The number of fringes is thus the same as the number of half wavelengths in 1.0000 cm, or

$$\frac{1.0000 \times 10^{-2} \text{ m}}{\frac{1}{2}(589.00 \times 10^{-9} \text{ m})} = 33,956 \text{ fringes.}$$

45-7   MICHELSON'S INTERFEROMETER AND LIGHT PROPAGATION *(Optional)*

In Chapter 21 we presented Einstein's hypothesis, now well verified, that in free space light travels with the same speed c no matter what the relative velocity of the source and the observer may be. We pointed out that this hypothesis contradicted the views of 19th-century physicists regarding wave propagation. It was difficult for these physicists, trained as they were in the classical physics of the time, to believe that a wave could be propagated without a medium. If such a medium could be established, the speed c of light would naturally be regarded as the speed *with respect to that medium,* just as the speed of sound always refers to a medium such as air. Although no medium for light propagation was obvious, physicists postulated one, called the *ether,* and hypothesized that its properties were such that it was undetectable by ordinary means such as weighing.

In 1881 (24 years before Einstein's hypothesis) A. A. Michelson set himself the task of direct physical verification of the existence of the ether. In particular, Michelson, later joined by E. W. Morley, tried to measure the speed u with which the Earth moves through the ether. Michelson's interferometer was their instrument of choice for this now-famous Michelson–Morley experiment.

The Earth together with the interferometer moving with velocity **u** through the ether is equivalent to the interferometer at rest with the ether streaming through it with velocity $-\mathbf{u}$, as shown in Fig. 20. Consider a wave moving along the path MM_1M and one moving along MM_2M. The first corresponds classically to a person rowing a boat a distance d downstream and the same distance upstream; the second corresponds to rowing a boat a distance d across a stream and back.

Based on the ether hypothesis the speed of light on the path MM_1 is $c + u$; on the return path M_1M it is $c - u$. The time required for the complete trip is

$$t_1 = \frac{d}{c+u} + \frac{d}{c-u} = d\frac{2c}{c^2 - u^2} = \frac{2d}{c}\frac{1}{1 - (u/c)^2}.$$

The speed of light, again *based on the ether hypothesis,* for path MM_2 is $\sqrt{c^2 - u^2}$, as Fig. 20 suggests. This same speed holds

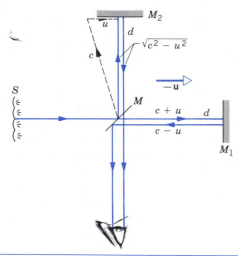

Figure 20 The "ether" streaming with velocity $-\mathbf{u}$ through Michelson's interferometer. The speeds shown are based on the (incorrect) ether hypothesis.

for the return path M_2M, so that the time required for this complete path is

$$t_2 = \frac{2d}{\sqrt{c^2 - u^2}} = \frac{2d}{c} \frac{1}{\sqrt{1 - (u/c)^2}}.$$

The difference of time for the two paths is

$$\Delta t = t_1 - t_2$$
$$= \frac{2d}{c} \left\{ \left[1 - \left(\frac{u}{c} \right)^2 \right]^{-1} - \left[1 - \left(\frac{u}{c} \right)^2 \right]^{-1/2} \right\}.$$

Assuming $u/c \ll 1$, we can expand the quantities in the square brackets by using the binomial theorem, retaining only the first two terms. This leads to

$$\Delta t = \frac{2d}{c} \left\{ \left[1 + \left(\frac{u}{c} \right)^2 + \cdots \right] - \left[1 + \frac{1}{2} \left(\frac{u}{c} \right)^2 + \cdots \right] \right\}$$

$$= \frac{2d}{c} \left\{ \frac{1}{2} \left(\frac{u}{c} \right)^2 \right\} = \frac{du^2}{c^3}. \quad (19)$$

Now let the entire interferometer be rotated through 90°. This interchanges the roles of the two light paths, MM_1M now being

the "cross-stream" path and MM_2M the "downstream and upstream" path. The time difference between the two waves entering the eye is also reversed; this changes the phase difference between the combining waves and alters the positions of the interference maxima. The experiment consists of looking for a shift of the interference fringes as the apparatus is rotated.

The *change* in time difference is $2\Delta t$, which corresponds to a phase difference of $\Delta \phi = \omega(2\Delta t)$, where $\omega (= 2\pi c/\lambda)$ is the angular frequency of the light wave. The expected maximum shift in the number of fringes on a 90° rotation is

$$\Delta N = \frac{\Delta \phi}{2\pi} = \frac{\omega(2\Delta t)}{2\pi} = \frac{2c\Delta t}{\lambda} = \frac{2d}{\lambda} \left(\frac{u}{c} \right)^2, \quad (20)$$

where we have used Eq. 19 for Δt.

In the Michelson–Morley interferometer let $d = 11$ m (obtained by multiple reflection in the interferometer) and $\lambda = 5.9 \times 10^{-7}$ m. If u is assumed to be roughly the orbital speed of the Earth, then $u/c \approx 10^{-4}$. The expected maximum fringe shift when the interferometer is rotated through 90° is then

$$\Delta N = \frac{2d}{\lambda} \left(\frac{u}{c} \right)^2 = \frac{(2)(11 \text{ m})}{5.9 \times 10^{-7} \text{ m}} (10^{-4})^2 = 0.4.$$

Even though a shift of only about 0.4 of a fringe was expected, Michelson and Morley were confident that they could observe a shift of 0.01 fringe. *They found from their experiment, however, that there was no observable fringe shift!*

The analogy between a light wave in the supposed ether and a boat moving in water, which seemed so evident in 1881, is simply incorrect. The derivation based on this analogy is incorrect for light waves. When the analysis is carried through based on Einstein's hypothesis, the observed null result is clearly predicted, the speed of light being c for all paths. The motion of the Earth around the Sun and the rotation of the interferometer have, in Einstein's view, no effect whatever on the speed of the light waves in the interferometer.

It should be made clear that although Einstein's hypothesis is completely consistent with the null result of the Michelson–Morley experiment, this experiment standing alone does not serve as a proof for Einstein's hypothesis. Einstein said that no number of experiments, however large, could prove him right but that a single experiment could prove him wrong. Our present-day belief in Einstein's hypothesis rests on consistent agreement in a large number of experiments designed to test it. The "single experiment" that might prove Einstein wrong has never been found. ■

QUESTIONS

1. Is Young's experiment an interference experiment or a diffraction experiment, or both?

2. In Young's double-slit interference experiment, using a monochromatic laboratory light source, why is screen A in Fig. 6 necessary? If the source of light is a laser beam, screen A is not needed. Why?

3. What changes, if any, occur in the pattern of interference fringes if the apparatus of Fig. 4 is placed under water?

4. Do interference effects occur for sound waves? Recall that sound is a longitudinal wave and that light is a transverse wave.

5. It is not possible to show interference effects between light from two separate sodium vapor lamps but you can show interference effects between sound from two loudspeakers that are driven by separate oscillators. Explain why this is so.

6. If interference between light waves of different frequencies is

possible, one should observe light beats, just as one obtains sound beats from two sources of sound with slightly different frequencies. Discuss how one might experimentally look for this possibility.

7. Why are parallel slits preferable to the pinholes that Young used in demonstrating interference?

8. Is coherence important in reflection and refraction?

9. Describe the pattern of light intensity on screen *C* in Fig. 4 if one slit is covered with a red filter and the other with a blue filter, the incident light being white.

10. If one slit in Fig. 4 is covered, what change would occur in the intensity of light at the center of the screen?

11. We are all bathed continuously in electromagnetic radiation, from the Sun, from radio and TV signals, from the stars and other celestial objects. Why do these waves not interfere with each other?

12. In calculating the disturbance produced by a pair of superimposed wavetrains, when should you add intensities and when amplitudes?

13. In Young's double-slit experiment suppose that screen *A* in Fig. 6 contained *two* very narrow parallel slits instead of one. (*a*) Show that if the spacing between these slits is properly chosen, the interference fringes can be made to disappear. (*b*) Under these conditions, would you call the beams emerging from slits S_1 and S_2 in screen *B* coherent? They do not produce interference fringes. (*c*) Discuss what would happen to the interference fringes in the case of a single slit in screen *A* if the slit width were gradually increased.

14. Defend this statement: Figure 7*a* is a sine (or cosine) wave but Fig. 7*b* is not. Indeed, you cannot assign a unique frequency to the curve of Fig. 7*b*. Why not? (*Hint*: Think of Fourier analysis.)

15. Most of us are familiar with rotating or oscillating radar antennas that produce rotating or oscillating beams of microwave radiation. It is also possible to produce an oscillating beam of microwave radiation *without* any mechanical motion of the transmitting antenna. This is done by periodically changing the phase of the radiation as it emerges from various sections of the (long) transmitting antenna. Convince yourself that, by constructive interference from various parts of the fixed antenna, an oscillating microwave beam can indeed be so produced.

16. What causes the fluttering of radio reception when an airplane flies overhead?

17. Is it possible to have coherence between light sources emitting light of different wavelengths?

18. An automobile directs its headlights onto the side of a barn. Why are interference fringes not produced in the region in which light from the two beams overlaps?

19. Suppose that the film coating in Fig. 14 had an index of refraction greater than that of the glass. Could it still be nonreflecting? If so, what difference would the coating make?

20. What are the requirements for maximum intensity when viewing a thin film by *transmitted* light?

21. Why does a film (for example, a soap bubble or an oil slick) have to be "thin" to display interference effects? Or does it? How thin is "thin"?

22. Why do coated lenses (see Sample Problem 5) look purple by reflected light?

23. Ordinary store windows or home windows reflect light from both their interior and exterior plane surfaces. Why then do we not see interference effects?

24. If you wet your eyeglasses to clean them you will notice that as the water evaporates the glasses become markedly less reflecting for a short time. Explain why.

25. A lens is coated to reduce reflection, as in Sample Problem 5. What happens to the energy that had previously been reflected? Is it absorbed by the coating?

26. Consider the following objects that produce colors when exposed to sunlight: (1) soap bubbles, (2) rose petals, (3) the inner surface of an oyster shell, (4) thin oil slicks, (5) nonreflecting coatings on camera lenses, and (6) peacock tail feathers. The colors displayed by all but one of these are purely interference phenomena, no pigments being involved. Which one is the exception? Why do the others seem to be "colored"?

27. A soap film on a wire loop held in air appears black at its thinnest portion when viewed by reflected light. On the other hand, a thin oil film floating on water appears bright at its thinnest portion when similarly viewed from the air above. Explain these phenomena.

28. Very small changes in the angle of incidence do not change the interference conditions much for "thin" films but they do change them for "thick" films. Why?

29. An *optical flat* is a slab of glass that has been ground flat to within a small fraction of a wavelength. How may it be used to test the flatness of a second slab of glass?

30. In a Newton's rings experiment, is the central spot, as seen by reflection, dark or light? Explain.

31. In connection with the phase change on reflection at an interface between two transparent media, do you think that phase shifts other than 0 or π are possible? Do you think that phase shifts can be calculated rigorously from Maxwell's equations?

32. The directional characteristics of a certain radar antenna as a receiver of radiation are known. What can be said about its directional characteristics as a transmitter?

33. A person in a dark room, looking through a small window, can see a second person standing outside in bright sunlight. The second person cannot see the first person. Is this a failure of the principle of optical reversibility? Assume no absorption of light.

34. Why is it necessary to rotate the interferometer in the Michelson–Morley experiment?

35. How is the negative result of the Michelson–Morley experiment interpreted according to Einstein's theory of relativity?

36. If the pathlength to the movable mirror in Michelson's interferometer (see Fig. 19) greatly exceeds that to the fixed mirror (say, by more than a meter) the fringes begin to disappear. Explain why. Lasers greatly extend this range. Why?

37. How would you construct an acoustical Michelson interferometer to measure wavelengths of sound? Discuss differences from the optical interferometer.

PROBLEMS

Section 45-1 Double-Slit Interference

1. Monochromatic green light, wavelength = 554 nm, illuminates two parallel narrow slits 7.7 μm apart. Calculate the angular deviation of the third-order, $m = 3$, bright fringe (a) in radians and (b) in degrees.

2. In a double-slit experiment to demonstrate the interference of light, the separation d of the two narrow slits is doubled. In order to maintain the same spacing of the fringes on the screen, how must the distance D of the screen from the slits be altered? (The wavelength of the light remains unchanged.)

3. A double-slit experiment is performed with blue-green light of wavelength 512 nm. The slits are 1.2 mm apart and the screen is 5.4 m from the slits. How far apart are the bright fringes as seen on the screen?

4. Find the slit separation of a double-slit arrangement that will produce bright interference fringes 1.00° apart in angular separation. Assume a wavelength of 592 nm.

5. A double-slit arrangement produces interference fringes for sodium light ($\lambda = 589$ nm) that are 0.23° apart. For what wavelength would the angular separation be 10% greater? Assume that the angle θ is small.

6. A double-slit arrangement produces interference fringes for sodium light ($\lambda = 589$ nm) that are 0.20° apart. What is the angular fringe separation if the entire arrangement is immersed in water ($n = 1.33$)?

7. In a double-slit experiment the distance between slits is 5.22 mm and the slits are 1.36 m from the screen. Two interference patterns can be seen on the screen, one due to light with wavelength 480 nm and the other due to light with wavelength 612 nm. Find the separation on the screen between the third-order interference fringes of the two different patterns.

8. In an interference experiment in a large ripple tank (see Fig. 3), the coherent vibrating sources are placed 120 mm apart. The distance between maxima 2.0 m away is 180 mm. If the speed of ripples is 25 cm/s, calculate the frequency of the vibrators.

9. If the distance between the first and tenth minima of a double-slit pattern is 18 mm and the slits are separated by 0.15 mm with the screen 50 cm from the slits, what is the wavelength of the light used?

10. A thin flake of mica ($n = 1.58$) is used to cover one slit of a double-slit arrangement. The central point on the screen is occupied by what used to be the seventh bright fringe. If $\lambda = 550$ nm, what is the thickness of the mica?

11. Sketch the interference pattern expected from using two pinholes, rather than narrow slits.

12. Two coherent radio point sources separated by 2.0 m are radiating in phase with $\lambda = 0.50$ m. A detector moved in a circular path around the two sources in a plane containing them will show how many maxima?

13. In the front of a lecture hall, a coherent beam of monochromatic light from a helium–neon laser ($\lambda = 632.8$ nm) illuminates a double slit. From there it travels a distance of 20.0 m to a mirror at the back of the hall, and returns the same distance to a screen. (a) In order that the distance between interference maxima be 10.0 cm what should be the distance between the two slits? (b) State what you will see if the lecturer slips a thin sheet of cellophane over one of the slits. The path through the cellophane contains 2.5 more waves than a path through air of the same geometric thickness.

14. One slit of a double-slit arrangement is covered by a thin glass plate of index of refraction 1.4, and the other by a thin glass plate of index of refraction 1.7. The point on the screen where the central maximum fell before the glass plates were inserted is now occupied by what had been the $m = 5$ bright fringe before. Assume $\lambda = 480$ nm and that the plates have the same thickness t and find the value of t.

15. Two point sources, S_1 and S_2 in Fig. 21, emit coherent waves. Show that curves, such as that given, over which the phase difference for rays r_1 and r_2 is a constant, are hyperbolas. (*Hint:* A constant phase difference implies a constant difference in length between r_1 and r_2.) The OMEGA system of sea navigation relies on this principle. S_1 and S_2 are phase-locked transmitters. The ship's navigator notes the received phase difference on an oscilloscope and locates the ship on a hyperbola. Reception of signals from a third transmitter is needed to determine the position on that hyperbola.

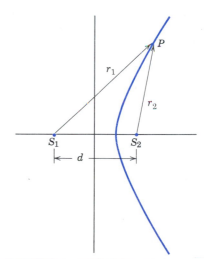

Figure 21   Problem 15.

16. Sodium light ($\lambda = 589$ nm) falls on a double slit of separation $d = 0.180$ mm. A thin lens ($f = 1.13$ m) is placed near the slit as in Fig. 5. What is the linear fringe separation on a screen placed in the focal plane of the lens?

17. Sodium light ($\lambda = 589$ nm) falls on a double slit of separation $d = 2.0$ mm. The slit–screen distance D is 40 mm. What fractional error is made by using Eq. 1 to locate the tenth bright fringe on the screen?

Section 45-2 Coherence

18. The *coherence length* of a wavetrain is the distance over which the phase constant is the same. (a) If an individual

atom emits coherent light for 1×10^{-8} s, what is the coherence length of the wavetrain? (b) Suppose this wavetrain is separated into two parts with a partially reflecting mirror and later reunited after one beam travels 5 m and the other 10 m. Do the waves produce interference fringes observable by a human eye?

Section 45-3   Intensity in Double-Slit Interference

19. Source A of long-range radio waves leads source B by 90°. The distance r_A to a detector is greater than the distance r_B by 100 m. What is the phase difference at the detector? Both sources have a wavelength of 400 m.

20. Find the phase difference between the waves from the two slits arriving at the mth dark fringe in a double-slit experiment.

21. Light of wavelength 600 nm is incident normally on two parallel narrow slits separated by 0.60 mm. Sketch the intensity pattern observed on a distant screen as a function of angle θ for the range of values $0 \le \theta \le 0.0040$ radians.

22. Find the sum of the following quantities (a) graphically, using phasors, and (b) using trigonometry:

$$y_1 = 10 \sin \omega t,$$
$$y_2 = 8.0 \sin (\omega t + 30°).$$

23. S_1 and S_2 in Fig. 22 are effective point sources of radiation, excited by the same oscillator. They are coherent and in phase with each other. Placed a distance $d = 4.17$ m apart, they emit equal amounts of power in the form of 1.06-m wavelength electromagnetic waves. (a) Find the positions of the first (that is, the nearest), the second, and the third maxima of the received signal, as the detector D is moved out along Ox. (b) Is the intensity at the nearest minimum equal to zero? Justify your answer.

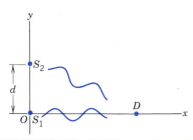

Figure 22   Problem 23.

24. Add the following quantities graphically, using the phasor method (see Sample Problem 3), and algebraically:

$$y_1 = 10 \sin \omega t,$$
$$y_2 = 14 \sin (\omega t + 26°),$$
$$y_3 = 4.7 \sin (\omega t - 41°).$$

25. Show that the half-width $\Delta\theta$ of the double-slit interference fringes is given by

$$\Delta\theta = \frac{\lambda}{2d}$$

if θ is small enough so that $\sin \theta \approx \theta$. The half-width is the angle between the two points in the fringe where the intensity is one-half that at the center of the fringe.

26. One of the slits of a double-slit system is wider than the other, so that the amplitude of the light reaching the central part of the screen from one slit, acting alone, is twice that from the other slit, acting alone. Derive an expression for the intensity I in terms of θ.

Section 45-4   Interference from Thin Films

27. We wish to coat a flat slab of glass ($n = 1.50$) with a transparent material ($n = 1.25$) so that light of wavelength 620 nm (in vacuum) incident normally is not reflected. What minimum thickness could the coating have?

28. A thin film in air is 410 nm thick and is illuminated by white light normal to its surface. Its index of refraction is 1.50. What wavelengths within the visible spectrum will be intensified in the reflected beam?

29. A disabled tanker leaks kerosene ($n = 1.20$) into the Persian Gulf, creating a large slick on top of the water ($n = 1.33$). (a) If you are looking straight down from an airplane onto a region of the slick where its thickness is 460 nm, for which wavelength(s) of visible light is the reflection the greatest? (b) If you are scuba-diving directly under this same region of the slick, for which wavelength(s) of visible light is the transmitted intensity the strongest?

30. In costume jewelry, rhinestones (made of glass with $n = 1.5$) are often coated with silicon monoxide ($n = 2.0$) to make them more reflective. How thick should the coating be to achieve strong reflection for 560-nm light, incident normally?

31. If the wavelength of the incident light is $\lambda = 572$ nm, the rays A and B in Fig. 23 are out of phase by 1.50λ. Find the thickness d of the film.

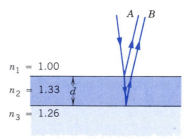

Figure 23   Problem 31.

32. Light of wavelength 585 nm is incident normally on a thin soapy film ($n = 1.33$) suspended in air. If the film is 0.00121 mm thick, determine whether it appears bright or dark when observed from a point near the light source.

33. A plane wave of monochromatic light falls normally on a uniformly thin film of oil that covers a glass plate. The wavelength of the source can be varied continuously. Complete destructive interference of the reflected light is observed for wavelengths of 485 and 679 nm and for no wavelengths between them. If the index of refraction of the oil is 1.32 and that of the glass is 1.50, find the thickness of the oil film.

34. White light reflected at perpendicular incidence from a soap film has, in the visible spectrum, an interference maximum at 600 nm and a minimum at 450 nm with no minimum

in-between. If $n = 1.33$ for the film, what is the film thickness, assumed uniform?

35. Two pieces of plate glass are held together in such a way that the air space between them forms a very thin wedge. Light of wavelength 480 nm strikes the upper surface perpendicularly and is reflected from the lower surface of the top glass and the upper surface of the bottom glass, thereby producing a series of interference fringes. How much thicker is the air wedge at the sixteenth fringe than it is at the sixth?

36. A sheet of glass having an index of refraction of 1.40 is to be coated with a film of material having an index of refraction of 1.55 such that green light (wavelength = 525 nm) is preferentially transmitted. (a) What is the minimum thickness of the film that will achieve the result? (b) Why are other parts of the visible spectrum not also preferentially transmitted? (c) Will the transmission of any colors be sharply reduced?

37. A thin film of acetone (index of refraction = 1.25) is coating a thick glass plate (index of refraction = 1.50). Plane light waves of variable wavelengths are incident normal to the film. When one views the reflected wave, it is noted that complete destructive interference occurs at 600 nm and contructive interference at 700 nm. Calculate the thickness of the acetone film.

38. An oil drop ($n = 1.20$) floats on a water ($n = 1.33$) surface and is observed from above by reflected light (see Fig. 24). (a) Will the outer (thinnest) regions of the drop correspond to a bright or a dark region? (b) How thick is the oil film where one observes the third blue region from the outside of the drop? (c) Why do the colors gradually disappear as the oil thickness becomes larger?

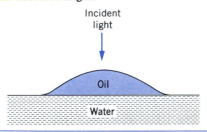

Figure 24   Problem 38.

39. A broad source of light ($\lambda = 680$ nm) illuminates normally two glass plates 120 mm long that touch at one end and are separated by a wire 0.048 mm in diameter at the other end (Fig. 25). How many bright fringes appear over the 120-mm distance?

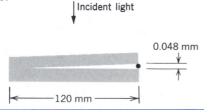

Figure 25   Problem 39.

40. A perfectly flat piece of glass ($n = 1.5$) is placed over a perfectly flat piece of black plastic ($n = 1.2$) as shown in Fig.

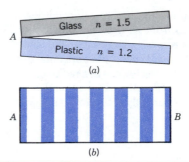

Figure 26   Problem 40.

26a. They touch at A. Light of wavelength 600 nm is incident normally from above. The location of the dark fringes in the reflected light is shown on the sketch of Fig. 26b. (a) How thick is the space between the glass and the plastic at B? (b) Water ($n = 1.33$) seeps into the region between the glass and plastic. How many dark fringes are seen when all the air has been displaced by water? (The straightness and equal spacing of the fringes is an accurate test of the flatness of the glass.)

41. Light of wavelength 630 nm is incident normally on a thin wedge-shaped film with index of refraction 1.50. There are ten bright and nine dark fringes over the length of film. By how much does the film thickness change over this length?

42. In an air wedge formed by two plane glass plates, touching each other along one edge, there are 4001 dark lines observed when viewed by reflected monochromatic light. When the air between the plates is evacuated, only 4000 such lines are observed. Calculate the index of refraction of the air from these data.

43. In a Newton's rings experiment the radius of curvature R of the lens is 5.0 m and its diameter is 20 mm. (a) How many rings are produced? (b) How many rings would be seen if the arrangement were immersed in water ($n = 1.33$)? Assume that $\lambda = 589$ nm.

44. The diameter of the tenth bright ring in a Newton's rings apparatus changes from 1.42 to 1.27 cm as a liquid is introduced between the lens and the plate. Find the index of refraction of the liquid.

45. A Newton's rings apparatus is used to determine the radius of curvature of a lens. The radii of the nth and $(n + 20)$th bright rings are measured and found to be 0.162 cm and 0.368 cm, respectively, in light of wavelength 546 nm. Calculate the radius of curvature of the lower surface of the lens.

46. In the Newton's rings experiment, show (a) that the difference in radius between adjacent rings (maxima) is given by

$$\Delta r = r_{m+1} - r_m \approx \tfrac{1}{2}\sqrt{\lambda R/m},$$

assuming $m \gg 1$, and (b) that the *area* between adjacent rings (maxima) is given by

$$A = \pi \lambda R,$$

assuming $m \gg 1$. Note that this area is independent of m.

47. In Sample Problem 5 assume that there is zero reflection for light of wavelength 550 nm at normal incidence. Calculate the factor by which the reflection is diminished by the coating at (a) 450 nm and (b) 650 nm. (*Hint*: Calculate ϕ in Eq. 13.)

48. A ship approaching harbor is transmitting at a wavelength of $\lambda = 3.43$ m from its antenna located $h = 23$ m above sea level. The receiving station antenna is located $H = 160$ m above sea level. What is the horizontal distance D between ship and receiving tower when radio contact is momentarily lost for the first time? Assume that the calm ocean reflects radio waves perfectly according to the law of reflection. See Fig. 27.

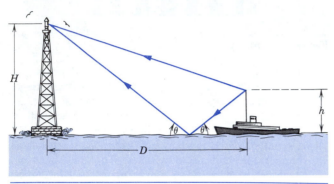

Figure 27   Problem 48.

Section 45-6 Michelson's Interferometer

49. If mirror M_2 in Michelson's interferometer is moved through 0.233 mm, 792 fringes are counted with a light meter. What is the wavelength of the light?

50. A thin film with $n = 1.42$ for light of wavelength 589 nm is placed in one arm of a Michelson interferometer. If a shift of 7.0 fringes occurs, what is the film thickness?

51. An airtight chamber 5.0 cm long with glass windows is placed in one arm of a Michelson interferometer as indicated in Fig. 28. Light of wavelength $\lambda = 500$ nm is used. The air is slowly evacuated from the chamber using a vacuum pump. While the air is being removed, 60 fringes are observed to pass through the view. From these data, find the index of refraction of air at atmospheric pressure.

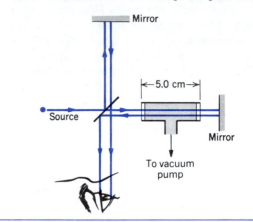

Figure 28   Problem 51.

52. Write an expression for the intensity observed in Michelson's interferometer (Fig. 19) as a function of the position of the movable mirror. Measure the position of the mirror from the point at which $d_1 = d_2$.

CHAPTER 46

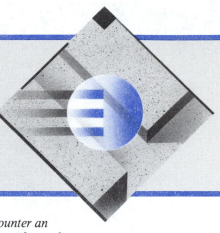

DIFFRACTION

Diffraction is the bending or spreading of waves that encounter an object (a barrier or an opening) in their path. This chapter considers only diffraction of light waves, but diffraction occurs for all types of waves. Sound waves, for example, are diffracted by ordinary objects, and as a result we can hear sounds even though we may not be in a direct line to their source. For diffraction to occur, the size of the object must be of the order of the wavelength of the incident waves; when the wavelength is much smaller than the size of the object, diffraction is ordinarily not observed and the object casts a sharp shadow.

Diffraction patterns consist of light and dark bands similar to the interference patterns discussed in Chapter 45. By studying these patterns, we can learn about the diffracting object. For example, diffraction of x rays is an important method for the study of the structure of solids, and diffraction of gamma rays is used to study nuclei. Diffraction also has unwanted effects, such as the spreading of light as it enters the aperture of a telescope, which limits its ability to resolve or separate stars that appear to be close to one another. These various effects of diffraction are considered in this chapter and the following one.

46-1 DIFFRACTION AND THE WAVE THEORY OF LIGHT

When light passes through a narrow slit (of width comparable to the wavelength of the light; see Fig. 1 of Chapter 43), the light beams not only flare out far beyond the geometrical shadow of the slit; they also give rise to a series of alternating light and dark bands that resemble interference fringes (Fig. 1). In Chapter 45, we argued that the appearance of interference fringes provides strong evidence for the wave nature of light. We can also argue that the appearance of diffraction patterns similarly requires that light must travel as waves.

Although diffraction was already known at the time of Huygens and Newton, neither of them believed that it provided evidence that light must be a wave. Newton in particular believed that light traveled as a stream of particles.

A strong proponent of the wave theory of light was the French engineer Augustin Fresnel (1788–1827). Fresnel explained diffraction based on the wave theory, which was not widely accepted even after Thomas Young's experiments on double-slit interference. In 1819, Fresnel submitted a paper on his theory of diffraction in a competition sponsored by the French Academy of Sciences. One of the members of the Academy, Simeon-Denis Poisson (a strong opponent of the wave theory of light), ridiculed

Figure 1   The diffraction pattern produced when light passes through a narrow slit.

967

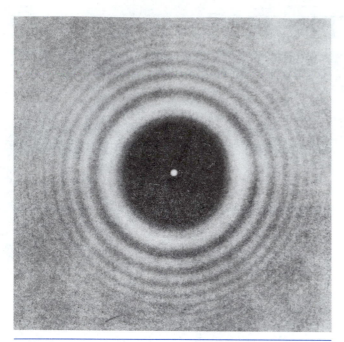

Figure 2 The diffraction pattern of a disk. Note the bright Poisson spot at the center of the pattern.

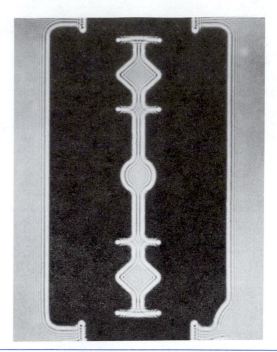

Figure 3 The diffraction pattern of a razor blade viewed in monochromatic light. Note the fringes near the edges.

Fresnel's theory because, as Poisson himself showed, Fresnel's diffraction theory led to the "absurd" prediction that the shadow of an opaque object should have a bright spot at its center. Figure 2 shows the diffraction pattern of a disk; the clearly visible bright spot at its center (known as the Poisson spot) supports Fresnel's interpretation.

Figure 3 shows the diffraction pattern produced when an ordinary object is illuminated by monochromatic light. Actually, you don't need special apparatus to observe diffraction. Hold two fingers so that there is a narrow slit between them, and look at a light bulb through the slit. The dark lines you see in the slit are caused by diffraction. Another common example of diffraction is the "floaters" that many people can observe in their field of view. Floaters are translucent dots or tiny chains that appear to float and drift. They can be seen by focusing the eyes at a distance while staring at a brightly illuminated piece of white paper. Floaters are caused by blood cells and other microscopic debris in the fluid of the eyeball; what we observe is the diffraction pattern on the retina.

Figure 4 shows the generalized diffraction situation. The curved surfaces on the left represent wavefronts of the incident light. The light falls on the diffracting object *B*, which we show in Fig. 4 as an opaque barrier containing an aperture of arbitrary shape. (Later, we consider an aperture that is a single narrow slit, which produced the diffraction pattern shown in Fig. 1.) *C* in Fig. 4 is a screen or photographic film that receives the light that passes through or around the diffracting object.

We can calculate the pattern of light intensity on screen *C* by subdividing the wavefront into elementary areas *d***A**, each of which becomes a source of an expanding

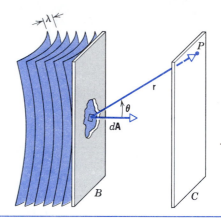

Figure 4 Diffraction occurs when coherent wavefronts of light fall on opaque barrier *B*, which contains an aperture of arbitrary shape. The diffraction pattern can be viewed on screen *C*.

Huygens wavelet. The light intensity at an arbitrary point *P* is found by superimposing the wave disturbances (that is, the **E** vectors) caused by the wavelets reaching *P* from all these elementary sources.

The wave disturbances reaching *P* differ in amplitude and in phase because (1) the elementary sources are at varying distances from *P*, (2) the light leaves the elementary sources at various angles to the normal to the wavefront, and (3) some sources are blocked by barrier *B*; others are not. Diffraction calculations, which are simple in principle, may become difficult in practice. The calculation must be repeated for every point on screen *C* at which we wish to know the light intensity. We followed

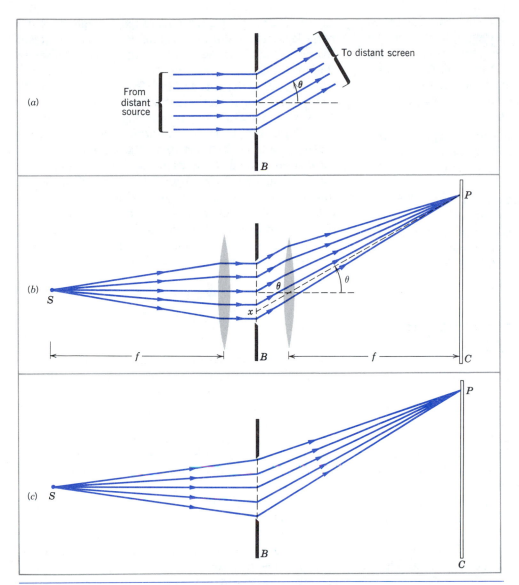

Figure 5 Light from point source S illuminates a slit in the opaque barrier B. The slit extends a long distance above and below the plane of the figure; this distance is much greater than the slit width a. The intensity at point P on screen C depends on the relative phases of the light received from various parts of the slit. (*a*) If source S and screen C are moved to large distances from the slit, both the incident and emergent light at B consist of nearly parallel rays. (*b*) Rather than using large distances, the source and the screen can each be placed in the focal plane of a lens; once again, parallel light rays enter and leave the slit. (*c*) Without the lens, the rays are not parallel.

exactly this program in calculating the double-slit intensity pattern in Section 45-3. The calculation there was simple because we assumed only two elementary sources, the two narrow slits.

Figure 5 shows another representation of Fig. 4, in the form of ray diagrams. The pattern formed on the screen depends on the separation between the screen C and the aperture B. In general, we can consider three cases:

1. *Very small separation.* When C is very close to B, the waves travel only a short distance after leaving the aper-

ture, and the rays diverge very little. The effects of diffraction are negligible, and the pattern on the screen is the geometric shadow of the aperture.

2. *Very large separation.* Figure 5a represents the situation when the screen is so far from the aperture that we can regard the rays as parallel or, equivalently, the wavefronts as planes. (In this case, we also assume the source to be far from the aperture, so that the incident wavefronts are also planes. The same effect can be achieved by illuminating the aperture with a laser.) One way of achieving this condition, which is known as *Fraunhofer diffraction,* in the laboratory is to use two converging lenses, as in Fig. 5b. The first lens converts the diverging light from the source into a plane wave, and the second lens focuses plane waves leaving the aperture to point P. All rays that reach P leave the aperture parallel to the dashed line Px drawn from P through the center of the second lens.

3. *Intermediate separation.* In the case shown in Fig. 5c, the screen can be at any distance from the aperture, and the rays entering and leaving the aperture are not parallel. This general case is called *Fresnel diffraction*.

Although Fraunhofer diffraction is a special limiting case of the more general Fresnel diffraction, it is an important case and is easier to handle mathematically. We assumed Fraunhofer diffraction in our analysis of double-slit interference in Chapter 45. In this book we deal only with Fraunhofer diffraction.

46-2   SINGLE-SLIT DIFFRACTION

The simplest diffraction pattern to analyze is that produced by a long narrow slit. In this section we discuss the locations of the minima and maxima in the pattern as shown in Fig. 1. In the next section we calculate the intensity of the pattern as a function of position on the screen.

Figure 6 shows a plane wave falling at normal incidence on a slit of width a. Let us first consider the central point P_0. Rays that leave the slit parallel to the central horizontal axis are brought to a focus at P_0. These rays are certainly in phase at the plane of the slit, and they remain in phase as they are brought to a focus by the lens (see, for example, Fig. 17a of Chapter 44). Since all rays arriving at P_0 are in phase, they interfere constructively and produce a maximum of intensity at P_0.

We now consider another point on the screen. Light rays that reach P_1 in Fig. 7 leave the slit at the angle θ, as

shown. The ray xP_1 passes undeflected through the center of the lens and therefore determines θ. Ray r_1 originates at the top of the slit and ray r_2 at its center. If θ is chosen so that the distance bb' in the figure is one-half wavelength, r_1 and r_2 are out of phase and interfere destructively at P_1. The same is true for a ray just below r_1 and another just below r_2. In fact, for every ray passing through the upper half of the slit, there is a corresponding ray passing through the lower half, originating at a point $a/2$ below the first ray, such that the two rays are out of phase at P_1. Every ray arriving at P_1 from the upper half of the slit interferes destructively with one coming from the bottom half of the slit. The intensity at P_1 is therefore zero, and P_1 is the first minimum of the diffraction pattern.

Since the distance bb' equals $(a/2) \sin \theta$, the condition for the first minimum can be written

$$\frac{a}{2} \sin \theta = \frac{\lambda}{2},$$

or

$$a \sin \theta = \lambda. \tag{1}$$

Equation 1 shows that the central maximum becomes wider as the slit is made narrower. If the slit width is as small as one wavelength ($a = \lambda$), the first minimum occurs at $\theta = 90°$ ($\sin \theta = 1$ in Eq. 1), which implies that the central maximum fills the entire forward hemisphere. We assumed a condition approaching this in our discussion of double-slit interference in Section 45-1.

In Fig. 8 the slit is divided into four equal zones, with a ray leaving the top of each zone. Let θ be chosen so that the distance bb' is one-half wavelength. Rays r_1 and r_2 then cancel at P_2. Rays r_3 and r_4 are also one-half wavelength out of phase and also cancel. Consider four other rays, emerging from the slit a given distance below these four rays. The two rays below r_1 and r_2 cancel, as do the two rays below r_3 and r_4. We can proceed across the entire slit and conclude again that no light reaches P_2; we have located a second point of zero intensity.

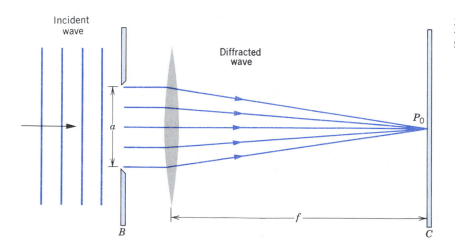

Figure 6   Conditions at the central maximum of the diffraction pattern.

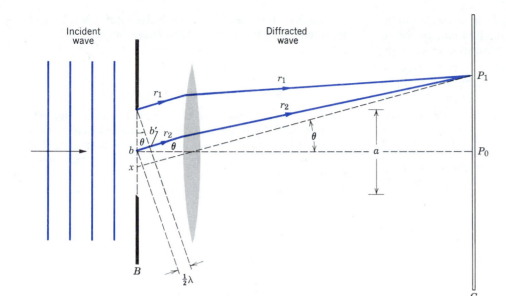

Figure 7   Conditions at the first minimum of the diffraction pattern. The angle θ is such that the distance bb' is one-half wavelength.

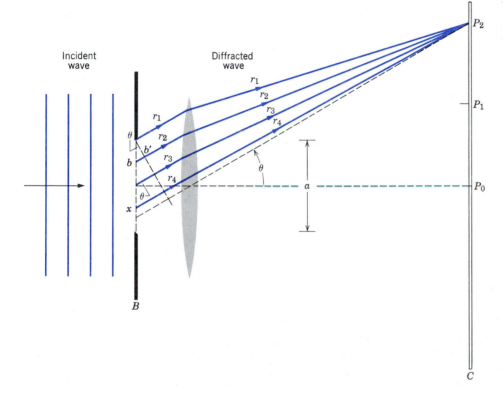

Figure 8   Conditions at the second minimum of the diffraction pattern. The angle θ is such that the distance bb' is one-half wavelength.

The condition described (see Fig. 8) requires that

$$\frac{a}{4}\sin\theta = \frac{\lambda}{2},$$

or

$$a\sin\theta = 2\lambda. \qquad (2)$$

For a given slit width a and wavelength λ, Eq. 2 gives the position on the screen of the second minimum in terms of the angle θ. By extension of Eqs. 1 and 2, the general formula for the minima in the diffraction pattern on screen C is

$$a\sin\theta = m\lambda \qquad m = 1, 2, 3, \ldots \quad \text{(minima)}. \quad (3)$$

There is a maximum approximately halfway between each adjacent pair of minima. Later in the chapter we derive a formula for the intensity of the diffracted light, from which the locations of the maxima can be found exactly. Note that Eq. 3 suggests *two* minima (and corresponding maxima) for each *m*, one at an angle θ above the central axis and one below (corresponding to $m < 0$). In deriving Eq. 3, consider how the assumption of parallel rays (Fraunhofer diffraction) has simplified the analysis.

Sample Problem 1 A slit of width *a* is illuminated by white light. For what value of *a* does the first minimum for red light ($\lambda = 650$ nm) fall at $\theta = 15°$?

Solution At the first minimum, $m = 1$ in Eq. 3. Solving for *a*, we then find

$$a = \frac{m\lambda}{\sin \theta} = \frac{(1)(650 \text{ nm})}{\sin 15°}$$

$$= 2510 \text{ nm} = 2.51 \text{ } \mu\text{m}.$$

For the incident light to flare out that much ($\pm 15°$) the slit must be very narrow indeed, amounting to about four times the wavelength (and far narrower than a fine human hair, which may only be about 100 μm in diameter).

Sample Problem 2 In Sample Problem 1, what is the wavelength λ' of the light whose first diffraction maximum (not counting the central maximum) falls at 15°, thus coinciding with the first minimum for red light?

Solution Maxima occur about halfway between minima, so

there is a maximum at 15° when the first minimum is at 10° and the second minimum is at 20°. In this case, for the second minimum,

$$a \sin \theta = 2\lambda',$$

or

$$\lambda' = \tfrac{1}{2}(2510 \text{ nm})(\sin 20°) = 430 \text{ nm}.$$

Light of this wavelength is violet. The second maximum for light of wavelength 430 nm always coincides with the first minimum for light of wavelength 650 nm, no matter what the slit width. If the slit is relatively narrow, the angle θ at which this overlap occurs is relatively large, and conversely.

46-3 INTENSITY IN SINGLE-SLIT DIFFRACTION

In Section 46-2, we located the positions of the minima of the single-slit diffraction pattern. We now wish to find an expression for the intensity of the *entire* pattern as a function of the diffraction angle θ. This expression will permit us to find the location and intensity of the maxima.

Figure 9 shows a slit of width *a* divided into *N* parallel strips, each of width Δx. The strips are very narrow, so that each strip can be regarded as a radiator of Huygens wavelets, and all the light from a given strip arrives at point *P* with the same phase. The waves arriving at *P*

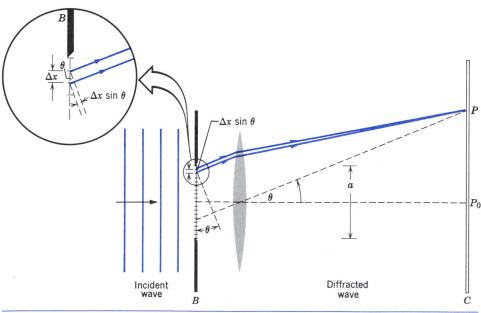

Figure 9 A slit of width *a* is divided into *N* strips of width Δx. The inset shows more clearly the conditions at the second strip. In the differential limit, the width *dx* of each strip becomes infinitesimally small and the number of strips becomes infinitely large. Here and in the next figure we take $N = 18$ for clarity.

from any pair of adjacent strips have the same (constant) phase difference $\Delta\phi$, which can be found from

$$\frac{\text{phase difference}}{2\pi} = \frac{\text{path difference}}{\lambda}$$

or

$$\Delta\phi = \frac{2\pi}{\lambda}\,\Delta x \sin\theta, \qquad (4)$$

where $\Delta x \sin\theta$, as the detail of Fig. 9 shows, is the path difference for rays originating from corresponding points of adjacent strips.

If the angle θ is not too large, each strip produces a wave of the same amplitude ΔE_0 at P. The net effect at P is due to the superposition of N vectors of the same amplitude, each differing in phase from the next by $\Delta\phi$. To find the intensity at P, we must first find the net electric field of the N vectors.

In Section 45-3, we introduced a graphical method for adding wave disturbances that enabled us to calculate the intensity in double-slit interference. That method is based on representing each wave disturbance as a phasor (a rotating vector) and finding the resultant phasor amplitude by vector addition, taking into account the relative phase given by Eq. 4. The resultant electric field E_θ varies with θ, because the phase difference $\Delta\phi$ varies with θ.

Let us consider some examples of the addition of phasors in single-slit diffraction. We first consider the resultant electric field at point P_0 (the center of the diffraction pattern on the screen). In this case $\theta = 0$, and Eq. 4 gives $\Delta\phi = 0$ as the phase difference between adjacent strips. According to the method of Section 45-3, we then lay N vectors of length ΔE_0 head to tail and parallel to one another ($\Delta\phi = 0$). The resultant E_θ is shown in Fig. 10a. This is clearly the maximum value that the resultant of these N vectors can take, so we label it E_m.

As we move away from $\theta = 0$, the phase difference $\Delta\phi$ assumes a definite nonzero value. Again laying the N vectors head to tail, each differing in direction from the previous one by $\Delta\phi$, we obtain the resultant shown in Fig. 10b. Note that E_θ is smaller than it was in Fig. 10a.

Now consider the first minimum of the diffraction pattern (point P_1 in Fig. 7). At this point the intensity is zero, so the resultant E_θ must be zero. This means that the N phasors, laid head to tail, must form a closed loop, as in Fig. 10c.

Beyond the first minimum, the phase shift $\Delta\phi$ is still larger, and the chain of vectors coils around through an angle greater than 360°. At a certain angle (corresponding to a certain phase shift, as in Fig 10d), the resultant E_θ has its greatest length within this loop, corresponding to the first maximum beyond the central one. Note that the intensity of this maximum is much smaller than the intensity of the central maximum, represented in Fig. 10a. Eventually, this loop closes on itself, giving a resultant of zero and corresponding to the second minimum.

Our goal in finding the intensity of the single-slit dif-

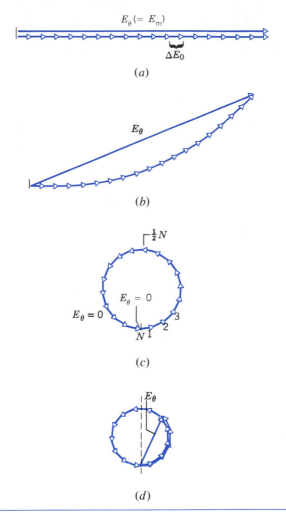

Figure 10 Phasors in single-slit diffraction, showing conditions at (a) the central maximum, (b) a direction slightly removed from the central maximum, (c) the first minimum, and (d) the first maximum beyond the central maximum. This figure corresponds to $N = 18$ in Fig. 9.

fraction pattern for any θ is to evaluate the phase shift according to Eq. 4 and find the resultant E_θ, as in Fig. 10b. The square of this resultant then gives the relative intensity, as in Section 45-3.

The light arriving at P from a given strip is in phase only if the strip is infinitesimally small and the number of strips is correspondingly large. The chain of phasors of Fig. 10b then approaches the arc of a circle, as drawn in Fig. 11. The length of the arc is E_m, while the amplitude we seek for the resultant field is indicated by the chord E_θ. The angle ϕ is the total phase difference between the rays from the top and bottom of the strip; as Fig. 11 shows, ϕ is also the angle between the two radii R.

From this figure we can write

$$E_\theta = 2R \sin\frac{\phi}{2}.$$

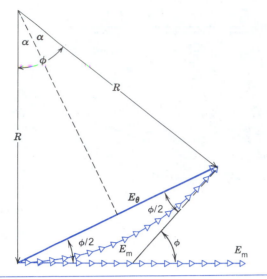

Figure 11   A construction used to calculate the intensity in single-slit diffraction. The situation corresponds to that of Fig. 10*b*.

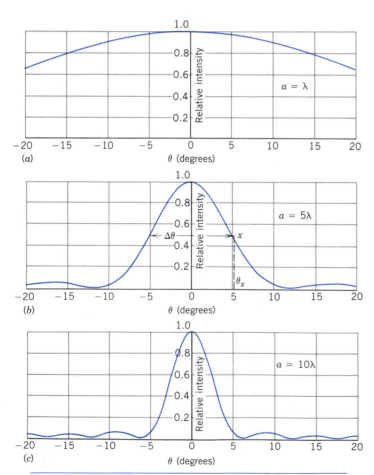

Figure 12   The intensity distribution in single-slit diffraction for three different values of the ratio a/λ. The wider the slit, the narrower is the central diffraction peak. As indicated in (*b*), $\Delta\theta$ gives a measure of the width of the central peak.

From Fig. 11, ϕ in radian measure is

$$\phi = \frac{E_m}{R}.$$

Combining yields

$$E_\theta = \frac{E_m}{\phi/2} \sin\frac{\phi}{2},$$

or

$$E_\theta = E_m \frac{\sin\alpha}{\alpha}, \qquad (5)$$

in which

$$\alpha = \frac{\phi}{2}. \qquad (6)$$

From Fig. 9, recalling that ϕ is the phase difference between rays from the top and the bottom of the slit and that the path difference for these rays is $a \sin\theta$, we have

$$\frac{\text{phase difference}}{2\pi} = \frac{\text{path difference}}{\lambda},$$

or

$$\phi = \frac{2\pi}{\lambda}(a \sin\theta).$$

Combining with Eq. 6 yields

$$\alpha = \frac{\phi}{2} = \frac{\pi a}{\lambda}\sin\theta. \qquad (7)$$

Equation 5, with α evaluated according to Eq. 7, gives the amplitude of the wave disturbance for a single-slit diffraction pattern at any angle θ. The intensity I_θ for the pattern is proportional to the square of the amplitude, so

$$I_\theta = I_m \left(\frac{\sin\alpha}{\alpha}\right)^2. \qquad (8)$$

Equation 8, combined with Eq. 7, gives the result we seek for the intensity of the single-slit diffraction pattern at any θ. Figure 12 shows plots of the relative intensity I_θ/I_m for several values of the ratio a/λ. Note that the pattern becomes narrower as we increase a/λ. (See also Fig. 1 of Chapter 43.)

Minima occur in Eq. 8 when

$$\alpha = m\pi \qquad m = \pm 1, \pm 2, \pm 3, \ldots . \qquad (9)$$

Combining with Eq. 7 leads to

$$a \sin\theta = m\lambda \qquad m = \pm 1, \pm 2, \pm 3, \ldots \quad \text{(minima)},$$

which is the result derived in the preceding section (Eq. 3). In that section, however, we derived *only* this result, obtaining no quantitative information about the intensity of the diffraction pattern at places in which it was not zero. Here (Eq. 8) we have complete intensity information.

Sample Problem 3   Calculate, approximately, the relative intensities of the secondary maxima in the single-slit Fraunhofer diffraction pattern.

Solution The secondary maxima lie approximately halfway between the minima and are roughly given by (see Problem 15)

$$\alpha \approx (m + \tfrac{1}{2}) \pi \qquad m = 1, 2, 3, \ldots \, ,$$

with a similar result for $m < 0$. Substituting into Eq. 8 yields

$$I_\theta = I_m \left[\frac{\sin (m + \tfrac{1}{2})\pi}{(m + \tfrac{1}{2})\pi} \right]^2 ,$$

which reduces to

$$\frac{I_\theta}{I_m} = \frac{1}{(m + \tfrac{1}{2})^2 \, \pi^2} \, .$$

This yields $I_\theta / I_m = 0.045$ $(m = 1)$, 0.016 $(m = 2)$, 0.0083 $(m = 3)$, and so forth. The successive maxima decrease rapidly in intensity.

Sample Problem 4 Derive the width $\Delta\theta$ of the central maximum in a single-slit Fraunhofer diffraction (see Fig. 12b). The width can be represented as the angle between the two points in the pattern where the intensity is one-half that at the center of the pattern.

Solution Point x in Fig. 12b is chosen so that $I_\theta = \tfrac{1}{2} I_m$, or, from Eq. 8,

$$\frac{1}{2} = \left(\frac{\sin \alpha_x}{\alpha_x} \right)^2 .$$

This equation cannot be solved analytically for α_x. Using a pocket calculator or a computer, we can find an approximate solution to any desired accuracy. Let us rewrite the equation as

$$\alpha_x = \sqrt{2} \sin \alpha_x. \qquad (10)$$

To solve this on your calculator, enter the "radian" mode. Pick any starting value for α_x, say $\alpha_x = 1$. Plug this value into the right side of Eq. 10 and solve, obtaining 1.19. Equation 10 requires that this value must then be equal to α_x, which it is clearly not ($1 \neq 1.19$). Take 1.19 as the new trial value, and again evaluate the right-hand side, obtaining 1.31. We still do not have a solution that satisfies Eq. 10 ($1.19 \neq 1.31$), but we are closer than we were on our first try. Continue in this way, using the result of one calculation as the starting point of the next, until the difference between the calculated value of the right-hand side of Eq. 10 and the starting value becomes as small as you like. (You can set this up as a program for a calculator or a computer and have it do the repetitions automatically. You can also have it stop when the difference between successive values becomes smaller than a limit you can set.) This method is called the *iterative* technique for solving equations. After 10 iterations, the result is

$$\alpha_x = 1.39156,$$

and additional iterations change only the fifth decimal place. Inserting this value into Eq. 7, we obtain

$$\theta_x = \sin^{-1} \left(\frac{\alpha_x \lambda}{\pi a} \right) = \sin^{-1} \left(\frac{1.39}{5\pi} \right) = 5.1°.$$

The width of the curve is then found from

$$\Delta\theta = 2\theta_x = 10.2°.$$

46-4 DIFFRACTION AT A CIRCULAR APERTURE

In focusing an image, a lens passes only the light that falls within its circular perimeter. From this point of view, a lens behaves like a circular aperture in an opaque screen. Such an aperture forms a diffraction pattern analogous to that of a single slit. Diffraction effects often limit the ability of telescopes and other optical instruments to form precise images.

The image formed by a lens can be distorted by other effects, including chromatic and spherical aberrations. These effects can be substantially reduced or eliminated by suitable shaping of the lens surfaces or by introducing correcting elements into the optical system. However, no amount of clever design can eliminate the effects of diffraction, which are determined only by the size of the aperture (the diameter of the lens) and the wavelength of the light. In diffraction, nature imposes a fundamental limitation on the precision of our instruments.

When we used geometrical optics to analyze lenses, we assumed diffraction not to occur. However, geometrical optics is itself an approximation, being the limit of wave optics. If we were to make a rigorous wave-optical analysis of the formation of an image by a lens, we would find that diffraction effects arise in a natural way.

Figure 13 shows the image of a distant point source of light (a star) formed on a photographic film placed in the focal plane of the converging lens of a telescope. It is not a

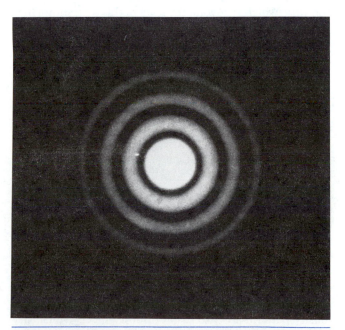

Figure 13 The diffraction pattern of a circular aperture. The central maximum is sometimes called the Airy disk (after Sir George Airy, who first solved the problem of diffraction by a circular aperture in 1835). Note the circular secondary maxima.

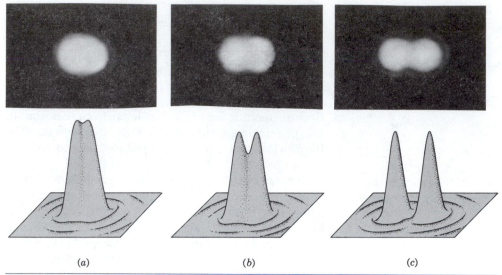

Figure 14   The images of two distant point sources (stars) formed by a converging lens. The diameter of the lens (which is the diffracting aperture) is 10 cm, so that $a/\lambda = 200{,}000$ if the effective wavelength is about 500 nm. In (*a*) the stars are so close together that their images can scarcely be distinguished, owing to the overlap of their diffraction patterns. In (*b*) the stars are farther apart and their separation meets Rayleigh's criterion for resolution of their images. In (*c*) the stars are still farther apart and their images are well resolved. Computer-generated profiles of the intensities are shown below the images.

point, as the (approximate) geometrical optics treatment suggests, but a circular disk surrounded by several progressively fainter secondary rings. Comparison with Fig. 1 leaves little doubt that we are dealing with a diffraction phenomenon.

The mathematical analysis of diffraction by a circular aperture, which is beyond the level of this text, shows that (under Fraunhofer conditions) the first minimum occurs at an angle from the central axis given by

$$\sin \theta = 1.22 \frac{\lambda}{d}, \tag{11}$$

where d is the diameter of the aperture. This is to be compared with Eq. 1,

$$\sin \theta = \frac{\lambda}{a},$$

which locates the first minimum of a slit of width a. These expressions differ by the factor 1.22, which arises when we divide the circular aperture into elementary Huygens sources and integrate over the aperture.

The fact that lens images are diffraction patterns is important when we wish to distinguish two distant point objects whose angular separation is small. Figure 14 shows the visual appearances and the corresponding intensity patterns for two distant point objects (stars, say) with small angular separations and approximately equal central intensities. In Fig. 14*a* the objects are not resolved; that is, they cannot be distinguished from a single point object. In Fig. 14*b* they are barely resolved, and in Fig. 14*c* they are fully resolved.

In Fig. 14*b* the angular separation of the two point sources is such that the central maximum of the diffraction pattern of one source falls on the first minimum of the diffraction pattern of the other. This is called *Rayleigh's criterion* for resolving images. From Eq. 11, two objects that are barely resolvable by Rayleigh's criterion must have an angular separation θ_R of

$$\theta_R = \sin^{-1}\left(\frac{1.22\lambda}{d}\right).$$

Since the angles involved are rather small, we can replace $\sin \theta_R$ by θ_R, so

$$\theta_R = 1.22 \frac{\lambda}{d}, \tag{12}$$

in which θ_R is expressed in radians. If the angular separation θ between the objects is greater than θ_R, we can resolve the two objects; if it is less, we cannot. The angle θ_R is the smallest angular separation for which resolution is possible, using Rayleigh's criterion.

When we wish to use a lens to resolve objects of small angular separation, it is desirable to make the central disk of the diffraction pattern as small as possible. This can be done (see Eq. 12) by increasing the lens diameter or by using a shorter wavelength. One reason for constructing large telescopes is to produce *sharper* images so that we can examine astronomical objects in finer detail. The images are also *brighter,* not only because the energy is concentrated into a smaller diffraction disk but because the larger lens collects more light. Thus fainter objects, for example, distant galaxies, can be seen.

Figure 15 An image of a chain of streptococcus bacteria (diameter 10^{-6} m) obtained with an electron microscope. Note the sharpness of the image, which would not be possible using visible light.

To reduce diffraction effects in *microscopes* we often use ultraviolet light, which, because of its shorter wavelength, permits finer detail to be examined than would be possible if the same microscope used visible light. We shall see in Chapter 50 that beams of electrons behave like waves under some circumstances. In the *electron microscope* such beams may have an effective wavelength of 4×10^{-3} nm, of the order of 10^5 times shorter than that of visible light. This permits the detailed examination of tiny objects such as bacteria or viruses (Fig. 15). If such a small object were examined with an optical microscope, its structure would be hopelessly concealed by diffraction.

Sample Problem 5   A converging lens 32 mm in diameter has a focal length f of 24 cm. (*a*) What angular separation must two distant point objects have to satisfy Rayleigh's criterion? Assume that $\lambda = 550$ nm. (*b*) How far apart are the centers of the diffraction patterns in the focal plane of the lens?

Solution   (*a*) From Eq. 12

$$\theta_R = 1.22 \, \frac{\lambda}{d} = \frac{(1.22)(550 \times 10^{-9} \text{ m})}{32 \times 10^{-3} \text{ m}}$$

$$= 2.10 \times 10^{-5} \text{ rad} = 4.3 \text{ arc seconds.}$$

(*b*) The linear separation is

$$\Delta x = f \theta_R = (0.24 \text{ m})(2.10 \times 10^{-5} \text{ rad})$$

$$= 5.0 \ \mu\text{m},$$

or about 9 wavelengths of the light.

46-5 DOUBLE-SLIT INTERFERENCE AND DIFFRACTION COMBINED

In our analysis of double-slit interference (Section 45-1) we assumed that the slits were arbitrarily narrow; that is, that $a \ll \lambda$. For such narrow slits, the central part of the screen on which the light falls is uniformly illuminated by the diffracted waves from each slit. When such waves interfere, they produce interference fringes of uniform intensity.

In practice, for visible light, the condition $a \ll \lambda$ is usually not met. For such relatively wide slits, the intensity of the interference fringes formed on the screen is *not* uniform. Instead, the intensity of the fringes varies within an envelope due to the diffraction pattern of a single slit.

The effect of diffraction on a double-slit interference pattern is illustrated in Fig. 16, which compares the double-slit pattern with the diffraction pattern produced by a single slit of the same width as each of the double slits. You can see from Fig. 16*a* that the diffraction does indeed provide an intensity envelope for the more closely spaced double-slit interference fringes.

Let us now analyze the combined interference and diffraction pattern of Fig. 16*a*. The interference pattern for two infinitesimally narrow slits is given by Eq. 11 of Chapter 45, or, with a small change in notation,

$$I_{\theta,\text{int}} = I_{\text{m,int}} \cos^2 \beta, \tag{13}$$

where

$$\beta = \frac{\pi d}{\lambda} \sin \theta \tag{14}$$

in which d is the distance between the center-lines of the slits.

Figure 16   (*a*) Interference fringes for a double-slit system in which the slit width is not negligible in comparison with the wavelength. (*b*) The diffraction pattern of a single slit of the same width. Note that the diffraction pattern modulates the intensity of the interference fringes, as shown in part (*a*).

(*a*)

(*b*)

The intensity for the diffracted wave from either slit is given by Eq. 8, or, again with a small change in notation,

$$I_{\theta,\text{dif}} = I_{\text{m,dif}} \left(\frac{\sin \alpha}{\alpha} \right)^2, \qquad (15)$$

where

$$\alpha = \frac{\pi a}{\lambda} \sin \theta. \qquad (16)$$

We find the combined effect by regarding $I_{\text{m,int}}$ in Eq. 13 as a variable amplitude, given in fact by $I_{\theta,\text{dif}}$ of Eq. 15. This assumption, for the combined pattern, leads to

$$I_\theta = I_\text{m}(\cos \beta)^2 \left(\frac{\sin \alpha}{\alpha} \right)^2, \qquad (17)$$

in which we have dropped all subscripts referring separately to interference and diffraction. Later in this section we derive this result using phasors.

Let us express this result in words. At any point on the screen the available light intensity from each slit, considered separately, is given by the diffraction pattern of that slit (Eq. 15). The diffraction patterns for the two slits, again considered separately, coincide because parallel rays in Fraunhofer diffraction are focused at the same spot. Because the two diffracted waves are coherent, they interfere.

The effect of interference is to redistribute the available energy over the screen, producing a set of fringes. In Section 45-1, where we assumed $a \ll \lambda$, the available energy was virtually the same at all points on the screen so that the interference fringes had virtually the same intensities (see Fig. 8 of Chapter 45). If we relax the assumption $a \ll \lambda$, the available energy is *not* uniform over the screen but is given by the diffraction pattern of a slit of width a. In this case the interference fringes have intensities that are determined by the intensity of the diffraction pattern at the location of a particular fringe. Equation 17 is the mathematical expression of this argument. This is especially clear in Fig. 17, which shows (*a*) the "interference factor" in Eq. 17 (that is, the factor $\cos^2 \beta$), (*b*) the "diffraction factor" $(\sin \alpha/\alpha)^2$, and (*c*) their product.

Figure 18 is a plot of the relative intensity I_θ/I_m given by Eq. 17 for $d = 50\lambda$ and for three values of a/λ. It shows clearly that for narrow slits ($a = \lambda$) the fringes are nearly uniform in intensity. As the slits are widened, the intensities of the fringes are markedly modulated by the "diffraction factor" in Eq. 17, that is, by the factor $(\sin \alpha/\alpha)^2$; compare with Fig. 12.

If we decrease the slit width a, the envelope of the fringe pattern becomes broader, and the central peak spreads out (compare Figs. 18*a* and 18*b*). As the slit width a approaches zero, $\alpha \to 0$ and $\sin \alpha/\alpha \to 1$. Thus Eq. 17 reduces to Eq. 13, which describes interference from a pair of vanishingly narrow slits. If we let the slit separation d approach zero, the two slits coalesce into a single slit of width a. From Eq. 14, $\beta \to 0$ as $d \to 0$, and Eq. 17 re-

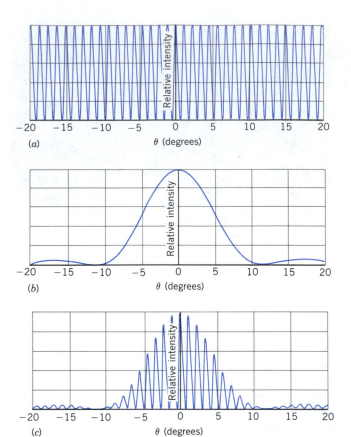

Figure 17 (*a*) Interference fringes that would be produced by a double slit of vanishingly narrow widths. (*b*) The diffraction pattern for a slit of finite width. (*c*) The pattern of interference fringes formed by two slits of the same width as that of (*b*). This pattern is equivalent to the product of the curves shown in (*a*) and (*b*). Compare Fig. 16*a*.

duces to Eq. 15, the diffraction equation for a single slit of width a.

If we increase the slit width a, the envelope of the fringe pattern becomes narrower, and the central peak becomes sharper (compare Figs. 18*b* and 18*c*). The separation between the fringes, which depends on d/λ, does not change. If we increase the slit separation d, the fringes are closer together, but the envelope of the fringe pattern, which depends on a/λ, does not change.

If we increase the wavelength of the incident light, both the diffraction and interference patterns broaden: the diffraction envelope becomes wider and the fringe separation increases. The reverse effect occurs as we decrease the wavelength. Put another way, the relationship between the diffraction envelope and the interference fringes (for example, the number of fringes in the central peak) depends on the ratio d/a and is independent of λ.

The double-slit pattern illustrated in Fig. 17 combines interference and diffraction in an intimate way. Both are superposition effects that depend on adding wave disturbances at a given point, taking phase differences properly

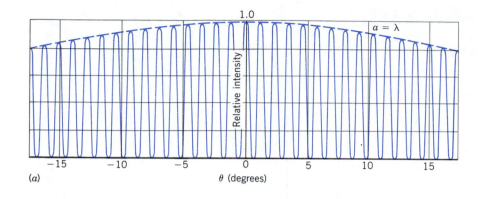

(a)

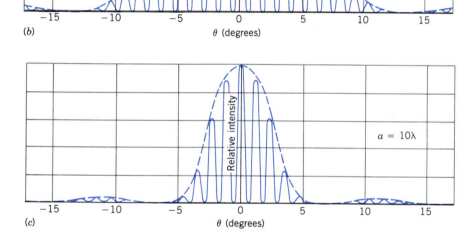

(b)

(c)

Figure 18   Interference fringes for a double slit with slit separation $d = 50\lambda$. Three different slit widths are shown.

into account. If the waves to be combined originate from a *finite* (and usually small) number of elementary coherent radiators, as in the double slit, we call the effect *interference.* If the waves to be combined originate by subdividing a wave into *infinitesimal* coherent radiators, as in our treatment of a single slit, we call the effect *diffraction.* This distinction between interference and diffraction is convenient and useful. However, it should not cause us to lose sight of the fact that both are superposition effects and that often both are present simultaneously, as in the double-slit experiment.

Sample Problem 6   In a double-slit experiment, the distance D of the screen from the slits is 52 cm, the wavelength λ is 480 nm, the slit separation d is 0.12 mm, and the slit width a is 0.025 mm. (*a*) What is the spacing between adjacent fringes? (*b*) What is the distance from the central maximum to the first minimum of the fringe envelope?

Solution   (*a*) The intensity pattern is given by Eq. 17, the fringe spacing being determined by the interference factor $\cos^2 \beta$. From Sample Problem 2, Chapter 45, we have

$$\Delta y = \frac{\lambda D}{d}.$$

Substituting yields

$$\Delta y = \frac{(480 \times 10^{-9} \text{ m})(52 \times 10^{-2} \text{ m})}{0.12 \times 10^{-3} \text{ m}}$$

$$= 2.1 \text{ mm}.$$

(*b*) The angular position of the first minimum follows from Eq. 1, or

$$\sin \theta = \frac{\lambda}{a} = \frac{480 \times 10^{-9} \text{ m}}{25 \times 10^{-6} \text{ m}} = 0.0192.$$

This is so small that, with little error, we can put $\sin \theta \approx \tan \theta \approx \theta$, so

$$y = D \tan \theta \approx D\theta = (52 \times 10^{-2} \text{ m})(0.0192)$$
$$= 10 \text{ mm}.$$

You can show that there are about 9 fringes in the central peak of the diffraction envelope.

Sample Problem 7   What requirements must be met for the central maximum of the envelope of the double-slit interference pattern to contain exactly 11 fringes?

Solution   The required condition will be met if the sixth minimum of the interference factor ($\cos^2 \beta$ in Eq. 17) coincides with the first minimum of the diffraction factor [$(\sin \alpha/\alpha)^2$ in Eq. 17].

The sixth minimum of the interference factor occurs when

$$\beta = (11/2)\pi$$

in Eq. 17. The first minimum in the diffraction term occurs for

$$\alpha = \pi$$

in Eq. 17. Dividing (see Eqs. 14 and 16) yields

$$\frac{\beta}{\alpha} = \frac{d}{a} = \frac{11}{2}.$$

This condition depends only on the ratio of the slit separation d to the slit width a and not at all on the wavelength. For larger λ the pattern is broader than for smaller λ, but there are always 11 fringes in the central peak of the envelope.

Phasor Derivation of Eq. 17   *(Optional)*

Figure 19 shows the geometry appropriate for the analysis of the double slit using phasors. Each of the two slits is divided into N zones, as was done for the single slit in Fig. 9. The net electric field at P is found from the superposition of the N electric field vectors from the upper slit and the N electric field vectors from the lower slit. The phasor method allows us to combine these contributions to the electric field at P, taking into account their relative phases.

Figure 20 shows the first N phasors (corresponding to the upper slit of Fig. 19) and their resultant E_1, as in Fig. 11. There is a phase difference $\Delta\phi = \phi/N$ between each of the N phasors. To add the second group of N phasors (corresponding to the lower slit) we must find the phase angle ξ between the last phasor from the upper slit and the first phasor from the lower slit. We than draw the N phasors from the lower slit and find their resultant, E_2. The sum of the E_1 and E_2 phasors gives the resultant E_θ that characterizes the double slit.

From Fig. 20, we see that E_θ is the base of an isosceles triangle whose sides have equal lengths E_1 and E_2, which are given by Eq. 5. From the geometry of Fig. 20,

$$E_\theta = 2E_1 \sin \frac{\delta}{2}, \tag{18}$$

where δ, the apex angle of the triangle, can be found from

$$\frac{\phi}{2} + \delta + \frac{\phi}{2} + \xi = \pi, \tag{19}$$

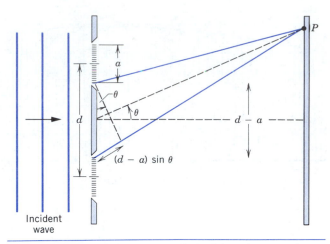

Figure 19   Each slit in a double slit is divided into N strips. In the differential limit, the strips become infinitesimally small and infinitely numerous. Here, as we did in Fig. 9, we show $N = 18$.

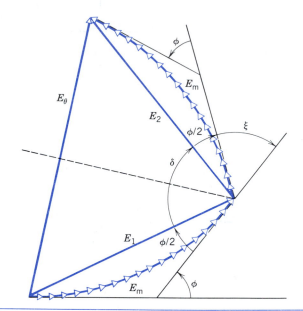

Figure 20   Phasor diagram used to calculate the resultant electric field in double-slit interference.

which gives

$$\delta = \pi - (\xi + \phi). \tag{20}$$

Using Eq. 20 to evaluate $\sin \delta/2$, we find

$$\sin \frac{\delta}{2} = \sin \left(\frac{\pi}{2} - \frac{\xi + \phi}{2} \right) = \cos \frac{\xi + \phi}{2}. \tag{21}$$

From the expression

$$\frac{\text{path difference}}{\lambda} = \frac{\text{phase difference}}{2\pi}$$

with the phase difference between the two rays (from the bottom of the upper slit and the top of the lower slit, as shown in Fig. 19) of ξ and the path difference of $(d - a) \sin \theta$, we have

$$\frac{\xi}{2} = \frac{\pi}{\lambda}(d - a)\sin\theta.$$

Combining this with Eq. 7, $\phi/2 = (\pi a/\lambda)\sin\theta$, we find

$$\frac{\xi + \phi}{2} = \frac{\pi}{\lambda}d\sin\theta,$$

which is just β, according to Eq. 14. Substituting into Eq. 21, we find

$$\sin\frac{\delta}{2} = \cos\beta.$$

Inserting this result into Eq. 18 and using Eq. 5 for the magnitude of E_1 (or E_2), we obtain

$$E_\theta = 2E_m \frac{\sin\alpha}{\alpha}\cos\beta. \tag{22}$$

Squaring Eq. 22 gives the intensities as

$$I_\theta = I_m(\cos\beta)^2 \left(\frac{\sin\alpha}{\alpha}\right)^2,$$

which is identical to Eq. 17. Note that, as was the case in Eq. 11 of Chapter 45, $I_m = 4I_0$. ■

QUESTIONS

1. Distinguish between Fresnel and Fraunhofer diffraction. Do different physical principles underlie them? If so, what are they? If the same broad principle underlies them, what is it?

2. In what way are interference and diffraction similar? In what way are they different?

3. Suppose that you hold a single narrow vertical slit in front of the pupil of your eye and look at a distant light source in the form of a long heated filament. Is the diffraction pattern that you see a Fresnel or a Fraunhofer pattern?

4. Do diffraction effects occur for virtual images as well as for real images? Explain.

5. Do diffraction effects occur for images formed by (a) plane mirrors and (b) spherical mirrors? Explain.

6. Comment on this statement: "Diffraction occurs in all regions of the electromagnetic spectrum." Consider the x-ray region and the microwave region, for example, and give arguments for believing the statement to be true or false.

7. We have claimed (correctly) that Maxwell's equations predict all the classical optical phenomena. Yet in Chapter 45 (Interference) and in this chapter (Diffraction), there is little mention of Maxwell's equations. Is there an inconsistency here? Where is the impact of Maxwell's equations felt? Discuss.

8. If we were to redo our analysis of the properties of lenses in Chapter 44 by the methods of geometrical optics but without restricting our consideration to paraxial rays and to "thin" lenses, would diffraction phenomena emerge from the analysis? Discuss.

9. Why is the diffraction of sound waves more evident in daily experience than that of light waves?

10. Sound waves can be diffracted. About what width of a single slit should you use if you wish to broaden the distribution of an incident plane sound wave of frequency 1 kHz?

11. Why do radio waves diffract around buildings, although light waves do not?

12. A loudspeaker horn, used at a rock music concert, has a rectangular aperture 1 m high and 30 cm wide. Will the pattern of sound intensity be broader in the horizontal plane or in the vertical?

13. A particular radar antenna is designed to give accurate measurements of the altitude of an aircraft but less accurate measurements of its direction in a horizontal plane. Must the height-to-width ratio of the radar antenna be less than, equal to, or greater than unity?

14. Describe what happens to a Fraunhofer single-slit diffraction pattern if the whole apparatus is immersed in water.

15. In single-slit diffraction, what is the effect of increasing (a) the wavelength and (b) the slit width?

16. While listening to the car radio, you may have noticed that the AM signal fades, but the FM signal doesn't, when you drive under a bridge. Could diffraction have anything to do with this?

17. What will the single-slit diffraction pattern look like if $\lambda > a$?

18. What would the pattern on a screen formed by a double slit look like if the slits did not have the same width? Would the location of the fringes be changed?

19. The shadow of a vertical flagpole cast by the Sun has clearly defined edges near its base, but less-well-defined edges near its top end. Why?

20. Sunlight falls on a single slit of width $1\,\mu$m. Describe qualitatively what the resulting diffraction pattern looks like.

21. In Fig. 8, rays r_1 and r_3 are in phase; so are r_2 and r_4. Why isn't there a maximum intensity at P_2 rather than a minimum?

22. When we speak of diffraction by a single slit we imply that the width of the slit must be much less than its length. Suppose that, in fact, the length was equal to twice the width. Make a rough guess at what the diffraction pattern would look like.

23. In Fig. 7 the optical path lengths from the slit to point P_0 are all the same. Why?

24. In Fig. 10d, why is E_θ, which represents the first maximum beyond the central maximum, not vertical? (*Hint*: Consider the effects of a slight winding or unwinding of the coil of phasors in this figure.) See Problem 16.

25. Give at least two reasons why the usefulness of large telescopes increases as we increase the lens diameter.

26. Are diffraction effects associated with reflecting telescopes, such as the Hubble Space Telescope, that use mirrors instead of lenses? If so, why do we go to the effort of putting such telescopes in space?

27. We have seen that diffraction limits the resolving power of optical telescopes (see Fig. 14). Does it also do so for large radio telescopes?

28. Diffraction is more of a nuisance in a telescope than in a camera. Why?

29. The double-slit pattern of Fig. 21a seen with a monochromatic light source is somehow changed to the pattern of Fig. 21b. Consider the following possible changes in conditions: (a) the wavelength of the light was decreased; (b) the wavelength of the light was increased; (c) the width of each slit was increased; (d) the separation of the slits was increased; (e) the separation of the slits was decreased; (f) the width of each slit was decreased. Which selection(s) of the above changes could explain the alteration of the pattern?

30. In double-slit interference patterns such as that of Fig. 16a, we said that the interference fringes were modulated in in-

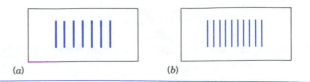

(a)   (b)

Figure 21 Question 29.

tensity by the diffraction pattern of a single slit. Could we reverse this statement and say that the diffraction pattern of a single slit is intensity-modulated by the interference fringes? Discuss.

PROBLEMS

Section 46-2 Single-Slit Diffraction

1. When monochromatic light is incident on a slit 0.022 mm wide, the first diffraction minimum is observed at an angle of 1.8° from the direction of the incident beam. Find the wavelength of the incident light.

2. Can you demonstrate the wave nature of x rays by diffracting them through a single slit? Determine the maximum slit width that could be used if a central maximum angular width of 0.12 mrad can just be detected and you guess the wavelength of the x rays to be 0.10 nm.

3. Monochromatic light of wavelength 441 nm falls on a narrow slit. On a screen 2.16 m away, the distance between the second minimum and the central maximum is 1.62 cm. (a) Calculate the angle of diffraction θ of the second minimum. (b) Find the width of the slit.

4. Light of wavelength 633 nm is incident on a narrow slit. The angle between the first minimum on one side of the central maximum and the first minimum on the other side is 1.97°. Find the width of the slit.

5. A single slit is illuminated by light whose wavelengths are λ_a and λ_b, so chosen that the first diffraction minimum of the λ_a component coincides with the second minimum of the λ_b component. (a) What relationship exists between the two wavelengths? (b) Do any other minima in the two patterns coincide?

6. A plane wave, with wavelength of 593 nm, falls on a slit of width 420 μm. A thin converging lens, having a focal length of 71.4 cm, is placed behind the slit and focuses the light on a screen. Find the distance on the screen from the center of the pattern to the second minimum.

7. In a single-slit diffraction pattern the distance between the first minimum on the right and the first minimum on the left is 5.20 mm. The screen on which the pattern is displayed is 82.3 cm from the slit and the wavelength is 546 nm. Calculate the slit width.

8. The distance between the first and fifth minima of a single-slit pattern is 0.350 mm with the screen 41.3 cm away from the slit, using light having a wavelength of 546 nm. (a) Calculate the diffraction angle θ of the first minimum. (b) Find the width of the slit.

9. A slit 1.16 mm wide is illuminated by light of wavelength 589 nm. The diffraction pattern is seen on a screen 2.94 m away. Find the distance between the first two diffraction minima on the same side of the central maximum.

10. Manufacturers of wire (and other objects of small dimensions) sometimes use a laser to continually monitor the thickness of the product. The wire intercepts the laser beam, producing a diffraction pattern like that of a single slit of the same width as the wire diameter; see Fig. 22. Suppose a He–Ne laser, wavelength 632.8 nm, illuminates a wire, the diffraction pattern being projected onto a screen 2.65 m away. If the desired wire diameter is 1.37 mm, what would be the observed distance between the two tenth-order minima on each side of the central maximum?

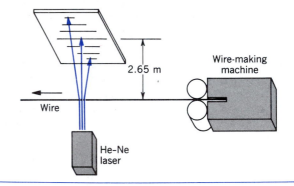

Figure 22 Problem 10.

Section 46-3 Intensity in Single-Slit Diffraction

11. Monochromatic light with wavelength 538 nm falls on a slit with width 25.2 μm. The distance from the slit to a screen is 3.48 m. Consider a point on the screen 1.13 cm from the central maximum. (a) Calculate θ. (b) Calculate α. (c) Calculate the ratio of the intensity at this point to the intensity at the central maximum.

12. If you double the width of a single slit, the intensity of the central maximum of the diffraction pattern increases by a factor of four, even though the energy passing through the slit only doubles. Explain this quantitatively.

13. Calculate the width of the central maximum in a single-slit diffraction pattern in which $a = 10\lambda$. Compare your result with Fig. 12c. See Sample Problem 4.

14. A monochromatic beam of parallel light is incident on a

"collimating" hole of diameter $a \gg \lambda$. Point P lies in the geometrical shadow region on a distant screen, as shown in Fig. 23a. Two obstacles, shown in Fig. 23b, are placed in turn over the collimating hole. A is an opaque circle with a hole in it and B is the "photographic negative" of A. Using superposition concepts, show that the intensity at P is identical for each of the two diffracting objects A and B (Babinet's principle). In this connection, it can be shown that the diffraction pattern of a wire is that of a slit of equal width. See "Measuring the Diameter of a Hair by Diffraction," by S. M. Curry and A. L. Schawlow, *American Journal of Physics,* May 1974, p. 412.

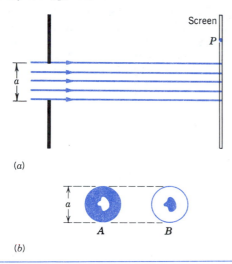

(a)

(b)

Figure 23 Problem 14.

15. (a) Show that the values of α at which intensity maxima for single-slit diffraction occur can be found exactly by differentiating Eq. 8 with respect to α and equating to zero, obtaining the condition

$$\tan \alpha = \alpha.$$

(b) Find the values of α satisfying this relation by plotting graphically the curve $y = \tan \alpha$ and the straight line $y = \alpha$ and finding their intersections or by using a pocket calculator to find an appropriate value of α by trial and error. (c) Find the (nonintegral) values of m corresponding to successive maxima in the single-slit pattern. Note that the secondary maxima do not lie exactly halfway between minima.

16. In Fig. 10d, calculate the angle E_θ makes with the vertical; see Question 24 and Problem 15.

Section 46-4 Diffraction at a Circular Aperture

17. The two headlights of an approaching automobile are 1.42 m apart. At what (a) angular separation and (b) maximum distance will the eye resolve them? Assume a pupil diameter of 5.00 mm and a wavelength of 562 nm. Also assume that diffraction effects alone limit the resolution.

18. An astronaut in a satellite claims to be able to just barely resolve two point sources on the Earth, 163 km below. Calculate their (a) angular and (b) linear separation, assuming ideal conditions. Take $\lambda = 540$ nm and the pupil diameter of the astronaut's eye to be 4.90 mm.

19. Find the separation of two points on the Moon's surface that can just be resolved by the 200-in. ($= 5.08$-m) telescope at

Mount Palomar, assuming that this distance is determined by diffraction effects. Assume a wavelength of 565 nm.

20. The wall of a large room is covered with acoustic tile in which small holes are drilled 5.20 mm from center to center. How far can a person be from such a tile and still distinguish the individual holes, assuming ideal conditions? Assume the diameter of the pupil of the observer's eye to be 4.60 mm and the wavelength to be 542 nm.

21. If Superman really had x-ray vision at 0.12-nm wavelength and a 4.3-mm pupil diameter, at what maximum altitude could he distinguish villains from heroes assuming the minimum detail required was 4.8 cm?

22. A navy cruiser employs radar with a wavelength of 1.57 cm. The circular antenna has a diameter of 2.33 m. At a range of 6.25 km, what is the smallest distance that two speedboats can be from each other and still be resolved as two separate objects by the radar system?

23. The paintings of Georges Seurat consist of closely spaced small dots (≈ 2 mm in diameter) of pure pigment, as indicated in Fig. 24. The illusion of color mixing occurs because the pupils of the observer's eyes diffract light entering them. Calculate the minimum distance an observer must stand from such a painting to achieve the desired blending of color. Take the wavelength of the light to be 475 nm and the diameter of the pupil to be 4.4 mm.

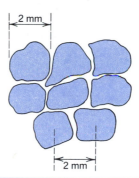

Figure 24 Problem 23.

24. A "spy in the sky" satellite orbiting at 160 km above the Earth's surface has a lens with a focal length of 3.6 m. Its resolving power for objects on the ground is 30 cm; it could easily measure the size of an aircraft's air intake. What is the effective lens diameter, determined by diffraction consideration alone? Assume $\lambda = 550$ nm. Far more effective satellites are reported to be in operation today.

25. (a) A circular diaphragm 60 cm in diameter oscillates at a frequency of 25 kHz in an underwater source of sound used for submarine detection. Far from the source the sound intensity is distributed as a diffraction pattern for a circular hole whose diameter equals that of the diaphragm. Take the speed of sound in water to be 1450 m/s and find the angle between the normal to the diaphragm and the direction of the first minimum. (b) Repeat for a source having an (audible) frequency of 1.0 kHz.

26. In June 1985 a laser beam was fired from the Air Force Optical Station on Maui, Hawaii, and reflected back from the shuttle *Discovery* as it sped by, 354 km overhead. The diameter of the central maximum of the beam at the shuttle position was said to be 9.14 m and the beam wavelength was

500 nm. What is the effective diameter of the laser aperture at the Maui ground station? (*Hint*: A laser beam spreads because of diffraction; assume a circular exit aperture.)

27. Millimeter-wave radar generates a narrower beam than conventional microwave radar. This makes it less vulnerable to antiradar missiles. (*a*) Calculate the angular width, from first minimum to first minimum, of the central "lobe" produced by a 220-GHz radar beam emitted by a 55-cm diameter circular antenna. (The frequency is chosen to coincide with a low-absorption atmospheric "window.") (*b*) Calculate the same quantity for the ship's radar described in Problem 22.

28. In a Soviet–French experiment to monitor the Moon's surface with a light beam, pulsed radiation from a ruby laser ($\lambda = 0.69 \, \mu$m) was directed to the Moon through a reflecting telescope with a mirror radius of 1.3 m. A reflector on the Moon behaved like a circular plane mirror with radius 10 cm, reflecting the light directly back toward the telescope on Earth. The reflected light was then detected by a photometer after being brought to a focus by this telescope. What fraction of the original light energy was picked up by the detector? Assume that for each direction of travel all the energy is in the central diffraction circle.

29. It can be shown that, except for $\theta = 0$, a circular obstacle produces the same diffraction pattern as a circular hole of the same diameter. Furthermore, if there are many such obstacles, such as water droplets located randomly, then the interference effects vanish leaving only the diffraction associated with a single obstacle. (*a*) Explain why one sees a "ring" around the Moon on a foggy night. The ring is usually reddish in color; explain why. (*b*) Calculate the size of the water droplets in the air if the ring around the Moon appears to have a diameter 1.5 times that of the Moon. The angular diameter of the Moon in the sky is 0.5°. (*c*) At what distance from the Moon might a bluish ring be seen? Sometimes the rings are white; why? (*d*) The color arrangement is opposite to that in a rainbow; why should this be so?

Section 46-5   Double-Slit Interference and Diffraction Combined

30. (*a*) Design a double-slit system in which the fourth fringe, not counting the central maximum, is missing. (*b*) What other fringes, if any, are also missing?

31. Two slits of width a and separation d are illuminated by a coherent beam of light of wavelength λ. What is the linear separation of the interference fringes observed on a screen that is a distance D away?

32. Suppose that, as in Sample Problem 7, the envelope of the central peak contains 11 fringes. How many fringes lie between the first and second minima of the envelope?

33. (*a*) For $d = 2a$ in Fig. 25, how many interference fringes lie in the central diffraction envelope? (*b*) If we put $d = a$, the two slits coalesce into a single slit of width $2a$. Show that Eq. 17 reduces to the diffraction pattern for such a slit.

34. (*a*) How many complete fringes appear between the first minima of the fringe envelope on either side of the central maximum for a double-slit pattern if $\lambda = 557$ nm, $d = 0.150$ mm, and $a = 0.030$ mm? (*b*) What is the ratio of the intensity of the third fringe to the side of the center to that of the central fringe?

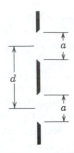

Figure 25   Problem 33.

35. Light of wavelength 440 nm passes through a double slit, yielding the diffraction pattern of intensity I versus deflection angle θ shown in Fig. 26. Calculate (*a*) the slit width and (*b*) the slit separation. (*c*) Verify the displayed intensities of the $m = 1$ and $m = 2$ interference fringes.

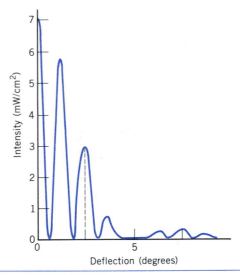

Figure 26   Problem 35.

36. An acoustic double-slit system (slit separation d, slit width a) is driven by two loudspeakers as shown in Fig. 27. By use of a variable delay line, the phase of one of the speakers may be varied. Describe in detail what changes occur in the intensity pattern at large distances as this phase difference is varied from zero to 2π. Take both interference and diffraction effects into account.

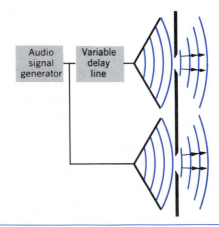

Figure 27   Problem 36.

CHAPTER 47

GRATINGS AND SPECTRA

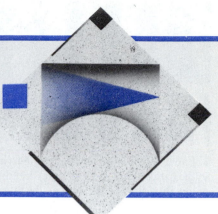

In Chapter 45 we discussed the interference pattern produced when monochromatic light is incident on a double slit: a pattern of bright and dark bands (interference fringes) is produced. Each of the slits is regarded as an elementary radiator. At first (in Chapter 45) we assumed that the slit width was much smaller than the wavelength of the light, so that light diffracted from each slit uniformly covered the observation screen. Later (in Chapter 46) we took the slit width into account and determined the "diffraction factor" that modulates the interference pattern.

In this chapter, we extend our discussion to cases in which the number of elementary radiators or diffraction centers is larger (often much larger) than two. We consider multiple arrays of slits in a plane and also three-dimensional arrays of atoms in a solid (for which we use x rays rather than visible light).

In both cases, we must distinguish carefully between the diffracting properties of a single radiator (one slit or atom) and the interference of the waves coherently diffracted from the assembly of radiators.

47-1 MULTIPLE SLITS

A logical extension of double-slit interference is to increase the number of slits from two to a larger number N. Figure 1 shows an example with five slits. Such a multiple-slit arrangement (in which N may be as large as 10^4) is called a *diffraction grating*. As in the case of the double slit, when monochromatic light falls on such a multiple-slit arrangement, a pattern of interference fringes is produced. For a given wavelength, the spacing of the fringes is determined by the slit separation d, while the relative intensities of the fringes are determined by diffraction effects associated with the slit width a.

In this chapter we analyze the interference patterns for multiple slits. We consider only the Fraunhofer region, in which there is an assumed infinite distance between the light source and the slits as well as between the slits and the screen. Equivalently, parallel light is incident on the slits, and parallel rays emerge from the slits (perhaps to be focused by a lens) to form an image on the screen.

Figure 2 shows a portion of the central maximum of the diffraction envelope for $N = 2$ and $N = 5$. We see that two important changes occur when the number of slits increases from two to five: (1) the bright fringes become

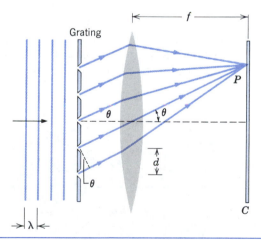

Figure 1 An idealized diffraction grating containing five slits. The slit width a is assumed to be much smaller than λ, although this condition may not be realized in practice. Also, the focal length f will in practice be much greater than d; the figure distorts these dimensions for clarity.

narrower, and (2) faint *secondary maxima* (three in Fig. 2b) appear between the bright fringes. Figure 3 shows the results of a theoretical calculation of the intensities (ne-

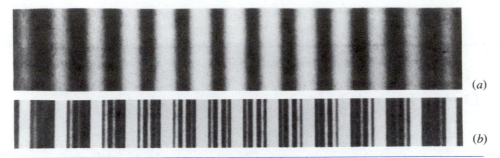

(a)

(b)

Figure 2 The diffraction pattern for a grating with (*a*) two slits and (*b*) five slits. Note that, in the case of the five-slit grating, the fringes are sharper (narrower), and secondary maxima of low intensity appear between the bright principal maxima.

glecting the effect of the diffraction envelope), in which the sharpening of the principal maxima is more apparent. As *N* increases, the number of secondary maxima increases and their brightness diminishes, until they become negligible; correspondingly, the principal maxima become sharper with increasing *N*. In the following discussion we ignore the secondary maxima and consider only the principal maxima.

A principal maximum occurs when the path difference between rays from any pair of adjacent slits, which is given by $d \sin \theta$, is equal to an integral number of wavelengths, or

$$d \sin \theta = m\lambda \qquad m = 0, \pm 1, \pm 2, \ldots, \qquad (1)$$

where *m* is called the *order number*. Equation 1 is identical with Eq. 1 of Chapter 45 for the maxima of the double slit. Note that if light passing through any pair of *adjacent* slits is in phase at a particular point on the screen, then light passing through any pair of slits, even nonadjacent ones, is also in phase at that point. For a given slit separation *d*, the locations of the principal maxima are deter-

mined by the wavelength, and so measuring their locations is a means for precise determination of wavelengths. The locations of the principal maxima are independent of the number of slits *N*, which, as we shall see, determines the width or sharpness of the principal maxima. The relative intensities of the principal maxima within the diffraction envelope are determined by the ratio a/λ, which does not affect their locations.

Width of the Maxima

The sharpening of the principal maxima as *N* is increased can be understood by a graphical argument, using phasors. Figures 4*a* and 4*b* show conditions at the central principal maximum for a two-slit and a five-slit grating. The small arrows represent the amplitudes of the wave disturbances arriving at the screen at the position of the central maximum, for which $m = 0$, and thus $\theta = 0$, in Eq. 1.

On either side of the central maximum there is a minimum of zero intensity, which lies at an angle $\delta\theta_0$ off the

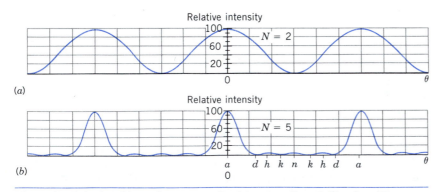

(a)

(b)

Figure 3 Calculated intensity patterns for (*a*) a two-slit and (*b*) a five-slit grating, having the same values of *d* and λ. Note the sharpening of the principal maxima and the appearance of faint secondary maxima in (*b*); compare with Fig. 2. The letters in (*b*) refer to Fig. 6. This calculation does not include diffraction effects due to the slit width; that is, we assume we are near the central region of Fig. 2 where the principal maxima have essentially equal intensities.

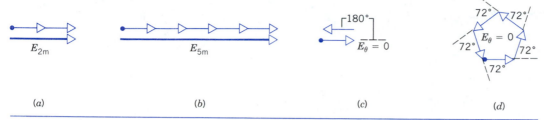

(a) (b) (c) (d)

Figure 4 (a,b) The conditions at the central maximum for a two-slit and a five-slit grating, respectively. (c,d) The corresponding conditions at a minimum of zero intensity that lies on either side of this central maximum.

central axis, as shown in Fig. 5. Figures 4c and 4d show the phasors at this point. The phase difference between waves from adjacent slits, which is zero at the central principal maximum, must increase by an amount $\Delta\phi$ chosen so that the array of phasors just closes on itself, yielding zero resultant intensity. For $N = 2$, $\Delta\phi = 2\pi/2$ (=180°); for $N = 5$, $\Delta\phi = 2\pi/5$ (=72°). In the general case it is given by

$$\Delta\phi = \frac{2\pi}{N}. \tag{2}$$

This increase in phase difference for adjacent waves corresponds to an increase in the path difference ΔL given by

$$\frac{\text{phase difference}}{2\pi} = \frac{\text{path difference}}{\lambda},$$

or

$$\Delta L = \left(\frac{\lambda}{2\pi}\right)\Delta\phi = \left(\frac{\lambda}{2\pi}\right)\left(\frac{2\pi}{N}\right) = \frac{\lambda}{N}. \tag{3}$$

From Fig. 1, however, the increase in path difference ΔL at the first minimum is also given by $d \sin \delta\theta_0$, so that we can write

$$d \sin \delta\theta_0 = \frac{\lambda}{N},$$

or

$$\sin \delta\theta_0 = \frac{\lambda}{Nd}. \tag{4}$$

Since $N \gg 1$ for actual gratings, $\sin \delta\theta_0$ is ordinarily quite small (that is, the lines are sharp), and to a good approximation we may replace $\sin \delta\theta_0$ by $\delta\theta_0$, expressed in radians, or

$$\delta\theta_0 = \frac{\lambda}{Nd}. \tag{5}$$

This equation shows specifically that if we increase N for a given λ and d, then $\delta\theta_0$ decreases, which means that the central principal maximum becomes sharper.

To obtain the result for *any* principal maximum, we consider the geometry of Fig. 5, in which the mth principal maximum occurs at an angle θ. We move away from the maximum through an angular displacement $\delta\theta$ to

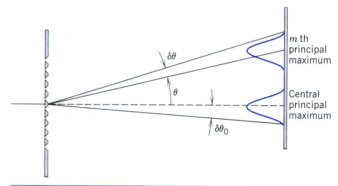

Figure 5 A principal maximum lies at the position given by the angle θ, and the first minimum occurs at the angle $\delta\theta$ from that maximum. The angle $\delta\theta$ can be taken as a measure of the width or sharpness of the maximum. The width of the central maximum is given by the angle $\delta\theta_0$.

arrive at the next minimum; we take this angle $\delta\theta$ to be a measure of the angular width of the maximum. At the maximum, the path difference between rays from adjacent slits is $m\lambda$ (see Eq. 1). At the next minimum, the path difference between rays from adjacent slits is $m\lambda + \lambda/N$, the additional path length of λ/N being given by Eq. 3. For example, consider the case of $N = 10$. The additional path length between adjacent slits at the minimum is 0.1λ. The path difference between slits 1 and 6 is therefore $5(m\lambda + 0.1\lambda) = 5m\lambda + 0.5\lambda$; the path lengths differ by a half-integral number of wavelengths, so the rays interfere destructively. The same is true for slits 2 and 7, slits 3 and 8, and so forth. If the additional path difference is λ/N, then rays from the lower $N/2$ slits undergo pairwise destructive interference with rays from the upper $N/2$ slits.

At the angle $\theta + \delta\theta$, the path difference between rays from adjacent slits is

$$d \sin (\theta + \delta\theta) = d(\sin \theta \cos \delta\theta + \cos \theta \sin \delta\theta)$$

$$\approx d \sin \theta + (d \cos \theta)\delta\theta,$$

where we assume $\delta\theta$ is small, which allows us to approximate $\cos \delta\theta \approx 1$ and $\sin \delta\theta \approx \delta\theta$. Setting this path differ-

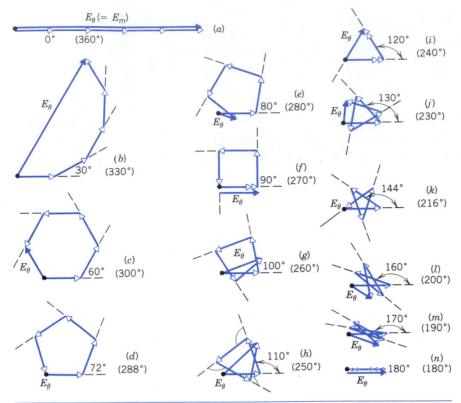

Figure 6   The figures taken in sequence from (*a*) to (*n*) and then from (*n*) to (*a*) show conditions as the intensity pattern of a five-slit grating is traversed from the central principal maximum to an adjacent principal maximum. Phase differences between waves from adjacent slits are shown directly; those in going from (*n*) to (*a*) are in parentheses. Principal maxima occur at (*a*), secondary maxima at or near (*h*) and (*n*), and minima of zero intensity at (*d*) and (*k*). Compare with Fig. 3*b*.

ence equal to $m\lambda + \lambda/N$, its value at the minimum, we obtain

$$d \sin \theta + (d \cos \theta)\delta\theta = m\lambda + \frac{\lambda}{N}$$

or, using Eq. 1,

$$(d \cos \theta)\delta\theta = \frac{\lambda}{N}.$$

Solving for $\delta\theta$ gives

$$\delta\theta = \frac{\lambda}{Nd \cos \theta}. \qquad (6)$$

This result gives the angular width* for the principal maximum that occurs at the angle θ, corresponding to the particular order *m*. Note that Eq. 6 reduces to Eq. 5 for the

---

* As defined by Eq. 6, the width is the angular interval from the peak to the first minimum. The usual definition of the width of a peak is the full interval covered by the peak at half its maximum height (see, for example, Fig. 12 of Chapter 46). These two measures of the width are roughly equal, and we take Eq. 6 to represent a measure of the width of the peak.

central maximum ($\theta = 0$). For a given N, d, and λ, the central maximum is the narrowest ($\cos \theta = 1$); the widths increase as we go to larger θ (and therefore to larger orders *m*). Equation 6 shows that $\delta\theta$ becomes smaller (the maxima become sharper) as the product Nd increases. This product (the number of slits times the distance between slits) gives the total width of the grating. Thus the peaks become sharper as the width of the grating increases.

The Secondary Maxima   *(Optional)*
The origin of the secondary maxima that appear for $N > 2$ can also be understood using the phasor method. Figure 6*a* shows conditions for the central principal maximum for a five-slit grating. The phasors are in phase. As we depart from the central maximum, θ in Fig. 1 increases from zero and the angle between adjacent phasors increases from zero to $\Delta\phi = (2\pi/\lambda)(d \sin \theta)$. Successive figures show how the resultant wave amplitude E_θ varies with $\Delta\phi$. Verify by graphical construction that a given figure represents conditions for both $\Delta\phi$ and $2\pi - \Delta\phi$. Thus we start at $\Delta\phi = 0$, proceed to $\Delta\phi = 180°$, and then trace backward through the sequence, following the phase differences shown in parentheses, until we reach $\Delta\phi = 360°$. This sequence corresponds to traversing the intensity pattern from the central principal maximum to an adjacent one. Figure 6, which should be

compared with Fig. 3*b*, shows that for *N* = 5 there are three secondary maxima, corresponding to $\Delta\phi = 110°$, $180°$, and $250°$. Make a similar analysis for *N* = 3 and show that only one secondary maximum occurs. In general, for a grating with *N* slits, there are *N* − 2 secondary maxima. In actual gratings, which commonly contain 10,000 to 50,000 "slits," the secondary maxima lie so close to the principal maxima or are so reduced in intensity that they cannot be observed experimentally. ∎

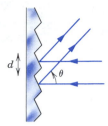

Figure 7 A cross section of a blazed grating viewed in reflected light. There is a path difference of $d \sin\theta$ between the two rays shown.

Sample Problem 1 A certain grating has 10^4 slits with a spacing of $d = 2.1\ \mu$m = 2100 nm. It is illuminated with yellow sodium light ($\lambda = 589$ nm). Find (*a*) the angular position of all principal maxima observed and (*b*) the angular width of the largest-order maximum.

Solution (*a*) From Eq. 1, we have

$$\sin\theta = \frac{m\lambda}{d} = \frac{m(589\text{ nm})}{2100\text{ nm}},$$

which gives

$$\theta = 16.3° \ (m = 1),\ 34.1° \ (m = 2),\ \text{and}\ 57.3° \ (m = 3),$$

with corresponding values at $\theta < 0$ for $m < 0$. For $m = 4$, $\sin\theta > 1$. Thus $m = 3$ is the highest order observed, which corresponds to a total of seven principal maxima (a central maximum and three on each side of center).

(*b*) For the $m = 3$ maximum, Eq. 6 gives

$$\delta\theta = \frac{\lambda}{Nd\cos\theta} = \frac{589\text{ nm}}{(10^4)(2100\text{ nm})(\cos 57.3°)}$$
$$= 5.2 \times 10^{-5}\text{ rad} = 0.0030°.$$

This is an exceedingly narrow principal maximum.

Note that Eq. 6, being a dimensionless ratio, gives its result in radian measure. This occurs because we derived Eq. 6 using the approximation $\sin\delta\theta \approx \delta\theta$, which is valid only in radian measure.

47-2 DIFFRACTION GRATINGS

A typical grating might contain *N* = 10,000 slits distributed over a width of a few centimeters, equivalent to a grating spacing *d* of a few micrometers. As we have seen in Sample Problem 1, when *Nd* is a few centimeters, the maxima are very narrow, which allows their position to be measured with great precision. Gratings are therefore used to determine wavelengths and to study the structure and intensity of the principal maxima.

Any regular periodic structure can serve as a diffraction grating, for example, the grooves of a compact disk, which produce a rainbow pattern when light is reflected from the surface of the disk. Gratings can produce their images by transmitted light, as in Fig. 1; there are also *reflection gratings*, which produce their images in reflected light. In the grating of Fig. 1, there is a periodic change in the *amplitude* (and no change in phase) of the light as a function of position across the grating. It is also possible to make gratings (of either the reflection or transmission type) that cause a periodic change in the *phase* (and a negligible change in the amplitude) of the light as a function of position across the grating. Most gratings used for visible light, whether of the reflection or transmission type, are phase gratings.

Gratings are made by ruling equally spaced parallel grooves in a thin layer of aluminum or gold deposited on a glass plate, using a diamond cutting point whose motion is automatically controlled by a ruling engine. Once such a master grating has been prepared, replicas can be formed by pouring a liquid plastic on the grating, allowing it to harden, and stripping it off. The stripped plastic, fastened to a flat piece of glass or other backing, forms a good grating.

Figure 7 shows a cross section of a common type of reflection phase grating. (If the grating were transparent, it could function as a transmission phase grating, since light passing through different thicknesses will have varying changes in phase.) The angles of the grooves are chosen so that light of a particular order is reflected in a particular direction. In this way the intensity of one particular order can be enhanced over that of other orders. Cutting gratings in this way is called *blazing*. Most gratings in use today are blazed gratings.

Figure 8 shows a simple grating spectroscope, used for viewing the spectrum of a light source, assumed to emit a number of discrete wavelengths. The light from source *S* is focused by lens L_1 on a slit S_1 placed in the focal plane of lens L_2. The parallel light emerging from collimator *C* falls on grating *G*. Parallel rays associated with a particular interference maximum occurring at angle θ fall on lens L_3, being brought to a focus in plane *FF'*. The image formed in this plane is examined, using a magnifying lens arrangement *E* (the eyepiece). The entire spectrum can be viewed by rotating telescope *T* through various angles. Instruments used for scientific research or in industry are more complex than the simple arrangement of Fig. 8. They invariably employ photographic or photoelectric

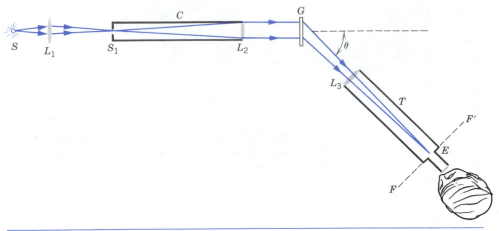

Figure 8 A simple type of grating spectroscope used to analyze the wavelengths of the light emitted by the source *S*.

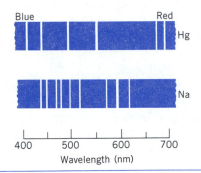

Figure 9 Examples of spectra of visible light emitted by gases of sodium (Na) and mercury (Hg).

recording and are called *spectrographs*. Figure 9 shows examples of spectra of visible light recorded by a spectrograph. Each line in the figure is in effect an image of the slit S_1 corresponding to one of the many individual wavelengths emitted from the source. For this reason, such images are called *spectral lines*; a "line" in a spectrum, no matter what technique is used to record the spectrum, is taken to mean a particular wavelength component.

In general, gratings may produce several images of spectral lines, corresponding to $m = \pm 1, \pm 2, \ldots$, in Eq. 1, and they can also separate wavelengths that are distributed continuously (as in Fig. 10) rather than as sharp spectral lines. The light from a hot, glowing object such as a lamp filament or the Sun gives a continuous spectrum. The Sun's spectrum also contains sharp spectral lines, which appear as dark lines superimposed on the continuous spectrum. These lines are caused by absorption of light by atoms of elements in the atmosphere surrounding the Sun. The element helium (from the Greek word *helios,* meaning the Sun) was discovered from an analysis of these lines.

Light can also be analyzed into its component wavelengths if the grating in Fig. 8 is replaced by a prism. In a *prism spectrograph* each wavelength in the incident beam is deflected through a definite angle θ, determined by the index of refraction of the prism material for that wavelength. Curves such as Fig. 4 of Chapter 43, which gives the index of refraction of fused quartz as a function of wavelength, show that the shorter the wavelength, the larger the angle of deflection θ. Such curves vary from substance to substance and must be found by measure-

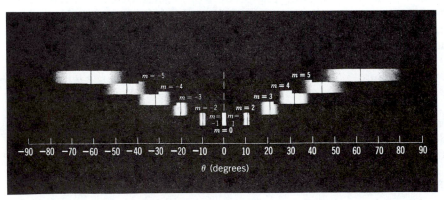

Figure 10 The spectrum of white light as viewed in a grating spectroscope such as that of Fig. 8. The different orders, identified by the index *m*, are shown separated vertically for clarity. As actually viewed, they would not be so displaced. The central line in each order corresponds to a wavelength of 550 nm. Diffraction gratings in common use today are designed to concentrate the intensity of the light in a particular order, and they do not show the ideal symmetrical patterns illustrated here.

ment. Prism instruments are not adequate for accurate *absolute* measurements of wavelength because the index of refraction of the prism material at the wavelength in question is usually not known precisely enough. Both prism and grating instruments make accurate *comparisons* of wavelength, using a suitable comparison spectrum such as that shown in Fig. 9, in which careful absolute determinations have been made of the wavelengths of the spectral lines.

Sample Problem 2   A diffraction grating has 10^4 rulings uniformly spaced over 25.0 mm. It is illuminated at normal incidence by yellow light from a sodium vapor lamp. This light contains two closely spaced lines (the well-known *sodium doublet*) of wavelengths 589.00 and 589.59 nm. (*a*) At what angle will the first-order maximum occur for the first of these wavelengths? (*b*) What is the angular separation between the first-order maxima for these lines?

Solution   (*a*) The grating spacing d is 2500 nm. The first-order maximum corresponds to $m = 1$ in Eq. 1. We thus have

$$\theta = \sin^{-1}\left(\frac{m\lambda}{d}\right) = \sin^{-1}\left(\frac{(1)(589 \text{ nm})}{2500 \text{ nm}}\right) = 13.6°.$$

(*b*) The straightforward way to find the angular separation is to repeat the calculation of part (*a*) for $\lambda = 589.59$ nm and to subtract the two angles. A difficulty, which can best be appreciated by doing the calculation, is that we must carry a large number of significant figures to obtain a meaningful value for the difference between the angles. To calculate the difference in angular positions *directly*, let us solve Eq. 1 for $\sin \theta$ and differentiate the result, treating θ and λ as variables:

$$\sin \theta = \frac{m\lambda}{d}$$

$$\cos \theta \, d\theta = \frac{m}{d} \, d\lambda.$$

If the wavelengths are close enough together, as in this case, $d\lambda$ can be replaced by $\Delta\lambda$, the actual wavelength difference; $d\theta$ then becomes $\Delta\theta$, the quantity we seek. This gives

$$\Delta\theta = \frac{m \, \Delta\lambda}{d \cos \theta} = \frac{(1)(0.59 \text{ nm})}{(2500 \text{ nm})(\cos 13.6°)}$$

$$= 2.4 \times 10^{-4} \text{ rad} = 0.014°.$$

Note that although the wavelengths involve five significant figures, our calculation, done this way, involves only two or three, with consequent reduction in numerical manipulation.

47-3 DISPERSION AND RESOLVING POWER

The ability of a grating to produce spectra that permit precise measurement of wavelengths is determined by two intrinsic properties of the grating: (1) the separation

$\Delta\theta$ between spectral lines that differ in wavelength by a small amount $\Delta\lambda$ and (2) the width or sharpness of the lines.

In Sample Problem 2, we calculated the angular separation between the closely spaced lines of the yellow sodium doublet, for which $\Delta\lambda = 0.59$ nm. We found in that case a separation of $\Delta\theta = 0.014°$ between the first-order principal maxima of these lines. The angular separation $\Delta\theta$ per unit wavelength interval $\Delta\lambda$ is called the *dispersion D* of the grating, or

$$D = \frac{\Delta\theta}{\Delta\lambda}. \qquad (7)$$

For lines of nearly equal wavelengths to appear as widely separated as possible, we would like our grating to have the largest possible dispersion.

To see what physical property of the grating determines its dispersion, we differentiate Eq. 1 ($d \sin \theta = m\lambda$), treating θ and λ as variables, which gives

$$d \cos \theta \, d\theta = m \, d\lambda,$$

or, in terms of small differences instead of differentials,

$$d \cos \theta \, \Delta\theta = m \, \Delta\lambda. \qquad (8)$$

The dispersion D is given by $\Delta\theta/\Delta\lambda$, or

$$D = \frac{m}{d \cos \theta}. \qquad (9)$$

The dispersion increases as the spacing between the slits decreases. We can also increase the dispersion by working at higher order (large m), as Fig. 10 illustrates. Note that the dispersion does not depend on the number of rulings.

Resolving Power

If a grating produces lines of large width, then the maxima of spectral lines of closely spaced wavelengths may overlap, making it difficult to determine whether such lines have one or more components and to measure the wavelengths of the lines to high precision. We therefore want to select a grating that produces the narrowest possible lines.

We obtain a reasonable measure of the ability to resolve nearby lines of different wavelengths by applying Rayleigh's criterion (see Section 46-4): if the maximum of one line falls on the first minimum of its neighbor, we should be able to resolve the lines. In Section 47-1, we defined the width of a spectral line in just that way, as the angular interval $\delta\theta$ from the maximum to the first minimum. The limit of resolution of the grating occurs when two lines in the spectrum are separated by a wavelength interval $\Delta\lambda$ such that the difference $\delta\theta$ between their angular positions is given by Eq. 6. We define the *resolving power R* of the grating as

$$R = \frac{\lambda}{\Delta\lambda}. \qquad (10)$$

If the lines are to be narrow ($\delta\theta$ is small), then the corre-

TABLE 1   PROPERTIES OF THREE GRATINGS[a]

Grating	N	d (nm)	θ	R	D (10^{-4} rad/nm)
A	5,000	10,000	2.9°	5,000	1.0
B	5,000	5,000	5.7°	5,000	2.0
C	10,000	10,000	2.9°	10,000	1.0

[a] For $\lambda = 500$ nm and $m = 1$.

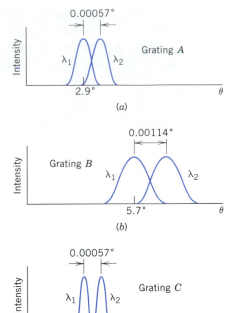

sponding wavelength interval $\Delta\lambda$ must be small, and the resolving power must be large. We should therefore choose a grating with the largest R.

To find the physical property of the grating that determines R, let us solve Eq. 8 for the spacing $\Delta\theta$ between nearby lines and (using Rayleigh's criterion) set this result equal to the width $\delta\theta$ of the line, given by Eq. 6 as the spacing between the maximum and first minimum. This gives

$$\frac{m\,\Delta\lambda}{d\cos\theta} = \frac{\lambda}{Nd\cos\theta},$$

and solving for $R\ (=\lambda/\Delta\lambda)$ gives

$$R = Nm. \qquad (11)$$

The resolving power, like the dispersion, increases with the order number. Unlike the dispersion, R depends on the number of lines N but is independent of their separation d. To maximize the resolving power, we choose a grating with the largest number of lines. For a given slit spacing d, the grating with the greatest total width has the greatest resolving power (that is, it produces the sharpest spectral lines).

Dispersion and resolving power measure different aspects of a diffraction grating's ability to produce cleanly separated lines. Consider, for example, three gratings A, B, and C whose properties are listed in Table 1. Suppose that the gratings are illuminated with light consisting of a doublet of lines at 500 nm separated by an interval $\Delta\lambda = 0.10$ nm. We have chosen the properties of grating A such that the two lines of the doublet in the first-order maximum are just at the limit of resolution; that is, the maximum of one line falls on the minimum of the other, as shown in Fig. 11a. Grating B has twice the dispersion of A but the same resolving power, and it produces the spectrum shown in Fig. 11b. In effect all angular intervals are scaled by a factor of 2, including the angular width and angular separation of the peaks. If our measurement with grating A had been limited by our ability to determine small angular intervals, changing to grating B would improve the measurement.

Grating C has twice the resolving power of A but the same dispersion. The peaks in Fig. 11c appear with the same angular separation as those in Fig. 11a, but with smaller widths. The maximum of one peak now clearly falls outside the first minimum of the other, and the two lines are more clearly distinguished from one another using grating C.

Figure 11   The intensity pattern of two lines at $\lambda = 500$ nm separated by $\Delta\lambda = 0.10$ nm, produced by the three gratings of Table 1. Grating B has the largest dispersion and grating C the largest resolving power.

The total widths of the three gratings, equal to the product Nd, are 50 mm for grating A, 25 mm for grating B, and 100 mm for grating C. Note from Fig. 11 that the peak widths depend inversely on the grating width, as suggested by Eq. 6.

Sample Problem 3   A grating has 9600 lines uniformly spaced over a width $W = 3.00$ cm and is illuminated by light from a mercury vapor discharge. (*a*) What is the expected dispersion, in the third order, in the vicinity of the intense green line ($\lambda = 546$ nm)? (*b*) What is the resolving power of this grating in the fifth order?

Solution   (*a*) The grating spacing is given by

$$d = \frac{W}{N} = \frac{3.00 \times 10^{-2}\text{ m}}{9600} = 3125\text{ nm}.$$

We must find the angle θ at which the line in question occurs. From Eq. 1, we have

$$\theta = \sin^{-1}\left(\frac{m\lambda}{d}\right) = \sin^{-1}\left(\frac{(3)(546\text{ nm})}{3125\text{ nm}}\right) = 31.6°.$$

We can now calculate the dispersion. From Eq. 9

$$D = \frac{m}{d\cos\theta} = \frac{3}{(3125\text{ nm})(\cos 31.6°)}$$

$$= 1.13 \times 10^{-3}\text{ rad/nm}$$

$$= 0.0646°/\text{nm} = 3.87\text{ arc min/nm}.$$

(*b*) From Eq. 11

$$R = Nm = (9600)(5) = 4.80 \times 10^4.$$

Thus, near $\lambda = 546$ nm and in fifth order, a wavelength difference given by (see Eq. 10)

$$\Delta\lambda = \frac{\lambda}{R} = \frac{546 \text{ nm}}{4.80 \times 10^4} = 0.011 \text{ nm}$$

can be resolved.

Sample Problem 4   A diffraction grating has 1.20×10^4 rulings uniformly spaced over a width $W = 2.50$ cm. It is illuminated at normal incidence by yellow light from a sodium vapor lamp. This light contains two closely spaced lines of wavelengths 589.00 and 589.59 nm. (*a*) At what angle does the first-order maximum occur for the first of these wavelengths? (*b*) What is the angular separation between these two lines (in first order)? (*c*) How close in wavelength can two lines be (in first order) and still be resolved by this grating? (*d*) How many rulings can a grating have and just resolve the sodium doublet lines?

Solution   (*a*) The grating spacing d is given by

$$d = \frac{W}{N} = \frac{2.50 \times 10^{-2} \text{ m}}{1.20 \times 10^4} = 2083 \text{ nm}.$$

The first-order maximum corresponds to $m = 1$ in Eq. 1. We thus have

$$\theta = \sin^{-1}\left(\frac{m\lambda}{d}\right) = \sin^{-1}\left(\frac{(1)(589.00 \text{ nm})}{2083 \text{ nm}}\right) = 16.4°.$$

(*b*) Here the *dispersion* of the grating comes into play. From Eq. 9, the dispersion is

$$D = \frac{m}{d \cos\theta} = \frac{1}{(2083 \text{ nm})(\cos 16.4°)}$$
$$= 5.01 \times 10^{-4} \text{ rad/nm}.$$

From Eq. 7, the defining equation for dispersion, we have

$$\Delta\theta = D\,\Delta\lambda$$

$$= (5.01 \times 10^{-4} \text{ rad/nm})(589.59 \text{ nm} - 589.00 \text{ nm})$$

$$= 2.95 \times 10^{-4} \text{ rad} = 0.0169° = 1.02 \text{ arc min}.$$

As long as the grating spacing d remains fixed, this result holds no matter how many lines there are in the grating.

(*c*) Here the *resolving power* of the grating comes into play. From Eq. 11, the resolving power is

$$R = Nm = (1.20 \times 10^4)(1) = 1.20 \times 10^4.$$

From Eq. 10, the defining equation for resolving power, we have

$$\Delta\lambda = \frac{\lambda}{R} = \frac{589 \text{ nm}}{1.20 \times 10^4} = 0.049 \text{ nm}.$$

This grating can easily resolve the two sodium lines, which have a wavelength separation of 0.59 nm. Note that this result depends only on the number of grating rulings and is independent of d, the spacing between adjacent rulings.

(*d*) From Eq. 10, the defining equation for R, the grating must have a resolving power of

$$R = \frac{\lambda}{\Delta\lambda} = \frac{589 \text{ nm}}{0.59 \text{ nm}} = 998.$$

From Eq. 11, the number of rulings needed to achieve this resolving power (in first order) is

$$N = \frac{R}{m} = \frac{998}{1} = 998 \text{ rulings.}$$

Since the grating has about 12 times as many rulings as this, it can easily resolve the sodium doublet lines, as we have already shown in part (*c*).

47-4 X-RAY DIFFRACTION

X rays are electromagnetic radiation with wavelengths of the order of 0.1 nm (compared with 500 nm for a typical wavelength of visible light). Figure 12 shows how x rays are produced when electrons from a heated filament F are accelerated by a potential difference V and strike a metal target.

For such small wavelengths a standard optical diffraction grating, as normally employed, cannot be used. For $\lambda = 0.10$ nm and $d = 3000$ nm, for example, Eq. 1 shows that the first-order maximum occurs at

$$\theta = \sin^{-1}\left(\frac{m\lambda}{d}\right) = \sin^{-1}\left(\frac{(1)(0.10 \text{ nm})}{3 \times 10^3 \text{ nm}}\right)$$
$$= 0.0019°.$$

This is too close to the central maximum to be practical. A grating with $d \approx \lambda$ is desirable, but, because x-ray wavelengths are about equal to atomic diameters, such gratings cannot be constructed mechanically.

In 1912 it occurred to physicist Max von Laue that a crystalline solid, consisting as it does of a regular array of atoms, might form a natural three-dimensional "diffraction grating" for x rays. Figure 13 shows that if a collimated beam of x rays, continuously distributed in wavelength, is allowed to fall on a crystal, such as sodium chloride, intense beams (corresponding to constructive interference from the many diffracting centers of which

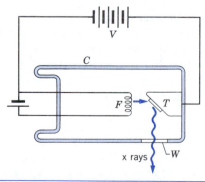

Figure 12   X rays are generated when electrons from heated filament F, accelerated through a potential difference V, strike a metal target T in the evacuated chamber C. Window W is transparent to x rays.

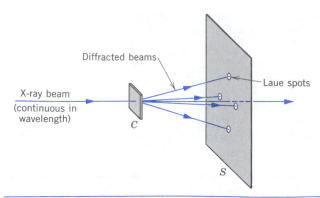

Figure 13 A beam of x rays strikes a crystal *C*. Strong diffracted beams appear in certain directions, forming a Laue pattern on the photographic film *S*.

Figure 14 A Laue x-ray diffraction pattern from a crystal of sodium chloride.

the crystal is made up) appear in certain sharply defined directions. If these beams fall on a photographic film, they form an assembly of "Laue spots." Figure 14, which shows an actual example of these spots, demonstrates that the hypothesis of Laue is indeed correct. The atomic arrangements in the crystal can be deduced from a careful study of the positions and intensities of the Laue spots in much the same way that we might deduce the structure of an optical grating (that is, the detailed profile of its slits) by a study of the positions and intensities of the lines in the interference pattern. Other experimental arrangements have supplanted the Laue technique to a considerable extent today, but the principle remains unchanged (see Question 25).

Figure 15 shows how sodium and chlorine atoms (strictly, Na⁺ and Cl⁻ ions) are stacked to form a crystal of sodium chloride. This pattern, which has *cubic* symmetry, is one of the many possible atomic arrangements exhibited by solids. The model represents the *unit cell* for

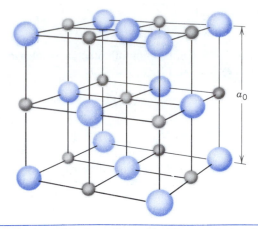

Figure 15 A model of a sodium chloride crystal, showing how the sodium ions Na⁺ (small spheres) and chloride ions Cl⁻ (large spheres) are stacked in the unit cell, whose edge a_0 has the length 0.563 nm.

sodium chloride. This is the smallest unit from which the crystal may be built up by repetition in three dimensions. You should verify that no smaller assembly of atoms possesses this property. For sodium chloride the length a_0 of the cube edge of the unit cell is 0.563 nm.

Each unit cell in sodium chloride has four sodium ions and four chlorine ions associated with it. In Fig. 15 the sodium ion in the center belongs entirely to the cell shown. Each of the other twelve sodium ions shown is shared with three adjacent unit cells so that each contributes one-fourth of an ion to the cell under consideration. The total number of sodium ions is then $1 + \frac{1}{4}(12) = 4$. By similar reasoning you can show that although there are fourteen chlorine ions in Fig. 15, only four are associated with the unit cell shown.

The unit cell is the fundamental repetitive diffracting unit in the crystal, corresponding to the slit (and its adjacent opaque strip) in the optical diffraction grating of Fig. 1. Figure 16a shows a particular plane in a sodium chloride crystal. If each unit cell intersected by this plane is represented by a small cube, Fig. 16b results. You may imagine each of these figures extended indefinitely in three dimensions.

Let us treat each small cube in Fig. 16b as an elementary diffracting center, corresponding to a slit in an optical grating. The *directions* (but not the intensities) of all the diffracted x-ray beams that can emerge from a sodium chloride crystal (for a given x-ray wavelength and a given orientation of the incident beam) are determined by the geometry of this three-dimensional lattice of diffracting centers. In exactly the same way the *directions* (but not the intensities) of all the diffracted beams that can emerge from a particular optical grating (for a given wavelength and orientation of the incident beam) are determined only by the geometry of the grating, that is, by the grating spacing *d*. Representing the unit cell by what is essentially

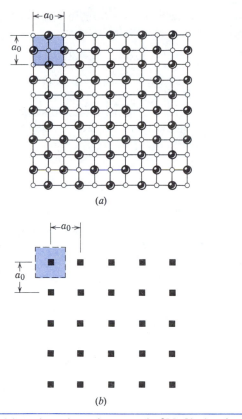

(a)

(b)

Figure 16 (a) A plane through a crystal of NaCl, showing the Na and Cl ions. (b) The corresponding unit cells in this section. Each cell is represented by a small black square.

a point, as in Fig. 16b, corresponds to representing the slits in a diffraction grating by lines, as we did in discussing the double-slit experiment in Section 45-1.

The *intensities* of the lines from an optical diffraction grating depend on the diffracting characteristics of a single slit, determined in particular by the slit width a; see, for example, Fig. 2 for a set of slits. The characteristics of actual optical gratings are determined by the profile of the grating rulings.

In exactly the same way the *intensities* of the diffracted beams emerging from a crystal depend on the diffracting characteristics of the unit cell. Fundamentally, the x rays are diffracted by electrons, diffraction by nuclei being negligible in most cases. Thus the diffracting characteristics of a unit cell depend on how the electrons are distributed throughout the volume of the cell. By studying the *directions* of diffracted x-ray beams, we can learn the basic symmetry of the crystal. By studying the *intensities* we can learn how the electrons are distributed in a unit cell. Figure 17 shows an example of this technique.

Bragg's Law

Bragg's law predicts the conditions under which diffracted x-ray beams from a crystal are possible. In deriving it, we ignore the structure of the unit cell, which is related only to the intensities of these beams. The dashed sloping lines in Fig. 18a represent the intersection with the

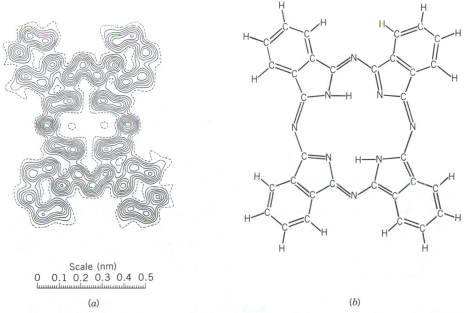

Scale (nm)
0   0.1 0.2 0.3 0.4 0.5

(a)   (b)

Figure 17 (a) Electron density contours for phthalocyanine ($C_{32}H_{18}N_8$) determined from the intensity distribution of scattered x rays. The dashed curves represent a density of one electron per 0.01 nm², and each adjacent curve represents an increase of one electron per 0.01 nm². (b) A structural representation of the molecule. Note that the greatest electron density occurs in (a) near the N atoms, which have the largest number of electrons (7). Note also that the H atoms, which contain only a single electron, are not prominent in (a).

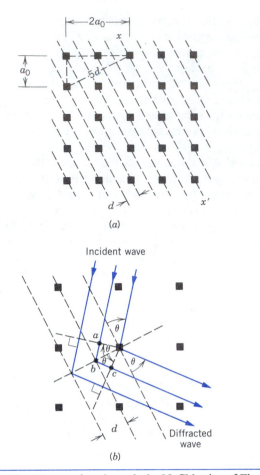

(a)

(b)

Figure 18 (*a*) A section through the NaCl lattice of Fig. 16. The dashed lines represent an arbitrary set of parallel planes connecting unit cells. The interplanar spacing is *d*. (*b*) An incident beam falls on a set of planes. A strong diffracted beam will be observed if Bragg's law is satisfied.

rays from adjacent planes (*abc* in Fig. 18*b*) must be an integral number of wavelengths or

$$2d \sin \theta = m\lambda \qquad m = 1, 2, 3, \ldots . \qquad (12)$$

This relation is called *Bragg's law* after W. L. Bragg who first derived it. The quantity *d* in this equation (the interplanar spacing) is the perpendicular distance between the planes. For the planes of Fig. 18*a* we see that *d* is related to the unit cell dimension a_0 by

$$d = \frac{a_0}{\sqrt{5}} . \qquad (13)$$

If an incident monochromatic x-ray beam falls at an arbitrary angle θ on a particular set of atomic planes, a diffracted beam will *not* result because Eq. 12 will not, in general, be satisfied. If the incident x rays are *continuous* in wavelength, diffracted beams will result when wavelengths given by

$$\lambda = \frac{2d \sin \theta}{m} \qquad m = 1, 2, 3, \ldots$$

are present in the incident beam (see Eq. 12).

X-ray diffraction is a powerful tool for studying both x-ray spectra and the arrangement of atoms in crystals. To study the spectrum of an x-ray source, a particular set of crystal planes, having a known spacing *d*, is chosen. Diffraction from these planes locates different wavelengths at different angles. A detector that can discriminate one angle from another can be used to determine the wavelength of radiation reaching it. On the other hand, we can study the crystal itself, using a monochromatic x-ray beam to determine not only the spacings of various crystal planes but also the structure of the unit cell. The DNA molecule and many other equally complex structures have been mapped by x-ray diffraction methods.

plane of the figure of an arbitrary set of planes passing through the elementary diffracting centers. The perpendicular distance between adjacent planes is *d*. Many other such families of planes, with different *interplanar spacings*, can be defined.

Figure 18*b* shows an incident wave striking the *family* of planes, the incident rays making an angle θ with the plane.* For a single plane, mirror-like "reflection" occurs for *any* value of θ. To have a constructive interference in the beam diffracted from the entire family of planes in the direction θ, the rays from the separate planes must reinforce each other. This means that the path difference for

---

* In x-ray diffraction it is customary to specify the direction of a wave by giving the angle between the ray and the plane (the *glancing angle*) rather than the angle between the ray and the normal.

Sample Problem 5 At what angles must an x-ray beam with $\lambda = 0.110$ nm fall on the family of planes represented in Fig. 18*b* if a diffracted beam is to exist? Assume the material to be sodium chloride ($a_0 = 0.563$ nm).

Solution The interplanar spacing *d* for these planes is given by Eq. 13, or

$$d = \frac{a_0}{\sqrt{5}} = \frac{0.563 \text{ nm}}{\sqrt{5}} = 0.252 \text{ nm}.$$

Equation 12 gives

$$\theta = \sin^{-1} \left(\frac{m\lambda}{2d} \right) = \sin^{-1} \left(\frac{(m)(0.110 \text{ nm})}{(2)(0.252 \text{ nm})} \right).$$

Diffracted beams are possible for $\theta = 12.6°$ ($m = 1$), $\theta = 25.9°$ ($m = 2$), $\theta = 40.9°$ ($m = 3$), and $\theta = 60.9°$ ($m = 4$). Higher order beams cannot exist because they require that $\sin \theta > 1$.

Actually, the unit cell in cubic crystals such as NaCl has symmetry properties that require the intensity of diffracted x-ray beams corresponding to odd values of m to be zero. (See Problem 42.) Thus the only beams that are expected are $\theta = 25.9°$ ($m = 2$) and $\theta = 60.9°$ ($m = 4$).

47-5   HOLOGRAPHY   *(Optional)*

The light emitted by an object contains the complete information on the size and shape of the object. We can consider that information to be stored in the wavefronts of the light from the object, specifically in the variation of intensity and phase of the electromagnetic fields. If we could record this information, we could reproduce a complete three-dimensional image of the object. However, photographic films record only the intensity variations; the films are not sensitive to phase variations. It is there-

fore not possible to use a photographic negative to reconstruct a three-dimensional image.

One exception to this restriction occurs in the case of x-ray diffraction from a crystal. Because of the regular spacing of the atoms of a crystal, we can easily deduce the relative phases of the diffracted waves reaching the film from different atoms. This possibility was realized by W. L. Bragg, who illuminated a photographic negative of an x-ray diffraction pattern and so reconstructed the image of a crystal. In this "double diffraction" method, diffraction of radiation from a diffraction pattern gives an image of the original object. For objects whose atoms are not arranged in such a periodic array, this simple method of image reconstruction does not work.

A scheme for recording the intensity *and* phase of the waves from objects was developed in 1948 by Dennis Gabor, who was awarded the 1971 Nobel Prize in physics for this discovery. This type of image formation is called *holography*, from the Greek words meaning "entire picture," and the image is called a *hologram*. The procedure is illustrated in Fig. 19. A wave diffracted from an object interferes on the photographic film with a reference wave. The interference between the two waves serves as the means for storing on the film information on the phase of the wave from the object. When the photographic image is viewed using light identical with the reference beam, a three-dimensional virtual image of the original object is reconstructed (Fig. 20). A second image (a real image), not shown in Fig. 20, is also produced by the hologram.

Because the film is illuminated uniformly by the diffracted light from the object and the reference beam, every piece of the film contains the information necessary to reproduce the three-dimensional image. The hologram itself (Fig. 21) shows only the interference fringes; in general, it is necessary to use a suitable monochromatic and coherent beam to reconstruct the image. For this reason, active development of holography did not occur until the early 1960s, when lasers became commonly available.

Some holograms can be viewed in white light. White-light holograms use a thick photographic emulsion, in which light is reflected by successive layers of grains in the film. Constructive interference occurs in the reflected light for the wavelength of the original reference beam, and destructive interference occurs for

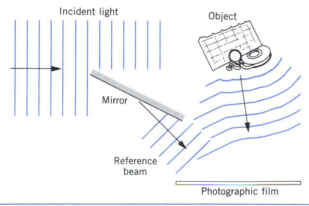

Figure 19 Apparatus for producing holograms. A portion of the beam from a source of coherent light (a laser, for instance) illuminates an object. The light diffracted by the object interferes on the film with a portion of the original beam, which serves as the reference.

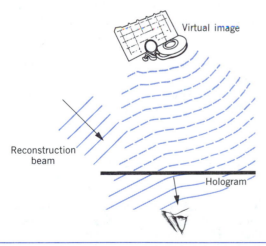

Figure 20 To view a hologram, it is illuminated with light identical to the reference beam. A three-dimensional virtual image can be seen, at the location of the original object.

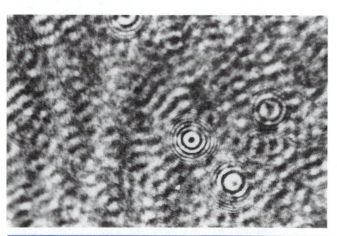

Figure 21 A close-up view of a hologram, showing the interference pattern.

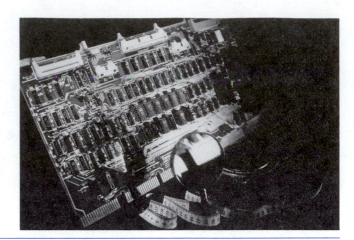

Figure 22 Two different views of the same hologram, taken from different directions. Note the relative movement of the objects in the images.

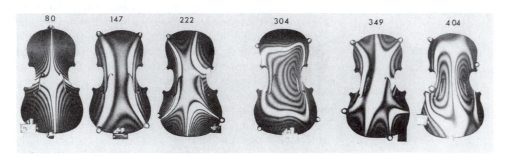

Figure 23 A holographic interference pattern of a top violin plate vibrating at different frequencies. The frequencies (in Hz) are shown above the plates.

other wavelengths. By using reference beams of several different colors, a full-color image can be produced.*

The hologram reconstructs a true three-dimensional image; for example, nearby objects appear "in front of" more distant objects, and by moving your head from side to side you can change the relative spatial orientation of the objects. Figure 22 shows two different views of the same hologram, illustrating the *parallax* effect of viewing the hologram from two different directions.

Holograms have a variety of applications in basic and applied

---

* See "White-Light Holograms," by Emmett N. Leith, *Scientific American,* October 1976, p. 80.

science. For example, in producing holograms the object must be kept absolutely still while the film is exposed; any small movement would change the relative phase between the diffracted and reference beams and thereby change the interference pattern stored on the film. If a hologram is made by superimposing on the film two successive exposures of a vibrating object, such as the top or bottom plate of a violin, locations on the object that moved between the exposures by an integral number of wavelengths will show constructive interference, while parts of the object that moved by a half-integral number of wavelengths ($\lambda/2$, $3\lambda/2$, . . .) will show destructive interference. Figure 23 shows an example of the use of this technique, called *holographic interferometry.* ∎

QUESTIONS

1. Discuss this statement: "A diffraction grating can just as well be called an interference grating."

2. How would the spectrum of an enclosed source that is formed by a diffraction grating on a screen change (if at all) when the source, grating, and screen are all submerged in water?

3. (*a*) For what kind of waves could a long picket fence be considered a useful grating? (*b*) Can you make a diffraction grating out of parallel rows of fine wire strung closely together?

4. Could you construct a diffraction grating for sound? If so, what grating spacing is suitable for a wavelength of 0.5 m?

5. A crossed diffraction grating is ruled in two directions, at right angles to each other. Predict the pattern of light inten-

sity on the screen if light is sent through such a grating. Is there any practical value to such a grating?

6. Suppose that, instead of a slit, a small circular aperture were placed in the focal plane of the collimating lens in the telescope of a spectrometer. What would be seen when the telescope is illuminated by sodium light? Why then do we usually call spectra *line* spectra?

7. In a grating spectrograph, several lines having different wavelengths and formed in different orders might appear near a certain angle. How could you distinguish between their orders?

8. You are given a photograph of a spectrum on which the angular positions and the wavelengths of the spectrum lines are marked. (*a*) How can you tell whether the spectrum was taken with a prism or a grating instrument? (*b*) What information could you gather about either the prism or the grating from studying such a spectrum?

9. A glass prism can form a spectrum. Explain how. How many "orders" of spectra will a prism produce?

10. For the simple spectroscope of Fig. 8 show (*a*) that θ increases with λ for a grating and (*b*) that θ decreases with λ for a prism.

11. According to Eq. 6 the principal maxima become wider (that is, $\delta\theta$ increases) the higher the order m (that is, the larger θ becomes). According to Eq. 11 the resolving power becomes greater the higher the order m. Explain this apparent paradox.

12. Explain in your own words why increasing the number of slits N in a diffraction grating sharpens the maxima. Why does decreasing the wavelength do so? Why does increasing the grating spacing d do so?

13. How much information can you discover about the structure of a diffraction grating by analyzing the spectrum it forms of a monochromatic light source? Let $\lambda = 589$ nm, for example.

14. Assume that the limits of the visible spectrum are 430 and 680 nm. How would you design a grating, assuming that the incident light falls normally on it, such that the first-order spectrum barely overlaps the second-order spectrum?

15. (*a*) Why does a diffraction grating have closely spaced rulings? (*b*) Why does it have a large number of rulings?

16. Two light beams of nearly equal wavelengths are incident on a grating of N rulings and are not quite resolvable. However, they become resolved if the number of rulings is increased. Formulas aside, is the explanation of this that: (*a*) more light can get through the grating? (*b*) the principal maxima be-

come more intense and hence resolvable? (*c*) the diffraction pattern is spread more and hence the wavelengths become resolved? (*d*) there is a large number of orders? or (*e*) the principal maxima become narrower and hence resolvable?

17. The relation $R = Nm$ suggests that the resolving power of a given grating can be made as large as desired by choosing an arbitrarily high order of diffraction. Discuss this possibility.

18. Show that at a given wavelength and a given angle of diffraction the resolving power of a grating depends only on its width $W (= Nd)$.

19. How would you experimentally measure (*a*) the dispersion D and (*b*) the resolving power R of a grating spectrograph?

20. For a given family of planes in a crystal, can the wavelength of incident x rays be (*a*) too large or (*b*) too small to form a diffracted beam?

21. If a parallel beam of x rays of wavelength λ is allowed to fall on a randomly oriented crystal of any material, generally no intense diffracted beams will occur. Such beams appear if (*a*) the x-ray beam consists of a continuous distribution of wavelengths rather than a single wavelength or (*b*) the specimen is not a single crystal but a finely divided powder. Explain each case.

22. Does an x-ray beam undergo refraction as it enters and leaves a crystal? Explain your answer.

23. Why cannot a simple cube of edge $a_0/2$ in Fig. 15 be used as a unit cell for sodium chloride?

24. In some respects Bragg reflection *differs* from plane grating diffraction. Of the following statements, which one is true for Bragg reflection but not true for grating diffraction? (*a*) Two different wavelengths may be superposed. (*b*) Radiation of a given wavelength may be sent in more than one direction. (*c*) Long waves are deviated more than short waves. (*d*) There is only one grating spacing. (*e*) Diffraction maxima of a given wavelength occur only for particular angles of incidence.

25. In Fig. 24*a* we show schematically the Debye–Scherrer experimental arrangement and in Fig. 24*b* a corresponding x-ray diffraction pattern. (*a*) Keeping in mind that the Laue method uses a large single crystal and an x-ray beam continuously distributed in wavelength, explain the origin of the spots in Fig. 14. (*Hint:* Each spot corresponds to the direction of scattering from a family of planes.) (*b*) Keeping in mind that the Debye–Scherrer method uses a large number of small single crystals randomly oriented and a monochromatic beam of x rays, explain the origin of the rings. (*Hint:* Because the small crystals are randomly oriented, all possible angles of incidence are obtained.)

Figure 24   Question 25.

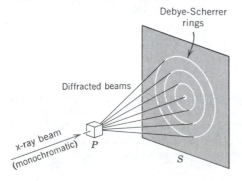

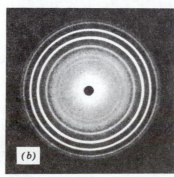

PROBLEMS

Section 47-1 Multiple Slits

1. A diffraction grating 21.5 mm wide has 6140 rulings. (a) Calculate the distance d between adjacent rulings. (b) At what angles will maximum-intensity beams occur if the incident radiation has a wavelength of 589 nm?

2. A diffraction grating 2.86 cm wide produces a deviation of $33.2°$ in the second order with light of wavelength 612 nm. Find the total number of rulings on the grating.

3. With light from a gaseous discharge tube incident normally on a grating with a distance 1.73 μm between adjacent slit centers, a green line appears with sharp maxima at measured transmission angles $\theta = \pm17.6°$, $37.3°$, $-37.1°$, $65.2°$, and $-65.0°$. Compute the wavelength of the green line that best fits the data.

4. A narrow beam of monochromatic light strikes a grating at normal incidence and produces sharp maxima at the following angles from the normal: $6° 40'$, $13° 30'$, $20° 20'$, $35° 40'$. No other maxima appear at any angle between $0°$ and $35° 40'$. The separation between adjacent ruling centers in the grating is 5040 nm. Find the wavelength of light used.

5. Light of wavelength 600 nm is incident normally on a diffraction grating. Two adjacent principal maxima occur at $\sin \theta = 0.20$ and $\sin \theta = 0.30$. The fourth order is missing. (a) What is the separation between adjacent slits? (b) What is the smallest possible individual slit width? (c) Name all orders actually appearing on the screen with the values derived in (a) and (b).

6. A diffraction grating is made up of slits of width 310 nm with a 930-nm separation between centers. The grating is illuminated by monochromatic plane waves, $\lambda = 615$ nm, the angle of incidence being zero. (a) How many diffraction maxima are there? (b) Find the width of the spectral lines observed in first order if the grating has 1120 slits.

7. Derive this expression for the intensity pattern for a three-slit "grating":

$$I = \tfrac{1}{9}I_m(1 + 4 \cos \phi + 4 \cos^2 \phi),$$

where

$$\phi = \frac{2\pi d \sin \theta}{\lambda}.$$

Assume that $a \ll \lambda$ and be guided by the derivation of the corresponding double-slit formula (Eq. 17 of Chapter 46).

8. (a) Using the result of Problem 7, show that the halfwidth of the fringes for a three-slit diffraction pattern, assuming θ small enough so that $\sin \theta \approx \theta$, is

$$\Delta\theta \approx \frac{\lambda}{3.2d}.$$

(b) Compare this with the expression derived for the two-slit pattern in Problem 25, Chapter 45 and show that these results support the conclusion that for a fixed slit spacing the interference maxima become sharper as the number of slits is increased.

9. (a) Using the result of Problem 7, show that a three-slit "grating" has only one secondary maximum. Find (b) its location and (c) its relative intensity.

10. A three-slit grating has separation d between adjacent slits. If the middle slit is covered up, will the halfwidth of the intensity maxima become broader or narrower and by what factor? See Problem 8.

11. A diffraction grating has a large number N of slits, each of width a. Let I_{max} denote the intensity at some principal maximum, and let I_k denote the intensity of the kth adjacent secondary maximum. (a) If $k \ll N$, show from the phasor diagram that, approximately, $I_k/I_{max} = 1/(k + \tfrac{1}{2})^2\pi^2$. (Compare this with the single-slit formula.) (b) For those secondary maxima that lie roughly midway between two adjacent principal maxima, show that roughly $I_k/I_{max} = 1/N^2$. (c) Consider the central principal maximum and those adjacent secondary maxima for which $k \ll N$. Show that this part of the diffraction pattern quantitatively resembles that for one single slit of width Na.

Section 47-2 Diffraction Gratings

12. A diffraction grating has 200 rulings/mm and a principal maximum is noted at $\theta = 28°$. (a) What are the possible wavelengths of the incident visible light? (b) What colors are they?

13. A grating has 315 rulings/mm. For what wavelengths in the visible spectrum can fifth-order diffraction be observed?

14. Show that in a grating with alternately transparent and opaque strips of equal width, all the even orders (except $m = 0$) are absent.

15. Given a grating with 400 rulings/mm, how many orders of the entire visible spectrum (400–700 nm) can be produced?

16. Assume that light is incident on a grating at an angle ψ as shown in Fig. 25. Show that the condition for a diffraction maximum is

$$d(\sin \psi + \sin \theta) = m\lambda \qquad m = 0, 1, 2, \ldots .$$

Only the special case $\psi = 0$ has been treated in this chapter (compare with Eq. 1).

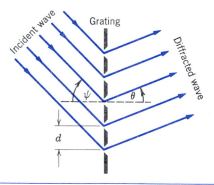

Figure 25 Problem 16.

17. A transmission grating with $d = 1.50$ μm is illuminated at various angles of incidence by light of wavelength 600 nm. Plot as a function of angle of incidence (0 to 90°) the angular deviation of the first-order diffracted beam from the incident direction. See Problem 16.

18. Assume that the limits of the visible spectrum are arbitrarily chosen as 430 and 680 nm. Calculate the number of rulings per mm of a grating that will spread the first-order spectrum through an angular range of 20.0°.

19. White light (400 nm < λ < 700 nm) is incident on a grating. Show that, no matter what the value of the grating spacing d, the second- and third-order spectra overlap.

20. A grating has 350 rulings/mm and is illuminated at normal incidence by white light. A spectrum is formed on a screen 30 cm from the grating. If a 10-mm square hole is cut in the screen, its inner edge being 50 mm from the central maximum and parallel to it, what range of wavelengths passes through the hole?

Section 47-3 Dispersion and Resolving Power

21. The "sodium doublet" in the spectrum of sodium is a pair of lines with wavelengths 589.0 and 589.6 nm. Calculate the minimum number of rulings in a grating needed to resolve this doublet in the second-order spectrum.

22. A grating has 620 rulings/mm and is 5.05 mm wide. (*a*) What is the smallest wavelength interval that can be resolved in the third order at $\lambda = 481$ nm? (*b*) How many higher orders can be seen?

23. A source containing a mixture of hydrogen and deuterium atoms emits light containing two closely spaced red colors at $\lambda = 656.3$ nm whose separation is 0.180 nm. Find the minimum number of rulings needed in a diffraction grating that can resolve these lines in the first order.

24. (*a*) How many rulings must a 4.15-cm-wide diffraction grating have to resolve the wavelengths 415.496 nm and 415.487 nm in the second order? (*b*) At what angle are the maxima found?

25. In a particular grating the sodium doublet (see Problem 21) is viewed in third order at 10.2° to the normal and is barely resolved. Find (*a*) the ruling spacing and (*b*) the total width of the grating.

26. Show that the dispersion of a grating can be written

$$D = \frac{\tan \theta}{\lambda}.$$

27. A grating has 40,000 rulings spread over 76 mm. (*a*) What is its expected dispersion D in °/nm for sodium light ($\lambda = 589$ nm) in the first three orders? (*b*) What is its resolving power in these orders?

28. Light containing a mixture of two wavelengths, 500 nm and 600 nm, is incident normally on a diffraction grating. It is desired (1) that the first and second principal maxima for each wavelength appear at $\theta \leq 30°$, (2) that the dispersion be as high as possible, and (3) that the third order for 600 nm be a missing order. (*a*) What should be the separation between adjacent slits? (*b*) What is the smallest possible individual slit width? (*c*) Name all orders for 600 nm that actually appear on the screen with the values derived in (*a*) and (*b*).

29. A diffraction grating has a resolving power $R = \lambda/\Delta\lambda = Nm$. (*a*) Show that the corresponding frequency range $\Delta\nu$ that can just be resolved is given by $\Delta\nu = c/Nm\lambda$.

(*b*) From Fig. 1, show that the "times of flight" of the two extreme rays differ by an amount $\Delta t = (Nd/c) \sin \theta$. (*c*) Show that $(\Delta\nu)(\Delta t) = 1$, this relation being independent of the various grating parameters. Assume $N \gg 1$.

Section 47-4 X-Ray Diffraction

30. X rays of wavelength 0.122 nm are found to reflect in the second order from the face of a lithium fluoride crystal at a Bragg angle of 28.1°. Calculate the distance between adjacent crystal planes.

31. A beam of x rays of wavelength 29.3 pm is incident on a calcite crystal of lattice spacing 0.313 nm. Find the smallest angle between the crystal planes and the beam that will result in constructive reflection of the x rays.

32. Monochromatic high-energy x rays are incident on a crystal. If first-order reflection is observed at Bragg angle 3.40°, at what angle would second-order reflection be expected?

33. An x-ray beam, containing radiation of two distinct wavelengths, is scattered from a crystal, yielding the intensity spectrum shown in Fig. 26. The interplanar spacing of the scattering planes is 0.94 nm. Determine the wavelengths of the x rays present in the beam.

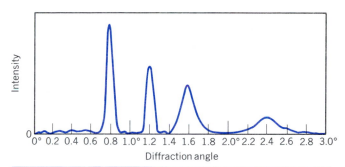

Figure 26   Problem 33.

34. In comparing the wavelengths of two monochromatic x-ray lines, it is noted that line A gives a first-order reflection maximum at a glancing angle of 23.2° to the face of a crystal. Line B, known to have a wavelength of 96.7 pm, gives a third-order reflection maximum at an angle of 58.0° from the same face of the same crystal. (*a*) Calculate the interplanar spacing. (*b*) Find the wavelength of line A.

35. Monochromatic x rays are incident on a set of NaCl crystal planes whose interplanar spacing is 39.8 pm. When the beam is rotated 51.3° from the normal, first-order Bragg reflection is observed. Find the wavelength of the x rays.

36. Show that, in Bragg diffraction by a monochromatic beam of x rays, no intense maxima will be obtained if the wavelength of the x rays is greater than twice the largest crystal plane separation. See Question 20.

37. Prove that it is not possible to determine both wavelength of radiation and spacing of Bragg reflecting planes in a crystal by measuring the angles for Bragg reflection in several orders.

38. Assume that the incident x-ray beam in Fig. 27 is not monochromatic but contains wavelengths in a band from 95.0 to 139 pm. Will diffracted beams, associated with the planes

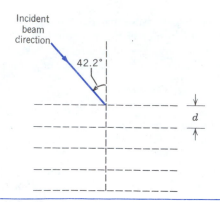

Figure 27   Problems 38 and 40.

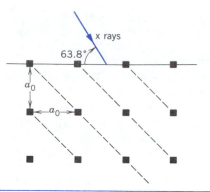

Figure 28   Problem 39.

shown, occur? If so, what wavelengths are diffracted? Assume $d = 275$ pm.

39. First-order Bragg scattering from a certain crystal occurs at an angle of incidence of 63.8°; see Fig. 28. The wavelength of the x rays is 0.261 nm. Assuming that the scattering is from the dashed planes shown, find the unit cell size a_0.

40. Monochromatic x rays ($\lambda = 0.125$ nm) fall on a crystal of sodium chloride, making an angle of 42.2° with the reference line shown in Fig. 27. The planes shown are those of Fig. 18a, for which $d = 0.252$ nm. Through what angles must the crystal be turned to give a diffracted beam associated with the planes shown? Assume that the crystal is turned about an axis that is perpendicular to the plane of the page.

41. Consider an infinite two-dimensional square lattice as in Fig. 16b. One interplanar spacing is obviously a_0 itself. (a) Calculate the next five smaller interplanar spacings by sketching figures similar to Fig. 18a. (b) Show that the general formula is

$$d = a_0/\sqrt{h^2 + k^2},$$

where h and k are both relatively prime integers that have no common factors other than unity.

42. In Sample Problem 5 the $m = 1$ beam, permitted by interference considerations, has zero intensity because of the diffracting properties of the unit cell for this geometry of beams and crystal. Prove this. (*Hint*: Show that the "reflection" from an atomic plane through the top of a layer of unit cells is canceled by a "reflection" from a plane through the middle of this layer of cells. All odd-order beams prove to have zero intensity.)

CHAPTER 48

POLARIZATION

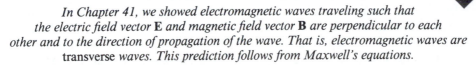

*In Chapter 41, we showed electromagnetic waves traveling such that the electric field vector **E** and magnetic field vector **B** are perpendicular to each other and to the direction of propagation of the wave. That is, electromagnetic waves are transverse waves. This prediction follows from Maxwell's equations.*

In many of the experiments we have described so far, light waves do not reveal their transverse nature. For example, reflection, refraction, interference, and diffraction can occur for longitudinal waves (such as sound) as well as for transverse waves. Thomas Young (whom we also remember for the double-slit experiment) in 1817 provided the experimental basis for believing that light waves are transverse, building on experiments by his contemporaries Arago and Fresnel on the phenomenon we now call double refraction (see Section 48-4).

*In this chapter, we consider the polarization of light and other electromagnetic waves. The direction of polarization refers to the direction of the **E** vector of the wave. We discuss different types of polarization, including linear and circular, and we consider the experimental techniques for producing and detecting polarized light.*

48-1 POLARIZATION

Consider the experimental arrangement shown in Fig. 1. A microwave transmitter on the left is connected to a dipole antenna. Charges surging up and down in the antenna produce an electromagnetic wave whose **E** vector is (at large distances from the dipole) parallel to its axis. When this wave is incident on the antenna of the microwave receiver at the right, the **E** vector of the wave causes charges to move up and down in the antenna. These moving charges produce a signal in the receiver.

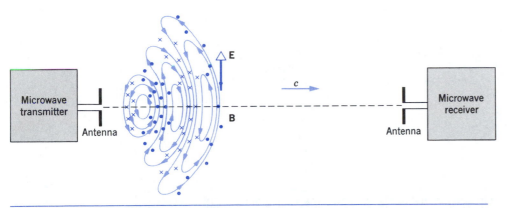

Figure 1   The electromagnetic wave generated by the transmitter is polarized in the plane of the page, its **E** vector being parallel to the axis of the transmitting antenna. The receiving antenna can detect this wave with maximum effectiveness if its antenna also lies in the plane and parallel to **E**. If the receiving antenna were rotated through 90° about the direction of propagation, no signal would be detected.

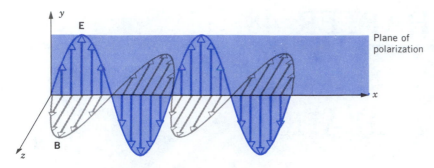

Figure 2   An instantaneous snapshot of a traveling electromagnetic wave showing the **E** and **B** vectors. The wave is linearly polarized, in this case in the *y* direction. The *plane of polarization* is defined to be the plane containing the **E** vector and the direction of propagation; in this case, the plane of polarization is the *xy* plane.

If the transmitter were rotated by 90° about the direction of propagation of the wave, the signal in the receiver would drop to zero. In this case, the **E** vector of the wave would be at right angles to the axis of the receiving antenna; the wave would produce no movement of charge along the antenna and thus no signal in the receiver. A similar result would be obtained if the receiver were rotated instead of the transmitter.

Figure 2 represents an electromagnetic wave such as that of Fig. 1. As is always the case, the **E** and **B** vectors are perpendicular to one another and to the direction of propagation of the wave, which is the basic picture of a transverse wave. By convention, we define the *direction of polarization* of the wave to be the direction of the **E** vector (the *y* direction in Fig. 2). The plane determined by the **E** vector and the direction of propagation of the wave (the *xy* plane in Fig. 2) is called the *plane of polarization* of the wave. Note that specifying two directions of an electromagnetic wave (the direction of propagation and the direction of **E**) completely specifies the wave, because the direction of **B** is fixed by these two directions.*

_____

* Recall the Poynting vector, $\mathbf{S} = (\mathbf{E} \times \mathbf{B})/\mu_0$, discussed in Section 41-4, where **S** is in the direction of propagation of the wave. Given **S** and **E**, we can find the magnitude and direction of **B**.

The wave illustrated in Fig. 2 is said to be *linearly polarized* (also called plane polarized). This means that the **E** field remains in a fixed direction (the *y* direction in Fig. 2) as the wave propagates.

As in the experiment shown in Fig. 1, linearly polarized electromagnetic waves in the microwave or radio regions can be produced by orienting the axis of a dipole transmitting antenna in a certain direction. For example, waves used to transmit television signals in the United States are polarized in a horizontal plane; for that reason, TV receiving antennas are mounted on the roofs of houses in a horizontal plane. (In England, TV signals are transmitted with a vertical plane of polarization, and so antennas are mounted in a vertical plane.)

The motions of the electrons in the microwave antenna of Fig. 1 are *coherent*; they act in unison to transmit a polarized electromagnetic wave (see Fig. 3a). In ordinary sources of light, such as an incandescent bulb or the Sun, the atoms behave independently and emit waves whose planes of polarization are randomly oriented about the direction of propagation (Fig. 3b). This light is transverse but *unpolarized*; that is, there is no preferred plane of polarization. The symmetry about the direction of propagation conceals the true transverse nature of the waves. Laser light, on the other hand, is coherent and polarized.

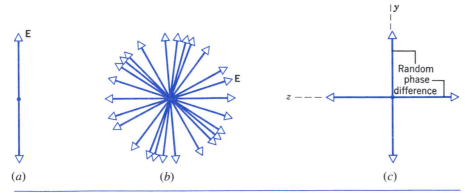

Figure 3   (a) A linearly polarized wave, such as that of Fig. 2, viewed from along the direction of propagation. The wave is moving out of the plane of the page. Only the direction of the **E** vector is shown. (b) An unpolarized wave, which can be considered to be a random superposition of many polarized waves. (c) An equivalent way of showing the unpolarized wave, as two waves linearly polarized at right angles to one another and with a random phase difference between them. The orientation of the *y* and *z* axes about the direction of propagation is completely arbitrary.

Figure 3c shows an alternative and useful way to represent an unpolarized wave. The random **E** vectors are represented by components on any two perpendicular axes (here y and z). For unpolarized waves, the components have equal amplitudes, and the phase difference between them varies randomly with time.

48-2   POLARIZING SHEETS

Figure 4 shows unpolarized light falling on a sheet of commercial polarizing material called *Polaroid*.† There exists in the sheet a certain characteristic polarizing direction, shown by the parallel lines. The sheet transmits only those wavetrain components whose electric field vectors vibrate parallel to this direction and absorbs those that vibrate at right angles to this direction. The light emerging from the sheet is linearly polarized. The polarizing direction of the sheet is established during the manufacturing process by embedding certain long-chain molecules in a flexible plastic sheet and then stretching the sheet so that the molecules are aligned parallel to each other. Radiation with its **E** vector parallel to the long molecules is strongly absorbed, while radiation with its **E** vector perpendicular to that direction passes through the sheet.

In Fig. 5 the polarizing sheet or *polarizer* lies in the plane of the page, and the direction of propagation is out of the page. The vector **E** shows the plane of vibration of a randomly selected wavetrain falling on the sheet. Two vector components, $\mathbf{E}_z$ (of magnitude $E \sin \theta$) and $\mathbf{E}_y$ (of magnitude $E \cos \theta$), can replace **E**, one parallel to the polarizing direction and one at right angles to it. Only the component $\mathbf{E}_y$ is transmitted; the component $\mathbf{E}_z$ is absorbed within the sheet.

When unpolarized light is incident on an ideal polarizing sheet, the intensity of the polarized light transmitted through the sheet is half the incident intensity, no matter what the orientation of the sheet. We can see this from the representation of the incident unpolarized light given in Fig. 3c, in which each of the components has, on the average, half the intensity of the incident light. Because the orientation of the axes in Fig. 3c is arbitrary, we are free to choose one of them to be along the direction of transmission of the polarizing sheet on which it is incident. Since this component of the light would be completely transmitted and the other completely absorbed, the sheet transmits 50% of the incident light. We can reach the same conclusion from Fig. 5, in which a wave

† There are other ways of producing polarized light without using this well-known commercial product. We mention some of them later. Also see "The Amateur Scientist," by Jearl Walker, *Scientific American*, December 1977, p. 172, for ways of making polarizing sheets and quarter-wave and half-wave plates and for various experiments that can be done with them.

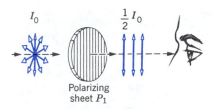

Figure 4   Unpolarized light is linearly polarized (and reduced in intensity by half) after passing through a single polarizing sheet. The parallel lines, which are not actually visible on the sheet, suggest its polarization direction.

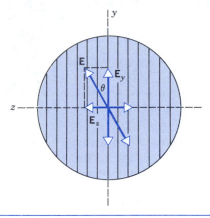

Figure 5   Another view of the action of a polarizing sheet. A linearly polarized wave (perhaps one of those shown in Fig. 3b) oriented in a random direction θ falls on the sheet. The y component of **E** is transmitted, and the z component is absorbed.

polarized in an arbitrary direction is incident on a polarizing sheet. The component E_y ($= E \cos \theta$) is transmitted, so the transmitted intensity is proportional to $E_y^2 = E^2 \cos^2 \theta$. If the incident light is unpolarized, we find the total transmitted intensity by averaging this expression over all possible orientations of the plane of polarization of the incident light, that is, over all possible values of θ. The average value of $\cos^2 \theta$ is $\frac{1}{2}$, so we again conclude that half the incident light is transmitted. Owing to reflection and partial absorption of the light along the polarizing direction, real polarizing sheets may transmit only 40% of the incident intensity. In our discussions, we assume ideal polarizers.

Let us place a second polarizing sheet P_2 (usually called, when so used, an *analyzer*) as in Fig. 6. If P_2 is rotated

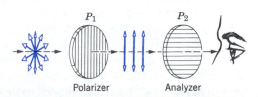

Figure 6   Unpolarized light is not transmitted by two polarizing sheets whose polarizing directions are perpendicular to one another.

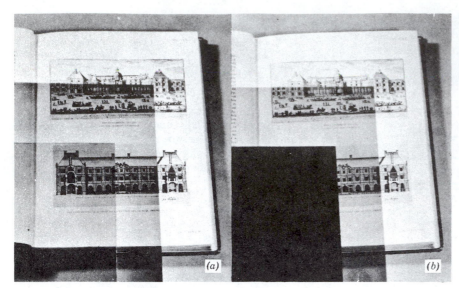

Figure 7   Two sheets of polarizing material are placed over an illustration from a book. In (*a*) the polarization directions of the two sheets are parallel, so that light passes through; in (*b*) the polarization directions are perpendicular, so that no light passes through. (The illustration shows the Luxembourg Palace in Paris. Malus discovered the phenomenon of polarization by reflection, using a calcite crystal to view sunlight reflected from the windows of this building.)

about the direction of propagation, there are two positions, 180° apart, at which the transmitted light intensity falls to zero; these are the positions in which the polarizing directions of P_1 and P_2 are at right angles.

If the amplitude of the linearly polarized light incident on P_2 is E_m, the amplitude of the light that emerges is $E_m \cos \theta$, where θ is the angle between the polarizing directions of P_1 and P_2. Recalling that the intensity of the light beam is proportional to the square of the amplitude, we see that the transmitted intensity I varies with θ according to

$$I = I_m \cos^2 \theta, \qquad (1)$$

in which I_m, the maximum value of the transmitted intensity, occurs when the polarizing directions of P_1 and P_2 are parallel, that is, when $\theta = 0$ or 180°. Figure 7*a*, in which two overlapping polarizing sheets are in the parallel position ($\theta = 0$ or 180° in Eq. 1), shows that the intensity of the light transmitted through the region of overlap has its maximum value. In Fig. 7*b* one or the other of the sheets has been rotated through 90° so that θ in Eq. 1 has the value 90° or 270°; the intensity of the light transmitted through the region of overlap is now a minimum. Equation 1, called the *law of Malus*, was discovered by Etienne Louis Malus (1775–1812) experimentally in 1809, using polarizing techniques other than those so far described (see Section 48-3).

Historically, polarization studies were made (by Young and by Malus, for example) to investigate the nature of light. Today we reverse the procedure and deduce something about the nature of an object from the polarization state of the light emitted by, or scattered from, that object. It has been possible to deduce, from studies of the polarization of light reflected from them, that the grains of cosmic dust present in our galaxy have been oriented in the weak galactic magnetic field (about 10^{-8} T) so that their long dimension is parallel to this field. Polarization studies

have suggested that Saturn's rings consist of ice crystals. The size and shape of virus particles can be determined by the polarization of ultraviolet light scattered from them. Information about the structure of atoms and nuclei is obtained from polarization studies of their emitted radiations in all parts of the electromagnetic spectrum. Thus we have a useful research technique for structures ranging in size from a galaxy (10^{+22} m) to a nucleus (10^{-14} m).

Polarized light also has many applications in industry and in engineering. Figure 8 shows a piece of plastic that

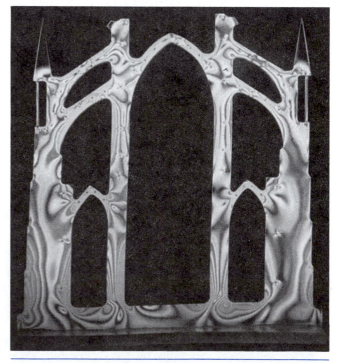

Figure 8   A piece of plastic is viewed between crossed polarizing sheets. The light and dark patterns show regions of stress in the structure.

Figure 9 A portable computer with a liquid crystal display.

has been stressed and placed between polarizing sheets. The stress pattern is revealed, allowing engineers to refine their designs to reduce stress at critical locations in the structure.* Figure 9 shows a common liquid crystal display, which uses polarized light to form letters and numbers, such as on watches and calculator displays. The liquid crystal is a material with stretched molecules like polarizing sheets; however, the long direction can be made to follow an applied electric field. The liquid crystal is arranged so that it normally transmits light through the polarizer and analyzer. When the electric field (from a battery) is applied to certain regions, the molecules line up in such a way that no light is transmitted through those regions, which form the dark patterns of the display.

Sample Problem 1 Two polarizing sheets have their polarizing directions parallel so that the intensity I_m of the transmitted light is a maximum. Through what angle must either sheet be turned if the intensity is to drop by one-half?

Solution From Eq. 1, since $I = \frac{1}{2}I_m$, we have

$$\tfrac{1}{2}I_m = I_m \cos^2 \theta,$$

or

$$\theta = \cos^{-1}\left(\pm\frac{1}{\sqrt{2}}\right) = \pm 45°, \pm 135°.$$

The same effect is obtained no matter which sheet is rotated or in which direction.

---

* For examples of how such models are used to study classical architecture, see "The Architecture of Christopher Wren," by Harold Dorn and Robert Mark, *Scientific American*, July 1981, p. 160, and "Gothic Structural Experimentation," by Robert Mark and William W. Clark, *Scientific American*, November 1984, p. 176.

48-3 POLARIZATION BY REFLECTION

Malus discovered in 1809 that light can be partially or completely polarized by reflection. Anyone who has watched the Sun's reflection in water, while wearing a pair of sunglasses made of polarizing material, has probably noticed the effect. It is necessary only to tilt the head from side to side, thus rotating the polarizing sheets, to observe that the intensity of the reflected sunlight passes through a minimum.

Figure 10 shows an unpolarized beam falling on a glass surface. The **E** vectors are resolved into two components (as in Fig. 3c), one perpendicular to the plane of incidence (the plane of Fig. 10) and one parallel to this plane. On the average, for completely unpolarized incident light, these two components are of equal amplitude.

For glass or other dielectric materials, there is a particular angle of incidence, called the *polarizing angle θ_p* (also known as *Brewster's angle*), at which the reflection coefficient for the polarization component in the plane of Fig. 10 is zero. This means that the beam reflected from the glass, although of low intensity, is linearly polarized, with its plane of polarization perpendicular to the plane of incidence. This polarization of the reflected beam can easily be verified by analyzing it with a polarizing sheet.

When light is incident at the polarizing angle, the component with polarization parallel to the plane of incidence is entirely refracted, while the perpendicular component is partially reflected and partially refracted. Thus the refracted beam, which is of high intensity, is partially polarized. If this refracted beam passed out of the glass into the air and were then incident on a second glass surface (again at angle θ_p), the perpendicular component would be reflected, and the refracted beam would have a slightly

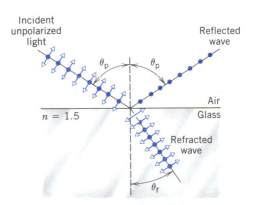

Figure 10 For a particular angle of incidence θ_p, the reflected light is completely polarized. The refracted light is partially polarized. The dots indicate polarization components perpendicular to the plane of the page, and the double arrows indicate polarization components parallel to the plane of the page.

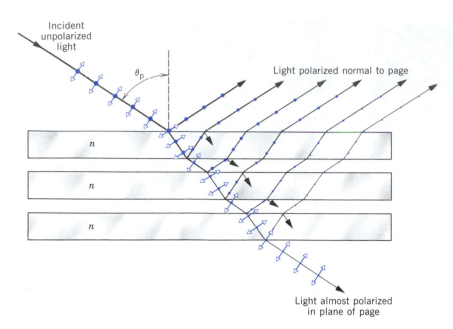

Incident
unpolarized
light

θ_p

Light polarized normal to page

Light almost polarized
in plane of page

n

n

n

Figure 11   Polarization of light by a stack of glass plates. Unpolarized light is incident at the angle θ_p. All reflected waves are polarized perpendicular to the plane of the page. After passing through several layers, the transmitted wave no longer contains any appreciable component polarized perpendicular to the page.

greater polarization. By using a stack of glass plates, we obtain reflections from successive surfaces, and we can increase the intensity of the emerging reflected beam (see Fig. 11). The perpendicular components are progressively removed from the transmitted beam, making it more completely polarized in the plane of Fig. 11.

At the polarizing angle it is found experimentally that the reflected and the refracted beams are at right angles, or (Fig. 10)

$$\theta_p + \theta_r = 90°.$$

From Snell's law,

$$n_1 \sin \theta_p = n_2 \sin \theta_r.$$

Combining these equations leads to

$$n_1 \sin \theta_p = n_2 \sin (90° - \theta_p) = n_2 \cos \theta_p,$$

or

$$\tan \theta_p = \frac{n_2}{n_1}, \tag{2}$$

where the incident beam is in medium 1 and the refracted beam in medium 2. If medium 1 is air ($n_1 = 1$), this becomes

$$\tan \theta_p = n, \tag{3}$$

where n is the index of refraction of the medium on which the light is incident. Equation 2 is known as *Brewster's law* after Sir David Brewster (1781–1868), who deduced it empirically in 1812. It is possible to prove this law rigorously from Maxwell's equations (see also Question 14).

Sample Problem 2   We wish to use a plate of glass ($n = 1.50$) in air as a polarizer. Find the polarizing angle and the angle of refraction.

Solution   From Eq. 3

$$\theta_p = \tan^{-1} 1.50 = 56.3°.$$

The angle of refraction follows from Snell's law:

$$\sin \theta_p = n \sin \theta_r,$$

or

$$\sin \theta_r = \frac{\sin 56.3°}{1.50} = 0.555 \quad \text{or} \quad \theta_r = 33.7°.$$

48-4   DOUBLE REFRACTION

In earlier chapters we assumed that the speed of light, and thus the index of refraction, is independent of the direction of propagation in the medium and of the state of polarization of the light. Liquids, amorphous solids such as glass, and crystalline solids having cubic symmetry normally show this behavior and are said to be *optically isotropic*. Many other crystalline solids are optically *anisotropic* (that is, not isotropic).* Optical anisotropy is responsible for the stress pattern illustrated in Fig. 8, although in this case the material is not crystalline.

Figure 12, in which a polished crystal of calcite ($CaCO_3$) is laid over a printed pattern, shows the optical anisotropy of this material; *the image appears double.* Furthermore,

---

* Solids may be anisotropic in many properties: mechanical (mica cleaves readily in one plane only), electric (a cube of crystalline graphite does not have the same electric resistance between all pairs of opposite faces), magnetic (a cube of crystalline nickel magnetizes more readily in certain directions than in others), and so forth.

Figure 12 A view through a birefringent crystal, showing the two images that result from the two different indices of refraction. The double images can be seen where there is no strip of polarizing material. The polarization axis of each strip is parallel to its long direction. Note that the two images have perpendicular polarizations.

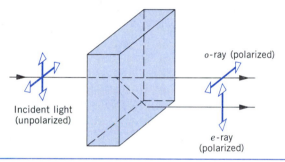

Figure 13 Unpolarized light falling on a birefringent material (such as a calcite crystal) splits into two components, the *o*-ray (which follows Snell's law of refraction) and the *e*-ray (which does not follow Snell's law). The two refracted rays have perpendicular polarizations, as shown.

the two images show perpendicular polarizations, as indicated in Fig. 13, which shows a beam of unpolarized light falling on a calcite crystal at right angles to one of its faces. The single beam splits into two at the crystal surface. The "double-bending" of a beam transmitted through calcite, exhibited in Figs. 12 and 13, is called *double refraction* or

birefringence. This phenomenon was studied by Huygens, who described it in his *Treatise on Light* published in 1678.

If the two emerging beams in Fig. 13 are analyzed with a polarizing sheet, they are found to be linearly polarized with their planes of vibration at right angles to each other. Figure 12 shows that each of the two crossed polarizers transmits only one of the two images (but not the other).

Some doubly refracting materials are strongly absorbing for one polarization component, while the other passes through with little absorption. Such materials are called *dichroic*. Polarizing sheets are examples of dichroic material.

If experiments are carried out at various angles of incidence, one of the beams in Fig. 13 (represented by the *ordinary ray*, or *o*-ray) is found to obey Snell's law of refraction at the crystal surface, just like a ray passing from one isotropic medium into another. The second beam (represented by the *extraordinary ray*, or *e*-ray) does not obey Snell's law. In Fig. 13, for example, the angle of incidence for the incident light is zero but the angle of refraction of the *e*-ray, contrary to the prediction of Snell's law, is nonzero. In general, the *e*-ray does not even lie in the plane of incidence.

This difference between the waves represented by the *o*- and *e*-rays with respect to Snell's law can be explained in these terms:

1. The *o*-wave travels in the crystal with the same speed v_o in all directions. In other words, the crystal has, for this wave, a single index of refraction n_o, just like an isotropic solid.

2. The *e*-wave travels in the crystal with a speed that varies with direction from v_o to v_e. In other words, the index of refraction, defined as c/v, varies with direction from n_o to n_e.

The quantities n_o and n_e are called the *principle indices of refraction* for the crystal. Problem 19 suggests how to measure them. Table 1 shows these indices for six doubly refracting cyrstals. For three of them the *e*-wave is slower; for the other three it is faster than the *o*-wave. Some doubly refracting crystals (such as mica and topaz) are more complex optically than calcite and require *three* principal

TABLE 1 **PRINCIPAL INDICES OF REFRACTION OF SEVERAL DOUBLY REFRACTING CRYSTALS**[a]

Crystal	Formula	n_o	n_e	$n_e - n_o$
Ice	H_2O	1.309	1.313	$+0.004$
Quartz	SiO_2	1.544	1.553	$+0.009$
Wurzite	ZnS	2.356	2.378	$+0.022$
Calcite	$CaCO_3$	1.658	1.486	-0.172
Dolomite	$CaO \cdot MgO \cdot 2CO_2$	1.681	1.500	-0.181
Siderite	$FeO \cdot CO_2$	1.875	1.635	-0.240

[a] For sodium light, $\lambda = 589$ nm.

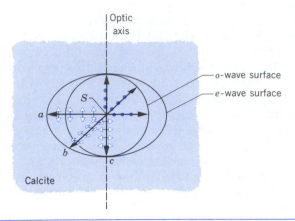

Figure 14   Huygens wave surfaces produced by a point source *S* imbedded in calcite. The polarization states of three *o*-rays and three *e*-rays are shown by the dots and arrows, respectively. In the general case (see ray *Sb*, for instance), the polarization direction is not perpendicular to the ray.

indices of refraction for a complete description of their optical properties. Crystals whose basic structure is cubic (such as NaCl; see Fig. 15 of Chapter 47) are optically isotropic, requiring only *one* index of refraction.

The behavior for the speeds of the two waves traveling in calcite is summarized by Fig. 14, which shows two wave surfaces spreading out from an imaginary point light source *S* imbedded in the crystal. The characteristic direc-

tion in the crystal in which $v_o = v_e$ is called the *optic axis*. The optic axis is a property of the crystal itself and is independent of the polarization or direction of propagation of the light.

The *o*-wave surface in Fig. 14 is a sphere, because the medium is isotropic for *o*-waves. The *e*-wave surface cannot be spherical, because the speed of the *e*-wave varies with direction relative to the optic axis. The *e*-wave surface is an ellipsoid of revolution about the optic axis. The two wave surfaces represent light having two different polarization states. If we consider for the present only rays lying in the plane of Fig. 14, then (1) the plane of polarization for the *o*-rays is perpendicular to the figure, as suggested by the dots, and (2) that for the *e*-rays coincides with the plane of the figure, as suggested by the double arrows. We describe the polarization states more fully at the end of this section.

We can use Huygens' principle to study the propagation of light waves in doubly refracting crystals. The most general situation may be quite complicated, with the *e*-wave emerging in a different plane than the *o*-wave. However, we may orient the crystal so the propagation directions for the incident wave, the *o*-wave, and the *e*-wave are all in the same plane. In the following discussion, we assume this has been done.

Figure 15*a* shows the special case in which unpolarized light falls at normal incidence on a calcite slab cut from a crystal in such a way that the optic axis is normal to the

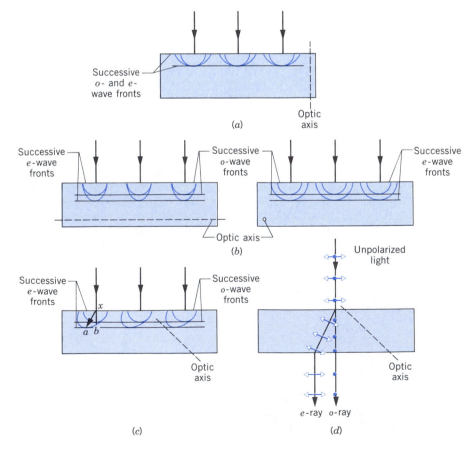

Figure 15   Unpolarized light falls at normal incidence on a slab cut from a calcite crystal. Huygens wavelets are shown, as in Fig. 14. (*a*) No double refraction or speed difference occurs. (*b*) No double refraction occurs, but there is a speed difference. (*c*) Both double refraction and a speed difference occur. (*d*) Same as (*c*), but showing the polarization states and the emerging rays.

surface. Consider a wavefront that, at time $t = 0$, coincides with the crystal surface. Following Huygens, we may let any point on this surface serve as a radiating center for a double set of Huygens wavelets, such as those in Fig. 14. The plane of tangency to these wavelets represents the new position of this wavefront at a later time t. The incident beam in Fig. 15a is propagated through the crystal without deviation at speed v_o. The beam emerging from the slab has the same polarization character as the incident beam. The calcite slab, in these special circumstances only, behaves like an isotropic material, and no distinction can be made between the o- and the e-waves. They both travel parallel to the optic axis and so have the same speed.

Figure 15b shows two views of another special case, namely, unpolarized incident light falling at right angles on a slab cut so that the optic axis is parallel to its surface. In this case too the incident beam is propagated without deviation. However, the propagation direction is perpendicular to the optic axis, and those waves that are polarized perpendicular to the axis have a different speed than those that are polarized along the axis. The first are o-waves and their speed is v_o; the second are e-waves and their speed is v_e. There will be a phase difference between the o-waves and e-waves as they emerge from the bottom of the slab.

Figure 15c shows unpolarized light falling at normal incidence on a calcite slab cut so that its optic axis makes an arbitrary angle with the crystal surface. Two spatially separated beams are produced, as in Fig. 13. They travel through the crystal at different speeds, that for the o-wave being v_o and that for the e-wave being intermediate between v_o and v_e. Note that ray *xa* represents the shortest *optical* path for the transfer of light energy from point *x* to the e-wavefront. Energy transferred along any other ray, in particular along ray *xb*, would have a longer transit time, a consequence of the fact that the speed of e-waves varies with direction. Figure 15d represents the same case as Fig. 15c. It shows the rays emerging from the slab, as in Fig. 13, and makes clear that the emerging beams are polarized at right angles to each other; that is, they are *cross-polarized*.

A Mechanical Analogy   *(Optional)*

We now seek to understand, in terms of the atomic structure of optically anisotropic crystals, how cross-polarized light waves with different speeds can exist. Light is propagated through a crystal by the action of the vibrating **E** vectors of the wave on the electrons in the crystal. These electrons, which experience electrostatic restoring forces if they are moved from their equilibrium positions, are set into forced periodic oscillation about these positions and pass along the transverse wave disturbance that constitutes the light wave. The strength of the restoring forces may be measured by a force constant k, as for the simple harmonic oscillator discussed in Chapter 15 (for which $F = -kx$).

In optically isotropic materials the force constant k is the same

for all directions of displacement of the electrons from their equilibrium positions. In doubly refracting crystals, however, k varies with direction. For electron displacements that lie in a plane at right angles to the optic axis, k has the constant value k_o, no matter how the displacement is oriented in this plane. For displacements parallel to the optic axis, k has the larger value (for calcite) k_e. Note carefully that the speed of a wave in a crystal is determined by the direction in which the **E** vectors vibrate and *not* by the direction of propagation. It is the transverse **E**-vector vibrations that call the restoring forces into play and thus determine the wave speed. Note too that the stronger the restoring force, that is, the larger k, the faster the wave. For waves traveling along a stretched cord, for example, the restoring force for the transverse displacements is determined by the tension F in the cord. Equation 18 of Chapter 19 ($v = \sqrt{F/\mu}$) shows that an increase in F means an increase in the wave speed v.

Figure 16, a long weighted "tire chain" supported at its upper end, provides a one-dimensional mechanical analogy for double refraction. It applies specifically to o- and e-waves traveling at right angles to the optic axis, as in Fig. 15b. If the supporting block is made to oscillate, as in Fig. 16a, a transverse wave travels along the chain with a certain speed. If the block oscillates lengthwise, as in Fig. 16b, another transverse wave is also propagated. The restoring force for the second wave is greater than for the first, the chain being more rigid for vibrations in its plane (Fig. 16b) than perpendicular to the plane (Fig. 16a). Thus the second wave travels along the chain with a greater speed.

In the language of optics we would say that the speed of a transverse wave in the chain depends on the orientation of the plane of vibration of the wave. If we oscillate the top of the chain in a random way, the wave disturbance at a point along the chain

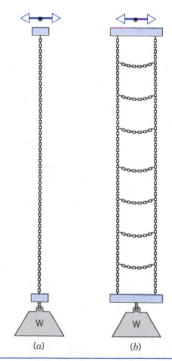

Figure 16 A one-dimensional mechanical model for double refraction. (*a*) Vibration perpendicular to the plane of the chain. (*b*) Vibration in the plane of the chain.

can be described as the sum of two waves, polarized at right angles and traveling with different speeds. This corresponds exactly to the optical situation of Fig. 15b.

For waves traveling parallel to the optic axis, as in Fig. 15a, or for waves in optically isotropic materials, the appropriate mechanical analogy is a single weighted hanging chain. Here there is only one speed of propagation, no matter how the upper end oscillates. The restoring forces are the same for all orientations of the plane of polarization of waves traveling along such a chain.

These considerations allow us to understand more clearly the polarization states of the light represented by the double-wave surface of Fig. 14. For the (spherical) o-wave surface, the E-vector vibrations must be everywhere at right angles to the optic axis. If this is so, the same force constant k_o always applies, and the o-waves travel with the same speed in all directions. More specifically, if we draw a ray in Fig. 14 from S to the o-wave surface, considered three-dimensionally (that is, as a sphere), the E-vector vibrations are always at right angles to the plane defined by this ray and the optic axis. Thus these vibrations are always at right angles to the optic axis.

For the (ellipsoidal) e-wave surface, the E-vector vibrations in general have a component parallel to the optic axis. For rays such as Sa in Fig. 14 or for the e-rays of Fig. 15b, the vibrations are completely parallel to this axis. Thus a relatively strong force constant (in calcite) k_e is operative, and the wave speed v_e is relatively high. For e-rays such as Sb in Fig. 14, the parallel component of the E-vector vibrations is less than 100%, so that the corresponding wave speed is less than v_e. For ray Sc in Fig. 14, the parallel component is zero, and the distinction between o- and e-rays disappears. ■

48-5 CIRCULAR POLARIZATION

Let linearly polarized light of angular frequency ω ($= 2\pi\nu$) fall at normal incidence on a slab of calcite cut so that the optic axis is parallel to the face of the slab, as in Fig. 17. The two waves that emerge are linearly polarized at right angles to each other, and, if the incident plane of

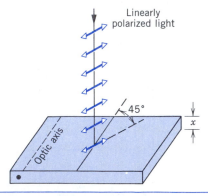

Figure 17 Linearly polarized light falls on a doubly refracting slab cut with its optic axis parallel to the surface. The plane of polarization makes an angle of 45° with the optic axis.

vibration is at 45° to the optic axis, they have equal amplitudes. Since the waves travel through the crystal at different speeds, there is a phase difference ϕ between them when they emerge from the crystal. If the crystal thickness is chosen so that (for a given frequency of light) $\phi = 90°$, the slab is called a *quarter-wave plate*. The emerging light is said to be *circularly polarized*.

In Section 15-7 we saw that two linearly polarized waves vibrating at right angles with a 90° phase difference can be represented as the projections on two perpendicular axes of a vector rotating with angular frequency ω about the propagation direction. This description applies to the emerging light in Fig. 17. These two descriptions of circularly polarized light are completely equivalent. Figure 18 clarifies the relationship between these two descriptions.

Suppose circularly polarized light, such as that of Fig. 18, is incident on a polarizing sheet. The emerging light is linearly polarized. Let us calculate its intensity. As it enters the sheet, the circularly polarized light can be represented by

$$E_y = E_m \sin \omega t \quad \text{and} \quad E_z = E_m \cos \omega t, \quad (4)$$

where y and z represent arbitrary perpendicular axes for a wave propagating in the x direction. These equations represent the equivalence between a circularly polarized wave and two linearly polarized waves with equal amplitudes E_m and a 90° phase difference. The resultant intensity in the incident circularly polarized wave is proportional to $E^2 = E_y^2 + E_z^2$, which equals E_m^2 when the components of the electric field are given by Eq. 4. Hence

$$I_{cp} \propto E_m^2 \quad (5)$$

Let the polarizing direction of the sheet make an arbitrary angle θ with the y axis as shown in Fig. 19. The instantaneous amplitude of the linearly polarized wave transmitted by the sheet is

$$\begin{aligned} E &= E_z \sin \theta + E_y \cos \theta \\ &= E_m \cos \omega t \sin \theta + E_m \sin \omega t \cos \theta \\ &= E_m \sin (\omega t + \theta). \end{aligned} \quad (6)$$

The intensity of the wave transmitted by the sheet is proportional to E^2, or

$$I \propto E_m^2 \sin^2(\omega t + \theta). \quad (7)$$

The eye and other measuring instruments respond only to the average intensity $\bar{I}$, which is found by replacing $\sin^2(\omega t + \theta)$ by its average value over one or more cycles ($= \frac{1}{2}$), so

$$\bar{I} \propto \frac{1}{2}E_m^2. \quad (8)$$

Comparison with Eq. 5 shows that inserting the polarizing sheet reduces the intensity by one-half. The orientation of the sheet makes no difference, since θ does not appear in this equation; this is to be expected if circularly polarized

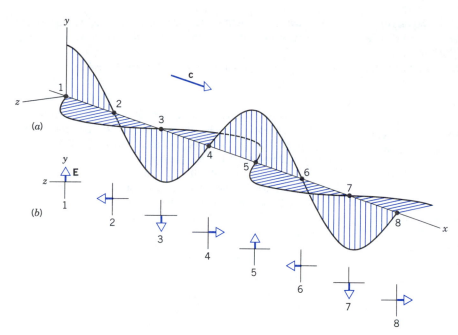

(a)

(b)

Figure 18 (a) Two waves of equal amplitude and linearly polarized in perpendicular directions move in the x direction. Only the **E** vectors are shown. The waves differ in phase by 90°, such that one reaches its maximum when the other is zero. (b) The resultant amplitude of the approaching wave as seen by observers at the numbered positions shown on the x axis. Note that, as the wave propagates, each observer will see at later times what the previous observer has seen. For instance, one-quarter cycle after the instant of this snapshot, the condition shown here for observer 7 will occur for observer 8. The resultant **E** vector thus appears to each observer to rotate clockwise with time.

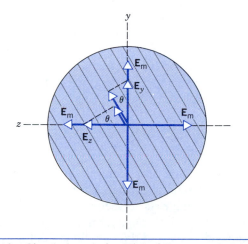

Figure 19 Circularly polarized light falls on a polarizing sheet. E_y and E_z are instantaneous values of the two components, which have maximum values E_m.

light is represented by a rotating vector, because all orientations about the propagation direction are equivalent. When unpolarized light is incident on a polarizing sheet, the intensity of the transmitted light is also reduced by $\frac{1}{2}$, independent of the orientation of the sheet, as we discussed in Section 48-2. A simple polarizing sheet therefore cannot be used to distinguish between unpolarized and circularly polarized light.

To distinguish between circularly polarized and unpolarized light, we can use a quarter-wave plate. Suppose circularly polarized light is incident on a quarter-wave plate whose optic axis has an arbitrary orientation. Com-

ponents of the incident light along and perpendicular to the direction of the optic axis differ in phase by 90°. After passing through the quarter-wave plate, an *additional* phase difference of 90° is introduced, which will either add to or subtract from the previous phase difference, depending on the orientation of the axis of the quarter-wave plate. The resulting phase difference is either 0° or 180°; that is, the polarization components along two perpendicular axes reach their maximum values at the same instant. The total **E** field is the sum of these two vectors and makes an angle of 45° with the two components. The emerging light is therefore linearly polarized in a direction at an angle of ±45° with the optic axis, which we could demonstrate by placing a polarizing sheet in the path of the light and rotating the sheet to show the extinction of the intensity.

This experiment is in effect the reverse of Fig. 17, in which circularly polarized light emerges when linearly polarized light is incident on a quarter-wave plate. Here we have linearly polarized light emerging when circularly polarized light is incident. This is an example of *time-reversal symmetry* in nature; if we reverse all motions in a physical situation, the result must also be an allowed physical situation. While certain very weak forces between elementary particles may not follow this symmetry, all other known forces, including electromagnetism and gravity, strictly follow the time-reversal symmetry.

Sample Problem 3 A quartz quarter-wave plate is to be used with sodium light ($\lambda = 589$ nm). What is the minimum thickness of such a plate?

Solution Two waves travel through the slab at speeds corresponding to the two principal indices of refraction given in Table 1 ($n_e = 1.553$ and $n_o = 1.544$). If the crystal thickness is x, the number of wavelengths of the first wave contained in the crystal is

$$N_e = \frac{x}{\lambda_e} = \frac{xn_e}{\lambda},$$

where λ_e is the wavelength of the *e*-wave in the crystal and λ is the wavelength in air. For the second wave the number of wavelengths is

$$N_o = \frac{x}{\lambda_o} = \frac{xn_o}{\lambda},$$

where λ_o is the wavelength of the *o*-wave in the crystal. The difference $N_e - N_o$ must be $m + \frac{1}{4}$, where $m = 0, 1, 2, \ldots$. The minimum thickness corresponds to $m = 0$, in which case

$$\frac{1}{4} = \frac{x}{\lambda}(n_e - n_o).$$

This equation yields

$$x = \frac{\lambda}{4(n_e - n_o)} = \frac{589 \text{ nm}}{(4)(1.553 - 1.544)} = 0.016 \text{ mm}.$$

This plate is rather thin. Most quarter-wave plates are made from mica; the sheet is split to the correct thickness by trial and error.

Sample Problem 4 A linearly polarized light wave of amplitude E_0 falls on a calcite quarter-wave plate with its plane of polarization at 45° to the optic axis of the plate, which is taken as the y axis; see Fig. 20. The emerging light will be circularly polarized. In what direction will the electric vector appear to rotate? The direction of propagation is out of the page.

Solution The wave component whose vibrations are parallel to the optic axis (the *e*-wave) can be represented as it emerges from the plate as

$$E_y = (E_0 \cos 45°) \sin \omega t = \frac{1}{\sqrt{2}} E_0 \sin \omega t = E_m \sin \omega t.$$

The wave component whose vibrations are at right angles to the optic axis (the *o*-wave) can be represented as

$$E_z = (E_0 \sin 45°) \sin (\omega t - 90°) = -\frac{1}{\sqrt{2}} E_0 \cos \omega t$$

$$= -E_m \cos \omega t,$$

the 90° phase shift representing the action of the quarter-wave plate. Note that E_z reaches its maximum value one-fourth of a cycle *later* than E_y does, for, in calcite, wave E_z (the *o*-wave) travels *slower* than wave E_y (the *e*-wave).

To decide the direction of rotation, let us locate the tip of the rotating electric vector at two instants of time, (Fig. 20a) $t = 0$ and (Fig. 20b) a short time t_1 later chosen so that ωt_1 is a small angle. At $t = 0$ the coordinates of the tip of the rotating vector (see Fig. 20a) are

$$E_y = 0 \quad \text{and} \quad E_z = -E_m.$$

At $t = t_1$ these coordinates become, approximately,

$$E_y = E_m \sin \omega t_1 \approx E_m \omega t_1$$
$$E_z = -E_m \cos \omega t_1 \approx -E_m.$$

Figure 20b shows that the vector representing the emerging circularly polarized light is rotating counterclockwise; by convention such light is called *left-circularly polarized*, the observer always being considered to face the light source.

You should verify that if the plane of vibration of the incident light in Fig. 20 is rotated through ±90°, the emerging light will be *right-circularly polarized*.

48-6 SCATTERING OF LIGHT

A light wave, falling on a transparent solid, causes the electrons in the solid to oscillate periodically in response to the time-varying electric vector of the incident wave. The wave that travels through the medium is the resultant of the incident wave and the radiations from the oscilla-

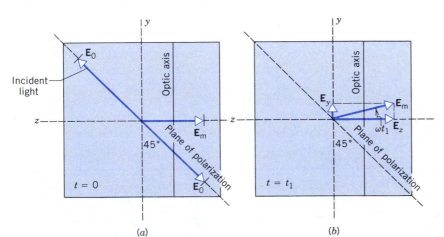

(a)   *(b)*

Figure 20 Sample Problem 4. Linearly polarized light falls (from behind the page) on a quarter-wave plate. The incident light is polarized at 45° with the y and z axes. (a) At a particular time $t = 0$, the emerging **E** vector points in the $-z$ direction. (b) A short interval of time t_1 later the vector has rotated to a new position. In this case the **E** vector rotates counterclockwise as seen by an observer on the x axis facing the light source.

ting electrons. The resultant wave has a maximum intensity in the direction of the incident beam, falling off rapidly on either side. The lack of sideways scattering, which would be essentially complete in a large "perfect" crystal, comes about because the oscillating charges in the medium act cooperatively or coherently.

When light passes through a liquid or a gas, we find much more sideways scattering. The oscillating electrons in this case, being separated by relatively large distances and not being bound together in a rigid structure, act independently rather than cooperatively. Thus a rigid cancellation of wave disturbances that are not in the forward direction is less likely to occur; there is more sideways scattering.

Light scattered sideways from a gas can be wholly or partially polarized, even though the incident light is unpolarized. Figure 21 shows an unpolarized beam moving upward on the page and striking a gas atom at O. The electrons at O oscillate in response to the electric components of the incident wave, their motion being equivalent to two oscillating dipoles whose axes are in the y and z directions at O. For transverse electromagnetic waves, an oscillating dipole does not radiate along its own axis. Thus an observer at O' would receive no radiation from the dipole at O oscillating in the z direction. The radiation reaching O' would come entirely from the dipole at O oscillating in the y direction and would be linearly polarized in the y direction.

As observer O' moves off the z axis, the radiation becomes less than fully polarized, because the dipole at O oscillating along the z axis can radiate somewhat in these directions. At points along the x axis, the transmitted ($x > 0$) or backscattered ($x < 0$) radiation is unpolarized,

because both dipoles can radiate equally well in the x direction.

A familiar example of this effect is the scattering of sunlight by the molecules of the Earth's atmosphere. If the atmosphere were not present, the sky would appear black except in the direction of the Sun, as observed by astronauts orbiting above the atmosphere. We can easily check with a polarizer that the light from the cloudless sky is at least partially polarized. This fact is used in polar exploration in the *solar compass*. In this device we establish direction by noting the nature of the polarization of the scattered sunlight. As is well known, magnetic compasses are not useful in these regions. It has been learned[*] that bees orient themselves in their flights between their hive and the pollen sources by means of polarization of the light from the sky; bees' eyes contain built-in polarization-sensing devices.

It still remains to be explained why the light scattered from the sky is predominantly blue and why the light received directly from the Sun—particularly at sunset when the length of the atmosphere that it must traverse is greatest—is red. The cross section of an atom or molecule for light scattering depends on the wavelength, blue light being scattered more effectively than red light. Since the blue light is more strongly scattered, the transmitted light has the color of normal sunlight with the blues largely removed; it is therefore more reddish in appearance.

The conclusion that the scattering cross section for blue light is higher than that for red light can be made reason-

---

[*] See "Polarized-Light Navigation by Insects," by Rudiger Wehner, *Scientific American*, July 1976, p. 106.

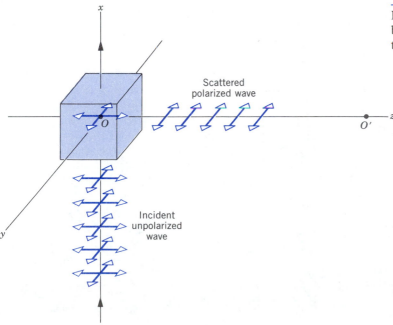

Figure 21 An unpolarized incident wave is scattered by an atom at O. The wave scattered toward O' on the z axis is linearly polarized.

able with a mechanical analogy. An electron in an atom or molecule is bound there by strong restoring forces. It has a definite natural frequency, like a small mass suspended in space by an assembly of springs. The natural frequency for electrons in atoms and molecules is usually in a region corresponding to violet or ultraviolet light.

When light is allowed to fall on such bound electrons, it sets up forced oscillations at the frequency of the incident light beam. In mechanical resonant systems it is possible to "drive" the system most effectively if we impress on it an external force whose frequency is as close as possible to the natural resonant frequency. In the case of light, the frequency of blue light is closer to the natural resonant frequency of the bound electron than is that of red light. We would expect the blue light to be more effective in causing the electron to oscillate, and it is more effectively scattered.

Double Scattering   *(Optional)*

Experiments similar to that shown in Fig. 21 can demonstrate that electromagnetic waves must be transverse; that is, there can be no component of the **E** vector parallel to the direction of propagation. Suppose there were such a component along the direction of the incident wave (the x direction in Fig. 21). Then the electrons at O would oscillate in all three directions, and the scattered wave directed toward O' would show all three possible polarization directions (two transverse and one longitudinal). This radiation would thus be unpolarized. If the incident radiation is only transverse, as in Fig. 21, the radiation propagated to O' is linearly polarized. The question as to the transverse nature

of the radiation is thus equivalent to determining whether the radiation traveling to O' is polarized or unpolarized.

There is another way to make this determination. Let us place a second scatterer at O'. A dipole at O' will oscillate in response to the incident (polarized) wave in only one direction (the y' direction, that of the incident **E** vector, as shown in Fig. 22). Radiation scattered by that dipole can travel in the $\pm x'$ directions, but (for transverse radiation) *not in the* y' *direction*. Thus a detector D measuring the *intensity* of the radiation should see a maximum in the $\pm x'$ directions and a mimimum of zero intensity in the y' direction. Such an experiment, as illustrated in Fig. 22, is called a *double scattering* experiment. Note that the *polarization* of the radiation scattered by the first target is determined through the *intensity* of the radiation scattered by the second target. If the radiation traveling to O' were not polarized (and not purely transverse), then the detector D would record the same intensity in all directions.

We can establish the transverse nature of electromagnetic radiation either by measuring the polarization of the radiation scattered from the first target (as shown in Fig. 21) or the intensity distribution of radiation scattered from the second target (as shown in Fig. 22). For some radiations (such as light), polarization measurements are relatively easy to make, and the double scattering method provides no great advantage. For other radiations (such as x rays or gamma rays), double scattering is usually the preferred method. Indeed, following the discovery of x rays in 1898, there was speculation whether they were waves or particles. A double scattering experiment, performed in 1906 by Charles Barkla, established that x rays, like visible light, were transverse in nature and helped to confirm that x rays are part of the electromagnetic spectrum. ■

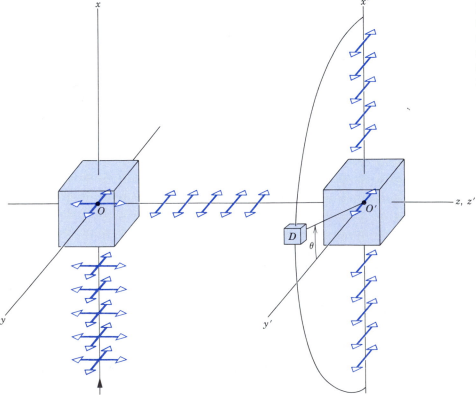

Figure 22   The polarized radiation scattered at O can be scattered by another atom at O'. A detector D measures the intensity of the radiation scattered by O' at various locations θ in the $x'y'$ plane.

48-7   TO THE QUANTUM LIMIT

In this chapter, we have described such properties of electromagnetic waves as polarization and scattering based on analysis in terms of the wave picture. As an alternative and complementary explanation, we can consider the quantum picture, in which the properties of the radiation are associated not with the fields but with individual quanta of radiation (photons).

As an example, we review the *linear* momentum carried by a monochromatic light wave. In Section 41-5, we showed that the absorption by an object of energy U from a light wave is accompanied by the transfer of momentum P to the object, where U and P are related by

$$P = \frac{U}{c}, \tag{9}$$

where c is the speed of light. In contrast to the wave picture, we can regard the light as a stream of photons, each of which carries an energy E. The photon is a massless particle, for which Eq. 32 of Chapter 21 gives $E = pc$, so the momentum p carried by each particle (photon) is given by

$$p = \frac{E}{c}. \tag{10}$$

Comparison of Eqs. 9 and 10 indicates the relationship between the photon and the wave pictures, or equivalently between the quantum and classical (nonquantum) domains. The absorption of energy U from a light wave is accomplished by the absorption of many individual photons of energy E by the atoms of the object. Similarly, the momentum P delivered to the object by the light wave can be analyzed in terms of the momentum p delivered to individual atoms by photons in the beam.

The absorption of a *circularly* polarized light wave can, in an analogous way, deliver *angular* momentum to an object. Classical electromagnetism gives the relationship between the energy U and the angular momentum L as

$$L = \frac{U}{\omega}, \tag{11}$$

where ω is the angular frequency of the wave. According to quantum mechanics, the energy E of a photon can be written (see Eq. 38 of Chapter 8, $\Delta E = h\nu$)

$$E = h\nu = \frac{h}{2\pi} \omega, \tag{12}$$

where h is the Planck constant. In Section 13-6, we showed that $h/2\pi$ is the basic quantum unit of angular momentum, which we write here as l. Hence Eq. 12 can be written

$$l = \frac{E}{\omega}. \tag{13}$$

In the quantum picture, when an atom absorbs a photon of energy E, its angular momentum changes by a definite amount l. Comparing Eqs. 11 and 13, we see the correspondence between the classical and quantum descriptions. The total angular momentum L absorbed by the object can be regarded as the net effect of the quanta of angular momentum l absorbed by individual atoms.

Classical physics, including the wave description of electromagnetic radiation, works perfectly well in analyzing a wide class of phenomena, including diffraction, polarization, and scattering. It is not necessary to invoke the quantum theory to explain these effects (although they can often be equally well explained based on quantum effects, as we have discussed in this section). For example, the Barkla x-ray double scattering experiment, discussed in the previous section, can also be interpreted in the quantum picture if we assign to each photon an intrinsic angular momentum ("spin") and demand that individual photons must have their spins aligned parallel or antiparallel to their direction of propagation. This is in fact the behavior that quantum theory predicts for photons.

This competition between particle and wave descriptions of phenomena associated with electromagnetic waves dates from the time of Newton, who sought to explain refraction based on a particle theory of light. Ultimately, it is interference and diffraction experiments, such as we discussed in Chapters 45 and 46, that lead us to favor the wave interpretation.

Beginning in the early 20th century, a new class of experiments was done that upset the conventional view of electromagnetic waves. The photoelectric effect (in which a metal surface irradiated with light emits electrons) and Compton scattering (in which the wavelength of the radiation scattered in the geometry of Fig. 21 is found to differ from the incident wavelength) cannot be accommodated in the wave picture. Further difficulties with classical physics arose when particles such as electrons were found to exhibit wavelike behavior under certain circumstances. The quantum theory, developed in the 1920s, offers an alternative explanation for all of these failures of classical physics and stresses the complementary roles of the wave and particle pictures. Chapters 49–56 in the extended version of this text present an introduction to the quantum theory and some of its many applications, ranging from the quark structure of elementary particles to the origin and evolution of the universe.

QUESTIONS

1. It is said that light from ordinary sources is unpolarized. Can you think of any common sources that emit polarized light?

2. Light from a laboratory gas discharge tube is unpolarized. How can this be made consistent with the fact that atoms and molecules radiate as electric dipoles whose radiation is linearly polarized?

3. Polarizing sheets contain long hydrocarbon chains that are made to line up in a parallel array during the production process. Explain how a polarizing sheet is able to polarize light. (*Hint*: Electrons are relatively free to move along these chains.)

4. As we normally experience them, radio waves are almost always polarized and visible light is almost always unpolarized. Why is this so?

5. What determines the desirable length and orientation of the rabbit ears on a portable TV set?

6. Why are sound waves unpolarized?

7. Suppose that each slit in Fig. 4 of Chapter 45 is covered with a polarizing sheet, the polarizing directions of the two sheets being at right angles. What is the pattern of light intensity on screen *C*? (The incident light is unpolarized.)

8. Why do sunglasses made of polarizing materials have a marked advantage over those that simply depend on absorption effects? What disadvantages might they have?

9. Unpolarized light falls on two polarizing sheets so oriented that no light is transmitted. If a third polarizing sheet is placed between them, can light be transmitted? If so, explain how.

10. Sample Problem 1 shows that, when the angle between the two polarizing directions is turned from 0° to 45°, the intensity of the transmitted beam drops to one-half its initial value. What happens to this "missing" energy?

11. You are given a number of polarizing sheets. Explain how you would use them to rotate the plane of polarization of a linearly polarized wave through any given angle. How could you do it with the least energy loss?

12. In the early 1950s, 3-D movies were very popular. Viewers wore polarizing glasses and a polarizing sheet was placed in front of each of the *two* projectors needed. Explain how the system worked. Can you suggest any problems that may have led to the early abandonment of the system?

13. A wire grid, consisting of an array of wires arranged parallel to one another, can polarize an incident unpolarized beam of electromagnetic waves that pass through it. Explain the facts that (*a*) the diameter of the wires and the spacing between them must be much less than the incident wavelength to obtain effective polarization and (*b*) the transmitted component is the one whose electric vector oscillates in a direction perpendicular to the wires.

14. Brewster's law, Eq. 2, determines the polarizing angle on reflection from a dielectric material such as glass; see Fig. 10. A plausible interpretation for zero reflection of the parallel component at that angle is that the charges in the dielectric are caused to oscillate parallel to the reflected ray by this component and produce no radiation in this direction. Explain this and comment on the plausibility.

15. Explain how polarization by reflection could occur if the light is incident on the interface from the side with the higher index of refraction (glass to air, for example).

16. Find a way to identify the polarizing direction of a polarizing sheet. No marks appear on the sheet.

17. Is the optic axis of a doubly refracting crystal simply a line or a direction in space? Has it a direction sense, like an arrow? What about the characteristic direction of a polarizing sheet?

18. If ice is doubly refracting (see Table 1), why don't we see two images of objects viewed through an ice cube?

19. Is it possible to produce interference effects between the *o*-beam and the *e*-beam, which are separated by the calcite crystal from the incident unpolarized beam in Fig. 13, by recombining them? Explain your answer.

20. From Table 1, would you expect a quarter-wave plate made from calcite to be thicker than one made from quartz?

21. Does the *e*-wave in doubly refracting crystals always travel at a speed given by c/n_e?

22. In Figs. 15*a* and 15*b* describe qualitatively what happens if the incident beam falls on the crystal with an angle of incidence that is not zero. Assume in each case that the incident beam remains in the plane of the figure.

23. Devise a way to identify the direction of the optic axis in a quarter-wave plate.

24. If linearly polarized light falls on a quarter-wave plate with its plane of vibration making an angle of (*a*) 0° or (*b*) 90° with the axis of the plate, describe the transmitted light. (*c*) If this angle is arbitrarily chosen, the transmitted light is called *elliptically polarized*; describe such light.

25. You are given an object that may be (*a*) a disk of grey glass, (*b*) a polarizing sheet, (*c*) a quarter-wave plate, or (*d*) a half-wave plate (see Problem 21). How could you identify it?

26. Can a linearly polarized light beam be represented as a sum of two circularly polarized light beams of opposite rotation? What effect has changing the phase of one of the circular components on the resultant beam?

27. Could a radar beam be circularly polarized?

28. How can a right-circularly polarized light beam be transformed into a left-circularly polarized beam?

29. A beam of light is said to be unpolarized, linearly polarized, or circularly polarized. How could you choose among them experimentally?

30. A parallel beam of light is absorbed by an object placed in its path. Under what circumstances will (*a*) linear momentum and (*b*) angular momentum be transferred to the object?

31. When observing a clear sky through a polarizing sheet, you find that the intensity varies on rotating the sheet. This does not happen when viewing a cloud through the sheet. Why?

32. In 1949 it was discovered that light from distant stars in our galaxy is slightly linearly polarized, with the preferred plane of vibration being parallel to the plane of the galaxy. This is probably due to nonisotropic scattering of the starlight by elongated and slightly aligned interstellar grains (see Problem 31 in Chapter 24). If the grains are oriented with their long axes parallel to the interstellar magnetic field lines, as discussed in Section 48-2, and they absorb and radiate elec-

tromagnetic waves like the oscillating electrons in a radio antenna, how must the magnetic field be oriented with respect to the galactic plane?

33. Verify that Eq. 11 is dimensionally correct.
34. Is polarization or interference a better test for identifying waves? Do they give the same information?

PROBLEMS

Section 48-1 Polarization

1. The magnetic field equations for an electromagnetic wave in free space are $B_x = B \sin (ky + \omega t)$, $B_y = B_z = 0$. (a) What is the direction of propagation? (b) Write the electric field equations. (c) Is the wave polarized? If so, in what direction?

2. Prove that two linearly polarized light waves of equal amplitude, their planes of vibration being at right angles to each other, cannot produce interference effects. (*Hint*: Prove that the intensity of the resultant light wave, averaged over one or more cycles of oscillation, is the same no matter what phase difference exists between the two waves.)

Section 48-2 Polarizing Sheets

3. A beam of unpolarized light of intensity 12.2 mW/m² falls at normal incidence upon a polarizing sheet. (a) Find the maximum value of the electric field of the transmitted beam. (b) Calculate the radiation pressure exerted on the polarizing sheet.

4. Unpolarized light falls on two polarizing sheets placed one on top of the other. What must be the angle between the characteristic directions of the sheets if the intensity of the transmitted light is one-third the intensity of the incident beam? Assume that each polarizing sheet is ideal, that is, that it reduces the intensity of unpolarized light by exactly 50%.

5. Three polarizing plates are stacked. The first and third are crossed; the one between has its axis at 45° to the axes of the other two. What fraction of the intensity of an incident unpolarized beam is transmitted by the stack?

6. A beam of linearly polarized light strikes two polarizing sheets. The characteristic direction of the second is 90° with respect to the incident light. The characteristic direction of the first is at angle θ with respect to the incident light. Find angle θ for a transmitted beam intensity that is 0.100 times the incident beam intensity.

7. A beam of unpolarized light is incident on a stack of four polarizing sheets that are lined up so that the characteristic direction of each is rotated by 30° clockwise with respect to the preceding sheet. What fraction of the incident intensity is transmitted?

8. A beam of light is linearly polarized in the vertical direction. The beam falls at normal incidence on a polarizing sheet with its polarizing direction at 58.8° to the vertical. The transmitted beam falls, also at normal incidence, on a second polarizing sheet with its polarizing direction horizontal. The intensity of the original beam is 43.3 W/m². Find the intensity of the beam transmitted by the second sheet.

9. Suppose that in Problem 8 the incident beam was unpolarized. What now is the intensity of the beam transmitted by the second sheet?

10. A beam of light is a mixture of polarized light and unpolarized light. When it is sent through a Polaroid sheet, we find that the transmitted intensity can be varied by a factor of five depending on the orientation of the Polaroid. Find the relative intensities of these two components of the incident beam.

11. At a particular beach on a particular day near sundown the horizontal component of the electric field vector is 2.3 times the vertical component. A standing sunbather puts on polaroid sunglasses; the glasses suppress the horizontal field component. (a) What fraction of the light energy received before the glasses were put on now reaches the eyes? (b) The sunbather, still wearing the glasses, lies on his side. What fraction of the light energy received before the glasses were put on reaches the eyes now?

12. It is desired to rotate the plane of vibration of a beam of polarized light by 90°. (a) How might this be done using only polarizing sheets? (b) How many sheets are required in order for the total intensity loss to be less than 5.0%?

Section 48-3 Polarization by Reflection

13. (a) At what angle of incidence will the light reflected from water be completely polarized? (b) Does this angle depend on the wavelength of the light?

14. Light traveling in water of index of refraction 1.33 is incident on a plate of glass of index of refraction 1.53. At what angle of incidence is the reflected light completely linearly polarized?

15. Calculate the range of polarizing angles for white light incident on fused quartz. Assume that the wavelength limits are 400 and 700 nm and use the dispersion curve of Fig. 4, Chapter 43.

16. When red light in vacuum is incident at the polarizing angle on a certain glass slab, the angle of refraction is 31.8°. What are (a) the index of refraction of the glass and (b) the polarizing angle?

Section 48-4 Double Refraction

17. Linearly polarized light of wavelength 525 nm strikes, at normal incidence, a wurzite crystal, cut with its faces parallel to the optic axis. What is the smallest possible thickness of the crystal if the emergent o- and e-rays combine to form linearly polarized light? See Table 1.

18. A narrow beam of unpolarized light falls on a calcite crystal cut with its optic axis as shown in Fig. 23. (a) For $t = 1.12$ cm and for $\theta_i = 38.8°$, calculate the perpendicular distance between the two emerging rays x and y. (b) Which is the o-ray and which the e-ray? (c) What are the states of polarization of the emerging rays? (d) Describe what happens if a polarizer is placed in the incident beam and rotated. (*Hint*: Inside the crystal the **E**-vector vibrations for one ray

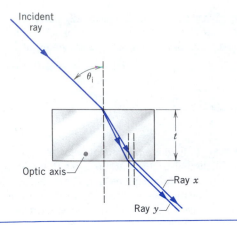

Figure 23 Problem 18.

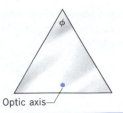

Figure 24 Problem 19.

are always perpendicular to the optic axis and for the other ray they are always parallel. The two rays are described by the indices n_o and n_e; *in this plane* each ray obeys Snell's law.)

19. A prism is cut from calcite so that the optic axis is parallel to the prism edge as shown in Fig. 24. Describe how such a prism might be used to measure the two principal indices of refraction for calcite. (*Hint*: See hint in Problem 18; see also Sample Problem 3, Chapter 43.)

Section 48-5 Circular Polarization

20. Find the greatest number of quarter-wave plates, to be used with light of wavelength 488 nm, that could be cut from a dolomite crystal 0.250 mm thick.

21. What would be the action of a *half-wave plate* (that is, a plate twice as thick as a quarter-wave plate) on (*a*) linearly polarized light (assume the plane of vibration to be at 45° to the optic axis of the plate), (*b*) circularly polarized light, and (*c*) unpolarized light?

22. A polarizing sheet and a quarter-wave plate are glued together in such a way that, if the combination is placed with face *A* against a shiny coin, the face of the coin can be seen when illuminated with light of appropriate wavelength. When the combination is placed with face *A* away from the coin, the coin cannot be seen. (*a*) Which component is on face *A* and (*b*) what is the relative orientation of the components?

Section 48-7 To the Quantum Limit

23. Assume that a parallel beam of circularly polarized light whose power is 106 W is absorbed by an object. (*a*) At what rate is angular momentum transferred to the object? (*b*) If the object is a flat disk of diameter 5.20 mm and mass 9.45 mg, after how long a time (assuming it is free to rotate about its axis) would it attain an angular speed of 1.50 rev/s? Assume a wavelength of 516 nm.

CHAPTER 49

LIGHT AND QUANTUM PHYSICS

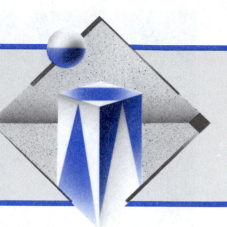

Thus far we have studied radiation—including not only light but all of the electromagnetic spectrum—through the phenomena of reflection, refraction, interference, diffraction, and polarization, all of which can be understood by treating radiation as a wave. The evidence in support of this wave behavior is overwhelming.

We now move off in a new direction and consider experiments that can be understood only by making quite a different assumption about electromagnetic radiation, namely, that it behaves like a stream of particles.

The concepts of wave and particle are so different that it is hard to understand how light (and other radiation) can be both. In a wave, for example, the energy and momentum are distributed smoothly over the wavefront, while they are concentrated in bundles in a stream of particles. We delay a discussion of this dual nature until Chapter 50. In the meantime, we ask that you not worry about this puzzle and that you consider the compelling experimental evidence that radiation has this particlelike nature. This begins our study of quantum physics, which leads eventually to our understanding of the fundamental structure of matter.

49-1 THERMAL RADIATION

We see most objects by the light that is reflected from them. At high enough temperatures, however, bodies become self-luminous, and we can see them glow in the dark. Incandescent lamp filaments and bonfires (see Fig. 1) are familiar examples. Although we see such objects by the visible light that they emit, we do not have to linger too long near a bonfire to believe that it also emits copiously in the infrared region of the spectrum. It is a curious fact that quantum physics, which dominates our modern view of the world around us, arose from the study—under controlled laboratory conditions—of the radiations emitted by hot objects.

Radiation given off by a body because of its temperature is called *thermal radiation*. All bodies not only emit such radiation but also absorb it from their surroundings. If a body is hotter than its surroundings it emits more radiation than it absorbs and tends to cool. Normally, it will come to thermal equilibrium with its surroundings, a condition in which its rates of absorption and emission of radiation are equal.

The spectrum of the thermal radiation from a hot solid body is continuous, its details depending strongly on the temperature. If we were steadily to raise the temperature of such a body, we would notice two things: (1) the higher the temperature, the more thermal radiation is emitted—at first the body appears dim, then it glows brightly; and (2) the higher the temperature, the shorter is the wavelength of that part of the spectrum radiating most intensely—the predominant color of the hot body shifts from dull red through bright yellow-orange to bluish "white heat." Since the characteristics of its spectrum depend on the temperature, we can estimate the temperature of a hot body, such as a glowing steel ingot or a star, from the radiation it emits. The eye sees chiefly the color corresponding to the most intense emission in the visible range.

The radiation emitted by a hot body depends not only on the temperature but also on the material of which the body is made, its shape, and the nature of its surface. For example, at 2000 K a polished flat tungsten surface emits radiation at a rate of 23.5 W/cm²; for molybdenum, however, the corresponding rate is 19.2 W/cm². In each case the rate increases somewhat if the surface is roughened.

Figure 1 Students contemplating thermal radiation. The study of such radiation, under controlled laboratory conditions, laid the foundations for modern quantum mechanics.

Other differences appear if we measure the distribution in wavelength of the emitted radiation. Such details make it hard to understand thermal radiation in terms of simpler physical ideas; it reminds us of the complications that arise in trying to understand the properties of real gases in terms of a simple atomic model. The "gas problem" was managed by introducing the notion of an ideal gas. In much the same spirit, the "radiation problem" can be made manageable by introducing an "ideal radiator" for which the spectrum of the emitted thermal radiation depends *only* on the temperature of the radiator and not on the material, the nature of the surface, or other factors.

We can make such an ideal radiator by forming a cavity within a body, the walls of the cavity being held at a uniform temperature. We must pierce a small hole through the wall so that a sample of the radiation inside the cavity can escape into the laboratory to be examined. It turns out that such thermal radiation, called *cavity radiation,** has a very simple spectrum whose nature is indeed determined only by the temperature of the walls and not

---

* Also known as *black-body radiation,* because an ideal black body (one that absorbs all radiation incident on it) would emit the same type of radiation. We assume that the dimensions of the cavity are much greater than the wavelength of the radiation.

in any way by the material of the cavity, its shape, or its size. Cavity radiation (radiation in a box) helps us to understand the nature of thermal radiation, just as the ideal gas (matter in a box) helped us to understand matter in its gaseous form.

Figure 2 shows a cavity radiator made of a thin-walled cylindrical tungsten tube about 1 mm in diameter and heated to incandescence by passing a current through it. A small hole has been drilled in its wall. It is clear from the figure that the radiation emerging from this hole is much more intense than that from the outer wall of the cavity, even though the temperatures of the outer and inner walls are more or less equal.

There are three interrelated properties of cavity radiation—all well verified in the laboratory—that any theory of cavity radiation must explain.

1. *The Stefan–Boltzmann law.* The total radiated power per unit area of the cavity aperture, summed over all wavelengths, is called its *radiant intensity $I(T)$* and is related to the temperature by

$$I(T) = \sigma T^4, \qquad (1)$$

in which $\sigma\,(= 5.670 \times 10^{-8}\ \text{W/m}^2 \cdot \text{K}^4)$ is a universal constant, called the *Stefan–Boltzmann constant.* Ordinary hot objects always radiate less efficiently than do cavity radiators. We express this by generalizing Eq. 1 to

$$I(T) = \epsilon \sigma T^4, \qquad (2)$$

in which ϵ, a dimensionless quantity, is called the *emissivity* of the surface material. For a cavity radiator, $\epsilon = 1$, but for the surfaces of ordinary objects, the emissivity is always less than unity and is almost always a function of temperature.

2. *The spectral radiancy.* The *spectral radiancy $R(\lambda)$* tells us how the intensity of the cavity radiation varies with

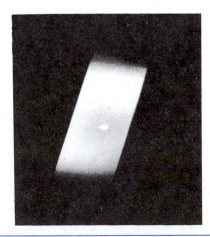

Figure 2 An incandescent tungsten tube with a small hole drilled in its wall. The radiation emerging from the hole is cavity radiation.

wavelength for a given temperature. It is defined so that the product $R(\lambda)\,d\lambda$ gives the radiated power per unit area that lies in the wavelength band that extends from λ to $\lambda + d\lambda$. $R(\lambda)$ is a statistical distribution function of the same type we considered in Chapter 24. We can find the radiant intensity $I(T)$ for any temperature by adding up (that is, by integrating) the spectral radiancy over the complete range of wavelengths. Thus

$$I(T) = \int_0^\infty R(\lambda)\,d\lambda \qquad \text{(fixed } T). \qquad (3)$$

Figure 3 shows the spectral radiancy for cavity radiation at four selected temperatures. Equation 3 shows that we can interpret the radiant intensity $I(T)$ as the area under the appropriate spectral radiancy curve. We see from the figure that, as the temperature increases, so does this area and thus the radiant intensity, as Eq. 1 predicts.

3. *The Wien displacement law.* We can see from the spectral radiancy curves of Fig. 3 that $\lambda_{\max}$, the wavelength at which the spectral radiancy is a maximum, decreases as the temperature increases. Wilhelm Wien (German, 1864–1928) deduced that $\lambda_{\max}$ varies as $1/T$ and that the product $\lambda_{\max} T$ is a universal constant. Its measured value is

$$\lambda_{\max} T = 2898 \ \mu\text{m}\cdot\text{K}. \qquad (4)$$

This relationship is called the *Wien displacement law;* Wien was awarded the 1911 Nobel prize in physics for his research into thermal radiation.

Sample Problem 1   How hot is a star? The "surfaces" of stars are not sharp boundaries like the surface of the Earth. Most of the radiation that a star emits is in thermal equilibrium with the hot gases that make up the star's outer layers. Without too much error, then, we can treat starlight as cavity radiation. Here are the wavelengths at which the spectral radiancies of three stars have their maximum values:

Star	λ_{max}	Appearance
Sirius	240 nm	Blue-white
Sun	500 nm	Yellow
Betelgeuse	850 nm	Red

(*a*) What are the surface temperatures of these stars? (*b*) What are the radiant intensities of these three stars? (*c*) The radius r of the Sun is 7.0×10^8 m and that of Betelgeuse is over 500 times larger, or 4.0×10^{11} m. What is the total radiated power output (that is, the *luminosity L*) of these stars?

Solution   (*a*) From Eq. 4 we find, for Sirius,

$$T = \frac{2898 \ \mu\text{m}\cdot\text{K}}{\lambda_{\max}}$$

$$= \left(\frac{2898 \ \mu\text{m}\cdot\text{K}}{240 \ \text{nm}}\right)\left(\frac{1000 \ \text{nm}}{1 \ \mu\text{m}}\right) = 12,000 \ \text{K}.$$

The temperatures for the Sun and for Betelgeuse work out in the same way to be 5800 K and 3400 K, respectively. At 5800 K, most of the radiation from the Sun's surface lies within the visible region of the spectrum. This suggests that over ages of

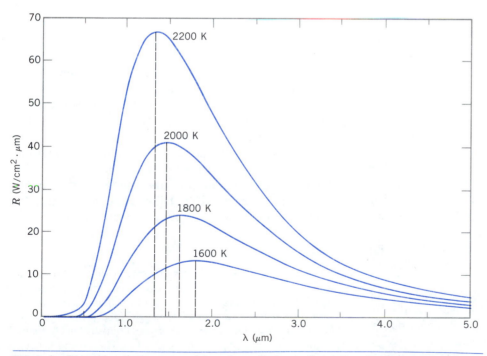

Figure 3   Spectral radiancy curves for cavity radiation at four selected temperatures. Note that as the temperature increases, the wavelength of the maximum spectral radiancy shifts to lower values.

evolution, eyes have adapted to the Sun to become most sensitive to those wavelengths that it radiates most intensely.

(*b*) For Sirius we have, from the Stefan–Boltzmann law (Eq. 1)

$$I = \sigma T^4 = (5.67 \times 10^{-8} \text{ W/m}^2 \cdot \text{K}^4)(12{,}000 \text{ K})^4$$
$$= 1.2 \times 10^9 \text{ W/m}^2.$$

The radiant intensities for the Sun and for Betelgeuse work out to be 6.4×10^7 W/m^2 and 7.7×10^6 W/m^2, respectively.

(*c*) We find the luminosity of a star by multiplying its radiant intensity by its surface area. Thus, for the Sun,

$$L = I(4\pi r^2) = (6.4 \times 10^7 \text{ W/m}^2)(4\pi)(7.0 \times 10^8 \text{ m})^2$$
$$= 3.9 \times 10^{26} \text{ W}.$$

For Betelgeuse the luminosity works out to be 1.5×10^{31} W, about 38,000 times larger. The enormous size of Betelgeuse, which is classified as a "red giant," much more than makes up for the relatively low radiant intensity associated with its low surface temperature.

The colors of stars are not strikingly apparent to the average observer because the retinal cones, which are responsible for color vision, do not function well in dim light. If this were not so, the night sky would be spangled with color.

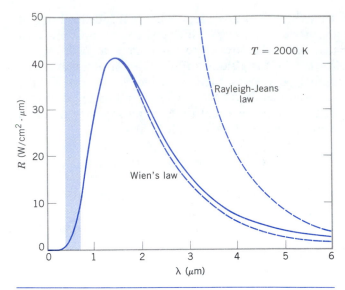

Figure 4   The solid curve shows the experimental spectral radiancy for radiation from a cavity at 2000 K. The predictions of the classical Rayleigh–Jeans law and Wien's law are shown as dashed lines. The shaded vertical bar represents the range of visible wavelengths.

49-2   PLANCK'S RADIATION LAW

Is there a simple formula, derivable from basic principles, that fits the experimental radiancy curves of Fig. 3? In September 1900 there were two suggested formulas, neither of which could fit the curves over the entire range of wavelengths.

The first, due originally to Lord Rayleigh but later derived independently by Einstein and modified by James Jeans, was developed rigorously from its classical base. Unfortunately, it completely fails to fit the curves, not even passing through a maximum. However, the Rayleigh–Jeans formula, as it is called, *does* fit the curves quite well in the limit of very long wavelengths. Figure 4 shows the spectral radiancy curve for cavity radiation at 2000 K, along with the Rayleigh–Jeans prediction. The good fit we speak of occurs for wavelengths much greater than 50 μm, far beyond the scale of that figure. The Rayleigh–Jeans formula, unsatisfactory though it may be, is the best that classical physics has to offer.

Wilhelm Wien also derived a theoretical expression for the spectral radiancy. His formula (see also Fig. 4) is much better. It fits the curves quite well at short wavelengths, passes through a maximum, but departs noticeably at the long-wavelength end of the scale. However, Wien's formula was not based on classical radiation theory but instead on a conjecture—it has been called a "guess"—that there is an analogy between the spectral radiancy curves and the Maxwell speed distribution curves for the molecules of an ideal gas.

Thus we have two formulas, one agreeing with experiment at long wavelengths and the other at short wavelengths. Max Planck,* seeking to reconcile these two radiation laws, made an inspired interpolation between them that turned out to fit the data at *all* wavelengths. Planck's radiation formula, announced to the Berlin Physical Society on October 19, 1900, is

$$R(\lambda) = \frac{a}{\lambda^5} \frac{1}{e^{b/\lambda T} - 1}, \tag{5}$$

in which a and b are empirical constants, chosen to give the best fit of Eq. 5 to the experimental data. Figure 5 shows how good the agreement is. Even though correct, Planck's formula was originally only empirical and did not constitute a true theory.

Planck set to work at once to derive his formula from simple assumptions and, in 2 months, he succeeded. In the process he recast his formula slightly, presenting the two arbitrary constants it contained in a different form. In this new notation, Planck's radiation law becomes

---

* Max Planck (1858–1947) was a German theoretical physicist whose specialization in thermodynamics led him to the study of thermal radiation and the discovery of the quantization of energy, for which he was awarded the 1918 Nobel prize in physics. Under his leadership, theoretical physics flourished in Germany in the 1920s; young physicists trained by Planck and his colleagues produced a complete mathematical formulation of the quantum theory. In his later life, Planck wrote extensively on religious and philosophical issues.

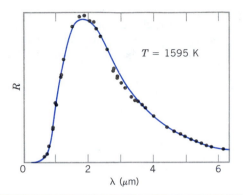

Figure 5 Planck's radiation law fitted to experimental data for a cavity radiator at 1595 K.

$$R(\lambda) = \frac{2\pi c^2 h}{\lambda^5} \frac{1}{e^{hc/\lambda kT} - 1}. \tag{6}$$

The two adjustable constants a and b in Eq. 5 are here replaced by quantities involving two different constants, the *Boltzmann constant k* (see Section 23-1) and a new constant, now called the *Planck constant h;* the quantity c is the speed of light.

By fitting Eq. 6 to the experimental data, Planck could find values for k and h. His values were within a percent or so of their presently accepted values, which are

$$k = 1.381 \times 10^{-23} \text{ J/K}$$

and

$$h = 6.626 \times 10^{-34} \text{ J} \cdot \text{s}.$$

Sample Problem 2 Figure 4 suggests that Planck's radiation law (Eq. 6) approaches the classical Rayleigh–Jeans law at long wavelengths. To what expression does Planck's law reduce as $\lambda \to \infty$?

Solution For algebraic convenience, we can write Eq. 6 in the form

$$R(\lambda) = \frac{2\pi c^2 h}{\lambda^5} \frac{1}{e^x - 1},$$

in which $x = hc/\lambda kT$. As $\lambda \to \infty$, we see that $x \to 0$. Recalling that

$$e^x = 1 + x + \frac{x^2}{2!} + \frac{x^3}{3!} + \cdots$$

(see Appendix H) allows us to make the approximation

$$e^x - 1 \approx x.$$

Thus we have

$$R(\lambda) \approx \frac{2\pi c^2 h}{\lambda^5} \frac{1}{x} = \frac{2\pi c^2 h}{\lambda^5} \left(\frac{\lambda kT}{hc} \right) = \frac{2\pi ckT}{\lambda^4}.$$

Note that the Planck constant h, a sure identifier of a quantum formula, conveniently cancels out as we approach the classical long-wave limit. The above result, in fact, is precisely the classical Rayleigh–Jeans expression for the spectral radiancy.

49-3   THE QUANTIZATION OF ENERGY

We turn now to the assumptions made by Planck in deriving his radiation law and to the significance of the constant h that appears in it. These assumptions and their consequences were not immediately clear to Planck's contemporaries or for that matter (as he confirmed later) to Planck himself. In what follows we describe the situation as it appeared some 6 or 7 years after Planck first advanced his theory. It seems to be true that the basic premise underlying Planck's radiation law — the quantization of energy — was not understood at any earlier date.

Planck derived his radiation law by analyzing the interplay between the radiation in the cavity volume and the atoms that make up the cavity walls. He assumed that these atoms behave like tiny oscillators, each with a characteristic frequency of oscillation. These oscillators radiate energy into the cavity and absorb energy from it. It should be possible to deduce the characteristics of the cavity radiation from the characteristics of the oscillators that generate it.

Classically, the energy of these tiny oscillators is a smoothly continuous variable. We certainly assume this for large-scale oscillators such as pendulums or mass–spring systems. It turns out, however, that in order to derive Planck's radiation law it is necessary to make a radical assumption; namely, *atomic oscillators may not emit or absorb any energy E but only energies chosen from a discrete set, defined by*

$$E = nhv, \qquad n = 1, 2, 3, \ldots, \tag{7}$$

in which v is the oscillator frequency. Here the Planck constant h is introduced into physics for the first time. We say that the energy of an atomic oscillator is *quantized* and that the integer n is a *quantum number*. Equation 7 tells us that the oscillator energy levels are evenly spaced, the interval between adjacent levels being hv; see Fig. 6.

The assumption of energy quantization is indeed a radical one, and Planck himself resisted accepting it for many years. In his words, "My futile attempts to fit the elementary quantum of action [that is, h] somehow into the classical theory continued for a number of years, and they cost me a great deal of effort." Max von Laue, the 1914 Nobel laureate in physics and a student of Planck's, has written: "After 1900 Planck strove for many years to bridge, if not to close, the gap between the older and the quantum physics. The effort failed, but it had value in that it provided the most convincing proof that the two could not be joined."

Let us look at energy quantization in the context of a large-scale oscillator such as a swinging pendulum. Our experience suggests that a pendulum can oscillate with *any* reasonable total energy and not only with certain selected energies. As friction causes the pendulum ampli-

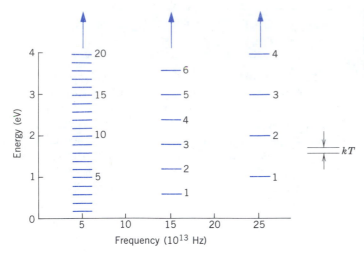

Figure 6 The energy levels for atomic oscillators at three selected frequencies. The quantum numbers of some of the levels are indicated. On the right is shown the energy kT for a classical oscillator at 2000 K.

tude to decay, it seems that the energy is dissipated in a perfectly continuous way and not in "jumps" or "quanta." However, because the Planck constant is so small, there is no basis in such everyday experience to dismiss energy quantization as a violation of "common sense." The "jumps" are there; they are just far too small for us to detect.

We would not apply quantum theory to a pendulum because classical theory works perfectly well in that case. We now see classical theory as a limiting case of quantum theory, the two being connected by the *correspondence principle,* which states that:

Quantum theory must agree with classical theory in the limit in which classical theory is known to agree with experiment.

Another way of stating the correspondence principle is:

Quantum theory must agree with classical theory in the limit of large quantum numbers.

It is in this way that the swinging pendulum and the oscillating atom relate to each other. The classical limit is illustrated in the following sample problem.

Sample Problem 3 A 300-g body, connected to a spring whose force constant k is 3.0 N/m, is oscillating with an amplitude A of 10 cm. Treat this system as a quantum oscillator and find (a) the energy interval between adjacent energy levels and (b) the quantum number that describes the oscillations.

Solution (a) The frequency of oscillation is found from

$$\nu = \frac{1}{2\pi} \sqrt{\frac{k}{m}} = \frac{1}{2\pi} \sqrt{\frac{3.0 \text{ N/m}}{0.3 \text{ kg}}} = 0.50 \text{ s}^{-1}.$$

The total mechanical energy E of the oscillating system is

$$E = \tfrac{1}{2}kA^2 = \tfrac{1}{2}(3.0 \text{ N/m})(0.10 \text{ m})^2 = 0.015 \text{ J}.$$

As the amplitude of the oscillations dies away by friction, quantum theory predicts that the energy E will fall in "jumps" whose size is

$$\Delta E = h\nu = (6.63 \times 10^{-34} \text{ J·s})(0.50 \text{ s}^{-1})$$
$$= 3.3 \times 10^{-34} \text{ J}.$$

Thus

$$\frac{\Delta E}{E} = \frac{3.3 \times 10^{-34} \text{ J}}{0.015 \text{ J}} = 2.2 \times 10^{-32}.$$

Energy measurements of such precision simply cannot be made. The quantum jumps of this oscillator are too small to be detected, and we are quite safe in treating the problem by the methods of classical physics.

(b) From the quantization relation, Eq. 7, we have

$$n = \frac{E}{h\nu} = \frac{0.015 \text{ J}}{3.3 \times 10^{-34} \text{ J}} = 4.6 \times 10^{31},$$

an enormous number! It is not surprising that we cannot detect energy quantization in the operation of an oscillator; we cannot measure changes of one unit out of 4.6×10^{31}.

The quantization of energy simply does not show up for large-scale oscillators. The smallness of the Planck constant h makes the graininess in the energy too fine to detect. In much the same way, we cannot tell by waving our hand through air that it is made up of molecules.

The Planck constant might as well be zero as far as classical systems are concerned and, indeed, one way to reduce quantum formulas to their limiting classical counterparts is to let $h \rightarrow 0$ in those formulas. In a similar way, we reduce relativistic formulas to their limiting classical counterparts by letting $c \rightarrow \infty$, where c is the speed of light.

This leaves the question: "Why should letting the wavelength increase mean that we are approaching a realm in which classical physics holds?" The answer is that as the wavelength increases, the frequency decreases and thus the basic energy quantum ($= h\nu$) becomes smaller. To tell whether we are in a classical or a quantum situation we must compare this quantity with kT, which is a (classical) measure of the mean translational energy of a particle at temperature T. If $hv \ll kT$, the "graininess" of the energy of the atomic oscillators (which is measured by hv) will not be noticed and we are in the classical realm. To sum up then, we approach classical situations as $v \rightarrow 0$, as $\lambda \rightarrow \infty$, or (for that matter) as $T \rightarrow \infty$; all three lead to the condition that $hv \ll kT$.

49-4   THE HEAT CAPACITY OF SOLIDS

Energy quantization was slow to be accepted, not an unusual fate for a radically new idea. It is not hard to see why. The systems whose energies were first quantized were the hypothetical "oscillators" that Planck assumed to form the walls of a cavity radiator. In fact, there are no such simple, one-dimensional, harmonic oscillators. The atoms that make up the wall are far more complex.

Energy quantization started to come into its own only after 1907, when Einstein showed that the same ideas that had worked so well for the cavity radiation problem could be used to solve another problem, that of the heat capacities of solids. Here, as we shall see, the systems whose energies are to be quantized are real and familiar atoms.

If you transfer heat Q to 1 mole of a solid and if a temperature rise ΔT results, the *molar heat capacity* is defined from

$$C_V = \frac{Q}{\Delta T} \qquad \text{(constant volume)}. \qquad (8)$$

We have chosen to transfer heat under conditions of constant volume, so that the distances between atoms remain constant and any added energy appears entirely as energy of oscillation of the atoms about their fixed lattice sites. We take the amount of the substance to be 1 mole, so that comparisons from element to element can be made on the basis of the same number of atoms. See Section 25-3 for more details on molar heat capacities and for the relationship between C_V, which is easier to calculate, and C_p, which is easier to measure.

Table 1 shows the molar heat capacities of some elemental solids at or near room temperature. A glance at the table shows a regularity known as the Dulong and Petit rule, after the investigators who first pointed it out in 1819. It asserts simply that, with a few exceptions, all solids have the same molar heat capacity, namely, about 25 J/mol·K. Values that were substantially less than this were called "anomalous" in those early days.

Figure 7, which shows the molar heat capacity of lead, aluminum, and carbon as a function of temperature, clarifies the situation. We see that C_V for all three elements approaches the same limiting value at high temperatures. That carbon appears "anomalous" in Table 1 simply re-

TABLE 1   MOLAR HEAT CAPACITIES OF SOME SOLIDS[a]

Solid	C_V (J/mol·K)	
Aluminum	23	
Beryllium		11
Bismuth	25	
Boron		13
Cadmium	25	
Carbon (diamond)		6
Copper	24	
Gold	25	
Lead	25	
Platinum	25	
Silver	24	
Tungsten	24	

[a] All measurements were made at room temperature; three "anomalous" values have been offset for emphasis.

flects the fact that, for this element, room temperature is not a very high temperature.

What does classical physics predict for the molar heat capacity of a solid?

The atoms in a solid are arranged in a three-dimensional lattice. Each atom, bound to its lattice site by electromagnetic forces, oscillates about that site with an amplitude that increases as the temperature increases. Each atom behaves like a tiny oscillator with three independent degrees of freedom, corresponding to three independent directional axes along which the atom is free to move.

The classical equipartition of energy theorem associates an energy of $\frac{1}{2}kT$ with each degree of freedom. The three-dimensional oscillator has six degrees of freedom, two for each direction (corresponding to the kinetic and potential energies for motion of the oscillation in that direction). The internal energy per mole of a solid is then

$$E_{int} = 6(\tfrac{1}{2}kT)N_A = 3RT, \qquad (9)$$

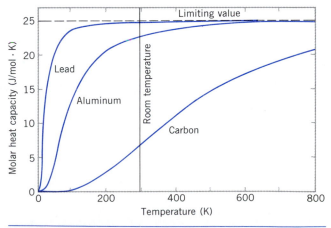

Figure 7   The molar heat capacities of three solids as a function of temperature.

in which N_A is the Avogadro constant and R is the universal gas constant.

If the solid is held at a constant volume, we can replace Q in Eq. 8 (the heat transferred per mole) by ΔE_{int}, the change in internal energy per mole. Doing so yields $C_V = \Delta E_{int}/\Delta T$, which becomes

$$C_V = \frac{dE_{int}}{dT} \qquad (10)$$

in the differential limit. Substituting from Eq. 9 yields finally

$$C_V = \frac{d}{dT}(3RT) = 3R. \qquad (11)$$

Classical theory predicts the molar heat capacity to be constant, the same for all substances, and independent of temperature. Substituting the value of R ($= 8.31$ J/mol·K) yields $C_V = 24.9$ J/mol·K. This agrees very well with the limiting value of C_V at high temperatures, as Fig. 7 and Table 1 show. However, there is no indication from this classical theory of the variation at lower temperatures that is shown in Fig. 7.

Quantum Theory of Heat Capacity

We turn now to the prediction of quantum theory. Einstein assumed that the energies of the atomic oscillators in the solid were quantized according to Eq. 7, and he assigned to each oscillator an average energy per direction, not of kT as in the classical case, but of

$$E = \frac{hv}{e^{hv/kT} - 1}. \qquad (12)$$

This is the same expression used by Planck for the average energy of the oscillators in the cavity radiation problem. In Eq. 12, v is the natural vibrational frequency of the oscillating atom, which Einstein left as an adjustable constant.

Multiplying Eq. 12 by the Avogadro constant and also by a factor of 3 to take account of the three directions, we obtain the internal energy per mole:

$$E_{int} = \frac{3N_A hv}{e^{hv/kT} - 1}. \qquad (13)$$

Differentiating, as in Eq. 10, gives eventually

$$C_V = \frac{dE_{int}}{dT} = 3R(hv/kT)^2 \frac{e^{hv/kT}}{(e^{hv/kT} - 1)^2} \qquad (14)$$

as Einstein's prediction for the molar heat capacity. There is only one adjustable parameter in Eq. 14, the oscillator frequency v. Commonly, this frequency v is expressed in terms of a characteristic *Einstein temperature* $T_E = hv/k$. This temperature can be chosen so that Einstein's equation fits the data rather well, although there are small deviations at low temperatures, deviations that had not yet been experimentally established when Einstein proposed his theory.

The failure to agree with experiment at low temperature can be traced to the fact that Einstein—perhaps deliberately—made an overly simple assumption, namely, that the oscillations of a particular atom are not influenced by those of its neighbors. In 1912 the Dutch physicist Peter Debye refined Einstein's theory by taking the interaction of the atomic oscillators with neighboring atoms into account. Figure 8 shows the excellent agreement of the Debye theory with experiment for a number of solids. The temperature scale in that figure is dimensionless, T_D being a constant that has a different value for each material. When these characteristic *Debye temperatures,* as they are called, are properly assigned, we see how nicely all the experimental points fall on the same theoretical curve. This agreement is a major triumph for quantum theory!

Figures 7 and 8 immediately suggest the explanation for the "anomalous" values of Table 1. For these substances, T_D is much greater than room temperature, so that the heat capacity has not yet reached its limiting value.

At high temperatures, you can show (see Problem 22) that Einstein's expression for the heat capacity (Eq. 14) reduces to the classical result (Eq. 11). This occurs for the same reasons that we discussed at the end of Sample

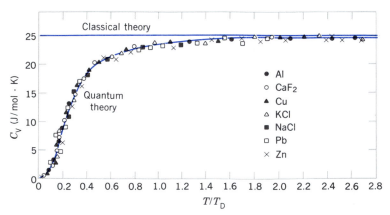

Figure 8 The quantum theory result for the heat capacity of solids is in excellent agreement with the experimental results. The horizontal scale is the dimensionless ratio of the temperature T to the Debye temperature T_D, the latter having a characteristic value for each substance.

Problem 3. In a solid, the frequency v of the atomic oscillators was assumed by Einstein to have a single constant value, characteristic of the substance. Thus as $T \to \infty$, we approach the condition in which $hv \ll kT$. This, as we have seen, means that the energy interval between adjacent levels for the atomic oscillators ($= hv$) is much less than the mean translational energy of the atoms (measured classically by kT). Under these conditions, the energy quantization of the atomic oscillators is not apparent, and classical conditions hold.

Sample Problem 4   In terms of the Einstein temperature T_E, find the temperature at which the heat capacity of a substance has half its classical value.

Solution   The classical value is $3R$, so we seek the temperature at which C_V in Eq. 14 has the value $3R/2$, or

$$3R \left(\frac{hv}{kT}\right)^2 \frac{e^{hv/kT}}{(e^{hv/kT} - 1)^2} = \frac{3R}{2}.$$

Letting $x = hv/kT = T_E/T$, we can write this as

$$x^2 \frac{e^x}{(e^x - 1)^2} = \frac{1}{2}.$$

There is no analytic technique for solving this equation. A numerical solution can be found on a calculator by trial and error or on a computer by calculating and displaying a table of values of the function on the left-hand side and noting the value of x when the function has the value $\frac{1}{2}$. The result is

$$x = 2.98.$$

Since $x = T_E/T$, we have $T/T_E = x^{-1} = 0.336$, or

$$T = 0.336 T_E.$$

49-5   THE PHOTOELECTRIC EFFECT

We were led to the idea of energy quantization by looking at the interplay between matter and radiation at the walls of a cavity radiator. Here we consider another example of a radiation–matter interaction, the *photoelectric effect*. It involves the Planck constant in a central way and extends the idea of quantization to the very nature of radiation itself.

Figure 9 shows a typical apparatus used to study the photoelectric effect. Light of frequency v falls on a metal surface (emitter E) and, if the frequency is high enough, the light will eject electrons out of the surface. If we set up a suitable potential difference V between E and the collector C, we can collect these *photoelectrons,* as we call them, and measure them as a photoelectric current i.

The potential difference V that acts between the emitter and the collector is not the same as the potential differ-

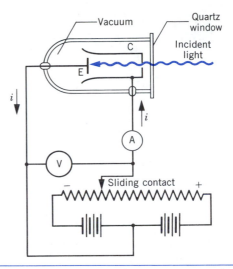

Figure 9   An apparatus for studying the photoelectric effect. The arrows show the direction of the current in the external circuit, which is opposite to the direction of motion of the electrons. The voltmeter V measures the externally applied voltage V_{ext}.

ence V_{ext} supplied by the external battery and read on the voltmeter in Fig. 9. There is also a second emf—a hidden battery, if you will—associated with the fact that the emitter and the collector are almost always made of different materials. If suitable precautions are taken, this *contact potential difference* V_{cpd} remains constant throughout the experiment. The potential difference V that the electrons "see" is the algebraic sum of these two quantities, or

$$V = V_{ext} + V_{cpd}. \tag{15}$$

In all that follows we shall assume that this contact potential difference has been measured and taken into account, and we shall express all our results in terms of V as defined by Eq. 15.

Figure 10 (curve a) shows the photoelectric current as a function of the potential difference V. We see that if V is positive and large enough, the photoelectric current reaches a constant saturation value, at which *all* photoelectrons ejected from E are collected by C.

If we reduce V to zero and then reverse it, the photoelectric current does not immediately drop to zero because the electrons emerge from emitter E with nonzero speeds. Some will reach the collector even though the potential difference opposes their motion. However, if we make the reversed potential difference large enough, we reach a value V_0, called the *stopping potential,* at which the photoelectric current does indeed drop to zero. This potential difference, multiplied by the electronic charge e, gives us the kinetic energy K_{max} of the most energetic of the emitted photoelectrons:

$$K_{max} = eV_0. \tag{16}$$

The stopping potential V_0, and thus K_{max}, is independent of the intensity of the incident light. Curve b in Fig.

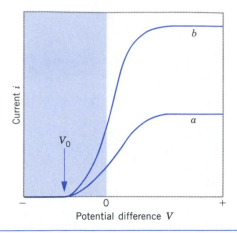

Figure 10   A plot (not to scale) of data taken with the apparatus of Fig. 9. The intensity of the incident light is twice as great for curve *b* as for curve *a*.

10, in which the light intensity has been doubled, shows this. Although the saturation current is also doubled, the stopping potential remains unchanged.

Figure 11 is a plot of the stopping potential as a function of the frequency of the incident light. We see by extrapolation that there is a sharp *cutoff frequency* ν_0 corresponding to a stopping potential of zero. For light of a lower frequency than this, no photoelectrons at all are emitted. There simply is no photoelectric effect.

Three major features of the photoelectric effect cannot be explained in terms of the classical wave theory of light. As for the cavity radiation and the heat capacity problems, the failure of classical wave theory in these cases is not a matter of a small numerical disagreement. The failure is total and indisputable. Here are the three problems:

1.   *The intensity problem.* Wave theory requires that the oscillating electric vector **E** of the light wave increases in amplitude as the intensity of the light beam is increased. Since the force applied to the electron is $e\mathbf{E}$, this suggests that the *kinetic energy* of the photoelectrons should also

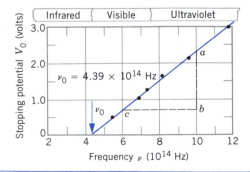

Figure 11   The stopping potential as a function of frequency for a sodium surface. The data come from Millikan's measurements in 1916.

increase as the light beam is made more intense. However, Fig. 10 shows that K_{max} ($= eV_0$) is *independent of the light intensity;* this has been tested over a range of intensities of 10^7.

2.   *The frequency problem.* According to the wave theory, the photoelectric effect should occur for any frequency of the light, provided only that the light is intense enough to supply the energy needed to eject the photoelectrons. However, Fig. 11 shows that there exists, for each surface, a characteristic cutoff frequency ν_0. *For frequencies less than ν_0, the photoelectric effect does not occur, no matter how intense the illumination.*

3.   *The time delay problem.* If the energy acquired by a photoelectron is absorbed directly from the wave incident on the metal plate, the "effective target area" for an electron in the metal is limited and probably not much more than that of a circle of diameter roughly equal to that of an atom. In the classical theory, the light energy is uniformly distributed over the wavefront. Thus, if the light is feeble enough, there should be a measurable time lag, which we shall estimate in Sample Problem 5, between the impinging of the light on the surface and the ejection of the photoelectron. During this interval the electron should be absorbing energy from the beam until it had accumulated enough to escape. However, *no detectable time lag has ever been measured.*

In the next section we see how quantum theory solves these problems in providing the correct interpretation of the photoelectric effect.

Sample Problem 5   A potassium foil is placed a distance r ($= 0.5$ m) from a light source whose output power P_0 is 1.0 W. How long would it take for the foil to soak up enough energy ($= 1.8$ eV) from the beam to eject an electron? Assume that the ejected photoelectron collected its energy from a circular area of the foil whose radius equals the radius of a potassium atom (1.3×10^{-10} m).

Solution   If the source radiates uniformly in all directions, the intensity I of the light at a distance r is given by

$$I = \frac{P_0}{4\pi r^2} = \frac{1.0 \text{ W}}{4\pi(0.5 \text{ m})^2} = 0.32 \text{ W/m}^2.$$

The target area A is $\pi(1.3 \times 10^{-10} \text{ m})^2$ or 5.3×10^{-20} m^2, so that the rate at which energy falls on the target is given by

$$P = IA = (0.32 \text{ W/m}^2)(5.3 \times 10^{-20} \text{ m}^2)$$
$$= 1.7 \times 10^{-20} \text{ J/s}.$$

If all this incoming energy is absorbed, the time required to accumulate enough energy for the electron to escape is

$$t = \left(\frac{1.8 \text{ eV}}{1.7 \times 10^{-20} \text{ J/s}}\right)\left(\frac{1.6 \times 10^{-19} \text{ J}}{1 \text{ eV}}\right) = 17 \text{ s}.$$

Our selection of a radius for the effective target area was somewhat arbitrary, but no matter what reasonable area we choose, we would still calculate a "soak-up time" within the range of easy measurement. However, no time delay has ever been observed under any circumstances, the early experiments setting an upper limit of about 10^{-9} s for such delays.

49-6   EINSTEIN'S PHOTON THEORY

In 1905 Einstein made a remarkable assumption about the nature of light; namely, that, under some circumstances, it behaves as if its energy is concentrated into localized bundles, later called *photons*. The energy E of a single photon is given by

$$E = h\nu, \tag{17}$$

where ν is the frequency of the light. This notion that a light beam behaves like a stream of particles is in sharp contrast to the notion that it behaves like a wave. In the wave theory of light, the energy is *not* concentrated into bundles but is spread out uniformly over the wavefronts.

When Planck, in 1900, derived his radiation law and first introduced the quantity h into physics, he made use of the relation $E = h\nu$. He applied it, however, not to the radiation within the cavity but to the atomic oscillators that made up its walls. Planck treated the cavity radiation on the basis of wave theory, but Einstein was later able to derive Planck's radiation law on the basis of his photon concept. His method was both clear and simple and avoided many of the special assumptions that Planck had found it necessary to make in his pioneering effort.

If we apply Einstein's photon concept to the photoelectric effect, we can write

$$h\nu = \phi + K_{max}, \tag{18}$$

where $h\nu$ is the energy of the photon. Equation 18 says that a *single* photon carries an energy $h\nu$ into the surface where it is absorbed by a *single* electron. Part of this energy (ϕ, called the *work function* of the emitting surface) is used in causing the electron to escape from the metal surface. The excess energy ($h\nu - \phi$) becomes the electron kinetic energy; if the electron does not lose energy by internal collisions as it escapes from the metal, it will still have this much kinetic energy after it emerges. Thus K_{max} represents the maximum kinetic energy that the photoelectron can have outside the surface.*

Consider how Einstein's photon hypothesis meets the three objections raised against the wave-theory interpre-

tation of the photoelectric effect. As for the first objection ("the intensity problem"), there is complete agreement of the photon theory with experiment. If we double the light intensity, we double the number of photons and thus double the photoelectric current; we do not change the energy of the individual photons or the nature of the individual photoelectric processes described by Eq. 18.

The second objection ("the frequency problem") is met by Eq. 18. If K_{max} equals zero, we have

$$h\nu_0 = \phi,$$

which asserts that the photon has just enough energy to eject the photoelectrons and none extra to appear as kinetic energy. If ν is reduced below ν_0, $h\nu$ will be smaller than ϕ and the individual photons, no matter how many of them there are (that is, no matter how intense the illumination), will not have enough energy to eject photoelectrons.

The third objection ("the time delay problem") follows from the photon theory because the required energy is supplied in a concentrated bundle. It is not spread uniformly over the beam cross section as in the wave theory.

Let us rewrite Einstein's photoelectric equation (Eq. 18) by substituting for K_{max} from Eq. 16. This yields, after rearrangement,

$$V_0 = (h/e)\nu - (\phi/e). \tag{19}$$

Thus Einstein's theory predicts a linear relationship between V_0 and ν, in complete agreement with experiment; see Fig. 11. The slope of the experimental curve in this figure should be h/e, so

$$\frac{h}{e} = \frac{ab}{bc} = \frac{2.30 \text{ V} - 0.68 \text{ V}}{(10 \times 10^{14} - 6.0 \times 10^{14}) \text{ Hz}}$$
$$= 4.1 \times 10^{-15} \text{ V} \cdot \text{s}.$$

We can find h by multiplying this ratio by the electron charge e,

$$h = (4.1 \times 10^{-15} \text{ V} \cdot \text{s})(1.6 \times 10^{-19} \text{ C})$$
$$= 6.6 \times 10^{-34} \text{ J} \cdot \text{s}.$$

From a more careful analysis of these and other data, including data taken with lithium surfaces, Millikan found the value $h = 6.57 \times 10^{-34}$ J·s with an accuracy of about 0.5%. This agreement with the value of h derived from Planck's radiation formula is a striking confirmation of Einstein's photon concept.

When Einstein first advanced his photon theory of light, the facts of photoelectricity were not nearly as well established experimentally as we have described. Precise photoelectric measurements are difficult, and it was not until 1916 that Millikan successfully subjected Einstein's photoelectric equation to rigorous experimental test. Although Millikan showed that this equation agreed with experiment in every detail, he himself remained unconvinced that Einstein's light particles were real. He wrote of Einstein's "bold, not to say reckless, hypothesis" and

---

* The work function represents the energy needed to remove the least tightly bound electrons from the surface. More tightly bound electrons require a larger energy and (for a fixed photon energy) emerge with a kinetic energy smaller than K_{max}.

wrote further that Einstein's photon concept "seems at present to be wholly untenable."

Planck, the very originator of the constant h, did not at once accept Einstein's photons either. In recommending Einstein for membership in the Royal Prussian Academy of Sciences in 1913, he wrote: "that he may sometimes have missed the target in his speculations, as for example in his theory of light quanta, cannot really be held against him." It is not unusual for truly novel ideas to be accepted only slowly, even by leading scientists such as Millikan and Planck. It was, incidentally, for his photon theory as applied to the photoelectric effect that Einstein received the Nobel prize in physics for 1921.

Sample Problem 6   Find the work function of sodium from the data plotted in Fig. 11.

Solution   The intercept of the straight line in Fig. 11 on the frequency axis is the cutoff frequency ν_0. Putting $V_0 = 0$ and $\nu = \nu_0$ in Eq. 19 yields

$$\phi = h\nu_0 = (6.63 \times 10^{-34} \text{ J·s})(4.39 \times 10^{14} \text{ Hz})$$
$$= 2.91 \times 10^{-19} \text{ J} = 1.82 \text{ eV}.$$

We note from Eq. 19 that a determination of the Planck constant h involves only the slope of the straight line in Fig. 11 and a determination of the work function ϕ involves only the intercept. Convince yourself that in the first case you need not take the contact potential difference V_{cpd} into account but in the second case you must do so.

Sample Problem 7   At what rate per unit area do photons strike the metal plate in Sample Problem 5? Assume a wavelength of 589 nm (yellow sodium light).

Solution   Recall our previous definition (see Section 41-4) of the intensity of light: energy per unit time per unit area (the area being taken as perpendicular to the direction of propagation of the light). Here we consider the intensity (for monochromatic light) in terms of photons as the energy per photon times the rate per unit area at which the photons strike a surface perpendicular to their motion. The two interpretations of intensity are equivalent.

The intensity of the light falling on the plate is, from Sample Problem 5,

$$I = (0.32 \text{ J/m}^2\text{·s})(1 \text{ eV}/1.6 \times 10^{-19} \text{ J})$$
$$= 2.0 \times 10^{18} \text{ eV/m}^2\text{·s}.$$

Each photon has an energy given by

$$E = h\nu = \frac{hc}{\lambda} = \frac{(6.63 \times 10^{-34} \text{ J·s})(3.00 \times 10^8 \text{ m/s})}{5.89 \times 10^{-7} \text{ m}}$$
$$= (3.4 \times 10^{-19} \text{ J})\left(\frac{1 \text{ eV}}{1.6 \times 10^{-19} \text{ J}}\right) = 2.1 \text{ eV}.$$

The rate per unit area r at which photons strike the plate is then the intensity divided by the energy per photon, or

$$r = \frac{I}{E} = \frac{2.0 \times 10^{18} \text{ eV/m}^2\text{·s}}{2.1 \text{ eV/photon}} = 9.5 \times 10^{17} \text{ photons/m}^2\text{·s}.$$

Even at this modest light intensity the photon rate is very great, with about 10^{12} photons falling on 1 mm^2 each second.

49-7   THE COMPTON EFFECT

Cavity radiation, which involved largely the infrared part of the spectrum, was our first example of the interaction of radiation with matter. The photoelectric effect, our second example, involved visible and ultraviolet light. Here we describe the *Compton* * *effect,* in which the key experiments occur in the x-ray and the gamma-ray regions of the electromagnetic spectrum.

The Compton effect, which involves the scattering of radiation from atoms, can readily be understood in terms of billiard-ball-like collisions between photons and electrons. In the explanation we must take into account not only the energy of the photons but also their linear momentum, a property that we have not needed to introduce so far. We have seen that Einstein's analysis of the heat capacity of a solid in quantum terms went far to convince people to accept the notion of energy quantization. In the same way, Compton's analysis of the effect that bears his name went far to convince people of the reality of photons.

In Compton's experiment, a beam of x rays with sharply defined wavelength λ falls on a graphite target T, as in Fig. 12. For various angles of scattering ϕ, the intensity of the scattered x rays is measured as a function of their wavelength. Figure 13 shows Compton's experimen-

---

* Arthur H. Compton (1892–1962) discovered in 1923 that the wavelengths of x rays change after they are scattered from electrons. He received the 1927 Nobel prize in physics for this discovery. Later he became the director of the laboratory at the University of Chicago where the first nuclear reactor was built.

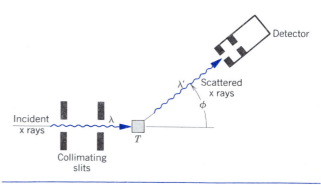

Figure 12   The experimental setup for observing the Compton effect. The detector can be moved to different angles ϕ.

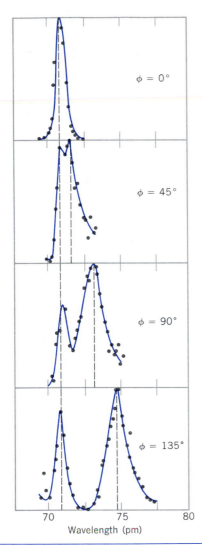

Figure 13 Compton's experimental results for four different values of the scattering angle ϕ.

Compton explained his experimental results by postulating that the incoming x-ray beam behaved not as a wave but as an assembly of photons of energy $E (= h\nu)$ and that these photons experienced billiard-ball-like collisions with the free electrons in the scattering target. In this view, the scattered radiation consists of the recoiling photons emerging from the target. Since the incident photon transfers some of its energy to the electron with which it collides, the scattered photon must have a lower energy E'. It must therefore have a lower frequency ν' $(= E'/h)$, which implies a larger wavelength λ' $(= c/\nu')$. This point of view accounts, at least qualitatively, for the wavelength shift $\Delta\lambda$. Note how different this particle model of x-ray scattering is from that based on the wave picture.

Now let us analyze a single photon–electron collision quantitatively. Figure 14 shows a collision between a photon and an electron. The electron is assumed to be initially at rest and essentially free, that is, not bound to the atoms of the scatterer. (This approximation holds for the loosely bound outer electrons, whose binding energy is much less than the energy of the x-ray photon.) Let us apply the law of conservation of energy to this collision. Since the recoil electrons may have a speed v that is comparable with that of light, we must use the relativistic expression for the kinetic energy of the electron. From the relativistic expression for the conservation of energy (see Section 21-9) we may write

$$E_i = E_f$$

or

$$h\nu + mc^2 = h\nu' + mc^2 + K, \qquad (20)$$

tal results. We see that although the incident beam consists essentially of a single wavelength λ, the scattered x rays have intensity peaks at two wavelengths; one of them is the same as the incident wavelength, but the other (λ') is larger by an amount $\Delta\lambda$. This *Compton shift* $\Delta\lambda$ varies with the angle at which the scattered x rays are observed.

The presence of a scattered wave of wavelength λ' cannot be understood if the incident x rays are regarded as an electromagnetic wave. In the wave picture, the incident wave of frequency ν causes electrons in the scattering target to oscillate at that same frequency. These oscillating electrons, like charges surging back and forth in a small radio transmitting antenna, radiate electromagnetic waves that again have this same frequency ν. Thus, according to this interpretation, the scattered wave should have the same frequency and the same wavelength as the incident wave. This conclusion disagrees with the experimental evidence (Fig. 13), which shows a variation in the wavelength of the scattered wave.

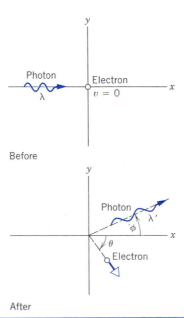

Figure 14 A photon of wavelength λ strikes an electron at rest. The photon is scattered at an angle ϕ with an increased wavelength λ'. The electron moves off with speed v at the angle θ.

where mc^2 is the rest energy of the struck electron and K is its (relativistic) kinetic energy. Substituting c/λ for v (and c/λ' for v') and using Eq. 25 of Chapter 7 for the relativistic kinetic energy, we have

$$\frac{hc}{\lambda} = \frac{hc}{\lambda'} + mc^2 \left(\frac{1}{\sqrt{1 - (v/c)^2}} - 1 \right). \qquad (21)$$

Now let us apply the (vector) law of conservation of linear momentum to the collision of Fig. 14. We first need an expression for the momentum of a photon. In Section 41-5 we saw that if an object completely absorbs an energy U from a parallel light beam that falls on it, the light beam, according to the wave theory of light, simultaneously transfers to the object a linear momentum U/c. In Section 48-7, we showed that we could consider this situation from the standpoint of a beam of photons of energy E, each delivering momentum $p = E/c$ to the absorbing object. In this case,

$$p = \frac{E}{c} = \frac{h\nu}{c} = \frac{h}{\lambda}. \qquad (22)$$

For the electron, the relativistic expression for the linear momentum is given by Eq. 22 of Chapter 9,

$$\mathbf{p} = \frac{m\mathbf{v}}{\sqrt{1 - (v/c)^2}}.$$

We can then write for the conservation of the x component of linear momentum

$$\frac{h}{\lambda} = \frac{h}{\lambda'} \cos \phi + \frac{mv}{\sqrt{1 - (v/c)^2}} \cos \theta \qquad (23)$$

and for the y component

$$0 = \frac{h}{\lambda'} \sin \phi - \frac{mv}{\sqrt{1 - (v/c)^2}} \sin \theta. \qquad (24)$$

Our aim is to find $\Delta\lambda \, (= \lambda' - \lambda)$, the wavelength shift of the scattered photons, so that we may compare it with the experimental results of Fig. 13. Compton's experiment did not involve observations of the recoil electron in the scattering block. Of the five variables (λ, λ', v, ϕ, and θ) that appear in the three equations (21, 23, and 24) we may eliminate two. We choose to eliminate v and θ, which deal only with the electron, thereby reducing the three equations to a single relation among the variables.

Carrying out the necessary algebraic steps (see Problem 64) leads to this simple result for the change in wavelength of the scattered photons:

$$\Delta\lambda = \lambda' - \lambda = \frac{h}{mc} (1 - \cos \phi). \qquad (25)$$

The Compton shift $\Delta\lambda$ depends only on the scattering angle ϕ and not on the initial wavelength λ. Equation 25 predicts within experimental error the observed Compton shifts of Fig. 13. Note from the equation that $\Delta\lambda$ varies from zero (for $\phi = 0$, corresponding to a "grazing" colli-

sion in Fig. 14, the incident photon being scarcely deflected) to $2h/mc$ (for $\phi = 180°$, corresponding to a "head-on" collision, the incident photon being reversed in direction).

Remember that the Compton shift $\Delta\lambda$ is a purely quantum effect, not expected to occur on the basis of classical physics. As in cavity radiation and the photoelectric effect, the presence of the Planck constant h in the expression for the Compton shift (Eq. 25) indicates a quantum phenomenon. Equation 25 shows that $\Delta\lambda \rightarrow 0$ as $h \rightarrow 0$. The method of letting the Planck constant approach zero is a formal way of testing quantum equations to see whether they predict what would happen if the laws of classical physics applied not only to large objects but also to atoms and electrons.

It remains to explain the presence of the peak in Fig. 13 for which the wavelength does not change on scattering. This peak results from collisions between photons and electrons that, instead of being nearly free, are tightly bound in an ionic core in the scattering target. During photon collisions the bound electrons behave like very heavy free electrons. This is because the ionic core as a whole recoils during the collision. Thus the effective mass M for a carbon scatterer is approximately the mass of a carbon nucleus. Since this nucleus contains six protons and six neutrons, we have approximately, $M = 12 \times 1840m = 22{,}000m$. If we replace m by M in Eq. 25, we see that the Compton shift for collisions with tightly bound electrons is immeasurably small.

Sample Problem 8   X rays with $\lambda = 100$ pm are scattered from a carbon target. The scattered radiation is viewed at 90° to the incident beam. (*a*) What is the Compton shift $\Delta\lambda$? (*b*) What kinetic energy is imparted to the recoiling electron?

Solution   (*a*) Putting $\phi = 90°$ in Eq. 25, we have, for the Compton shift,

$$\Delta\lambda = \frac{h}{mc} (1 - \cos \phi)$$

$$= \frac{6.63 \times 10^{-34} \text{ J} \cdot \text{s}}{(9.11 \times 10^{-31} \text{ kg})(3.00 \times 10^8 \text{ m/s})} (1 - \cos 90°)$$

$$= 2.43 \times 10^{-12} \text{ m} = 2.43 \text{ pm}.$$

(*b*) Using $\nu = c/\lambda$, we can write Eq. 20 as

$$\frac{hc}{\lambda} = \frac{hc}{\lambda'} + K.$$

Substituting $\lambda' = \lambda + \Delta\lambda$ and solving for K, we obtain

$$K = \frac{hc \, \Delta\lambda}{\lambda(\lambda + \Delta\lambda)}$$

$$= \frac{(6.63 \times 10^{-34} \text{ J} \cdot \text{s})(3.00 \times 10^8 \text{ m/s})(2.43 \times 10^{-12} \text{ m})}{(100 \times 10^{-12} \text{ m})(100 \text{ pm} + 2.43 \text{ pm})(10^{-12} \text{ m/pm})}$$

$$= 4.72 \times 10^{-17} \text{ J} = 295 \text{ eV}.$$

You can show that the initial photon energy E in this case

$(=h\nu = hc/\lambda)$ is 12.4 keV so that the photon lost about 2.4% of its energy in this collision. A photon whose energy was ten times as large $(= 124$ keV$)$ can be shown to lose 20% of its energy in a similar collision. This is consistent with the fact that $\Delta\lambda$ does not depend on the intial wavelength. More energetic x rays, which have smaller wavelengths, will experience a larger percent increase in wavelength and thus a larger percent loss in energy.

49-8   LINE SPECTRA

Experimental results from the photoelectric effect and the Compton effect give indisputable evidence for the existence of the photon or the particlelike nature of electromagnetic radiation. Historically, however, the photon concept emerged from the study of thermal radiation, which has a continuous spectrum of energies. The discrete (quantized) nature of the photon energies in this case is hidden in this broad distribution of energies. A more direct verification of the quantized nature of the radiation would result from detecting individual photons and measuring their energies.

The complication in the case of thermal radiation occurs because the atoms in the cavity walls behave cooperatively and must be analyzed using statistical considerations. If we analyze absorbing or emitting systems that are isolated from one another, we find that the radiation spectrum is not continuous but *discrete,* consisting of individual wavelengths separated by gaps where no radiation occurs. In the case of visible light, spectra are often displayed and analyzed using spectroscopes with prisms or diffraction gratings, such as that of Fig. 8 of Chapter 47, which give spectra such as Fig. 9 of Chapter 47.

These spectra are called *line spectra,* and the individual components are called *spectral lines.* Line spectra can result from the emission or absorption of radiation by any isolated system, including molecules, atoms, nuclei, or subnuclear particles. (Line spectra of atoms and molecules were available in Planck's time, but they were not interpreted in terms of energy quantization until after Planck and Einstein developed the photon concept.)

Analysis of radiation as photons of a definite energy strongly suggests that the system emitting or absorbing the radiation has discrete energy states, such that the difference in energy between the states equals the photon energy, as indicated in Figs. 18 and 19 of Chapter 8. (Here we are neglecting the small "recoil" energy needed to conserve linear momentum in the absorption or emission process.) In effect, this is a consequence of Einstein's interpretation of the spectrum of cavity radiation: quantization of the radiation implies quantization of the sources of the radiation. By studying line spectra, we can learn about the energy states of the atoms or other systems that emit the radiation. In the next chapter we discuss *quantum mechanics,* a mathematical procedure for calculating the energies of these discrete states, based on assuming a particular force to act between the components of the system (for example, the electrostatic force between the electron and the nucleus in an atom). The calculated energies of the states can then be compared with those deduced from experiment to see if the assumption about the force that acts in the system is reasonable.

Figure 15 shows examples of line spectra from molecules, atoms, nuclei, and particles, which have contributed to our understanding of their internal structure. These spectra, whose origins are listed in Table 2, indicate the great variety of line spectra that can be measured in the laboratory and the corresponding range of wavelengths. In Chapter 51, we discuss the line spectrum of atomic hydrogen, which provided the insight that led to the quantum theory of atomic structure.

TABLE 2   SOME SELECTED LINE SPECTRA

Figure	Entity	Wavelength Region (m)	Spectrum Region	Mode
15a	H_2 molecule	40	Radio	Absorption
15b	NH_3 molecule	1×10^{-2}	Microwave	Absorption
15c	HCl molecule	3×10^{-6}	Infrared	Absorption
15d	Fe atom	3×10^{-7}	Ultraviolet	Emission
15e	Mo atom	6×10^{-11}	X ray	Emission
15f	^{198}Hg nucleus	3×10^{-12}	Gamma ray	Emission
15g	Proton	4×10^{-15}	Gamma ray	Absorption

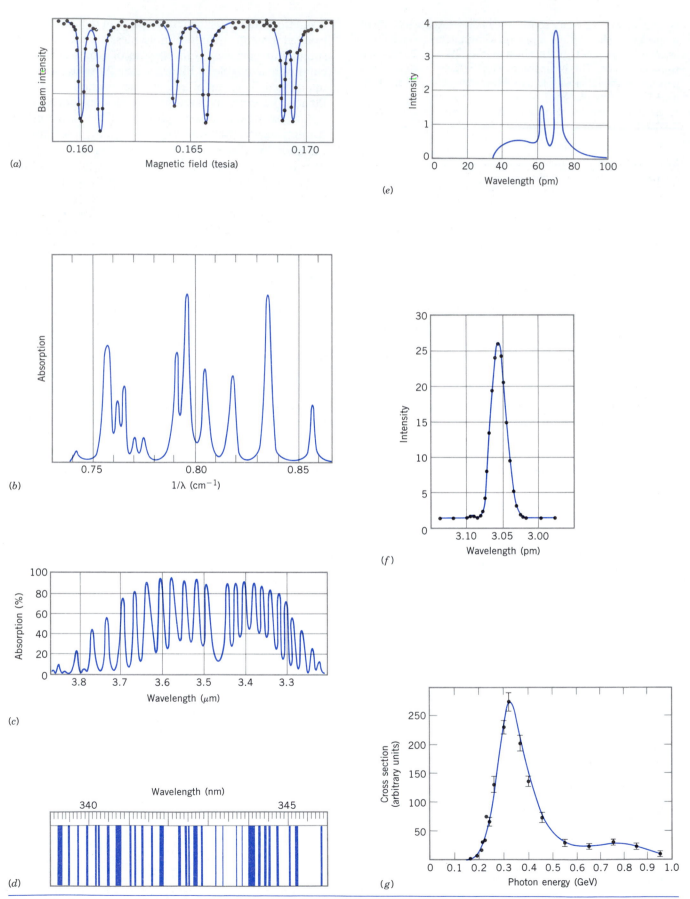

Figure 15   Some selected line spectra. See Table 2 for their identifications.

QUESTIONS

1. "Pockets" formed by the coals in a coal fire seem brighter than the coals themselves. Is the temperature in such pockets appreciably higher than the surface temperature of an exposed glowing coal? Explain this common observation.

2. The relation $I(T) = \sigma T^4$ (Eq. 1) is exact for true cavities and holds for all temperatures. Why don't we use this relation as the basis of a definition of temperature at, say, 100°C?

3. Do all incandescent solids obey the fourth-power law of temperature, as Eq. 2 seems to suggest?

4. A hole in the wall of a cavity radiator is sometimes called a *black body*. Why?

5. If we look into a cavity whose walls are maintained at a constant temperature, no details of the interior are visible. Explain.

6. By simply looking at the sky at night, can you pick out stars that are hotter than the Sun? Cooler than the Sun? What do you look for? Is the star's brightness a clue?

7. Betelgeuse, the prominent red star in the constellation Orion, has a surface temperature that is much lower than that of the Sun, yet it radiates energy into space at a much higher rate than the Sun does. How can that be?

8. Less than a few percent of the energy supplied to a 100-W lamp appears in the form of visible light. What happens to the rest of it? What could be done to increase this percentage? Why hasn't it already been done?

9. Your skin temperature is about 300 K. In what region of the electromagnetic spectrum do you emit thermal radiation most intensely?

10. Spectral radiancy curves for cavity radiators at different temperatures do not intersect; see Fig. 3. Suppose, however, that they did. Can you show that this would violate the second law of thermodynamics?

11. We claim that all objects radiate energy by virtue of their temperature and yet we cannot see all objects in the dark. Why not?

12. Is energy quantized in classical physics?

13. Show that the Planck constant has the dimensions of angular momentum. Does this necessarily mean that angular momentum is a quantized quantity?

14. For quantum effects to be "everyday" phenomena in our lives, what order of magnitude value would h need to have? [See G. Gamow, *Mr. Tompkins in Wonderland* (Cambridge University Press, Cambridge, 1957), for a delightful popularization of a world in which the physical constants c, G, and h make themselves obvious.]

15. For you to be able to detect energy quantization by watching a swinging pendulum, what order of magnitude would the Planck constant have to be? (*Hint*: See Sample Problem 3.)

16. Explain in your own words just why assuming that the energy of the atomic oscillators in a solid is quantized leads to a reduction in the heat capacity at low temperatures.

17. The classical law of equipartition of energy (see Section 23–6) leads to the Rayleigh–Jeans radiation law when applied to cavity radiation and to the Dulong–Petit law when applied to the heat capacities of solids. In both cases there is serious disagreement with experiment. Can you relate these two failures of the equipartition law and explain why energy quantization leads, in each case, to theories that *do* agree with experiment?

18. "For the cavity radiation problem and the heat capacity of solids problem, the disagreements between experiment and classical theory in certain ranges of the variables are not small but are total, and beyond all dispute." Can you identify, in each case, the specific disagreements to which this statement refers?

19. Explain why a tube used to examine photoelectric emission is (*a*) evacuated and (*b*) fitted with a window made of quartz rather than glass.

20. Determine whether or not relativistic mechanics is needed to verify the photoelectric equation (Eq. 18) with a 1% uncertainty. Note that typical stopping potentials are a few volts.

21. In Fig. 10, why doesn't the photoelectric current rise vertically to its maximum (saturation) value when the applied potential difference is slightly more positive than V_0?

22. In the photoelectric effect, why does the existence of a cutoff frequency speak in favor of the photon theory and against the wave theory?

23. Why are photoelectric measurements so sensitive to the nature of the photoelectric surface?

24. An insulated metal plate yields photoelectrons when first illuminated by ultraviolet light but then doesn't give up any more. Explain.

25. Why is it that even for incident radiation that is monochromatic the photoelectrons are emitted with a spread of velocities?

26. We claim that all the energy of an absorbed photon is given to an emitted photoelectron. Why can we neglect the energy taken by the lattice?

27. Is the Compton effect more supportive of the photon theory of light than is the photoelectric effect? Explain your answer.

28. Consider the following procedures: (*a*) bombard a metal with electrons; (*b*) place a strong electric field near a metal; (*c*) illuminate a metal with light; (*d*) heat a metal to a high temperature. Which of the above procedures can result in the emission of electrons?

29. A certain metal plate is illuminated by light of a definite frequency. Whether or not photoelectrons are emitted as a result depends on which of the following features: (*a*) intensity of illumination; (*b*) length of time of exposure to the light; (*c*) thermal conductivity of the plate; (*d*) area of the plate; (*e*) material of the plate?

30. Does Einstein's theory of photoelectricity, in which light is postulated to be a stream of photons, invalidate the double-slit interference experiment in which light is postulated to be a wave?

31. Explain the statement that one's eyes could not detect faint starlight if light were not particlelike.

32. How can a photon energy be given by $E = h\nu$ when the very presence of the frequency ν in the formula implies that light is a wave?

33. Distinguish between the Planck relation $E = nh\nu$ (Eq. 7) and the Einstein relation $E = h\nu$ (Eq. 17).

34. A photon has no rest mass since it can never be at rest with respect to any observer. If energy equals mc^2, how can a photon have any energy?

35. The momentum p of a photon is given by $p = h/\lambda$. Why is it that c, the speed of light, does not appear in this expression?

36. In discussing the propagation of light we sometimes use straight rays, sometimes waves, and still other times discrete photons. To what extent, if at all, are these views compatible with one another? Are there cases in which one view is clearly superior to the others?

37. Given that $E = h\nu$ for a photon, the Doppler shift in frequency of radiation from a receding light source would seem to indicate a reduced energy for the emitted photons. Is this in fact true? If so, what happened to the conservation of energy principle? (See "Questions Students Ask," *The Physics Teacher,* December 1983, p. 616.)

38. Photon A has twice the energy of photon B. What is the ratio of the momentum of A to that of B?

39. How does a photon differ from a material particle?

40. What is the direction of a Compton scattered electron with maximum kinetic energy compared with the direction of the incident monochromatic photon beam?

41. Why, in the Compton scattering picture (Fig. 14), would you expect $\Delta\lambda$ to be independent of the materials of which the scatterer is composed?

42. Why don't we observe a Compton effect with visible light?

43. Light from distant stars is Compton scattered many times by free electrons in outer space before reaching us. This shifts the light toward the red. How can this shift be distinguished from the Doppler red shift due to the motion of receding stars?

44. In both the photoelectric effect and the Compton effect there is an incident photon and an ejected electron. What is the difference between these two effects?

45. List and discuss the assumptions made by Planck in connection with the cavity radiation problem, by Einstein in connection with the photoelectric effect, and by Compton in connection with the Compton effect.

46. Describe several experimental methods that can be used to determine the value of the Planck constant h.

PROBLEMS

Section 49-1 Thermal Radiation

1. In 1983 the Infrared Astronomical Satellite (IRAS) detected a cloud of solid particles surrounding the star Vega, radiating maximally at a wavelength of 32 μm. What is the temperature of this cloud of particles? Assume an emissivity of unity.

2. Low-temperature physicists would not consider a temperature of 2.0 mK (0.0020 K) to be particularly low. At what wavelength is the spectral radiancy of a cavity at this temperature a maximum? To what region of the electromagnetic spectrum does this radiation belong? What are some of the practical difficulties of operating a cavity radiator at such a low temperature?

3. Calculate the wavelength of maximum spectral radiancy and identify the region of the electromagnetic spectrum to which it belongs for each of the following: (a) The 2.7-K cosmic background radiation, a remnant of the primordial fireball. (b) Your body, assuming a skin temperature of 34°C. (c) A tungsten lamp filament at 1800 K. (d) The Sun, at an assumed surface temperature of 5800 K. (e) An exploding thermonuclear device, at an assumed fireball temperature of 10^7 K. (f) The universe immediately after the Big Bang, at an assumed temperature of 10^{38} K. Assume cavity radiation conditions throughout.

4. (a) The effective surface temperature of the Sun is 5800 K. At what wavelength would you expect the Sun to radiate most strongly? In what region of the spectrum is this? Why then does the Sun appear yellow? (b) At what temperature is cavity radiation most visible to the human eye? See Fig. 1 in Chapter 42.

5. A cavity whose walls are held at 1900 K has a small hole, 1.00 mm in diameter, drilled in its wall. At what rate does energy escape through this hole from the cavity interior?

6. Calculate the thermal power radiated from a fireplace assuming an emissivity of 0.90, an effective radiating surface of 0.50 m², and a radiating temperature of 500°C. Does your answer seem reasonable?

7. (a) Show that a human body of area 1.80 m², emissivity $\epsilon = 1.0$, and temperature 34°C emits radiation at the rate of 910 W. (b) Why, then, do people not glow in the dark?

8. A cavity at absolute temperature T_1 radiates energy at a power level of 12.0 mW. At what power level does the same cavity radiate at temperature $2T_1$?

9. A cavity radiator has its maximum spectral radiancy at a wavelength of 25.0 μm, in the infrared region of the spectrum. The temperature of the body is now increased so that the radiant intensity $I(T)$ of the body is doubled. (a) What is this new temperature? (b) At what wavelength will the spectral radiancy now have its maximum value?

10. A 100-W incandescent lamp has a coiled tungsten filament whose diameter is 0.42 mm and whose extended length is 33 cm. The effective emissivity under operating conditions is 0.22. Find the operating temperature of the filament.

11. An oven with an inside temperature $T_o = 215$°C is in a room with a temperature of $T_r = 26.2$°C. There is a small opening of area $A = 5.20$ cm² in one side of the oven. How much net power is transferred from the oven to the room? (*Hint:* Consider both oven and room as cavities with $\epsilon = 1$.)

12. A *thermograph* is a medical instrument used to measure radiation from the skin. For example, normal skin radiates at a temperature of about 34°C and the skin over a tumor radiates at a slightly higher temperature. (a) Derive an approximate expression for the fractional difference $\Delta I/I$ in the radiant intensity between adjacent areas of the skin that are at slightly different temperatures T and $T + \Delta T$. (b) Evaluate this expression for a temperature difference

of 1.3 C°. Assume that the skin radiates with a constant emissivity.

13. A convex lens 3.8 cm in diameter and of focal length 26 cm produces an image of the Sun on a thin black screen the same size as the image. Find the highest temperature to which the screen can be raised. The effective temperature of the Sun is 5800 K.

14. The filament of a particular 100-W light bulb is a cylindrical wire of tungsten 0.280 mm in diameter and 1.80 cm long. See Appendix D for needed data on tungsten. Assume an emissivity of unity and ignore absorption of energy by the filament from the surroundings. (a) Calculate the operating temperature of the filament. (b) How long does it take for the filament to cool by 500 C° after the bulb is switched off?

15. Consider a planet, with radius R, revolving about the Sun in a circular orbit of radius r. Suppose that the planet has no atmosphere (and therefore no "greenhouse effect" on its surface temperature). (a) Show that the surface temperature T of the planet is given from the relation $T^4 = P_{Sun}/16\pi\sigma r^2$, where P_{Sun} is the radiant power output of the Sun. (b) Evaluate the temperature numerically for the Earth.

Section 49-2 Planck's Radiation Law

16. Show that the wavelength λ_{max} at which Planck's spectral radiation law, Eq. 6, has its maximum is given by Eq. 4:

$$\lambda_{max} = (2898 \; \mu m \cdot K)/T.$$

(*Hint*: Set $dR/d\lambda = 0$; an equation will be encountered whose numerical solution is 4.965.)

17. (a) By integrating the Planck radiation law, Eq. 6, over all wavelengths, show that the power radiated per square meter of a cavity surface is given by

$$I(T) = \left(\frac{2\pi^5 k^4}{15h^3c^2}\right) T^4 = \sigma T^4.$$

(*Hint*: Make a change in variables, letting $x = hc/\lambda kT$. The definite integral

$$\int_0^\infty \frac{x^3 \, dx}{e^x - 1}$$

will be encountered, which has the value $\pi^4/15$.) (b) Verify that the numerical value of the constant σ is 5.67×10^{-8} W/(m²·K⁴).

18. (a) An ideal radiator has a spectral radiancy at 400 nm that is 3.50 times its spectral radiancy at 200 nm. What is its temperature? (b) What would be its temperature if its spectral radiancy at 200 nm were 3.50 times its spectral radiancy at 400 nm?

Section 49-4 The Heat Capacity of Solids

19. In terms of the Einstein temperature T_E, at what temperature will the molar internal energy of a solid achieve one-half its classical value of $3RT$?

20. (a) Show that the molar internal energy E_{int} of a solid can be written, according to Einstein's theory of heat capacities, as

$$E_{int} = 3RT_E \left(\frac{1}{e^x - 1}\right),$$

in which $x = T_E/T$, where T_E is the *Einstein temperature*

$h\nu/k$. (b) Verify that E_{int} approaches its classical value of $3RT$ as $T \rightarrow \infty$.

21. In terms of Einstein's theory of heat capacity, (a) what is the molar heat capacity at constant volume of a solid at its Einstein temperature? Express your answer as a percentage of its classical value of $3R$. (b) What is the molar internal energy at the Einstein temperature? Express your answer as a percentage of its classical value of $3RT$.

22. Show that, at high enough temperatures, Einstein's expression for the heat capacity of a solid, Eq. 14, reduces to the classical formula, Eq. 11.

23. The Einstein temperatures of lead, aluminum, and beryllium may be taken as 68 K, 290 K, and 690 K, respectively. For each of these elements, find (a) the frequency ν of its atomic oscillators, (b) the spacing ΔE between adjacent oscillator levels, and (c) the effective spring constant k.

24. The Einstein temperature of aluminum may be taken as 290 K. According to Einstein's theory of heat capacity, what are (a) its molar internal energy (see Problem 20) at 150 K and (b) its molar heat capacity, under constant-volume conditions, at 150 K?

25. A 12.0-g block of aluminum is heated from 80 K up to 180 K, under constant-volume conditions. How much heat is required according to (a) the classical theory of heat capacity and (b) Einstein's quantum theory of heat capacity? The Einstein temperature for aluminum may be taken to be 290 K.

26. Assume that 25.0 g of aluminum at 80.0 K are mixed thoroughly with 12.0 g of aluminum at 200 K in an insulated container. What is the final temperature of the mixture? Assume that Einstein's theory of heat capacities is valid and that, at these relatively low temperatures, the differences between the heat capacity at constant volume and that at constant pressure may be neglected. Assume further that there are no energy exchanges between the two aluminum specimens and the container. The Einstein temperature of aluminum may be taken to be 290 K.

Section 49-6 Einstein's Photon Theory

27. (a) By using the "best" values of the fundamental constants, as found in Appendix B, show that the energy E of a photon is related to its wavelength λ by

$$E = \frac{1240 \; eV \cdot nm}{\lambda}.$$

This result can be useful in solving many problems. (b) The orange-colored light from a highway sodium lamp has a wavelength of 589 nm. How much energy is possessed by an individual photon from such a lamp?

28. Consider monochromatic light falling on a photographic film. The incident photons will be recorded if they have enough energy to dissociate a AgBr molecule in the film. The minimum energy required to do this is about 0.60 eV. Find the cutoff wavelength greater than which the light will not be recorded. In what region of the spectrum does this wavelength fall?

29. An atom absorbs a photon having a wavelength of 375 nm and immediately emits another photon having a wavelength

of 580 nm. What was the net energy absorbed by the atom in this process?

30. (a) A spectral emission line of hydrogen, important in radioastronomy, has a wavelength of 21.11 cm. What is its corresponding photon energy? (b) At one time the meter was defined as 1,650,763.73 wavelengths of the orange light emitted by a light source containing krypton-86 atoms. What is the corresponding photon energy of this radiation?

31. Most gaseous ionization processes require energy changes of 1.0×10^{-18} to 1.0×10^{-16} J. What region then of the Sun's electromagnetic spectrum is chiefly responsible for creating the ionosphere in the Earth's atmosphere?

32. Under ideal conditions the normal human eye will record a visual sensation at 540 nm if incident photons are absorbed at a rate as low as 100 s^{-1}. To what power level does this correspond?

33. You wish to pick a substance for a photocell operable with visible light. Which of the following will do (work function in parentheses): tantalum (4.2 eV), tungsten (4.5 eV), aluminum (4.2 eV), barium (2.5 eV), lithium (2.3 eV), cesium (1.9 eV)?

34. Satellites and spacecraft in orbit about the Earth can become charged due, in part, to the loss of electrons caused by the photoelectric effect induced by sunlight on the space vehicle's outer surface. Suppose that a satellite is coated with platinum, a metal with one of the largest work functions: $\phi = 5.32$ eV. Find the smallest-frequency photon that can eject a photoelectron from platinum. (Satellites must be designed to minimize such charging.)

35. (a) The energy needed to remove an electron from metallic sodium is 2.28 eV. Does sodium show a photoelectric effect for red light, with $\lambda = 678$ nm? (b) What is the cutoff wavelength for photoelectric emission from sodium and to what color does this wavelength correspond?

36. Find the maximum kinetic energy in eV of photoelectrons if the work function of the material is 2.33 eV and the frequency of the radiation is 3.19×10^{15} Hz.

37. Incident photons strike a sodium surface having a work function of 2.28 eV, causing photoelectric emission. When a stopping potential of 4.92 V is imposed, there is no photocurrent. Find the wavelength of the incident photons.

38. Light of wavelength 200 nm falls on an aluminum surface. In aluminum, 4.2 eV is required to remove an electron. What is the kinetic energy of (a) the fastest and (b) the slowest emitted photoelectrons? (c) Find the stopping potential. (d) Calculate the cutoff wavelength for aluminum.

39. (a) If the work function for a metal is 1.85 eV, what would be the stopping potential for light having a wavelength of 410 nm? (b) What would be the maximum speed of the emitted photoelectrons at the metal's surface?

40. The stopping potential for photoelectrons emitted from a surface illuminated by light of wavelength 491 nm is 710 mV. When the incident wavelength is changed to a new value, the stopping potential is found to be 1.43 V. (a) What is this new wavelength? (b) What is the work function for the surface?

41. Millikan's photoelectric data for lithium are:

Wavelength (nm)	433.9	404.7	365.0	312.5	253.5
Stopping potential (V)	0.55	0.73	1.09	1.67	2.57

Make a plot like Fig. 11, which is for sodium, and find (a) the Planck constant and (b) the work function for lithium.

42. A lithium surface for which the work function is 2.49 eV is irradiated with light of frequency 6.33×10^{14} Hz. The loss of electrons causes the metal to acquire a positive potential. What must this potential have become by the time its value prevents further loss of electrons from the surface?

43. A satellite in Earth orbit maintains a panel of solar cells at right angles to the direction of the Sun's rays. Assume that the solar radiation is monochromatic with a wavelength of 550 nm and arrives at the rate of 1.38 kW/m². What must be the area of the panels in order that "one mole of photons" arrives each minute?

44. In the photon picture of radiation, show that if two parallel beams of light of different wavelengths are to have the same intensity, then the rates per unit area at which photons pass through any cross section of the beams are in the same ratio as the wavelengths.

45. An ultraviolet light bulb, emitting at 400 nm, and an infrared light bulb, emitting at 700 nm, each are rated at 130 W. (a) Which bulb radiates photons at the greater rate? (b) How many more photons does it generate per second than does the other bulb?

46. To remove an inner, most tightly bound, electron from an atom of molybdenum requires an energy of 20 keV. If this is to be done by allowing a photon to strike the atom, (a) what must be the associated wavelength of the photon? (b) In what region of the spectrum does the photon lie? (c) Could this process be called a photoelectric effect? Discuss your answers.

47. X rays with a wavelength of 71.0 pm eject photoelectrons from a gold foil, the electrons originating from deep within the gold atoms. The ejected electrons move in circular paths of radius r in a region of uniform magnetic field **B**. Experiment shows that $rB = 188 \ \mu\text{T} \cdot \text{m}$. Find (a) the maximum kinetic energy of the photoelectrons and (b) the work done in removing the electrons from the gold atoms that make up the foil.

48. A special kind of light bulb emits monochromatic light at a wavelength of 630 nm. It is rated at 70.0 W and is 93.2% efficient in converting electrical energy to light. How many photons will the bulb emit over its 730-h lifetime?

49. Assume that a 100-W sodium-vapor lamp radiates its energy uniformly in all directions in the form of photons with an associated wavelength of 589 nm. (a) At what rate are photons emitted from the lamp? (b) At what distance from the lamp will the average flux of photons be 1.00 photon/(cm²·s)? (c) At what distance from the lamp will the average density of photons be 1.00 photon/cm³? (d) Calculate the photon flux and the photon density 2.00 m from the lamp.

50. Show, by analyzing a collision between a photon and a free electron (using relativistic mechanics), that it is impossible for a photon to give all its energy to the free electron. In other words, the photoelectric effect cannot occur for completely

free electrons; the electrons must be bound in a solid or in an atom.

Section 49-7 The Compton Effect

51. A particular x-ray photon has a wavelength of 41.6 pm. Calculate the photon's (a) energy, (b) frequency, and (c) momentum.

52. Find (a) the frequency, (b) the wavelength, and (c) the momentum of a photon whose energy equals the rest energy of the electron.

53. By how much does a sodium atom slow down upon absorbing a photon of wavelength 589 nm with which it collides head-on?

54. The quantity h/mc in Eq. 25 is often called the Compton wavelength, λ_C, of the scattering particle and that equation is written

$$\Delta\lambda = \lambda_C(1 - \cos\phi).$$

(a) Calculate the Compton wavelength of an electron. Of a proton. (b) What is the energy of a photon whose wavelength is equal to the Compton wavelength of the electron? Of the proton? (c) Show that in general the energy of a photon whose wavelength is equal to the Compton wavelength of a particle is just the rest energy of that particle.

55. Photons of wavelength 2.17 pm are incident on free electrons. (a) Find the wavelength of a photon that is scattered 35.0° from the incident direction. (b) Do the same if the scattering angle is 115°.

56. A 511-keV gamma-ray photon is Compton-scattered from a free electron in an aluminum block. (a) What is the wavelength of the incident photon? (b) What is the wavelength of the scattered photon? (c) What is the energy of the scattered photon? Assume a scattering angle of 72.0°.

57. Show that $\Delta E/E$, the fractional loss of energy of a photon during a Compton collision, is given by

$$\frac{\Delta E}{E} = \frac{h\nu'}{mc^2}(1 - \cos\phi).$$

58. What fractional increase in wavelength leads to a 75% loss of photon energy in a Compton collision with a free electron?

59. Find the maximum wavelength shift for a Compton collision between a photon and a free proton.

60. A 6.2-keV x-ray photon falling on a carbon block is scattered by a Compton collision and its frequency is shifted by 0.010%. (a) Through what angle is the photon scattered? (b) How much kinetic energy is imparted to the electron?

61. An x-ray photon of wavelength $\lambda = 9.77$ pm is back-scattered by an electron ($\phi = 180°$). Determine (a) the change in wavelength of the photon, (b) the change in energy of the photon, and (c) the final kinetic energy of the electron.

62. Calculate the fractional change in photon energy for a Compton collision with ϕ in Fig. 14 equal to 90° for radiation in (a) the microwave range, with $\lambda = 3.00$ cm, (b) the visible range, with $\lambda = 500$ nm, (c) the x-ray range, with $\lambda = 0.10$ nm, and (d) the gamma-ray range, with $\lambda = 1.30$ pm. What are your conclusions about the importance of the Compton effect in these various regions of the electromagnetic spectrum, judged solely by the criterion of energy loss in a single Compton encounter?

63. Through what angle must a 215-keV photon be scattered by a free electron so that it loses 10.0% of its energy?

64. Carry out the necessary algebra to eliminate v and θ from Eqs. 21, 23, and 24 to obtain the Compton shift relation, Eq. 25.

65. (a) Show that when a photon of energy E scatters from a free electron, the maximum recoil kinetic energy of the electron is given by

$$K_{max} = \frac{E^2}{E + mc^2/2}.$$

(b) Find the maximum kinetic energy of the Compton-scattered electrons knocked out of a thin copper foil by an incident beam of 17.5-keV x rays.

CHAPTER 50

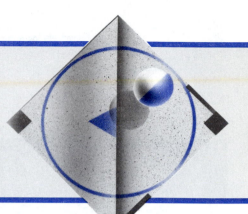

THE WAVE NATURE
OF MATTER

*Physicists have rarely gone wrong in relying on the underlying
symmetries of nature. For example, after learning that a changing magnetic
field produces an electric field, it is a good bet to guess (and it turns out to be true) that a
changing electric field produces a magnetic field. The electron was known to have an
antiparticle (a particle of the same mass but opposite charge), and one might guess that the
proton also has an antiparticle. To confirm this guess, a proton accelerator of the proper
energy (see Sample Problem 9 of Chapter 21) was built, and the antiproton was discovered.*

*In the previous chapter, we discussed the particlelike properties of light and other radiation,
which we traditionally analyze as waves. On the basis of symmetry, we are led to ask the
following question: Does matter, which we traditionally analyze as particles, also have
wavelike properties? In this chapter we show that this hypothesis turns out to be correct, and
we discuss the mechanics of this wavelike behavior, which is known as* quantum mechanics.
*As we shall see in the remaining chapters, quantum mechanics provides the means to
understand the fundamental behavior of physical systems from solids to quarks.*

50-1 THE WAVE BEHAVIOR OF PARTICLES

Before we discuss the analysis of the wave behavior of particles, let us try to persuade you that they really do show this kind of behavior. As we have seen, electromagnetic radiation can under most circumstances be treated as waves; its particlelike properties are directly revealed only in a few special experiments. We have also had a great deal of success at treating matter as composed of entities with only particlelike properties; for example, there is no need to take the wave nature into account in a game of billiards or in the construction of a building. However, there are a number of experiments that can be understood only if the entities that we have ordinarily analyzed as particles behave as waves.

In Chapters 45 and 46, we discussed interference and diffraction experiments, and we pointed out that the appearance of an interference or diffraction pattern is a definite signal of wave-type behavior. If we are going to search for direct evidence of the wave-type behavior of particles, interference and diffraction experiments are a logical place to begin.

Figure 1a shows a beam of electrons incident on a double slit. The electrons are accelerated to a chosen energy in a potential difference V, and after passing through the double slit they strike a fluorescent screen (like a TV screen). The resulting pattern on the screen leaves an image on the photographic film. The experimental setup of Fig. 1a looks very much like that for double-slit interference with light waves.

The results of this experiment are shown in Fig. 1b. There should be no doubt that we are observing an interference pattern. If the electrons had no wavelike behavior, we should expect to see bright regions on the film only in front of the two slits; clearly there is more to this outcome, and the wavelike behavior of the electrons is responsible for it.

We can replace the double slit with a circular aperture, which produces the diffraction pattern shown in Fig. 2. Once again, there is clearly more to this result than the passage of particles through an aperture.

Figure 3 compares the results of diffraction at a straight edge for beams of electrons and light. The comparison is convincing evidence that electrons have wavelike behavior.

Is there something unique about electrons that causes

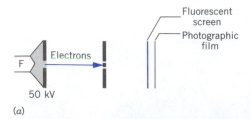

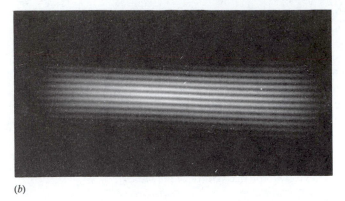

(a)

(b)

Figure 1 (a) Apparatus for producing double-slit interference with electrons. A filament F produces a spray of electrons, which are accelerated through 50 kV, pass through the single slit, and strike the double slit. They produce a visible pattern when they strike a fluorescent screen, which can be photographed. (b) The resulting electron double-slit interference pattern, showing the interference fringes.

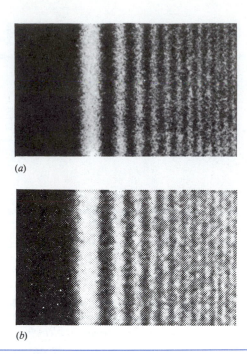

(a)

(b)

Figure 3 Diffraction of (a) light and (b) electrons from a straight edge.

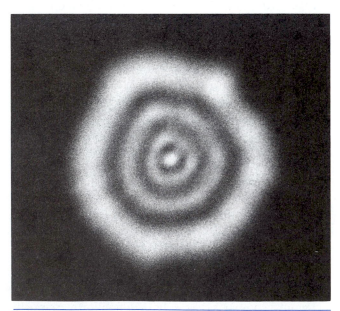

Figure 2 An electron diffraction pattern using a circular aperture of diameter 30 μm and 100-keV electrons. Compare with the optical pattern (Fig. 2 of Chapter 46).

this behavior? Let us instead try the experiment with neutrons, which differ from electrons in several respects: neutrons are more massive (by a factor of about 2000), they are uncharged, and they are composite particles (as opposed to electrons, which are fundamental "point" particles). If neutrons also show wavelike behavior, we should suspect that this behavior has nothing to do with the special properties of electrons but instead may be characteristic of particles in general.

To show interference with neutrons, a wire made of material (boron, for example) that is highly absorbing of neutrons is placed in a gap (of width slightly larger than the wire) in a similarly absorptive material, creating in effect a double slit. A detector scans across the transmitted neutron beam and measures the intensity as a function of location. The results are shown in Fig. 4. Once again, if neutrons behaved like traditional particles, we should expect to find peaks in the transmitted intensity only directly in front of the slits. Here, on the other hand, we see definite evidence of interference effects.

Figure 5 shows the results of an experiment in which a beam of helium atoms was passed through a double slit. Although the results are not as dramatic as those for the electrons and neutrons, there is again evidence for interference fringes similar to those obtained with light.

These experiments use very different types of particles and different types of slit systems and detector systems, yet they all have one feature in common: the particles

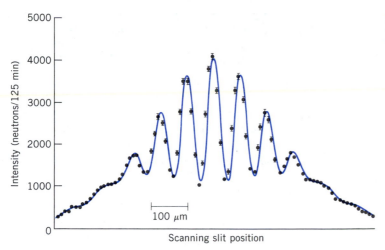

Figure 4   Intensity pattern of neutrons passing through a double slit.

seem to be undergoing some sort of interference. Such experiments provide direct evidence for the wave nature of particles.

At this point, you are probably wondering how a particle can produce an interference pattern. Our analysis of the double-slit experiment in terms of waves in Chapter 45 was based on parts of a single wavefront passing through each slit and then recombining on the screen. Is it possible for parts of an electron or a neutron or a helium atom to pass through each slit and then to recombine? This is a difficult question, but one that is essential to the understanding of quantum behavior. The answer is yes, and we discuss it further in the final section of this chapter.

Sample Problem 1   In the data shown in Fig. 5, the slit separation d was 8 μm and the detector was a distance $D = 64$ cm from the slits. From the observed spacing between the fringes, find the wavelength of the helium atoms.

Solution   In Sample Problem 2 of Chapter 45, we found the spacing between adjacent interference fringes to be

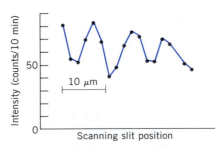

Figure 5   Intensity pattern of helium atoms passing through a double slit.

$$\Delta y = \frac{\lambda D}{d} \, .$$

From Fig. 5, we estimate about 8 μm for the spacing between the minima (which is the same as the spacing between the maxima), and so

$$\lambda = \frac{d \, \Delta y}{D} = \frac{(8 \times 10^{-6} \text{ m})(8 \times 10^{-6} \text{ m})}{0.64 \text{ m}} = 1.0 \times 10^{-10} \text{ m}.$$

50-2   THE DE BROGLIE WAVELENGTH

The experiments discussed in the previous section were done in recent years, when the precise apparatus needed to produce narrow slits or stable beams was available. However, the proposal that particles have a wavelike nature was made long before these results were obtained, on the basis of indirect arguments based partly on the symmetry of nature.

In 1924 Louis de Broglie, a physicist and a member of a distinguished French aristocratic family, puzzled over the fact that radiation seemed to have a dual wave–particle aspect, but matter (at that time) seemed entirely particle-like. On the other hand, matter and radiation had other aspects in common: both are forms of energy, each can be transformed into the other, and both are governed by the spacetime symmetries of the theory of relativity. De Broglie began to think that matter should also have a dual character and that particles such as electrons should have wavelike properties.

Equation 22 of Chapter 49 provided a connection between a wavelike property of radiation, the wavelength, and a particlelike property, the momentum: $p = h/\lambda$, where h is Planck's constant. De Broglie suggested that this same relationship connects the particlelike and wavelike properties of matter. That is, associated with a free particle moving with linear momentum p there is a sinusoidal wave having a wavelength λ given by

$$\lambda = \frac{h}{p}. \tag{1}$$

The wavelength of a particle computed according to Eq. 1 is called its *de Broglie wavelength*. Note that Planck's constant provides the connecting link between the wave and particle natures of both matter and radiation.

Equation 1 immediately shows why we don't observe the wave behavior of ordinary objects. Planck's constant is so small ($\approx 10^{-33}$ J·s) that the wavelengths of ordinary objects are many orders of magnitude smaller than the size of a nucleus! No double slit could possibly be constructed on this scale to reveal the wave nature. In the atomic or subatomic realm, however, the momentum p can be sufficiently small to bring the de Broglie wavelength into the range in which wave properties can be observed, as we illustrated in the previous section.

De Broglie's relationship provides us with a means to calculate the wavelength associated with the wave behavior of matter. It does not indicate anything about the amplitude of the wave, nor does it suggest the physical variable that is oscillating as the wave travels. We deal with these questions in Section 50-6.

Sample Problem 2   Calculate the de Broglie wavelength of (*a*) a virus particle of mass 1.0×10^{-15} kg moving at a speed of 2.0 mm/s, and (*b*) an electron whose kinetic energy is 120 eV.

Solution   Using Eq. 1, we find

(*a*) $\lambda = \dfrac{h}{p} = \dfrac{h}{mv} = \dfrac{6.6 \times 10^{-34} \text{ J·s}}{(1.0 \times 10^{-15} \text{ kg})(2.0 \times 10^{-3} \text{ m/s})}$

   $= 3.3 \times 10^{-16}$ m,

(*b*) $\lambda = \dfrac{h}{p} = \dfrac{h}{\sqrt{2mK}}$

   $= \dfrac{6.6 \times 10^{-34} \text{ J·s}}{\sqrt{2(9.11 \times 10^{-31} \text{ kg})(120 \text{ eV})(1.6 \times 10^{-19} \text{ J/eV})}}$

   $= 1.1 \times 10^{-10}$ m.

Even for so small an object as a virus particle moving slowly, the de Broglie wavelength is too small for observation (smaller than an atomic nucleus). For larger objects, the wave behavior is entirely unobservable. For the electron in part (*b*), however, the de Broglie wavelength is about the same size as an atom, and (as we shall see) by using atoms as diffracting objects for electrons we can verify that the de Broglie wavelength does indeed characterize the wave behavior of electrons.

50-3   TESTING DE BROGLIE'S HYPOTHESIS

If you want to prove that you are dealing with a wave, a convincing thing to do is to measure the wavelength. In 1801, for example, Thomas Young made a strong case for the wave nature of visible light when he measured its wavelength, using double-slit interference.

To measure a wavelength using the double-slit (or any similar) method, we need two or more diffracting centers (slits) separated by a distance that is of the order of magnitude of the wavelength itself. Sample Problem 2*a* shows at once that it is hopeless to try to measure the wavelength of even so small a particle as a virus; we would need two "slits" separated by 10^{-16} m. That is why our daily experiences with large moving objects gives no clues to the wave nature of matter. Sample Problem 2*b*, however, suggests that we *should* be able to measure the wavelength of a moving electron. We now describe two ways to do so.

1. *The Davisson–Germer experiment*. Sample Problem 2*b* suggests that we should be able to use a crystal as a diffraction grating to measure the de Broglie wavelengths of electrons with kinetic energies of a few hundred electron volts. Figure 6 shows the apparatus used for this purpose by C. J. Davisson and L. H. Germer of what is now the AT&T Bell Laboratories. In 1937 Davisson shared the Nobel prize for this work.

In the Davisson–Germer apparatus of Fig. 6, electrons are accelerated from a heated filament F by an adjustable potential difference V. The beam, made up of electrons whose kinetic energy is eV, is then allowed to fall on a crystal C, which, in their experiment, was of nickel. They set detector D at an arbitrary angle ϕ and read the current I of electrons entering D for various values of the potential

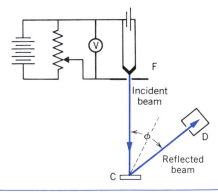

Figure 6   The apparatus used in the Davisson–Germer experiment. Electrons are emitted from the filament F and accelerated by the adjustable potential difference V. After reflection from the crystal C, they are recorded by the detector D, which can be moved to various angular positions ϕ.

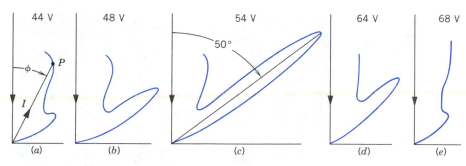

Figure 7 The results obtained by Davisson and Germer for five different accelerating voltages, shown as polar plots of current I as a function of the angle ϕ. A strong diffraction peak is observed in (c) at $\phi = 50°$ for $V = 54$ V.

difference V. Figure 7 shows the results of five such runs. We see that there is a strong diffracted beam for $\phi = 50°$ and $V = 54$ V. If either the angle or the accelerating potential are changed, the intensity of the diffracted beam drops.

Figure 8 is a simplified representation of the nickel crystal C of Fig. 6. Because this low-energy electron beam does not penetrate very far into the crystal, it is sufficient to consider the diffraction to take place in the plane of atoms on the surface. The situation is very similar to light reflected from a diffraction grating. In this case the grating lines are the parallel rows of atoms lying on the crystal surface, and the grating spacing is the interval D in Fig. 8. The principal maxima for such a grating must satisfy Eq. 1 of Chapter 47,

$$m\lambda = D \sin \phi \qquad (m = 1, 2, 3, \ldots). \qquad (2)$$

For their crystal Davisson and Germer knew that $D = 215$ pm. For $m = 1$, which corresponds to a first-order diffraction peak, Eq. 2 leads to

$$\lambda = D \sin \phi = (215 \text{ pm})(\sin 50°) = 165 \text{ pm}.$$

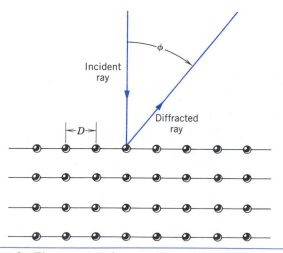

Figure 8 The crystal surface acts like a diffraction grating with spacing D.

The expected de Broglie wavelength for a 54-eV electron, calculated as in Sample Problem 2b, is 167 pm, in good agreement with the measured value. De Broglie's prediction is confirmed.

2. *G. P. Thomson's experiment.* In 1927 George P. Thomson, working at the University of Aberdeen in Scotland, independently confirmed de Broglie's equation, using a somewhat different method. As in Fig. 9a, he directed a monoenergetic beam of 15-keV electrons through a thin metal target foil. The target was specifically *not* a single crystal (as in the Davisson–Germer experiment) but was made up of a large number of tiny, randomly oriented crystallites.

If a photographic film is placed parallel to the target, as shown in Fig. 9a, the central beam spot will be surrounded by diffraction rings. Figure 9b shows this pattern for an x-ray beam incident on an aluminum target. Figure 9c shows the pattern for aluminum when an electron beam of the same wavelength replaces the x-ray beam. A simple glance at these two diffraction patterns leaves no doubt that both originate in the same way. Numerical analysis of the patterns confirms de Broglie's hypothesis in every detail.

Thomson shared the 1937 Nobel prize with Davisson for his electron diffraction experiments. George P. Thomson was the son of J. J. Thomson, who won the Nobel prize in 1906 for his discovery of the electron and for his measurement of its charge-to-mass ratio. It has been said that "Thomson, the father, was awarded the Nobel prize for having shown that the electron is a particle, and Thomson, the son, for having shown that the electron is a wave."

Today the wave nature of matter is taken for granted, and diffraction studies by beams of electrons or neutrons are used routinely to study the atomic structures of solids or liquids. Matter waves serve as a valuable supplement to x rays as tools for structure analysis. Electrons, for example, are less penetrating than x rays and so are particularly useful for studying surfaces. X rays interact largely with the electrons in a target, and for that reason it is not easy to

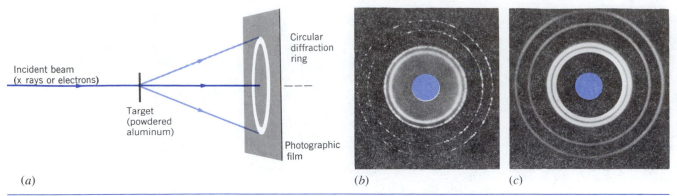

(a)   (b)   (c)

Figure 9 (a) An arrangement for producing a diffraction pattern using a powdered or crystalline target. (b) The x-ray diffraction pattern of a powdered aluminum target. (c) The electron diffraction pattern of the same target. The electron energy has been chosen so that the de Broglie wavelength is the same as the x-ray wavelength used in (b).

use them to locate light atoms—particularly hydrogen—which have few electrons. Neutrons, on the other hand, interact largely with the nucleus of the atom and can be used to fill this gap. Figure 10, for example, shows the structure of solid benzene as deduced from neutron diffraction studies.

Sample Problem 3 Suppose a beam of atoms emerges from an oven at a temperature T. The beam has a Maxwellian distribution of speeds (see Section 24-3). Based on this distribution, it can be shown (see Problem 20) that the most probable value of the de Broglie wavelength of the atoms in the beam is

$$\lambda_p = \frac{h}{\sqrt{5mkT}}. \qquad (3)$$

The data of Fig. 5 were obtained for helium atoms ($m = 4.0026$ u) emerging from an oven at a temperature $T = 83$ K. Find the most probable de Broglie wavelength for these helium

atoms and compare with the wavelength estimated in Sample Problem 1.

Solution From Eq. 3,

$$\lambda_p = \frac{6.63 \times 10^{-34} \text{ J} \cdot \text{s}}{\sqrt{5(4.0026 \text{ u})(1.66 \times 10^{-27} \text{ kg/u})(1.38 \times 10^{-23} \text{ J/K})(83 \text{ K})}}$$
$$= 1.07 \times 10^{-10} \text{ m}.$$

This value agrees very well with the estimate obtained in Sample Problem 1, verifying once again that the de Broglie wavelength characterizes the wave behavior of material particles.

Sample Problem 4 Nuclear reactors are often designed so that a beam of low-energy neutrons emerges after passing through a graphite cylinder in the shielding wall (see Fig. 11). After many collisions with the carbon atoms, the neutrons are in thermal equilibrium with them at room temperature (293 K). Such neutrons are called *thermal neutrons*. (a) Find the most probable de Broglie wavelength in a beam of thermal neutrons. (b) Let a beam of these neutrons be incident on a crystal C in which the spacing between the Bragg planes is $d = 0.304$ nm. An intense first-order Bragg diffraction is observed for neutrons of wavelength λ_p when the Bragg scattering angle θ is as shown in Fig. 11. Find the angle θ.

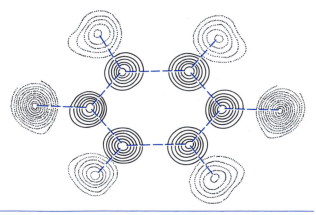

Figure 10 The atomic structure of solid benzene as deduced from neutron diffraction. The solid circles show the location of the six carbon atoms that form the familiar benzene ring. The dotted circles show the locations of the hydrogen atoms.

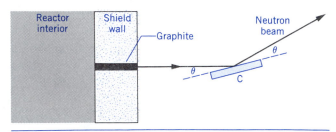

Figure 11 Sample Problem 4. An arrangement for observing neutron diffraction. The neutrons emerging from the graphite thermalizing column have a distribution of energies. After Bragg reflection from the crystal C, the beam at the angle θ is monoenergetic.

Solution   (*a*) Using Eq. 3, we have

$$\lambda_p = \frac{6.63 \times 10^{-34} \text{ J} \cdot \text{s}}{\sqrt{5(1.67 \times 10^{-27} \text{ kg})(1.38 \times 10^{-23} \text{ J/K})(293 \text{ K})}}$$

$$= 1.14 \times 10^{-10} \text{ m} = 0.114 \text{ nm}.$$

(*b*) The Bragg formula for x-ray diffraction was given as Eq. 12 of Chapter 47,

$$2d \sin \theta = m\lambda \qquad (m = 1, 2, 3, \ldots).$$

The same formula can be applied to the diffraction of particles, if we use the de Broglie wavelength λ. Solving for the angle θ, we obtain

$$\theta = \sin^{-1}\left(\frac{m\lambda_p}{2d}\right) = \sin^{-1}\left(\frac{(1)(0.114 \text{ nm})}{(2)(0.304 \text{ nm})}\right) = 10.8°.$$

By diffracting the neutrons in this way, a monoenergetic beam can be obtained. This beam can then be diffracted by other materials to study their structure, as was done to obtain Fig. 10.

50-4   WAVES, WAVE PACKETS, AND PARTICLES

As we have just seen, the evidence that matter is wavelike is very strong. Still, we cannot forget that the evidence that matter is particlelike is just as strong. The basic difference between these two points of view is that the position of a particle can be localized in both space and time but a wave can not, being spread out in both of these dimensions. Let us begin to reconcile these two approaches by seeing whether we can put together an assembly of waves in such a way that we end up with something that reminds us of a particle. What we will have to say will hold for all kinds of waves, whether they be water waves, sound waves, electro-

magnetic waves, or de Broglie waves. We discuss, in sequence, localizing a wave in space and in time.

1.   *Localizing a wave in space.* Figure 12*a* is a "snapshot" of a wave taken at an arbitrary time, say, $t = 0$. The wave extends from $x = -\infty$ to $x = +\infty$ and has a sharply defined wavelength λ_0 and a correspondingly sharply defined wave number $k_0 (= 2\pi/\lambda_0)$, as Fig. 12*b* shows. However, there is nothing about this wave that suggests the localization in space that we associate with the word "particle." Put another way, if the wave of Fig. 12*a* is to represent a particle, the uncertainty Δx of its position along the x axis is infinite: it could be anywhere along the axis.

We recall that it is possible (see Section 19-7) to create almost any wave shape we want by adding sine waves with properly chosen wave numbers, amplitudes, and phases. Figure 13*a* shows a *wave packet* that can be put together in this way. This collection of infinite waves adds to make a sine wave over a certain region of width Δx and, by destructive interference, adds to zero everywhere else. We now have a localization in space, measured by Δx, the length of the packet. The price we have paid is the sacrifice of the "purity" of our original wave because our packet now no longer contains a single wave number k_0 but rather a spread of wave numbers centered about k_0; see Fig. 13*b*.

Let Δk in Fig. 13*b* be a rough measure of the spread of wave numbers that forms the packet of Fig. 13*a*. It is reasonable to believe that the sharper (that is, the more particlelike) we wish the wave packet to be, the broader the range of wave numbers that we must use to build it up. In Fig. 12*a*, for example, the "packet" was not sharp at all ($\Delta x \to \infty$) but, on the other hand, we needed only a single wave number to "build it up" ($\Delta k = 0$). At the other extreme, we could build a very sharp wave packet ($\Delta x \to 0$), but we would need to combine waves with a very large

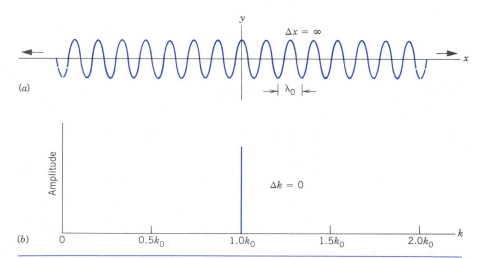

Figure 12   (*a*) A harmonic wave viewed at $t = 0$. (*b*) The distribution of wave numbers, shown as a plot of the amplitude of the harmonic component as a function of its wave number. In this plot, all waves with $k \neq k_0$ have an amplitude of zero.

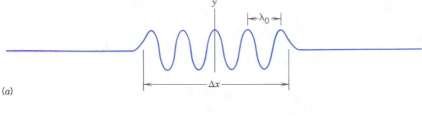

(a)

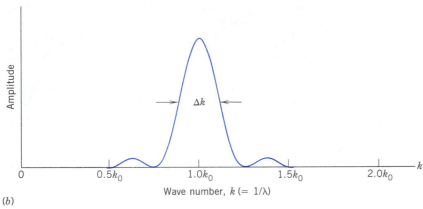

(b)

Figure 13   (*a*) A wave packet of length Δx, viewed at $t = 0$. (*b*) The relative amplitudes of the various harmonic components that combine to make up the packet. The central peak has a width Δk.

spread in wave number ($\Delta k \rightarrow \infty$) to do so. In general, as Δx decreases, Δk increases, and conversely. The relationship between them proves to be very simple; namely,

$$\Delta k \cdot \Delta x \sim 1. \qquad (4)$$

The symbol $\sim$ in Eq. 4 should be taken to mean "is of the order of," because so far we have not defined Δx or Δk very precisely.*

2. *Localizing a wave in time.* A particle is localized in time as well as in space. If we replaced the space variable x in Fig. 12*a* by the time variable t (and the wavelength λ_0 by the period T_0), that figure would then show how our wave would vary with time as it passes a particular fixed point, say, $x = 0$. As before, there is nothing at all about this wave that suggests the localization in time that we associate with the word "particle," because a particle would pass our observation point at a particular time, rather than spread over an infinite time interval.

We can build up a wave packet in time as well as in space. Figure 13*a* can illustrate this, provided we replace the space variables by the corresponding time variables, as above, and also replace the wave number k_0 by the angular frequency ω_0. By analogy with Eq. 4, the duration Δt of our new wave packet is related to the spread $\Delta \omega$ of angular frequencies needed to make up the wave packet by

$$\Delta \omega \cdot \Delta t \sim 1. \qquad (5)$$

---

* The estimate given by Eq. 4 represents the best we can do in constructing a wave packet. It is possible to do much worse ($\Delta k \cdot \Delta x \gg 1$), but it is *not* possible to do much better ($\Delta k \cdot \Delta x$ can never be $\ll 1$).

This equation has many practical applications. For example, much of our information in today's society, including telephone communication, radar, and computer data storage, is sent from point to point in the form of pulses. The electronic amplifiers through which these pulses pass should be sensitive over the full range of frequencies included in the pulses they are designed to handle. Equation 5 tells us that the shorter the time duration of the pulse, the greater must be the frequency bandwidth (as it is called) of the amplifier, and conversely.

Sample Problem 5   A radar transmitter emits pulses 0.15 μs long at a wavelength of 1.2 cm. (*a*) To what central frequency should the radar receiver be tuned? (*b*) What is the length of the wave packet? (*c*) What should be the frequency bandwidth of the receiver? That is, to what range of frequencies should it be able to respond?

Solution   (*a*) The central frequency $\nu_0 \; (= \omega_0/2\pi)$ is given by

$$\nu_0 = \frac{c}{\lambda_0} = \frac{3.00 \times 10^8 \text{ m/s}}{0.012 \text{ m}} = 2.5 \times 10^{10} \text{ Hz} = 25 \text{ GHz}.$$

(*b*) The length of the wave packet is

$$\Delta x = c \, \Delta t = (3.00 \times 10^8 \text{ m/s})(0.15 \times 10^{-6} \text{ s}) = 45 \text{ m}.$$

(*c*) The receiver's bandwidth is given approximately by Eq. 5, or

$$\Delta \nu = \frac{\Delta \omega}{2\pi} \sim \frac{1}{2\pi \Delta t} = \frac{1}{2\pi (0.15 \times 10^{-6} \text{ s})}$$

$$= 1.1 \times 10^6 \text{ Hz} = 1.1 \text{ MHz}.$$

If the receiver cannot respond to frequencies throughout this range, it will not be able to reproduce faithfully the shape of the transmitted radar pulse.

50-5   HEISENBERG'S UNCERTAINTY RELATIONSHIPS

Equation 4 applies to all kinds of waves. Let us apply it to de Broglie waves. We write, for the quantity Δk that appears in that equation,

$$\Delta k = \Delta\left(\frac{2\pi}{\lambda}\right) = \Delta\left(\frac{2\pi p_x}{h}\right) = \frac{2\pi}{h}\Delta p_x.$$

Here we have identified λ with the de Broglie wavelength of the particle and substituted h/p_x for it. The subscript on the momentum reminds us that we are dealing with motion along the x axis only. Substituting this result into Eq. 4 leads to

$$\Delta k \cdot \Delta x = \frac{2\pi}{h}\Delta p_x \cdot \Delta x \sim 1$$

or

$$\Delta p_x \cdot \Delta x \sim h/2\pi.$$

Taking into account the fact that momentum is a vector, we can generalize this relationship to

$$\begin{align}
\Delta p_x \cdot \Delta x &\sim h/2\pi, \\
\Delta p_y \cdot \Delta y &\sim h/2\pi, \qquad\qquad (6) \\
\Delta p_z \cdot \Delta z &\sim h/2\pi.
\end{align}$$

Equations 6 are the *Heisenberg uncertainty relationships,* first derived by Werner Heisenberg in 1927. They can be regarded as the mathematical formulation of *Heisenberg's uncertainty principle:*

It is not possible to determine both the position and the momentum of a particle with unlimited precision.

It is our goal in quantum mechanics to represent a particle by a wave packet that has large amplitude where the particle is likely to be found and small amplitude elsewhere. The width Δx of the wave packet indicates something about the probable location of the particle. However, as we have seen, construction of such a wave packet requires the superposition of waves with a range Δk in wave number or, equivalently, a range Δp_x in momentum. Hence, another way of stating the uncertainty principle is: *a particle cannot be described by a wave packet in which both the position and the momentum have arbitrarily small ranges.* As you make the range of one of them smaller, the range of the other becomes larger, according to Eq. 6.

Even though an individual measurement of the momentum of a particle can yield an arbitrarily precise value, that value can be anywhere in a range Δp_x about the "true" value p_x. (In effect, quantum mechanics tells us that we cannot determine the "true" value p_x except to within a range Δp_x.) If we repeat the measurement a large number of times on identically prepared systems, our re-sults will cluster about p_x with a statistical distribution characterized by the width Δp_x. Similarly, measurements of position will cluster about the position x with a statistical distribution characterized by the width Δx.

These limitations have nothing whatever to do with the practical problems of measurement. Equations 6, in fact, *assume* ideal instruments. In practice, you will always do worse. Sometimes these relationships are written with a " $\geq$ " symbol replacing the " $\sim$ " symbol, reminding us of this fact.

When we use the word "particle" to describe objects such as electrons, it conjures up in our mind the image of a tiny dot moving along a path, its position and velocity being well-defined at every moment. This way of thinking is a natural extension of familiar experiences with objects like baseballs and pebbles that we can see and touch. We must, however, accept the fact that this picture simply does not hold up experimentally beyond the limits set by the uncertainty principle. The quantum world is a world beyond our direct experience, and we must be prepared for new ways of thinking.

In Sample Problem 6 we shall see that the uncertainty principle does *not* limit our precision of measurement when we are dealing with large objects such as golf balls. Here ordinary instrumental errors overwhelm the fundamental limits set by this principle. When we deal with electrons and other elementary particles, however, the situation is quite different, as Sample Problem 6 shows.

Sample Problem 6   (*a*) A free 10-eV electron moves in the x direction with a speed of 1.88×10^6 m/s. Assume that you can measure this speed to a precision of 1%. With what precision can you simultaneously measure its position? (*b*) A golf ball has a mass of 45 g and a speed of 40 m/s, which you can measure with a precision of 1%. What limits does the uncertainty principle place on your ability to measure its position?

Solution   (*a*) The electron's momentum is

$$p_x = mv_x = (9.11 \times 10^{-31}\text{ kg})(1.88 \times 10^6\text{ m/s})$$
$$= 1.71 \times 10^{-24}\text{ kg} \cdot \text{m/s}.$$

The uncertainty Δp_x in momentum is 1% of this or 1.71×10^{-26} kg·m/s. The uncertainty in position is then, from Eq. 5,

$$\Delta x \sim \frac{h/2\pi}{\Delta p_x} = \frac{1.06 \times 10^{-34}\text{ J}\cdot\text{s}}{1.71 \times 10^{-26}\text{ kg}\cdot\text{m/s}} = 6.2\text{ nm},$$

which is about 30 atomic diameters. Given your measurement of the electron's momentum, there is simply no way that you can simultaneously pin down its position to a better precision than this.

(*b*) This example is exactly like part (*a*) except that the golf ball is much more massive and much slower than the electron. The same calculation yields, in this case, $\Delta x \sim 6 \times 10^{-31}$ m. This is a very small distance, about 10^{16} times smaller than the diameter of a typical atomic nucleus. Where large objects are concerned the uncertainty principle sets no meaningful limit to the precision of measurement. You could never have discovered this principle by studying flying golf balls!

The Uncertainty Principle and Single-Slit Diffraction

Here we learn more about the uncertainty principle by seeing how it works in a particular case. Consider a beam of electrons of speed v_0, moving upward as in Fig. 14. We set ourselves this task: measure simultaneously and with unlimited precision the horizontal position x and the velocity component v_x for an electron in this beam. As we shall see, this task (which violates the uncertainty principle) cannot be accomplished.

To measure x let us block the beam with a screen A in which we put a slit of width Δx. If an electron passes through this slit, we can claim to know its horizontal position to this precision. By narrowing the slit we can improve the precision of this measurement as much as we wish.

So far so good. However, something happens that perhaps we hadn't counted on. The electron beam—being wavelike—flares out by diffraction as it passes through the slit. If we put a suitably sensitive screen B in Fig. 14, a typical single-slit diffraction pattern shows up. Electrons that form the left half of this pattern must have been moving to the left (some faster, some slower) as they emerged from the slit. Those that form the right half must have been moving to the right. Even though—as the sym-

metry of the arrangement requires—the *average* value of v_x for the emerging electrons is zero, individual electrons can have nonzero values.

There is a particular value of v_x that will cause the electron to land at the first minimum of the diffraction pattern, identified by the angle θ_1 in Fig. 14. We take this value of v_x—somewhat arbitrarily—as a rough measure of the uncertainty of our knowledge of v_x and we call it Δv_x.

The location of the first minimum of the diffraction pattern is given by Eq. 1 of Chapter 46 ($\sin \theta_1 = \lambda/\Delta x$). If we assume that θ_1 is small enough, we can replace $\sin \theta_1$ by θ_1, obtaining

$$\theta_1 \approx \lambda/\Delta x.$$

To reach the first minimum it must be true that

$$\theta_1 \approx \Delta v_x/v_0.$$

Combining these two relations leads to

$$\Delta v_x \cdot \Delta x \approx \lambda v_0.$$

Now λ, the de Broglie wavelength of the electron, is equal to h/p or h/mv_0; putting this into the above and rearranging, we find

$$\Delta p_x \cdot \Delta x \approx h.$$

This is certainly consistent with Eq. 6; minor differences (the factor 2π) result from the arbitrary way we have defined Δx and Δp_x.

See how the uncertainty principle operates in this case. If we want to pin down the horizontal position of the electron, we must narrow the slit. This, however, broadens the diffraction pattern so that Δp_x increases. On the other hand, if we want to pin down the horizontal momentum component of the electron, we must somehow reduce the angular width of the diffraction pattern. The only way to do this is to widen the slit but that, in turn, means that we no longer know the horizontal position of the electron as precisely as we did. As we try to increase our knowledge about one variable, we simultaneously reduce our knowledge of the other. The uncertainty principle is not a statement about electrons (or other particles); it is a statement about our ability simultaneously to determine certain properties of those particles.

The Energy–Time Uncertainty Relationship

Thus far we have considered only the *wavelengths* of matter waves and have said nothing about their *frequencies*.

By analogy with Einstein's photon equation ($E = h\nu$), the uncertainty in the frequency of a matter wave is related to the uncertainty in the energy E of the corresponding particle by $\Delta \nu = \Delta E/h$. Substituting this into Eq. 5 yields, with $\Delta \omega = 2\pi \, \Delta \nu$,

$$\Delta E \cdot \Delta t \sim h/2\pi, \tag{7}$$

which is the mathematical relationship of the uncertainty

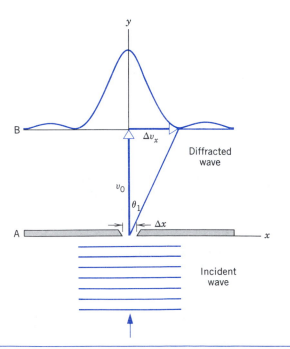

Figure 14   An incident beam of electrons is diffracted at the slit in screen A. If the slit is made narrower, the diffraction pattern becomes wider.

principle expressed in terms of different parameters. In words, it says:

> *It is not possible to determine both the energy and the time coordinate of a particle with unlimited precision.*

All energy measurements carry an inherent uncertainty unless you have an infinite time available for the measurement. In an atom, for example, the lowest energy state (the so-called ground state) has a well-defined energy because the atom normally exists indefinitely in that state. The energies of all states of higher energy (the excited states) are less precisely defined because the atom — sooner or later — will move spontaneously to a state of lower energy. On average, you have only a certain time Δt available so that your energy measurement will be uncertain by an amount ΔE given by $(h/2\pi)/\Delta t$.

Sample Problem 7   In 1974 an important new particle, more than three times as massive as the proton, was discovered simultaneously and independently by two groups of physicists, using the high-energy accelerators at the Brookhaven National Laboratory and at Stanford University. The rest energy of this particle was measured to be 3097 MeV, the uncertainty of the measurement being only 0.063 MeV. Such a massive particle was expected to decay extremely rapidly into particles of smaller mass. What is the mean time interval between production and decay for these short-lived particles?

Solution   The answer follows from the uncertainty principle, in the form of Eq. 7. Solving for Δt yields

$$\Delta t \sim \frac{h/2\pi}{\Delta E} = \frac{1.06 \times 10^{-34} \text{ J} \cdot \text{s}}{(0.063 \text{ MeV})(1.60 \times 10^{-13} \text{ J/MeV})}$$
$$= 1.1 \times 10^{-20} \text{ s}.$$

This time interval can be identified with the lifetime of the particle. By ordinary standards this seems to be a short lifetime but, in fact, the experimenters were astonished by how *long* it was or — equivalently — by how sharply the rest energy of the particle was defined. Theory had predicted that the decay of this massive particle (called ψ) would be very much faster. One observer remarked that the observed slowing down of the decay was as if Cleopatra, floating on the Nile in her Royal Barge, had dropped a pebble over the side and the falling pebble, as of today, had not yet hit the water! This new particle proved to be so significant that the leaders of the two groups, Burton Richter and Samuel Ting, were awarded the Nobel prize in 1976 for its discovery.

The uncertainty principle is used in this way to deduce the lifetimes of the excited states of molecules, atoms, and unstable fundamental particles of all kinds.

50-6   THE WAVE FUNCTION

By this time you should be comfortable with the fact that a moving particle can be viewed as a wave, and you should know how to measure its wavelength. It remains to ask:

"What is the quantity whose variation in time and space makes up this wave?" To put it loosely: "What is waving?"

For a wave in a string we can represent the wave disturbance by the transverse displacement y. For sound waves we use the differential pressure Δp and for electromagnetic waves the electric field vector $\mathbf{E}$. For waves representing particles, we introduce the *wave function* Ψ. The problem at hand may be that of a proton traveling along the axis of an evacuated tube in a particle accelerator, a conduction electron moving through a copper wire, or an electron moving about the nucleus of a hydrogen atom. Whatever the case may be, if we know the wave function $\Psi(x, y, z, t)$ for every point of space and for every instant of time, we know all that can be known about the behavior of the particle.

Before we look into the physical meaning of the wave function, let us consider a problem that involves radiation rather than matter: a plane electromagnetic wave traveling through free space. We can think of such a wave (following Maxwell) as an arrangement of electric and magnetic fields that varies in space and time or (following Einstein) as a beam of photons, each moving with the speed of light. On the first picture, the rate per unit area at which energy is transported by the wave (see Section 41-4) is proportional to E^2, where E is the amplitude of the electric field vector. On the second picture, this rate is proportional to the average number of photons per unit volume of the beam, each photon having an energy $h\nu$. We see here a connection between the wave and particle pictures of radiation, namely, the notion — first advanced by Einstein — that the square of the electric field intensity is a direct measure of the average density of photons.

Max Born proposed that the wave function Ψ for a beam of particles be interpreted in this same way, namely, that its square is a direct measure of the average density of particles in the beam. In many problems, however, such as the structure of the hydrogen atom, there is only a single electron present. What can it then mean to speak of the "average density of particles"? Born proposed that in such cases we should interpret the square of the wave function at any point as the probability (per unit volume) that the particle will be at that point.* Specifically, if dV is a volume element located at a point whose coordinates are x, y, z, then the probability that the particle will be found in that volume element at time t is proportional to $\Psi^2 \, dV$. Perhaps by analogy with ordinary mass density (a mass per unit volume), we call the square of the wave

---

* The wave function Ψ is usually a complex quantity; that is, it involves $\sqrt{-1}$, which we represent by the symbol i. By Ψ^2 (more properly written as $|\Psi|^2$) we mean the square of the absolute value of the wave function. This is always a real quantity. We give a physical interpretation only to the square of the wave function, not to the wave function itself.

function a *probability density*, that is, a probability per unit volume.

Note that the relationship between the wave function and its associated particle is *statistical,* involving only the *probability* that the particle will find itself within a specified volume element. In classical physics we also deal with particles on a statistical basis (see Chapters 23 and 24), but in those cases statistical methods are just a handy way of dealing with large numbers of particles. In quantum mechanics, however, the statistical nature is inherent and is dictated by the uncertainty principle, which, as we have seen, sets limits to the meaning that we can attach to the word "particle."

The probability that our particle will be *somewhere* must be equal to unity (corresponding to a 100% chance to find it) so that we have

$$\int \Psi^2 \, dV = 1 \qquad \text{(normalization condition),} \qquad (8)$$

the integration being taken over all space. To *normalize* a wave function means to multiply it by a constant factor, chosen so that Eq. 8 is satisfied.

We save for last an obvious question: In any given problem, how do we know what the wave function is? Waves on strings and sound waves are governed by Newton's laws of mechanics. Electromagnetic waves are predicted and described by Maxwell's equations. From where do the wave functions come?

In 1926 Erwin Schrödinger, inspired by de Broglie's concept, thought along these lines: geometrical optics deals with rays and with the motion of light in a straight line; it turned out to be a special case of a much more general wave optics. Newtonian mechanics also has "rays" (trajectories) and straight-line motion (of free particles). Could it turn out to be a special case of a much more general — but as yet undiscovered — wave mechanics?

Schrödinger derived a remarkably successful theory based on this analogy. Its central feature is a differential equation, now known as *Schrödinger's equation,* which governs the variation in space and time of the wave function Ψ for a wide range of problems. We obtain solutions to problems in classical mechanics by manipulating Newton's laws of motion; we obtain solutions to problems in electromagnetism by manipulating Maxwell's equations; in exactly the same spirit, we obtain solutions to atomic problems by manipulating Schrödinger's equation.

In the next section we shall study an important problem from the point of view of wave mechanics, that of a particle trapped in a region from which it can never escape (or can it?).

50-7   TRAPPED PARTICLES AND PROBABILITY DENSITIES

Before we consider the situation for matter waves, let us review two analogous examples involving mechanical waves and electromagnetic waves.

In Sections 19-9 and 19-10, we considered the standing waves that can occur in a string of length L that is fixed at both ends. The fixed ends of the string are constrained by the supports to be *vibrational nodes,* that is, locations where the amplitude is zero at all times. Only a limited set of wavelengths can occur for these standing waves. As shown in Section 19-10, these allowed wavelengths can be written

$$\lambda_n = \frac{2L}{n} \qquad (n = 1, 2, 3, \ldots) \qquad (9)$$

or, in terms of wave number,

$$k_n = \frac{2\pi}{\lambda_n} = \frac{n\pi}{L} \qquad (n = 1, 2, 3, \ldots). \qquad (10)$$

At any point along the string, the amplitude of vibration is

$$y_n(x) = y_{max} \sin k_n x = y_{max} \sin \frac{n\pi x}{L}$$
$$(n = 1, 2, 3, \ldots), \qquad (11)$$

where y_{max} is the maximum displacement of the string. Figure 15 shows examples of the vibrational patterns of these standing waves, which might characterize some of the lower vibrational modes of a guitar string or a violin string.

In electromagnetism, a similar situation results when a plane electromagnetic wave oscillates back and forth (in one dimension) between two perfectly reflecting surfaces

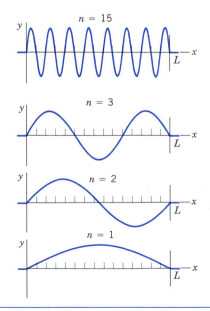

Figure 15 Four standing wave patterns for a stretched string of length L, clamped rigidly at each end. These patterns are determined from Eq. 11.

(mirrors, for instance) separated by a distance L. Such a situation might occur for light waves in a laser. Just as in the mechanical case, a standing wave is established in the cavity. This standing electromagnetic wave can be considered as the superposition of two similar waves traveling in opposite directions. At the ends of the cavity, where the reflection occurs from a conducting material such as the silvering on a mirror, the electric field must drop to zero (which is true for all ideal conductors under conditions of electrostatics). Imposing these conditions that $E = 0$ at $x = 0$ and $x = L$, we find that only certain wavelengths are permitted for the standing wave; the permitted wavelengths are given by Eq. 9, and the amplitude of the electric field oscillations can be written

$$E_n(x) = E_{\max} \sin \frac{n\pi x}{L} \qquad (n = 1, 2, 3, \ldots). \quad (12)$$

Figure 16a shows a plot of $E_n(x)$ as a function of x for the lowest modes of oscillation ($n = 1$ and 2). Note the similarity between this figure and Fig. 15, which showed the mechanical standing wave on a string.

Figure 16b shows a plot of $E_n^2(x)$, which is proportional to the energy density of the wave, as we discussed in Section 41-4. In terms of the photon picture, we can regard the standing wave as a collection of photons, and Fig. 16b represents the density of photons as a function of x. That is, in the mode of oscillation with $n = 1$ you would find the greatest density of photons at $x = L/2$ and the smallest density next to the walls. In the $n = 2$ mode you would find a minimum in density at $x = L/2$ and maxima at $x = L/4$ and $3L/4$.

Suppose we reduce the intensity of light in the cavity until it contains only a single photon. Figure 16b would still apply, but we must change its interpretation slightly, since it is no longer appropriate to speak of the density of photons. Instead, we use a related statistical concept: *the square of the electric field amplitude at a particular coordinate gives the probability to find the photon at that location.* Figure 16b shows locations where the probability to

find the photon is large and where it is zero. Note that we do not consider the actual location of the photon, but instead its probability to be found in a certain location.

These characteristics of mechanical and electromagnetic standing waves in one dimension can be directly carried over to matter waves. Consider a particle confined to move between two perfectly reflecting walls a distance L apart. Figure 17 shows a device that might be used to trap an electron. Although this is a large-scale device, it is possible to construct microscopic devices that accomplish the same result. For example, "quantum well" structures are built from a few atomic layers of semiconducting material surrounded by insulating material; such devices are used for optical communication and logic gates.

In the apparatus shown in Fig. 17a, the electron can move freely in the central section, where no forces act on it. When it reaches either end, it encounters a region in which the potential changes rapidly from 0 (which we take to be that of the central section) to $-V_0$, the potential associated with either battery. Equivalently, the potential energy of the electron is 0 in the central section and U_0 ($= eV_0$) in the outer sections (Fig. 17b). If the kinetic energy of the electron in the central region is less than U_0, then classically it has insufficient energy to escape from the well and it oscillates back and forth between the walls.

We seek a description of the motion of the electron in the potential well using the language of wave mechanics. While this may seem like a trivial problem far removed from, say, the structure of atoms, it turns out to demonstrate the important features of wave-mechanical behavior in a way that avoids the mathematical complexity of more complicated systems.

In describing mechanical and electromagnetic standing waves, we used functions $y(x)$ and $E_n(x)$, which lack the time dependence that must be present to describe a wave. However, as we showed above, for our analysis we were more interested in the variations of amplitude with position, and so it was not necessary to consider the time dependence. We shall do the same in the case of matter

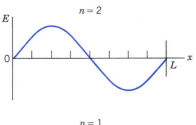

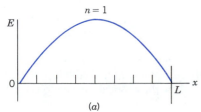

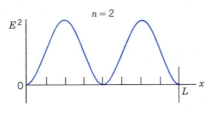

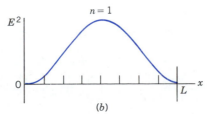

(a)   (b)

Figure 16   (a) The electric field for two electromagnetic standing wave patterns in a cavity of length L. These patterns are determined from Eq. 12. (b) The density of photons in the cavity can be found from the square of the fields.

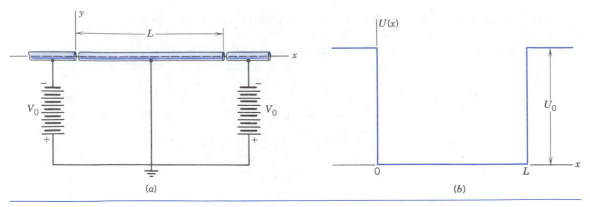

Figure 17   (*a*) An arrangement that can be used to confine an electron to a region of length *L* along the *x* axis. (*b*) The potential energy of the electron. In any real device, the potential energy would not change immediately from 0 to U_0; the graph of the potential energy would have rounded corners and nonvertical sides.

waves. Instead of seeking the general wave function $\Psi(x, t)$, we shall consider only the spatial part, which we write as $\psi(x)$.

We begin by assuming the walls to be perfectly reflecting for all particles; that is, we consider an infinitely deep well, such that $U_0 \rightarrow \infty$. The Schrödinger equation in this case turns out to be identical with the wave equations that describe mechanical or electromagnetic waves. Its solution is

$$\psi_n(x) = A \sin k_n x = A \sin \frac{n\pi x}{L}$$
$$(n = 1, 2, 3, \ldots), \quad (13)$$

where the allowed set of wave numbers or wavelengths is given by Eqs. 9 and 10. The constant *A* must be determined by the normalization condition (see Sample Problem 10).

The function $\psi_n(x)$ has no physical interpretation. However, the square of the wave function does have a physical meaning—it gives the *probability density* $P_n(x)$:

$$P_n(x) = |\psi_n(x)|^2 = A^2 \sin^2 \frac{n\pi x}{L}$$
$$(n = 1, 2, 3, \ldots). \quad (14)$$

Just as was the case for the square of the electric field in Fig. 16, the square of the wave function at a particular location indicates the probability to find the electron at that location. Some of these probability densities are plotted in Fig. 18.

The energy of the particle (which is entirely kinetic inside the well, since $U = 0$) is restricted to a certain set of values; we say that the energy is *quantized*. Let us see how this comes about. The allowed de Broglie wavelengths of the particle are given by Eq. 9, and so the magnitude of its momentum is restricted to the values

$$p_n = \frac{h}{\lambda_n} = \frac{nh}{2L} \quad (n = 1, 2, 3, \ldots). \quad (15)$$

The (kinetic) energy is then

$$E_n = \frac{p_n^2}{2m} = n^2 \frac{h^2}{8mL^2} \quad (n = 1, 2, 3, \ldots), \quad (16)$$

where *m* is the mass of the particle. When we write the energy in this way, the index *n* is called a *quantum number*. The allowed energies are plotted in Fig. 19. The electron is permitted to occupy only those states of motion corresponding to this set of energies; no other energies are permitted for the particle.

Once we have found the permitted energies and wave functions, we have solved the problem of the trapped particle using the techniques of quantum mechanics. The quantum solution shows a number of unexpected features that are not part of the classical solution for a trapped particle. Let us consider some of these.

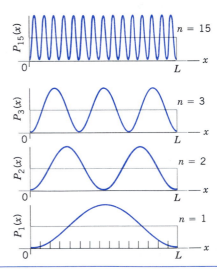

Figure 18   The probability density $P_n(x)$, computed according to Eq. 14, for four different values of the quantum number *n*. The horizontal lines show the classical expectations for the probability density.

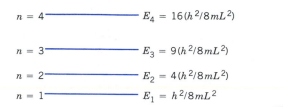

$n = 4$ ———————— $E_4 = 16(h^2/8mL^2)$

$n = 3$ ———————— $E_3 = 9(h^2/8mL^2)$

$n = 2$ ———————— $E_2 = 4(h^2/8mL^2)$

$n = 1$ ———————— $E_1 = h^2/8mL^2$

Figure 19   The allowed energy levels, calculated from Eq. 16, for an electron trapped in a one-dimensional region.

1. *The electron cannot be at rest in the well.* The lowest energy state, called the *ground state,* corresponds to $n = 1$ in Eq. 16 and Fig. 18. This lowest energy is not zero. We cannot reduce the energy of the particle (that is, its kinetic energy) to zero. The minimum energy, given by

$$E_1 = \frac{h^2}{8mL^2}, \tag{17}$$

is called the *zero-point energy* for the infinite well. In other quantum systems, the zero-point energy may take different forms, but the phenomenon exists for all quantum systems. Even at the absolute zero of temperature, where a particle is in the lowest possible energy state, it still has motion and energy.

In effect, the zero-point motion occurs as a result of confining the particle to a region of space. Let us see how the uncertainty principle helps us understand this effect. If we confine a particle in the well, we know its position to within an uncertainty of approximately L. The corresponding uncertainty in its momentum is, from Eq. 6,

$$\Delta p_x \sim \frac{h}{2\pi L}.$$

The smaller the region in which we confine the particle, the larger is the uncertainty in its momentum.

If we know the kinetic energy of the particle, we know the magnitude of its momentum precisely, but we don't know the direction. An uncertainty in momentum of Δp_x suggests that the particle may be moving to the right with momentum $p_x = \frac{1}{2}\Delta p_x = h/4\pi L$ or to the left with momentum $p_x = -\frac{1}{2}\Delta p_x = -h/4\pi L$. This is comparable to the momentum of this state given by Eq. 15; that is, $p_1 = h/2L$, the difference of a factor of 2π having to do with the arbitrary way we have defined the uncertainties. The estimate does indicate that the zero-point motion is consistent with the uncertainty relationship, and that the motion is a result of our confining the particle to a region of space.

2. *The electron spends more time in certain parts of the trap than in others.* For this one-dimensional problem a "volume" element becomes a line element so that the product $P_n(x)\,dx$ is the probability that the electron will be found in the interval x to $x + dx$. A glance at the probability density curves of Fig. 18 shows that in the ground state ($n = 1$), the electron is much more likely to be found near the center of the trap than near its ends. Once again we

have a prediction in sharp contrast with the prediction of classical mechanics. According to classical theory all positions between the walls of the trap are equally likely, as the horizontal lines in Fig. 18 suggest. For both the quantum and the classical curves, the area under the curve is unity, as the normalization condition (Eq. 8) requires.

For states of higher quantum number—and thus of higher energy—the distribution of the electron probability density across the trap becomes more uniform and the quantum prediction merges with the classical prediction. This agreement between classical and quantum physics for large quantum numbers is called the *correspondence principle* and is discussed in Section 50-9.

3. *The electron can escape from its trap.* So far we have dealt with a well of infinite depth. A major quantum surprise awaits us if we relax this requirement and deal with the more realistic case of a well of finite depth. In Fig. 20 we compare two wells of the same width ($= 2 \times 10^{-10}$ m, about the size of a large atom), one of the wells being infinitely deep and the other having a depth of only 20 eV. To find the allowed energies and the corresponding probability densities for a *finite* well, we need the full power of the Schrödinger equation. Here we simply give the results, without proof.* We consider the ground state only.

Figure 20 shows that the ground state energy for the finite well ($= 4.45$ eV) is substantially less than the ground state energy for the infinite well ($= 9.41$ eV, calculated from Eq. 16). We can tell that this should be the case simply by inspecting the probability density curves of Fig. 20. For the infinite well, half a de Broglie wavelength fits neatly and exactly between the rigid walls of the well. For the finite well, however, the de Broglie wavelength is too large to fit in this way; it spills over beyond the walls. Now if the de Broglie wavelength is *larger* for the finite well, the momentum ($= h/\lambda$) must be *smaller,* which means the energy must also be smaller, just as we observe. Thus the reduced energy for the finite well is consistent with the form of its probability density curve.

The spilling over of the exponential tails of the probability curve (Fig. 20b) beyond the walls means that there is a finite probability that the electron will be found *outside* the well! This curve has been normalized (see Eq. 8) so that the area under the curve is unity. The area under the two exponential tails is 0.074, which means that, if you measured the position of the trapped electron, you would find it outside the well 7.4% of the time.

How can an electron whose energy is only 4.45 eV escape from a well that is 20 eV deep? It is clearly a classical impossibility. It is as if you put a jelly bean into a closed box and sometimes (but not always) the jelly bean

---

* See, for example, Robert Eisberg and Robert Resnick, *Quantum Physics of Atoms, Molecules, Solids, Nuclei, and Particles,* 2nd edition (Wiley, 1985), Appendices G and H.

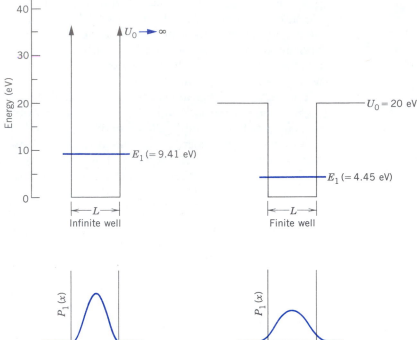

Figure 20   A potential well of (*a*) infinite depth and (*b*) finite depth (20 eV) are compared. The wells have the same width (0.2 nm). The energies E_1 and the probability densities $P_1(x)$ are compared.

materialized outside. This is so unlikely for jelly beans that we can safely use the word "impossible." However, electrons are not jelly beans. They are governed by quantum laws and not at all by classical Newtonian laws. How are we to understand this behavior of the electron?

Once again the uncertainty principle provides the answer. Recall that, when we applied the uncertainty principle to an electron trapped in an infinite well, we assumed that $\Delta x \approx L$, the width of the well, and $\Delta p_x \approx 2p_x$. For the finite well the electron's momentum, as we have just seen, is *smaller* than it is for the infinite well. Therefore, for our finite well, the uncertainty in position in the ground state must be *larger* than it is for the infinite well. Thus, for the finite well, *the uncertainty in position is larger than the width of the well!* We should not then be surprised to find the electron outside the well from time to time.

Sample Problem 8   Consider an electron confined by electrical forces to an infinitely deep potential well whose length L is 100 pm, which is roughly one atomic diameter. What are the energies of its three lowest allowed states and of the state with $n = 15$?

Solution   From Eq. 16, with $n = 1$, we have

$$E_1 = \frac{h^2}{8mL^2} = \frac{(6.63 \times 10^{-34} \text{ J} \cdot \text{s})^2}{(8)(9.11 \times 10^{-31} \text{ kg})(100 \times 10^{-12} \text{ m})^2}$$
$$= 6.03 \times 10^{-18} \text{ J} = 37.7 \text{ eV}.$$

The energies of the remaining states ($n = 2$, 3, and 15) are $2^2 \times 37.7$ eV, $3^2 \times 37.7$ eV, and $15^2 \times 37.7$ eV or 151 eV, 339 eV, and 8480 eV, respectively.

Sample Problem 9   Consider a 1-μg speck of dust moving back and forth between two rigid walls separated by 0.1 mm. It moves so slowly that it takes 100 s for the particle to cross this gap. What quantum number describes this motion?

Solution   The energy of the particle is

$$E (=K) = \tfrac{1}{2}mv^2 = \tfrac{1}{2}(1 \times 10^{-9} \text{ kg})(1 \times 10^{-6} \text{ m/s})^2$$
$$= 5 \times 10^{-22} \text{ J}.$$

Solving Eq. 16 for n yields

$$n = \frac{L}{h}\sqrt{8mE} = \frac{1 \times 10^{-4} \text{ m}}{6.63 \times 10^{-34} \text{ J} \cdot \text{s}}\sqrt{(8)(10^{-9} \text{ kg})(5 \times 10^{-22} \text{ J})}$$
$$\approx 3 \times 10^{14}.$$

This is a very large number. It is experimentally impossible to distinguish between $n = 3 \times 10^{14}$ and $n = 3 \times 10^{14} + 1$, so that the quantized nature of this motion would never reveal itself. If you compare this example with the previous one, you will see that, although its mass and its kinetic energy are both extremely small, our speck of dust is still a gross macroscopic object when compared to an electron. Quantum mechanics gives the correct answers but, since these answers coincide in this case with the answers given by classical physics, the complications of the quantum calculations are not needed.

Sample Problem 10   Evaluate the normalization constant A in Eq. 13, which gives the probability density for a particle trapped in an infinitely deep well of width L.

Solution   For this one-dimensional problem, the "volume" element is a length element and the normalization equation (Eq. 8) becomes

$$\int_0^L P_n(x)\,dx = 1,$$

in which L is the width of the well. Substituting for $P_n(x)$ from Eq. 14 yields

$$A^2 \int_0^L \sin^2 \frac{n\pi x}{L}\,dx = 1.$$

This integral is carried out most easily by introducing a new variable θ, defined from

$$\theta = \frac{n\pi x}{L}.$$

With this change, the integral becomes

$$\frac{A^2 L}{n\pi} \int_0^{n\pi} \sin^2 \theta\,d\theta = \frac{A^2 L}{n\pi}\left(\frac{1}{2}\theta - \frac{1}{4}\sin 2\theta\right)\Big|_0^{n\pi} = 1.$$

Evaluating and solving for A lead eventually to

$$A = \sqrt{\frac{2}{L}}. \qquad (18)$$

Note that the normalization constant A does not involve the quantum number n and is thus the same for all states of the system.

Sample Problem 11   An electron is trapped in an infinitely deep well of width L. If the electron is in its ground state, what fraction of its time does it spend in the central third of the well?

Solution   In the preceding example we showed that the normalization constant A that appears in Eq. 13 is $\sqrt{2/L}$, so that the probability density for the ground state, which corresponds to $n = 1$, is given from Eq. 14 by

$$P_1(x) = \frac{2}{L}\sin^2 \frac{\pi x}{L}.$$

The integral of this quantity over the entire well is unity, and the fraction that we seek is given by

$$f = \int_{L/3}^{2L/3} P_1(x)\,dx = \frac{2}{L}\int_{L/3}^{2L/3} \sin^2 \frac{\pi x}{L}\,dx.$$

Evaluating this integral as in the previous sample problem leads to

$$f = 0.61.$$

Thus the electron in its ground state spends 61% of its time in the central third of its trap, and about 19.5% in each of the outer two-thirds ($0.195 + 0.61 + 0.195 = 1$). If the electron obeyed the laws of classical physics, it would spend exactly one-third of its time in each of these regions of its trap. The probability density curve for the ground state, displayed in Fig. 18, supports graphically the calculation that we have made in this example.

50-8   BARRIER TUNNELING

In Section 50-7 we saw that an electron trapped in a well from which—classically—it could not escape has nevertheless a finite probability of being found outside the well.

We pointed out that it was as if you put a jelly bean in a closed box and found it outside the box a certain fraction of the times that you checked. Things like this don't happen to massive objects like jelly beans, but they do happen to electrons and to other light particles.

Here we discuss a related quantum phenomenon, the penetration of classically impenetrable barriers. In this case it is as if you tossed a jelly bean at a window pane and—to your surprise—it materialized on the other side with the glass unbroken. Again, don't expect this to happen for jelly beans. *Barrier tunneling*, as it is called, certainly *does* happen for electrons and is, as we shall see, a phenomenon of great practical importance.

Figure 21a shows such a barrier, of height U and thickness L. An electron of total energy E approaches the barrier from the left. Classically, because $E < U$, the electron would be reflected at the barrier and would move back in the direction from which it came. In wave mechanics, however, there is a finite chance that the electron will penetrate the barrier and continue its motion to the right.

We can describe the situation by assigning a reflection coefficient R and a transmission coefficient T, the sum of these two quantities being unity. Thus, for example, if $T = 0.05$, 5 of every 100 electrons fired at the barrier will, on the average, get through and 95 will be reflected.

Figure 21b shows the probability density for the situation. To the left of the barrier the reflected matter wave has a smaller amplitude than the incident wave so that, although there is interference, there are no points at which the cancellation is total. Within the barrier the wave decays exponentially, just as it did outside the potential well in Fig. 20b. On the far side of the barrier we have a

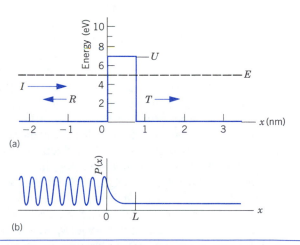

(a)

(b)

Figure 21   A particle of total energy E is incident from the left on a barrier of height U. I represents the incident beam of particles, R the reflected beam, and T the beam transmitted through the barrier. (b) The probability density for the wave describing this particle. The incident and reflected beams combine to produce standing waves to the left of the barrier.

traveling matter wave of reduced amplitude, which gives a uniform probability density.

From the Schrödinger equation we can show that the transmission coefficient T is given by*

$$T \approx e^{-2kL}, \qquad (19)$$

in which

$$k = \sqrt{\frac{8\pi^2 m(U - E)}{h^2}}.$$

This formula is an approximation that holds only for barriers that are either high enough and/or thick enough so that the transmission coefficient T is small ($T \ll 1$). Nevertheless, Eq. 19 displays well enough the main features of the barrier-tunneling phenomenon.

The value of the transmission coefficient is very sensitive to the thickness of the barrier L and to the factor k, which, in turn, depends on the mass m of the particle and the height U of the barrier. Equation 19 shows us that the transmission coefficient decreases if we increase either the thickness L or the height U of the barrier. This is just what we expect from the correspondence principle. The transmission coefficient also decreases as the mass of the particle increases, becoming vanishingly small very rapidly indeed as we proceed from electrons to jelly beans. Again, this is just what we expect from the correspondence principle. Sample Problem 12 shows some numerical predictions of Eq. 19.

Barriers and Waves

The penetration of barriers by waves of all kinds is not uncommon in classical physics. It is only when the wave is a matter wave and when, in addition, we choose to focus our attention on its associated particle, that nonclassical behavior presents itself.

Consider Fig. 22*a*, which represents an incident electromagnetic wave (visible light) falling on a glass–air interface at an angle of incidence such that total internal reflection occurs. When we treated this subject in Section 43-6, we assumed that there was no penetration of the incident ray into the air space beyond the interface. However, that treatment was based on geometrical optics, which, as we know, is always an approximation, being a limiting case of the more general wave optics. In much the same way, Newtonian mechanics (with its raylike trajectories) is a limiting case of the more general wave mechanics.

If we analyze total internal reflection from the wave optics point of view, we learn that there *is* a penetration of the wave, for a distance of the order of a few wavelengths, beyond the interface. Speaking very loosely, we can say that such a penetration is necessary because the incident

wave must "feel out" the situation locally before it can "know for sure" that an interface is present.

In Fig. 22*b*, we place the face of a second glass prism parallel to the interface, the gap between them being no more than a few wavelengths. The incident wave can then "tunnel" through this narrow "barrier" and generate a transmitted wave T. The energy in the transmitted wave comes at the expense of the reflected wave R, which is now reduced in intensity. The comparison with the barrier tunneling of a matter wave is direct. In one case we deal with an electromagnetic wave (governed by Maxwell's equations) and in the other with a matter wave (governed by Schrödinger's equation).

You can check out the phenomenon shown in Fig. 22*b* using a glass of water. Look down into the glass at the side wall, at such an angle that the light entering your eye has been totally reflected from the wall. The wall will look silvery when this condition holds. Then press your (moistened) fingertip against the outside of the glass. You will be able to see the ridges of your fingerprint because, at those points, you have interfered with the total reflection process, as in Fig. 22*b*. The valleys between the ridges of your prints are still far enough away from the glass surface that the reflection here remains total, and you see simply a silvery whorl.

It is also possible to demonstrate the phenomenon of Fig. 22*b* on a large scale by using incident microwaves and large paraffin prisms. In this case the wavelength may be a few centimeters so that the gap between the prisms can also be of this order of magnitude.

Barrier Tunneling: Some Examples

The barrier tunneling of matter waves is an important phenomenon in the natural world and has many practical applications. For a simple example consider a bare copper wire that has been cut and the two ends twisted together. It still conducts electricity readily, in spite of the fact that the wires are coated with a thin layer of copper oxide, an

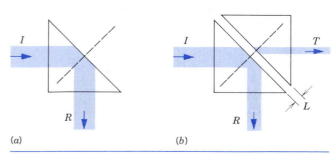

Figure 22   (*a*) An incident light beam I undergoes total internal reflection at the glass–air interface. (*b*) The beam tunnels through the narrow air gap, and as a result there is a transmitted beam T in the second glass. This condition is called *frustrated total internal reflection*. In these drawings, the width of the beam represents its intensity.

---

* See, for example, Robert Eisberg and Robert Resnick, *Quantum Physics of Atoms, Molecules, Solids, Nuclei, and Particles* 2nd edition (Wiley, 1985), Section 6-5.

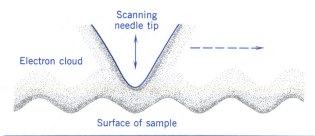

Figure 23   A needle is scanned over the surface of a sample in a scanning tunneling microscope.

insulating material. How do the electrons get through this (extremely thin) oxide layer? By barrier tunneling.

For a more exotic example, consider the core of the Sun, where the Sun's energy is being generated by thermonuclear fusion processes. Such processes involve the fusing together of light nuclei to form heavier ones, with the release of energy. Suppose that two protons are rushing together at high speed. They must get extremely close before their strong attractive nuclear forces can take effect and cause them to fuse. Meanwhile, they are slowed down by the repulsive Coulomb force that tends to drive them apart. They are, we can say, separated by a Coulomb barrier. The likelihood of fusion depends critically on the ability of the protons to tunnel through this mutual barrier. Without barrier tunneling the solar furnace would shut down and the Sun would collapse into itself.

The emission of (positively charged) alpha particles by radioactive nuclei and the spontaneous fission of heavy nuclei into two large fragments are among other natural processes in which tunneling plays a role.

Among practical applications we may list the tunnel diode, in which the flow of electrons (by tunneling) through a device can be rapidly turned on or off by controlling the height of the barrier (by varying an externally applied voltage, for instance). This can be done with a very short response time (of the order of 10^{-11} s or 10 ps)

so that the device is suitable for applications where speed of response is critical. The 1973 Nobel prize was shared by three "tunnelers," Leo Esaki (tunneling in semiconductors), Ivar Giaver (tunneling in superconductors), and Brian Josephson (the Josephson junction, a quantum switching device based on tunneling).

In a *scanning tunneling microscope,* a fine needle tip is scanned mechanically (in a TV-like raster pattern) over the surface of the sample being investigated, as in Fig. 23. Electrons from the sample tunnel through the gap between the sample and the needle and are recorded as a "tunnel current." Normally, this tunnel current would vary widely as the gap between the sample and the needle changes during the scan.

However, a mechanism is provided that automatically moves the needle up or down during the scan, so as to keep the tunnel current—and thus the gap—constant. The needle's vertical position can then be displayed on a screen as a function of its location, producing a three-dimensional plot of the surface. The 1986 Nobel prize was awarded to Gerd Binnig and Heinrich Rohrer for the development of the scanning tunneling microscope.*

Figure 24 shows the result of a scan over a graphite surface. The "bumps" suggest individual carbon atoms. Features as small as 1/100 of an atomic diameter can be resolved with this remarkable device.

Sample Problem 12   Consider an electron whose total energy E is 5.0 eV approaching a barrier whose height U is 6.0 eV, as in Fig. 21a. Let the barrier thickness L be 0.70 nm. (*a*) What is the de Broglie wavelength of the incident electron? (*b*) What transmission coefficient follows from Eq. 19? (*c*) What would be the transmission coefficient if the barrier thickness were reduced to

---

* See "The Scanning Tunneling Microscope," by Gerd Binnig and Heinrich Rohrer, *Scientific American,* August 1985, p. 50.

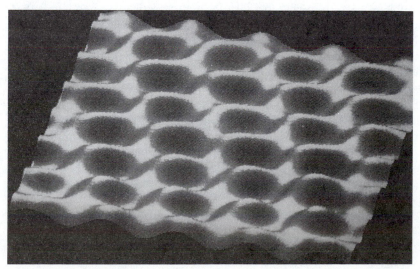

Figure 24   The regular arrangement of carbon atoms on the surface of graphite is revealed in this image made with a scanning tunneling microscope.

0.30 nm? If its height were increased to 7.0 eV? If the incident particle were a proton?

Solution   (a) Before the electron reaches the barrier, its total energy E is entirely kinetic, the potential energy in that region being zero. Proceeding as in Sample Problem 2b, we find $\lambda = 0.55$ nm. Thus the barrier is about 0.70 nm/0.55 nm or about 1.3 de Broglie wavelengths thick.

(b) We have

$$k = \sqrt{\frac{8\pi^2 m(U - E)}{h^2}}$$

$$= \sqrt{\frac{8\pi^2(9.11 \times 10^{-31} \text{ kg})(6.0 \text{ eV} - 5.0 \text{ eV})(1.60 \times 10^{-19} \text{ J/eV})}{(6.63 \times 10^{-34} \text{ J} \cdot \text{s})^2}}$$

$$= 5.12 \times 10^9 \text{ m}^{-1}.$$

The quantity kL is then $(5.12 \times 10^9 \text{ m}^{-1})(700 \times 10^{-12} \text{ m}) = 3.58$, and the transmission coefficient is

$$T = e^{-2kL} = e^{-2(3.58)} = 7.7 \times 10^{-4}.$$

Of every 100,000 electrons that strike the barrier, only 77 will tunnel through it.

(c) Making the appropriate changes in the solution to part (b), we find:

$$L = 0.30 \text{ nm}: \quad T = 0.10$$
$$U = 7.0 \text{ eV}: \quad T = 5.9 \times 10^{-5}$$
$$m = 1836 m_e: \quad T = 10^{-130}.$$

It is easier for the electron to penetrate the thinner barrier, but more difficult to penetrate the higher one. The more massive proton penetrates hardly at all. (Imagine how small T would be for a jelly bean!)

50-9   THE CORRESPONDENCE PRINCIPLE

In several cases in this chapter and the previous one, we have tried to make comparisons between classical and quantum behaviors. For example, in Sample Problem 3 of Chapter 49, we showed that the quantized behavior of an oscillator of ordinary size is too small to be observable; we are therefore safe in treating that oscillator using classical (nonquantum) techniques and in regarding the energy of the oscillator as a continuous (rather than a quantized) variable. In this chapter, we showed in Sample Problem 2 that the de Broglie wavelength of a virus particle is unobservably small; in Sample Problem 6b that the uncertainty principle should not affect your golf game; and in Sample Problem 9 that the energy quantization of a trapped dust particle cannot be observed.

We seem to have two sets of rules for analyzing mechanical behavior: we use quantum mechanics for "small" particles and classical mechanics for "large" particles. Clearly a golf ball is a large particle, but even dust particles and viruses can be regarded as "large." Perhaps you are wondering where to draw the line between classical particles and quantum particles.

Niels Bohr was similarly puzzled when he attempted (before de Broglie's bold hypothesis led to the development of quantum theory) to work out the structure of atoms based on the nuclear model of electrons orbiting a central nucleus. Bohr's model was based on discrete energy levels (and discrete transitions as the electron jumped from one level to another) for the atom, but he knew that a "classical" atom would be characterized by a continuous spectrum of radiation as the electron spiraled in toward the nucleus. As in the case of the quantum system, Bohr faced the dilemma of one set of rules for systems on one scale and a different set of rules on another scale.

Bohr resolved his problem by proposing the *correspondence principle*, which can be stated in general terms as:

Quantum theory must agree with classical theory in the limit of large quantum numbers.

This avoids the problem of having to find a boundary between the two different systems. The predictions of quantum mechanics must be identical to those of classical mechanics as the quantum system grows to classical dimensions. For example, consider the probability densities for a particle trapped in a well (Fig. 18). The behavior for $n = 1$ or $n = 2$ differs markedly from the classical behavior of a uniform probability density in the well. However, for $n = 15$, the probability density has become much more uniform. As n increases, the oscillations of P are packed closer and closer together, so that if we examine the probability in an interval of length Δx greater than L/n, we find no change as the interval is moved throughout the well. Here we are approaching the classical situation of a uniform probability density for large quantum numbers.

The correspondence principle tells us that we do not need to draw a line between classical and quantum behaviors. If we are in doubt whether to apply classical or quantum laws to a virus or a dust particle, we now know that we are safe in applying the quantum laws, the results of which must duplicate the classical laws if we are in a region where classical behavior is expected. Indeed, by using the probability densities calculated from the Schrödinger equation and doing the appropriate averaging, it can be shown that the average force on a particle in the quantum regime equals the mass times the average acceleration: $\overline{F} = m\overline{a}$. In the limit of large quantum numbers, the fluctuations from the average become negligible, and $F = ma$ becomes exact. Even though the Schrödinger equation looks very different from Newton's second law, its outcomes reduce to Newton's second law in the limit of large systems. This justifies our use of Newton's second law, which is easier to apply in the large limit, to bodies composed of atoms that are individually governed by the Schrödinger equation.

50-10 WAVES AND PARTICLES

On several earlier occasions we have promised to address the question of how an electron (or a photon) can be wavelike under some circumstances and particlelike under others. We now keep that promise. First, we remind you in Table 1 of the clear experimental evidence that both matter and radiation do indeed have this dual character.

Our mental images of "wave" and "particle" are drawn from our familiarity with large-scale objects such as ocean waves and tennis balls. In a way it is fortunate that we are able to extend these concepts into the atomic domain and to apply them to entities such as the electron, which we can neither see nor touch. We say at once, however, that no single concrete mental image, combining the features of *both* wave and particle, is possible in the quantum world. As Paul Davies, physicist and science writer, has written: "It is impossible to visualize a wave-particle, so don't try." What then are we to do?

Niels Bohr, who not only played a major role in the development of quantum mechanics but also served as its major philosopher and interpreter, has shown the way with his *principle of complementarity*, which states:

The wave and the particle aspects of a quantum entity are both necessary for a complete description. However, the two aspects cannot be revealed simultaneously in a single experiment. The aspect that is revealed is determined by the nature of the experiment being done.

Consider a beam of light, perhaps from a laser, that passes across a laboratory table. What is the nature of the light beam? Is it a wave or a stream of particles?

You cannot answer this question unless you interact with the beam in some way. If you put a diffraction grating in the path of the beam you reveal it as a wave. If you interpose a photoelectric apparatus (Section 49-5), you will need to regard the beam as a stream of particles (photons) if you are to interpret your measurements in a satisfactory way. Try as you will, there is no single experiment that you can carry out with the beam that will require you to interpret it as a wave *and* as a particle *at the same time.*

Complementarity: A Case Study

Let us see how complementarity works by trying to set up an experiment that will force nature to reveal both the wave and the particle aspects of electrons at the same time. In Fig. 25 a beam of electrons falls on a double-slit arrangement in screen *A* and sets up a pattern of interference fringes on screen *B*. This is convincing proof of the wave nature of the incident electron beam.

Suppose now that we replace screen *B* with a small electron detector, designed to generate and record a "click" every time an electron hits it. We find that such clicks do indeed occur. If we move the detector up and down in Fig. 25, we can, by plotting the click rate against the detector position, trace out the pattern of interference fringes. Have we not succeeded in demonstrating both wave and particle? We see the fringes (wave) and we hear the clicks (particle).

We have not. The "click" shows that the electron is localized (like a particle) at the detector, but it does not indicate how it got there. The concept of "particle" involves the concept of "trajectory" and a mental image of a dot following a path. As a minimum, we want to be able to know which of the two slits in screen *A* the electron passed through on its way to generating a click in the detector. Can we find out?

We can, in principle, by putting a very thin detector in front of each slit, designed so that, if an electron passes through it, it will generate an electronic signal. We can then try to correlate each click, or "screen arrival signal," with a "slit passage signal," thus identifying the path of the electron involved.

If we succeed in modifying the apparatus to do this, we find a surprising thing. The interference fringes have disappeared! In passing through the slit detectors, the electrons were affected in ways that destroyed the interference pattern. Although we have now shown the particle nature of the electron, the evidence for its wave nature has vanished.

The converse to our thought experiment is also true. If we start with an experiment that shows that electrons are particles and if we tinker with it to bring out the wave aspect, we will always find that the evidence for particles has vanished. Also, our experiment would work in precisely the same way if we substituted a light beam for the incident electron beam in Fig. 25.

TABLE 1 SELECTED EXPERIMENTS SHOWING THE DUAL WAVE–PARTICLE NATURE OF MATTER AND OF RADIATION

	Matter	*Radiation*
Wave nature	Davisson–Germer electron diffraction experiments (Section 50-3)	Young's double-slit interference experiment (Section 45-1)
Particle nature	J. J. Thomson's measurement of e/m for the electron (Section 34-2)	The Compton effect (Section 49-7)

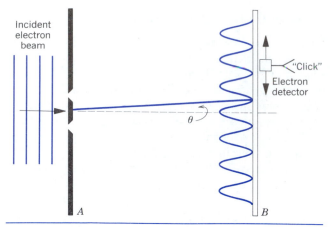

Figure 25 An electron beam falls on a double slit in screen A and produces interference fringes on screen B. Screen B can be replaced by an electron detector, which can be moved along the location occupied by the screen.

A Quantum Puzzle Resolved

In Section 50-1, we asked how it is possible for particles to undergo double-slit interference. A particle, after all, must travel a definite trajectory. What is the source of the interference?

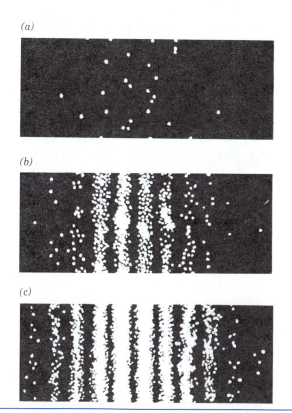

Figure 26 The buildup of interference fringes as electrons fall on screen B of Fig. 25. In (*a*), about 30 electrons have landed on the screen, in (*b*) about 1000, and in (*c*) about 10,000. The probability density of the wave describing the electron determines where the electrons will land on the screen.

As the beginning of an answer we look again at the thought experiment of Fig. 25, in which the pattern of fringes on screen *B* is neatly accounted for by the alternating constructive and destructive interference of matter wavelets radiating from each of the two slits in screen *A*. The connection of these waves with the particle is that the square of their associated wave function at any point gives the probability (per unit volume) that the particle will be found at that point. Thus, on screen *B*, electrons will pile up at those places where this probability amplitude is large, and they will be found in lesser abundance at those places where it is small. Figure 26, a computer simulation, shows how the fringes build up with time for a weak incident beam.

These considerations apply even if the incident beam is deliberately made so weak that, by calculation, there should be—on average—only a single electron in the apparatus at any given time. You might think that, because the single electron that happens to be in the apparatus must go through one slit or the other, the fringes must vanish; after all, you may reason, the electron cannot interfere with itself and there is nothing else for it to interfere with. However, experiment shows that the fringes will *still* be formed, built up slowly as electron after electron falls on screen *B*. Even under these conditions the associated wave always passes through *both* slits, and *it* is what determines where the electrons are likely to fall on screen *B*.

To get a better idea of the role of the wave in the motion of a particle, consider Fig. 27, in which a particle (an electron, say) is generated at point *I* and detected at point *F*. How does it travel this straight-line path?

The quantum answer is that the wave explores all possible paths, as the figure suggests, assigning an equal probability to each. However, only for the straight line connecting the two points do the waves add constructively, yielding a high probability that the particle will be found there if sought. For points not near this straight line, the waves cancel each other by destructive interference, the cancellation being more severe the more massive the particle. It is in this way that the trajectories of particles in Newtonian mechanics are related to their associated waves.

It is not hard to imagine the effect of inserting a double slit in Fig. 27 between *I* and *F*. In the process of exploring

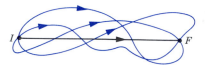

Figure 27 An electron moves from *I* to *F*. The waves that describe its journey interfere constructively along the straight path and destructively along all other paths. The wave explores all possible paths between *I* and *F*.

the possible paths, only those waves passing through the slits survive. The particle is more likely to be found in regions on the screen of high probability density, and in this way the interference pattern is formed.

QUESTIONS

1. How can the wavelength of an electron be given by $\lambda = h/p$? Doesn't the very presence of the momentum p in this formula imply that the electron is a particle?

2. In a repetition of Thomson's experiment for measuring e/m for the electron (see Section 34-2), a beam of electrons is collimated by passage through a slit. Why is the beamlike character of the emerging electrons not destroyed by diffraction of the electron wave at this slit?

3. Why is the wave nature of matter not more apparent in our daily observations?

4. Considering the wave behavior of electrons, we should expect to be able to construct an "electron microscope" using short-wavelength electrons to provide high resolution. This, indeed, has been done. (a) How might an electron beam be focused? (b) What advantages might an electron microscope have over a light microscope? (c) Why not make a proton microscope? A neutron microscope?

5. How many experiments can you recall that support the wave theory of light? The particle theory of light? The wave theory of matter? The particle theory of matter?

6. Is an electron a particle? Is it a wave? Explain your answer, citing relevant experimental evidence.

7. If the particles listed below all have the same energy, which has the shortest wavelength: electron; α particle; neutron; proton?

8. What common expression can be used for the momentum of either a photon or a particle?

9. Discuss the analogy between (a) wave optics and geometrical optics and (b) wave mechanics and classical mechanics.

10. Does a photon have a de Broglie wavelength? Explain.

11. Discuss similarities and differences between a matter wave and an electromagnetic wave.

12. Can the de Broglie wavelength associated with a particle be smaller than the size of the particle? Larger? Is there any relation necessarily between such quantities?

13. If, in the de Broglie formula $\lambda = h/mv$, we let $m \rightarrow \infty$, do we get the classical result for particles of matter?

14. Considering electrons and photons as particles, how are they different from each other?

15. Is Eq. 1 for the de Broglie wavelength, $\lambda = h/p$, valid for a relativistic particle? Justify your answer.

16. How could Davisson and Germer be sure that the "54-V" peak of Fig. 7 was a first-order diffraction peak, that is, that $m = 1$ in Eq. 2?

17. Do electron diffraction experiments give different information about crystals than can be obtained from x-ray diffraction experiments? From neutron diffraction experiments? Give examples.

18. Why are the hydrogen atoms clearly visible in Fig. 10 but not in Fig. 17 of Chapter 47?

19. In Fig. 9b (made with x rays) the diffraction circles are speckled, but in Fig. 9c (made with electrons) they are smooth. Can you explain why?

20. Electromagnetic waves will penetrate seawater to a certain extent if their frequency is low enough. This is the basis of one plan to communicate with submerged submarines. A difficulty with this plan is that the lower the frequency, the longer the time it takes to transmit a message (in Morse code pulses, say). Can you explain why this should be?

21. Why is the Heisenberg uncertainty principle not more readily apparent in our daily observations?

22. (a) Give examples of how the process of measurement disturbs the system being measured. (b) Can the disturbances be taken into account ahead of time by suitable calculations?

23. You measure the pressure in a tire, using a pressure gauge. The gauge, however, bleeds a little air from the tire in the process, so that the act of measuring changes the property that you are trying to measure. Is this an example of the Heisenberg uncertainty principle? Explain.

24. "The energy of the ground state of an atomic system can be precisely known, but the energies of its excited states are always subject to some uncertainty." Can you explain this statement on the basis of the uncertainty principle?

25. "If an electron is localized in space, its momentum becomes uncertain. If it is localized in time, its energy becomes uncertain." Explain this statement.

26. The quantity $\psi(x)$, the amplitude of a matter wave, is called a *wave function*. What is the relationship between this quantity and the particles that form the matter wave?

27. In Section 50-7 we solved the wave mechanical problem of a particle trapped in an infinitely deep well without ever using (or even writing down) Schrödinger's equation. How were we able to do that?

28. A standing wave can be viewed as the superposition of two traveling waves. Can you apply this view to the problem of a particle confined between rigid walls, giving an interpretation in terms of the motion of the particle?

29. The allowed energies for a particle confined between rigid walls are given by Eq. 16. First, convince yourself that, as n increases, the energy levels become farther apart. How can this possibly be? The correspondence principle would seem to require that they move closer together as n increases, approaching a continuum.

30. How can the predictions of wave mechanics be so exact if the only information we have about the positions of the electrons in atoms is statistical?

31. In the $n = 1$ state, for a particle confined between rigid walls, what is the probability that the particle will be found in a small-length element at the surface of either wall?

32. What are the dimensions of $P_n(x)$ in Fig. 18? What is the value of the classically expected probability density, represented by the horizontal lines? What value do the areas under the curves have? How does the area under any curve compare with the area under the horizontal line? All these questions can be answered by inspection of the figure.

33. In Fig. 18 what do you imagine the curve for $P_n(x)$ for $n = 100$ looks like? Convince yourself that these curves approach classical expectations as $n \rightarrow \infty$.

34. We have seen that barrier tunneling works for matter waves and for electromagnetic waves. Do you think that it also works for water waves? For sound waves?

35. Comment on the statement: "A particle can't be detected while tunneling through a barrier, so that it doesn't make sense to say that such a thing actually happens."

36. List examples of barrier tunneling occurring in nature and in manufactured devices.

37. A proton and a deuteron, each having 3 MeV of energy, attempt to penetrate a rectangular potential barrier of height 10 MeV. Which particle has the higher probability of succeeding? Explain in qualitative terms.

38. A laser projects a beam of light across a laboratory table. If you put a diffraction grating in the path of the beam and observe the spectrum, you declare the beam to be a wave. If instead you put a clean metal surface in the path of the beam and observe the ejected photoelectrons, you declare this same beam to be a stream of particles (photons). What can you say about the beam if you don't put anything in its path?

39. State and discuss (a) the correspondence principle, (b) the uncertainty principle, and (c) the complementarity principle.

40. In Fig. 25, why would you expect the electrons from each slit to arrive at the screen over a range of positions? Shouldn't they all arrive at the same place? How does your answer relate to the complementarity principle?

41. Several groups of experimenters are trying to detect gravity waves, perhaps coming from our galactic center, by measuring small distortions in a massive object through which the hypothesized waves pass. They seek to measure displacements as small as 10^{-21} m. (The radius of a proton is $\sim 10^{-15}$ m, a million times larger!) Does the uncertainty principle put any restriction on the precision with which this measurement can be carried out?

42. Figure 18 shows that for $n = 3$ the probability function $P_n(x)$ for a particle confined between rigid walls is zero at two points between the walls. How can the particle ever move across these positions? (*Hint*: Consider the implications of the uncertainty principle.)

43. In Sample Problem 8, the electron's energy is determined *exactly* by the size of the box. How do you reconcile this with the fact that the uncertainty in the location of the electron cannot exceed 100 pm and, if the uncertainty principle is to be obeyed, the electron's momentum must be correspondingly uncertain?

PROBLEMS

Section 50-2 The De Broglie Wavelength

1. A bullet of mass 41 g travels at 960 m/s. (a) What wavelength can we associate with it? (b) Why does the wave nature of the bullet not reveal itself through diffraction effects?

2. Using the classical relation between momentum and kinetic energy, show that the de Broglie wavelength of an electron can be written (a) as

$$\lambda = \frac{1.226 \text{ nm}}{\sqrt{K}},$$

in which K is the kinetic energy in electron volts, or (b) as

$$\lambda = \sqrt{\frac{1.50}{V}},$$

where λ is in nm, and V is the accelerating potential in volts. (Use the best values of the needed constants as found in Appendix B.)

3. Calculate the wavelength of a 1.00-keV (a) electron, (b) photon, and (c) neutron.

4. The wavelength of the yellow spectral emission line of sodium is 589 nm. At what kinetic energy would an electron have the same de Broglie wavelength?

5. If the de Broglie wavelength of a proton is 0.113 pm, (a) what is the speed of the proton and (b) through what electric potential would the proton have to be accelerated from rest to acquire this speed?

6. Singly charged sodium ions are accelerated through a potential difference of 325 V. (a) What is the momentum acquired by the ions? (b) Calculate their de Broglie wavelength.

7. The existence of the atomic nucleus was discovered in 1911 by Ernest Rutherford, who properly interpreted some experiments in which a beam of alpha particles was scattered from a foil of atoms such as gold. (a) If the alpha particles had a kinetic energy of 7.5 MeV, what was their de Broglie wavelength? (b) Should the wave nature of the incident alpha particles have been taken into account in interpreting these experiments? The distance of closest approach of the alpha particle to the nucleus in these experiments was about 30 fm. (The wave nature of matter was not postulated until more than a decade after these crucial experiments were first performed.)

8. The highest achievable resolving power of a microscope is limited only by the wavelength used; that is, the smallest detail that can be separated is about equal to the wavelength. Suppose one wishes to "see" inside an atom. Assuming the atom to have a diameter of 100 pm this means that we wish to resolve detail of separation about 10 pm. (a) If an electron microscope is used, what minimum energy of electrons is needed? (b) If a light microscope is used, what minimum energy of photons is needed? (c) Which microscope seems more practical for this purpose? Why?

9. The 32-GeV electron accelerator at Stanford provides an electron beam of small wavelength, suitable for probing the fine details of nuclear structure by scattering experiments. What is this wavelength and how does it compare with the size of an average nucleus? (At these energies it is sufficient to use the extreme relativistic relationship between momentum and energy; namely, $p = E/c$. This is the same relationship used for light and is justified when the kinetic energy of a particle is much greater than its rest energy, as in this case. The radius of a middle-mass nucleus is about 5.0 fm.)

10. Consider a balloon filled with (monatomic) helium gas at 18°C and 1.0 atm pressure. Calculate (*a*) the average de Broglie wavelength of the helium atoms and (*b*) the average distance between the atoms. Can the atoms be treated as particles under these conditions?

11. A nonrelativistic particle is moving three times as fast as an electron. The ratio of their de Broglie wavelengths, particle to electron, is 1.813×10^{-4}. By calculating its mass, identify the particle. See Appendix B.

12. (*a*) A photon in free space has an energy of 1.5 eV and an electron, also in free space, has a kinetic energy of that same amount. What are their wavelengths? (*b*) Repeat for an energy of 1.5 GeV.

13. In an ordinary color television set, electrons are accelerated through a potential difference of 25.0 kV. Find the de Broglie wavelength of such electrons (*a*) using the classical expression for momentum and (*b*) taking relativity into account.

14. What accelerating voltage would be required for electrons in an electron microscope to obtain the same ultimate resolving power as that which could be obtained from a gamma-ray microscope using 136-keV gamma rays? (*Hint*: See Problem 8.)

Section 50-3 Testing De Broglie's Hypothesis

15. A neutron crystal spectrometer utilizes crystal planes of spacing $d = 73.2$ pm in a beryllium crystal. What must be the Bragg angle θ so that only neutrons of energy $K = 4.2$ eV are reflected? Consider only first-order reflections.

16. A beam of thermal neutrons from a nuclear reactor falls on a crystal of calcium fluoride, the beam direction making an angle θ with the surface of the crystal. The atomic planes parallel to the crystal surface have an interplanar spacing of 54.64 pm. The de Broglie wavelength of neutrons in the incident beam is 11.00 pm. For what values of θ will the first three orders of Bragg-reflected neutron beams occur? (*Hint*: Neutrons, which carry no charge and are thus not subject to electrical forces, are not refracted as they pass through a crystal surface. Thus neutron diffraction can be treated in strict analogy with x-ray diffraction.)

17. In the experiment of Davisson and Germer (*a*) at what angles would the second- and third-order diffracted beams corresponding to a strong maximum in Fig. 7 occur, provided they are present? (*b*) At what angle would the first-order diffracted beam occur if the accelerating potential were changed from 54 to 60 V?

18. A potassium chloride (KCl) crystal is cut so that the layers of atomic planes parallel to its surface have a spacing of 314 pm between adjacent lines of atoms. A beam of 380-eV elec-

trons is incident normally on the crystal surface. Calculate the angles ϕ at which the detector must be positioned to record strongly diffracted beams of all orders present.

19. A beam of low-energy neutrons emerges from a reactor and is diffracted from a crystal. The kinetic energies of the neutrons are contained in a band of width ΔK centered on kinetic energy K. Show that the angles for a given order of diffraction are spread over a range $\Delta\theta$ given in degrees by

$$\Delta\theta = \left(\frac{90}{\pi}\right)(\tan\theta)\frac{\Delta K}{K},$$

where θ is the diffraction angle for a neutron with kinetic energy K.

20. A beam of atoms emerges from an oven that is at a temperature T. The distribution of the speeds of the atoms in the beam is proportional to $v^3 e^{-mv^2/2kT}$ (see Section 24-3). (*a*) Show that the distribution of de Broglie wavelengths of the atoms is proportional to $\lambda^{-5} e^{-h^2/2mkT\lambda^2}$, and (*b*) that the most probable de Broglie wavelength is

$$\lambda_p = \frac{h}{\sqrt{5mkT}}.$$

Section 50-4 Waves, Wave Packets, and Particles

21. Using a rotating shutter arrangement, you listen to a 540-Hz standard tuning fork for 0.23 s. What approximate spread of frequencies is contained in this acoustic pulse?

22. The signal from a television station contains pulses of full width $\Delta t \approx 10$ ns. Is it feasible to transmit television in the AM broadcasting band, which runs from about 500 to 1600 kHz?

Section 50-5 Heisenberg's Uncertainty Relationships

23. A nucleus in an excited state will return to its ground state, emitting a gamma ray in the process. If its mean lifetime is 8.7 ps in a particular excited state of energy 1.32 MeV, find the uncertainty in the energy of the corresponding emitted gamma-ray photon.

24. An atom in an excited state has a lifetime of 12 ns; in a second excited state the lifetime is 23 ns. What is the uncertainty in energy for a photon emitted when an electron makes a transition between these two states?

25. A microscope using photons is employed to locate an electron in an atom to within a distance of 12 pm. What is the minimum uncertainty in the momentum of the electron located in this way?

26. Imagine playing baseball in a universe where Planck's constant was 0.60 J·s. What would be the uncertainty in the position of a 0.50-kg baseball moving at 20 m/s with an uncertainty in velocity of 1.2 m/s? Why would it be hard to catch such a ball?

27. Find the uncertainty in the location of a particle, in terms of its de Broglie wavelength λ, so that the uncertainty in its velocity is equal to its velocity.

Section 50-7 Trapped Particles and Probability Densities

28. What must be the width of an infinite well such that a trapped electron in the $n = 3$ state has an energy of 4.70 eV?

29. (a) Calculate the smallest allowed energy of an electron confined to an infinitely deep well with a width equal to the diameter of an atomic nucleus (about 1.4×10^{-14} m). (b) Repeat for a neutron. (c) Compare these results with the binding energy (several MeV) of protons and neutrons inside the nucleus. On this basis, should we expect to find electrons inside nuclei?

30. The ground-state energy of an electron in an infinite well is 2.6 eV. What will the ground-state energy be if the width of the well is doubled?

31. An electron, trapped in an infinite well of width 253 pm, is in the ground ($n = 1$) state. How much energy must it absorb to jump up to the third excited ($n = 4$) state?

32. (a) Calculate the fractional difference between two adjacent energy levels of a particle confined in a one-dimensional well of infinite depth. (b) Discuss the result in terms of the correspondence principle.

33. (a) Calculate the separation in energy between the lowest two energy levels for a container 20 cm on a side containing argon atoms. (b) Find the ratio with the thermal energy of the argon atoms at 300 K. (c) At what temperature does the thermal energy equal the spacing between these two energy levels? Assume, for simplicity, that the argon atoms are trapped in a one-dimensional well 20 cm wide. The molar mass of argon is 39.9 g/mol.

34. Consider a conduction electron in a cubical crystal of a conducting material. Such an electron is free to move throughout the volume of the crystal but cannot escape to the outside. It is trapped in a three-dimensional infinite well. The electron can move in three dimensions, so that its total energy is given by (compare with Eq. 16),

$$E = \frac{h^2}{8mL^2}(n_1^2 + n_2^2 + n_3^2),$$

in which n_1, n_2, n_3 each take on the values 1, 2, ... Calculate the energies of the lowest five distinct states for a conduction electron moving in a cubical crystal of edge length $L = 250$ nm.

35. Consider an electron trapped in an infinite well whose width is 98.5 pm. If it is in a state with $n = 15$, what are (a) its energy? (b) The uncertainty in its momentum? (c) The uncertainty in its position?

36. Repeat Sample Problem 11, but assume now that the electron is in the $n = 2$ state.

37. Where are the points of (a) maximum and (b) minimum probability for a particle trapped in an infinitely deep well of length L if the particle is in the state n?

38. A particle is confined between rigid walls separated by a distance L. (a) Show that the probability P that it will be found within a distance $L/3$ from one wall is given by

$$P = \frac{1}{3}\left(1 - \frac{\sin(2\pi n/3)}{2\pi n/3}\right).$$

Evaluate the probability for (b) $n = 1$, (c) $n = 2$, (d) $n = 3$, and (e) under the assumption of classical physics.

39. A particle is confined between rigid walls located at $x = 0$ and $x = L$. For the $n = 4$ energy state, (a) sketch the probability density curve for the particle's location. Calculate the approximate probabilities of finding the particle within a region $\Delta x = 0.0003L$ when (b) Δx is located at $x = L/8$ and (c) at $x = 3L/16$. Refer to your figure to see whether or not your results seem reasonable. (*Hint*: No integration is necessary.)

Section 50-8 Barrier Tunneling

40. In Sample Problem 12, suppose that you can vary the thickness L of the barrier. To what value should the thickness be adjusted so that 1 electron out of 100 striking the barrier will tunnel through it?

41. (a) A proton and (b) a deuteron (which has the same charge as a proton but twice the mass) are incident on a barrier of thickness 10 fm and height 10 MeV. Each particle has a kinetic energy of 3.0 MeV. Find the transmission probabilities for them.

42. Consider a barrier such as that of Fig. 21, but whose height U is 6.00 eV and whose thickness L is 700 pm. Calculate the energy of an incident electron such that its transmission probability is 1 in 1000.

43. Suppose that an incident beam of 5.0-eV protons fell on a barrier of height 6.0 eV and thickness 0.70 nm, and at a rate equivalent to a current of 1.0 kA. How long would you have to wait — on the average — for one proton to be transmitted?

44. Consider the barrier tunneling situation defined by Sample Problem 12. What fractional change in the transmission coefficient occurs for a 1% increase in (a) the barrier height, (b) the barrier thickness, and (c) the incident energy of the electron?

CHAPTER 51

THE STRUCTURE OF ATOMIC HYDROGEN

*Ever since it has been known that matter is made up of atoms, the
fundamental question has been: "What is an atom like?" Our aim in this
chapter is to answer this question from the point of view of wave mechanics. Understanding
the structure of atoms is essential if we hope to understand how atoms join to form
molecules and solids. Chemistry and solid-state physics both depend on knowledge of
atomic structure acquired from wave mechanics.*

*We start in this chapter with hydrogen, which is both the simplest atom and the most
abundant atom in the universe. Understanding how the principles of wave mechanics
account for the structure of hydrogen leads us to apply similar considerations to explaining
the structure of more complex atoms, which we do in the next chapter.*

*Because of its simplicity, hydrogen has the advantage that its properties can be calculated
exactly and without approximation, which has permitted comparison between prediction
and experiment for a variety of physical theories from quantum mechanics in the 1920s to
quantum electrodynamics in the 1940s and 1950s.*

51-1 THE BOHR THEORY

Most of our knowledge about atoms, molecules, and nuclei comes from studying the radiation emitted or absorbed by them, as we illustrated by the line spectra in Fig.
15 of Chapter 49. This also is the case with atomic hydrogen. A spectrum of atomic hydrogen in the visible region
is illustrated in Fig. 1. This spectrum, which might be
obtained with a prism or diffraction grating in a spectrograph such as that of Fig. 8 of Chapter 47, had been measured with great precision in the late 1800s, and its interpretation was puzzling for scientists of that era. The initial
approach in analyzing this spectrum was to find an empirical formula that fit the data. It took another 30 years for a
theory to be developed that could explain the formula.

The spectrum in Fig. 1 shows several regularities. The
spacing of the lines decreases as we go to shorter wavelengths, while the wavelengths themselves approach a
limit called the *series limit*. An empirical formula for the
wavelengths of the lines of atomic hydrogen was developed in 1885 by Johannes Balmer, a Swiss high school
teacher. Balmer's formula for the wavelength λ in nano-
meters is

$$\lambda = 364.6 \, \frac{n^2}{n^2 - 4}, \qquad n = 3, 4, 5, \ldots \qquad (1)$$

This series of lines of hydrogen in the visible region is
called the *Balmer series*.

In 1890, J. J. Rydberg modified Balmer's formula and
wrote it as

$$\frac{1}{\lambda} = R \left(\frac{1}{4} - \frac{1}{n^2} \right), \qquad n = 3, 4, 5, \ldots, \qquad (2)$$

where R, called the *Rydberg constant,* has a value of
$1.097 \times 10^7 \text{ m}^{-1}$. Recognizing that 4 can be written as 2^2,
Rydberg rewrote the formula in a more general form as

$$\frac{1}{\lambda} = R \left(\frac{1}{m^2} - \frac{1}{n^2} \right), \qquad (3)$$

$$n = m + 1, m + 2, m + 3, \ldots,$$

where $m = 2$ for the Balmer series. The obvious question
that occurred was whether there were other series of lines,
corresponding to other fixed values of m. Soon searchers
turned up a series in the infrared corresponding to $m = 3$
and another in the ultraviolet with $m = 1$. All these series
could be fit by Eq. 3 (called the *Balmer–Rydberg formula*) with a given value of m and a series of values of n

1069

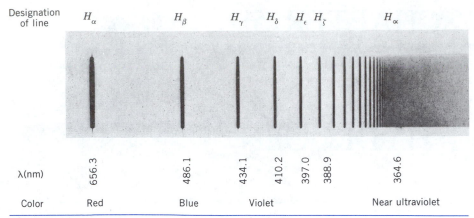

Figure 1   A photograph of the spectral lines of the Balmer series in hydrogen.

starting with $m + 1$ and ending with the series limit as $n \rightarrow \infty$. Figure 2 shows the series for $m = 1$ (called the Lyman series), $m = 2$ (the Balmer series), and $m = 3$ (the Paschen series). The wavelengths of some of these lines are listed in Table 1.

The key to understanding this empirical formula was provided by the Danish physicist Niels Bohr in 1913. After completing his Ph.D., Bohr went to England, where he worked first with J. J. Thomson and then with Ernest Rutherford (see the discussions of the Thomson and Rutherford models of the atom in Section 29-7). Bohr immediately recognized the importance of the Ruther-ford nuclear atom in understanding the structure of atoms. He was led to propose a model in which the elec-tron circulates about the nucleus like a planet about the Sun (Fig. 3). However, he recognized that such a model would violate one of the predictions of classical physics, namely, that an accelerated electron (even centripetally accelerated) would emit a continuous spectrum of radia-tion as it loses energy and spirals into the nucleus. Clearly this does not happen; if Bohr's planetary structure of the atom is correct, the classical physics of Newton and Max-well must be suspended! (Keep in mind that Bohr's work occurred 10 years before de Broglie's bold hypothesis of matter waves.)

Realizing that classical physics had come to a dead end on the hydrogen atom problem, Bohr put forward two bold postulates. Both turned out to be enduring features that carry over in full force to our modern point of view.

Moreover, both turned out to be quite general, applying not only to the hydrogen atom but to atomic, molecular, and nuclear systems of all kinds. We discuss each postu-late in turn.

1. *The postulate of stationary states.* Bohr assumed that the hydrogen atom can exist for a long time *without ra-diating* in any one of a number of *stationary states* of well-defined energy. This assumption contradicts classi-cal theory, but Bohr's attitude was: "Let's assume it any-way and see what happens." Note that this postulate says nothing at all about what these states look like. There is, for example, no mention of orbits.

2. *The frequency postulate.* Bohr assumed that the hy-drogen atom can emit or absorb radiation *only* when the atom changes from one of its stationary states to another. The energy of the emitted (or absorbed) photon is equal to the difference in energy between these two states. Thus if an atom changes from an initial state of energy E_n to a final state of (lower) energy E_m, the energy of the emitted photon is given by

$$h\nu_{nm} = E_n - E_m \quad \text{(Bohr's frequency postulate)}. \quad (4)$$

This postulate ties together two new ideas (the photon hypothesis and energy quantization) with one familiar old idea (the conservation of energy).

Bohr now sought to interpret the empirical Balmer–Rydberg formula in terms of his postulates. We start by

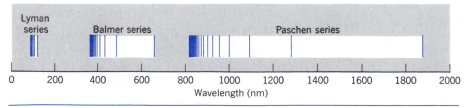

Figure 2   The Lyman, Balmer, and Paschen series of atomic hydrogen. The series limit is at the short wavelength (left) end of each series.

TABLE 1   THE HYDROGEN SPECTRUM (SOME SELECTED LINES)

Name of Series	Quantum Number		Wavelength (nm)
	m (Lower State)	n (Upper State)	
Lyman	1	2	121.6
	1	3	102.6
	1	4	97.0
	1	∞ (series limit)	91.2
Balmer	2	3	656.3
	2	4	486.1
	2	5	434.1
	2	∞ (series limit)	364.6
Paschen	3	4	1875.1
	3	5	1281.8
	3	6	1093.8
	3	∞ (series limit)	822.0

recasting this formula (Eq. 3) in the general format of Bohr's frequency postulate (Eq. 4). If we multiply each side of Eq. 3 by hc and if we replace c/λ by ν_{nm}, we can write

$$h\nu_{nm} = \left(-\frac{hcR}{n^2}\right) - \left(-\frac{hcR}{m^2}\right).$$

A term by term comparison with Eq. 4 allows us to infer

$$E_n = -\frac{hcR}{n^2}, \qquad n = 1, 2, 3, \ldots \qquad (5)$$

for the energies of the stationary states of the hydrogen atom. The energy is negative because the atom is in a bound state; that is, work must be done by some external agent to pull it apart. (The potential energy, which is zero at infinite separation of the proton and electron, is negative and larger in magnitude than the kinetic energy.) In the same way, the Earth–Sun system is a bound state; work must be done by an external agent to tear *this* system apart against the gravitational force that holds it together.

Figure 4 shows an energy level diagram for the hydrogen atom, the energies being calculated from Eq. 5. Each level is marked with its *quantum number n*. A downward-pointing arrow connecting two levels represents the emission of a photon, in accord with Bohr's frequency postulate (Eq. 4). Table 1 displays the wavelengths of some of the lines shown in this figure.

Equation 5 is not yet the end of the path that Bohr followed. The value of the Rydberg constant in that formula—at this stage—can be found only from experiment. What is needed is a way of expressing this constant in terms of other, known physical constants. This, as we shall see, is precisely what Bohr did next.

Sample Problem 1   Calculate the binding energy of the hydrogen atom, that is, the energy that must be added to the atom to remove the electron from its lowest energy state.

Solution   The energy of the atom when the electron has been removed from it, found by letting $n \to \infty$ in Eq. 5, is zero. The

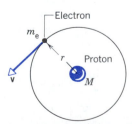

Figure 3   In the Bohr model of the hydrogen atom, the electron moves in a circular orbit about the central proton.

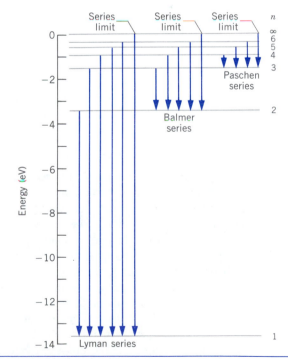

Figure 4   Energy levels and transitions in the spectrum of atomic hydrogen. Compare with the spectral lines represented in Fig. 2.

binding energy E_b is therefore numerically equal to the energy of the atom in its lowest energy state, found by putting $n = 1$ in Eq. 5. That is,

$$E_b = -E_1 = -\left(-\frac{hcR}{1^2}\right)$$

$$= (6.63 \times 10^{-34} \text{ J} \cdot \text{s})(3.00 \times 10^8 \text{ m/s})(1.097 \times 10^7 \text{ m}^{-1})$$

$$= 2.18 \times 10^{-18} \text{ J} = 13.6 \text{ eV}.$$

This calculated value agrees with the experimentally observed binding energy for the hydrogen atom.

Sample Problem 2 (*a*) What is the wavelength of the least energetic photon in the Balmer spectrum? (*b*) What is the wavelength of the series limit for the Balmer series?

Solution (*a*) We identify the Balmer series (see Fig. 4) by putting $m = 2$ in Eq. 3. From the relation $E = h\nu$, the least energetic photon has the smallest frequency and thus the greatest wavelength. This means that we must put $n = 3$ (the smallest possible value) in Eq. 3; any higher value of n would yield a smaller wavelength. With these substitutions we have

$$\frac{1}{\lambda} = R\left(\frac{1}{m^2} - \frac{1}{n^2}\right)$$

$$= (1.097 \times 10^7 \text{ m}^{-1})\left(\frac{1}{2^2} - \frac{1}{3^2}\right) = 1.524 \times 10^6 \text{ m}^{-1}$$

or

$$\lambda = 6.563 \times 10^{-7} \text{ m} = 656.3 \text{ nm}.$$

(*b*) Again we put $m = 2$ in Eq. 3. To find the series limit (see Fig. 4) we let $n \rightarrow \infty$. Equation 3 then becomes

$$\frac{1}{\lambda} = (1.097 \times 10^7 \text{ m}^{-1})\left(\frac{1}{2^2} - 0\right) = 2.743 \times 10^6 \text{ m}^{-1}$$

or

$$\lambda = 3.646 \times 10^{-7} \text{ m} = 364.6 \text{ nm}.$$

Note that both of these numerical results appear in Table 1.

Derivation of the Bohr Theory

So far everything we have done has been empirical, that is, based on measured values rather than on derived values. Our goal now is to derive an expression that gives the Rydberg constant or, equivalently, the energy levels (Eq. 5). We shall do this following Bohr's calculation by invoking the correspondence principle (see Section 50-9): the classical theory (which holds for macroscopic orbits) and the quantum theory must agree where they overlap in the region of large quantum numbers (large values of n).

We begin by analyzing the properties of an atom, such as that of Fig. 3, using classical principles. We shall then compare the results with those of the quantum calculation in the limit of large n.

Let us apply Newton's second law ($F = m_e a$) to the motion of the electron in the classical orbit shown in Fig.

3. For generality, we take the central charge to be Ze rather than e, where Z is the atomic number, $Z = 1$ identifying hydrogen. We assume further that $M \gg m_e$, where M is the nuclear mass and m_e is the mass of the orbiting electron.

Combining Coulomb's force law with Newton's second law gives

$$\frac{1}{4\pi\epsilon_0}\frac{(Ze)(e)}{r^2} = m_e\frac{v^2}{r}, \tag{6}$$

in which v is the speed of the electron in its orbit. Solving for v yields

$$v(r) = \sqrt{\frac{Ze^2}{4\pi\epsilon_0 m_e r}}, \tag{7}$$

which tells us the orbital speed if we know the orbit radius.

From this result we can write an expression for the frequency of revolution of the electron in its orbit:

$$\nu(r) = \frac{v}{2\pi r} = \sqrt{\frac{Ze^2}{16\pi^3\epsilon_0 m_e r^3}}. \tag{8}$$

The kinetic energy follows from

$$K(r) = \tfrac{1}{2}m_e v^2 = \frac{Ze^2}{8\pi\epsilon_0 r}. \tag{9}$$

The potential energy is given by

$$U(r) = -\frac{Ze^2}{4\pi\epsilon_0 r} \tag{10}$$

so that the total mechanical energy $E(r)$ follows from

$$E(r) = K(r) + U(r) = -\frac{Ze^2}{8\pi\epsilon_0 r}. \tag{11}$$

Finally, the angular momentum follows directly from Eq. 7:

$$L(r) = m_e vr = \sqrt{\frac{Ze^2 m_e r}{4\pi\epsilon_0}}. \tag{12}$$

Thus, if we knew the orbit radius, we could find the orbital linear speed, the frequency of revolution, the kinetic energy, the potential energy, the total mechanical energy, and the angular momentum. We see from their interconnections that if any one of these quantities turns out to be quantized, all of them will be. There is, however, no quantization of anything in these purely classical calculations.

We continue by eliminating the radius r between Eqs. 8 and 11 to find a relation between the frequency and the energy:

$$\nu_{cm} = \left(-\frac{32\epsilon_0^2 E^3}{Z^2 m_e e^4}\right)^{1/2}, \tag{13}$$

in which we have added the subscript cm to remind us that this expression is derived on the basis of classical mechanics.

Substituting for E from Eq. 5 gives an expression for the frequency calculated from classical mechanics in the region of large quantum numbers:

$$\nu_{cm} = \left(\frac{32\epsilon_0^2 h^3 c^3 R^3}{Z^2 m_e e^4}\right)^{1/2} \frac{1}{n^3}. \tag{14}$$

In classical physics, this frequency of revolution is also the frequency of the emitted radiation.

We turn now to the quantum point of view. In quantum terms (that is, now using Eq. 4, the second quantum postulate), the frequency ν_{qm} that corresponds to the classical frequency we have just calculated is the lowest emitted frequency, which is associated with a transition from a state with quantum number n to the next lower state, whose quantum number is $n - 1$. Putting $m = n - 1$ in Eq. 3 gives

$$\nu_{qm} = \frac{c}{\lambda} = cR\left(\frac{1}{(n-1)^2} - \frac{1}{n^2}\right) = cR\frac{(2n-1)}{(n-1)^2 n^2}. \tag{15}$$

This expression should agree with the classical expression in the limit of large quantum numbers. When $n \gg 1$, Eq. 15 can be written

$$\nu_{qm} \approx \frac{2cR}{n^3} \quad \text{for } n \gg 1, \tag{16}$$

which is the relationship we seek.

We are ready at last to apply the correspondence principle. This principle tells us that, in the limit of large quantum numbers, the frequency ν_{qm} calculated from Eq. 16 (a quantum expression) must equal the frequency ν_{cm} calculated from Eq. 14 (a classical expression). Table 2 shows this principle in action.

Equating Eqs. 14 and 16 and solving for the Rydberg constant R, we find

$$R = \frac{m_e Z^2 e^4}{8\epsilon_0^2 h^3 c}, \tag{17}$$

a *theoretically predicted value* for the Rydberg constant in terms of other fundamental constants: the charge e and the mass m_e of the electron, the speed c of light, and the Planck constant h. Bohr, using data available in his time for these constants, obtained good agreement with the experimentally determined value of R, the agreement today being within extremely narrow limits of experimental error.

We can now regard the constant R as theoretically determined and, by substituting Eq. 17 into Eq. 5, obtain

$$E_n = -\frac{m_e Z^2 e^4}{8\epsilon_0^2 h^2}\frac{1}{n^2}, \tag{18}$$

a purely quantum expression for the energies of the stationary states of the hydrogen atom. This expression is Bohr's triumph. Everything that he has done so far, including the postulate of stationary states, the frequency postulate, the correspondence principle, and Eq. 18, the expression for the energy of the hydrogen atom states, carries over unchanged into modern quantum mechanics.

By eliminating the energy E between the classical (Eq. 11) and the quantum (Eq. 18) expressions, we can find the radii of the quantized Bohr orbits. They are given by

$$r_n = \left(\frac{\epsilon_0 h^2}{Ze^2\pi m_e}\right)n^2 = a_0 n^2, \qquad n = 1, 2, 3, \ldots. \tag{19}$$

The quantity a_0, called the *Bohr radius,* has the value

$$a_0 = \frac{\epsilon_0 h^2}{Ze^2\pi m_e} = 5.292 \times 10^{-11} \text{ m} = 52.92 \text{ pm}.$$

In a formal sense, a_0 is the radius of the Bohr orbit corresponding to $n = 1$, which defines the ground state of the hydrogen atom in Bohr's semiclassical planetary model of the one-electron atom, where we visualize the electron moving in planetary orbits. Today we do not believe in such orbits but, based on experiment, we do have some notion of the size of the atoms. They are all

TABLE 2   THE CORRESPONDENCE PRINCIPLE AND THE HYDROGEN ATOM

Quantum Number n	Frequency of Revolution in Orbit ν_{cm} (Hz)	Frequency of Transition to Next Lowest State ν_{qm} (Hz)	Difference (%)
2	8.22×10^{14}	24.7×10^{14}	67
5	5.26×10^{13}	7.40×10^{13}	29
10	6.58×10^{12}	7.72×10^{12}	15
50	5.26×10^{10}	5.43×10^{10}	3.1
100	6.580×10^{9}	6.680×10^{9}	1.5
1,000	6.5797×10^{6}	6.5896×10^{6}	0.15
10,000	6.5797×10^{3}	6.5807×10^{3}	0.015
25,000	4.2110×10^{2}	4.2113×10^{2}	0.007
100,000	6.5798	6.5799	0.0007

roughly of the order of magnitude of the Bohr radius! It is amazing that, although Bohr put no specific assumption into his theory concerning the size of atoms, it nevertheless generated a number that gave just about the right size. Today we use the Bohr radius as a convenient unit in which to measure lengths on the scale of atomic dimensions.

The fact that the energy (see Eq. 18) and the radius (see Eq. 19) of the Bohr semiclassical atom are quantized means that other mechanical properties are also quantized in his planetary model. The quantization of the angular momentum of the orbiting electron turns out to be particularly simple. It is (see Problem 23)

$$L_n = n\left(\frac{h}{2\pi}\right) = n\hbar, \qquad n = 1, 2, 3, \ldots, \quad (20)$$

in which we have written $\hbar$ (pronounced "h-bar") as a convenient shorthand for $h/2\pi$.

In 1924, 11 years after Bohr presented his theory, de Broglie gave a satisfying physical interpretation of the Bohr rule for the quantization of angular momentum. If we represent the circulating electron in terms of its de Broglie wave, then the stationary states are those in which the electron's de Broglie wave joins onto itself with the same phase after each revolution; otherwise, the wave would destroy itself by destructive interference. Put another way, the de Broglie wavelength must fit around the circumference of the orbit an integral number of times, or

$$n\lambda = 2\pi r, \qquad n = 1, 2, 3, \ldots,$$

as suggested by Fig. 5. Substituting h/p for the de Broglie wavelength in this expression leads directly to Eq. 20.

Like the Bohr model itself, Fig. 5 is not consistent with modern quantum theory. Although the quantization of angular momentum plays a central role, it differs somewhat from Eq. 20. For the ground state of the hydrogen atom, for example, this equation predicts $L = \hbar$; modern quantum theory, on the other hand, predicts $L = 0$, in agreement with experiment. It is to Bohr's credit that he foresaw the crucial importance of angular momentum quantization and, indeed, proposed Eq. 20 as an alternative basic hypothesis from which his theory could be developed; see Problem 27.

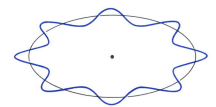

Figure 5   A Bohr orbit with the electron represented as a de Broglie wave.

51-2   THE HYDROGEN ATOM AND SCHRÖDINGER'S EQUATION

The Bohr theory was surprisingly successful in analyzing the radiations emitted by hydrogen, but it is a very incomplete theory. For example, it doesn't provide any basis to calculate which among the many permitted radiations are more likely to be emitted, nor does it provide us with the information we need to understand how hydrogen forms molecular bonds with other atoms. To obtain a complete analysis, we must use methods of wave mechanics.

As we discussed in Chapter 50, the proper treatment of the electron in any particular dynamical situation must take into account its wave nature. In the case of an electron confined to a region in which no force acts on it (Section 50-7), we saw that the wave behavior was similar to that of a classical standing wave on a string. In the case of an electron subject to a force, especially a force that varies with location, the wave behavior is more complicated, as the classical string would be if the tension varied with location along its length.

To analyze the wave behavior of the electron, we require a mathematical procedure in which we can specify the interaction of the electron with its environment and then solve for its motion. This is of course just what we did in classical physics using Newton's laws, in which the interactions were described in terms of forces.

The wave-mechanical procedure for studying the behavior of electrons (and other particles) is based on an equation proposed by the Austrian physicist Erwin Schrödinger (1887–1961) in 1926, just 2 years after de Broglie's hypothesis concerning matter waves. Schrödinger's equation, which we shall not present in detail, is for wave mechanics what Newton's second law is for classical mechanics. We begin by specifying the interaction of the particle with its environment, which we do in terms of potential energy rather than force. (The two descriptions are of course equivalent, as suggested by the one-dimensional expression $F = -dU/dx$; we can find the force from the potential energy or the potential energy from the force.) We then carry out the mathematical procedure specified by Schrödinger's equation, and the results include the wave functions that describe the particle, the quantized energy levels that the particle is permitted to occupy, and a set of *quantum numbers* that specify the allowed states of motion of the particle. Figure 6 represents this procedure symbolically. The box in Fig. 6 might in fact represent a computer, for we currently solve most quantum-mechanical problems of practical interest on computers using numerical methods.

The wave functions corresponding to the allowed states of motion encompass all the information about the behavior of the particle that can be known. Using those wave

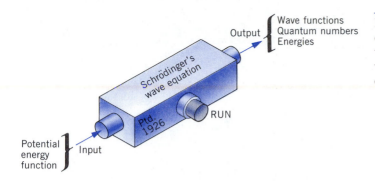

Figure 6   A schematic representation of Schrödinger's wave equation as a "machine" in which the potential energy function must be supplied as input, and the output consists of the wave functions, quantum numbers, and energy levels that characterize the quantum behavior of the system.

functions, we can calculate anything we can know about the particle. In the case of the hydrogen atom, we can use the wave functions resulting from Schrödinger's equation to find the mean radius of the atom, the probability to find the electron at any specified location, the probability for the electron to make a transition from any specified initial state to any specified final state (emitting or absorbing a photon in the process), the magnetic moment of the atom, and so on. By combining two wave functions, we can even study bonds formed between the two atoms in *molecular* hydrogen, H_2.

The potential energy that serves as our starting point results from the Coulomb force between the electron and the proton:

$$U(r) = -\frac{1}{4\pi\epsilon_0}\frac{e^2}{r}. \qquad (21)$$

Figure 7 is a plot of this familiar potential energy function on a scale appropriate for an atom of hydrogen. We specify the distance between the electron and proton in terms of the Bohr radius a_0 defined in Eq. 19. We shall not discuss the mathematical procedure for finding the wave functions,* examples of which are given in Sections 51-7 and 51-8. The energy levels that result from this procedure are

$$E_n = -\frac{m_e e^4}{8\epsilon_0^2 h^2}\frac{1}{n^2}, \qquad n = 1, 2, 3, \ldots , \qquad (22)$$

which are exactly those obtained from the Bohr model (Eq. 18). This agreement should not surprise us because we have seen that the Bohr theory provided a perfect match with the observed wavelengths of the hydrogen spectral lines.

From the Schrödinger wave functions, we can calculate the most probable distance between the electron and proton. This turns out to be $n^2 a_0$. That is, in the lowest energy state, the most likely place to find the electron is at a distance of one Bohr radius from the proton. Here the

Bohr model (which has the electron moving in a fixed orbit at a unique distance from the nucleus) gives an incomplete interpretation. As we shall see in Section 51-7, the electron can be found anywhere from $r = 0$ to $r = \infty$, but $r = a_0$ is its most probable location.

The hydrogen atom is a three-dimensional system, and the Schrödinger equation must be solved in three dimensions. Because of the form of the potential energy (Eq. 21), it is most convenient to solve this problem in spherical coordinates, using as coordinates the radius r and two angles θ and ϕ to fix the direction. When we solve the Schrödinger equation for this system, we find that three quantum numbers are necessary to describe the states of the electron. These quantum numbers are defined and displayed in Table 3. We discuss these quantum numbers, along with a fourth one based on the electron spin (a relativistic effect that is not predicted by the Schrödinger equation, which is nonrelativistic), later in this chapter.

The Bohr theory is of only limited usefulness in understanding the structure of atomic hydrogen and ions with a

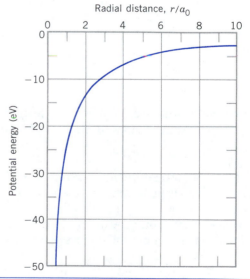

Figure 7   The potential energy function $U(r)$ for the hydrogen atom. The radial distance between the electron and proton is measured in terms of the Bohr radius a_0.

---

* For a full treatment, see Robert Eisberg and Robert Resnick, *Quantum Physics of Atoms, Molecules, Solids, Nuclei, and Particles,* 2nd edition (Wiley, 1985), Chapter 7.

TABLE 3   THE HYDROGEN ATOM QUANTUM NUMBERS[a]

Symbol	Name	Associated with	Allowed Values
n	Principal quantum number	Energy, mean radius	1, 2, 3, . . .
l	Orbital quantum number	Magnitude of orbital angular momentum	0, 1, 2, . . . , $n-1$
m_l	Magnetic quantum number	Direction of orbital angular momentum	0, ±1, ±2, . . . , ±l

[a] A fourth quantum number, associated with the spin, will be introduced later.

single electron (He⁺, Li⁺⁺, and so forth), and it is of no help at all in understanding details beyond the wavelengths of spectral lines. It provides only a very limited basis for understanding atoms more complex than hydrogen, which can be studied in detail with the Schrödinger equation and the exclusion principle, which is explained in the next chapter. The Bohr theory does not show how to calculate the properties of systems more complex than a single atom, such as a molecule or a solid. Today we regard the Bohr theory as an important and ingenious step toward understanding the atom, and we should remember that two principles developed by Bohr to make his theory work (the correspondence principle and the existence of stationary states) are essential parts of the complete quantum theory.

51-3   ANGULAR MOMENTUM

The energy of a state is a scalar and, in the hydrogen atom, it is specified by a single quantum number n. The angular momentum of a state, however, is a vector and we see from Table 3 that it takes two quantum numbers, l and m_l, to describe it. The angular momentum is *doubly* quantized, in both magnitude and direction. We discuss each in turn.

The Magnitude of L

In solving the Schrödinger equation, we learn that the angular momentum is quantized. Its allowed values are

$$L = \sqrt{l(l+1)}\ \hbar \qquad (23)$$

in which l is the orbital quantum number. For convenience, we have again introduced the symbol $\hbar$ (pronounced "h-bar") as an abbreviation for $h/2\pi$.

The values that l can have in Eq. 23 depend on the value of the principal quantum number n and are given by

$$l = 0, 1, 2, . . . , n-1. \qquad (24)$$

For example, the ground state of the hydrogen atom, which has $n = 1$, must have $l = 0$ (and thus $L = 0$), no other value being permitted by Eq. 24. For another exam-

ple, a state with $n = 2$ can have $l = 0$ or $l = 1$. These two states of the hydrogen atom share the same principal quantum number n and have the same energy, even though they represent very different states of motion.

The Direction of L

Let us choose a direction in space, which we arbitrarily label as the z axis, and let us determine the direction of **L** with respect to this axis.

It turns out that the angular momentum vector **L** cannot take *any* position with respect to the z axis, but only those positions that have a component along the z axis given by

$$L_z = m_l \hbar, \qquad (25)$$

in which m_l, the *magnetic quantum number*, may have only the values

$$m_l = 0, \pm 1, \pm 2, . . . , \pm l. \qquad (26)$$

This restriction on the direction of **L** is called *space quantization*.

We see from Eqs. 23–26 that, for a hydrogen atom state with $l = 2$, the magnitude of **L** is $\sqrt{2(2+1)}\hbar$ or $2.45\hbar$. L_z, the component of **L** along the z axis, may have the values $0, \pm 1\hbar$, and $\pm 2\hbar$, five components in all. *No other orientations of the angular momentum vector with respect to the z axis are allowed.* Note that the maximum value of L_z ($=2\hbar$ in this case) is less than the magnitude of **L** ($=2.45\hbar$). This will always be the case; the angular momentum vector **L** can never be fully lined up with the z axis.

Figure 8 shows the allowed values of L_z for $l = 1, 2$, and 10. In the latter case we begin to approach the classical situation, in which space quantization has faded away and *any* orientation of the angular momentum vector is allowed. Sample Problem 3 gives further details of how the correspondence principle operates in this case.

The quantization of L_z means that the angle θ between **L** and the z axis (see Fig. 8) is quantized, its values being restricted to

$$\theta = \cos^{-1} \frac{L_z}{L} = \cos^{-1} \frac{m_l}{\sqrt{l(l+1)}}, \qquad (27)$$

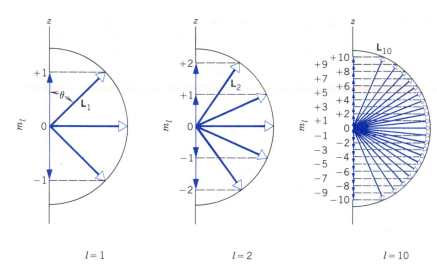

Figure 8 The allowed values of L_z for $l = 1, 2,$ and 10. The numbers on the z axis show values of m_l. The figures are drawn to different scales.

$l = 1$ $l = 2$ $l = 10$

where we have used Eq. 23 for L and Eq. 25 for L_z. The minimum value of θ occurs when m_l has its greatest value, which is l. For $l = 2$, for example, you can easily show that this minimum value is

$$\theta_{min} = \cos^{-1}[2/\sqrt{2(2+1)}] = \cos^{-1} 0.817 = 35.3°.$$

For hydrogen atom states with $l = 2$ it is simply not possible for the (orbital) angular momentum vector **L** to make any smaller angle with the z axis.

Once we have selected an axis and determined the component of **L** along that axis, the components of **L** on all other axes are completely uncertain. That is, we can have an exact knowledge of only one selected component of **L** (which we arbitrarily assume to be the z component).

Figure 9 suggests a classical vector model that helps in visualizing the space quantization of **L**. It shows the vector precessing about the z direction, like a spinning top or a gyroscope precessing about a vertical axis in the Earth's gravitational field. The component L_z remains constant

as the motion proceeds, but the x and y components of **L** do not have definite values.

Heisenberg's uncertainty principle helps us to understand the space quantization of the angular momentum vector. In its angular form (compare Eq. 6 of Chapter 50) this principle is

$$\Delta L_z \cdot \Delta\phi \sim h/2\pi \quad (z \text{ component}), \tag{28}$$

in which ϕ is the angle of rotation about the z axis in Fig. 9. Equation 25 tells us that L_z is precisely known, once we have specified the quantum number m_l. It follows that ΔL_z, the uncertainty in L_z, must be zero. Equation 28 then requires that $\Delta\phi \to \infty$, which means that we have no information at all about the angular position about the z axis of the precessing angular momentum vector **L**. We know the magnitude of **L** and its projection L_z on the z axis, and *nothing else*.

Recall that our original choice for the direction of the z axis was completely arbitrary. There is no special feature that singles out this particular direction in space, and we could just as easily have labeled our choice as the x axis or the y axis. What *is* significant is that we can choose *any* direction in space, and we will observe space quantization with respect to that direction. By convention, we usually refer to our choice of the quantization axis as the z axis.

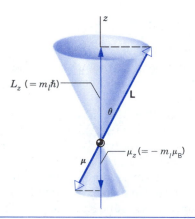

Figure 9 A vector representation of the space quantization of the angular momentum **L** and the magnetic dipole moment **μ**.

Sample Problem 3 Find the minimum value of θ in Fig. 9 for $l = 1, 10^2, 10^3, 10^4,$ and 10^9.

Solution The minimum value of θ occurs when we put $m_l = l$ in Eq. 27. Doing so and rearranging lead to

$$\theta_{min} = \cos^{-1} \frac{l}{\sqrt{l(l+1)}} = \cos^{-1}\left[1 + \frac{1}{l}\right]^{-1/2}.$$

We see by inspection that if we let $l \to \infty$, then $\theta \to \cos^{-1} 1 = 0$. This is just what we expect from the correspondence principle.

Substituting for l in this equation leads to these results:

l	θ_{min}
1	45.0°
10^2	5.7°
10^3	1.8°
10^4	0.57°
10^9	0.0018°

For a macroscopic object like a spinning top or a phonograph record, l would be enormously larger than 10^9 and θ_{min} would be so close to zero that the difference would be beyond the possibility of measurement. Thus as the angular momentum of a spinning object gets larger and larger, the space quantization of wave mechanics merges gently into the continuous distribution of classical mechanics. We see once again how the correspondence principle works.

Computational note: If you use the above formula to calculate θ_{min} for $l = 10^9$, your calculator will probably overflow. Take advantage of the fact that $(1/l) \ll 1$ and develop an approximate formula. You will need to use both the binomial expansion and the series expansion for $\cos \theta$; see Appendix H.

Sample Problem 4   (*a*) For $n = 4$, what is the largest allowed value of l? (*b*) What is the magnitude of the corresponding angular momentum? (*c*) How many different components on the z axis may this angular momentum vector have? (*d*) What is the magnitude of the largest projected component? (*e*) What is the smallest angle that the angular momentum vector can make with the z axis?

Solution   (*a*) From Eq. 24, the largest allowed value of l is $n - 1$, so $l = 3$.

(*b*) From Eq. 23 we have

$$L = \sqrt{l(l+1)}\,(h/2\pi)$$
$$= \sqrt{3(3+1)}\,(6.63 \times 10^{-34}\text{ J·s})/(2\pi)$$
$$= 3.66 \times 10^{-34}\text{ J·s}.$$

In practice, atomic angular momenta are rarely reported in SI units. It would be more customary to report the magnitude of the angular momentum in this case as simply $\sqrt{12}\hbar$ or $3.46\hbar$. See part (*d*).

(*c*) The number of components that the angular momentum vector may have on the z axis is equal to the number of allowed values of the magnetic quantum number m_l. From Eq. 26 this number is $2l + 1$ or $2 \times 3 + 1 = 7$.

(*d*) The largest projected component is found from Eq. 25, in which the magnetic quantum number m_l is given its largest possible value. From Eq. 26 this largest value is just l, so we have

$$L_z = l(h/2\pi)$$
$$= (3)(6.63 \times 10^{-34}\text{ J·s})/(2\pi)$$
$$= 3.17 \times 10^{-34}\text{ J·s}.$$

Note from part (*b*) that this is smaller than the magnitude of the angular momentum vector, as it must be. As we remarked in part (*b*), the maximum projected angular momentum component would be reported as simply $3\hbar$. When we refer to angular momentum, in fact, we almost always mean this maximum projected value. The magnitude of the angular momentum rarely enters quantum calculations and is seldom given.

(*e*) The smallest angle that the angular momentum vector can make with the z axis follows from Eq. 27, with $m_l = l$. We then have

$$\theta = \cos^{-1}\,[l/\sqrt{l(l+1)}]$$
$$= \cos^{-1}\,[3/\sqrt{3(3+1)}]$$
$$= \cos^{-1}\,0.866 = 30°.$$

The angular momentum vector can make no smaller angle with the z axis than this.

Orbital Angular Momentum and Magnetism

The Bohr model also suggests that the orbiting electron — a tiny current loop — should have an (orbital) magnetic dipole moment associated with it. Both the angular momentum **L** and the magnetic dipole moment μ are vectors and share a common axis. Because the electron has a negative charge, however, these vectors point in opposite directions along this axis. In Section 37-2 we showed that these two vectors are related by

$$\mu = -\frac{e}{2m_e}\mathbf{L}, \qquad (29)$$

the minus sign showing the opposite directions of **L** and μ. Although Eq. 29 was derived on a semiclassical basis, it remains true in wave mechanics.

Consider a state in which the z component of the angular momentum is $h/2\pi$. Substituting this value for **L** in Eq. 29 yields

$$\mu_z = \frac{eh}{4\pi m_e}.$$

This quantity is called the *Bohr magneton* μ_B and has the value

$$\mu_B = \frac{eh}{4\pi m_e}$$
$$= 9.274 \times 10^{-24}\text{ J/T} = 5.788 \times 10^{-5}\text{ eV/T}. \quad (30)$$

The Bohr magneton is a convenient unit in which to measure atomic magnetic moments, much as we took the Bohr radius a_0 as a convenient unit in which to measure atomic distances.

Bohr theory predicts that the magnetic dipole moment of the hydrogen atom in its ground state will be one Bohr magneton. The theory is simply not correct on this point. Experiment shows that, in accordance with the predictions of wave mechanics, both the (orbital) angular momentum and the (orbital) magnetic dipole moment of the hydrogen atom in its ground state are zero. This failure of Bohr theory, however, does not stop us from using the Bohr magneton as a convenient unit of measure.

If the angular momentum is quantized, the magnetic dipole moment must also be quantized, and in the same fashion. Combining Eq. 29 (*z* components only) and Eq. 25 allows us to write

$$\mu_z = \frac{e}{-2m_e} L_z = -\frac{e}{2m_e}(m_l \hbar) = -m_l \frac{eh}{4\pi m_e}.$$

From Eq. 30, we see that the *z* component of the magnetic dipole moment is given by

$$\mu_z = -m_l \mu_B, \tag{31}$$

in which μ_B is the Bohr magneton. As shown by Fig. 9, the classical vector model accounts for the space quantization of $\boldsymbol{\mu}$ as well as **L**. Both vectors precess about the *z* direction, and both are characterized by their *z* components.

The magnetic dipole moment of the atom—much like a compass needle—can respond to an external magnetic field. This gives the atom a convenient "handle" by means of which we can explore its inner workings by probing from the outside. Because the magnetic dipole moment is rigidly coupled to the angular momentum, keeping track of the former automatically keeps track of the latter as well.

We do not have to look far to find evidence that atoms can be the carriers of magnetism. An ordinary iron bar shows no external magnetic properties because its elementary atomic magnets are randomly arranged, their effects canceling at all external points. When these elementary atomic magnets are lined up, however, as they are in a bar magnet, their combined magnetic strength is there for all to see.

When the magnetic dipole moments of an assembly of atoms are lined up, their angular momenta—to which they are rigidly coupled—must also be lined up. In 1915

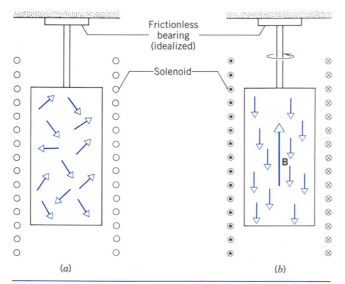

(*a*)   (*b*)

Figure 10   The Einstein–de Haas effect. (*a*) The atomic angular momentum vectors in an iron cylinder are randomly oriented. (*b*) When an axial magnetic field is applied, the atomic angular momenta line up as shown and the cylinder as a whole starts to rotate in the opposite sense.

Einstein, working with W. J. de Haas (the son-in-law of the great Dutch physicist H. A. Lorentz), carried out an experiment to explore this phenomenon. If an iron bar is suddenly magnetized, perhaps by switching on a current in a solenoid as in Fig. 10, the angular momenta of its atoms suddenly become lined up. Because angular momentum must be conserved, the bar as a whole must start to rotate in the opposite sense. This *Einstein–de Haas effect,* as it is called, is small and the measurements are difficult. Bear in mind that in 1915, when this experiment was carried out, wave mechanics had not been discovered, the Bohr theory was only 2 years old, and the intrinsic spin of the electron had yet to be discovered.

It turned out later that the Einstein–de Haas effect (and also, for that matter, the ferromagnetism of a bar magnet) is due largely to the intrinsic angular momentum (spin) of the electrons rather than to their orbital angular momentum. This, however, does not alter the fact that this experiment demonstrates, in a macroscopic way, that atoms can be the carriers of both magnetism and angular momentum.

Sample Problem 5   An unmagnetized iron cylinder, whose radius *R* is 5 mm, hangs from a frictionless bearing so that it can rotate freely about its axis; see Fig. 10. A magnetic field is suddenly applied parallel to this axis, causing the magnetic dipole moments of the atoms to align themselves parallel to the field. The atomic angular momentum vectors, which are coupled back-to-back with the magnetic dipole moment vectors, also become aligned and the cylinder will start to rotate in the opposite sense. Find *T*, the period of rotation of the cylinder. Assume that each iron atom has an angular momentum of $h/2\pi$. The molar mass *M* of iron is 55.8 g/mol (=0.0558 kg/mol).

Solution   The angular momentum of the rotating cylinder ($=\mathbf{L}_{cyl}$) must be equal in magnitude (though opposite in direction) to the angular momentum associated with the aligned atoms ($=\mathbf{L}_{atoms}$). If *N* is the number of atoms in the cylinder, N_A is the Avogadro constant, and *m* is the mass of the cylinder, we can write

$$L_{atoms} = N(h/2\pi) = (N_A m/M)(h/2\pi).$$

For the rotating cylinder we have

$$L_{cyl} = I\omega = (\tfrac{1}{2}mR^2)(2\pi/T),$$

in which *I* is the rotational inertia of the cylinder about its rotational axis and ω is its angular speed.

Equating these two expressions and solving for *T* yields

$$T = \frac{2\pi^2 R^2 M}{N_A h}$$

$$= \frac{(2\pi^2)(5 \times 10^{-3}\text{ m})^2(0.0558\text{ kg/mol})}{(6.02 \times 10^{23}\text{ mol}^{-1})(6.63 \times 10^{-34}\text{ J·s})}$$

$$= 6.90 \times 10^4\text{ s} = 19.2\text{ h}.$$

This is indeed a slowly rotating cylinder! Actually, Einstein and de Haas suspended their cylinder from a torsion fiber and used more refined techniques in their experiment dealing with this effect.

51-4 THE STERN–GERLACH EXPERIMENT

Space quantization, that is, the notion that an atomic angular momentum vector **L** or an atomic magnetic dipole moment vector μ can have only a certain discrete set of projections on a selected axis, is not an easy concept for the classically oriented mind to accept. Nevertheless, it was predicted theoretically (by Wolfgang Pauli) and verified experimentally (in 1922, by Otto Stern and Walther Gerlach) several years before the development of wave mechanics.

Figure 11 shows the apparatus of Stern and Gerlach. Silver is vaporized in an electrically heated "oven," and silver atoms spray into the external vacuum of the apparatus from a small hole in the oven wall. The atoms (which are electrically neutral but which have a magnetic dipole moment) are formed into a narrow beam as they pass through a slit in an interposed screen. The collimated beam then passes between the poles of an electromagnet and, finally, deposits itself on a glass detector plate.

Often in laboratory experiments we want the magnetic field to be uniform, but in this case the pole faces are shaped to make the field as *nonuniform* as possible. The atomic beam passes very close to the sharp V-shaped ridge in the upper pole piece, where the nonuniformity of the magnetic field is greatest.

A Dipole in a Nonuniform Field

Figure 12*a* shows a dipole of magnetic moment μ, making an angle θ with a uniform magnetic field. We can imagine the dipole to be a tiny bar magnet, with the magnetic dipole moment vector μ pointing (by convention) from its south pole to its north pole. We may imagine the forces to be concentrated at the poles as shown in the figure. We

see that, for a uniform field, there is no net force on the dipole. The upward and downward forces on the poles are of the same magnitude and they cancel, no matter what the orientation of the dipole.

Figures 12*b* and 12*c* show the situation in a nonuniform field. Here the upward and downward forces do *not* have the same magnitude because the two poles are immersed in fields of different strengths. In this case there *is* a net force, both its magnitude and direction depending on the orientation of the dipole, that is, on the value of θ. In Fig. 12*b* this net force is up and in Fig. 12*c* it is down. Thus the silver atoms in the beam of Fig. 11, as they pass through the electromagnet, are deflected up or down, and in greater or lesser amounts, depending on the orientation of their magnetic moment dipole vectors with respect to the magnetic field.

Now let us calculate the deflecting force quantitatively. The magnetic potential energy of a dipole in a magnetic field **B** is given by Eq. 38 of Chapter 34,

$$U(\theta) = -\mu \cdot \mathbf{B} = -(\mu \cos \theta)B.$$

In our mind's eye let us follow the silver atoms in the beam as they move through the electromagnet of Fig. 11 parallel to the sharp edge. From symmetry (see also Fig. 12*b*, *c*), the magnetic field at this central position has no x or y components. Thus $B = B_z$. Because $\mu \cos \theta = \mu_z$, we can write the potential energy as

$$U(\theta) = -\mu_z B_z. \tag{32}$$

The net force F_z on the atom is $-(dU/dz)$ or, from Eq. 32,

$$F_z = \mu_z \frac{dB_z}{dz}. \tag{33}$$

Note that the deflecting force is determined by the *derivative* of the magnetic field and does not depend on the magnitude of the field itself. In Fig. 12*b*, *c*, B_z increases as z increases so that the derivative is positive. Thus the sign of the deflecting force F_z in Eq. 33 depends on the sign of μ_z. If μ_z is positive (as in Fig. 12*b*) the atom is deflected upward; if it is negative (as in Fig. 12*c*) the deflection is downward.

One troublesome point remains. If the individual

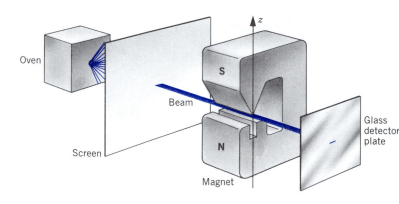

Oven

Screen

Beam

Magnet

Glass detector plate

Figure 11 The apparatus of Stern and Gerlach.

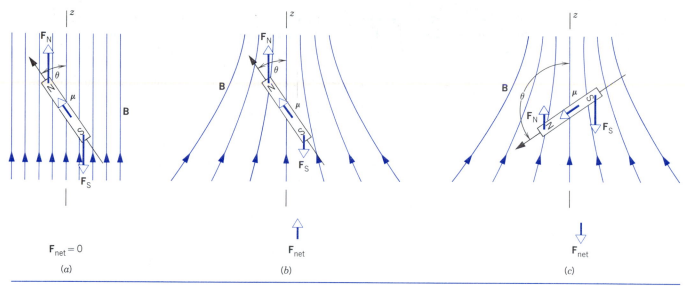

Figure 12   A magnetic dipole, represented as a small bar magnet with two poles, in (*a*) a uniform field and (*b,c*) a nonuniform field. The net force acting on the magnet is zero in (*a*), points up in (*b*), and points down in (*c*).

atoms in the beam behave like tiny bar magnets, why don't they all simply line up with the magnetic field? Why should any of them point, even partially, in the opposite direction? The answer is that the atoms not only have a magnetic dipole moment; they also have angular momentum. The result is that they precess around the field direction (see Fig. 9) rather than line up with it. In the same way a top that is not spinning will simply fall over if you place it at an angle with the Earth's gravitational field. We saw in Section 13-5, however, that if the top is spinning, it will precess about this direction. It is the angular momentum that does it!

The Experimental Results

When the electromagnet in Fig. 11 is turned off (or is operating at very low power), there will be no deflections of the atoms and the beam will form a narrow line on the detecting plate.

When the electromagnet is turned on, however, strong deflecting forces come into play. Then there are two possibilities, depending on whether space quantization exists or not. (Don't forget that the entire object of this experiment is to find out!) If there is no space quantization, the atomic magnetic dipole moment vectors have a continuous distribution of values, some positive and some negative; the beam will simply broaden.

On the other hand, if space quantization *does* exist, there is only a discrete set of values of μ_z. This means that there is only a discrete set of values for the deflecting force F_z in Eq. 33, and the beam splits up into a number of discrete components.

Figure 13 shows what happens. The beam does not broaden but splits cleanly into two subbeams. Space

quantization exists! Stern and Gerlach ended the published report of their work with the words: "We view these results as direct experimental verification of space quantization in a magnetic field." Physicists everywhere agree.

Sample Problem 6   In an experiment of the Stern–Gerlach type, the magnetic field gradient dB_z/dz at the beam position was 1.4 T/mm and the length h of the beam path through the magnet was 3.5 cm. The temperature of the "oven" in which the silver was evaporated was adjusted so that the most probable speed v for the atoms in the beam was 750 m/s. Find the separa-

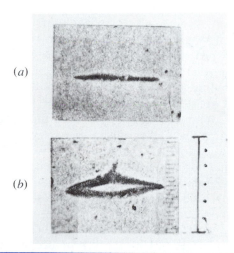

Figure 13   The results of the Stern–Gerlach experiment, showing the silver deposit on the glass detector plate of Fig. 11, with the magnetic field (*a*) turned off and (*b*) turned on. The beam has been split into two subbeams by the magnetic field. The vertical bar at the right in (*b*) represents 1 mm.

tion d between the two deflected subbeams as they emerge from the magnet; see Fig. 13b. The mass m of a silver atom is 1.8×10^{-25} kg, and its magnetic moment μ_z is 1 Bohr magneton ($= 9.28 \times 10^{-24}$ J/T).

Solution   The acceleration of a silver atom as it passes through the electromagnet is given (see Eq. 33) by

$$a = \frac{F_z}{m} = \frac{\mu_z(dB_z/dz)}{m}.$$

The vertical deflection Δz of either of the subbeams as it clears the magnet is

$$\Delta z = \tfrac{1}{2}at^2 = \tfrac{1}{2}\frac{\mu_z(dB_z/dz)}{m}\left(\frac{h}{v}\right)^2.$$

The separation d of the two beams is $2\Delta z$, or

$$
\begin{aligned}
d &= \frac{\mu_z(dB_z/dz)h^2}{mv^2} \\
&= \frac{(9.28 \times 10^{-24}\ \text{J/T})(1.4 \times 10^3\ \text{T/m})(3.5 \times 10^{-2}\ \text{m})^2}{(1.8 \times 10^{-25}\ \text{kg})(750\ \text{m/s})^2} \\
&= 1.6 \times 10^{-4}\ \text{m} = 0.16\ \text{mm}.
\end{aligned}
$$

This is the order of magnitude of the separation displayed in Fig. 13b; note the scale in that figure.

51-5 THE SPINNING ELECTRON

The Stern–Gerlach experiment clearly demonstrates that the magnetic moment vector of an atom can have only a finite number of discrete directions in space, as opposed to the infinite number allowed by classical physics. However, there is a curious feature of the experiment. Figure 13 shows the beam of silver atoms splitting into two components, corresponding to two different orientations of the magnetic moment vector of the atom (or, equivalently, of its angular momentum vector, since the two vectors are related by Eq. 29). Yet a glance at Table 3 or Fig. 8 shows that there is always an *odd* number of possible orientations of the **L** vector. Put another way, the number of possible orientations of **L** is $2l + 1$, and for this to equal 2 we must have $l = \tfrac{1}{2}$; however, this contradicts the restriction that l takes only integer values.

The solution to this dilemma was proposed in 1924 by the Austrian-born physicist Wolfgang Pauli (see Fig. 14). He suggested that there is yet another quantum number that describes the state of an electron in an atom, and that this quantum number can take the values $+\tfrac{1}{2}$ or $-\tfrac{1}{2}$. In the following year two Dutch graduate students, Samuel Goudsmit and George Uhlenbeck, proposed the notion of *electron spin* as the physical interpretation of Pauli's proposed new quantum number.

Spin is often called *intrinsic angular momentum*, and it is often useful (although strictly not correct) to visualize the spin as the angular momentum of a particle rotating

on its axis. There are many parallels between spin and orbital angular momentum. The spin quantum number s is analogous to the orbital quantum number l; however, unlike l, the value of s does not change with the electron's state of motion. All electrons, no matter what their state of motion, have $s = \tfrac{1}{2}$. In fact, we usually consider s to be a fundamental property of a particle, along with its mass and electric charge.

The spin of the electron can be represented by a vector **S** of magnitude (compare Eq. 23)

$$S = \sqrt{s(s + 1)}\ \hbar. \tag{34}$$

The component of this vector in the z direction can be written (compare Eq. 25)

$$S_z = m_s\hbar. \tag{35}$$

Just like the components of **L**, the permitted components of **S** differ by one unit of $\hbar$. We therefore see that the permitted values of m_s are

$$m_s = \pm\tfrac{1}{2}. \tag{36}$$

Associated with the spin angular momentum there is a magnetic moment, which is given by (compare Eq. 29)

$$\mu_s = -\frac{e}{m_e}\mathbf{S}. \tag{37}$$

Note the difference in Eqs. 37 and 29 by a factor of 2. This suggests that the spin angular momentum is twice as effective as the orbital angular momentum in producing magnetic effects. For further details on orbital and spin magnetic moments, see Section 37-2.

The quantum numbers for the orbital angular momentum l and its magnetic projection m_l arise in a natural way from solving the Schrödinger equation for the hydrogen atom. The spin angular momentum and its magnetic projection seem to be introduced arbitrarily with no theoretical justification. The English mathematical physicist Paul A. M. Dirac developed a relativistic wave equation similar to the nonrelativistic Schrödinger equation, and Dirac showed that solutions to his equation for the hydrogen atom gave the electron spin as a fourth quantum number. To obtain the complete solution for the hydrogen atom, we must replace the Schrödinger machine of Fig. 6 by a Dirac machine! This is another great triumph for relativity theory, without which we would have no theoretical basis for understanding this fundamental part of the structure of atoms.

We can now explain the appearance of two beams in the Stern–Gerlach experiment. The electron in a silver atom happens to occupy a state in which $l = 0$, so the total angular momentum of the electron is due only to its spin. This spin vector has only two possible orientations relative to the magnetic field, hence the two components of the beam.

Figure 14 Wolfgang Pauli (left) and Niels Bohr watching a "tippy top," a top that spins for a while on one end and then turns upside down. They are waiting for the "spin flip."

Every fundamental particle has its characteristic spin and magnetic moment. The proton and neutron, like the electron, have a spin of $\frac{1}{2}$. Their magnetic moments are discussed in Section 37-2. In Section 51-8 we consider other observable consequences of the existence of electron spin.

Consequences of proton and neutron spin have proved to be of great practical value through the phenomenon of *nuclear magnetic resonance* (see Section 13-6 and Sample Problem 1 of Chapter 37). When a proton is placed in a magnetic field B, an energy change of $2\mu_s B$ occurs when the spin changes direction or "flips." This spin flip can be caused by subjecting the protons to an electromagnetic wave whose frequency is selected such that $h\nu = 2\mu_s B$. The field B consists of an external field B_{ext} (perhaps due to an electromagnet) and an internal field B_{int} (due to the chemical environment in which the proton is found). For example, in a molecule of ethanol, whose formula we may write as $CH_3 — CH_2 — OH$, each hydrogen nucleus experiences a different internal field because of its different location in the molecule. By keeping B_{ext} fixed and varying the frequency ν, we can find several frequencies at which the spin flips occur, each corresponding to a particular environment of a hydrogen nucleus in an ethanol molecule. Equivalently, as in Fig. 15, we can keep ν fixed and vary B_{ext}. Either way, we get a unique signature that identifies ethanol. In this way nuclear magnetic resonance

proves to be an important analytical tool in organic chemistry. Other applications include measuring B_{int} in various molecular or solid environments and measuring nuclear magnetic dipole moments.

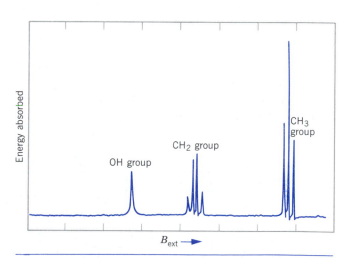

Figure 15 A nuclear magnetic resonance spectrum of ethanol. All the lines are due to absorption of the incident radiation when the proton spin flips. The groups of lines correspond to different groupings of hydrogen within the molecule. The entire horizontal scale is considerably less than 10^{-4} T.

51-6 COUNTING THE HYDROGEN ATOM STATES

We have now described the four quantum numbers that define the stationary states of the hydrogen atom, and we have shown how each of them can be interpreted physically. Although we did not prove it, it is nevertheless true that they emerge from Schrödinger's wave equation, with an important assist from Dirac's equation in the case of the electron spin. In solving the Schrödinger wave equation, the quantum numbers and the other information come tumbling out in a natural way.

Our next task is to see whether we can arrange the hydrogen atom states in some orderly fashion. Consider first the principal quantum number n. All states with the same value of n have the same energy, and we say that the assembly of such states forms a *shell*. Equation 24 tells us that the number of different values of l that are possible for a given value of n is just equal to n. Thus for $n = 3$ we can have three values of l (=0, 1, and 2).

The shells can be further subdivided. Within a given shell, all states with the same value of l have the same angular momentum and are said to form a *subshell*. For example, the shell defined by $n = 3$ contains three subshells, each with the same energy but a different angular momentum.

Within the subshells, the electrons may have different states of motion because of the different ways the angular momentum vector can be oriented. For a given l, Eq. 26 tells us that there are $2l + 1$ values of m_l. In our example, then, the subshell with $l = 2$ contains 5 ($= 2 \times 2 + 1$) states, and those with $l = 1$ and 0 contain 3 states and 1 state, respectively. This adds up to a total of 9 ($= 5 + 3 + 1$) states in the shell with $n = 3$.

The effect of the spin quantum number m_s is simply to double the number of states. Each combination of n, l, and m_l that we have identified can now be associated with either $m_s = +\frac{1}{2}$ or $m_s = -\frac{1}{2}$, thus producing two states where one existed before. To continue with our example, there are not 9 but 18 states in the shell with $n = 3$.

Table 4 summarizes this classification of hydrogen atom states into shells and subshells.

Do the numbers of states in the shells (that is, 2, 8, and 18) in the bottom row of Table 4 seem familiar? They are the lengths of the horizontal rows (periods) in the periodic table of the elements! As Appendix E shows, period 1 contains 2 elements, periods 2 and 3 contain 8 elements each, and periods 4 and 5 contain 18. In Chapter 52 we shall see in detail just how the order of the elements in the periodic table arises from wave-mechanical principles.

Sample Problem 7   A certain shell has a principal quantum number n of 4. (a) How many subshells does this shell contain? (b) What is the number of states in each of these subshells? (c) What is the number of states in the shell?

Solution   (a) If $n = 4$, we know from Eq. 24 that the allowed values of l are 0, 1, 2, and 3. This is a total of four values, in agreement with the fact that the number of allowed values of l for a given n is just equal to n. Each value of n defines a shell and each value of l defines a subshell within that shell. Thus there are four subshells in the $n = 4$ shell.

(b) The number of states in a subshell is given by $2(2l + 1)$, the factor of 2 coming from the two allowed values of the spin magnetic quantum number. For the numbers of states in the various subshells in the $n = 4$ shell we then have

l	$2(2l + 1)$
0	2
1	6
2	10
3	14

(c) The number of states in the $n = 4$ shell is found by adding up the numbers in the subshells that it contains. From the table above we have $2 + 6 + 10 + 14$ or 32. Note that 32 is the number of elements in horizontal row 6 of the periodic table of the elements. See Appendix E.

Can you prove that, in general, the number of states in a shell defined by principal quantum number n is given by $2n^2$? That works out in this case because $2 \times 4^2 = 32$, as we found by explicit counting.

TABLE 4   STATES OF THE HYDROGEN ATOM[a]

n	1	2		3		
l	0	0	1	0	1	2
m_l	0	0	$0, \pm 1$	0	$0, \pm 1$	$0, \pm 1, \pm 2$
m_s	$\pm\frac{1}{2}$	$\pm\frac{1}{2}$	$\pm\frac{1}{2}$	$\pm\frac{1}{2}$	$\pm\frac{1}{2}$	$\pm\frac{1}{2}$
Number of states in the subshells	2	2	6	2	6	10
Number of states in the shells	2	8		18		

[a] Complete to $n = 3$ only.

51-7   THE GROUND STATE OF HYDROGEN

In this section, without going into the mathematical details, we present the results of using the Schrödinger equation to study the ground state of hydrogen. The procedure involves solving the Schrödinger equation in spherical polar coordinates when the electron and the proton interact through the Coulomb force, given by the potential energy

$$U(r) = -\frac{1}{4\pi\epsilon_0}\frac{e^2}{r}. \qquad (38)$$

If we insert this potential energy function into the Schrödinger equation and carry out the needed mathematical manipulations, we are able to derive an expression for the energies of the allowed stationary states of the atom and also for the wave functions that describe those states.

The expression for the energies of the stationary states turns out to be exactly Eq. 18, the expression derived from Bohr theory. We focus our attention here on the ground state of the hydrogen atom, that is, on the state of lowest energy.

The wave function for the ground state also emerges from the Schrödinger equation and turns out to depend only on the single variable r. It is given by

$$\psi(r) = \frac{1}{\sqrt{\pi a_0^3}}e^{-r/a_0}, \qquad (39)$$

in which a_0 is the Bohr radius.

This wave function has spherical symmetry, by which we mean that it depends only on the magnitude of the vector **r** (which defines the point at which the wave function is evaluated), but not on its direction. This is perhaps not surprising. The potential energy function (Eq. 38) is also spherically symmetric, so that the atom has no built-in preferred direction. Like a billiard ball, the atom in its ground state looks the same in all directions.

The square of the wave function, which we have called the *probability density,* has the property that $\psi^2(r)\,dV$ gives the probability of finding the electron in a volume element dV located at a position defined by the position vector **r**. From Eq. 39 we have

$$\psi^2(r) = \frac{1}{\pi a_0^3}e^{-2r/a_0}. \qquad (40)$$

Figure 16a is a "dot plot" representation of Eq. 40, the density of the dots suggesting the probabilistic nature of the electron's location. A circle of one Bohr radius has been drawn to show the scale.

Another useful quantity for representing the electron's position is the *radial probability density* $P_r(r)$. This is defined so that $P_r(r)\,dr$ gives the probability that, regardless of direction, the electron will be found to lie between two spherical shells whose radii are r and $r + dr$. The volume between those shells is $(4\pi r^2)(dr)$ so that we can write

$$P_r(r)\,dr = \psi^2(r)\,dV = \psi^2(r)(4\pi r^2)(dr),$$

which, combined with Eq. 40, leads to

$$P_r(r) = \psi^2(r)(4\pi r^2) = \frac{4}{a_0^3}r^2 e^{-2r/a_0}. \qquad (41)$$

Figure 16b shows a plot of Eq. 41. We note (see Sample Problem 8) that the maximum value of $P_r(r)$ occurs at $r = a_0$. In wave mechanics we do not say that the electron in the ground state of the hydrogen atom moves in an orbit of one Bohr radius. We say instead that the electron is more likely to be found in a thin shell at this distance from the central nucleus than in a shell of equal thickness at any other distance, either larger or smaller. The so-

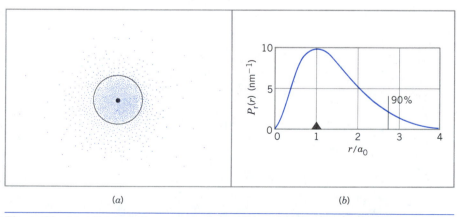

(a)	(b)

Figure 16   (a) A "dot plot" representation of the probability density for the ground state of the hydrogen atom, given by Eq. 40. A circle has been drawn at the radius $r = a_0$. (b) The radial probability density, given by Eq. 41. The filled triangle marks the maximum probability at $r = a_0$. The line marked "90%" shows the radius of a sphere containing 90% of the probability density.

called 90% radius is also indicated in the figure. It defines a sphere such that the probability that the electron will be found inside is 90%. The probability that it will be found outside is, of course, 10%.

We see that the answer to the question, "How big is the hydrogen atom?" is not so simple. You can say that it is one Bohr radius, but (see Problem 56) 68% of the times that you measure it, you will find that the electron is farther away than this. A more reasonable answer is to give the radius of the 90% probability sphere, which turns out to be 2.7 Bohr radii. The inherent quantum fuzziness of the atom simply does not permit us to answer the question any more precisely.

We have said nothing so far about the role of the spin of the electron in the ground state of the hydrogen atom. As Table 4 reminds us, the shell corresponding to $n = 1$ in the hydrogen atom contains two states, corresponding to the two allowed values ($= \pm \frac{1}{2}$) of the spin quantum number m_s. These two states, however, have exactly the same energy and, for an isolated atom, there is no way to tell them apart experimentally. The atom in its ground state can be either in the state with $m_s = +\frac{1}{2}$ or in the state with $m_s = -\frac{1}{2}$. The energy of the atom and the probability density curve of Fig. 16b are the same for each of these states.

If you *really* want to distinguish between these two spin orientation states you can do so by putting the atom in an external magnetic field. This not only provides a natural reference axis with respect to which the electron's spin angular momentum vector (and its spin magnetic dipole moment vector) can orient itself, but it also separates the two states in energy. This, in fact, is exactly what was done in the Stern–Gerlach experiment.

Sample Problem 8   Verify that the maximum of the radial probability density curve of Fig. 16b falls at $r = a_0$, where a_0 is the Bohr radius.

Solution   Differentiating $P_r(r)$ in Eq. 41 with respect to r, we obtain

$$\frac{dP_r}{dr} = \frac{4}{a_0^3}(2r)e^{-2r/a_0} + \frac{4}{a_0^3}r^2\left(-\frac{2}{a_0}\right)e^{-2r/a_0}$$

$$= \frac{8r}{a_0^3}e^{-2r/a_0}\left(1 - \frac{r}{a_0}\right).$$

At the maximum of the curve we must have $dP_r/dr = 0$ and, as inspection of this equation shows, this does indeed occur at $r = a_0$. Note that we also have $dP_r/dr = 0$ at $r = 0$ and as $r \rightarrow \infty$. These conditions are quite consistent with Fig. 16b and correspond to minima rather than maxima.

51-8   THE EXCITED STATES OF HYDROGEN

The state next highest in energy above the ground state is called the first excited state. Its energy ($= -3.40$ eV) is found by putting $n = 2$ in the energy equation (Eq. 18). As Table 4 reminds us, the $n = 2$ shell contains two subshells, corresponding to $l = 0$ and to $l = 1$. We deal with each in turn.

The $n = 2,\ l = 0$ Subshell

The wave function for this state is

$$\psi_{200}(r) = \frac{1}{4\sqrt{2\pi a_0^3}}(2 - r/a_0)e^{-r/2a_0}, \qquad (42)$$

in which the subscripts 200 represent the quantum number sequence $n = 2$, $l = 0$, and $m_l = 0$. This state, just like the ground state, has spherical symmetry in that it is a function of r only and involves no angles as variables.

The probability density $\psi^2(r)$ and the radial probability density $P_r(r)$ are given by

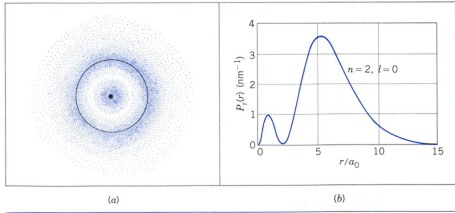

(a)   (b)

Figure 17   (a) A "dot plot" representation of the probability density of atomic hydrogen for the excited state with $n = 2$ and $l = 0$. A circle has been drawn at the radius $r = 4a_0$. (b) The radial probability density.

$$\psi^2(r) = \frac{1}{32\pi a_0^3}(2 - r/a_0)^2 e^{-r/a_0} \qquad (43)$$

and

$$P_r(r) = \psi^2(r)(4\pi r^2)$$
$$= \frac{1}{8a_0}\left(\frac{r}{a_0}\right)^2\left(2 - \frac{r}{a_0}\right)^2 e^{-r/a_0}. \qquad (44)$$

Figure 17a is a "dot plot" of Eq. 43, the probability density $\psi^2(r)$. Figure 17b is a plot of Eq. 44, the radial probability density. Note that the latter curve has two maxima and goes to zero at $r = 2a_0$, as simple inspection of Eq. 44 makes clear.

The remarks about spin at the end of Section 51-7 apply with equal force here. This state also has complete spherical symmetry. Its angular momentum is $\frac{1}{2}$, in units of $\hbar$, due (as before) entirely to the spin of the electron.

The $n = 2$, $l = 1$ Subshell

The states that comprise this subshell *do* have orbital angular momentum, its z-axis projections being given by $m_l\hbar$, where m_l can take on the values of 0 or ± 1. The wave functions for these three states are *not* spherically symmetric, being functions not only of r but also of the polar angle θ, defined in Fig. 18a.

Figure 18a shows "dot plots" of the three probability densities, $\psi_{21-1}^2(r,\theta)$, $\psi_{210}^2(r,\theta)$, and $\psi_{21+1}^2(r,\theta)$. The subscripts represent the quantum number sequence n, l, and m_l. All three plots have rotational symmetry about the z axis, the plots for $m_l = -1$ and for $m_l = +1$ being identical.

You are entitled to be a little puzzled about the lack of spherical symmetry shown in Fig. 18a. After all, the potential energy function that we inserted into the Schrödinger equation depended only on r. Does the electron in a state with $n = 2$, $l = 1$, and $m_l = 0$ *really* like to cluster about the z axis, avoiding the equatorial plane? How is the direction of this axis chosen?

The answer to this puzzle comes when we realize that the three states in question have the same energy and, in the absence of an external magnetic field, there is no way to isolate them experimentally. If we assume that—on average—the atom spends one-third of its time in each of the three states shown in Fig. 18a, we can calculate a *weighted average probability density* for the subshell as a whole. The result is

$$\psi_{21}^2(r) = \frac{1}{3}\psi_{21-1}^2(r,\theta) + \frac{1}{3}\psi_{210}^2(r,\theta) + \frac{1}{3}\psi_{21+1}^2(r,\theta)$$
$$= \frac{1}{96\pi a_0^5} r^2 e^{-r/a_0}. \qquad (45)$$

The subscript on the probability density gives the values of $n (= 2)$ and $l (= 1)$. Note that the angular variable θ has disappeared from the final result! The probability density for the subshell as a whole depends only on r and has the spherical symmetry that we expect it to have. This means that if you superimpose the three cylindrically symmetrical dot plots of Fig. 18a, the resulting dot plot (imagined in three dimensions) will be spherically symmetric.

We now find the radial probability density for this subshell, proceeding as we did in Section 51-7 for the ground state; namely,

$$P_r(r) = \psi_{21}^2(r)(4\pi r^2)$$
$$= \frac{1}{24a_0^5} r^4 e^{-r/a_0}. \qquad (46)$$

Figure 18b shows a plot of this radial probability density. Note that the maximum of the distribution occurs at $r = 4a_0$, which (see Eq. 19) is just the radius of the second Bohr orbit.

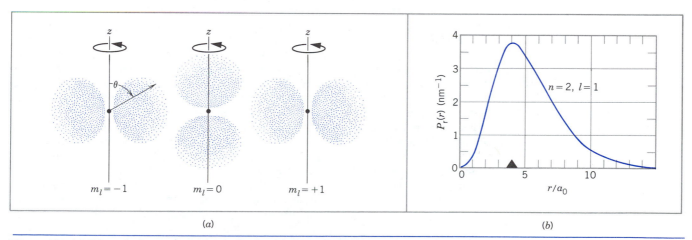

(a)   (b)

Figure 18 (a) "Dot plot" representations of the probability density of atomic hydrogen for the excited state with $n = 2$ and $l = 1$. To obtain the full three-dimensional picture, imagine each plot rotated about the z axis. (b) The radial probability density. The filled triangle shows the location of the maximum at $r = 4a_0$.

51-9 DETAILS OF ATOMIC STRUCTURE *(Optional)*

So far in this chapter we have outlined basic aspects of the quantum theory applied to the structure of atomic hydrogen, which allows us to understand such details of its properties as the Balmer series (and other series of emitted radiations). Here we mention briefly some additional details of atomic structure that we can similarly understand.

Fine Structure

When we study the spectral lines under high resolution, we find that what appears to be a single line is often a pair of very closely spaced lines (a doublet). This is called the *fine structure* of the spectrum. The effect is usually very small. In the case of the transition between the first excited state and the ground state in hydrogen ($E = 10.2$ eV), the energy difference between the two components due to the fine structure is 4.5×10^{-5} eV. The fine structure splitting increases rapidly with atomic number, however; in sodium, it is responsible for the splitting of the yellow D-lines, which differ in wavelength by about 0.6 nm out of 590 nm, or about 1 part in 10^3.

The fine structure splitting is usually analyzed in terms of the total angular momentum of the electron, obtained from the sum of the orbital and spin contributions. In the excited $l = 1$ state, the possible values of the total angular momentum quantum number are, according to the rules for adding angular momentum in quantum mechanics,

$$j = l \pm s = 1 \pm \tfrac{1}{2} = \tfrac{3}{2} \text{ or } \tfrac{1}{2}.$$

Loosely speaking, these two possibilities correspond respectively to the **L** and **S** vectors being parallel or antiparallel. These two different orientations have slightly different energies, which gives an energy splitting in the atom between the states corresponding to these combinations. (You can think of the parallel combination as corresponding to two tiny bar magnets aligned side by side and parallel to each other, with their like poles repelling one another; in the antiparallel configuration, the magnets are aligned in the opposite direction, with neighboring N and S poles attracting each other. The latter arrangement lowers the energy of the atom, that is, makes it more tightly bound.)

The total angular momentum has properties similar to the orbital and spin angular momenta. Specifically, there are $2j + 1$ different orientations of the **J** vector, corresponding to the different possible values of its z component $J_z = m_j \hbar$, where m_j ranges from $+j$ to $-j$ in integer steps. That is, for $j = \tfrac{3}{2}$, we have $m_j = +\tfrac{3}{2}, +\tfrac{1}{2}, -\tfrac{1}{2}$, or $-\tfrac{3}{2}$, while for $j = \tfrac{1}{2}$ we have $m_j = +\tfrac{1}{2}$ or $-\tfrac{1}{2}$.

Figure 19 shows a representation of the fine structure splitting in sodium. Note that, not considering the fine structure splitting, the $l = 1$ excited state includes six substates (three corresponding to $m_l = +1, 0$, and -1, each of which can have $m_s = +\tfrac{1}{2}$ or $-\tfrac{1}{2}$). Considering the fine structure, there are still six substates, four associated with $j = \tfrac{3}{2}$ and two with $j = \tfrac{1}{2}$. The ground state, with $l = 0$, can have only $j = \tfrac{1}{2}$.

The Zeeman Effect

Michael Faraday had the intuitive idea that the light from a source would change if you put the source in a strong magnetic field. Faraday was not successful in attempting to do this experi-

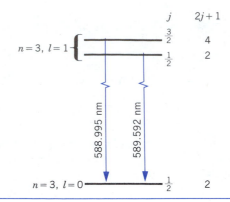

Figure 19   The fine structure splitting of the energy levels in sodium that emit the light of the familiar doublet. The drawing is not to scale; the actual splitting of the upper levels is about 1/1000 of the energy difference between the upper and lower levels. To the right is shown the number of different orientations of the **J** vector (that is, the number of different values of m_j for each level).

ment, because the equipment available to him did not have sufficient resolution to observe this small effect. About 30 years later in 1896, the Dutch physicist Pieter Zeeman repeated the experiment with more sensitive apparatus and observed that the spectral lines were measurably broadened in a strong magnetic field. With higher fields and better resolution, it is possible to see the lines dividing into components whose splitting increases in proportion to the field. For this work, Zeeman shared the 1902 Nobel prize in physics. Figure 20 shows an example of the Zeeman effect. In an intense magnetic field, the splitting of the lines in the Zeeman effect may be about 1 part in 10^4 of the energy or wavelength of the lines.

To understand the origin of the Zeeman effect, consider the energy-level diagram for sodium shown in Fig. 21, which shows only the $j = \tfrac{3}{2}$ excited state and the $j = \tfrac{1}{2}$ ground state. When the magnetic field is turned on, the four possible orientations of the **J** vector of the excited state (corresponding to the four different m_j values) give different energies, which can be calculated from the magnetic moment of the state. Similarly, the $j = \tfrac{1}{2}$ ground state splits into two substates. With the field off, there is only one possible transition between the excited state and the ground state; with the field on, there might be transitions from any substate of the excited state to any substate of the ground state, giving eight possible transitions. Two of these transitions are forbidden to occur by the rules of quantum mechanics, leaving six individual components.

The number of components and their separations observed in the Zeeman effect can be calculated using wave mechanics. As Fig. 20 illustrates, different lines in a given spectrum may show different patterns of splitting. It is a triumph of wave mechanics that these details of the Zeeman effect, along with the relative intensities of the components and even their polarizations, can be calculated and agree precisely with experiment.

Reduced Mass

Our derivation of the energy levels of hydrogen assumed that the electron revolves about a stationary nucleus. Actually, the elec-

Figure 20 The Zeeman effect in rhodium. The bottom spectrum shows the splitting of the spectral lines when the magnet is turned on.

tron and proton each orbit about the center of mass of the system. One convenient way of taking this into account is to assign the electron an effectively smaller mass called the *reduced mass* (see Section 15-10), defined according to

$$m = \frac{m_e M}{M + m_e} = \frac{m_e}{1 + m_e/M}, \qquad (47)$$

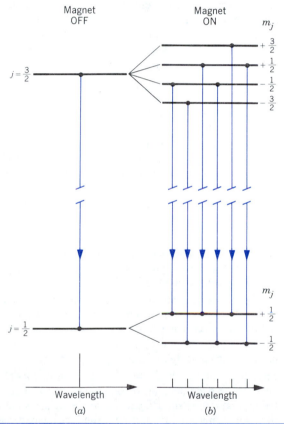

Figure 21 An energy-level diagram illustrating the Zeeman splitting of one member of the fine-structure doublet in sodium. When the magnetic field is applied, the single spectral line of (*a*) splits into the six closely spaced components shown in (*b*).

where M is the mass of the nucleus. That is, in all expressions involving the electron mass, we should replace it with its reduced mass.

In the case of hydrogen, Eq. 47 gives

$$m = \frac{m_e}{1 + \dfrac{1}{1836.15}} = \frac{m_e}{1.000545},$$

and the corresponding value of the ground-state energy changes from -13.6057 eV (corresponding to an infinitely massive nucleus) to -13.5983 eV. All spectral lines scale correspondingly. For example, the first line in the Lyman series would have an energy of 10.2043 eV for an infinitely massive nucleus, while in hydrogen the observed energy is 10.1972 eV. These differences, amounting to about 1 part in 10^3, are easily observable with spectroscopes.

About one hydrogen atom in 6000 is deuterium, or "heavy hydrogen," whose nucleus contains one proton and one neutron, making it about twice as massive as ordinary hydrogen. Most of the chemical and physical properties of heavy hydrogen are identical to those of ordinary hydrogen, except for those properties that specifically depend on mass. As we have seen, the energy of the spectral lines depends slightly on the mass of the nucleus. In an atom of heavy hydrogen, the reduced mass of the electron is

$$m = \frac{m_e}{1 + \dfrac{1}{3670.48}} = \frac{m_e}{1.000272},$$

and the corresponding ground-state energy is -13.6020 eV. The first line of the Lyman series would have an energy of 10.2015 eV. If we examined the spectral lines from a sample of hydrogen, we would find that each line consisted of a doublet, separated by an interval of 0.027% of the energy (or wavelength) and with relative intensities of about 6000 to 1. By increasing the concentration of heavy hydrogen (by distillation, for example) the intensity ratio can be varied. Using this procedure, deuterium was discovered in 1932 by H. C. Urey, who was awarded the 1934 Nobel prize in chemistry for his discovery. ∎

QUESTIONS

1. Discuss the analogy between the Kepler – Newton relationship in the development of Newton's law of gravitation and the Balmer – Bohr relationship in developing the Bohr theory of atomic structure.

2. Why was the Balmer series, rather than the Lyman or Paschen series, the first to be detected and analyzed in the hydrogen spectrum?

3. Any series of atomic hydrogen yet to be observed will probably be found to be in what region of the spectrum?

4. In Bohr's theory for the hydrogen atom orbits, what is the implication of the fact that the potential energy is negative and is greater in magnitude than the kinetic energy?

5. Can a hydrogen atom absorb a photon whose energy exceeds its binding energy (13.6 eV)?

6. On emitting a photon, an isolated hydrogen atom recoils to conserve momentum. Explain the fact that the energy of the emitted photon is slightly less than the energy difference between the energy levels involved in the emission process.

7. Why are some lines in the hydrogen spectrum brighter than others?

8. Radioastronomers observe lines in the hydrogen spectrum that originate in hydrogen atoms that are in states with $n = 350$ or so. Why can't hydrogen atoms in states with such high quantum numbers be produced and studied in the laboratory?

9. Only a relatively small number of Balmer lines can be observed from laboratory discharge tubes, whereas a large number are observed in stellar spectra. Explain this in terms of the small density, high temperature, and large volume of gases in stellar atmospheres.

10. According to classical mechanics, an electron moving in an orbit should be able to do so with any angular momentum whatever. According to Bohr's theory of the hydrogen atom, however, the angular momentum is quantized according to $L = nh/2\pi$. Reconcile these two statements, using the correspondence principle.

11. Why does the concept of Bohr orbits violate the uncertainty principle?

12. Consider a hydrogen-like atom in which a positron (a positively charged electron) circulates about a (negatively charged) antiproton. In what way, if any, would the emission spectrum of this "antimatter atom" differ from the spectrum of a normal hydrogen atom?

13. If Bohr's theory and Schrödinger's wave mechanics predict the same result for the energies of the hydrogen atom states, then why do we need wave mechanics, with its greater complexity?

14. Compare Bohr's theory and wave mechanics. In what respects do they agree? In what respects do they differ?

15. How would you show in the laboratory that an atom has angular momentum? That it has a magnetic dipole moment?

16. Why don't we observe space quantization for a spinning top?

17. The angular momentum of the electron in the hydrogen atom is quantized. Why isn't the linear momentum also quantized? (*Hint*: Consider the implications of the uncertainty principle.)

18. Angular momentum is a vector and you might expect that it would take three quantum numbers to describe it, corresponding to the three space components of a vector. Instead, in an atom, only two quantum numbers characterize the angular momentum. Explain why.

19. Justify the statement that, in the Einstein – de Haas effect, the angular momentum of the iron bar as a whole must be conserved when the bar is suddenly magnetized.

20. In the Einstein – de Haas experiment (see Sample Problem 5), can you justify the fact that the predicted period of rotation of the cylinder depends only on the cylinder radius and not, for example, on its height? What assumptions were made in deriving the expression for the period of rotation?

21. Convince yourself that the directions of the arrows in Fig. 10*b* representing the current in the solenoid, the magnetic field, the atomic angular momenta, and the direction of rotation of the cylinder are consistent with each other.

22. Does the Einstein – de Haas effect provide any evidence that angular momentum is quantized?

23. A beam of circularly polarized light, viewed as a beam of photons whose spins are aligned, can exert a torque on an absorbing screen. Develop the analogy to the Einstein – de Haas experiment.

24. A beam of neutral silver atoms is used in a Stern – Gerlach experiment. What is the origin of both the force and the torque that act on the atom? How is the atom affected by each?

25. What determines the number of subbeams into which a beam of neutral atoms is split in a Stern – Gerlach experiment?

26. If in a Stern – Gerlach experiment an ion beam is resolved into five component beams, then what angular momentum quantum number does each ion have?

27. In a Stern – Gerlach apparatus, is it possible to have a magnetic field configuration in which the magnetic field itself is zero along the beam path but the field gradient is not? If your answer is yes, can you design an electromagnet that will produce such a field configuration?

28. The silver atoms in the Stern – Gerlach experiment of Sample Problem 6 are uncharged. Suppose that a silver atom in the apparatus were suddenly to lose an electron, becoming a silver ion. What would be the nature and the relative magnitude of the forces acting on it (*a*) before and (*b*) after this event?

29. How do we arrive at the conclusion that the spin magnetic quantum number m_s can have only the values $\pm \frac{1}{2}$? What kinds of experiments support this conclusion?

30. Why is the magnetic moment of the spinning electron directed opposite to its spin angular momentum?

31. Discuss how good an analogy the rotating Earth revolving about the Sun is to a spinning electron moving about a proton in the hydrogen atom.

32. An atom in a state with zero angular momentum has spherical symmetry as far as its interaction with other atoms is concerned. It is sometimes called a "billiard-ball atom." Explain.

33. "If the angular momentum of electrons in atoms were not quantized, the periodic table of the elements would not be what it is." Discuss this statement.

34. How would the properties of helium differ if the electron had no spin, that is, if the only operative quantum numbers were n, l, and m_l?

35. We assert that the number of quantum numbers needed for a complete description of the motion of the electron in the hydrogen atom is equal to the number of degrees of freedom that the electron possesses. What is this number? How can you justify it?

36. Define and distinguish among the terms wave function, probability density, and radial probability density.

37. What are the dimensions and the SI units of a wave function, a probability density, and a radial probability density? Are the dimensions what you expect?

38. In the hydrogen atom state with $l = 1$, the spin and the orbital angular momentum vectors can be aligned either parallel or antiparallel. Which arrangement has the greater energy and why?

39. How can you account for the fact that in the state of the hydrogen atom with $n = 2$ and $l = 0$, the probability density is a maximum at $r = 0$ but the radial probability density is zero there? See Fig. 17.

40. Figure 18a shows the three probability densities for the hydrogen atom states with $n = 2$ and $l = 1$. What determines the direction in space that we choose for the z axis?

41. Consider the three probability density "dot plots" of Fig. 18a, each of which is a figure of revolution about the z axis. Do you see any connection between these figures and the semiclassical vector model of the atom (Fig. 9) for the case of $l = 1$?

42. Use Heisenberg's uncertainty principle to show that the probability densities in an $l = 2$ state have cylindrical symmetry about the z axis.

43. Explain how the interaction between the spin and the orbital motions of the valence electron in sodium leads to the splitting of the spectral lines of sodium, producing the familiar sodium doublet. See Fig. 19.

PROBLEMS

Section 51-1 The Bohr Theory

1. (a) By direct substitution of numerical values of the fundamental constants, verify that the energy of the ground state of the hydrogen atom is -13.6 eV; see Eq. 18. (b) Similarly, from Eq. 17 show that the value of the Rydberg constant R is 0.01097 nm^{-1}. (c) Also verify the numerical value of a_0 by direct computation of its expression given in Eq. 19.

2. Answer the questions of Sample Problem 2, but for the Lyman series.

3. Using the Balmer–Rydberg formula, Eq. 3, calculate the five longest wavelengths of the Balmer series.

4. What are the (a) wavelength, (b) momentum, and (c) energy of the photon that is emitted when a hydrogen atom undergoes a transition from the state $n = 3$ to $n = 1$?

5. Show, on an energy-level diagram for hydrogen, the quantum numbers corresponding to a transition in which the wavelength of the emitted photon is 121.6 nm.

6. (a) If the angular momentum of the Earth due to its motion around the Sun were quantized according to Bohr's relation $L = nh/2\pi$, what would the quantum number be? (b) Could such quantization be detected if it existed?

7. Calculate the binding energy of the hydrogen atom in the first excited state.

8. Find the value of the quantum number for a hydrogen atom that has an orbital radius of 847 pm.

9. Light with a wavelength of 1281.8 nm is emitted by a hydrogen atom. (a) What transition of the hydrogen atom is responsible for this radiation? (b) To what series does this radiation belong? (*Hint*: See Fig. 2.)

10. A hydrogen atom is excited from a state with $n = 1$ to one with $n = 4$. (a) Calculate the energy that must be absorbed by the atom. (b) Calculate and display on an energy-level diagram the different photon energies that may be emitted if the atom returns to the $n = 1$ state.

11. The lifetime of an electron in the state $n = 2$ in hydrogen is about 10 ns. What is the uncertainty in the energy of the $n = 2$ state? Compare this with the energy of this state.

12. A diatomic gas molecule consists of two atoms of mass m separated by a fixed distance d rotating about an axis as indicated in Fig. 22. Assuming that its angular momentum is quantized as in the Bohr atom, determine (a) the possible angular velocities and (b) the possible rotational energies. (c) Calculate, according to this model, the ground-state energy, in eV, of an O_2 molecule for which $d = 121$ pm and $m = 16.0$ u.

13. If an electron is revolving in an orbit at frequency ν_0, classical electromagnetism predicts that it will radiate energy not

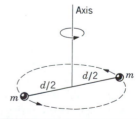

Figure 22 Problem 12.

only at this frequency but also at $2\nu_0$, $3\nu_0$, $4\nu_0$, and so on. Show that this is also predicted by Bohr's theory of the hydrogen atom in the limiting case of large quantum numbers.

14. In Table 2 show that the quantity in the last column is given by

$$\frac{100(\nu - \nu_0)}{\nu} \approx \frac{150}{n}$$

for large quantum numbers.

15. A neutron, with kinetic energy of 6.0 eV, collides with a resting hydrogen atom in its ground state. Show that this collision must be elastic (that is, energy must be conserved). (*Hint:* Show that the atom cannot be raised to a higher excitation state as a result of the collision.)

16. (*a*) Calculate, according to the Bohr model, the speed of the electron in the ground state of the hydrogen atom. (*b*) Calculate the corresponding de Broglie wavelength. (*c*) Comparing the answers to (*a*) and (*b*), find a relation between the de Broglie wavelength λ and the radius a_0 of the ground-state Bohr orbit.

17. According to the correspondence principle, as $n \to \infty$ we expect classical results in the Bohr atom. Hence the de Broglie wavelength associated with the electron (a quantum result) should get smaller compared with the radius of the Bohr orbit as n increases. Indeed, we expect that $\lambda/r \to 0$ as $n \to \infty$. Show that this is the case.

18. A hydrogen atom in a state having a *binding energy* (the energy required to remove an electron) of 0.85 eV makes a transition to a state with an *excitation energy* (the difference in energy between the state and the ground state) of 10.2 eV. (*a*) Find the energy of the emitted photon. (*b*) Show this transition on an energy-level diagram for hydrogen, labeling with the appropriate quantum numbers.

19. From the energy-level diagram for hydrogen, explain the observation that the frequency of the second Lyman-series line is the sum of the frequencies of the first Lyman-series line and the first Balmer-series line. This is an example of the empirically discovered *Ritz combination principle*. Use the diagram to find some other valid combinations.

20. Calculate the recoil speed of a hydrogen atom, assumed initially at rest, if the electron makes a transition from the $n = 4$ state directly to the ground state. (*Hint:* Apply conservation of linear momentum.)

21. (*a*) How much energy is required to remove the electron from a He^+ ion in its ground state? (*b*) From a Li^{2+} ion in a state with $n = 3$? (*Hint:* See Eq. 18.)

22. In stars the Pickering series is found in the He^+ spectrum. It is emitted when the electron in He^+ jumps from higher levels to the level with $n = 4$. (*a*) Show that the wavelengths of the lines in this series are given by

$$\lambda = \frac{1}{R} \frac{4n^2}{n^2 - 16},$$

in which $n = 5, 6, 7, \ldots$ (*b*) Calculate the wavelength of the first line in this series and of the series limit. (*c*) In what region(s) of the spectrum does this series occur?

23. Show that in Bohr's semiclassical one-electron model of the atom (*a*) the orbital speeds are quantized as $v_n =$ $(Ze^2/2\epsilon_0 h)(1/n)$ and (*b*) the orbital angular momenta are quantized as $L_n = (h/2\pi)n$.

24. In the ground state of the hydrogen atom, according to Bohr's theory, what are (*a*) the quantum number, (*b*) the orbit radius, (*c*) the angular momentum, (*d*) the linear momentum, (*e*) the angular velocity, (*f*) the linear speed, (*g*) the force on the electron, (*h*) the acceleration of the electron, (*i*) the kinetic energy, (*j*) the potential energy, and (*k*) the total energy?

25. How do the quantities (*b*) to (*k*) in Problem 24 vary with the quantum number n?

26. Suppose that we wish to test the possibility that electrons in atoms move in orbits by "viewing" them with photons with sufficiently short wavelength, say 10.0 pm. (*a*) What would be the energy of such photons? (*b*) How much energy would such a photon transfer to a free electron in a head-on Compton collision? (*c*) What does this tell you about the possibility of confirming orbital motion by "viewing" an atomic electron at two or more points along its path? Assume that the speed of the electron is $0.10c$.

27. Bohr proposed that, as an alternative to the correspondence principle, the quantization expression for the angular momentum (Eq. 20) could be taken as a basic postulate. Starting from this point, and using only classical results, derive Bohr's expression for the quantized energies of the stationary states of the hydrogen atom (Eq. 18).

28. (*a*) Calculate the wavelength intervals over which the Lyman, the Balmer, and the Paschen series extend. (The interval extends from the longest wavelength to the series limit.) (*b*) Find the corresponding frequency intervals.

Section 51-3 Angular Momentum

29. Verify that $\mu_B = 9.274 \times 10^{-24}$ J/T $= 5.788 \times 10^{-5}$ eV/T, as reported in Eq. 30.

30. If an electron in a hydrogen atom is in a state with $l = 5$, what is the smallest possible angle between $\mathbf{L}$ and $\mathbf{L}_z$?

31. For a hydrogen atom in a state with $l = 3$, calculate the allowed values of (*a*) L_z, (*b*) μ_z, and (*c*) θ. Find also the magnitudes of (*d*) $\mathbf{L}$ and (*e*) μ. Where appropriate, express answers in units of $\hbar$ and μ_B.

32. (*a*) Show that the magnetic moments of the electrons in the various Bohr orbits are given, according to the Bohr theory, by

$$\mu = n\mu_B$$

in which μ_B is the Bohr magneton and $n = 1, 2, 3, \ldots$. (*b*) How does this expression compare with the actual values?

33. (*a*) Show, by reanalyzing Problem 12 for a diatomic molecule with the angular momentum quantized by Eqs. 23 and 24, that the energy levels can be written as

$$E_l = \frac{h^2 l(l + 1)}{4\pi^2 m d^2}, \qquad l = 1, 2, 3, \ldots$$

(*b*) Calculate the energies of the lowest three levels of the O_2 molecule, for which the two atoms are 121 pm apart. The mass of the oxygen atom is 16.0 u. Compare your result with Problem 12.

34. Show that Eq. 28 is a plausible version of the uncertainty

principle $\Delta p \cdot \Delta x = h/2\pi$. (*Hint*: Multiply by r/r; associate p with mv, and L with mvr.)

Section 51-4 The Stern–Gerlach Experiment

35. Of the three scalar components of **L**, one, L_z, is quantized, according to Eq. 25. In view of the restrictions imposed by Eqs. 23 and 24, taken together, show that the most that can be said about the other two components of **L** is

$$\sqrt{L_x^2 + L_y^2} = \sqrt{l(l+1) - m_l^2}\,\hbar \ .$$

Note that these two components are not separately quantized. Show also that

$$\sqrt{l}\,\hbar \leq \sqrt{L_x^2 + L_y^2} \leq \sqrt{l(l+1)}\,\hbar \ .$$

Correlate these results with Fig. 9.

36. Suppose a hydrogen atom (in its ground state) moves 82 cm in a direction perpendicular to a magnetic field that has a gradient, in the vertical direction, of 16 mT/m. (*a*) What is the force on the atom due to the magnetic moment of the electron, which we take to be 1 Bohr magneton? (*b*) Find its vertical displacement if its speed is 970 m/s.

37. Calculate the acceleration of the silver atom as it passes through the deflecting magnet in the Stern–Gerlach experiment of Sample Problem 6.

38. Assume that in the Stern–Gerlach experiment described for neutral silver atoms the magnetic field **B** has a magnitude of 520 mT. (*a*) What is the energy difference between the orientations of the silver atoms in the two subbeams? (*b*) What is the frequency of the radiation that would induce a transition between these two states? (*c*) What is its wavelength, and to what part of the electromagnetic spectrum does it belong? The magnetic moment of a neutral silver atom is 1 Bohr magneton.

Section 51-6 Counting the Hydrogen Atom States

39. Write down the quantum numbers for all the hydrogen atom states belonging to the subshell for which $n = 4$ and $l = 3$.

40. A hydrogen atom state is known to have the quantum number $l = 3$. What are the possible n, m_l, and m_s quantum numbers?

41. A hydrogen atom state has a maximum m_l value of $+4$. What can you say about the rest of its quantum numbers?

42. How many hydrogen atom states are there with $n = 5$? How are they distributed among the subshells?

43. What are the quantum numbers n, l, m_l, m_s for the two electrons of the helium atom in its ground state?

44. Calculate the two possible angles between the electron spin angular momentum vector and the magnetic field in Sample Problem 6. Bear in mind that the *orbital* angular momentum of the valence electron is zero.

45. Label as true or false these statements involving the quantum numbers n, l, m_l. (*a*) One of these subshells cannot exist: $n = 2$, $l = 1$; $n = 4$, $l = 3$; $n = 3$, $l = 2$; $n = 1$, $l = 1$. (*b*) The number of values of m_l that are allowed depends only on l and not on n. (*c*) The $n = 4$ shell contains four subshells. (*d*) The smallest value of n that can go with a given l is $l + 1$. (*e*) All states with $l = 0$ also have $m_l = 0$, regardless of the value of n. (*f*) Every shell contains n subshells.

46. What is the wavelength of a photon that will induce a transition of an electron spin from parallel to antiparallel orientation in a magnetic field of magnitude 190 mT? Assume that $l = 0$.

47. The proton as well as the electron has spin $\frac{1}{2}$. In the hydrogen atom in its ground state, with $n = 1$ and $l = 0$, there are two energy levels, depending on whether the electron and the proton spins are in the same direction or in opposite directions. The state with the spins in the opposite direction has the higher energy. If an atom is in this state and one of the spins "flips over," the small energy difference is released as a photon of wavelength 21 cm. This spontaneous spin-flip process is very slow, the mean life for the process being about 10^7 y. However, radio astronomers observe this 21-cm radiation from interstellar space, where the density of hydrogen is so small that an atom can flip before being disturbed by collisions with other atoms. Calculate the effective magnetic field (due to the magnetic dipole moment of the proton) experienced by the electron in the emission of this 21-cm radiation.

48. Show that the number of states in any shell is given by $2n^2$.

Section 51-7 The Ground State of Hydrogen

49. In the ground state of the hydrogen atom, evaluate the square of the wave function, $\psi^2(r)$, and the radial probability density $P_r(r)$ for the positions (*a*) $r = 0$ and (*b*) $r = a_0$. Explain what these quantities mean.

50. In Fig. 16*b*, verify the plotted values of $P_r(r)$ at (*a*) $r = 0$, (*b*) $r = a_0$, and (*c*) $r = 2a_0$.

51. Find the ratio of the probabilities of finding the electron in the hydrogen atom in a thin shell at the Bohr radius to that of finding it in a shell of the same thickness at twice that distance.

52. A spherical region of radius $0.05a_0$ is located a distance a_0 from the nucleus of a hydrogen atom in its ground state. Calculate the probability that the electron will be found inside this sphere. (Assume that ψ is constant inside the sphere.)

53. For a hydrogen atom in its ground state, calculate the probability of finding the electron between two spheres of radii $r = 1.00a_0$ and $r = 1.01a_0$.

54. In atoms there is a finite, though very small, probability that, at some instant, an orbital electron will actually be found inside the nucleus. In fact, some unstable nuclei use this occasional appearance of the electron to decay by *electron capture*. Assuming that the proton itself is a sphere of radius 1.1×10^{-15} m and that the hydrogen atom electron wave function holds all the way to the proton's center, use the ground-state wave function to calculate the probability that the hydrogen atom electron is inside its nucleus. (*Hint*: When $x \ll 1$, $e^{-x} \approx 1$.)

55. Repeat Problem 54 for an electron in the state $n = 2$, $l = 0$; that is, calculate the probability that the electron will be found inside the proton, radius = 1.1 fm, that constitutes the nucleus of the hydrogen atom.

56. (*a*) In the ground state of the hydrogen atom show that the probability P that the electron lies within a sphere of radius r is given by

$$P = 1 - e^{-2x}(1 + 2x + 2x^2),$$

in which $x = r/a_0$. (b) Evaluate the probability that, in the ground state, the electron lies within a sphere of radius a_0.

57. Use the result of Problem 56 to calculate the probability that the electron in a hydrogen atom, in the ground state, will be found between the spheres $r = a_0$ and $r = 2a_0$.

58. For an electron in the ground state of the hydrogen atom, calculate the radius of a sphere for which the probability that the electron will be found inside the sphere equals the probability that the electron will be found outside the sphere. (*Hint*: See Problem 56.)

Section 51-8 The Excited States of Hydrogen

59. For the state $n = 2, l = 0$, (a) locate the two maxima for the radial probability density curve of Fig. 17b, and (b) calculate the values of the radial probability density at the two maxima; compare with Fig. 17b.

60. Using Eq. 46, show that, for the hydrogen atom state with $n = 2$ and $l = 1$,

$$\int_0^\infty P_r(r) \, dr = 1.$$

What is the physical interpretation of this result?

61. For a hydrogen atom in a state with $n = 2$ and $l = 0$, calculate the probability of finding the electron between two spheres of radii $r = 5.00a_0$ and $r = 5.01a_0$.

62. For a hydrogen atom in a state with $n = 2$ and $l = 0$, what is the probability of finding the electron somewhere within the smaller of the two maxima of its radial probability density function? See Fig. 17b.

Section 51-9 Details of Atomic Structure

63. Potassium ($Z = 19$), like sodium ($Z = 11$), is an alkali metal, its single valence electron moving around a filled 18-electron argon-like core. As in sodium, there is a potassium doublet, its wavelengths being 764.5 nm and 769.9 nm. The quantum numbers of the levels that give rise to these lines are just the same as for sodium (see Fig. 19) except that $n = 4$. Calculate (a) the energy splitting between the two upper states and (b) the energy difference between the uppermost state and the ground state.

64. The wavelengths of the lines of the sodium doublet (see Fig. 19) are 588.995 nm and 589.592 nm. (a) What is the difference in energy between the two upper levels in that figure? (b) This energy difference comes about because the electron's spin magnetic dipole moment can be oriented either parallel or antiparallel to the internal magnetic field associated with the electron's orbital motion. Use the result you have just calculated to find the strength of this internal magnetic field. The electron's spin magnetic dipole moment has a magnitude of 1 Bohr magneton.

65. Apply Bohr's model to a muonic atom, which consists of a nucleus of charge Ze with a negative muon (an elementary particle with a charge $q = -e$ and a mass $m = 207m_e$, where m_e is the electron mass) circulating about it. Calculate (a) the muon–nucleus separation in the first Bohr orbit, (b) the ionization energy, and (c) the wavelength of the most energetic photon that can be emitted. Assume that the muon is circulating about a hydrogen nucleus ($Z = 1$). See "The Muonium Atom," by Vernon W. Hughes, *Scientific American*, April 1966, p. 93.

66. Apply Bohr's model to the positronium atom. This consists of a positive and a negative electron revolving around their center of mass, which lies halfway between them. (a) What relationship exists between this spectrum and the hydrogen spectrum? (b) What is the radius of the ground-state orbit? (*Hint*: Calculate the reduced mass of the atom.) See "Exotic Atoms," by E. H. S. Burhop, *Contemporary Physics*, July 1970, p. 335.

CHAPTER 52

ATOMIC PHYSICS

*In the preceding three chapters, we have developed the foundations of
wave mechanics and used its principles to understand the structure of the
hydrogen atom. In this chapter, we broaden the development by considering
the structure of atoms beyond hydrogen.*

*We begin by considering the emission of x rays by atoms, which historically provided
the first definitive means to measure the number of electrons in an atom. We then consider
the rules for determining how to construct atoms with more than one electron, and we
consider how those rules and the resulting structure determine the arrangement of elements
in the familiar periodic table. We use information from atomic structure to analyze the
operation of the helium–neon laser, and we conclude with a brief look at how we can extend
our knowledge of atomic structure and wave functions to learn about
the structure of molecules.*

52-1 THE X-RAY SPECTRUM

So far we have dealt with the behavior of single electrons
in atoms, either the lone electron of hydrogen or the single
valence electron of sodium. We now shift our attention to
the behavior of electrons deep within the atom. We move
from a region of relatively low binding energy (5 eV for
the work required to remove the valence electron from
sodium, for example) to a region of higher energy (70 keV
for the work required to remove an innermost electron
from tungsten, for example). The radiations we deal with,
though of course still part of the electromagnetic spectrum, differ drastically in wavelength, for example, from
6×10^{-7} m for the sodium doublet lines to 2×10^{-11} m
for one of the tungsten characteristic radiations, a ratio of
about 30,000. We are now speaking of x rays.

The usefulness of these penetrating radiations in medical and dental diagnostics and in therapy is well known, as
are their many industrial applications, such as examining
welded joints in pipe lines. In Section 47-4 we described
how x rays can be used to deduce the atomic structures of
crystalline materials. The structures of such complex substances as insulin and DNA have been worked out by
these methods. In astronomy, x-ray satellites have shown
us an entirely new view of our universe through images of
the x rays emitted by stars and galaxies.

We saw in Section 47-4 that x rays are produced when
energetic electrons strike a solid target and are brought to
rest in it. Figure 1 shows the wavelength spectrum of the x
rays produced when 35-keV electrons strike a molybdenum target.

The Continuous X-Ray Spectrum

We first examine the continuous spectrum of Fig. 1,
ignoring—for the time being—the two prominent peaks
that rise from it. Consider an electron of kinetic energy K
that scatters from the nucleus of one of the molybdenum
atoms in the target, as in Fig. 2. In such a collision, momentum is transferred to the atom, and the electron loses
kinetic energy. (Because the atom is so massive, the momentum imparted to it by the electron results in a negligible kinetic energy.) The energy lost by the electron appears as the energy $h\nu$ of an x-ray photon that radiates
away from the site of the encounter. This process is called
bremsstrahlung (German, "braking radiation"), and it
accounts for the continuous x-ray spectrum.

Suppose electrons are accelerated through a potential
difference V and fall on a thick target. Due to bremsstrahlung processes in the target, the electrons can lose any
amount of energy from 0 to their maximum energy of eV.
The bremsstrahlung photons have a corresponding continuous spectrum from 0 to eV.

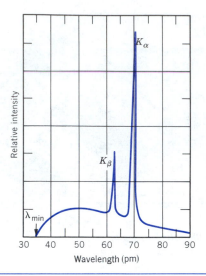

Figure 1   The wavelength spectrum of x rays produced when 35-keV electrons strike a molybdenum target (1 pm = 10^{-12} m).

A prominent feature of the continuous spectrum of Fig. 1 is the sharply defined cutoff wavelength λ_{min} below which the continuous spectrum does not exist. This minimum wavelength corresponds to a decelerating event in which one of the incident electrons (with initial kinetic energy eV) loses *all* this energy in a single encounter, radiating it away as a single photon. Thus

$$eV = h\nu_{max} = \frac{hc}{\lambda_{min}},$$

or

$$\lambda_{min} = \frac{hc}{eV}. \tag{1}$$

Equation 1 shows that if $h \to 0$, then $\lambda_{min} \to 0$, which is the prediction of classical theory. The existence of a minimum wavelength is a quantum phenomenon.

Note that as you change the target material, perhaps from molybdenum to copper, the general shape and intensity of the continuous spectrum may change but the cutoff wavelength will not change. This wavelength depends only on the kinetic energy of the electrons that bombard the target and not at all on the target material.

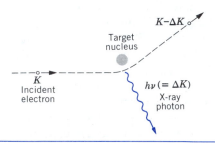

Figure 2   An electron passing near the nucleus of a target atom is accelerated and radiates a photon, losing part of its kinetic energy in the process.

The Characteristic X-Ray Spectrum

We now turn our attention to the two peaks of Fig. 1, labeled K_α and K_β. These peaks are characteristic of the target material and, together with other peaks that appear at longer wavelengths, form the *characteristic x-ray spectrum* of the element in question.

Here is how these x-ray photons arise. (1) An energetic incoming electron strikes an atom in the target and knocks out one of its deep-lying electrons. If the electron is in the shell with $n = 1$ (called, for historical reasons, the K shell) there remains a vacancy, or a "hole" as we shall call it, in this shell. (2) One of the outer electrons moves in to fill this hole and, in the process, the atom emits a characteristic x-ray photon. If the electron falls from the shell with $n = 2$ (called the L shell), we have the K_α line of Fig. 1; if it falls from the next outermost shell (called the M shell) we have the K_β line, and so on. Of course, such a transition will leave a hole in either the L or the M shell, but this will be filled by an electron from still further out in the atom; in the process, yet another characteristic x-ray spectrum line is emitted.

Figure 3 shows an x-ray atomic energy-level diagram for molybdenum, the element to which Fig. 1 refers. The base line ($E = 0$) represents the energy of a neutral molybdenum atom in its ground state. The level marked K ($E = 20$ keV) represents the energy of a molybdenum atom with a hole in its K shell. Similarly, the level marked L ($E = 2.7$ keV) represents the energy of an atom with a hole in its L shell, and so on. Note that in representing the

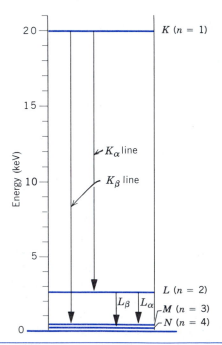

Figure 3   An atomic energy-level diagram for molybdenum, showing the transitions that give rise to the characteristic x rays of that element. (All levels, except the K level, contain a number of close-lying components, which are not shown in the figure.)

energy levels for the hydrogen atom (see Fig. 4 of Chapter 51), we chose a different base line. Rather than the atom in its ground state, there we selected the atom with its electron removed to infinity as our $E = 0$ configuration. Actually, the atomic configuration for which we choose to put $E = 0$ does not matter. Only differences in energy are physically significant, and these are the same no matter what our choice of an $E = 0$ base line is.

The transitions K_α and K_β in Fig. 3 show the origin of the two peaks in Fig. 1. The K_α line, for example, originates when an electron from the L shell of molybdenum — moving *upward* on the energy-level diagram — fills the hole in the K shell. This is the same as saying that a hole — moving *downward* on the diagram — moves from the K shell to the L shell. It is easier to keep track of a single hole than of the 41 electrons in ionized molybdenum that are potentially available to fill it. We have drawn the arrows in Fig. 3 from the point of view of hole transitions.

Sample Problem 1   Calculate the wavelength λ_{min} for the continuous spectrum of x rays emitted when 35-keV electrons fall on a molybdenum target, as in Fig. 1.

Solution   From Eq. 1, we have

$$\lambda_{min} = \frac{hc}{eV} = \frac{(4.14 \times 10^{-15}\ eV \cdot s)(3.00 \times 10^8\ m/s)}{35.0 \times 10^3\ eV}$$

$$= 3.54 \times 10^{-11}\ m = 35.4\ pm.$$

This is in agreement with the experimental result shown by the vertical arrow in Fig. 1. Note that Eq. 1 contains no reference to the target material. For a given accelerating potential all targets, no matter what they are made of, exhibit the same cutoff wavelength.

52-2   X RAYS AND THE NUMBERING OF THE ELEMENTS

In this section, we consider what x rays can teach us about the structure of the atoms that emit or absorb them. We focus on the work of the British physicist H. G. J. Moseley,* who, by x-ray studies, developed the concept of atomic number and gave physical meaning in terms of

---

* Henry G. J. Moseley (1887–1915) joined Ernest Rutherford's laboratory at the University of Manchester in 1910. Through a brilliant series of experiments, Moseley showed that characteristic x-ray frequencies increased regularly with the atomic number of the element, and he was able to locate gaps in the sequences corresponding to elements not yet discovered in his time. Moseley's promising research career was cut short when he died at age 27 at the battle of Gallipoli in World War I.

atomic structure to the ordering of the elements in the periodic table.

In his investigation of the atomic number concept, Moseley generated characteristic x rays by using as many elements as he could find — he found 38 — as targets for electron bombardment in a special evacuated x-ray tube of his own design. He measured the wavelengths of a number of the lines of the characteristic x-ray spectrum by the crystal diffraction method described in Section 47-4. He then sought, and readily found, regularities in the spectra as he moved from element to element in the periodic table. In particular, he noted that if, for a given spectrum line such as K_α, he plotted the square root of its frequency $(= \sqrt{\nu} = \sqrt{c/\lambda})$ against the position of the associated element in the periodic table, a straight line resulted. Figure 4 shows a portion of his data. We shall see below why it is logical to plot the data in this way and why a straight line is to be expected. Moseley's conclusion from the full body of his data was:

> We have here a proof that there is in the atom a fundamental quantity, which increases by regular steps as we pass from one element to the next. This quantity can only be the charge on the central atomic nucleus.

Moseley's achievement can be appreciated all the more when we realize the status of understanding of atomic structure at that time (1913). The nuclear model of the atom had been proposed by Rutherford only 2 years earlier. Little was known about the magnitude of the nuclear charge or the arrangement of the atomic electrons; Bohr published his first paper on atomic structure only in that same year. An element's place in the periodic table was at that time assigned by atomic *mass,* although there were several cases in which it was necessary to invert this order to fit the demands of the chemical evidence. The table had several empty squares, and a surprisingly large number of claims for the discovery of new elements had been advanced; the rare earth elements, because of the problems caused by their similar chemical properties, had not yet been properly sorted out.

Due to Moseley's work, the characteristic x-ray spectrum became the universally accepted signature of an element. Through such studies it became possible to string the elements in a line, so to speak, and to assign consecutive numbers to them, all without the slightest need to know anything about their chemical properties.

It is not hard to see why the characteristic x-ray spectrum shows such impressive regularities from element to element and the optical spectrum does not. The key to the identity of an element is the charge on its nucleus. This determines the number of its atomic electrons and thus its chemical and physical properties. Gold, for example, is what it is because its atoms have a nuclear charge of $+79e$. If it had just one more unit of charge, it would not be gold but mercury; if it had one fewer, it would be platinum. The K electrons, which play such a large role in

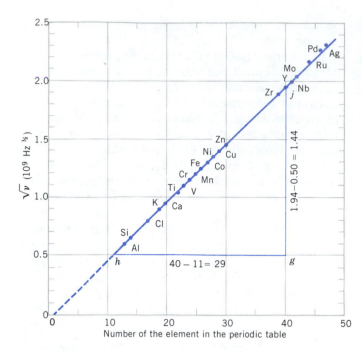

Figure 4   A Moseley plot of the K_α line of the characteristic x-ray spectra of 21 elements. The frequency is determined from the measured wavelength.

the production of the characteristic x-ray spectrum, lie very close to the nucleus and are sensitive probes of its charge. The optical spectrum, on the other hand, is associated with transitions of the outermost or valence electrons, which are heavily screened from the nucleus by the remaining $Z - 1$ electrons of the atom; they are not sensitive probes of the nuclear charge.

Bohr Theory and the Moseley Plot

Bohr's theory works well for hydrogen but fails for atoms with more than one electron (in part because it does not include the repulsive interaction between the electrons). Nevertheless, it provides an excellent first approximation in accounting for the Moseley plot of Fig. 4.

Consider an electron in the L shell of an atom that is about to move into a hole in the K shell, emitting a K_α x-ray photon in the process. Using Gauss' law (see Eq. 17 of Chapter 29), we find that the electric field at the location of the L electron is determined by the charge enclosed in an imaginary sphere of radius equal to the radial coordinate of the L electron. This sphere encloses a charge $+Ze$ from the nucleus and a charge $-e$ from the single remaining K electron. We say that the K electron "screens" the charge of the nucleus. In part because of this screening and in part because of readjustments that take place in the electron cloud as a whole, the effective atomic number for the transition turns out to be $Z - b$, where $b \approx 1$.

Bohr's formula for the frequency of the radiation corresponding to a transition in a hydrogen-like atom between any two atomic levels differing in energy by ΔE is

$$\nu = \frac{\Delta E}{h} = \frac{m_e e^4 Z^2}{8\epsilon_0^2 h^3}\left(\frac{1}{m^2} - \frac{1}{n^2}\right), \tag{2}$$

using Eq. 18 of Chapter 51 for the energy levels. For the K_α transition of Fig. 3, we can replace Z by $Z - b$ and substitute 1 for m and 2 for n. Doing so yields

$$\nu = \frac{m_e e^4}{8\epsilon_0^2 h^3}\left(\frac{1}{1^2} - \frac{1}{2^2}\right)(Z - b)^2.$$

Taking the square root of each side leads to

$$\sqrt{\nu} = \left(\frac{3 m_e e^4}{32 \epsilon_0^2 h^3}\right)^{1/2}(Z - b), \tag{3}$$

which we can write in the form

$$\sqrt{\nu} = aZ - ab, \tag{4}$$

where a is the indicated constant and $b \approx 1$.

Equation 4 represents a straight line, in full agreement with the experimental data of Fig. 4. If this plot is extended to higher atomic numbers, however, it turns out to be not quite straight but somewhat concave upward. Nevertheless, the quantitative agreement with Bohr theory is surprisingly good, as Sample Problem 2 shows.

Sample Problem 2   Calculate the value of the quantity a in Eq. 4 and compare it with the measured slope of the straight line in Fig. 4.

Solution   Comparing Eqs. 3 and 4 allows us to write

$$a = \left(\frac{3 m_e e^4}{32 \epsilon_0^2 h^3}\right)^{1/2}$$

$$= \frac{\sqrt{3}\,(9.11 \times 10^{-31}\text{ kg})^{1/2}(1.60 \times 10^{-19}\text{ C})^2}{4\sqrt{2}\,(8.85 \times 10^{-12}\text{ F/m})(6.63 \times 10^{-34}\text{ J}\cdot\text{s})^{3/2}}$$

$$= 4.95 \times 10^7 \text{ Hz}^{1/2}.$$

Careful measurement of Fig. 4, using the triangle *hgj*, yields

$$a = \frac{gj}{hg} = \frac{(1.94 - 0.50) \times 10^9 \text{ Hz}^{1/2}}{(40 - 11)} = 4.96 \times 10^7 \text{ Hz}^{1/2},$$

which is in agreement with the value predicted by Bohr theory within the uncertainty of the graphical measurement. Note also that the intercept in Fig. 4 is in fact close to 1, as expected from our screening argument.

The agreement with Bohr theory is not nearly as good for other lines in the x-ray spectrum, corresponding to the transitions of electrons farther from the nucleus; here we must rely on calculations based on wave mechanics.

Sample Problem 3   A cobalt target is bombarded with electrons, and the wavelengths of its characteristic spectrum are measured. A second, fainter, characteristic spectrum is also found, because of an impurity in the target. The wavelengths of the K_α lines are 178.9 pm (cobalt) and 143.5 pm (impurity). What is the impurity?

Solution   Let us apply Eq. 4 both to cobalt and to the impurity. Putting c/λ for ν (and assuming $b = 1$), we obtain

$$\sqrt{\frac{c}{\lambda_{\text{Co}}}} = aZ_{\text{Co}} - a \quad \text{and} \quad \sqrt{\frac{c}{\lambda_{\text{x}}}} = aZ_{\text{x}} - a.$$

Dividing yields

$$\sqrt{\frac{\lambda_{\text{Co}}}{\lambda_{\text{x}}}} = \frac{Z_{\text{x}} - 1}{Z_{\text{Co}} - 1}.$$

Substituting gives us

$$\sqrt{\frac{178.9 \text{ pm}}{143.5 \text{ pm}}} = \frac{Z_{\text{x}} - 1}{27 - 1}.$$

Solving for the unknown, we find $Z_{\text{x}} = 30.0$; a glance at the periodic table identifies the impurity as zinc.

52-3 BUILDING ATOMS

In the preceding section we saw how, by measuring the wavelengths of the characteristic x-ray spectrum of an element, we could assign an *atomic number Z* to each element and thus string them in a line according to a logical principle.

Here we go a step further. We try to see how far the principles of wave mechanics can take us in breaking up this line into a series of segments, corresponding to the horizontal periods of the periodic table of the elements.

The attempt meets with essentially total success. Every detail of the table (see Appendix E) can be accounted for, including (1) the numbers of elements in the seven horizontal periods into which the table is divided, (2) the similarity of the chemical properties of the elements in the various vertical columns—the alkali metals and the inert gases, for example—and (3) the existence of the rare earth, or lanthanide, series of elements, all crammed into one square of the table. In short, wave mechanics, supplemented by certain guiding principles that we discuss in this section, accounts for every feature of this table and thus, essentially, for all of chemistry.

Let us imagine that—in Tinker Toy fashion—we are going to construct a typical atom for each of the more than 100 elements that make up the periodic table. Our starting materials will be a supply of nuclei, each characterized by a charge $+ Ze$, with Z ranging by integers from 1 to over 100. We also need an ample supply of electrons. Our plan is to add Z electrons to each nucleus in such a way as to produce a neutral atom in its ground state.

Success follows only if we observe these three principles of atom building:

1. *The quantum number principle.* The electron in a hydrogen atom may—to mention one possibility—be in a state described by the quantum numbers $n = 2$, $l = 1$, $m_l = +1$, and $m_s = -\frac{1}{2}$. It turns out that a particular electron in *any* atom may also be fully identified by this same set of quantum numbers. That is not to say that the electrons in these different cases will move in the same way, because they will not. Put another way, although these electrons may share the same set of quantum numbers, the potentials in which they move—and thus their wave functions—will be quite different. Specifically, the quantum number principle asserts:

The hydrogen atom quantum numbers can be used to describe electron states and to assign electrons to shells and subshells, in any atom, no matter how many electrons it contains. Furthermore, the restrictions among the quantum numbers discussed in Section 51-6 remain in force.

2. *The Pauli exclusion principle.* This powerful principle was put forward by the Austrian-born physicist Wolfgang Pauli in 1925. Speaking generally, it tells us that no two electrons can be in the same state of motion at the same time. More specifically, it asserts that:

In a multielectron atom there can never be more than one electron in any given quantum state. That is, no two electrons in an atom can have the same set of quantum numbers.

If this principle did not hold, all the electrons in an atom would pile up in its *K* shell, and chemistry as we know it would not exist. You would not be here to read this sentence, and we would not have been here to have written it. Pauli's exclusion principle is no trivial assertion!

3. *The minimum energy principle.* As we fill subshells with electrons in the course of atom building, the question arises: In what order shall we fill them? The answer is:

When one subshell is filled, put the next electron in whichever vacant subshell will lead to an atom lowest in energy. To do otherwise would be to depart from our stated aim of building atoms in their ground states.

The lowest-energy subshell can be identified with the help of the following rule, which we first state and then try to make reasonable:

For a given principal quantum number n in a multielectron atom, the order of increasing energy of the subshells is the order of increasing l.

Table 1 helps to clarify this rule. Consider first a hydrogen atom whose single electron is in a state with $n = 4$. There are four allowed values of l, namely, 0, 1, 2, and 3. For electrons in true one-electron atoms—such as hydrogen—the energy does not depend on l at all but only on n, being given by Eq. 18 of Chapter 51,

$$E = -\frac{m_e e^4}{8\epsilon_0 h^2}\frac{Z^2}{n^2}, \qquad n = 1, 2, 3, \ldots . \quad (5)$$

Recall that this relation is predicted not only by Bohr theory but also by wave mechanics. Putting $Z = 1$ and $n = 4$ in this relation yields, for hydrogen, $E = -0.85$ eV, as Table 1 shows.

Consider now a lead nucleus ($Z = 82$) around which only a single electron circulates, again in a state with $n = 4$. Equation 5 also applies to this (admittedly rather unlikely) one-electron atom. For $Z = 82$ and $n = 4$, the table shows that we have $E = -5720$ eV, once more independent of l. The electron moves in the field of a nucleus with a charge of $+82e$; furthermore, it is drawn in very close to this nucleus, the equivalent Bohr orbit radius (see Eq. 19 of Chapter 51) being 82 times smaller than for hydrogen.

Finally, let us construct a normal, neutral lead atom by "sprinkling in" the missing 81 electrons. The outermost or valence electrons in lead have $n = 6$, so that an electron with $n = 4$ would lie somewhere in the middle of the smeared-out electron cloud surrounding the lead nucleus.

Equation 5 no longer holds for this multielectron atom, but we can find the energies of the four $n = 4$ subshells experimentally from x-ray studies. Their approximate values are shown in the last column of Table 1. We see at once that they lie higher in energy (that is, the binding energies are smaller) than for our hypothetical one-electron lead "atom" and that they vary with l just as the minimum energy rule predicts.

That the electrons in lead become more loosely bound when the entire electron cloud is present follows because some substantial fraction of this cloud screens the nucleus electrically. A typical $n = 4$ electron no longer "sees" the full positive nuclear charge, but rather sees this charge reduced by the negative charge of that part of the electron cloud that lies between the nucleus and the effective radius of the electron in question.

As for the variation of energy with l, let us ask ourselves what an $l = 0$ orbit would have to look like under the mechanical constraints of the Bohr picture. Truly to have no angular momentum, the electron would have to oscillate back and forth on a straight line segment passing directly through the nucleus. This does not happen, of course. The equivalent wave-mechanical statement is that an electron with $l = 0$ must spend a larger fraction of its time near the nucleus than do electrons with higher values of l. Such electrons would then, on the average, "see" a higher effective nuclear charge and would be more tightly bound; they would lie lower in energy, just as the minimum energy principle and Table 1 predict. It is interesting to compare Figs. 17 and 18 of Chapter 51, which show the $n = 2$, $l = 0$ and $n = 2$, $l = 1$ states of hydrogen. In the $l = 0$ state there is indeed a marked tendency for the electron to cluster near the nucleus—note the close-in secondary maximum—just as our qualitative argument suggests that it would.

52-4 THE PERIODIC TABLE

Figure 5 shows how the periodic table is put together, using the three rules for atom building that we have described in the previous section. Energy increases upward

TABLE 1 ENERGY LEVELS FOR ELECTRONS WITH $n = 4$, IN THREE DIFFERENT ATOMS

Orbital Quantum Number l	Energy (eV)		
	Hydrogen[a] $Z = 1$	*"Lead"*[b] $Z = 82$	*Lead*[c] $Z = 82$
0	−0.85	−5720	−890
1	−0.85	−5720	−710
2	−0.85	−5720	−420
3	−0.85	−5720	−140

[a] A neutral hydrogen atom; see Eq. 5.
[b] A hypothetical one-electron atom with $Z = 82$; see Eq. 5.
[c] An actual neutral lead atom ($Z = 82$); data from experiment.

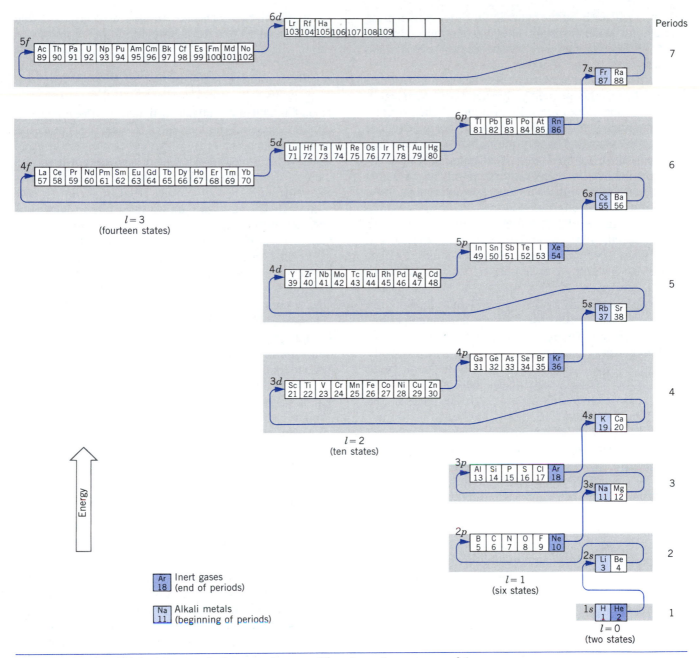

Figure 5   Starting with hydrogen at the bottom, the curved line shows the sequence of the seven horizontal periods of the periodic table. Each period starts with an alkali metal and ends with an inert gas.

in this figure. States with the same value of l have been displaced to the left for clarity and grouped into columns according to their l value.

Before we look more closely at this table, we introduce a new notation for the angular momentum quantum number l. For historical reasons* the values of l have been

given letter equivalents, according to this scheme:

l	0	1	2	3	4	5 . . .
Symbol	s	p	d	f	g	h . . .

In this notation, a state with $n = 1$ and $l = 0$ is called a "$1s$" state. Similarly, a state with $n = 4$ and $l = 3$ is called a "$4f$" state, and so on. These states are also known as *subshells*.

The dependence of energy on l is a dominant feature of Fig. 5. Look, for example, at the sequence of states $4s$, $4p$,

---

* The letters s, p, d, f stand for *sharp, principal, diffuse,* and *fundamental,* which were early spectroscopic designations of spectral lines. Beyond f, the states are labeled in alphabetic order.

4*d*, and 4*f*. They lie in the figure in the order of increasing energy, just as the minimum energy rule requires. In fact, the 4*f* states lie so high that they are above the 5*s* and the 5*p* states, which have a higher value of *n*.

The term *shell* is used to designate a group of states, lying close together in energy, that has a particular stability when those states are fully occupied. When dealing with the hydrogen atom (for which the energy depends only on the principal quantum number *n*), we identified the shells by giving the value of that quantum number. We now see that, in many-electron atoms, the principal quantum number alone is no longer a good indicator. As Fig. 5 shows, the shell we have labeled "6," corresponding to the sixth horizontal period of the periodic table, does indeed contain all the 6*s* and 6*p* states, corresponding to *n* = 6. However, it also contains all the 4*f* and 5*d* states. Moreover, the 6*d* states do not lie in this shell at all but in the shell above.

By starting with hydrogen in Fig. 5 and following the curved line, we can see how the seven horizontal periods of the periodic table are built up, each starting with an alkali metal and ending with an inert gas. Consider again the long sixth period, which starts with the alkali metal cesium (*Z* = 55) and ends with the inert gas radon (*Z* = 86). The order in which the subshells are filled, as the curved line indicates, is 6*s*, 4*f*, 5*d*, and 6*p*.

The sixth period contains a run of 15 elements (*Z* = 57 to *Z* = 71), listed separately at the bottom of the periodic table in Appendix E. These elements are called the *rare earths* or *lanthanides* (after the element lanthanum that begins the series). Their chemical properties are so similar that they are all grouped into a single square of the table. This similarity arises because, while the 4*f* state is being filled deep within the electron cloud, an outer screen of one or two 6*s* electrons remains in place. It is these outermost electrons that determine the chemical properties of the atom. A similar series (the *actinides*) occurs in the seventh period.

The maximum number of electrons permitted in any subshell is 2(2*l* + 1). This follows from the Pauli principle; for any value of *l*, there are 2*l* + 1 different m_l values, and for each of those there are 2 m_s values. There are thus 2(2*l* + 1) different possible labels for electrons in any subshell, and by the Pauli principle each electron in an atom must have a different label. If you count the number of elements in each of the labeled subshells of Fig. 5, you will find there to be 2(2*l* + 1); that is, 2 for *s* subshells (*l* = 0), 6 for *p* subshells (*l* = 1), 10 for *d* subshells (*l* = 2), and 14 for *f* subshells (*l* = 3).

Electron Configurations

We can describe the lowest energy state of an atom by giving its *electron configuration,* that is, by specifying the number of electrons in each occupied state. For example, for lithium (*Z* = 3) we have, from Fig. 5, $1s^2 2s^1$, where

the superscript indicates the number of electrons in that state. Consider these three configurations:

$$\text{F } (Z = 9): \qquad 1s^2 2s^2 2p^5$$
$$\text{Ne } (Z = 10): \qquad 1s^2 2s^2 2p^6$$
$$\text{Na } (Z = 11): \qquad 1s^2 2s^2 2p^6 3s^1$$

Neon has a filled 2*p* state, a particularly stable configuration. It takes considerable energy to break this configuration, and therefore neon does not readily give up electrons. There is a particularly large gap between neon and the next element (sodium), which indicates that neon is also reluctant to accept another electron. Neon is correspondingly an *inert gas*; under most circumstances, it does not form compounds with other elements. The elements in the column above Ne in Fig. 5 (or below Ne in Appendix E) are also inert gases. (This property depends on the filling of an entire shell, not just a particular subshell. The elements Zn, Cd, and Hg, for example, all have filled *d* states, but all form compounds readily.)

Fluorine, on the other hand, lacks one electron from a filled 2*p* state. Since the filled state is a stable configuration, fluorine readily accepts an electron from another atom to form compounds. Elements in the column above fluorine in Fig. 5 (or below F in Appendix E) behave similarly; they are collectively known as *halogens*. For another example, sodium has a single electron in the 3*s* state. This electron is not particularly tightly bound, and sodium can give up that electron in forming chemical compounds with other atoms (as in NaCl, for example). The elements with a single *s* electron, called the *alkali* elements, have similar properties.

Ionization Energy

The energy needed to remove the outermost electron from an atom is called its *ionization energy*. Figure 6 shows the ionization energies of the elements. Note the regular behavior that is consistent with the electron configurations. For each shell, the ionization energy rises gradually and reaches a maximum at an inert gas, and there is a sharp drop for the alkali element that follows. The element Na, for example, has an ionization energy of 5.14 eV. To remove a *second* electron from Na, however, takes nearly an order of magnitude more energy (47.3 eV); with one electron removed, an Na ion has an electron configuration similar to inert Ne, consisting of a filled shell, in which the electrons are more tightly bound.

There are occasionally small irregularities in the order in which the states are filled. For example, from Fig. 5 we would expect copper (*Z* = 29) to have the outer configuration $4s^2 3d^9$. However, it is energetically favorable for one of the 4*s* electrons to complete the filling of the 3*d* state; as a result, the outer configuration of Cu is $4s^1 3d^{10}$. The single 4*s* electron is responsible for the large electrical conductivity of copper. A similar situation occurs for silver (*Z* = 47) and gold (*Z* = 79).

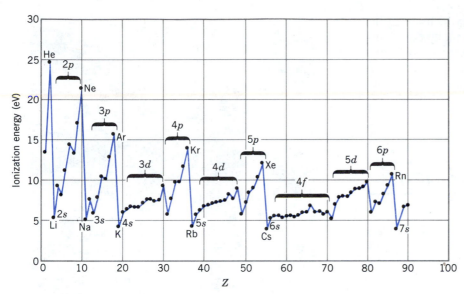

Figure 6 The ionization energies of the elements plotted against their atomic number. Subshell labels are indicated.

Excited States and Optical Transitions

So far we have discussed only the minimum-energy or ground-state configuration of atoms. When we add energy to the atom, such as when we place it in an electric discharge tube or illuminate it with radiation, we can cause the electrons to move to higher states. If we supply sufficient energy, it is possible to remove an electron completely, thereby ionizing the atom. If an inner electron is removed, the ensuing filling of levels gives the characteristic x rays, as we have discussed in Section 52-1.

The energy differences are typically of the order of eV between the minimum-energy state of the outer electron and the next higher-lying states to which it can be excited. When the electron drops back to its lowest energy, the atom emits radiation of energy in the eV range, that is,

visible light. For this reason, such changes in the state of the electron are called *optical transitions*. Using the wave functions corresponding to the various states, it is possible to calculate the relative probability for different transitions to occur. When we do so, we find that transitions that change *l* by one unit are strongly favored over transitions that change *l* by any other amount. Such a conclusion is called a *selection rule*. For all electromagnetic transitions in atoms (optical, x-ray, and so forth), the selection rule is

$$\Delta l = \pm 1. \tag{6}$$

Selection rules are often not absolute. In atoms it is possible to observe transitions corresponding to other changes in *l*; they are just far less likely to occur.

Figure 7 shows some excited states in sodium and some

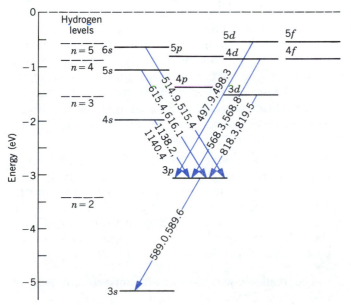

Figure 7 The excited states of sodium. Some emitted radiations are indicated. Note the operation of the $\Delta l = \pm 1$ selection rule.

transitions that may occur (not all of which are in the optical region). Most of the states are actually close-lying doublets, which give two spectral lines of nearly equal energies, such as the familiar sodium doublet, which, as you can see from Fig. 7, corresponds to the electron making a transition from the first excited state ($3p$) back to the ground state.

52-5   LASERS AND LASER LIGHT

In the late 1940s and again in the early 1960s quantum physics made two enormous contributions to technology, the transistor and the laser. The first stimulated the growth of *electronics,* which deals with the interaction (at the quantum level) between electrons and bulk matter. The laser has led to a new field—sometimes called *photonics*—which deals with the interaction (again at the quantum level) between photons and bulk matter.

To see the importance of lasers, let us look at some of the characteristics of laser light (see Fig. 8). We shall compare it as we go along with the light emitted by such sources as a tungsten filament lamp (continuous spectrum) or a neon gas discharge tube (line spectrum). We shall see that referring to laser light as "the light fantastic" goes far beyond whimsy.

1. *Laser light is highly monochromatic.* Tungsten light, spread over a continuous spectrum, gives us no basis for comparison. The light from selected lines in a gas discharge tube, however, can have wavelengths in the visible region that are precise to about 1 part in 10^6. The sharp-

Figure 9   The NOVA laser room at the Lawrence Livermore National Laboratory. These lasers, with a power of about 10^{14} W, are used in controlled thermonuclear fusion research (see Section 55-10).

ness of definition of laser light can easily be a thousand times greater, or 1 part in 10^9.

2. *Laser light is highly coherent.* Wavetrains for laser light may be several hundred kilometers long. Interference fringes can be set up by combining two beams that have followed separate paths whose lengths differ by as much as this amount. The corresponding coherence length for light from a tungsten filament lamp or a gas discharge tube is typically considerably less than 1 m.

3. *Laser light is highly directional.* A laser beam departs from strict parallelism only because of diffraction effects, determined (see Section 46-4) by the wavelength and the diameter of the exit aperture. Light from other sources can be made into an approximately parallel beam by a lens or a mirror, but the beam divergence is much greater than for laser light. For example, focused light from a tungsten filament source forms a beam, the angular divergence of which is determined by the spatial extent of the filament.

4. *Laser light can be sharply focused.* This property is related to the parallelism of the laser beam. As for star light, the size of the focused spot for a laser beam is limited only by diffraction effects and not by the size of the source. Flux densities for focused laser light of 10^{15} W/cm² are readily achieved. An oxyacetylene flame, by contrast, has a flux density of only 10^3 W/cm².

The smallest lasers, used for telephone communication over optical fibers, have as their active medium a semi-

Figure 8   Laser beams light up the sky.

Figure 10   A Michelson interferometer using lasers at the National Institute of Standards and Technology, used to measure the *x*, *y*, *z* coordinates of a point in space with extreme precision.

conducting gallium arsenide crystal about the size of a pinhead. The largest lasers, used for laser fusion research (see Fig. 9), fill a large building. They can generate pulses of laser light of 10^{-10} s duration, which have a power level of 10^{14} W during the pulse. This is about 100 times the total power-generating capacity of all the electric power stations on Earth.

Other laser uses include spot-welding detached retinas, drilling tiny holes in diamonds for drawing fine wires, cutting cloth (50 layers at a time, with no frayed edges) in the garment industry, precision surveying, precise length measurements by interferometry, precise fluid-flow velocity measurements using the Doppler effect, and the generation of holograms (see Section 47-5).

Figure 10 shows another example of laser technology, namely, a facility at the National Institute of Standards and Technology used to measure the *x*, *y*, and *z* coordinates of a point, by laser interference techniques, with a precision of $\pm 2 \times 10^{-8}$ m ($= \pm 20$ nm). It is used for measuring the dimensions of special three-dimensional gauges, which, in turn, are used in industry to check the dimensional accuracy of complicated machined parts. A number of laser beams, visible by scattered light, appear in the figure.

52-6   EINSTEIN AND THE LASER

In 1917 Einstein introduced into physics a new concept, that of *stimulated emission,* which we shall define and discuss below. Even though the first operating laser did not appear until 1960, the groundwork for its invention was put in place by Einstein's work. The importance of stimulated emission is indicated by the name "laser," which is an acronym for **l**ight **a**mplification by the **s**timu**l**ated **e**mission of **r**adiation.

What was Einstein working on when the concept of stimulated emission occurred to him? Nothing other than the cavity radiation problem, which in the hands of Planck and others established the new science of quantum mechanics. In 1917 Einstein succeeded in deriving the Planck radiation law in terms of beautifully simple assumptions and in a way that made quite clear the role of energy quantization and the photon concept.*

It is interesting that Einstein was also thinking deeply about this same fundamental cavity radiation problem when, in 1905, he first proposed the concept of the photon and realized that the photoelectric effect could be explained with its use. We learn from both of these examples that practical devices of major importance can flow from a concern over problems that seem to have no relevance to technology. When you next see a photoelectric elevator door opener or listen to a compact-disc stereo system, think of Einstein.

Now let us take a look at three processes that involve the interaction between matter and radiation. Two of them, absorption and spontaneous emission, have long been familiar; the third is stimulated emission.

---

* See Robert Resnick and David Halliday, *Basic Concepts in Relativity and Early Quantum Theory,* 2nd edition (Wiley, 1985), Supplementary Topic E.

1. *Absorption.* Figure 11a suggests an atomic system in the lower of two possible states, of energies E_1 and E_2. A continuous spectrum of radiation is present. Let a photon from this radiation field approach the two-level atom and interact with it, and let the associated frequency v of the photon be such that

$$hv = E_2 - E_1. \tag{7}$$

The result is that the photon vanishes and the atomic system moves to its upper energy state. We call this process *absorption*.

2. *Spontaneous emission.* In Fig. 11b the atomic system is in its upper state and there is no radiation nearby. After a mean time τ, this (isolated) atomic system moves of its own accord to the state of lower energy, emitting a photon of energy $hv (= E_2 - E_1)$ in the process. We call this process *spontaneous emission,* in that no outside influence triggered the emission.

Normally the mean life τ for spontaneous emission by excited atoms is of the order of 10^{-8} s. However, there are some states for which τ is much longer, perhaps 10^{-3} s. We call such states *metastable;* they play an essential role in laser operation. (They have such long lifetimes because they can emit radiation only through processes that violate the selection rule of Eq. 6.)

The light from a glowing lamp filament is generated by spontaneous emission. Photons produced in this way are totally independent of each other. In particular, they have different directions and phases. Put another way, the light they produce has a low degree of coherence.

3. *Stimulated emission.* In Fig. 11c the atomic system is again in its upper state, but this time radiation of frequency given by Eq. 7 is present. As in absorption, a photon of energy hv interacts with the system. The result is

that the system is driven down to its lower state, and there are now *two* photons where only one existed before. We call this process *stimulated emission.*

The emitted photon in Fig. 11c is in every way identical with the "triggering" or "stimulating" photon. It has the same energy, direction, phase, and state of polarization. Furthermore, each of these two photons can cause another stimulated emission event, giving a total of four photons, which can cause additional stimulated emissions, and so on. We can see how a chain reaction of similar processes could be triggered by one such event. This is the "amplification" of the laser acronym. The photons have identical energies, directions, phases, and states of polarization. This is how laser light acquires its characteristics.

Figure 11 refers to the interaction with radiation of a single atom. In the usual case, however, we find ourselves dealing with a large number of atoms. For the two-level system of Fig. 11, how many of these atoms will be in level E_1 and how many in level E_2? In any system at thermal equilibrium, the number occupying a state at energy E is determined by the exponential factor $e^{-E/kT}$ in the Maxwell–Boltzmann distribution (see Eqs. 27 and 32 of Chapter 24). The ratio of the number of atoms in the upper level to the number in the lower level is

$$n(E_2)/n(E_1) = e^{-(E_2 - E_1)/kT}. \tag{8}$$

Figure 12a illustrates this situation. The quantity kT is the mean energy of agitation of an atom at temperature T, and we see that the higher the temperature the more atoms — on long-term average — will be "bumped up" by thermal agitation to the level E_2. Because $E_2 > E_1$, the ratio $n(E_2)/n(E_1)$ will always be less than unity, which means that there will always be fewer atoms in the higher

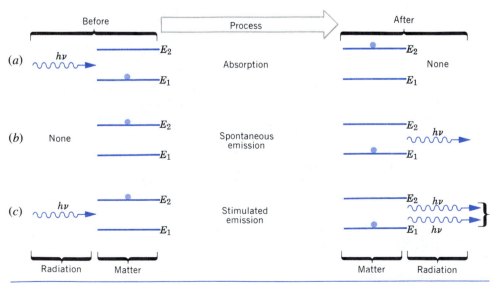

Figure 11 The interaction of matter and radiation for the processes of (a) absorption, (b) spontaneous emission, and (c) stimulated emission.

Figure 12 (*a*) The normal thermal equilibrium distribution of atomic systems occupying one of two possible states. (*b*) An inverted population distribution, which can be obtained using special techniques.

energy level than in the lower. This is what we would expect if the level populations are determined only by the action of thermal agitation.

If we expose a system like that of Fig. 12*a* to radiation, the dominant process—by sheer weight of numbers—will be absorption. However, if the level populations were inverted, as in Fig. 12*b*, the dominant process in the presence of radiation would be stimulated emission, and with it the generation of laser light. A *population inversion* like that of Fig. 12*b* is not a situation that is obtained by thermal processes; we must use clever tricks to bring it about.

52-7 HOW A LASER WORKS

Figure 13 shows schematically how a population inversion can be achieved so that laser action—or "lasing" as it is called—can occur. Atoms from the ground state E_1 are "pumped" up to an excited state E_3, for example by the absorption of light energy from an intense, continuous-spectrum source that surrounds the lasing material.

From E_3 the atoms decay rapidly to a state of energy E_2. For lasing to occur this state must be metastable; that is, it must have a relatively long mean life against decay by spontaneous emission. If conditions are right, state E_2 can then become more heavily populated than state E_1, thus providing the needed population inversion. A stray photon of the right energy can then trigger an avalanche of stimulated emission events, resulting in the production of laser light. A number of lasers using crystalline solids

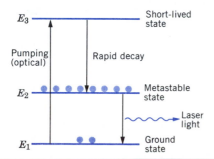

Figure 13 The basic three-level scheme for laser operation. Metastable state E_2 has a greater population than the ground state E_1.

(such as ruby) as a lasing material operate in this three-level mode.

Figure 14 shows the elements of a type of laser that is often found in student laboratories. The glass discharge tube is filled with an 80%–20% mixture of the inert gases helium and neon, the helium being the "pumping" medium and the neon the "lasing" medium. Figure 15 is a simplified version of the level structures for these two atoms. Note that four levels, labeled E_0, E_1, E_2, and E_3, are involved in this lasing scheme, rather than three levels as in Fig. 13.

Pumping is accomplished by setting up an electrically induced gas discharge in the helium–neon mixture. Electrons and ions in this discharge occasionally collide with helium atoms, raising them to level E_3 in Fig. 15. This level is metastable, spontaneous emission to the ground state (level E_0) being very infrequent. Level E_3 in helium (= 20.61 eV) is, by chance, very close to level E_2 in neon (= 20.66 eV), so that, during collisions between helium and neon atoms, the excitation energy of the helium can readily be transferred to the neon. In this way level E_2 in Fig. 15 can become more highly populated than level E_1 in that figure. This population inversion is maintained because (1) the metastability of level E_3 ensures a ready supply of neon atoms in level E_2 and (2) level E_1 decays rapidly (through intermediate stages not shown) to the neon ground state, E_0. Stimulated emission from level E_2 to level E_1 predominates, and red laser light of wavelength 632.8 nm is generated.

Most stimulated emission photons initially produced in the discharge tube of Fig. 14 will not happen to be parallel to the tube axis and will be quickly stopped at the walls. Stimulated emission photons that *are* parallel to the axis, however, can move back and forth through the discharge tube many times by successive reflections from mirrors M_1 and M_2. These photons can in turn cause other stimulated emissions to occur. A chain reaction thus builds up rapidly in this direction, and the inherent parallelism of the laser light results.

Rather than thinking in terms of the photons bouncing back and forth between the mirrors, it is perhaps more useful to think of the entire arrangement of Fig. 14 as an optical resonant cavity that, like an organ pipe for sound waves, can be tuned to be sharply resonant at one (or more) wavelengths.

The mirrors M_1 and M_2 are concave, with their focal points nearly coinciding at the center of the tube. Mirror M_1 is coated with a dielectric film whose thickness is carefully adjusted to make the mirror as close as possible to totally reflective at the wavelength of the laser light; see Section 45-4. Mirror M_2, on the other hand, is coated so as to be slightly "leaky," so that a small fraction of the laser light can escape at each reflection to form the useful beam.

The windows W in Fig. 14, which close the ends of the discharge tube, are slanted so that their normals make an

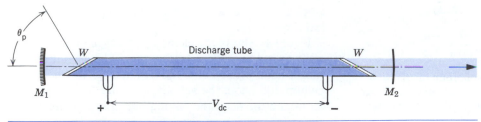

Figure 14 The basic elements of a helium–neon gas laser.

angle θ_p, the Brewster angle, with the tube axis, where

$$\tan \theta_p = n, \qquad (9)$$

n being the index of refraction of the glass at the wavelength of the laser light. In Section 48-3 we showed that such windows transmit light without loss by reflection, provided only that the light is polarized with its plane of polarization in the plane of Fig. 14. If the windows were square to the tube ends, beam loss by reflection (about 4% from each surface of each window) would make laser operation impossible.

Sample Problem 4 A three-level laser of the type shown in Fig. 13 emits laser light at a wavelength of 550 nm, near the center of the visible band. (*a*) If the optical pumping mechanism is shut off, what will be the ratio of the population of the upper level (energy E_2) to that of the lower level (energy E_1)? Assume that $T = 300$ K. (*b*) At what temperature for the conditions of (*a*) would the ratio of populations be $\frac{1}{2}$?

Solution (*a*) From the Bohr frequency condition, the energy difference between the two levels is given by

$$E_2 - E_1 = h\nu = \frac{hc}{\lambda}$$

$$= \frac{(6.63 \times 10^{-34} \text{ J} \cdot \text{s})(3.00 \times 10^8 \text{ m/s})}{(550 \times 10^{-9} \text{ m})(1.60 \times 10^{-19} \text{ J/eV})}$$

$$= 2.26 \text{ eV}.$$

The mean energy of thermal agitation is equal to

$$kT = (8.62 \times 10^{-5} \text{ eV/K})(300 \text{ K}) = 0.0259 \text{ eV}.$$

From Eq. 8 we have then, for the desired ratio,

$$n(E_2)/n(E_1) = e^{-(E_2 - E_1)/kT}$$

$$= e^{-(2.26 \text{ eV})/(0.0259 \text{ eV})} = e^{-87.3} = 1.3 \times 10^{-38}.$$

This is an incredibly small number. It is not unreasonable, however. An atom whose mean thermal agitation energy is only 0.0259 eV will not often impart an energy of 2.26 eV (87 times as great) to another atom in a collision.

(*b*) Setting the ratio in Eq. 8 equal to $\frac{1}{2}$, taking the natural logarithm of each side, and solving for T yields

$$T = \frac{E_2 - E_1}{k(\ln 2)} = \frac{2.26 \text{ eV}}{(8.62 \times 10^{-5} \text{ eV/K})(0.693)}$$

$$= 37,800 \text{ K}.$$

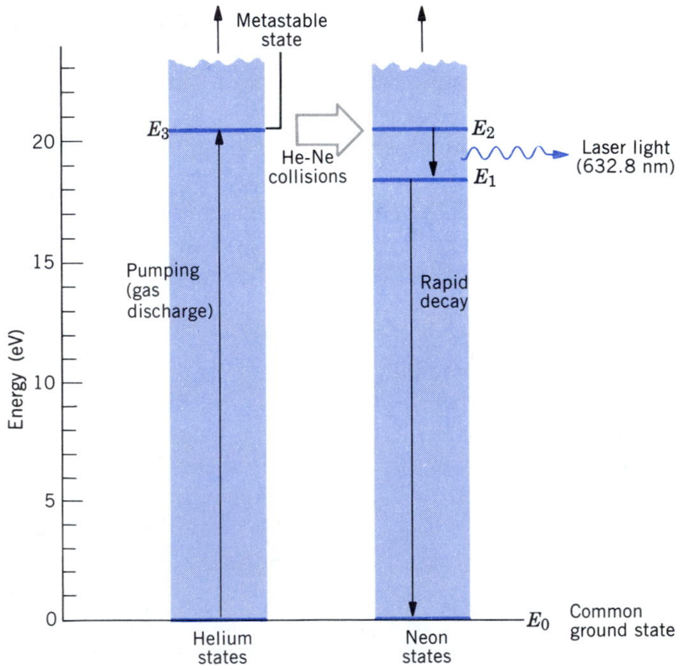

Figure 15 The atomic levels involved in the operation of a He–Ne gas laser.

This is much hotter than the surface of the Sun. It is clear that, if we are to invert the populations of these two levels, a special mechanism is needed. Without population inversion, lasing is not possible.

Sample Problem 5   A pulsed ruby laser has as its active element a synthetic ruby crystal in the form of a cylinder 6 cm long and 1 cm in diameter. Ruby consists of Al_2O_3 in which — in this case — one aluminum ion in every 3500 has been replaced by a chromium ion, Cr^{3+}. It is in fact the optical absorption properties of this small chromium "impurity" that account for the characteristic color of ruby. These same ions also account for the lasing ability of ruby, which occurs — by the three-level mechanism of Fig. 13 — at a wavelength of 694.4 nm.

Suppose that *all* the Cr^{3+} ions are in a metastable state corresponding to state E_2 of Fig. 13 and that *none* are in the ground state E_1. How much energy is available for release in a single pulse of laser light if *all* these ions revert to the ground state in a single stimulated emission chain reaction episode? Our answer will be an upper limit only because the conditions postulated cannot be realized in practice. The density ρ of Al_2O_3 is 3700 kg/m^3, and its molar mass M is 0.102 kg/mol.

Solution   The number of Al^{3+} ions is

$$N_{Al} = \frac{2N_A m}{M} = \frac{2N_A \rho V}{M},$$

where m is the mass of the ruby cylinder and the factor 2 accounts for there being two aluminum ions in each "molecule" of Al_2O_3. The volume V is

$$V = (\pi/4)(1.0 \times 10^{-2}\ m)^2(6.0 \times 10^{-2}\ m)$$
$$= 4.7 \times 10^{-6}\ m^3.$$

Thus

$$N_{Al} = \frac{(2)(6.0 \times 10^{23}/mol)(3.7 \times 10^3\ kg/m^3)(4.7 \times 10^{-6}\ m^3)}{0.102\ kg/mol}$$
$$= 2.1 \times 10^{23}.$$

The number of Cr^{3+} ions is then

$$N_{Cr} = \frac{N_{Al}}{3500} = 6.0 \times 10^{19}.$$

The energy of the stimulated emission photon is

$$E = h\nu = \frac{hc}{\lambda} = \frac{(4.1 \times 10^{-15}\ eV \cdot s)(3.0 \times 10^8\ m/s)}{694 \times 10^{-9}\ m} = 1.8\ eV,$$

and the total available energy per laser pulse is

$$U = N_{Cr}E = (6.0 \times 10^{19})(1.8\ eV)(1.6 \times 10^{-19}\ J/eV) = 17\ J.$$

Such large pulse energies have indeed been achieved, but only by much more elaborate laser arrangements than that described here.

In this example we have postulated an ideal circumstance, namely, a total population inversion, in which the ground state remains virtually unpopulated. The actual population inversion in a working ruby laser will be very much less than total. For this and other reasons the pulse energy in practice will be very much less than the upper limit calculated above.

52-8 MOLECULAR STRUCTURE

Understanding the structure of atoms is the first step in the process that eventually leads to the understanding of the structure of macroscopic objects. The next step is to understand how atoms join together to form molecules.

The force responsible for binding atoms together in molecules is the same electrostatic force that binds electrons in atoms. However, atoms are ordinarily electrically neutral and would thus exert no electrostatic force on one another. To have molecular bonds between atoms, there must therefore occur some readjustment of the electronic structure of the atoms.

Consider, for example, a molecule of hydrogen, H_2 (Fig. 16a). The separation between the two protons in a molecule of H_2 is measured to be 0.074 nm, which is comparable to the radius of the lowest electronic orbit in *atomic* hydrogen, 0.0529 nm. The single electron in atomic hydrogen would like to acquire a partner to fill the 1s shell, and therefore the electron from one H atom can be regarded as pairing with the other electron in the same shell. (Of course, the electrons are identical, and in accordance with the quantum rules for indistinguishable particles, we can no longer speak of the electrons from the two original atoms as having separate identities. Instead, a pair of electrons, represented by a two-electron wave func-

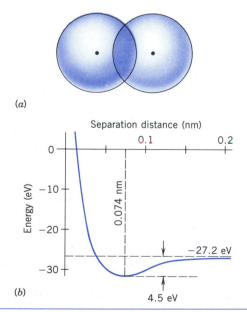

Figure 16   (a) The overlap of the s electrons in H is responsible for the formation of the H_2 molecule. (b) The total energy of the two electrons in the bound state of the H_2 molecule, as a function of the atomic separation distance. When the separation is large, the energy is -27.2 eV (twice the energy of the single electron in atomic hydrogen, -13.6 eV). The minimum energy of the bound molecule is -31.7 eV when the separation is 0.074 nm.

tion, moves in the combined electrostatic field of the two protons.)

We can regard this binding as originating from a sharing of electrons. The electrons "belong to" neither proton. In effect, each proton attracts the pair of electrons, and this attraction is sufficient to overcome the Coulomb repulsion of the protons. (A similar effect, involving a very different type of shared particle, is responsible for the binding of two protons in the nucleus of an atom.)

This type of molecular bonding, based on shared electrons, is called *covalent* bonding. Molecules containing two atoms of the same element are common examples of those having covalent bonds. A measure of the strength of the bond is the *dissociation energy,* the energy necessary to break the molecule into two neutral atoms. In H_2, the dissociation energy is 4.5 eV, as indicated in Fig. 16b. It is also possible to have covalent bonding in atoms in which the outer electrons are in the *p* shell, such as in the case of N_2 or O_2. In the case of N_2, the sharing of the three $2p$ electrons from each atom gives a total of six $2p$ electrons, a configuration that would (in a single atom) correspond to a filled shell; the N_2 molecule is therefore very stable (the energy needed to dissociate the molecule is 9.8 eV). In O_2, on the other hand, there are eight $2p$ electrons, which is a less stable configuration; the dissociation energy of O_2 is only 5.1 eV. In practical terms, this difference makes O_2 more reactive than N_2. Molecules of O_2 can be broken by relatively modest chemical reactions, as, for example, the oxidation of metals exposed to air. The F_2 molecule (ten $2p$ electrons) is even less stable than O_2; its dissociation energy is only 1.6 eV, less than the energy of photons of visible light, and as a result, F_2 can be dissociated by exposure to light.

Covalent molecular bonds can also be formed between dissimilar atoms, even in cases in which the shared electrons originate from different atomic shells. Bonds between *s* and *p* electrons are common, such as in H_2O (*s* electrons from H and *p* electrons from O, as illustrated in Fig. 17a) and in hydrocarbon compounds (*s* electrons from H and *p* electrons from C). We can regard the *p* electrons as having wave functions with lobes of high probability along the coordinate axes (see Fig. 18 of Chapter 51, for example). An *s* electron can be attached at each of these lobes. In H_2O, for instance, an *s* electron from each H atom attaches to two of the different *p* electrons. We would therefore expect the angle between the bonds in H_2O to be 90°; the measured angle is 104°, indicating that there is some Coulomb repulsion of the H atoms that spreads the bond angle. Ammonia (NH_3) is another example of this type of structure, illustrated in Fig. 17b.

In carbon, the $2s$ and $2p$ electrons are mixed, giving C an effective valence of 4. These four electrons can form a variety of covalent bonds with other atoms, which is responsible for the diversity of organic compounds, from simple molecules such as methane (CH_4) to the complex molecules that form the basis of living things.

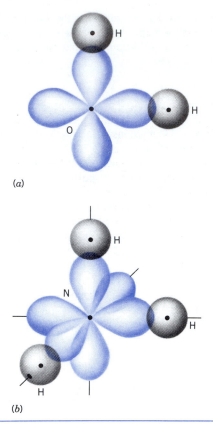

(a)

(b)

Figure 17   (a) The overlap of *s* electrons from H and *p* electrons from O in a molecule of H_2O. (b) The overlap of *s* electrons from H and *p* electrons from N in a molecule of NH_3.

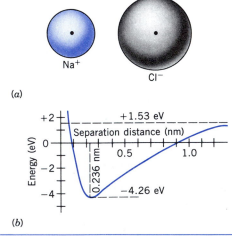

(a)

(b)

Figure 18   (a) Ionic bonding in NaCl. Note that there is no appreciable overlap of the electron distributions. (b) Binding energy in NaCl. The zero of energy corresponds to Na and Cl *atoms* separated by a large distance. The dashed line represents the energy of Na^+ and Cl^- *ions* separated by a large distance.

At the other extreme from covalent bonds are those in which the electrons are not shared but belong to one atom or another. In a molecule (*not* a solid crystal) of NaCl, the Cl lacks one electron from a complete *p* shell, while the Na has a single valence electron in the *s* shell. As neutral atoms of Na and Cl are brought close together, it becomes energetically favorable for the valence electron from Na to be transferred to Cl, thereby filling its *p* shell. As a result, we have ions Na$^+$ and Cl$^-$, which then exert electrostatic forces on one another and bind together in a molecule of NaCl (Fig. 18*a*). The atoms are prevented from approaching too close to one another, because the Pauli principle does not allow the filled *p* shells to overlap. The stable equilibrium separation is 0.236 nm, and the binding energy (the energy needed to split the molecule into its neutral atoms) is 4.26 eV, as shown in Fig. 18*b*. Molecules

of this type, based on the bonding of ions, are called *ionic molecules.*

Molecular bonding has analogs in the bonding of atoms in solids. There are ionic solids (such as NaCl), which we can regard as being made of assemblies of positive and negative ions. There are also covalent solids, such as diamond, whose structure depends on the overlap of electron wave functions. Other types include molecular solids (such as ice), in which the molecules retain their electronic structure and are bound by much weaker forces based on electric dipole interactions, and metallic solids, in which each atom contributes one or more electrons to a "sea" of electrons that are shared throughout the entire solid. In the next chapter, we consider some solids whose properties can be understood on the basis of this structure.

QUESTIONS

1. What is the origin of the cutoff wavelength λ_{min} of Fig. 1? Why is it an important clue to the photon nature of x rays?

2. In Fig. 2, why is the emitted photon shown moving off in the direction that it is? Could it be shown moving off in any other direction? Explain.

3. What are the characteristic x rays of an element? How can they be used to determine the atomic number of an element?

4. Compare Figs. 1 and 3. How can you be sure that the two prominent peaks in Fig. 1 do indeed correspond numerically with the two transitions similarly labeled in Fig. 3?

5. Can atomic hydrogen be caused to emit x rays? If so, describe how. If not, why not?

6. How does the x-ray energy-level diagram of Fig. 3 differ from the energy-level diagram for hydrogen, displayed in Fig. 4 of Chapter 51? In what respects are the two diagrams similar?

7. When extended to higher atomic numbers, the Moseley plot of Fig. 4 is not a straight line but is concave upward. Does this affect the ability to assign atomic numbers to the elements?

8. Why is it that Bohr theory, which does not work very well even for helium ($Z = 2$), gives such a good account of the characteristic x-ray spectra of the elements, or at least of that portion that originates deep within the atom?

9. Why does the characteristic x-ray spectrum vary in a systematic way from element to element but the spectrum in the visible range does not?

10. Why do you expect the wavelengths of radiations generated by transitions deep within the atom to be shorter than those generated by transitions occurring in the outer fringes of the atom?

11. Given the characteristic x-ray spectrum of a certain element, containing a number of lines, how would you go about identifying and labeling them?

12. On what quantum numbers does the energy of an electron in (*a*) a hydrogen atom and (*b*) a vanadium atom depend?

13. The periodic table of the elements was based originally on atomic mass, rather than on atomic number, the latter concept having not yet been developed. Why were such early tables as successful as they proved to be? In other words, why is the atomic mass of an element (roughly) proportional to its atomic number?

14. How does the structure of the periodic table support the need for a fourth quantum number, corresponding to electron spin?

15. If there were only three quantum numbers (that is, if the electron had no spin), how would the chemical properties of helium be different?

16. Explain why the effective radius of a helium atom is less than that of a hydrogen atom.

17. Why does it take more energy to remove an electron from neon ($Z = 10$) than from sodium ($Z = 11$)?

18. What can Fig. 5 tell you about why the inert gases are so chemically stable?

19. Does it make any sense to assign quantum numbers to a vacancy in an otherwise filled subshell?

20. Why do the lanthanide series of elements (see Appendix E) have such similar chemical properties? How can we justify putting them all into a single square of the periodic table? Why is it that, in spite of their similar chemical properties, they can be so easily sorted out by measuring their characteristic x-ray spectra?

21. In your own words, state the minimum energy principle for atom building and give a physical argument in support of it.

22. Figure 5 shows that the 2*s* state is lower in energy than the 2*p* state. Can you explain why this should be so, basing your argument on the radial probability densities of the two states (see Figs. 17 and 18 of Chapter 51)?

23. If you start with a bare nucleus and fill in the electrons to form an atom in its ground state, the energies of the unfilled levels change as you proceed. Why do they change? Do they increase or decrease in energy as electrons are added?

24. Why is focused laser light inherently better than focused light from a tiny incandescent lamp filament for delicate surgical jobs such as spot-welding detached retinas?

25. Laser light forms an almost parallel beam. Does the intensity of such light fall off as the inverse square of the distance from the source?

26. In what ways are laser light and star light similar? In what ways are they different?

27. Arthur Schawlow, one of the laser pioneers, invented a typewriter eraser, based on focusing laser light on the unwanted character. Can you imagine what its principle of operation is?

28. In what ways do spontaneous emission and stimulated emission differ?

29. We have spontaneous emission and stimulated emission. From symmetry, why don't we also have spontaneous and stimulated absorption? Discuss in terms of Fig. 11.

30. Why is a population inversion necessary between two atomic levels for laser action to occur?

31. What is a metastable state? What role do such states play in the operation of a laser?

32. Comment on this statement: "Other things being equal, a four-level laser scheme such as that of Fig. 15 is preferable to a three-level scheme such as that of Fig. 13 because, in the three-level scheme, one-half of the population of atoms in level E_1 must be moved to state E_2 before a population inversion can even begin to occur."

33. Comment on this statement: "In the laser of Fig. 14, only light whose plane of polarization lies in the plane of that figure is transmitted through the right-hand window. Therefore, half of the energy potentially available is lost." (*Hint*: Is this second statement really true? Consider what happens to photons whose effective plane of polarization is at right angles to the plane of Fig. 14. Do such photons participate fully in the stimulated emission amplification process?)

34. A beam of light emerges from an aperture in a "black box" and moves across your laboratory bench. How could you test this beam to find out the extent to which it is coherent over its cross section? How could you tell (without opening the box) whether or not the concealed light source is a laser?

35. Why is it difficult to build an x-ray laser?

PROBLEMS

Section 52-1 The X-Ray Spectrum

1. Show that the short-wavelength cutoff in the continuous x-ray spectrum is given by

$$\lambda_{\min} = 1240 \text{ pm}/V,$$

where V is the applied potential difference in kilovolts.

2. Determine Planck's constant from the fact that the minimum x-ray wavelength produced by 40.0-keV electrons is 31.1 pm.

3. What is the minimum potential difference across an x-ray tube that will produce x rays with a wavelength of 0.126 nm?

4. In Fig. 1, the x rays shown are produced when 35.0-keV electrons fall on a molybdenum target. If the accelerating potential is maintained at 35.0 kV but a silver target ($Z = 47$) is substituted for the molybdenum target, what values of (a) $\lambda_{\min}$, (b) λ_{K_β}, and (c) λ_{K_α} result? The K, L, and M atomic x-ray levels for silver (compare with Fig. 3) are 25.51, 3.56, and 0.53 keV.

5. Electrons bombard a molybdenum target, producing both continuous and characteristic x rays as in Fig. 1. In that figure the energy of the incident electrons is 35.0 keV. If the accelerating potential applied to the x-ray tube is increased to 50.0 kV, what values of (a) $\lambda_{\min}$, (b) λ_{K_α}, and (c) λ_{K_β} result?

6. The wavelength of the K_α line from iron is 19.3 pm. (a) Find the energy difference between the two states of the iron atom (see Fig. 3) that give rise to this transition. (b) Find the corresponding energy difference for the hydrogen atom. Why is the difference so much greater for iron than for hydrogen? (*Hint*: In the hydrogen atom the K shell corresponds to $n = 1$ and the L shell to $n = 2$.)

7. From Fig. 1, calculate approximately the energy difference

$E_L - E_M$ for the x-ray atomic energy levels of molybdenum. Compare with the result that may be found from Fig. 3.

8. Find the minimum potential difference that must be applied to an x-ray tube to produce x rays with a wavelength equal to the Compton wavelength of the electron. (See Problem 54 of Chapter 49.)

9. X rays are produced in an x-ray tube by a target potential of 50.0 kV. If an electron makes three collisions in the target before coming to rest and loses one-half of its remaining kinetic energy on each of the first two collisions, determine the wavelengths of the resulting photons. Neglect the recoil of the heavy target atoms.

10. A tungsten target ($Z = 74$) is bombarded by electrons in an x-ray tube. (a) What is the minimum value of the accelerating potential that will permit the production of the characteristic K_β and K_α lines of tungsten? (b) For this same accelerating potential, what is the value of $\lambda_{\min}$? (c) Calculate λ_{K_β} and λ_{K_α}. The K, L, and M atomic x-ray levels for tungsten (see Fig. 3) are 69.5, 11.3, and 2.3 keV, respectively.

11. A molybdenum target ($Z = 42$) is bombarded with 35.0-keV electrons and the x-ray spectrum of Fig. 1 results. Here $\lambda_{K_\beta} = 63$ pm and $\lambda_{K_\alpha} = 71$ pm. (a) What are the corresponding photon energies? (b) It is desired to filter these radiations through a material that will absorb the K_β line much more strongly than it will absorb the K_α line. What substance(s) would you use? The K ionization energies for molybdenum and for four neighboring elements are as follows:

Z	40	41	42	43	44
Element	Zr	Nb	Mo	Tc	Ru
E_K (keV)	18.00	18.99	20.00	21.04	22.12

(*Hint*: A substance will selectively absorb one of two x radiations more strongly if the photons of one have enough energy to eject a *K* electron from the atoms of the substance but the photons of the other do not.)

12. The binding energies of *K*-shell and *L*-shell electrons in copper are 8.979 keV and 0.951 keV, respectively. If a K_α x ray from copper is incident on a sodium chloride crystal and gives a first-order Bragg reflection at 15.9° when reflected from the alternating planes of the sodium atoms, what is the spacing between these planes?

13. A 20.0-keV electron is brought to rest by undergoing two successive bremsstrahlung events, thus transferring its kinetic energy into the energy of two photons. The wavelength of the second photon is 130 pm greater than the wavelength of the first photon to be emitted. (*a*) Find the energy of the electron after its first deceleration. (*b*) Calculate the wavelengths and energies of the two photons.

14. In an x-ray tube an electron moving initially at a speed of 2.73×10^8 m/s slows down in passing near a nucleus. A single photon of energy 43.8 keV is emitted. Find the final speed of the electron. (Relativity must be taken into account; ignore the energy imparted to the nucleus.)

15. Show that a moving electron cannot spontaneously emit an x-ray photon in free space. A third body (atom or nucleus) must be present. Why is it needed? (*Hint*: Examine conservation of total energy and of momentum.)

Section 52-2 X Rays and the Numbering of the Elements

16. Using the Bohr theory, calculate the ratio of the wavelengths of the K_α line for niobium (Nb) to that of gallium (Ga). Take needed data from the periodic table.

17. Here are the K_α wavelengths of a few elements:

Ti	27.5 pm	Co	17.9 pm
V	25.0	Ni	16.6
Cr	22.9	Cu	15.4
Mn	21.0	Zn	14.3
Fe	19.3	Ga	13.4

Make a Moseley plot (see Fig. 4) and verify that its slope agrees with the value calculated in Sample Problem 2.

Section 52-4 The Periodic Table

18. If a uranium nucleus ($Z = 92$) had only a single electron, what would be the radius of its ground-state orbit, according to Bohr's theory?

19. Two electrons in lithium ($Z = 3$) have as their quantum numbers n, l, m_l, m_s, the values $1, 0, 0, \pm\frac{1}{2}$. (*a*) What quantum numbers can the third electron have if the atom is to be in its ground state? (*b*) If the atom is to be in its first excited state?

20. By inspection of Fig. 5, what do you think might be the atomic number of the next higher inert gas above radon ($Z = 86$)?

21. If the electron had no spin, and if the Pauli exclusion principle still held, how would the periodic table be affected? In particular, which of the present elements would be inert gases?

22. Suppose there are two electrons in the same system, both of which have $n = 2$ and $l = 1$. (*a*) If the exclusion principle did not apply, how many combinations of states are conceivably possible? (*b*) How many states does the exclusion principle forbid? Which ones are they?

23. In the alkali metals there is one electron outside a closed shell. (*a*) Using the Bohr theory, calculate the effective charge number of the nucleus as seen by the valence electron in sodium (ionization energy = 5.14 eV) and potassium (ionization energy = 4.34 eV). (*b*) For each element, what fraction is this of the actual nuclear charge *Z*? Needed quantum numbers can be found on Fig. 5.

Section 52-7 How a Laser Works

24. A ruby laser emits light at wavelength 694.4 nm. If a laser pulse is emitted for 12.0 ps and the energy release per pulse is 150 mJ, (*a*) what is the length of the pulse, and (*b*) how many photons are in each pulse?

25. Lasers have become very small as well as very large. The active volume of a laser constructed of the semiconductor GaAlAs has a volume of only 200 $(\mu m)^3$ (smaller than a grain of sand) and yet it can continuously deliver 5.0 mW of power at 0.80-μm wavelength. Calculate the production rate of photons.

26. A He–Ne laser emits light at a wavelength of 632.8 nm and has an output power of 2.3 mW. How many photons are emitted each minute by this laser when operating?

27. It is entirely possible that techniques for modulating the frequency or amplitude of a laser beam will be developed so that such a beam can serve as a carrier for television signals, much as microwave beams do now. Assume also that laser systems will be available whose wavelengths can be precisely "tuned" to anywhere in the visible range, that is, in the range 400 nm $< \lambda <$ 700 nm. If a television channel occupies a bandwidth of 10 MHz, how many channels could be accommodated with this laser technology? Comment on the intrinsic superiority of visible light to microwaves as carriers of information.

28. A hypothetical atom has energy levels evenly spaced by 1.2 eV in energy. For a temperature of 2000 K, calculate the ratio of the number of atoms in the 13th excited state to the number in the 11th excited state.

29. A particular (hypothetical) atom has only two atomic levels, separated in energy by 3.2 eV. In the atmosphere of a star there are 6.1×10^{13} of these atoms per cm³ in the excited (upper) state and 2.5×10^{15} per cm³ in the ground (lower) state. Calculate the temperature of the star's atmosphere.

30. A population inversion for two levels is often described by assigning a negative Kelvin temperature to the system. Show that such a negative temperature would indeed correspond to an inversion. What negative temperature would describe the system of Sample Problem 4 if the population of the upper level exceeds that of the lower by 10.0%?

31. An atom has two energy levels with a transition wavelength of 582 nm. At 300 K, 4.0×10^{20} atoms are in the lower state. (*a*) How many occupy the upper state, under conditions of thermal equilibrium? (*b*) Suppose, instead, that 7.0×10^{20} atoms are pumped into the upper state, with

4.0×10^{20} in the lower state. How much energy could be released in a single laser pulse?

32. The mirrors in the laser of Fig. 14 form a cavity in which standing waves of laser light are set up. In the vicinity of 533 nm, how far apart in wavelength are the adjacent allowed operating modes? The mirrors are 8.3 cm apart.

33. A high-powered laser beam ($\lambda = 600$ nm) with a beam diameter of 11.8 cm is aimed at the Moon, 3.82×10^5 km distant. The spreading of the beam is caused only by diffraction effects. The angular location of the edge of the central diffraction disk (see Eq. 11 in Chapter 46) is given by

$$\sin \theta = \frac{1.22 \, \lambda}{d},$$

where d is the diameter of the beam aperture. Find the diameter of the central diffraction disk at the Moon's surface.

34. The beam from an argon laser ($\lambda = 515$ nm) has a diameter d of 3.00 mm and a power output of 5.21 W. The beam is focused onto a diffuse surface by a lens of focal length $f = 3.50$ cm. A diffraction pattern such as that of Fig. 13 in Chapter 46 is formed. (a) Show that the radius of the central disk is given by

$$R = \frac{1.22 \, f\lambda}{d}.$$

The central disk can be shown to contain 84% of the incident power. Calculate (b) the radius R of the central disk, and the average power flux density (c) in the incident beam and (d) in the central disk.

35. The use of lasers for defense against ballistic missiles is being studied. A laser beam of intensity 120 MW/m² would probably burn into and destroy a hardened (nonspinning) missile in about 1 s. (a) If the laser has a power output of 5.30 MW, a wavelength of 2.95 μm, and a beam diameter of 3.72 m (a very powerful laser indeed), would it destroy a missile at a distance of 3000 km? (b) If the wavelength could be changed, what minimum value would work? (c) If the wavelength of the laser could not be changed, what would be the destructive range of the laser in (a)? Use the equation for central disk given in Problem 34 and take the focal length to be the distance to the target.

36. The active medium in a particular ruby laser ($\lambda = 694$ nm) is a synthetic ruby crystal 6.00 cm long and 1.0 cm in diameter. The crystal is silvered at one end and — to permit the formation of an external beam — only partially silvered at the other. (a) Treat the crystal as an optical resonant cavity in analogy to a closed organ pipe and calculate the number of standing-wave nodes there are along the crystal axis. (b) By what amount $\Delta \nu$ would the beam frequency have to shift to increase this number by one? Show that $\Delta \nu$ is just the inverse of the travel time of light for one round trip back and forth along the crystal axis. (c) What is the corresponding fractional frequency shift $\Delta \nu / \nu$? The appropriate index of refraction is 1.75.

CHAPTER 53

ELECTRICAL CONDUCTION IN SOLIDS

We have seen in the previous two chapters how well quantum theory works when we apply it to individual atoms. In this chapter we show that this powerful theory works equally well when we apply it to collections of atoms in the form of solids.

Every solid has an enormous range of properties that we can choose to examine: Is it soft or hard? Can it be hammered into a thin sheet or drawn into a fine wire? Is it transparent? What kind of waves travel through it and at what speeds? Does it conduct heat? What are its magnetic properties? What is its crystal structure? And so on. In each case, we should like to use quantum theory to understand the measured properties.

In this chapter, we focus on one particular property of solids: conduction of electricity. We discuss the classification of solids into conductors, insulators, semiconductors, and superconductors, and we show how quantum theory provides the framework for understanding why some materials behave one way and some another.

53-1 CONDUCTION ELECTRONS IN A METAL

An isolated copper atom has 29 electrons. In solid copper, 28 of these are held close to their lattice sites by electromagnetic forces and are not free to move throughout the volume of the solid. The remaining electron *is* free to so move and, if we apply an emf between the ends of a copper wire, it is these *conduction electrons* (one per atom) that constitute the current that is set up in the wire.

In Section 32-5 we looked at this problem from the point of view of classical physics, comparing the conduction electrons in a metal cube to the atoms of a gas confined to a cubical box. Using this (classical) *free electron gas* model, we derived an expression for the resistivity of the metal. It is (see Eq. 20 of Chapter 32)

$$\rho = \frac{m}{ne^2\tau},\qquad(1)$$

in which m is the mass and e the charge of the electron, n is the number of conduction electrons per unit volume, and τ is the average time between collisions of the electrons with the lattice.

We showed in Section 32-5 that τ is essentially constant, independent of whether or not an electric field has

been set up inside the cube by an externally applied emf. Thus the resistivity ρ is independent of the applied electric field, which is another way of saying that metals obey Ohm's law.

Although this derivation of the form of Ohm's law is a fine achievement for classical physics, it is no simple matter to go much further. Also, there is one problem—the heat capacities of metals—about which this classical theory *does* have something to say, but unfortunately its predictions do not agree with experiment. Looking beyond this level of concern, it is hard to imagine how we would explain something as complicated as a transistor on the basis of the classical free electron gas model. We had better see what wave mechanics has to offer.

The first step in solving any wave mechanical problem is to specify the potential energy of the particle—which we take to be a single conduction electron—as a function of its position. As Fig. 6 of Chapter 51 reminds us, we need this information to substitute into the Schrödinger equation. We start with the simplest reasonable assumption, namely, that the potential energy is zero for all points within the cubical metal sample and that it approaches an infinitely great value for all points outside. We are still dealing with a free electron gas, but it is now one that is governed by quantum—rather than classical—rules.

This potential energy reminds us of the problem of an

electron trapped in an infinite well that we solved in Section 50-7. Note, however, two differences: the present problem is three-dimensional, and it involves a well of macroscopic, rather than atomic, dimensions.

We represent a single conduction electron trapped in its metal cube by a (standing) matter wave $\psi(\mathbf{r})$, in which $\mathbf{r}$ is a position vector, and we impose the condition that the probability density $\psi^2(\mathbf{r})$ be zero both at the surface of the cube and at all outside points. This is our way of recognizing that the electron is truly trapped inside the metal cube. Figure 18 of Chapter 50 reminds us that we proceeded in just the same way in the one-dimensional case.

If we impose these boundary conditions on the wave function, the Schrödinger equation tells us that the total energy E of the electron will be quantized, just as it was for an electron trapped in a one-dimensional well. There is a big difference, however. Because our metal cube is so very large on the scale of atomic dimensions, the number of standing matter waves that we can fit into the volume of the cube and still satisfy the boundary requirements is enormous, and the allowed electron energies are extremely close together. Sample Problem 1 shows that, for a cube 1 cm on edge, there are about 10^{20} quantized states that lie between $E = 5$ eV and $E = 5.01$ eV! Compare this with the limited array of well-spaced levels shown, for example, for the hydrogen atom in Fig. 4 of Chapter 51.

We cannot possibly deal with this vast number of states one at a time; we must use statistical methods. Instead of asking, "What is the energy of this state?" we must ask, "How many states have energies that lie in the range E to $E + dE$?"

We have met situations like this before. For example, in describing the speeds of the molecules of an ideal gas in Section 24-3, we saw that the only way to proceed was to pose the question: "How many molecules have speeds that lie in the range v to $v + dv$?"

For the conduction electrons, the number of states (per unit volume of the solid) whose energies lie in the range E to $E + dE$ can be written as $n(E)dE$, where $n(E)$ is a function called the *density of states*. For our (quantum) free electron gas it can be shown to be*

$$n(E) = \frac{8\sqrt{2}\pi m^{3/2}}{h^3}E^{1/2}. \qquad (2)$$

At this stage we are simply counting the states that are available to a single conduction electron. Note that there is nothing in Eq. 2 that depends on the material of which our sample is made. Fitting patterns of standing waves into a cubical box is a purely geometrical problem. In the following section we shall see how to go about filling those states.

** See *Quantum Physics of Atoms, Molecules, Solids, Nuclei, and Particles*, by Robert Eisberg and Robert Resnick (Wiley, 1985) 2nd ed., Section 11-11.*

You may well ask, "If the energies of the allowed states are so close together, why don't we just forget about the quantization and assume a continuous distribution in energy?" The answer, as we shall see in the next section, rests on the fact that the Pauli exclusion principle applies to electrons wherever we find them, whether as orbital electrons in atoms or as conduction electrons in metals. Even though there are many states in our problem, there are also many conduction electrons available to occupy them, and the Pauli principle allows us to put only one electron in each of these states. Thus, even though we cannot easily detect directly the quantized nature of the energies of the conduction electrons, the fact of quantization remains an absolutely central feature and has important consequences.

Sample Problem 1 A cube of copper is 1 cm on edge. How many states are available for its conduction electrons in the energy interval between $E = 5.00$ eV and 5.01 eV? Assume that the conduction electrons behave like a (quantum) free electron gas.

Solution These energy limits are so close together that we can safely say that the answer, on a per unit volume basis, is $n(E)\Delta E$, where $E = 5$ eV and $\Delta E = 0.01$ eV. From Eq. 2 we have

$$n(E) = \frac{8\sqrt{2}\pi m^{3/2}}{h^3}E^{1/2}$$

$$= \frac{(8\sqrt{2}\pi)(9.11 \times 10^{-31}\text{ kg})^{3/2}}{(6.63 \times 10^{-34}\text{ J·s})^3}[(5\text{ eV})(1.6 \times 10^{-19}\text{ J/eV})]^{1/2}$$

$$= 9.48 \times 10^{46}\text{ m}^{-3}\text{J}^{-1} = 1.52 \times 10^{28}\text{ m}^{-3}\text{eV}^{-1}.$$

Note that we must express the energy E in joules before substituting into Eq. 2, even though we wish our final result to be given in terms of electron volts.

The actual number N of states that lie in the range from $E = 5.00$ eV to $E = 5.01$ eV in our cube is, if a is the length of the cube edge,

$$N = n(E)\Delta E\, a^3$$

$$= (1.52 \times 10^{28}\text{ m}^{-3}\text{eV}^{-1})(0.01\text{ eV})(1 \times 10^{-2}\text{ m})^3$$

$$= 1.52 \times 10^{20}.$$

That is, there are 1.52×10^{20} individual energy states between 5.00 eV and 5.01 eV. The average energy interval ΔE_{adj} between adjacent levels in this interval follows readily from

$$\Delta E_{adj} = \frac{\Delta E}{N} = \frac{0.01\text{ eV}}{1.52 \times 10^{20}} \approx 7 \times 10^{-23}\text{ eV}.$$

We conclude that, even in this narrow energy band, there are very many states and they lie exceedingly close together in energy.

Our conclusions are completely independent of the material of the sample. Nor is it important that the sample is cubical; any other shape enclosing the same volume would give the same final result. What we *have* assumed to be true is that the conduction electrons behave like a (quantum) free electron gas. That is, we have assumed that their potential energy is constant (which we have taken to be zero) for all points within the sample. For actual metals this assumption is never strictly true. Nevertheless, our central conclusion holds: many states are available to a conduction electron in a metal and they lie very close together in energy.

53-2   FILLING THE ALLOWED STATES

Now that we have seen how many states there are, we are ready to start filling them with electrons. We went through this process in Section 52-3 in connection with building up the periodic table of the elements. There we saw the central importance of Pauli's exclusion principle, which tells us that we can allocate only one electron to a given state. This powerful principle is just as important for our present problem.

Figure 1*a* shows the density of states given by Eq. 2. This function gives the number of possible states in any energy interval. However, not all those states are occupied. We fill the available states in a metal just as we filled the available states in an atom: we add electrons, one per quantum state, starting at the lowest energy and ending when we have added all the necessary electrons to the metal.

Let us first consider conditions at the absolute zero of temperature. This represents the lowest energy state of our sample, and we achieve it by placing the conduction electrons into the unfilled states that lie lowest in energy. This process is suggested by Fig. 1*b*, which shows the *probability function p(E)*. This function gives the probability of the state at the energy E to be occupied. At $T = 0$, all states below a certain energy are filled ($p = 1$) and all states above that energy are vacant ($p = 0$). The highest occupied state under these conditions is called the *Fermi level*, and its energy, marked E_F in Fig. 1*b*, is called the *Fermi energy*. The Fermi energy for copper, for example, is 7.06 eV.

If we multiply the density $n(E)$ of *available* states by the probability $p(E)$ that those states are occupied, the result is the *density of occupied states, $n_o(E)$*, or

$$n_o(E) = n(E)p(E). \qquad (3)$$

This quantity is plotted in Fig. 1*c*.

The shaded area in Fig. 1*c* represents the total number of occupied states (per unit volume). Finding this area and equating it to the density n of conduction electrons in the metal gives a means to find the Fermi energy. Integrating between the limits of $E = 0$ and $E = E_F$ to find the area, we obtain

$$n = \int_0^{E_F} n_o(E)dE, \qquad (4)$$

or, carrying out the integral,

$$n = \frac{8\sqrt{2}\pi m^{3/2}}{h^3} \int_0^{E_F} E^{1/2}\, dE = \left(\frac{8\sqrt{2}\pi m^{3/2}}{h^3}\right)\left(\tfrac{2}{3}E_F^{3/2}\right).$$

Solving for E_F gives

$$E_F = \frac{h^2}{8m}\left(\frac{3n}{\pi}\right)^{2/3}. \qquad (5)$$

A glance at Fig. 1*c* should be enough to shatter the popular misconception that all motion ceases at the absolute zero of temperature. We see that, entirely because of Pauli's exclusion principle, the electrons are stacked up in energy from zero to the Fermi energy. The *average* energy for the conditions of Fig. 1*c* turns out to be about 4.2 eV. By comparison, the average translational kinetic energy of a molecule of an ideal gas at room temperature is only 0.025 eV. The conduction electrons in a metal have plenty of energy at the absolute zero!

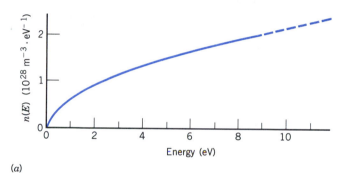

(*a*)

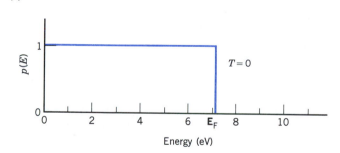

(*b*)

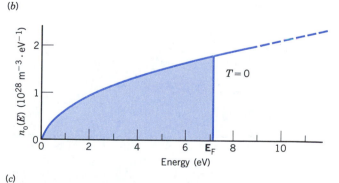

(*c*)

Figure 1   (*a*) The density of states $n(E)$ plotted as a function of the energy E. (*b*) The probability function $p(E)$ at $T = 0$. (*c*) The density of occupied states $n_o(E)$, equal to the product of $n(E)$ and $p(E)$. All states below E_F are occupied and all states above E_F are vacant.

It seems clear that the molecules of a gas at ordinary temperatures and the conduction electrons in a metal behave in quite different ways. Formally, we say that the gas molecules obey the (classical) Maxwell–Boltzmann statistics and that the conduction electrons obey the (quantum) Fermi–Dirac statistics. The word "statistics" here refers to the formal rules for counting particles. In Maxwell–Boltzmann statistics, for example, we assume that we can tell identical particles apart, but in Fermi–Dirac statistics we assume that we cannot. Again, in Maxwell–Boltzmann statistics, Pauli's exclusion principle plays no role, but in Fermi–Dirac statistics its role, as we have seen, is vital. See Section 24-6 for a discussion of these statistical distributions.

What happens to the electron distribution of Fig. 1 as we raise the temperature? Only a small change occurs in the distribution, but that small change has important con-

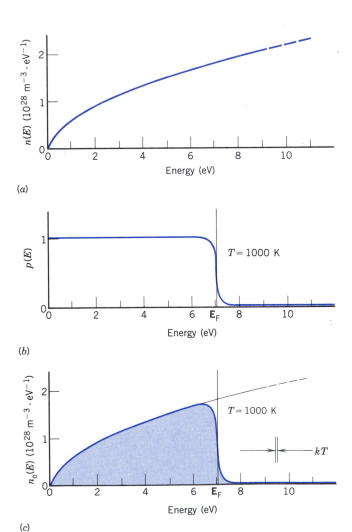

(a)

(b)

(c)

Figure 2 Same as Fig. 1, but for $T = 1000$ K. Note how little the plots differ from those of Fig. 1. (These plots are somewhat idealized in that they assume the electrons move in a region of uniform potential. Measured density of states plots in real metals do not have this simple shape.)

sequences. Figure 2 shows what the distributions of Fig. 1 would look like at $T = 1000$ K, a temperature at which a metal sample would glow brightly in a dark room.

The striking feature of Fig. 2 is how little it differs from Fig. 1, the distributions at absolute zero. At $T = 0$, the probability function $p(E)$ was strictly unity below E_F and strictly zero above E_F. As Fig. 2b shows, at $T = 1000$ K there is a small probability to have a few vacant states below E_F and a few occupied states above E_F. The density $n_o(E)$ of occupied states, again given by Eq. 3 as the product of $n(E)$ and $p(E)$, is shown in Fig. 2c. Because a few states above E_F are occupied, the average energy is a little larger than it was at the absolute zero but it is not much larger. This is again in striking contrast with the behavior of an ideal gas, for which the average kinetic energy of the molecules is proportional to the temperature.

By comparing conditions at $T = 0$ and at $T = 1000$ K, we see that all the "action" takes place for conduction electrons whose energies are close to the Fermi energy. The motion of most of the electrons remains unchanged as the temperature is raised, their large store of energy being effectively locked up.

Let us see why this is so. Figure 2c displays the magnitude of kT, a measure of the energy available from thermal agitation; its value at 1000 K is only 0.086 eV. No electron can hope to have its energy changed by more than a few times this relatively small amount by thermal agitation alone. Because of the exclusion principle, only electrons whose energies are near the Fermi energy have vacant states close enough to them for such thermal transitions to occur. An electron with an energy of, say, 2 eV can neither gain nor lose energy because all states close enough to it in energy are already filled; it simply has nowhere to go. In analogy with waves on the ocean, thermal agitation of the electrons normally causes only ripples on the surface of the "Fermi sea"; the vast depths of that sea lie undisturbed.

The probability function $p(E)$ plotted in Figs. 1b and 2b is called the *Fermi–Dirac probability function* and can be shown to be

$$p(E) = \frac{1}{e^{(E - E_F)/kT} + 1}, \tag{6}$$

in which E_F is the Fermi energy, now defined (see Fig. 2b) as the energy corresponding to $p = \frac{1}{2}$.

Note that Eq. 6 yields the rectangular plot in Fig. 1b for $T = 0$. As $T \to 0$, the exponent $(E - E_F)/kT$ in Eq. 6 approaches $-\infty$ if $E < E_F$ and $+\infty$ if $E > E_F$. In the first case we have $p(E) = 1$ and in the second case $p(E) = 0$, just as required.

Equation 6 also shows us that the important quantity is not the energy E but rather $E - E_F$, the energy interval between E and the Fermi energy. We see further that, because of the exponential nature of the term in the denominator of Eq. 6, $p(E)$ is very sensitive to small changes in $E - E_F$. This confirms our assertion that electrons whose energies are close to the Fermi energy are the only ones that play an active role. As we shall see, the first question in dealing with the electrons in a solid — be it a conductor, a semiconductor, or an insulator — is likely to be: "On an energy scale, where is the Fermi level?"

Sample Problem 2   Calculate the Fermi energy for copper, given that the number of conduction electrons per unit volume (see Sample Problem 2, Chapter 32) is 8.49×10^{28} m^{-3}.

Solution   From Eq. 5 we obtain

$$
\begin{aligned}
E_F &= \frac{h^2}{8m}\left(\frac{3n}{\pi}\right)^{2/3} \\
&= \frac{(6.63 \times 10^{-34}\ \text{J·s})^2}{(8)(9.11 \times 10^{-31}\ \text{kg})}\left[\frac{(3)(8.49 \times 10^{28}\ \text{m}^{-3})}{\pi}\right]^{2/3} \\
&= 1.13 \times 10^{-18}\ \text{J} = 7.06\ \text{eV}.
\end{aligned}
$$

Sample Problem 3   What is the probability of occupancy for a state whose energy is (a) 0.1 eV above the Fermi energy, (b) 0.1 eV below the Fermi energy, and (c) equal to the Fermi energy? Assume a temperature of 800 K.

Solution   (a) The (dimensionless) exponent in Eq. 6 is

$$
\frac{E - E_F}{kT} = \frac{0.1\ \text{eV}}{(8.62 \times 10^{-5}\ \text{eV/K})(800\ \text{K})} = 1.45.
$$

Inserting this exponent into Eq. 6 yields

$$
p(E_+) = \frac{1}{e^{1.45} + 1} = 0.19.
$$

Thus the occupancy probability for this state is 19%.

(b) For an energy 0.1 eV *below* the Fermi energy, the exponent in Eq. 6 has the same numerical value as above but is negative. Thus, from Eq. 6,

$$
p(E_-) = \frac{1}{e^{-1.45} + 1} = 0.81.
$$

The occupancy probability for this state is 81%.

(c) For $E = E_F$ the exponent in Eq. 6 is zero and that equation becomes

$$
p(E_F) = \frac{1}{e^0 + 1} = \frac{1}{1 + 1} = 0.50.
$$

Note that this result does not depend on the temperature. Note also that none of these three results depends on the actual value of the Fermi energy, only on the energy interval between the Fermi energy and the energy of the state in question.

Sample Problem 4   (a) For copper at 1000 K, find the energy at which the probability $p(E)$ that a conduction electron state will be occupied is 90%. (Assume that the conduction electrons in copper behave like a free electron gas, with a Fermi energy of 7.06 eV.) (b) For this energy, what is $n(E)$, the distribution in

energy of the available states? (c) For this same energy, what is $n_o(E)$, the distribution in energy of the occupied states?

Solution   (a) Substitution into Eq. 6 yields

$$
p(E) = \frac{1}{e^{\Delta E/kT} + 1} = 0.9,
$$

in which $\Delta E = E - E_F$. A little algebra leads to $\Delta E/kT = -2.20$ so that

$$
\begin{aligned}
\Delta E &= -2.20 kT = -(2.20)(8.62 \times 10^{-5}\ \text{eV/K})(1000\ \text{K}) \\
&= -0.19\ \text{eV}.
\end{aligned}
$$

For copper, assuming that $E_F = 7.06$ eV, we have

$$
E = E_F + \Delta E = 7.06\ \text{eV} - 0.19\ \text{eV} = 6.87\ \text{eV}.
$$

(b) Carrying out a calculation just like that of Sample Problem 1 for $E = 6.87$ eV yields $n(E) = 1.78 \times 10^{28}$ m^{-3} eV^{-1}.

(c) From Eq. 3 we have, again for $E = 6.87$ eV,

$$
\begin{aligned}
n_o(E) &= n(E)p(E) \\
&= (1.78 \times 10^{28}\ \text{m}^{-3}\ \text{eV}^{-1})(0.90) \\
&= 1.60 \times 10^{28}\ \text{m}^{-3}\ \text{eV}^{-1}.
\end{aligned}
$$

53-3   ELECTRICAL CONDUCTION IN METALS

Figure 3 represents the Fermi distribution of *velocities* in a metal. The *Fermi speed* v_F is the speed of an electron whose kinetic energy equals E_F, the Fermi energy. With no applied electric field, electrons have speeds ranging from 0 to approximately v_F, corresponding to energies ranging from 0 to approximately E_F. The distribution in Fig. 3 represents a typical *velocity* component, rather than the speed. This illustrates that there are equal numbers of electrons moving in opposite directions, so that the net current is zero in the absence of an electric field.

When an electric field is applied, the electrons are accelerated by the field and acquire a small increase in velocity in a direction opposite to the field. (Because electrons are

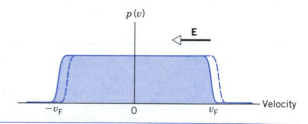

Figure 3   The Fermi distribution of velocities. With no electric field (solid line) states up to the Fermi speed v_F are filled. When an electric field **E** is applied in the direction shown, the distribution shifts to the right (dashed line) as the electrons are accelerated by the field.

negatively charged, the force on an electron is $\mathbf{F} = -e\mathbf{E}$, which is in a direction opposite to $\mathbf{E}$.) The entire velocity distribution in the presence of a field is shifted slightly to the right in Fig. 3. However, most of the electrons still add pairwise to zero velocity and do not contribute to the conduction.

The electrons that contribute to the conduction are those in a small group of velocities near v_F. The electric field causes states having velocities just below v_F in the direction of $\mathbf{E}$ to become unoccupied, while states having velocities just above v_F in a direction opposite to $\mathbf{E}$ become occupied. You can see from Fig. 3 why the drift velocity (the average velocity of all the electrons) is much smaller than v_F, because in the averaging process many positive and negative velocities will cancel one another. The drift speed is determined primarily by the small number of electrons moving from states below speed v_F to states above speed v_F under the action of the electric field.

The resistivity of the metal to the flow of these electrons is determined by collisions made by the electrons with the ion cores of the lattice. In Sample Problem 5 we show that, for copper at room temperature, the Fermi speed, which is the average speed of the conduction electrons between collisions, is 1.6×10^6 m/s, a significant fraction of the speed of light. The mean time between collisions is 2.5×10^{-14} s and the mean free path is 41 nm, which is about 150 nearest-neighbor distances in the copper lattice.

You may be surprised that, at room temperature, a conduction electron can move so far through a copper lattice without hitting an ion core. At lower temperatures —where the resistivity is lower—it can move even much further. It is, in fact, a perhaps unexpected prediction of wave mechanics that a perfectly periodic lattice at the absolute zero of temperature would be totally transparent to conduction electrons. There would never be any collisions!

There are, however, no perfectly periodic lattices. Vacant lattice sites and impurity atoms are always present, no matter how hard we try to eliminate them. Furthermore, at temperatures above the absolute zero the lattice is vibrating, and these motions also spoil the periodicity of the lattice. At room temperature the "collisions" of which we have spoken are largely interactions between the conduction electrons and the vibrations of the lattice.

Sample Problem 5   Take the Fermi energy of copper to be 7.06 eV. (*a*) What is the speed of a conduction electron with this kinetic energy? (*b*) The resistivity of copper at room temperature is $1.7 \times 10^{-8} \, \Omega \cdot$m. What is the average time τ between collisions? (*c*) What mean free path λ may be calculated from the results of (*a*) and (*b*)?

Solution   (*a*) Throughout this section we have assumed that the conduction electrons are moving in a region in which their potential energy is zero. Thus their total energy E is all kinetic and

we can write, if $E = E_F$,

$$E_F = \tfrac{1}{2} m v_F^2,$$

in which v_F is the Fermi speed. Solving for v_F yields

$$v_F = \left(\frac{2E_F}{m} \right)^{1/2} = \left[\frac{(2)(7.06 \text{ eV})(1.6 \times 10^{-19} \text{ J/eV})}{9.11 \times 10^{-31} \text{ kg}} \right]^{1/2}$$

$$= 1.6 \times 10^6 \text{ m/s}.$$

You should not confuse this speed with the *drift speed* of the conduction electrons, which is of the order of 10^{-4} m/s and is thus smaller by about a factor of 10^{10}. As we explained in Section 32-5, the drift speed is the average speed at which electrons actually drift through a conductor when an electric field is applied; the Fermi speed is their average speed between collisions.

(*b*) Solving Eq. 1 for τ yields

$$\tau = \frac{m}{ne^2 \rho}$$

$$= \frac{9.11 \times 10^{-31} \text{ kg}}{(8.49 \times 10^{28} \text{ m}^{-3})(1.60 \times 10^{-19} \text{ C})^2(1.7 \times 10^{-8} \, \Omega \cdot \text{m})}$$

$$= 2.5 \times 10^{-14} \text{ s}.$$

(*c*) To find the mean free path, we have

$$\lambda = v_F \tau = (1.6 \times 10^6 \text{ m/s})(2.5 \times 10^{-14} \text{ s})$$

$$= 4.1 \times 10^{-8} \text{ m} = 41 \text{ nm}.$$

In the copper lattice the centers of neighboring ion cores are 0.26 nm apart. Thus a typical conduction electron can move a substantial distance through a copper lattice at room temperature without making a collision.

53-4   BANDS AND GAPS

Figure 4*a* suggests the potential energy variation that we have been using to describe a conduction electron in a metal. The potential energy is zero inside the metal, and it rises to infinity at the surface. However, there are problems with this model. For example, it tells us that, because of the infinite potential barrier, an electron could never escape from inside the sample through its surface. We know that this isn't true, because electrons can be "boiled out" of a metal by raising its temperature, as in the heated filament of a vacuum tube (thermionic emission). They can also be "kicked out" if we shine light of high enough frequency on the metal surface (photoelectric effect).

Figure 4*b* shows that we can take care of this difficulty easily enough by making the potential energy at the surface finite. We made the same realistic adjustment (see Fig. 20 of Chapter 50) for the electron trapped in an atom-sized, one-dimensional well. The quantity ϕ in Fig. 4*b* is the *work function* of the metal, defined as the least amount of energy that must be supplied to an electron to remove it from the sample.

Figure 4*b* has been drawn to bring our energy scale into agreement with that used for the hydrogen atom. That is,

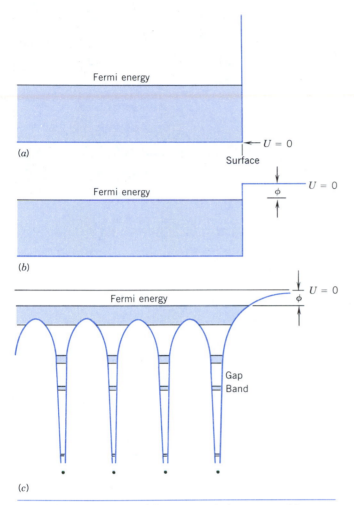

Figure 4   (*a*) The potential energy variation assumed for a metal in the free electron gas model. (*b*) A more realistic variation, showing a finite change in potential energy at the surface of the sample. (*c*) A still more realistic variation, taking the lattice of ion cores into account. This curve is a one-dimensional cut along a line of ion cores (shown as dots at the bottom of the figure). The shaded regions are the energy bands permitted for the electrons. Electrons are not permitted to have energies corresponding to the gaps.

we have chosen the $E = 0$ configuration to represent an electron at rest far outside the sample. It is possible to make this change because the potential energy always contains an arbitrary additive constant and we are more concerned, in any case, with changes in the total energy E than with E itself. On our new energy scale, the total energies of electrons trapped in the sample are negative, just as they are for the hydrogen atom.

By far the largest difficulty remaining in Fig. 4*b* is that it assumes that the potential energy of a conduction electron is constant throughout the volume of the sample. This ignores the fact that the conduction electrons move about among an array of positively charged ion cores. It is rather remarkable, in fact, that we have been able to learn as much as we have about the resistivity of metals without

taking into account the potential energy variations caused by the ion cores of the lattice. We have not, however, been able to answer such questions as: "Why is copper a conductor and diamond a nonconductor?" If we take the lattice periodicity into account, we shall be able to answer this question and to go far beyond.

Figure 4*c* shows a potential energy curve that takes the ion cores into account. Substituting this potential energy (or some approximation to it) into the Schrödinger equation brings out an interesting new phenomenon. As Fig. 4*c* shows, *the allowed states are now grouped into bands, with energy gaps between them in which no states exist.* Note that electrons just below the Fermi level are free to move throughout the lattice, but electrons with lower energies, the *core electrons,* are not. Let us see if we can understand these bands and gaps in physical terms.

The distance between nearest neighbors in a copper lattice is 0.26 nm. Consider, however, two copper atoms separated by a much greater distance, say, 50 nm, so that we may describe them as "isolated"; see Fig. 5*a*. In each atom the 29 electrons are assigned to the levels shown in Fig. 5*b*.

Now let us bring the two atoms closer together so that an outer electron in either atom can be influenced, however slightly, by forces exerted on it by the other atom. In the language of wave mechanics we say that their wave functions begin to overlap. We state without proof that the two overlapping wave functions can be combined in two independent ways, describing two states having (slightly) different energies, as shown in the second column of Fig. 6. Because the overlap is greater for the outer electrons, the energy splitting will be greater for them than for the inner electrons.

By extension, if we bring N copper atoms together to form a copper lattice, each level of the isolated atom becomes N levels of the solid. Thus the $1s$ level of the atom becomes the $1s$ band of the solid and so on. Figure 6 suggests the process.

From this point of view the forbidden gaps are not so hard to understand, being familiar from the level structure of the isolated atom. Indeed, we can say that Niels Bohr, even before wave mechanics, "invented" energy gaps when he said, in effect: "I assume that atoms can exist without radiating in a discrete set of stationary states of definite energy, *states of intermediate energy being forbidden.*"

53-5   CONDUCTORS, INSULATORS, AND SEMICONDUCTORS

Figure 7*a* represents the band structure of a conductor, such as copper. Its central feature is that the most energetic band that contains any electrons at all is only par-

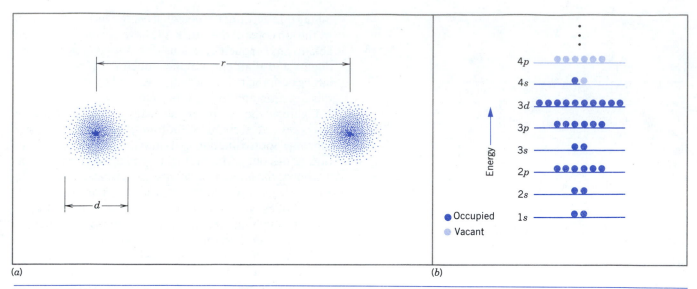

(a)   (b)

Figure 5   (a) Two neutral copper atoms of diameter d, separated by a distance r, with $r \gg d$. (b) The atoms are independent systems, and in its ground state each has the same quantum number assignments for its electrons, as shown. The energy scale is symbolic only.

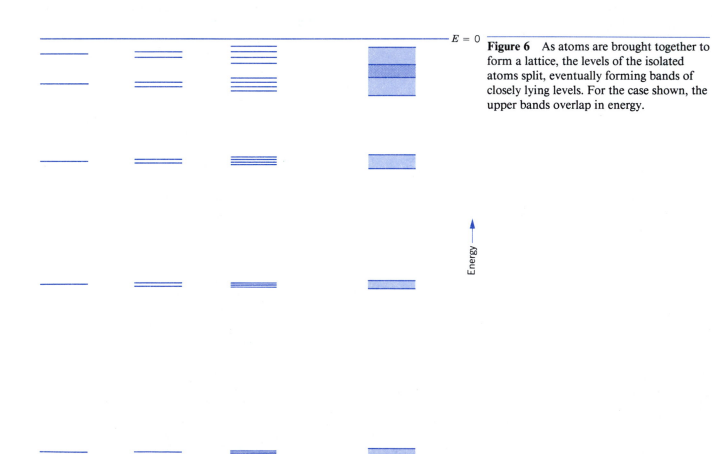

Figure 6   As atoms are brought together to form a lattice, the levels of the isolated atoms split, eventually forming bands of closely lying levels. For the case shown, the upper bands overlap in energy.

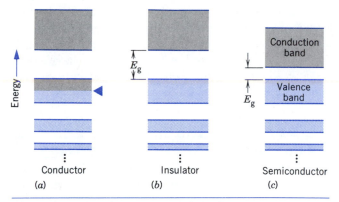

Figure 7   An idealized representation of the energy bands for (*a*) a conductor, (*b*) an insulator, and (*c*) a semiconductor. Filled bands are shown in colored shading, and empty bands in gray shading. The black triangle marks the Fermi level for the conductor.

tially filled. There are vacant states above the Fermi level so that, if you apply an electric field **E**, every electron in this band is able to increase its momentum in the $-\mathbf{E}$ direction, and there will be a current. The lower energy bands are completely filled and cannot contribute to the conduction process, all velocities adding pairwise to zero.

Figure 7*b* represents an insulator. Its central feature is that the most energetic band that contains any electrons at all is completely filled, and the forbidden energy gap lying immediately above it, marked E_g in the figure, is substantial. By "substantial" we mean that $E_g \gg kT$, so that the probability that an electron will be lifted, by thermal agitation, into the empty band that lies above the gap is negligible. If you set up an electric field within an insulator, there is no way for any of the electrons to respond to it so that there will be no current.

Carbon in its diamond form is an excellent insulator, its energy gap being 5.5 eV, more than 200 times the value of kT at room temperature.

Figure 7*c* represents a semiconductor. It differs from an insulator in that its energy gap is small enough so that thermal excitation of electrons across it can occur to a useful extent at room temperature. This puts some electrons into the (nearly empty) band labeled *conduction band* in the figure and leaves an equal number of vacant states, or *holes,* in the (nearly filled) *valence band.* In a band that is nearly full, it turns out to be more convenient to analyze its contribution to the electrical conduction in terms of the motion of holes, which behave like *positively* charged particles.

Silicon is our prototype semiconductor. It has the same crystal structure as diamond but its gap width ($= 1.1$ eV) is considerably smaller. At the absolute zero of temperature, where thermal agitation is absent, all semiconductors are insulators. At any higher temperature the probability that an electron will be raised across the gap is very sensitive to the gap width. Thus the distinction between

an insulator and a semiconductor is somewhat arbitrary. There is no hesitation, however, in calling diamond ($E_g = 5.5$ eV) an insulator and silicon ($E_g = 1.1$ eV) a semiconductor.

Sample Problem 6   What is the probability at room temperature that a state at the bottom of the conduction band is occupied in diamond and in silicon? Take the Fermi energy to be at the middle of the gap between the conduction and valence bands.

Solution   For a state at the bottom of the conduction band, the energy difference $E - E_F$ is $0.5E_g$, if the Fermi energy is at the middle of the gap. At room temperature (300 K), $kT = 0.026$ eV. We therefore have $E - E_F \gg kT$, and so we can approximate the Fermi–Dirac probability function (Eq. 6) as

$$p(E) = \frac{1}{e^{(E-E_F)/kT} + 1} \approx e^{-(E-E_F)/kT} = e^{-E_g/2kT}.$$

For diamond, $E_g = 5.5$ eV and so

$$p(E) = e^{-(5.5 \text{ eV})/2(0.026 \text{ eV})} = 1.2 \times 10^{-46}.$$

For silicon, $E_g = 1.1$ eV and so

$$p(E) = e^{-(1.1 \text{ eV})/2(0.026 \text{ eV})} = 6.5 \times 10^{-10}.$$

In a cubic centimeter of material, containing roughly 10^{23} atoms, there will be a negligible probability to find even one electron in the conduction band of diamond, while there may be roughly 10^{13} electrons in the conduction band (and an equal number of holes in the valence band) available for electrical conduction in silicon. This calculation illustrates the extreme difference in conductivity that results from small variations in the gap energy, and it clearly shows the distinction between insulators and semiconductors. In a cubic centimeter of a conductor, on the other hand, there might be 10^{23} electrons available for electrical conduction.

Semiconductors

In the previous sample problem, we compared a property of a semiconductor with that of an insulator. Table 1 compares some properties of a typical semiconductor (silicon) and a typical conductor (copper). Let us now discuss these properties in more detail.

TABLE 1   SOME ELECTRICAL PROPERTIES OF COPPER AND SILICON[a]

	Copper	*Silicon*
Type of material	Conductor	Semiconductor
Density of charge carriers[b] n (m^{-3})	9×10^{28}	1×10^{16}
Resistivity ρ ($\Omega \cdot$m)	2×10^{-8}	3×10^{3}
Temperature coefficient of resistivity α (K^{-1})	$+4 \times 10^{-3}$	-70×10^{-3}

[a] All values refer to room temperature.
[b] Includes, for silicon, both electrons and holes.

1. *The density of charge carriers, n.* Copper has many more charge carriers than does silicon, by a factor of about 10^{13}. For copper the carriers are the conduction electrons, about one per atom. Figure 7c shows that, at the absolute zero of temperature, silicon would have no charge carriers at all. At room temperature, to which Table 1 refers, charge carriers arise only because, at thermal equilibrium, thermal agitation has caused a certain (very small) number of electrons to be raised to the conduction band, leaving an equal number of vacant states (holes) in the valence band.

The holes in the valence band of a semiconductor also serve effectively as charge carriers because they permit a certain freedom of movement to the electrons in that band. If an electric field is set up in a semiconductor, the electrons in the valence band, being negatively charged, drift in the direction of $-\mathbf{E}$. The holes drift in the direction of the field and behave like particles carrying a charge $+e$, which is exactly how we shall regard them. Conduction by holes is an important characteristic of semiconductors.

2. *The resistivity, ρ.* At room temperature the resistivity of silicon is considerably higher than that of copper, by a factor of about 10^{11}. For both elements, the resistivity is determined by Eq. 1. As that equation shows, the resistivity increases as n, the density of charge carriers, decreases. The vast difference in resistivity between copper and silicon can be accounted for by the vast difference in n. (The mean collision time τ will also be different for copper and for silicon, but the effect of this on the resistivity is overwhelmed by the enormous difference in the density of charge carriers.)

For completeness, we mention that the resistivity of a good insulator (fused quartz or diamond, for example) may be as high as $10^{20}\ \Omega \cdot m$, about 10^{28} times higher than that of copper at room temperature. Few physical properties have as wide a range of measurable values as the electrical resistivity.

3. *The temperature coefficient of resistivity, α.* This quantity (see Eq. 16 of Chapter 32) is the fractional change in the resistivity ρ per unit change of temperature, or

$$\alpha = \frac{1}{\rho}\frac{d\rho}{dT}.$$

The resistivity of copper and other metals *increases* with temperature ($d\rho/dT > 0$). This happens because collisions occur more frequently the higher the temperature, thus reducing τ in Eq. 1. For metals, the density of charge carriers n in that equation is independent of temperature.

On the other hand, the resistivity of silicon (and other semiconductors) *decreases* with temperature ($d\rho/dT < 0$). This happens because the density of charge carriers n in Eq. 1 increases rapidly with temperature. The decrease in τ mentioned above for metals also occurs for semiconduc-

tors, but its effect on the resistivity is overwhelmed by the very rapid increase of the density of charge carriers.

In the laboratory, you can identify a semiconductor by its large resistivity ρ and, especially, by its large—and negative—temperature coefficient of resistivity α, both quantities being compared to values for a typical metal.

53-6 DOPED SEMICONDUCTORS

The performance of semiconductors can be substantially changed by deliberately introducing a small number of suitable replacement atoms as impurities into the semiconductor lattice, a process called *doping*. We describe the semiconductor that results as *extrinsic*, to distinguish it from the pure undoped or *intrinsic* material. Essentially all semiconducting devices today are based on extrinsic material.

Figure 8a is a two-dimensional representation of a lattice of pure silicon. Each silicon atom has 4 valence electrons and forms a two-electron bond with each of its four nearest neighbors, the electrons involved in the bonding making up the valence band of the sample.

In Fig. 8b one of the silicon atoms has been replaced by an atom of phosphorus, which has 5 valence electrons. Four of these electrons form bonds with the 4 neighboring silicon atoms, but the fifth electron is loosely bound to the phosphorus ion core, as Fig. 8b suggests. It is far easier for *this* electron to be thermally excited into the conduction band than it is for one of the silicon valence electrons to be so excited.

The phosphorus atom is called a *donor* atom because it readily *donates* an electron to the conduction band. The "extra" electron in Fig. 8b can be said to lie in a localized donor level, as Fig. 9a shows. This level is separated from the bottom of the conduction band by an energy gap E_d, where usually $E_d \ll E_g$. Adding donor atoms can greatly increase the density of electrons in the conduction band.

Semiconductors doped with donor atoms are called *n*-type semiconductors, the "*n*" standing for "negative" because the negative charge carriers (electrons) greatly outnumber the positive charge carriers (holes). In *n*-type semiconductors, the electrons in the conduction band are called the *majority carriers*, while the holes in the valence band are called the *minority carriers*.

Figure 8c shows a silicon lattice in which a silicon atom has been replaced by an aluminum atom, which has 3 valence electrons. In this case there is a "missing" electron, and it is easy for the aluminum ion core to "steal" a valence electron from a nearby silicon atom, thus creating a hole in the valence band.

The aluminum atom is called an *acceptor* atom because it so readily *accepts* an electron from the valence band. The electron so accepted moves into a localized acceptor level, as Fig. 9b shows. This level is separated from the top

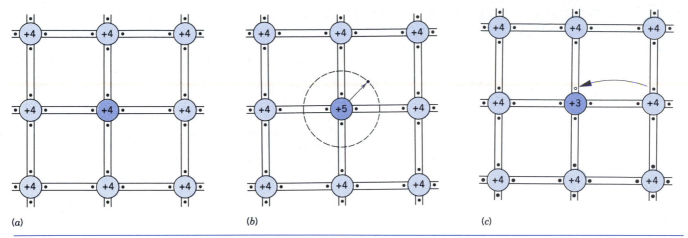

Figure 8 (*a*) A two-dimensional representation of the silicon lattice. Each silicon ion (core charge $= +4e$) is bonded to each of its four nearest neighbors by a shared two-electron bond. The dots show those valence electrons. (*b*) A phosphorus atom (valence $= 5$) is substituted for a silicon atom, creating a donor site. (*c*) An aluminum atom (valence $= 3$) is substituted for a silicon atom, creating an acceptor site.

of the valence band by an energy gap E_a, for which $E_a \ll E_g$. Adding acceptor atoms can greatly increase the number of holes in the valence band.

Semiconductors doped with acceptor atoms are called *p-type* semiconductors, the "p" standing for "positive" because the positive charge carriers (holes) greatly outnumber the negative carriers (electrons). In *p*-type semiconductors the majority carriers are the holes in the valence band and the minority carriers are the electrons in the conduction band.

Table 2 summarizes the properties of a typical *n*-type and a typical *p*-type semiconductor. Note particularly that the donor and acceptor ion cores, although they are charged, are not charge carriers because, at normal temperatures, they remain fixed in their lattice sites.

Sample Problem 7   The density of conduction electrons in pure silicon at room temperature is about 10^{16} m^{-3}. Assume that, by doping the lattice with phosphorus, you want to increase this number by a factor of 10^6. What fraction of the silicon atoms must you replace by phosphorus atoms? (Assume that, at room

temperature, the thermal agitation is effective enough so that essentially every phosphorus atom donates its "extra" electron to the conduction band.)

Solution   The density n_P of the phosphorus atoms must be about $(10^{16}$ m$^{-3})(10^6)$ or 10^{22} m^{-3}. The density of silicon atoms in a pure silicon lattice may be found from

$$n_{Si} = \frac{N_A d}{M},$$

in which N_A is the Avogadro constant, d ($= 2330$ kg/m^3) is the density of silicon, and M ($= 28.1$ g/mol) is the molar mass of silicon. Substituting yields

$$n_{Si} = \frac{(6.02 \times 10^{23} \text{ mol}^{-1})(2330 \text{ kg/m}^3)}{0.0281 \text{ kg/mol}} = 5 \times 10^{28} \text{ m}^{-3}.$$

The ratio of these two number densities is the quantity we are looking for. Thus

$$\frac{n_{Si}}{n_P} = \frac{5 \times 10^{28} \text{ m}^{-3}}{10^{22} \text{ m}^{-3}} = 5 \times 10^6.$$

We see that if only *one silicon atom in five million* is replaced by a phosphorus atom, the number of electrons in the conduction band will be increased by a factor of 10^6.

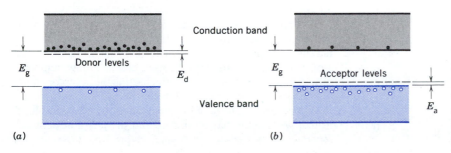

(*a*)   (*b*)

Figure 9 (*a*) An *n*-type semiconductor, showing donor levels that have contributed electrons (majority carriers) to the conduction band. The small number of holes (minority carriers) in the valence band is also shown. (*b*) A *p*-type semiconductor, showing acceptor levels that have contributed holes (majority carriers) to the valence band. The small number of electrons (minority carriers) in the conduction band is also shown.

TABLE 2 TWO TYPICAL EXTRINSIC SEMICONDUCTORS

	n-Type	p-Type
Matrix material	Silicon	Silicon
Dopant	Phosphorus	Aluminum
Type of dopant	Donor	Acceptor
Dopant valence	5 (= 4 + 1)	3 (= 4 − 1)
Dopant energy gap	0.045 eV	0.057 eV
Majority carriers	Electrons	Holes
Minority carriers	Holes	Electrons
Dopant ion core charge	+e	−e

How can such a tiny admixture of phosphorus atoms have such a big effect? The answer is that, for pure silicon at room temperature, there were not many conduction electrons there to start with! The density of conduction electrons was 10^{16} m^{-3} before doping and 10^{22} m^{-3} after doping. For copper, however, the conduction electron density (see Table 1) is about 10^{29} m^{-3}. Thus, even *after* doping, the conduction electron density of silicon remains much less than that of a typical conductor such as copper.

Sample Problem 8 Assume that the "extra" electron in a phosphorus donor atom moves in a Bohr orbit around the central phosphorus ion core, as in Fig. 8b. Calculate (a) the binding energy and (b) the orbit radius for this electron.

Solution (a) The Bohr theory expression for the binding energy E_b of the $n = 1$ state is (see Eq. 18 of Chapter 51)

$$E_b = -E_1 = \frac{mZ^2e^4}{8\epsilon_0^2h^2}. \tag{7}$$

Here we put $Z = 1$ because the orbiting electron "sees" a net central charge of $+e$.

We derived the Bohr energy by considering a hydrogen-like atom, its orbiting electron moving in a vacuum. In this case, however, the electron moves through a silicon lattice. One effect of this is to reduce the electrostatic force by a factor of κ_e, the dielectric constant of silicon. To realize this force reduction quantitatively, we must replace ϵ_0 in Coulomb's law by $\kappa_e\epsilon_0$. Making the same replacement in Eq. 7 leads to

$$E_b = \frac{1}{\kappa_e^2}\left(\frac{me^4}{8\epsilon_0^2h^2}\right),$$

in which the factor in parentheses is just 13.6 eV, the binding energy of the hydrogen atom. For silicon we have $\kappa_e = 12$, so that

$$E_b = \frac{13.6 \text{ eV}}{(12)^2} = 0.094 \text{ eV}.$$

This result is in rough order-of-magnitude agreement with the value of 0.045 eV listed in Table 2.

(b) The orbit radius follows from Eq. 19 of Chapter 51. Substituting as in part (a) leads to

$$r = \kappa_e\left(\frac{\epsilon_0h^2}{\pi me^2}\right). \tag{8}$$

The factor in parentheses is just the Bohr radius (= 52.9 pm).

Thus

$$r = (12)(52.9 \text{ pm}) = 630 \text{ pm}.$$

This is comparable to the atomic spacing in the silicon lattice (540 pm).

We should also replace the electron mass m in Eqs. 7 and 8 by an effective electron mass m_{eff}, to take partially into account the periodic nature of the silicon lattice potential. Doing so reduces the estimated binding energy and increases the estimated orbit radius, both changes being in the direction of improving the agreement with experiment.

53-7 THE *pn* JUNCTION

In the next few sections we describe some commonly used semiconducting devices, such as diode rectifiers, light-emitting diodes, and transistors. There is no end to the number of such devices that we could have chosen to describe. With today's technology, in fact, it is possible to tailor-make complex semiconducting devices to meet specific needs.

Essentially all semiconducting devices involve one or more *pn junctions*. Consider a hypothetical plane through a rod of a pure crystalline semiconducting material such as silicon. On one side of the plane, the rod is doped with donor atoms (thus creating *n*-type material), and on the other side it is doped with acceptor atoms (thus creating *p*-type material). This combination gives a *pn* junction.*

Figure 10a represents a *pn* junction at the imagined moment of its creation. There is an abundance of electrons in the *n*-type material and of holes in the *p*-type material.

Electrons close to the junction plane will tend to diffuse across it, for much the same reason that gas molecules will diffuse through a permeable membrane into a vacuum beyond it. The diffusing electrons in the conduction band, which move from right to left in Fig. 10a, will readily combine with the holes in the valence band on the other side of the junction plane. Similarly, the holes in the *p*-type region diffuse across the junction plane from left to right and combine with electrons in the *n*-region.

For every such diffusion–recombination event, the portion of the bar on the right side of this plane acquires a

---

* In common practice, to make a *pn* junction one starts with, say, *p*-type material, made by adding acceptor atoms to the molten silicon from which the solid silicon crystal is drawn. Donor atoms are then diffused into the solid sample at high temperature in a special furnace, overcompensating the acceptor atoms to a certain (controllable) depth below the surface and creating the *n*-type region. The junction that we analyze here is idealized in that we assume that the *n*-type and the *p*-type regions are separated by a well-defined plane; in practice, these regions blend into each other gradually.

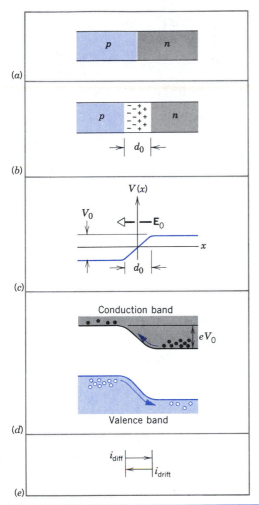

Figure 10 (a) A pn junction at the imagined moment of its creation. (b) Diffusion of majority carriers across the junction plane causes a space charge of fixed donor and acceptor ions to appear. (c) The space charge establishes a potential difference V_0 and a corresponding electric field $\mathbf{E}_0$ across the junction plane. (d) The electron energy bands near the junction. The arrows show the diffusion of the majority carriers. (e) In equilibrium, the diffusion of majority carriers across the junction plane is just balanced by the drift of minority carriers in the opposite direction.

positive charge and the portion on the left side a negative charge. These charges† cause a potential difference V_0 to build up across the junction, as Fig. 10c shows. Related to the potential difference (by the equation $E = -dV/dx$) is an internal electric field E_0 that appears across the junc-

† The fixed charges, which are close to — and separated by — the junction plane, are those of the donor and the acceptor ion cores, which, we recall, are not mobile. Normally the charges of these ion cores are compensated by the (opposite) charges of the mobile charge carriers. However, when charge carriers cross the junction plane, the ion cores are no longer fully compensated and are, so to speak, uncovered.

tion plane, pointing as shown in Fig. 10c. This field exerts a force on the electrons, opposing their motion of diffusion. Put another way, for an electron to succeed in diffusing from right to left or a hole from left to right in Fig. 10b, it must be energetic enough to overcome the potential barrier represented by Fig. 10c. This is represented in Fig. 10d, which shows the electron energy bands. To diffuse from the n-type region to the p-type region, an electron must "climb" the hill of height eV_0. A hole must also "climb" a hill of this same height to diffuse from left to right. The diffusion of both electrons and holes gives a current whose direction, in the usual conventional sense, is from left to right in Fig. 10. We call this current the *diffusion current* i_{diff}.

It is, of course, not possible to have an isolated silicon rod resting on a shelf with a current flowing indefinitely along its length. Something must happen to stop, or to compensate, this current. To find out what it is we turn our attention to the minority carriers.

As Fig. 9 and Table 2 show, although the majority carriers in n-type material are electrons, there are nevertheless also a few holes, the minority carriers. Likewise in p-type material, although the majority carriers are holes, there are also a few electrons. The minority carriers are shown in Fig. 10d.

Although the electric field in Fig. 10c acts to retard the motions of the majority carriers—being a barrier for them—it is a downhill trip for the minority carriers, be they electrons or holes. When, by thermal agitation, an electron close to the junction plane is raised from the valence band to the conduction band of the p-type material, it drifts steadily from left to right across the junction plane, swept along by the electric field E_0. Similarly, if a hole is created in the n-type material, it too drifts across to the other side. The space-charge region shown in Fig. 10b is effectively swept free of charge carriers by this process and, for that reason, we call it the *depletion zone*. The current represented by the motions of the minority carriers, called the *drift current* i_{drift}, is in the opposite direction to the diffusion current and just compensates it at equilibrium, as Fig. 10e shows.

Thus, at equilibrium, a pn junction resting on a shelf develops a contact potential difference V_0 between its ends. The diffusion current i_{diff} that moves through the junction plane from the direction p to n is just balanced by a drift current i_{drift} that moves in the opposite direction. An electric field E_0 acts across the depletion layer, whose width is d_0.

Sample Problem 9 A silicon-based pn junction has an equal concentration n_0 of donor and acceptor atoms. Its depletion zone, of width d, is symmetrical about the junction plane, as Fig. 11a shows. (a) Derive an expression for E_{max}, the maximum intensity of the electric field in the depletion zone. (b) Derive an expression for V_0, the potential difference that exists across the

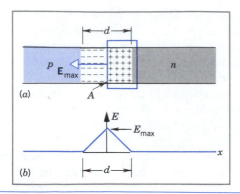

(a)

(b)

Figure 11   Sample Problem 9. (a) The depletion zone of a *pn* junction. The rectangle represents the cross section of a Gaussian surface with end caps of area A. (b) The variation of the electric field in the depletion zone.

depletion zone; see Fig. 10c. (c) Assume that $n_0 = 3 \times 10^{22} \text{ m}^{-3}$ and that V_0 is measured to be 0.6 V. Calculate the width of the depletion zone. (d) Using this value of d, calculate the value of E_{max}.

Solution   (a) The electric field may be taken as zero in the *n*-type and the *p*-type material outside the depletion zone. The field points from right to left within the depletion zone and, from symmetry, has its maximum value in the center of this zone; see Fig. 11b.

Let us apply Gauss' law to the closed "box" (Gaussian surface) shown in Fig. 11a. This law is

$$\epsilon_0 \oint \kappa_e \mathbf{E} \cdot d\mathbf{A} = q,$$

in which κ_e (=12) is the dielectric constant of silicon and q [$= n_0 e A(d/2)$] is the free charge contained within the box. The integral is to be taken over the surface of the box.

The only contribution to the integral comes from the face of the box that lies in the junction plane so that the integral has the value $\kappa_e E_{max} A$. Making these substitutions and solving for E_{max} yields

$$E_{max} = \frac{n_0 e d}{2\kappa_e \epsilon_0}, \qquad (9)$$

the relationship we seek.

(b) As Fig. 11b shows, the electric field drops linearly from its central value of E_{max} to zero at each edge of the depletion zone. Its *average* value throughout the zone is thus $\frac{1}{2}E_{max}$. The potential difference V_0 is equal to the work per unit charge required to carry a test charge q_0 from one face of the depletion zone to the other. Thus if F is the average force acting on the test charge,

$$V_0 = \frac{W}{q_0} = \frac{Fd}{q_0} = \frac{(\frac{1}{2}E_{max}q_0)d}{q_0} = \frac{1}{2}dE_{max} .$$

Substituting for E_{max} from Eq. 9 above leads to

$$V_0 = \frac{n_0 e d^2}{4\kappa_e \epsilon_0} . \qquad (10)$$

(c) Solving Eq. 10 for d and substituting the given values, we find

$$d = \left(\frac{4\kappa_e \epsilon_0 V_0}{e n_0}\right)^{1/2}$$

$$= \left[\frac{(4)(12)(8.85 \times 10^{-12} \text{ F/m})(0.60 \text{ V})}{(1.60 \times 10^{-19} \text{ C})(3 \times 10^{22} \text{ m}^{-3})}\right]^{1/2}$$

$$= 2.3 \times 10^{-7} \text{ m} = 230 \text{ nm}.$$

(d) Substituting in Eq. 9 leads to

$$E_{max} = \frac{n_0 e d}{2\kappa_e \epsilon_0}$$

$$= \frac{(3 \times 10^{22} \text{ m}^{-3})(1.60 \times 10^{-19} \text{ C})(2.3 \times 10^{-7} \text{ m})}{(2)(12)(8.85 \times 10^{-12} \text{ F/m})}$$

$$= 5.2 \times 10^6 \text{ V/m}.$$

What assumptions were made in this problem that might lead to different values of the calculated quantities under practical laboratory conditions?

The Diode Rectifier

Although a *pn* junction can be used in many ways, it is basically a rectifier. That is, if you connect it across the terminals of a battery, the current (a few picoamperes) in the circuit will be very much smaller for one polarity of the battery connection than for the other. Figure 12 shows that, for a typical silicon-based *pn* junction diode, the current for the *reverse-biased* connection ($V < 0$) is negligible by comparison with the current for the forward-biased connection ($V > 0$).

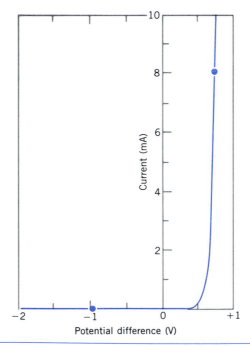

Figure 12   A current–voltage plot for a typical *pn* junction, showing that it conducts easily in the forward direction but is essentially nonconducting in the reverse direction. The dots refer to Problem 43.

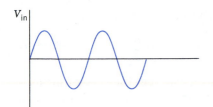

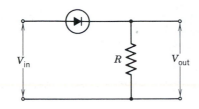

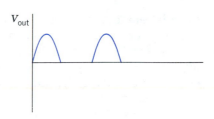

Figure 13   A *pn* junction diode connected as a rectifier. The diode conducts easily in the forward direction (positive sections of input wave) but not at all in the reverse direction (negative sections of input wave).

Figure 13 shows one of many possible applications of a diode rectifier. A sine wave input potential generates a half-wave output potential, the diode rectifier acting as essentially a short circuit for one polarity of the input potential and as essentially an open circuit for the other. An ideal diode rectifier, in fact, has only these two modes of operation. It is either on (zero resistance) or off (infinite resistance).

Figure 13 displays the conventional symbol for a diode rectifier. The arrow head corresponds to the *p*-type terminal of the device and points in the direction of "easy" conventional current flow. That is, the diode is on when the terminal with the arrow head is (sufficiently) positive with respect to the other terminal.

Figure 14 shows details of the two connections. In Fig.

14*a* (the reverse-biased arrangement) the battery emf simply adds to the contact potential difference, thus increasing the height of the barrier that the majority carriers must surmount. Fewer of them can do so and, as a result, the diffusion current decreases markedly.

The drift current, however, senses no barrier and thus is independent of the magnitude or direction of the external potential. The current balance that existed at zero bias (see Fig. 10*e*) is thus upset and, as shown in Fig. 14*a*, a current—but a very small one—appears in the circuit.

Another effect of reverse bias is to widen the depletion zone, as a comparison of Figs. 10*b* and 14*a* shows. This seems reasonable because the positive battery terminal, connected to the *n*-type end of the junction, tends to pull electrons out of the depletion zone back into the *n*-type

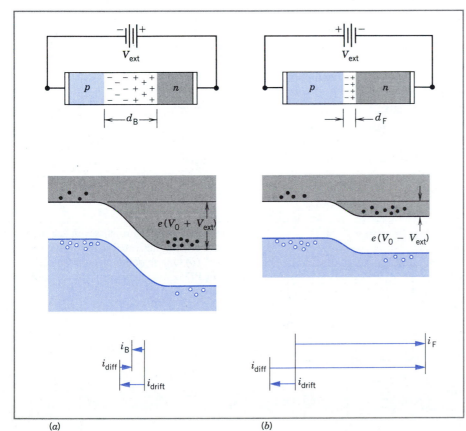

(a)   (b)

Figure 14   (*a*) The reverse-biased connection of a *pn* junction, showing the wide depletion zone, the energy bands, and the corresponding small back current i_B. (*b*) The forward-biased connection, showing the narrow depletion zone, the energy bands, and the large forward current i_F. Note that the drift current is the same in each case.

material and to repel holes back into the *p*-type material. Because the depletion zone contains very few charge carriers, it is a region of high resistivity. Thus its substantially increased width means a substantially increased resistance, consistent with the small value of the current in a reverse-biased diode.

Figure 14*b* shows the forward-biased connection, the positive terminal of the battery being connected to the *p*-type end of the *pn* junction. Here the applied emf *subtracts* from the contact potential, the diffusion current *rises* substantially, and a relatively *large* net forward current results. The depletion zone becomes *narrower*, its low resistance being consistent with the large forward current.

53-8   OPTICAL ELECTRONICS

We are all familiar with the brightly colored numbers that flash and glow at us from cash registers, gasoline pumps, and electronic equipment. In nearly all cases this light is emitted from an assembly of *pn* junctions operating as *light-emitting diodes* (LEDs).

Figure 15*a* shows the familiar seven-segment display from which the numbers are formed. Figure 15*b* shows that each element of this display is the end of a flat plastic lens, at the other end of which is a small LED, possibly about 1 mm² in area. Figure 15*c* shows a typical circuit, in which the LED is forward biased.

How can a *pn* junction emit light? When an electron at the bottom of the conduction band of a semiconductor falls into a hole at the top of the valence band, an energy E_g is released, where E_g is the gap width. What happens to this energy? There are at least two possibilities. It might be transformed into internal energy of the vibrating lattice and, with high probability, that is exactly what happens in a silicon-based semiconductor.

In some semiconducting materials, however, the emitted energy can also appear as electromagnetic radiation, the wavelength being given by

$$\lambda = \frac{c}{\nu} = \frac{c}{E_g/h} = \frac{hc}{E_g}. \tag{11}$$

Commercial LEDs designed for the visible region are usually based on a semiconducting material that is a gallium–arsenic–phosphorus compound. By adjusting the ratio of phosphorus to arsenic, the gap width—and thus the wavelength of the emitted light—can be varied.

If light is emitted when an electron falls from the conduction band to the valence band, then light of that same wavelength will be *absorbed* when an electron moves in the other direction, that is, from the valence band to the conduction band. To avoid having all the emitted photons absorbed, it is necessary to have a great surplus of both electrons and holes present in the material, in much greater numbers than would be generated by thermal agitation in the intrinsic semiconducting material. These are precisely the conditions that result when majority carriers—be they electrons or holes—are injected across the central plane of a *pn* junction by the action of an external potential difference. That is why a simple intrinsic semiconductor will not serve as an LED. You need a *pn* junction! To provide lots of majority carriers—and thus lots of photons—it should be heavily doped and strongly forward biased.

Sample Problem 10   An LED is constructed from a *pn* junction based on a certain semiconducting material whose energy gap is 1.9 eV. What is the wavelength of its emitted light?

Solution   From Eq. 11 we have

$$\lambda = \frac{hc}{E_g} = \frac{(6.63 \times 10^{-34} \text{ J} \cdot \text{s})(3.00 \times 10^8 \text{ m/s})}{(1.9 \text{ eV})(1.60 \times 10^{-19} \text{ J/eV})}$$

$$= 6.53 \times 10^{-7} \text{ m} = 653 \text{ nm}.$$

Light of this wavelength is red.

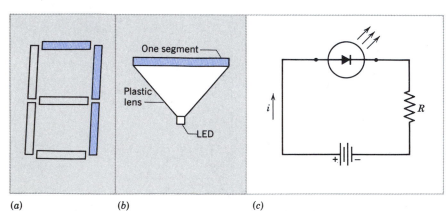

(a)   (b)   (c)

Figure 15   (*a*) The familiar seven-segment number display, activated to show the number "7." (*b*) One segment of such a display. (*c*) An LED connected to an external source of emf.

The Diode Laser

The dropping down of an electron from the conduction band to fill a hole in the valence band, with the emission of a photon, bears a strong resemblance to the dropping down of electrons in transitions between atomic states we considered in Chapter 52. There is an important application based on this similarity: by injecting electrons into the conduction band and holes into the valence band, it is possible to create a population inversion analogous to that considered in our discussion of lasers in Section 52-6. In this way it is possible to make a *diode laser,* in which the lasing medium is not a gas but a solid semiconductor. Diode lasers are commonly used in compact disk players and other optical data retrieval systems.

Figure 16 shows a representation of the energy levels in a diode laser. The lasing material is sandwiched between layers of *p*-type and *n*-type material, which have slightly larger gap energies. Electrons are injected by an external circuit into the *n*-type material; some of these excess electrons drift into the lasing layer, where they are prevented from drifting into the *p*-type material by a potential barrier. Similarly, holes are injected into the *p*-type material, drift into the lasing layer, and are trapped there. The excess of electrons (and holes) in the active region gives the lasing action.

The physical construction of the device is illustrated schematically in Fig. 17, and Fig. 18 shows a photograph of a diode laser. The lasing material is a narrow (0.2 μm) layer of a material such as GaAs (gallium arsenide), and the *p*-type and *n*-type material on each side may be layers of GaAlAs (gallium aluminum arsenide) a few micrometers in thickness. The ends of the material are cleaved to create mirror-like surfaces that reflect a portion of the light wave to enable stimulated emission in the active region. The device illustrated in Fig. 18 emits at 840 nm (in the infrared region). Diode lasers at this wavelength are commonly used in communication to send signals

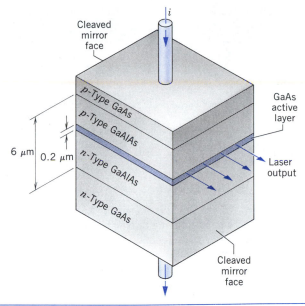

Figure 17 The physical construction of a diode laser. The lasing action occurs in the narrow GaAs layer.

along optical fibers. Other materials can be used in similar fashion to give visible radiation.

Among the advantages of diode lasers are their small size and low power input (in the range of 10 milliwatts, compared with the standard HeNe laser that may require several watts of electrical power). Like other semiconductor devices, the diode laser can be powered by batteries. Efficiencies of the order of 20% are possible (that is, 20% of the electrical power supplied to the device appears in the laser beam), compared with 0.1% in the HeNe laser. The light signal can easily be modulated by controlling the injection current, and thus we have an optical device that

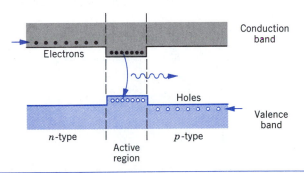

Figure 16 The energy bands in a diode laser. The active region has a smaller energy gap than the *n*-type and *p*-type materials on either side. When electrons in the conduction band of the active region drop down to fill holes in the valence band, light is emitted.

Figure 18 A diode laser, compared in size with a grain of table salt on the right.

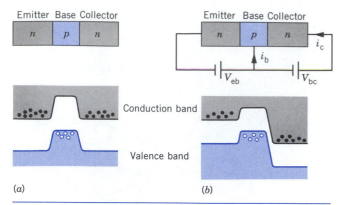

Figure 19 (a) An *npn* junction transistor. At the bottom are shown the energy bands and majority carriers in the three regions. (b) The emitter–base junction is forward biased, and the base–collector junction is reverse biased. Electrons that move from the emitter to the base either recombine with holes or (far more likely) continue to the collector.

can respond at the rapid switching times (< 100 ps) characteristic of electronic circuits.*

53-9 THE TRANSISTOR

The junction diodes we have considered so far are two-terminal devices. Here we consider a device with three (or more) terminals, called a *transistor*.† A transistor often operates in the following mode: a current established between two of the terminals is regulated by a current or voltage at the third terminal.

One common variety of transistor is the *junction transistor,* which consists of three layers of doped semiconductors, such as *npn* or *pnp*. Figure 19a shows a typical configuration for a *npn* junction transistor. The three sections are called the *emitter, base,* and *collector*. The conduction and valence bands are shown, and only the majority carriers are indicated. The emitter–base and base–collector junctions behave much like ordinary *pn* junctions.

In normal operation, as illustrated in Fig. 19b, the emitter–base junction is forward biased and the base–collector junction is reverse biased, which gives the energy bands shown in the figure.

Electrons flow from the heavily doped *n*-type emitter into the base. Because the base is very narrow, most of these electrons reach the collector, but a few recombine

* See "Applications of Lasers," by Elsa Garmire, in *Fundamentals of Physics* by David Halliday and Robert Resnick (Wiley, 1988), essay 19.

† The transistor was invented in 1947 at what is now the AT&T Bell Laboratories by John Bardeen, Walter Brattain, and William Shockley, who shared the 1956 Nobel prize in physics for their discovery.

with holes in the *p*-type region. To replenish the holes in the base, electrons from the valence band in the base must leave the transistor through the external circuit as the small base current i_b. A small change in the base current i_b can result in a large change in the collector current i_c. In this configuration, the transistor serves as a current amplifier, and the current gain i_c/i_b can have typical values in excess of 100.

A second type of transistor is called a *field-effect transistor* (FET). Figure 20 illustrates the basic geometry. Electrons flow through the *n*-type region from the *source* to the *drain* when there is an external potential difference V_{ds} between the drain and the source. The *p*-type regions are heavily doped, and the depletion layers formed at the two *pn* junctions determine the width of the *n*-type channel. An external voltage V_g applied to the *p*-type region (the *gate*) changes the width of the depletion region and consequently changes the width of the *n*-type channel. This in turn changes the current through the device, because the ability of current to flow along the *n*-channel depends on the width of the channel. A small change in the gate voltage changes the width of the channel and causes a large change in the current through the *n*-channel, so that the device can operate as an amplifier.

If the gate voltage is made large enough, the *n*-channel width can become zero, and the FET stops conducting. Here the transistor is acting like a switch: it is either conducting (on) or not conducting (off). The current can be switched on or off very rapidly by the signal applied to the gate; switching times smaller than 1 ns (10^{-9} s) are common.

A common type of FET widely used in digital circuits is the metal-oxide-semiconductor FET (MOSFET), which is fabricated by depositing and etching successive layers in a *p*-type substrate. A cross section of a MOSFET is shown in Fig. 21. The *n*-region and the *n*-channel are made by etching a mask onto the *p*-type substrate and diffusing donor atoms a known distance into the substrate. An oxide layer (SiO$_2$) is then deposited, and a metal layer is then deposited to form the contacts for the *n*-region and the gate.

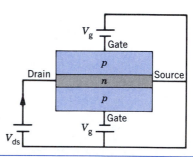

Figure 20 The basic structure of a field-effect transistor. Electrons travel along the narrow *n*-channel from the source to the drain. The width of the channel can be controlled by varying the voltage V_g at the gate.

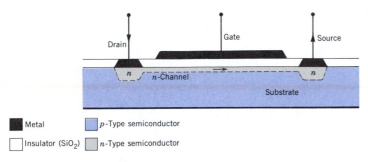

Figure 21   The structure of a MOSFET.

| Metal | | p-Type semiconductor |
| Insulator (SiO_2) | | n-Type semiconductor |

53-10 SUPERCONDUCTORS

The resistivity of a typical metallic conductor decreases as the temperature decreases. However, the resistivity does not fall to zero, even as T approaches 0 K. As we have seen in Chapter 32, the resistivity of a conductor originates with collisions made by the conducting electrons as they move through the lattice. Impurities and lattice defects increase the chances for electrons to have collisions, and collisions of electrons with atoms displaced from their lattice sites by vibrational motion contribute to the resistivity.

In certain materials called *superconductors* (see Section 32-8), the resistance falls gradually with decreasing temperature, as expected; however, at a certain *critical temperature* T_c the resistivity drops suddenly to zero (Fig. 22). Below T_c, electrons move unimpeded through the material. Table 3 shows a selection of some superconductors and their critical temperatures.

Superconductivity has been observed in 27 elements and in numerous compounds, but it has not been observed for the best metallic conductors (Cu, Ag, Au). We conclude that a superconductor is not merely a good conductor getting better, and we are led to suspect that the mechanism that causes superconductivity may differ from the mechanism that causes the conductivity of ordi-

nary metals. As we shall see, superconductivity results from a *strong* coupling between conduction electrons and the lattice. Normal conduction in the best conductors occurs when there is a *weak* coupling between the valence electron and the lattice.

Consider an electron moving through a lattice. As it moves, it pulls the positive ion cores toward it and changes the charge density in its vicinity. It leaves a somewhat higher positive charge density in its wake than would otherwise be there. This positive charge attracts other electrons. The electrons interact with one another through the intermediary of the lattice, somewhat like two boats on a lake interacting through their wakes. The net result is a slight attraction of the electrons for each other.

The BCS (Bardeen–Cooper–Schrieffer) theory* of superconductivity shows that the electron system has the lowest possible energy if the electrons are bound together in pairs, called *Cooper pairs.* When no current exists in a superconductor, the two electrons of a Cooper pair have momenta of equal magnitude but exactly opposite directions, so that the total momentum and the electric current both vanish. When a current is generated, both electrons in a pair acquire the same increase in momentum, resulting in a motion of the center of mass of the pair. All Cooper pairs acquire the same momentum.

---

* This theory of superconductivity was developed in 1957 by John Bardeen, Leon N. Cooper, and J. Robert Schrieffer, who were awarded the 1972 Nobel prize in physics for their work. Bardeen also shared the 1956 Nobel prize for his research on semiconductors and the discovery of the transistor.

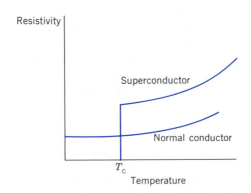

Figure 22   Comparison of the dependence of resistivity on temperature for a normal conductor and a superconductor. The resistivity of a normal conductor falls gradually with decreasing temperature. In superconducting materials, the resistivity drops suddenly to zero at the critical temperature T_c.

TABLE 3   PROPERTIES OF SOME SUPERCONDUCTORS

Material	T_c (K)	Pairing Energy (meV)
Cd	0.56	0.27
Al	1.19	0.34
Sn	3.75	1.15
Hg	4.16	1.65
Pb	7.22	2.73
Nb	9.46	3.05
Nb_3Sn	18.1	
$YBa_2Cu_3O_7$	90	
$Tl_2Ba_2Ca_2Cu_3O_{10}$	125	

Superconductivity is a cooperative phenomenon. If some Cooper pairs have been formed, the reduction in energy that occurs for the next pair is greater than if no pairs were previously formed. Once the temperature drops below T_c and some pairs are formed, a small additional reduction in temperature causes many more pairs to form. The change from the normal to the superconducting state is quite precipitous. The cooperative motions of the Cooper pairs also force every pair to have the same momentum.

The Cooper pairs have a binding energy Δ, called the *pairing energy,* which is typically in the range of 10^{-4} to 10^{-3} eV, as shown in Table 3. Note that critical temperatures of $1-10$ K (typical for most of the superconductors shown in Table 3) correspond to energies kT_c in the same range of 10^{-4} to 10^{-3} eV. The critical temperature of a superconductor is directly related to the pairing energy. Above T_c, the pairs are broken and the material has normal electrical resistance.

The binding energy of a Cooper pair introduces a *pairing gap* 2Δ into the density of states $n(E)$ near the Fermi energy. (Figure $1a$ shows an example of the density of states for a normal conductor.) It is energetically favorable for electrons near the Fermi energy in a superconductor to bind together in Cooper pairs. As a result, the density of states decreases to zero within an interval of $\pm\Delta$ of E_F, with a corresponding increase in $n(E)$ just above and below E_F. Figure 23 shows the resulting density of states and the pairing gap 2Δ. Above T_c, the density of states of a superconductor might be as shown in Fig. $1a$. The gap begins to open as the superconductor is cooled below T_c; the gap energy increases as the temperature decreases, reaching its maximum as T approaches 0 K.

The occupation probability of electron states in a superconductor can be found from the product of the density of states, shown in Fig. 23, and a Fermi–Dirac distribution function, as was shown in Fig. $2b$. This leads to a high occupation probability of the superconducting states just below the gap. Above the gap, a small density of normal (unpaired) states occurs for $T > 0$.

Beginning in 1986, a new class of superconductors was discovered* with unusually high values of T_c. The last two

---

* See "Superconductors Beyond 1-2-3," by Robert J. Cava, *Scientific American,* August 1990, p. 42.

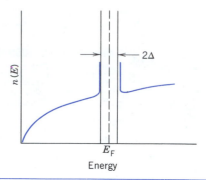

Figure 23   The density of states in a superconductor below its transition temperature. There is an energy gap of 2Δ, within which the density of states is zero. The scale of this drawing has been exaggerated; typically the Fermi energy E_F is a few electron-volts, while the pairing gap is 10^{-4} to 10^{-3} eV.

entries in Table 3 are examples of these compounds, which are ceramic materials that (unlike the more familiar types of ceramics) are conductors at room temperature. Since the highest temperature at which superconductivity had previously been observed was about 20 K, these new materials represent a substantial leap in technology. In particular, they allow superconductivity to be attained at temperatures characteristic of cooling with liquid nitrogen (77 K) rather than at those characteristic of the more expensive and less convenient liquid helium (4 K). This jump of a factor of 6 in T_c holds out the hope that, with another jump of a factor of less than 3, it might be possible to achieve superconductivity at room temperature.

These high-temperature superconductors are oxides of copper in combination with various other elements. The theory of operation of these materials is not yet understood; it is not clear whether there is a BCS-type mechanism involved. It seems apparent that the superconductivity resides with the copper oxides; although elemental copper is not superconducting, the copper oxide combinations are. The crystal structure of these compounds places the copper and oxygen in planes anchored between the other elements, and it is likely that these planes provide the pathway for the electrons that carry the superconducting current.

QUESTIONS

1. Do you think that any of the properties of solids listed in the opening to this chapter are related to each other? If so, which?

2. The conduction electrons in a metallic sphere occupy states of quantized energy. Does the average energy interval between adjacent states depend on (*a*) the material of which the sphere is made, (*b*) the radius of the sphere, (*c*) the energy of the state, or (*d*) the temperature of the sphere?

3. What role does the Pauli exclusion principle play in accounting for the electrical conductivity of a metal?

4. In what ways do the classical model and the quantum mechanical model for the electrical conductivity of a metal differ?

5. If we compare the conduction electrons of a metal with the atoms of an ideal gas, we are surprised to note (see Fig. 1c) that so much kinetic energy is locked into the conduction electron system at absolute zero. Would it be better to compare the conduction electrons, not with the atoms of a gas, but with the inner electrons of a heavy atom? After all, a lot of kinetic energy is also locked up in this case, and we don't seem to find that surprising. Discuss.

6. What features of Fig. 2 make it specific for copper, for which it was drawn? What features are independent of the identity of the metal?

7. Why do the curves in Figs. 1c and 2c differ so little from each other?

8. Distinguish carefully among the density of states function $n(E)$, the density of occupied states function $n_o(E)$, and the Fermi–Dirac probability function $p(E)$, all of which appear in Eq. 3.

9. Does the Fermi energy for a given metal depend on the volume of the sample? If, for example, you compare a sample whose volume is 1 cm³ with one whose volume is twice that, the latter sample has just twice as many available conduction electrons; it might seem that you would have to go to higher energies to fill its available levels. Do you?

10. In Section 25-4 we showed that the (molar) heat capacity of an ideal monatomic gas is $\frac{3}{2}R$. If the conduction electrons in a metal behaved like such a gas, we would expect them to make a contribution of about this amount to the measured specific heat of a metal. However, this measured specific heat can be accounted for quite well in terms of energy absorbed by the vibrations of the ion cores that form the metallic lattice. The electrons do not seem to absorb much energy as the temperature of the specimen is increased. How does Fig. 2 provide an explanation of this prequantum-days puzzle?

11. Give a physical argument to account qualitatively for the existence of allowed and forbidden energy bands in solids.

12. Is the existence of a forbidden energy gap in an insulator any harder to accept than the existence of forbidden energies for an electron in, say, the hydrogen atom?

13. On the band theory picture, what are the *essential* requirements for a solid to be (a) a metal, (b) an insulator, or (c) a semiconductor?

14. What can band theory tell us about solids that the classical model (see Section 32-5) cannot?

15. Distinguish between the drift speed and the Fermi speed of the conduction electrons in a metal.

16. Why is it that, in a solid, the allowed bands become wider as one proceeds from the inner to the outer atomic electrons?

17. Do pure (undoped) semiconductors obey Ohm's law?

18. At room temperature a given applied electric field will generate a drift speed for the conduction electrons of silicon that is about 40 times as great as that for the conduction electrons of copper. Why isn't silicon a better conductor of electricity than copper?

19. Consider these two statements: (a) At low enough temperatures silicon ceases to be a semiconductor and becomes a rather good insulator. (b) At high enough temperatures silicon ceases to become a semiconductor and becomes a rather good conductor. Discuss the extent to which each statement is either true or not true.

20. Does the electrical conductivity of an intrinsic (undoped) semiconductor depend on the temperature? On the energy gap E_g between the full and empty bands?

21. How do you account for the fact that the resistivity of metals increases with temperature but that of semiconductors decreases?

22. The energy gaps for the semiconductors silicon and germanium are 1.1 eV and 0.67 eV, respectively. Which substance do you expect would have the higher density of charge carriers at room temperature? At the absolute zero of temperature?

23. Discuss this sentence: "The distinction between a metal and a semiconductor is sharp and clear-cut, but that between a semiconductor and an insulator is not."

24. The Hall effect is much greater in semiconductors than in metals. Why? What practical use can be made of this result?

25. Does a slab of *n*-type material carry a *net* negative charge?

26. Suppose that a semiconductor contains equal numbers of donor and acceptor impurities. Do they cancel each other in their electrical effects? If so, what is the mechanism? If not, why not?

27. Why does an *n*-type semiconductor have so many more electrons than holes? Why does a *p*-type semiconductor have so many more holes than electrons? Explain in your own words.

28. What elements other than phosphorus are good candidates to use as donor impurities in silicon? What elements other than aluminum are good candidates to use as acceptor impurities? Consult the periodic table.

29. Does one distinguish between majority and minority carriers for an intrinsic semiconductor such as silicon or germanium? If not, why not? If so, what criterion do you use?

30. In preparing *n*-type or *p*-type semiconductors by doping, why is it extremely important to avoid contamination of the sample with even very small concentrations of unwanted impurities?

31. Would you expect doping to change the resistivity of silicon by very much?

32. When a current flows through a *p*-type material, positive holes move toward the negative terminal of the battery and combine with electrons in the ohmic electrode connected to the boundary of the crystal. Why doesn't the crystal become negatively charged?

33. Why is silicon often preferred to germanium for making semiconductor devices?

34. Germanium and silicon are similar semiconducting materials whose principal distinction is that the gap width E_g is 0.67 eV for the former and 1.1 eV for the latter. If you wished to construct a *pn* junction (see Fig. 10) in which the back current is to be kept as small as possible, which material would you choose and why?

35. In a *pn* junction (see Fig. 10) we have seen that electrons and holes may diffuse, in opposite directions, through the junction region. What is the eventual fate of each such particle as it diffuses into the material on the opposite side of the junc-

tion? Why is it that the electrons and positive holes do not *all* recombine, thus removing the possibility of conduction?

36. Consider two possible techniques for fabricating a *pn* junction (see Fig. 10). (*a*) Prepare separately an *n*-type and a *p*-type sample and join them together, making sure that their abutting surfaces are planar and highly polished. (*b*) Prepare a single *n*-type sample and diffuse an excess acceptor impurity into it from one face, at high temperature. Which method is preferable and why?

37. The *pn* junction shown in Fig. 11*a* has equal dopant concentrations on each side of its junction plane. Suppose, however, that the donor concentration were significantly greater than the acceptor concentration. Would the depletion zone still be symmetrically located about the junction plane? If not, would the central plane of the zone move toward the *n*-type or toward the *p*-type face of the junction? Give your reason.

38. Why can't you measure the contact potential difference generated at a *pn* junction by simply connecting a voltmeter across it?

39. In Fig. 10*b*, why does the depletion zone build up close to the junction plane? Why does it not spread out throughout the volume of the sample?

40. What does it mean to say that a *pn* junction is biased in the forward direction?

41. (*a*) Discuss the motions of the majority carriers (both electrons and holes) in a forward-biased *pn* junction. (*b*) Discuss the motions of the minority carriers in this same junction.

42. Explain in your own words how the thickness of the depletion zone of a *pn* junction can be decreased by (*a*) increasing the forward-bias voltage and (*b*) increasing the dopant concentration.

43. If you increase the temperature of a reverse-biased *pn* junction, what happens to the current (see Fig. 14*a*)? Is the effect larger for silicon or for germanium? (The intrinsic energy gap E_g for silicon is larger than that for germanium.)

44. Does the diode rectifier whose characteristics are shown in Fig. 12 obey Ohm's law? What is your criterion for deciding?

45. We have seen that a simple intrinsic (undoped) semiconductor cannot be used as a light-emitting diode. Why not? Would a heavily doped *n*-type or *p*-type semiconductor work?

46. Explain in your own words how the MOSFET device of Fig. 21 works.

47. Do you think that there is a correlation between the critical temperature of a superconductor (Table 3) and its electrical conductivity (inverse of resistivity) at room temperature?

PROBLEMS

Section 53-1 Conduction Electrons in a Metal

1. (*a*) Show that Eq. 2 can be written as

$$n(E) = CE^{1/2},$$

where $C = 6.81 \times 10^{27}$ m$^{-3}\cdot$eV$^{-3/2}$. (*b*) Use this relation to verify a calculation of Sample Problem 1, namely, that for $E = 5.00$ eV, $n(E) = 1.52 \times 10^{28}$ m$^{-3}\cdot$eV^{-1}.

2. Calculate the density $n(E)$ of conduction electron states in a metal for $E = 8.00$ eV and show that your result is consistent with the curve of Fig. 1*c*.

3. Gold is a monovalent metal with a molar mass of 197 g/mol and a density of 19.3 g/cm^3 (see Appendix D). Calculate the density of charge carriers.

4. At what pressure would an ideal gas have a density of molecules equal to that of the density of the conduction electrons in copper ($= 8.49 \times 10^{28}$ m^{-3})? Assume that $T = 297$ K.

5. The density and molar mass of sodium are 971 kg/m^3 and 23.0 g/mol, respectively; the radius of the ion Na$^+$ is 98 pm. (*a*) What fraction of the volume of metallic sodium is available to its conduction electrons? (*b*) Carry out the same calculation for copper. Its density, molar mass, and ionic radius are, respectively, 8960 kg/m^3, 63.5 g/mol, and 96 pm. (*c*) For which of these two metals do you think the conduction electrons behave more like a free electron gas?

Section 53-2 Filling the Allowed States

6. Calculate the probability that a state 0.0730 eV above the Fermi energy is occupied at (*a*) $T = 0$ K and (*b*) $T = 320$ K.

7. The Fermi energy of silver is 5.5 eV. (*a*) At $T = 0°$C, what are the probabilities that states at the following energies are occupied: 4.4 eV, 5.4 eV, 5.5 eV, 5.6 eV, 6.4 eV? (*b*) At what temperature will the probability that a state at 5.6 eV is occupied be 0.16?

8. Prove that the occupancy probabilities for two states whose energies are equally spaced above and below the Fermi energy add up to one.

9. The density of gold is 19.3 g/cm^3. Each atom contributes one conduction electron. Calculate the Fermi energy of gold. See Appendix D for the molar mass of gold.

10. Figure 2*c* shows the density of occupied states $n_o(E)$ of the conduction electrons in copper at 1000 K. Calculate $n_o(E)$ for copper for the energies $E = 4.00, 6.75, 7.00, 7.25,$ and 9.00 eV. The Fermi energy of copper is 7.06 eV.

11. In Section 50-7 we considered the situation of an electron trapped in an infinitely deep well. Suppose that 100 electrons are placed in a well of width 120 pm, two to a level with opposite spins. Calculate the Fermi energy of the system. (*Note*: The Fermi energy is the energy of the highest occupied level at the absolute zero of temperature.)

12. The conduction electrons in a metal behave like an ideal gas if the temperature is high enough. In particular, the temperature must be such that $kT \gg E_F$, the Fermi energy. What temperatures are required for copper ($E_F = 7.06$ eV) to satisfy this requirement? Compare your answer with the boiling point of copper; see Appendix D. Study Fig. 2*c* in this connection and note that we have $kT \ll E_F$ for the conditions of that figure. This is just the reverse of the requirement cited above.

13. Show that Eq. 5 can be written as

$$E_F = An^{2/3},$$

where the constant A has the value 3.65×10^{-19} m$^2 \cdot$eV.

14. The Fermi energy of copper is 7.06 eV. (a) For copper at 1050 K, find the energy at which the occupancy probability is 0.910. For this energy, evaluate (b) the density of states and (c) the density of occupied states.

15. Show that the density of states function given by Eq. 2 can be written in the form

$$n(E) = \tfrac{3}{2}nE_F^{-3/2}E^{1/2}.$$

Explain how it can be that $n(E)$ is independent of material when the Fermi energy E_F ($=7.06$ eV for copper, 9.44 eV for zinc, etc.) appears explicitly in this expression.

16. Show that if $E \gg E_F$, the distribution in energy of the occupied states $n_o(E)$ can be written as

$$n_o(E) \approx CE^{1/2}e^{-E/kT},$$

in which C is a constant. Compare this result with that calculated for the Maxwell-Boltzmann distribution in Section 24-4. What do you conclude?

17. Show that the probability p_h that a *hole* exists at a state of energy E is given by

$$p_h = \frac{1}{e^{-(E-E_F)/kT} + 1}.$$

(*Hint:* The existence of a hole means that the state is unoccupied; convince yourself that this implies that $p_h = 1 - p$.)

18. The Fermi energy of aluminum is 11.66 eV; its density is 2.70 g/cm^3 and its molar mass is 27.0 g/mol (see Appendix D). From these data, determine the number of free electrons per atom.

19. White dwarf stars represent a late stage in the evolution of stars like the Sun. They become dense enough and hot enough that we can analyze their structure as a solid in which all Z electrons per atom are free. For a white dwarf with a mass equal to that of the Sun and a radius equal to that of the Earth, calculate the Fermi energy of the electrons. Assume the atomic structure to be represented by iron atoms, and $T = 0$ K.

20. A neutron star can be analyzed by techniques similar to those used for ordinary metals. In this case the neutrons (rather than electrons) obey the probability function, Eq. 6. Consider a neutron star of 2.00 solar masses with a radius of 10.0 km. Calculate the Fermi energy of the neutrons.

21. Estimate the number N of conduction electrons in a metal that have energies greater than the Fermi energy as follows. Strictly, N is given by

$$N = \int_{E_F}^{\infty} n(E)p(E)dE.$$

By studying Fig. 2c, convince yourself that, to a good degree of approximation, this expression can be written as

$$N = \int_{E_F}^{E_F + 4kT} n(E_F)(\tfrac{1}{4})dE.$$

By substituting the density of states function, evaluated at the Fermi energy, show that this yields for the fraction f of conduction electrons excited to energies greater than the Fermi energy,

$$f = \frac{N}{n} = \frac{3kT/2}{E_F}.$$

Why not evaluate the first integral above directly without resorting to an approximation?

22. Use the result of Problem 21 to calculate the fraction of excited electrons in copper at temperatures of (a) absolute zero, (b) 300 K, and (c) 1000 K.

23. At what temperature will the fraction of excited electrons in lithium equal 0.0130? The Fermi energy of lithium is 4.71 eV. See Problem 21.

24. Silver melts at 962°C. At the melting point, what fraction of the conduction electrons are in states with energies greater than the Fermi energy of 5.5 eV? See Problem 21.

25. Show that, at the absolute zero of temperature, the average energy $\overline{E}$ of the conduction electrons in a metal is equal to $\tfrac{3}{5}E_F$, where E_F is the Fermi energy. (*Hint:* Note that, by definition of average, $\overline{E} = (1/n) \int En_o(E)dE$.)

26. (a) Using the result of Problem 25, estimate how much energy would be released by the conduction electrons in a penny (assumed all copper; mass $= 3.1$ g) if we could suddenly turn off the Pauli exclusion principle. (b) For how long would this amount of energy light a 100-W lamp? Note that there is no known way to turn off the Pauli principle!

Section 53-3 Electrical Conduction in Metals

27. Silver is a monovalent metal. Calculate (a) the number of conduction electrons per cubic meter, (b) the Fermi energy, (c) the Fermi speed, and (d) the de Broglie wavelength corresponding to this speed. Extract needed data from Appendix D.

28. Zinc is a bivalent metal. Calculate (a) the number of conduction electrons per cubic meter, (b) the Fermi energy, (c) the Fermi speed, and (d) the de Broglie wavelength corresponding to this speed. See Appendix D for needed data on zinc.

29. For silver, calculate (a) the mean free path of conduction electrons and (b) the ratio of the mean free path to the distance between neighboring ion cores. Silver has a Fermi energy of 5.51 eV and a resistivity of 1.62×10^{-8} $\Omega \cdot$m. See Problem 27.

Section 53-5 Conductors, Insulators, and Semiconductors

30. Repeat the calculation of Sample Problem 6 for a temperature of (a) 1000 K and (b) 4.0 K.

31. The Fermi–Dirac distribution function can be applied to semiconductors as well as to metals. In semiconductors, E is the energy above the top of the valence band. The Fermi level for an intrinsic semiconductor is nearly midway between the top of the valence band and the bottom of the conduction band. For germanium these bands are separated by a gap of 0.67 eV. Calculate the probability that (a) a state at the bottom of the conduction band is occupied and (b) a state at the top of the valence band is unoccupied at 290 K.

32. The band gap in pure germanium is 0.67 eV. Assume that the Fermi level is at the middle of the gap. (a) Calculate the probability that a state at the bottom of the conduction band

is occupied at 16°C. (*b*) At what temperature will the occupation probability of this state be 3.0 times the probability at 16°C?

33. In a simplified model of an intrinsic semiconductor (no doping), the actual distribution in energy of states is replaced by one in which there are N_v states in the valence band, all these states having the same energy E_v, and N_c states in the conduction band, all these states having the same energy E_c. The number of electrons in the conduction band equals the number of holes in the valence band. (*a*) Show that this last condition implies that

$$\frac{N_c}{e^{(E_c - E_F)/kT} + 1} = \frac{N_v}{e^{-(E_v - E_F)/kT} + 1}.$$

(*Hint*: See Problem 17.) (*b*) If the Fermi level is in the gap between the two bands and is far from both bands compared to kT, then the exponentials dominate in the denominators. Under these conditions, show that

$$E_F = \tfrac{1}{2}(E_c + E_v) + \tfrac{1}{2}kT \ln(N_v/N_c),$$

and therefore that, if $N_v \approx N_c$, the Fermi level is close to the center of the gap.

Section 53-6 Doped Semiconductors

34. Identify the following as *p*-type or *n*-type semiconductors: (*a*) Sb in Si; (*b*) In in Ge; (*c*) Al in Ge; (*d*) As in Si.

35. Pure silicon at 300 K has an electron density in the conduction band of 1.5×10^{16} m^{-3} and an equal density of holes in the valence band. Suppose that one of every 1.0×10^7 silicon atoms is replaced by a phosphorus atom. (*a*) What charge carrier density will the phosphorus add? Assume that all the donor electrons are in the conduction band. (See Appendix D for needed data on silicon.) (*b*) Find the ratio of the charge carrier density in the doped silicon to that for the pure silicon.

36. What mass of phosphorus would be needed to dope a 1.0-g sample of silicon to the extent described in Sample Problem 7?

37. A silicon crystal is doped with phosphorus to a concentration of 10^{22} phosphorus atoms per cubic meter. On average, how far apart are these atoms? See Sample Problem 7.

38. A sample of very pure germanium has one impurity atom to 1.3×10^9 atoms of germanium. Calculate the distance between impurity atoms.

39. In Fig. 24 two energy bands of a hypothetical solid are represented. The bands are filled to level E_x, which may be in either band 1 or band 2. There may be an impurity level at E_i. Indicate whether the solid is a conductor, insulator, intrinsic semiconductor, or extrinsic semiconductor. The impurity type may be donor, acceptor, or none, and extrinsic semiconductors may be either *p*-type or *n*-type. Complete the table.

E_x (eV)	E_i (eV)	E_b (eV)	Solid	Impurity	Extrinsic Semiconductor
3.00	—	9.00			
3.00	4.06	4.10			
3.00	—	4.10			
1.49	—	9.00			
4.40	—	4.10			
3.00	3.04	4.10			

(with header *Type* spanning Solid, Impurity, Extrinsic Semiconductor)

40. Doping changes the Fermi energy of a semiconductor. Consider silicon, with a gap of 1.1 eV between the valence and conduction bands. At 290 K the Fermi level of the pure material is nearly at the midpoint of the gap. Suppose that it is doped with donor atoms, each of which has a state 0.15 eV below the bottom of the conduction band, and suppose further that doping raises the Fermi level to 0.084 eV below the bottom of that band. (*a*) For both the pure and doped silicon, calculate the probability that a state at the bottom of the conduction band is occupied. (*b*) Also calculate the probability that a donor state in the doped material is occupied. See Fig. 25.

41. A silicon sample is doped with atoms having a donor state 0.11 eV below the bottom of the conduction band. (*a*) If each of these states is occupied with probability 4.8×10^{-5} at temperature 290 K, where is the Fermi level relative to the top of the valence band? (*b*) What then is the probability that a state at the bottom of the conduction band is occupied? The energy gap in silicon is 1.1 eV.

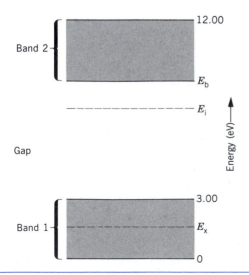

Figure 24 Problem 39.

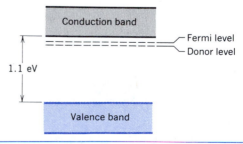

Figure 25 Problem 40.

Section 53-7 The pn Junction

42. When a photon enters the depletion region of a *pn* junction, electron–hole pairs can be created as electrons absorb part of the photon's energy and are excited from the valence band to the conduction band. These junctions are thus often used as detectors for photons, especially for x rays and nuclear gamma rays. When a 662-keV gamma-ray photon is totally absorbed by a semiconductor with an energy gap of 1.1 eV, on the average how many electron–hole pairs are created?

43. Calculate and compare the resistances of the diode rectifier for the two points shown on the characteristic curve of Fig. 12. The current for the left-hand dot (too small to show in the figure) is 50 pA.

44. For an ideal *pn*-junction diode, with a sharp boundary between the two semiconducting materials, the current *i* is related to the potential difference *V* across the diode by

$$i = i_0(e^{eV/kT} - 1),$$

where i_0, which depends on the materials but not on the current or potential difference, is called the *reverse saturation current*. *V* is positive if the junction is forward biased and negative if it is reverse biased. (*a*) Verify that this expression predicts the behavior expected of a diode by sketching *i* as a function of *V* over the range $-0.12 \text{ V} < V < +0.12 \text{ V}$. Take $T = 290$ K and $i_0 = 5.0$ nA. (*b*) For the same temperature, calculate the ratio of the current for a 0.50-V forward bias to the current for a 0.50-V reverse bias.

45. A drop of lead (work function = 3.4 eV) is in close contact with a sheet of copper (work function = 4.5 eV). Find the contact potential difference that appears across the lead–copper interface. How might you measure it? Draw an energy diagram, showing (in the style of Fig. 4*b*) the relative Fermi levels both before and after the two metals are joined together. Can such a junction serve as a diode rectifier?

46. (*a*) A capacitance is associated with a *pn* junction. Explain why. (*b*) Derive an expression for the capacitance of the *pn* junction of Sample Problem 9.

Section 53-8 Optical Electronics

47. (*a*) Calculate the maximum wavelength that will produce photoconduction in diamond, which has a band gap of 5.5 eV. (*b*) In what part of the electromagnetic spectrum does this wavelength lie?

48. In a particular crystal, the highest occupied band of states is full. The crystal is transparent to light of wavelengths longer than 295 nm but opaque at shorter wavelengths. Calculate the width, in electron-volts, of the gap between the highest occupied band and the next (empty) band.

49. The KCl crystal has a band gap of 7.6 eV above the topmost occupied band, which is full. Is this crystal opaque or transparent to radiation of wavelength 140 nm?

50. (*a*) Fill in the seven-segment display shown in Fig. 15*a* to show how all 10 numbers may be generated. (*b*) If the numbers are displayed randomly, in what fraction of the displays will each of the seven segments be used?

51. Section 53-8 discussed the mode of operation of a light-emitting diode, in which light is emitted when charge carriers are injected across the central plane of a *pn* junction by an external potential. The reverse device, a *photodiode,* is also a possibility. That is, you can shine light on a *pn* junction and a current will develop across the junction plane. Discuss how such a device might operate. Would it be best to operate it in a forward- or a reverse-biased mode?

CHAPTER 54

NUCLEAR PHYSICS

Deep within the atom lies its nucleus, occupying only 10^{-15} of the volume of the atom but providing most of its mass as well as the force that holds it together. The next goal in our study of physics is to understand the structure of the nucleus and the substructure of its components.

Our task is made easier by the many similarities between the study of atoms and the study of nuclei. Both systems are governed by the laws of quantum mechanics. Like atoms, nuclei have excited states that can decay to the ground state through the emission of photons (gamma rays). In certain circumstances, as we shall see, nuclei can exhibit shell effects that are very similar to those of atoms. We shall also see that there are differences between the study of atoms and the study of nuclei that keep us from achieving as complete an understanding of nuclei as we have of atoms.

In this chapter we study the structure of nuclei and their constituents. We consider some experimental techniques for studying their properties, and we conclude with a description of the theoretical basis for understanding the structure of nuclei.

54-1 DISCOVERING THE NUCLEUS

In the first years of the 20th century not much was known about the structure of atoms beyond the fact that they contained electrons. This particle had been discovered (by J. J. Thomson) only in 1897, and its mass was unknown in those early days. Thus it was not possible even to say just how many electrons a given atom contained. Atoms are electrically neutral so they must also contain some positive charge, but at that time nobody knew what form this compensating positive charge took. How the electrons moved within the atom and how the mass of the atom was divided between the electrons and the positive charge were also open questions.

In 1911, Ernest Rutherford, interpreting some experiments carried out in his laboratory, was led to propose that the positive charge of the atom was densely concentrated at the center of the atom and that, furthermore, it was responsible for most of the mass of the atom. He had discovered the atomic nucleus!

Until this step had been taken, all attempts to under-

stand the motions of the electrons within the atom were doomed to failure. Only 2 years after Rutherford's proposal, Niels Bohr used the concept of the nuclear atom to develop the semiclassical theory of atomic structure that we described in Chapter 51. This early work by Rutherford and Bohr marks the beginning of our understanding of the structure of atoms.

How did Rutherford come to make this proposal? It was not an idle conjecture but was based firmly on the results of an experiment suggested by him and carried out by his collaborators, Hans Geiger (of Geiger counter fame) and Ernest Marsden, a 20-year-old student who had not yet earned his bachelor's degree.

Rutherford's idea was to probe the forces acting within an atom by firing energetic alpha (α) particles through a thin target foil and measuring the extent to which they were deflected as they passed through the foil. Alpha particles, which are about 7300 times more massive than electrons, carry a charge of $+2e$ and are emitted spontaneously (with energies of a few MeV) by many radioactive materials. We now know that these useful projectiles are the nuclei of the atoms of ordinary helium. Figure 1 shows the experimental arrangement of Geiger and Mars-

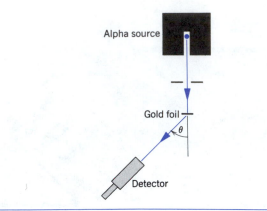

Figure 1 The experimental arrangement used in Rutherford's laboratory to study the scattering of α particles by thin metal foils. The detector can be rotated to various scattering angles θ.

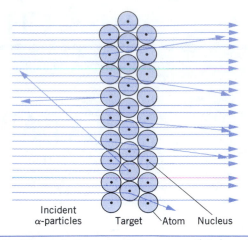

Figure 3 The angle through which an α particle is scattered depends on how close its extended incident path lies to the nucleus of an atom. Large deflections result only from very close encounters.

den. The experiment consists in counting the number of α particles deflected through various scattering angles θ. (See Section 29-7.)

Figure 2 shows their results. Note especially that the vertical scale is logarithmic. We see that most of the α particles are scattered through rather small angles, but—and this was the big surprise—a very small fraction of them is scattered through very large angles, approaching 180°. In Rutherford's words: "It was quite the most incredible event that ever happened to me in my life. It was almost as incredible as if you had fired a 15-inch shell at a piece of tissue paper and it came back and hit you."

Why was Rutherford so surprised? At the time of these

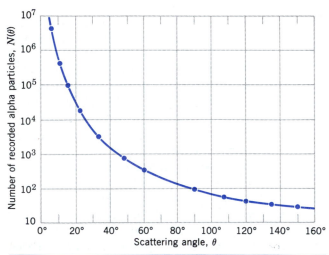

Figure 2 The dots show the α-particle scattering results from the experiments of Geiger and Marsden, and the solid curve is computed according to Rutherford's theory of the nucleus. Note that the vertical axis is marked in powers of 10.

experiments, many physicists believed in a model of the atom that had been proposed by J. J. Thomson. In Thomson's model, the positive charge of the atom was thought to be spread out through the entire volume of the atom. The electrons were thought to be distributed throughout this volume, somewhat like seeds in a watermelon, and to vibrate about their equilibrium positions within this sphere of charge.

The maximum deflecting force acting on the α particle as it passes through such a positive sphere of charge proves to be far too small to deflect the α particle by even as much as one degree. The electrons in the atom would also have very little effect on the massive, energetic α particle. They would, in fact, be themselves strongly deflected, much as a swarm of gnats would be brushed aside by a stone thrown through them. There is simply no mechanism in Thomson's atom model to account for the backward deflection of an α particle.

Rutherford saw that to produce such a large deflection there must be a large force, which could be provided if the positive charge were concentrated tightly at the center of the atom, instead of being spread throughout its volume. On this model the incoming α particle can get very close to the center of the positive charge without penetrating it, resulting in a large deflecting force; see Sample Problem 1.

Figure 3 shows the paths taken by typical α particles as they pass through the atoms of the target foil. As we see, most are deflected only slightly or not at all, but a few (those whose extended incoming paths pass, by chance, close to a nucleus) are deflected through large angles. From an analysis of the data, Rutherford concluded that the dimensions of the nucleus must be smaller than the diameter of an atom by a factor of about 10^4. The atom is mostly empty space! It is not often that the piercing insight of a gifted scientist, supported by a few simple calculations,* leads to results of such importance.

Sample Problem 1   A 5.30-MeV α particle happens, by chance, to be headed directly toward the nucleus of an atom of gold ($Z = 79$). How close does it get before it comes momentarily to rest and reverses its course? Neglect the recoil of the (relatively massive) gold nucleus.

Solution   Initially the total mechanical energy of the two interacting particles is just equal to K_α ($= 5.30$ MeV), the initial kinetic energy of the α particle. At the moment the α particle comes to rest, the total energy is the electrostatic potential energy of the system of two particles. Because energy must be conserved, these two quantities must be equal, or

$$K_\alpha = \frac{1}{4\pi\epsilon_0}\frac{qQ}{d},$$

in which q ($= 2e$) is the charge of the α particle, Q ($= 79e$) is the charge of the gold nucleus, and d is the distance between the centers of the two particles.

Substituting for the charges and solving for d yield

$$d = \frac{qQ}{4\pi\epsilon_0 K_\alpha}$$

$$= (8.99 \times 10^9 \text{ N}\cdot\text{m}^2/\text{C}^2) \frac{(2)(79)(1.60 \times 10^{-19} \text{ C})^2}{(5.30 \text{ MeV})(1.60 \times 10^{-13} \text{ J/MeV})}$$

$$= 4.29 \times 10^{-14} \text{ m} = 42.9 \text{ fm}.$$

This is a small distance by atomic standards but not by nuclear standards. As we shall see in the following section, it is considerably larger than the sum of the radii of the gold nucleus and the α particle. Thus the α particle reverses its course without ever "touching" the gold nucleus.

If the positive charge associated with the gold atom had been spread uniformly throughout the volume of the atom, the maximum retarding force acting on the α particle would have occurred at the moment the α particle began to touch the surface of the atom. This force (see Problem 2) would have been far too weak to have had much effect on the motion of the α particle, which would have gone barreling right through such a "spongy" atom.

54-2   SOME NUCLEAR PROPERTIES

The nucleus, tiny as it may be, has a structure that is every bit as complex as that of the atom. Nuclei are made up of protons and neutrons. These particles (unlike the electron) are not true elementary particles, being made up of

---

* For an analysis of this scattering experiment, see Kenneth S. Krane, *Modern Physics* (Wiley, 1983), Chapter 6.

other particles, called *quarks*. However, nuclear physics —the subject of this chapter—is concerned primarily with studies of the nucleus that do not involve the internal structure of the protons and neutrons themselves. The fundamental nature of these two particles is a topic in the field of elementary particle physics, which we consider in Chapter 56.

Nuclear Systematics

Nuclei are made up of protons and neutrons. The number of protons in the nucleus is called the *atomic number* and is represented by Z. The number of neutrons is called the *neutron number,* and we represent it by N. Aside from the difference in their electric charges ($q = +e$ for the proton, $q = 0$ for the neutron), the proton and the neutron are very similar particles: they have nearly equal masses and experience identical nuclear forces inside nuclei. For this reason, we classify the proton and neutron together as *nucleons*. The total number of nucleons ($= Z + N$) is called the *mass number,* and we represent it by A.

By specifying Z and A (and therefore N) we uniquely identify a particular nuclear species or *nuclide*. We use A, the total number of nucleons, as an identifying superscript in labeling nuclides. In ^{81}Br, for example, there are 81 nucleons. The symbol "Br" tells us that we are dealing with bromine, for which $Z = 35$. The remaining 46 nucleons are neutrons, so that, for this nuclide, $Z = 35$, $N = 46$, and $A = 81$. Two nuclides with the same Z but different N and A, such as ^{81}Br and ^{82}Br, are called *isotopes*.

Figure 4 shows a chart of the known nuclides as a plot of Z against N. The dark shading represents stable nuclides; the lighter shading represents known radioactive nuclides, or *radionuclides*. Table 1 shows some properties of a few selected nuclides.

Note that there is a reasonably well-defined zone of stability in Fig. 4. Unstable radionuclides lie on either side of the stability zone.

The Nuclear Force

The force that controls the electronic structure and properties of the atom is the familiar Coulomb force. To bind the nucleus together, however, there must be a strong attractive force of a totally new kind acting between the neutrons and the protons. This force must be strong enough to overcome the repulsive Coulomb force between the (positively charged) protons and to bind both neutrons and protons into the tiny nuclear volume. Experiments suggest that this *strong force*, as it is simply called, has the same character between any pair of nuclear constituents, be they neutrons or protons.

The "strong force" has a short range, roughly equal to 10^{-15} m. This means that the attractive force between pairs of nucleons drops rapidly to zero for nucleon separations greater than a certain critical value. This in turn

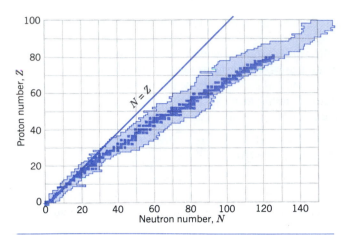

Figure 4 A plot of the known nuclides. The dark shading indicates stable nuclides and the light shading shows radioactive nuclides. Note that light stable nuclides have essentially equal numbers of protons and neutrons, while $N > Z$ for heavy nuclei.

means that, except in the smallest nuclei, a given nucleon cannot interact through the strong force with all the other nucleons in the nucleus but only with a few of its nearest neighbors. By contrast, the Coulomb force is not a short-range force. A given proton in a nucleus exerts a Coulomb repulsion on all the other protons, no matter how large their separation; see Problem 12.

Figure 4 shows that the lightest stable nuclides tend to lie on or close to the line $Z = N$. The heavier stable nuclides lie well below this line and thus typically have many more neutrons than protons. The tendency to an excess of neutrons at large mass numbers is a Coulomb repulsion effect. Because a given nucleon interacts with only a small number of its neighbors through the strong force, the amount of energy tied up in strong-force bonds between nucleons increases just in proportion to A. The energy tied up in Coulomb-force bonds between protons increases more rapidly than this because each proton interacts with all other protons in the nucleus. Thus the Coulomb energy becomes increasingly important at high mass numbers.

Consider a nucleus with 238 nucleons. If it were to lie on the $Z = N$ line, it would have $Z = N = 119$. However, such a nucleus, if it could be assembled, would fly apart at once because of Coulomb repulsion. Relative stability is found only if we replace 27 of the protons by neutrons, thus greatly diluting the Coulomb repulsion effect. We then would have the nuclide ^{238}U, which has $Z = 92$ and $N = 146$, a neutron excess of 54.

Even in ^{238}U, Coulomb effects are evident in that (1) this nuclide is radioactive and emits α particles, and (2) it can easily break up (fission) into two fragments. Both of these processes reduce the Coulomb energy more than they do the energy in the strong-force bonds.

Nuclear Radii

We have used the Bohr radius a_0 ($= 5.29 \times 10^{-11}$ m) as a convenient unit for measuring the dimensions of atoms. Nuclei are smaller by a factor of about 10^4, and a convenient unit for measuring distances of this scale is the *femtometer* ($= 10^{-15}$ m). This unit is often called the *fermi* and shares the same abbreviation. Thus

$$1 \text{ fermi} = 1 \text{ femtometer} = 1 \text{ fm} = 10^{-15} \text{ m}.$$

We can learn about the size and structure of nuclei by doing scattering experiments, much as suggested by Fig. 1, using an incident beam of high-energy electrons. The energy of the incident electrons must be high enough (> 200 MeV) so that their de Broglie wavelength will be small enough for them to act as structure-sensitive nuclear probes. In effect, these experiments measure the diffraction pattern of the scattered particles and so deduce the shape of the scattering object (the nucleus).

From a variety of scattering experiments, the nuclear density has been deduced to be of the form shown in Fig. 5. We see that the nucleus does not have a sharply defined surface. It does, however, have a characteristic mean radius R. The density $\rho(r)$ has a constant value in the nuclear interior and drops to zero through the fuzzy surface zone. From these experiments it has been found that

TABLE 1 SOME PROPERTIES OF SELECTED NUCLIDES

Nuclide	Z	N	A	Stability[a]	Atomic Mass (u)	Radius (fm)	Binding Energy per Nucleon (MeV)	Spin ($h/2\pi$)	Magnetic Moment (μ_N)
^{7}Li	3	4	7	92.5%	7.016003	2.30	5.61	$\frac{3}{2}$	+3.26
^{14}N	7	7	14	99.6%	14.003074	2.89	7.48	1	+0.403
^{31}P	15	16	31	100%	30.973762	3.77	8.48	$\frac{1}{2}$	+1.13
^{88}Rb	37	51	88	18 m	87.911326	5.34	8.68	2	+0.508
^{120}Sn	50	70	120	32.4%	119.902199	5.92	8.50	0	0
^{157}Gd	64	93	157	15.7%	156.923956	6.47	8.20	$\frac{3}{2}$	−0.340
^{197}Au	79	118	197	100%	196.966543	6.98	7.92	$\frac{3}{2}$	+0.146
^{239}Pu	94	145	239	24,100 y	239.052158	7.45	7.56	$\frac{1}{2}$	+0.203

[a] For stable nuclides the isotopic abundance is given; for radionuclides, the half-life.

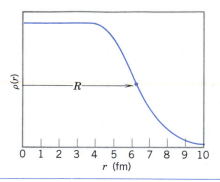

Figure 5   The variation with radial distance of the density of a nucleus of ^{197}Au.

R increases with A approximately as

$$R = R_0 A^{1/3}, \qquad (1)$$

in which A is the mass number and R_0 is a constant with a value of about 1.2 fm. For ^{63}Cu, for example,

$$R = (1.2 \text{ fm})(63)^{1/3} = 4.3 \text{ fm}.$$

By comparison, the mean radius of a copper ion in a lattice of solid copper is 1.8 Bohr radii, about 2×10^4 times larger.

Nuclear Masses and Binding Energies

Atomic masses can be measured with great precision using modern mass spectrometer and nuclear reaction techniques. We recall that such masses are measured in *unified atomic mass units* (abbreviation u), chosen so that the atomic mass (*not* the nuclear mass) of ^{12}C is exactly 12 u. The relation of this unit to the SI mass standard is

$$1 \text{ u} = 1.6605 \times 10^{-27} \text{ kg}.$$

Note that the mass number (symbol A) identifying a nuclide is so named because this number is equal to the atomic mass of the nuclide, rounded to the nearest integer. Thus the mass number of the nuclide ^{137}Cs is 137; this nuclide contains 55 protons and 82 neutrons, a total of 137 particles; its atomic mass is 136.907073 u, which rounds off numerically to 137.

In nuclear physics, as contrasted with atomic physics, the energy changes per event are commonly so great that Einstein's well-known mass–energy relation $E = \Delta m c^2$ is an indispensable work-a-day tool. We shall often need to use the energy equivalent of 1 atomic mass unit, and we find it from

$$E = \Delta m \, c^2 = \frac{(1.6605 \times 10^{-27} \text{ kg})(2.9979 \times 10^8 \text{ m/s})^2}{1.6022 \times 10^{-13} \text{ J/MeV}}$$

$$= 931.5 \text{ MeV}.$$

This means that we can write c^2 as 931.5 MeV/u and can thus easily find the energy equivalent (in MeV) of any mass or mass difference (in u), or conversely.

As an example, consider the deuteron, the nucleus of the heavy hydrogen atom. A deuteron consists of a proton and a neutron bound together by the strong force. The energy E_B that we must add to the deuteron to tear it apart into its two constituent nucleons is called its *binding energy*. In effect, the binding energy is the total internal energy of the nucleus, due in part to the strong force between the nucleons, the Coulomb force between the nucleons, and the kinetic energies of the nucleons relative to the center of mass of the entire nucleus. From the conservation of energy we can write, for this pulling-apart process,

$$m_d c^2 + E_B = m_n c^2 + m_p c^2. \qquad (2)$$

If we add $m_e c^2$, the energy equivalent of one electron mass, to each side of this equation, we have

$$(m_d + m_e)c^2 + E_B = m_n c^2 + (m_p + m_e)c^2,$$

or

$$m(^2\text{H})c^2 + E_B = m_n c^2 + m(^1\text{H})c^2. \qquad (3)$$

Here $m(^2\text{H})$ and $m(^1\text{H})$ are the masses of the neutral *heavy* hydrogen atom and the neutral *ordinary* hydrogen atom, respectively. They are atomic masses, not nuclear masses. Solving Eq. 3 for E_B yields

$$E_B = [m_n + m(^1\text{H}) - m(^2\text{H})]c^2 = \Delta m c^2, \qquad (4)$$

in which Δm is the mass difference. In making calculations of this kind we always use atomic, rather than nuclear, masses, as this is what is normally tabulated. As in this example, the electron masses conveniently cancel.*

For the deuteron calculation the needed masses are

$$m_n = 1.008665 \text{ u}, \qquad m(^1\text{H}) = 1.007825 \text{ u},$$
$$m(^2\text{H}) = 2.014102 \text{ u}.$$

Substituting into Eq. 4 and replacing c^2 by its equivalent, 931.5 MeV/u, we find the binding energy to be

$$E_B = (1.008665 \text{ u} + 1.007825 \text{ u}$$
$$- 2.014102 \text{ u})(931.5 \text{ MeV/u})$$
$$= (0.002388 \text{ u})(931.5 \text{ MeV/u}) = 2.224 \text{ MeV}.$$

Compare this with the binding energy of the hydrogen atom in its ground state, which is 13.6 eV, about five orders of magnitude smaller.

If we divide the binding energy of a nucleus by its mass number, we get the average binding energy per nucleon, a property we have listed in Table 1. Figure 6 shows a plot of this quantity as a function of mass number. The fact that this *binding energy curve* "droops" at both high and low mass numbers has practical consequences of the greatest importance.†

---

* See, however, Problem 51, for an exception.
† *The Curve of Binding Energy* has even been adopted as the title of a book (by John McPhee) about the possibilities of nuclear terrorism!

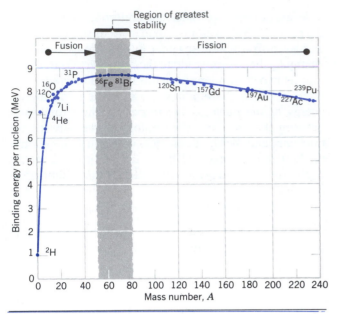

Figure 6   The binding energy per nucleon over the range of mass numbers. Some of the nuclides of Table 1 are identified, along with a few others. The region of greatest stability corresponds to mass numbers from about 50 to 80.

The drooping of the binding energy curve at high mass numbers tells us that nucleons are more tightly bound when they are assembled into two middle-mass nuclei rather than into a single high-mass nucleus. In other words, energy can be released in the *nuclear fission* of a single massive nucleus into two smaller fragments.

The drooping of the binding energy curve at low mass numbers, on the other hand, tells us that energy will be released if two nuclei of small mass number combine to form a single middle-mass nucleus. This process, the reverse of fission, is called *nuclear fusion*. It occurs inside our Sun and other stars and is the mechanism by which the Sun generates the energy it radiates to us.

Nuclear Spin and Magnetism

Nuclei, like atoms, have an intrinsic angular momentum whose maximum component along any chosen z axis is given by $J\hbar$. Here J is a quantum number, which may be integral or half-integral, called the *nuclear spin*; some values for selected nuclides are shown in Table 1.

Again as for atoms, a nuclear angular momentum has a nuclear magnetic moment associated with it. Recall that, in atomic magnetism, the *Bohr magneton* μ_B, defined as

$$\mu_B = \frac{eh}{4\pi m_e} = 5.79 \times 10^{-5} \text{ eV/T},$$

is a unit of convenience. In nuclear physics the corresponding unit of convenience is the *nuclear magneton* μ_N, defined similarly to the Bohr magneton except that the

electron mass m_e is replaced by the proton mass m_p. That is,

$$\mu_N = \frac{eh}{4\pi m_p} = 3.15 \times 10^{-8} \text{ eV/T}.$$

Because the magnetic moment of the free electron is (very closely) one Bohr magneton, it might be supposed that the magnetic moment of the free proton would be (very closely) one nuclear magneton. It is not very close, however, the measured value being $+2.7929\ \mu_N$. To understand the magnetic moments of the proton and neutron, it is necessary to consider their internal structure. The magnetic moments of heavier nuclei can in turn be analyzed in terms of the magnetic moments of the constituent protons and neutrons.

Sample Problem 2   What is the approximate density of the *nuclear matter* from which all nuclei are made?

Solution   We know that this density is large, because virtually all the mass of the atom resides in its tiny nucleus. The volume of the nucleus, approximated as a uniform sphere of radius R, is given by Eq. 1 as

$$V = \tfrac{4}{3}\pi R^3 = \tfrac{4}{3}\pi(R_0^3 A).$$

The density ρ_n of nuclear matter, expressed in nucleons per unit volume, is then

$$\rho_n = \frac{A}{V} = \frac{A}{(4\pi/3)R_0^3 A}$$

$$= \frac{1}{(4\pi/3)(1.2 \text{ fm})^3} = 0.14 \text{ nucleons/fm}^3.$$

The mass of a nucleon is 1.7×10^{-27} kg. The *mass* density ρ_m of nuclear matter is then

$$\rho_m = (0.14 \text{ nucleons/fm}^3)(1.7 \times 10^{-27} \text{ kg/nucleon})$$
$$\times (1 \text{ fm}/10^{-15} \text{ m})^3$$
$$= 2.4 \times 10^{17} \text{ kg/m}^3,$$

or 2.4×10^{14} times the density of water! Unlike the orbital electrons, the nuclides have a density nearly independent of the number of their nucleons. To some extent nucleons are packed in like marbles in a bag.

Sample Problem 3   Imagine that a typical middle-mass nucleus such as ^{120}Sn is picked apart into its constituent protons and neutrons. Find (a) the total energy required and (b) the energy per nucleon. The atomic mass of ^{120}Sn is 119.902199 u; see Table 1.

Solution   (a) ^{120}Sn contains 50 protons and $120 - 50 = 70$ neutrons. The combined atomic mass of these free particles is

$$M = Zm_p + Nm_n = 50 \times 1.007825 \text{ u} + 70 \times 1.008665 \text{ u}$$
$$= 120.997800 \text{ u}.$$

This exceeds the atomic mass of ^{120}Sn by

$$\Delta m = 120.997800 \text{ u} - 119.902199 \text{ u} = 1.095601 \text{ u}.$$

Converting this to a rest energy yields the total binding energy,

$$E_B = \Delta mc^2 = (1.0956 \text{ u})(931.5 \text{ MeV/u}) = 1020.6 \text{ MeV}.$$

(b) The binding energy E per nucleon is

$$E = \frac{E_B}{A} = \frac{1020.6 \text{ MeV}}{120} = 8.50 \text{ MeV/nucleon}.$$

This agrees with the value that may be read from the curve of Fig. 6.

54-3 RADIOACTIVE DECAY

As Fig. 4 shows, most of the nuclides that have been identified are radioactive. That is, they spontaneously emit a particle, transforming themselves in the process into a different nuclide. In this chapter we discuss the two most common situations, the emission of an α particle (alpha decay) and the emission of an electron (beta decay).

No matter what the nature of the decay, its main feature is that it is statistical. Consider, for example, a 1-mg sample of uranium metal. It contains 2.5×10^{18} atoms of the very long-lived alpha emitter ^{238}U. The nuclei of these atoms have existed without decaying since they were created (before the formation of our solar system) in the explosion of a supernova.

During any given second about 12 of the nuclei in our sample will decay, emitting an α particle in the process. We have absolutely no way of predicting, however, whether any given nucleus in the sample will be among those that do so. Every single ^{238}U nucleus has exactly the same probability as any other to decay during any 1-s observation period, namely, $12/(2.5 \times 10^{18})$, or one chance in 2×10^{17}.

In general, if a sample contains N radioactive nuclei, we can express the statistical character of the decay process by saying that the ratio of the decay rate $R\ (=-dN/dt)$ to the number of nuclei in the sample is equal to a constant, or

$$\frac{-dN/dt}{N} = \lambda, \tag{5}$$

in which λ, the *disintegration constant*, has a different characteristic value for each radioactive nuclide. We can rewrite Eq. 5 as

$$\frac{dN}{N} = -\lambda \, dt,$$

which integrates readily to

$$N = N_0 e^{-\lambda t}. \tag{6}$$

Here N_0 is the number of radioactive nuclei in the sample at $t = 0$. We see that the decrease of N with time follows a simple exponential law.

We are often more interested in the *activity* or decay rate $R\ (=-dN/dt)$ of the sample than we are in N. Differentiating Eq. 6 yields

$$R = R_0 e^{-\lambda t}, \tag{7}$$

in which $R_0\ (=\lambda N_0)$ is the decay rate at $t = 0$. Note also that $R = \lambda N$ at any time t. We assumed initially that the ratio of R to N is constant, so we are not surprised to confirm that they both decrease with time according to the same exponential law.

A quantity of interest is the time $t_{1/2}$, called the *half-life*, after which both N and R are reduced to one-half of their initial values. Putting $R = \frac{1}{2}R_0$ in Eq. 7 gives

$$\tfrac{1}{2}R_0 = R_0 e^{-\lambda t_{1/2}},$$

which leads readily to

$$t_{1/2} = \frac{\ln 2}{\lambda}, \tag{8}$$

a relationship between the half-life and the disintegration constant.

The following two sample problems show how λ can be measured for decay processes with relatively short half-lives and also with relatively long half-lives.

Sample Problem 4 In short-lived decays, it is possible to measure directly the decrease in the decay rate R with time. The following table gives some data for a sample of ^{128}I, a radionuclide often used medically as a tracer to measure the iodine uptake rate of the thyroid gland. Find (a) the disintegration constant λ and (b) the half-life $t_{1/2}$ from these data.

Time (min)	R (counts/s)	Time (min)	R (counts/s)
4	392.2	132	10.9
36	161.4	164	4.56
68	65.5	196	1.86
100	26.8	218	1.00

Solution (a) If we take the natural logarithm of each side of Eq. 7, we find that

$$\ln R = \ln R_0 - \lambda t. \tag{9}$$

Thus if we plot the natural logarithm of R against t, we should obtain a straight line whose slope is $-\lambda$. Figure 7 shows such a plot. Equating the slope of the line to $-\lambda$ yields

$$-\lambda = -\frac{(6.06 - 0)}{(220 \text{ min} - 0)},$$

or

$$\lambda = 0.0275 \text{ min}^{-1}.$$

(b) Equation 8 yields for $t_{1/2}$:

$$t_{1/2} = \frac{\ln 2}{\lambda} = \frac{0.693}{0.0275 \text{ min}^{-1}} = 25.2 \text{ min}.$$

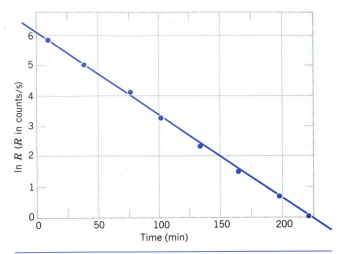

Figure 7   Sample Problem 4. A logarithmic plot of the decay data is fitted by a straight line, showing the exponential nature of the decay. The disintegration constant λ can be found from the slope of the line.

Sample Problem 5   A 1.00-g sample of pure KCl from the chemistry stock room is found to be radioactive and to decay at an absolute rate R of 1600 counts/s. The decays are traced to the element potassium and in particular to the isotope ^{40}K, which constitutes 1.18% of normal potassium. What is the half-life for this decay?

Solution   In the case of long-lived decays it is not possible to wait long enough to observe a measurable decrease in the decay rate R with time. We must find λ by measuring both N and $-dN/dt$ in Eq. 5. The molar mass of KCl is 74.9 g/mol, so the number of potassium atoms in the sample is

$$N_K = \frac{(6.02 \times 10^{23} \text{ mol}^{-1})(1.00 \text{ g})}{74.9 \text{ g/mol}} = 8.04 \times 10^{21}.$$

The number of ^{40}K atoms is 1.18% of N_K, or

$$N_{40} = (0.0118)(8.04 \times 10^{21}) = 9.49 \times 10^{19}.$$

From Eq. 5 we have

$$\lambda = \frac{-dN/dt}{N} = \frac{R}{N_{40}} = \frac{1600 \text{ s}^{-1}}{9.49 \times 10^{19}} = 1.69 \times 10^{-17} \text{ s}^{-1},$$

and the half-life, from Eq. 8, is

$$t_{1/2} = \frac{\ln 2}{\lambda} = \left(\frac{0.693}{1.69 \times 10^{-17} \text{ s}^{-1}}\right)\left(\frac{1 \text{ y}}{3.16 \times 10^{7} \text{ s}}\right)$$
$$= 1.30 \times 10^{9} \text{ y}.$$

This is of the order of magnitude of the age of the universe. No wonder we cannot measure the half-life of this nuclide by waiting around for its decay rate to decrease! (Interestingly, the potassium in our own bodies has its normal share of the ^{40}K isotope. We are all slightly radioactive.)

54-4   ALPHA DECAY

The radionuclide ^{238}U, a typical alpha emitter, decays spontaneously according to the scheme

$$^{238}\text{U} \rightarrow ^{234}\text{Th} + ^{4}\text{He}, \tag{10}$$

with a half-life of 4.47×10^9 y. In Sample Problem 6 we show that, in every such decay, an energy of 4.27 MeV is emitted, appearing as kinetic energy shared between the α particle (^{4}He) and the recoiling residual nucleus (^{234}Th).

We now ask ourselves: "If energy is released in every such decay event, why did the ^{238}U nuclei not decay shortly after they were created?" The creation process is believed to have occurred in the violent explosions of ancestral stars (supernovas), predating the formation of our solar system. Why did these nuclei wait so very long before getting rid of their excess energy by emitting an α particle? To answer this question, we must study the detailed mechanism of alpha decay.

We choose a model in which the α particle is imagined

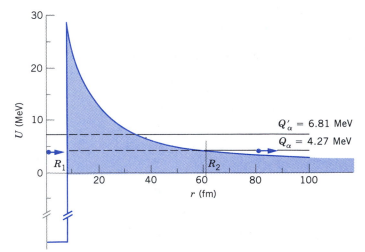

Figure 8   A potential energy function representing the emission of α particles by ^{238}U. The shaded area represents the potential barrier that inhibits the decay process. The horizontal lines represent the decay energies of ^{238}U (4.27 MeV) and ^{228}U (6.81 MeV).

to exist preformed inside the nucleus before it escapes. Figure 8 shows the approximate potential energy function $U(r)$ for the α particle and the residual ^{234}Th nucleus as a function of their separation. It is a combination of a potential well associated with the (attractive) strong nuclear force that acts in the nuclear interior ($r < R_1$) and a Coulomb potential associated with the (repulsive) electrostatic force that acts between the two particles after the decay has occurred ($r > R_1$).

The horizontal line marked $Q_\alpha = 4.27$ MeV shows the disintegration energy for the process, as calculated in Sample Problem 6. Note that this line intersects the potential energy curve at two points, R_1 and R_2. We now see why the α particle is not immediately emitted from the ^{238}U nucleus! That nucleus is surrounded by an impressive potential barrier, shown by the shaded area in Fig. 8. Visualize this barrier as a spherical shell whose inner radius is R_1 and whose outer radius is R_2, its volume being forbidden to the α particle under the laws of classical physics. If the α particle found itself in that region, its potential energy U would exceed its total energy E, which would mean, classically, that its kinetic energy K ($= E - U$) would be negative, an impossible situation.

Indeed, we now change our question and ask: "How can the ^{238}U nucleus *ever* emit an α particle?" The α particle seems permanently trapped inside the nucleus by the barrier.

The answer is that, as we learned in Section 50-8, in wave mechanics there is always a chance (described by Eq. 19 of Chapter 50) that a particle can tunnel through a barrier that is classically insurmountable. In fact, the explanation of alpha decay by wave mechanical barrier tunneling was one of the very first applications of the new quantum physics.

For the long-lived decay of ^{238}U the barrier is actually not very "leaky." We can show that the α particle, presumed to be rattling back and forth within the nucleus, must present itself at the inner surface of the barrier about 10^{38} times before it succeeds in tunneling through. This is about 10^{20} times per second for about 10^9 years! We, of course, are waiting on the outside, taking note of only those α particles that *do* manage to escape.

We can test this barrier tunneling explanation of alpha decay by looking at other alpha emitters, for which the barrier would be different. For an extreme contrast, consider the alpha decay of another uranium nuclide, ^{228}U, which has a disintegration energy Q'_α of 6.81 MeV, as shown in Fig. 8. The barrier in this case is both thinner (compare the lengths of the dashed lines in Fig. 8) and lower (compare the heights of the barrier above the dashed lines); if our barrier tunneling notions are correct, we would expect alpha decay to occur more readily for ^{228}U than for ^{238}U. Indeed it does. As Table 2 shows, the half-life of ^{228}U is only 550 s! We recall from Section 50-8 that the transmission coefficient of a barrier—because of the exponential nature of Eq. 19 of Chapter 50—is very

TABLE 2   THE ALPHA DECAY OF ^{238}U AND ^{228}U

Nuclide	Q_α	Half-life
^{238}U	4.27 MeV	4.5×10^9 y
^{228}U	6.81 MeV	550 s

sensitive to small changes in the dimensions of the barrier. We see that an increase in Q_α by a factor of only 1.6 produces a decrease in half-life (that is, in the effectiveness of barrier tunneling) by a factor of 3×10^{14}.

Sample Problem 6 (a) Find the energy released during the alpha decay of ^{238}U. (b) Show that this nuclide cannot spontaneously emit a proton. The needed atomic masses are

$$^{238}\text{U} \quad 238.050785 \text{ u} \qquad ^4\text{He} \quad 4.002603 \text{ u}$$

$$^{234}\text{Th} \quad 234.043593 \text{ u} \qquad ^1\text{H} \quad 1.007825 \text{ u}$$

$$^{237}\text{Pa} \quad 237.051143 \text{ u}.$$

Solution (a) In the alpha decay process of Eq. 10 the total atomic mass of the decay products ($= 238.046196$ u) is less than the atomic mass of ^{238}U by $\Delta m = 0.004589$ u, whose energy equivalent is

$$Q_\alpha = \Delta m \, c^2 = (0.004589 \text{ u})(931.5 \text{ MeV/u}) = 4.27 \text{ MeV}.$$

This *disintegration energy* is available to share as kinetic energy between the α particle and the recoiling ^{234}Th atom.

(b) If ^{238}U were to emit a proton, the decay process would be

$$^{238}\text{U} \rightarrow \,^{237}\text{Pa} + \,^1\text{H}.$$

In this case the mass of the decay products *exceeds* the mass of ^{238}U by $\Delta m = 0.008183$ u, the energy equivalent Q_p being -7.622 MeV. The minus sign means that we must *add* energy to split ^{238}U into ^{237}Pa plus a proton. Thus ^{238}U is stable against spontaneous proton emission.

54-5   BETA DECAY

A nucleus that decays spontaneously by emitting an electron (either positive or negative) is said to undergo *beta decay*.* Here are two examples:

$$^{32}\text{P} \rightarrow \,^{32}\text{S} + e^- + \bar{\nu} \qquad (t_{1/2} = 14.3 \text{ d}) \qquad (11)$$

and

$$^{64}\text{Cu} \rightarrow \,^{64}\text{Ni} + e^+ + \nu \qquad (t_{1/2} = 12.7 \text{ h}). \qquad (12)$$

The symbols ν and $\bar{\nu}$ represent the *neutrino* and its antiparticle, the *antineutrino*, neutral particles that are emit-

---

* Beta decay also includes electron capture, in which a nucleus decays by absorbing one of its orbital electrons. We do not consider that process here.

ted from the nucleus along with the electron or positron (positive electron) during the decay process. Neutrinos interact only very weakly with matter and—for that reason—are so extremely difficult to detect that, for many years, their presence went unnoticed. We consider the fundamental nature and importance of these elusive particles in Chapter 56.

It may seem surprising that nuclei can emit electrons (and neutrinos) in view of the fact that we have said that nuclei are made up of neutrons and protons only. However, we saw earlier that atoms emit photons, and we certainly do not say that atoms "contain" photons. We say that the photons are created during the emission process.

So it is with the electrons and the neutrinos emitted from nuclei during beta decay. They are both created during the emission process, a neutron transforming itself into a proton within the nucleus (or conversely) according to

$$n \rightarrow p + e^- + \bar{\nu} \quad (\beta^- \text{ decay}) \qquad (13)$$

or

$$p \rightarrow n + e^+ + \nu \quad (\beta^+ \text{ decay}). \qquad (14)$$

These are the basic beta-decay processes.

In any decay process, the amount of energy released is uniquely determined by the difference in rest energy between the initial nucleus and the final nucleus plus decay products (see, for example, Sample Problem 6). In a particular alpha-decay process, such as that of ^{238}U, every emitted α particle carries the same kinetic energy. In beta decay, however, the kinetic energy of the emitted electrons is not uniquely determined. Instead, the emitted electrons have a continuous spectrum of energies, from zero up to a maximum K_{max}, as Fig. 9 illustrates for the beta decay of ^{64}Cu (Eq. 12).

For many years, before the neutrino was identified, curves such as that of Fig. 9 were a challenging puzzle. They suggested that some energy was "missing" in the decay process and led many reputable physicists, including Niels Bohr, to speculate that perhaps the law of conservation of energy might be valid only statistically in such decays.

The answer to this puzzle lies in the emission of the neutrino or antineutrino, which carries a share of the decay energy. If we were to measure the energies of both particles (electron and antineutrino or positron and neutrino) in a particular decay process and add them up, we would come out every time with the same fixed value, equal to the disintegration energy. Energy is indeed conserved in each individual decay process.

The existence of an undetected particle as a solution to the missing energy problem was proposed by Pauli in 1931, and the neutrino was made a part of a formal theory of beta decay by Fermi in 1934. Nevertheless, it took another 20 years before neutrinos were detected in the laboratory. The difficulty in their measurement results from their exceedingly weak interactions with matter—their mean free path through solid matter is of the order of several thousand light years. Today neutrino physics is an important subfield of nuclear and particle physics, and its practitioners study not only neutrinos from radioactive sources but also those emitted in great quantities by the Sun and those that were created during the formation of the universe (which have a present density of about 100 per cm³). Figure 10 shows evidence of a burst of neutrinos detected on Earth resulting from the 1987 supernova in the nearby Large Magellanic Cloud (see Fig. 17 of Chapter 8). Because the detector was located in Japan and the supernova occurred in the southern sky, the neutrinos had to travel completely through the Earth to reach the detector.

Our study of alpha and beta decay permits us to look at the nuclidic chart of Fig. 4 in a new way. Let us construct a three-dimensional surface by plotting the mass of each nuclide in a direction at right angles to the NZ plane of that figure. The surface so formed gives a graphic representation of nuclear stability. As Fig. 11 shows (for the light nuclides), it describes a "valley of the nuclides," the stability zone of Fig. 4 running along its bottom. Nuclides on

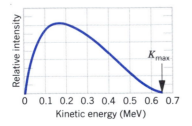

Figure 9   The kinetic energy distribution of the positrons emitted in the beta decay of ^{64}Cu. The maximum kinetic energy is 0.653 MeV.

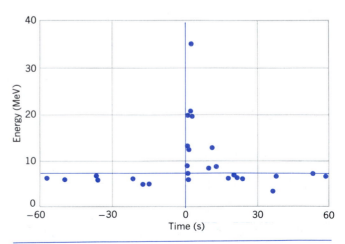

Figure 10   Evidence for a burst of neutrinos from the supernova SN 1987A.

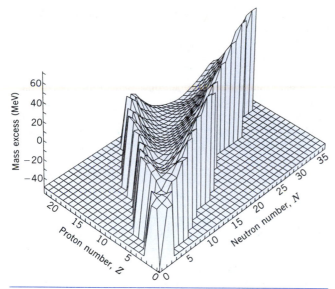

Figure 11 A portion of the valley of the nuclides, showing only the lightest nuclides. The quantity plotted on the vertical axis is the mass excess, defined as $(m - A)c^2$, where m is the atomic mass in u.

the headwall of the valley (a region not displayed in Fig. 11) decay into it largely by chains of alpha decay and by spontaneous fission. Nuclides on the proton-rich side of the valley decay into it by emitting positive electrons and those on the neutron-rich side do so by emitting negative electrons.

Sample Problem 7 Calculate the disintegration energy Q in the beta decay of ^{32}P, as described by Eq. 11. The needed atomic masses are 31.973907 u for ^{32}P and 31.972071 u for ^{32}S.

Solution Because of the presence of the emitted electron, we must be especially careful to distinguish between nuclear and atomic masses. Let m' represent the nuclear masses of ^{32}P and ^{32}S, and let m represent their atomic masses. We take the disintegration energy Q to be $\Delta m\, c^2$, where

$$\Delta m = m'(^{32}\text{P}) - [m'(^{32}\text{S}) + m_e],$$

m_e being the mass of the electron and the neutrino being assumed to be massless. If we add and subtract $15m_e$ on the right-hand side, we have

$$\Delta m = [m'(^{32}\text{P}) + 15m_e] - [m'(^{32}\text{S}) + 16m_e].$$

The quantities in brackets are the atomic masses. Thus we have

$$\Delta m = m(^{32}\text{P}) - m(^{32}\text{S}).$$

If we subtract the atomic masses in this way, the mass of the emitted electron is automatically taken into account.*

---

* This is not the case for positron decay or for electron capture; see Problems 51 and 52. Note also that in this sample problem we neglect the (small) difference in the binding energies of the atomic electrons before and after the beta decay.

The disintegration energy for the ^{32}P decay is then

$$Q = \Delta m\, c^2 = (31.973907 \text{ u} - 31.972071 \text{ u})(931.5 \text{ MeV/u})$$
$$= 1.71 \text{ MeV}.$$

This is just equal to the measured value of K_{max}, the maximum energy of the emitted electrons. Thus although 1.71 MeV is released every time a ^{32}P nucleus decays, in essentially every case the electron carries away less energy than this. The neutrino gets the rest, carrying it away from the laboratory undetected. (A negligible share, of the order of eV, also goes to the ^{32}S nucleus in order to conserve momentum in the decay.)

54-6 MEASURING IONIZING RADIATION*

When radiations such as x rays, gamma rays, beta particles, or alpha particles encounter an atom, they can cause the atom to eject electrons and to become ionized. Because ionization can damage individual cells of living tissue, the effects of ionizing radiations have become a matter of general public interest. Such radiations arise in nature from the cosmic rays and also from radioactive elements in the Earth's crust. Artificially produced radiations also contribute, including diagnostic and therapeutic x rays and radiations from radionuclides used in medicine and in industry. The disposal of radioactive waste and the evaluation of the probabilities of nuclear accidents continue to be dealt with at the level of national policy.

It is not our task here to explore the various sources of ionizing radiations but simply to describe the units in which the properties and effects of these radiations are expressed. There are four such units, and they are often used loosely or incorrectly in popular reporting.

1. *The curie (abbreviation Ci).* This is a measure of the *activity* or rate of decay of a radioactive source. It was originally defined as the activity of 1 g of radium in equilibrium with its by-products, but it is now defined simply as

1 curie $= 3.7 \times 10^{10}$ disintegrations per second.

This definition says nothing about the nature of the decays. Note also that this unit is not appropriate to describe the ionizing effects of x rays from, say, a medical x-ray machine. The radiations must be emitted from a radionuclide.

An example of the proper use of the curie is the statement: "One milligram of ^{239}Pu has an activity of 62 μCi." The fact that ^{239}Pu is an alpha emitter does not enter.

---

* See "Radiation Exposure in Our Daily Lives," by Stewart C. Bushong, *The Physics Teacher,* March 1977, p. 135.

2. *The roentgen (abbreviation R).* This is a measure of *exposure,* that is, of the ability of a beam of x rays or gamma rays to produce ions in a particular substance. Specifically, one roentgen is defined as that exposure that would produce 1.6×10^{12} ion pairs per gram of air, the air being dry and at standard temperature and pressure. We might say, for example: "In 0.1 s, this dental x-ray beam provides an exposure of 30 mR." This says nothing about whether ions are actually produced or whether or not there is a patient in the chair.

3. *The rad.* This is an acronym for radiation absorbed dose and is a measure, as its name suggests, of the dose actually delivered to a specific object, in terms of the energy transferred to it. An object, which might be a person (whole body) or a specific part of the body (the hands, say) is said to have received an absorbed dose of 1 rad when 10^{-5} J/g have been delivered to it by ionizing radiations. A typical statement to show the usage is: "A whole-body gamma-ray dose of 300 rad will cause death in 50% of the population exposed to it." By way of comfort we note that the present average exposure to radiation from both natural and artificial sources is about 0.2 rad ($= 200$ mrad) per year.

4. *The rem.* This is an acronym for roentgen equivalent in man and is a measure of *dose equivalent.* It takes account of the fact that, although different types of radiation (gamma rays and neutrons, say) may deliver the same energy per unit mass to the body, they do not have the same biological effect. The dose equivalent (in rems) is found by multiplying the absorbed dose (in rads) by a *quality factor* QF, which may be found tabulated in various reference sources. For x rays and electrons, QF = 1. For slow neutrons, QF = 5, and so on. Personnel monitoring devices such as film badges are designed to register the dose equivalent in rems.

An example of correct usage of the rem is: "The recommendation of the National Council on Radiation Protection is that no individual who is (nonoccupationally) exposed to radiations should receive a dose equivalent greater than 500 mrem ($= 0.5$ rem) in any one year." This includes radiations of all kinds, using the appropriate quality factors.

Sample Problem 8 A dose of 300 rad is lethal to 50% of the population that receives it. If the equivalent amount of energy were absorbed directly as heat, what temperature increase would result? Assume that c, the specific heat capacity of the human body, is the same as that of water, namely, 4180 J/kg·K.

Solution An absorbed dose of 300 rad corresponds to an absorbed energy per unit mass of

$$(300 \text{ rad}) \left(\frac{10^{-2} \text{ J/kg}}{1 \text{ rad}} \right) = 3 \text{ J/kg}.$$

The temperature increase that would result from such an influx of heat is found from

$$\Delta T = \frac{Q/m}{c} = \frac{3 \text{ J/kg}}{4180 \text{ J/kg·K}} = 7.2 \times 10^{-4} \text{ K}.$$

We see from this tiny temperature increase that the damage done by ionizing radiation has very little to do with thermal heating. The harmful effects arise because the ionizing radiation succeeds in breaking molecular bonds and thus interfering with the normal functioning of the tissue in which it has been absorbed.

54-7 NATURAL RADIOACTIVITY

All the elements beyond hydrogen and helium were made in nuclear reactions in the interiors of stars or in explosive supernovas. Both radioactive and stable nuclides are created in these processes. The solar system is composed of nuclides that were formed about 4.5×10^9 years ago. (How this is determined is discussed later in this section.) Most of the radioactive nuclides that were formed at that time have half-lives that are far shorter than a billion years, and so they have long since decayed to stable nuclides through alpha or beta emission. A few of the original radioactive nuclides, however, have half-lives that are not short in comparison with the age of the solar system. The decay of these nuclides can still be observed, and these decays form part of the background of natural radioactivity in our environment.

Some of these radioactive species are part of decay chains that start with heavy nuclides, such as ^{232}Th ($t_{1/2} = 1.4 \times 10^{10}$ y) or ^{238}U ($t_{1/2} = 4.5 \times 10^9$ y). These nuclides decay through a sequence of alpha and beta decays, eventually reaching stable end products (respectively, ^{208}Pb and ^{206}Pb). The intermediate nuclei in these decay chains have much shorter half-lives; the rate at which the original nuclide disappears and is replaced with the stable end product is determined by the longest-lived member of the chain. These decay processes have presumably been going on since the solar system was formed, and so (as we discuss later) the relative amounts of the initial nuclide and stable decay products present in a material can give a measure of the age of the material. These decays are also thought to contribute to the internal heating of the planets.

Normally, the products of these decays remain in place in the rocks or minerals containing the parent nuclide. However, one of the intermediate substances produced in these decay chains, radon, is a gas. Natural decays that occur near the surface of the Earth (and in building materials, such as concrete) release radioactive radon gas into the atmosphere. The hazards of breathing this radon gas are currently the subject of active research. Radon gas can

TABLE 3 SOME NATURAL RADIOACTIVE ISOTOPES

Isotope	$t_{1/2}$ (y)
^{40}K	1.28×10^9
^{87}Rb	4.8×10^{10}
^{113}Cd	9×10^{15}
^{115}In	4.4×10^{14}
^{138}La	1.3×10^{11}
^{176}Lu	3.6×10^{10}
^{187}Re	5×10^{10}

also be released from the fracture of rocks beneath the surface; therefore the detection of radon gas has been used as a way of predicting earthquakes.

In addition to the heavy elements, other long-lived radioactive nuclides are present in natural substances. Some of these are listed in Table 3.

Other radioactive nuclides are continually produced by natural processes, generally in the Earth's atmosphere by reactions of molecules of the air with cosmic rays (high-energy protons from space). Notable among these is ^{14}C ($t_{1/2} = 5730$ y), which has important applications in radioactive dating of organic materials.

Radioactive Dating

Suppose we have an initial radionuclide I that decays to a final product F with a known half-life $t_{1/2}$. At a particular time $t = 0$, we start with N_0 initial nuclei and none of the final product nuclei. At a later time t, we find N_I of the original nuclei remain, while $N_F (= N_0 - N_I)$ of the product nuclei have appeared. The initial nuclei decay according to

$$N_I = N_0 e^{-\lambda t},$$

and thus

$$t = \frac{1}{\lambda} \ln \frac{N_0}{N_I} = \frac{t_{1/2}}{\ln 2} \ln \frac{N_0}{N_I}$$

or, substituting $N_I + N_F$ for N_0,

$$t = \frac{t_{1/2}}{\ln 2} \ln \left(1 + \frac{N_F}{N_I} \right). \tag{15}$$

That is, a measurement of the present ratio of product and original nuclei can determine the age of the sample.

This calculation has been based on the assumption that none of the product nuclei were present at $t = 0$. This assumption may not always be valid, but there are techniques for radioactive dating that can correct for the presence of these original product nuclei.

This method can be used to determine the time since the formation of the solar system; examples include the ratios of ^{238}U to ^{206}Pb, ^{87}Rb to ^{87}Sr, and ^{40}K to ^{40}Ar. Terrestrial rocks, Moon rocks, and meteorites analyzed

by these methods all seem to have common ages of around 4.5×10^9 y, which we take to be the age of the solar system.

The radioactive isotope ^{14}C is present in the atmosphere; about 1 carbon atom in 10^{12} is radioactive ^{14}C. Each gram of carbon has an activity of about 12 decays per minute due to ^{14}C. Living organisms can absorb this activity by aspiration of CO_2 or by eating plants that have done so. When the organism dies, it stops absorbing ^{14}C, and the ^{14}C present at its death begins to decay. By measuring the decay rate of ^{14}C, we can determine the age of the sample. For example, if we examine a sample and it shows 6 decays per minute per gram of carbon, we know that the original activity has been reduced by half, and the sample must be one half-life (5730 y) old.

This method of *radiocarbon dating* (which was developed in 1947 by Willard Libby, who was awarded the 1960 Nobel prize in chemistry for this work) is useful for samples of organic matter that are less than about 10 half-lives in age. In 10 half-lives, the activity of a sample drops by a factor of 2^{-10}, or about 10^{-3}, and the decay rate becomes too small to be determined with precision. The practical upper limit on the age of samples that can be dated by this method is about 50,000 y. In recent years, a new technique has been developed in which an accelerator is used as a mass spectrometer to determine the $^{14}C/^{12}C$ ratio to high precision. In this way the usefulness of radiocarbon dating has been extended to samples as old as 100,000 y.

Sample Problem 9 In a sample of rock, the ratio of ^{206}Pb to ^{238}U nuclei is found to be 0.65. What is the age of the rock?

Solution From Eq. 15, using 4.5×10^9 y for the half-life of ^{238}U, we have

$$t = \frac{4.5 \times 10^9 \text{ y}}{0.693} \ln (1 + 0.65) = 3.3 \times 10^9 \text{ y}.$$

This rock is somewhat younger than the maximum age of 4.5×10^9 y that we determine for rocks in the solar system. This may suggest that the rock did not solidify until 3.3×10^9 y ago. The ^{206}Pb that was formed prior to that time was "boiled off" from the molten rock. Only after the rock solidified could the ^{206}Pb begin to accumulate.

54-8 NUCLEAR REACTIONS

We can represent a nuclear reaction by

$$X + a \rightarrow Y + b \tag{16}$$

or, in more compact notation,

$$X(a,b)Y. \tag{17}$$

Usually, particle *a* is the *projectile nucleus* and particle *X* is the *target nucleus*, which is often at rest in the laboratory. If the projectile is a charged particle, it may be raised to its desired energy in a Van de Graaff accelerator (see Section 30-11) or a cyclotron (see Section 34-3). The projectile may also be a neutron from a nuclear reactor. It is customary to designate product particle *Y* as the heavier *residual nucleus* and *b* as the lighter *emerging nucleus*.

The reaction energy *Q* is defined as

$$Q = (m_X + m_a)c^2 - (m_Y + m_b)c^2. \quad (18)$$

Using energy conservation, we can write Eq. 18 as

$$Q = (K_Y + K_b) - (K_X + K_a), \quad (19)$$

in which *K* represents the kinetic energy.

Equations 18 and 19 are valid only when *Y* and *b* are in their ground states. As we discuss later in this section, if either nuclide is produced in an excited state, the reaction energy is reduced by the excitation energy.

A typical reaction is

$$^{19}F(p,\alpha)^{16}O,$$

for which $Q = 8.13$ MeV. Equations 18 and 19 tell us that, in this reaction, the system loses rest energy and gains kinetic energy, in amount 8.13 MeV per event. Reactions, like this one, for which $Q > 0$ are called *exothermic*. Reactions for which $Q < 0$ are called *endothermic*; such reactions will not "go" unless a certain minimum kinetic energy (the *threshold energy*) is carried into the system by the projectile.

If *a* and *b* are identical particles, which requires that *X* and *Y* also be identical, we describe the reaction as *scattering*. If the kinetic energy of the system is the same both before and after the event (which means that $Q = 0$ and all nuclides remain in their ground states), we have *elastic scattering*. If these energies are different ($Q \neq 0$), we have *inelastic scattering*, in which case *Y* or *b* may be left in an excited state.

We can easily keep track of nuclear reactions by plotting them on a nuclidic chart like that of Fig. 4. Figure 12 shows an enlarged portion of such a chart, centered arbitrarily on the nuclide ^{197}Au. Stable nuclides are shaded, and their isotopic abundances are shown. The unshaded squares represent radionuclides, their half-lives being shown.

Figure 13 suggests a transparent overlay that we can place over a nuclidic chart such as that of Fig. 12. If the shaded central square of Fig. 13 overlays a particular target on the chart of Fig. 12, the residual nuclides resulting from the various reactions printed on the overlay are identified.

Thus if we chose ^{197}Au as a target, a (p,α) reaction will produce (stable) ^{194}Pt, and either an (n,γ) or a (d,p) reaction will produce the radionuclide ^{198}Au, whose half-life is 2.70 d.

82	^{197}Pb 8 min	^{198}Pb 2.4 h	^{199}Pb 1.5 h	^{200}Pb 21.5 h	^{201}Pb 9.42 h	^{202}Pb 5250 y	^{203}Pb 52.0 h
81	^{196}Tl 1.84 h	^{197}Tl 2.83 h	^{198}Tl 5.3 h	^{199}Tl 7.4 h	^{200}Tl 26.1 h	^{201}Tl 73.6 h	^{202}Tl 12.2 d
80	^{195}Hg 9.5 h	^{196}Hg 0.15%	^{197}Hg 64.1 h	^{198}Hg 10.0%	^{199}Hg 16.9%	^{200}Hg 23.1%	^{201}Hg 13.2%
79	^{194}Au 39.5 h	^{195}Au 183 d	^{196}Au 6.18 d	^{197}Au 100%	^{198}Au 2.70 d	^{199}Au 3.14 d	^{200}Au 48.4 min
78	^{193}Pt 50 y	^{194}Pt 32.9%	^{195}Pt 33.8%	^{196}Pt 25.3 %	^{197}Pt 18.3 h	^{198}Pt 7.2%	^{199}Pt 30.8 min
77	^{192}Ir 74.2 d	^{193}Ir 62.7%	^{194}Ir 19.2 h	^{195}Ir 2.5 h	^{196}Ir 52 s	^{197}Ir 5.8 min	^{198}Ir 8 s
76	^{191}Os 15.4 d	^{192}Os 41.0%	^{193}Os 30.5 h	^{194}Os 6.0 y	^{195}Os 6.5 min	^{196}Os 35 min	—
	115	116	117	118	119	120	121

Atomic number, *Z* (vertical axis) · Neutron number, *N* (horizontal axis)

Figure 12 An expanded portion of the chart of the nuclides (Fig. 4).

Nuclei, like atoms, have stationary states of definite energy, and reaction studies can be used to identify them. Consider, for example, the reaction

$$^{27}Al(d,p)^{28}Al, \qquad Q = 5.49 \text{ MeV},$$

in which a thin aluminum target foil is bombarded with 2.10-MeV deuterons. In the laboratory the emerging protons are seen to come off with a number of well-defined discrete energies and are accompanied by gamma rays. Figure 14 shows the energy distribution of the emerging protons.

In every reaction event we know that an energy equal to the kinetic energy of the incident deuteron ($= 2.10$ MeV)

			α,n	α,γ
	p,n	p,γ d,n	d,γ α,d	α,p
	γ,n p,d		n,γ d,p	
p,α	γ,d d,α	n,d γ,p	n,p	
γ,α	n,α			

Figure 13 Placing this as an overlay on Fig. 12, with the shaded central square over a particular target nuclide, shows the residual nuclides that result from the indicated reactions.

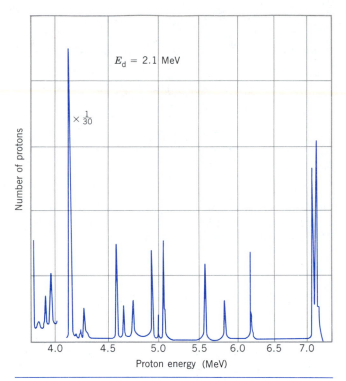

Figure 14   The energy distribution of protons resulting from the reaction ^{27}Al(d,p)^{28}Al. The incident deuteron has an energy of 2.10 MeV. The protons are detected as they emerge from the target at right angles to the incident beam.

plus the reaction energy $Q (= 5.49$ MeV) is available to be shared between the two reaction products, that is, between the residual nucleus ^{28}Al and the emerging proton p. How is this total energy (2.10 MeV + 5.49 MeV = 7.59 MeV) to be shared between these two particles?

It all depends on whether the residual nucleus ^{28}Al is produced in its ground state or in one of its excited stationary states. In the former case, the emerging proton will have the maximum possible energy, corresponding to the peak on the extreme right of the proton spectrum in Fig. 14. If, however, the residual nucleus is formed in an excited state, that nucleus will retain more of the available energy and there will be less energy left for the emerging proton. The residual nucleus will not remain in its excited state very long but will get rid of its excess energy, such as by emitting a gamma ray.

Every proton peak in the spectrum of Fig. 14 corresponds to a stationary state of the residual nucleus ^{28}Al. Figure 15 shows the energy levels that may be deduced by analyzing this spectrum. You can see the correspondence between the peaks of Fig. 14 and the energy levels of Fig. 15. We have seen that our understanding of the way atoms are put together rests on the measured energies of the hydrogen atom states as its firm foundation. In the same way, we can learn how nuclei are put together by

studying the energies and other properties of their stationary states.

Sample Problem 10   In the reaction

$$^1H + {}^3H \rightarrow {}^2H + {}^2H,$$

protons (^{1}H) with kinetic energy 5.70 MeV are incident on ^{3}H at rest. (*a*) What is the Q value for this reaction? (*b*) Find the kinetic energies of the deuterons emitted along the direction of the incident proton.

Solution   (*a*) From Eq. 18 we have

$$Q = [m(^1H) + m(^3H) - m(^2H) - m(^2H)]c^2$$
$$= (1.007825 \text{ u} + 3.016049 \text{ u} - 2.014102 \text{ u}$$
$$\quad - 2.014102 \text{ u})(931.5 \text{ MeV/u})$$
$$= -4.03 \text{ MeV}.$$

This reaction is endothermic; the final products have the greater mass and correspondingly the smaller kinetic energy by Eq. 19.

(*b*) Using Eq. 19, with $K = 0$ for the initial ^{3}H, we have

$$K_1 + K_2 = Q + K_p = -4.03 \text{ MeV} + 5.70 \text{ MeV}$$
$$= 1.67 \text{ MeV}. \tag{20}$$

Here the subscripts 1 and 2 refer to the two ^{2}H product nuclei. Conservation of momentum along the direction of the incident protons gives

$$p_1 + p_2 = p_p = \sqrt{2m(^1H)K_p} = \sqrt{2(938 \text{ MeV}/c^2)(5.70 \text{ MeV})}$$
$$= 103.4 \text{ MeV}/c. \tag{21}$$

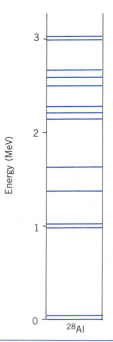

Figure 15   Energy levels of ^{28}Al, deduced from data such as those of Fig. 14.

Equations 20 and 21 can be solved as two equations in two unknowns (either p_1 and p_2 or K_1 and K_2). The results are

$$K_1 = 0.24 \text{ MeV} \quad \text{and} \quad K_2 = 1.43 \text{ MeV}.$$

Note that we have used nonrelativistic dynamics in solving this problem. Is this a good approximation?

54-9   NUCLEAR MODELS   *(Optional)*

The structure of atoms is now well understood. The Coulomb force is exerted by the massive center (the nucleus) on the electrons, and (given enough computer time) we can use the methods of quantum mechanics to calculate properties of the atom.

Things are not quite so well understood in the case of nuclei. The force law is complicated and cannot be written down explicitly in full detail. Nor is there a natural force center to simplify the calculations. To understand nuclear structure, we face a many-body problem of great complexity.

In the absence of a comprehensive theory of nuclear structure, we try instead to construct *nuclear models.* Physicists use models as simplified ways of looking at a complex system to give physical insight into its properties. The usefulness of a model is tested by its ability to make predictions that can be verified experimentally in the laboratory.

Two models of the nucleus have proved useful. One model describes situations in which we can consider all the protons and neutrons to behave cooperatively, while the other model neglects all but one proton or neutron in determining the properties of the nucleus. These two models represent quite opposing views of nuclear structure, but they can be combined to create a single unified model of the nucleus.

The Collective Model

In the collective model, we ignore the motions of individual nucleons and treat the nucleus as a single entity. This model, originally called the "liquid drop model," was developed by Niels Bohr to explain nuclear fission. We imagine the nucleus as a body analogous to a liquid drop, in which the nucleons interact with each other like molecules in the liquid.

The equilibrium shape of the liquid drop is determined by the interactions of its molecules, and similarly the equilibrium shape of a nucleus is determined by the interactions of its nucleons. Many nuclei have spherical equilibrium shapes, while others may be ellipsoidal.

Like a liquid drop, a nucleus can absorb energy by the entire nucleus rotating about an axis or vibrating about its equilibrium shape. Through radioactive decay or nuclear reaction experiments, it is possible to study the spectra of these excited states. Figure 16 shows examples of the two kinds of situations. The rotational energy $\frac{1}{2}I\omega^2$ can be written in terms of the angular momentum L ($=I\omega$) as $L^2/2I$. Writing the quantized angular momentum according to Eq. 23 of Chapter 51 as $L = \sqrt{J(J+1)}\hbar$, where J is the rotational angular momentum quantum number of the entire nucleus, we obtain

$$E_J = \frac{\hbar^2}{2I} J(J+1). \tag{22}$$

Note that the spacing between the states grows as the angular momentum increases.

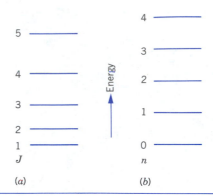

Figure 16   (*a*) Rotational excited states, labeled with the angular momentum quantum number *J*. (*b*) Vibrational states, labeled with the vibrational quantum number *n*.

The vibrational states have energies given by

$$E_n = n\hbar\omega = nh\nu, \qquad n = 1, 2, 3, \ldots, \tag{23}$$

where ν is the vibrational frequency. This is the same expression that was used by Planck to describe the quantized vibrations of the atomic oscillators in the theory of cavity radiation (see Eq. 7 and Fig. 6 of Chapter 49). Note that the vibrational states in Fig. 16*b* are equally spaced, as given by Eq. 23.

Evidence for collective structure can also be found in nuclear reactions. In a certain class of reactions $X + a \rightarrow Y + b$, an intermediate state is formed when X and a coalesce into a single entity C^*, called a *compound nucleus,* which then breaks apart into $Y + b$. The energy carried by projectile a into target X is quickly shared more or less equally in the random motion of the nucleons of the compound nucleus. (In the context of the liquid drop model, think of two drops coming together to form a larger drop whose molecules have a higher mean kinetic energy, corresponding to a higher temperature for the combined drop.)

The compound nucleus may exist for as long as 10^{-16} s, a very long time by the standards of nuclear reactions, which may typically last for only 10^{-22} s. Eventually, a nucleon or a group of nucleons will, by a statistical fluctuation, acquire enough energy to break free of the compound nucleus. We observe this outgoing particle b and the residual nucleus Y.

A central feature of this hypothesis is that, once the energy of the projectile is shared among the nucleons, the compound nucleus "forgets" how it was formed and decays purely according to statistical considerations. Figure 17 represents this process, in which a compound nucleus $^{20}\text{Ne}^*$ is formed in any of three ways

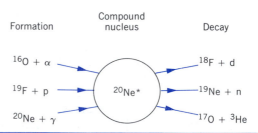

Figure 17   A few of the many possible formation and decay modes of the compound nucleus $^{20}\text{Ne}^*$.

and decays in any of three different ways. Experimentally, we observe that the relative probability of the different decay modes has the same value for any of the combinations of projectile and target. This confirms the compound nucleus interpretation and provides another example of the collective behavior of nucleons in the nucleus.

The Independent Particle Model

In the liquid drop model, the nucleons move around at random and bump into each other frequently. The independent particle model, however, considers that each nucleon moves in a well-defined orbit within the nucleus and hardly makes any collisions at all! The nucleus—unlike the atom—has no fixed center of charge, and we assume in this model that each nucleon moves in a potential that is determined by the smeared-out motions of all the other nucleons.

A nucleon in a nucleus, like an electron in an atom, has a set of quantum numbers that defines its state of motion. Also nucleons, again like electrons, obey the Pauli exclusion principle. That is, no two nucleons may occupy the same state at the same time. In considering nucleon states, the neutrons and the protons are treated separately, each having its own array of available quantized states.

The fact that nucleons obey the Pauli principle helps us to understand the relative stability of the nucleon states. If two nucleons within the nucleus are to collide, the energy of each of them after the collision must correspond to the energy of an unoccupied stationary state. If these states (or even just one of them) are filled, the collision simply cannot occur. In time, any given nucleon will find it possible to collide, but meanwhile it will have made enough revolutions in its orbit to give meaning to the notion of a stationary nucleon state with a quantized energy.

In the atomic realm, the essence of the periodic table of the elements is that it is periodic. That is, certain properties of the elements repeat themselves in a regular fashion as one proceeds through the table. These repetitions are associated with the fact that the atomic electrons arrange themselves in shells and subshells that have a special stability when they are fully occupied. We can take the atomic numbers of the inert gases,

$$2, 10, 18, 36, 54, 86, \ldots ,$$

as *magic electron numbers* that mark the completion of such shells.

Nuclei also show shell effects, associated with certain *magic nucleon numbers:*

$$2, 8, 20, 28, 50, 82, 126, \ldots .$$

Any nuclide whose proton number Z or neutron number N has one of these values turns out to have a special stability that may be made apparent in a variety of ways.

Examples of "magic" nuclides are ^{18}O ($Z = 8$), ^{40}Ca ($Z = 20$, $N = 20$), ^{92}Mo ($N = 50$), and ^{208}Pb ($Z = 82$, $N = 126$). Both ^{40}Ca and ^{208}Pb are said to be "doubly magic" because they contain filled shells of both protons *and* neutrons.

The magic number 2 shows up in the exceptional stability of the α particle (^{4}He), which, with $Z = N = 2$, is doubly magic. For example, the binding energy per nucleon for this nuclide stands well above that of its neighbors on the binding energy curve of Fig. 6. The α particle is so tightly bound, in fact, that it is impossible to add another particle to it; there is no stable nuclide with $A = 5$.

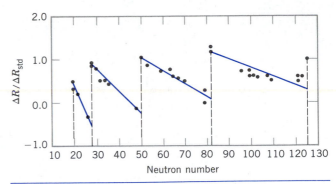

Figure 18 The variation in nuclear radius as a function of neutron number. The variation is expressed relative to the "standard" variation expected from the "collective" structure of $R = R_0 A^{1/3}$. The sudden jumps indicate shell structure.

Evidence for *atomic* shell structure can be found, for example, from measurements of the ionization energies or mean radii of atoms. Figure 6 of Chapter 52 shows the variation in the ionization energy of atoms as a function of the number of electrons. If we plot the atomic radii as a function of electron number, we find a gradual decrease as one shell is filled and then a sudden jump as we begin filling the next shell, because the radius depends primarily on the principal quantum number n. These sudden jumps in the radius and in the ionization energy occur when the number of electrons is equal to one of the magic electron numbers.

In the case of nuclei, we can gather similar evidence for *nuclear* shell structure. Sample Problem 12 gives an example of the change in "ionization energy" (the energy needed to remove a single proton or neutron from the nucleus) at closed shells. Figure 18 shows the variation in the nuclear radius as a function of neutron number. Just as in the atomic case, the radius gradually decreases within a shell and then increases suddenly as we begin filling the next shell. The sudden jumps occur when either the proton number or neutron number is equal to one of the magic nucleon numbers. Similar evidence for shell structure can be found in other nuclear properties, including alpha-decay half-lives, magnetic dipole moments, cross sections for capture of neutrons and scattering of electrons, and energies of excited states. ∎

Sample Problem 11 Consider the neutron-capture reaction

$$^{109}Ag + n \rightarrow {}^{110}Ag^* \rightarrow {}^{110}Ag + \gamma.$$

Figure 19 shows its cross section as a function of the energy of the incident neutron. Analyze this figure in terms of the compound nucleus concept and the uncertainty principle.

Solution The cross-section curve of Fig. 19 is sharply peaked, reaching a maximum cross section of 12,500 barns.† This "reso-

---

† The *cross section* for a reaction is a measure of the probability for the reaction to occur. A common unit for expressing cross section is the *barn,* which is equivalent to 10^{-28} m².

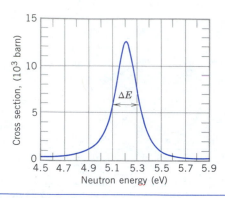

Figure 19   Sample Problem 11. The cross section for the reaction $^{109}\text{Ag}(n,\gamma)^{110}\text{Ag}$ as a function of the energy of the incident neutron. The width of the peak at half its maximum is about 0.20 eV.

nance peak" suggests that we are dealing with a single excited level in the compound nucleus $^{110}\text{Ag}^*$. When the available energy just matches the energy of this level above the ^{110}Ag ground state, we have "resonance," and the reaction really "goes."

However, the resonance peak is not infinitely sharp. From the figure we can measure that it has an approximate width at half maximum (that is, at 6250 barns) of 0.20 eV. We account for this by saying that ^{110}Ag in its excited state is not sharply defined in energy; it is "fuzzy," with an energy uncertainty ΔE of 0.20 eV.

We can use the uncertainty principle, written in the form

$$\Delta E \cdot \Delta t \sim h/2\pi \qquad (24)$$

to tell us something about any state of an atomic or nuclear system. We have seen that ΔE is a measure of the uncertainty of our knowledge of the energy of the state. The quantity Δt is interpreted as the time available to measure the energy of the state; it is in fact the mean life of the state before it decays.

For the excited state $^{110}\text{Ag}^*$ we have, from Eq. 24,

$$\Delta t \sim \frac{h/2\pi}{\Delta E} = \frac{6.58 \times 10^{-16}\ \text{eV} \cdot \text{s}}{0.20\ \text{eV}} = 3.3 \times 10^{-15}\ \text{s}.$$

This is the order of magnitude of the lifetime that is characteristic of a compound nucleus.

Sample Problem 12   The nuclide ^{120}Sn ($Z = 50$) has a filled proton shell, 50 being one of the magic nucleon numbers. The nuclide ^{121}Sb ($Z = 51$) has an "extra" proton outside this shell. According to the shell concept, this extra proton should be easier to remove than a proton from the filled shell. Verify this by calculating the required energy in each case. Use the following mass data:

Nuclide	Z	N	Atomic Mass (u)
^{121}Sb	$50 + 1$	70	120.903821
^{120}Sn	50	70	119.902199
^{119}In	$50 - 1$	70	118.905819

The proton atomic mass is 1.007825 u.

Solution   Removing the "extra" proton corresponds to the process

$$^{121}\text{Sb} \rightarrow {}^{120}\text{Sn} + \text{p}.$$

The required energy E follows from

$$E = [m(^{120}\text{Sn}) + m(^{1}\text{H}) - m(^{121}\text{Sb})]c^2$$
$$= (119.902199\ \text{u} + 1.007825\ \text{u} - 120.903821\ \text{u})$$
$$\times (931.5\ \text{MeV/u})$$
$$= 5.8\ \text{MeV}.$$

Removing the proton from the filled shell corresponds to

$$^{120}\text{Sn} \rightarrow {}^{119}\text{In} + \text{p}.$$

The required energy follows from

$$E = [m(^{119}\text{In}) + m(^{1}\text{H}) - m(^{120}\text{Sn})]c^2$$
$$= (118.905819\ \text{u} + 1.007825\ \text{u} - 119.902199\ \text{u})$$
$$\times (931.5\ \text{MeV/u})$$
$$= 10.7\ \text{MeV}.$$

This is considerably greater than the energy required to remove an "extra" proton ($= 5.8$ MeV), just as the shell model predicts. In much the same way the energy needed to remove an *electron* from a filled *electron shell* ($= 22$ eV for the filled shell of neon) is much greater than that needed to remove an "extra" electron from outside such a filled shell ($= 5$ eV for the "extra" electron from sodium).

QUESTIONS

1. When a thin foil is bombarded with α particles, a few of them are scattered back toward the source. Rutherford concluded from this that the positive charge of the atom—and also most of its mass—must be concentrated in a very small "nucleus" within the atom. What was his line of reasoning?

2. In what ways do the so-called strong force and the electrostatic or Coulomb force differ?

3. Why does the *relative* importance of the Coulomb force compared to the strong nuclear force increase at large mass numbers?

4. In your body, are there more neutrons than protons? More protons than electrons? Discuss.

5. Why do nuclei tend to have more neutrons than protons at high mass numbers?

6. Why do we use atomic rather than nuclear masses in analyzing most nuclear decay and reaction processes?

7. How might the equality 1 u $= 1.6605 \times 10^{-27}$ kg be arrived at in the laboratory?

8. The atoms of a given element may differ in mass, have different physical characteristics, and yet not vary chemically. Why is this?

9. The deviation of isotopic masses from integer values is due to many factors. Name some. Which is most responsible?

10. How is the mass of the neutron determined?

11. The most stable nuclides have a mass number A near 60 (see Fig. 6). Why don't *all* nuclides have mass numbers near 60?

12. If we neglect the very lightest nuclides, the binding energy per nucleon in Fig. 6 is roughly constant at 7 to 8 MeV/nucleon. Do you expect the mean electronic binding energy per electron in atoms also to be roughly constant throughout the periodic table?

13. Why is the binding energy per nucleon (Fig. 6) low at low mass numbers? At high mass numbers?

14. In the binding energy curve of Fig. 6, what is special or notable about the nuclides ^{2}H, ^{4}He, ^{62}Ni, and ^{239}Pu?

15. The magnetic moment of the neutron is $-1.9130\,\mu_N$. What is a nuclear magneton and how does it differ from a Bohr magneton? What does the minus sign mean? How can the neutron, which carries no net charge, have a magnetic moment in the first place?

16. A particular ^{238}U nucleus was created in a massive stellar explosion, perhaps 10^{10} y ago. It suddenly decays by α emission while we are observing it. After all those years, why did it decide to decay at this particular moment?

17. Can you justify this statement: "In measuring half-lives by the method of Sample Problem 4, it is not necessary to measure the absolute decay rate R; any quantity proportional to it will suffice. However, in the method of Sample Problem 5 an absolute rate *is* needed."

18. Does the temperature affect the rate of decay of radioactive nuclides? If so, how?

19. You are running longevity tests on light bulbs. Do you expect their "decay" to be exponential? What is the essential difference between the decay of light bulbs and of radionuclides?

20. Generally clocks exhibit complete regularity of some periodic process. Considering that radioactive decay is completely random, how can it nevertheless be used for the measurement of time?

21. Can you give a justification, even a partial one, for the barrier tunneling phenomenon in terms of basic ideas about the wave nature of matter?

22. Explain why, in alpha decay, short half-lives correspond to large disintegration energies, and conversely.

23. A radioactive nucleus can emit a positron, e$^+$. This corresponds to a proton in the nucleus being converted to a neutron. The mass of a neutron, however, is greater than that of a proton. How then can positron emission occur?

24. In beta decay the emitted electrons form a continuous spectrum, but in alpha decay they form a discrete spectrum. What difficulties did this cause in the explanation of beta decay, and how were these difficulties finally overcome?

25. How do neutrinos differ from photons? Each has zero charge and (presumably) zero rest mass and travels at the speed of light.

26. The decay of radioactive elements produces helium, which eventually passes into the Earth's atmosphere. The amount of helium actually present in the atmosphere, however, is very much less than the amount released in this way. Explain.

27. The half-life of ^{238}U is 4.5×10^9 y, about the age of the solar system. How can such a long half-life be measured?

28. In radioactive dating with ^{238}U, how do you get around the fact that you don't know how much ^{238}U was present in the rocks to begin with? (*Hint:* What is the ultimate decay product of ^{238}U?)

29. Make a list of the various sources of ionizing radiation encountered in our environment, whether natural or artificial.

30. Which of these conservation laws apply to all nuclear reactions? Conservation of (*a*) charge, (*b*) mass, (*c*) total energy, (*d*) rest energy, (*e*) kinetic energy, (*f*) linear momentum, (*g*) angular momentum, and (*h*) total number of nucleons.

31. Small temperature changes have a large effect on the rate of chemical reactions but generally have a negligible effect on the rate of nuclear reactions. Explain.

32. In the development of our understanding of the atom, did we use atomic models as we now use nuclear models? Is Bohr's theory such an atomic model? Are models now used in atomic physics? What is the difference between a model and a theory?

33. What are the basic assumptions of the liquid drop and the independent particle models of nuclear structure? How do they differ? Are there similarities between them?

34. Does the collective model of the nucleus give us a picture of the following phenomena: (*a*) acceptance by the nucleus of a colliding particle; (*b*) loss of a particle by spontaneous emission; (*c*) fission; (*d*) dependence of stability on energy content?

35. What is so special ("magic") about the magic nucleon numbers?

36. Why aren't the magic nucleon numbers and the magic electron numbers the same? What accounts for each?

37. The average number of stable (or very long-lived) isotopes of the inert gases is 3.7. The average number of stable nuclides for the four magic neutron numbers, however, is 5.8, considerably greater. If the inert gases are so stable, why were not more stable isotopes of them created when the elements were formed?

PROBLEMS

Section 54-1 Discovering the Nucleus

1. Calculate the distance of closest approach for a head-on collision between a 5.30-MeV α particle and the nucleus of a copper atom.

2. (*a*) Calculate the electric force on an α particle at the surface of a gold atom, presuming that the positive charge is spread uniformly throughout the volume of the atom. Ignore the atomic electrons. A gold atom has a radius of 0.16 nm; treat

the α particle as a point particle. (b) Through what distance would this force, presumed constant, have to act to bring a 5.30-MeV α particle to rest? Express your answer in terms of the diameter of a gold atom.

3. Assume that a gold nucleus has a radius of 6.98 fm (see Table 1), and an α particle has a radius of 1.8 fm. What energy must an incident α particle have to just touch the gold nucleus?

4. When an α particle collides elastically with a nucleus, the nucleus recoils. A 5.00-MeV α particle has a head-on elastic collision with a gold nucleus, initially at rest. What is the kinetic energy (a) of the recoiling nucleus and (b) of the rebounding α particle? The mass of the α particle may be taken to be 4.00 u and that of the gold nucleus 197 u.

Section 54-2 Some Nuclear Properties

5. Locate the nuclides displayed in Table 1 on the nuclidic chart of Fig. 4. Which of these nuclides are within the stability zone?

6. The radius of a nucleus is measured, by electron-scattering methods, to be 3.6 fm. What is the likely mass number of the nucleus?

7. Arrange the 25 nuclides given below in squares as a section of the nuclidic chart similar to Fig. 4. Draw in and label (a) all isobaric (constant A) lines and (b) all lines of constant neutron excess, defined as $N - Z$. Consider nuclides $^{118-122}$Te, $^{117-121}$Sb, $^{116-120}$Sn, $^{115-119}$In, and $^{114-118}$Cd.

8. A neutron star is a stellar object whose density is about that of nuclear matter, as calculated in Sample Problem 2. Suppose that the Sun were to collapse into such a star without losing any of its present mass. What would be its expected radius?

9. Verify that the binding energy per nucleon given in Table 1 for ^{239}Pu is indeed 7.56 MeV/nucleon. The needed atomic masses are 239.052158 u (^{239}Pu), 1.007825 u (^{1}H), and 1.008665 u (neutron).

10. Calculate the average binding energy per nucleon of ^{62}Ni, which has an atomic mass of 61.928346 u. This nucleus has the greatest binding energy per nucleon of all the known stable nuclei.

11. The atomic masses of ^{1}H, ^{12}C, and ^{238}U are 1.007825 u, 12.000000 u (by definition), and 238.050785 u, respectively. (a) What would these masses be if the mass unit were defined so that the mass of ^{1}H was (exactly) 1.000000 u? (b) Use your result to suggest why this perhaps obvious choice was not made.

12. (a) Convince yourself that the energy tied up in nuclear, or strong-force, bonds is proportional to A, the mass number of the nucleus in question. (b) Convince yourself that the energy tied up in Coulomb-force bonds between the protons is proportional to $Z(Z - 1)$. (c) Show that, as we move to larger and larger nuclei (see Fig. 4), the importance of (b) increases more rapidly than does that of (a).

13. In the periodic table, the entry for magnesium is:

$$\boxed{\begin{array}{c} 12 \\ \text{Mg} \\ 24.305 \end{array}}$$

There are three isotopes:

$$^{24}\text{Mg, atomic mass} = 23.985042 \text{ u.}$$
$$^{25}\text{Mg, atomic mass} = 24.985837 \text{ u.}$$
$$^{26}\text{Mg, atomic mass} = 25.982594 \text{ u.}$$

The abundance of ^{24}Mg is 78.99% by mass. Calculate the abundances of the other two isotopes.

14. Because a nucleon is confined to a nucleus, we can take its uncertainty in position to be approximately the nuclear radius R. What does the uncertainty principle yield for the kinetic energy of a nucleon in a nucleus with, say, $A = 100$? (*Hint*: Take the uncertainty in momentum Δp to be the actual momentum p.)

15. You are asked to pick apart an α particle (^{4}He) by removing, in sequence, a proton, a neutron, and a proton. Calculate (a) the work required for each step, (b) the total binding energy of the α particle, and (c) the binding energy per nucleon. Needed atomic masses are

^{4}He	4.002603 u	^{2}H	2.014102 u
^{3}H	3.016049 u	^{1}H	1.007825 u

$$\text{n} \quad 1.008665 \text{ u.}$$

16. To simplify calculations, atomic masses are sometimes tabulated, not as the actual atomic mass m but as $(m - A)c^2$, where A is the mass number expressed in mass units. This quantity, usually reported in MeV, is called the *mass excess*, symbol Δ. Using data from Sample Problem 3, find the mass excesses for (a) ^{1}H, (b) the neutron, and (c) ^{120}Sn.

17. (a) Show that the total binding energy of a nuclide can be written as

$$E_B = Z\Delta_H + N\Delta_n - \Delta,$$

where Δ_H, Δ_n, and Δ are the appropriate mass excesses; see Problem 16. (b) Using this method calculate the binding energy per nucleon for ^{197}Au. Compare your result with the value listed in Table 1. The needed mass excesses are $\Delta_H = +7.289$ MeV, $\Delta_n = +8.071$ MeV, and $\Delta_{197} = -31.17$ MeV. Δ_H is the mass excess of ^{1}H. Note the economy of calculation that results when mass excesses are used in place of the actual masses.

18. A penny has a mass of 3.00 g. Calculate the nuclear energy that would be required to separate all the neutrons and protons in this coin. Ignore the binding energy of the electrons. For simplicity assume that the penny is made entirely of ^{63}Cu atoms (mass = 62.929599 u). The atomic masses of the proton and the neutron are 1.007825 u and 1.008665 u, respectively.

19. Nuclear radii may be measured by scattering high-energy electrons from nuclei. (a) What is the de Broglie wavelength for 480-MeV electrons? (b) Are they suitable probes for this purpose? Relativity must be taken into account.

20. Because the neutron has no charge, its mass must be found in some way other than by using a mass spectrometer. When a resting neutron and a proton meet, they combine and form a deuteron, emitting a gamma ray whose energy is 2.2233 MeV. The atomic masses of the proton and the deuteron are 1.007825 u and 2.014102 u, respectively. Find the

mass of the neutron from these data, to as many significant figures as the data warrant.

21. The spin and the magnetic moment (maximum z component) of ^{7}Li in its ground state (see Table 1) are $\frac{3}{2}$ and $+3.26$ nuclear magnetons, respectively. A free ^{7}Li nucleus is placed in a magnetic field of 2.16 T. (*a*) Into how many levels will the ground state split because of space quantization? (*b*) What is the energy difference between adjacent pairs of levels? (*c*) What is the wavelength that corresponds to a transition between such a pair of levels? (*d*) In what region of the electromagnetic spectrum does this wavelength lie?

22. (*a*) Show that the electrostatic potential energy of a uniform sphere of charge Q and radius R is given by

$$U = \frac{3Q^2}{20\pi\epsilon_0 R}.$$

(*Hint*: Assemble the sphere from thin spherical shells brought in from infinity.) (*b*) Find the electrostatic potential energy for the nuclide ^{239}Pu, assumed spherical; see Table 1. (*c*) Compare its electrostatic potential energy per particle with its binding energy per nucleon of 7.56 MeV. (*d*) What do you conclude?

Section 54-3 Radioactive Decay

23. The half-life of a radioactive isotope is 140 d. How many days would it take for the activity of a sample of this isotope to fall to one-fourth of its initial decay rate?

24. The half-life of a particular radioactive isotope is 6.5 h. If there are initially 48×10^{19} atoms of this isotope in a particular sample, how many atoms of this isotope remain after 26 h?

25. A radioactive isotope of mercury, ^{197}Hg, decays into gold, ^{197}Au, with a decay constant of 0.0108 h^{-1}. (*a*) Calculate its half-life. (*b*) What fraction of the original amount will remain after three half-lives? (*c*) After 10 days?

26. From data presented in the first few paragraphs of Section 54-3, deduce (*a*) the disintegration constant λ and (*b*) the half-life of ^{238}U.

27. ^{67}Ga, atomic mass $= 66.93$ u, has a half-life of 78.25 h. Consider an initially pure 3.42-g sample of this isotope. (*a*) Find its activity (decay rate). (*b*) Find its activity 48.0 h later.

28. Show that the law of radioactive decay (Eq. 6) can be written in the form

$$N = N_0(\tfrac{1}{2})^{t/t_{1/2}}.$$

29. ^{223}Ra decays by alpha decay with a half-life of 11.43 d. How many helium atoms are created in 28 d from an initially pure sample of ^{223}Ra containing 4.70×10^{21} atoms?

30. The radionuclide ^{64}Cu has a half-life of 12.7 h. How much of an initially pure 5.50-g sample of ^{64}Cu will decay during the 2-h period beginning 14.0 h later?

31. The radionuclide ^{32}P (half-life = 14.28 d) is often used as a tracer to follow the course of biochemical reactions involving phosphorus. (*a*) If the counting rate in a particular experimental setup is 3050 counts/s, after what time will it fall to 170 counts/s? (*b*) A solution containing ^{32}P is fed to the root system of an experimental tomato plant and the ^{32}P activity in a leaf is measured 3.48 d later. By what factor must this

reading be multiplied to correct for the decay that has occurred since the experiment began?

32. A 1.00-g sample of samarium emits α particles at a rate of 120 particles/s. ^{147}Sm, whose natural abundance in bulk samarium is 15.0%, is the responsible isotope. Calculate the half-life of this isotope.

33. ^{239}Pu, atomic mass $= 239$ u, decays by alpha decay with a half-life of 24,100 y. How many grams of helium are produced by an initially pure 12.0-g sample of ^{239}Pu after 20,000 y? (Recall that an α particle is a helium nucleus, with an atomic mass of 4.00 u.)

34. A source contains two phosphorus radionuclides, ^{32}P ($t_{1/2} = 14.3$ d) and ^{33}P ($t_{1/2} = 25.3$ d). Initially 10.0% of the decays come from ^{33}P. How long must one wait until 90.0% do so?

35. After a brief neutron irradiation of silver, two activities are present: ^{108}Ag ($t_{1/2} = 2.42$ min) with an initial decay rate of 3.1×10^5/s, and ^{110}Ag ($t_{1/2} = 24.6$ s) with an initial decay rate of 4.1×10^6/s. Make a plot similar to Fig. 7 showing the total combined decay rate of the two isotopes as a function of time from $t = 0$ until $t = 10$ min. In Fig. 7, the extraction of the half-life for simple decays was illustrated. Given only the plot of total decay rate, can you suggest a way to analyze it in order to find the half-lives of both isotopes?

36. As of this writing there is speculation that the free proton may not actually be a stable particle but may be radioactive, with a half-life of about 1×10^{32} y. If this turns out to be true, about how long would you have to wait to be reasonably sure that one proton in your body has decayed? Assume that you are made of water and have a mass of 70 kg.

37. A certain radionuclide is being manufactured in a cyclotron, at a constant rate P. It is also decaying, with a disintegration constant λ. Let the production process continue for a time that is long compared to the half-life of the radionuclide. Convince yourself that the number of radioactive nuclei present at such times will be constant and will be given by $N = P/\lambda$. Convince yourself further that this result holds no matter how many of the radioactive nuclei were present initially. The nuclide is said to be in *secular equilibrium* with its source; in this state its decay rate is just equal to its production rate.

38. The radionuclide ^{56}Mn has a half-life of 2.58 h and is produced in a cyclotron by bombarding a manganese target with deuterons. The target contains only the stable manganese isotope ^{55}Mn and the reaction that produces ^{56}Mn is

$$^{55}\text{Mn} + \text{d} \rightarrow {}^{56}\text{Mn} + \text{p}.$$

After being bombarded for a time $\gg 2.58$ h, the activity of the target, due to ^{56}Mn, is 8.88×10^{10} s^{-1}; see Problem 37. (*a*) At what constant rate P are ^{56}Mn nuclei being produced in the cyclotron during the bombardment? (*b*) At what rate are they decaying (also during the bombardment)? (*c*) How many ^{56}Mn nuclei are present at the end of the bombardment? (*d*) What is their total mass? The atomic mass of ^{56}Mn is 55.94 u.

39. A radium source contains 1.00 mg of ^{226}Ra, which decays with a half-life of 1600 y to produce ^{222}Rn, an inert gas. This radon gas in turn decays by alpha decay with a half-life of 3.82 d. (*a*) Calculate the decay rate of ^{226}Ra in the source.

(*b*) At what rate is the radon decaying when it has come to secular equilibrium with the radium source? See Problem 37. (*c*) How much radon is in secular equilibrium with the radium source?

Section 54-4 Alpha Decay

40. Generally, heavier nuclides tend to be more unstable to alpha decay. For example, the most stable isotope of uranium, ^{238}U, has an alpha decay half-life of 4.5×10^9 y. The most stable isotope of plutonium is ^{244}Pu with a 8.2×10^7 y half-life, and for curium we have ^{248}Cm and 3.4×10^5 y. When half of an original sample of ^{238}U has decayed, what fractions of the original isotopes of (*a*) plutonium and (*b*) curium are left?

41. Consider a ^{238}U nucleus to be made up of an α particle (^{4}He) and a residual nucleus (^{234}Th). Plot the electrostatic potential energy $U(r)$, where r is the distance between these particles. Cover the range 10 fm $< r < 100$ fm and compare your plot with that of Fig. 8.

42. A ^{238}U nucleus emits an α particle of energy 4.196 MeV. Calculate the disintegration energy Q for this process, taking the recoil energy of the residual ^{234}Th nucleus into account. The atomic mass of an α particle is 4.0026 u and that of the ^{234}Th is 234.04 u. Compare your result with that of Sample Problem 6*a*.

43. Heavy radionuclides emit an α particle rather than other combinations of nucleons because the α particle is such a stable, tightly bound structure. To confirm this, calculate the disintegration energies for these hypothetical decay processes and discuss the meaning of your findings:

$$^{235}\text{U} \rightarrow {}^{232}\text{Th} + {}^3\text{He}, \quad Q_3;$$

$$^{235}\text{U} \rightarrow {}^{231}\text{Th} + {}^4\text{He}, \quad Q_4;$$

$$^{235}\text{U} \rightarrow {}^{230}\text{Th} + {}^5\text{He}, \quad Q_5.$$

The needed atomic masses are

^{232}Th	232.038051 u	^{3}He	3.016029 u
^{231}Th	231.036298 u	^{4}He	4.002603 u
^{230}Th	230.033128 u	^{5}He	5.01222 u .
^{235}U	235.043924 u		

44. Consider that a ^{238}U nucleus emits (*a*) an α particle or (*b*) a sequence of neutron, proton, neutron, proton. Calculate the energy released in each case. (*c*) Convince yourself both by reasoned argument and also by direct calculation that the difference between these two numbers is just the total binding energy of the α particle. Find that binding energy. Needed atomic masses are

^{238}U	238.050785 u	^{4}He	4.002603 u
^{237}U	237.048725 u	^{1}H	1.007825 u
^{236}Pa	236.048890 u	n	1.008665 u.
^{235}Pa	235.045430 u		
^{234}Th	234.043593 u		

45. Under certain circumstances, a nucleus can decay by emitting a particle heavier than an α particle. Such decays are very rare. Consider the decays

$$^{223}\text{Ra} \rightarrow {}^{209}\text{Pb} + {}^{14}\text{C}$$

and

$$^{223}\text{Ra} \rightarrow {}^{219}\text{Rn} + {}^4\text{He}.$$

(*a*) Calculate the Q-values for these decays and determine that both are energetically possible. (*b*) The Coulomb barrier height for α particles in this decay is 30 MeV. What is the barrier height for ^{14}C decay? Atomic masses are

^{223}Ra	223.018501 u	^{14}C	14.003242 u
^{209}Pb	208.981065 u	^{4}He	4.002603 u.
^{219}Rn	219.009479 u		

Section 54-5 Beta Decay

46. A certain stable nuclide, after absorbing a neutron, emits a negative electron and then splits spontaneously into two α particles. Identify the nuclide.

47. ^{137}Cs is present in the fallout from above-ground detonations of nuclear bombs. Because it beta decays with a slow 30.2-y half-life into ^{137}Ba, releasing considerable energy in the process, it is an environmental concern. The atomic masses of the Cs and Ba are 136.907073 u and 136.905812 u, respectively. Calculate the total energy released in one decay.

48. A free neutron decays according to Eq. 13. Calculate the maximum energy K_{max} of the beta spectrum. Needed atomic masses are:

$$\text{n} \quad 1.008665 \text{ u}; \quad ^1\text{H} \quad 1.007825 \text{ u}.$$

49. An electron is emitted from a middle-mass nuclide ($A = 150$, say) with a kinetic energy of 1.00 MeV. (*a*) Find its de Broglie wavelength. (*b*) Calculate the radius of the emitting nucleus. (*c*) Can such an electron be confined as a standing wave in a "box" of such dimensions? (*d*) Can you use these numbers to disprove the argument (long since abandoned) that electrons actually exist in nuclei?

50. The radionuclide ^{32}P decays to ^{32}S as described by Eq. 11. In a particular decay event, a 1.71-MeV electron is emitted, the maximum possible value. Find the kinetic energy of the recoiling ^{32}S atom in this event. The atomic mass of ^{32}S is 31.97 u. (*Hint:* For the electron it is necessary to use the relativistic expressions for the kinetic energy and the linear momentum. Newtonian mechanics may safely be used for the relatively slow-moving ^{32}S atom.)

51. The radionuclide ^{11}C decays according to

$$^{11}\text{C} \rightarrow {}^{11}\text{B} + e^+ + \nu, \quad t_{1/2} = 20.3 \text{ min.}$$

The maximum energy of the positron spectrum is 960.8 keV. (*a*) Show that the disintegration energy Q for this process is given by

$$Q = (m_\text{C} - m_\text{B} - 2m_e)c^2,$$

where m_C and m_B are the atomic mass of ^{11}C and ^{11}B, respectively and m_e is the electron (positron) mass. (*b*) Given that $m_\text{C} = 11.011433$ u, $m_\text{B} = 11.009305$ u, and $m_e = 0.0005486$ u, calculate Q and compare it with the maximum energy of the positron spectrum, given above. (*Hint:* Let m'_C and m'_B be the nuclear masses and proceed as in Sample Problem 7 for beta decay. Note that positron decay is an exception to the general rule that, if atomic masses are used in nuclear decay processes, the mass of the emitted electron is automatically taken care of.)

52. Some radionuclides decay by capturing one of their own atomic electrons, a K-electron, say. An example is

$$^{49}V + e^- \rightarrow \, ^{49}Ti + \nu, \qquad t_{1/2} = 331 \text{ d}.$$

Show that the disintegration energy Q for this process is given by

$$Q = (m_V - m_{Ti})c^2 - E_K,$$

where m_V and m_{Ti} are the atomic masses of ^{49}V and ^{49}Ti, respectively, and E_K is the binding energy of the vanadium K-electron. (*Hint*: Put m'_V and m'_{Ti} as the corresponding nuclear masses and proceed as in Sample Problem 7; see the footnote in that sample problem.)

53. Find the disintegration energy Q for the decay of ^{49}V by K-electron capture, as described in Problem 52. The needed data are $m_V = 48.948517$ u, $m_{Ti} = 48.947871$ u, and $E_K = 5.47$ keV.

Section 54-6 Measuring Ionizing Radiation

54. A Geiger counter records 8722 counts in 1 min. Calculate the activity of the source in Ci, assuming that the counter records all decays.

55. A typical chest x-ray radiation dose is 25 mrem, delivered by x rays with a quality factor of 0.85. Assuming that the mass of the exposed tissue is one-half the patient's mass of 88 kg, calculate the energy absorbed in joules.

56. A 75-kg person receives a whole-body radiation dose of 24 mrad, delivered by α particles for which the quality factor is 12. Calculate (*a*) the absorbed energy in joules and (*b*) the equivalent dose in rem.

57. An activity of 3.94 μCi is needed in a radioactive sample to be used in a medical procedure. One week before treatment, a nuclide sample with a half-life of 1.82×10^5 s is prepared. What should be the activity of the sample at the time of preparation in order that it have the required activity at the time of treatment?

58. The plutonium isotope ^{239}Pu, atomic mass 239.05 u, is produced as a by-product in nuclear reactors and hence is accumulating in reactor fuel elements. It is radioactive, decaying by alpha decay with a half-life of 2.411×10^4 y. But plutonium is also one of the most toxic chemicals known; as little as 2.00 mg is lethal to a human. (*a*) How many nuclei constitute a chemically lethal dose? (*b*) What is the decay rate of this amount? (*c*) Its activity in curies?

59. Cancer cells are more vulnerable to x and gamma radiation than are healthy cells. Though linear accelerators are now replacing it, in the past the standard source for radiation therapy has been radioactive ^{60}Co, which beta decays into an excited nuclear state of ^{60}Ni, which immediately drops into the ground state, emitting two gamma-ray photons, each of approximate energy 1.2 MeV. The controlling beta-decay half-life is 5.27 y. How many radioactive ^{60}Co nuclei are present in a 6000-Ci source used in a hospital? The atomic mass of ^{60}Co is 59.93 u.

60. An airline pilot spends an average of 20 h per week flying at 12,000 m, at which altitude the dose equivalent rate due to cosmic and solar radiation is 12 μSv/h (1 Sv = 1 sievert = 100 rem; the sievert is the SI unit of dose equivalent). Calculate the annual equivalent dose in mrem.

61. After long effort, in 1902, Marie and Pierre Curie succeeded in separating from uranium ore the first substantial quantity of radium, 1 decigram (dg) of pure $RaCl_2$. The radium was the radioactive isotope ^{226}Ra, which decays with a half-life of 1600 y. (*a*) How many radium nuclei had they isolated? (*b*) What was the decay rate of their sample, in Bq? (1 Bq = 1 becquerel = 1 decay/s.) (*c*) In curies? The molar mass of Cl is 35.453 g/mol; the atomic mass of the radium isotope is 226.03 u.

62. Calculate the mass of 4.60 μCi of ^{40}K, which has a half-life of 1.28×10^9 y and an atomic mass of 40.0 u.

63. One of the dangers of radioactive fallout from a nuclear bomb is ^{90}Sr, which beta decays with a 29-y half-life. Because it has chemical properties much like calcium, the strontium, if eaten by a cow, becomes concentrated in its milk and ends up in the bones of whoever drinks the milk. The energetic decay electrons damage the bone marrow and thus impair the production of red blood cells. A 1-megaton bomb produces approximately 400 g of ^{90}Sr. If the fallout spreads uniformly over a 2000-km^2 area, what area would have radioactivity equal to the allowed bone burden for one person of 0.002 mCi? The atomic mass of ^{90}Sr is 89.9 u.

64. The nuclide ^{198}Au, half-life = 2.693 d, is used in cancer therapy. Calculate the mass of this isotope required to produce an activity of 250 Ci.

65. An 87-kg worker at a breeder reactor plant accidentally ingests 2.5 mg of ^{239}Pu dust. ^{239}Pu has a half-life of 24,100 y, decaying by alpha decay. The energy of the emitted α particles is 5.2 MeV, with a quality factor of 13. Assume that the plutonium resides in the worker's body for 12 h, and that 95% of the emitted α particles are stopped within the body. Calculate (*a*) the number of plutonium atoms ingested, (*b*) the number that decay during the 12 h, (*c*) the energy absorbed by the body, (*d*) the resulting physical dose in rad, and (*e*) the equivalent biological dose in rem.

Section 54-7 Natural Radioactivity

66. A rock is found to contain 4.20 mg of ^{238}U and 2.00 mg of ^{206}Pb. Assume that the rock contained no lead at formation, all the lead now present arising from the decay of the uranium. Find the age of the rock. The half-life of ^{238}U is 4.47×10^9 y.

67. Two radioactive materials that are unstable to alpha decay, ^{238}U and ^{232}Th, and one that is unstable to beta decay, ^{40}K, are sufficiently abundant in granite to contribute significantly to the heating of the Earth through the decay energy produced. The alpha-unstable isotopes give rise to decay chains that stop at stable lead isotopes. ^{40}K has a single beta decay. Decay information follows:

Parent Nuclide	Decay Mode	Half-life (y)	Stable Endpoint	Q (MeV)	f (ppm)
^{238}U	α	4.47×10^9	^{206}Pb	51.7	4
^{232}Th	α	1.41×10^{10}	^{208}Pb	42.7	13
^{40}K	β	1.28×10^9	^{40}Ca	1.32	4

Q is the total energy released in the decay of one parent nucleus to the final stable endpoint and f is the abundance of the isotope in kilograms per kilogram of granite; ppm means parts per million. (*a*) Show that these materials give rise to a total heat production of 987 pW for each kilogram of granite. (*b*) Assuming that there is 2.7×10^{22} kg of granite in a 20-km thick, spherical shell around the Earth, estimate the power this will produce over the whole Earth. Compare this with the total solar power intercepted by the Earth, 1.7×10^{17} W.

68. A particular rock is thought to be 260 million years old. If it contains 3.71 mg of ^{238}U, how much ^{206}Pb should it contain?

69. A rock, recovered from far underground, is found to contain 860 μg of ^{238}U, 150 μg of ^{206}Pb, and 1.60 mg of ^{40}Ca. How much ^{40}K will it very likely contain? Needed half-lives are listed in Problem 67.

Section 54-8 Nuclear Reactions

70. Fill in the missing nuclide in each of the following reactions: (*a*) ^{116}Sn(?,p)^{117}Sn; (*b*) ^{40}Ca(α,n)?; and (*c*) ?(p,n)^{7}Be.

71. Calculate Q for the reaction ^{59}Co(p,n)^{59}Ni. Needed atomic masses are

^{59}Co   58.933198 u ^{1}H   1.007825 u

^{59}Ni   58.934349 u   n   1.008665 u.

72. Making mental use of the overlay of Fig. 13 applied to Fig. 12, write down the reactions by which the radionuclide ^{197}Pt ($t_{1/2} = 18.3$ h) can be prepared, at least in principle. Except in special circumstances, only stable nuclides can serve as practical targets for nuclear reactions.

73. The radionuclide ^{60}Co ($t_{1/2} = 5.27$ y) is much used in cancer therapy. Tabulate possible reactions that might be used in preparing it. Limit the projectiles to neutrons, protons, and deuterons. Limit the targets to stable nuclides. The stable nuclides suitably close to ^{60}Co are ^{63}Cu, ^{60,61,62}Ni, ^{59}Co, and ^{57,58}Fe. (Commercially, ^{60}Co is made by bombarding elemental cobalt, which consists only of the isotope ^{59}Co, with neutrons in a reactor.)

74. A beam of deuterons falls on a copper target. Copper has two stable isotopes, ^{63}Cu (69.2%) and ^{65}Cu (30.8%). Tabulate the residual nuclides that can be produced by the reactions (d,n), (d,p), (d,α), and (d,γ). By inspection of Fig. 4, indicate which residual nuclides are stable and which are radioactive.

75. Prepare an overlay like that of Fig. 13 in which that figure is extended to include reactions involving the light nuclides ^{3}H (tritium) and ^{3}He, considered both as projectiles and as emerging particles.

76. A platinum target is bombarded with cyclotron-accelerated deuterons for several hours and then the element iridium ($Z = 77$) is separated chemically from it. What radioisotopes of iridium are present and by what reactions are they formed? (*Note*: ^{190}Pt and ^{192}Pt, not shown in Fig. 12, are stable platinum isotopes, but their isotopic abundances are so small that we may ignore their presence.)

77. Consider the reaction $X(a,b)Y$, in which X is taken to be at rest in the laboratory reference frame. The initial kinetic energy in this frame is

$$K_{\text{lab}} = \tfrac{1}{2}m_a v_a^2.$$

(*a*) Show that the initial velocity of the center of mass of the system in the laboratory frame is

$$V = v_a \left(\frac{m_a}{m_X + m_a} \right).$$

Is this quantity changed by the reaction? (*b*) Show that the initial kinetic energy, viewed now in a reference frame attached to the center of mass of the two particles, is given by

$$K_{\text{cm}} = K_{\text{lab}} \left(\frac{m_X}{m_X + m_a} \right).$$

Is this quantity changed by the reaction? (*c*) In the reaction ^{90}Zr(d,p)^{91}Zr the kinetic energy of the deuteron, measured in the laboratory frame, is 15.9 MeV. Find $v_a \,(= v_d)$, V, and K_{cm}. Ignore the small relativistic effects.

78. In an endothermic reaction ($Q < 0$), the interacting particles a and X must have a kinetic energy, measured in the center-of-mass reference frame, of at least $|Q|$ if the reaction is to "go." Show, using the result of Problem 77, that the *threshold energy* for particle a, measured in the laboratory reference frame, is

$$K_{\text{th}} = |Q| \frac{m_X + m_a}{m_X}.$$

Is it reasonable that K_{th} should be greater than $|Q|$?

79. Prepare an overlay like that of Fig. 13 in which two nucleons or light nuclei may appear as emerging particles. The reaction ^{63}Cu(α,pn)^{65}Zn is an example. Consider the combinations nn, np, and pd as possibilities.

Section 54-9 Nuclear Models

80. A typical kinetic energy for a nucleon in a middle-mass nucleus may be taken as 5 MeV. To what effective nuclear temperature does this correspond, using the assumptions of the liquid drop model of nuclear structure? (*Hint*: See Eq. 30 in Chapter 24.)

81. An intermediate nucleus in a particular nuclear reaction decays within 1.2×10^{-22} s of its formation. (*a*) What is the uncertainty ΔE in our knowledge of this intermediate state? (*b*) Can this state be called a compound nucleus? See Sample Problem 11.

82. From the following list of nuclides, identify (*a*) those with filled nucleon shells, (*b*) those with one nucleon outside a filled shell, and (*c*) those with one vacancy in an otherwise filled shell. Nuclides: ^{13}C, ^{18}O, ^{40}K, ^{49}Ti, ^{60}Ni, ^{91}Zr, ^{92}Mo, ^{121}Sb, ^{143}Nd, ^{144}Sm, ^{205}Tl, and ^{207}Pb.

83. As Table 1 shows, the nuclide ^{197}Au has a nuclear spin of $\tfrac{3}{2}$. (*a*) If we regard this nucleus as a spinning rigid sphere with a radius given in Table 1, what rotational frequency results? (*b*) What rotational kinetic energy? Note that this picture is overly mechanistic.

84. Consider the three formation modes shown for the compound nucleus ^{20}Ne* in Fig. 17. What energies must (*a*) the α particle, (*b*) the proton, and (*c*) the gamma-ray photon have to provide 25.00 MeV of excitation energy to the compound nucleus? Needed atomic masses are

^{20}Ne   19.992435 u α 4.002603 u

^{19}F   18.998403 u ^{1}H   1.007825 u.

^{16}O   15.994915 u

85. Consider the three decay modes shown for the compound nucleus ^{20}Ne* in Fig. 17. If the compound nucleus is initially at rest and has an excitation energy of 25.0 MeV, what kinetic energies, measured in the laboratory, will (a) the deuteron, (b) the neutron, and (c) the ^{3}He nuclide have when the nucleus decays? Needed atomic masses are

^{20}Ne   19.992435 u   d   2.014102 u

^{19}Ne   19.001879 u   n   1.008665 u

^{18}F   18.000937 u ^{3}He 3.016029 u.

^{17}O   16.999131 u

86. The nuclide ^{208}Pb is "doubly magic" in that both its proton number $Z\,(=82)$ and its neutron number $N\,(=126)$ represent filled nucleon shells. An additional proton would yield ^{209}Bi and an additional neutron ^{209}Pb. These "extra" nucleons should be easier to remove than a proton or a neutron from the filled shells of ^{208}Pb. (a) Calculate the energy required to move the "extra" proton from ^{209}Bi and compare it with the energy required to remove a proton from the filled proton shell of ^{208}Pb. (b) Calculate the energy required to remove the "extra" neutron from ^{209}Pb and compare it with the energy required to remove a neutron from the filled neutron shell of ^{208}Pb. Do your results agree with expecta-

tion? Use these atomic mass data:

Nuclide	Z	N	Atomic Mass (u)
^{209}Bi	$82 + 1$	126	208.980374
^{208}Pb	82	126	207.976627
^{207}Tl	$82 - 1$	126	206.977404
^{209}Pb	82	$126 + 1$	208.981065
^{207}Pb	82	$126 - 1$	206.975872

The atomic masses of the proton and the neutron are 1.007825 u and 1.008665 u, respectively.

87. The nucleus ^{91}Zr ($Z = 40$, $N = 51$) has a single neutron outside a filled 50-neutron core. Because 50 is a magic number, this neutron should perhaps be especially loosely bound. (a) Calculate its binding energy. (b) Calculate the binding energy of the next neutron, which must be extracted from the filled core. (c) Find the binding energy per particle for the entire nucleus. Compare these three numbers and discuss. Needed atomic masses are

^{91}Zr   90.905644 u   n   1.008665 u

^{90}Zr   89.904703 u ^{1}H   1.007825 u.

^{89}Zr   88.908890 u

CHAPTER 55

ENERGY FROM THE NUCLEUS

*In a system of interacting particles, we can extract useful energy when
the system moves to a lower energy state (that is, a more tightly bound state).
In an atomic system, we can extract this energy through chemical reactions, such as
burning. In a nuclear system, we can extract energy in a variety of ways. For example, the
energy released in radioactive decay has been used to provide electrical power to cardiac
pacemakers and to space probes.*

*In this chapter, we consider the two primary methods that are used to extract energy from
the nucleus and convert it to useful purposes. In* nuclear fission, *a heavy nucleus is split into
two fragments. In* nuclear fusion, *two light nuclei are combined into a heavier nucleus.
Figure 6 of Chapter 54 showed that either of these processes can result in more tightly
bound nuclei and therefore can release excess nuclear binding energy to be converted into
other forms of energy. Reactors based on nuclear fission today provide a significant share of
the world's electrical power. Research and engineering are actively underway to develop
reactors based on nuclear fusion.*

55-1 THE ATOM AND THE NUCLEUS

When we get energy from coal by burning it in a furnace, we are tinkering with *atoms* of carbon and oxygen, rearranging their outer *electrons* in more stable combinations. When we get energy from uranium by consuming it in a nuclear reactor, we are tinkering with its *nucleus*, rearranging its *nucleons* in more stable combinations.

Electrons are held in atoms by the Coulomb force, and it takes a few electron volts to remove one of the outer electrons. On the other hand, nucleons are held in nuclei by the strong nuclear force, and it takes a few *million* electron volts to pull one of *them* out. This factor is also reflected in our ability to extract about a million times more energy from a kilogram of uranium than from a kilogram of coal.

In both the atomic and nuclear cases, the appearance of energy is accompanied by a decrease in the rest energy of the fuel. The only difference between consuming uranium and burning coal is that, in the former case, a much larger fraction of the available rest energy (again, by a factor of several million) is converted to other forms of energy.

We must be clear about whether our concern is for the quantity of energy or for the rate at which the energy is delivered, that is, the *power*. In the nuclear case will the kilogram of uranium burn slowly in a power reactor or explosively in a bomb? In the atomic case, are we thinking about exploding a stick of dynamite or digesting a jelly doughnut? (Surprisingly, the energy release is greater in the second case than in the first!)

Table 1 shows how much energy can be extracted from 1 kg of matter by doing various things to it. Instead of reporting the energy directly, we measure it by showing how long the extracted energy could operate a 100-W light bulb. Row 6, the total mutual annihilation of matter and antimatter, is the ultimate in extracting energy from matter. When you have used up all the available mass you can do no more. (However, no one has yet figured out an economical way to produce and store 1 kg of antimatter to use for energy production.)

Keep in mind that the comparisons of Table 1 are on a per-unit-mass basis. Kilogram for kilogram we get several million times more energy from uranium than we do from coal or from falling water. On the other hand, there is a lot of coal in the Earth's crust and there is a lot of water backed up behind the Bonneville Dam on the Columbia River.

TABLE 1 ENERGY FROM 1 kg OF MATTER

Form of Matter	Process	Time[a]
Water	A 50-m waterfall	5 s
Coal	Burning	8 h
Enriched UO_2 (3%)	Fission in a reactor	680 y
^{235}U	Complete fission	3×10^4 y
Hot deuterium gas	Complete fusion	3×10^4 y
Matter and antimatter	Complete annihilation	3×10^7 y

[a] These numbers show how long the energy generated could power a 100-W light bulb.

55-2 NUCLEAR FISSION: THE BASIC PROCESS

In 1932 the English physicist James Chadwick discovered the neutron. A few years later Enrico Fermi and his collaborators in Rome discovered that, if various elements are bombarded by these new projectiles, new radioactive elements are produced. Fermi had predicted that the neutron, being uncharged, would be a useful nuclear projectile; unlike the proton or the α particle, it experiences no repulsive Coulomb force when it approaches a nuclear surface. Because there is no Coulomb barrier for it, the slowest neutron can penetrate and interact with even the most massive, highly charged nucleus. *Thermal neutrons,* which are neutrons in equilibrium with matter at room temperature, are convenient and effective bombarding particles. At 300 K, the mean kinetic energy of such neutrons is

$$\overline{K} = \tfrac{3}{2}kT = \tfrac{3}{2}(8.62 \times 10^{-5} \text{ eV/K})(300 \text{ K}) = 0.04 \text{ eV}.$$

In 1939 the German chemists Otto Hahn and Fritz Strassmann, following work initiated by Fermi and his collaborators, bombarded uranium with thermal neutrons. They found by chemical analysis that after the bombardment a number of new radioactive elements were present, among them one whose chemical properties were remarkably similar to those of barium. Repeated tests finally convinced these chemists that this "new" element was not new at all; it really *was* barium. How could this middle-mass element ($Z = 56$) be produced by bombarding uranium ($Z = 92$) with neutrons?

The riddle was solved within a few weeks by the physicists Lise Meitner and her nephew Otto Frisch. They showed that a uranium nucleus, having absorbed a neutron, could split, with the release of energy, into two roughly equal parts, one of which might well be barium. They named this process *nuclear fission.*† Figure 1 shows the tracks left in the gas of a cloud chamber by the two energetic fission fragments that result from a fission event occurring near the center of the chamber.

The fission of ^{235}U by thermal neutrons, a process of great practical importance, can be represented by

$$^{235}U + n \rightarrow {}^{236}U^* \rightarrow X + Y + bn, \tag{1}$$

in which ^{236}U, as the asterisk indicates, is a compound nucleus. Here X and Y stand for *fission fragments,* middle-mass nuclei that are usually highly radioactive. The factor b, which has the average value 2.47 for fission events of this type, is the number of neutrons released in such events.

Figure 2 shows the distribution by mass number of the fission fragments X and Y. We see that in only about 0.01% of the events will the fragments have equal mass. The most probable mass numbers, occurring in about 7% of the events, are $A = 140$ and $A = 95$. We can also tell from the difference in the length of the two fission fragment tracks in Fig. 1 that the two fragments in this particular fission event do not have the same mass.

The nuclide ^{236}U, which is the fissioning nucleus in Eq. 1, has 92 protons and $236 - 92$ or 144 neutrons, a

---

† See "The Discovery of Fission," by Otto Frisch and John Wheeler, *Physics Today,* November 1967, p. 43, for a fascinating account of the early days of discovery.

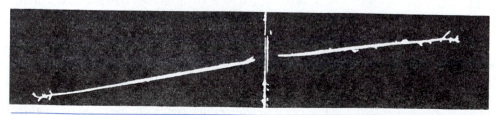

Figure 1 When a fast charged particle passes through a cloud chamber, it leaves a track of liquid droplets. The two back-to-back tracks represent fission fragments, produced by a fission event that took place in a thin vertical uranium foil in the center of the chamber.

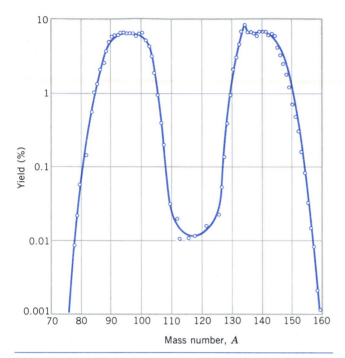

Figure 2 The distribution in mass of the fission fragments X and Y (see Eq. 1) from the fission of ^{235}U by thermal neutrons. Note that the vertical scale is logarithmic.

neutron/proton ratio of about 1.6. The primary fragments formed immediately after fission will retain this same neutron/proton ratio. Study of the stability curve of Fig. 4 of Chapter 54, however, shows that stable nuclides in the middle-mass region ($75 < A < 150$) have a neutron/proton ratio of only 1.2 to 1.4. The primary fragments will thus be excessively neutron-rich and will "boil off" a small number of neutrons, 2.47 of them on the average for the reaction of Eq. 1. The fragments X and Y that remain are still too neutron-rich and approach the stability line by a chain of successive beta decays.

A specific example of the generalized fission process of Eq. 1 is

$$^{235}\text{U} + \text{n} \rightarrow {}^{236}\text{U}^* \rightarrow {}^{140}\text{Xe} + {}^{94}\text{Sr} + 2\text{n}. \qquad (2)$$

The fission fragments ^{140}Xe and ^{94}Sr decay until each reaches a stable end product, as follows:

$$^{140}\text{Xe} \xrightarrow[14\text{ s}]{} {}^{140}\text{Cs} \xrightarrow[65\text{ s}]{} {}^{140}\text{Ba} \xrightarrow[13\text{ d}]{} {}^{140}\text{La} \xrightarrow[40\text{ h}]{} {}^{140}\text{Ce (stable)}$$

$$^{94}\text{Sr} \xrightarrow[75\text{ s}]{} {}^{94}\text{Y} \xrightarrow[19\text{ min}]{} {}^{94}\text{Zr (stable)}$$

The decays are β^- events, the half-lives being indicated at each stage. As for all beta decays, the mass numbers (140 and 94) remain unchanged as the decays continue.

The disintegration energy Q for fission is very much larger than for chemical processes. We can support this by a rough calculation. From the binding energy curve of Fig. 6 of Chapter 54, we see that for heavy nuclides ($A =$ 240, say) the binding energy per nucleon is about 7.6

MeV. In the intermediate range ($A = 120$, say), it is about 8.5 MeV. The difference in total binding energy between a single nucleus ($A = 240$) and two fragments (assumed equal) into which it may be split is then

$$Q = 2(8.5\text{ MeV})\frac{A}{2} - (7.6\text{ MeV})A \approx 200\text{ MeV}.$$

Sample Problem 1 shows a more careful calculation, which agrees very well with this rough estimate.

Sample Problem 1 Calculate the disintegration energy Q for the fission event of Eq. 2, taking into account the decay of the fission fragments. Needed atomic masses are

^{235}U	235.043924 u	^{140}Ce	139.905433 u
n	1.008665 u	^{94}Zr	93.906315 u.

Solution If we replace the fission fragments in Eq. 2 by their stable end products, we see that the overall transformation is

$$^{235}\text{U} \rightarrow {}^{140}\text{Ce} + {}^{94}\text{Zr} + \text{n}.$$

The single neutron comes about because the (initiating) neutron on the left side of Eq. 2 cancels one of two neutrons on the right side of that equation.

The mass difference for this reaction is

$$\Delta m = 235.043924\text{ u} - (139.905433\text{ u} + 93.906315\text{ u}$$
$$+ 1.008665\text{ u})$$
$$= 0.223511\text{ u},$$

and the corresponding energy is

$$Q = \Delta m\, c^2 = (0.223511\text{ u})(931.5\text{ MeV/u}) = 208.2\text{ MeV},$$

in good agreement with our previous rough estimate of 200 MeV.

About 80% of the disintegration energy is in the form of the kinetic energy of the two fragments, the remainder going to the neutron and the radioactive decay products.

If the fission event takes place in a bulk solid, most of the disintegration energy appears as an increase in the internal energy of the solid, which shows a corresponding rise in temperature. Five percent or so of the disintegration energy, however, is associated with neutrinos that are emitted during the beta decay of the primary fission fragments. This energy is carried out of the system and does not contribute to the increase in its internal energy.

55-3 THEORY OF NUCLEAR FISSION

Soon after the discovery of fission, Niels Bohr and John Wheeler developed a theory, based on the analogy between a nucleus and a charged liquid drop, that explained its main features.

Figure 3 suggests how the fission process proceeds.

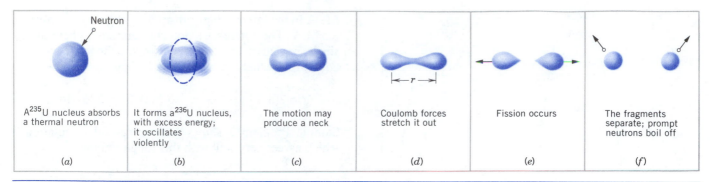

Figure 3   The stages in a fission process, according to the liquid-drop fission model.

When a heavy nucleus such as ^{235}U absorbs a slow neutron, as in Fig. 3a, that neutron falls into the potential well associated with the strong nuclear forces that act in the nuclear interior. Its potential energy is then transformed into internal excitation energy, as Fig. 3b suggests.

The amount of excitation energy that a slow neutron carries into the nucleus that absorbs it is equal to the work required to pull the neutron back out of the nucleus, that is, to the binding energy E_n of the neutron. In much the same way, the amount of excitation energy delivered to a well when a stone is dropped into it is equal to the work required to pull the stone back out of the well, that is, to the "binding energy" E_s of the stone. In Sample Problem 2 we show that the binding energy E_n of a neutron in ^{236}U is 6.5 MeV.

Figure 3c shows that the nucleus, behaving like an energetically oscillating charged liquid drop, will sooner or later develop a short "neck" and will begin to separate into two charged "globs." If conditions are right, the electrostatic repulsion between these two globs will force them apart, breaking the neck. The two fragments, each still carrying some residual excitation energy, then fly apart. Fission has occurred.

So far this model gives a good qualitative picture of the fission process. It remains to be seen, however, whether it can answer a hard question: "Why are some heavy nuclides (^{235}U and ^{239}Pu, say) readily fissionable by slow neutrons but other, equally heavy, nuclides (^{238}U and ^{243}Am, say) are not?"

Bohr and Wheeler were able to answer this question. Figure 4 shows the potential energy curve for the fission process that they derived from their model. The horizontal axis displays the *distortion parameter r*, which is a rough measure of the extent to which the oscillating nucleus departs from a spherical shape. Figure 3d suggests how this parameter is defined before fission occurs. When the fragments are far apart, this parameter is simply the distance between their centers.

The energy interval between the initial state and the final state of the fissioning nucleus—that is, the disintegration energy Q—is displayed in Fig. 4. The central feature of that figure, however, is that the potential energy

curve passes through a maximum at a certain value of r. There is a *potential barrier* of height E_b that must be surmounted (or tunneled) before fission can occur. This reminds us of alpha decay (see Fig. 8 of Chapter 54), which also is a process that is inhibited by a potential barrier. We see then that fission will occur *only* if the absorbed neutron provides an excitation energy E_n great enough to overcome the barrier or to have a reasonable probability of tunneling through it.

Table 2 shows a test of fissionability by thermal neutrons applied to four heavy nuclides, chosen from dozens of candidates that might have been considered. For each nuclide both the barrier height E_b and the excitation energy E_n are given. E_b was calculated from the theory of Bohr and Wheeler, and E_n was computed (as in Sample Problem 2) from the known masses.

For ^{235}U and ^{239}Pu we see that $E_n > E_b$. This means that fission by absorbing a thermal neutron is predicted to occur for these nuclides. This is confirmed by noting, in the table, the large measured cross sections (that is, reaction probabilities) for the process.

For the other two nuclides (^{238}U and ^{243}Am), we have $E_n < E_b$, so that there is not enough energy to surmount the barrier or to tunnel through it effectively. The excited nucleus (Fig. 3b) prefers to get rid of its excitation energy by emitting a gamma ray instead of by breaking into two large fragments. Table 2 shows, as we expect, that the cross sections for thermal neutron fission in these cases

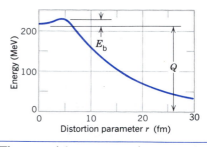

Figure 4   The potential energy at various stages in the fission process, showing the disintegration energy Q and the barrier height E_b.

TABLE 2   TEST OF THE FISSIONABILITY OF FOUR NUCLIDES

Target Nuclide	Nuclide Being Fissioned	E_b (MeV)	E_n (MeV)	$E_n - E_b$ (MeV)	Fission Cross Section[a] (barns)
^{235}U	^{236}U	5.2	6.5	+1.3	584
^{238}U	^{239}U	5.7	4.8	−0.9	2.7×10^{-6}
^{239}Pu	^{240}Pu	4.8	6.4	+1.6	742
^{243}Am	^{244}Am	5.8	5.5	−0.3	<0.08

[a] The cross section is a measure of the probability for a nuclear reaction to occur. The cross section is measured in units of barns, where 1 barn $= 10^{-28}$ m^2.

are exceedingly small. These nuclides *can* be made to fission, however, if they absorb a substantially energetic (rather than a thermal) neutron. For ^{238}U, for example, the absorbed neutron must have an energy of at least 1.3 MeV for the fission process to "go" with reasonable probability.

Sample Problem 2   Consider a ^{236}U nucleus in its ground state. How much energy is required to remove a neutron from it, leaving a ^{235}U nucleus behind? The needed atomic masses are

^{235}U 235.043924 u;   n   1.008665 u; ^{236}U 236.045563 u.

Solution   The increase in mass of the system as the neutron is pulled out is

$$\Delta m = 1.008665 \text{ u} + 235.043924 \text{ u} - 236.045563 \text{ u}$$
$$= 0.007026 \text{ u}.$$

This means that an energy equal to

$$E_n = \Delta m \, c^2 = (0.007026 \text{ u})(931.5 \text{ MeV/u}) = 6.545 \text{ MeV}$$

must be expended. This, by definition, is the binding energy of the neutron in the ^{236}U nucleus.

When a ^{235}U nucleus absorbs a thermal neutron, 6.545 MeV is the amount of excitation energy that the thermal neutron brings into the ^{236}U nucleus. In effect, the ^{236}U nucleus is formed in an excited state 6.545 MeV above the ground state. The excited nucleus can get rid of this energy either by emitting gamma rays (which leaves a ^{236}U nucleus in its ground state) or by fission (see Eq. 1). It turns out that fission is about six times more likely than gamma-ray emission.

55-4   NUCLEAR REACTORS: THE BASIC PRINCIPLES

Energy releases per atom in individual nuclear events such as alpha emission are roughly a million times larger than those of chemical events. To make large-scale use of nuclear energy, we must arrange for one nuclear event to trigger another until the process spreads throughout bulk matter like a flame through a burning log. The fact that more neutrons are generated in fission than are consumed

(see Eq. 1) raises just this possibility; the neutrons that are produced can cause fission in nearby nuclei and in this way a chain of fission events can propagate itself. Such a process is called a *chain reaction*. It can either be rapid and uncontrolled as in a nuclear bomb or controlled as in a nuclear reactor.

Suppose that we wish to design a nuclear reactor based, as most present reactors are, on the fission of ^{235}U by slow neutrons. The fuel in such reactors is almost always artificially "enriched," so that ^{235}U makes up a few percent of the uranium nuclei rather than the 0.7% that occurs in natural uranium; the remaining 99.3% of natural uranium is ^{238}U, which is not fissionable by thermal neutrons. Although on the average 2.47 neutrons are produced in ^{235}U fission for every thermal neutron consumed, there are serious difficulties in making a chain reaction "go." Here are three of the difficulties, together with their solutions:

1.   *The neutron leakage problem.* A certain percentage of the neutrons produced will simply leak out of the reactor core and be lost to the chain reaction. If too many do so, the reactor will not work. Leakage is a surface effect, its magnitude being proportional to the *square* of a typical reactor core dimension (surface area $= 4\pi r^2$ for a sphere). Neutron production, however, is a volume effect, proportional to the *cube* of a typical dimension (volume $= \frac{4}{3}\pi r^3$ for a sphere). The fraction of neutrons lost by leakage can be made as small as we wish by making the reactor core large enough, thereby decreasing its surface-to-volume ratio ($= 3/r$ for a sphere).

2.   *The neutron energy problem.* Fission produces *fast* neutrons, with kinetic energies of about 2 MeV, but fission is induced most effectively by *slow* neutrons. The fast neutrons can be slowed down by mixing the uranium fuel with a substance that has these properties: (a) it is effective in causing neutrons to lose kinetic energy by collisions and (b) it does not absorb neutrons excessively, thereby removing them from the fission chain. Such a substance is called a moderator. Most power reactors in this country are moderated by water, in which the hydrogen nuclei (protons) are the effective moderating element.

3.   *The neutron capture problem.* Neutrons may be captured by nuclei in ways that do not result in fission. The most common possibility is capture followed by the emission of a gamma ray. In particular, as the fast (MeV) neutrons generated in the fission processes are slowed down in the moderator to thermal equilibrium (0.04 eV), they must pass through an energy interval (1 – 100 eV) in which they are particularly susceptible to nonfission capture by ^{238}U.

To minimize such *resonance capture,* as it is called, the uranium fuel and the moderator (water, say) are not intimately mixed but are "clumped," remaining in close contact with each other but occupying different regions of the reactor volume. The hope is that a fast fission neutron, produced in a uranium "clump" (which might be a fuel rod), will with high probability find itself in the moderator as it passes through the "dangerous" resonance energy range. Once it has reached thermal energies, it will very likely wander back into a clump of fuel and produce a fission event. The task for reactor designers is to produce the most effective geometrical arrangement of fuel and moderator.

Figure 5 shows the neutron balance in a typical power reactor operating with a steady output. Let us trace the behavior of a sample of 1000 thermal neutrons in the reactor core. They produce 1330 neutrons by fission in the ^{235}U fuel and 40 more by fast fission in the ^{238}U, making a total of 370 new neutrons, all of them fast. Exactly this same number of neutrons is then lost to the chain by leakage from the core and by nonfission capture, leaving 1000 thermal neutrons to continue the chain. What has been gained in this cycle, of course, is that each of the 370 neutrons produced by fission has deposited

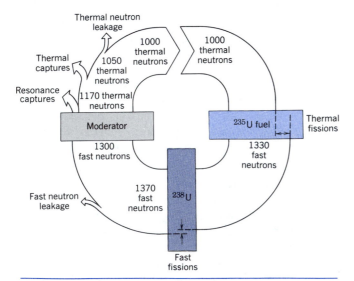

Figure 5   A generation of 1000 thermal neutrons is followed through various stages in a reactor. At a steady operating level, the loss of neutrons due to captures (in the fuel, moderator, and structural elements) and leakage through the surface is exactly balanced by the production of neutrons in the fission processes.

about 200 MeV of energy in the reactor core, heating it up.

An important reactor parameter is the *multiplication factor k,* the ratio of the number of neutrons present at the beginning of a particular generation to the number present at the beginning of the next generation. For the situation of Fig. 5, the multiplication factor is exactly 1. For $k = 1$, the operation of the reactor is said to be exactly *critical,* which is what we wish it to be for steady power production. Reactors are designed so that they are inherently *supercritical* ($k > 1$); the multiplication factor is then adjusted to critical operation ($k = 1$) by inserting *control rods* into the reactor core. These rods, containing a material such as cadmium that absorbs neutrons readily, can then be withdrawn as needed to compensate for the tendency of reactors to go subcritical as (neutron-absorbing) fission products build up in the core during continued operation.

If you pulled out one of the control rods, how fast would the reactor power level increase? This *response time* is controlled by the fascinating circumstance that a small fraction of the neutrons generated by fission is not emitted promptly from the newly formed fission fragments but is emitted from these fragments later, as they decay by beta emission. Of the 370 "new" neutrons analyzed in Fig. 5, for example, about 16 are delayed, being emitted from fragments following beta decays whose half-lives range from 0.2 to 55 s. These delayed neutrons are few in number but they serve the useful purpose of slowing down the reactor response time to match human reaction times.

Figure 6 shows the broad outlines of an electric power plant based on a *pressurized-water reactor* (PWR), a type in common use in the United States. In such a reactor, water is used both as the moderator and as the heat transfer medium. In the *primary loop,* water at high temperature and pressure (possibly 600 K and 150 atm) circulates through the reactor vessel and transfers heat from the reactor core to the steam generator, which provides high-pressure steam to operate the turbine that drives the generator. To complete the *secondary loop,* low-pressure steam from the turbine is condensed to water and forced back into the steam generator by a pump. To give some idea of scale, a typical reactor vessel for a 1000-MW (electric) plant may be 10 m high and weigh 450 tons. Water flows through the primary loop at a rate of about 300,000 gal/min.

An unavoidable feature of reactor operation is the accumulation of radioactive wastes, including both fission products and heavy "transuranic" nuclides such as plutonium and americium. One measure of their radioactivity is the rate at which they release energy in thermal form. Figure 7 shows the variation with time of the thermal power generated by such wastes from one year's operation of a typical large nuclear plant. Note that both scales are logarithmic. The total activity of the waste 10 years after its removal from the reactor is about 3×10^7 Ci.

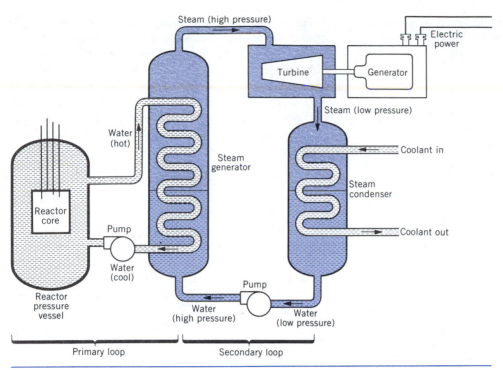

Figure 6 A simplified layout of a nuclear power plant based on a pressurized-water reactor.

Sample Problem 3 A large electric generating station is powered by a pressurized-water nuclear reactor. The thermal power in the reactor core is 3400 MW, and 1100 MW of electricity is generated. The fuel consists of 86,000 kg of uranium, in the form of 110 tons of uranium oxide, distributed among 57,000 fuel rods. The uranium is enriched to 3.0% ^{235}U. (*a*) What is the plant efficiency? (*b*) At what rate R do fission events occur in the

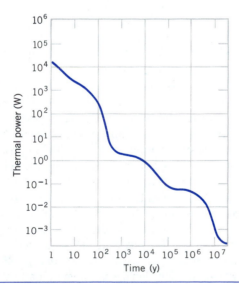

Figure 7 Thermal power released by the radioactive wastes of one year's operation of a typical large nuclear power plant, as a function of time after the fuel is removed. The curve represents the effect of many radionuclides with a range of half-lives. Note that both scales are logarithmic.

reactor core? (*c*) At what rate is the ^{235}U fuel disappearing? Assume conditions at start-up. (*d*) At this rate of fuel consumption, how long would the fuel supply last? (*e*) At what rate is mass being lost in the reactor core?

Solution (*a*) The efficiency e is the ratio between the power output (in the form of electric energy) to the power input (in the form of thermal energy), or

$$e = \frac{\text{electric output}}{\text{thermal input}} = \frac{1100 \text{ MW}}{3400 \text{ MW}}$$

$$= 0.32 \text{ or } 32\%.$$

As for all power plants, whether based on fossil fuel or nuclear fuel, the efficiency is controlled by the second law of thermodynamics. In this plant, 3400 MW − 1100 MW or 2300 MW of power must be discharged as thermal energy to the environment.

(*b*) If P (= 3400 MW) is the thermal power in the core and Q (= 200 MeV) is the average energy released per fission event, then, in steady-state operation,

$$R = \frac{P}{Q} = \frac{3.4 \times 10^9 \text{ J/s}}{(200 \text{ MeV/fission})(1.6 \times 10^{-13} \text{ J/MeV})}$$

$$= 1.06 \times 10^{20} \text{ fissions/s.}$$

(*c*) ^{235}U disappears by fission at the rate calculated in (*b*). It is also consumed by (nonfission) neutron capture at a rate about one-fourth as large. The total ^{235}U consumption rate is then $(1.25)(1.06 \times 10^{20} \text{ s}^{-1})$ or $1.33 \times 10^{20} \text{ s}^{-1}$. We recast this as a mass rate as follows:

$$\frac{dM}{dt} = (1.33 \times 10^{20} \text{ s}^{-1}) \left(\frac{0.235 \text{ kg/mol}}{6.02 \times 10^{23} \text{ atoms/mol}} \right)$$

$$= 5.19 \times 10^{-5} \text{ kg/s} = 4.5 \text{ kg/d.}$$

(*d*) From the data given, we can calculate that, at start-up, about (0.03)(86,000 kg) or 2600 kg of ^{235}U were present. Thus a somewhat simplistic answer would be

$$T = \frac{2600 \text{ kg}}{4.5 \text{ kg/d}} = 580 \text{ d}.$$

In practice, the fuel rods are replaced (often in batches) before their ^{235}U content is entirely consumed.

(*e*) From Einstein's $E = \Delta m \, c^2$ relation, we can write

$$\frac{dM}{dt} = \frac{dE/dt}{c^2} = \frac{3.4 \times 10^9 \text{ W}}{(3.00 \times 10^8 \text{ m/s})^2}$$
$$= 3.8 \times 10^{-8} \text{ kg/s} = 3.3 \text{ g/d}.$$

The mass loss rate is about the mass of one penny every day! This mass loss rate (reduction in rest energy) is quite a different quantity than the fuel consumption rate (loss of ^{235}U) calculated in part (*c*).

55-5 A NATURAL REACTOR

On December 2, 1942, when the reactor assembled by Enrico Fermi and his associates first went critical, they had every right to expect that they had put into operation the first fission reactor that had ever existed on this planet. About 30 years later it was discovered that, if they did in fact think that, they were wrong.

Some two billion years ago, in a uranium deposit now being mined in Gabon, West Africa, a natural fission reactor went into operation and ran for perhaps several hundred thousand years before shutting itself off.

The story of this discovery is fascinating at the level of the best detective thriller. More important, it provides a first-class example of the nature of the scientific evidence needed to back up what may seem at first to be an improbable claim. It set a high standard for all who speculate about past events. We consider here only two points.*

1. *Was there enough fuel?* The fuel for a uranium-based fission reactor must be the easily fissionable isotope ^{235}U, which constitutes only 0.72% of natural uranium. This isotopic ratio has been measured not only for terrestrial samples but also in Moon rocks and in meteorites, in which the same value is always found. The initial clue to the discovery in Gabon was that the uranium from this deposit was deficient in ^{235}U, some samples having an abundance as low as 0.44%. Investigation led to the speculation that this deficit in ^{235}U could be accounted for if, at some time in the past, this isotope was partially consumed by the operation of a natural fission reactor.

The serious problem remains that, with an isotopic abundance of only 0.72%, a reactor can be assembled (as Fermi and his team learned) only with the greatest of difficulty. There seems no chance at all that it could have happened naturally.

However, things were different in the distant past. Both ^{235}U and ^{238}U are radioactive, with half-lives of 0.704×10^9 y and 4.47×10^9 y, respectively. Thus the half-life of the readily fissionable ^{235}U is about 6.4 times shorter than that of ^{238}U. Because ^{235}U decays faster, there must have been more of it, relative to ^{238}U, in the past. Two billion years ago, in fact, this abundance was not 0.72%, as it is now, but 3.8%. This abundance happens to be just about the abundance to which natural uranium is artificially enriched to serve as fuel in modern power reactors.

With this amount of readily fissionable fuel available in the distant past, the presence of a natural reactor (providing certain other conditions are met) is much less surprising. The fuel was there. Two billion years ago, incidentally, the highest order of life forms that had evolved were the blue-green algae.

2. *What is the evidence?* The mere depletion of ^{235}U in an ore deposit is not enough evidence on which to base a claim for the existence of a natural fission reactor. More convincing proof is needed.

If there were a reactor, there must also be fission products; see Fig. 2. Of the 30 or so elements whose stable isotopes are produced in this way, some must still remain. Study of their isotopic ratios could provide the convincing evidence we need.

Of the several elements investigated, the case of neodymium is spectacularly convincing. Figure 8*a* shows the isotopic abundances of the seven stable neodymium isotopes as they are normally found in nature. Figure 8*b* shows these abundances as they appear among the ultimate stable products of the fission of ^{235}U. The clear differences are not surprising, considering their totally different origins. The isotopes shown in Fig. 8*a* were formed in supernova explosions that occurred before the formation of our solar system. The isotopes of Fig. 8*b* were cooked up in a reactor by totally different processes. Note particularly that ^{142}Nd, the dominant isotope in the natural element, is totally absent from the fission products.

The big question is: "What do the neodymium isotopes found in the uranium ore body in Gabon look like?" We must expect that, if a natural reactor operated there, isotopes from *both* sources (that is, natural isotopes as well as fission-produced isotopes) might be present. Figure 8*c* shows the results after this and other corrections have been made to the raw data. Comparison of Figs. 8*b* and 8*c* certainly suggests that there was indeed a natural fission reactor at work!

---

* For the complete story, see "A Natural Fission Reactor," by George A. Cowan, *Scientific American,* July 1976, p. 36.

Sample Problem 4 The isotopic ratio of ^{235}U to ^{238}U in natural uranium deposits today is 0.0072. What was this ratio $2.0 \times$

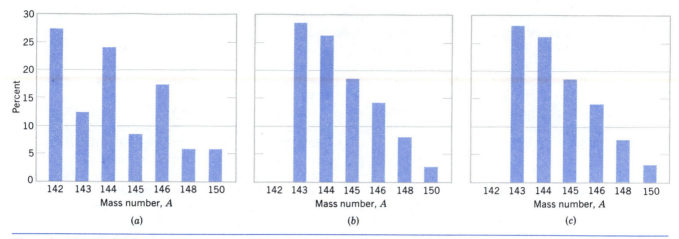

Figure 8   The distribution by mass number of the isotopes of neodymium as they occur in (a) natural terrestrial deposits, (b) the spent fuel of a power reactor, and (c) the uranium mine in Gabon, West Africa. Note that (b) and (c) are virtually identical and quite different from (a).

10^9 y ago? The half-lives of the two isotopes are 0.704×10^9 y and 4.47×10^9 y, respectively.

Solution   Consider two samples that, at a time t in the past, contained $N_5(0)$ and $N_8(0)$ atoms of ^{235}U and ^{238}U, respectively. The numbers of atoms remaining at the present time are

$$N_5(t) = N_5(0)e^{-\lambda_5 t} \quad \text{and} \quad N_8(t) = N_8(0)e^{-\lambda_8 t},$$

respectively, in which λ_5 and λ_8 are the corresponding disintegration constants. Dividing gives

$$\frac{N_5(t)}{N_8(t)} = \frac{N_5(0)}{N_8(0)} e^{-(\lambda_5 - \lambda_8)t}.$$

Expressed in terms of the isotopic ratio $R = N_5/N_8$, this becomes

$$R(0) = R(t)e^{(\lambda_5 - \lambda_8)t}.$$

The disintegration constants are related to the half-lives by Eq. 8 of Chapter 54, or

$$\lambda_5 = \frac{\ln 2}{t_{1/2,\,5}} = \frac{0.693}{7.04 \times 10^8 \text{ y}} = 0.984 \times 10^{-9} \text{ y}^{-1}$$

and

$$\lambda_8 = \frac{\ln 2}{t_{1/2,\,8}} = \frac{0.693}{4.47 \times 10^9 \text{ y}} = 0.155 \times 10^{-9} \text{ y}^{-1}.$$

Substituting in the expression for the isotopic ratio gives

$$R(0) = R(t)e^{(\lambda_5 - \lambda_8)t}$$
$$= (0.0072)e^{(0.984 - 0.155)(10^{-9}\text{ y}^{-1})(2.00 \times 10^9 \text{ y})}$$
$$= (0.0072)e^{1.65} = 0.0378 \text{ or } 3.78\%.$$

We see that, two billion years ago, the ratio of ^{235}U to ^{238}U in natural uranium deposits was much higher than it is today. When the Earth was formed (4.5 billion years ago) this ratio was 30%.

55-6   THERMONUCLEAR FUSION: THE BASIC PROCESS

We pointed out in connection with the binding energy curve of Fig. 6 of Chapter 54 that energy can be released if light nuclei are combined to form nuclei of somewhat larger mass number, a process called *nuclear fusion*. However, this process is hindered by the mutual Coulomb repulsion that tends to prevent two such (positively) charged particles from coming within range of each other's attractive nuclear forces and "fusing." This reminds us of the potential barrier that inhibits nuclear fission (see Fig. 4) and also of the barrier that inhibits alpha decay (see Fig. 8 of Chapter 54).

In the case of alpha decay, two charged particles—the α particle and the residual nucleus—are initially *inside* their mutual potential barrier. For alpha decay to occur, the α particle must leak through this barrier by the barrier-tunneling process and appear on the *outside*. In nuclear fusion the situation is just reversed. Here the two particles must penetrate their mutual barrier from the *outside* if a nuclear interaction is to occur.

The interaction between two deuterons is of particular importance in fusion. Sample Problem 5 gives a rough calculation of the potential barrier between two deuterons, which works out to be about 200 keV. The corresponding barrier for two interacting ^{3}He nuclei (charge $= +2e$) is about 1 MeV. For more highly charged particles the barrier, of course, is correspondingly higher.

One way to arrange for light nuclei to penetrate their mutual Coulomb barrier is to use one light particle as a target and to accelerate the other by means of a cyclotron or a similar device. To generate power in a useful way from the fusion process, however, we must have the interaction of matter in bulk, just as in the combustion of coal.

The cyclotron technique holds no promise in this direction. The best hope for obtaining fusion in bulk matter in a controlled fashion is to raise the temperature of the material so that the particles have sufficient energy to penetrate the barrier due to their thermal motions alone. This process is called *thermonuclear fusion.*

The mean thermal kinetic energy $\overline{K}$ of a particle in equilibrium at a temperature T is given, as we have seen in Chapter 23, by

$$\overline{K} = \tfrac{3}{2}kT, \tag{3}$$

where $k (= 8.62 \times 10^{-5}\,\text{eV/K})$ is the Boltzmann constant. At room temperature ($T \approx 300$ K), $\overline{K} = 0.04$ eV, which is, of course, far too small for our purpose.

Even at the center of the Sun, where $T \approx 1.5 \times 10^7$ K, the mean thermal kinetic energy calculated from Eq. 3 is only 1.9 keV. This still seems hopelessly small in view of the magnitude of the Coulomb barrier of 200 keV calculated in Sample Problem 5. Yet we know that thermonuclear fusion not only occurs in the solar interior but is its central and dominant feature.

The puzzle is solved with the realization that (1) the energy calculated from Eq. 3 is a *mean* kinetic energy; particles with energies much greater than this mean value constitute the high-energy "tails" of the Maxwellian speed distribution curves (see Fig. 10 of Chapter 24). Also, (2) the barrier heights that we have quoted represent only the *peaks* of the barriers. Barrier tunneling can occur to a significant extent at energies well below these peaks, as we saw in Section 54-4 in the case of alpha decay.

Figure 9 summarizes the situation by a quantitative example. The curve marked $n(K)$ in this figure is a Maxwell energy distribution curve drawn to correspond to the Sun's central temperature, 1.5×10^7 K. Although the same curve holds no matter what particle is under consideration, we focus our attention on protons, bearing in

mind that hydrogen forms about 35% of the mass of the Sun's central core.

For $T = 1.5 \times 10^7$ K, Eq. 3 yields $\overline{K} = 1.9$ keV, and this value is indicated by a vertical line in Fig. 9. Note that there are many particles whose energies exceed this mean value.

The curve marked $p(K)$ in Fig. 9 is the probability of barrier penetration for two colliding protons. At $K = 6$ keV, for example, we have $p = 2.4 \times 10^{-5}$. This is the probability that two colliding protons, each with $K = 6$ keV, will succeed in penetrating their mutual Coulomb barrier and coming within range of each other's strong nuclear forces. Put another way, on the average, one of every 42,000 such encounters will succeed.

It turns out that the most probable energy for proton–proton fusion events to occur at the Sun's central temperature is about 6 keV. If the energy is much higher, the barrier is more easily penetrated (that is, p is greater), but there are too few protons in the Maxwellian "tail" (n is smaller). If the energy is much lower, there are plenty of protons but the barrier is now too formidable.

Sample Problem 5   The deuteron (^{2}H) has a charge $+e$, and its radius has been measured to be 2.1 fm. Two such particles are fired at each other with the same initial kinetic energy K. What must K be if the particles are brought to rest by their mutual Coulomb repulsion when the two deuterons are just "touching"?

Solution   Because the two deuterons are momentarily at rest when they "touch" each other, their kinetic energy has all been transformed into electrostatic potential energy associated with the Coulomb repulsion between them. If we treat them as point charges separated by a distance $2R$, we have

$$2K = \frac{1}{4\pi\epsilon_0}\frac{q_1 q_2}{r} = \frac{1}{4\pi\epsilon_0}\frac{e^2}{2R},$$

which yields

$$K = \frac{e^2}{16\pi\epsilon_0 R}$$

$$= \frac{(1.6 \times 10^{-19}\,\text{C})^2}{(16\pi)(8.9 \times 10^{-12}\,\text{C}^2/\text{J}\cdot\text{m})(2.1 \times 10^{-15}\,\text{m})}\left(\frac{1\,\text{keV}}{1.6 \times 10^{-16}\,\text{J}}\right)$$

$$\approx 200\,\text{keV}.$$

This quantity provides a reasonable measure of the height of the Coulomb barrier between two deuterons.

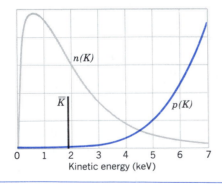

Figure 9   The curve marked $n(K)$ gives the distribution in energy of protons in the core of the Sun, corresponding to a temperature of 1.5×10^7 K. The vertical line indicates the mean kinetic energy per particle at that temperature. The curve marked $p(K)$ gives the probability of barrier penetration in proton–proton collisions. The two curves are drawn to different arbitrary vertical scales.

55-7   THERMONUCLEAR FUSION IN STARS

Here we consider in more detail the thermonuclear fusion processes that take place in our Sun and in other stars. In the Sun's deep interior, where its mass is concentrated and where most of the energy production takes place, the (central) temperature is 1.5×10^7 K and the central density is on the order of 10^5 kg/m^3, about 13 times the density of lead. The central temperature is so high that, in spite of the high central pressure (2×10^{11} atm), the Sun remains gaseous throughout.

The present composition of the Sun's core is about 35% hydrogen by mass, about 65% helium, and about 1% other elements. At these temperatures the light elements are essentially totally ionized, so that our picture is one of an assembly of protons, electrons, and α particles in random motion.

The Sun radiates at the rate of 3.9×10^{26} W and has been doing so for as long as the solar system has existed, which is about 4.5×10^9 y. It has been known since the 1930s that thermonuclear fusion processes in the Sun's interior account for its prodigious energy output. Before analyzing this further, however, let us dispose of two other possibilities that had been put forward earlier. Consider first chemical reactions such as simple burning. If the Sun, whose mass is 2.0×10^{30} kg, were made of coal and oxygen in just the right proportions for burning, it would last only about 10^3 y, which of course is far too short (see Problem 47). The Sun, as we shall see, does not burn coal but hydrogen, and in a nuclear furnace, not an atomic or chemical one.

Another possibility is that, as the core of the Sun cools and the pressure there drops, the Sun will shrink under the action of its own strong gravitational forces. By transferring gravitational potential energy to internal energy (just as we do when we drop a stone onto the Earth's surface), the temperature of the Sun's core will rise so that radiation may continue. Calculation shows, however, that the Sun could radiate from this cause for only about 10^8 y, too short by a factor of 30 (see Problem 51).

The Sun's energy is generated by the thermonuclear "burning" (that is, "fusing") of hydrogen to form helium. Figure 10 shows the *proton–proton cycle* by means of which this is accomplished. Note that each reaction shown is a fusion reaction, in that one of the products (^{2}H, ^{3}He, or ^{4}He) has a higher mass number than any of the reacting particles that form it. The reaction energy Q for each reaction shown in Fig. 10 is positive. This characterizes an exothermic reaction, with the net production of energy.

The cycle is initiated by the collision of two protons (^{1}H $+$ ^{1}H) to form a deuteron (^{2}H), with the simultaneous creation of a positron (e^+) and a neutrino (ν). The positron very quickly encounters a free electron (e^-) in the Sun and both particles annihilate, their rest energies appearing as two gamma-ray photons (γ), as we discussed in Section 8-7. In Fig. 10 we follow the consequences of two such events, as indicated in the top row of the figure. Such events are extremely rare. In fact, only once in about 10^{26} proton–proton collisions is a deuteron formed; in the vast majority of cases the colliding protons simply scatter from each other. It is the slowness of this process that regulates the rate of energy production and keeps the Sun from exploding. In spite of this slowness, there are so very many protons in the huge volume of the Sun's core that deuterium is produced there in this way at the rate of about 10^{12} kg/s!

Once a deuteron has been produced, it quickly (within a few seconds) collides with another proton and forms a ^{3}He nucleus, as the second row of Fig. 10 shows. Two such ^{3}He nuclei may then eventually (within about 10^5 y) collide, forming an α particle (^{4}He) and two protons, as the third row of the figure shows. There are other variations of the proton–proton cycle, involving other light elements, but we concentrate on the principal sequence as represented in Fig. 10.

Taking an overall view of the proton–proton cycle, we see that it amounts to the combination of four protons and two electrons to form an α particle, two neutrinos, and six gamma rays:

$$4\,^1\text{H} + 2e^- \rightarrow \,^4\text{He} + 2\nu + 6\gamma. \qquad (4)$$

Now, in a formal way, let us add two electrons to each side of Eq. 4, yielding,

$$4(^1\text{H} + e^-) \rightarrow (^4\text{He} + 2e^-) + 2\nu + 6\gamma. \qquad (5)$$

The quantities in parentheses then represent *atoms* (not bare nuclei) of hydrogen and of helium.

The energy release in the reaction of Eq. 5 is, using the atomic masses of hydrogen and helium,

$$Q = \Delta m\, c^2 = [4m(^1\text{H}) - m(^4\text{He})]c^2$$
$$= [4(1.007825 \text{ u}) - 4.002603 \text{ u}](931.5 \text{ MeV/u})$$
$$= 26.7 \text{ MeV}.$$

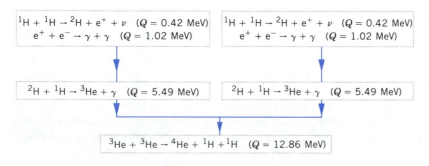

^{1}H $+$ ^{1}H $\rightarrow$ ^{2}H $+$ e^+ $+$ ν (Q = 0.42 MeV)
e^+ $+$ e^- $\rightarrow$ γ $+$ γ (Q = 1.02 MeV)

^{1}H $+$ ^{1}H $\rightarrow$ ^{2}H $+$ e^+ $+$ ν (Q = 0.42 MeV)
e^+ $+$ e^- $\rightarrow$ γ $+$ γ (Q = 1.02 MeV)

^{2}H $+$ ^{1}H $\rightarrow$ ^{3}He $+$ γ (Q = 5.49 MeV)

^{2}H $+$ ^{1}H $\rightarrow$ ^{3}He $+$ γ (Q = 5.49 MeV)

^{3}He $+$ ^{3}He $\rightarrow$ ^{4}He $+$ ^{1}H $+$ ^{1}H (Q = 12.86 MeV)

Figure 10 The proton–proton cycle that primarily accounts for energy production in the Sun.

Neutrinos and gamma-ray photons have no mass and thus do not enter into the calculation of the disintegration energy. This same value of Q follows (as it must) by adding up the Q values for the separate steps of the proton–proton cycle in Fig. 10.

Not quite all this energy is available as internal energy inside the Sun. About 0.5 MeV is associated with the two neutrinos that are produced in each cycle. Neutrinos are so penetrating that in essentially all cases they escape from the Sun, carrying this energy with them. Some are intercepted by the Earth, bringing us our only direct information about the Sun's interior.

Subtracting the neutrino energy leaves 26.2 MeV per cycle available within the Sun. As we show in Sample Problem 6, this corresponds to a "heat of combustion" for the nuclear burning of hydrogen into helium of 6.3×10^{14} J/kg of hydrogen consumed. By comparison, the heat of combustion of coal is about 3.3×10^7 J/kg, some 20 million times lower, reflecting roughly the general ratio of energies in nuclear and chemical processes.

We may ask how long the Sun can continue to shine at its present rate before all the hydrogen in its core has been converted into helium. Hydrogen burning has been going on for about 4.5×10^9 y, and calculations show that there is enough available hydrogen left for about 5×10^9 y more. At that time major changes will begin to happen. The Sun's core, which by then will be largely helium, will begin to collapse and to heat up while the outer envelope will expand greatly, perhaps so far as to encompass the Earth's orbit. The Sun will become what astronomers call a *red giant*.

If the core temperature heats up to about 10^8 K, energy can be produced by burning helium to make carbon. Helium does not burn readily, the only possible reaction being

$$^4\text{He} + {}^4\text{He} + {}^4\text{He} \rightarrow {}^{12}\text{C} + \gamma \qquad (Q = +7.3 \text{ MeV}).$$

Such a three-body collision of three α particles must occur within 10^{-16} s if the reaction is to go. Nevertheless, if the density and temperature of the helium core are high enough, carbon will be manufactured by the burning of helium in this way.

As a star evolves and becomes still hotter, other elements can be formed by other fusion reactions. However, elements beyond $A = 56$ cannot be manufactured by further fusion processes. The elements with $A = 56$ (^{56}Fe, ^{56}Co, ^{56}Ni) lie near the peak of the binding energy curve of Fig. 6 of Chapter 54, and fusion between nuclides beyond this point involves the consumption, and not the production, of energy. The production of the elements in fusion processes is discussed in Chapter 56.

Sample Problem 6 At what rate is hydrogen being consumed in the core of the Sun, assuming that all the radiated energy is generated by the proton–proton cycle of Fig. 10?

Solution We have seen that 26.2 MeV appears as internal energy in the Sun for every four protons consumed, a rate of 6.6 MeV/proton. We can express this as

$$\left(\frac{6.6 \text{ MeV/proton}}{1.67 \times 10^{-27} \text{ kg/proton}} \right) (1.6 \times 10^{-13} \text{ J/MeV})$$

$$= 6.3 \times 10^{14} \text{ J/kg,}$$

which tells us that the Sun radiates away 6.3×10^{14} J for every kilogram of protons consumed. The hydrogen consumption rate is then just the output power ($= 3.9 \times 10^{26}$ W) divided by this quantity, or

$$\frac{dm}{dt} = \frac{3.9 \times 10^{26} \text{ W}}{6.3 \times 10^{14} \text{ J/kg}} = 6.2 \times 10^{11} \text{ kg/s.}$$

To keep this number in perspective, keep in mind that the Sun's mass is 2.0×10^{30} kg.

55-8 CONTROLLED THERMONUCLEAR FUSION

Thermonuclear reactions have been going on in the universe since its creation in the presumed cosmic "big bang" of some 15 billion years ago. Such reactions have taken place on Earth, however, only since October 1952, when the first fusion (or hydrogen) bomb was exploded. The high temperatures needed to initiate the thermonuclear reaction in this case were provided by a fission bomb used as a trigger.

A sustained and controllable thermonuclear power source—a fusion reactor—is proving much more difficult to achieve. The goal, however, is being vigorously pursued because many look to the fusion reactor as the ultimate power source of the future, at least as far as the generation of electricity is concerned.

The proton–proton interaction displayed in Fig. 10 is not suitable for use in a terrestrial fusion reactor because the process displayed in the first row is hopelessly slow. The reaction cross section is in fact so small that it cannot be measured in the laboratory. The reaction succeeds under the conditions that prevail in stellar interiors only because of the enormous number of protons available in the high-density stellar cores.

The most attractive reactions for terrestrial use appear to be the deuteron–deuteron (d-d) and the deuteron–triton (d-t) reactions:

d-d: $^2\text{H} + {}^2\text{H} \rightarrow {}^3\text{He} + \text{n} \qquad (Q = +3.27 \text{ MeV}),$ (6)

d-d: $^2\text{H} + {}^2\text{H} \rightarrow {}^3\text{H} + {}^1\text{H} \qquad (Q = +4.03 \text{ MeV}),$ (7)

d-t: $^2\text{H} + {}^3\text{H} \rightarrow {}^4\text{He} + \text{n} \qquad (Q = +17.59 \text{ MeV}).$ (8)

Here *triton* indicates ^{3}H, the nucleus of hydrogen with $A = 3$. Note that each of these reactions is indeed a fusion reaction and has a positive Q value. Deuterium, whose isotopic abundance in normal hydrogen is 0.015%, is available in unlimited quantities as a component of seawater. Tritium (atomic ^{3}H) is radioactive and is not normally found in naturally occurring hydrogen.

There are three basic requirements for the successful operation of a thermonuclear reactor.

1. *A high particle density n.* The number of interacting particles (deuterons, say) per unit volume must be great enough to ensure a sufficiently high deuteron–deuteron collision rate. At the high temperatures required, the deuterium gas would be completely ionized into a neutral *plasma* consisting of deuterons and electrons.

2. *A high plasma temperature T.* The plasma must be hot. Otherwise the colliding deuterons will not be energetic enough to penetrate the mutual Coulomb barrier that tends to keep them apart. In fusion research, temperatures are often reported by giving the corresponding value of kT (not $\frac{3}{2}kT$). A plasma temperature of 33 keV, corresponding to 2.8×10^8 K, has been achieved in the laboratory. This is much higher than the Sun's central temperature (1.3 keV, or 1.5×10^7 K).

3. *A long confinement time τ.* A major problem is containing the hot plasma long enough to ensure that its density and temperature remain sufficiently high. It is clear that no actual solid container can withstand the high temperatures necessarily involved, so that special techniques, to be described later, must be employed. By use of one such technique, confinement times greater than 1 s have been achieved.

It can be shown that, for the successful operation of a thermonuclear reactor, it is necessary to have

$$n\tau \geq 10^{20} \text{ s} \cdot \text{m}^{-3}, \qquad (9)$$

a condition called *Lawson's criterion.* Equation 9 tells us, loosely speaking, that we have a choice between confining a lot of particles for a relatively short time or confining fewer particles for a somewhat longer time. Beyond meeting this criterion, it is also necessary that the plasma temperature be sufficiently high.

There are two techniques that have been used to attempt to achieve the combination of temperature T and Lawson's parameter $n\tau$ that are necessary to produce fusion reactions. *Magnetic confinement* uses magnetic fields to confine the plasma while its temperature is increased. In *inertial confinement,* on the other hand, a small amount of fuel is compressed and heated so rapidly that fusion occurs before the fuel can expand and cool. These techniques are discussed in the following two sections.

Deriving Lawson's Criterion   (Optional)
Let us see how Lawson's criterion comes about. To raise a plasma to a suitably high temperature and to maintain it there against losses, energy must be added to the plasma at a rate per unit volume P_h, where the subscript stands for "heating." The heating may be done by passing an electric current through the

plasma, by firing a beam of energetic neutral particles into it, or in other ways. The denser the plasma, the greater the heating power required, in direct proportion, or

$$P_h = C_h n, \qquad (10)$$

where C_h is a suitable constant.

If thermonuclear fusion occurs in the plasma, there will be a certain rate of energy generation per unit volume P_f, where the subscript now stands for "fusion." P_f is proportional to the confinement time τ. It is also proportional to n^2, where n is the particle density. To see this, suppose that we double the particle density. Not only will a given particle make twice as many collisions as it wanders through the plasma, but there will be twice as many wandering particles, giving an overall factor of four. Thus

$$P_f = C_f n^2 \tau. \qquad (11)$$

To have a net production power, we must have

$$P_f > P_h$$

or, from Eqs. 10 and 11,

$$n\tau > C_h / C_f,$$

which leads directly to Eq. 9 if the constants C_h and C_f are suitably evaluated. The condition in which $P_f = P_h$ is called *breakeven.* ■

55-9   MAGNETIC CONFINEMENT

Because a plasma consists of charged particles, its motion can be controlled with magnetic fields. For example, charged particles spiral about the direction of a uniform magnetic field. By suitably varying the field strength, it is possible to design a "magnetic mirror" (see Fig. 14 of Chapter 34) from which particles can be reflected. Another design makes use of the toroidal geometry, in which the particles spiral around the axis of a toroid inside a "doughnut-shaped" vacuum chamber. The type of fusion reactor based on this principle, which was first developed in Russia, is called a *tokamak,* which comes from the Russian-language acronym for "toroidal magnetic chamber." Several large machines of this type have been built and tested.

In a tokamak, there are two components to the magnetic field, as illustrated in Fig. 11. The *toroidal* field B_t is the one we usually associate with a toroidal winding of wires; Fig. 11 shows one small section of an external coil that contributes to the toroidal field. Because the toroidal field decreases with increasing radius, it is necessary to add a second field component to confine the particles. This *poloidal* component B_p of the field adds to the toroidal component to give the total field a helical structure, as illustrated in Fig. 11. The poloidal field is produced by a current i' in the plasma itself, which is induced by a set of windings not illustrated in the figure. This current also serves to heat the plasma. Additional means of heating, such as by firing neutral beams of particles into the

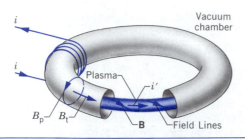

Figure 11   The toroidal chamber that forms the basis of the tokamak. Note the plasma, the helical magnetic field **B** that confines it, and the induced current i' that heats it.

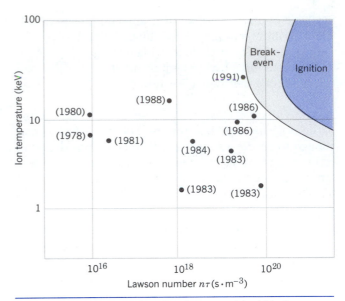

Figure 13   The approach to breakeven and ignition in controlled fusion reactors, shown as a plot of Lawson number against temperature.

plasma, are also necessary to achieve the desired plasma temperature.

Figure 12 shows a worker in the interior of the toroidal vacuum chamber of the Tokamak Fusion Test Reactor at the Princeton Plasma Physics Laboratory. The interior radius of the vacuum chamber is about 2 m, and the major radius of the toroid is 2.5 m.

In designing magnetic confinement devices such as the tokamak, the goal is to increase both the Lawson confinement parameter $n\tau$ and the temperature T of the plasma. At sufficiently high values of these parameters, fusion reactions in the plasma will produce enough energy to equal the energy that must be supplied to heat the plasma. This condition is called "breakeven." At still higher values of these parameters, the device will achieve "igni-

tion," where self-sustaining fusion reactions will occur. Figure 13 illustrates the steady progress toward these goals that has been made. Despite the approach to the breakeven condition, many formidable engineering problems remain to be solved, and the production of electric power from fusion is likely many decades away.

Sample Problem 7   The Tokamak Fusion Test Reactor at Princeton has achieved a confinement time of 400 ms. (*a*) What must be the density of particles in the plasma if Lawson's criterion is to be satisfied? (*b*) How does this number compare with the particle density of the atoms of an ideal gas at standard conditions? (*c*) If a next-generation tokamak could achieve ignition, with a plasma temperature of 10 keV and confinement time of 1 s, what would the particle density of its plasma have to be?

Solution   (*a*) Using Lawson's criterion (Eq. 9), we must have

$$n = \frac{10^{20} \text{ s} \cdot \text{m}^{-3}}{0.40 \text{ s}} = 2.5 \times 10^{20} \text{ m}^{-3}.$$

(*b*) The number density of atoms in an ideal gas at standard conditions is given by $n' = N_A/V_m$, where N_A is the Avogadro constant and $V_m (= 2.24 \times 10^{-2} \text{ m}^3/\text{mol})$ is the molar volume of an ideal gas at standard conditions, which gives

$$n' = \frac{N_A}{V_m} = \frac{6.02 \times 10^{23} \text{ mol}^{-1}}{2.24 \times 10^{-2} \text{ m}^3/\text{mol}} = 2.7 \times 10^{25} \text{ m}^{-3}.$$

The particle density of the plasma we found in part (*a*) is smaller than that of an ideal gas by a factor of about 10^5.

(*c*) From Fig. 13 we see that the 10-keV temperature line intersects the curve marked "ignition" at a value of the Lawson number of about 1×10^{21} s·m^{-3}. (In making this last estimate,

Figure 12   A worker inside the toroidal chamber of the Tokamak Fusion Test Reactor at Princeton University.

bear in mind that the scale is logarithmic.) The necessary particle density is then

$$n = \frac{1 \times 10^{21} \text{ s} \cdot \text{m}^{-3}}{1 \text{ s}} = 1 \times 10^{21} \text{ m}^{-3}.$$

55-10 INERTIAL CONFINEMENT

A second technique for confining plasma so that thermonuclear fusion can take place is called *inertial confinement*. In terms of Lawson's criterion (Eq. 9), it involves working with extremely high particle densities n for extremely short confinement times τ. These times are arranged to be so short that the fusion episode is over before the particles of the plasma have time to move appreciably from the positions they occupy at the onset of fusion. The interacting particles are confined by their own inertia.

Laser fusion, which relies on the inertial-confinement principle, is being investigated in laboratories throughout the world. At the Lawrence Livermore Laboratory, for example, in the NOVA laser fusion project (see Fig. 14) deuterium–tritium fuel pellets, each smaller than a grain of sand (see Fig. 15), are to be "zapped" by 10 synchronized, high-powered laser pulses, symmetrically arranged around the pellet. The laser pulses are designed to deliver in total some 200 kJ of energy to each fuel pellet in less than a nanosecond. This is a delivered power of 2×10^{14} W during the pulse, which is roughly 100 times the total installed electric power generating capacity of the world!

The laser pulse energy serves to heat the fuel pellet, ionizing it to a plasma and — it is hoped — raising its tem-

Figure 14 The target chamber of the NOVA inertial confinement fusion facility at the Lawrence Livermore National Laboratory. The photo shows several of the 10 laser beam tubes.

Figure 15 The tiny spheres, shown resting on a dime, are deuterium–tritium fuel pellets for use in inertial confinement experiments.

perature to around 10^8 K. As the surface layers of the pellet evaporate at these high thermal speeds, the reaction force of the escaping particles compresses the core of the pellet, increasing its density by a factor of perhaps 10^3. If all these things happened, then conditions would be right for thermonuclear fusion to occur in the core of the highly compressed pellet of plasma, the fusion reaction being the d-t reaction given in Eq. 8.

In an operating thermonuclear reactor of the laser fusion type, it is visualized that fuel pellets would be exploded, like miniature hydrogen bombs, at the rate of perhaps 10–100 per second. The energetic emerging particles of the fusion reaction (^{4}He and n) might be absorbed in a "blanket" consisting of a moving stream of molten lithium, heating it up. Internal energy would then be extracted from the lithium stream at another location and used to generate steam, just as in a fission reactor or a fossil-fuel power plant. Lithium would be a suitable choice for a heat-transfer medium because the energetic neutron would, with high probability, deliver up its energy to the "blanket" by the reaction

$$^6\text{Li} + \text{n} \rightarrow {}^4\text{He} + {}^3\text{H}.$$

The two charged particles would readily be brought to rest in the lithium. The tritium produced in the reaction can be extracted for use as fuel in the reactor. The feasibility of laser fusion as the basis of a fusion reactor has not been conclusively demonstrated as of 1991, but vigorous research is continuing.

Sample Problem 8   Suppose that a fuel pellet in a laser fusion device is made of a liquid deuterium–tritium mixture containing equal numbers of deuterium and tritium atoms. The density $d\,(= 200$ kg/m³) of the pellet is increased by a factor of 10^3 by the action of the laser pulses. (a) How many particles per unit volume (either deuterons or tritons) does the pellet contain in its compressed state? (b) According to Lawson's criterion, for how long must the pellet maintain this particle density if breakeven operation is to take place?

Solution   (a) We can write, for the density d' of the compressed pellet,

$$d' = 10^3 d = m_d \frac{n}{2} + m_t \frac{n}{2},$$

in which n is the number of particles per unit volume (either deuterons or tritons) in the compressed pellet, m_d is the mass of a

deuterium atom, and m_t is the mass of a tritium atom. These atomic masses are related to the Avogadro constant N_A and to the corresponding molar masses (M_d and M_t) by

$$m_d = M_d/N_A \quad \text{and} \quad m_t = M_t/N_A.$$

Combining these equations and solving for n lead to

$$
\begin{aligned}
n &= \frac{2d'N_A}{M_d + M_t} \\
&= \frac{(2)(10^3 \times 200 \text{ kg/m}^3)(6.02 \times 10^{23} \text{ mol}^{-1})}{2.0 \times 10^{-3} \text{ kg/mol} + 3.0 \times 10^{-3} \text{ kg/mol}} \\
&= 4.8 \times 10^{31} \text{ m}^{-3}.
\end{aligned}
$$

(b) From Lawson's criterion (Eq. 9), we have

$$\tau > \frac{10^{20} \text{ s} \cdot \text{m}^{-3}}{n} = \frac{10^{20} \text{ s} \cdot \text{m}^{-3}}{4.8 \times 10^{31} \text{ m}^{-3}} = 2 \times 10^{-12} \text{ s}.$$

The pellet must remain compressed for at least this long if breakeven operation is to occur. (It is also necessary for the effective temperature to be suitably high.)

A comparison with Sample Problem 7 shows that, unlike tokamak operation, laser fusion seeks to operate in the realm of very high particle densities and correspondingly very short confinement times.

QUESTIONS

1. If it's so much harder to get a nucleon out of a nucleus than to get an electron out of an atom, why try?

2. Can you say, from examining Table 1, that one source of energy, or of power, is better than another? If not, what other considerations enter?

3. To which of the processes in Table 1 does the relationship $E = \Delta m\, c^2$ apply?

4. Of the two fission fragment tracks shown in Fig. 1, which fragment has the larger (a) momentum, (b) kinetic energy, (c) speed, (d) mass?

5. In the generalized equation for the fission of ^{235}U by thermal neutrons, ^{235}U + n → X + Y + bn, do you expect the Q of the reaction to depend on the identity of X and Y?

6. Is the fission fragment curve of Fig. 2 necessarily symmetrical about its central minimum? Explain your answer.

7. In the chain decays of the primary fission fragments (see Eq. 2) why do no β^+ decays occur?

8. The half-life of ^{235}U is 7.0×10^8 y. Discuss the assertion that if it had turned out to be shorter by a factor of 10 or so, there would not be any atomic bombs today.

9. ^{238}U is not fissionable by thermal neutrons. What minimum neutron energy do you think would be necessary to induce fission in this nuclide?

10. The half-life for the decay of ^{235}U by alpha emission is 7×10^8 y; by spontaneous fission, acting alone, it would be 3×10^{17} y. Both are barrier-tunneling processes, as Fig. 8 in Chapter 54 and Fig. 4 in Chapter 55 reveal. Why this enormous difference in barrier-tunneling probability?

11. Compare fission with alpha decay in as many ways as possi-

ble. How can a thermal neutron deliver several million electron-volts of excitation energy to a nucleus that absorbs it, as in Fig. 3a? The neutron has essentially no energy to start with!

12. The binding energy curve of Fig. 6 in Chapter 54 tells us that any nucleus more massive than $A \approx 56$ can release energy by the fission process. Only very massive nuclides seem to do so, however. Why can't lead, for example, release energy by the fission process?

13. By bombardment of heavy nuclides in the laboratory it is possible to prepare other heavy nuclides that decay, at least in part, by *spontaneous fission*. That is, after a certain mean life they spontaneously break up into two major fragments. Can you explain this on the basis of the theory of Bohr and Wheeler?

14. Slow neutrons are more effective than fast ones in inducing fission. Can you make that plausible? (*Hint*: Consider how the de Broglie wavelength of a neutron might be related to its capture cross section in ^{235}U.)

15. Compare a nuclear reactor with a coal fire. In what sense does a chain reaction occur in each? What is the energy-releasing mechanism in each case?

16. Not all neutrons produced in a reactor are destined to initiate a fission event. What happens to those that do not?

17. Explain just what is meant by the statement that in a reactor core neutron leakage is a surface effect and neutron production is a volume effect.

18. Explain the purpose of the moderator in a nuclear reactor. Is it possible to design a reactor that does not need a modera-

tor? If so, what are some of the advantages and disadvantages of such a reactor?

19. Describe how to operate the control rods of a nuclear reactor (*a*) during initial start-up, (*b*) to reduce the power level, and (*c*) on a long-term basis, as fuel is consumed.

20. A reactor is operating at full power with its multiplication factor k adjusted to unity. If the reactor is now adjusted to operate stably at half power, what value must k now assume?

21. Separation of the two isotopes ^{238}U and ^{235}U from natural uranium requires a physical method, such as diffusion, rather than a chemical method. Explain why.

22. A piece of pure ^{235}U (or ^{239}Pu) will spontaneously explode if it is larger than a certain "critical size." A smaller piece will not explode. Explain.

23. What can you say, if anything, about the value of the multiplication factor k in an atomic (fission) bomb?

24. The Earth's core is thought to be mostly iron because, during the formation of the Earth, heavy elements such as iron would have sunk toward the Earth's center and lighter elements, such as silicon, would have floated upward to form the Earth's crust. However, iron is far from the heaviest element. Why isn't the Earth's core made of uranium?

25. From information given in the text, collect and write down the approximate heights of the Coulomb barriers for (*a*) the alpha decay of ^{238}U, (*b*) the fission of ^{235}U by thermal neutrons, and (*c*) the head-on collision of two deuterons.

26. The Sun's energy is assumed to be generated by nuclear reactions such as the proton–proton cycle. What alternative ways of generating solar energy were proposed in the past, and why were they rejected?

27. Elements up to mass number ≈ 56 are created by thermonuclear fusion in the cores of stars. Why are heavier elements not also created by this process?

28. Do you think that the thermonuclear fusion reaction controlled by the two curves plotted in Fig. 9 necessarily has its maximum effectiveness for the energy at which the two curves cross each other? Explain your answer.

29. In Fig. 9, are you surprised that, as judged by the areas under the curve marked $n(K)$, the number of particles with $K > \overline{K}$ is smaller than the number with $K < \overline{K}$, where $\overline{K}$ is the average thermal energy?

30. The uranium nuclides present in the Earth today were originally built up and spewed into space during the explosion of stars, so-called supernova events. These explosions, which occurred before the formation of our solar system, represent the collapse of stars under their own gravity. Can you then say that the energy derived from fission was once stored in a gravitational field? Does fission energy then, in this limited sense, have something in common with energy derived from hydroelectric sources?

31. Why does it take so long ($\sim 10^6$ y!) for gamma-ray photons generated by nuclear reactions in the Sun's central core to diffuse to the surface? What kinds of interactions do they have with the protons, α particles, and electrons that make up the core?

32. The primordial matter of the early universe is thought to have been largely hydrogen. Where did all the silicon in the Earth come from? All the gold?

33. Do conditions at the core of the Sun satisfy Lawson's criterion for a sustained thermonuclear fusion reaction? Explain.

34. To achieve ignition in a tokamak, why do you need a high plasma temperature? A high density of plasma particles? A long confinement time?

35. Which would generate more radioactive waste products, a fission reactor or a fusion reactor?

36. Does Lawson's criterion hold both for tokamaks and for laser fusion devices?

PROBLEMS

Section 55-2 Nuclear Fission: The Basic Process

1. You wish to produce 1.0 GJ of energy. Calculate and compare (*a*) the amount of coal needed if you obtain the energy by burning coal and (*b*) the amount of natural uranium needed if you obtain the energy by fission in a reactor. Assume that the combustion of 1.0 kg of coal releases 2.9×10^7 J; the fission of 1.0 kg of uranium in a reactor releases 8.2×10^{13} J.

2. In the United States, coal commonly contains about 3 parts per million (3 ppm) of fissionable uranium and thorium. Calculate and compare (*a*) the energy derived from burning 100 kg of coal and (*b*) the energy that could be derived from the fission of the fissionable impurities that remain in its ashes. Assume that the combustion of 1 kg of coal releases 2.9×10^7 J; the fission of 1 kg of uranium or thorium in a reactor releases 8.2×10^{13} J.

3. (*a*) How many atoms are contained in 1.00 kg of pure ^{235}U? (*b*) How much energy, in joules, is produced by the complete fissioning of 1.00 kg of ^{235}U? Assume $Q = 200$ MeV. (*c*) For how many years would this energy light a 100-W lamp?

4. At what rate must ^{235}U nuclei undergo fission by neutrons to generate 2.00 W? Assume that $Q = 200$ MeV.

5. Verify that, as reported in Table 1, the fission of the ^{235}U in 1.0 kg of UO_2 (enriched so that ^{235}U is 3.0% of the total uranium) could keep a 100-W lamp burning for 680 y.

6. The fission properties of the plutonium isotope ^{239}Pu are very similar to those of ^{235}U. The average energy released per fission is 180 MeV. How much energy, in joules, is liberated if all the atoms in 1.00 kg of pure ^{239}Pu undergo fission?

7. Very occasionally a ^{233}U nucleus, having absorbed a neutron, breaks up into *three* fragments. If two of these fragments are identified chemically as isotopes of chromium and gallium and if no prompt neutrons are involved, what is at least one possibility for the identity of the fragments? Consult a nuclidic chart or table.

8. Show that, in Sample Problem 1, there is no need to take the masses of the electrons emitted during the beta decay of the primary fission fragments explicitly into account.

9. ^{235}U decays by alpha emission with a half-life of 7.04×10^8 y. It also decays (rarely) by spontaneous fission, and if the alpha decay did not occur, its half-life due to this process alone would be 3.50×10^{17} y. (a) At what rate do spontaneous fission decays occur in 1.00 g of ^{235}U? (b) How many alpha-decay events are there for every spontaneous fission event?

Section 55-3 Theory of Nuclear Fission

10. Fill in the following table, which refers to the generalized fission reaction

$$^{235}U + n \rightarrow X + Y + bn.$$

X	Y	b
^{140}Xe	—	1
^{139}I	—	2
—	^{100}Zr	2
^{141}Cs	^{92}Rb	—

11. Calculate the distintegration energy Q for the spontaneous fission of ^{52}Cr into two equal fragments. The needed masses are ^{52}Cr, 51.940509 u; and ^{26}Mg, 25.982593 u. Discuss your result.

12. Calculate the disintegration energy Q for the fission of ^{98}Mo into two equal parts. The needed masses are ^{98}Mo, 97.905406 u; and ^{49}Sc, 48.950022 u. If Q turns out to be positive, discuss why this process does not occur spontaneously.

13. Calculate the energy released in the fission reaction

$$^{235}U + n \rightarrow ^{141}Cs + ^{92}Rb + 3n.$$

Needed atomic masses are

^{235}U	235.043924 u	^{92}Rb	91.919661 u
^{141}Cs	140.920006 u	n	1.008665 u.

14. ^{238}Np has a barrier energy for fission of 4.2 MeV. To remove a neutron from this nuclide requires an energy expenditure of 5.0 MeV. Is ^{237}Np fissionable by thermal neutrons?

15. Consider the fission of ^{238}U by fast neutrons. In one fission event no neutrons were emitted and the final stable end products, after the beta decay of the primary fission fragments, were ^{140}Ce and ^{99}Ru. (a) How many beta-decay events were there in the two beta-decay chains, considered together? (b) Calculate Q. The relevant atomic masses are

^{238}U	238.050784 u	^{140}Ce	139.905433 u
n	1.008665 u	^{99}Ru	98.905939 u.

16. In a particular fission event of ^{235}U by slow neutrons, it happens that no neutron is emitted and that one of the primary fission fragments is ^{83}Ge. (a) What is the other fragment? (b) How is the disintegration energy $Q = 170$ MeV split between the two fragments? (c) Calculate the initial speed of each fragment.

17. Assume that just after the fission of ^{236}U* according to Eq. 2, the resulting ^{140}Xe and ^{94}Sr nuclei are just touching at their

surfaces. (a) Assuming the nuclei to be spherical, calculate the Coulomb potential energy (in MeV) of repulsion between the two fragments. (*Hint*: Use Eq. 1 in Chapter 54 to calculate the radii of the fragments.) (b) Compare this energy with the energy released in a typical fission process. In what form will this energy ultimately appear in the laboratory?

18. A ^{236}U* nucleus undergoes fission and breaks up into two middle-mass fragments, ^{140}Xe and ^{96}Sr. (a) By what percentage does the surface area of the ^{236}U nucleus change during this process? (b) By what percentage does its volume change? (c) By what percentage does its electrostatic potential energy change? The potential energy of a uniformly charged sphere of radius r and charge Q is given by

$$U = \frac{3}{5}\left(\frac{Q^2}{4\pi\epsilon_0 r}\right).$$

Section 55-4 Nuclear Reactors: The Basic Principles

19. Many fear that helping additional nations develop nuclear power reactor technology will increase the likelihood of nuclear war because reactors can be used not only to produce energy but, as a by-product through neutron capture with inexpensive ^{238}U, to make ^{239}Pu, which is a "fuel" for nuclear bombs (*breeder* reactors). What simple series of reactions involving neutron capture and beta decay would yield this plutonium isotope?

20. A 190-MW fission reactor consumes half its fuel in 3 years. How much ^{235}U did it contain initially? Assume that all the energy generated arises from the fission of ^{235}U and that this nuclide is consumed only by the fission process. See Sample Problem 3.

21. Repeat Problem 20 taking into account nonfission neutron capture by the ^{235}U. See Sample Problem 3.

22. (a) A neutron with initial kinetic energy K makes a head-on elastic collision with a resting atom of mass m. Show that the fractional energy loss of the neutron is given by

$$\frac{\Delta K}{K} = \frac{4m_n m}{(m + m_n)^2},$$

in which m_n is the neutron mass. (b) Find $\Delta K/K$ if the resting atom is hydrogen, deuterium, carbon, or lead. (c) If $K = 1.00$ MeV initially, how many such collisions would it take to reduce the neutron energy to thermal values (0.025 eV) if the material is deuterium, a commonly used moderator? (*Note*: In actual moderators, most collisions are not "head-on.")

23. The neutron generation time t_{gen} in a reactor is the average time between one fission and the fissions induced by the neutrons emitted in that fission. Suppose that the power output of a reactor at time $t = 0$ is P_0. Show that the power output a time t later is $P(t)$, where

$$P(t) = P_0 k^{t/t_{gen}},$$

where k is the multiplication factor. Note that for constant power output $k = 1$.

24. The neutron generation time (see Problem 23) of a particular power reactor is 1.3 ms. It is generating energy at the rate of 1200 MW. To perform certain maintenance checks, the power level must be temporarily reduced to 350 MW. It is desired that the transition to the reduced power level take

2.6 s. To what (constant) value should the multiplication factor be set to effect the transition in the desired time?

25. The neutron generation time t_{gen} (see Problem 23) in a particular reactor is 1.0 ms. If the reactor is operating at a power level of 500 MW, about how many free neutrons (neutrons that will subsequently induce a fission) are present in the reactor at any moment?

26. A reactor operates at 400 MW with a neutron generation time of 30 ms. If its power increases for 5.0 min with a multiplication factor of 1.0003, find the power output at the end of the 5.0 min. See Problem 23.

27. The thermal energy generated when radiations from radio-nuclides are absorbed in matter can be used as the basis for a small power source for use in satellites, remote weather stations, and so on. Such radionuclides are manufactured in abundance in nuclear power reactors and may be separated chemically from the spent fuel. One suitable radionuclide is ^{238}Pu ($t_{1/2} = 87.7$ y) which is an alpha emitter with $Q = 5.59$ MeV. At what rate is thermal energy generated in 1.00 kg of this material?

28. Among the many fission products that may be extracted chemically from the spent fuel of a nuclear power reactor is ^{90}Sr ($t_{1/2} = 29$ y). It is produced in typical large reactors at the rate of about 18 kg/y. By its radioactivity it generates thermal energy at the rate of 2.3 W/g. (a) Calculate the effective disintegration energy Q_{eff} associated with the decay of a ^{90}Sr nucleus. (Q_{eff} includes contributions from the decay of the ^{90}Sr daughter products in its decay chain but not from neutrinos, which escape totally from the sample.) (b) It is desired to construct a power source generating 150 W (electric) to use in operating electronic equipment in an underwater acoustic beacon. If the source is based on the thermal energy generated by ^{90}Sr and if the efficiency of the thermal–electric conversion process is 5.0%, how much ^{90}Sr is needed? The atomic mass of ^{90}Sr is 89.9 u.

29. In an atomic bomb (A-bomb), energy release is due to the uncontrolled fission of plutonium ^{239}Pu (or ^{235}U). The magnitude of the released energy is specified in terms of the mass of TNT required to produce the same energy release (bomb "rating"). One megaton (10^6 tons) of TNT produces 2.6×10^{28} MeV of energy. (a) Calculate the rating, in tons of TNT, of an atomic bomb containing 95 kg of ^{239}Pu, of which 2.5 kg actually undergoes fission. For plutonium, the average Q is 180 MeV. (b) Why is the other 92.5 kg of ^{239}Pu needed if it does not fission?

30. A 66-kiloton A-bomb (see Problem 29) is fueled with pure ^{235}U, 4.0% of which actually undergoes fission. (a) How much uranium is in the bomb? (b) How many primary fission fragments are produced? (c) How many neutrons generated in the fissions are released to the environment? (On the average, each fission produces 2.47 neutrons.)

31. One possible method for revealing the presence of concealed nuclear weapons is to detect the neutrons emitted in the spontaneous fission of ^{240}Pu in the warhead. In an actual trial, a neutron detector of area 2.5 m^2, carried on a helicopter, measured a neutron flux of 4.0 s^{-1} at a distance of 35 m from a missile warhead. Estimate the mass of ^{240}Pu in the warhead. The half-life for spontaneous fission in ^{240}Pu is 1.34×10^{11} y and 2.5 neutrons, on the average, are emitted in each fission.

Section 55-5 A Natural Reactor

32. The natural fission reactor discussed in Section 55-5 is estimated to have generated 15 gigawatt-years of energy during its lifetime. (a) If the reactor lasted for 200,000 y, at what average power level did it operate? (b) How much ^{235}U did it consume during its lifetime?

33. Some uranium samples from the natural reactor site described in Section 55-5 were found to be slightly *enriched* in ^{235}U, rather than depleted. Account for this in terms of neutron absorption by the abundant isotope ^{238}U and the subsequent beta and alpha decay of its products.

34. How far back in time would natural uranium have been a practical reactor fuel, with a ^{235}U/^{238}U ratio of 3.00%? See Sample Problem 4.

Section 55-6 Thermonuclear Fusion: The Basic Process

35. Calculate the height of the Coulomb barrier for the head-on collision of two protons. The effective radius of a proton may be taken to be 0.80 fm. See Sample Problem 5.

36. The equation of the curve $n(K)$ in Fig. 9 is

$$n(K) = \frac{2N}{\sqrt{\pi}} \frac{K^{1/2}}{(kT)^{3/2}} e^{-K/kT},$$

where N is the total density of particles. At the center of the Sun the temperature is 1.5×10^7 K and the mean proton energy $\bar{K}$ is 1.9 keV. Find the ratio of the density of protons at 5.0 keV to that at the mean proton energy.

37. Methods other than heating the material have been suggested for overcoming the Coulomb barrier for fusion. For example, one might consider using particle accelerators. If you were to use two of them to accelerate two beams of deuterons directly toward each other so as to collide "head-on," (a) what voltage would each require to overcome the Coulomb barrier? (b) Would this voltage be difficult to achieve? (c) Why do you suppose this method is not presently used?

38. Calculate the Coulomb barrier height for two ^{7}Li nuclei, fired at each other with the same initial kinetic energy K. See Sample Problem 5. (*Hint*: Use Eq. 1 in Chapter 54 to calculate the radii of the nuclei.)

39. For how long could the fusion of 1.00 kg of deuterium by the reaction

$$^2H + {}^2H \rightarrow {}^3He + n \qquad (Q = +3.27 \text{ MeV})$$

keep a 100-W lamp burning? The atomic mass of deuterium is 2.014 u.

Section 55-7 Thermonuclear Fusion in Stars

40. We have seen that Q for the overall proton–proton cycle is 26.7 MeV. How can you relate this number to the Q values for the three reactions that make up this cycle, as displayed in Fig. 10?

41. Show that the energy released when three alpha particles fuse to form ^{12}C is 7.27 MeV. The atomic mass of ^{4}He is 4.002603 u, and of ^{12}C is 12.000000 u.

42. At the central core of the Sun the density is 1.5×10^5 kg/m^3 and the composition is essentially 35% hydrogen by mass and 65% helium. (a) What is the density of protons at the Sun's core? (b) What is the ratio of this to the density of

particles for an ideal gas at standard conditions of temperature and pressure?

43. Calculate and compare the energy in MeV released by (*a*) the fusion of 1.0 kg of hydrogen deep within the Sun and (*b*) the fission of 1.0 kg of ^{235}U in a fission reactor.

44. The Sun has a mass of 2.0×10^{30} kg and radiates energy at the rate of 3.9×10^{26} W. (*a*) At what rate does the mass of the Sun decrease? (*b*) What fraction of its original mass has the Sun lost in this way since it began to burn hydrogen, about 4.5×10^9 y ago?

45. Let us assume that the core of the Sun has one-eighth the Sun's mass and is compressed within a sphere whose radius is one-fourth of the solar radius. We assume further that the composition of the core is 35% hydrogen by mass and that essentially all of the Sun's energy is generated there. If the Sun continues to burn hydrogen at the rate calculated in Sample Problem 6, how long will it be before the hydrogen is entirely consumed? The Sun's mass is 2.0×10^{30} kg.

46. Verify the Q values reported for the reactions in Fig. 10. The needed atomic masses are

^1H	1.007825 u	^3He	3.016029 u
^2H	2.014102 u	^4He	4.002603 u
$e^{\pm}$	0.0005486 u.		

(*Hint*: Distinguish carefully between atomic and nuclear masses, and take the positrons properly into account.)

47. Coal burns according to

$$C + O_2 \rightarrow CO_2.$$

The heat of combustion is 3.3×10^7 J/kg of atomic carbon consumed. (*a*) Express this in terms of energy per carbon atom. (*b*) Express it in terms of energy per kilogram of the initial reactants, carbon and oxygen. (*c*) Suppose that the Sun (mass = 2.0×10^{30} kg) were made of carbon and oxygen in combustible proportions and that it continued to radiate energy at its present rate of 3.9×10^{26} W. How long would it last?

48. After converting all its hydrogen to helium, a particular star is 100% helium in composition. It now proceeds to convert the helium to carbon via the triple–alpha process

$$^4He + {}^4He + {}^4He \rightarrow {}^{12}C + \gamma;$$

$Q = 7.27$ MeV. The mass of the star is 4.6×10^{32} kg, and it generates energy at the rate of 5.3×10^{30} W. How long will it take to convert all the helium to carbon?

49. In certain stars the *carbon cycle* is more likely than the proton–proton cycle to be effective in generating energy. This cycle is

$^{12}C + {}^1H \rightarrow {}^{13}N + \gamma,$	$Q_1 = 1.95$ MeV,
$^{13}N \rightarrow {}^{13}C + e^+ + \nu,$	$Q_2 = 1.19$ MeV,
$^{13}C + {}^1H \rightarrow {}^{14}N + \gamma,$	$Q_3 = 7.55$ MeV,
$^{14}N + {}^1H \rightarrow {}^{15}O + \gamma,$	$Q_4 = 7.30$ MeV,
$^{15}O \rightarrow {}^{15}N + e^+ + \nu,$	$Q_5 = 1.73$ MeV,
$^{15}N + {}^1H \rightarrow {}^{12}C + {}^4He,$	$Q_6 = 4.97$ MeV.

(*a*) Show that this cycle of reactions is exactly equivalent in

its overall effects to the proton–proton cycle of Fig. 10. (*b*) Verify that both cycles, as expected, have the same Q.

50. (*a*) Calculate the rate at which the Sun is generating neutrinos. Assume that energy production is entirely by the proton–proton cycle. (*b*) At what rate do solar neutrinos impinge on the Earth?

51. The gravitational potential energy of a uniform spherical object of mass M and radius R is

$$U = -3GM^2/5R,$$

in which G is the gravitational constant. (*a*) Demonstrate the consistency of this expression with that of Problem 22 in Chapter 54. (*b*) Use this expression to find the maximum energy that could be released by a spherical object, initially of infinite radius, in shrinking to the present size of the Sun. (*c*) Assume that during this shrinking, the Sun radiated energy at its present rate and calculate the age of the Sun based on the hypothesis that the Sun derives its energy from gravitational contraction.

Section 55-8 Controlled Thermonuclear Fusion

52. Verify the Q values reported in Eqs. 6, 7, and 8. The needed masses are

^1H	1.007825 u	^3He	3.016029 u
^2H	2.014102 u	^4He	4.002603 u
^3H	3.016049 u	n	1.008665 u.

53. Suppose we had a quantity of N deuterons (^2H nuclei). (*a*) Which of the following procedures for fusing these N nuclei releases more energy, and how much more? (A) $N/2$ fusion reactions of the type $^2H + {}^2H \rightarrow {}^3H + {}^1H$, or (B) $N/3$ fusion reactions of the type $^2H + {}^3H \rightarrow {}^4He + n$, using $N/3$ nuclei of ^3H that are first made in $N/3$ reactions of type A. (*b*) List the ultimate product nuclei resulting from the two procedures and the quantity of each.

54. Ordinary water consists of roughly 0.015% by mass of "heavy water," in which one of the two hydrogens is replaced with deuterium, ^2H. How much average fusion power could be obtained if we "burned" all the ^2H in 1 liter of water in 1 day through the reaction $^2H + {}^2H \rightarrow {}^3He + n + 3.27$ MeV?

55. In the deuteron–triton fusion reaction of Eq. 8, how is the reaction energy Q shared between the α particle and the neutron (that is, calculate the kinetic energies K_α and K_n)? Neglect the relatively small kinetic energies of the two combining particles.

56. Figure 16 shows an idealized representation of a hydrogen bomb. The fusion fuel is lithium deuteride (LiD). The high temperature, particle density, and neutrons to induce fusion are provided by an atomic (fission) bomb "trigger." The fusion reactions are

$$^6Li + n \rightarrow {}^3H + {}^4He + Q$$

and

$$^2H + {}^3H \rightarrow {}^4He + n + 17.59 \text{ MeV},$$

the tritium (^3H) produced in the first reaction fusing with the deuterium (D) in the fuel; see Eq. 8. By calculating Q for the first reaction, find the mass of LiD required to produce a

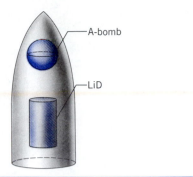

A-bomb

LiD

Figure 16   Problem 56.

fusion yield of 1 megaton of TNT ($=2.6 \times 10^{28}$ MeV). Needed atomic masses are

^{6}Li	6.015121 u	^{4}He	4.002603 u
^{3}H	3.016049 u	n	1.008665 u.

Section 55-10 Inertial Confinement

57. Assume that a plasma temperature of 1.3×10^8 K is reached in a laser-fusion device. (*a*) What is the most probable speed of a deuteron at this temperature? (*b*) How far would such a deuteron move in the confinement time calculated in Sample Problem 8?

58. The uncompressed radius of the fuel pellet of Sample Problem 8 is 20 μm. Suppose that the compressed fuel pellet "burns" with an efficiency of 10%. That is, only 10% of the deuterons and 10% of the tritons participate in the fusion reaction of Eq. 8. (*a*) How much energy is released in each such microexplosion of a pellet? (*b*) To how much TNT is each such pellet equivalent? The heat of combustion of TNT is 4.6 MJ/kg. (*c*) If a fusion reactor is constructed on the basis of 100 microexplosions per second, what power would be generated? (Note that part of this power must be used to operate the lasers.)

CHAPTER 56

PARTICLE PHYSICS AND COSMOLOGY

Research in particle physics is often done at accelerators where a beam of particles moving at speeds close to the speed of light (and thus having kinetic energies many times their rest energies) is incident on a target, usually consisting of protons. In other accelerators, two high-energy particle beams moving in opposite directions may be brought together. Collisions of individual particles cause reactions in which dozens or perhaps hundreds of new particles are produced. Some of these particles live for unimaginably short times, often less than 10^{-20} s. Nevertheless, physicists can track these particles and study their properties. This is our primary means for learning about the fundamental constituents of matter.

Astrophysicists use a very different method to unlock the secrets of the universe. From observations with telescopes and detectors that are sensitive to radiations from all parts of the electromagnetic spectrum, they try to look backward in time to learn about the universe when it was very young, and they also project their conclusions into the future to try to understand the subsequent evolution of the universe. These investigations are part of cosmology, the study of the origin and evolution of the universe.

It may seem surprising that we have grouped these two very different studies in a single chapter. As we shall see, measurements by particle physicists can tell us about the structure of the universe just after its birth, and conclusions by cosmologists can set limits on the variety of fundamental particles and the interactions between them. Although they are at opposite ends of the scale of observations, particle physics and cosmology go hand-in-hand in providing an understanding of the structure of the universe.

56-1 PARTICLE INTERACTIONS

There are tens of thousands of chemical compounds of varying degrees of complexity. Understanding this huge number of systems would be a hopeless task if it were not for the underlying simplicity of the 109 fundamental units (elements) of which these compounds are made and the relatively small number of types of bonds through which they can interact. In order to understand chemistry, we need not study the properties of tens of thousands of compounds, but only those of about 100 elements, along with a few basic types of bonds between them.

In fact, the task is even simpler. The 109 known elements can be classified into groups with similar properties: inert gases, halogens, alkali metals, transition metals, rare earths, and so forth. If we understand the properties of one member of a group, we can infer the properties of the other members of that group.

The subatomic world can be understood in a similar way. We know that the 109 different kinds of atoms are not fundamental units, but instead that they are in turn composed of three different particles: protons, neutrons, and electrons. When we look still further, by smashing particles together at high energy and studying the debris of the collisions (see Fig. 1), we find what appears at first glance to be a complexity approaching that of chemistry: hundreds of different particles are produced. Yet when we look carefully we find that we can classify those particles into a few groups whose members have similar properties. Eventually we find that this classification leads to clues about the underlying substructure that is based again on a small number of truly fundamental particles and a small number of possible interactions among them.

(a)

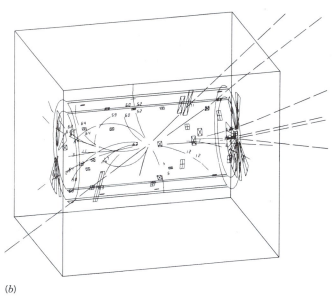

(b)

Figure 1 (*a*) The UA1 detector at the proton–antiproton collider of the European Organization for Nuclear Research (CERN) accelerator near Geneva, Switzerland. Oppositely moving beams of protons and antiprotons are made to collide in the central region of this detector, which is designed to record the trajectories of all electromagnetically or strongly interacting particles that leave the reaction region. (*b*) A computer reconstruction of the trajectories leaving the central region after one collision. A magnetic field causes the curvature of the paths that permits the momentum of the particles to be determined and helps to identify the particles. Events of this type were responsible for the discovery of the W and Z particles at CERN in 1983.

The Four Basic Forces

All of the known forces in the universe can be grouped into four basic types. In order of increasing strength, these are: gravitation, the weak force, electromagnetism, and the strong force. These forces have important roles not only in the interactions between particles, but also in the decay of one particle into other particles.

1. *The gravitational force.* Gravity is of course exceedingly important in our daily lives, but on the scale of fundamental interactions between particles in the sub-atomic realm, it is of no importance at all. To give a relative figure, the gravitational force between two protons just touching at their surfaces is about 10^{-38} of the strong force between them. The principal difference between gravitation and the other forces is that, on the practical scale, gravity is cumulative and infinite in range. For example, your weight is the cumulative effect of the gravitational force exerted by each atom of the Earth on each atom of your body.

2. *The weak force.* The weak force is responsible for nuclear beta decay (see Section 54-5) and other similar decay processes involving fundamental particles. It does not play a major role in the binding of nuclei. The weak force between two neighboring protons is about 10^{-7} of the strong force between them, and the range of the weak force is smaller than 1 fm. That is, at separations greater than about 1 fm, the weak force between particles is negligible. Nevertheless, the weak force is important in understanding the behavior of fundamental particles, and it is critical in understanding the evolution of the universe.

3. *The electromagnetic force.* Electromagnetism is important in the structure and the interactions of the fundamental particles. For example, some particles interact or decay primarily through this mechanism. Electromagnetic forces are of infinite range, but shielding generally diminishes their effect for ordinary objects. The properties of atoms and molecules are determined by electromagnetic forces, and many common macroscopic forces (such as friction, air resistance, drag, and tension) are ultimately due to the electromagnetic force. The electromagnetic force between neighboring protons is about 10^{-2} of the strong force, but within the nucleus the electromagnetic forces can act cumulatively because there is no shielding. As a result, the electromagnetic force can compete with the strong force in determining the stability and the structure of nuclei.

4. *The strong force.* The strong force, which is responsible for the binding of nuclei, is the dominant one in the reactions and decays of most of the fundamental particles. However, as we shall see, some particles (such as the electron) do not feel this force at all. It has a relatively short range, on the order of 1 fm.

The relative strength of a force determines the time scale over which it acts. If we bring two particles close enough together for any of these forces to act, then a longer time is required for the weak force to cause a decay or reaction than for the strong force. As we shall see, the mean lifetime of a decay process is often a signal of the type of interaction responsible for the process, with strong forces being at the shortest end of the time scale (often down to 10^{-23} s). Table 1 summarizes the four forces and some of their properties. The characteristic time for each force gives a typical range of time intervals observed for systems in which each force acts. Usually this is the typical lifetime of a particle that decays through that force.

Unification of Forces

One of the landmark achievements in the history of physics was the 19th century theory of electromagnetism, based on experiments by Faraday and Oersted showing that magnetic effects could produce electric fields and electrical effects could produce magnetic fields. The previously separate sciences of electricity and magnetism became linked under the common designation of electromagnetism. This linking was later shown to be a fundamental part of the special theory of relativity, according to which electric fields and magnetic fields can be transformed into one another due entirely to the relative motion of the observer.

In the 20th century, it has been attempted to carry this linking further to include other forces. First it was shown that electromagnetism and the weak force can be understood as two different aspects of the same force, called the *electroweak* force. If we study particle interactions at a high enough energy, these two forces behave similarly. It is convenient for us to regard them as separate forces for many of the effects we shall discuss, just as we often find it convenient to speak separately of electric and magnetic forces when we discuss electromagnetic phenomena. The theory of the electroweak force, which was proposed independently in 1967 by Stephen Weinberg and Abdus Salam (and for which they, along with Sheldon Glashow, another originator of the theory, received the 1979 Nobel prize in physics), suggested that, just as the photon is the carrier of the electromagnetic force, there should be heavy particles that carry the weak force, and these new particles should, on an energy scale of 100 GeV (about 100 times the rest energy of the proton), behave similarly to a high-energy photon. In 1983, a research team at the European Center for Nuclear Physics (CERN), led by Carlo Rubbia and using experimental techniques developed by Simon van der Meer, discovered the predicted particles, for

TABLE 1 THE FOUR BASIC FORCES

Type	Range	Relative Strength	Characteristic Time
Strong	1 fm	1	10^{-23} s
Electromagnetic	∞	10^{-2}	10^{-14}–10^{-20} s
Weak	$\ll$ 1 fm	10^{-7}	10^{-8}–10^{-13} s
Gravitational	∞	10^{-38}	Years

which they were awarded the 1984 Nobel prize in physics. The discovery of these particles provided the evidence for the unification of the electromagnetic and weak interactions into the electroweak interaction.

Next it was attempted to combine the strong and electroweak forces at a new higher level of unification. Theories that do so are called *grand unified theories* (GUTs), and at the present time there are many candidates for GUTs but none has as yet emerged as the correct one. Because the energy at which the forces merge is immense, perhaps 10^{15} GeV (10^{11} times the energy of the largest particle accelerator yet built or even contemplated), we cannot do experiments to test the GUTs directly. We must therefore rely on tests at obtainable energies, where the effects are exceedingly small. One prediction of these theories is that the proton should not be a stable particle but should decay on a time scale greater than 10^{31} years. (Compare this number with the age of the universe, about 10^{10} years.) Searches for proton decay have so far been unsuccessful and have excluded certain of the GUTs from consideration, but as yet there has been no verification of any of the theories.

The final step in the unification would be to include gravity in the scheme to create a *theory of everything* (TOE). There is not yet a quantum theory of gravity, so it is difficult to anticipate the form that these theories might take, but they nevertheless provide challenges for theoretical speculation.

Sample Problem 1 Suppose the half-life of the proton were 10^{31} y, as predicted by certain GUTs. (*a*) On the average, how long must we observe a liter of water before we would see one of its protons decay? (*b*) What volume of water would be required to have a proton decay rate of one per day?

Solution (*a*) A liter of water (approximately 1000 g) contains a number of molecules given by

$$\frac{(1000 \text{ g})(6.02 \times 10^{23} \text{ molecules/mole})}{18 \text{ g/mole}} = 3.3 \times 10^{25} \text{ molecules.}$$

Each molecule contains 10 protons (2 from the hydrogens and 8 from the oxygen), so that the number of protons in a liter of water is $N = 3.3 \times 10^{26}$. The decay rate R is given by Eq. 5 of Chapter 54 as

$$R = \lambda N = \frac{\ln 2}{t_{1/2}} N = \frac{0.693}{10^{31} \text{ y}} 3.3 \times 10^{26}$$
$$= 2.3 \times 10^{-5} \text{ y}^{-1}$$
$$= \frac{1}{43,000 \text{ y}}.$$

That is, on the average we must wait for 43,000 years before a proton decay occurs in a liter of water.

(*b*) If $R = 1 \text{ d}^{-1}$, we obtain

$$N = \frac{R}{\lambda} = \frac{1 \text{ d}^{-1}}{0.693/(10^{31} \text{ y})} = 5.3 \times 10^{33} \text{ protons}$$

Figure 2 An underground chamber, lined with plastic, in the Morton salt mine near Cleveland. Its size can be inferred from the worker standing in the corner. This chamber was later filled with 10,000 tons of water in which were suspended 2048 detectors that respond to the tiny flashes of light that would be emitted in the decay of one of the protons in the water.

or 5.3×10^{32} molecules of water. This works out to be 1.6×10^7 L, equivalent to a cube of water measuring 25 m on a side! (See Fig. 2.)

56-2 FAMILIES OF PARTICLES

We can learn a lot about things by classifying them. This is a technique used commonly by biologists; by grouping plants or animals into categories based on certain obvious features of their structure, a basis can be found for studying their behavior. From the scientific standpoint, for example, it may be more enlightening to compare one spider with another spider than with a fly or a moth. Part of the training of a scientist is concerned with learning how to make and to use these classifications.

The earliest classification scheme for particles was based on their masses. The lightest particles, including the electron ($m_e c^2 = 0.511$ MeV), were called *leptons* (from the Greek word for "small"). The heaviest particles, including the proton ($m_p c^2 = 938$ MeV), were called *baryons* (from the Greek word for "heavy"). In between were particles, including the pion ($m_\pi c^2 = 140$ MeV), called *mesons* (from the Greek word for "middle"). Today these classifications based on mass are no longer valid; for ex-

TABLE 2 THREE FAMILIES OF PARTICLES

Family	Structure	Interactions	Spin	Examples
Leptons	Fundamental	Weak, electromagnetic	Half integral	e, ν
Mesons	Composite	Weak, electromagnetic, strong	Integral	π, K
Baryons	Composite	Weak, electromagnetic, strong	Half integral	p, n

ample, one lepton and many mesons are more massive than the proton. However, we retain these three names as descriptive of particles with similar properties, even though the classification based only on mass is no longer valid. Table 2 summarizes these three families of particles and some of their properties.

Leptons

The leptons are fundamental particles that interact only through the weak and electromagnetic interactions; even though the strong force can exceed the weak or electromagnetic force in strength by many orders of magnitude, the leptons do not feel this force at all. The leptons are true fundamental particles; they have no internal structure and are not composed of other still smaller particles. We can consider the leptons to be point particles with no finite dimensions. All known leptons have a spin of $\frac{1}{2}$.

Table 3 shows the six leptons, which appear as three pairs of particles. Each pair includes a charged particle (e^-, μ^-, τ^-) and an uncharged neutrino $(\nu_e, \nu_\mu, \nu_\tau)$. We discussed the electron neutrino previously in connection with beta decay (Section 54-5). Both the charged leptons and the neutrinos have antiparticles.

According to some theories of the structure of fundamental particles, the neutrinos are massless (and correspondingly travel at the speed of light) and stable. Other theories predict that the neutrinos should have a small but definitely nonzero mass and should transform into one another. So far no experiment has revealed a mass inconsistent with zero, but only for the electron neutrino is the upper limit very small (rest energy < 20 eV). The neutrinos and their possible masses have important implications for cosmology, as we discuss later in this chapter.

The electron is a stable particle, but the muon and tau decay to other leptons, according to

$$\mu^- \rightarrow e^- + \bar{\nu}_e + \nu_\mu \quad \text{(mean life} = 2.2 \times 10^{-6} \text{ s)},$$

$$\tau^- \rightarrow \mu^- + \bar{\nu}_\mu + \nu_\tau \quad \text{(mean life} = 3.0 \times 10^{-13} \text{ s)}.$$

These decays are caused by the weak interaction, as we can conclude from the presence of neutrinos (which *always* indicates a weak interaction process) among the decay products and as we infer from the typical decay lifetimes listed in Table 1. The form of these decays can be understood based on a conservation law for leptons discussed in Section 56-3.

Mesons

Mesons are strongly interacting particles having integral spin. A partial list of some mesons is given in Table 4. Generally, mesons are produced in reactions by the strong interaction; they decay, usually to other mesons or leptons, through the strong, electromagnetic, or weak interactions. For example, pions can be produced in reactions of nucleons, such as

$$p + n \rightarrow p + p + \pi^- \quad \text{or} \quad p + n \rightarrow p + n + \pi^0,$$

and the pions can decay according to

$$\pi^- \rightarrow \mu^- + \bar{\nu}_\mu \quad \text{(mean life} = 2.6 \times 10^{-8} \text{ s)},$$

$$\pi^0 \rightarrow \gamma + \gamma \quad \text{(mean life} = 8.4 \times 10^{-17} \text{ s)},$$

where the first decay occurs due to the weak interaction (indicated by the neutrinos and suggested by the mean life) and the second due to the electromagnetic interaction (indicated by the photons and suggested by the mean life).*

---

* While neutrinos always indicate a weak-interaction decay, not all weak-interaction decays produce neutrinos. The same is true for photons in electromagnetic decays.

TABLE 3 THE LEPTON FAMILY

Particle	Antiparticle	Particle Charge (e)	Spin ($h/2\pi$)	Rest Energy (MeV)	Mean Life (s)	Typical Decay Products
e^-	e^+	-1	$\frac{1}{2}$	0.511	∞	—
ν_e	$\bar{\nu}_e$	0	$\frac{1}{2}$	< 20 eV	∞	—
μ^-	μ^+	-1	$\frac{1}{2}$	105.7	2.2×10^{-6}	$e^- + \bar{\nu}_e + \nu_\mu$
ν_μ	$\bar{\nu}_\mu$	0	$\frac{1}{2}$	< 0.3	∞	—
τ^-	τ^+	-1	$\frac{1}{2}$	1784	3.0×10^{-13}	$\mu^- + \bar{\nu}_\mu + \nu_\tau$
ν_τ	$\bar{\nu}_\tau$	0	$\frac{1}{2}$	< 40	∞	—

TABLE 4 SOME SELECTED MESONS

Particle	Antiparticle	Charge[a] (e)	Spin (h/2π)	Strangeness[a]	Rest Energy (MeV)	Mean Life (s)	Typical Decay Products
π^+	π^-	+1	0	0	140	2.4×10^{-8}	$\mu^+ + \nu_\mu$
π^0	π^0	0	0	0	135	8.4×10^{-17}	$\gamma + \gamma$
K^+	K^-	+1	0	+1	494	1.2×10^{-8}	$\mu^+ + \nu_\mu$
K^0	$\overline{K}^0$	0	0	+1	498	0.9×10^{-10}	$\pi^+ + \pi^-$
η	η	0	0	0	549	8.0×10^{-19}	$\gamma + \gamma$
ρ^+	ρ^-	+1	1	0	769	4.5×10^{-24}	$\pi^+ + \pi^0$
η'	η'	0	0	0	958	2.2×10^{-21}	$\eta + \pi^+ + \pi^-$
D^+	D^-	+1	0	0	1869	1.1×10^{-12}	$K^- + \pi^+ + \pi^+$
ψ	ψ	0	1	0	3097	1.0×10^{-20}	$e^+ + e^-$
B^+	B^-	+1	0	0	5278	1.2×10^{-12}	$D^- + \pi^+ + \pi^+$
Y	Y	0	1	0	9460	1.3×10^{-20}	$e^+ + e^-$

[a] The charge and strangeness are those of the particle. Values for the antiparticle have the opposite sign. The spin, rest energy, and mean life are the same for a particle and its antiparticle.

Baryons

Baryons are strongly interacting particles having half-integral spins ($\frac{1}{2}$, $\frac{3}{2}$, $\frac{5}{2}$, . . .). A partial listing of some baryons is given in Table 5. The familiar members of the baryon family are the proton and neutron. Baryons have distinct antiparticles, for example, the antiproton ($\overline{p}$) and antineutron ($\overline{n}$).

We can produce heavier baryons in reactions between nucleons, such as

$$p + p \rightarrow p + \Lambda^0 + K^+,$$

which produces the Λ^0 baryon and the K^+ meson. The Λ^0 decays according to

$$\Lambda^0 \rightarrow p + \pi^- \quad \text{(mean life} = 2.6 \times 10^{-8} \text{ s)}.$$

Although there are no neutrinos produced in the decay, the mean life indicates that the decay is governed by the weak interaction. We shall learn the reason for this "slow" decay in Section 56-3.

Field Particles and Exchange Forces

There is one additional small family of particles that cannot be classified among the leptons, mesons, or baryons. These are the *field particles*, those responsible for carrying the forces with which the particles interact.

Newton's law of gravitation and Coulomb's law of electrostatics were originally based on the concept of action-at-a-distance. Later, in the nineteenth century, this concept was replaced by the notion of a field. Two particles interact through the fields that they establish: one particle sets up a field and the other interacts with that field, rather than directly with the first particle. Quantum field theory takes this notion one step further by supposing that the fields are carried by quanta. In this view, instead of the first particle setting up the field, we say that it emits quanta of the field. The second particle then absorbs these quanta. For example, the electromagnetic interaction between two particles can be viewed in terms of the emission and absorption of photons, which are quanta of the elec-

TABLE 5 SOME SELECTED BARYONS

Particle	Antiparticle	Charge[a] (e)	Spin (h/2π)	Strangeness[a]	Rest Energy (MeV)	Mean Life (s)	Typical Decay Products
p	$\overline{p}$	+1	$\frac{1}{2}$	0	938	∞	
n	$\overline{n}$	0	$\frac{1}{2}$	0	940	889	$p + e^- + \overline{\nu}_e$
Λ^0	$\overline{\Lambda}^0$	0	$\frac{1}{2}$	−1	1116	2.6×10^{-10}	$p + \pi^-$
Σ^+	$\overline{\Sigma}^+$	+1	$\frac{1}{2}$	−1	1189	0.8×10^{-10}	$p + \pi^0$
Σ^0	$\overline{\Sigma}^0$	0	$\frac{1}{2}$	−1	1192	5.8×10^{-20}	$\Lambda^0 + \gamma$
Σ^-	$\overline{\Sigma}^-$	−1	$\frac{1}{2}$	−1	1197	1.5×10^{-10}	$n + \pi^-$
Ξ^0	$\overline{\Xi}^0$	0	$\frac{1}{2}$	−2	1315	2.9×10^{-10}	$\Lambda^0 + \pi^0$
Ξ^-	$\overline{\Xi}^-$	−1	$\frac{1}{2}$	−2	1321	1.6×10^{-10}	$\Lambda^0 + \pi^-$
Δ^*	$\overline{\Delta}^*$	+2, +1, 0, −1	$\frac{3}{2}$	0	1232	6×10^{-24}	$p + \pi$
Σ^*	$\overline{\Sigma}^*$	+1, 0, −1	$\frac{3}{2}$	−1	1385	2×10^{-23}	$\Lambda^0 + \pi$
Ξ^*	$\overline{\Xi}^*$	−1, 0	$\frac{3}{2}$	−2	1530	6×10^{-23}	$\Xi + \pi$
Ω^-	$\overline{\Omega}^-$	−1	$\frac{3}{2}$	−3	1672	8.2×10^{-11}	$\Lambda^0 + K^-$

[a] The charge and strangeness are those of the particle. Values for the antiparticle have the opposite sign. The spin, rest energy, and mean life are the same for a particle and its antiparticle.

TABLE 6   THE FIELD PARTICLES

Particle	Symbol	Interaction	Charge (e)	Spin ($h/2\pi$)	Rest Energy (GeV)
Graviton		Gravitation	0	2	0
Weak boson	W^+, W^-	Weak	± 1	1	80.6
Weak boson	Z^0	Weak	0	1	91.2
Photon	γ	Electromagnetic	0	1	0
Gluon	g	Strong (color)	0	1	0

tromagnetic field. Each type of field has its characteristic field particles. A list of the particles associated with the four basic forces can be found in Table 6.

A force accomplished through the exchange of particles is called an *exchange force*. For example, the force between two nucleons in a nucleus takes place through the exchange of pions. In this case the pions, along with other mesons, can act as field particles associated with the strong force between nucleons.

How is it possible for a particle, such as a proton, to emit another particle with nonzero mass and still remain a proton? This process seems to violate conservation of energy. The solution to this dilemma lies in the energy–time form of the uncertainty relationships. The uncertainty principle is a fundamental limitation on our ability to measure a system. That is, if we observe a system for a time interval Δt, there is a corresponding uncertainty ΔE in its energy, according to Eq. 7 of Chapter 50, given at minimum by

$$\Delta E = \frac{h}{2\pi \, \Delta t} . \tag{1}$$

We cannot know the energy of a system more precisely than this ΔE unless we measure for a time longer than Δt. If we observe only for a very short time, the uncertainty in the rest energy of a proton can be at least as large as the rest energy of a pion, as the following sample problem demonstrates.

Sample Problem 2   (*a*) What is the longest interval of time for which we can observe a proton for its rest energy to be uncertain by the pion rest energy? (*b*) What is the greatest distance the pion can travel in that time?

Solution   (*a*) For the proton's rest energy to be uncertain by an amount $\Delta E = m_\pi c^2$, the observation time interval can, according to Eq. 1, be at most

$$\Delta t = \frac{h}{2\pi \, \Delta E} = \frac{h}{2\pi m_\pi c^2}$$
$$= \frac{4.14 \times 10^{-15} \text{ eV} \cdot \text{s}}{(2\pi)(140 \text{ MeV})} = 4.7 \times 10^{-24} \text{ s}.$$

In a time interval shorter than 4.7×10^{-24} s, a proton can emit and absorb a pion without our observing a violation of conservation of energy.

(*b*) If the pion travels at nearly the speed of light, the maximum distance d it can travel in this time interval is

$$d = c \, \Delta t = (3.00 \times 10^8 \text{ m/s})(4.7 \times 10^{-24} \text{ s})$$
$$= 1.4 \times 10^{-15} \text{ m} = 1.4 \text{ fm}.$$

This distance defines the *range* of the nuclear force. Two nucleons closer than about 1.4 fm can interact through the exchange of pions. If the nucleons are separated by a greater distance, pion exchange cannot operate, and there is no nuclear force.

56-3   CONSERVATION LAWS

We would have a difficult time analyzing physical processes without the laws of conservation of energy and linear and angular momentum. These conservation laws help us understand why certain outcomes occur (such as in the case of the collisions that we considered in Chapter 10). They also help us understand why certain processes (those that violate the conservation laws) are never observed. In one sense they are empirical laws, deduced from observing physical processes and carefully tested in the laboratory. In another sense they reveal to us fundamental aspects of the laws of nature.

An example of a conservation law is the conservation of electric charge. By observing the outcomes of many processes, we are led to propose this law: the net amount of electric charge must not change in any process. Equivalently, we may say that the net charge before a particular reaction or decay must equal the net charge after the reaction or decay. No violation of this law has ever been observed, even though it has been carefully tested (see Section 27-6).

Conservation of Lepton Number

In reactions and decays of fundamental particles, we often find a certain set of outcomes but fail to observe a set of related outcomes that would otherwise be expected to occur. When this happens, we suspect that there is some unknown conservation law at work that permits the first set and forbids the second. For example, we can produce an electron neutrino when a proton captures an electron:

$$e^- + p \rightarrow n + \nu_e.$$

We always find neutrinos in this process, but we never observe antineutrinos. Furthermore, the reaction always produces electron neutrinos and never muon or tau neutrinos.

We account for the failure to observe certain processes by proposing a conservation law for *lepton number* that is similar to the conservation law for electric charge. To each lepton we assign a lepton number +1 and to each antilepton we assign a lepton number −1. All other particles have lepton numbers of 0. The law of conservation of lepton number then states:

In any process, the lepton number for electron-type leptons, muon-type leptons, and tau-type leptons must each remain constant.

As far as we know, the law of lepton conservation is strictly valid; despite precise experimental searches for violations, none has yet been found.

In the electron capture process, we assign an electron lepton number L_e of +1 to the electron and to the electron neutrino, while $L_e = 0$ for the proton and neutron. This process then has $L_e = +1$ on both sides and upholds the law of conservation of lepton number. If an electron *anti*neutrino were produced, the right side would have $L_e = -1$, and the law would be violated. This accounts for our failure to observe this process. If another type of neutrino, for example, a muon neutrino, were produced, the process would have $L_e = +1$ on the left and $L_e = 0$ on the right. Furthermore, it would have $L_\mu = 0$ on the left and $L_\mu = +1$ on the right. The process would therefore violate conservation of both electron and muon lepton numbers, and it has never been observed.

Through the law of lepton conservation, we can account for many experimental observations. Like other conservation laws, this law proves to be of great value in analyzing decays and reactions.

Sample Problem 3   Analyze the decay of the muon from the standpoint of conservation of lepton number.

Solution   Let us assign lepton numbers to each particle in the decay as follows:

$$\mu^- \rightarrow e^- + \bar{v}_e + v_\mu$$

$$L_e: \quad 0 \quad +1 \quad -1 \quad 0$$

$$L_\mu: \quad +1 \quad 0 \quad 0 \quad +1$$

Note that electron-type leptons are assigned $L_\mu = 0$ and muon-type leptons are assigned $L_e = 0$. We see that $L_e = 0$ and $L_\mu = +1$ both before and after the decay, so the process is allowed by conservation of lepton number. Because of this conservation law, we can understand why there must be an electron antineutrino and a muon neutrino among the decay products, rather than, for example, an electron neutrino and a muon antineutrino.

Conservation of Baryon Number

A similar conservation law occurs in the case of baryons. To each baryon, such as the proton or neutron, we assign a baryon number B of +1, and we assign $B = -1$ to antibaryons such as the antiproton. The law of conservation of baryon number then states:

In any process, the total baryon number must remain constant.

No violation of this law has yet been observed. (However, certain speculative theories, the GUTs discussed in Section 56-1, suggest that the proton can decay into nonbaryons, which would violate the law of conservation of baryon number. This decay has never been observed; if it were observed, the law of conservation of baryon number would need to be changed accordingly.)

Consider, for example, the reaction in which antiprotons are produced when a proton beam is incident on a target of protons:

$$p + p \rightarrow p + p + p + \bar{p}$$

$$B: \quad +1 + 1 \quad +1 + 1 + 1 - 1$$

In this reaction, the net baryon number is +2 on both the left and the right sides.

Contrary to the case of lepton number, there is only one type of baryon number. The law of conservation of baryon number is a more general version of the rule we used in analyzing nuclear processes in Chapters 54 and 55; there we kept the total of neutrons plus protons constant in all decays and reactions, which, because neutrons and protons are baryons, is equivalent to conserving the total number of baryons.

Even though there are conservation laws for two types of particles (leptons and baryons), there is no conservation law for mesons. For example, in a reaction of protons on protons, any number of mesons can be produced (as long as the incident particles have enough kinetic energy):

$$p + p \rightarrow p + n + \pi^+,$$

$$p + p \rightarrow p + p + \pi^+ + \pi^-,$$

$$p + p \rightarrow p + n + \pi^+ + \pi^0 + \pi^0.$$

Note the conservation of electric charge in these processes.

Strangeness

There are still other processes that are difficult to understand based only on the conservation laws we have discussed so far. For example, consider the group of kaons (K mesons), which in many respects are similar to the pions. Because there is no conservation law for mesons, we might expect that any number of kaons can be produced in reactions. What we instead find is that kaons are either produced in pairs, for example,

$$p + p \rightarrow p + p + K^+ + K^-,$$

$$p + p \rightarrow p + n + K^+ + \overline{K^0},$$

or if a single kaon is produced, it is always accompanied by another "strange" particle, for example, a Λ^0,

$$p + p \rightarrow p + \Lambda^0 + K^+.$$

We account for these processes (and the failure to observe others that appear to be permitted by the previously known conservation laws) by assigning to particles a new quantum number called *strangeness,* which is found to follow a new conservation law, called *conservation of strangeness.* Two kaons (K^+ and K^0) are assigned to have strangeness $S = +1$, and the other two (K^- and $\overline{K^0}$) are assigned $S = -1$. All nonstrange particles (such as p, n, and e) have $S = 0$. The reactions in which two kaons are produced then have $S = 0$ on the left (only nonstrange particles) and also $S = 0$ on the right. The Λ^0 baryon is assigned $S = -1$, so the reaction in which $\Lambda^0 + K^+$ is produced also has $S = 0$ on both sides.

When we analyze the decays of the strange particles, the conservation of strangeness appears to break down. The kaons can decay into two (nonstrange) pions, for example,

$$K^+ \rightarrow \pi^+ + \pi^0.$$

Here we have $S = +1$ on the left and $S = 0$ on the right, a clear violation of the conservation of strangeness. We get a clue about how to resolve this difficulty when we measure the lifetime for this decay, which turns out to be about 10^{-8} s. The kaons and pions are strongly interacting particles, and we would expect this decay to occur with a typical strong interaction lifetime in the range of 10^{-23} s (see Table 1). Instead, it is slowed by 15 orders of magnitude! What could be responsible for slowing this decay?

Another clue comes from the more likely decay mode of the K^+:

$$K^+ \rightarrow \mu^+ + \nu_\mu,$$

a weak-interaction process, for which the mean life of 10^{-8} s would not be unusual. It appears that *the weak interaction can change strangeness* by one unit. In either of these kaon decay modes, we have S changing by one unit. Even though it does not produce the neutrinos that usually characterize a weak-interaction process, the decay $K^+ \rightarrow \pi^+ + \pi^0$ is governed by the weak interaction. In this case, the strangeness violation is a clue that it cannot be a strong-interaction process (strangeness is conserved in all strong interactions), and it must therefore be a weak-interaction decay.

Does the electromagnetic interaction conserve strangeness? To answer this question, we look for strangeness-violating electromagnetic decays, such as $\Lambda^0 \rightarrow n + \gamma$. This decay apparently does not occur, and so we conclude that *the electromagnetic interaction conserves strangeness.*

We can summarize these results in the law of conservation of strangeness:

In processes governed by the strong or electromagnetic interactions, the total strangeness must remain constant. In processes governed by the weak interaction, the total strangeness either remains constant or changes by one unit.

Sample Problem 4   The Ω^- baryon has $S = -3$. (*a*) It is desired to produce the Ω^- using a beam of K^- incident on protons. What other particles are produced in this reaction? (*b*) How might the Ω^- decay?

Solution   (*a*) Reactions usually proceed only through the strong interaction, which conserves strangeness. We consider the reaction

$$K^- + p \rightarrow \Omega^- + ?$$

On the left side, we have $S = -1$, $B = +1$, and $Q = 0$. On the right side, we have $S = -3$, $B = +1$, and $Q = -1$. We must therefore add to the right side particles with $S = +2$, $B = 0$, and $Q = +1$. Scanning through the tables of mesons and baryons, we find that we can satisfy these criteria with K^+ and K^0, so the reaction is

$$K^- + p \rightarrow \Omega^- + K^+ + K^0.$$

(*b*) The Ω^- cannot decay by the strong interaction, because no $S = -3$ final states are available. It must therefore decay to particles having $S = -2$ through the weak interaction, which can change S by one unit. One of the product particles must be a baryon in order to conserve baryon number. Two possibilities are

$$\Omega^- \rightarrow \Lambda^0 + K^- \quad \text{and} \quad \Omega^- \rightarrow \Xi^0 + \pi^-.$$

56-4   THE QUARK MODEL

Decays and reactions involving mesons and baryons are subject to conservation laws involving two quantities: the electric charge Q and the strangeness S. It then makes sense to ask whether there is any connection between the electric charge and the strangeness of a particle. In a particular group of similar particles (the spin-0 mesons or the spin-$\frac{1}{2}$ baryons, for example), do we find all possible combinations of Q and S or only certain ones? Finding only a restricted set of combinations suggests that the particles are built according to a set of rules out of more fundamental units whose electric charge and strangeness have certain values.

To answer this question, we make a plot showing electric charge on one axis and strangeness on another. We locate particles on this grid according to their values of electric charge and strangeness. Figure 3 shows this kind

TABLE 7   PROPERTIES OF THREE QUARKS

Quark	Symbol	Antiquark	Charge[a] (e)	Spin ($h/2\pi$)	Baryon Number[a]	Strangeness[a]
Up	u	$\bar{u}$	$+\frac{2}{3}$	$\frac{1}{2}$	$+\frac{1}{3}$	0
Down	d	$\bar{d}$	$-\frac{1}{3}$	$\frac{1}{2}$	$+\frac{1}{3}$	0
Strange	s	$\bar{s}$	$-\frac{1}{3}$	$\frac{1}{2}$	$+\frac{1}{3}$	-1

[a] The values for charge, baryon number, and strangeness refer to the quarks. Values for the antiquarks have opposite signs.

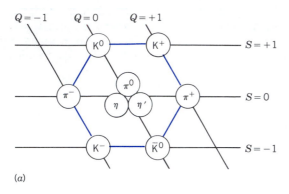

(a)

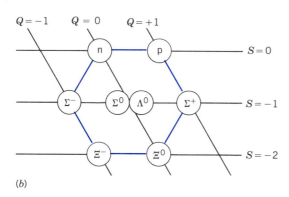

(b)

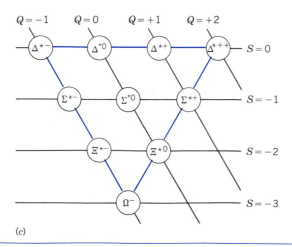

(c)

Figure 3   A chart showing (a) the spin-0 mesons, (b) the spin-$\frac{1}{2}$ baryons, and (c) the spin-$\frac{3}{2}$ baryons. Each particle is located on a grid according to its strangeness S and electric charge Q. The grid lines for electric charge have been drawn obliquely so that the patterns appear more symmetric.

of plot for the spin-0 mesons, the spin-$\frac{1}{2}$ baryons, and the spin-$\frac{3}{2}$ baryons. The regularity of these patterns suggests that these particles are composed out of more basic units.

In 1964, it was realized independently by Murray Gell-Mann and George Zweig that these regular patterns could be explained if it were assumed that the baryons and mesons are composed of three fundamental units, which soon became known as *quarks*. Table 7 shows the properties of the three quarks, which are called up (u), down (d), and strange (s).

According to this model, the mesons are composed of a quark and an antiquark, while the baryons are composed of three quarks. Consider the combination $u\bar{d}$ of an up quark and an antidown quark, such that their two spins add to 0. The charge of the up quark (in units of e) is $+\frac{2}{3}$, while the charge of the antidown quark is $+\frac{1}{3}$ (the charge of an antiparticle is opposite to that of the particle). The combination $u\bar{d}$ has $Q = +1$, $S = 0$ (because both quarks have $S = 0$), and $B = 0$ (because the quark has $B = +\frac{1}{3}$ and the antiquark has $B = -\frac{1}{3}$). This combination has the same quantum numbers as the π^+ meson. Continuing in this way, we find nine possible combinations of a quark and an antiquark, which are listed in Table 8. These nine combinations exactly reproduce the electric charge and strangeness combinations of the spin-0 mesons, as indicated by Fig. 4a.

Baryons are composed of three quarks, the simplest combination that gives $B = +1$. There are ten different

TABLE 8   QUARK–ANTIQUARK COMBINATIONS

Combination	Charge (e)	Spin ($h/2\pi$)	Baryon Number	Strangeness
$u\bar{u}$	0	0,1	0	0
$u\bar{d}$	$+1$	0,1	0	0
$u\bar{s}$	$+1$	0,1	0	$+1$
$d\bar{u}$	-1	0,1	0	0
$d\bar{d}$	0	0,1	0	0
$d\bar{s}$	0	0,1	0	$+1$
$s\bar{u}$	-1	0,1	0	-1
$s\bar{d}$	0	0,1	0	-1
$s\bar{s}$	0	0,1	0	0

combinations that can be made from three quarks, as listed in Table 9. Plotting the allowed spin-$\frac{1}{2}$ and spin-$\frac{3}{2}$ combinations, we obtain Figs. 4b and 4c.

The similarities between Figs. 3 and 4 are remarkable. Based on only three quarks, we are able to account for the Q, S, and B quantum numbers of all these particles. However, the quark model does far more than produce these

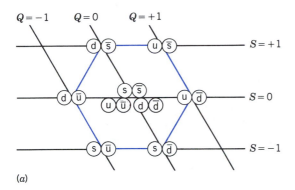

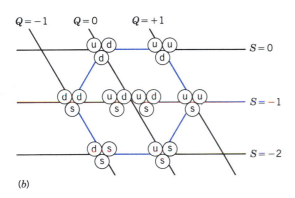

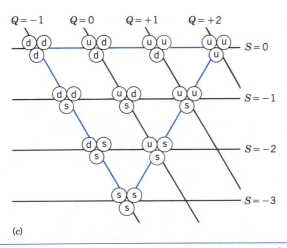

(c)

Figure 4   A chart showing (a) the spin-0 combinations of a quark and an antiquark, (b) the spin-$\frac{1}{2}$ combinations of three quarks, and (c) the spin-$\frac{3}{2}$ combinations of three quarks. Compare with Fig. 3.

simple geometrical patterns. You should think of these patterns as ways of organizing particles with similar properties, just as the periodic table allows us to organize atoms with similar properties. Underlying the periodic table is atomic theory, which can be used to calculate properties of atoms beyond their geometrical arrangements. In a similar way, the quark model allows us to calculate properties of particles, including masses, magnetic dipole moments, decay modes, lifetimes, and reaction products. The agreement between the measured and calculated properties has been a spectacular success for the model. In fact, all known particles (hundreds of them!) have been accounted for based on this model, with a few additional quarks that we describe later.

The most unusual aspect of the quark model is the fractional electric charges of the quarks. All particles yet discovered have electric charges that can be expressed as integral multiples of the basic unit of charge e. No particle with a fractional electric charge has ever been seen. In fact, no one has ever seen a free quark, despite heroic experimental efforts to search for one. It is possible that our particle accelerators do not yet have enough energy to produce a free quark. It has also been suggested that free quarks may be forbidden to exist, so that we may only observe quarks bound in mesons and baryons.

Even though free quarks have never been seen, individual *bound* quarks have been observed. Scattering experiments that probe deep inside the nucleon have revealed three pointlike objects that appear to have a spin of $\frac{1}{2}$ and a charge of $+\frac{2}{3}$ or $-\frac{1}{3}$. These experiments give direct proof of the existence of quarklike particles within the nucleon.

The Force Between Quarks

What holds the quarks together inside a meson or a nucleon? This force is the most fundamental version of the strong force, brought about through the exchange of particles called *gluons*. Just as the electromagnetic force between charged particles can be regarded as an exchange of photons, the strong force between quarks is accomplished through the exchange of gluons. We therefore picture a nucleon as composed of three quarks mutually exchanging gluons. It is possible, through indirect means, to measure the fraction of the momentum of the internal structure of a nucleon that is due to the quarks. This fraction turns out to be only around 50%. The rest must be due to the exchanged gluons. The resulting picture of the nucleon is of three quarks "swimming in a sea" of exchanged gluons.

The force between quarks has two unusual properties. (1) It takes a large (perhaps infinite) energy to separate two quarks to a distance greater than the size of a nucleon or a meson (about 1 fm). This may be the reason that no free quarks have yet been seen. When we try to pump energy into a nucleon to separate one of its quarks, the energy

TABLE 9 COMBINATIONS OF THREE QUARKS

Combination	Charge (e)	Spin ($h/2\pi$)	Baryon Number	Strangeness
uuu	$+2$	$\frac{3}{2}$	$+1$	0
uud	$+1$	$\frac{3}{2},\frac{1}{2}$	$+1$	0
udd	0	$\frac{3}{2},\frac{1}{2}$	$+1$	0
uus	$+1$	$\frac{3}{2},\frac{1}{2}$	$+1$	-1
uss	0	$\frac{3}{2},\frac{1}{2}$	$+1$	-2
uds	0	$\frac{3}{2},\frac{1}{2}$	$+1$	-1
ddd	-1	$\frac{3}{2}$	$+1$	0
dds	-1	$\frac{3}{2},\frac{1}{2}$	$+1$	-1
dss	-1	$\frac{3}{2},\frac{1}{2}$	$+1$	-2
sss	-1	$\frac{3}{2}$	$+1$	-3

actually creates a quark–antiquark pair. The antiquark combines with one of the quarks to form a meson, which agrees with our observations: when we smash two nucleons together at high energies, we get our nucleons (or other baryons) back, plus some additional mesons. The more energy we put in, the more mesons we get out, but no free quark emerges. (2) Paradoxically, inside the nucleon or the meson, the quarks appear to move freely. At very short distances (less than the size of a nucleon), the force between quarks approaches zero.

This unusual behavior of quarks and gluons can be understood by comparison with electromagnetism. Two charged particles interact with one another through the exchange of photons. However, the photon itself carries no electric charge, and so the interaction between the charged particle and the exchanged photon does not result in the exchange of additional photons. A quark, on the other hand, can emit a gluon and interact with it. This force between the quark and the gluon can create additional gluons. When it interacts with another electron, an electron can emit a photon and still remain an electron. It does not sacrifice its "electricness" (that is, its electric charge) to emit the photon. A quark, however, gives its emitted gluon a share of its "strongness," which physicists call "color." In the interaction of quarks, color plays the same role as electric charge in the interaction of charged particles. A photon carries no electric charge, but a gluon carries color, and in doing so it changes the residual color left behind in the quark that emitted the gluon. In effect, the quark is spreading its color over a sphere the size of a nucleon (the range of the gluons), and as a result the interaction between quarks is considerably weakened at these distances.

Particle physicists have chosen amusing and whimsical names to describe the fundamental particles and their properties. Names such as quark, strangeness, gluon, or color have meaning only as labels. Gluons do provide the "glue" that binds quarks together, but it has no similarity to any other "glue" in our experience. The "color" carried by quarks and gluons has nothing to do with our ordinary use of color. It is simply easier to remember and discuss these properties if we give them familiar names.

More Quarks

In simultaneous 1974 experiments at the Brookhaven National Laboratory in New York and the Stanford Linear Accelerator Center in California, investigators discovered an unusual meson with a rest energy about three times that of the proton. This new meson, called ψ (psi), was expected to decay into lighter mesons in a strong interaction time of perhaps 10^{-23} s. Instead, it was observed to decay in a time of about 10^{-20} s, which is more characteristic of the electromagnetic interaction (see Table 1). Moreover, its decay products were not mesons but an electron and a positron, another signal of an electromagnetic process.

Why is the rapid, strong-interaction decay path blocked for this particle, slowing its decay by three orders of magnitude? We discussed a similar effect in the case of strangeness, a new quantum number that was introduced partly to explain certain slow decays. We accounted for those decays through a violation of the conservation of strangeness.

In a similar fashion, we assume that the decay of ψ is slowed by the violation of another conservation law, called *charm*. According to this interpretation, the ψ meson is composed of a new quark c (for charm) and its antiquark $\bar{c}$. The c quark has an electric charge of $+\frac{2}{3}$. Just as the strange quark is assigned a strangeness quantum number of $S = -1$, the charmed quark is assigned a charm of $C = +1$. The decay of the ψ meson is slowed, because the c quark must decay into other quarks (u, d, or s), all of which have $C = 0$. The decay thus involves a violation of the conservation of charm and therefore cannot occur through the strong interaction, which conserves charm.

In 1977 a similar discovery was made at the Fermi National Accelerator Laboratory near Chicago. Again, a heavy meson (in this case, ten times the proton rest energy) was discovered, which was expected to decay to other mesons in a time characteristic of the strong interaction, but instead it decayed into $e^- + e^+$ in about 10^{-20} s. In this case, the decay was again slowed by the violation of yet another conservation rule, involving yet another new quark, called b (for bottom) and having an electric charge of $-\frac{1}{3}$. This new meson, called Υ (upsilon), is assumed to be composed of the combination $b\bar{b}$. If we assign to the b quark a new quantum number that represents bottomness, then the decay is slowed because the b quark must change into lighter quarks that lack this property; this violation of the conservation of bottomness is responsible for slowing the decay.

A New Symmetry

Ordinary matter is composed of protons and neutrons, which are in turn made up only of u and d quarks. Ordinary matter is also composed of electrons, and in the conversion of protons to neutrons or neutrons to protons in the beta decay of ordinary matter, we find electron-type neutrinos emitted along with the positron or electron.

We can therefore construct our entire world and all the phenomena we commonly observe out of two pairs of fundamental particles: u and d quarks, and e^- and ν_e leptons. Within each pair, the charges differ by one unit $(+\frac{2}{3}$ and $-\frac{1}{3}; -1$ and $0)$.

If we do experiments at a somewhat higher energy, we find new types of particles: a new pair of leptons (μ^- and its neutrino ν_μ) and a new pair of quarks (c and s). Once again, within each pair the electric charges differ by one unit.

At still higher energy, we find a new pair of leptons (τ and ν_τ) and a new quark (b). It is assumed that the b quark has a partner, called t (for top), and if the t quark has a charge of $+\frac{2}{3}$, this latest pair of quarks will be similar to the other pairs. Searches for the t quark have been made by looking for new mesons up to about 30 times the proton's rest energy, but as yet no evidence for this quark has been found. Nevertheless, physicists are sure of its existence and confident it will be found if enough energy is available.

It therefore seems that the truly fundamental particles, the quarks and leptons, appear in pairs, and that a pair of quarks and a pair of leptons can be combined into a "generation," as follows:

1st generation: $\begin{pmatrix} e \\ \nu_e \end{pmatrix}$ and $\begin{pmatrix} u \\ d \end{pmatrix}$

2nd generation: $\begin{pmatrix} \mu \\ \nu_\mu \end{pmatrix}$ and $\begin{pmatrix} c \\ s \end{pmatrix}$

3rd generation: $\begin{pmatrix} \tau \\ \nu_\tau \end{pmatrix}$ and $\begin{pmatrix} t \\ b \end{pmatrix}$

Properties of these six quarks and leptons are summarized in Appendix F.

It probably now occurs to you that we may be headed in the same direction all over again. That is, might we someday have hundreds of "fundamental" quarks and leptons, so that instead of simplicity we have a new layer of com-

plexity? It is possible to suppose that bigger and bigger accelerators will reveal new generations of ever more massive leptons and quarks, and the only limit on their number may appear to be imposed by the amount of energy we have available. To answer this question, we turn to discoveries at the opposite end of the scale from accelerator laboratories: we look to the earliest moments after the birth of the universe, and we shall see that the previous list of quarks and leptons may be complete.

Sample Problem 5   Analyze these processes in terms of their quark content:

(a) $p \rightarrow n + e^+ + \nu_e$,
(b) $\Omega^- \rightarrow \Lambda^0 + K^-$,
(c) $K^- + p \rightarrow \Omega^- + K^+ + K^0$.

Solution (a) Using Figs. 3 and 4 to find the quark content of each of the particles, we can rewrite the decay as

$$uud \rightarrow udd + e^+ + \nu_e.$$

Canceling the common pair of ud quarks from each side, we find

$$u \rightarrow d + e^+ + \nu_e.$$

The u quark changes to a d quark by beta decay.

(b) The quark content is

$$sss \rightarrow uds + s\bar{u}.$$

Canceling the common pair of s quarks from each side, we find the net process to be

$$s \rightarrow u + d + \bar{u}.$$

That is, the s quark is transformed into a d quark, and a $u\bar{u}$ pair is created from the decay energy.

(c) Again replacing the particles by their quark content, we can write the reaction as

$$s\bar{u} + uud \rightarrow sss + u\bar{s} + d\bar{s},$$

and removing the common quarks of u, d, and s from each side we are left with

$$u\bar{u} \rightarrow s\bar{s} + s\bar{s}.$$

The net process consists of the annihilation of the $u\bar{u}$ pair and the production of two $s\bar{s}$ pairs from the reaction energy.

These examples are typical of quark processes: the weak interaction can change one type of quark into another. The strong interaction can create or destroy quark–antiquark pairs, but it cannot change one type of quark into another.

56-5   THE BIG BANG COSMOLOGY

Since the beginnings of recorded history, human beings have speculated about the origin and future of the universe, a branch of science now called *cosmology*. Until the 20th century, these speculations were done mostly by phi-

Figure 5   Edwin Hubble (1889–1953) at the controls of the 100-in. telescope on Mount Wilson, where he did much of the research that led him to propose that the universe is expanding.

losophers and theologians, because there was no experimental evidence of any sort that would form the basis of any scientific theory. In this century, two major experimental discoveries have pointed the way to a coherent theory that is now accepted by nearly all physicists.

The Expansion of the Universe

The first of the two great discoveries was made by astronomer Edwin Hubble (see Fig. 5) in the 1920s. Hubble was studying the wispy objects known previously as *nebulae.*

By eventually resolving individual stars in the nebulae, Hubble was able to show that they are galaxies just like our Milky Way, composed of hundreds of billions of stars. More startlingly, Hubble deduced that the galaxies are moving away from one another and from us, and that the greater their distance from us, the greater is their recessional speed. That is, if *d* is the distance of the galaxy from Earth (or from any other point of reference in the universe) and *v* is the speed with which the galaxy appears to be moving away from us, Hubble's law gives

$$v = Hd, \qquad (2)$$

where *H* is a proportionality constant known as the *Hubble parameter.*

The Hubble parameter has the dimensions of inverse time. Its value can be learned only by experiment: we must independently deduce the distance of a galaxy from Earth and its speed relative to Earth. The recessional speeds can be measured in a straightforward way using the Doppler shift of the light from the galaxy (see Fig. 6 of Chapter 42), but the distance scale is difficult to determine (in fact, Hubble's early estimates were off by a factor of 10). Nevertheless, today we have a consistent set of data (Fig. 6) that confirms Hubble's law and gives a value of the Hubble parameter of about

$$H = 67 \, \frac{\text{km/s}}{\text{Mpc}},$$

where the Mpc (megaparsec) is a commonly used unit of distance on the cosmic scale:

$$1 \text{ Mpc} = 10^6 \text{ pc} = 3.26 \times 10^6 \text{ light-years}$$
$$= 3.084 \times 10^{19} \text{ km}.$$

Because of uncertainties in the estimates of the cosmic

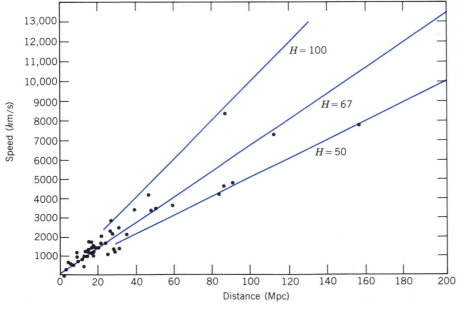

Figure 6   The relationship between speed and distance for groups and clusters of galaxies. The straight lines show the Hubble relationships for various values of the Hubble parameter *H*.

scale of distance, the Hubble parameter is uncertain, with possible values in the range of 50–100 (km/s)/Mpc.

If the universe has been expanding forever at the same rate, then H^{-1} is the age of the universe. Using the accepted value of the Hubble parameter, we would estimate the age of the universe as 15×10^9 y, with the range of uncertainty of H permitting values in the range of $10–19 \times 10^9$ y. However, as we shall see later, the expansion of the universe has not been constant, so the true age is less than the currently deduced value of H^{-1}.

The Cosmic Microwave Background Radiation

Although there were other explanations of the expansion of the universe, the one that gained favor was based on the assumption that, if the galaxies are presently rushing apart, they must have been closer together in the distant past. If we run the cosmic clock back far enough, we find that in its early state the universe consisted of unimaginably high densities of matter and radiation. As the universe expanded, both the matter and the radiation cooled; you can think of the wavelengths of the radiant photons being stretched in the expansion. The radiation filled the entire universe in its compact state, and it continues to fill the entire universe in the expansion. We should still find that radiation present today, cooled to the extent that its most intense component is in the microwave region of the electromagnetic spectrum. This is known as the *cosmic microwave background radiation.*

This radiation was discovered in 1965 by Arno Penzias and Robert Wilson of the Bell Laboratories in New Jersey, who were testing a microwave antenna used for satellite communications (see Fig. 7). No matter where they pointed their antenna, they found the same annoying background "hiss." Eventually they realized that they were indeed seeing a remnant of the early universe, and they were awarded the 1978 Nobel prize in physics for their discovery.

The microwave background radiation has a true thermal spectrum of the type we discussed in Sections 49-1 and 49-2. Figure 8 shows measurements of the intensity of the background radiation at various wavelengths, and you can see how well it is fit by Planck's radiation law with a temperature of 2.735 K. The data points include recent measurements made from a satellite in Earth orbit, thereby eliminating atmospheric absorption.

Measurements of the intensity of the microwave background radiation in various directions show that the radiation has a uniform intensity in all directions; it does not appear to come from any particular source in the sky, but instead fills the entire universe uniformly, as would be expected for radiation that likewise filled the early universe. Recent observations, however, show that there are temperature fluctuations of about 10^{-5} K between different regions of the sky. These results have been interpreted

Figure 7   Arno Penzias (right) and Robert Wilson, standing in front of the large horn antenna with which they first detected the microwave background radiation.

as evidence for the nonuniform distribution of matter in the early universe that led ultimately to the condensation of stars and galaxies.

The energy density of the radiation can be found from Planck's radiation law (Eq. 6 of Chapter 49). The number density of these background photons is about 400 per cm³, and the energy density is about 0.25 eV/cm³ (roughly corresponding to half the rest energy of an electron per m³). The mean energy per photon is about 0.00063 eV, which suggests why we are not ordinarily aware of the presence of these photons.

The Big Bang Cosmology

The cosmological theory that is in best agreement with these two experimental findings (the Hubble law and the background radiation) is the *Big Bang cosmology.* According to this theory, the universe began some 10–20 billion years ago in a state of extreme density and temperature. There were no galaxies or even clumped matter as we now know it; the "stuff" of the universe at early times was a great variety of particles and antiparticles, plus radiation. The density of radiation and matter is related to the temperature of the universe. As the universe expands, it cools (just as an expanding thermodynamic system cools). If we make some reasonable assumptions about the expansion rate, we can find a relationship between the temperature and the time after the formation of the universe:

$$T = \frac{1.5 \times 10^{10} \text{ s}^{1/2} \cdot \text{K}}{t^{1/2}}, \tag{3}$$

where the temperature T is in K and the time t is in seconds.

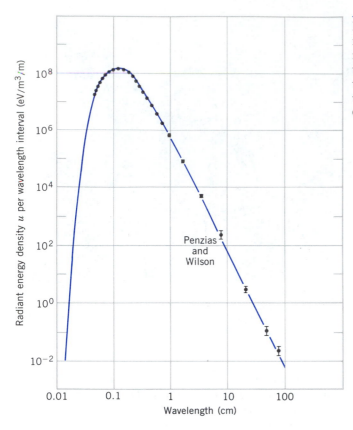

Figure 8 The spectrum of the cosmic microwave background radiation. The dots represent observations, and the solid line represents the Planck spectrum for the radiant energy corresponding to a temperature of 2.735 K. Note the excellent agreement between the data points and the theoretical curve. The data between 0.05 cm and 1.0 cm come from observations made by the COBE (COsmic Background Explorer) satellite launched in 1989.

The radiation in the early universe consisted of high-energy photons, whose typical energy can be roughly estimated as kT, where k is the Boltzmann constant and T is the temperature at a particular time t, determined from Eq. 3. The dominant processes in the early universe can be represented as

$$\text{photons} \leftrightarrows \text{particle} + \text{antiparticle}.$$

That is, photons can engage in *pair production* and produce a particle–antiparticle pair, for example, an electron and a positron or a proton and an antiproton. Conversely, a particle and its antiparticle can annihilate into photons. In each case, the total energy of the photons must be at least as large as the rest energy of the particle and the antiparticle.

Our goal in describing the early universe is to understand the formation of ordinary matter from the particles and radiation produced in the Big Bang. Since ordinary matter is composed of nucleons, let us consider the formation and annihilation of protons and neutrons:

$$\gamma + \gamma \leftrightarrows p + \bar{p} \quad \text{and} \quad \gamma + \gamma \rightleftarrows n + \bar{n},$$

where we represent the photons as gamma rays. For photons to produce nucleon–antinucleon pairs, the photon energy kT must be at least as large as the rest energy mc^2 of a nucleon (940 MeV). The minimum temperature of the universe that will permit production of nucleons and antinucleons is

$$T = \frac{mc^2}{k} = \frac{940 \text{ MeV}}{8.62 \times 10^{-5} \text{ eV/K}} = 1.1 \times 10^{13} \text{ K}.$$

According to Eq. 3, the universe cooled below this temperature at the time

$$t = \left(\frac{1.5 \times 10^{10} \text{ s}^{1/2} \cdot \text{K}}{T} \right)^2 = \left(\frac{1.5 \times 10^{10} \text{ s}^{1/2} \cdot \text{K}}{1.1 \times 10^{13} \text{ K}} \right)^2$$

$$= 2 \times 10^{-6} \text{ s}.$$

That is, at times earlier than 2 μs, the universe was hot enough for the photons to produce nucleon–antinucleon pairs, but after 2 μs the photons were (on the average) not energetic enough to produce nucleon–antinucleon pairs.

At earlier times (corresponding to higher temperatures), the radiation may have been able to create quark–antiquark pairs. We can regard the universe at these earliest times as consisting only of fundamental particles (quarks and leptons) and radiation. The quarks and antiquarks combined to form mesons and baryons, which were disassociated by the radiation as rapidly as they could form. As the universe expanded and cooled, the radiation eventually became too feeble to blast apart the mesons and baryons. Because the details of the quark model (and the existence of free quarks) are not yet confirmed, we will begin the story at a time of about 10^{-6} s, when we can regard the universe as being composed of protons, neutrons, antiprotons, antineutrons, mesons, leptons, antileptons, and photons. The rates of production and disassociation are roughly equal, so that the numbers of particles and their corresponding antiparticles are roughly equal. Furthermore, the number of photons is roughly equal to the number of protons, which is in turn roughly equal to the number of electrons. Before this time, the strong interaction played a prominent role in determining the structure and composition of the universe, through such processes as the combinations of quarks and antiquarks into baryons or mesons or collisions of baryons to form mesons or new baryons. After about 10^{-6} s (corresponding to $T = 1.5 \times 10^{13}$ K or $kT = 1300$ MeV), the particles and radiation have too little energy to induce these reactions, and the era of the strong interaction ends at about this time.

Electromagnetic and weak-interaction processes continue to take place. Electromagnetic processes are represented by the production of particles and antiparticles (for example, electrons and positrons) by photons, while weak interactions occur through such processes as

$$n + \nu_e \leftrightarrows p + e^- \quad \text{and} \quad p + \bar{\nu}_e \leftrightarrows n + e^+$$

and similar processes, in which neutrinos are being created and destroyed at the same rate. As long as the leptons and neutrinos have enough energy, the forward and reverse reaction rates are equal, which maintains the balance between the number of charged leptons (e^+ and e^-) and neutrinos. Since these reactions convert neutrons to protons and protons to neutrons with equal ease, the very early universe contained roughly equal numbers of protons and neutrons.

The difference in rest energy between protons and neutrons is about $\Delta E = 1.3$ MeV (neutrons being slightly more massive). At any temperature T, the difference between the relative numbers of protons and neutrons is determined in part by the Boltzmann factor $e^{-\Delta E/kT}$ (see Section 24-4). When $t < 10^{-6}$ s (corresponding to $T > 1.5 \times 10^{13}$ K or $kT > 1300$ MeV), the Boltzmann factor is very nearly equal to 1 and has a negligible influence on the relative number of protons and neutrons.

At times after about 10^{-6} s, the radiation has on the average too little energy to create nucleon–antinucleon pairs (that is, we no longer have $\gamma + \gamma \rightarrow n + \bar{n}$ or $p + \bar{p}$), but nucleon–antinucleon annihilation continues to occur ($n + \bar{n} \rightarrow \gamma + \gamma$ and $p + \bar{p} \rightarrow \gamma + \gamma$). From 10^{-6} s until about 10^{-2} s ($T = 1.5 \times 10^{11}$ K or $kT = 13$ MeV), weak-interaction processes maintain a rough balance between the numbers of protons and neutrons.

Between 10^{-2} s and 1 s ($T = 1.5 \times 10^{10}$ K or $kT = 1.3$ MeV), the Boltzmann factor begins to upset the balance between protons and neutrons, and at $t = 1$ s the ratio between the number of neutrons and the number of protons is about e^{-1}, so that the nucleons consist of about 73% protons and 27% neutrons. During this period, the influence of the neutrinos has been decreasing, and by about $t = 1$ s the neutrinos (which are cooling along with the rest of the universe as it expands) have too little energy to cause proton–neutron transformations, which diminishes the role of the weak interactions in the evolution of the universe. This is known as the time of "neutrino decoupling," when the interactions between matter and neutrinos no longer occur. From this time on, the neutrinos continue to fill the universe, cooling along with the expansion of the universe. These primordial neutrinos today have roughly the same density as the microwave background photons but a slightly lower temperature (about 2 K). Because neutrinos interact only feebly with matter, detection of energetic neutrinos ($E > 1$ MeV) requires equipment of great size and sophistication. Detection of these primordial neutrinos ($E < 10^{-3}$ eV) seems a hopeless task, but observing them, measuring their energy distribution, and deducing their temperature would provide another dramatic confirmation of the Big Bang cosmology.

At a time of about 6 s ($T = 6 \times 10^9$ K or $kT = 0.5$ MeV), the radiation has cooled to a temperature at which it no longer has enough energy to produce even the lightest particle–antiparticle pairs (electrons and positrons), so no new particles are formed by pair production. Particle–antiparticle annihilation continues to occur, and the resulting photons slow the rate of cooling somewhat. Furthermore, the electrons have too little energy to cause protons to transform into neutrons ($p + e^- \rightarrow n + \nu_e$ no longer occurs). The only weak-interaction process that continues to occur is the decay of the neutron ($n \rightarrow p + e^- + \bar{\nu}_e$). At this time the nucleons consist of about 83% protons and 17% neutrons.

During this period, particle–antiparticle annihilation has continued to occur, so that no positrons or antinucleons remain in the universe. The universe now consists of a number N of protons, an equal number N of electrons (to make it electrically neutral), and about $0.2N$ neutrons. Because particle–antiparticle annihilation has substantially decreased the number of particles while maintaining the amount of radiation, there are far more photons (perhaps $10^8 - 10^9 \, N$) than nucleons or electrons. There are about as many neutrinos as photons.

As far as we know, the present universe contains no stars or galaxies made of antimatter. What happened to all the antimatter, which represented 50% of the particles in the early universe? According to the Big Bang cosmology, in an early epoch of the evolution of the universe one of the forces that acted between the particles caused a very slight imbalance of matter over antimatter, perhaps 1 part in 10^8 or 10^9. The exact nature of this force is not yet well understood, although experiments testing this force and distinguishing matter from antimatter have been duplicated in the laboratory. During the subsequent stages in the evolution of the universe, all the antimatter annihilated with all but about 1 part in 10^8 or 10^9 of the matter. That is, for every 1,000,000,000 positrons there may originally have been 1,000,000,001 electrons, but following the annihilation of 2,000,000,000 particles the remainder is just 1 electron.

This description of the evolution of the universe, illustrated in Fig. 9, has taken us from the formation of the universe at the Big Bang, through hot and turbulent eras dominated by nuclear reactions, to a time of a few seconds when the composition became identical with the particles that now make up our universe. How these particles combined to form the nuclei and atoms that we observe today is discussed in the next section.

Sample Problem 6   When did the universe become too cool to permit the radiation to create $\mu^+ \mu^-$ pairs?

Solution   The rest energy of the muon is 105.7 MeV. Photons have this mean energy at a temperature determined by

$$T = \frac{E}{k} = \frac{105.7 \text{ MeV}}{8.62 \times 10^{-5} \text{ eV/K}} = 1.23 \times 10^{12} \text{ K}.$$

The corresponding time is found from Eq. 3:

$$t = \left(\frac{1.5 \times 10^{10} \text{ s}^{1/2} \cdot \text{K}}{1.23 \times 10^{12} \text{ K}} \right)^2 = 1.5 \times 10^{-4} \text{ s}.$$

56-6   NUCLEOSYNTHESIS

At an age of a few seconds, the universe consisted of protons, neutrons, and electrons. Today, the composition of the universe is mostly hydrogen and helium, with a small abundance of heavier elements. How were the present nuclei and atoms produced from the Big Bang? The formation of the elements of the present universe is known as *nucleosynthesis*. As we shall see, observing the present abundances of the elements can give us clues about the processes that occurred during the Big Bang.

Big Bang Nucleosynthesis

The first step in building up complex atoms is the formation of deuterium nuclei (deuterons) from the combination of a proton and a neutron, according to

$$n + p \rightarrow d + \gamma.$$

The binding energy of the deuteron (see Section 54-2) is 2.2 MeV, which is the energy of the γ ray that is given off during the formation. The reverse reaction,

$$d + \gamma \rightarrow n + p,$$

can break apart the deuterium nuclei into their constituent protons and neutrons, if the γ ray energy is at least 2.2 MeV.

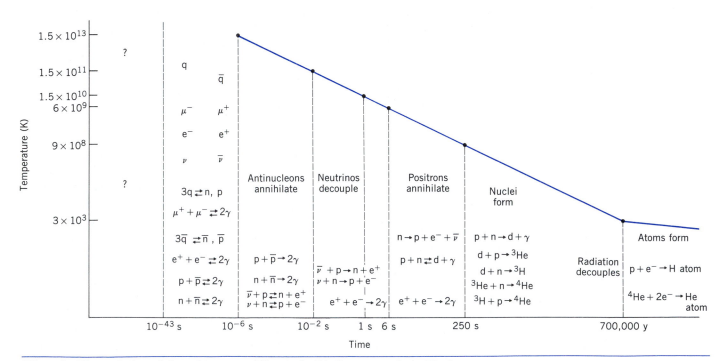

Figure 9   The evolution of the universe according to the Big Bang cosmology. The heavy solid line shows the relationship between temperature and time according to Eq. 3. The important reactions in each era are shown. (Here q and $\bar{q}$ stand for quark and antiquark, respectively.)

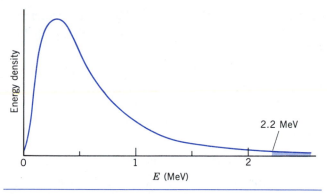

Figure 10   The energy spectrum of photons at a particular time in the evolution of the universe. Photons with energy above 2.2 MeV, which constitute a tiny fraction of the total number of photons, can dissociate deuterons.

If the universe is filled with energetic photons, the two reactions will take place at the same rate, and deuterium will be disassociated as quickly as it is formed. However, if the universe is sufficiently old, the photons will not have enough energy to accomplish the disassociation reaction, and deuterium can start to build up.

When we ended our story in the previous section, the universe was about 6 s old, and the mean energy of the radiation was about 0.5 MeV, which is less than what is needed to keep deuterium from forming. However, it must be remembered that the radiation has a Planck distribution of energies (see Fig. 10, which was discussed in Section 49-2) and that there are $10^8 - 10^9$ photons for every proton or neutron. There is a high-energy tail in the Planck distribution, which suggests that no matter what the temperature of the radiation, there will always be some photons of energy above 2.2 MeV that can break apart deuterium nuclei. If, on the average, the number of these energetic photons is less than the number of protons and neutrons, deuterium can start to build up.

The neutron-to-proton ratio is about 0.2 at this point in the evolution of the universe, and there are roughly 10^9 photons per nucleon, so that the ratio of neutrons to photons is about 0.2×10^{-9}. If the fraction of photons with energies above 2.2 MeV is less than 0.2×10^{-9} of the total number of photons, there will be less than one energetic photon per neutron, and deuterium formation can proceed. From the expression for the Planck distribution (obtained from Eq. 6 of Chapter 49), we find that the fraction of photons of energy greater than 2.2 MeV will be less than 0.2×10^{-9} when the temperature has fallen to 9×10^8 K. Equation 3 shows that this temperature occurs at a time of 250 s.

At a time of 250 s, the formation of deuterium nuclei begins. Because the deuterons are less abundant than protons or neutrons, the deuterons will readily react with protons and neutrons, according to the reactions

$$d + n \rightarrow {}^3H + \gamma \quad \text{and} \quad d + p \rightarrow {}^3He + \gamma.$$

Finally, the ${}^3H$ and ${}^3He$ will also react with protons and neutrons, as given by

$${}^3H + p \rightarrow {}^4He + \gamma \quad \text{and} \quad {}^3He + n \rightarrow {}^4He + \gamma.$$

For all four of these reactions, the binding energy of the final particle is greater than that of the deuteron. Thus if the radiation is too feeble to prevent the formation of deuterons, it will certainly be too feeble to prevent the succeeding reactions. We can therefore assume that nearly all the deuterons are eventually converted into ${}^4He$, so that the end products of this stage of the evolution of the universe are protons and α particles. Because there are no stable nuclei with a mass number of 5, these reactions cannot continue beyond ${}^4He$.

To find the relative number of α particles, we must find the number of available neutrons at $t = 250$ s, when deuterons begin to form. At $t = 6$ s, about 17% of the nucleons are neutrons, but as a result of the radioactive decay of the neutron, some neutrons will be converted into protons between $t = 6$ s and $t = 250$ s. Using the half-life of the neutron (about 11 min), we find that at $t = 250$ s the nucleons will consist of about 12.5% neutrons and 87.5% protons. That is, out of every 10,000 nucleons there will be 1250 neutrons and 8750 protons. These neutrons will combine with 1250 protons to form 625 α particles, leaving $8750 - 1250 = 7500$ protons. Of the total number of nuclei in the universe at this time, 7.7% are α particles and 92.3% are protons. In terms of mass, the α particles constitute a fraction of the total mass of the universe given by

$$\frac{4 \times 625}{7500 + 4 \times 625} = 0.25 \text{ or } 25\%.$$

The abundance of ${}^4He$ in the present universe should equal this value, if we neglect the burning of hydrogen to helium that takes place in stars. The measured helium abundance in a variety of systems, including stars, gaseous nebulae, and planetary nebulae, turns out to be $24 \pm 1\%$, which agrees with our estimate and indicates that our description is certainly reasonable.

The final step in the production of matter in the Big Bang is the formation of neutral atoms of hydrogen and helium when the protons and α particles combine with electrons. As in the case of deuteron formation, this cannot occur when there are enough photons in the high-energy tail of the Planck distribution to break apart any neutral atoms that may form. In this case, we want the relative fraction of photons with energies above 13.6 eV (the binding energy of atomic hydrogen) to be less than about 10^{-9}. This occurs for a temperature of about 6000 K, which corresponds to an age of the universe of around 200,000 y. (As the radiation cools, the energy density of the universe becomes less dominated by radiation and more by matter. In this case Eq. 3, which assumes a radiation-dominated universe, is not quite correct. Taking this effect into account, the temperature of the universe when hydrogen atoms begin to form is closer to 3000 K, corresponding to an age of around 700,000 y.)

Once neutral atoms have formed, there are essentially no free charged particles left in the universe. This is the time of decoupling of the matter and the radiation field. The universe becomes transparent to the radiation, which can travel long distances without interacting with matter. This radiation, which has been traveling since the decoupling time, is observed today as the microwave background. The expansion of the universe has reduced the radiation temperature by a factor of 1000 since the decoupling time.

The story of the evolution of the universe as described by the Big Bang cosmology is a remarkable one. It integrates modern experiments in nuclear and particle physics with quantum physics and classical thermodynamics. It yields results that can be tested in the present universe, including the helium abundance, the microwave background radiation, and a small abundance of left-over deuterium that did not get "cooked" into mass-3 nuclei. It is a story that depends in critical ways on the strengths of nuclear or subnuclear forces and on the variety of particles that took part in the early universe. For example, if there were a fourth generation of leptons, the reaction rates of weak-interaction processes would be greater, and more neutrons would be formed, thereby increasing the abundance of ^{4}He. Although this conclusion is subject to interpretation, the observed present abundance of ^{4}He is regarded by many cosmologists as limiting the number of generations of leptons to three.

Nucleosynthesis in Fusion Reactions

After the decoupling of matter and radiation, the matter (consisting of hydrogen and helium) was subject only to the gravitational force. Recent precise observations of the microwave background have shown that the distribution of matter at the decoupling time was slightly nonuniform. Regions of slightly higher density began to condense into clouds of ever increasing density. As each cloud contracted under its own gravity, its temperature rose until it became hot enough to initiate fusion reactions. This is how first-generation stars formed.

We have seen in Chapter 55 that stars convert hydrogen into helium by means of fusion reactions. After a star has used up its supply of hydrogen and become mostly helium, it can again begin to contract, which increases its temperature. (This increase in temperature causes an increased radiation pressure, which causes the radius of the star to increase. The surface area increases more rapidly than the temperature, so that the energy per unit area of the surface actually decreases, and the color of the star goes from bright yellow to red. This is the *red giant* phase of the evolution of the star.) Eventually, the temperature is high enough that the Coulomb barrier between two ^{4}He

nuclei can be successfully breached by their thermal motion, and helium fusion can occur. The simple helium fusion reaction

$$^4\text{He} + {}^4\text{He} \rightarrow {}^8\text{Be}$$

does not contribute to the fusion in a star, because ^{8}Be is unstable and breaks apart as rapidly as it forms. Helium fusion requires a third α particle to participate, so that the net reaction is

$$^4\text{He} + {}^4\text{He} + {}^4\text{He} \rightarrow {}^{12}\text{C} + \gamma.$$

Once ^{12}C forms, we can have additional α-particle reactions, such as

$$^{12}\text{C} + {}^4\text{He} \rightarrow {}^{16}\text{O} + \gamma,$$

$$^{16}\text{O} + {}^4\text{He} \rightarrow {}^{20}\text{Ne} + \gamma,$$

$$^{20}\text{Ne} + {}^4\text{He} \rightarrow {}^{24}\text{Mg} + \gamma,$$

and so on. These reactions have increasingly high Coulomb barriers and therefore require increasing temperatures.

When the helium fuel is exhausted, contraction sets in again to increase the temperature, so that other reactions can occur, such as carbon burning:

$$^{12}\text{C} + {}^{12}\text{C} \rightarrow {}^{24}\text{Mg} + \gamma.$$

Eventually, these reactions reach the peak of the binding energy curve (Fig. 6 of Chapter 54) at mass 56. Beyond this point, energy is no longer released in fusion reactions.

Figure 11 shows the abundance of nuclei in this mass range. The relative abundances support this scenario for producing the elements in fusion reactions. Note that C is more than five orders of magnitude more abundant than Li, Be, and B, which are not produced in these processes. Also note that the even-Z nuclei are, on the average, more than an order of magnitude more abundant than their odd-Z neighbors. These fusion reactions with α particles produce only even-Z products, so the observed higher abundances of these products are consistent with this explanation of their formation.

Note also the last point in Fig. 11, which indicates that the *total* abundance of the 50 elements beyond the nuclei in the mass-56 range is less than the abundance of all but one of the individual elements in the region from C to Zn. It certainly appears that most of the matter we know about was produced in fusion processes.

Nucleosynthesis by Neutron Capture

The elements beyond mass 56 were produced either slowly in stars or suddenly in supernovas by neutron-capture reactions. There will be a small density of neutrons in stars, produced through sequences such as

$$^{12}\text{C} + {}^1\text{H} \rightarrow {}^{13}\text{N} + \gamma$$

$$^{13}\text{N} \rightarrow {}^{13}\text{C} + e^+ + \nu_e$$

$$^{13}\text{C} + {}^4\text{He} \rightarrow {}^{16}\text{O} + \text{n}.$$

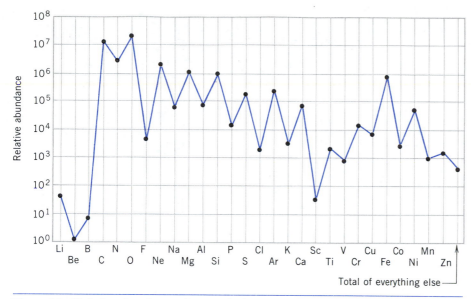

Figure 11   Relative abundances (by mass) of the elements beyond helium in the solar system.

Suppose we have ^{56}Fe, which has been produced through fusion processes. A sequence of neutron-capture reactions can then occur:

$$^{56}\text{Fe} + n \rightarrow {}^{57}\text{Fe} + \gamma$$

$$^{57}\text{Fe} + n \rightarrow {}^{58}\text{Fe} + \gamma$$

$$^{58}\text{Fe} + n \rightarrow {}^{59}\text{Fe} + \gamma.$$

Both ^{57}Fe and ^{58}Fe are stable, but ^{59}Fe is radioactive and decays to ^{59}Co by negative beta decay with a half-life of 45 days. Whether the ^{59}Fe captures another neutron, thereby forming ^{60}Fe, or decays to ^{59}Co depends on the density of neutrons. If the density of neutrons is so low that the ^{59}Fe is unlikely to encounter a neutron within a time of the order of 45 days, then it will decay to ^{59}Co. If the chance of encountering a neutron is high, the ^{59}Fe will be converted into ^{60}Fe and then possibly to ^{61}Fe, ^{62}Fe, and so forth. These nuclei are very rich in neutrons and are thus mov-

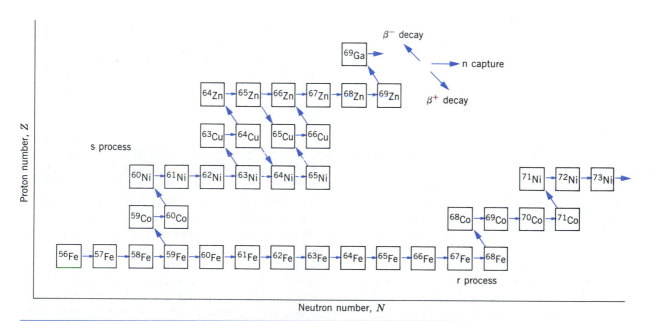

Figure 12   A section of the chart of the nuclides (Fig. 4 of Chapter 54), showing the s- and r-process paths from ^{56}Fe. Many r-process paths are possible, as the short-lived nuclei beta decay; only one such path is shown. All the nuclei in the r-process path are unstable and may beta decay toward the stable nuclei.

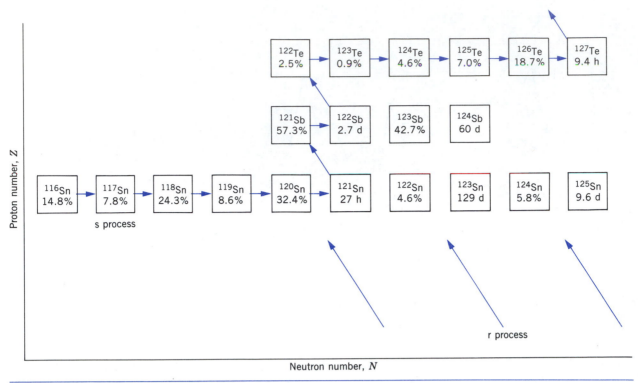

Figure 13   The r- and s-process paths leading to Sn, Sb, and Te isotopes.

ing further from the realm of the stable nuclei shown in Fig. 4 of Chapter 54. They correspondingly have ever shorter half-lives. For example, ^{64}Fe has a half-life of only 2 s. Eventually the half-life becomes so short that no neutron is encountered before the decay occurs, and finally there is a beta decay to the corresponding nuclide of Co, such as is indicated in Fig. 12.

The branch that proceeds through ^{59}Co, which permits time for even long-lived beta decays to occur before the next neutron is captured, is called the *s process* (s for slow). The process in which the density of neutrons is so great that many captures can occur before the beta decay is called the *r process* (r for rapid). These two processes are indicated in Fig. 12.

Of course, the highly unstable nuclei produced in the r process will eventually beta decay toward the stable nuclei, but as you can see from Fig. 12, the decays follow a path different from that of the s process. Consider, for example, the Sn nuclides illustrated in Fig. 13. The s process proceeds through ^{120}Sn, ^{121}Sn, and ^{121}Sb. Because of the beta decay of ^{121}Sn, it cannot capture a neutron through the s process. Therefore ^{122}Sn, which has a natural abundance of 4.6%, cannot be produced in the s process; it must be produced in the r process. The nuclide ^{120}Sn can be produced through both the s and r process, but the nuclide ^{124}Te can be produced *only* through the s process, because the r-process beta decays along the mass-124 line are stopped at stable ^{124}Sn.

In a red giant star, the density of neutrons may be of the

order of 10^{14} per m^3, which is sufficient to maintain the s process. In a supernova explosion the neutron density may be 10–20 orders of magnitude larger, but the high density lasts for a time of only a few seconds. In that time, the r-process chains occur all the way up to the heaviest nuclei, which are then hurled into space and gradually decay to form the stable r-process nuclei. The elements found on Earth beyond mass 56 were produced in first-generation stars, perhaps through the s process or the r process, and the planets of our solar system (and in fact we ourselves) are made of the recycled ashes of burnt-out stars.

56-7   THE AGE OF THE UNIVERSE

In Section 54-7, we discussed the use of radioactive dating methods to determine the age of the Earth. By examining the relative amounts of parent and daughter isotopes in certain radioactive decay processes having half-lives in the range 10^8–10^9 y (for example, ^{238}U $\rightarrow$ ^{206}Pb, ^{87}Rb $\rightarrow$ ^{87}Sr, and ^{40}K $\rightarrow$ ^{40}Ar), it has been determined that the age of the oldest rocks on Earth is about 4.5×10^9 y. An identical value is obtained for meteorites and for rocks from the Moon. We can therefore be fairly certain that this value represents the time since the condensation of the solar system.

We know that the universe must be much older than

this value, because the solar system formed out of elements that were created in the interiors of stars or in supernovas. The present chemical composition of the solar system was determined during a previous era of nucleosynthesis, which occurred in a previous generation of stars. To find the true age of the universe, we must determine the time interval needed for the elements to be produced.

The total time from the Big Bang to the present can be divided into four periods: (1) from the Big Bang until the formation of neutral H and He atoms (t_1); (2) the condensation of galaxies and the formation of first-generation stars (t_2); (3) nucleosynthesis in stars and supernovas, leading to the present chemical elements (t_3); and (4) formation and evolution of the solar system from the debris of earlier stars (t_4). The present age of the universe is just the sum of these four terms:

$$t = t_1 + t_2 + t_3 + t_4. \tag{4}$$

We know from our discussion of the Big Bang cosmology that the time t_1 from the Big Bang until neutral atoms formed is no more than 10^6 y. The time t_2 for galaxies to condense from hydrogen and helium produced in the Big Bang is not precisely known but has been estimated to be in the range $1-2 \times 10^9$ y. Since t_4 is known to be 4.5×10^9 y, the age of the universe can be determined if we can find the time t_3 associated with nucleosynthesis.

This time must be estimated from the relative abundances of the products that remain at the end of nucleosynthesis. For example, consider the isotopes ^{235}U and ^{238}U, which at present have a relative abundance of 0.72% (see the discussion in connection with the natural fission reactor in Section 55-5). Both ^{235}U and ^{238}U have been decaying during the interval since the formation of the solar system. Their ratio 4.5×10^9 years ago is (see Sample Problem 4 of Chapter 55)

$$R(0) = R(t)e^{(\lambda_5 - \lambda_8)t}$$
$$= (0.0072)e^{(0.984 - 0.155)(10^{-9}\,\text{y}^{-1})(4.5 \times 10^9\,\text{y})} = 0.30.$$

During the interval t_3, both isotopes were being formed

more-or-less continuously through the r process, while the relative decay of course also took place. Because of the production of both isotopes during this time, the ratio of their abundances in this period did not change as rapidly as it did during the free decay in the interval t_4; see Fig. 14. Evidence from the uranium abundance suggests that t_3 is in the range $4-9 \times 10^9$ y; similar values result from the analysis of the abundances of other r-process nuclei.

An independent estimate of t_3 comes from the r- and s-process production of the isotope ^{187}Os, which is illustrated in Fig. 15. The isotope ^{187}Re is formed only in the r process, and it decays to ^{187}Os with a half-life of 40×10^9 y, which is in the proper range to serve as a measure of the age of the universe. Comparing the relative amounts of parent ^{187}Re and daughter ^{187}Os should give a measure of the duration of r-process nucleosynthesis. However, ^{187}Os can also be formed through the s process, as shown in Fig. 15. Correcting its observed abundance for the fraction produced in the s process (which can be determined from the abundance of ^{186}Os), we find from the relative amounts of ^{187}Re and ^{187}Os that t_3 is in the range $9-12 \times 10^9$ y. The lower end of this range, 9×10^9 y, is consistent with the upper end of the range for t_3 determined from the uranium abundances, so we choose this value as our estimate for t_3.

Combining these results, we have as our estimate for the age of the universe

$$t = t_1 + t_2 + t_3 + t_4$$
$$= 10^6\,\text{y} + 1-2 \times 10^9\,\text{y} + 9 \times 10^9\,\text{y} + 4.5 \times 10^9\,\text{y}$$
$$= 15 \times 10^9\,\text{y}.$$

This number is somewhat uncertain as a result of the range of values in the estimate for t_3. Taking this uncertainty into account, we obtain

$$t = 10-18 \times 10^9\,\text{y}.$$

Consider the enormous amount of physics contained in this simple result. To determine t_1, we used the cumulative knowledge of particle physics, electromagnetism,

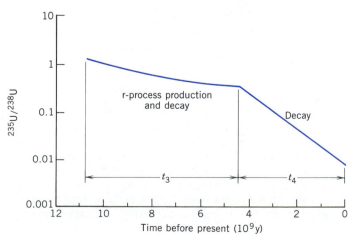

Figure 14 The change in the ^{235}U/^{238}U ratio with time. During the life of the solar system (the time t_4), the ratio changes due only to the relative decays, eventually reaching the present value of 0.0072. During the interval t_3, production by the r process occurs along with the decay. The duration deduced for the interval t_3 depends on the value that we take for the initial ratio, which must be determined from calculation.

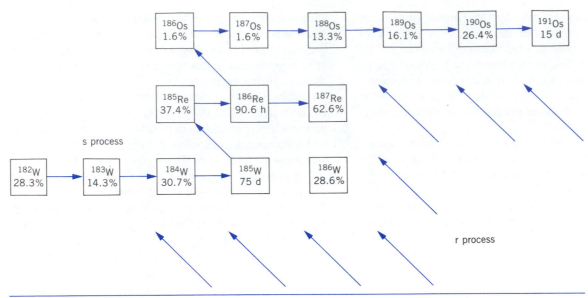

Figure 15   The r- and s-process formation of Re and Os isotopes.

thermal physics, and atomic and nuclear physics to trace the formation of matter as we know it. The interval t_2 is determined from calculations using thermodynamics and gravitational theory to analyze the condensation of cold matter into hot stars. Our estimate for t_3 is based on our knowledge of r-process and s-process nucleosynthesis based on nuclear physics studies in laboratories on Earth, and the interval t_4 is based on further experiments in nuclear physics and research in geochemistry.

Cosmological Determination of the Age

If we make the rough but not quite correct assumption that the universe has been expanding at the same rate since its formation, than the separation d between typical galaxies should be related to the age of the universe roughly according to

$$d = vt,$$

where v is the (assumed constant) speed of separation. Comparing this result with Eq. 2 shows that the age t of the universe is just the inverse of the Hubble parameter:

$$t = H^{-1}. \tag{5}$$

The present best estimate for the Hubble parameter, $H = 67$ (km/s)/Mpc, gives a value for the age of the universe of

$$t = 15 \times 10^9 \text{ y},$$

in remarkably good agreement with the value obtained from the nucleosynthesis calculation. However, the range of uncertainty of the Hubble parameter, 50–100 (km/s)/Mpc, is very large and gives a corresponding range in ages of

$$t = 10 - 19 \times 10^9 \text{ y},$$

which corresponds with the range determined from nucleosynthesis.

Our assumption about the constant separation speed of the galaxies is almost certainly incorrect. The mutual gravitational attraction of the galaxies has been slowing their separation since the Big Bang, so that at earlier times the speed of separation was greater than it is at present. Figure 16 shows a representation of a typical intergalactic separa-

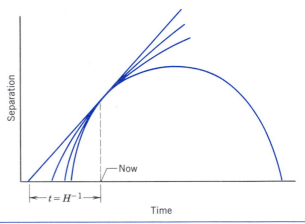

Figure 16   The dependence of a typical galactic separation distance on time during the evolution of the universe according to different models. If the universe has been expanding at a constant rate (straight line), we can extrapolate backward to zero separation (the Big Bang) at a time of H^{-1} before the present. If the expansion has been slowing due to the gravitational interaction (a more reasonable scenario), the Big Bang occurred at a time less than H^{-1} before the present. If the gravitational interaction is strong enough, the expansion may eventually become a contraction.

tion distance as a function of the time. If the "constant-speed" model were valid, the age of the universe would be $t = H^{-1}$. If the speed has been decreasing since the Big Bang, the deduced age depends on the rate of deceleration. Since humans have not been observing long enough to detect any change in the separation rate, we must rely on two indirect methods to determine the deceleration: (1) we can measure the red shifts and thereby deduce the speeds of the most distant (and therefore the oldest) objects we can observe with telescopes, or (2) we can calculate the deceleration based on the gravitational effects of the total amount of matter in the universe.

If the galaxies are decelerating, the most distant objects would be observed to have larger recessional speeds than we would deduce from the Hubble relationship. Unfortunately, our observations of these distant objects are not yet precise enough to indicate any definite deceleration. The second method is also unsuccessful: the amount of matter observed with telescopes (that is, matter that emits some type of electromagnetic radiation) is not even enough to account for the gravitational attraction within galaxies and clusters of galaxies. It is possible that as much as 90% of the matter in the universe is in a form unknown to us, possibly neutrinos (if they have nonzero mass) or other elementary particles left over from the Big Bang or perhaps burnt-out stars. Because of this unknown "dark matter," we cannot make a reliable calculation of the deceleration of the universe.

Various cosmological models have been proposed that give curves in Fig. 16 of differing curvatures and therefore different ages of the universe. For example, some of these give an age that is one-half or two-thirds of H^{-1}, or $5-12 \times 10^9$ y. Although we don't know which (if any) of these models is correct, it seems clear that both the nucleosynthesis and cosmological estimates for the age of the universe are consistent with values in the range $10-15 \times 10^9$ y.

It is a source of great frustration for physicists not to be able to view the history of the universe with more certainty, because our ability to look forward is similarly limited. Will the expansion continue forever, or is there enough matter present to reverse the expansion? Figure 16 shows several possible outcomes. Perhaps cosmologists of later eras will observe the galaxies rushing together as the universe "heats up" and the galaxies come together, eventually reaching a single point (a "Big Crunch") that may be followed by another Big Bang. Or perhaps the expansion continues forever, until the universe is cold and dark. If the solution to this fundamental problem is to be found, it will require vigorous investigations at the forefronts of astrophysics, nuclear physics, and particle physics.

QUESTIONS

1. The ratio of the gravitational force between the electron and the proton in the hydrogen atom to the magnitude of the electromagnetic force of attraction between them is about 10^{-40}. If the gravitational force is so very much weaker than the electromagnetic force, how was it that the gravitational force was discovered first and is so much more apparent to us?

2. What is really meant by an elementary particle? In arriving at an answer, consider such properties as lifetime, mass, size, decays into other particles, fusion to make other particles, and reactions.

3. Why do particle physicists want to accelerate particles to higher and higher energies?

4. Name two particles that have neither mass nor charge. What properties do these particles have?

5. Why do neutrinos leave no tracks in detecting chambers?

6. Neutrinos have no mass (presumably) and travel with the speed of light. How, then, can they carry varying amounts of energy?

7. Do all particles have antiparticles? What about the photon?

8. In the beta decay of an antineutron to an antiproton, is a neutrino or an antineutrino emitted?

9. Photons and neutrinos are alike in that they have zero charge, zero mass, and travel with the speed of light. What are the differences between these particles? How would you produce them? How do you detect them?

10. Explain why we say that the π^0 meson is its own antiparticle.

11. Why can't an electron decay by disintegrating into two neutrinos?

12. Why is the electron stable? That is, why does it not decay spontaneously into other particles?

13. Why cannot a resting electron emit a single gamma-ray photon and disappear? Could a moving electron do so?

14. A neutron is massive enough to decay by the emission of a proton and two neutrinos. Why does it not do so?

15. A positron invariably finds an electron and they annihilate each other. How then can we call a positron a stable particle?

16. What is the mechanism by which two electrons exert forces on each other?

17. Why are particles not grouped into families on the basis of their mass?

18. A particle that responds to the strong force is either a meson or a baryon. You can tell which it is by allowing the particle to decay until only stable end products remain. If there is a proton among these products, the original particle was a baryon. If there is no proton, the original particle was a meson. Explain this classification rule.

19. How many kinds of stable leptons are there? Stable mesons? Stable baryons? In each case, name them.

20. Most particle physics reactions are endothermic, rather than exothermic. Why?

21. What is the lightest strongly interacting particle? What is the heaviest particle unaffected by the strong interaction?

22. For each of the following particles, state which of the four basic forces are influential: (*a*) electron; (*b*) neutrinos; (*c*) neutron; (*d*) pion.

23. Just as x rays are used to discover internal imperfections in a metal casting caused by gas bubbles, so cosmic-ray muons have been used in an attempt to discover hidden burial chambers in Egyptian pyramids. Why were muons used?

24. Are strongly-interacting particles affected by the weak interaction?

25. Do all weak-interaction decays produce neutrinos?

26. Mesons and baryons are each sensitive to the strong force. In what ways are they different?

27. By comparing Tables 3 and 7, point out as many similarities between leptons and quarks as you can and also as many differences.

28. What is the experimental evidence for the existence of quarks?

29. We can explain the "ordinary" world around us with two leptons and two quarks. Name them.

30. The neutral pion has the quark structure $u\bar{u}$ and decays with a mean life of only 8.3×10^{-17} s. The charged pion, on the other hand, has a quark structure of $u\bar{d}$ and decays with a mean life of 2.6×10^{-8} s. Explain, in terms of their quark structure, why the mean life of the neutral pion should be so much shorter than that of the charged pion. (*Hint*: Think of annihilation.)

31. Do leptons contain quarks? Do mesons? Do photons? Do baryons?

32. The Δ^* baryon can have an electric charge of $+2e$ (see Table 5). Based on the quark model, do we expect to find mesons with charge $+2e$? Baryons with charge $-2e$?

33. The Σ^+ baryon decays with a mean life characteristic of the weak interaction (see Table 5). It should be able to decay to the Λ^0 baryon by the strong interaction without changing strangeness. Why doesn't it?

34. Why can't we find the center of the expanding universe? Are we looking for it?

35. Due to the effect of gravity, the rate of expansion of the universe must have decreased in time following the Big Bang. Show that this implies that the age of the universe is less than $1/H$.

36. It is not possible, using telescopes that are sensitive in any part of the electromagnetic spectrum, to "look back" any farther than about 500,000 y from the Big Bang. Why?

37. How does one arrive at the conclusion that visible matter may account for only about 10% of the matter in the universe?

38. Are we always looking back in time as we observe a distant galaxy? Does the direction in which we look make a difference?

39. Can you think of any possible explanation for the expanding universe other than a Big Bang?

PROBLEMS

Section 56-1 Particle Interactions

1. (*a*) An electron and a positron are separated by a distance *r*. Find the ratio of the gravitational force to the electrostatic force between them. What do you conclude from the result concerning the forces acting between particles detected in a bubble chamber or similar detector? (*b*) Repeat for a proton-antiproton pair.

2. Some of the GUTs predict the following possible decay schemes for the proton:

$$p \rightarrow e^+ + \gamma,$$

$$p \rightarrow e^+ + \pi^0.$$

(*a*) Calculate the *Q*-values for these decays. (*b*) Show that the decays do not violate the conservation laws of charge, relativistic energy, or linear momentum. The rest energy of a proton is 938.27 MeV, of an electron is 0.511 MeV, and of a neutral pion is 135 MeV.

3. An electron and a proton are placed a distance apart equal to one Bohr radius a_0. Find the radius *R* of a lead sphere that must be placed directly behind the electron so that the gravitational force on the electron just overcomes the electrostatic attraction between the proton and the electron; see Fig. 17. Assume that Newton's law of gravitation holds, and that the density of the sphere equals the density of lead on the Earth.

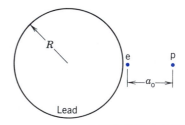

Figure 17   Problem 3.

Section 56-2 Families of Particles

4. A neutral pion decays into two gamma rays: $\pi^0 \rightarrow \gamma + \gamma$. Calculate the wavelengths of the gamma rays produced by the decay of a neutral pion at rest.

5. The rest energy of many short-lived particles cannot be measured directly, but must be inferred from the measured momenta and known rest energies of the decay products. Consider the ρ^0 meson, which decays by $\rho^0 \rightarrow \pi^+ + \pi^-$. Calculate the rest energy of the ρ^0 meson given that each of the oppositely directed momenta of the created pions has magnitude 358.3 MeV/*c*. See Table 4 for the rest energies of the pions.

6. Observations of neutrinos emitted by the supernova SN1987a in the Large Magellanic Cloud, see Fig. 18, place an upper limit on the rest energy of the electron neutrino of

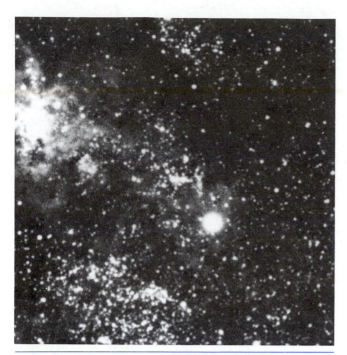

Figure 18 Problem 6.

20 eV. Suppose that the rest energy of the neutrino, rather than being zero, is in fact equal to 20 eV. How much slower than light is a 1.5-MeV neutrino, emitted in a β-decay, moving?

7. A neutral pion has a rest energy of 135 MeV and a mean life of 8.4×10^{-17} s. If it is produced with an initial kinetic energy of 80 MeV and it decays after one mean lifetime, what is the longest possible track that this particle could leave in a bubble chamber? Take relativistic time dilation into account.

8. A positive tau (τ^+, rest energy = 1784 MeV) is moving with 2200 MeV of kinetic energy in a circular path perpendicular to a uniform 1.2-T magnetic field. (a) Calculate the momentum of the tau in kg · m/s. Relativistic effects must be considered. (b) Find the radius of the circular path. (*Hint:* See Section 34-3.)

9. Calculate the range of the weak force between two neighboring protons. Assume that the Z^0 boson is the field particle; see Table 6.

10. Identify the interaction responsible for each of the following decays: (a) $\eta \rightarrow \gamma + \gamma$; (b) $K^+ \rightarrow \mu^+ + \nu_\mu$; (c) $\eta' \rightarrow \eta + \pi^+ + \pi^-$; (d) $K^0 \rightarrow \pi^+ + \pi^-$.

Section 56-3 Conservation Laws

11. What conservation law is violated in each of these proposed decays? (a) $\mu^- \rightarrow e^+ + \nu_\mu + \bar{\nu}_e$; (b) $\mu^+ \rightarrow \pi^+ + \bar{\nu}_\mu$.

12. The reaction $\pi^+ + p \rightarrow p + p + \bar{n}$ proceeds by the strong interaction. By applying the conservation laws, deduce the charge, baryon number, and strangeness of the antineutron.

13. By examining strangeness, determine which of the following decays or reactions proceed via the strong interaction. (a) $K^0 \rightarrow \pi^+ + \pi^-$; (b) $\Lambda^0 + p \rightarrow \Sigma^+ + n$; (c) $\Lambda^0 \rightarrow p + \pi^-$; (d) $K^- + p \rightarrow \Lambda^0 + \pi^0$. See Tables 4 and 5 for values of S.

14. What conservation law is violated in these proposed reactions and decays? (a) $\Lambda^0 \rightarrow p + K^-$; (b) $K^- + p \rightarrow \Lambda^0 + \pi^+$.

15. Use the conservation laws to identify the particle labeled x in the following reactions, which proceed by means of the strong interaction. (a) $p + p \rightarrow p + \Lambda^0 + x$; (b) $p + \bar{p} \rightarrow n + x$; (c) $\pi^- + p \rightarrow \Xi^0 + K^0 + x$.

Section 56-4 The Quark Model

16. Show that, if instead of plotting S versus Q for the spin-$\frac{1}{2}$ baryons in Fig. 3b and for the spin-0 mesons in Fig. 3a, the quantity $Y = B + S$ is plotted against the quantity $T_z = Q - \frac{1}{2}B$, then the hexagonal patterns emerge with the use of nonsloping (perpendicular) axes. (The quantity Y is called *hypercharge* and T_z is related to a quantity called *isospin*.)

17. The quark composition of the proton and the neutron are uud and udd, respectively. What are the quark compositions of (a) the antiproton and (b) the antineutron?

18. From Tables 5 and 7, determine the identity of the baryons formed from the following combinations of quarks. Check your answers with the baryon octet shown in Fig. 3b. (a) ddu; (b) uus; (c) ssd.

19. What quark combinations form (a) Λ^0; (b) Ξ^0?

20. Using the up, down, and strange quarks only, construct, if possible, a baryon (a) with $Q = +1$ and $S = -2$. (b) With $Q = +2$ and $S = 0$.

21. There is no known meson with $Q = +1$ and $S = -1$ or with $Q = -1$ and $S = +1$. Explain why, in terms of the quark model.

22. Analyze the following decays or reactions in terms of the quark content of the particles: (a) $\Sigma^- \rightarrow n + \pi^-$; (b) $K^0 \rightarrow \pi^+ + \pi^-$; (c) $\pi^+ + p \rightarrow \Sigma^+ + K^+$; (d) $\gamma + n \rightarrow \pi^- + p$.

Section 56-5 The Big Bang Cosmology

23. By choosing two points on each line of Fig. 6 and calculating the slopes, verify the given numerical values of the Hubble parameter.

24. If Hubble's law can be extrapolated to very large distances, at what distance would the recessional speed become equal to the speed of light?

25. What is the observed wavelength of the 656.3-nm H_α line of hydrogen emitted by a galaxy at a distance of 2.4×10^8 pc?

26. In the laboratory, one of the lines of sodium is emitted at a wavelength of 590.0 nm. When observing the light from a particular galaxy, however, this line is seen at a wavelength of 602.0 nm. Calculate the distance to the galaxy, assuming that Hubble's law holds.

27. The wavelength of the photons at which a radiation field of temperature T radiates most intensely is given by $\lambda_{max} = (2898 \ \mu m \cdot K)/T$ (see Eq. 4 in Chapter 49). (a) Show that the energy E in MeV of such a photon can be computed from

$$E = (4.28 \times 10^{-10} \ \text{MeV/K})T.$$

(b) At what minimum temperature can this photon create an electron-positron pair?

28. The recessional speeds of galaxies and quasars at great distances are close to the speed of light, so that the relativistic Doppler shift formula (see Eq. 10 in Chapter 42) must be

used. The redshift is reported as z, where $z = \Delta\lambda/\lambda_0$ is the (fractional) red shift. (a) Show that, in terms of z, the recessional speed parameter $\beta = v/c$ is given by

$$\beta = \frac{z^2 + 2z}{z^2 + 2z + 2}.$$

(b) The most distant quasar detected (as of 1990) has $z = 4.43$. Calculate its speed parameter. (c) Find the distance to the quasar, assuming that Hubble's law is valid to these distances.

29. Due to the presence everywhere of the microwave radiation background, the minimum temperature possible of a gas in interstellar or intergalactic space is not 0 K but 2.7 K. This implies that a significant fraction of the molecules in space that possess excited states of low excitation energy may, in fact, be in those excited states. Subsequent de-excitation leads to the emission of radiation that could be detected. Consider a (hypothetical) molecule with just one excited state. (a) What would the excitation energy have to be in order that 23% of the molecules be in the excited state? (*Hint*: see Section 52-6.) (b) Find the wavelength of the photon emitted in a transition to the ground state.

30. Will the universe continue to expand forever? To attack this question, make the (reasonable?) assumption that the recessional speed v of a galaxy a distance r from us is determined only by the matter that lies inside a sphere of radius r centered on us; see Fig. 19. If the total mass inside this sphere is M, the escape speed v_e is given by $v_e = \sqrt{2GM/r}$ (see Sample

Figure 19   Problem 30.

Problem 6 in Chapter 16). (a) Show that the average density ρ inside the sphere must be at least equal to the value given by

$$\rho = 3H^2/8\pi G$$

to prevent unlimited expansion. (b) Evaluate this "critical density" numerically; express your answer in terms of H-atoms/m³. Measurements of the actual density are difficult and complicated by the presence of dark matter.

31. (a) What is the minimum temperature of the universe necessary for the photons to produce $\pi^+ - \pi^-$ pairs? (b) At what age did the universe have this temperature?

Section 56-7   *The Age of the Universe*

32. The existence of dark (i.e., nonluminous) matter in a galaxy (such as our own) can be inferred by determining through observation the variation with distance in the orbital period of revolution of stars about the galactic center. This is then compared with the variation derived on the basis of the distribution of matter as indicated by the luminous material (mostly stars). Any significant deviation implies the existence of dark matter. For example, suppose that the matter (stars, gas, dust) of a particular galaxy, total mass M, is distributed uniformly throughout a sphere of radius R. A star, mass m, is revolving about the center of the galaxy in a circular orbit of radius $r < R$. (a) Show that the orbital speed v of the star is given by

$$v = r\sqrt{GM/R^3},$$

and therefore that the period T of revolution is

$$T = 2\pi\sqrt{R^3/GM},$$

independent of r. (b) What is the corresponding formula for the orbital period assuming that the mass of the galaxy is strongly concentrated toward the center of the galaxy, so that essentially all of the mass is at distances from the center less than r? These considerations applied to our own Milky Way galaxy indicate that substantial quantities of dark matter are present.

33. (a) Show that the number N of photons radiated, per unit area per unit time, by a cavity radiator at temperature T is given by

$$N = \int_0^\infty \frac{R(\lambda)}{hc/\lambda}d\lambda \approx \frac{30\sigma}{\pi^4 k}T^3.$$

(*Hint*: In evaluating the integral, ignore the "1" in the denominator of $R(\lambda)$; see Eq. 6 in Chapter 49. Use the change of variables given in Problem 17(a) in Chapter 49.) (b) To the same approximation, show that the fraction of photons, by number, with energies greater than 2.2 MeV at a temperature of 9×10^8 K is 2.1×10^{-10}.

APPENDIX A

THE INTERNATIONAL SYSTEM OF UNITS (SI)*

THE SI BASE UNITS

Quantity	Name	Symbol	Definition
length	meter	m	". . . the length of the path traveled by light in vacuum in 1/299,792,458 of a second." (1983)
mass	kilogram	kg	". . . this prototype [a certain platinum-iridium cylinder] shall henceforth be considered to be the unit of mass." (1889)
time	second	s	". . . the duration of 9,192,631,770 periods of the radiation corresponding to the transition between the two hyperfine levels of the ground state of the cesium-133 atom." (1967)
electric current	ampere	A	". . . that constant current which, if maintained in two straight parallel conductors of infinite length, of negligible circular cross section, and placed 1 meter apart in vacuum, would produce between these conductors a force equal to 2×10^{-7} newton per meter of length." (1946)
thermodynamic temperature	kelvin	K	". . . the fraction 1/273.16 of the thermodynamic temperature of the triple point of water." (1967)
amount of substance	mole	mol	". . . the amount of substance of a system which contains as many elementary entities as there are atoms in 0.012 kilogram of carbon 12." (1971)
luminous intensity	candela	cd	". . . the luminous intensity, in the perpendicular direction, of a surface of 1/600,000 square meter of a blackbody at the temperature of freezing platinum under a pressure of 101.325 newton per square meter." (1967)

* Adapted from "The International System of Units (SI)," National Bureau of Standards Special Publication 330, 1972 edition. The definitions above were adopted by the General Conference of Weights and Measures, an international body, on the dates shown. In this book we do not use the candela.

SOME SI DERIVED UNITS

Quantity	Name of Unit	Symbol	Equivalent
area	square meter	m^2	
volume	cubic meter	m^3	
frequency	hertz	Hz	s^{-1}
mass density (density)	kilogram per cubic meter	kg/m^3	
speed, velocity	meter per second	m/s	
angular velocity	radian per second	rad/s	
acceleration	meter per second squared	m/s^2	
angular acceleration	radian per second squared	rad/s^2	
force	newton	N	$kg \cdot m/s^2$
pressure	pascal	Pa	N/m^2
work, energy, quantity of heat	joule	J	$N \cdot m$
power	watt	W	J/s
quantity of electricity	coulomb	C	$A \cdot s$
potential difference, electromotive force	volt	V	$N \cdot m/C$
electric field	volt per meter	V/m	N/C
electric resistance	ohm	Ω	V/A
capacitance	farad	F	$A \cdot s/V$
magnetic flux	weber	Wb	$V \cdot s$
inductance	henry	H	$V \cdot s/A$
magnetic field	tesla	T	Wb/m^2, $N/A \cdot m$
entropy	joule per kelvin	J/K	
specific heat capacity	joule per kilogram kelvin	$J/(kg \cdot K)$	
thermal conductivity	watt per meter kelvin	$W/(m \cdot K)$	
radiant intensity	watt per steradian	W/sr	

THE SI SUPPLEMENTARY UNITS

Quantity	Name of Unit	Symbol
plane angle	radian	rad
solid angle	steradian	sr

APPENDIX B

SOME FUNDAMENTAL CONSTANTS OF PHYSICS

Constant	Symbol	Computational Value	Best (1986) value Value[a]	Best (1986) value Uncertainty[b]
Speed of light in a vacuum	c	3.00×10^8 m/s	2.99792458	exact
Elementary charge	e	1.60×10^{-19} C	1.60217733	0.30
Electron rest mass	m_e	9.11×10^{-31} kg	9.1093897	0.59
Permittivity constant	ϵ_0	8.85×10^{-12} F/m	8.85418781762	exact
Permeability constant	μ_0	1.26×10^{-6} H/m	1.25663706143	exact
Electron rest mass[c]	m_e	5.49×10^{-4} u	5.48579902	0.023
Neutron rest mass[c]	m_n	1.0087 u	1.008664904	0.014
Hydrogen atom rest mass[c]	$m(^1H)$	1.0078 u	1.007825035	0.011
Deuterium atom rest mass[c]	$m(^2H)$	2.0141 u	2.014101779	0.012
Helium atom rest mass[c]	$m(^4He)$	4.0026 u	4.00260324	0.012
Electron charge-to-mass ratio	e/m_e	1.76×10^{11} C/kg	1.75881962	0.30
Proton rest mass	m_p	1.67×10^{-27} kg	1.6726231	0.59
Proton-to-electron mass ratio	m_p/m_e	1840	1836.152701	0.020
Neutron rest mass	m_n	1.67×10^{-27} kg	1.6749286	0.59
Muon rest mass	m_μ	1.88×10^{-28} kg	1.8835327	0.61
Planck constant	h	6.63×10^{-34} J·s	6.6260755	0.60
Electron Compton wavelength	λ_e	2.43×10^{-12} m	2.42631058	0.089
Universal gas constant	R	8.31 J/mol·K	8.314510	8.4
Avogadro constant	N_A	6.02×10^{23} mol^{-1}	6.0221367	0.59
Boltzmann constant	k	1.38×10^{-23} J/K	1.3806513	1.8
Molar volume of ideal gas at STP[d]	V_m	2.24×10^{-2} m³/mol	2.2413992	1.7
Faraday constant	F	9.65×10^4 C/mol	9.6485309	0.30
Stefan-Boltzmann constant	σ	5.67×10^{-8} W/m²·K⁴	5.670399	6.8
Rydberg constant	R	1.10×10^7 m^{-1}	1.0973731571	0.00036
Gravitational constant	G	6.67×10^{-11} m³/s²·kg	6.67259	128
Bohr radius	a_0	5.29×10^{-11} m	5.29177249	0.045
Electron magnetic moment	μ_e	9.28×10^{-24} J/T	9.2847700	0.34
Proton magnetic moment	μ_p	1.41×10^{-26} J/T	1.41060761	0.34
Bohr magneton	μ_B	9.27×10^{-24} J/T	9.2740154	0.34
Nuclear magneton	μ_N	5.05×10^{-27} J/T	5.0507865	0.34
Fine structure constant	α	1/137	1/137.0359895	0.045
Magnetic flux quantum	Φ_0	2.07×10^{-15} Wb	2.06783461	0.30
Quantized Hall resistance	R_H	25800 Ω	25812.8056	0.045

[a] Same unit and power of ten as the computational value.
[b] Parts per million.
[c] Mass given in unified atomic mass units, where 1 u = $1.6605402 \times 10^{-27}$ kg.
[d] STP—standard temperature and pressure = 0°C and 1.0 bar.

APPENDIX C

SOME ASTRONOMICAL DATA

THE SUN, THE EARTH, AND THE MOON

Property	Sun[a]	Earth	Moon
Mass (kg)	1.99×10^{30}	5.98×10^{24}	7.36×10^{22}
Mean radius (m)	6.96×10^{8}	6.37×10^{6}	1.74×10^{6}
Mean density (kg/m³)	1410	5520	3340
Surface gravity (m/s²)	274	9.81	1.67
Escape velocity (km/s)	618	11.2	2.38
Period of rotation[c] (d)	26–37[b]	0.997	27.3
Mean orbital radius (km)	2.6×10^{17d}	1.50×10^{8e}	3.82×10^{5f}
Orbital period	2.4×10^{8} y[d]	1.00 y[e]	27.3 d[f]

[a] The Sun radiates energy at the rate of 3.90×10^{26} W; just outside the Earth's atmosphere solar energy is received, assuming normal incidence, at the rate of 1380 W/m².

[b] The Sun—a ball of gas—does not rotate as a rigid body. Its rotational period varies between 26 d at the equator and 37 d at the poles.

[c] Measured with respect to the distant stars.

[d] About the galactic center.

[e] About the Sun.

[f] About the Earth.

SOME PROPERTIES OF THE PLANETS

	Mercury	Venus	Earth	Mars	Jupiter	Saturn	Uranus	Neptune	Pluto
Mean distance from Sun (10^6 km)	57.9	108	150	228	778	1,430	2,870	4,500	5,900
Period of revolution (y)	0.241	0.615	1.00	1.88	11.9	29.5	84.0	165	248
Period of rotation[a] (d)	58.7	243[b]	0.997	1.03	0.409	0.426	0.451[b]	0.658	6.39
Orbital speed (km/s)	47.9	35.0	29.8	24.1	13.1	9.64	6.81	5.43	4.74
Inclination of axis to orbit	0.0°	2.6°	23.5°	24.0°	3.08°	26.7°	82.1°	28.8°	65°
Inclination of orbit to Earth's orbit	7.00°	3.39°	—	1.85°	1.30°	2.49°	0.77°	1.77°	17.2°
Eccentricity of orbit	0.206	0.0068	0.0167	0.0934	0.0485	0.0556	0.0472	0.0086	0.250
Equatorial diameter (km)	4,880	12,100	12,800	6,790	143,000	120,000	51,800	49,500	3,400
Mass (Earth = 1)	0.0558	0.815	1.000	0.107	318	95.1	14.5	17.2	0.002
Mean density (g/cm³)	5.60	5.20	5.52	3.95	1.31	0.704	1.21	1.67	0.5(?)
Surface gravity[c] (m/s²)	3.78	8.60	9.78	3.72	22.9	9.05	7.77	11.0	0.03
Escape speed (km/s)	4.3	10.3	11.2	5.0	59.5	35.6	21.2	23.6	1.3
Known satellites	0	0	1	2	16 + rings	19 + rings	15 + rings	8 + rings	1

[a] Measured with respect to the distant stars.

[b] The sense of rotation is opposite to that of the orbital motion.

[c] Measured at the planet's equator.

APPENDIX D

PROPERTIES OF THE ELEMENTS

Element	Symbol	Atomic number, Z	Molar mass (g/mol)	Density (g/cm³) at 20°C	Melting point (°C)	Boiling point (°C)	Specific heat (J/g·C°) at 25°C
Actinium	Ac	89	(227)	—	1050	3200	0.092
Aluminum	Al	13	26.9815	2.699	660	2467	0.900
Americium	Am	95	(243)	13.7	994	2607	—
Antimony	Sb	51	121.75	6.69	630.5	1750	0.205
Argon	Ar	18	39.948	1.6626×10^{-3}	−189.2	−185.7	0.523
Arsenic	As	33	74.9216	5.72	817 (28 at.)	613	0.331
Astatine	At	85	(210)	—	302	337	—
Barium	Ba	56	137.33	3.5	725	1640	0.205
Berkelium	Bk	97	(247)	—	—	—	—
Beryllium	Be	4	9.0122	1.848	12.78	2970	1.83
Bismuth	Bi	83	208.980	9.75	271.3	1560	0.122
Boron	B	5	10.811	2.34	20.79	2550	1.11
Bromine	Br	35	79.909	3.12 (liquid)	−7.2	58	0.293
Cadmium	Cd	48	112.41	8.65	320.9	765	0.226
Calcium	Ca	20	40.08	1.55	839	1484	0.624
Californium	Cf	98	(251)	—	—	—	—
Carbon	C	6	12.011	2.25	3550	—	0.691
Cerium	Ce	58	140.12	6.768	798	3443	0.188
Cesium	Cs	55	132.905	1.873	28.40	6.69	0.243
Chlorine	Cl	17	35.453	3.214×10^{-3} (0°C)	−101	−34.6	0.486
Chromium	Cr	24	51.996	7.19	1857	2672	0.448
Cobalt	Co	27	58.9332	8.85	1495	2870	0.423
Copper	Cu	29	63.54	8.96	1083.4	2567	0.385
Curium	Cm	96	(247)	—	1340	—	—
Dysprosium	Dy	66	162.50	8.55	1412	2567	0.172
Einsteinium	Es	99	(252)	—	—	—	—
Erbium	Er	68	167.26	9.07	1529	2868	0.167
Europium	Eu	63	151.96	5.245	822	1527	0.163
Fermium	Fm	100	(257)	—	—	—	—
Fluorine	F	9	18.9984	1.696×10^{-3} (0°C)	−219.6	−188.2	0.753
Francium	Fr	87	(223)	—	(27)	(677)	—
Gadolinium	Gd	64	157.25	7.90	1313	3273	0.234
Gallium	Ga	31	69.72	5.907	29.78	2403	0.377
Germanium	Ge	32	72.61	5.323	937.4	2830	0.322
Gold	Au	79	196.967	19.32	1064.43	2808	0.131
Hafnium	Hf	72	178.49	13.31	2227	4602	0.144
Helium	He	2	4.0026	0.1664×10^{-3}	−272.2	−268.9	5.23
Holmium	Ho	67	164.930	8.79	1474	2700	0.165
Hydrogen	H	1	1.00797	0.08375×10^{-3}	−259.34	−252.87	14.4
Indium	In	49	114.82	7.31	156.6	2080	0.233
Iodine	I	53	126.9044	4.94	113.5	184.35	0.218
Iridium	Ir	77	192.2	22.5	2410	4130	0.130
Iron	Fe	26	55.847	7.87	1535	2750	0.447
Krypton	Kr	36	83.80	3.488×10^{-3}	−156.6	−152.3	0.247
Lanthanum	La	57	138.91	6.145	918	3464	0.195
Lawrencium	Lr	103	(260)	—	—	—	—

Element	Symbol	Atomic number, Z	Molar mass (g/mol)	Density (g/cm³) at 20°C	Melting point (°C)	Boiling point (°C)	Specific heat (J/g·C°) at 25°C
Lead	Pb	82	207.19	11.36	327.50	1740	0.129
Lithium	Li	3	6.939	0.534	180.54	1342	3.58
Lutetium	Lu	71	174.97	9.84	1663	3402	0.155
Magnesium	Mg	12	24.305	1.74	649	1090	1.03
Manganese	Mn	25	54.9380	7.43	1244	1962	0.481
Mendelevium	Md	101	(258)	—	—	—	—
Mercury	Hg	80	200.59	13.55	−38.87	357	0.138
Molybdenum	Mo	42	95.94	10.22	2617	4612	0.251
Neodymium	Nd	60	144.24	7.00	1021	3074	0.188
Neon	Ne	10	20.180	0.8387×10^{-3}	−248.67	−246.0	1.03
Neptunium	Np	93	(237)	20.25	640	3902	1.26
Nickel	Ni	28	58.69	8.902	1453	2732	0.444
Niobium	Nb	41	92.906	8.57	2468	4742	0.264
Nitrogen	N	7	14.0067	1.1649×10^{-3}	−210	−195.8	1.03
Nobelium	No	102	(259)	—	—	—	—
Osmium	Os	76	190.2	22.57	3045	5027	0.130
Oxygen	O	8	15.9994	1.3318×10^{-3}	−218.4	−183.0	0.913
Palladium	Pd	46	106.4	12.02	1554	3140	0.243
Phosphorus	P	15	30.9738	1.83	44.25	280	0.741
Platinum	Pt	78	195.09	21.45	1772	3827	0.134
Plutonium	Pu	94	(244)	19.84	641	3232	0.130
Polonium	Po	84	(209)	9.24	254	962	—
Potassium	K	19	39.098	0.86	63.25	760	0.758
Praseodymium	Pr	59	140.907	6.773	931	3520	0.197
Promethium	Pm	61	(145)	7.264	1042	(3000)	—
Protactinium	Pa	91	(231)	—	1600	—	—
Radium	Ra	88	(226)	5.0	700	1140	—
Radon	Rn	86	(222)	9.96×10^{-3} (0°C)	−71	−61.8	0.092
Rhenium	Re	75	186.2	21.04	3180	5627	0.134
Rhodium	Rh	45	102.905	12.44	1965	3727	0.243
Rubidium	Rb	37	85.47	1.53	38.89	686	0.364
Ruthenium	Ru	44	101.107	12.2	2310	3900	0.239
Samarium	Sm	62	150.35	7.49	1074	1794	0.197
Scandium	Sc	21	44.956	2.99	1541	2836	0.569
Selenium	Se	34	78.96	4.79	217	685	0.318
Silicon	Si	14	28.086	2.33	1410	2355	0.712
Silver	Ag	47	107.68	10.49	961.9	2212	0.234
Sodium	Na	11	22.9898	0.9712	97.81	882.9	1.23
Strontium	Sr	38	87.62	2.54	769	1384	0.737
Sulfur	S	16	32.066	2.07	112.8	444.6	0.707
Tantalum	Ta	73	180.948	16.6	2996	5425	0.138
Technetium	Tc	43	(98)	11.46	2172	4877	0.209
Tellurium	Te	52	127.60	6.24	449.5	990	0.201
Terbium	Tb	65	158.924	8.25	1357	3230	0.180
Thallium	Tl	81	204.38	11.85	304	1457	0.130
Thorium	Th	90	(232)	11.72	1750	(3850)	0.117
Thulium	Tm	69	168.934	9.31	1545	1950	0.159
Tin	Sn	50	118.71	7.31	231.97	2270	0.226
Titanium	Ti	22	4788	4.54	1660	3287	0.523
Tungsten	W	74	183.85	19.3	3410	5660	0.134
Uranium	U	92	(238)	19.07	1132	3818	0.117
Vanadium	V	23	50.942	6.1	1890	3380	0.490
Xenon	Xe	54	131.30	5.495×10^{-3}	−111.79	−108	0.159
Ytterbium	Yb	70	173.04	6.966	819	1196	0.155
Yttrium	Y	39	88.905	4.469	1552	5338	0.297
Zinc	Zn	30	65.37	7.133	419.58	907	0.389
Zirconium	Zr	40	91.22	6.506	1852	4377	0.276

The values in parentheses in the column of atomic masses are the mass numbers of the longest-lived isotopes of those elements that are radioactive.
Melting points and boiling points in parentheses are uncertain.
All the physical properties are given for a pressure of one atmosphere except where otherwise specified.
The data for gases are valid only when these are in their usual molecular state, such as H_2, He, O_2, Ne, etc. The specific heats of the gases are the values at constant pressure.

Source: Handbook of Chemistry and Physics, 71st edition (CRC Press, 1990).

APPENDIX E

PERIODIC TABLE
OF THE
ELEMENTS

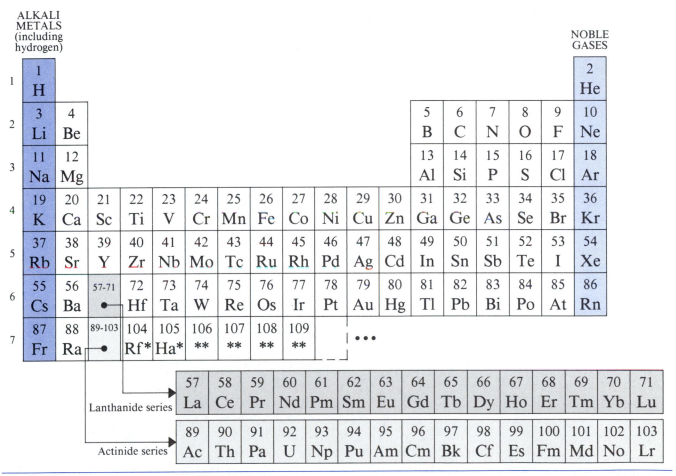

	ALKALI METALS (including hydrogen)																NOBLE GASES

1 1 H

2 3 Li | 4 Be | | | | | | | | | | | 5 B | 6 C | 7 N | 8 O | 9 F | 10 Ne

3 11 Na | 12 Mg | | | | | | | | | | | 13 Al | 14 Si | 15 P | 16 S | 17 Cl | 18 Ar

4 19 K | 20 Ca | 21 Sc | 22 Ti | 23 V | 24 Cr | 25 Mn | 26 Fe | 27 Co | 28 Ni | 29 Cu | 30 Zn | 31 Ga | 32 Ge | 33 As | 34 Se | 35 Br | 36 Kr

5 37 Rb | 38 Sr | 39 Y | 40 Zr | 41 Nb | 42 Mo | 43 Tc | 44 Ru | 45 Rh | 46 Pd | 47 Ag | 48 Cd | 49 In | 50 Sn | 51 Sb | 52 Te | 53 I | 54 Xe

6 55 Cs | 56 Ba | 57-71 | 72 Hf | 73 Ta | 74 W | 75 Re | 76 Os | 77 Ir | 78 Pt | 79 Au | 80 Hg | 81 Tl | 82 Pb | 83 Bi | 84 Po | 85 At | 86 Rn

7 87 Fr | 88 Ra | 89-103 | 104 Rf* | 105 Ha* | 106 ** | 107 ** | 108 ** | 109 ** | · · ·

Lanthanide series: 57 La | 58 Ce | 59 Pr | 60 Nd | 61 Pm | 62 Sm | 63 Eu | 64 Gd | 65 Tb | 66 Dy | 67 Ho | 68 Er | 69 Tm | 70 Yb | 71 Lu

Actinide series: 89 Ac | 90 Th | 91 Pa | 92 U | 93 Np | 94 Pu | 95 Am | 96 Cm | 97 Bk | 98 Cf | 99 Es | 100 Fm | 101 Md | 102 No | 103 Lr

* The names of these elements (Rutherfordium and Hahnium) have not been accepted because of conflicting claims of discovery. A group in the USSR has proposed the names Kurchatovium and Nielsbohrium.

** Discovery of these elements has been reported but names for them have not yet been adopted.

APPENDIX F

ELEMENTARY PARTICLES

1. THE FUNDAMENTAL PARTICLES

LEPTONS

Particle	Symbol	Anti-particle	Charge (e)	Spin $(h/2\pi)$	Rest energy (MeV)	Mean life (s)	Typical decay products
Electron	e^-	e^+	-1	$1/2$	0.511	∞	
Electron neutrino	ν_e	$\bar{\nu}_e$	0	$1/2$	<0.00002	∞	
Muon	μ^-	μ^+	-1	$1/2$	105.7	2.2×10^{-6}	$e^- + \bar{\nu}_e + \nu_\mu$
Muon neutrino	ν_μ	$\bar{\nu}_\mu$	0	$1/2$	<0.3	∞	
Tau	τ^-	τ^+	-1	$1/2$	1784	3.0×10^{-13}	$\mu^- + \bar{\nu}_\mu + \nu_\tau$
Tau neutrino	ν_τ	$\bar{\nu}_\tau$	0	$1/2$	<40	∞	

QUARKS

Flavor	Symbol	Antiparticle	Charge (e)	Spin $(h/2\pi)$	Rest energy[a] (MeV)	Other property
Up	u	$\bar{u}$	$+2/3$	$1/2$	300	$C = S = T = B = 0$
Down	d	$\bar{d}$	$-1/3$	$1/2$	300	$C = S = T = B = 0$
Charm	c	$\bar{c}$	$+2/3$	$1/2$	1500	Charm $(C) = +1$
Strange	s	$\bar{s}$	$-1/3$	$1/2$	500	Strangeness $(S) = -1$
Top[b]	t	$\bar{t}$	$+2/3$	$1/2$	$>40,000$	Topness $(T) = +1$
Bottom	b	$\bar{b}$	$-1/3$	$1/2$	4700	Bottomness $(B) = -1$

FIELD PARTICLES

Particle	Symbol	Interaction	Charge (e)	Spin $(h/2\pi)$	Rest energy (GeV)
Graviton[b]		Gravity	0	2	0
Weak boson	W^+, W^-	Weak	± 1	1	80.6
Weak boson	Z^0	Weak	0	1	91.2
Photon	γ	Electromagnetic	0	1	0
Gluon	g	Strong (color)	0	1	0

2. SOME COMPOSITE PARTICLES

BARYONS

Particle	Symbol	Quark content	Anti-particle	Charge (e)	Spin ($h/2\pi$)	Rest energy (MeV)	Mean life (s)	Typical decay
Proton	p	uud	$\bar{\text{p}}$	+1	1/2	938	$>10^{40}$	$\pi^0 + e^+$ (?)
Neutron	n	udd	$\bar{\text{n}}$	0	1/2	940	889	$p + e^- + \bar{\nu}_e$
Lambda	Λ^0	uds	$\overline{\Lambda^0}$	0	1/2	1116	2.6×10^{-10}	$p + \pi^-$
Omega	Ω^-	sss	$\overline{\Omega^-}$	−1	3/2	1673	8.2×10^{-11}	$\Lambda^0 + K^-$
Delta	Δ^{++}	uuu	$\overline{\Delta^{++}}$	+2	3/2	1232	5.7×10^{-24}	$p + \pi^+$
Charmed lambda	Λ_c^+	udc	$\overline{\Lambda_c^+}$	+1	1/2	2285	1.9×10^{-13}	$\Lambda^0 + \pi^+$

MESONS

Particle	Symbol	Quark content	Anti-particle	Charge (e)	Spin ($h/2\pi$)	Rest energy (MeV)	Mean life (s)	Typical decay
Pion	π^+	$u\bar{d}$	π^-	+1	0	140	2.6×10^{-8}	$\mu^+ + \nu_\mu$
Pion	π^0	$u\bar{u} + d\bar{d}$	π^0	0	0	135	8.4×10^{-17}	$\gamma + \gamma$
Kaon	K^+	$u\bar{s}$	K^-	+1	0	494	1.2×10^{-8}	$\mu^+ + \nu_\mu$
Kaon	K^0	$d\bar{s}$	$\overline{K^0}$	0	0	498	0.9×10^{-10}	$\pi^+ + \pi^-$
Rho	ρ^+	$u\bar{d}$	ρ^-	+1	1	768	4.5×10^{-24}	$\pi^+ + \pi^-$
D-meson	D^+	$c\bar{d}$	D^-	+1	0	1869	1.1×10^{-12}	$K^- + \pi^+ + \pi^+$
Psi	ψ	$c\bar{c}$	ψ	0	1	3097	1.0×10^{-20}	$e^+ + e^-$
B-meson	B^+	$u\bar{b}$	B^-	+1	0	5278	1.2×10^{-12}	$D^- + \pi^+ + \pi^+$
Upsilon	Y	$b\bar{b}$	Y	0	1	9460	1.3×10^{-20}	$e^+ + e^-$

[a] The rest energies listed for the quarks are not those associated with free quarks; since no free quarks have yet been observed, measuring their rest energies in the free state has not yet been possible. The tabulated values are effective rest energies corresponding to *constituent* quarks, those bound in composite particles.

[b] Particles expected to exist but not yet observed.

Source: "Review of Particle Properties," *Physics Letters B,* vol. 239 (April 1990).

APPENDIX G

CONVERSION FACTORS

Conversion factors may be read directly from the tables. For example, 1 degree = 2.778×10^{-3} revolutions, so $16.7° = 16.7 \times 2.778 \times 10^{-3}$ rev. The SI quantities are capitalized. Adapted in part from G. Shortley and D. Williams, *Elements of Physics,* Prentice-Hall, Englewood Cliffs, NJ, 1971.

PLANE ANGLE

	°	′	″	RADIAN	rev
1 degree =	1	60	3600	1.745×10^{-2}	2.778×10^{-3}
1 minute =	1.667×10^{-2}	1	60	2.909×10^{-4}	4.630×10^{-5}
1 second =	2.778×10^{-4}	1.667×10^{-2}	1	4.848×10^{-6}	7.716×10^{-7}
1 RADIAN =	57.30	3438	2.063×10^{5}	1	0.1592
1 revolution =	360	2.16×10^{4}	1.296×10^{6}	6.283	1

SOLID ANGLE

1 sphere = 4π steradians = 12.57 steradians

LENGTH

	cm	METER	km	in.	ft	mi
1 centimeter =	1	10^{-2}	10^{-5}	0.3937	3.281×10^{-2}	6.214×10^{-6}
1 METER =	100	1	10^{-3}	39.37	3.281	6.214×10^{-4}
1 kilometer =	10^{5}	1000	1	3.937×10^{4}	3281	0.6214
1 inch =	2.540	2.540×10^{-2}	2.540×10^{-5}	1	8.333×10^{-2}	1.578×10^{-5}
1 foot =	30.48	0.3048	3.048×10^{-4}	12	1	1.894×10^{-4}
1 mile =	1.609×10^{5}	1609	1.609	6.336×10^{4}	5280	1

1 angström = 10^{-10} m
1 nautical mile = 1852 m
  = 1.151 miles = 6076 ft
1 fermi = 10^{-15} m

1 light-year = 9.460×10^{12} km
1 parsec = 3.084×10^{13} km
1 fathom = 6 ft
1 Bohr radius = 5.292×10^{-11} m

1 yard = 3 ft
1 rod = 16.5 ft
1 mil = 10^{-3} in.
1 nm = 10^{-9} m

AREA

	METER^2	cm^2	ft^2	in.^{2}
1 SQUARE METER =	1	10^{4}	10.76	1550
1 square centimeter =	10^{-4}	1	1.076×10^{-3}	0.1550
1 square foot =	9.290×10^{-2}	929.0	1	144
1 square inch =	6.452×10^{-4}	6.452	6.944×10^{-3}	1

1 square mile = 2.788×10^{7} ft^2 = 640 acres
1 barn = 10^{-28} m^2

1 acre = 43,560 ft^2
1 hectare = 10^{4} m^2 = 2.471 acre

VOLUME

	METER^3	cm^3	L	ft^3	in.^{3}
1 CUBIC METER =	1	10^6	1000	35.31	6.102×10^4
1 cubic centimeter =	10^{-6}	1	1.000×10^{-3}	3.531×10^{-5}	6.102×10^{-2}
1 liter =	1.000×10^{-3}	1000	1	3.531×10^{-2}	61.02
1 cubic foot =	2.832×10^{-2}	2.832×10^4	28.32	1	1728
1 cubic inch =	1.639×10^{-5}	16.39	1.639×10^{-2}	5.787×10^{-4}	1

1 U.S. fluid gallon = 4 U.S. fluid quarts = 8 U.S. pints = 128 U.S. fluid ounces = 231 in.^{3}
1 British imperial gallon = 277.4 in^3 = 1.201 U.S. fluid gallons

MASS

	g	KILOGRAM	slug	u	oz	lb	ton
1 gram =	1	0.001	6.852×10^{-5}	6.022×10^{23}	3.527×10^{-2}	2.205×10^{-3}	1.102×10^{-6}
1 KILOGRAM =	1000	1	6.852×10^{-2}	6.022×10^{26}	35.27	2.205	1.102×10^{-3}
1 slug =	1.459×10^4	14.59	1	8.786×10^{27}	514.8	32.17	1.609×10^{-2}
1 u =	1.661×10^{-24}	1.661×10^{-27}	1.138×10^{-28}	1	5.857×10^{-26}	3.662×10^{-27}	1.830×10^{-30}
1 ounce =	28.35	2.835×10^{-2}	1.943×10^{-3}	1.718×10^{25}	1	6.250×10^{-2}	3.125×10^{-5}
1 pound =	453.6	0.4536	3.108×10^{-2}	2.732×10^{26}	16	1	0.0005
1 ton =	9.072×10^5	907.2	62.16	5.463×10^{29}	3.2×10^4	2000	1

1 metric ton = 1000 kg
Quantities in the colored areas are not mass units but are often used as such. When we write, for example, 1 kg "=" 2.205 lb this means that a kilogram is a *mass* that *weighs* 2.205 pounds under standard condition of gravity ($g = 9.80665$ m/s^2).

DENSITY

	slug/ft^3	KILOGRAM/METER^3	g/cm^3	lb/ft^3	lb/in.^{3}
1 slug per ft^3	1	515.4	0.5154	32.17	1.862×10^{-2}
1 KILOGRAM per METER^3 =	1.940×10^{-3}	1	0.001	6.243×10^{-2}	3.613×10^{-5}
1 gram per cm^3 =	1.940	1000	1	62.43	3.613×10^{-2}
1 pound per ft^3 =	3.108×10^{-2}	16.02	1.602×10^{-2}	1	5.787×10^{-4}
1 pound per in.^{3} =	53.71	2.768×10^4	27.68	1728	1

Quantities in the colored areas are weight densities and, as such, are dimensionally different from mass densities. See note for mass table.

TIME

	y	d	h	min	SECOND
1 year =	1	365.25	8.766×10^3	5.259×10^5	3.156×10^7
1 day =	2.738×10^{-3}	1	24	1440	8.640×10^4
1 hour =	1.141×10^{-4}	4.167×10^{-2}	1	60	3600
1 minute =	1.901×10^{-6}	6.944×10^{-4}	1.667×10^{-2}	1	60
1 SECOND =	3.169×10^{-8}	1.157×10^{-5}	2.778×10^{-4}	1.667×10^{-2}	1

SPEED

	ft/s	km/h	METER/SECOND	mi/h	cm/s
1 foot per second =	1	1.097	0.3048	0.6818	30.48
1 kilometer per hour =	0.9113	1	0.2778	0.6214	27.78
1 METER per SECOND =	3.281	3.6	1	2.237	100
1 mile per hour =	1.467	1.609	0.4470	1	44.70
1 centimeter per second =	3.281×10^{-2}	3.6×10^{-2}	0.01	2.237×10^{-2}	1

1 knot = 1 nautical mi/h = 1.688 ft/s   1 mi/min = 88.00 ft/s = 60.00 mi/h

FORCE

	dyne	NEWTON	lb	pdl	gf	kgf
1 dyne =	1	10^{-5}	2.248×10^{-6}	7.233×10^{-5}	1.020×10^{-3}	1.020×10^{-6}
1 NEWTON =	10^5	1	0.2248	7.233	102.0	0.1020
1 pound =	4.448×10^5	4.448	1	32.17	453.6	0.4536
1 poundal =	1.383×10^4	0.1383	3.108×10^{-2}	1	14.10	1.410×10^{-2}
1 gram-force =	980.7	9.807×10^{-3}	2.205×10^{-3}	7.093×10^{-2}	1	0.001
1 kilogram-force =	9.807×10^5	9.807	2.205	70.93	1000	1

Quantities in the colored areas are not force units but are often used as such. For instance, if we write 1 gram-force "=" 980.7 dynes, we mean that a gram-mass experiences a force of 980.7 dynes under standard conditions of gravity ($g = 9.80665$ m/s²).

ENERGY, WORK, HEAT

	Btu	erg	ft·lb	hp·h	JOULE	cal	kW·h	eV	MeV	kg	u
1 British thermal unit =	1	1.055×10^{10}	777.9	3.929×10^{-4}	1055	252.0	2.930×10^{-4}	6.585×10^{21}	6.585×10^{15}	1.174×10^{-14}	7.070×10^{12}
1 erg =	9.481×10^{-11}	1	7.376×10^{-8}	3.725×10^{-14}	10^{-7}	2.389×10^{-8}	2.778×10^{-14}	6.242×10^{11}	6.242×10^5	1.113×10^{-24}	670.2
1 foot-pound =	1.285×10^{-3}	1.356×10^7	1	5.051×10^{-7}	1.356	0.3238	3.766×10^{-7}	8.464×10^{18}	8.464×10^{12}	1.509×10^{-17}	9.037×10^9
1 horsepower-hour =	2545	2.685×10^{13}	1.980×10^6	1	2.685×10^6	6.413×10^5	0.7457	1.676×10^{25}	1.676×10^{19}	2.988×10^{-11}	1.799×10^{16}
1 JOULE =	9.481×10^{-4}	10^7	0.7376	3.725×10^{-7}	1	0.2389	2.778×10^{-7}	6.242×10^{18}	6.242×10^{12}	1.113×10^{-17}	6.702×10^9
1 calorie =	3.969×10^{-3}	4.186×10^7	3.088	1.560×10^{-6}	4.186	1	1.163×10^{-6}	2.613×10^{19}	2.613×10^{13}	4.660×10^{-17}	2.806×10^{10}
1 kilowatt-hour =	3413	3.6×10^{13}	2.655×10^6	1.341	3.6×10^6	8.600×10^5	1	2.247×10^{25}	2.247×10^{19}	4.007×10^{-11}	2.413×10^{16}
1 electron volt =	1.519×10^{-22}	1.602×10^{-12}	1.182×10^{-19}	5.967×10^{-26}	1.602×10^{-19}	3.827×10^{-20}	4.450×10^{-26}	1	10^{-6}	1.783×10^{-36}	1.074×10^{-9}
1 million electron volts =	1.519×10^{-16}	1.602×10^{-6}	1.182×10^{-13}	5.967×10^{-20}	1.602×10^{-13}	3.827×10^{-14}	4.450×10^{-20}	10^6	1	1.783×10^{-30}	1.074×10^{-3}
1 kilogram =	8.521×10^{13}	8.987×10^{23}	6.629×10^{16}	3.348×10^{10}	8.987×10^{16}	2.146×10^{16}	2.497×10^{10}	5.610×10^{35}	5.610×10^{29}	1	6.022×10^{26}
1 unified atomic mass unit =	1.415×10^{-13}	1.492×10^{-3}	1.101×10^{-10}	5.559×10^{-17}	1.492×10^{-10}	3.564×10^{-11}	4.146×10^{-17}	9.32×10^8	932.0	1.661×10^{-27}	1

Quantities in the colored areas are not properly energy units but are included for convenience. They arise from the relativistic mass-energy equivalence formula $E = mc^2$ and represent the energy equivalent of a mass of one kilogram or one unified atomic mass unit (u).

PRESSURE

	atm	dyne/cm²	inch of water	cm Hg	PASCAL	lb/in.²	lb/ft²
1 atmosphere =	1	1.013×10^6	406.8	76	1.013×10^5	14.70	2116
1 dyne per cm² =	9.869×10^{-7}	1	4.015×10^{-4}	7.501×10^{-5}	0.1	1.405×10^{-5}	2.089×10^{-3}
1 inch of water[a] at 4°C =	2.458×10^{-3}	2491	1	0.1868	249.1	3.613×10^{-2}	5.202
1 centimeter of mercury[a] at 0°C =	1.316×10^{-2}	1.333×10^4	5.353	1	1333	0.1934	27.85
1 PASCAL =	9.869×10^{-6}	10	4.015×10^{-3}	7.501×10^{-4}	1	1.450×10^{-4}	2.089×10^{-2}
1 pound per in.² =	6.805×10^{-2}	6.895×10^4	27.68	5.171	6.895×10^3	1	144
1 pound per ft² =	4.725×10^{-4}	478.8	0.1922	3.591×10^{-2}	47.88	6.944×10^{-3}	1

[a] Where the acceleration of gravity has the standard value 9.80665 m/s².

1 bar = 10^6 dyne/cm² = 0.1 MPa   1 millibar = 10^3 dyne/cm² = 10^2 Pa   1 torr = 1 millimeter of mercury

POWER

	Btu/h	ft·lb/s	hp	cal/s	kW	WATT
1 British thermal unit per hour =	1	0.2161	3.929×10^{-4}	6.998×10^{-2}	2.930×10^{-4}	0.2930
1 foot-pound per second =	4.628	1	1.818×10^{-3}	0.3239	1.356×10^{-3}	1.356
1 horsepower =	2545	550	1	178.1	0.7457	745.7
1 calorie per second =	14.29	3.088	5.615×10^{-3}	1	4.186×10^{-3}	4.186
1 kilowatt =	3413	737.6	1.341	238.9	1	1000
1 WATT =	3.413	0.7376	1.341×10^{-3}	0.2389	0.001	1

MAGNETIC FLUX

	maxwell	WEBER
1 maxwell =	1	10^{-8}
1 WEBER =	10^8	1

MAGNETIC FIELD

	gauss	TESLA	milligauss
1 gauss =	1	10^{-4}	1000
1 TESLA =	10^4	1	10^7
1 milligauss =	0.001	10^{-7}	1

1 tesla = 1 weber/meter²

APPENDIX H

MATHEMATICAL FORMULAS

GEOMETRY

Circle of radius r: circumference $= 2\pi r$; area $= \pi r^2$.

Sphere of radius r: area $= 4\pi r^2$; volume $= \frac{4}{3}\pi r^3$.

Right circular cylinder of radius r and height h: area $= 2\pi r^2 + 2\pi rh$; volume $= \pi r^2 h$.

Triangle of base a and altitude h: area $= \frac{1}{2}ah$.

QUADRATIC FORMULA

If $ax^2 + bx + c = 0$, then $x = \dfrac{-b \pm \sqrt{b^2 - 4ac}}{2a}$.

TRIGONOMETRIC FUNCTIONS OF ANGLE θ

$$\sin\theta = \frac{y}{r} \quad \cos\theta = \frac{x}{r}$$

$$\tan\theta = \frac{y}{x} \quad \cot\theta = \frac{x}{y}$$

$$\sec\theta = \frac{r}{x} \quad \csc\theta = \frac{r}{y}$$

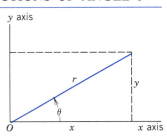

PYTHAGOREAN THEOREM

$a^2 + b^2 = c^2$

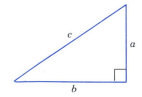

TRIANGLES

Angles A, B, C
Opposite sides a, b, c

$A + B + C = 180°$

$\dfrac{\sin A}{a} = \dfrac{\sin B}{b} = \dfrac{\sin C}{c}$

$c^2 = a^2 + b^2 - 2ab \cos C$

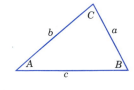

MATHEMATICAL SIGNS AND SYMBOLS

$=$ equals
$\approx$ equals approximately
$\sim$ is of the order of magnitude of
$\neq$ is not equal to
$\equiv$ is identical to, is defined as
$>$ is greater than ($\gg$ is much greater than)
$<$ is less than ($\ll$ is much less than)
$\geq$ is greater than or equal to (or, is no less than)
$\leq$ is less than or equal to (or, is no more than)
$\pm$ plus or minus ($\sqrt{4} = \pm 2$)
$\propto$ is proportional to
Σ the sum of
$\bar{x}$ the average value of x

PRODUCTS OF VECTORS

Let $\mathbf{i}$, $\mathbf{j}$, $\mathbf{k}$ be unit vectors in the x, y, z directions. Then

$$\mathbf{i\cdot i} = \mathbf{j\cdot j} = \mathbf{k\cdot k} = 1, \quad \mathbf{i\cdot j} = \mathbf{j\cdot k} = \mathbf{k\cdot i} = 0,$$

$$\mathbf{i\times i} = \mathbf{j\times j} = \mathbf{k\times k} = 0,$$

$$\mathbf{i\times j} = \mathbf{k}, \quad \mathbf{j\times k} = \mathbf{i}, \quad \mathbf{k\times i} = \mathbf{j}.$$

Any vector $\mathbf{a}$ with components a_x, a_y, a_z along the x, y, z axes can be written

$$\mathbf{a} = a_x\mathbf{i} + a_y\mathbf{j} + a_z\mathbf{k}.$$

Let $\mathbf{a}$, $\mathbf{b}$, $\mathbf{c}$ be arbitrary vectors with magnitudes a, b, c. Then

$$\mathbf{a}\times(\mathbf{b}+\mathbf{c}) = (\mathbf{a}\times\mathbf{b}) + (\mathbf{a}\times\mathbf{c})$$

$$(s\mathbf{a})\times\mathbf{b} = \mathbf{a}\times(s\mathbf{b}) = s(\mathbf{a}\times\mathbf{b}) \quad (s = \text{a scalar}).$$

Let θ be the smaller of the two angles between $\mathbf{a}$ and $\mathbf{b}$. Then

$$\mathbf{a\cdot b} = \mathbf{b\cdot a} = a_xb_x + a_yb_y + a_zb_z = ab\cos\theta$$

$$\mathbf{a}\times\mathbf{b} = -\mathbf{b}\times\mathbf{a} = \begin{vmatrix} \mathbf{i} & \mathbf{j} & \mathbf{k} \\ a_x & a_y & a_z \\ b_x & b_y & b_z \end{vmatrix}$$

$$= (a_yb_z - b_ya_z)\mathbf{i} + (a_zb_x - b_za_x)\mathbf{j} + (a_xb_y - b_xa_y)\mathbf{k}$$

$$|\mathbf{a}\times\mathbf{b}| = ab\sin\theta$$

$$\mathbf{a\cdot(b\times c)} = \mathbf{b\cdot(c\times a)} = \mathbf{c\cdot(a\times b)}$$

$$\mathbf{a}\times(\mathbf{b}\times\mathbf{c}) = (\mathbf{a\cdot c})\mathbf{b} - (\mathbf{a\cdot b})\mathbf{c}$$

TRIGONOMETRIC IDENTITIES

$\sin(90° - \theta) = \cos \theta$

$\cos(90° - \theta) = \sin \theta$

$\sin \theta / \cos \theta = \tan \theta$

$\sin^2 \theta + \cos^2 \theta = 1 \quad \sec^2 \theta - \tan^2 \theta = 1 \quad \csc^2 \theta - \cot^2 \theta = 1$

$\sin 2\theta = 2 \sin \theta \cos \theta$

$\cos 2\theta = \cos^2 \theta - \sin^2 \theta = 2 \cos^2 \theta - 1 = 1 - 2 \sin^2 \theta$

$\sin(\alpha \pm \beta) = \sin \alpha \cos \beta \pm \cos \alpha \sin \beta$

$\cos(\alpha \pm \beta) = \cos \alpha \cos \beta \mp \sin \alpha \sin \beta$

$\tan(\alpha \pm \beta) = \dfrac{\tan \alpha \pm \tan \beta}{1 \mp \tan \alpha \tan \beta}$

$\sin \alpha \pm \sin \beta = 2 \sin \frac{1}{2}(\alpha \pm \beta) \cos \frac{1}{2}(\alpha \mp \beta)$

BINOMIAL THEOREM

$(1 \pm x)^n = 1 \pm \dfrac{nx}{1!} + \dfrac{n(n-1)x^2}{2!} + \cdots \quad (x^2 < 1)$

$(1 \pm x)^{-n} = 1 \mp \dfrac{nx}{1!} + \dfrac{n(n+1)x^2}{2!} + \cdots \quad (x^2 < 1)$

EXPONENTIAL EXPANSION

$e^x = 1 + x + \dfrac{x^2}{2!} + \dfrac{x^3}{3!} + \cdots$

LOGARITHMIC EXPANSION

$\ln(1 + x) = x - \frac{1}{2}x^2 + \frac{1}{3}x^3 - \cdots \quad (|x| < 1)$

TRIGONOMETRIC EXPANSIONS (θ in radians)

$\sin \theta = \theta - \dfrac{\theta^3}{3!} + \dfrac{\theta^5}{5!} - \cdots$

$\cos \theta = 1 - \dfrac{\theta^2}{2!} + \dfrac{\theta^4}{4!} - \cdots$

$\tan \theta = \theta + \dfrac{\theta^3}{3} + \dfrac{2\theta^5}{15} + \cdots$

DERIVATIVES AND INTEGRALS

In what follows, the letters u and v stand for any functions of x, and a and m are constants. To each of the indefinite integrals should be added an arbitrary constant of integration. The *Handbook of Chemistry and Physics* (CRC Press Inc.) gives a more extensive tabulation.

1. $\dfrac{dx}{dx} = 1$

2. $\dfrac{d}{dx}(au) = a\dfrac{du}{dx}$

3. $\dfrac{d}{dx}(u + v) = \dfrac{du}{dx} + \dfrac{dv}{dx}$

4. $\dfrac{d}{dx}x^m = mx^{m-1}$

5. $\dfrac{d}{dx}\ln x = \dfrac{1}{x}$

6. $\dfrac{d}{dx}(uv) = u\dfrac{dv}{dx} + v\dfrac{du}{dx}$

7. $\dfrac{d}{dx}e^x = e^x$

8. $\dfrac{d}{dx}\sin x = \cos x$

9. $\dfrac{d}{dx}\cos x = -\sin x$

10. $\dfrac{d}{dx}\tan x = \sec^2 x$

11. $\dfrac{d}{dx}\cot x = -\csc^2 x$

12. $\dfrac{d}{dx}\sec x = \tan x \sec x$

13. $\dfrac{d}{dx}\csc x = -\cot x \csc x$

14. $\dfrac{d}{dx}e^u = e^u \dfrac{du}{dx}$

15. $\dfrac{d}{dx}\sin u = \cos u \dfrac{du}{dx}$

16. $\dfrac{d}{dx}\cos u = -\sin u \dfrac{du}{dx}$

1. $\displaystyle\int dx = x$

2. $\displaystyle\int au\, dx = a\int u\, dx$

3. $\displaystyle\int (u + v)\, dx = \int u\, dx + \int v\, dx$

4. $\displaystyle\int x^m\, dx = \dfrac{x^{m+1}}{m + 1} \quad (m \neq -1)$

5. $\displaystyle\int \dfrac{dx}{x} = \ln |x|$

6. $\displaystyle\int u\dfrac{dv}{dx}\, dx = uv - \int v\dfrac{du}{dx}\, dx$

7. $\displaystyle\int e^x\, dx = e^x$

8. $\displaystyle\int \sin x\, dx = -\cos x$

9. $\displaystyle\int \cos x\, dx = \sin x$

10. $\displaystyle\int \tan x\, dx = \ln |\sec x|$

11. $\displaystyle\int \sin^2 x\, dx = \frac{1}{2}x - \frac{1}{4}\sin 2x$

12. $\displaystyle\int e^{-ax}\, dx = -\dfrac{1}{a}e^{-ax}$

13. $\displaystyle\int xe^{-ax}\, dx = -\dfrac{1}{a^2}(ax + 1)e^{-ax}$

14. $\displaystyle\int x^2 e^{-ax}\, dx = -\dfrac{1}{a^3}(a^2x^2 + 2ax + 2)e^{-ax}$

15. $\displaystyle\int_0^\infty x^n e^{-ax}\, dx = \dfrac{n!}{a^{n+1}}$

16. $\displaystyle\int_0^\infty x^{2n} e^{-ax^2}\, dx = \dfrac{1 \cdot 3 \cdot 5 \cdots (2n - 1)}{2^{n+1}a^n}\sqrt{\dfrac{\pi}{a}}$

APPENDIX I

COMPUTER PROGRAMS

Here we give examples of programs that can be used to calculate the trajectories of charged particles moving in electric and magnetic fields. They are based on the programs for kinematic calculations involving nonconstant forces given in Appendix I of Volume 1. The programs are written in the BASIC language and can easily be adapted for most personal computers. By making appropriate modifications, these programs can be used for any electric or magnetic field configuration. In both programs, all quantities are in SI units.

In using both programs, special care should be taken to determine that the time interval DT (specified in line 130 in the electric-field program and in line 120 in the mag-

netic-field program) is small enough that the approximations used in the integrations do not introduce significant errors into the calculation. As the interval is made smaller, the number of intervals becomes larger, thereby increasing both the length of time it takes to run the program and the quantity of data it produces. To reduce the amount of output, each program has a provision for limiting the output to a certain number of points. This limits only the output data and does not affect the calculation. The total length of time for which the program follows the motion is equal to the product of the number of intervals (NT) and the size of each interval (DT).

1. ELECTRIC FIELDS

This program was used in Section 28–6 to find the motion of a particle moving along the axis of a ring of charge. The program calculates motion in the xz plane only. The x and z components of the electric field are specified in lines 180 and 190. As given here, the electric field compo-

nents are $E_x = 0$, $E_z = zR\lambda/2\epsilon_0(z^2 + R^2)^{3/2}$ (see Eq. 22 of Chapter 28), with $R = 0.02$ m and $\lambda = +2 \times 10^{-7}$ C/m. The output of the program is plotted in Fig. 16a of Chapter 28.

PROGRAM LISTING

```
10 'CALCULATIONS OF MOTION OF CHARGED PARTICLE
20 '  IN ELECTRIC FIELD IN XZ PLANE
30 E0 = 8.85E-12
40 'SPECIFY MASS, CHARGE, INITIAL POSITION, AND INITIAL VELOCITY
50 M = 1.67E-27
60 Q = 1.6E-19
70 X = 0
80 Z = .5
90 VX = 0
100 VZ = -700000!
110 'SPECIFY STARTING TIME, TIME INTERVAL, NUMBER OF INTERVALS
```

(Continued)

A-16

```
120 T = 0
130 DT = 5E-10
140 NT=3000
150 'SPECIFY NUMBER OF INTERVALS TO PRINT
160 N=15
170 'SPECIFY ELECTRIC FIELD COMPONENTS
180 DEF FNEX(X,Z) = 0
190 DEF FNEZ(X,Z) = Z*.03*.0000002/2/E0/(Z^2+(.03)^2)^1.5
200 PRINT " TIME       X        Z        VX        VZ"
210 LPRINT " TIME         X          Z          VX          VZ"
220 PRINT USING "##.##^^^^ ";T,X,Z,VX,VZ
230 LPRINT USING "##.##^^^^ ";T,X,Z,VX,VZ
240 FOR I = 1 TO NT
250 T = T+DT
260 AX = Q*FNEX(X,Z)/M
270 AZ = Q*FNEZ(X,Z)/M
280 X = X + VX*DT + .5*AX*DT*DT
290 Z = Z + VZ*DT + .5*AZ*DT*DT
300 VX = VX + AX*DT
310 VZ = VZ + AZ*DT
320 IF (NT/N)*INT(I/NT*N) < > I THEN 350
330 PRINT USING "##.##^^^^ ";T,X,Z,VX,VZ
340 LPRINT USING "##.##^^^^ ";T,X,Z,VX,VZ
350 NEXT I
400 END
```

SAMPLE OUTPUT

TIME	X	Z	VX	VZ
0.00E+00	0.00E+00	5.00E-01	0.00E+00	-7.00E+05
1.00E-07	0.00E+00	4.31E-01	0.00E+00	-6.85E+05
2.00E-07	0.00E+00	3.63E-01	0.00E+00	-6.64E+05
3.00E-07	0.00E+00	2.98E-01	0.00E+00	-6.35E+05
4.00E-07	0.00E+00	2.37E-01	0.00E+00	-5.90E+05
5.00E-07	0.00E+00	1.81E-01	0.00E+00	-5.16E+05
6.00E-07	0.00E+00	1.35E-01	0.00E+00	-3.89E+05
7.00E-07	0.00E+00	1.06E-01	0.00E+00	-1.78E+05
8.00E-07	0.00E+00	1.02E-01	0.00E+00	9.89E+04
9.00E-07	0.00E+00	1.25E-01	0.00E+00	3.37E+05
1.00E-06	0.00E+00	1.66E-01	0.00E+00	4.86E+05
1.10E-06	0.00E+00	2.20E-01	0.00E+00	5.72E+05
1.20E-06	0.00E+00	2.80E-01	0.00E+00	6.24E+05
1.30E-06	0.00E+00	3.44E-01	0.00E+00	6.58E+05
1.40E-06	0.00E+00	4.11E-01	0.00E+00	6.81E+05
1.50E-06	0.00E+00	4.80E-01	0.00E+00	6.97E+05

2. MAGNETIC FIELDS

This program, which was used in Section 34-3, calculates the motion of a particle confined to the xy plane and subject to a magnetic field in the z direction. The z component of the magnetic field (in units of tesla) is specified in line 170. As given here, the field is uniform. The output of the program is plotted in Fig. 17a of Chapter 34.

PROGRAM LISTING

```
10 'CALCULATION OF MOTION OF CHARGED PARTICLE IN XY PLANE
20 ' WITH MAGNETIC FIELD IN Z DIRECTION
30 'SPECIFY MASS, CHARGE, INITIAL POSITION, INITIAL VELOCITY
40 M = 6.645E-27
50 Q = 3.2E-19
60 X = 0
70 Y = 0
80 VX = 3000000!
90 VY = 0
100 'SPECIFY STARTING TIME, TIME INTERVAL, NUMBER OF INTERVALS
110 T = 0
120 DT = 1E-10
130 NT = 10000
140 'SPECIFY NUMBER OF INTERVALS TO PRINT
150 N = 20
160 'SPECIFY Z COMPONENT OF MAGNETIC FIELD
170 DEF FNBZ(X,Y) = .15
180 PRINT " TIME        X       Y      VX        VY"
190 LPRINT " TIME        X       Y        VX            VY"
200 PRINT USING "##.##^^^^ ";T,X,Y,VX,VY
210 LPRINT USING "##.##^^^^ ";T,X,Y,VX,VY
220 FOR I = 1 TO NT
230 T = T + DT
240 AX = Q*VY*FNBZ(X,Y)/M
250 AY = -Q*VX*FNBZ(X,Y)/M
260 X = X + VX*DT + .5*AX*DT*DT
270 Y = Y + VY*DT + .5*AY*DT*DT
280 VX = VX + AX*DT
290 VY = VY + AY*DT
300 IF (NT/N)*INT(I/NT*N) < > I THEN 330
310 PRINT USING "##.##^^^^ ";T,X,Y,VX,VY
320 LPRINT USING "##.##^^^^ ";T,X,Y,VX,VY
330 NEXT I
400 END
```

SAMPLE OUTPUT

TIME	X	Y	VX	VY
0.00E+00	0.00E+00	0.00E+00	3.00E+06	0.00E+00
5.00E-08	1.47E-01	−2.68E-02	2.81E+06	−1.06E+06
1.00E-07	2.75E-01	−1.04E-01	2.25E+06	−1.98E+06
1.50E-07	3.67E-01	−2.21E-01	1.41E+06	−2.65E+06
2.00E-07	4.12E-01	−3.63E-01	3.77E+05	−2.98E+06
2.50E-07	4.04E-01	−5.12E-01	−6.99E+05	−2.92E+06
3.00E-07	3.44E-01	−6.49E-01	−1.69E+06	−2.48E+06
3.50E-07	2.39E-01	−7.55E-01	−2.46E+06	−1.73E+06
4.00E-07	1.03E-01	−8.18E-01	−2.91E+06	−7.49E+05
4.50E-07	−4.55E-02	−8.29E-01	−2.99E+06	3.27E+05
5.00E-07	−1.89E-01	−7.86E-01	−2.68E+06	1.36E+06
5.50E-07	−3.08E-01	−6.95E-01	−2.02E+06	2.22E+06
6.00E-07	−3.87E-01	−5.69E-01	−1.11E+06	2.79E+06
6.50E-07	−4.16E-01	−4.22E-01	−5.14E+04	3.00E+06

(Continued)

```
7.00E-07     -3.92E-01     -2.75E-01     1.01E+06      2.83E+06
7.50E-07     -3.17E-01     -1.45E-01     1.95E+06      2.29E+06
8.00E-07     -2.01E-01     -5.08E-02     2.63E+06      1.45E+06
8.50E-07     -5.94E-02     -3.27E-03     2.98E+06      4.29E+05
9.00E-07      9.00E-02     -8.86E-03     2.94E+06     -6.50E+05
9.50E-07      2.28E-01     -6.69E-02     2.52E+06     -1.65E+06
1.00E-06      3.36E-01     -1.70E-01     1.77E+06     -2.43E+06
```

APPENDIX J

NOBEL PRIZES IN PHYSICS*

1901	Wilhelm Konrad Röntgen	1845–1923	for the discovery of x-rays
1902	Hendrik Antoon Lorentz	1853–1928	for their researches into the influence of magnetism upon radiation
	Pieter Zeeman	1865–1943	phenomena
1903	Antoine Henri Becquerel	1852–1908	for his discovery of spontaneous radioactivity
	Pierre Curie	1859–1906	for their joint researches on the radiation phenomena discovered by
	Marie Sklowdowska-Curie	1867–1934	Professor Henri Becquerel
1904	Lord Rayleigh (John William Strutt)	1842–1919	for his investigations of the densities of the most important gases and for his discovery of argon
1905	Philipp Eduard Anton von Lenard	1862–1947	for his work on cathode rays
1906	Joseph John Thomson	1856–1940	for his theoretical and experimental investigations on the conduction of electricity by gases
1907	Albert Abraham Michelson	1852–1931	for his optical precision instruments and metrological investigations carried out with their aid
1908	Gabriel Lippmann	1845–1921	for his method of reproducing colors photographically based on the phenomena of interference
1909	Guglielmo Marconi	1874–1937	for their contributions to the development of wireless
	Carl Ferdinand Braun	1850–1918	telegraphy
1910	Johannes Diderik van der Waals	1837–1932	for his work on the equation of state for gases and liquids
1911	Wilhelm Wien	1864–1928	for his discoveries regarding the laws governing the radiation of heat
1912	Nils Gustaf Dalén	1869–1937	for his invention of automatic regulators for use in conjunction with gas accumulators for illuminating lighthouses and buoys
1913	Heike Kamerlingh Onnes	1853–1926	for his investigations of the properties of matter at low temperatures which led, *inter alia,* to the production of liquid helium
1914	Max von Laue	1879–1960	for his discovery of the diffraction of Röntgen rays by crystals
1915	William Henry Bragg	1862–1942	for their services in the analysis of crystal structure by means of
	William Lawrence Bragg	1890–1971	x-rays
1917	Charles Glover Barkla	1877–1944	for his discovery of the characteristic x-rays of the elements
1918	Max Planck	1858–1947	for his discovery of energy quanta
1919	Johannes Stark	1874–1957	for his discovery of the Doppler effect in canal rays and the splitting of spectral lines in electric fields
1920	Charles-Édouard Guillaume	1861–1938	for the service he has rendered to precision measurements in Physics by his discovery of anomalies in nickel steel alloys
1921	Albert Einstein	1879–1955	for his services to Theoretical Physics, and especially for his discovery of the law of the photoelectric effect
1922	Niels Bohr	1885–1962	for the investigation of the structure of atoms, and of the radiation emanating from them
1923	Robert Andrews Millikan	1868–1953	for his work on the elementary charge of electricity and on the photoelectric effect
1924	Karl Manne Georg Siegbahn	1888–1979	for his discoveries and research in the field of x-ray spectroscopy
1925	James Franck	1882–1964	for their discovery of the laws governing the impact of an electron
	Gustav Hertz	1887–1975	upon an atom
1926	Jean Baptiste Perrin	1870–1942	for his work on the discontinuous structure of matter, and especially for his discovery of sedimentation equilibrium

* See *Nobel Lectures, Physics,* 1901–1970, Elsevier Publishing Company for biographies of the awardees and for lectures given by them on receiving the prize.

1927	Arthur Holly Compton	1892–1962	for his discovery of the effect named after him
	Charles Thomson Rees Wilson	1869–1959	for his method of making the paths of electrically charged particles visible by condensation of vapor
1928	Owen Willans Richardson	1879–1959	for his work on the thermionic phenomenon and especially for the discovery of the law named after him
1929	Prince Louis-Victor de Broglie	1892–1987	for his discovery of the wave nature of electrons
1930	Sir Chandrasekhara Venkata Raman	1888–1970	for his work on the scattering of light and for the discovery of the effect named after him
1932	Werner Heisenberg	1901–1976	for the creation of quantum mechanics, the application of which has, among other things, led to the discovery of the allotropic forms of hydrogen
1933	Erwin Schrödinger	1887–1961	for the discovery of new productive forms of atomic theory
	Paul Adrien Maurice Dirac	1902–1984	
1935	James Chadwick	1891–1974	for his discovery of the neutron
1936	Victor Franz Hess	1883–1964	for the discovery of cosmic radiation
	Carl David Anderson	1905–1991	for his discovery of the positron
1937	Clinton Joseph Davisson	1881–1958	for their experimental discovery of the diffraction of electrons
	George Paget Thomson	1892–1975	by crystals
1938	Enrico Fermi	1901–1954	for his demonstrations of the existence of new radioactive elements produced by neutron irradiation, and for his related discovery of nuclear reactions brought about by slow neutrons
1939	Ernest Orlando Lawrence	1901–1958	for the invention and development of the cyclotron and for results obtained with it, especially for artificial radioactive elements
1943	Otto Stern	1888–1969	for his contribution to the development of the molecular ray method and his discovery of the magnetic moment of the proton
1944	Isidor Isaac Rabi	1898–1988	for his resonance method for recording the magnetic properties of atomic nuclei
1945	Wolfgang Pauli	1900–1958	for the discovery of the Exclusion Principle (Pauli Principle)
1946	Percy Williams Bridgman	1882–1961	for the invention of an apparatus to produce extremely high pressures, and for the discoveries he made therewith in the field of high-pressure physics
1947	Sir Edward Victor Appleton	1892–1965	for his investigations of the physics of the upper atmosphere, especially for the discovery of the so-called Appleton layer
1948	Patrick Maynard Stuart Blackett	1897–1974	for his development of the Wilson cloud chamber method, and his discoveries therewith in nuclear physics and cosmic radiation
1949	Hideki Yukawa	1907–1981	for his prediction of the existence of mesons on the basis of theoretical work on nuclear forces
1950	Cecil Frank Powell	1903–1969	for his development of the photographic method of studying nuclear processes and his discoveries regarding mesons made with this method
1951	Sir John Douglas Cockcroft	1897–1967	for their pioneer work on the transmutation of atomic nuclei by
	Ernest Thomas Sinton Walton	1903–	artificially accelerated atomic particles
1952	Felix Bloch	1905–1983	for their development of new methods for nuclear magnetic precision
	Edward Mills Purcell	1912–	methods and discoveries in connection therewith
1953	Frits Zernike	1888–1966	for his demonstration of the phase-contrast method, especially for his invention of the phase-contrast microscope
1954	Max Born	1882–1970	for his fundamental research in quantum mechanics, especially for his statistical interpretation of the wave function
	Walther Bothe	1891–1957	for the coincidence method and his discoveries made therewith
1955	Willis Eugene Lamb	1913–	for his discoveries concerning the fine structure of the hydrogen spectrum
	Polykarp Kusch	1911–	for his precision determination of the magnetic moment of the electron
1956	William Shockley	1910–1989	for their researches on semiconductors and their discovery of the
	John Bardeen	1908–1991	transistor effect
	Walter Houser Brattain	1902–1987	
1957	Chen Ning Yang	1922–	for their penetrating investigation of the parity laws which has led to
	Tsung Dao Lee	1926–	important discoveries regarding the elementary particles
1958	Pavel Aleksejecič Čerenkov	1904–	for the discovery and the interpretation of the Cerenkov effect
	Il' ja Michajlovič Frank	1908–1990	
	Igor' Evgen' evič Tamm	1895–1971	
1959	Emilio Gino Segrè	1905–1989	for their discovery of the antiproton
	Owen Chamberlain	1920–	
1960	Donald Arthur Glaser	1926–	for the invention of the bubble chamber
1961	Robert Hofstadter	1915–1990	for his pioneering studies of electron scattering in atomic nuclei and for his thereby achieved discoveries concerning the structure of the nucleons
	Rudolf Ludwig Mössbauer	1929–	for his researches concerning the resonance absorption of γ-rays and his discovery in this connection of the effect which bears his name

1962	Lev Davidoviĉ Landau	1908–1968	for his pioneering theories of condensed matter, especially liquid helium
1963	Eugene P. Wigner	1902–	for his contribution to the theory of the atomic nucleus and the elementary particles, particularly through the discovery and application of fundamental symmetry principles
	Maria Goeppert Mayer	1906–1972	for their discoveries concerning nuclear shell structure
	J. Hans D. Jensen	1907–1973	
1964	Charles H. Townes	1915–	for fundamental work in the field of quantum electronics which has led to the construction of oscillators and amplifiers based on the maser-laser principle
	Nikolai G. Basov	1922–	
	Alexander M. Prochorov	1916–	
1965	Sin-itiro Tomonaga	1906–1979	for their fundamental work in quantum electrodynamics, with deep-ploughing consequences for the physics of elementary particles
	Julian Schwinger	1918–	
	Richard P. Feynman	1918–1988	
1966	Alfred Kastler	1902–1984	for the discovery and development of optical methods for studying Hertzian resonance in atoms
1967	Hans Albrecht Bethe	1906–	for his contributions to the theory of nuclear reactions, especially his discoveries concerning the energy production in stars
1968	Luis W. Alvarez	1911–1988	for his decisive contribution to elementary particle physics, in particular the discovery of a large number of resonance states, made possible through his development of the technique of using hydrogen bubble chamber and data analysis
1969	Murray Gell-Mann	1929–	for his contribution and discoveries concerning the classification of elementary particles and their interactions
1970	Hannes Alvén	1908–	for fundamental work and discoveries in magneto-hydrodynamics with fruitful applications in different parts of plasma physics
	Louis Néel	1904–	for fundamental work and discoveries concerning antiferromagnetism and ferrimagnetism which have led to important applications in solid state physics
1971	Dennis Gabor	1900–1979	for his discovery of the principles of holography
1972	John Bardeen	1908–1991	for their development of a theory of superconductivity
	Leon N. Cooper	1930–	
	J. Robert Schrieffer	1931–	
1973	Leo Esaki	1925–	for his discovery of tunneling in semiconductors
	Ivar Giaever	1929–	for his discovery of tunneling in superconductors
	Brian D. Josephson	1940–	for his theoretical prediction of the properties of a super-current through a tunnel barrier
1974	Antony Hewish	1924–	for the discovery of pulsars
	Sir Martin Ryle	1918–1984	for his pioneering work in radioastronomy
1975	Aage Bohr	1922–	for the discovery of the connection between collective motion and particle motion and the development of the theory of the structure of the atomic nucleus based on this connection
	Ben Mottelson	1926–	
	James Rainwater	1917–1986	
1976	Burton Richter	1931–	for their (independent) discovery of an important fundamental particle
	Samuel Chao Chung Ting	1936–	
1977	Philip Warren Anderson	1923–	for their fundamental theoretical investigations of the electronic structure of magnetic and disordered systems
	Nevill Francis Mott	1905–	
	John Hasbrouck Van Vleck	1899–1980	
1978	Peter L. Kapitza	1894–1984	for his basic inventions and discoveries in low-temperature physics
	Arno A. Penzias	1926–	for their discovery of cosmic microwave background radiation
	Robert Woodrow Wilson	1936–	
1979	Sheldon Lee Glashow	1932–	for their unified model of the action of the weak and electromagnetic forces and for their prediction of the existence of neutral currents
	Abdus Salam	1926–	
	Steven Weinberg	1933–	
1980	James W. Cronin	1931–	for the discovery of violations of fundamental symmetry principles in the decay of neutral K mesons
	Val L. Fitch	1923–	
1981	Nicolaas Bloembergen	1920–	for their contribution to the development of laser spectroscopy
	Arthur Leonard Schawlow	1921–	
	Kai M. Siegbahn	1918–	for his contribution of high-resolution electron spectroscopy
1982	Kenneth Geddes Wilson	1936–	for his method of analyzing the critical phenomena inherent in the changes of matter under the influence of pressure and temperature
1983	Subrehmanyan Chandrasekhar	1910–	for his theoretical studies of the structure and evolution of stars
	William A. Fowler	1911–	for his studies of the formation of the chemical elements in the universe
1984	Carlo Rubbia	1934–	for their decisive contributions to the large project, which led to the discovery of the field particles W and Z, communicators of the weak interaction
	Simon van der Meer	1925–	
1985	Klaus von Klitzing	1943–	for his discovery of the quantized Hall resistance
1986	Ernst Ruska	1906–	for his invention of the electron microscope
	Gerd Binnig	1947–	for their invention of the scanning-tunneling electron microscope
	Heinrich Rohrer	1933–	

1987	Karl Alex Müller	1927–	for their discovery of a new class of superconductors
	J. Georg Bednorz	1950–	
1988	Leon M. Lederman	1922–	for experiments with neutrino beams and the discovery of the muon neutrino
	Melvin Schwartz	1932–	
	Jack Steinberger	1921–	
1989	Hans G. Dehmelt	1922–	for their development of techniques for trapping individual atoms
	Wolfgang Paul	1913–	
	Norman F. Ramsey	1915–	for his discoveries in atomic resonance spectroscopy, which led to hydrogen masers and atomic clocks
1990	Richard E. Taylor	1929–	for their experiments on the scattering of electrons from nuclei, which revealed the presence of quarks inside nucleons
	Jerome I. Friedman	1930–	
	Henry W. Kendall	1926–	
1991	Pierre-Gilles de Gennes	1932–	for discoveries about the ordering of molecules in substances such as liquid crystals, superconductors, and polymers
1992	George Charpak	1924-	for his invention of fast electronic detectors for high energy particles

ANSWERS TO ODD NUMBERED PROBLEMS

CHAPTER 27

1. 2.74 N on each charge. **3.** 0.50 C. **5.** (a) 1.77 N.
(b) 3.07 N. **7.** $q_1 = -4q_2$.
9. 24.5 N, along the angle bisector.
11. 1.00 μC and 3.00 μC, of opposite sign.
13. (a) A charge $-4q/9$ must be located on the line segment joining the two positive charges, a distance $L/3$ from the $+q$ charge.
15. $q = Q/2$. **17.** (b) 2.96 cm. **19.** $a/\sqrt{2}$.
23. $\sqrt{\pi^3 m \epsilon_0 d^3/2qQ}$. **25.** 2.89×10^{-9} N. **27.** 3.8 N.
29. 5.08 m below the electron. **31.** 13.4 MC.
33. (a) 57.1 TC; no. (b) 598 metric tons. **35.** (a) Boron.
(b) Nitrogen. (c) Carbon.

CHAPTER 28

1. 10.5 mN/C, westward. **3.** 203 nN/C, up. **5.** 144 pC.
7. 19.5 kN/C. **9.** 9:30. **11.** (b) Parallel to **p**.
15. (a) $\dfrac{1}{4\pi\epsilon_0} \dfrac{qz}{(R^2 + z^2)^{3/2}}$. (b) $\dfrac{1}{2\pi^2\epsilon_0} \dfrac{(q_1 - q_2)R}{(R^2 + z^2)^{3/2}}$.
19. To the right. **25.** $R/\sqrt{3}$. **27.** (a) 104 nC. (b) 1.31×10^{17}.
(c) 4.96×10^{-6}. **29.** (a) 6.50 cm. (b) 4.80 μC.
35. $q/8\pi\epsilon_0 R^2$. **37.** (a) 6.53 cm. (b) 26.9 ns. (c) 0.121.
39. (a) 585 kN/C, toward the negative charge.
(b) 93.6 fN, toward the positive charge.
41. $5e$. **43.** 1.64×10^{-19} C ($\approx 2.5\%$ high). **45.** 1.2 mm.
47. The upper plate; 4.06 cm. **49.** (a) Zero.
(b) 8.50×10^{-22} N·m. (c) Zero. **51.** $2pE \cos \theta_0$.
53. (a) $8q/\pi\epsilon_0 a^3$.

CHAPTER 29

1. -0.0078 N·m²/C. **3.** (a) $-\pi R^2 E$. (b) $\pi R^2 E$.
5. 208 kN·m²/C. **7.** $q/6\epsilon_0$. **9.** 4.6 μC.
13. (a) 22.3 N·m²/C. (b) 197 pC. **15.** (a) 452 nC/m².
(b) 51.1 kN/C. **17.** (a) $-Q$. (b) $-Q$. (c) $-(Q + q)$. (d) Yes.
19. (a) 53 MN/C. (b) 60 N/C. **21.** (a) 322 nC. (b) 143 nC.
23. (a) Zero. (b) σ/ϵ_0, to the left. (c) Zero. **25.** 5.11 nC/m².
27. 5.09 μC/m³. **29.** (a) $q/2\pi\epsilon_0 Lr$, radially inward.
(b) $-q$ on both inner and outer surfaces.
(c) $q/2\pi\epsilon_0 Lr$, radially outward. **31.** -1.13 nC.
33. (a) $\lambda/2\pi\epsilon_0 r$. (b) Zero. **35.** 270 eV.
37. (a) 2.19 MN/C, radially out. (b) 436 kN/C, radially in.
39. 97.9 cm. **41.** $0.557R$. **45.** (b) $\rho R^2/2\epsilon_0 r$.

CHAPTER 30

1. (a) 484 keV. (b) Zero. **3.** (a) 27.2 fJ = 170 keV.
(b) 3.02×10^{-31} kg, in error by a factor of about three.
5. (a) 3.0 kN. (b) 240 MeV. **7.** (a) 30 GJ. (b) 7.1 km/s.
(c) 9.0×10^4 kg. **9.** (a) 256 kV. (b) $0.745c$. **11.** 2.6 km/s.
13. $\dfrac{qQ}{8\pi\epsilon_0}\left[\dfrac{1}{r_1} - \dfrac{1}{r_2}\right]$. **15.** 2.17 d. **17.** (a) 24.4 kV/m.
(b) 2.93 kV. **19.** (a) 132 MV/m. (b) 8.43 kV/m.
21. (a) 32 MeV. **23.** (a) -3.85 kV. (b) -3.85 kV.
25. -1.1 nC. **27.** (a) 0.562 mm. (b) 813 V. **29.** 637 MV.
31. (a) $qd/2\pi\epsilon_0 a(a + d)$. **33.** (a) -5.40 nm. (b) 9.00 nm.
(c) No. **37.** 186 pJ. **41.** (a) 4.5 m. (b) No. **45.** 746 V/m.
47. -2.3×10^{21} V/m. **49.** -39.2 V/m.
51. (a) $\dfrac{k}{4\pi\epsilon_0}[\sqrt{L^2 + y^2} - y]$. (b) $\dfrac{k}{4\pi\epsilon_0}\left[1 - \dfrac{y}{\sqrt{L^2 + y^2}}\right]$.
(d) $3L/4$. **53.** (a) $V_1 = V_2$. (b) $q_1 = q/3$; $q_2 = 2q/3$.
55. 840 V. **57.** 2.0×10^{-8}. **59.** (a) Zero. (b) Zero. (c) Zero.
(d) Zero. (e) No. **63.** (a) 1.75 kV. (b) 7.40 cm. **65.** 9.65 kW.

CHAPTER 31

1. 7.5 pC. **3.** 3.25 mC. **5.** 0.546 pF. **7.** (a) 84.5 pF.
(b) 191 cm². **11.** 9090. **13.** 7.17 μF. **15.** (a) 2.4 μF.
(b) $q_4 = q_6 = 480$ μC. (c) $V_4 = 120$ V; $V_6 = 80$ V. **17.** (a) $d/3$.
(b) $3d$. **19.** (a) 942 μC. (b) 91.4 V. **23.** (a) 45.4 V.
(b) 52.7 μC. (c) 146 μC. **25.** (a) 50 V. (b) Zero.
27. (a) $q_1 = 9.0$ μC, $q_2 = 16$ μC, $q_3 = 9.0$ μC, $q_4 = 16$ μC.
(b) $q_1 = 8.40$ μC, $q_2 = 16.8$ μC, $q_3 = 10.8$ μC, $q_4 = 14.4$ μC.
29. 200 nJ. **31.** (a) 28.6 pF. (b) 17.9 nC. (c) 5.59 μJ.
(d) 482 kV/m. (e) 1.03 J/m³. **33.** 74.1 mJ/m³.
35. (a) 2.0 J. **37.** (a) $2V$. (b) $U_i = \epsilon_0 AV^2/2d$; $U_f = \epsilon_0 AV^2/d$.
(c) $\epsilon_0 AV^2/2d$. **39.** (a) $e^2/32\pi^2\epsilon_0 r^4$. (b) $e^2/8\pi\epsilon_0 R$.
(c) 1.40 fm. **43.** 3.89. **45.** The mica sheet. **47.** 86.3 nF.
49. (a) 730 pF. (b) 28 kV. **51.** (a) $\epsilon_0 A/(d - b)$. (b) $d/(d - b)$.
(c) $q^2 b/2A\epsilon_0$; pulled in. **53.** 1.63 kV. **57.** (a) 13.4 kV/m.
(b) 6.16 nC. (c) 5.02 nC. **59.** (a) 6.53. (b) 754 nC.
61. (a) 85.6 pF. (b) 119 pF. (c) 10.3 nC; 10.3 nC.
(d) 9.86 kV/m. (e) 2.05 kV/m. (f) 86.6 V. (g) 170 nJ.

CHAPTER 32

1. (a) 1.33 kC. (b) 8.31×10^{21}. **3.** (a) 9.41 A/m² north.
5. 0.400 mm. **7.** 0.67 A, toward the negative terminal.
9. 7.1 ms. **11.** (a) 654 nA/m². (b) 83.4 MA. **13.** 52.5 min.
15. (a) 95.0 μC. (b) 158 C°. **17.** 0.59 Ω. **19.** (a) 1.5 kA.
(b) 53 MA/m². (c) 110 nΩ·m; platinum. **23.** (a) 250°C.
25. (a) 380 μV. (b) Negative. (c) 4.3 min. **27.** 54 Ω. **29.** 3.

31. (*a*) 6.00 mA. (*b*) 15.9 nV. (*c*) 21.2 nΩ.
33. 1190 (Ω·m)$^{-1}$. **35.** (*a*) Cu: 55.3 A/cm^2; Al: 34.0 A/cm^2.
(*b*) Cu: 1.01 kg; Al: 0.495 kg. **37.** (*a*) Silver. (*b*) 60.8 nΩ.
39. 0.036. **41.** (*a*) 8.52 kΩ. (*b*) 4.51 μA. **43.** 7.16 fs.
45. 18 kC. **47.** (*a*) 1.03 kW. (*b*) 34.5 cents. **49.** (*a*) $4.46.
(*b*) 144 Ω. (*c*) 833 mA. **51.** (*a*) 2.88 × 10^{11}. (*b*) 24.0 μA.
(*c*) 1.14 kW; 23.1 MW. **53.** (*a*) 6.1 m. (*b*) 13 m.
55. 27.4 cm/s. **57.** 311 nJ. **59.** (*a*) 37.0 min. (*b*) 122 min.
61. (*a*) 1.37L. (*b*) 0.730A.

CHAPTER 33

1. 10.6 kJ. **3.** 13 h 38 min. **5.** −10 V. **7.** (*a*) 14 Ω.
(*b*) 35 mW. **9.** (*a*) 44.2 V. (*b*) 21.4 V. (*c*) Left.
11. The cable. **13.** (*a*) 1.5 kΩ. (*b*) 400 mV. (*c*) 0.26%.
15. (*a*) 3.4 A. (*b*) 0.29 V. *And:* (*a*) 0.59 A. (*b*) 1.7 V.
17. 4.0 Ω; 12 Ω. **19.** 7.5 V. **21.** 262 Ω or 38.2 Ω.
23. (*a*) In parallel. (*b*) 72.0 Ω; 144 Ω.
25. (*a*) ρ_A = 16.3 nΩ·m; ρ_B = 7.48 nΩ·m.
(*b*) $j_A = j_B$ = 62.3 kA/cm^2.
(*c*) E_A = 10.2 V/m; E_B = 4.66 V/m.
(*d*) V_A = 435 V; V_B = 195 V.
27. (*a*) $R/2$. (*b*) 5$R/8$. **29.** (*a*) 3$R/4$. (*b*) 5$R/6$. **31.** (*a*) R_2.
(*b*) R_1. **33.** $\mathcal{E}/7R$.
35. (50 kW)$\left(\dfrac{x}{2000 + 10x - x^2}\right)^2$, x in cm.
37. (*a*) i_1 = 668 mA, down; i_2 = 85.7 mA, up; i_3 =
582 mA, up. (*b*) −3.60 V. **39.** (*a*) 0.45 A. **41.** 0.90%.
45. (*a*) Top: 70.9 mA; 4.91 V; bottom: 55.2 mA; 4.86 V.
(*b*) Top: 69.3 Ω; bottom: 88.0 Ω. **49.** 4.61. **51.** (*a*) 2.20 s.
(*b*) 44 mV. **53.** 2.35 MΩ. **55.** (*a*) 955 pC/s. (*b*) 1.08 μW.
(*c*) 2.74 μW. (*d*) 3.82 μW.

CHAPTER 34

1. 1: +; 2: −; 3: 0; 4: −. **3.** (*a*) 3.4 km/s. **5.** 8.2 × 10^9.
7. 0.75k, T. **9.** (*a*) To the East. (*b*) 6.27 × 10^{14} m/s^2.
(*c*) 2.98 mm. **11.** (*a*) 0.34 mm. (*b*) 2.6 keV.
13. (*a*) 1.11 × 10^7 m/s. (*b*) 0.316 mm. **15.** (*a*) 2600 km/s.
(*b*) 110 ns. (*c*) 140 keV. (*d*) 70 kV. **19.** (*a*) K_p. (*b*) $K'_p/2$.
21. (*a*) $r_p\sqrt{2}$. (*b*) r_p. **23.** (*a*) $B(qm/2V)^{1/2}\Delta x$. (*b*) 7.91 mm.
25. (*a*) −q. (*b*) $\pi m/qB$. **27.** (*a*) 0.999928c.
29. An alpha particle. **31.** (*a*) 78.6 ns. (*b*) 9.16 cm.
(*c*) 3.20 cm. **33.** 240 m. **39.** 37 cm/s.
41. 467 mA; left to right. **43.** (*a*) 330 MA.
(*b*) 1.1 × 10^{17} W. **45.** 4.2 C. **47.** −0.414k, N.
49. (*a*) 0; 138 mN; 138 mN.
53. 2πaiB sin θ, normal to plane of ring, up. **55.** 1.63 A.
57. 2.1 GA. **59.** (*a*) −2.86k, A·m^2. (*b*) 1.10k, A·m^2.

CHAPTER 35

1. 7.7 mT. **3.** 12 nT. **5.** (*a*) 0.324 fN, parallel to current.
(*b*) 0.324 fN, radially outward. (*c*) Zero.
7. 30.0 A, antiparallel. **9.** (*a*) 4. (*b*) ½. **11.** (*a*) 2.43 A·m^2.
(*b*) 46 cm. **13.** 2 rad. **15.** $\dfrac{\mu_0 i\theta}{4\pi}\left(\dfrac{1}{b} - \dfrac{1}{a}\right)$, out of figure.
19. (*b*) ia^2. **21.** $\left(\dfrac{\mu_0 i}{2\pi w}\right)\ln\left(1 + \dfrac{w}{d}\right)$; up. **25.** ½$\pi i(a^2 + b^2)$.
29. (*c*) ½nia^2 sin(2π/n). **31.** (*a*) (2$\mu_0 i/3\pi L$)(2 $\sqrt{2}$ + $\sqrt{10}$).
(*b*) Greater. **35.** 606 μN, toward the center of the square.
37. (*b*) 2.3 km/s. **39.** (*a*) −2.5 μT·m. (*b*) Zero.
41. 6.0 μT·m. **45.** (*a*) $\dfrac{\mu_0 ir}{2\pi c^2}$. (*b*) $\dfrac{\mu_0 i}{2\pi r}$. (*c*) $\dfrac{\mu_0 i}{2\pi r}\dfrac{a^2 - r^2}{a^2 - b^2}$.

(*d*) Zero. **47.** $\mu_0 ir^2/2\pi a^3$. **49.** 3$i_0/8$, into the page.
51. 109 m. **53.** 272 mA. **55.** (*a*) Negative. (*b*) 9.7 cm.

CHAPTER 36

1. 57 μWb. **3.** (*a*) 31 mV. (*b*) Right to left. **5.** (*a*) 1.12 mΩ.
(*b*) 1.27 T/s. **7.** (*b*) 58 mA. **9.** 4.97 μW. **11.** (*b*) No.
15. (*a*) 28.2 μV. (*b*) From *c* to *b*. **17.** 80 μV; clockwise.
19. Zero. **21.** $iLBt/m$, away from G. **23.** 455 mV.
27. (*b*) Design it so that Nab = (5/2π) m^2. **29.** 6.3 rev/s.
31. 25 μC. **33.** (*a*) 253 μV. (*b*) 610 μA. (*c*) 154 nW.
(*d*) 31.7 nN. (*e*) 154 nW. **35.** (*a*) $\dfrac{\mu_0 ia}{2\pi}\ln\left(1 + \dfrac{b}{D}\right)$.
(*b*) $\dfrac{\mu_0 iabv}{2\pi RD(D + b)}$. **39.** $(Bar)^2\omega\sigma t$. **41.** (*a*) −1.20 mV.
(*b*) −2.79 mV. (*c*) 1.59 mV. **47.** (*a*) 34 V/m.
(*b*) 6.0 × 10^{12} m/s^2. **49.** (*a*) 0.15°.

CHAPTER 37

1. +3 Wb. **3.** (*a*) Stable. (*b*) Unstable. (*c*) Stable.
(*d*) Unstable. **5.** $\left(\dfrac{\mu_0 iL}{\pi}\right)$ ln 3. **7.** (*a*) 514 GV/m.
(*b*) 19.0 mT. **11.** 24 mJ/T. **13.** (*a*) 0.86 μT. (*b*) 0.68 A/m.
15. 0.58 K. **17.** (*a*) 150 T. (*b*) 600 T. **19.** Yes.
23. (*a*) 3.0 μT. (*b*) 9.0 × 10^{-29} J. **27.** (*a*) 630 MA.
31. 1660 km. **33.** 61 μT; 84°.

CHAPTER 38

1. 100 nWb. **3.** 261 μH/m. **5.** (*a*) 600 μH. (*b*) 120.
7. 7.87 H. **15.** $\left(\dfrac{\mu_0 l}{2\pi}\right)\ln\dfrac{b}{a}$. **17.** 29.8 Ω. **19.** (*a*) 4.78 mH.
(*b*) 2.42 ms. **21.** 42 + 20t, V. **23.** 12 A/s.
25. (*a*) $i_1 = i_2$ = 3.33 A. (*b*) i_1 = 4.55 A; i_2 = 2.73 A.
(*c*) i_1 = 0; i_2 = 1.82 A. (*d*) $i_1 = i_2$ = 0. **29.** (*a*) 13.2 H.
(*b*) 124 mA. **31.** 63.2 MJ/m^3. **33.** 150 MV/m.
35. (*a*) 78 kJ. (*b*) 3.7 kg. **37.** (*a*) 117 H. (*b*) 225 μJ.
39. (*a*) $\mu_0 i^2 N^2/8\pi^2 r^2$. (*b*) $(\mu_0 N^2 hi^2/4\pi)$ ln $\dfrac{b}{a}$. **41.** 12 PJ.
45. 123 mA. **47.** 38 μH. **51.** (*a*) 6.08 μs. (*b*) 164 kHz.
(*c*) 3.04 μs. **53.** (*a*) No. (*b*) 6.1 kHz. (*c*) 16 nF.
55. (*a*) 5800 rad/s. (*b*) 1.1 ms. **57.** (*a*) $q_m/\sqrt{3}$.
(*b*) t/T = 0.152. **59.** (*a*) 6.0 : 1. (*b*) 36 pF; 220 μH.
61. (*a*) 180 μC. (*b*) $T/8$. (*c*) 67 W. **63.** (*a*) Zero. (*b*) 2i.
65. (L/R) ln 2. **67.** 8.7 mΩ. **69.** 2.96 Ω.

CHAPTER 39

1. 377 rad/s. **3.** (*a*) 3.75 krad/s. (*b*) 23.4 Ω. **5.** (*a*) 39.1 mA.
(*b*) Zero. (*c*) 32.6 mA. (*d*) Taking energy. **7.** (*a*) 6.73 ms.
(*b*) 2.24 ms. (*c*) Capacitor. (*d*) 56.6 μF. **13.** 1.0 kV > $\mathcal{E}_m$.
15. (*a*) 36.0 V. (*b*) 27.4 V. (*c*) 17.0 V. (*d*) 8.4 V.
17. (*a*) 39.1 Ω. (*b*) 21.7 Ω. (*c*) Capacitive. **19.** (*a*) 45°.
(*b*) 76.0 Ω. **23.** 177 Ω. **25.** (*a*) 1.82 W. (*b*) 3.13 W.
27. 100 V. **31.** (*a*) 0.74. (*b*) Leads. (*c*) Capacitive. (*d*) No.
(*e*) Yes; no; yes. (*f*) 33 W. **33.** (*a*) 2.49 A.
(*b*) 37.4 V; 153 V; 218 V; 65.0 V; 75.0 V.
(*c*) $P_C = P_L$ = 0; P_R = 93.0 W.
37. 166 Ω; 315 mH; 14.8 μF. **39.** (*a*) 2.4 V.
(*b*) 3.2 mA; 160 mA. **43.** 10.

CHAPTER 40

3. $r = 2.5$ m; 10 m.
5. Change the potential across the plates at 1.0 kV/s.
7. (a) 1.84 A. (b) 140 GV/m·s. (c) 460 mA. (d) 578 nT·m.
9. (a) 840 mA. (b) Zero. (c) 1.3 A. 11. (a) 623 nT.
(b) 2.11 TV/m·s. 13. 2.27 pT.
19. 1900 km in radius, independent of its length.

CHAPTER 41

3. (a) 4.5×10^{24} Hz. (b) 10,000 km. 5. 5.0×10^{-21} H.
7. 1.07 pT. 11. 100 kJ. 13. 4.62×10^{-29} W/m².
15. 78 cm. 17. (a) 883 m. (b) No. 19. (a) 6.53 nT.
(b) 5.10 mW/m². (c) 8.04 W.
21. (a) $\pm EBa^2/\mu_0$ for faces parallel to the xy plane; zero through each of the other four faces. (b) Zero.
23. (a) 9.14 mW/cm². (b) 1.68 MW. 25. (a) 76.8 mV/m.

(b) 256 pT. (c) 12.6 kW. 29. (a) $\dfrac{\omega}{k} = c$; $E_m = cB_m$.

(b) $S = \left(\dfrac{E_m^2}{4\mu_0 c}\right) \sin 2\omega t \sin 2kx$.

31. (a) $E = \mathcal{E}/r \ln(b/a)$; $B = \mu_0 \mathcal{E}/2\pi Rr$.
(b) $S = \mathcal{E}^2/2\pi Rr^2 \ln(b/a)$. 33. 0.043 kg·m/s. 35. 7.7 MPa.
37. (a) 586 MN. (b) 1.66×10^{-14}. 39. (a) 94.3 MHz.
(b) $+z$; 960 nT. (c) 1.98 m⁻¹; 593 Mrad/s. (d) 110 W/m².
(e) 678 nN; 367 nPa. 41. $I(2-f)/c$. 45. (a) 3.60 GW/m².
(b) 12.0 Pa. (c) 16.7 pN. (d) 2.78 km/s². 47. 1.06 km².
49. (b) 585 nm.

CHAPTER 42

1. (a) 515 nm; 610 nm. (b) 555 nm; 541 THz; 1.85 fs.
3. (a) 8.68 y. (b) 4.4 My. 5. 67 ps. 11. Yellow-orange.
13. (b) $0.80c$. 15. ± 0.0036 nm. 17. (a) 1.66×10^{-5}.
(b) 0.83×10^{-5}. 19. (a) 6 min. (b) 12 min. (c) 6 min.
23. (a) 0.067. (b) 10°; 7.0°; 2.2°. 25. 4.43 nm. 27. 78.9°.

CHAPTER 43

1. (a) 38.0°. (b) 52.9°. 3. 1.56. 5. 1.95×10^8 m/s.
7. 1.25. 9. 1.5. 13. 74 m. 15. (b) 0.60 mm.
19. 43 mm. 21. 750 m. 23. $1.24 < n < 1.37$. 27. (a) $2v$.
(b) v. 33. 390 cm beneath the mirror surface.
35. $I_{new} = (10/9)I_{old}$. 37. Six. 39. (a) 405 nm. (b) 2.37 μm.
(c) 112°. 41. (a) 72.07°. (b) From A to B.
43. (a) $n_{liquid} < n_{glass}$. 45. 187 cm. 47. (b) 0.170.
49. (b) 60.2 μs. 51. (a) Yes. (b) No. (c) 43°.

CHAPTER 44

1. 11.0 cm. 3. (a) $+$, $+40$, -20, $+2$, no, yes.
(b) Plane, ∞, ∞, -10, yes.
(c) Concave, $+40$, $+60$, -2, yes, no.
(d) Concave, $+20$, $+40$, $+30$, yes, no.
(e) Convex, -20, $+20$, $+0.50$, no, yes.
(f) Convex, $-$, -40, -18, $+180$, no, yes.
(g) -20, $-$, $-$, $+5.0$, $+0.80$, no, yes.
(h) Concave, $+8.0$, $+16$, $+12$, $-$, yes. 9. (b) 2.0. (c) None.
11. 12 cm to the left of the lens. 13. 2.5 mm.
17. (a) 40 cm. (b) 80 cm. (c) 240 cm. (d) -40 cm.
(e) -80 cm. (f) -240 cm. 23. 22 cm. 27. 30 cm to the left of the diverging lens; virtual; upright; $m = 0.75$. 29. (b) No.
(c) Light passes undeviated.
31. (a) 73.6 cm on the side of the lens away from the mirror.
(b) Real. (c) Upright. (d) 0.289. 35. 2.0 mm.

CHAPTER 45

37. (a) 2.34 cm. (b) Smaller. 39. (a) 5.3 cm. (b) 3.0 mm.
41. 103. 43. 25 ms.

1. (a) 0.22 rad. (b) 12°. 3. 2.3 mm. 5. 650 nm.
7. 0.103 mm. 9. 600 nm. 13. (a) 0.253 mm.
(b) Maxima and minima are interchanged. 17. 3.2×10^{-4}.
19. 0°. 23. (a) 1.21 m; 3.22 m; 8.13 m. 27. 124 nm.
29. (a) 552 nm. (b) 442 nm. 31. 215 nm. 33. 643 nm.
35. 2.4 μm. 37. 840 nm. 39. 141. 41. 1.89 μm.
43. (a) 34. (b) 45. 45. 1.00 m. 47. (a) 88%. (b) 95%.
49. 588 nm. 51. 1.0003.

CHAPTER 46

1. 690 nm. 3. (a) 0.430°. (b) 118 μm. 5. (a) $\lambda_a = 2\lambda_b$.
(b) Minima coincide when $m_b = 2m_a$. 7. 173 μm.
9. 1.49 mm. 11. (a) 0.186°. (b) 0.478 rad. (c) 0.926.
13. 5.07°. 15. (b) 0; 4.493 rad; 7.725 rad; . . .
(c) -0.50; 0.93; 1.96; . . . 17. (a) 137 μrad. (b) 10.4 km.
19. 51.8 m. 21. 1400 km. 23. 15 m. 25. (a) 6.8°.
(b) No answer. 27. (a) 0.35°. (b) 0.94°. 29. (b) 70 μm.
(c) Three times the lunar diameter.
31. $\lambda D/d$. 33. (a) 3. 35. (a) 5.0 μm. (b) 20 μm.

CHAPTER 47

1. (a) 3.50 μm. (b) 9.69°; 19.7°; 30.3°; 42.3°; 57.3°.
3. 523 nm. 5. (a) 6.0 μm. (b) 1.5 μm.
(c) $m = 0, 1, 2, 3, 5, 6, 7, 9$.
9. (b) Halfway between principal maxima. (c) $I_m/9$.
13. 400 nm $< \lambda < 635$ nm. 15. 3. 21. 491. 23. 3650.
25. (a) 9.98 μm. (b) 3.27 mm.
27. (a) 0.032 °/nm; 0.077 °/nm; 0.25 °/nm.
(b) 40,000; 80,000; 120,000. 31. 2.68°. 33. 26 pm; 39 pm.
35. 49.8 pm. 39. 0.206 nm.
41. (a) $a_0/\sqrt{2}$; $a_0/\sqrt{5}$; $a_0/\sqrt{10}$; $a_0/\sqrt{13}$; $a_0/\sqrt{17}$.

CHAPTER 48

1. (a) $-y$. (b) $E_x = 0$, $E_y = 0$, $E_z = -cB \sin(ky + \omega t)$.
(c) Linearly polarized; z direction. 3. (a) 2.14 V/m.
(b) 20.3 pPa. 5. $\frac{1}{8}$. 7. 27/128. 9. 15.8 W/m².
11. (a) 0.16. (b) 0.84. 13. (a) 53.1°. (b) Yes, slightly.
15. 55°31′ to 55°46′. 17. 12 μm. 21. (a) Turns plane of polarization by 90°. (b) Reverses handedness of circular polarization. (c) Light remains unpolarized.
23. (a) 2.90×10^{-14} kg·m²/s². (b) 2.88 h.

CHAPTER 49

1. 91 K. 3. (a) 1.06 mm; microwave. (b) 9.4 μm; infrared.
(c) 1.6 μm; infrared. (d) 500 nm; visible. (e) 0.29 nm; x ray.
(f) 2.9×10^{-41} m; hard gamma ray. 5. 580 mW.
9. (a) 138 K. (b) 21.0 μm. 11. 1.44 W. 13. 780 K.
15. (b) 6°C. 19. $0.796 T_E$. 21. (a) 92.1%. (b) 58.2%.
23. (a) 1.4×10^{12}, 6.0×10^{12}, 1.4×10^{13}, Hz.
(b) 5.9, 25, 60, meV. (c) 27, 64, 120, N/m. 25. (a) 1110 J.
(b) 713 J. 27. (a) 2.11 eV. 29. 1.17 eV. 31. Ultraviolet.
33. Cesium, lithium, barium. 35. (a) No.
(b) 544 nm; green. 37. 172 nm. 39. (a) 1.17 V. (b) 641 km/s. 43. 2.63 m². 45. (a) The infrared bulb.
(b) 1.97×10^{20}. 47. (a) 3.10 keV. (b) 14.4 keV.
49. (a) 2.97×10^{20} s⁻¹. (b) 48,600 km. (c) 281 m.
(d) 5.91×10^{18}/m² · s; 1.97×10^{10} m⁻³. 51. (a) 29.8 keV.

(b) 7.19×10^{18} Hz. (c) 1.59×10^{-23} kg · m/s $= 29.8$ keV/c.
53. 2.95 cm/s. **55.** (a) 2.87 pm. (b) 5.89 pm. **59.** 2.64 fm.
61. (a) 4.86 pm. (b) -42.1 keV. (c) 42.1 keV. **63.** 42.6°.
65. (b) 1.12 keV.

CHAPTER 50

1. (a) 1.7×10^{-35} m. **3.** (a) 38.8 pm. (b) 1.24 nm.
(c) 907 fm. **5.** (a) 3.51×10^6 m/s. (b) 64.4 kV.
7. (a) 5.3 fm. **9.** 3.9×10^{-17} m. **11.** A neutron.
13. (a) 7.77 pm. (b) 7.68 pm. **15.** 5.5°.
17. (a) The beams are not present. (b) 47°. **21.** 690 mHz.
23. 76 μeV. **25.** 8.8×10^{-24} kg · m/s. **27.** $\lambda/2\pi$.
29. (a) 1900 MeV. (b) 1.0 MeV. **31.** 88.3 eV.
33. (a) 6.2×10^{-41} J. (b) 1.0×10^{-20}. (c) 3.0×10^{-18} K.
35. (a) 8.74 keV. (b) 1.01×10^{-22} kg · m/s. (c) 98.5 pm.
37. (a) $x = NL/2n$, $N = 1,3,5, \ldots, (n-1)$. (b) $x = NL/n$,
$N = 0,1,2, \ldots, n$. **39.** (b) 0.0006. (c) 0.0003.
41. (a) 9.2×10^{-6}. (b) 7.5×10^{-8}. **43.** 1.1×10^{104} y.

CHAPTER 51

3. 656.3, 486.1, 434.1, 410.2, 397.0 nm. **7.** 3.40 eV.
9. (a) $n = 5 \rightarrow 3$. (b) Paschen. **11.** 66 neV; $E_2 = -3.4$ eV.
21. (a) 54.4 eV. (b) 13.6 eV. **25.** (b) n^2. (c) n. (d) $1/n$.
(e) $1/n^3$. (f) $1/n$. (g) $1/n^4$. (h) $1/n^4$. (i) $1/n^2$. (j) $1/n^2$. (k) $1/n^2$.
31. (a) 3, 2, 1, 0, $-1, -2, -3\hbar$.
(b) $-3, -2, -1, 0, 1, 2, 3$ μ_B.
(c) 30.0°, 54.7°, 73.2°, 90°, 107°, 125°, 150°. (d) $\sqrt{12}\hbar$.
(e) $\sqrt{12}\mu_B$. **33.** (b) 0.358 meV; 1.07 meV; 2.15 meV.
37. 72 km/s^2.
39. $n = 4$; $l = 3$; $m_l = 3, 2, 1, 0, -1, -2, -3$; $m_s = \pm\frac{1}{2}$.
41. $n \geq 5$; $l = 4$; $m_s = \pm\frac{1}{2}$. **43.** 1, 0, 0, $\frac{1}{2}$; 1, 0, 0, $-\frac{1}{2}$.
45. All the statements are true. **47.** 51 mT.
49. (a) 2150 nm^{-3}; zero. (b) 291 nm^{-3}; 10.2 nm^{-1}.
51. 1.85. **53.** 5.41×10^{-3}. **55.** 1.5×10^{-15}. **57.** 0.439.
59. (a) 0.764a_0; 5.236a_0. (b) 0.981 nm^{-1}; 3.61 nm^{-1}.
61. 1.90×10^{-3}. **63.** (a) 11.4 meV. (b) 1.62 eV.
65. (a) 0.284 pm. (b) 2.53 keV. (c) 490 pm.

CHAPTER 52

3. 9.84 kV. **5.** (a) 24.8 pm. (b) Unchanged. (c) Unchanged.
7. $\approx$2.1 keV. **9.** 49.6 pm; 99.2 pm; 99.2 pm.
11. (a) 19.7 keV; 17.5 keV. (b) Zr or Nb. **13.** (a) 5.72 keV.
(b) 86.8 pm, 14.3 keV; 217 pm, 5.72 keV.
19. (a) 2, 0, 0, $\pm\frac{1}{2}$. (b) $n = 2$; $l = 1$; $m_l = 1, 0, -1$; $m_s = \pm\frac{1}{2}$.
21. Only argon would remain an inert gas.
23. (a) 1.84; 2.26. (b) 0.167; 0.119. **25.** 2.0×10^{16} s^{-1}.
27. 3.2×10^7. **29.** 10,000 K. **31.** (a) None. (b) 51.1 J.
33. 4.74 km. **35.** (a) No. (b) 0.11 μm. (c) 110 km.

CHAPTER 53

3. 5.90×10^{28} m^{-3}. **5.** (a) 0.90. (b) 0.69. (c) Sodium.
7. (a) 1.00; 0.986; 0.500; 0.014; zero. (b) 700 K.
9. 5.53 eV. **11.** 65.4 keV. **19.** 234 keV. **23.** 201°C.
27. (a) 5.86×10^{28} m^{-3}. (b) 5.51 eV. (c) 1390 km/s.
(d) 524 pm. **29.** (a) 52.1 nm. (b) 202. **31.** (a) 1.5×10^{-6}.
(b) 1.5×10^{-6}. **35.** (a) 5.0×10^{21} m^{-3}. (b) 1.7×10^5.

37. 46 nm.
39.

insulator	none	—
ext. semi.	donor	n
int. semi.	none	—
conductor	none	—
conductor	none	—
ext. semi.	acceptor	p

41. (a) 0.74 eV above it. (b) 5.6×10^{-7}. **43.** 20 GΩ; 90 Ω.
45. 1.1 eV; no. **47.** (a) 230 nm. (b) Ultraviolet. **49.** Opaque.

CHAPTER 54

1. 15.7 fm. **3.** 26 MeV.
11. (a) 1.000000 u; 11.906830 u; 236.202500 u.
13. ^{25}Mg: 10.01%; ^{26}Mg: 11.00%.
15. (a) 19.81 MeV; 6.258 MeV; 2.224 MeV. (b) 28.30 MeV.
(c) 7.075 MeV. **17.** (b) 7.92 MeV. **19.** (a) 2.59 fm.
(b) Yes. **21.** (a) 4. (b) 148 neV. (c) 8.38 m.
(d) Radio region. **23.** 280 d. **25.** (a) 64.2 h. (b) 0.125.
(c) 0.0749. **27.** (a) 7.57×10^{16} s^{-1}. (b) 4.95×10^{16} s^{-1}.
29. 3.84×10^{21}. **31.** (a) 59.5 d. (b) 1.18. **33.** 87.8 mg.
39. (a) 3.65×10^7 s^{-1}. (b) 3.65×10^7 s^{-1}. (c) 6.41 ng.
43. $Q_3 = -9.460$ MeV; $Q_4 = 4.679$ MeV; $Q_5 =$
-1.326 MeV. **45.** (a) 31.85 MeV; 5.979 MeV. (b) 73 MeV.
47. 1.17 MeV. **49.** (a) 874 fm. (b) 6.4 fm. (c) No.
51. (b) 960.2 keV. **53.** 596 keV. **55.** 13 mJ.
57. 39.4 μCi. **59.** 5.33×10^{22}. **61.** (a) 2.03×10^{20}.
(b) 2.78×10^9 Bq. (c) 75.1 mCi. **63.** 730 cm^2.
65. (a) 6.3×10^{18}. (b) 2.5×10^{11}. (c) 200 mJ. (d) 230 mrad.
(e) 3.0 rem. **67.** (b) 27 TW. **69.** 1.78 mg.
71. -1.855 MeV. **77.** (c) 3.9×10^7 m/s; 8.8×10^5 m/s;
15.6 MeV. **81.** (a) 5.5 MeV. **83.** (a) 5.10×10^{18} Hz.
(b) 20.5 keV. **85.** (a) 3.55 MeV. (b) 7.72 MeV.
(c) 3.26 MeV. **87.** (a) 7.19 MeV. (b) 12.0 MeV. (c) 8.69 MeV.

CHAPTER 55

1. (a) 34 kg. (b) 12 mg. **3.** (a) 2.56×10^{24}. (b) 81.9 TJ.
(c) 25,900 y. **9.** (a) 13.9 d^{-1}. (b) 4.97×10^8.
11. -23.0 MeV. **13.** 174 MeV. **15.** 231 MeV.
17. (a) 253 MeV.
19. ^{238}U $+ n \rightarrow$ ^{239}U $\rightarrow$ ^{239}Np $+$ e; ^{239}Np $\rightarrow$ ^{239}Pu $+$ e.
21. 548 kg. **25.** 1.6×10^{16}. **27.** 566 W. **29.** (a) 44 kton.
31. 24 g. **35.** 450 keV. **37.** (a) 170 kV. **39.** 24,800 y.
43. (a) 4.0×10^{27} MeV. (b) 5.1×10^{26} MeV. **45.** 4.5 Gy.
47. (a) 4.1 eV/atom. (b) 9.0 MJ/kg. (c) 1500 y.
51. (b) 2.28×10^{42} J. (c) 1.85×10^8 y.
53. (a) B; 5.19N, MeV. (b) A: $\frac{1}{2}N$ ^{3}H, $\frac{1}{2}N$ n; B: $\frac{1}{3}N$ ^{1}H, ^{4}He, n.
55. $K_\alpha = 3.52$ MeV; $K_n = 14.07$ MeV. **57.** (a) 1000 km/s.
(b) 2.0 μm.

CHAPTER 56

1. (a) 2.4×10^{-43}. (b) 8.1×10^{-37}. **3.** 2.84×10^{28} m.
5. 769 MeV. **7.** 31 nm. **9.** 2.2×10^{-18} m.
11. (a) Charge; electron lepton number.
(b) Relativistic energy. **13.** b, d. **15.** (a) K$^+$. (b) $\bar{n}$. (c) π^0.
17. (a) $\overline{u}\,\overline{u}\,\overline{d}$. (b) $\overline{u}\,\overline{u}\,\overline{d}$. **19.** (a) sud. (b) uss. **25.** 690 nm.
27. (b) 2.39 GK. **29.** (a) 280 μeV. (b) 4.4 mm.
31. (a) 1.6×10^{12} K. (b) 88 μs.

PHOTO CREDITS

CHAPTER 27
Figure 2: Courtesy Xerox Corporation. Figure 9: *Seattle Times.*

CHAPTER 28
Figure 9: Courtesy Educational Services, Inc.

CHAPTER 30
Figure 23: Courtesy High Voltage Engineering Company. Figure 30: Courtesy NASA.

CHAPTER 31
Figure 2: Courtesy Spague Electric Company. Figure 8: Courtesy Lawrence Livermore Laboratory. Figure 19: Courtesy Pasco Scientific.

CHAPTER 34
Figures 1 and 2: D. C. Heath and Company with Education Development Center. Figure 3: Courtesy Varian Associates. Figure 10: Courtesy Professor J. le P. Webb, University of Sussex, Brighton, England. Figure 11: Courtesy Argonne National Laboratory. Figure 13: Courtesy Fermi National Accelerator Laboratory.

CHAPTER 37
Figure 5: Courtesy GE Medical Systems. Figure 10: Courtesy R. W. De Blois. Figure 12: Dr. Syun Akasofu/Geophysical Institute, University of Alaska, Copyright © 1977.

CHAPTER 40
Figure 6: Courtesy Stanford Linear Accelerator Laboratory.

CHAPTER 41
Figure 2a: Courtesy NASA. Figure 2b: Astronomical Society of the Pacific. Figure 3: Courtesy AT&T Bell Labs. Figure 4: Courtesy NASA. Figure 5: Astronomical Society of the Pacific. Figure 17: Courtesy NASA.

CHAPTER 42
Figure 2: Oregon State University. Figure 3: Copyright © Fotocentre Ltd. Oamuaru, New Zealand. Figure 6: Courtesy Mount Wilson and Mount Palomar Observatories.

CHAPTER 43
Figure 2: Education Development Center, Inc. Figure 3a: PSSC, *Physics, 2nd Ed.,* D. C. Heath and Co. with Education Development Center, 1965, Newton, Mass. Figure 20: Science Photo Library/Photo Researchers. Figure 21: Bell System.

CHAPTER 44
Figure 27: Courtesy NASA.

CHAPTER 45
Figure 2: from *Atlas of Optical Phenomena* by Cagnet et al., Springer-Verlag, Prentice-Hall, 1962. Figure 3: Education Development Center, Newton, Mass. Figure 16: Courtesy Bausch & Lomb Optical, Co. Figure 18b: Courtesy Robert Guenther.

CHAPTER 46
Figures 1 and 2: from *Atlas of Optical Phenomena* by Cagnet et al., Springer-Verlag, Prentice-Hall, 1962. Figure 3: from Sears, Zemansky, and Young, *University Physics,* 5th ed., Addison-Wesley, Reading, Mass., 1976. Figure 13: from *Atlas of Optical Phenomena,* by Cagnet et al., Springer-Verlag, Prentice-Hall, 1962. Figure 15: Courtesy Dr. G. D. Shockman from D. C. Shingo, J. B. Cornett, G. D. Shockman, *J. Bacteriology,* 138: 598–608, 1979. Figure 16: from *Atlas of Optical Phenomena,* by Cagnet et al., Springer-Verlag, Prentice-Hall, 1962.

CHAPTER 47
Figure 2: from *Atlas of Optical Phenomena,* by Cagnet et al., Springer-Verlag, Prentice-Hall, 1962. Figure 14: W. Arrington and J. L. Katz, X-Ray Laboratory, Rensselaer Polytechnic Institute. Figures 21 and 22: Ronald R. Erickson and Museum of Holography. Figure 23: from Rigden, "Physics and the Sound of Music," *Scientific American,* John Wiley & Sons, Inc., 1985.

CHAPTER 48
Figure 8: Copyright © R. Mark, *Experiments in Gothic Structure,* MIT Press, Cambridge, Mass., 1982. Figure 9: Courtesy Apple Computer, Inc. and Paul Matsuda. Figure 12: from Robert Guenther, *Modern Optics,* John Wiley & Sons, Inc., 1990.

CHAPTER 49
Figure 1: Courtesy Alice Halliday

CHAPTER 50
Figure 1: Professor C. Jonsson, University Tübingen, Germany. Figure 2: Courtesy G. Matteucci. Figure 24: Philippe Plailly/SPL/Photo Researchers.

CHAPTER 51
Figure 1: W. Finkelnburg, *Structure of Matter,* Springer-Verlag, 1964. Figure 14: American Institute of Physics, Neils Bohr Library, Margaret Bohr Collection.

CHAPTER 52
Figure 8: Dave Roback/AP/Wide World Photos. Figure 9: Roger Ressmeyer/Starlight Pictures. Figure 10: National Bureau of Standards.

CHAPTER 53
Figure 18: Courtesy AT&T

CHAPTER 55
Figure 12: Princeton University Plasma Physics Lab. Figure 14: Courtesy Lawrence Livermore Laboratory. Figure 15: Los Alamos National Laboratory.

CHAPTER 56
Figure 1a: Courtesy CERN. Figure 5: American Institute of Physics, Neils Bohr Library, Margaret Bohr Collection. Figure 7: Courtesy AT&T.

INDEX

SOME MATHEMATICAL SYMBOLS

$=$	equals	$\sim$	is of the order of magnitude of
$\approx$	equals approximately	$\propto$	is proportional to
$\neq$	is not equal to	lim	the limit of
$\equiv$	is identical to, is defined as	Σ	the sum of
$>$	is greater than	$\int$	the integral of
$\gg$	is much greater than	Δx	the change or difference in x
$\geq$	is greater than or equal to	$|x|$	the absolute value or magnitude of x
$<$	is less than	$\bar{x}$	the average value of x
$\ll$	is much less than	df/dx	the derivative of f with respect of x
$\leq$	is less than or equal to	$\partial f/\partial x$	the partial derivative of f with respect to x

SI PREFIXES

Factor	Prefix	Symbol	Factor	Prefix	Symbol
10^{18}	exa	E	10^{-1}	deci	d
10^{15}	peta	P	10^{-2}	centi	c
10^{12}	tera	T	10^{-3}	milli	m
10^{9}	giga	G	10^{-6}	micro	μ
10^{6}	mega	M	10^{-9}	nano	n
10^{3}	kilo	k	10^{-12}	pico	p
10^{2}	hecto	h	10^{-15}	femto	f
10^{1}	deka	da	10^{-18}	atto	a

THE GREEK ALPHABET

Alpha	A	α	Iota	I	ι	Rho	P	ρ
Beta	B	β	Kappa	K	κ	Sigma	Σ	σ
Gamma	Γ	γ	Lambda	Λ	λ	Tau	T	τ
Delta	Δ	δ	Mu	M	μ	Upsilon	Υ	υ
Epsilon	E	ϵ	Nu	N	ν	Phi	Φ	ϕ, φ
Zeta	Z	ζ	Xi	Ξ	ξ	Chi	X	χ
Eta	H	η	Omicron	O	o	Psi	Ψ	ψ
Theta	Θ	θ	Pi	Π	π	Omega	Ω	ω